PHYSICAL CHEMISTRY THROUGH PROBLEMS

PHYSICAL CHEMISTRY THROUGH PROBLEMS

SECOND EDITION

S K DOGRA
INSA Honorary Scientist
Ex-Professor, Department of Chemistry
Indian Institute of Technology Kanpur, Kanpur

S DOGRA
Ex-Reader, Department of Chemistry
Maitreyi College
University of Delhi, Delhi

Publishing Globally

NEW AGE INTERNATIONAL (P) LIMITED, PUBLISHERS
LONDON • NEW DELHI • NAIROBI
Bangalore • Chennai • Cochin • Guwahati • Hyderabad • Kolkata • Lucknow • Mumbai
Visit us at **www.newagepublishers.com**

Published by New Age International (P) Ltd., Publishers
First Edition: 1984
Second Edition: 2015
Reprint: 2018

GLOBAL OFFICES

- **New Delhi** NEW AGE INTERNATIONAL (P) LIMITED, PUBLISHERS
7/30 A, Daryaganj, New Delhi-110002, (INDIA)
Tel.: (011) 23253771, 23253472, **Telefax:** 23267437, 43551305
E-mail: contactus@newagepublishers.com • Visit us at www.newagepublishers.com
- **London** NEW AGE INTERNATIONAL (UK) LTD.
27 Old Gloucester Street, London, WC1N 3AX, UK
E-mail: info@newacademicscience.co.uk • Visit us at www.newacademicscience.co.uk
- **Nairobi** NEW AGE GOLDEN (EAST AFRICA) LTD.
Ground Floor, Westlands Arcade, Chiromo Road (Next to Naivas Supermarket)
Westlands, Nairobi, KENYA, **Tel.:** 00-254-713848772, 00-254-725700286
E-mail: kenya@newagepublishers.com

BRANCHES

- **Bangalore** 37/10, 8th Cross (Near Hanuman Temple), Azad Nagar, Chamarajpet, Bangalore- 560 018
Tel.: (080) 26756823, **Telefax:** 26756820, **E-mail: bangalore@newagepublishers.com**
- **Chennai** 26, Damodaran Street, T. Nagar, Chennai-600 017, **Tel.:** (044) 24353401, **Telefax:** 24351463
E-mail: chennai@newagepublishers.com
- **Cochin** CC-39/1016, Carrier Station Road, Ernakulam South, Cochin-682 016
Tel.: (0484) 2377303, **Telefax:** 4051304, **E-mail: cochin@newagepublishers.com**
- **Guwahati** Hemsen Complex, Mohd. Shah Road, Paltan Bazar, Near Starline Hotel, Guwahati-781 008
Tel.: (0361) 2513881, **Telefax:** 2543669
E-mail: guwahati@newagepublishers.com
- **Hyderabad** 105, 1st Floor, Madhiray Kaveri Tower, 3-2-19, Azam Jahi Road, Near Kumar Theater, Nimboliadda Kachiguda, Hyderabad-500 027, **Tel.:** (040) 24652456, **Telefax:** 24652457
E-mail: hyderabad@newagepublishers.com
- **Kolkata** RDB Chambers (Formerly Lotus Cinema) 106A, 1st Floor, S N Banerjee Road, Kolkata-700 014
Tel.: (033) 22273773, **Telefax:** 22275247
E-mail: kolkata@newagepublishers.com
- **Lucknow** 16-A, Jopling Road, Lucknow-226 001, **Tel.:** (0522) 2209578, 4045297, **Telefax:** 2204098
E-mail. lucknow@newagepublishers.com
- **Mumbai** 142C, Victor House, Ground Floor, N.M. Joshi Marg, Lower Parel, Mumbai-400 013
Tel.: (022) 24927869, **Telefax:** 24915415, **E-mail: mumbai@newagepublishers.com**
- **New Delhi** 22, Golden House, Daryaganj, New Delhi-110 002, **Tel.:** (011) 23262368, 23262370
Telefax: 43551305, **E-mail: sales@newagepublishers.com**

ISBN: 978-81-224-3805-5
C-18-01-10991

Printed in India at Rashtriya Printers, Delhi.
Typeset at J R Enterprises, Delhi.

NEW AGE INTERNATIONAL (P) LIMITED, PUBLISHERS
7/30 A, Daryaganj, New Delhi-110002
Visit us at **www.newagepublishers.com**
(CIN: U74899DL1966PTC004618)

To
Our Parents

Preface to the Second Edition

Since this book came into market three decades ago, a lot of changes have taken place in the undergraduate and postgraduate courses of the universities and colleges. Major revision in their course contents was suggested by University Grants Commission (UGC) in the end of the Nineties and a majority of universities have adopted these changes. Significant changes in the course contents have been suggested in the areas of Quantum Chemistry and Spectroscopy. Keeping this in mind, it was necessary to revise this book so that the students can get comprehensive benefits when they use this book.

An important topic not covered in the earlier edition of the book was "Symmetry and Group Theory". This chapter has been added and is named as 0 (zero) chapter. The chapter starts with the basic idea of Symmetry and its applications to the different molecules. It is followed by the basic idea of Group Theory, method depicting the construction of character tables and molecular orbitals, followed by applications to find out whether the important integrals are zero or not. Chapters 2 (Elementary Quantum Mechanics), 3 (Atomic Structure and Atomic Spectrum), 4 (Molecular Structure) and 5 (Molecular Spectroscopy) have been revised tremendously. In Chapter 3, topics like Atomic Spectrum have been introduced and Zeeman effect has been revised. In Chapter 4, a new topic of Intermolecular Forces has been introduced. In Chapter 5, a new section on Intensity, which includes Lambert-Beer Law and Classical and Quantum Mechanical Theories have been introduced. Sections on Electronic Spectrum and NMR have been revised drastically. Similarly a lot of changes have been incorporated in Chapter 9 on Statistical Mechanics. A number of new solved examples and unsolved examples have added. We are hopeful that these changes will help the students and will keep them in touch with the modern developments. The answers to the additional problems have not been given. If the demands will come, we will add the answers to these problems at a later date. One of the authors (Prof. S. K. Dogra) is very thankful to the Indian National Academy of Sciences, New Delhi for providing the financial help to improve the academics of this book. Further the faculty and students are requested to suggest any changes required for the improvement of this book so these can be added in the next addition.

S. K. Dogra
S. Dogra

New Delhi

Preface to the First Edition

A proper understanding of the principles of physical chemistry is required for a study of pure, industrial and biological sciences. It is the authors opinion that these principles can be learnt properly only through their applications to solving problems and that the existing texts do not prepare a student sufficient in the diverse applications of the fundamental principles of physical chemistry. This book has been written to fill this gap.

The topics in this book have been put in the time tested arrangement of first structure, then dynamics and lastly synthesis. The first two fall in the domain of the physical chemistry proper and last in that of organic and inorganic chemistry. The first four chapters are devoted to the understanding of the elementary quantum mechanics and atomic and molecular structure. The care has been taken to give problems which illuminate the principles of atomic and molecular structures. So that the students can grasp the more involved details that follow. Chapter five illustrates the physical methods used in establishing the structures and parameters of molecules. Chapters six, seven and eight give the preliminary ideas about the three states of matter, i.e. solids, liquids and gases. Chapter nine serves as a bridge between the results of microsystems and their applications to the macrosystems, i.e. statistical mechanics. This chapter illustrates the methods of calculating the properties of bulk system from those of microsystems. Chapters 10 to 17 deal with the principles of thermodynamics and electrochemistry as applicable to ideal systems. These help in determining whether a particular process, physical or chemical, is feasible or not. Chapter 18 deals with quantities that measure the departure of real systems from ideal behaviour. The basic principles of chemical kinetics are dealt in Chapter 19. This helps in following the progress of a process.

Each section of a chapter starts with a short summary of the basic fundamentals to refresh the memory of the students and to place in perspective the results required for solving problems. This is followed by illustrations. At the end of each chapter there are number of unsolved problems, arranged in the order in which concepts have been introduced. The answers of these problems are given at the end. Care has been taken to provide all the data needed for a particular problem but a hand-book may be consulted wherever proper data has not been given.

The SI units have been used predominantly, though some non-SI units like those for pressure etc. have been used along side. This is because a large amount of information is still available in non-SI units and the students should be familiar with both the systems.

A large number of problems enclosed in this book have been tested on first and second year students of the B. Tech. programme and on the M.Sc. students, studying at I.I.T. Kanpur. The data needed in these problems have been taken from International critical tables, hand-book for physics and chemistry, journals and physical chemistry textbooks. The authors acknowledge their debt. We also express our thanks to Mr. R.D. Singh and Mr. Anil Kumar for typing the manuscript and to Mr. B.S. Panesar for the preparation of drawings. The authors also acknowledge the financial grant from the curriculum development group of quality improvement programme at Indian Institute of Technology, Kanpur. Finally we would like to express our sincere thanks to Mr. S.N. Mahey (late) whose constant encouragement has made possible to complete this work.

***January*, 1984**

S K Dogra

S Dogra

CONTENTS

CHAPTER 0

SYMMETRY, POINT GROUP AND GROUP THEORY

0.1 INTRODUCTION

Understanding of symmetry is fundamental to many areas of electronic structure, spectroscopy and chemical reactivity. It has become customary:

(*i*) to represent the structure of molecules in terms of symmetry that the molecules possess.

(*ii*) for spectroscopists to describe molecular vibrations in terms of symmetry.

(*iii*) knowledge of symmetry properties is very important for the modern applications of spectroscopy methods to the problem of structure determination.

In this chapter, we will be considering the symmetry of an isolated molecule and thus is called *point symmetry*. In other words point symmetry refers to the set of operations transforming a system about a common point, which turns out to be the centre of gravity of a molecule.

Symmetry: If a molecule possesses two or more orientations in three dimensional space which are indistinguishable, the molecule is supposed to possess symmetry.

Symmetry elements and symmetry operations are used in such a general manner that we do not attach separate meaning, but these are two different identities. Symmetry element is the line, point, or plane about which symmetry operations are carried out. Symmetry operation can be defined only with respect to the symmetry element and existence of the symmetry element can be shown only by carrying out the symmetry operation.

Before we define the symmetry elements, we would like to define two things.

Identical or *original structure* is that structure in which even if the similar atoms are labelled with prime or double prime marks or any other notations no change would be detected if any symmetry operation is carried out. The other structure is called *equivalent one* and when the symmetry operation is carried out to this structure, one gets the structure which is similar to the identical one but is not the original or identical structure.

0.2 SYMMETRY ELEMENTS

There are five different types of symmetry elements.

0.2.1. *Identity*

When this operation is carried out, no change is observed in the molecule, *i.e.* the result of this operation is not only to produce an equivalent structure but an identical one. This is denoted by the letter *E* or *I*, we will use the former.

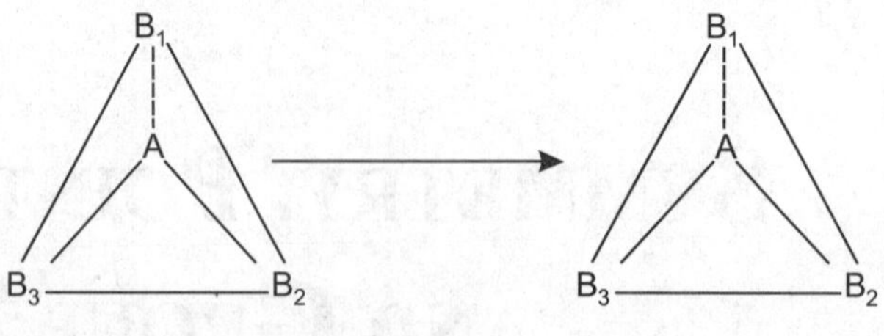

Fig. 0.1 *The identity operation.*

0.2.2. Centre of Symmetry or Inversion Centre

This element is denoted by *i*. A molecule is said to possess a centre of symmetry or centre of inversion when every atom in the molecule is joined to the centre by a straight line and if we move further in the same direction by the same distance and encounters a similar atom. Mathematically this operation corresponds to placing this centre at the centre of coordinate system and assigning every atom with coordinates (x, y, z) and changing its coordinates to $(-x, -y, -z)$. *The presence of this symmetry element requires that atoms in the molecule must occur in pairs and if there is an odd atom in the molecule it must be present at the centre of inversion*. Here we can also make very important deduction pertaining to symmetry elements present in a molecule with centre of symmetry. *The molecules having centre of symmetry must possess at least one reflection plane passing through the centre of symmetry but molecules having other reflection planes not passing through the centre of inversion do not reflect on the centre of inversion*. For example, O_2 possesses the centre of inversion. In case of CO_2, oxygen atoms occur in pairs, whereas C-atom is an odd one and is present at the centre of inversion, whereas OCS is a linear molecule does not possess the centre of inversion. Similarly although CH_4 (*i*) molecule contains even number of hydrogen atoms but structure of methane is such that it does not contain the centre of inversion. This operation generates only one operation as second operation will lead to identical structure, *i.e.*, $i^2 = i \times i = E$. The other examples can be (*ii*) trans-dichloroethene and tetrachloroplatinate (II).

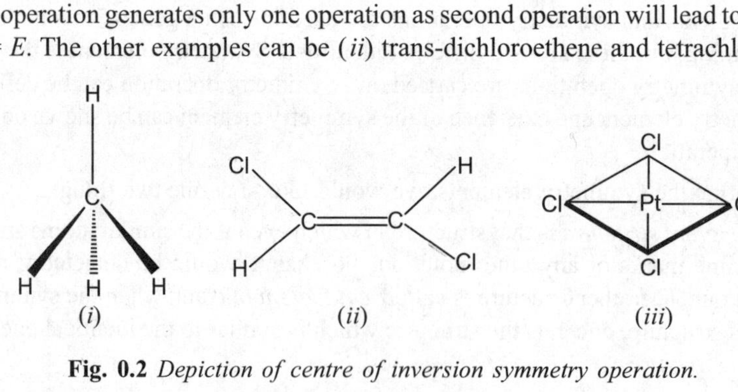

Fig. 0.2 *Depiction of centre of inversion symmetry operation.*

0.2.3. *Plane of Symmetry or Mirror Plane*

This symmetry operation involves selecting a plane, drawing a line perpendicular to the plane from every atom and placing an atom at the end of the line at equal distance to the opposite side of the plane. This symmetry element may also be defined as follows, *i.e.*, if a plane exists in a molecule that separates the molecule into two halves that are mirror images of each other. If the equivalent configuration is obtained after this is done to all the atoms, the plane selected is a mirror plane and this plane cannot lie outside the molecule but must pass through it. *The presence of this symmetry element also demands the presence of even number of similar atoms in the molecule and if any odd atom is present it must lie on the plane of symmetry*.

For example, molecule SFClO (*i*) does not contain any symmetry plane. In case of water molecule (H_2O, *ii*), the vertical plane passes through the oxygen atom and bisect the angle HOH as well vertical plane containing the molecule. Similarly $CHBrCl_2$ (*iii*) contains a vertical plane passing through CHBr atoms and bisecting the angle ClCCl, in other words atoms C, H and Br lie on the plane. NH_3 molecule (*iv*) contains three vertical planes, passing through the nitrogen atom and one of the hydrogen atoms, *i.e.*, NH^1 atoms lying in the plane and bisecting the angle H^2NH^3 and so on.

In general the presence of symmetry plane is denoted by the symbol σ. In molecules which contain more than one mirror plane (*e.g.* AX_3, BF_3, v), the horizontal plane σ_h is taken as horizontal plane and is perpendicular to the principal axis (see next section, highest fold rotation) and three vertical planes, (σ_v, $\sigma_{v'}$, $\sigma_{v''}$), but all the vertical planes contain boron atom and are perpendicular to σ_h.

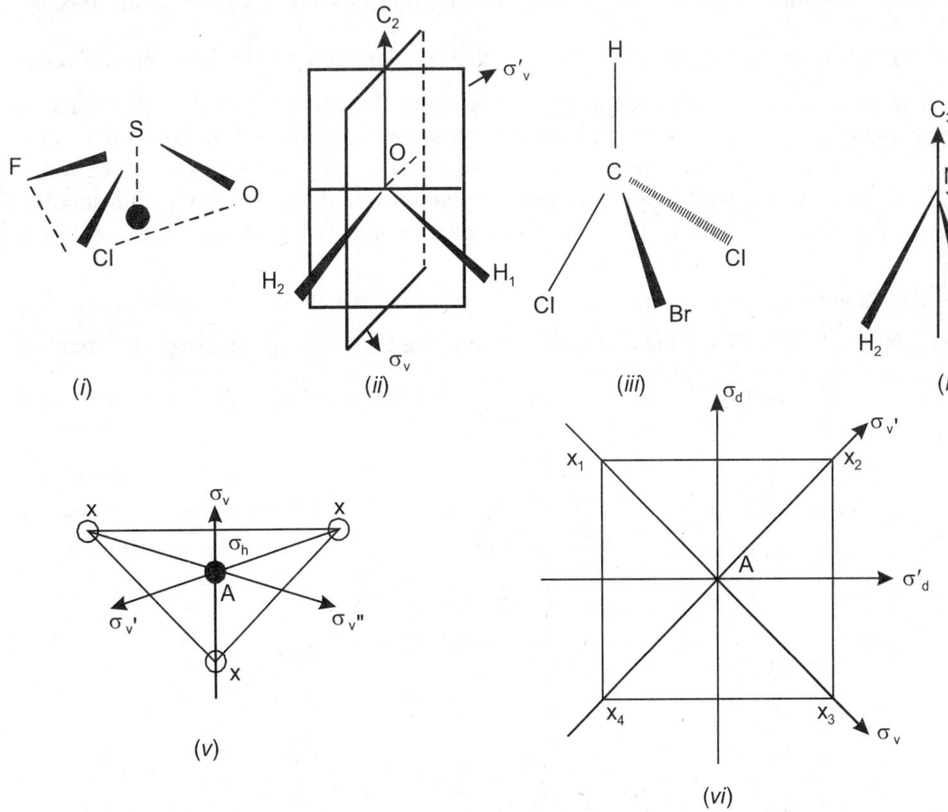

Fig. 0.3 *Depiction of symmetry plane operation.*

When a molecule is located in a coordinate system, the *Z*-axis always lies in the vertical plane. Some molecules have mirror planes containing the principal axis but none of the perpendicular C_2 axis. These planes bisect the angle between two of the C_2 axes (in XY-plane), they are referred to as dihedral planes and is abbreviated as σ_d and is shown in Fig. 0.3. (*vi*) AX_4. In AX_4 planar molecules, *Z*-axis is perpendicular to *XY*-plane and coincides with principal axis and X_1AX_3 and X_2AX_4 bond axes as the *X*- and *Y*-coordinates. One set of vertical planes are *XZ*- and *YZ*- planes and the other set bisect the angle between *X*- and *Y*- axes. The former set including the *X* atoms is called σ_v and the latter set is called σ_d. The diagonal plane is always taken as that bisects *X*-, *Y*- and *Z*- or twofold axes in the molecule. Summarising, *the horizontal plane (σ_h) by definition must be perpendicular to highest-fold rotation axis and vertical plane is the plane which contains the highest-fold rotation axis* .

0.2.4. *n-Fold Rotation or Symmetry Axis (Proper Axis) of Rotation*

Molecule is said to have *n*-fold *symmetry axis* or *proper axis of rotation* if rotation of a molecule by a least angle of $\theta = \frac{2\pi}{n}$ (*or* $\frac{360}{n}$) about such an axis produces a configuration equivalent to the original one. It is denoted by the symbol $\hat{C}_n$ or simply by C_n. The subscript *n* denotes the order of the axis. By the order of the axis we mean the largest value of *n* such that rotation through $\frac{2\pi}{n}$ gives an equivalent configuration. The rotation by 2π angle about any axis passing through a molecule will produce a configuration identical to the original one so the molecule possesses one fold axis of rotation, *i.e.*, C_1. Here the order of C_1 axis will be 1. The rotation can be clockwise as well as anti-clockwise, but in this section, we will use the anti-clockwise rotation. It may be mentioned here that the convention adopted here for rotation is similar to that used for the sign of an angular momentum vector along the *Z*-axis in a species.

This has been elaborated here by taking example of a planar AB_3 type of molecules (*i.e.*, BF_3, BCl_3, NO_3, SO_3, NO_3^-, CO_3^{2-}). *B*'s in the molecule are numbered just for convenience. The effect of rotation through $\frac{2\pi}{3}$, $\frac{2x2\pi}{3}$ and $\frac{3x2\pi}{3}$ can be seen as indicated below in Fig. 0.4 (*a*). Configurations

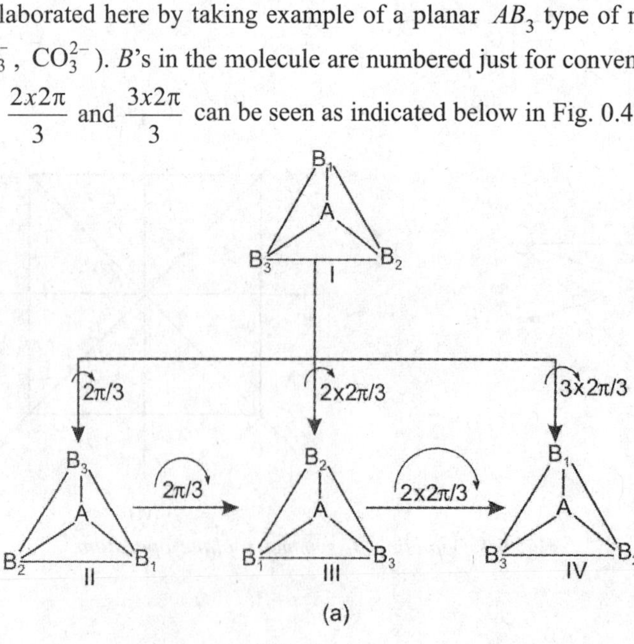

(a)

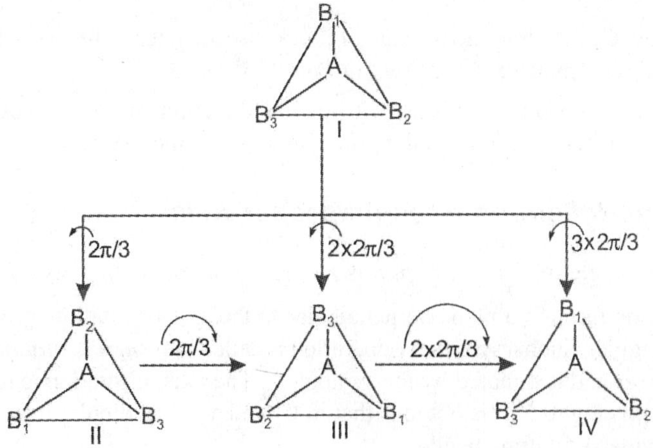

(b)

Fig. 0.4 *Effect of rotation on planar AB_3 type molecule: (a) C_3 principal axis carried out in clockwise manner and (b) carried out in anticlockwise manner.*

II and III are equivalent to I, without the labels these are identical, although with labels these are distinguishable. However configuration IV is identical to I both with and without the labels. Hence, it is not merely equivalent, it is identical to the original configuration. In this example since the least angle of rotation is $\frac{2\pi}{3}$, so the C_3 axis is called the three-fold rotation axis. C_3 or $\hat{C}_3$ will represent the operation of rotation by $\frac{2\pi}{3}$ around C_3 axis. The rotation by $\frac{2x2\pi}{3}$ is denoted by the symbol C_3^2 and by $\frac{3x2\pi}{3}$ by the symbol C_3^3. Since $C_3^4 = C_3^1$ so only C_3, C_3^2 and C_3^3 are the separate and distinct operations. Since the operation C_3^3 produces a configuration identical to the original one hence we may write $C_3^3 = E$. Thus following results can be obtained and are listed below:

(*i*) *n*-fold axis is denoted by C_n and minimum angle of rotation will be $\frac{2\pi}{n}$. Similarly rotations carried out *k* times will be denoted by C_n^k. The element $C_n^n = E$ and $C_n^{n+1} = C_n^1$, $C_n^{n+2} = C_n^2$ and so on.

(*ii*) Unlike, reflection in a symmetry plane and centre of inversion (where both elements generate only one symmetry operation), C_n axis of rotation generate *n* symmetry operations, *e.g.*, $C_n, C_n^2, C_n^3, \ldots C_n^{n-1}, C_n^n\ (E)$.

(*iii*) There is no restrictions on the number of atoms of each species, whether even or odd lying on the proper axis of rotation. This is because on carrying out rotation around this axis will not shift the position of these kinds of atoms in a molecule. But *the atoms of the species which are displaced on rotation must be equal to the order of the rotation axis*. For example, C_n axis will require the presence of '*n*' operations and in each operation equivalent configuration will be only generated if the similar atoms displace each other.

(*iv*) The clockwise rotation of the system or of a vector when it is symbolised as C_n (sometimes as C_n^+) as the clockwise rotation and C_n^{-1} (or C_n^-) as anti-clockwise rotation. Figure 0.4 *a* gives the C_3 rotation carried out in clockwise manner, whereas when carried out in an anticlockwise manner it is shown in Fig. 0-I.4 *b*.

(*v*) There are many molecules which possess different order of axes, but the highest fold rotation axis is referred to as the principal axis in a molecule.

0.2.5. *Rotation–Reflection Axes (Improper Axis)*

Rotation through an angle of $\frac{2\pi}{n}$, (where *n* is an order), about such an axis (called proper rotation) followed by reflection through a plane perpendicular to the axis of rotation, produces a configuration equivalent to the original. Such a symmetry operation is called the *mirror rotation* or *rotation-reflection axes* or *improper axes* and is denoted by the symbol S_n. The axis of rotation is referred to as the *n*-fold rotation reflection axis Fig. 0.5. It is obvious that if C_n and a perpendicular reflection plane are present independently, S_n will exist automatically.

Since this operation combines both rotation and reflection it can be represented as *a* multiplication of operations, *i.e.*, rotation followed by reflection through a plane perpendicular to the axis of rotation or vice versa Both will result into the same configuration.

$$S_n = C_n \sigma_h = \sigma_h C_n$$

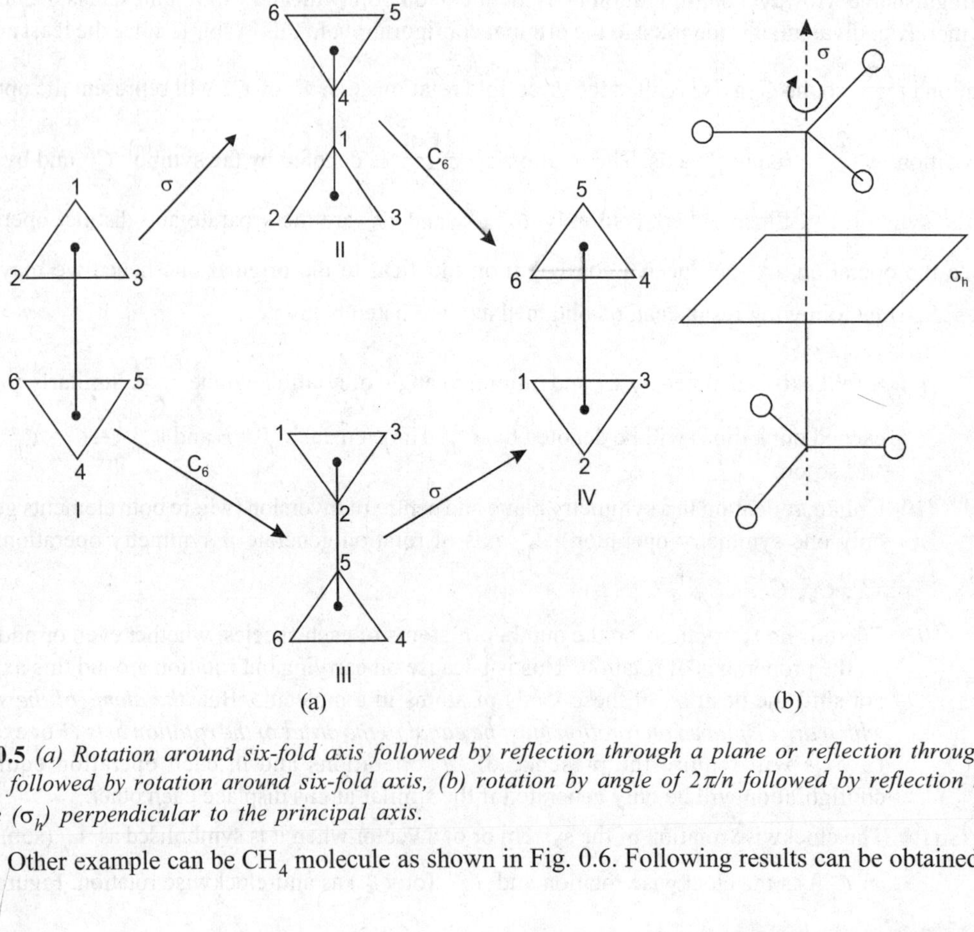

Fig. 0.5 *(a) Rotation around six-fold axis followed by reflection through a plane or reflection through a plane followed by rotation around six-fold axis, (b) rotation by angle of 2π/n followed by reflection in a plane (σ_h) perpendicular to the principal axis.*

Other example can be CH_4 molecule as shown in Fig. 0.6. Following results can be obtained:

(*i*) S_n, when n is even, leads to the presence of $C_{\frac{n}{2}}$ axis of rotation. Further a σ reflection plane does not co-exist as a symmetry element with in that system.

(*ii*) S_n, when n is odd, leads to the (*a*) presence of C_n and σ operations independently and (*b*) generates $2n$ symmetry operations.

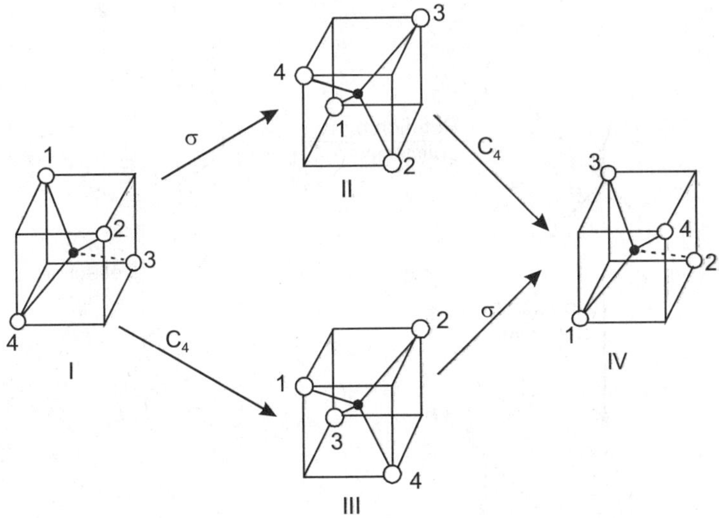

Fig. 0.6 *Reflection through a plane followed by four-fold axis of rotation or rotation around four-fold axes of rotation followed by reflection through a plane.*

For easy reference all the five symmetry operations can be summarised in Table 0.1.

TABLE 0.1. The five symmetry elements and their associated operations

Symmetry elements		Symmetry operations	
Symbol	**Description**	**Symbol**	**Description**
E	Identity	$\hat{E}$	No change
i	Centre of symmetry	$\hat{i}$	Reflection through the centre
σ	Plane of symmetry	$\hat{\sigma}$	Reflection through the plane
C_n	n-Fold rotation-reflection	$\hat{C}_n$	Rotation about the axis by 360°/n
S_n	n-fold rotation-reflection axis of symmetry, also called improper rotation	$\hat{S}_n$	Rotation about the axis by 360°/n followed by reflection through a plane perpendicular to the axis

***Exercise* 0.1:** By taking the example of staggered ethane (Fig. 0.7) show that centre of inversion operation (the operation *i*) converts every point (x, y, z) to $(-x, -y, -z)$. Also examine the effects of C_2 operation on the sign of the coordinates.

(*b*) Fill the vacant spaces marked as? in the following conformations of staggered ethane.

Let us place the staggered ethane Fig. 0.7 (*a*) in the coordinate system as shown in Fig. 0.7 (*b*). Reflection of H^3 (x, y, z) through the centre of the molecule gives H^2 $(-x, -y, -z)$. Similarly reflection of: H^1 (x, y, z) gives H^4 $(-x, -y, -z)$; H^5 (x, y, z) gives H^6 $(-x, -y, -z)$. Thus H^5 and H^6 exchanged by *i* are in the same plane, *i.e.*, the *YZ* plane. Since we have taken atoms as points it may seem that H^5 and H^6 do

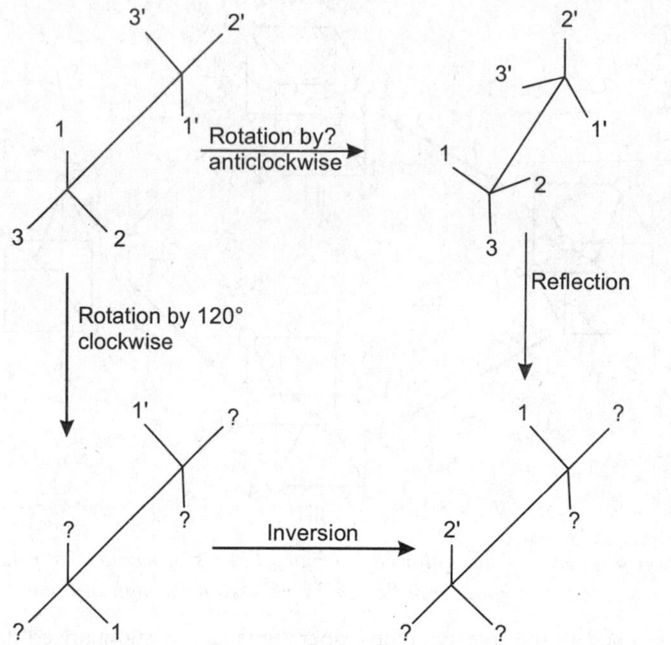

not have different signs for the *x* coordinate since if $-x = 0$, $x = 0$. Since for our purpose, assuming atom to be a sphere, then on H^5 towards the rear of H^3 has $-x$ positive and inversion would invert it to $-x$, an equivalent point. Now coming to the effect of C_2 operation on the sign of the coordinates, we

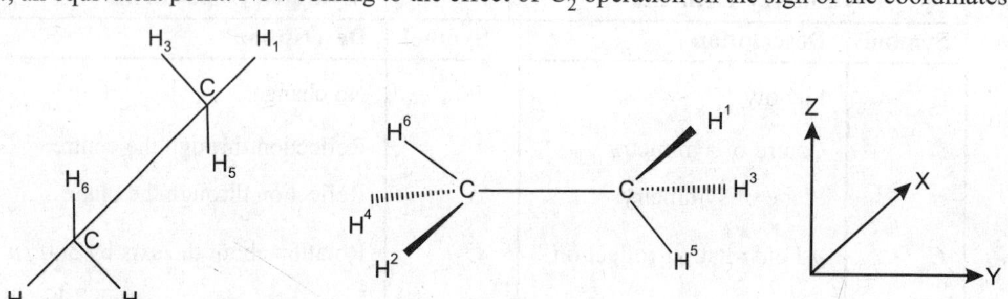

(a) (b)

Fig. 0.7 *The staggered form of ethane and applying the different symmetry operations on the front of H* ^{6}.

find that staggered form for ethane has three C_2axes, one of which is coincident with *x*-axis.

C_2^x operation transforms:

$$H^1\ (x, y, z) \rightarrow H^2\ (-x, -y, -z)$$

$H^3 (x, y, z) \rightarrow H^4 (-x, -y, -z)$

$H^5 (x, y, z) \rightarrow H^6 (-x, -y, -z)$

In general C_2 operations around x-axis changes the sign of y and z coordinates leaving x coordinate unaffected.

Similarly C_2^z inverts the sign of x and y and leaves z unaffected and C_2^y inverts the sign of x and z and leaves y unaffected. All points on the symmetry element are left unaffected by the symmetry operation

(*c*) This is left an exercise for the students.

***Exercise* 0.2:** Find out the symmetry elements present in the following molecules:

(*i*) $POClBr_2$, (*ii*) ClF_3, (*iii*) N_2O_2

(*i*) Structure of $POClBr_2$ molecule is shown in Fig. 0.8, (*a*). Identity element, *E* will always be present in every molecule. This molecule is having a tetrahedral shape. So rotation axis can be C_2, passing through phosphorous atom and bisecting the angles OPCl and BrPBr angles. But this symmetry element is not present. On the other hand, a plane containing O, P and Cl is present. So it contains one vertical plane.

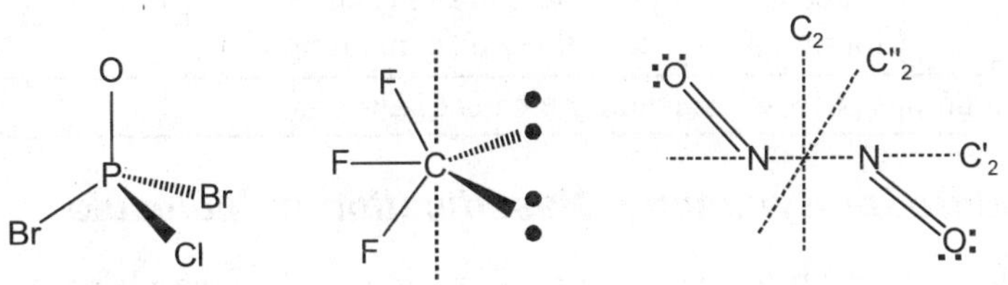

(a) (b) (c)

Fig. 0.8 *Structures of (a) dibromophosphorous oxychloride, (b) chlorine trifluoride and (c) dimer of nitrogen oxide.*

(*ii*) ClF_3 structure is shown in Fig. 0.8 (*b*). The geometry of ClF_3 is basically a triagonal bipyramid, with two lone pairs of electrons and one fluorine atom in the equatorial plane and with two fluorine atoms bent towards equatorial fluorine. The structure can be approximated as *T*-shaped structure, *i.e.*, two lone pairs and one of the fluorine atoms form the planar triangle and rest of the two fluorine atoms are present on the vertical axis passing through Cl-atom. Principle axis in trigonal bilyramidal structure is C_3 but is absent as is clear from Fig. 0.8 (*b*). The only axis of rotation present in the molecule ClF_3 is C_2 symmetry axis of rotation, passing through F, Cl and bisecting the angle lone pair Cl and lone pair, as well as plane containing the equilateral triangle. The student should verify the absence of symmetry along the axes indicated by dashed line.

(*iii*) Structure of N_2O_2 molecule is shown in Fig. 0.8 (*c*). It is a planar molecule. Three C_2 axes have been drawn. As is clear from the diagram C_2 and C_2^2 rotation axes are absent. The only rotation axis as an independent element is present in the molecule. Similarly centre of inversion is also present. No independent symmetry molecular plane is present. S_2 symmetry element is present *i.e.*, rotation C_2 followed by reflection in symmetry plane perpendicular to the C_2 axis is present.

0.3 POINTS GROUPS

A point group is a collection of all the symmetry operations that can be carried out on a molecule belonging to this group. It is possible to classify any given molecule into one of the point groups. Symmetry elements of some common point groups are listed in Table 0.2.

TABLE 0.2. Symmetry elements of some common point groups

Point Group	Symmetry elements	Examples
C_1	No symmetry	SiBrClFI
C_2	One C_2 axis	H_2O_2
C_{nh}	One *n*-fold rotation axis and a horizontal plane must be perpendicular to the *n*-fold rotation axis	Trans-$C_2H_2Cl_2$ (C_{2h})
C_{nv}	One *n*-fold rotation axis and *n*-vertical plane containing the *n*-fold axis of rotation.	H_2O, SO_2Cl_2, $SiCl_2Br_2$ (C_{2v}) NH_3, CH_3Cl, $POCl_3$ (C_{3v})
D_{nh}	One *n*-fold axis, *n*-C_2 axis perpendicular to C_n axis, *n* vertical planes and one σ_h.	N_2O_4 (planar) C_{2h} AX_3 type (planar) C_{3h} AX_4 type (planar) C_{4h}
D_{nd}	One *n*-fold axis, *n*-C_2 axes perpendicular to C_n axis, *n* vertical planes	$CH_2=C=CH_2$ (D_{2d})
T_d	Three C_2 axes perpendicular to each other, four C_3 axes, six σ, three S_4 containing C_2	CH_4, SiH_4
All point groups possess the identity element, *E*		

0.3.1. *Procedure for Symmetry Classification of Molecule*

Five step procedures to establish the symmetry point groups of a molecule are shown in Fig. 0.9 as a flow sheet diagram (only basic point groups are shown in Fig. 0.9).

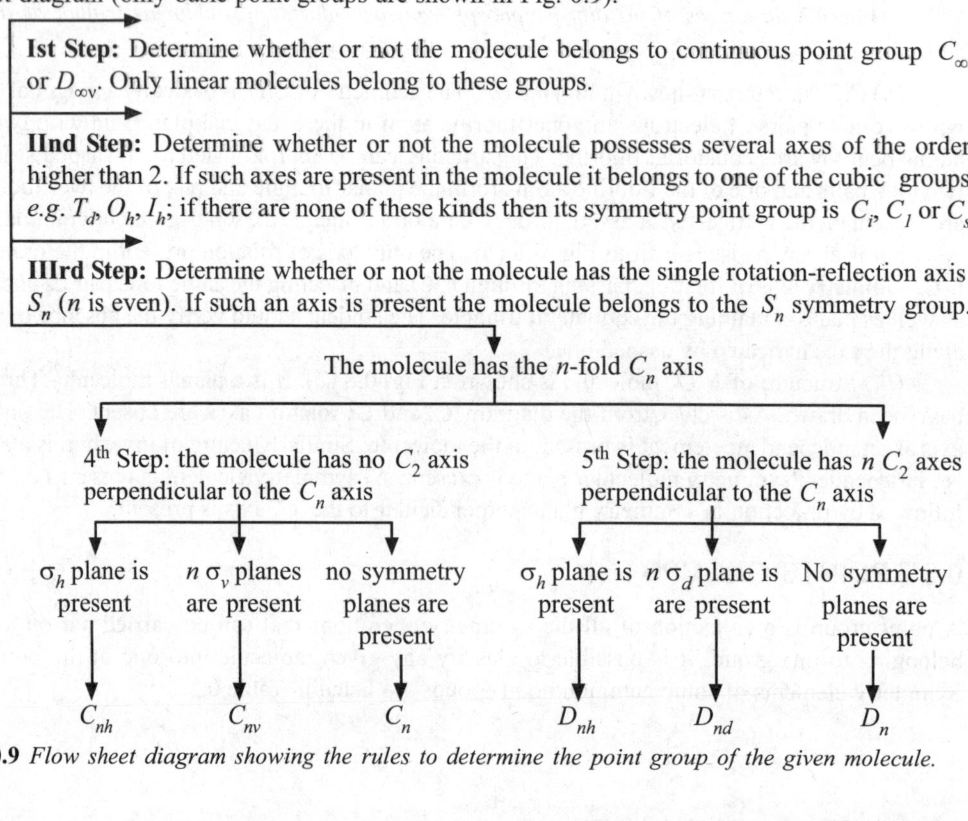

Fig. 0.9 *Flow sheet diagram showing the rules to determine the point group of the given molecule.*

***Exercise* 0.3:** Assign the point group to (*a*) Ferrocene, (*b*) benzene and (*c*) PCl_5.

(*a*) Ferrocene (Fig. 0.10 *a*)

(*i*) Since the ferrocene molecule is nonlinear, it cannot belong to either $C_{\infty v}$ or $D_{\infty v}$ point group.

(*ii*) Molecule has only one proper or unique axis of order higher than 2, *i.e.*, the C_5 axis. So the molecule does not belong to cubic symmetry groups either.

(*iii*) An even order improper axis S_{10} in the Fig. 0.10*a*, cannot be present in the molecule as there are other unrelated symmetries present in the molecule. So the molecule cannot belong to S_n group either.

(*iv*) Let us now look for two fold axis of symmetry in the molecule. There are five C_2-axes, perpendicular to the unique C_5 axis. Thus the molecule may belong to either of D_{nh}, D_{nd} or D_n groups.

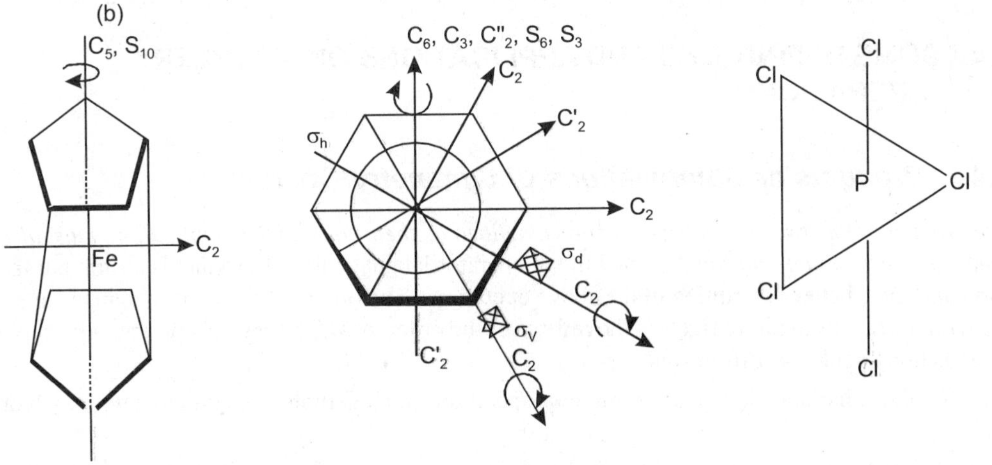

(a) (b) (c)

Fig. 0.10 *Structures of (a) ferrocene, (b) benzene and (c) PCl_5.*

(*v*) Because of the staggered relationship of the rings, the presence of σ_h plane is ruled out, but we can draw five vertical planes which pass between any two C_2–axes. Thus the molecule belongs to D_{5d} point group. On the other hand, the eclipsed configuration of ferrocene has all the symmetry elements of staggered form but due to the non-staggered nature of the rings, the horizontal plane σ_h is also present in the molecule. So the eclipsed form belongs to D_{5h} point group.

(*b*) Benzene (Fig. 0.10 *b*)

(*i*) Benzene being nonlinear molecule can not belong to either $C_{\infty v}$ or $D_{\infty v}$ point group.

(*ii*) Molecule has C_6–axis perpendicular to the ring plane, C_3–axis (order greater than 2) is also present in the molecule. So the molecule does not belong to cubic groups either.

(*iii*) Benzene ring has also improper axis S_6 perpendicular to the plane of benzene ring, but there are other symmetry elements independent of it. So the molecule can not belong to any S_n group.

(*iv*) Let us now look for two-fold axis of symmetry in the molecule. There are six C_2–axis lying in the plane of the ring. So the molecule may belong to either of the D_{nh}, D_{nd} or D_n group.

(*v*) Since there is a σ_h plane in the molecule which is the plane of the molecule itself, so the molecule belongs to D_{6h}. There are also six vertical planes ($3\ \sigma_v + 3\ \sigma_d$) but they contain the C_2–axis.

(c) PCl_5 type (Trigonal Bipyramidal Molecules, Fig. 0.10c)

(*i*) Molecule is nonlinear, so it does not belong to either $C_{\infty v}$ or $D_{\infty v}$ point group.

(*ii*) Molecule has unique three fold axis of symmetry (*i.e.*, C_3), so the possibility of cubic group is ruled out.

(*iii*) There is no even order S_n axis present in the molecule. Thus the possibility of S_n group is ruled out.

(*iv*) Molecule has three two-fold axis of symmetry which are perpendicular to the C_3–axis. Thus the molecule may belong to either D_{nh}, or D_{nd} or D_n group.

(*v*) Presence of σ_h plane in the molecule shows that it belongs to D_{3h} point group.

0.4 SOME DEFINITIONS AND APPLICATIONS OF SYMMETRY OPERATIONS

0.4.1. *Products or Combinations of Symmetry Operations*

The product of any two symmetry operations, defined as their consecutive application, must also be a symmetry operation of the same group. In the operator language use of product is alright but the term *combination* is better description of the above operations. The order in which the symmetry operations are written (*i.e.*, from left to right) is the reverse of the order in which they are applied, *i.e.*, application of operation should start from right.

***Exercise* 0.4:** Find out whether the symmetry operations in H_2O molecule commute with each other or not.

H_2O belongs to C_{2v} point group and contains E, C_2, σ_v and σ'_v symmetry elements. *Symmetry element E always commutes with every symmetry element* (*i.e.*, $E \times C_2 = C_2 \times E$). Consider the combination of C_2 and σ_v which can be written as

$$C_2 \times \sigma_v = \sigma'_v$$

or

$$\sigma_v \times C_2 = \sigma'_v$$

Here C_2 axis is the two-fold rotation axis containing the *Z*-axis and molecular plane is *XZ* (σ'_v) and σ_v is the plane bisecting the angle HOH. This means that the combination of C_2 and σ_v will lead to configuration of H_2O which can be obtained by the symmetry operation σ'_v. This is shown in Fig. 0.11. It is clear that these symmetry operations commute with each other. Although in the Example 0.4, described above, it is immaterial whether we write C_2 and σ_v or σ_v and C_2, the symmetry operation generated is the same, *i.e.*, σ'_v. This result is not universally applicable, i.e. order of following operations can give different results. This is supported by taking the example of C_{3v}. It is shown in Exercise 0.5 and is depicted in Fig. 0.12.

$$C_3 \times \sigma_v \neq N$$

If the order of symmetry operations in the combination of the operations gives the similar results (as observed in case of C_{2v} symmetry point group) the two symmetry operations are said to commute with each other.

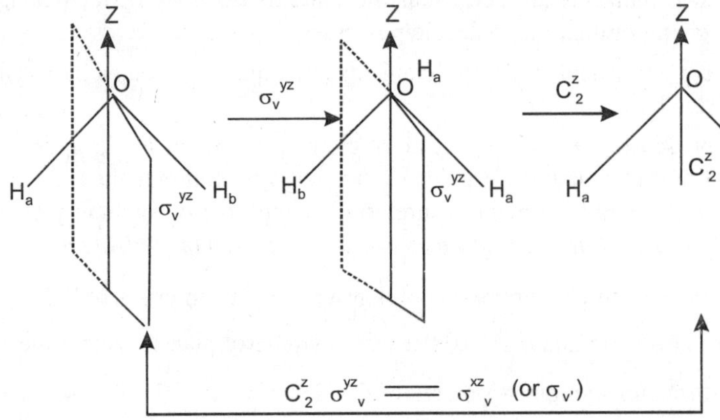

Fig. 0.11 *Combination of symmetry operations C_2 and σ_v leads to σ'_v N.*

***Exercise* 0.5:** Find out whether symmetry operations in NH_3 molecule commute with each other.

***Solution*:** NH_3 molecule belongs to C_{3v} symmetry point group. The symmetry elements present in C_{3v} symmetry point group are E, $2C_3$, and $3\sigma_v$ symmetry planes. Consider C_3 rotation and σ_v symmetry plane operations as shown in Fig. 0.12. Symmetry operations have been carried out as shown in the Fig. 0.12. It is clear from the final results that the resultant configurations obtained in both cases are equivalent but not identical. This clearly suggests that the two operations listed do not commute with each other.

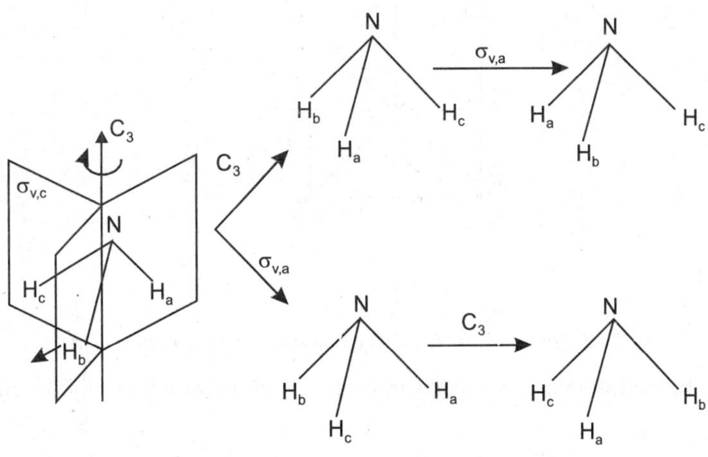

Fig. 0.12 *Effect of sequence of operations on the final positions of the atoms in an ammonia molecule. N is out of the plane of the molecule and C_3-axis is perpendicular to the plane of paper passing through it.*

0.4.2. *Equivalent Symmetry and Equivalent Atoms*

If a symmetry element A can be moved into the element B by a symmetry operation corresponding to symmetry element X, then A and B are said to be *equivalent*. Furthermore, if A can be carries into C, then there must be a symmetry operation (or a sequence of symmetry operations) that carries B into C. Since B can be carried into A, Then symmetry elements A, B and C are said to form an equivalent set.

Equivalent atoms in a molecule are defined as those atoms that may be interchanged with one another by a symmetry operation that the molecule possesses.

***Exercise* 0.6:** Find out the equivalent symmetry elements and equivalent atoms in the molecules: (*i*) H_2O, (*ii*) NH_3 and (*iii*) $[PtCl_4]^{2-}$.

(*i*) H_2O molecule possesses a C_{2v} symmetry point group and contain E, C_2, σ_v and σ'_v symmetry elements. From Fig. 0.11, it is clear that if we apply C_2 symmetry operation to the H_2O molecule it will not generate either σ_v or σ'_v symmetry planes. Similarly by applying σ_v symmetry element will not generate either C_2 or σ'_v Thus, *all the three symmetry operations are not equivalent*.

(*ii*) NH_3 molecule belongs to C_{3v} symmetry point group, which comprises of E, 2 C_3 (C_3^1 and C_3^2) and $3\sigma_v$ symmetry planes (σ_v^1, σ_v^2 and σ_v^3). All the three symmetry planes contain the C_3 symmetry rotation element. Now if rotation symmetry operation C_3^1 is applied to NH_3, σ_v^1 symmetry plane will generate the σ_v^2 symmetry plane and similarly applying C_3^2 will generate σ_v^3. Thus three σ_v symmetry planes (σ_v^1, σ_v^2 and σ_v^3) are equivalent. Similarly two axes of rotation are equivalent symmetry elements.

(*iii*) $[PtCl_4]^{2-}$ is a square planar complex ion and belongs to D_{4h} symmetry point group and contains 16 symmetry elements, *i.e.*, E, $2C_4$, C_2, $2C'$, $2C''$, i, $2S_4$, σ_h, $2\sigma_v$ and $2\sigma_d$ symmetry elements. These are shown in Fig. 0.13. Using the similar procedure it can be shown that C'_2 and C''_2; C''_2 and C''_2; σ_d and σ_d; σ_v and σ_v; σ_h form the set of equivalent symmetry elements.

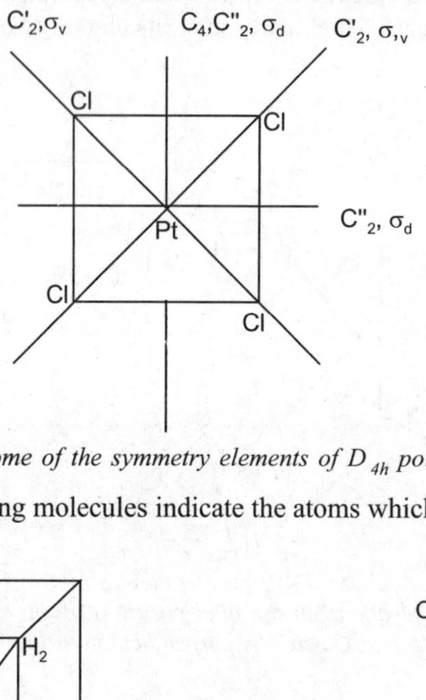

Fig. 0.13 *Some of the symmetry elements of* D_{4h} *point group.*

***Exercise* 0.7:** In the following molecules indicate the atoms which are equivalent?

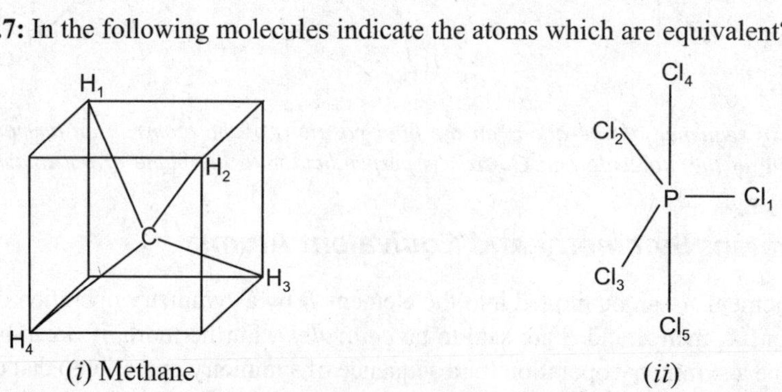

(*i*) Methane (*ii*)

Fig. 0.14 *Geometries of (i) methane and (ii) phosphorous pentachloride.*

By definition equivalent atoms are those which may be interchanged with one another by a symmetry operation that the molecule possesses.

(*i*) Methane belong to tetrahedral point group, *i.e.* cubic symmetry as shown above and carbon atom is the point. By applying C_2 rotation operation, H_1 will go to H_2 and H_3 will go to H_4, suggesting that H_1 and H_2 as well as H_3 and H_4 are equivalent. Consider C_3 rotation symmetry operation, passing through H_1 and centre of the triangle formed by H_2, H_3 and H_4. Now by applying C_3, H_2 will go to H_3, H_3 will go to H_4 and H_4 will go to H_2. This means that H_2, H_3 and H_4 are equivalent. Since H_1 is equivalent to H_2, and in turn equivalent to H_3 and H_4. This suggests all the four hydrogen atoms are equivalent.

(*ii*) PCl_5 have trigonal bipyramidal structure and D_{3h} symmetry point group. Three chlorine atoms (Cl_1, Cl_2 and Cl_3 are called equatorial Chlorines and Cl_4 and Cl_5 are called axial chlorines. By applying C_3 rotation symmetry operation, passing through Cl_4, *P* and Cl_5 atoms replace Cl_1 by Cl_3; Cl_2 by Cl_1 and Cl_3 by Cl_2. This suggests that Cl_1, Cl_2 and Cl_3 are equivalent. D_{3h} symmetry point group possesses σ_h symmetry plane and thus replaces Cl_4 by Cl_5 and vice versa. Thus Cl_4 and Cl_5 are equivalent. Now applying any other symmetry operation it is not possible to replace equatorial positions by axial positions. In other words, PCl_5 possesses two set of chlorine atoms, *i.e.*, Cl_1, Cl_2 and Cl_3 as well as Cl_4 and Cl_5.

0.4.3. Dipole Moments

Molecules consist of nuclei (containing positive charge) and electrons (containing negative charge). Thus the positions of nuclei give rise to centre of gravity of the positive charge. Similarly the positions of electrons give rise to the centre of gravity of the negative charge. If centres of gravities of the positive and negative charges are at different positions, the molecules possess the dipole moments, whose magnitude is given by the product of the magnitude of the charge and distance between the charges. Further dipole moment is a vector quantity. For a fixed geometry (*i.e.*, non-oscillating molecules) the dipole moment is non-fluctuating property (*i.e.*, non-changing) of the molecule. Similar to total energy of the molecules, dipole moment must remain unchanged by the operation of symmetry element of the molecule. *In order for this to occur, the dipole moment vector must be coincident with each of the symmetry elements*. Thus the molecules with following symmetry elements will not have the dipole moment

(*i*) Molecules having centre of inversion do not possess the dipole moment.

(*ii*) Molecules having more than one type of C_n axes do not show dipole moment.

(*iii*) Molecules with one C_n ($n > 1$), those with one σ plane and no C_n, those with C_n and symmetry planes that include C_n axes and those that have no symmetry elements.

***Exercise* 0.8:** Find out whether H_2O, NH_3 and $[PtCl_4]^{2-}$ molecules possess permanent dipole moment.

(*i*) Water molecule consists of two polar bonds and thus will have two bond moments as vectors. The directions are depicted in the Fig. 0.15, (*i*). Data given in the Exercise 0.6 (*i*), can be used to find out whether the water possesses the dipole moment. It is clear from the data of Fig. 0.12, (*i*) that all the symmetry elements contain the *Z*-axis and *Z*-axis passing through the oxygen atom and bisecting the angle HOH. Thus from the results of the Section 0.4.3. it is clear that water molecule will have the permanent dipole moment.

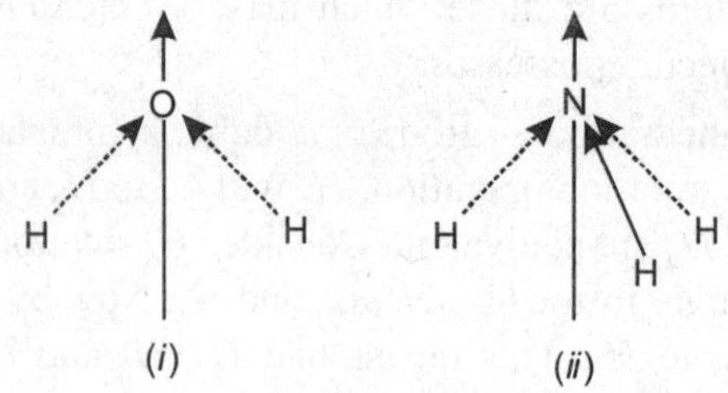

Fig. 0.15 *The dipole moments and bond moments are shown.*

(*ii*) In the similar manner as carried out in Exercise 0.6 (*i*), we can also say that all the six symmetry elements of C_{3v} symmetry point group contain *Z*-axis and thus if the dipole moment coincides with the *Z*-axis the molecule NH_3 will have the dipole moment.

(*iii*) $[PtCl_4]^{2-}$ is square planar complex. It belongs to D_{4h} symmetry point group and symmetry elements are given in Exercise 0.6. There are reasons to say that this complex has zero dipole moment. First, the complex contains centre of symmetry, secondly the molecule contains more than one C_n rotation axes ($n > 1$) and there is horizontal plane.

0.4.4. Optical Activity

It is commonly accepted that molecule is optically active if the mirror image of a molecule cannot be superimposed on the original molecule and if it can be superimposed, the molecule is optically inactive. In case of simple molecules, it is easy to find out whether the mirror image can be superimposed on the original or not, but in case of complex molecules it is not easy. Symmetry elements or symmetry operation play a major role in finding whether the molecule is optically active or not. *Any molecule that has no improper axis of rotation (i.e. S_n) is said to be dissymmetric and optically active molecules must be dissymmetric*. Instead of dissymmetric, it is often mentioned that optically active molecules should be asymmetric, which is not correct. The reason being is that *a dissymmetric molecule differs from an asymmetric one because the latter type is completely lacking symmetric*. In summary, then, we can state that if a molecule possesses only C_n symmetric element, it is dissymmetric and optically active. If $n = 1$, the molecule is asymmetric as well as dissymmetric; and if $n > 1$, the molecule is dissymmetric. If a molecule possesses S_n with any n it cannot be optically active.

***Exercise* 0.9:** Indicate whether *trans*- or *cis*-1, 2-dichlorocyclopropane is optically active?

The structures of *trans*- and *cis*-1,2-dichlorocyclopropane as well as their mirror images are shown in Fig. 0.16. It is clear from the figure that mirror image of *cis*-form is superimposable over its original one, whereas the mirror image of *trans*-form is not superimposable over its original.

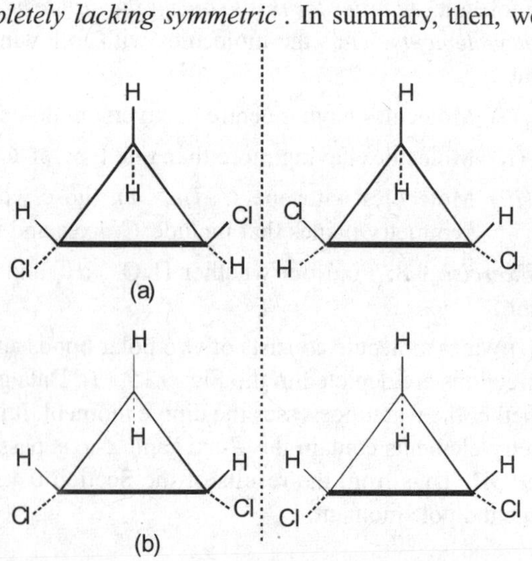

Fig. 0.16 *Trans–(a) and cis–1,2–dichlorocyclopropane (b) and their mirror images.*

In other words, *trans*-form is optically active whereas *cis*-form is optically inactive. Further *trans*-form also possesses C_2 axis of rotation and thus this form is not asymmetric but dissymmetric.

0.5. CONCEPT OF SYMMETRY POINT GROUP

Group theory is a topic in abstract algebra that can be applied to certain systems if they meet specific requirements. *Mathematicians define a group as a set of elements (i.e., rational numbers between 0 and ∞) in which any two group elements are associated with the same group*. But in case of group theory the members of the group are the collections of symmetry operations which are intrinsic to the given molecule constitute its symmetry point group. It is called the *point group* because whatever symmetry operations we perform on the given molecule at least one atom remains stationary. Symmetry point group is a mathematical group in the sense that its symmetry operations obey the rules of a mathematical group given below.

0.5.1. *Identity Number*

A group must possess an element (E) such that

$$AE = EA = A \qquad ...(0.1)$$

where A is an element in the group. According to this rule, *one element in the group must commute with all other elements of the group and leave them unchanged*.

0.5.2. *Laws of Combination*

The symmetry elements must obey the combinative law, *i.e.*, the product C of the multiplication of any two elements say A and B of the group is also an element of the group. This group property is known as *closure*. If the law of combination is multiplication, then the group is said to be closed with respect to multiplication.

$$AB = C \qquad ...(0.2)$$

If $A\ B = B\ A$, the elements A and B are said to commute with each other. Elements of C_{2v} type groups commute with each other. *A group in which all the members commute is called an Abellian or commutative group*. It is possible that the group can be non-abellian (although some of its elements may commute), *i.e.*, product of symmetry operations does not always commute, *i.e.*, $A\ B \neq B\ A$. In other words, the order in which symmetry operation is applied is very important. The common procedure followed is that application of symmetry operations is starting from right and proceed towards left, symmetry operation B will be applied first and then followed by symmetry operation A. Non-commuting group represents molecules which have at least one C_n $(n > 2)$ and which show degenerate vibrations.

0.5.3. *Law of Associative Combination*

The symmetry elements of the point group obey the law of associative multiplication.

$$A\ B\ (C) = (A\ B)\ C \qquad ...(0.3)$$

In this case the operation performed first is written to the right.

0.5.4. *Inverse Number*

Every element A in the group has inverse in the group such that

$$A\ A^{-1} = A^{-1}\ A = E, \text{ where } A^{-1} \text{ is inverse of element } A. \qquad ...(0.4)$$

An important theorem, which we will be encountering later on, can be proved before proceeding further, *i.e.*, *the reciprocal of a product of two or more than two elements is equal to the product of the reciprocals in the reverse order*, *i.e.*,

$$(A\,B\,C.........Y\,Z)^{-1} = Z^{-1}\,Y^{-1}.........C^{-1}\,B^{-1}\,A^{-1} \quad ...(0.5)$$

***Exercise* 0.10:** Prove the equation (0.5).

Let us start with product of three elements, *i.e.*, *X*, *Y* and *Z*. The theorem states that if *X*, *Y* and Z are the three elements of a group, then the product of these elements will also be an element (*W*) of the group, viz.

$$X\,Y\,Z = W \quad ...(0.6)$$

Multiplying each side of the equation (4-I.6) by $Z^{-1}\,Y^{-1}\,X^{-1}$ from the left hand side, we get

$$Z^{-1}\,Y^{-1}\,X^{-1}\,X\,Y\,Z = Z^{-1}\,Y^{-1}\,X^{-1}\,W$$

$$Z^{-1}\,Y^{-1}\,E\,Y\,Z = Z^{-1}\,Y^{-1}\,X^{-1}\,W$$

$$Z^{-1}\,Y^{-1}\,Y\,Z = Z^{-1}\,Y^{-1}\,X^{-1}\,W$$

$$Z^{-1}\,E\,Z = Z^{-1}\,Y^{-1}\,X^{-1}\,W$$

$$E = Z^{-1}\,Y^{-1}\,X^{-1}\,W$$

Since $Z^{-1}\,Y^{-1}\,X^{-1}$ times *W* is equal to *E*, it means that *W* is a reciprocal of $Z^{-1}\,Y^{-1}\,X^{-1}$ and *W* should be equal to *X Y Z* as defined in the beginning. Thus

$$W^{-1} = (X\,Y\,Z)^{-1} = Z^{-1}\,Y^{-1}\,X^{-1} \quad ...(0.7)$$

Although we have carried out the proof for the product of three elements of a group, it can be extended to the product of any number of elements of a group. This law is also called the law of *reciprocality* with itself.

***Exercise* 0.11:** Determine the order of D_{4h} symmetry point group.

According to the symmetry point group tables, given in Appendix-1, the symmetry elements present in D_{4h} symmetry point group are 12. However C_4 symmetry elements generate C_4^2, C_4^3 symmetry operations besides C_4^1 and similarly S_4 also generates S_4^3 symmetry element besides S_4^1. Thus, the total symmetry operations will be 16. Thus the order of the D_{4h} symmetry point groups is 16.

0.6. GROUP MULTIPLICATION TABLES

If we complete non-redundant list of all the elements in a finite group, and we know all the products, then the group is completely defined. This information can be summarised with a *group multiplication table*. The number of elements in the group is referred as the *order of the group*, ***h***. The group of multiplication table will consist of ***h*** rows and ***h*** columns.

***Exercise* 0.12:** Prove the multiplication table of C_{2v} symmetry point group given below:

C_{2v}	E	C_2	σ_v	σ'_v
E	E	C_2	σ_v	σ'_v
C_2	C_2	E	σ'_v	σ_v
σ'_v	σ_v	σ'_v	E	C_2
σ'_v	σ'_v	σ_v	C_2	E

Let us start with the table having only the symmetry elements, *i.e.*,

C_{2v}	E	C_2	σ_v	σ'_v
E				
C_2				
σ_v				
σ'_v				

As mentioned above the elements of the multiplication tabel can be obtained by writing row x column, *i.e.*, *RC* (column operation will be applied first followed by row operation). Since we know that identity operation commute with every operation of the point group, the above table can be written as, *i.e.*, $E \times E = E$; $E \times C_2 = = C_2$ and so on.

C_{2v}	E	C_2	σ_v	σ'_v
E	E	C_2	σ_v	σ'_v
C_2	C_2			
σ_v	σ_v			
σ'_v	σ'_v			

There are two ways to complete the above table. These involve defining either $C_2 \times C_2 = E$ or $\sigma_v \times \sigma_v = E$ or $\sigma'_v \times \sigma'_v = E$. If $C_2 \times C_2 = E$, then according to rule 0.5.4, C_2 is its own reciprocal and so other two relations, *i.e.*, $\sigma_v \times \sigma_v = E$ or $\sigma'_v \times \sigma'_v = E$ (C_2 cannot be reciprocal to C_2 and σ_v or σ'_v unless all the three operations are the same, *i.e.*, σ_v is reciprocal to σ_v and σ'_v is reciprocal to σ'_v). Thus the above table can be written as

C_{2v}	E	C_2	σ_v	σ'_v
E	E	C_2	σ_v	σ'_v
C_2	C_2	E		
σ_v	σ_v		E	
σ_v	σ'_v			E

Thus each row requires two more elements to complete the table. Now if $C_2 \times \sigma_v = C_2$ or σ_v, we will not be able to complete the table without repeating the element in a row or column. For example if $C_2 \times \sigma_v = C_2$, this is being repeated in the row and if $C_2 \times \sigma_v = \sigma_v$, it is being repeated in column 3. Thus the combination of C_2 and σ_v will generate $= \sigma'_v$. Thus the last element of the row 2 will be σ_v, *i.e.*, $C_2 \times \sigma'_v = \sigma_v$. Thus other elements can be generated in the similar manner, knowing that in C_2 symmetry point group each symmetry element commute with each other. So the multiplication table is the same as given.

C_{2v}	E	C_2	σ_v	σ'_v
E	E	C	σ_v	σ'_v
C_2	C_2	E	σ'_v	σ_v
σ_v	σ_v	σ'_v	E	C_2
σ_v	σ'_v	σ_v	C_2	E

***Exercise* 0.13:** Generate the multiplication table of C_{3v} symmetry point group.

C_{3v}	E	C_1^3	C_3^2	σ_{v1}	σ_{v2}	σ_{v2}
E						
C_1^3						
C_3^2						
σ_{v1}						
σ_{v2}						
σ_{v3}						

Using the result of Exercise 0.12 that symmetry element E commutes with each symmetry element of the point group, we can write the partial table as

C_{3v}	E	C_3^1	C_3^2	σ_{v1}	σ_{v2}	σ_{v3}
E	E	C_3^1	C_3^2	σ_{v1}	σ_{v2}	σ_{v3}
C_3^1	C_3^1					
C_3^2	C_3^2					
σ_{v1}	σ_{v1}					
σ_{v2}	σ_{v2}					
σ_{v3}	σ_{v3}					

The multiplication table is little more complicated as compared to C_{2v} because as we have learnt in Exercise 0.5, the symmetry operations of C_{3v} point group do not commute with each other. So have to do step by step. For Example, $C_3^1 \times C_3^1$ will be same as we carry out the step twice, *i.e.*, C_3^2. Similarly $C_3^1 \times C_3^2 = C_3^3 = E$. Similarly $C_3^2 \times C_3^1 = C_3^3 = E. C_3^2 \times C_3^2 = C_3^4 = E \times C_3^1 = C_3^1$. So let us fill these results to the above table and we get

C_{3v}	E	C_3^1	C_3^2	σ_{v1}	σ_{v2}	σ_{v3}
E	E	C_3^1	C_3^2	σ_{v1}	σ_{v2}	σ_{v3}
C_3^1	C_3^1	C_3^2	E			
C_3^2	C_3^2	E	C_3^1			
σ_{v1}	σ_{v1}					
σ_{v2}	σ_{v2}					
σ_{v3}	σ_{v3}					

Similarly proceeding with symmetry planes and knowing $\sigma^2 = \sigma \times \sigma = E$, so $\sigma_{v1} \times \sigma_{v1} = E$

C_{3v}	E	C_3^1	C_3^2	σ_{v1}	σ_{v2}	σ_{v3}
E	E	C_3^1	C_3^2	σ_{v1}	σ_{v2}	σ_{v3}
C_3^1	C_3^1	C_3^2	E			

C_3^2	C_3^2	E	C_3^1			
σ_{v1}	σ_{v1}	σ_{v2}	σ_{v3}	E		
σ_{v2}	σ_{v2}	σ_{v3}	σ_{v1}		E	
σ_{v3}	σ_{v3}	σ_{v1}	σ_{v2}			E

Now product of $C_3^1 \times \sigma_{v1}$ cannot be either C_3^1 or C_3^2 or σ_{v1} as these operations will be the similar in the same row or same column. So the multiplication will lead to either σ_{v2} or σ_{v3}. Carrying out these symmetry operations will lead to σ_{v3}. The only symmetry elements left in row 2 are σ_{v1} and σ_{v2}. It cannot be σ_{v2}, so it has to be σ_{v1}.

Now using the third row, we can easily guess that $C_3^2 \times \sigma_{v1}$ will be σ_{v2}, $C_3^2 \times \sigma_{v2}$ will be σ_{v3} and $C_3^2 \times \sigma_{v3}$ will be σ_{v1}. Knowing from earlier discussion that the symmetry elements do not commute with each other, and using the earlier procedure, we can write rest of elements left.

Table 0.13: Multiplication table for C_{3v} symmetry point group

C_{3v}	E	C_3^1	C_3^2	σ_{v1}	σ_{v2}	σ_{v3}
E	E	C_3^1	C_3^2	σ_{v1}	σ_{v2}	σ_{v3}
C_3^1	C_3^1	C_3^2	E	σ_{v3}	σ_{v1}	σ_{v2}
C_3^2	C_3^2	E	C_3^1	σ_{v2}	σ_{v3}	σ_{v1}
σ_{v1}	σ_{v1}	σ_{v2}	σ_{v3}	E	C_3^1	C_3^2
σ_{v2}	σ_{v2}	σ_{v3}	σ_{v1}	C_3^2	E	C_3^1
σ_{v3}	σ_{v3}	σ_{v1}	σ_{v2}	C_3^1	C_3^2	E

0.7. SIMILARITY TRANSFORM AND CONJUGATE ELEMENTS

If A and C are two elements of a group and

$$C^{-1} A C = B \quad ...(0.8)$$

where B is also an element of the same group, then we say that B is the similarity transform of A by C. In other words, if two elements A and B are similar, (*i.e.,* the A and B are related to each other by the above relation), we say that A and B are *conjugate* to each other. The following are the characteristics of the conjugate elements

(*i*) *Every element is conjugate with itself*. That is

$$E^{-1} A E = A \quad ...(0.9)$$

where E is the identity element. In place of E, it may be some other element say C which commutes with the chosen element, A. Thus

$$C^{-1} A C = A \quad ...(0.10)$$

This can be proven as follows. Multiply both sides by A^{-1} from left hand side

Left hand side: $A^{-1} C^{-1} A C = (C A)^{-1} (A C) = E$ $\quad [(CA)^{-1} = A^{-1} C^{-1}]$

This relation can hold goods only if A and C commute with each other. Thus the element C may always be E or it can be any other symmetry element if it commutes with chosen element A.

Right hand side: $A^1 A = E$ ...(0.11)

(*ii*) If A is conjugate with B, then B is conjugate with A, *i.e.*, using similarity transform on A we get

$$A = X^{-1}BX \quad ...(0.12)$$

Similarly there will be another symmetry element in the point group to satisfy the following relation, *i.e.*,

$$B = Y^{-1}AY \quad ...(0.13)$$

Multiply equation (0.12) from left hand side by X, *i.e.*,

$$XA = XX^{-1}BX = BX \quad ...(0.14)$$

Multiply equation (0.14) from right by X^{-1}, *i.e.*, $XAX^{-1} = BXX^{-1} = B$

Thus if $Y = X^{-1}$ (and also $Y^{-1} = X$), we have

$$B = Y^{-1}AY \quad ...(0.15)$$

The similarity transform is applicable to the symmetry group. For example in C_{3v} point group, the symmetry elements are: E, $2C_2$, $3\sigma_v$. Now consider $A = C_3^1$, $C = \sigma_{v'}$. We know that C^1 is also σ_v. Thus

$$\sigma_v C_3^1 \sigma_{v1}' = \sigma_{v1}\sigma_{v3} = C_3^2$$

Taking $A = C_3^1$, we can use any operation of C_{3v} point group and see that using the similarity transform we get either C_3^2 or C_3^1, *i.e.*, C_3^1 and C_3^2 are similar. In group theory language these belong to the same class as defined below. Similarly we can also show that σ_{v1}, σ_{v2}, and σ_{v3} belong to the same class in C_{3v} symmetry point group.

0.8. CLASSES

A complete set of any elements of a group which are conjugate to one another constitutes a *class of the group*. This means that the elements of the point group can be separated into smaller groups called *classes*. There can be more than one class in a group. *The order of all classes must be integral factors of the order of the group*. From above discussion we can say that although the C_{3v} point group has 6 symmetry operations but have three classes, *i.e.*, E, $2C_3$, $3\sigma_v$.

The simple method to find out the number of elements in a class of the symmetry point group is: take one element and work out all the similarities transform using all the elements of the point group including itself, then take a second element and carry out the similar procedure, which is not one of the conjugate to the first one. Carry out till all the elements in a group has been placed in one class or another. However, it is not always necessary to carry out the whole process of similarity transformation to know the number of classes in a group. It can be also done by inspection.

***Exercise:* 0.14:** Show that the six symmetry elements of C_{3v} symmetry point group belong to only three classes.

Symmetry elements of C_{3v} are E, C_3^1, C_3^2, σ_{v1}, σ_{v2}, σ_{v3}. Now let us take each symmetry element one by one. E will always constitute a class by itself of order 1, $E^{-1} \times E \times E = E$. Similarly $C_3^2 \times E \times C_3^1$

(because C_3^2 is inverse of C_3^1) *i.e.*, $C_3^2\, C_3^1 = E$. Similar procedure can be used for other symmetry elements

Now let us try rotation operations, *i.e.*, C_3^1, and C_3^2. It can be shown below that both these symmetry elements belong to the same class.

$$E \times C_3^1 \times = C_3^1$$

$$C_3^2 \times C_3^1 \times C_3^1 = C_3^4 = E \times C_3^1 = C_3^1$$

$$C_3^1 \times C_3^1 \times C_3^2 = C_3^4 = E \times C_3^1 = C_3^1$$

$$\sigma_{v1} \times C_3^1 \times \sigma_{v1} = C_3^2$$

$$\sigma_{v2} \times C_3^1 \times \sigma_{v2} = C_3^2$$

$$\sigma_{v3} \times C_3^1 \times \sigma_{v3} = C_3^2$$

On the same lines it can be shown that all symmetry plans belong to the same class, *i.e.*

$$E \times \sigma_{v_1} \times E = \sigma_{v_1}$$

$$C_3^2 \times \sigma_{v_1} \times C_3^1 = \sigma_{v_3}$$

$$C_3^1 \times \sigma_{v_1} \times C_3^2 = \sigma_{v_2}$$

$$\sigma_{v_1} \times \sigma_{v_1} \times \sigma_{v_1} = \sigma_{v_1}$$

$$\sigma_{v_2} \times \sigma_{v_1} \times \sigma_{v_2} = \sigma_{v_3}$$

$$\sigma_{v_3} \times \sigma_{v_1} \times \sigma_{v_3} = \sigma_{v_2}$$

The other examples can be

C_{2h}: E, C_2, σ_h, i = 4 classes

D_{3h}: $E, C_3, C_3^2, C_2, C_2', C_2'', \sigma_h, S_3, S_3^5, \sigma_v, \sigma_v', \sigma_v''$ = 12 symmetry operations.

= E, $2C_3$, $3C_2$, σ_h, $2S_3$, $3\sigma_v$ = 6 classes.

0.9. REPRESENTATIONS: GEOMETRIC TRANSFORMATIONS

We have learnt in the earlier sections that whenever a symmetry operation is carried out to the molecules, the positions of the atoms in the molecule are changed although equivalent configuration is achieved. We also know that each atom in the molecule is assigned with coordinates based on the some coordinate system considering the point as the origin. Relation between the old coordinates and the new coordinates can be described with the help of matrix notation. The various symmetry operations inherent of molecules can be described in terms of matrices.

0.9.1. *The identity*

The coordinates (x, y, z) of every atom in a molecule remain unchanged when it is subjected to identity operation. Simultaneous equations relating the old and new coordinates are given as follows.

$$x' = x + 0y + 0z$$
$$y' = 0\,x + y + 0z$$
$$z' = 0\,x + 0y + z$$

This set of equations can be expressed in a matrix equation as follows

$$\begin{bmatrix} 1 & 0 & 0 \\ 0 & 1 & 0 \\ 0 & 0 & 1 \end{bmatrix} \begin{bmatrix} x \\ y \\ z \end{bmatrix} = \begin{bmatrix} x' \\ y' \\ z' \end{bmatrix} \qquad ...(0.16)$$

Thus the identity operation is described by the unit matrix, *i.e.*, $\begin{bmatrix} 1 & 0 & 0 \\ 0 & 1 & 0 \\ 0 & 0 & 1 \end{bmatrix}$

0.9.2. *Proper Rotation*

When *Z*–axis is considered as the rotation axis, the *z*-coordinate will remain unchanged by rotation about the *Z*–axis. This may be expressed in a matrix form as follows

$$\begin{bmatrix} & & 0 \\ & & 0 \\ 0 & 0 & 1 \end{bmatrix}$$

The four missing elements in the matrix can be generated by considering the rotation of AB_2 type molecule (C_{2v}) in a two dimensional space, *i.e.*, in the *XY* plane. Let the molecule AB_2 is placed in the *XY* plane of an arbitrary Cartesian system of coordinates so that the atom '*A*' lies at the origin (Fig. 0.17). For lucidity the *B* atoms are labelled and the coordinates of atom B_1 atom are denoted by (x_1, y_1) and those of B_2 by (x_2, y_2). When the molecule is rotated through an angle ö in the *XY* plane in the anticlockwise direction the coordinates of atom B_1 changes to (x_1', y_1') and those of B_2 to (x_2', y_2'). The relationship between the old and the new coordinates is deduced as follows

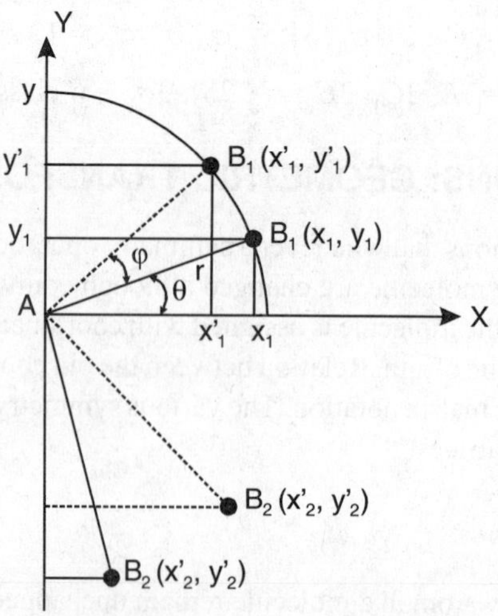

Fig. 0.17 *Rotation operation on AB_2 type molecule in XY plane in the anticlockwise direction.*

The components of distance r (AB_1) at point (x_1, y_1)

$$x_1 = r \cos \theta$$
$$y_1 = r \sin \theta$$

and those at point (x'_1, y'_1)

$$x'_1 = r \cos (\theta + \varphi)$$
$$y'_1 = r \sin (\theta + \varphi)$$

or
$$x'_1 = r \cos \theta \cos \varphi - \sin \theta \sin \varphi$$
$$y'_1 = r \sin\theta \cos \varphi + r \cos \theta \sin \varphi$$

or
$$x'_1 = x_1 \cos \varphi - y_1 \sin \varphi$$
$$y'_1 = x_1 \sin \varphi + y_1 \cos \varphi$$

or simply removing the subscripts

$$x' = x \cos \varphi - y \sin \varphi$$
$$y' = x \sin \varphi + y \cos \varphi$$

The transformation equation can be written in a matrix form equation as follows

$$\begin{bmatrix} \cos\varphi & -\sin\varphi \\ \sin\varphi & \cos\varphi \end{bmatrix} \begin{bmatrix} x \\ y \end{bmatrix} = \begin{bmatrix} x' \\ y' \end{bmatrix} \quad ...(0.17)$$

Thus the rotation is described by a matrix $\begin{bmatrix} \cos\varphi & -\sin\varphi \\ \sin\varphi & \cos\varphi \end{bmatrix}$. This result is of an anticlockwise rotation. Since cos φ = cos (–φ) and sin φ = – sin (– φ), the matrix for a clockwise rotation of a molecule through an angle φ in the *XY* plane is

$$\begin{bmatrix} \cos\varphi & \sin\varphi \\ -\sin\varphi & \cos\varphi \end{bmatrix}$$

and the total matrix equation for the clockwise rotation through an angle φ about the *Z*-axis is

$$\begin{bmatrix} \cos\varphi & \sin\varphi & 0 \\ -\sin\varphi & \cos\varphi & 0 \\ 0 & 0 & 1 \end{bmatrix} \begin{bmatrix} x \\ y \\ z \end{bmatrix} = \begin{bmatrix} x' \\ y' \\ z' \end{bmatrix}$$

Thus the matrix notation for clockwise rotation of a molecule through an angle φ about the *Z*-axis is

$$\begin{bmatrix} \cos\varphi & \sin\varphi & 0 \\ -\sin\varphi & \cos\varphi & 0 \\ 0 & 0 & 1 \end{bmatrix} \quad ...(0.18)$$

Similarly the matrix notation for anticlockwise rotation of a molecule through an angle φ about the *Z*–axis is

$$\begin{bmatrix} \cos\varphi & -\sin\varphi & 0 \\ \sin\varphi & \cos\varphi & 0 \\ 0 & 0 & 1 \end{bmatrix} \quad ...(0.18a)$$

The value of φ depicts the magnitude of angle through which the rotation around the principal axis is carried out. The magnitude of φ can be given as 360/*n*, where *n* is the number of times we get the equivalent structures when the rotation is carried out by 360º.

0.9.3. *Improper Rotation*

The matrix notation for the improper rotation of the molecule is obtained by the matrix for the proper rotation multiplied by the transformation matrix of the reflection through the plane perpendicular to the rotation axis.

Since the effect of improper rotation through an angel φ about the *Z*–axis is identical to that of proper rotation through the same angle except that the sign of the *z*-coordinate changes because of the reflection in *XY* plane. So the matrix notation for the improper rotation of molecule through an angle φ about the *Z*–axis in a clockwise direction is:

$$\begin{bmatrix} \cos\varphi & \sin\varphi & 0 \\ -\sin\varphi & \cos\varphi & 0 \\ 0 & 0 & -1 \end{bmatrix} \quad ...(0.19)$$

and

$$\begin{bmatrix} \cos\varphi & -\sin\varphi & 0 \\ \sin\varphi & \cos\varphi & 0 \\ 0 & 0 & -1 \end{bmatrix} \quad ...(0.19a)$$

in anticlockwise direction.

0.9.4. *Reflection Through a Plane*

Let the AB_2 molecule again is placed in the *XY* plane which coincides with the symmetry plane σ_{xy} of the molecule, of an arbitrary Cartesian system of coordinates so that the atom *A* coincides with the origin (Fig. 0.18). For the sake of lucidity *B* atoms are labelled and the coordinates of B_1 are denoted by (x_1, y_1, z_1) and those of atom B_2 are denoted by (x_2, y_2, z_2). The atom *A* does not change its position when reflected in this plane of the molecule. While the atom B_1 with coordinates (x_1, y_1, z_1) switches places with atom B_2 with change in the sign of the coordinate measured perpendicular to the plane, *i.e.*, z_1 coordinate here and leaving unchanged the two coordinates, *i.e.*, x_1, y_1 whose axes define the plane. Now denoting the new coordinates of the atom B_1 by x'_1, y'_1 and z'_1 as before and $x'_1 = x_1, y'_1 = y_1, z'_1 = -z_1$, we may write the following matrix equation for the reflection of a molecule in the σ_{XY} plane.

$$x'_1 = 1\,x_1 + 0\,y_1 + 0\,z_1$$
$$y'_1 = 0\,x_1 + 1\,y_1 + 0\,z_1$$
$$z'_1 = 0\,x_1 + 0\,y_1 + (-1)\,z_1$$

or

$$\sigma_{XY} = \begin{bmatrix} 1 & 0 & 0 \\ 0 & 1 & 0 \\ 0 & 0 & -1 \end{bmatrix} \text{ or}$$

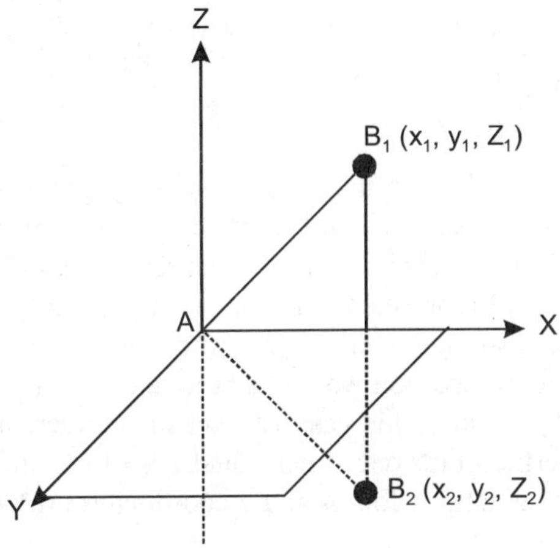

Fig. 0.18 *Reflection operation on molecule AB_2 in XY plane.*

$$\sigma_{YZ} = \begin{bmatrix} -1 & 0 & 0 \\ 0 & 1 & 0 \\ 0 & 0 & 1 \end{bmatrix} \text{ or } \sigma_{ZX} = \begin{bmatrix} 1 & 0 & 0 \\ 0 & -1 & 0 \\ 0 & 0 & 1 \end{bmatrix}$$

0.9.5. *Inversion*

Inversion operation simply changes signs of all the coordinates. Thus inversion may be denoted by a negative unit matrix, *i.e.*,

$$i = \begin{bmatrix} -1 & 0 & 0 \\ 0 & -1 & 0 \\ 0 & 0 & -1 \end{bmatrix}$$

and the matrix equation for the inversion operation as

$$\begin{bmatrix} -1 & 0 & 0 \\ 0 & -1 & 0 \\ 0 & 0 & -1 \end{bmatrix} \begin{bmatrix} x \\ y \\ z \end{bmatrix} = \begin{bmatrix} -x \\ -y \\ -z \end{bmatrix}$$

0.10. REPRESENTATION

A representation of a point group is defined as a set of matrices, each of which corresponds to a single symmetry operation in the group. By using such a method we can express the results in a very precise manner. In other words, we can use the molecular symmetry to arrive at the practical consideration such as whether or not a given chemical species is active in the infrared spectral region. In the coming section we will take the example of AB_2 molecule having C_{2v} symmetry point group.

0.10.1. *Group Representation*

Each symmetry operation of the given point group may be associated with a square matrix. For example, the C_{2v} point group consists of symmetry elements E, C_2, σ_v and σ'_v. Each of these elements has the corresponding symmetry matrix.

$$E = \begin{bmatrix} 1 & 0 & 0 \\ 0 & 1 & 0 \\ 0 & 0 & 1 \end{bmatrix}, C_2 = \begin{bmatrix} -1 & 0 & 0 \\ 0 & -1 & 0 \\ 0 & 0 & 1 \end{bmatrix}, \sigma'_v = \begin{bmatrix} 1 & 0 & 0 \\ 0 & -1 & 0 \\ 0 & 0 & 1 \end{bmatrix}, \sigma_v = \begin{bmatrix} -1 & 0 & 0 \\ 0 & 1 & 0 \\ 0 & 0 & 1 \end{bmatrix}$$

It can be noticed that each of these matrices is block diagonalised, *i.e.* the total matrix can be broken up of smaller matrices with no off-diagonal elements of the blocks. The fact that it is block diagonalised suggests that the total representation must consist of a set of so-called one dimensional representation, *i.e.*, if we were to be concerned only with operations on a point that had only x-coordinate [the column matrix $(x, 0, 0)$] then only the first row of the total representation would be required (*i.e.*, 1, –1, 1, –1). This is an irreducible representation, which in this case is a set of one-dimensional matrices describing the symmetry properties of the one dimensional x-vector in the specified group. Similarly the irreducible representations for y- and z-coordinates are (1, –1, –1, 1) and (1, 1, 1, 1) respectively.

Fig. 0.19

The total representation, the four 3 × 3 matrices is reducible representation. It is reducible into sets of 1 by 1 matrices, each set a representation by itself. *The trace or character of each of the total representation matrices is the sum of the character of each of the component irreducible representations.* In this case it is 3, –1, 1, 1. The block diagonalised 3 × 3 matrices for the effect of symmetry operations on our point resulted because the x-, y- and z-axes of our coordinate system were selected as the set of vectors for the symmetry operations, *i.e.*, z for rotation and xy for reflection. There is another representation which is not block diagonalised, *i.e.*, rotation around the R_z described by the sketch on the right. The C_2 rotation do not change the direction in which the arrow points, but the reflection by σ_{xz} σ_{yz} do. None of the irreducible representations we now have, nor any combination of these irreducible representations, describes the effect of the C_{2v} symmetry operations on this arrow representing rotation around the z-axis. The result is another irreducible representation, *i.e.*, 1, 1, –1, –1.

0.10.2. *Transformation Matrices and Group Multiplication Tables*

It can easily be seen that these matrices would reproduce the table of group multiplication for C_{2v} point group.

***Exercise* 0.15:** Using transformation matrices for different symmetry operations, show that σ_v' $\sigma_v = C_2$ for H_2O molecule.

In Exercise 0.4, we have shown the equality relation by carrying out the geometrical representation. The analogous relation is also valid by using for the matrices representation. Using the transformation matrices derived in Section 0.8, the transformation matrices for σ'_v and σ_v are

$$\sigma'_v = \begin{bmatrix} 1 & 0 & 0 \\ 0 & -1 & 0 \\ 0 & 0 & 1 \end{bmatrix} \text{ and } \sigma_v = \begin{bmatrix} -1 & 0 & 0 \\ 0 & 1 & 0 \\ 0 & 0 & 1 \end{bmatrix}$$

and substituting the respective transformation matrices in given relation, we get

$$\begin{bmatrix} 1 & 0 & 0 \\ 0 & -1 & 0 \\ 0 & 0 & 1 \end{bmatrix} \begin{bmatrix} -1 & 0 & 0 \\ 0 & 1 & 0 \\ 0 & 0 & 1 \end{bmatrix} = \begin{bmatrix} -1 & 0 & 0 \\ 0 & -1 & 0 \\ 0 & 0 & 1 \end{bmatrix}$$

which is same as derived for C_2 symmetry operation.

Thus *any collection of matrices each of which corresponds to a symmetry element in the group and obeys the laws of table of group multiplication is called the representation of the group*. The above set of matrices can be used to represent the C_{2v} point group.

Questions may come in mind about the number of representations present and the answer will be infinite number. It may be mentioned here that there are some special type of representations which can be used to represent all other representations. The former is called *irreducible representations* and the latter are called *reducible representations*.

0.10.3. *Character or Trace (or Spur in German) of Set of Irreducible Matrices*

In representing the results obtained above we do not need to use the whole matrix, but merely its character (*i.e.*, trace or spur in German) *since this is invariant under a similarity transformation*. *It is the sum of the elements of the leading diagonal*. Mathematically

$$\chi(a) = \sum_i a_{ii} \qquad ...(0.20)$$

0.10.4. *Contribution to Trace of the Representations*

The trace of the representation is defined as the sum of the elements present on the diagonal of the transformation matrix. The contributions to the diagonal elements have been from those vectors or the atoms which have not been displaced from their original positions in the molecule. Thus the following rules can be generalised.

(*i*) *If a symmetry operation R shifts an atom or one or more of its attached orthogonal vectors, then the contribution to the diagonal term is zero.*

(*ii*) *The contribution of an unshifted vector component or atom is* + 1 *and if the sign of the function representing vector or atom is reversed its contribution is* – 1.

For example, taking the *rotation axis* C_n *around the Z-axis, the contribution to the trace can be generalised as* $\chi(C_n) = [2\cos(\frac{360^o}{n}) + 1]$. For C_2 rotational axis, when $n = 2$, the $\chi(C_2) = [2\cos 180^o + 1] = 2 \times -1 + 1 = -1$.

***Exercise* 0.16:** Find the trace of the reducible representation of the NH_3 molecule. NH_3 molecule belongs to C_{3v} symmetry point group and possesses the E, $2C_3$ and $3\sigma_v$ symmetry elements. Let us apply each symmetry operations.

(1) We know that identity operation does not change the positions of each atom, so the trance will be 4 as there are four atoms in the NH_3 molecule.

(2) C_3 rotation axis passes through *N*-atom and the centre of the equilateral triangle formed by the three hydrogen atoms. Rotation whether carried out anticlockwise or clockwise changes the position of each hydrogen atom except that of *N*-atom. Thus the trace will be 1.

(3) Each symmetry reflection plane contains the *N*-atom and one of the hydrogen atom. Thus these two atoms will not change their positions but those of other two hydrogen atoms will be changed. Thus the trace will be 2.

So, the complete trace will be 4, 1, 2.

***Exercise* 0.17:** Consider two dimensional model of NH_3 molecule, *i.e.*, look the NH_3 molecule from the top of the nitrogen atom and it can be shown in below:

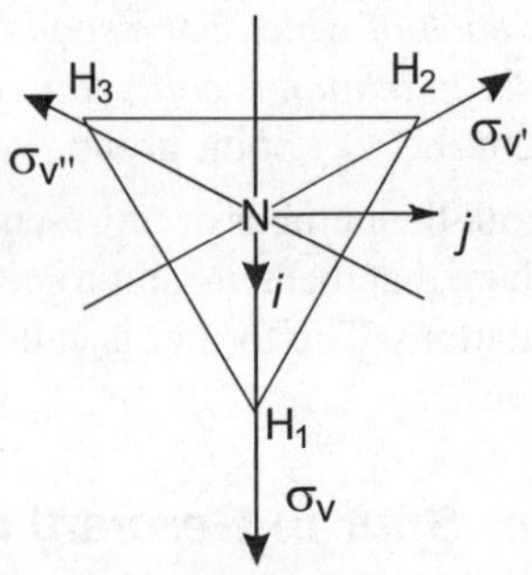

Fig. 0.20

The symmetry operations are drawn and also depicted the vector component of x- (i) and y- (j) axis. Construct the matrices representation of each symmetry operation.

We have taken the two dimensional model and thus have shown only the$x$- (i) and y- (j) components of the vectors. We have also taken x-axis combining with the σ_v symmetry plane and i vector lies on it, whereas the j component is shown perpendicular to the i component. Now considering the E symmetry operation where there is no change in position of any vector, the transformation matrix will be

$$E\begin{bmatrix} 1 & 0 \\ 0 & 1 \end{bmatrix},$$

In C_3^1. The angle of rotation, φ = 120°. Sin120 = $\frac{\sqrt{3}}{2}$, cos $120^0 = -\frac{1}{2}$, so the transformation matrix will be

$$C_3^1\begin{bmatrix} \cos\varphi & -\sin\varphi \\ \sin\varphi & \cos\varphi \end{bmatrix} = C_3^1\begin{bmatrix} \cos 120 & \sin 120 \\ -\sin 120 & \cos 120 \end{bmatrix} = \begin{bmatrix} -\frac{1}{2} & -\frac{\sqrt{3}}{2} \\ \frac{\sqrt{3}}{2} & -\frac{1}{2} \end{bmatrix}$$

In C_3^2, the angle of rotation is 240°. Sin 240 = $-\frac{\sqrt{3}}{2}$, cos 240° = $-\frac{1}{2}$ so the transformation matrix is

$$C_3^2\begin{bmatrix} \cos\varphi & -\sin\varphi \\ \sin\varphi & \cos\varphi \end{bmatrix} = C_3^2\begin{bmatrix} \cos 240 & -\sin 240 \\ \sin 240 & \cos 240 \end{bmatrix} = \begin{bmatrix} -\frac{1}{2} & \frac{\sqrt{3}}{2} \\ -\frac{\sqrt{3}}{2} & -\frac{1}{2} \end{bmatrix}$$

In σ_v symmetry of reflection, vector i makes an angle of 0 degree and vector j makes an angle of 90°. So, angle φ for i will be zero and for j it will be 180°. Thus

$$\sigma_v\begin{bmatrix} \cos\varphi & -\sin\varphi \\ \sin\varphi & \cos\varphi \end{bmatrix} = \sigma_v\begin{bmatrix} \cos 0 & -\sin 0 \\ \sin 180 & \cos 180 \end{bmatrix} = \sigma_v\begin{bmatrix} 1 & 0 \\ 0 & -1 \end{bmatrix}$$

In σ'_v symmetry of reflection, vector i makes an angle of 120° and vector j makes an angle of 30°. So, angle φ for i will be 240° and for j it will be 60°. Thus

$$\sigma_v'\begin{bmatrix}\cos\varphi & -\sin\varphi\\ \sin\varphi & \cos\varphi\end{bmatrix} = \sigma_v'\begin{bmatrix}\cos 240 & -\sin 240\\ \sin 60 & \cos 60\end{bmatrix} = \sigma_v'\begin{bmatrix}-\frac{1}{2} & \frac{\sqrt{3}}{2}\\ \frac{\sqrt{3}}{2} & \frac{1}{2}\end{bmatrix}$$

In σ''_v symmetry of reflection, vector i makes an angle of 240° and vector j makes an angle of 150°. So, angle φ for i will be 480° and for j it will be 300°. Thus

$$\sigma_v''\begin{bmatrix}\cos\varphi & -\sin\varphi\\ \sin\varphi & \cos\varphi\end{bmatrix} = \sigma_v''\begin{bmatrix}\cos 480 & -\sin 480\\ \sin 300 & \cos 300\end{bmatrix} = \sigma_v''\begin{bmatrix}-\frac{1}{2} & -\frac{\sqrt{3}}{2}\\ -\frac{\sqrt{3}}{2} & \frac{1}{2}\end{bmatrix}$$

So the transformation matrix for each symmetry element can be written as

$$E\begin{bmatrix}1 & 0\\ 0 & 1\end{bmatrix},\ C_3^1\begin{bmatrix}-\frac{1}{2} & -\frac{\sqrt{3}}{2}\\ \frac{\sqrt{3}}{2} & -\frac{1}{2}\end{bmatrix},\ C_3^2\begin{bmatrix}-\frac{1}{2} & \frac{\sqrt{3}}{2}\\ -\frac{\sqrt{3}}{2} & -\frac{1}{2}\end{bmatrix},$$

$$\sigma_v\begin{bmatrix}1 & 0\\ 0 & -1\end{bmatrix},\ \sigma_v'\begin{bmatrix}-\frac{1}{2} & \frac{\sqrt{3}}{2}\\ \frac{\sqrt{3}}{2} & \frac{1}{2}\end{bmatrix},\ \sigma_v''\begin{bmatrix}-\frac{1}{2} & -\frac{\sqrt{3}}{2}\\ -\frac{\sqrt{3}}{2} & \frac{1}{2}\end{bmatrix}.$$

From the above transformation matrices it is clear that the trace of C_3^1 and C_3^2 as well as of σ_v, σ'_v and σ''_v symmetry operations are identical, *i.e.*, –1 and 0. As mentioned earlier that trace of the symmetry operations belonging to the same class are the same, the above transformation matrices can be written as

$$E\begin{bmatrix}1 & 0\\ 0 & 1\end{bmatrix},\ 2C_3^1\begin{bmatrix}-\frac{1}{2} & -\frac{\sqrt{3}}{2}\\ \frac{\sqrt{3}}{2} & -\frac{1}{2}\end{bmatrix},\ 3\sigma_v\begin{bmatrix}1 & 0\\ 0 & -1\end{bmatrix}$$

0.11. CHARACTER TABLES

In the last section, we have empirically generated all the irreducible representations of the C_{2v} symmetry point group. All though irreducible representations for all the symmetry point groups are available and thus does not need to learn how they have been generated but it is always better to know this procedure. In this section, before we describe the procedure, we will just mention few theorems without proof.

0.11.1. *The Great Orthogonality Theorem*

Consider a group of order 'h' and let the dimensions of the ith, irreducible representation $\Gamma^{(i)}$ be l_i. Let

R represents any symmetry operation of the group and let an element in the mth row and nth column of the matrix representing R in the ith irreducible representation is denoted by $\Gamma^{(i)}(R)_{mn}$.

According to orthogonality theorem the collection of matrices constituting any given irreducible representation possess sets of elements (each matrix contributing one element to the set) which behave like the components of a vector in h dimensional space. Mathematically this is expressed as

$$\sum_R \Gamma^{(i)*}(\text{R})_{mn}\,\Gamma^{(j)}(\text{R})_{pq} = \frac{h}{\sqrt{l_i l_j}}\delta_{ij}\delta_{mp}\delta_{nq} \qquad ...(0.21)$$

where $\Gamma^{(i)*}$ represents the complex conjugate of $\Gamma^{(i)}$ and the sum extends over all the symmetry operations, each δ refers to Kronecker delta, *i.e.*, $\delta_{ij} = 0$ when $i \neq j$ and 1 when $i = j$. When all the characters are real the term $\Gamma^{(i)}$ can be used in place of its conjugate.

This theorem can be used as follows, *i.e.*, combine the matrices $\Gamma^{(i)}$ and $\Gamma^{(j)}$ and select the particular values for i, j and p, q. For the ijth term in $\Gamma^{(i)}$ and the pqth term in $\Gamma^{(j)}$, sum of the products of $\Gamma^{(i)*}(R)_{ij}\,\Gamma^{(j)}(R)_{pq}$ over all the symmetry operations R in the point group are formed and compare the result with the value of the right hand side equation (0.21).

0.11.2. *Relationship Between any Reducible Representation of a Group and the Irreducible Representations of that Group*

We know that for any reducible representation it is possible to find a similarity transformation which will reduce it to a partitioned matrix in which the matrix corresponding to the jth irreducible representation may appear a_j times. Without deriving the relation the value of a_j can be obtained from equation (0.22)

$$a_j = \frac{1}{h}\sum_R \chi(R)\,\chi^{(j)}(R) \qquad ...(0.22)$$

and

$$a_j = \frac{1}{h}\sum_{\text{classes}} \text{N}_\text{R}\,(j)\chi(R)\,\chi^{(j)}(R) \qquad ...(0.23)$$

where $N_R(j)$ is the number of symmetry operations in the class containing operator R. Thus

$$a_j = \frac{1}{h}\sum_{\text{classes}} \textit{number of operations in the class} \times \textit{character of operation of reducible representation} \times \textit{character of operation of irreducible representation.}$$

Equations (0.22) and (0.23) enable us to determine how many times the jth irreducible representation is contained in a reducible representation.

***Exercise* 0.18:** The reducible representation for the case of C_{2v} symmetry point group is found to be 2, –2, 0, 0. Find out the number and kinds of irreducible representations.

The total number of operations $h = 4$. Using the character table for C_{2v} symmetry point group and equation (0.22), we get

$$a\,(\Gamma^{(1)}) = \frac{1}{4}[1 \times 1 \times 2 + 1 \times 1 \times (-2) + 1 \times 1 \times 0 + 1 \times 1 \times 0] = 0$$

$$a(\Gamma^{(2)}) = \frac{1}{4}[1 \times 1 \times 2 + 1 \times 1 \times (-2) + 1 \times (-1) \times 0 + 1 \times (-1) \times 0] = 0$$

$$a(\Gamma^{(3)}) = \frac{1}{4}[1 \times 1 \times 2 + 1 \times (-1) \times (-2) + 1 \times 1 \times 0 + 1 \times (-1) \times 0] = 1$$

$$a(\Gamma^{(4)}) = \frac{1}{4}[1 \times 1 \times 2 + 1 \times (1) \times (-2) + 1 \times (-1) \times 0 + 1 \times (-1) \times 0] = 1$$

Thus $\Gamma = \Gamma^{(3)} + \Gamma^{(4)}$

In general we need not to write the matrices, only their characters would suffice.

0.11.3. *Mulliken Symbols*

The labelling of irreducible representations as $\Gamma^{(i)}$ is not very informative. The normal labelling employs the Mulliken symbols.

(*i*) All single dimensional representations ($l = 1$) are denoted by either the symbol A or B, the two dimensional ($l = 2$) by the symbol (E/I) and all the three dimensional ($l = 3$) by T (or F).

(*ii*) If the sign of the function Ψ (*i.e.*, sign of dimension of irreducible representation) does not change after rotation through an angle of $\frac{2\pi}{n}$ or $\frac{360}{n}$ about the principal axis C_n (*i.e.*, $\Psi \xrightarrow{C_n} \Psi$ or $\chi(C_n) = +1$), the single dimensional representation is denoted by A and if the sign changes (*i.e.*, $\Psi \xrightarrow{C_n} -\Psi$ or $\chi(C_n) = -1$), the irreducible representation is denoted by the letter B.

(*iii*) If the molecule possesses C_2 axis perpendicular to the principal axis or σ_v (and σ_d) planes, the function Ψ may either change sign after the corresponding rotations or reflections or not. If the function Ψ changes sign (*i.e.* $\Psi \xrightarrow{C_2/\sigma_v} -\Psi$ if C_2 itself is the principal axis, $\chi(C_2)$ or $\chi(\sigma_v) = 1$ as the case may be), $\Psi \xrightarrow{C_2/\sigma_v} +\Psi$ the letter A or B is supplemented with the subscript 2 (A_2 or B_2) if the function Ψ does not change sign, the letter A or B is supplemented (*i.e.* Ψ remains symmetrical), with the subscript 1 (A_1 or B_1). Similar subscripts may be seen with symbols E/I or T/F, but in that case the symbol rules are more complicated. But for our purpose it is sufficient to regard these subscripts as some labels making it possible to discern one irreducible representation from another.

(*iv*) The subscripts g and u by the irreducible symbol indicate how the function Ψ behaves in the inversion operation. When it changes sign (*i.e.*, $\Psi \xrightarrow{i} -\Psi$ subscript u for ungerade (German for odd) is used to supplement the letter A (A_u) or B (B_u). Otherwise (*i.e.*, $\Psi \xrightarrow{i} \Psi$) subscript g for gerade (German for even) is used to supplement the letter A (A_g) or B (B_g).

(*v*) The irreducible representations of the same dimension in groups such as C_s, D_{nh} and C_{nh} with odd h, are denoted by one stroke (′) over the letter symbol if the function Ψ changes sign (*i.e.*, $\Psi \rightarrow -\Psi$) or by two strokes (″) over the letter symbol if Ψ does not change sign (*i.e.*, $\Psi \rightarrow \Psi$).

(*vi*) A special system of numerical subscripts exists for the one dimensional representations of the groups D_2 and D_{2h}. In these groups there are three mutually perpendicular two-fold axes; the B representations are symmetrical with respect to rotation about only one such axis, *i.e.*, say

$\Psi \xrightarrow{C_2(Z)} \Psi$. By convention, the B_1, B_2, B_3 representations are symmetric with respect to rotation about the Z–, –Y and X–axes respectively.

(*vii*) As a rule different notation is used for the continuous group $D_{\infty h}$ and $C_{\infty h}$. The notations used are indicated below.

Dimension of irreducible representation	One	Two
Symbol	Σ	Π, Δ, Φ, etc.

Besides the symbols are supplemented with a superscript + or –. If the function Ψ does not change sign on reflection in σ_v planes, the symbol is supplemented by superscript + and if it changes sign on reflection through σ_v planes, the symbol is supplemented by the superscript –. The number of such planes in these groups is infinite. The irreducible symbols of the $D_{\infty h}$ group also have the parity subscripts g and u.

In scientific publications the same symbols are sometimes used (though not good enough but still used) both for these groups and for the finite groups. The relationship between two notations– "*strict*" and "*nonstrict*" is presented below for the $D_{\infty h}$ group

$$\sum\nolimits_g^+ \equiv A_{1g}; \sum\nolimits_u^+ \equiv A_{1u}; \Pi_g \equiv E_{1g} \Delta_g \equiv E_{2g}$$

$$\sum\nolimits_g^- \equiv A_{2g}; \sum\nolimits_u^- \equiv A_{2u}; \Pi_u \equiv E_{1u} \Delta_u \equiv E_{2u}$$

0.11.4. *Rules for Constructing the Character Tables*

We have summarised the rules as follows.

Rule 1: Number of irreducible representations (k) in a point group is equal to the number of classes in the point group, *i.e.*,

$$k = \text{Number of classes} \qquad ...(0.24)$$

There will be one irreducible representation in any group whose elements will be 1.

Rule 2: Sum of the squares of the dimensions of the irreducible representations of a point group is equal to the order of the point group, *i.e.*,

$$\sum_{i=1}^{k} l_i^2 = h \qquad ...(0.25)$$

Rule 3: Sum of the squares of the characters of identity matrix (E) of the irreducible representations is equal to the order of the point group, *i.e.*,

$$\sum_{i=1}^{k} \left|\chi^{(i)}(E)\right|^2 = h \qquad ...(0.26)$$

Rule 4: Any two rows of a character table of a point group, summed over the number of symmetry elements form a orthonormal set, *i.e.*,

$$\sum_R \chi^i(R)\, \chi^j(R) = h\, \delta_{ij} \quad (i \neq j) \qquad ...(0.27)$$

or $$\sum_{\text{classes}} N_R \chi^i(R)\, \chi^j(R) = h\, \delta_{ij} \qquad ...(0.27a)$$

Rule 5: The sum of the squares of characters in any irreducible representation is equal to the order of the group, *i.e.*,

$$\sum_R [\chi^i(R)]^2 = h \qquad ...(0.28)$$

or $$\sum_{\text{classes}} N_R [\chi^i(R)]^2 = h$$

Rule 6: Sum of the squares of the characters for any operation follows the rule

$$\sum_i [\chi^i(R)_m]^2 = \frac{h}{N_R} \qquad ...(0.29)$$

i.e., columns of the character table obey this rule.

Rule 7: First row of characters (an A, A', A_g, A_1, A_{1g} or A'_1, irreducible representation) in the character table of a point group are each equal to +1, *i.e., each character table must contain at least one irreducible representation which should be totally symmetric* .

Rule 8: Dimensions of the reducible representation are equal to the sum of the dimensions of the total number of irreducible representations it comprises of, *i.e.*,

$$l_{\text{red}} = \sum_{i=1}^{k} l_i \qquad ...(0.30)$$

In the following sections, we will construct the character table of the few point groups.

***Exercise* 0.19:** Using the above rules construct the character table for the Point Group C_{3v}

The multiplication table for the point group C_{3v} is

C_{3v}	E	C_3^1	C_3^2	σ_{v1}	σ_{v2}	σ_{v3}
E	E	C_3^1	C_3^2	σ_{v1}	σ_{v2}	σ_{v3}
C_3^1	C_3^1	C_3^2	E	σ_{v3}	σ_{v1}	σ_{v2}
C_3^2	C_3^2	E	C_3^1	σ_{v2}	σ_{v3}	σ_{v1}
σ_{v1}	σ_{v1}	σ_{v2}	σ_{v3}	E	C_3^1	C_3^2
σ_{v2}	σ_{v2}	σ_{v3}	σ_{v1}	C_3^2	E	C_3^1
σ_{v3}	σ_{v3}	σ_{v1}	σ_{v2}	C_3^1	C_3^2	E

We will follow the procedure adopted for the point group C_{2v} to find the elements of the character table:

(1) The six symmetry operations of this group, *i.e.*, E, C_3^1, C_3^2, σ_{v1}, σ_{v2}, σ_{v3} are divided among three classes, namely E, C_3 and σ_v.

By rule (1) the number of irreducible representations is three and for the time being we can label these as $\Gamma^{(1)}$, $\Gamma^{(2)}$, $\Gamma^{(3)}$. The table can now be written as

C_{3v}	E	$2C_3$	$3\sigma_v$
$\Gamma^{(1)}$			
$\Gamma^{(2)}$			
$\Gamma^{(3)}$			

(2) By rule (3), we write

$$l_1^2 + l_2^2 + l_3^2 = 6$$

Since $l_1 = 1$, so by inspection $l_2 = 1$ and $l_3 = 2$. Thus, the dimensions of three irreducible representations are 1, 1, 2. The elements of each irreducible representation corresponding to E will be positive as the symmetry operation E leaves each atom remains unchanged. There will always be one irreducible representation in any group whose characters are 1. Thus the elements of $\Gamma^{(1)}$ will be 1, 1, 1.

The information gathered is put in the table and the partially filled table is

C_{3v}	E	$2C_3$	$3\sigma_v$
$\Gamma^{(1)}$	1	1	1
$\Gamma^{(2)}$	1		
$\Gamma^{(3)}$	2		

Rule (5), *i.e.*, $\sum_R N_R[\chi^{(i)}(R)]^2 = h$ is obeyed by $\Gamma^{(1)}$, *i.e.*, $1 \times (1)^2 + 2 \times (1)^2 + 3 \times (1)^2 = 6$.

(3) In order to find the characters of the representation $\Gamma^{(2)}$, use is made of rule (4) or orthogonal vectors for the two different irreducible repres entations, *i.e.*, $\Gamma^{(1)}$ and $\Gamma^{(2)}$.

Accordingly we write

$1 \times 1 \times 1 + 2 \times 1 \times \chi^{(2)}(C_3) + 3 \times 1 \times \chi^{(2)}(\sigma_v) = 0$

or $1 + 2\chi^{(2)}(C_3) + 3\chi^{(2)}(\sigma_v) = 0$

By inspection; $\chi^{(2)}(C_3) = 1$ and $\chi^{(2)}(\sigma_v) = -1$. Thus the elements of $\Gamma^{(2)}$ are 1, 1, –1. The partially filled character table now appears as

C_{3v}	E	$2C_3$	$3\sigma_v$
$\Gamma^{(1)}$	1	1	1
$\Gamma^{(2)}$	1	1	–1
$\Gamma^{(3)}$	2		

(4) The remaining two elements of ro w $\Gamma^{(3)}$ can be determined by applying orthogonality rule to rows $\Gamma^{(1)}$ and $\Gamma^{(3)}$ and to rows $\Gamma^{(2)}$ and $\Gamma^{(3)}$ (Rule 4), *i.e.*,

$\Gamma^{(1)}\ \Gamma^{(3)}: 1 \times 1 \times 2 + 2 \times 1 \times \chi^{(3)}(C_3) + 3 \times (+1) \times \chi^{(3)}(\sigma_v) = 0$

$\Gamma^{(2)}\ \Gamma^{(3)}: 1 \times 1 \times 2 + 2 \times 1 \times \chi^{(3)}(C_3) + 3 \times (1) \times \chi^{(3)}(\sigma_v) = 0$

or

$\Gamma^{(1)}\ \Gamma^{(3)}: 2 + 2\chi^{(3)}(C_3) + 3\chi^{(3)}(\sigma_v) = 0$ ...(0.31)

$\Gamma^{(2)}\ \Gamma^{(3)}$: $2 + 2\ \chi^{(3)}\ (C_3)$ " $3\ \chi^{(3)}\ (\sigma_v) = 0$ ...(0.32)

Solving equations (0.31) and (0.32) for $\chi^{(3)}\ (C_3)$ and $\chi^{(3)}\ (\sigma_v)$ we obtain

$$\chi^{(3)}\ (C_3) = 1$$

$$\chi^{(3)}\ (\sigma_v) = 0$$

Thus the elements of the irreducible representation $\Gamma^{(3)}$ are 2, –1, 0 and the complete character table is

C_{3v}	E	$2C_3$	$3\ \sigma_v$
$\Gamma^{(1)}$	1	1	1
$\Gamma^{(2)}$	1	1	–1
$\Gamma^{(3)}$	2	–1	0

The correctness of characters of $\Gamma^{(3)}$ can be judged by rule (5), *i.e.*, square of the length of the vector defined by $\Gamma^{(3)}$ should be equal to h (= 6) and has been found to be so.

$1 \times (2)^2 + 2 \times (-1)^2 + 3 \times (0)^2 = 4 + 2 + 0 = 6.$

5. Since the representations $\Gamma^{(1)}$ and $\Gamma^{(2)}$ are single dimensional so they either be denoted by symbol A or symbol B. Since the representations $\Gamma^{(1)}$ and $\Gamma^{(2)}$ are symmetric with respect to the operation $\chi(C_3)$, *i.e.*, rotation by $\frac{2\pi}{3}$ about the principal axis C_3, so both representations are denoted by letter A. Since the dimension of representation $\Gamma^{(3)}$ is two, so it is denoted letter E. These transformations can be summarized as follows:

C_{3v}	E	**Symmetry**	**Mulliken Symbol**
$\Gamma^{(1)}$	$1 \xrightarrow{C_3} 1$	Does not change	A
$\Gamma^{(2)}$	$1 \xrightarrow{C_3} 1$	Does not change	A
$\Gamma^{(3)}$	$2 \xrightarrow{C_3} 2$	Does not change	E_2 or simply E

Further in order to differentiate between A's operator σ_v is applied on the column one, *i.e.*,

C_{3v}	E	**Symmetry**	**Mulliken Symbol**
A	$1 \xrightarrow{\sigma_v} 1$	Does not change	A_1
A	$1 \xrightarrow{\sigma_v} -1$	Change	A_2

So the final character table for the point group C_{3v} in terms of Mulliken symbols is

C_{3v}	**1**	**2C3**	$3\ \sigma_v$
A_1	1	1	1
A_2	1	1	–1
E	2	–1	0

An expanded form of C_{3v} point group is

C_{3v}	E	C_3	C_3^2	σ_{v_1}	σ_{v_2}	σ_{v_3}
A_1	1	1	1	1	1	1
A_2	1	1	1	–1	–1	–1
E	2	–1	–1	0	0	0

***Exercise* 0.20:** Construct the character table for the symmetry point group *Td*.

The multiplication table for the point group T_d is given below:

T_d	E	$(C_3^1, C_3^2, ...C_3^8)$	(C_2^1, C_2^2, C_2^3)	$(S_4^1 ....S_4^6)$	$(\sigma_d^1 ....\sigma_d^6)$
E					
C_3^1					
C_3^2					
.					
.					
C_3^8					
C_2^1					
C_2^2					
C_2^3					
S_4^1					
S_4^2					
.					
.					
S_4^6					
σ_d^1					
σ_d^2					
.					
σ_d^6					

In order to determine the elements of the character table we proceed as follows:

1. The twenty-four symmetry operations of this group have been divided into *five classes*, viz. *E*, C_3, C_2, S_4 and σ_d. By rule (1) the number of irreducible representations is '*five*' and for the time being these are labelled as $\Gamma^{(1)}$, $\Gamma^{(2)}$, $\Gamma^{(3)}$, $\Gamma^{(4)}$, $\Gamma^{(5)}$. The table can be now written as

T_d	E	8 C_3	3 C_2	6 S_4	6 σ_d
$\Gamma^{(1)}$					
$\Gamma^{(2)}$					
$\Gamma^{(3)}$					
$\Gamma^{(4)}$					
$\Gamma^{(5)}$					

2 By rule (3) we write:

$$l_1^2 + l_2^2 + l_3^2 + l_4^2 + l_5^2 = 24$$

Since $l_1 = 1$, so by inspection $l_2 = 1$, $l_3 = 2$, $l_4 = l_5 = 3$. We also know that there will be one irreducible representation in any group (Rule 1) whose elements will be 1, *i.e.*, the elements of $\Gamma^{(1)}$ are 1, 1, 1, 1, 1. The information gathered is stored in the tabular form is as follows:

T_d	E	8 C_3	3 C_2	6 S_4	6 σ_d
$\Gamma^{(1)}$	1	1	1	1	1
$\Gamma^{(2)}$	1				
$\Gamma^{(3)}$	2				
$\Gamma^{(4)}$	3				
$\Gamma^{(5)}$	3				

Rule (5), *i.e.*, $\sum_R N_R[\chi^{(i)}(R)]^2 = h$ is obeyed by $\Gamma^{(1)}$, *i.e.*,

$$1 \times (1)^2 + 8 \times (1)^2 + 3 \times (1)^2 + 6 \times (1)^2 + 6 \times (1)^2 = 24.$$

3. In order to find the elements of $\Gamma^{(2)}$, rule (4) of orthogonality of vectors for two different irreducible representations is applied, *i.e.*,

$$\sum_R \chi^{(1)}(R)\chi^{(2)}(R) = 0 \text{ or } \sum_{\text{classes}} N_R\chi^{(1)}(R)\chi^{(2)}(R) = 0$$

$$1 \times 1 \times 1 + 8 \times 1 \times \chi^{(2)}(C_3) + 3 \times 1 \times \chi^{(2)}(C_2) + 6 \times 1 \times \chi^{(2)}(S_4) + 6 \times 1 \times \chi^{(2)}(\sigma_d) = 0$$

$$1 + 8\,\chi^{(2)}(C_3) + 3\,\chi^{(2)}(C_2) + 6\,\chi^{(2)}(S_4) + 6\,\chi^{(2)}(\sigma_d) = 0$$

By inspection $\chi^{(2)}(C_3) = 1$, $\chi^{(2)}(C_2) = 1$, $\chi^{(2)}(S_4) = -1$, $\chi^{(2)}(\sigma_d) = -1$

Thus the elements of the $\Gamma^{(2)}$ row are 1, 1, 1, –1, –1 and the partially filled table is

T_d	E	8 C_3	3 C_2	6 S_4	6 σ_d
$\Gamma^{(1)}$	1	1	1	1	1
$\Gamma^{(2)}$	1	1	1	–1	–1
$\Gamma^{(3)}$	2				
$\Gamma^{(4)}$	3				
$\Gamma^{(5)}$	3				

Rule (5) is obeyed by $\Gamma^{(2)}$ also, *i.e.*,

$$1 \times (1)^2 + 8 \times (1)^2 + 3 \times (1)^2 + 6 \times (1)^2 + 6 \times (1)^2 = 24.$$

4. Apply the orthogonality to $\Gamma^{(1)}$, $\Gamma^{(3)}$ and $\Gamma^{(2)}$, $\Gamma^{(3)}$, *i.e.*,

$$\Gamma^{(1)} - \Gamma^{(3)} : 1 \times 1 \times 2 + 8 \times 1 \times \chi^{(3)}(C_3) + 3 \times 1 \times \chi^{(3)}(C_2) + 6 \times 1 \times \chi^{(3)}(S_4) + 6 \times 1 \times \chi^{(3)}(\sigma_d) = 0$$

$$\Gamma^{(2)} - \Gamma^{(3)} : 1 \times 1 \times 2 + 8 \times 1 \times \chi^{(3)}(C_3) + 3 \times 1 \times \chi^{(3)}(C_2) + 6 \times (1) \times \chi^{(3)}(S_4) + 6 \times (1) \times \chi^{(3)}(\sigma_d) = 0$$

or

$$2 + 8\,\chi^{(3)}(C_3) + 3\,\chi^{(3)}(C_2) + 6\,\chi^{(3)}(S_4) + 6\,\chi^{(3)}(\sigma_d) = 0 \qquad ...(0.33)$$

$$2 + 8\,\chi^{(3)}(C_3) + 3\,\chi^{(3)}(C_2) - 6\,\chi^{(3)}(S_4) - 6\,\chi^{(3)}(\sigma_d) = 0 \qquad ...(0.34)$$

Subtracting equation (0.34) from equation (0.33), we get

$$12\,\chi^{(3)}(S_4) + 12\,\chi^{(3)}(\sigma_d) = 0$$

$$\chi^{(3)}(S_4) = -\chi^{(3)}(\sigma_d) \qquad ...(A)$$

and adding equations (0.33) and (0.34), we get

$$4 + 16\,\chi^{(3)}(C_3) + 6\,\chi^{(3)}(C_2) = 0$$

By inspection we can assign $\chi^{(3)}(C_3) = -1$ and $\chi^{(3)}(C_2) = 2$

Further we know that $\sum_R N_R[\chi^{(i)}(R)]^2 = 24$ and substituting the values of $\chi^{(3)}(R_i)$ and $\chi^{(3)}(S_4) = -\chi^{(3)}(\sigma_d)$, we get

$$1 \times 2^2 + 8\,[-1]^2 + 3\,[2]^2 + 6[\chi^{(3)}(S_4)]^2 + 6[-\chi^{(3)}(S_4)]^2 = 24$$

$$4 + 8 + 12 + 12\,[\chi^{(3)}(S_4)]^2 = 24$$

or $\chi^{(3)}(S_4)] = \chi^{(3)}(S_4) = 0$

Consequently the elements of row three are 2, –1, 2, 0, 0 and the partially filled table is

T_d	E	8 C_3	3 C_2	6 S_4	6 σ_d
$\Gamma^{(1)}$	1	1	1	1	1
$\Gamma^{(2)}$	1	1	1	–1	1
$\Gamma^{(3)}$	2	–1	2	0	0
$\Gamma^{(4)}$	3				
$\Gamma^{(5)}$	3				

4. Applying rule (6) to columns 2, 3, 4 and 5, we will get four relations of type

$$\sum_{i=1}^{5} [\chi(i)(R)]^2 = \frac{h}{N_R}$$

Since columns 2, 3, 4 and 5 refer to classes C_3, C_2, S_4 and σ_d which contains 8. 3, 6, and 6 operations, $\frac{h}{N_R}$ will be 3, 8, 4 and 4 respectively. Apply this rule to column 2, we get

$$[\Gamma^{(1)}(C_3)]^2 + [\Gamma^{(2)}(C_3)]^2 + [\Gamma^{(3)}(C_3)]^2 + [\Gamma^{(4)}(C_3)]^2 + [\Gamma^{(5)}(C_3)]^2 = 3$$

or $[1]^2 + [1]^2 + [-1]^2 + [\Gamma^{(4)}(C_3)]^2 + [\Gamma^{(5)}(C_3)]^2 = 3$

$$[\Gamma^{(4)}(C_3)]^2 + [\Gamma^{(5)}(C_3)]^2 = 0$$

The only possible solution will be that $\Gamma^{(4)}(C_3) = [\Gamma^{(5)}(C_3) = 0$

5. Applying the orthogonal relation to $\Gamma^{(1)}\Gamma^{(4)}$ and $\Gamma^{(3)}\Gamma^{(4)}$, we write

$\Gamma^{(1)}–\Gamma^{(4)}: 1 \times 1 \times 3 + 8 \times 1 \times 0 + 3 \times 1 \times \chi^{(4)}(C_2) + 6 \times 1\ \chi^{(4)}(S_4) + 6 \times 1 \times \chi^{(4)}(\sigma_d) = 0$

$\Gamma^{(3)}–\Gamma^{(4)}: 1 \times 2 \times 3 + 8 \times (-1) \times 0 + 3 \times 2 \times \chi^{(4)}(C_2) + 6 \times 0\ \chi^{(4)}(S_4) + 6 \times 0 \times \chi^{(4)}(\sigma_d) = 0$

or

$$3 + 3\,\chi^{(4)}(C_2) + 6\,\chi^{(4)}(S_4) + 6\,\chi^{(4)}(\sigma_d) = 0 \quad ...(0.35)$$

$$6 + 6\,\chi^{(4)}(C_2) = 0 \quad ...(0.36)$$

Using equations (0.35) and (0.36) we obtain

$$\chi^{(4)}(C_2) = -1$$

and

$$3 - 3 + \chi^{(4)}(S_4) + \chi^{(4)}(\sigma_d) = 0$$

or

$$\chi^{(4)}(S_4) + \chi^{(4)}(\sigma_d) = 0$$

or

$$\chi^{(4)}(S_4) = -\chi^{(4)}(\sigma_d)$$

This in conjunction with orthogonal relation gives: $\chi^{(4)}(S_4) = 1$ and $\chi^{(4)}(\sigma_d) = -1$

Now applying rule (6)

$$[\Gamma^{(1)}(C_2)]^2 + [\Gamma^{(2)}(C_2)]^2 + [\Gamma^{(3)}(C_2)]^2 + [\Gamma^{(4)}(C_2)]^2 + [\Gamma^{(5)}(C_2)]^2 = 8$$

or $[1]^2 + [1]^2 + [2]^2 + [-1]^2 + [\Gamma^{(5)}(C_2)]^2 = 8$

$$[\Gamma^{(5)}(C_2)]^2 = 1 \text{ or } \Gamma^{(5)}(C_2) = \pm 1$$

By applying orthogonal condition to $\Gamma^{(3)}$ and $\Gamma^{(4)}$ we obtain $\chi^{(5)}(C_2) = 1$. Accordingly the partially filled table appears as

T_d	E	8 C_3	3 C_2	6 S_4	6 σ_d
$\Gamma^{(1)}$	1	1	1	1	1
$\Gamma^{(2)}$	1	1	1	–1	1
$\Gamma^{(3)}$	2	–1	2	0	0
$\Gamma^{(4)}$	3	0	–1	1	–1
$\Gamma^{(5)}$	3	0	–1		

The orthogonality relation when applied to $\Gamma^{(4)}$ and $\Gamma^{(5)}$ gives

$$\Gamma^{(4)} - \Gamma^{(5)}: 1 \times 3 \times 3 + 8 \times 0 \times 0 + 3 \times (-1)(-1) + 6 \times 1 \times \chi^{(5)}(S_4) + 6 \times (-1) \times \chi^{(5)}(\sigma_d) = 0$$

or $9 + 3 + 6\,\chi^{(5)}(S_4)\ \ 6\,\chi^{(5)}(\sigma_d) = 0$

or $\chi^{(5)}(S_4) - \chi^{(5)}(\sigma_d) = -2$

or $\chi^{(5)}(\sigma_d) - \chi^{(5)}(S_4) = -2$

or $\chi^{(5)}(\sigma_d) > \chi^{(5)}(S_4)$

$\therefore$ $\chi^{(5)}(S_4) = 1$ and $\chi^{(5)}(\sigma_d) = +1$

The complete character table for the point group T_d is

T_d	E	$8\,C_3$	$3\,C_2$	$6\,S_4$	$6\,\sigma_d$
$\Gamma^{(1)}$	1	1	1	1	1
$\Gamma^{(2)}$	1	1	1	–1	–1
$\Gamma^{(3)}$	2	–1	2	0	0
$\Gamma^{(4)}$	3	0	–1	1	–1
$\Gamma^{(5)}$	3	0	–1	–1	1

Since the irreducible representation s $\Gamma^{(1)}$ and $\Gamma^{(2)}$ are single dimensional and both are symmetric with respect to the operation C_3 but $\Gamma^{(2)}$ is asymmetric with respect to σ_d, so $\Gamma^{(1)}$ is denoted by the letter A_1, while $\Gamma^{(2)}$ is denoted by the letter A_2. $\Gamma^{(3)}$ being two dimensional is designated by the Mulliken symbol E. Both $\Gamma^{(4)}$ and $\Gamma^{(5)}$ being three dimensional are denoted by the letter F/T. $\Gamma^{(4)}$ by the symbol F_1/T_1 and $\Gamma^{(5)}$ by the symbol F_2. Thus the character table for the point group T_d in terms of Mulliken symbols is

T_d	E	$8\,C_3$	$3\,C_2$	$6\,S_4$	$6\,\sigma_d$
A_1	1	1	1	1	1
A_2	1	1	1	–1	–1
E	2	–1	2	0	0
F_1/T_1	3	0	–1	1	–1
F_2/T_2	3	0	–1	–1	1

0.11.5. *Examining the Character Table*

If we look at the Appendix A–1, we observe that there are four columns (or areas) in each character table. We have labelled only two in last two sections.

(*i*) First column gives the notations for the irreducible representations and notations have been explained in Section 0.10.3.

(*ii*) Second column gives the characters of each symmetry operations belonging to each irreducible representation.

(*iii*) Third column lists the transformation properties of vectors along the x, y and z axes and rotations R_x, R_y and R_z about the X–, Y– and Z-axes.

(*iv*) This column indicates the transformation properties of the squares and binary products of the coordinates. Although there are six possible square and binary products ($x^2, y^2, z^2, xy, yz. zx$), only five of them are generally indicated. The reason being that $x^2 + y^2 + z^2 = r^2$ and as a result, one of the linear combinations is redundant ($r^2 = x^2 + y^2 + z^2$, where r is the radius of the sphere). Further the effect of the symmetry operations can also be obtained on the functions like $x^2, y^2, z^2, xy, yz, zx$ etc. Although not given, but all the square functions transform in the same manner as z (obtained by the direct product of x and x and so on). Similarly the functions yz and zx transform in the same manner as x and y, whereas xy transform in a different manner. So it may be concluded that all the functions will transform only in four different ways under C_{2v} symmetry point group. It may be further tested by taking few examples, *i.e.* x^3, can be written as x into x^2. In other words, x^3 will transform like x and similarly x^4 will transform like x^2. Thus the four transformations A_1, A_2, B_1 and B_2 are the four irreducible transformations in the C_{2v} symmetry point group and the numbers in the column 2 gives the characters of the irreducible representations of C_{2v} symmetry point group. We also know that p orbitals are named as p_x, p_y and p_z because these are lying along these axes. Thus it may be mentioned that these three p orbitals will also transform in the same manner as the x, y and z coordinates, according to B_1, B_2 and A_1 respectively under the C_{2v} symmetry point group. The beauty of the character table is that d orbitals which Column 4 represents are the binary products of the coordinates.

It may be mentioned that the information provided in the third column of the character table will be used in discussing the p orbitals and vibrational spectra of the polyatomic molecules, whereas the information given in the fourth column is used in discussing about the d orbitals and the Raman spectra.

0.11.6. *Direct Product*

We have learnt earlier that multiplication of two or more than two symmetry operations can be achieved by multiplying the respective transformation matrices in the same order as we apply the symmetry operations. Great orthoganility theorem also tells that instead of using transformation matrices, we can use the trace of the respective matrices and results obtained are similar. Thus it is always convenient to use the trace or character of the respective symmetry operation. Similar results can be applied if we want to determine the direct product of the irreducible representations.

Two points are worth mentioning from the above results.

(*i*) *Square of the one dimensional irreducible representation will always belong to totally symmetric irreducible representation, i.e.,* A_1.

(*ii*) *In case of one dimensional irreducible representations, the direct product of totally symmetric irreducible representation and any other one dimensional irreducible representation gives back the same irreducible representation.*

(*iii*) *Square of the two dimensional or three dimensional irreducible representation will always contain at least one representation which will be totally symmetric,* A_1.

***Exercise* 0.21:** In C_{2v} symmetry point group, x, y and z transform as B_1, B_2 and A_1 irreducible representations. Show that the binary product xy, xz and x^2, y^2 and z^2 belong to A_2, B_1 and A_1 irreducible representations respectively. For ready reference the charcter table for C_{2v} is given below:

C_{2v}	E	C_2	σ_v	σ'_v		
A_1	1	1	1	1	T_z	T_{x2}, T_{y2}, T_{z2}
A_2	1	1	–1	–1	R_z	xy
B_1	1	–1	1	1	$T_{z'}, R_y$	zx
B_2	1	–1	–1	1	T_y, R_x	yz

xy can be written as: $(x) \times (y)$, *i.e.*, direct product of x and y. x belongs to B_1 irreducible representation and y belongs to B_2 irreducible representation. So the direct product of xy can be written as direct product of the characters of B_1 and B_2, *i.e.*,

$1 \times 1; -1 \times -1, 1 \times -1, -1 \times 1 = 1, 1, -1, -1$ is equivalent to A_2. It agrees with results given in the 4th column.

Similarly, $xz \equiv B_1 \times A_1 = 1 \times 1, -1 \times 1, 1 \times 1, -1 \times 1 \equiv 1, -1, 1 -1 = B_1$

$x^2 = (x) \times (x) = B_1 \times B_1 = 1 \times 1, -1 \times -1, 1 \times 1, -1 \times -1 \equiv 1, 1, 1, 1 \equiv A_1$

$y^2 = (y) \times (y) = B_2 \times B_2 = 1 \times 1, -1 \times -1, -1 \times -1, 1 \times 1 \equiv 1, 1, 1, 1 \equiv A_1$

$z^2 = (z) \times (z) = A_1 \times A_1 = 1 \times 1, 1 \times 1, 1 \times 1, 1 \times 1 \equiv 1, 1, 1, 1 \equiv A_1$

***Exercise* 0.22:** Consider the function $x^2 + y^3 + z$. Determine the irreducible representation for this function under the C_{2v} symmetry point group.

The function $x^2 + y^3 + z$ does not represent any irreducible representation. But it can be analysed as the individual elements. As mentioned above x^2 transform as A_1, y^3 will transform as y and thus B_2 and z will transform like A_1. Thus the complete function will span the irreducible representations $A_1 + B_2 + A_1$, *i.e.*, $2A_1 + B_2$.

Lastly it may be mentioned here that the construction of the character table is based on the selection of the axes. In C_{2v} symmetry point group, σ'_v has been chosen as ZX and σ_v as YZ planes and character table has been constructed. Now if we interchange the X and Y axes, X will span B_2 and y will span B_1 irreducible representation and so on. Since based on Mulliken notation, the character of B_1 under σ_v plane remains as + 1, so that to retain right hand axes the direction of Z would have to be reversed.

***Exercise* 0.23:** In the C_{2v} symmetry point group, show that (*a*) A_1 and A_2 are orthogonal, (*b*) B_1 is normalised.

(*a*) We know that the condition for orthogonality of characters of irreducible representations A_1 and A_2 is:

$$\sum_R \chi_{A_1}^{(i)}(R)\chi_{A_1}^{(j)}(R) = 0$$

So from the character table for the C_{2v} symmetry point group we write:

$[(1 \times 1) + (1 \times 1) + (1 \times -1) + (1 \times -1)] = (1 + 1 - 1 - 1) = 0$

The normalised results is also $\frac{1}{4}(0) = 0$, *i.e.*, two irreducible representations are orthogonal.

(*b*) The condition for the normalisation of character of any irreducible representation is

$$\sum_R [\chi_{A_1}^{(i)}(R)]^2 = h, \text{ i.e., } \frac{1}{h}\sum_R [\chi_{A_1}^{(i)}(R)]^2 = 1$$

Since $\frac{1}{4}[(1 \times 1) + (-1 \times -1) + (1 \times 1) + (-1 \times -1)] = \frac{1}{4}(1 + 1 + 1 + 1) = 1$

Therefore B_1 is normalised.

***Exercise* 0.24:** Consider the C_{3v} symmetry point group, show that $E \times E$ (dot product of E with E) will contain at least one totally symmetric irreducible representation.

The character of the E irreducible representation is 2, –1, 0. So

$E \times E : 2 \times 2, -1 \times -1, 0 \times 0; 4, 1, 0$

which is a reducible representation. Using equation (0.23) we can find the irreducible representations as

$$a(A_1) = \frac{1}{6}[1 \times 1 \times 4 + 2 \times 1 \times 1 + 3 \times 0 \times 0] = 1$$

$$a(A_2) = \frac{1}{6}[1 \times 1 \times 4 + 2 \times 1 \times 1 + 3 \times 1 \times 0] = 1$$

$$a(E) = \frac{1}{6}[2 \times 1 \times 4 + 2 \times 1 \times -1 + 3 \times 1 \times 0] = 1$$

Thus $E \times E = A_1 + A_2 + E$, *i.e.*, it contains at least one completely symmetric irreducible representation.

0.11.7. *Representation of an Atom*

After discussing the transformation of vector assigned to a point, followed by the bond vector in two dimensional model let us try to go little more in finer detail. Taking again the simplest example of H_2O, belonging to C_{2v} symmetry point group, assign a set of three orthogonal vectors $\boldsymbol{x}$, $\boldsymbol{y}$ and $\boldsymbol{z}$ to each atom. Thus we will have nine such variables. Let us designate them as x_O, y_O, z_O to oxygen atom and x_a, y_a, z_a and x_b, y_b, z_b to each of the two hydrogen atoms. We have chosen these set of axes for convenience and are parallel to the symmetry operations. Following the similar procedure as carried out earlier, each operation in C_{2v} point group will give 9×9 transformation matrix and thus we will get four such sets. Before we write such sets, let us try to apply couple of symmetry operations to one of the atoms of H_2O, say oxygen. Take C_2 as symmetry operations and the results obtained are

$$x_O \rightarrow x_O, \; y_O \rightarrow y_O, \; z_O \rightarrow z_O$$

where as under σ'_v as YZ–plane, we get

$$x_O \rightarrow -x_O, \; y_O \rightarrow y_O, \; z_O \rightarrow z_O$$

In similar manner as learnt earlier we can write four 9×9 transformation matrices as given below:

$$\begin{bmatrix} x'_O \\ y'_O \\ z'_O \\ x'_a \\ y'_a \\ z'_a \\ x'_b \\ y'_b \\ z'_b \end{bmatrix} = \begin{bmatrix} 1 & 0 & 0 & 0 & 0 & 0 & 0 & 0 & 0 \\ 0 & 1 & 0 & 0 & 0 & 0 & 0 & 0 & 0 \\ 0 & 0 & 1 & 0 & 0 & 0 & 0 & 0 & 0 \\ 0 & 0 & 0 & 1 & 0 & 0 & 0 & 0 & 0 \\ 0 & 0 & 0 & 0 & 1 & 0 & 0 & 0 & 0 \\ 0 & 0 & 0 & 0 & 0 & 1 & 0 & 0 & 0 \\ 0 & 0 & 0 & 0 & 0 & 0 & 1 & 0 & 0 \\ 0 & 0 & 0 & 0 & 0 & 0 & 0 & 1 & 0 \\ 0 & 0 & 0 & 0 & 0 & 0 & 0 & 0 & 1 \end{bmatrix} \begin{bmatrix} x_O \\ y_O \\ z_O \\ x_a \\ y_a \\ z_a \\ x_a \\ y_b \\ z_b \end{bmatrix} \quad \text{...(0.37)}$$

$$M(E): \chi(E) = 9$$

$$\begin{bmatrix} x'_O \\ y'_O \\ z'_O \\ x'_a \\ y'_a \\ z'_a \\ x'_b \\ y'_b \\ z'_b \end{bmatrix} = \begin{bmatrix} -1 & 0 & 0 & 0 & 0 & 0 & 0 & 0 & 0 \\ 0 & -1 & 0 & 0 & 0 & 0 & 0 & 0 & 0 \\ 0 & 0 & 1 & 0 & 0 & 0 & 0 & 0 & 0 \\ 0 & 0 & 0 & 0 & 0 & 0 & -1 & 0 & 0 \\ 0 & 0 & 0 & 0 & 0 & 0 & 0 & -1 & 0 \\ 0 & 0 & 0 & 0 & 0 & 0 & 0 & 0 & 1 \\ 0 & 0 & 0 & -1 & 0 & 0 & 0 & 0 & 0 \\ 0 & 0 & 0 & 0 & -1 & 0 & 0 & 0 & 0 \\ 0 & 0 & 0 & 0 & 0 & 1 & 0 & 0 & 0 \end{bmatrix} \begin{bmatrix} x_O \\ y_O \\ z_O \\ x_a \\ y_a \\ z_a \\ x_b \\ y_b \\ z_b \end{bmatrix} \quad \text{...(0.38)}$$

$$M(C_2): \chi(C_2) = -1$$

$$\begin{bmatrix} x'_O \\ y'_O \\ z'_O \\ x'_a \\ y'_a \\ z'_a \\ x'_b \\ y'_b \\ z'_b \end{bmatrix} = \begin{bmatrix} 1 & 0 & 0 & 0 & 0 & 0 & 0 & 0 & 0 \\ 0 & -1 & 0 & 0 & 0 & 0 & 0 & 0 & 0 \\ 0 & 0 & 1 & 0 & 0 & 0 & 0 & 0 & 0 \\ 0 & 0 & 0 & 0 & 0 & 0 & 1 & 0 & 0 \\ 0 & 0 & 0 & 0 & 0 & 0 & 0 & -1 & 0 \\ 0 & 0 & 0 & 0 & 0 & 0 & 0 & 0 & 1 \\ 0 & 0 & 0 & 1 & 0 & 0 & 0 & 0 & 0 \\ 0 & 0 & 0 & 0 & -1 & 0 & 0 & 0 & 0 \\ 0 & 0 & 0 & 0 & 0 & 1 & 0 & 0 & 0 \end{bmatrix} \begin{bmatrix} x_O \\ y_O \\ z_O \\ x_a \\ y_a \\ z_a \\ x_b \\ y_b \\ z_b \end{bmatrix} \quad \text{...(0.39)}$$

$$M(\sigma_v): \chi(\sigma_v) = 1$$

$$\begin{bmatrix} x_O' \\ y_O' \\ z_O' \\ x_a' \\ y_a' \\ z_a' \\ x_b' \\ y_b' \\ z_b' \end{bmatrix} = \begin{bmatrix} -1 & 0 & 0 & 0 & 0 & 0 & 0 & 0 & 0 \\ 0 & 1 & 0 & 0 & 0 & 0 & 0 & 0 & 0 \\ 0 & 0 & 1 & 0 & 0 & 0 & 0 & 0 & 0 \\ 0 & 0 & 0 & -1 & 0 & 0 & 0 & 0 & 0 \\ 0 & 0 & 0 & 0 & 1 & 0 & 0 & 0 & 0 \\ 0 & 0 & 0 & 0 & 0 & 1 & 0 & 0 & 0 \\ 0 & 0 & 0 & 0 & 0 & 0 & -1 & 0 & 0 \\ 0 & 0 & 0 & 0 & 0 & 0 & 0 & 1 & 0 \\ 0 & 0 & 0 & 0 & 0 & 0 & 0 & 0 & 1 \end{bmatrix} \begin{bmatrix} x_O \\ y_O \\ z_O \\ x_a \\ y_a \\ z_a \\ x_b \\ y_b \\ z_b \end{bmatrix} \qquad \text{...(0.40)}$$

$$M(\sigma'_v): \chi(\sigma'_v) = 3$$

From above system it can be mentioned here that 9 × 9 matrix is known as *reducible representation and the representation formed from the blocks are called irreducible representation of the symmetry point group*. The trace obtained in case of H_2O molecule is

C_{2v}	E	C_2	σ_v	σ'_v
$\Gamma_{red}(C_{2v})$	9	1	1	3

***Exercise* 0.25:** The trace for the reducible representation for the H_2O is given in the above table. Calculate the number and different types of irreducible representations.

***Solution*:** Using equation (0.22), *i.e.*,

$$a_j = \frac{1}{h}\sum_R \chi(R)\, \chi^{(j)}(R)$$

we get

$$a\,(\Gamma^{(1)}) = \frac{1}{4}\,[1 \times 1 \times 9 + 1 \times 1 \times (-1) + 1 \times 1 \times 1 + 1 \times 1 \times 3] = 3$$

$$a\,(\Gamma^{(2)}) = [1 \times 1 \times 9 + 1 \times 1 \times (-1) + 1 \times (-1) \times 1 + 1 \times (-1) \times 3] = 1$$

$$a\,(\Gamma^{(3)}) = [1 \times 1 \times 9 + 1 \times (-1) \times (-1) + 1 \times 1 \times 1 + 1 \times (-1) \times 3] = 2$$

$$a\,(\Gamma^{(4)}) = [1 \times 1 \times 9 + 1 \times (-1) \times (-1) + 1 \times (-1) \times 1 + 1 \times (-1) \times 3] = 3$$

The irreducible representations are $3A_1 + A_2 + 2B_1 + 3B_2$.

The other point to be noted is that *total number of irreducible representations is equal to the number of basis vector*, because each irreducible representation derives from one dimensional matrix.

Having discussed from the point to vector and atoms as individual in a molecule, the effect of symmetry operations on the wave functions (or atomic orbitals) can be extended to the systems mentioned in Section 0.11.

0.12. APPLICATIONS

Symmetry and group theory find many applications but here we will divide the applications into two broad categories.

1. The ideas of group theory will be used to generate the symmetry combinations. Physically this means that if we are given some basic set, *i.e.*, a set of orbitals or mathematical functions pertaining to a molecule, we can use group theory to *construct linear combinations* of the things in the basic set that *reflect* the symmetry of the molecule. For example, if we have a set of atomic orbitals for a molecule, we can use group theory to assist us in mapping out the shape of the molecular orbitals for the given molecule. Below we will give some areas where the group theory is very helpful.

(*i*) to determine hybrid orbitals used in the bonding of atoms in the molecule.

(*ii*) to determine which atomic orbitals can contribute to the various molecular orbitals in a molecule.

(*iii*) to determine the number and symmetries of molecular vibrations.

(*iv*) to predict how the degeneracies of the *d*-orbitals are removed by crystal fields of various symmetries.

(*v*) to predict and generate the spin functions to be sued in the Hamiltonian for EPR, NQR and Mossbauer spectroscopy.

(*vi*) to construct symmetry adapted combinations of nuclear spin functions to work out NMR equations.

2. Second broad range of application deals with to ascertain which quantum mechanical integrals are zero, *i.e.*, $\int \psi^* \hat{O} \psi d\tau$ equal to zero? Here ψ and ψ^* are wave functions of the given state and $\hat{O}$ is the operator for the observable quantity. Such integral find utility in the determination of:

(*i*) the allowedness of electronic transitions

(*ii*) the activity of infrared and Raman vibrations

(*iii*) the allowedness of any given transition in NMR, ESR etc.

(*iv*) the average property of any observable quantity for which operator is used, *e.g.* total energy, orbital angular momentum, spin angular momentum etc.

To discuss all the applications and give examples are beyond the scope of this book. The respective applications will be described at the respective chapter.

0.12.1. *Tri-atomic Molecules (AB_2 Type Molecules)*

The simplest tri-atomic molecule is water (H_2O), consists of two hydrogen atoms bonded to central oxygen atom. The number of valence electrons is six in oxygen atom and one each from two hydrogen atoms. The symmetry point group to which water molecule belongs is C_{2v}. The character table for C_{2v} symmetry point group and traces of the reducible representations are given below:

C_{2v}	E	C_2	σ_v	σ'_v	
A_1	1	1	1	1	z
A_2	1	1	–1	–1	
B_1	1	–1	1	–1	x
B_2	1	–1	–1	1	y
Γ_{r1}	2	0	0	2	$A_1 + B_1$
Γ_{r2}	3	–1	1	–1	$A_1 + B_1 + B_2$

Let us first consider the s_1 and s_2 orbitals belonging to different hydrogen atoms as our basis set and examine the effect of symmetry operations of C_{2v} symmetry point group.

$E(s_1, s_2) = (s_1, s_2)$ = both the s-orbitals does not change their positions, thus each orbital contribute +1; thus $\chi(E) = 2$

$C_2(s_1, s_2) = (s_2, s_1)$ = both the s-orbitals change their positions, thus contribute zero to $\chi(C_2) = 0$

$\sigma_v\ (YZ)\ (s_1, s_2) = (s_2, s_1)$ = both the s-orbitals change their positions, thus contribute zero to $\chi(\sigma_v) = 0$

$\sigma'_v\ (XZ)\ (s_1, s_2) = (s_1, s_2)$ = both the s-orbitals remain as such, thus each orbital contributes +1, to $\chi(\sigma_v) = 2$

This trace is shown at the bottom of the character table, which is a reducible representation. These set of characters by inspection can be easily resolved into A_1 and B_2, *i.e.*,

$$\Gamma_{r1} = A_1 + B_2 \qquad ...(0.41)$$

Let us now consider the p-orbitals of the oxygen atom as basis set and apply the symmetry operations. The character of each symmetry operation can be obtained as

$E(p_x, p_y, p_z) = (p_x, p_y, p_z)$ = orbitals remain unchanged, so $\chi(E) = 3$

$C_2(p_x, p_y, p_z) = (-p_x, -p_y, p_z) = p_x$ and p_y orbitals are changed into their negatives and thus contribute –1, whereas p_z remains unchanged, contribute +1, $\therefore\ \chi(C_2) = -1$

$\sigma_v\ (yz)(p_x, p_y, p_z) = (-p_x, p_y, p_z) = p_x$ orbital is inverted while p_y and p_z remain unchanged,

$\therefore\quad \chi(\sigma_v, yz) = -1 + 1 + 1 = 1$

$\sigma_v(xz)(p_x, p_y, p_z) = (p_x, -p_y, p_z) = p_y$ orbital is inverted while p_x and p_z remain unchanged,

$$\chi(\sigma_v, xz) = \therefore\ 1 - 1 + 1 = 1$$

The trace has also been shown as Γ_{r2} at the bottom of the character table. As shown above this reducible representation can also be resolved into A_1, B_1 and B_2 such that

$$\Gamma_{r2} = A_1 + B_1 + B_2 \qquad ...(0.42)$$

Equation (0.41) only tells us that one of the s-orbitals belongs to A_1 and one to B_2. Which one belongs to A_1 and which one to B_2, cannot be decided simply by inspection of the character table because all the s-orbitals are totally symmetric. Thus one requires some technique which will convert this set of 's' orbitals (s_1, s_2) into set of another equivalent set of functions which will belong to A_1 and B_2. Such equivalent functions are known as *Symmetry Adopted Linear Combination* (SALC) functions and are constructed by taking the suitable linear combinations of the original functions. These functions are also having the similar properties as the original ones, *i.e.*, each SALC function belong to one of the irreducible representation. Like irreducible representation, each of the SALC functions is also orthogonal.

These SALC functions are generated by using the method known as Projection Operators. Here it is reproduced as follows:

$$\psi_i = \left[\frac{l_i}{h}\sum_R \chi_i(R).R\right]\varphi_j \qquad ...(0.43)$$

where ψ_i is a SALC function belonging to Γ_i irreducible representation, the terms in the square bracket is known as projection operator, φ_j is one of the members of the basis set used for generating the reducible representation, $\chi_i(R)$ is the character of the transformation matrix for the symmetry operation R in the irreducible representation Γ_i whose dimensions are l_i and h is the order of the group.

Now select any one member of 's' orbitals of the water molecule (say s_1) as the basis set and apply the projection operator given in (0.43) to get the SALC $\psi(A_1)$, belonging to A_1. The expression becomes

$$\psi(A_1) = \frac{1}{4}[1 \times Es_1 + 1 \times C_2 s_1 + 1 \times \sigma_{yz} s_1 + 1 \times \sigma_{xz} s_1]$$

$$= \frac{1}{4}[s_1 + s_2 + s_2 + s_1] = \frac{1}{2}[s_1 + s_2] \quad ...(0.44)$$

The equation (0.44) has been obtained by applying the respective symmetry operations , knowing that E and σ_{xz} will not change the positions of s_1 orbital, whereas C_2 and σ_{yz} will replace s_1 by s_2.

The normalised wave function will be

$$\psi(A_1) = \frac{1}{\sqrt{2}}(s_1 + s_2) \quad ...(0.45)$$

Similarly the SALC function $\psi(B_2)$ belonging to B_2 representation can be obtained as

$$\psi(B_2) = \frac{1}{4}[1 \times Es_1 - 1 \times C_2 s_1 - 1 \times \sigma_{yz} s_1 + 1 \times \sigma_{xz} s_1]$$

$$= \frac{1}{4}[s_1 - s_2 - s_2 + s_1] = \frac{1}{2}[s_1 - s_2] \quad ...(0.46)$$

and its normalised wave function form will be

$$\psi(B_2) = \frac{1}{\sqrt{2}}(s_1 - s_2) \quad ...(0.47)$$

After forming the SALC functions (equations 0.45 and 0.46) we can decide which function will overlap with which of the orbitals of the oxygen atom. From the character table of C_{2v} symmetry point group, it is clear that p_z orbital of the oxygen atom belongs to A_1, p_x belongs to B_1 and p_y belongs to B_2 irreducible representations respectively. *Knowing that atomic orbitals of the same symmetry will combine to give rise to molecular orbitals, p_z will combine according SALC given by equation* (0.45) *and p_y will combine with SALC function given by equation* (0.47). $2s$ and $2p_x$ pairs of electrons of oxygen atom will act as the nonbonding pairs. The wave functions for the bonding molecular orbitals will be

$$\Psi_1 = C_1 2p_z(\text{O}) + C_2\psi(A_1) = C_1 2p_z(\text{O}) + C_2(s_1 + s_2) \quad ...(0.48)$$

$$\Psi_2 = C_3 2p_y(\text{O}) + C_4\psi(B_2) = C_3 2p_y(\text{O}) + C_4(s_1 - s_2) \quad ...(0.49)$$

The coefficients $C_1 - C_4$ can be determined using the normalisation conditions. Anti-bonding molecular orbitals will be obtained by taking the negative combinations. It may be mentioned here that group theory does not decide whether $2s$ or $2p_z$ orbital of oxygen combine with equation (0.45) as both these orbitals belongs to A_1 irreducible representation. It is the energy criterion that decides this. The energy of $2s$ atomic orbital of oxygen is located much below the $2p$ orbitals of oxygen and thus will not take enough participation in the overlap of LCAO to form the molecular orbitals.

0.12.2. *Ammonia Molecule (AB_3 Pyramidal Type Molecules)*

Ammonia molecule consists of one nitrogen atom and three hydrogen atoms. The nitrogen atom with ground state electronic configuration as $1s^2 2s^2 2p^3$ possesses five valence electrons. The three hydrogen atoms contribute one electron each. Further ammonia molecule belongs to C_{3v} symmetry point group with the following character table, along the character of the trace to C_{3v}:

C_{3v}	E	$2C_3$	$3\sigma_v$	Bases
A_1	1	1	1	z
A_2	1	1	–1	
E	2	–1	0	x, y
Γ_{r1}	3	0	1	$A_1 + E$

Proceeding on the same lines as we have carried out in case of water, consider $1s$ atomic orbitals of the three H-atoms as bases to generate a reducible representation and find the components of the irreducible representation in it.

$$E(s_1, s_2, s_3) = (s_1, s_2, s_3);\ \chi(E) = +3$$
$$C_3^1(s_1, s_2, s_3) = (s_3, s_1, s_2);\ \chi(C_3^1) = 0$$
$$C_3^2(s_1, s_2, s_3) = (s_2, s_3, s_1);\ \chi(C_3^2) = 0$$
$$\sigma_v^1(s_1, s_2, s_3) = (s_1, s_3, s_2);\ \chi(\sigma_v) = 1$$
$$\sigma_v^2(s_1, s_2, s_3) = (s_3, s_2, s_1);\ \chi(\sigma_v') = 1$$
$$\sigma_v^3(s_1, s_2, s_3) = (s_2, s_1, s_3);\ \chi(\sigma_v'') = 1$$

C_{3v}	E	$2C_3$	$3\sigma_v$	Bases
A_1	1	1	1	z
A_2	1	1	–1	
E	2	–1	0	x, y
Γ_{r1}	3	0	1	$A_1 + E$

Thus the trace of the reducible representation as Γ_{r1} is given in the character table. The reducible representation can be resolved on simple inspection as $A_1 + E$. Although in the above two molecules the reducible representations have been resolved into irreducible representations by simple inspection but sometimes we do come across some complicated cases. In these complex cases, the reducible representation can be resolved into irreversible representation using the equation (0.22).

Thus using equation (0.22), let us try to resolve the reducible representation into irreducible representation.

$$a(A_1) = \frac{1}{6}[(3\times1\times1)+(0\times1\times2)+(1\times1\times3)] = \frac{1}{6}[3+3] = 1$$

$$a(A_2) = \frac{1}{6}[(3\times1\times1)+(0\times1\times2)+(1\times-1\times3)] = \frac{1}{6}[3-3] = 0$$

$$a(E) = \frac{1}{6}[(3\times2\times1)+(0\times-1\times2)+(1\times0\times3)] = \frac{1}{6}[6] = 1$$

As found earlier, $\Gamma_{r1} = A_1 + E$, which is same as mentioned above. Similar to the treatment in water, here also we cannot say that which one of s orbital belongs to A_1 irreducible representation and which one to E. We will use the projection operator method to find out which combination of the s orbitals will correspond to A_1 and which one to E. The procedure is similar to that used earlier and it is as follows

$$\Psi(A_1) = \frac{1}{6}[1\times Es_1 + 1\times C_3^1 s_1 + 1\times C_3^2 s_1 + 1\times \sigma_v s_1 + 1\times \sigma_v' s_1 + 1\times \sigma_v'' s_1]$$

$$= \frac{1}{6}[s_1 + s_2 + s_3 + s_1 + s_2 + s_3] = \frac{1}{3}[s_1 + s_2 + s_3]$$

After normalisation

$$\Psi(A_1) = \frac{1}{\sqrt{3}}[s_1 + s_2 + s_3] \qquad ...(0.50)$$

Since A_1 is a one dimensional irreducible representation, there will be two other SALC functions which will correspond to E irreducible representation.

$$\Psi(E_1) = \frac{1}{6}[2\times Es_1 - 1\times C_3^1 s_1 - 1\times C_3^2 s_1 + 0\times \sigma_v s_1 + 0\times \sigma'_v s_1 + 0\times \sigma''_v s_1]$$

$$= \frac{1}{6}[2s_1 - s_2 - s_3] \qquad ...(0.51)$$

Its normalised form is

$$\Psi(E_1) = \frac{1}{\sqrt{6}}[2s_1 - s_2 - s_3] \qquad ...(0.51a)$$

To obtain the other SALC function belonging to E, like $\Psi(E_2)$, we have to apply the following rules

1. *$\Psi(E_2)$ must be orthogonal to $\Psi(E_1)$.*

2. *The second function (or symmetry adopted orbitals) in the degenerate representation in a given symmetry point group can be constructed by employing an appropriate rotation operation subgroup of that point group. Thus when appropriate rotation operation is applied to $\Psi(E_1)$ it will convert it into a linear combination of $\Psi(E_1)$ and $\Psi(E_2)$, i.e.,*

$$\hat{R}\Psi(E_1) = C_1\Psi(E_1) + C_2\Psi(E_2) \qquad ...(0.52)$$

where C_1 and C_2 are the coefficients to be determined. Let us take $\hat{R}$ to be C_3^2 and replace it in equation (0.52), *i.e.*,

$$C_3^2\Psi(E_1) = C_1\Psi(E_1) + C_2\Psi(E_2) \qquad ...(0.52a)$$

The effect of C_3^2 on ΨE_1 can be written as

$$C_3^2\Psi(E_1) = C_3^2[\frac{1}{6}(2s_1 - s_2 - s_3)] = \frac{1}{6}(C_3^2 2s_1 - C_3^2 s_2 - C_3^2 s_3)$$

$$= \frac{1}{6}(2s_2 - s_3 - s_1) \qquad ...(0.53)$$

According to rule 2, the function given in equation (0.53) will be a linear combination of type given in equation (0.52*a*). Multiply equation (0.52*a*) by $\Psi(E_1)$ and on integrating the resultant equation we obtain

$$\int\Psi(E_1)[C_3^2\Psi(E_1)]d\tau = C_1\int\Psi^2(E_1)d\tau + C_2\int\Psi(E_1)\Psi(E_2)d\tau = C_1\int\Psi^2(E_1)d\tau \qquad ...(0.54)$$

Second integral is equal to zero as $\Psi(E_1)$ and $\Psi(E_2)$ are orthogonal to each other. The equation (0.54) can be written as

$$C_1 = \frac{\int\Psi(E_1)(C_3^2\Psi(E_1))d\tau}{\int\Psi^2(E_1)d\tau} \qquad ...(0.55)$$

Combining equations (0.51), (0.53) and (0.55), we get

$$C_1 = \frac{\int\frac{1}{6}(2s_1 - s_2 - s_3)\frac{1}{6}(2s_2 - s_3 - s_1)d\tau}{\int[\frac{1}{6}(2s_1 - s_2 - s_3)]^2 d\tau} \qquad ...(0.56)$$

The integral in the numerator can be expanded as

$$\frac{1}{36}[4\int s_1s_2d\tau - 2\int s_1s_3d\tau - 2\int s_1s_1d\tau - 2\int s_2s_2d\tau + \int s_2s_3d\tau + \int s_2s_1d\tau - 2\int s_2s_3d\tau$$

$$+\int s_3s_3d\tau + \int s_1s_3d\tau]$$

Using the following relations

$$\int s_is_jd\tau = 0 \text{ when } i \neq j$$

$$= 1 \text{ when } i = j$$

above expression becomes

$$\frac{1}{36}(0 - 0 - 2 - 2 + 0 - 0 + 0 + 1 + 0] = -\frac{1}{12}$$

whereas the denominator becomes: $\frac{1}{36}(4 + 1 + 1 - 0 - 0 - 0 + 0 - 0 + 0] = \frac{1}{6}$

Substituting the values of numerator and denominator in equation (0.56), we get

$$C_1 = \frac{-\frac{1}{12}}{\frac{1}{6}} = -\frac{1}{2} \qquad ...(0.57)$$

The equation (0.52*a*) can be written as

$$C_2\Psi(E_2) = C_3^2\Psi(E_1) - C_1\Psi(E_1) = \frac{1}{6}(2s_2 - s_3 - s_1) - \left(-\frac{1}{2}\right)\left[\frac{1}{6}(2s_1 - s_2 - s_3)\right]$$

$$= \frac{1}{4}(s_2 - s_3)$$

Thus

$$\Psi(E_2) = \frac{1}{4C_2}(s_2 - s_3) \quad ...(0.58)$$

Thus the equations (0.51) and (0.58) represent the SALC functions belonging to E irreducible representation. These functions can be normalised using the normal relation, *i.e.*, $\int \psi^* \psi d\tau = 1$. So

$$1 = \int \frac{C_1}{6}(2s_1 - s_2 - s_3)\frac{C_1}{6}(2s_1 - s_2 - s_3)\, d\tau = \frac{1}{36}C_1^2(4+1+1) = \frac{1}{6}C_1^2$$

or

$$C_1 = \sqrt{6}$$

Thus

$$\Psi(E_1) = \frac{1}{\sqrt{6}}(2s_1 - s_2 - s_3) \quad ...(0.59)$$

which corresponds to equation (0.51 *a*).

and similarly normalised $\Psi(E_2) = \frac{1}{\sqrt{2}}(s_2 - s_3)$ ...(0.60)

Thus to summarise, three SALC functions of three hydrogen atoms in NH_3 are:

$$\Psi(A_1) = \frac{1}{\sqrt{3}}[s_1 + s_2 + s_3] \quad ...(0.50)$$

$$\Psi(E_1) = \frac{1}{\sqrt{6}}(2s_1 - s_2 - s_3) \quad ...(0.51)$$

$$\Psi(E_2) = \frac{1}{\sqrt{2}}(s_2 - s_3) \quad ...(0.60)$$

Following the same procedure as followed in case of oxygen atom in H_2O, the three p orbitals of nitrogen, p_z belong to A_1 irreducible representation and p_x and p_y belong to E irreducible representation. The molecular orbitals can be written on the basis that orbitals of similar symmetry will overlap, we get

$$\Psi_1(m.o.) = C_1p_z(N) + C_2\psi_1(A_1) = C_1p_z(N) + C_2[s_1 + s_2 + s_3] \quad ...(0.61)$$

$$\Psi_2(m.o.) = C_3p_x(N) + C_4\psi_2(E) = C_3p_x(N) + C_4[2s_1 - s_2 - s_3] \quad ...(0.62)$$

$$\Psi_3(m.o.) = C_5p_y(N) + C_6\psi_3(E) = C_5p_y(N) + C_6[s_2 - s_3] \quad ...(0.63)$$

0.12.3. Hybrid Orbitals

One of the most important classes of orbitals used in explaining the geometry of molecules in chemistry is the *hybrid atomic orbitals*. These orbitals give strong directional properties to the chemical bonds and lead to the molecular geometry which can be determined experimentally. Involvement of hybrid orbitals adds specificity in the formation of the bonds. The *hybrid atomic orbitals are obtained by taking the linear combination of the proper atomic orbitals, belonging to the same atom*. We have divided this discussion into two sections. One involves the formation of σ bonds with other atoms and the other involves the formation of π–bonds.

0.12.3.1. σ–ordbital

(*i*) The first step is to consider the set of equivalent chemical bonds as the vectors and assign the reducible representation under the symmetry point group of the molecule.

(*ii*) Reduce this reducible representation into the irreducible representations.

(*iii*) Atomic orbitals which transform according to the various irreducible representations of the particular symmetry point group are then selected from the character table of that symmetry point group and are combined to form the hybrid orbitals.

Let us consider an example to understand this method.

Consider methane (shown in Fig. 0.21), which belongs to tetrahedral geometry and T_d symmetry point group. First step is to consider four carbon-hydrogen bonds as the basis set and get the reducible representation obtained by performing the symmetry operations of T_d symmetry point group. The

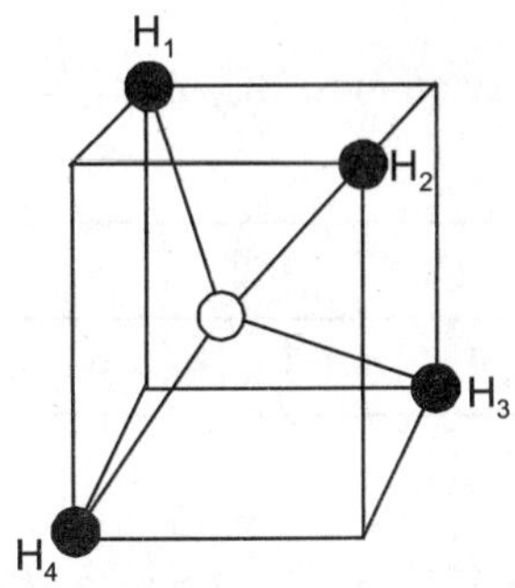

Fig. 0.21 *Methane molecule. $C_{3,1}$, $C_{3,2}$, $C_{3,3}$ and $C_{3,4}$ axes are three-fold axes passing through corners 1, 2, 3 and 4 respectively, C_{2x}, C_{2y} and C_{2z} axes are two-fold axes coinciding with X, Y and Z coordinates, σ_{d1}, σ_{d2}, σ_{d3}, σ_{d4}, σ_{d5} and σ_{d6} are the reflection planes passing through Cs_1s_2, Cs_1s_3, Cs_1s_4, Cs_2s_3, Cs_2s_4 and Cs_3s_4 respectively.*

reducible representation obtained under the symmetry point group T_d is given below. Using the character table of T_d (Appendix *A*–1) reducible representation can be reduced to $A_1 + T_2$.

T_d	E	$8C_3$	$3C_2$	$6S_4$	$6\sigma_D$
Γ_{C-H}	4	1	0	0	2

A simple look at the character table reveals that *s*-orbital belongs to the completely symmetric irreducible representation (A_1) and the set (*x*, *y*, *z*, *i.e.*, p_x, p_y and p_z atomic orbitals or *xy*, *yz* and *zx*, d_{xy}, d_{yz} and d_{zx} atomic orbitals) to T_2 irreducible representation. So for the formation of CH_4, the central carbon will use one *s*-orbital and three *p*-orbitals or three *d*-orbitals, thus forming sp^3 or sd^3 bonding. In other words, it will yield four sp^3 or sd^3 hybrid orbitals. It may be mentioned that on the basis of symmetry arguments we cannot differentiate whether to use three *p*-orbitals or three *d*-orbitals. But from energetic point of view the energy of 3 *d* orbitals of carbon ((962.3 kJ mol^{-1}) is at a much higher level than that of 2*p* orbitals of carbon atom. In other words, if we assume that sd^3 orbitals are involved for the formation of the hybrid orbitals on carbon atom rather than sp^3 each sd^3 bond would have to release an energy of somewhat $3 \times \frac{962.3}{4} = 721.7$ kJ mol^{-1}, which is not possible. Thus the combination of 2*s* and three 2*p* orbitals of carbon will be preferred, leading to sp^3 hybridisation over the sd^3 hybridisation. Thus it may be concluded that atoms present in the second short period (lithium to fluorine) will involve $2s2p^3$ combination of atomic orbitals to form four tetrahedral directed bonds. On the other hand, if silicon is the central atom, the choice whether to use three 3-*p* orbitals or three 3-*d* orbitals, is not very clear. But the *d*-orbitals are extensively used in bonding since they differ little in energy from 3*p* orbitals. It may be mentioned that CH_4 is the representative of a number of tetrahedral molecules which belong to T_d symmetry point group and the hybrid orbitals constructed for this molecule will be equally valid for other tetrahedral molecules, *i.e.,* the symmetry of the atomic orbital does not depend on the principal quantum number, *n*. So the types of orbitals to be selected for the hybridisation depend on the atomic number of the central atom. For example, carbon atom will use 2 *p* orbitals whereas the silicon may use 3 *d* orbitals.

Different combinations of atomic orbitals used in constructing the σ-atomic hybrid orbitals can be discussed in a similar manner. The useful information required is given below in a summarised manner along with few examples.

Trigonal planar molecules (AB_3): Examples; BF_3, AlR_3, CO_3^{2-}, NO_3^-, SO_3

Hybridisation: sp^2, sd^2, dp^2, d^3

D_{3h}	E	$2C_3^1$	$3C_2^1$	σ_h	$2S_3^1$	$3\sigma_v$
$\Gamma_r(\sigma)$	3	0	1	3	0	1

$$\Gamma_r(\sigma) = A'_1 + E_2$$

Possible combinations: (s, p_x, p_y), $(s, d_{xy}, d_{x^2-y^2})$, (d_{z^2}, p_x, p_y), $(d_{z^2}, d_{xy}, d_{x^2-y^2})$,

Square planar molecule (AB_4): Examples: XeF_4, $[Ni(CN)_4]^2$, $[AuCl_4]^-$

Hybridisation: dsp^2, d^2p^2

D_{3h}	E	$2C_4$	$2C'_2$	$2C''_2$	i	$2S_4$	σ_h	$2\sigma_v$	$2\sigma_d$
$\Gamma_r(\sigma)$	4	0	0	2	0	0	4	2	0

$$\Gamma_r(\sigma) = A_{1g} + B_{1g} + E_u$$

Possible combinations: (s, d_{z^2}, p_x, p_y), $(d_{x^2-y^2}, d_{z^2}, p_x, p_y)$,

Square pyramidal: AB_5

Possible combinations: $(d_{x^2-y^2}, s, p_x, p_y, p_z)$;$(s, d_{z^2}, d_{x^2-y^2}, d_{xz}, d_{yz})$;

$(s, p_x, d_{z^2}, d_{x^2-y^2}, d_{xz}); (p_z, d_{z^2}, d_{x^2-y^2}, d_{xz}, d_{yz}); (p_z; d_{x^2-y^2}, p_x, p_y, d_{z^2})$

Octahedral (AB_6); d^2sp^3 Examples $[Fe(CN)_6]^4$, SF_6, PE_6^-

O_h	E	$8C_3$	$6C_2$	$3C_2$	i	$2S_4$	$8S_6$	$3\sigma_h$	$6\sigma_d$
$\Gamma_r(\sigma)$	4	0	0	2	0	0	0	2	0

$$\Gamma_r(\sigma) = A_{1g} + E_g + T_u$$

Possible combinations: $s, p_x, p_y, p_z, d_{z^2}, d_{x^2-y^2}$ only.

The next job is to determine the coefficients of these atomic orbitals involved in the formation of hybrid orbitals, which is beyond the scope of this book. This procedure can be looked from the book "Symmetry and Group Theory in Chemistry", New Age Publishers, New Delhi, 2014.

0.12.4. *Prediction About Certain Integral in Quantum Mechanics*

There are two types of integrals which are used very often in quantum mechanics. Although it is essential to get quantitative information about these integrals, but many times if we know that whether these integrals are zero or finite it is more than enough. These integrals are:

$$S_{ij} = \int \varphi_i^* \varphi_j d\tau \text{ and } H_{ij} = \int \varphi_i^* \hat{H} \varphi_j d\tau \quad ...(0.64)$$

Although the given integrals are specific, *i.e.*, overlap integral and the total energy integral, the procedure to be discussed below is valid for general type to be present in quantum mechanics.

We know that *if an even function is integrated over the complete space, the integral is non-zero, whereas if an odd function is integrated over the complete space the integral is zero*. This can be generalised as: *If the product of the traces of the irreducible representations to which these functions belong to is totally symmetric or its reducible representation contains at least one of the irreducible representation as totally symmetric*, the integral will be nonzero.

The most important integral of second type is to find out whether the given transition is allowed or not. The transition moment integral (M) which decides the validity of the transition is given as

$$M = \int_{\text{all space}} \psi_i \hat{M} \psi_f d\tau \quad ...(0.65)$$

where ψ_i and ψ_f are initial state and final state wave functions between which the transition are talking place, $\hat{M} = e\hat{r}$ is the dipole moment operator. Since transition moment integral is a vector quantity, it can have magnitude and direction. The relation between the component of the transition moment integral and transition moment is given as:

$$\mathbf{M}^2 = \mathbf{M}_x^2 + \mathbf{M}_y^2 + \mathbf{M}_z^2 \quad ...(0.66)$$

where

$$M_x = \int_{\text{all space}} \psi_i \hat{M}_x \psi_f d\tau = \int_{\text{all space}} \psi_i e x \psi_f d\tau \quad ...(0.67)$$

$$M_y = \int_{\text{all space}} \psi_i \hat{M}_y \psi_f d\tau = \int_{\text{all space}} \psi_i ey \psi_f d\tau \quad ...(0.68)$$

$$M_z = \int_{\text{all space}} \psi_i \hat{M}_z \psi_f d\tau = \int_{\text{all space}} \psi_i ez \psi_f d\tau \quad ...(0.69)$$

Transition is known to be *x*-axis, *y*-axis or *z*-axis polarised if M_x, M_y or M_z is not zero.

***Exercise* 0.26:** Formaldehyde molecule belongs to C_{2v} symmetry point group. In the $n \rightarrow \pi^*$ transition *n*-orbital belongs to A_1 and π^* to A_2. Predict whether this transition is allowed or not?

The initial state is A_1 and the final state is A_2. The direct product of $\psi_i\ \psi_f\ (A_1A_2)$ will be A_2. Looking at the C_{2v} point group to which formaldehyde belongs to, $\hat{M}_x$, $\hat{M}_y$ *and* $\hat{M}_z$ the components of $\hat{M}$ belong to B_1, B_2 and A_1 respectively. Since none of the components of $\hat{M}$ belongs to A_2, it can be concluded that this transition is forbidden. This result can also be arrived at by considering the integrals (0.67, 0.68, 0.69), $\psi_i \hat{M}_x \psi_f$ or $\psi_i \hat{M}_y \psi_f$ or $\psi_i \hat{M}_z \psi_f$ ($A_1B_1A_2$ or $A_1B_2A_2$ or $A_1A_1A_2$) will give rise to B_2 or B_1 or A_2. None of them belongs to totally symmetric configuration, *i.e.*, A_1. Thus, the weak transition observed experimentally justifies this result.

***Exercise* 0.27:** The HOMO to LUMO electronic transition in benzene is 4 fold degenerate. After taking into account the configuration interactions, this splits into three energy levels, two belong to singly degenerate (B_{2u} and B_{1u}) and one doubly degenerate (E_{2u}), as shown in Fig. 0.22. Even though all the transitions involve $\pi \rightarrow \pi^*$, show that 256 nm transition is forbidden.

Using the integrals (0.67, 0.68 and 0.69), the direct product of $^1A_{1g}$ and $^1B_{2u}$ will lead to $^1B_{2u}$. Looking at the D_{6h} point group to which benzene belongs to, $\hat{M}_x$, $\hat{M}_y$ *and* $\hat{M}_z$ the components of $\hat{M}$ belong to E_{1u} A_{2u} respectively. Since none of the components of $\hat{M}$ belong to B_{2u}, it can be

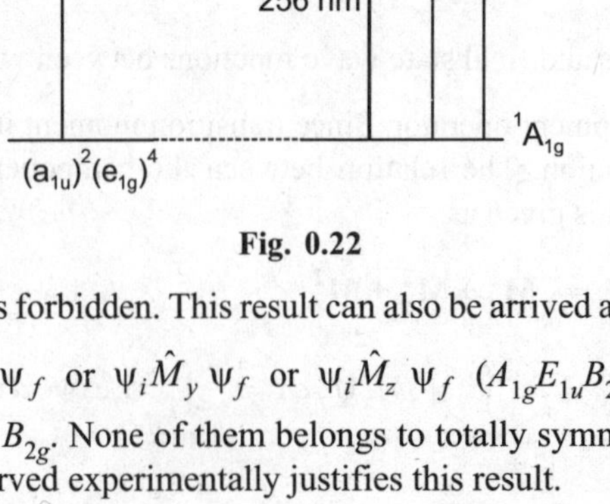

Fig. 0.22

concluded that this transition is forbidden. This result can also be arrived at by considering the integrals (0.67, 0.68 and 0.69), $\psi_i \hat{M}_x \psi_f$ or $\psi_i \hat{M}_y \psi_f$ or $\psi_i \hat{M}_z \psi_f$ ($A_{1g}E_{1u}B_{2u}$ or $A_{1g}E_{1u}B_{2u}$ or $A_{1g}A_{2u}B_{2u}$) will give rise to E_{2g} or E_{2g} or B_{2g}. None of them belongs to totally symmetric configuration, *i.e.*, A_{1g}. Thus the weak transition observed experimentally justifies this result.

PROBLEMS

0.1 Assign the symmetry point group to the molecules shown below in the configuration shown

(a) O = C = S (b) 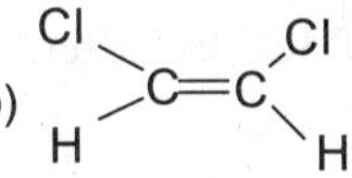 (c) 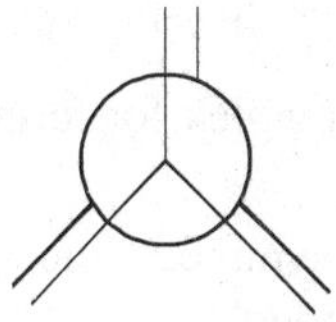

(e) 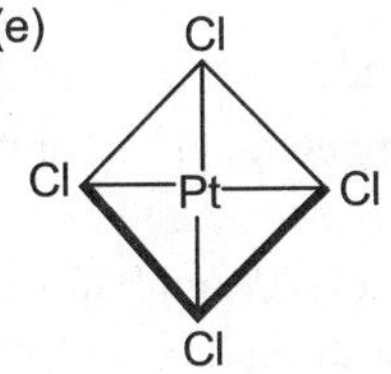

(f) 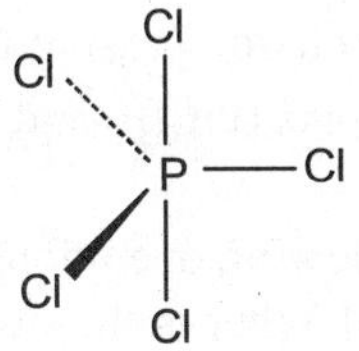

(g) 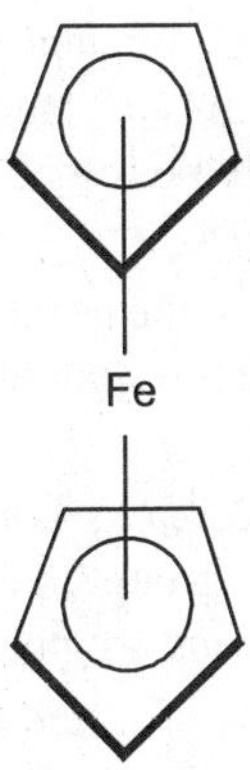

(h) 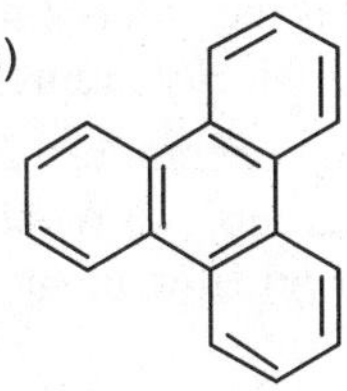

(i) 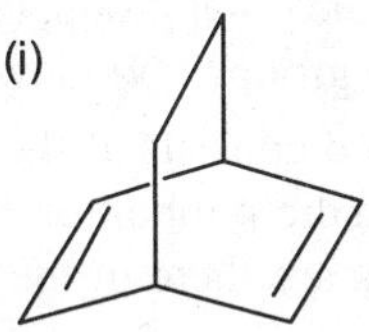

(j) 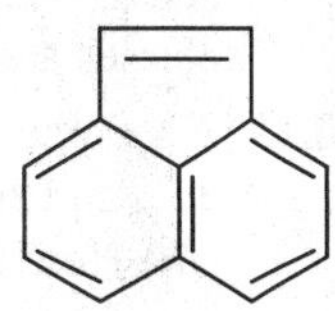

(k) 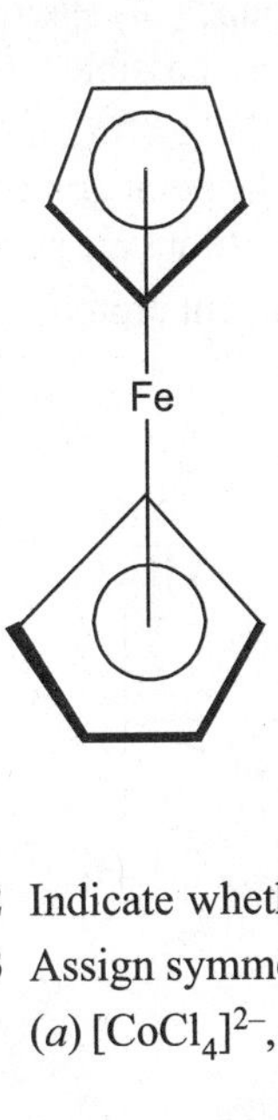

(l) 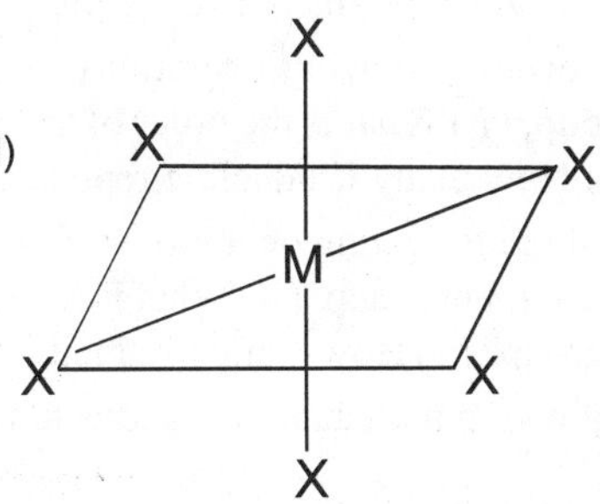

(m) 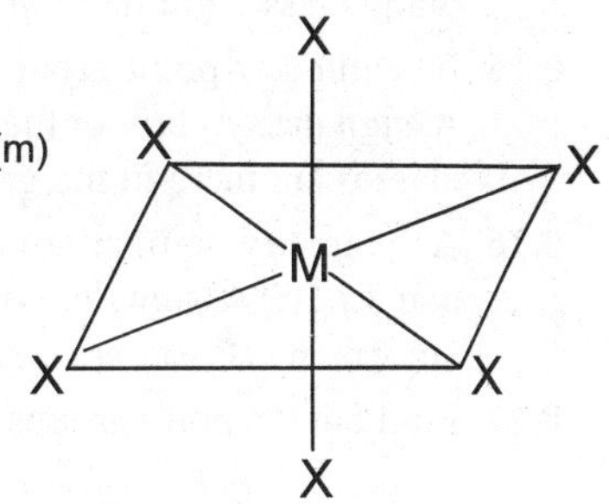

[**Ans:** (*a*) C_{∞}, (*b*) C_{2v}, (*c*) D_{3h}, (*d*) D_2, (*e*) D_{4h}, (*f*) D_{3h}, (*g*) D_{5h}, (*h*) D_{3h}, (*i*) C_{2v}, (*j*) C_{2v}, (*k*) D_{5d}, (*l*) O_h, (*m*) D_{4h},

0.2 Indicate whether each molecule possesses the dipole moment or not.

0.3 Assign symmetry point group to the following molecules:

(*a*) $[CoCl_4]^{2-}$, (*b*) $[Ni(CN)_4]^{2-}$, (*c*) [*cis*-$COCl_4(NH_3)_2]^-$ (ignore the hydrogen atoms of ammonia),

0.4 (*d*) PF_3, (*e*) C-I-Cl–, (*f*) planar [*trans*-$PdCl_2B_2$] (*B* is a base).
List the different kind of symmetry elements present in (*i*) SO_3, (*iii*) CH_4 and (*iii*) C_6H_6.

0.5 Determine the symmetry point group of xenon tetrafluoride molecule

0.6 Show that $C_3^1 \neq C_3^{-1}$ but $C_3^2 = C_3^{-1}$.

0.7 How does a D_{2d} complex for formula $[MCl_4]^{2-}$ differ from a T_d complex (*i.e.*, which symmetry elements differ).

0.8 Which important symmetry element is absent in the PF_3 molecule that is present in the D_{3h} symmetry point group?

0.9 Does $POCl_2Br$ possesses improper axis of rotation passing through *PO* bond?

0.10 Assign the symmetry point group to the complex $[PtCl_4]^{2-}$ and list the symmetry element belonging to each class.

0.11 A symmetry point group consists of the following operations: E, $2C_3$, $3C_2$, σ_h, $2S_3$ and $3\sigma_v$. (*i*) Assign the symbol for the point group, (*ii*) What is the order of the point group?, (*iii*) How many classes are there in the group? (*iv*) How many irreducible representations are possible?

0.12 A symmetry point group consists of the following operations: E, $2C_6$, $2C_3$, C_2 $3\sigma_V$ and $3\sigma_d$. (*i*) Assign the symbol for the point group, (*ii*) What is the order of the point group? (*iii*) How many classes are there in the group? (*iv*) How many irreducible representations are possible?

0.13 A symmetry point group consists of the following operations: E, $2C_4$, C_2, $2C'_2$, C''_2, i, $2S_4$, σ_h, $2\sigma_V$, and $2\sigma_d$. (*i*) Assign the symbol for the point group, (*ii*) What is the order of the point group? (*iii*) How many classes are there in the group? (*iv*) How many irreducible representations are possible?

0.14 A symmetry point group consists of the following operations: E, $2S_8$, $2C_4$, $2S_8^3$, C_2, 4, and $4\sigma_d$. (*i*) Assign the symbol for the point group, (*ii*) What is the order of the point group?, (*iii*) How many classes are there in the group? (*iv*) How many irreducible representations are possible?

0.15 A symmetry point group consists of the following operations: E, $8C_3$, $3C_2$, $6S_4$, and $6\sigma_d$. (*i*) Assign the symbol for the point group, (*ii*) What is the order of the point group? (*iii*) How many classes are there in the group? (*iv*) How many irreducible representations are possible?

0.16 A symmetry point group consists of the following operations: E, $4C_3$, $4C_3^2$, $3C_2$, i, $4S_6$, $6S_4$, $4S_6^5$ and $3\sigma_h$. (*i*) Assign the symbol for the point group, (*ii*) What is the order of the point group? (*iii*) How many classes are there in the group? (*iv*) How many irreducible representations are possible?

0.17 Find out the point groups of the following molecules and indicate the equivalent protons:

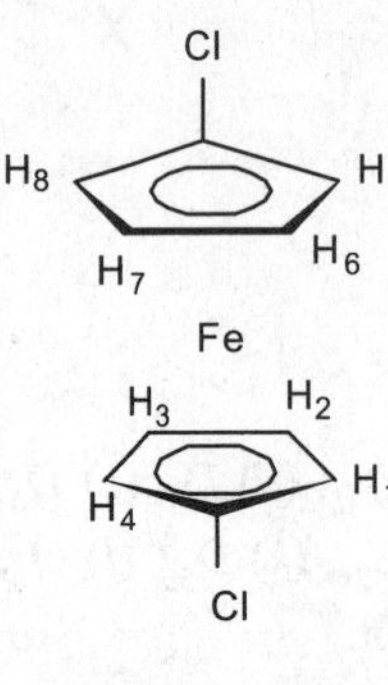

(*i*)

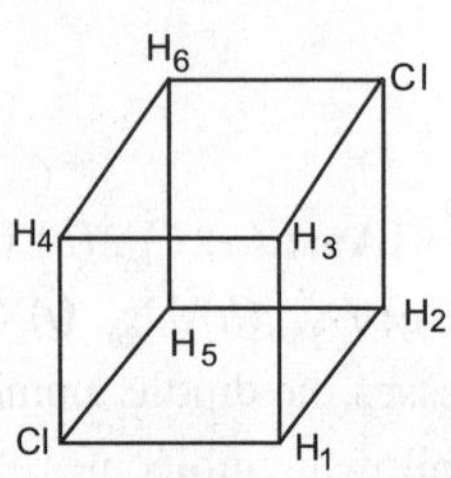

(*ii*)

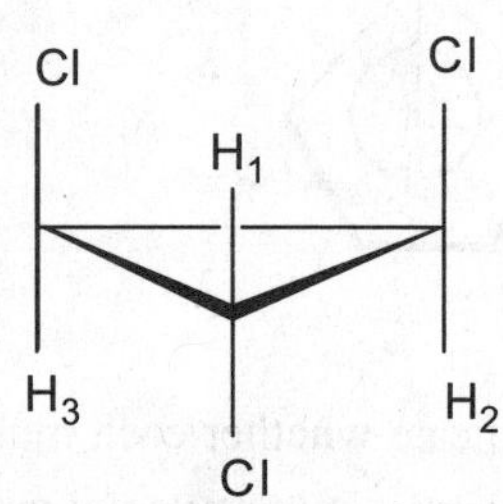

(*iii*)

0.18 Construct the combination table for the symmetry operations associated with molecule (*e*), (*f*) and (*h*) of Question 0.1.

0.19 List the symmetry operations associated with molecules listed in Problem (0.14).

0.20 Using the following combination of the operations find the element or elements arise. Assume that the both operations in each pair pass through the same point. (*i*) C_6 and σ_d, (*ii*) C_3 and σ_v, (*iii*) S_3 and σ_h, (*iv*) C_4 and *i*, (*v*) C_2 and S_4, (*vi*) C_6 and C''_2.

0.21 Using the Exercises 0.4 and 0.5, construct the multiplication tables for the C_{2v} and C_{3v} symmetry point groups.

0.22 With the help of character tables given in Appendix A-1, find out the number of symmetry elements present in C_{4v}, D_{4h}, T_d and O_h symmetry points groups.

0.23 Evaluate $\sigma_v\sigma'_v$, $C_2\sigma'_v$, $\sigma'_v\sigma_v$ and σ'_vC_2 in the C_{2v} symmetry point group.

0.24 Similarly evaluate, $\sigma_v\sigma'_v$, $C_3\sigma'_v$, $\sigma'_v\sigma_v$ and σ'_vC_3 in the C_{3v} symmetry point group compare the results.

0.25 Prove the law of associative multiplication for the following symmetry operations forC_{3v} symmetry point group by using the geometrical descriptions and making use of matrices description of these symmetry operations from taking the matrices from Exercise 0.17.

$$(i)\ C_3^1(\sigma_v\sigma'_v)=(C_3^1\sigma_v)\sigma'_v\ ,\ (ii)\ C_3^2(\sigma_v\sigma''_v)=(C_3^2\sigma_v)\sigma''_v\ ,\ (iii)\ C_3^1(C_3^2\sigma'_v)=(C_3^1C_3^2)\sigma'_v$$

0.26 Using equations 0.20, *i.e.*, $\sum_R[\chi^i(R)]^2 = h$ or $\sum_{\text{classes}} N_R[\chi^i(R)]^2 = h$

Show that the above equations hold good for the irreducible representations (*i*) B_{2u}, (*ii*) E_g and (*iii*) B_{1g} of the symmetry point group D_{4h}.

0.27 Repeat the Problem 0.26 for the irreducible representations (*i*) A_{1u}, (*ii*) E_{1u}, (*iii*) E_{2g} and (*iv*) B_{2u} of D_{6h} symmetry point group.

0.28 For a given column of any symmetry point group follows the equation: $\sum_i[\chi^i(R)_m]^2 = \dfrac{h}{N_R}$,

where N_R is the number of times the particular symmetry element occur and h is the order of the symmetry point group. Prove this relation for (*i*) C_2 in the D_{3h}, (*ii*) σ_v and σ_h for D_{4h} and (*iii*) S_3 and σ_d for D_{6h} symmetry point groups.

0.29 Write the transformation matrices for NH_3 molecule for each symmetry operation. Also determine the irreducible representations present in the reducible representation.

0.30 Consider a symmetry point group having the symmetry elements as: E, $2C_6$, $2C_3$, C_2, $3C'_2$, $3C''_2$, *i*, $2S_3$, $2S_6$, σ_h, $3\sigma_d$, and $3\sigma_v$. (*i*) Name the symmetry point group, (*ii*) calculate the order of the group, (*iii*) find the number of classes and (*iv*) number of irreducible representations.

0.31 Calculate the direct product of E_g with E_g and E_g with E_u of the D_{6h} symmetry point group. Also determine the irreducible representations present in reducible representations of the above direct product.

0.32 Are the following molecule optically active?

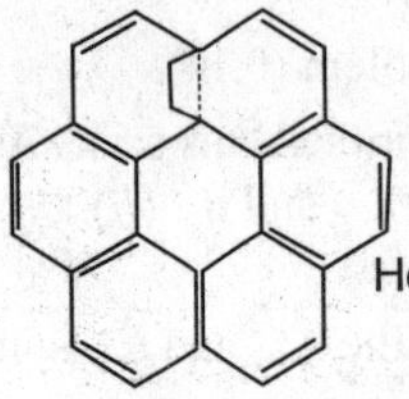

Hexahelicene

(*i*)

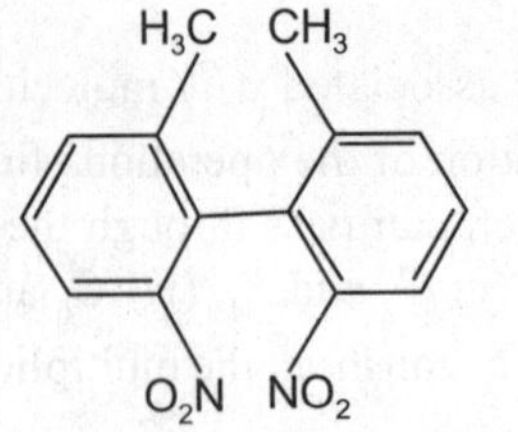

(*ii*)

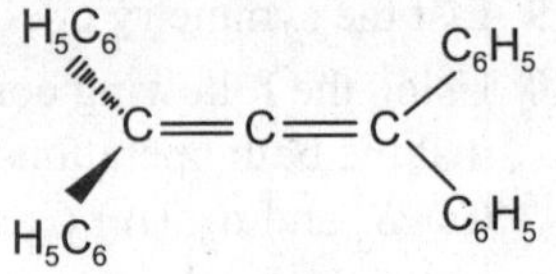

(*iii*)

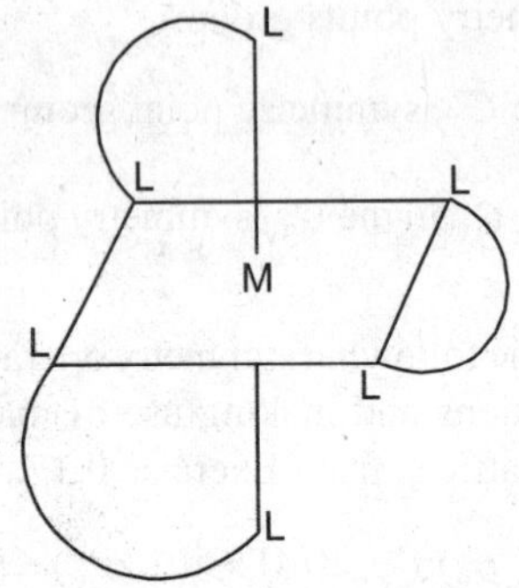

M = Ni^{2+}

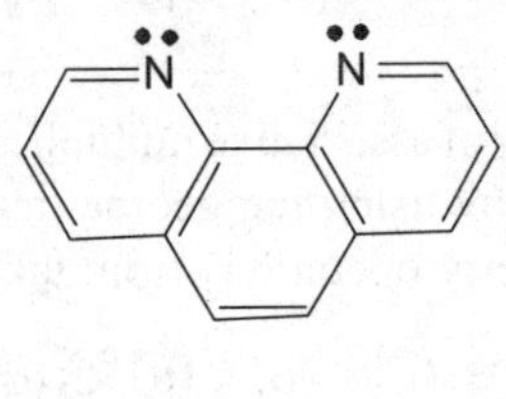

(*iv*) (*v*)

0.33 The square pyradimal molecule (AX_4) belongs to C_{4v} point group. Find the transformation matrix of C_4 and σ_v symmetry operations.

0.34 Construct the character table for the symmetry point groups: (*i*) C_{4v}, (*ii*) D_{4h} and D_{4d}.

0.35 The transformation matrix for the rotation symmetry element can be given as

$$\begin{bmatrix} \cos\theta & -\sin\theta & 0 \\ \sin\theta & \cos\theta & 0 \\ 0 & 0 & 1 \end{bmatrix}$$

when the rotation is carried out by an angle θ in an anticlockwise direction and similarly for S_θ the transformation matrix is

$$\begin{bmatrix} \cos\theta & -\sin\theta & 0 \\ \sin\theta & \cos\theta & 0 \\ 0 & 0 & -1 \end{bmatrix}$$

0.36 For a given column in the symmetry point group (*i*) D_{4h}, (*ii*) C_{4v} and D_{6h} equation (0.29) is obeyed.

0.37 Show that (*i*) the overlap integral between the $2p_x$ atomic orbital and [$1s\,(H_1) + 1s(H_2) + 1s\,(H_3)$] is zero in case of NH_3 molecule, (*ii*) although the symmetry of the 2*s* orbital of nitrogen and [$1s\,(H_1) + 1s(H_2) + 1s\,(H_3)$] is the same but still we try to avoid this overlap.

0.38 For the second group elements, suggest the possible hybrid atomic orbitals which can give trigonal planar structure. (*i*) (s, p_x, p_y), (*ii*) (s, d_{xy}, $d_{x^2-y^2}$), (*iii*) (d_{z^2}, p_x, p_y), (*iv*) d_{z^2}, d_{xy}, $d_{x^2-y^2}$.

0.39 Find the trace of the square planar molecules and irreducible representation of the D_{4h} symmetry point group. Suggest the possible combination of atomic orbitals on the central atom which can lead to hybrid orbitals giving this structure. (*i*) dsp^2 and (*ii*) d_2p^2. Label these p and d orbitals.

0.40 Repeat the Problem 0.36 for AB_5 structure. (*i*) dsp^3, (*ii*) d^4s, (*iii*) spd^3, (*iv*) pd^4 and (v) p^3d^2.

0.41 Benzene molecule belonging to D_{6h} symmetry point group can be treated by C_6 symmetry point group and the trace obtained by performing all the symmetry operations of C_6 symmetry point group is:

$\Gamma_r(\pi)$	6	0	0	0	–2	0

Determine the irreducible representations present in this reducible representation.

0.42 The electronic configuration for methane is $(1\,a_1)^2\,(1t_2)^6\,(2a_1)^0\,(2t_2)^0$. Since all the orbitals are paired, thus the ground state in methane is ^1A_g. Assign the irreducible representations for the transitions $1t_2 \rightarrow 2a_1$ and $1t_2 \rightarrow 2t_2$. Find out the number of allowed $\sigma \rightarrow \sigma$ transitions.

0.43 Energy levels for the cyclobutene in order of increasing energy belong to the following irreducible representations A, E and B of D_{4h} symmetry point group. Suggest the possible electronic transitions and find out which ones are allowed and which one are forbidden? Also find out the direction of polarisation.

0.44 1,3-Butadiene has two geometric structures, *i.e. cis*-butadiene and *trans*-butadiene. The former belongs to C_{2v} and the latter belongs to C_{2h} symmetry point groups respectively. In case of *cis* butadiene the energy levels in increasing order are designated as B_1, A_2, B_1 and A_2. Write the ground and excited states electronic configurations. In the first two transitions of lowest energies, *i.e.* $E_2 \rightarrow E_3$ and $E_2 \rightarrow E_4$, find out whether both are allowed?

[**Ans.** $(b_1)^2(a_2)^2$; $(b_1)^2(a_2)(b_1)$, $(b_1)^2(a_2)(a_2)$, both these transitions are allowed]

0.45 Repeat the problem 0.49 for tran-sbutadiene. See if there is any difference in the observation of spectrum.

[**Ans.** $(a_u)^2(b_g)^2$; $(a_u)^2(b_g)(a_u)$, $(a_u)^2(b_g)(b_g)$, $E_2 \rightarrow E_3$ is allowed and $E_2 \rightarrow E_4$ is forbidden]

0.46 Consider the $–NO_2$ group. Construct its π molecular orbitals. Assign the symmetry point group and irreducible representations to the π molecular orbitals. Find out whether the first-three transitions are allowed or forbidden?

0.47 Naphthalene molecule belongs to D_{2h} symmetry point group. Ten $p\pi$ orbitals can be clubbed as ψ_1, ψ_2 as (A_u), ψ_3, ψ_4, ψ_5 as (B_{1u}), ψ_6, ψ_7 as (B_{2g}), ψ_8 ψ_9 ψ_{10} ψ_2 as (B_{3g}). The ten π energy levels in terms of β (increasing order of energy) are: $(1B_{1u}, 2.303)$, $(1B_{2g}, 1.618)$, $(1B_{3g}, 1.303)$, $(2B_{1u}, 1.000)$, $(1A_u, –0.618)$, $(2B_{2g}, 0.618)$, $(2B_{3g}, 1.00)$, $(3B_{1u}, –1.303)$, $(2A_u, –1.618)$, $(3B_{3g}, –2.303)$. Write the ground state electronic configuration. What is the degeneracy for t he first excited state? Calculate the transition energy for the first-three electronic transitions in terms of β and find out whether these are allowed or forbidden, as well as direction of polarisation?

0.48 Three vibrational modes in water (C_{2v} symmetry point group) belong to $2A_1 + B_2$ irreducible representations. It is also obvious from the C_{2v} character table that the x, y and z components of the oscillating dipole moment are spanned by the B_1, B_2 and A_1 respectively. Find out whether $A_1 \rightarrow A_1$, and $A_1 \rightarrow B_2$ transitions ate allowed or not?

0.49 Continuing with the Problem 0.44, indicate whether $A_1 \rightarrow A_1$ and $A_1 \rightarrow B_2$ transitions are Raman active or not. Given, components of polarisability α_{ii} (as x^2, y^2 and z^2) are spanned by A_1, and α_{ij} (xy, yz, zx) are spanned by A_2, B_1 and B_2 respectively.

0.50 Molecules of type XY_3 belongs C_{3v} to symmetry point group. There are 6 vibrational modes belong to 2 A_1 + 2 E irreducible representations. Consulting the C_{3v} symmetry point group shows that all the transitions in these type of molecules are both IR and Raman active. And hence the number of vibrational modes, belonging to each irreducible representation.

0.51 Determine (*a*) infrared and (*b*) Raman activity as well as (*c*) the number of vibrations corresponding to irreducible representation in XY_4 molecule belonging to point group T_d.

0.52 We know that benzene has 30 degrees of vibrational modes. Two of these, given below, belong to A_{1g} and E_g irreducible representations of the D_{6h} symmetry point group. Find out whether these are infrared active or Raman active.

0.53 In Problem 0.4, suppose one of the hydrogen atoms is replaced by an atom X. Assign the symmetry point group to the new molecule (C_6H_5X). Indicate the differences observed in the vibrational spectra observed in pure benzene and mono-substituted benzene.

0.54 Following transitions are observed in the infrared and Raman spectrum of CCl_4 molecule, *i.e.* 459 cm^{-1}, 315 cm^{-1}, 217 cm^{-1}, 762 cm^{-1}, 791 cm^{-1}. Analyse these bands using group theory. Indicate whether there are any bands belonging to Fermi resonance or not?

CHAPTER 1

SI UNITS

1.1. INTRODUCTION

In normal day-to-day life, although variety of units have been used but in scientific world mainly two systems of units are commonly used. One is F.P.S. (Foot, Pound and Second), originated in U.K. and other C.G.S. (Centimetre, Gram and Second) originated in France. As the definitions stand today, foot or cm is a distance between the two fixed marks on a specific kind of metal, pound is a weight of a certain mass of metal whereas a gm used to be (originally specified) as a mass of one cubic centimetre of water at 0°C. The kilogram was later specified as being the mass of a particular piece of metal.

For a common man units of measure do not matter much and he will prefer to use the same units which he is using for centuries. But in the scientific world, if there is no common language it is not easy to understand the discoveries or an achievement of scientists of different countries, since the knowledge of science does not remain in domain of one person or of one country. The other major disadvantage of the present system of units is that they are not in any way derivable from the scientific formulas or natural constants. So the need was felt to arrive at a common system of units. The meeting was held under the auspicious of Conference Generale des Poids et Messues (C.G.P.M.) and agreed to adopt common units of measurements known as Systeme Internationale d′ Units or the International System of Units, the abbreviation of which is SI. Many countries have adopted this system of units.

According to SI, the unit for distance is 'metre (m)', defined as the length equal to 1 650 763.73 wavelengths in vacuum of the radiation corresponding to the transition between the levels $2\,p_{10}$ and $5d_5$ of the krypton-86 atom. The SI unit of area is the square metre (m^2). Land is often measured by the hectare (10000 m^2 or approximately 2.5 acres). The SI unit of volume is the cubic metre (m^3). Fluid volume is often measured by the litre (0.001 m^3 or dm^3).

The unit of time is second(s), defined as the duration of 9 192 631 770 periods of the radiation corresponding to the transition between two hyperfine levels of the ground state of the cesium-133 atom. The SI unit for frequency is the hertz (Hz) which equals one cycle per second.

The standard unit of mass, the kilogram (kg) is a cylinder of platinum-iridium alloy kept by the International Bureau of Weights and Measures at Paris. It is important to note that unit of mass is not defined in terms of physical constants and is quite possible that the base unit of mass now known as kg will be renamed as giorgi, symbol G. The SI unit of force is newton (N), which is the force when applied for one second will give to a 1 kg mass a speed of 1 m per second. One newton equals nearly 0.2 lb of force. The SI unit for work and any kind of energy is the Joule (J), 1 J = 1 N m. The SI unit for power of any kind is the Watt (W), 1W = 1 J s^{-1}.

The Kelvin or thermodynamic scale of temperature used in SI has its origin or zero point at absolute zero and has a fixed point at the triple point of water defined as 273.16 K. The triple point defined on Celsius scale is 0.01°C and on Fahrenheit scale as 32.02°F.

The basic unit of electric current is ampere (A), defined as the magnitude of the current that, when flowing through each of two long parallel wires separated by 1 m in free space, results in a force between the two wires (due to these magnetic fields) of 2×10^{-7} N for each metre of length. The SI unit of voltage is volt (V), which is defined 1 V = 1 W A^{-1}. The SI unit of electrical resistance is the ohm (Ω) which is defined 1 Ω = 1 V A^{-1}.

It is good if the standard units are used everywhere. Certain units, *e.g.* minute, hour, day, litre, degree Celsius, degree (angle), minute (angle) etc. will be used simultaneously along with SI units for sometimes though these units have no place in scientific world. Of course units like scurie, electron volt, gauss, bar and barn will be used along with SI units but we hope that with time they will be phased out.

In the following pages the usages of SI units will be discussed by taking certain examples. It will be shown that SI units are much better and one does not need to remember the units of many quantities.

1.2. SPECIAL NOTES ON THE USE OF SI UNITS

1. Only the singular form of units is to be used, *e.g.* km and not kms.
2. Full stop at the end of abbreviations is to be omitted, *e.g.* km not km.
3. The decimal sign is to be a point or a comma on the line, *e.g.* 2.5 or 2, 5 not 2.5.
4. Digits should be grouped in threes about the decimal point in order to facilitate the reading of long numbers. Commas should not be used to space digits in numbers, *e.g.* 16 543 121.133 45 and not 16, 543, 121.133, 45.
5. The degree sign is to be omitted when the Kelvin scale is employed, *e.g.* 273 K and not 273°K.
6. The solidus is not to be encouraged, but where used, no more than one should be employed, *e.g.* J kg^{-1} K^{-1} or J/kg K but not J/Kg/K.

1.3. BASIC UNITS

The basic units of the SI system are shown in Table 1.1.

TABLE 1.1. Base Units

Physical quantity	Unit	Symbol
Length	metre	m
Mass	kilogram	kg
Time	second	s

Electric current	ampere	A
Temperature	kelvin	K
Luminous intensity of light	Candela	Cd
Plane angle	radian	rad
Solid angle	Steradian	srn

The mol has been recommended by the ISO (International Organisation for Standarisation) as a basic unit but this remains to be accepted by the CGPM.

1.4. DERIVED UNITS

Derived units with special names are given in Table 1.2.

TABLE 1.2. Symbols for certain SI derived units

Physical Quantity	Name of SI Unit	Symbol of SI Unit	Definition of SI Unit
Force	newton	N	$kg\ m\ s^{-2}$
Pressure*	pascal	Pa	$kg\ m^{-1}\ s^{-2}\ (= N\ m^{-2})$
Energy	joule	J	$kg\ m^2\ s^{-2}\ (= N\ m)$
Power	watt	W	$kg\ m^2\ s^{-3}\ (J\ s^{-1})$
Electric charge	Coulomb	C	A s
Electric potential difference	volt	V	$kg\ m^2\ s^{-3}\ A^{-1}\ (= J\ A^{-1}\ s^{-1})$
Electric resistance	ohm	Ω	$kg\ m^2\ s^{-3}\ A^{-2}\ (= V\ A^{-1})$
Electric conductance*	Seimens	S	$kg^{-1}\ m^{-2}\ s^3 A^2 (= A\ V^{-1} = \Omega^{-1})$
Electric capacitance	farad	F	$A^2\ s^2\ kg^{-1}\ m^{-2}\ (= A\ s\ V^{-1})$
Magnetic flux	Weber	Wb	$kg\ m^2\ s^{-2}\ A^{-1}\ (= V\ s)$
Inductance	henry	H	$kg\ m^2\ s^{-2}\ A^{-2} (= V\ A^{-1}\ s)$
Magnetic flux density	Tesla	T	$kg\ s^{-2}\ A^{-1}\ (= V\ s\ m^{-2})$
Luminous flux	lumen	Lm	Cd sr
Illumination	lux	Lx	$Cd\ sr\ m^{-2}$
Frequency	hertz	Hz	s^{-1}

*Recommended by CIPM.

1.5. PREFIXES

It is recommended that prefixes to be used should differ from a unit in steps of 10^3 (Table 1.3).

Units to be eventually abandoned are listed in Table 1.4.

1.6. EQUATIONS OF ELECTROMAGNETISM FROM CGS TO SI

Table 1.5 gives the conversion factors for converting the equations representing the electromagnetic quantities from cgs to SI units, *i.e.* for each electromagnetic quantity expressed in cgs units, substitute the required value and rearrangement gives the desired quantities in SI units (see Exercises 1.9 and 1.10).

TABLE 1.3. Prefixes for fractions and multiples of SI units

Prefix	Symbol	Factor	Prefix	Symbol	Factor
tera	T	10^{12}	*centi	c	10^{-2}
giga	G	10^{9}	milli	m	10^{-3}
mega	M	10^{6}	micro	μ	10^{-6}
kilo	k	10^{3}	nano	n	10^{-9}
*hecto	h	10^{2}	pico	p	10^{-12}
*deca	da	10	femto	f	10^{-15}
*deci	d	10^{-1}	atto	a	10^{-18}

*An asterisk indicates that the use of these prefixes is limited to occasions where recommended prefixes are inconvenient.

TABLE 1.4

Physical quantity	Unit	SI equivalent
Length	*Ångstrom (Å)	10^{-10} m
	*Inch (in)	0.0254 m
	mile	1.609 34 km
	*nautical mile international	1.852 km
Area	*barn (b)	10^{-23} m^2
	*square inch (in^2)	645.16 mm^2
Volume	cubic inch	$1.638\ 71 \times 10^{-5}$ m^3
Mass	*pound (lb)	0.453 592 37 kg
Force	*dyne (dyn)	10^{-5} N
	poundal	0.138 255 N
	pound force	4.448 22 N
Pressure	*atmosphere (atm)	101.325 kN m^{-2}
	*bar (bar)	105 N m^{-2}
	torr (torr)	133.322 N m^{-2}
	p.s.i.	6 894.76 N m^{-2}
Energy	*calorie international	4.186 8 J
	calorie 15°C (cal_{IT})	4.185 5 J
	*calorie thermochemical (cal_{th})	4.184 J
	B.t.u.	1 055.06 J
	electron volt (eV)	$1.602\ 1 \times 10^{-19}$ J
	*erg (erg)	10^{-7} J
Power	horse power	745.700 W
Radioactivity	*curie (Ci)	3.7×10^{10} s^{-1}

Exposure	roentgen (R)	2.5×10^{-4} C kg^{-1}
Magnetic flux density	*gauss (G)	10^{-4} T
Kinematic viscosity	*stoke (St)	10^{-4}.m^2 s^{-1}
Dynamic viscosity	poise (P)	10^{-1} kg m^{-1} s^{-1}

*An asterisk indicates that the conversions are exact.

TABLE 1.5. Expressions for physical quantities

Physical quantity	CGS symbol	SI symbol	Physical quantity	CGS symbol	SI symbol
Electric charge	Q	$(4\pi\varepsilon_0)^{-1/2}$ Q	Resistivity	ρ	$4\pi\varepsilon_0\, \rho$
			Conductance	G	$G/4\pi\varepsilon_0$
Charge density	ρ	$(4\pi\varepsilon_0)^{-1/2}\rho$	Conductivity	K	$K/4\pi\varepsilon_0$
Surface charge density	σ	$(4\pi\varepsilon_0)^{1/2}$ σ	Impedance	Z	$4\pi\varepsilon_0 Z$
			Self-impedance	L	$4\pi\varepsilon_0$ L
Electric dipole moment	p	$(4\pi\varepsilon_0)^{-1/2}$ p	Capacitance	C	$C/4\pi\varepsilon_0$
			speed of light	c	$(\mu_0\varepsilon_0)^{-1/2}$
Electric current	i	$(4\pi\varepsilon_0)^{-1/2}$ i	Magnetic induction	B	$(4\pi\mu_0)^{1/2}$ B
Electric current density	j	$(4\pi\varepsilon_0)^{-1/2}$ j	Magnetic field strength	H	$(4\pi\mu_0)^{1/2}$ H
Electric field strength	E	$(4\pi\varepsilon_0)^{1/2}$ E	Magnetisation	M	$(\mu_0/4\pi)^{1/2}$ M
			Electromagnetic	m	$(\mu_0/4\pi)^{1/2}$ m
Electric displacement	D	$(4\pi\varepsilon_0)^{1/2}$ D			
			Moment relative	μ	μ/μ_0
Dielectric polarisation	P	$(4\pi\varepsilon_0)^{-1/2}$ P	permeability		
			Magnetic susceptibility	χ	$\chi/4\pi$
Electric potential	V	$(4\pi\varepsilon_0)^{1/2}$ V			
Relative permittivity	ε	$\varepsilon/\varepsilon_0$			
(Dielectric constant)	D				
Electric polarisability	α	$\alpha/4\pi\varepsilon_0$			
Resistance	R	$4\pi\varepsilon_0$ R			

TABLE 1.6. Revised list of general physical constants and their values

Constant	Symbol	Value	Units (SI)	Units (cgs)
Speed of light	c	2.997 923	10^8 m s^{-1}	10^{10} cm s^{-1}
Elementary charge	e	1.602 191 7	10^{-19} C	10^{-20} cm g$^{-1/2}$
		4.803 250		10^{-10} cm C^{-1} g^{-1} s

Avogadro constant	N_A	6.022 169	10^{23} mol^{-1}	10^{23} mol^{-1}
Atomic mass unit	u	1.660 531	10^{-27} kg	10^{-24} g
Electron rest mass	m_e	9.109 558	10^{-31} kg	10^{-28} g
Proton rest mass	m_p	1.672 614	10^{-27} kg	10^{-24} g
Faraday constant	F	9.648 670	10^{4} C mol^{-1}	10^{3} cm g^{-1} mol^{-1}
Planck constant	h	6.626 196	10^{-34} J s	10^{-27} erg s
Boltzmann constant	k_b	1.380 622	10^{-23} J K^{-1}	10^{-16} erg K^{-1}
Fine structure constant	α	7.297 351	10^{-3}	10^{-3}
Bohr magneton	μ_B	9.274 096	10^{-24} J T^{-1}	10^{-21} erg G 1
Nuclear magneton	μ_n	5.054 04	10^{-27} J T^{-1}	10^{-24} erg G^{-1}
Permittivity	ε_0	8.854 19	10^{-12} A^2 S^4 kg^{-1} m^{-3}	
Permeability of vacuum	μ_0	4π	10^{-7} kg s^{-2} A^{-2}	
Gas constant	R	8.314 33	J K^{-1} mol^{-1}	8.314 33 × 10^{7} erg K^{-1} mol^{-1} 1.987 17cal K^{-1} mol^{-1} 0.082 056 atm L K^{-1} mol^{-1}
Stefan-Boltzmann constant	σ	5.669 61	10^{-8} W m^{-2} K^{-4}	10^{-5} erg cm^{-2} s^{-1} K^{-4}
Gravitational	G	6.673 2	10^{-11} N m^2 kg	10^{-8} dyn cm^2 g^{-2}

1.7. ATOMIC UNITS

Till now we have been using SI units, so that the potential energy $V(r)$ is expressed in joules (J). However this is not a satisfactory procedure for atomic and molecular-level calculations. This can be seen from Exercise 1.11 which involves the calculation of centrifugal distortion coefficient (D) used in rotational spectrum (see Section 5.3). From this exercise it is clear that its value is ~ 10^{-3} cm^{-1}. It is very small. Now if we substitute the value of h^4, in SI units its value is 1.9276 × 10^{-132} J^4 s^4. The magnitude of this quantity is so small that many computers will incur an underflow error that causes $h^4/16\pi^4$ to be set equal to zero. Of course, we avoid this error first dividing $h^4/16\pi^4$ by r_e^6 before computing $h^4/16\pi^4$, but when we are executing computationally intensive procedures on a computer, the problem is too complex to allow us to foresee all the overflow and underflow danger point. Therefore, we need to employ a system of units in which the quantities of interest are much closer to unity. Such units give us the maximum protection against both underflows and overflows. These units are called atomic units. The conversion factors from each other are compiled in Table 1.7.

TABLE 1.7. Conversion factors between atomic, SI and electrostatic units

Atomic units	SI unit	Electrostatic unit
1 charge unit	1.602 177 × 10^{-19} C	4.803 21 × 10^{-10} statC
1 mass unit	9.109 39 × 10^{-31} kg	9.109 39 × 10^{-28} g
1 distance unit (1 bohr)	0.529 177 × 10^{-10} m	0.529 177 × 10^{-10} cm
1 time unit	2.418 88 × 10^{-17} s	2.418 88 × 10^{-17} s
1 energy unit (1 hartree)	4.359 75 × 10^{-18} J	4.359 75 × 10^{-18} ergs

TABLE 1.8. Fundamental constants and other conversion factors

Constant	Atomic unit	SI value	Electrostatic value
e	1	$1.602\ 177 \times 10^{-9}$ C	$4.803\ 21 \times 10^{-10}$ statC
$\hbar$	1	$1.054\ 57 \times 10^{-34}$ J s	$1.054\ 57 \times 10^{-34}$ erg s
h	2π	$6.626\ 08 \times 10^{-34}$ J s	$6.626\ 08 \times 10^{-34}$ erg s
m_e	1	$9.109\ 39 \times 10^{-31}$ kg	$9.109\ 39 \times 10^{-28}$ g
m_p	1,836.1	$1.672\ 62 \times 10^{-27}$ kg	$1.672\ 62 \times 10^{-24}$ g
a_0	1	$0.529\ 177 \times 10^{-10}$ m	$0.529\ 177 \times 10^{-8}$ cm
c Speed of light in a vacuum	137.036	$2.997\ 924 \times 10^{8}$ m s^{-1}	$2.997\ 924 \times 10^{10}$ cm s^{-1}

Note: 1 hartree = 27.2114 eV = 2 625.5 kJ mol^{-1} = 627.51 Kcal mol^{-1}.
1 hartree is equivalent to 2 19 475 cm^{-1}.
1 eV is equivalent to 8 065. 6 cm^{-1}.

***Exercise* 1.1:** The unit of pressure is atmosphere. Calculate its value in SI unit.

$$\begin{aligned} 1 \text{ atmosphere} &= \text{Density} \times \text{Length} \times \text{Acceleration} \\ &= (\text{kg m}^{-3})\ (\text{m})\ (\text{m s}^{-2}) \\ &= 13\ 595.1 \times 0.760 \times 9.806\ 65 \\ &= 101\ 325 \text{ kg m s}^{-2} \text{ m}^{-2} \\ &= 101\ 325 \text{ N m}^{-2} \end{aligned}$$

since kg m s^{-2} = 1 Newton

***Exercise* 1.2:** Root mean square velocity $\sqrt{\bar{c^2}}$ of a molecule is given by the relation:

$$\sqrt{\bar{c^2}} = (3RT/M)^{1/2}$$

where R is the gas constant, T is temperature and M is for molecular weight. What is the value in SI units?

Units of $T = \text{K}$

$R = \text{J K}^{-1} \text{ mol}^{-1}$

$M = (\text{g mol}^{-1})\ (10^{-3} \text{ kg g}^{-1})$

$$\sqrt{\bar{c^2}} = \left(\frac{\text{J K}^{-1} \text{ mol}^{-1} \text{ K}}{\text{kg mol}^{-1}}\right)^{1/2} = \left(\frac{\text{Joule}}{\text{kg}}\right)^{1/2}$$

Now Joule $= \text{kg m}^2 \text{ s}^{-2}$

$$\sqrt{\bar{c^2}} = \left(\frac{\text{kg m}^2 \text{ s}^{-2}}{\text{kg}}\right)^{1/2} = \text{m s}^{-1}$$

whereas traditional unit is cm s^{-1}.

***Exercise* 1.3:** In the limiting density method, the molecular weight is related to density (d) and pressure (p) by

$$M = (d/p)_0 RT$$

Derive the SI unit for molecular weight.

$$M = (d/p)_0 RT$$

$$d = \text{density} = \text{kg m}^{-3}$$

$$p = \text{N m}^{-2} = \text{kg m s}^{-2}\ \text{m}^{-2}$$

$$R = \text{J K}^{-1}\ \text{mol}^{-1} = \text{kg m}^2\ \text{s}^{-2}\ \text{K}^{-1}\ \text{mol}^{-1}$$

$$T = \text{K}$$

$$M = \frac{\text{kg m}^{-3}}{\text{kg m s}^{-2}\ \text{m}^{-2}}\ \text{kg m}^2\ \text{K}^{-1}\ \text{mol}^{-1}\ \text{K}$$

$$= \text{kg mol}^{-1}$$

***Exercise* 1.4:** Chapman's equation gives the value of viscosity by the equation

$$\eta = 0.499\ d\bar{c}l$$

where d is density, $\bar{c}$ mean velocity and l the mean free path. Calculate the SI unit for η.

$$\text{Viscosity } \eta = 0.499\ dcl$$

$$d = \text{density} = \text{kg m}^{-3}$$

$$\bar{c} = \text{mean velocity} = \text{m s}^{-1}$$

$$l = \text{mean free length} = \text{m}$$

$$\eta = (\text{kg m}^{-3})\ (\text{m s}^{-1})(\text{m})$$

$$= \text{kg m}^{-1}\ \text{s}^{-1}$$

$$= \text{kg m s}^{-2}\ \text{m}^{-2}\ \text{s} = \text{N s m}^{-2}$$

Traditional unit 'poise' but now it is (Newton second)/metre ^{2}.

***Exercise* 1.5:** Difference in heat capacities at constant pressure and at constant volume is

$$C_p - C_v = \frac{TV_0\alpha^2}{\beta}$$

If α is the coefficient of linear expansion and β is coefficient of compressibility, SI unit for $C_p - C_v$ is J K^{-1} mol^{-1}.

$$C_p - C_v = \frac{TV_0\alpha^2}{\beta}$$

$$\text{Unit of } T = \text{K}$$

$$V_0 = \text{m}^3$$

Coefficient of expansion $= \left(\frac{dV}{dT} \times \frac{1}{V^0}\right)$

$$\frac{\text{m}^3}{\text{K}} \cdot \frac{1}{\text{m}^3} = \text{K}^{-1}$$

Coefficient of compressibility

$$= -\frac{dV}{dp}\cdot\frac{1}{V} = \frac{m^3}{N\,m^{-2}}\cdot\frac{1}{m^3}$$

$$= N^{-1}m^2$$

$$C_P - C_v = \frac{K\,m^3(K^{-1})^2}{N^{-1}\,m^2} = N\,K\,m\,K^{-2}$$

$$= kg\,m\,s^{-2}\,m\,K^{-1} = kg\,m^2\,s^{-2}\,K^{-1}$$

$$= J\,K^{-1}\quad \text{since } J = kg\,m^2\,s^{-2}$$

***Exercise* 1.6:** Joule-Thomson coefficient is expressed as

$$\mu = \left(\frac{\Delta T}{\Delta P}\right)_H = \frac{(2a/RT - b)}{C_p}$$

where a and b are the van der Waals gas constant and the others have usual meaning. Express μ in SI unit. Joule-Thomson coefficient

$$\mu = \left(\frac{\Delta T}{\Delta P}\right)_H = \frac{(2a/RT - b)}{C_p}$$

Units of

$$a = m^4\,N\,mol^{-2} = kg\,m\,s^{-2}\,m^4\,mol^{-2}$$

$$b = m^3\,mol^{-1}$$

$$R = J\,mol^{-1}\,K^{-1} = kg\,m^2\,s^{-2}\,mol^{-1}\,K^{-1}$$

$$T = K$$

Molar heat capacity $= \bar{C}_p = J\,K^{-1}\,mol^{-1} = kg\,m^2\,s^{-2}\,K^{-1}\,mol^{-1}$

$$\mu = \frac{(2a/RT - b)}{C_p}$$

$$\mu = \frac{(kg\,m\,s^{-2}\,m^4\,mol^{-2}/kg\,m^2\,s^{-2}\,mol^{-1}\,K^{-1}K) - m^3\,mol^{-1}}{kg\,m^2\,s^{-2}\,K^{-1}\,mol^{-1}}$$

$$= \frac{m^3\,mol^{-1}}{kg\,m^2\,s^{-2}\,K^{-1}\,mol^{-1}} = \frac{K}{kg\,m^{-1}\,s^{-2}}$$

$$= \frac{K}{kg\,m\,s^{-2}\,m^{-2}} = \frac{K}{N\,m^{-2}} \qquad \because N = kg\,m\,s^{-2}$$

whereas from $\mu = \dfrac{\Delta T}{\Delta p} = \dfrac{K}{N\,m^{-2}} = K\,N^{-1}\,m^2$

***Exercise* 1.7:** The vapour pressure at different temperature is given by the equation

$$\log p = A - B/T$$

where p is in mm of Hg, A $(\log p)_{1/T = 0}$ and B is a constant. Express the above equation in SI unit.

Vapour pressure is related to temperature by the relation

$$\log p = A - B/T$$

Here p is mm of Hg. To express in N m^{-2}, we must multiply by a factor 133.3 as 1 mm Hg = 133.3 N m^{-2}. So

$$\log (p \times 133.3) = A - B/T$$

$$\log p + 2.124\,9 = A - B/T$$

***Exercise* 1.8:** In collision theory, the collision frequency (Z) is given by the relation

$$Z = 2\sigma^2 (\pi RT/M)^{1/2}\, n^2$$

where σ is the collision diameter, M molecular weight, n is number of molecules per unit volume. What will be the SI unit for Z?

Substituting the respective units for each quantity in above equation:

$$Z = \text{m}^2 \left[\frac{(\text{J K}^{-1}\ \text{mol}^{-1})\ (\text{K})}{(\text{g mol}^{-1})\ (10^{-3}\ \text{kg g}^{-1})}\right]^{1/2} (\text{molecules m}^{-3})^2$$

Expressing J as kg m s^{-2} m and neglecting the constant terms the units of Z, will be

$$(\text{m}^2)\left(\frac{\text{kg m}^2\ \text{s}^{-2}}{\text{kg}}\right)^{1/2} \text{molecules}^2\ \text{m}^{-6}$$

$$= \text{molecules}^2\ \text{s}^{-1}\ \text{m}^{-3}$$

***Exercise* 1.9:** Express Coulomb's law in SI unit.

The Coulomb's law between two electronic charges is given as

$$F = \frac{e^2}{r^2}$$

where e is cgs unit, r in cm and F in dynes. Replace e^2 by $e^2/4\pi\varepsilon_0$ as mentioned in Table 1.5 and express r in m, the equation becomes,

$$F = \frac{e}{(4\pi\varepsilon_0)^{1/2} r^2} \cdot \frac{e}{(4\pi\varepsilon_0)^{1/2}} = \frac{e^2}{4\pi\varepsilon_0 r^2} \text{ in N}$$

***Exercise* 1.10:** Find the relation between expressions for specific conductance (K) and molar conductance expressed as in SI and C.G.S. units.

In C.G.S. unit $\left(K = L\dfrac{l}{A}\right)$

where L is the conductance, l is the distance between two electrodes and A the area of the electrode. The unit for K in cgs unit is mho cm^{-1}. Replacing the values of l, A in SI units

$$\text{K} = L\ (\Omega^{-1})\ \frac{(1\ \text{cm})\ (10^{-2}\ \text{m cm}^{-1})}{(A\ \text{cm}^2)\ (10^{-2}\ \text{m cm}^{-1})^2} = 100\ L\ \frac{l}{A}\ \Omega^{-1}\ \text{m}^{-1}$$

i.e. the value of K in cgs unit should be multiplied by 100 to get its value in SI units.

Similarly

$$\Lambda = \frac{K \times 1000}{C}\ \Omega^{-1}\ cm^2\ mol^{-1}$$

where C is the concentration mol L^{-1}, K in $\Omega^{-1}\ cm^{-1}$. Now expressing in SI units

$$\Lambda = \frac{(K\ \Omega^{-1}\ cm^{-1})\ (10^2\ cm\ m^{-1})\ (1000)}{(C\ mol\ cm^{-3})\ (10^2\ m\ cm^{-1})^3}$$

$$= \frac{(K \times 10^{-5}\ \Omega^{-1} m^{-1})}{c \times 10^6\ mol\ m^{-3}} = \frac{K}{10C}\ \Omega^{-1}\ m^2\ mol^{-1}$$

$$= \Lambda\ (\text{cgs units}) \times 10^{-4}$$

***Exercise* 1.11:** The electrostatic force between the two charges Q_{1e} and Q_{2e} are given as

$$F(r) = \frac{Q_{1e}Q_{2e}}{r^2}$$

Find the value of Q_{1e}.

***Solution*:** Assume that the charges on the two particles are the same, say Q. The electrostatic force between the two particles is then given as

$$F(r) = 1 dyne = \frac{QQ}{r^2} = \frac{Q^2}{(4\pi\varepsilon_0)r^2} = 10^{-5}\ \text{Newton}\ (1N)$$

Knowing that 1 dyne is equal to 10^{-5} N. Substituting the values of the parameters we get

$Q = [(10^{-5}\ kg\ m\ s^{-2})(4)(3.14159927)(8.85419 \times 10^{-12}\ kg^{-1}\ m^{-2}\ s^2\ C^2\ m^{-1})(10^{-4}\ m^2)]^{1/2}$

$= 3.335641 \times 10^{-10}$ C

Therefore 1 statC is equivalent to 3.335641×10^{-10} C or 1 C is equivalent to 2.99792×10^9 statC.

***Exercise* 1.12:** Calculate the value of potential energy between two charged particles having $Q = e$ and at a distance $r = a_0$.

***Solution*:** The expression for the potential energy between two particles with $Q = e$ and at $r = a_0$ in terms of SI units is

$$V(r = a_0) = -\frac{e^2}{4\pi\varepsilon_0 r}$$

$$= -\frac{(1.602\ 177 \times 10^{-19}\ C)^2}{(4)(3.141\ 592\ 7)(8.854\ 19 \times 10^{-12}\ kg^{-1}\ m^{-2}\ s^2\ C^2\ m^{-1})(0.529\ 177 \times 10^{-10}\ m)}$$

$$= 4.359\ 75 \times 10^{-18}\ J \qquad ...(A)$$

The expression for the potential energy between two particles with $Q = e$ and at $r = a_0$ in terms of electrostatic units is

$$V\ (r = a_0) = -\frac{e^2}{r} = -\frac{(4.803\ 21 \times 10^{-10})^2}{(0.529\ 177 \times 10^{-8}\ m)} = -4.359\ 75 \times 10^{-11}\ ergs \qquad ...(B)$$

These figures obtained in SI units and electrostatic are too small and if we work with energies of this magnitude in atomic and molecular calculations, there is a danger of underflows and overflows. So it is better if we choose some other appropriate unit.

Now let us try to find the value of potential energy in atomic units (au) we can arrive at

$$V(r = a_0) = -\frac{e^2}{r} = -\frac{e^2}{a_0} = \frac{(1)^2}{1} = -1 \text{ energy unit } (-1 \text{ hartree}) \quad \text{...(C)}$$

So comparing (*A*), (*B*) and (*C*), we can say that the new unit, atomic unit (au) one hartree equivalent to of = 4.359 75 × 10^{-18} J in SI units and equivalent to = – 4.359 75 × 10^{-11} ergs in the electrostatic units.

Lastly, the atomic unit for mass is the mass of electron as the atomic unit, *i.e.* 1 au = 9.109 39 × 10^{-31} kg.

PROBLEMS

1.1 van der Waals equation of state for one mol of gas is $\left(p + \frac{a}{\bar{V}^2}\right)(\bar{V} - b) = RT,$ where *a* and *b* are the constants with units atm L^2 mol^{-2} and *L* mol^{-1} respectively. The critical pressure is related to the above constants as $p_c = \frac{a}{27b^2}$. Show that the SI unit for pressure is N m^{-2}.

1.2 In Clapeyron equation the change in pressure with temperature is $\frac{dp}{dT} = \frac{\Delta H}{T\Delta V}$ or $\Delta V = \frac{\Delta H}{T\left(\frac{dp}{dT}\right)}$, where ΔH is the enthalpy of vapourisation, *T* temperature and ΔV is the difference in molar volumes of vapours and liquid. Find the SI unit for volume.

1.3 (*a*) The frequency factor *A* in the transition state theory is $A = PZ = (e^n k_b T/h) \exp(\Delta S_0^{\neq}/R)$ where *P* is the steric factor, *e* the exponential, *n* the molecularity of the reaction, k_b Boltzmann constant and $\Delta S_0^{\neq}$, the standard entropy change which accompanies the formation of the activated complex and the reactants. What will be the SI unit for *A*?

(*b*) How the numerical value of $\Delta S_0^{\neq}$ will change if *A* is (*i*) 10^{14} cm^3 mol^{-1} s^{-1}, (*ii*) 10^{11} L. mol^{-1} s^{-1} and (*iii*) 10^8. m^3 mol^{-1}.

1.4 What will be the energy of a photon having wavelength (*a*) 200 nm; (*b*) 250 nm and (*c*) 10 μm?

1.5 Einstein's equation relating the vibrational frequency (ν) of solids to their compressibility (β) is

$$\nu = \frac{\sqrt{3}N_A^{1/3}}{2\pi} \frac{V^{1/6}}{A^{1/3}\beta^{1/3}}$$

where N_A is Avogadro's number, V, the specific volume, A, atomic weight. Show that the above relation gives the SI units for v as s^{-1}.

1.6 The change in the free energy (ΔG) of a chemical reaction in an electrochemical cell is given by the relation $\Delta G = -\,n\,FE$, where n is the number of electrons, F the Faraday constant and E, the electromotive force. From above relation calculate the SI unit for ΔG.

1.7 In Debye-Hückel theory of electrolytical conductance, the Debye length (1/K) is defined as

$$\mathrm{K} = \left(\frac{8\pi\, N_A e^2}{1000\, \varepsilon\, k_b T}\mu\right)^{1/2}$$

where ε is the permittivity in Faraday m^{-1} ($C^2\ N^{-1}\ m^{-2}$) e is charge in C, k_b, the Boltzmann constant, N_A, Avogadro's number, T, temperature (K), μ the ionic strength (mol dm^{-3}). Find the unit for κ.

1.8 In cgs unit Curie's law for the effective electromagnetic moment (m_{eff}) is

$$m_{\text{eff}} = \left(\frac{3k_b T\chi M}{N_A\, d}\right)^{1/2}$$

where χ, M and d are magnetic susceptibility, molar mass and density of the substance. Derive the above expression in SI units.

1.9 Find the relation between the expressions for Bohr magneton expressing in SI and cgs units.

1.10 Find the multiplication factor to convert molar susceptibility from cgs units to SI units:

$$\chi_M = \frac{\chi_M}{d}.$$

Addition Problems

1.11 Calculate the value of atomic unit for time in terms of either SI or electrostatic units.

1.12 Relate Planck's constant expressed in atomic unit to SI and electrostatic units.

1.13 Bohr's radius is expressed as $a_0 = [\frac{(4\pi\varepsilon_0)h^2}{4\pi^2 m_e e^2}]$. Using the SI unit system show that its value comes out to 1 bohr.

1.14 Show that 1 hartree is equivalent to 8065.6 cm^{-1}.

1.15 Show that 1 eV = 0.036 749 3 hartree or 1 hartree = 27.211 4 eV

1.16 The potential energy curve for the vibrational motion can be described by the Morse potential equation. The centrifugal distortion coefficient for this distortion is given as

$$D = \frac{h^4}{64\pi^4\mu^2\beta^2 D_e R_e^6}$$

where D is the centrifugal distortion constant, μ is the reduced mass (consider H_2 molecule), β is the Morse constant = $1.941 \times 10^{10}\ m^{-1}$, D_e is the depth of the potential well, also known as the equilibrium dissociation energy = 7.607×10^{-19} J and R_e is the equilibrium bond distance and its value for hydrogen molecule = 0.074 nm. Calculate its value in terms of SI units and also in atomic units. Compare the two values.

1.17 In Chapter 5, the energy of the rigid rotator is given by the expression $E_J = \frac{h^2}{8\pi^2 \mu r^2} J(J+1)$, where μ is the reduced mass of the diatomic molecule, *r* is the equilibrium bond length of the diatomic molecule. Considering the diatomic molecule to be H_2, calculate the value of $\frac{h^2}{8\pi^2 \mu r^2}$ in (*i*) in SI units and (*ii*) in atomic units.

1.18 In Chapter 1, we found that the energy of Bohr model for hydrogen atom, the energy of different levels are given by $E_n = -\frac{2\pi^2 \mu Z e^4}{(4\pi\varepsilon_0)^2 h^2 n^2}$, where *e* is the charge in SI units? (*i*) What is the equivalent expression in electrostatic units? (*ii*) What system of units used in this expression? (*iii*). What is the value of the ground state energy of Bohr hydrogen atom in hartrees and in eV?

LITERATURE CITED

1. G. Socrates *J. Chem. Ed.*, **46**, 710 (1969).
2. M.L. McGlashan, *Pure Applied Chem.*, **21**, 1 (1970).
3. N.H. Davies, *Chem. in Britain,* **6**, 344 (1970).
4. N.B.S. Tech. *News Bull.*, Jan., 1971; reprinted *J. Chem. Ed.*, **48**, 569 (1971).
5. T.I. Quickenden and R.C. Marshal, *J. Chem. Ed.*, **49**, 114 (1972).
6. A.C. Norris, *J. Chem. Ed.*, **48**, 797 (1971).
7. R.I. Holliday, *J. Chem. Ed.*, **53**, 21 (1976).
8. T. Coitas and N. Kallay, *J. Chem. Ed.*, **54**, 530 (1977).

CHAPTER 2

ELEMENTARY QUANTUM MECHANICS

2.1. DUAL NATURE OF RADIATION

The light rays have both the electrical and magnetic properties and are commonly called electromagnetic radiation. It has been shown that the eletromagnetic radiation has a dual character, namely that of a wave and that of a particles. (*i*) The frequency (ν) and wavelength (λ) are related to each other by the relation

$$\nu\lambda = c \qquad ...(2.1)$$

ν has the units Hertz (Hz), λ = metre and the velocity c of m s^{-1}. The other commonly used unit (specially in spectroscopy) for frequency is cm^{-1}, *i.e.* reciprocal of wavelength, called wave number $\bar{\nu}$

$$\bar{\nu} = \frac{1}{\lambda}, \qquad \nu = \bar{\nu}c \qquad ...(2.2)$$

(*ii*) Electromagnetic radiation consists of particles having small packets of energies called quanta. Planck was the first to relate the wave and particle characteristics by an equation

$$\varepsilon = h\nu \qquad ...(2.3)$$

ε is the energy of the particle, ν the frequency and h the Planck constant with unit J s.

***Exercise* 2.1:** Calculate the energy of a photon associated with light of wavelength 200 nm.

$$\lambda = 200 \text{ nm}$$

$$\nu = \frac{c}{\lambda} = \frac{(3.0 \times 10^8 \text{ m s}^{-1})}{200 \times 10^{-9} \text{ m}} = 1.5 \times 10^{15} \text{ s}^{-1}$$

$$\varepsilon = h\nu = (6.63 \times 10^{-34} \text{ J s})(1.5 \times 10^{15} \text{ s}^{-1}) = 9.94 \times 10^{-19} \text{ J}$$

***Exercise* 2.2:** The energy required to carry out the following reaction $H_2 \rightarrow 2\,H$ is 104 kcal mol^{-1}. What is the corresponding value of the wavelength of this photon?

$$E = N_{av}\, h\nu = N_{av}\, h\frac{c}{\lambda}$$

$$\lambda = \frac{(6.02 \times 10^{23}\ \text{mol}^{-1})\,(6.63 \times 10^{-34}\ \text{J s})\,(3.0 \times 10^{8}\,\text{m s}^{-1})}{(104\ \text{kcal mol}^{-1})\,(4.184 \times 10^{3}\ \text{J kcal}^{-1})}$$

$$= 274.5\ \text{nm}$$

2.2. CONCEPT OF QUANTUM IN ENERGY

2.2.1. *Blackbody Radiation*

When a body is heated it emits radiations of different frequencies (or wavelengths). The exact frequency spectrum emitted by this body depends on its *nature* at a given temperature. An ideal body that emits and absorbs all the frequencies with equal rate is called *blackbody*. This serves as the ideal material for any radiating body. The radiation emitted by a blackbody is called blackbody radiation. The intensity distribution of a blackbody at different temperatures as a function of frequency is given in Fig. 2.1.

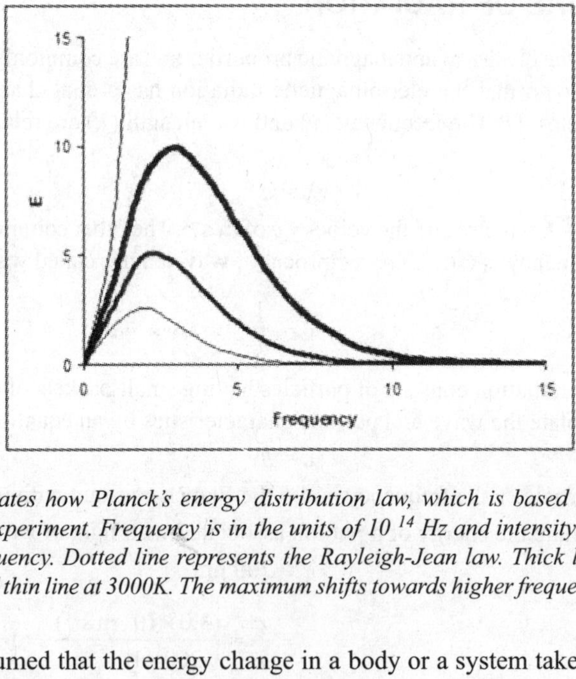

Fig. 2.1 *The graph indicates how Planck's energy distribution law, which is based on the idea of quantized oscillators, agreed with experiment. Frequency is in the units of 10^{14} Hz and intensity in the units of 10^{-16} J per unit volume per unit frequency. Dotted line represents the Rayleigh-Jean law. Thick line represents at 5000K, medium line at 4000K and thin line at 3000K. The maximum shifts towards higher frequency (energy) with increase of temperature.*

Earlier it was assumed that the energy change in a body or a system takes place in a continuous manner. Few relations available at that time to explain the spectra distribution, given in Fig. 2.1 are:

***(i)* Rayleigh-Jean Law**

This law states

$$\varepsilon(\nu,T)d\nu = \frac{8\pi k_b T}{c^3}\ -\nu^2 d\nu \qquad ...(2.4)$$

where $\varepsilon(\nu, T)$ is the density of radiative energy (J m^{-3} Hz^{-1}) between the frequency ν and $\nu + d\nu$, k_b is the

Boltzmann constant (= R/N_{av}, R is a gas constant and N_{av} is an Avogadro number), c is the velocity of light and T is the temperature in Kelvin. The curve predicted by equation (2.4) is given as a dashed line in Fig. 2.1. Agreement between the predicted curve and the experimental one is good at the lower frequencies, but departs at high frequencies. This departure from the experimental curves at higher frequencies is known as *ultraviolet catastrophe*.

Earlier it was assumed that energy change in a system takes place in a continuous manner. Hence, it was not possible to explain the energy distribution of radiation emitted by a black body at a certain temperature. Planck introduced the concept that increase or decrease in the energy does not take place continuously but in a discrete manner. The separation between energies of various groups of oscillators being given by the relation (2 3). Thus the average energy of the oscillator having frequency of oscillation ν is given by

$$\bar{\varepsilon} = \frac{h\nu}{\exp(h\nu / k_b T) - 1} \qquad \text{...(2.5)}$$

where k_b is the Boltzmann constant. The intensity of spectral distribution in the range λ and $\lambda + d\lambda$ is given by the expression

$$d\varepsilon(\lambda) = \left(\frac{8\pi hc}{\lambda^5}\right) \frac{\exp(-hc / k_b \lambda T)}{1 - \exp(-hc / k_b \lambda T)} d\lambda \qquad \text{...(2.6)}$$

Equation (2.6) expresses Planck's radiation distribution law in terms of wavelength. This can also be expressed in terms of frequency as

$$d\varepsilon(\nu, T) = \frac{8\pi h\nu^3}{c^3} \frac{d\nu}{e^{\frac{h\nu}{k_b T}} - 1} \qquad \text{...(2.7)}$$

A very useful relation between the wavelength (λ_{max}) at which $\varepsilon(\lambda, T)$ is maximum and temperature T can be derived from equation (2.7),

$$\lambda_{max} T = \frac{hc}{4.965 k_b} = 2.90 \times 10^{-3} \text{ m K} \qquad \text{...(2.8)}$$

Equation (2.8) is known as Wien's displacement law. Using this relation one can predict the temperature of the body if λ_{max} is known or one can predict the λ_{max} if temperature is known.

***Exercise* 2.3:** Calculate the average energy of an oscillator with frequency 10^{14} Hz at 1000 K.

$$\frac{h\nu}{k_b T} = \frac{(6.63 \times 10^{-34} \text{ J s})(10^{14} \text{ s}^{-1})}{(1.38 \times 10^{-23} \text{ J K}^{-1})(1000 \text{ K})} = 4.8$$

$$\exp(h\nu/k_b T) - 1 = e^{4.8} - 1 = 120.51$$

Using relation (2.5)

$$\bar{\varepsilon} = \frac{h\nu}{\exp(h\nu / k_b T) - 1} = \frac{(6.63 \times 10^{-34} \text{ J s})(10^{14} \text{ } s^{-1})}{120.51} = 5.5 \times 10^{-22} \text{ J}$$

***Exercise* 2.4:** A small cavity of volume 4×10^{-6} m^3 is heated at 2000 K. Calculate the amount of energy corresponding to wavelength in the range 400–430 nm.

The equation (2.6) is used to calculate the energy of photon emitted by the cavity. The middle of the wavelength is 415 nm and $d\lambda = 30$ nm.

$$\frac{hc}{k_b T\lambda} = \frac{(6.63\times10^{-34}\text{ J s})(3.0\times10^{8}\text{ m s}^{-1})}{(1.38\times10^{-23}\text{ J K}^{-1})(2000\text{ K})(415\times10^{-9}\text{ m})} = 17.36$$

$$1 - e^{-17.36} = 1$$

$$d\epsilon = \frac{\left(8\times\frac{22}{7}\right)(6.63\times10^{-34}\text{ J s})(3.0\times10^{8}\text{ m s}^{-1})(e^{-17.36})(30\times10^{-9}\text{ m})}{(415\times10^{-9}\text{ m})^{5}}$$

$$= 3.52\times10^{-7}\text{ J m}^{-3}$$

Therefore, the energy emitted by the 4×10^{-6} m^3 cavity in the range 400–430 nm is 14.08×10^{-13} J.

***Exercise* 2.5:** (*a*) Derive Equation (2.8) from Equation (2.6), (*b*) if the maximum in the radiation distribution of Sun radiation occurs at 500 nm calculate the temperature of Sun.

***Solution*:** Equation (2.6) can also be written as

$$d\varepsilon(\lambda, T)d\lambda = \frac{8\pi hc}{\lambda^5}\frac{d\lambda}{e^{\frac{hc}{\lambda k_b T}}-1}$$

Removing $d\lambda$ from both side, differentiate both sides with respect to λ and equating this equal to zero, we get

$$\frac{d\varepsilon}{d\lambda} = 0 = 8\pi hc\left[-\frac{5}{\lambda^6}\frac{1}{e^{\frac{hc}{\lambda k_b T}}-1} + \left(-\frac{1}{\lambda^5}\right)\frac{1}{\left(e^{\frac{hc}{\lambda k_b T}}-1\right)^2}e^{\frac{hc}{\lambda k_b T}}\frac{hc}{k_b T}\left(-\frac{1}{\lambda^2}\right)\right]$$

$$\frac{8\pi hc}{\lambda^6\left(e^{\frac{hc}{\lambda k_b T}}-1\right)}\left[-5 + \frac{e^{\frac{hc}{\lambda k_b T}}}{\left(e^{\frac{hc}{\lambda k_b T}}-1\right)}\frac{hc}{\lambda k_b T}\right] = 0$$

Then $$\lambda_{max} = \left(\frac{hc}{5k_b T}\right)\frac{e^{\frac{hc}{\lambda k_b T}}}{\left(e^{\frac{hc}{\lambda k_b T}}-1\right)}$$

For $\frac{hc}{\lambda} >> k_b T$, $\frac{hc}{\lambda k_b T} >> 1$, thus one (1) in the denominator can be neglected as compared to the exponential term. Thus above equation reduces to

$$\lambda_{max} = \frac{hc}{5k_b T} = \frac{C}{T}, \text{ where } C = \frac{hc}{5k_b}$$

Substituting the values of the constants in C, we get

$$C = \frac{hc}{5k_b} = \frac{(6.626\times10^{-34}\text{ J s})(3\times10^{8}\text{ m s}^{-1})}{5\times1.38\times10^{-23}\text{ J K}^{-1}} = 2.90\times10^{-3}\text{ m K}$$

So, we get

$$\lambda_{max} = \frac{hc}{5k_bT} = \frac{2.90 \times 10^{-3} \text{ m K}}{T(\text{K})}, \text{ equation (2.8)}$$

(*b*) Using equation (2.8) and substituting the value of λ_{max} = 500 nm, we get

$$T = \frac{C}{\lambda_{max}} = \frac{2.9 \times 10^{-3} \text{ m K}}{500 \times 10^{-9} \text{m}} = 5800 \text{ K}$$

***Exercise* 2.6:** Express equation (2.6) in term equation (2.7).

Equation (2.6) is

$$d\varepsilon(\lambda, T)d\lambda = \frac{8\pi hc}{\lambda^5}\frac{d\lambda}{e^{\frac{hc}{\lambda k_bT}} - 1} \qquad ...(2.6)$$

$$\lambda = \frac{c}{\nu}, \; d\lambda = -\frac{c}{\nu_2}d\nu$$

Substitute λ and $d\lambda$ in equation (2.6), we get

$$d\varepsilon(\nu, T)d\nu = \left(8\pi hc\frac{\nu^5}{c^5}\right)\left(\frac{1}{e^{\frac{h\nu}{k_bT}} - 1}\right)\left(-\frac{c}{\nu^2}\right)d\nu = 8\pi h\frac{\nu^3}{c^3}\frac{1}{e^{\frac{h\nu}{k_bT}} - 1}d\nu$$

Another beautiful law, known as *Stefan-Boltzmann law*, based on classical mechanics or on thermodynamic argument can also be devised from equation (2.7) as

$$E_V = \frac{4}{c}\sigma T^4 \qquad ...(2.9)$$

where σ is the Stefan-Boltzmann constant having unit as J m^2 K^{-4} s^{-1}. Its experimental value is 5.6697 $\times 10^{-8}$ J m^2 K^{-4} s^{-1}.

***Exercise* 2.7:** Derive the expression for Stefan-Boltzmann constant from equation (2.7).

Equation (2.7) gives the energy density distribution fro m blackbody radiation in the frequency region ν and $\nu + d\nu$

$$d\varepsilon(\nu, T) = \frac{8\pi h\nu^3}{c^3}\frac{d\nu}{e^{\frac{h\nu}{k_bT}} - 1}$$

Integrating equation (2.7) overall the frequencies, *i.e.*

$$\varepsilon\nu = \int_0^\infty \frac{8\pi h\nu^3}{c^3}\frac{d\nu}{e^{\frac{h\nu}{k_bT}} - 1} = \frac{8\pi h}{c^3}\int_0^\infty \frac{\nu^3 d\nu}{e^{\frac{h\nu}{k_bT}} - 1}$$

Substituting $\quad x = \frac{h\nu}{k_bT}$ and $dx = \frac{h}{k_bT}d\nu$

or $$\nu = \frac{k_b Tx}{h} \text{ and } d\nu = \frac{k_b T}{h}dx$$

$$E_\nu = \frac{8\pi h}{c^3}\int_0^\infty \left(\frac{k_b T}{h}\right)^3 x^3 \frac{k_b T}{h}\frac{\text{dx}}{e^x - 1} = \frac{8\pi h}{c^3}\left(\frac{k_b T}{h}\right)^4 \int_0^\infty x^3 \frac{\text{dx}}{e^x - 1}$$

Using standard tables

$$\int_0^\infty x^3 \frac{dx}{e^x - 1} = \frac{\pi^4}{15}$$

So $$E_\nu = \frac{8\pi h}{c^3}\left(\frac{k_b T}{h}\right)^4\left(\frac{\pi^4}{15}\right) = \frac{8\pi^5 k_b^4 T^4}{15c^3}$$

Comparing this expression with equation (2.9), we get

$$\sigma = \frac{2\pi^5 k_b^4}{15h^3c^2}$$

$$\sigma = \frac{2\pi^5 k_b^4}{15h^3c^2} = \frac{(2)(22/7)^5(1.38 \times 10^{-23}\ \text{J K}^{-1})^4}{(15)(6.626 \times 10^{-34}\ \text{J s})^3(3.0 \times 10^8\ \text{m s}^{-1})^2} = 5.67 \times 10^{-8}\ \text{J m}^{-2}\ \text{K}^{-4}\ \text{s}^{-1}$$

***Exercise* 2.8:** Average surface temperature of Sun is nearly 6000 K and its diameter is about 1.4×10^9 m. Calculate the total energy emitted by the Sun.

Equation (2.9) gives the expression for the radiation rate m^{-2}. The surface area of Sun can be calculated using its diameter and then the total energy emitted.

Energy emitted = $\sigma T^4 = (5.67 \times 10^{-8}\ \text{J m}^{-2}\ \text{K}^{-4}\ \text{s}^{-1})(6000\ \text{K})^4 = 7.35 \times 10^7\ \text{J m}^{-2}\ \text{s}^{-1}$

Surface area of Sun= $4\pi r^2 = (4)(22/7)(1.4 \times 10^9\ \text{m})^2 = 6.16 \times 10^{18}\ \text{m}^2$

Energy emitted s^{-1} = luminosity = $(7.35 \times 10^7\ \text{J m}^{-2}\ \text{s}^{-1})(6.16 \times 10^{18}\ \text{m}^2) = 4.53 \times 10^{26}\ \text{J s}^{-1}$

2.2.2. *The Photoelectric Effect*

Einstein stated the photoelectric effect, *i.e.* if a radiation is shone on a metal in an evacuated system, the metal plate emits electrons. The results are: (*i*) The electrons ejected are only function of frequency of light falling on it whereas the number of electrons being ejected is proportional to the intensity of light (*ii*) There is no time lag between the light falling on the metal plate and ejection of the electrons provided the energy of the photon is greater than the minimum energy required to eject the electron, called threshold energy (ε_0) or threshold frequency (v_0) or work function.

The relation between the energy of the radiation (ε), kinetic energy of the ejected electron and the threshold energy is

$$\varepsilon = \varepsilon_0 + \frac{1}{2}mv^2 \qquad ...(2.10a)$$

or in Planck's notation

$$h\nu = h\nu_0 + \frac{1}{2} mv^2 \qquad ...(2.10b)$$

v is the speed of the electron.

***Exercise* 2.9:** Calculate the velocity of electron ejected from platinum surface when radiation of wavelength 200 nm falls on it. The work function for platinum is 5 eV.

$$5 \text{ eV} = (5 \text{ eV}) (8068.3 \text{ cm}^{-1} \text{ eV}^{-1}) = 40341.5 \text{ cm}^{-1}$$

$$\nu_0 = \frac{c}{\lambda_0} = c\bar{\nu}_0 = (3.0 \times 10^8 \text{ m s}^{-1}) (10^2 \text{ cm m}^{-1}) (40341.5 \text{ cm}^{-1})$$

$$= 1.21 \times 10^{15} \text{ s}^{-1}$$

$$\nu = \frac{c}{\lambda} = \frac{(3.0\times10^8 \text{ ms}^{-1})}{(200\times10^{-9} \text{ m})} = 1.5 \times 10^{15} \text{ s}^{-1}$$

Substituting the values of ν, ν_0, h and m_e in equation (2.10b)

$$v^2 = \frac{2h(\nu - \nu_0)}{m_e} = \frac{(2)(6.63\times10^{-34} \text{ J s})(1.5-1.21)\times10^{15} \text{ s}^{-1}}{9.11\times10^{-31} \text{ kg}}$$

$$v = 6.5 \times 10^5 \text{ m s}^{-1}$$

***Exercise* 2.10:** The voltage required to stop the ejection of electrons from copper plate is 0.24 V, when the radiation of wavelength 253.7 nm strikes the plate. Calculate the work function.

$$\varepsilon = h\nu = \frac{hc}{\lambda} = \frac{(6.63\times10^{-34} \text{ J s})(3.0\times10^8 \text{ m s}^{-1})}{(253.7\times10^{-9} \text{ m})}$$

$$= 7.834 \times 10^{-19} \text{ J} = (7.834 \times 10^{-19} \text{ J}) (6.242 \times 10^{18} \text{ eV J}^{-1})$$

$$= 4.89 \text{ eV}$$

$$\text{K.E.} = 1/2\, mv^2$$

$$= \text{(charge of electron esu) (voltage volts)}$$

$$= 0.24 \text{ eV}$$

Using equation (2.10a)

$$\varepsilon_0 = \varepsilon - \frac{1}{2} mv^2 = 4.89 - 0.24$$

$$= 4.65 \text{ eV.}$$

2.3. RYDBERG EQUATION

The Rydberg equation is an empirical relation which completely describes the spectrum of hydrogen and hydrogen like atoms. The relation is

$$\bar{\nu} = Z^2 R \left[\frac{1}{n_2^2} - \frac{1}{n_1^2}\right] \qquad ...(2.11)$$

where $\bar{\nu}$ is the difference between the two energy levels expressed in cm^{-1}, Z is the atomic number of atom, R is the Rydberg constant, given as

$$R = (109737.32 \text{ cm}^{-1}) \left(\frac{m_{\text{nuc}}}{m_e + m_{\text{nuc}}} \right) \quad ...(2.12)$$

and n_i's are the integers with the condition $n_1 > n_2$. Modified forms of Rydberg equation can be used to describe the spectrum of all hydrogen-like atoms.

***Exercise* 2.11:** Calculate the energy of transition involving $n_1 = 6$ to $n_2 = 3$ in an hydrogen atom.

Since, the mass of nucleus is much larger than that of electron, the relation (2.12) gives

$$R = 109737.32 \text{ cm}^{-1}$$

$$\bar{\nu} = (1)^2 (109737.32 \text{ cm}^{-1}) \left(\frac{1}{3^2} - \frac{1}{6^2} \right)$$

$$= 9144.78 \text{ cm}^{-1}$$

$$\varepsilon = h\nu = hc\bar{\nu}$$

$$= (6.63 \times 10^{-3} \text{ J s}) (3.0 \times 10^8 \text{ m s}^{-1}) (10^2 \text{ cm m}^{-1}) (9144.78 \text{ cm}^{-1})$$

$$= 1.82 \times 10^{-19} \text{ J}$$

2.4. BOHR'S THEORY OF ATOMIC STRUCTURE

Bohr's theory of atomic structure is based on three postulates: (*i*) The electron moves around the nucleus of electrical charge Ze in a circular orbit of fixed energy, (*ii*) The electron can have only discrete values of angular momentum, given by the relation

$$L = \frac{nh}{2\pi} \quad ...(2.13)$$

where *n,* called quantum numbers may have values 1, 2, 3, ... (*iii*) The electron will move from one orbit to another when it absorbs those frequencies which are related by the equation (2.14)

$$\nu = \frac{E_2 - E_1}{h} = \frac{\Delta E}{h} \quad ...(2.14)$$

where E_2 is the energy of the upper orbit, E_1 that of lower orbit.

With these principles in mind one can calculate the radius, velocity of an electron in any orbit and also the energy of the orbit:

$$r_n = \frac{(4\pi\varepsilon_0)\, h^2}{4\pi^2 \mu e^2 Z}\, n^2 \quad ...(2.15)$$

$$v_n = \frac{2\pi\, Ze^2}{(4\pi\varepsilon_0)\, hn} \quad ...(2.16)$$

$$\varepsilon_n = -\frac{2\pi^2 \mu Z^2 e^4}{h^2\, (4\pi\varepsilon_0)^2\, n^2} \quad ...(2.17)$$

where $4\pi\varepsilon_0$ is the permittivity constant and has the value $1.113 \times 10^{-10}\, \text{C}^2 \text{N}^{-1} \text{m}^{-2}$, μ is the reduced mass of the system, *i.e.*

$$\mu = \frac{m_e m_{\text{nuc}}}{m_e + m_{\text{nuc}}}$$

In some books $4\pi\varepsilon_0$ is treated as (4) (π) (ε_0) *i.e.* separate terms. In this case $\in_0 = 8.854\ 19 \times 10^{-12}$ kg^{-1} m^{-2} s^2 C^2 m^{-1}.

The Bohr's theory explains the Rydberg equation but it cannot explain the fine structure of atomic spectra and cannot describe the structure of multielectron atoms. The frequency of spectral line due to a transition between levels with quantum numbers n_1 and n_2 is

$$\nu = \frac{\varepsilon_{n_1} - \varepsilon_{n_2}}{h} = \frac{2\pi^2 \mu Z^2 e^4}{(4\pi\varepsilon_0)^2 h^3}\left[\frac{1}{n_2^2} - \frac{1}{n_1^2}\right]$$

or

$$\bar{\nu} = \frac{\nu}{c} = \frac{2\pi^2 \mu Z^2 e^4}{(4\pi\varepsilon_0)^2 h^3 c}\left[\frac{1}{n_2^2} - \frac{1}{n_1^2}\right] \quad ...(2.18)$$

This relation is same as the Rydberg relation (2.11) and Rydberg constant is

$$R = \frac{2\pi^2 \mu e^4}{(4\pi\varepsilon_0)^2 h^3 c} = 109677.58 \text{ cm}^{-1}$$

after substituting the values of constants in SI units. The equation (2.18) can be rewritten as

$$\bar{\nu} = RZ^2\left[\frac{1}{n_2^2} - \frac{1}{n_1^2}\right] \quad ...(2.19)$$

Number of series has been observed in the emission spectrum of hydrogen atoms and has been analysed by different scientists, labelling them after their names. These series, their spectral regions, series formula and series limits have been compiled in Table 2.1.

TABLE 2.1. Atomic spectrum of hydrogen atom

Series	Spectral region	Series formula	Series limit
Layman (1906)	Ultraviolet	$\frac{1}{\lambda} = R\,[\frac{1}{1^2} - \frac{1}{n^2}]$, n = 2, 3, 4...	91.127 nm
Ballmer (1885)	Visible	$\frac{1}{\lambda} = R\,[\frac{1}{2^2} - \frac{1}{n^2}]$, n = 3, 4, 5...	364.51 nm
Paschen (1908)	Infrared	$\frac{1}{\lambda} = R\,[\frac{1}{3^2} - \frac{1}{n^2}]$, n = 4, 5, 6...	820.14 nm
Bracket (1922)	Infrared	$\frac{1}{\lambda} = R\,[\frac{1}{4^2} - \frac{1}{n^2}]$, n = 5, 6, 7...	1458.0 nm
Pfund (1924)	Infrared	$\frac{1}{\lambda} = R\,[\frac{1}{5^2} - \frac{1}{n^2}]$, n = 6, 7, 8...	2278.2 nm

***Exercise* 2.12:** Calculate the radii and energies of the first two Bohr orbits of hydrogen atom. Find also the frequency of the energy of photon emitted/absorbed during transition of electron from one to the other orbit.

Substituting the respective values in (2.15)

$$r_n = \frac{(1.113\times10^{-10}\ \text{C}^2\text{N}^{-1}\ \text{m}^{-2})\ (6.63\times10^{-34}\ \text{J s})^2 n^2}{(9.105\times10^{-31}\ \text{kg})\ (1.60\times10^{-19}\ \text{C})^2 (1)\ (4\pi)^2} = (5.295 \times 10^{-11}\ n^2)\ \text{m}$$

so $$r_1 = 52.95\ \text{pm},\ r_2 = (2^2)\ (52.95\ \text{pm}) = 211.8\ \text{pm}$$

Using equation (2.17)

$$E_n = -\frac{RZ^2}{n^2} = -\frac{109677.6}{n^2}\ \text{cm}^{-1}$$

$$E_1 = -\ 109677.6\ \text{cm}^{-1},\ E_2 = -\frac{109677.6}{2^2}$$

$$= -\ 27419.4\ \text{cm}^{-1}$$

$\therefore$ $$E_2 - E_1 = -27419.4 - (-109677.6)$$
$$= 82258.2\ \text{cm}^{-1}.$$

2.5. PARTICLES AND WAVES

de Broglie suggested that dual character (*i.e.* wave nature and particle behaviour) is not only possessed by the radiation but by all the particles. The wavelength (λ) of the moving particle can be calculated from the relation

$$\lambda = \frac{h}{mv} \qquad \text{...(2.20)}$$

where m is the mass of the moving particle and v is its velocity.

***Exercise* 2.13:** Calculate the velocity of an electron in the first Bohr orbit and the de Broglie wavelength associated with it.

The velocity is given by the expression (2.16)

$$v = \frac{(1.6\times10^{-19}\ \text{C})^2\ (1)\ (2\pi)}{(1.113\times10^{-10}\text{C}^2\ \text{N}^{-1}\text{m}^{-2})(1)\ (6.63\times10^{-34}\ \text{J s})} = 2.19 \times 10^6\ \text{m s}^{-1}$$

$$\lambda = \frac{h}{mv} = \frac{(6.63\times10^{-34}\ \text{J s})}{(9.11\times10^{-31}\ \text{kg})\ (2.19\times10^6\ \text{m s}^{-1})} = 3.32 \times 10^{-10}\ \text{m}.$$

2.6. HEISENBERG UNCERTAINTY PRINCIPLE

Heisenberg uncertainty principle states that both position as well as momentum associated with the particle cannot be measured accurately and simultaneously. Whereas according to classical mechanics both the position and momentum of the particle can be observed precisely. In Cartesian coordinates the uncertainty principle becomes

$$\Delta x\ .\Delta p_x \geq h/4\pi,\ \Delta y\ \Delta\ p_y \geq h/4\pi,\ \Delta\ z\ \Delta\ p_z \geq h/4\pi \qquad \text{...(2.21)}$$

Similar relation also exists between time and energy, *i.e.*

$$\Delta\ t\ .\Delta\ \varepsilon \geq h/4\pi \qquad \text{...(2.22)}$$

***Exercise* 2.14:** A car weighing 3.0×10^3 kg is moving on a highway. Its speed can be measured with an accuracy of ± 0. 0025 mile hr^{-1} and its position with an accuracy of ± 0.01 mile. Is the Heisenberg uncertainty principle valid?

$$m = 3.0 \times 10^3 \text{ kg}$$

$$\Delta v = \frac{(0.0025 \text{ mile hr}^{-1})(1.6\times10^3 \text{ m mile}^{-1})}{(3600 \text{ s hr}^{-1})} = 1.1 \times 10^{-3} \text{ m s}^{-1}$$

$$\Delta p = (3.0 \times 10^3 \text{ kg})(1.1 \times 10^{-3} \text{ m s}^{-1}) = 3.3 \text{ kg m s}^{-1}$$

$$\Delta x = (0.01 \text{ mile})(1.6 \times 10^3 \text{ m mile}^{-1}) = 16 \text{ m}$$

$$\Delta x.\Delta p = (16 \text{ m})(3.3 \text{ kg m s}^{-1}) = 52.8 \text{ J}$$

Since $\Delta x.\Delta p \gg \frac{h}{4\pi}$, the uncertainty principle is not vaild here.

2.7. WAVE MECHANICS

Further understanding in the structure of atoms and thus molecules was obtained from the two independent studies made by Heisenberg and Schrödinger. The former is called "matrix mechanics" and the latter is called "wave mechanics". But both the approaches lead to the same final results. In Chemistry wave mechanics is more popular, easy to understand and will be discussed briefly.

If a system consists of N particles then the physical description of the particles can be very well defined or its properties can be calculated from the wave function, ψ. The wave function is a function of $3N$ position coordinates and time, *i.e.* $\psi(x_1, x_2, \ldots x_{3N}, t)$ and the probability (P) of finding a particle in a volume element $d\tau$, which is more physically observable quantity, is given by the expression

$$P = \psi^* \psi d\tau \tag{2.23}$$

where $d\tau$ is the small volume element and ψ^* is the complex conjugate of ψ. The integration is carried out over the entire domain of wave function. The maximum value of P in the complete space is 'unity'.

It may be mentioned here that wave function is purely a mathematical function but can be used to calculate any physically observable quantities. Further ø is well behaved only for certain particular value of E and it is these values that are allowed energy levels.

There are certain restrictions imposed on the nature of wave function used in wave mechanics which are:

(*a*) ψ as well as its first and second derivatives should be continuous, finite and single valued at each point.

(*b*) The square of the wave function should be integrable. If the square of the wave function or the product of the wave function and its complex conjugate can be integrated, then the equations (2.23) and (2.37) can be evaluated and so the observable quantities like probability, momentum, energy, etc.

Other quantities commonly used in the wave mechanics are:

(*a*) Normalised wave functions: A function is normalised if the probability of finding the particle over the whole space is unity, *i.e.*

$$\int_{\text{all space}} \psi^* \psi d\tau = 1 \qquad \ldots(2.24)$$

(*b*) Orthogonal wave function: The wave functions are orthogonal if the relation

$$\int_{\text{all space}} \psi_i{}^* \psi_j d\tau = 0 \text{ is satisfied } (i \neq j) \qquad \ldots(2.25)$$

(c) Ortho-normal set: Wave functions form a ortho-normal set if the relation

$$\int_{\text{all space}} \psi_i{}^*\psi_j d\tau = 1 \text{ if } i = j$$

$$= 0 \text{ if } i \neq j \qquad \text{...(2.26)}$$

is satisfied.

***Exercise* 2.15:** Which of the following wave functions are acceptable in quantum mechanics? (*a*) sin *x*, (*b*) tan *x*, (*c*) cosec *x*, (*d*) cos *x* + sin *x*, over the range when *x* goes from 0 to 2π.

The wave functions (*a*) and (*d*) are acceptable wave functions whereas (*b*) and (*c*) are not, since (*b*) tends to infinite at $x = \pi/2$ and (*c*) tends to infinite at $x = 0$.

***Exercise* 2.16:** Normalise the wave function $\psi = A \sin \dfrac{n\pi}{a} x$ for a particle in one dimensional box of length *a*.

A wave function is normalised if the probability is unity, *i.e.*

$$\int_{\text{all space}} \psi^* \psi d\tau = 1$$

Substituting the values of limits and the wave function, the integral becomes

$$1 = A^2 \int_0^a \sin \frac{n\pi}{a} x \sin \frac{n\pi}{a} x dx = A^2 \int_0^a \sin^2 \frac{n\pi}{a} x dx$$

$$= (A^2/2)a$$

$$A = \sqrt{2/a}$$

$$\text{Thus } \psi = \sqrt{\frac{2}{a}}.\sin \frac{n\pi}{a} x$$

***Exercise* 2.17:** The Hermite polynomial for v = 2 and v = 1 are, $H_2(\xi) = 4\xi^2 - 2$ and $H_1(\xi) = 2\xi$. Show that the wave functions of harmonic oscillator in the stationary states 1 and 2 are orthogonal.

The wave functions for harmonic oscillator are

$$\psi_1 = N_1 \exp(-\xi^2/2)2\xi = 2N_1\xi \exp(-\xi^2/2)$$

$$\psi_2 = N_2(4\xi^2 - 2) \exp(-\xi^2/2)$$

where $x = \xi/\sqrt{a}$.

The above wave functions are orthogonal if the integral

$$\int_{-\infty}^{+\infty} \psi_1{}^* \psi_2 dx = 0$$

Substituting the values of wave functions and *x*

$$\int_{-\infty}^{+\infty} \psi_1{}^* \psi_2 dx = \frac{2N_1N_2}{\sqrt{a}} \int_{-\infty}^{+\infty} \xi(4\xi^2 - 2) \exp(-\xi^2) d\xi$$

$$= \frac{8N_1N_2}{\sqrt{a}} \int_{-\infty}^{+\infty} \xi^3 \exp(-\xi^2) d\xi - \frac{4N_1N_2}{\sqrt{a}} \int_{-\infty}^{+\infty} \exp(-\xi^2)\xi\, d\xi$$

Both the integrands are odd functions, so the values of both the integrals are zero. Thus the wave functions $\psi_1(x)$ and $\psi_2(x)$ are orthogonal.

***Exercise* 2.18:** The normalised wave functions describing the behaviour of rigid rotator is given as

$$\psi_{J,m}(\theta, \phi) = \left(\frac{1}{2\pi}\right)^{1/2} \Theta_{J,m}(\theta) \exp(\pm im\phi) \qquad ...(2.27)$$

where $J = 0, 1, 2, ...$ and $m = -J, -J+1, ... 0, ... J-1, J$ (J and m are called rotational quantum numbers) and

$$\Theta_{J,m}(\theta) = \left[\frac{(2J+1)(J-|m|)!}{2(J+|m|)!}\right]^{1/2} P_J^{|m|}(\cos\theta) \qquad ...(2.28)$$

$P_J^{|m|}(x)$ are the associated Legendre functions and given by

$$P_J^{|m|}(x) = \frac{1}{2^J J!}(1-x^2)^{|m|/2} \frac{d^{J+|m|}}{d^{J+|m|}x}(x^2-1)^J \qquad ...(2.29)$$

Find the wave function $\psi_{J,m}(\theta, \phi)$ for $J = 3$ and $m = \pm 2$. Show that these are normalised.

Now first find the associated Legendre function for $J = 3$ and $m = \pm 2$:

$$P_3^2(x) = \frac{1}{2^3.3!}(1-x^2)^{2/2}\frac{d^{3+2}}{dx^5}(x^2-1)^3$$

$$= \frac{1}{8\times 6}(1-x^2)\frac{d^4}{dx^4}(2x)(x^2-1)^2(3)$$

$$= \frac{1}{48}(1-x^2)\frac{d^4}{dx^4}6x(x^4-2x^2-1) = 15x(1-x^2)$$

Now,
$$\Theta_{3,2}(\theta) = \left[\frac{(7)(3-2)!}{2\times(3+2)!}\right]^{1/2} P_3^2(\cos\theta)$$

$$= \left[\frac{7\times 1}{(2)(5!)}\right]^{1/2} 15\cos\theta\sin^2\theta = \frac{\sqrt{105}}{4}\cos\theta\sin^2\theta$$

$$\psi_{3,+2}(\theta, \phi) = \frac{1}{\sqrt{2\pi}}\left(\frac{105}{16}\right)^{1/2}\cos\theta\sin^2\theta\exp(2i\phi)$$

This can be also written as

$$\psi_{3,+2}(\theta, \phi) = \left(\frac{105}{\pi}\right)^{1/2}.\frac{1}{4}\cos\theta\sin^2\theta.\cos 2\phi$$

and

$$\psi_{3,-2}(\theta, \phi) = \left(\frac{105}{\pi}\right)^{1/2}.\frac{1}{4}\cos\theta\sin^2\theta.\sin 2\phi$$

To see that the above function is normalised, the integral

$$\int \psi^2_{3,=2}\, d\tau = 1$$

Substitute the value of $\psi_{3,+2}$ in equation (2.24)

$$\int \psi^2_{3,+2}\, d\tau = \int_0^{\pi}\int_0^{2\pi} \frac{105}{16\pi} \cos^2\theta \sin^4\theta \sin^2 2\phi \sin\theta\, d\theta\, d\phi$$

$$= \frac{105}{16\pi}\int_0^{\pi} \cos^2\theta \sin^5\theta\, d\theta^a \int_0^{2\pi} 4\sin^2\phi \cos^2\phi\, d\phi^b \quad [\sin^2 2\phi = 2\sin\phi\cdot\cos\phi]$$

Using the following mathematical relations:

$$\psi^2_{3,+2}\, d\tau = \frac{105}{4}\left[\left(\left\{\frac{\cos\theta\sin^6\theta}{2+5}\right\}_0^{\pi} + \frac{2-1}{2+5}\int_0^{\pi}\sin^5\theta\, d\theta\right)\left(-\frac{1}{8}\left\{\frac{1}{4}\sin^4\phi\right\}_0^{2\pi}\right)\right]$$

$$= \frac{105}{4\pi}\times\frac{\pi}{4}\left[\frac{1}{7}\int_0^{\pi}\sin^5\theta\, d\theta\right]^c = \frac{105}{16\times 7}\left(\left[\frac{\sin^4\theta\cos\theta}{5}\right]_0^{\pi} + \frac{4}{5}\int_0^{\pi}\sin^3\theta\, d\theta\right)$$

$$= \frac{105\times 4}{7\times 5\times 16}\int_0^{\pi}\sin^3\theta\, d\theta = \frac{3}{4}\left[-\frac{\sin^2\theta\cos\theta}{3} + \frac{2}{3}\int \sin\theta\, d\theta\right]_0^{\pi}$$

$$= \frac{3}{4}\int_0^{\pi}\frac{2}{3}\sin\theta\, d\theta = \frac{3}{4}\times\frac{2}{3}[-\cos\theta]_0^{\pi}$$

$$= -\frac{1}{2}\,[-1-1] = 1$$

So the wave function $\psi_{3,+2}$ is normalised. Similarly it can also be proved that wave function $\psi_{3,-2}$ is also normalised.

***Exercise* 2.19:** The eigen values, *i.e.* energy of each rotational level, of the rigid rotator problem is given by

$$E_J = \frac{h^2}{8\pi^2 I}\, J(J+1) \qquad ...(2.30)$$

$$a \quad \int \cos^m x \sin^n x\, dx = \frac{\cos^{m-1} x \sin^{n+1} x}{m+n} + \frac{m-1}{m+n}\int \cos^{m-2} x \sin^n x\, dx$$

$$b \quad \int \sin^2 x \cos^2 x\, dx = -\frac{1}{8}\,[\sin 4x - x]$$

$$c \quad \int \sin^n x\, dx = -\frac{\sin^{n-1} x \cos x}{n} + \frac{n-1}{n}\int \sin^{n-2} x\, dx$$

where I is the moment of inertia and $J = 0, 1, 2, ...$ are rotational quantum numbers. Calculate the energies of the first four levels, find the energy difference between each level:

$$E_J = \frac{h^2}{8\pi^2 I} J(J+1)$$

$$E_0 = 0, E_1 = 2\frac{h^2}{8\pi^2 I}$$

$$E_2 = 6\frac{h^2}{8\pi^2 I}, E_3 = 12\frac{h^2}{8\pi^2 I}$$

$$E_4 = 20\frac{h^2}{8\pi^2 I}$$

$$E_{1-0} = 2\frac{h^2}{8\pi^2 I}, E_{2-1} = 4\frac{h^2}{8\pi^2 I}$$

$$E_{3-2} = 6\frac{h^2}{8\pi^2 I}, E_{4-3} = 8\frac{h^2}{8\pi^2 I}$$

2.8. OPERATOR

As the name suggests, an operator is an instruction given to carry out a mathematical operation on a given function, called operand. Similar to wave functions, the following rules are observed by the operators:

(*a*) Operators should be linear, *i.e.*

$$\hat{O}(A + B) = \hat{O}A + \hat{O}B \quad ...(2.31)$$

For example, d/dx is a linear operator but '√' is not.

(*b*) The product of the two operators is defined as

$$\hat{O}_1 \hat{O}_2 f(x) = \hat{O}_1 (\hat{O}_2 f(x)) \quad ...(2.32)$$

(*c*) Operators should be Hermitian, *i.e.*

$$\bar{O} = \int \psi^* \hat{O} \psi d\tau = \int \psi (\hat{O} \psi^*) d\tau \quad ...(2.33)$$

$$= \int \psi \hat{O} \psi^* d\tau$$

The advantages of Hermitian operators are that their eigen values are real.

(*d*) The law of associative multiplication should be followed, *i.e.*

$$\hat{A}(\hat{B}\hat{C}) = (\hat{A}\hat{B})\hat{C} \quad ...(2.34)$$

A quantum mechanical operator for any physical quantity can be obtained by first writing the classical expression in terms of variables like x, p_x, t, E, etc. and then converting the expression to an operator form with the help of rules mentioned in the Table 2.2.

Finally, it may not be necessary that $\hat{O}_1 \hat{O}_2$ be the same as $\hat{O}_2 \hat{O}_1$. If $\hat{O}_1 \hat{O}_2 = \hat{O}_2 \hat{O}_1$, the two operators are said to "*commute*" with each other.

It must be kept in mind that operator and operand must be compatible; the operation and the result must be mathematically well defined. Here we will denote the operator by a capital letter with a cap (or carrot) over it, *e.g.* $\hat{O}f(x) = g(x)$, where $\hat{O}$ is an operator and operates on operand $f(x)$ to give a function $g(x)$.

***Exercise* 2.20:** Write the Hamiltonian operator of a free particle moving in one direction under the influence of zero potential energy.

In classical mechanics, Hamiltonian is the sum of kinetic energy and potential energy, *i.e.*

$$H = \frac{1}{2} m\dot{x}^2 + \text{PE}$$

Since PE = 0.

$$H = \frac{1}{2} m\dot{x}^2 = \frac{1}{2m} p_x^2$$

where p_x is the linear momentum along x-axis. For linear momentum the operator is $\frac{h}{2\pi i}\frac{d}{dx}$, so

$$\hat{H} = \frac{1}{2m}\left[\frac{h}{2\pi i}\frac{d}{dx}\right]^2 = \frac{1}{2m}\left[\frac{h}{2\pi i}\frac{d}{dx}\right]\left[\frac{h}{2\pi i}\frac{d}{dx}\right]$$

$$= -\frac{h^2}{8\pi^2 m}\frac{d^2}{dx^2}$$

***Exercise* 2.21:** Do the operators x and $\frac{d}{dx}$ commute with each other? The condition for the two commuting operators is

$$\hat{O}_1 \hat{O}_2 f(x) = \hat{O}_2 \hat{O}_1 f(x) \qquad \text{...(2.35a)}$$

TABLE 2.2

Classical variable	Q.M. operator	Expression for operator	Operation
x	$\hat{x}$	x	Multiply by x
p_x	$\hat{p}_x$	$\frac{h}{2\pi i}\frac{d}{dx}$	Take derivative w.r.t. x and multiply by $h/2\pi i$
t	$\hat{t}$	t	Multiply by t
E	$\hat{E}$	$-\frac{h}{2\pi i}\frac{\partial}{\partial t}$	Take derivative w.r.t. t and multiply by $-h/2\pi i$

L_x	$\hat{L}_x$	$-\frac{ih}{2\pi}\left(y\frac{\partial}{\partial z}-z\frac{\partial}{\partial y}\right)$	
L_y	$\hat{L}_y$	$-\frac{ih}{2\pi}\left(z\frac{\partial}{\partial x}-x\frac{\partial}{\partial z}\right)$	
L_z	$\hat{L}_z$	$-\frac{ih}{2\pi}\left(x\frac{\partial}{\partial y}-y\frac{\partial}{\partial x}\right)$	

Let $f(x)$ be a function

Now,
$$\hat{O}_1\,\hat{O}_2 f(x) = x\,\frac{d}{dx}\,f(x)$$

$$= x\,\frac{df(x)}{dx} = xf'(x) \qquad ...(2.35b)$$

Similarly,
$$\hat{O}_2\,\hat{O}_1 f(x) = \left[\frac{d}{dx}, x\right] f(x)$$

$$= \left(\frac{d}{dx}\right) xf(x) = f(x) + xf'(x) \qquad ...(2.35c)$$

From equations (2.35*b*) and (2.35*c*), it is clear that relation (2.35*a*) is not satisfied, *i.e.* these two operators do not commute.

From equations (2.35*a*) and (2.35*b*), the commutator is

$$[\hat{O}_1\hat{O}_2]f(x) = \hat{O}_1\hat{O}_2 f(x) - \hat{O}_2\hat{O}_1 f(x) = xf'(x) - f(x) - xf'(x) = -f(x)$$

$\hat{O}_1\hat{O}_2 - \hat{O}_2\hat{O}_1 = -1$. Because this equation applies to all functions $f(x)$, it is valid to delete $f(x)$ here. *It may be mentioned here that this is not a general relation and thus may not be applicable to all the functions*. See Problem 2.22.

So the commutator is equal to – 1.

***Exercise* 2.22:** Prove the following relation

$$\left[\hat{L}_x.\hat{L}_y\right] = \frac{ih}{2\pi}L_z$$

Consider the function f. Now

$$\hat{L}_y f = -\frac{ih}{2\pi}\left[z\frac{\partial f}{\partial x} - x\frac{\partial f}{\partial z}\right]$$

$$\hat{L}_x\hat{L}_y f = -\frac{h^2}{4\pi^2}\left(y\frac{\partial}{\partial z} - z\frac{\partial}{\partial y}\right)\left(z\frac{\partial f}{\partial x} - x\frac{\partial f}{\partial z}\right)$$

$$= -\frac{h^2}{4\pi^2}\left(y\frac{\partial f}{\partial x} + yz\frac{\partial^2 f}{\partial z \partial x} - yx\frac{\partial^2 f}{\partial z^2} - z^2\frac{\partial^2 f}{\partial y \partial x} + zx\frac{\partial^2 f}{\partial y \partial z}\right) \quad ...(2.36a)$$

Similarly,
$$\hat{L}_x f = -\frac{ih}{2\pi}\left(y\frac{\partial f}{\partial z} - z\frac{\partial f}{\partial y}\right)$$

and
$$\hat{L}_y \hat{L}_x f = -\frac{h^2}{4\pi^2}\left(zy\frac{\partial^2 f}{\partial x \partial z} - z^2\frac{\partial^2 f}{\partial x \partial y} - xy\frac{\partial^2 f}{\partial z^2} + x\frac{\partial f}{\partial y} + xy\frac{\partial^2 f}{\partial z \partial y}\right) \quad ...(2.36b)$$

Subtracting (2.36*b*) from (2.36*a*) and making use of

$$\frac{\partial^2 f}{\partial x \partial y} = \frac{\partial^2 f}{\partial y \partial x} \text{ as } f \text{ is a state function}$$

$$\hat{L}_x \hat{L}_y f - \hat{L}_y \hat{L}_x f = -\frac{h^2}{4\pi^2}\left[y\frac{\partial f}{\partial x} - x\frac{\partial f}{\partial y}\right]$$

$$i.e.\ \left[\hat{L}_x \hat{L}_y\right] = \frac{ih}{2\pi}\hat{L}_z .$$

2.9. POSTULATES OF QUANTUM MECHANICS

In this section, we will summarise the postulates of quantum mechanics that will suffice for all the systems with which we encounter and are interested in chemistry. Results predicted using these postulates will be compared with the experimental results. The ultimate test of these axioms or postulates will be whether these results agree with the experimental ones or not.

2.9.1. *Postulate 1*

The state of a quantum mechanical system is completely defined by a function that depends on the coordinates of the system and time, *i.e.* $\psi(x_1, x_2, x_3, ...x_{3N}, t)$. Further other explanation or details can be obtained from Section 2.7.

2.9.2. *Postulate 2*

For every observable in classical mechanics there corresponds a linear operator in quantum mechanics. Some of the classical observable and their quantum mechanical operators are compiled in Table 2.2. This postulate states that all the quantum mechanical operators are linear. Consider an eigenvalue problem with a two fold degeneracy, *i.e.* consider the two equations

$$\hat{O}\varphi_1 = O\varphi_1 \text{ and } \hat{O}\varphi_2 = O\varphi_2$$

Both the functions φ_1 and φ_2 have the same eigen value O. In this case, then any linear combination of φ_1 and φ_2, *i.e.* $c_1\ \varphi_1 + c_2\ \varphi_2$ is an eigen function of $\hat{O}$. It can be shown as follows:

$$\hat{O}(c_1\varphi_1 + c_2\varphi_2) = \hat{O}c_1\varphi_1 + \hat{O}c_2\varphi_2 = c_1\hat{O}\varphi_1 + c_2\hat{O}\varphi_2 = c_1 O\varphi_1 + c_2 O\varphi_2 = O(c_1\varphi_1 + c_2\varphi_2)$$

That is linear combination of degenerate wave functions is also an eigen function of the same operator.

Postulate 3: Let $\hat{O}$ be an operator, corresponding to physically observable quantity and the system can be represented by a function ψ. The precise value of this observable can only be calculated if the following equation is followed.

$$\hat{O}\psi = O\psi \qquad \text{...(2.37)}$$

Here O is a constant and is the value of observable. The above equation imposes the restriction that ψ is an eigen function of the operator $\hat{O}$ and O is its eigen value. In other words, the precise value of the observable can only be obtained or the values obtained for observable will be the same for each measurement if ψ is an eigen function to the operator $\hat{O}$. In real sense this problem reduces to finding ψ and O that satisfies the eigen value–eigen function equation (2.37). It *may be mentioned here that if a function is an eigen function to the operator then one gets only one unique eigen value and its value remain identical even a number of measurements are carried out*.

Postulate 4: If in postulate III, ψ is not an eigen function to the operator, $\hat{O}$, the measurement carried out every time will give a different value of the observable. In this case the average value or expectation value of the observable is given by:

$$<O> = \bar{O} = \frac{\int \psi^* \hat{O} \psi \, d\tau}{\int \psi^* \psi \, d\tau} \qquad \text{...(2.38)}$$

Here $<O>$ or $\bar{O}$ is the average or expectation value of the observable corresponding to the operator $\hat{O}$ and the above relation is called *mean value theorem*. If the wave function ψ is a normalized wave function then expression (2.38) reduces to (2.39).

$$<O> = \bar{O} = \int \psi^* \hat{O} \psi \, d\tau \qquad \text{...(2.39)}$$

The average value of the observable obtained by the expression (2.38) is the average of the number of observations made to calculate the property corresponding to the operator $\hat{O}$ and it should not be related to the average in classical mechanics. *If ψ is an eigen function of operator, $\hat{O}$, the average value will be the same as the eigen value*. For example:

$$<O> = \frac{\int \psi^* \hat{O} \psi \, d\tau}{\int \psi^* \psi \, d\tau} = \frac{\int \psi^* O \psi \, d\tau}{\int \psi^* \psi \, d\tau} = O\frac{\int \psi^* \psi \, d\tau}{\int \psi^* \psi \, d\tau} = O \qquad \text{...(2.40)}$$

One postulate which we have not mentioned deliberately is regarding the time dependence part of the Schrödinger wave equation. It is simply because in the present book we are not going to deal any problem which is an explicit function of time, *i.e.* we will be dealing mainly the steady state functions and their corresponding wave functions.

***Exercise* 2.23:** If ψ_n is an eigen function of operator, $\hat{O}$, then the following relation holds true, *i.e.* $\sigma_O^2 = <O^2> - (<O>)^2 = 0$, where σ is the variance of measurements,

Using equation (2.39), assuming ψ_n is a normalised function.

$$\hat{O}\psi_n = O_n\psi_n \quad ...(2.41)$$

Then $$<O> = \int_{-\infty}^{+\infty} \psi_n^* \hat{O}\psi_n d\tau = \int_{-\infty}^{+\infty} \psi_n^* O_n \psi_n d\tau = O_n \int_{-\infty}^{+\infty} \psi_n^* \psi_n d\tau = O_n \quad ...(1)$$

Furthermore if $\hat{O}\psi_n = O_n\psi_n$, then

$$\hat{O}^2\psi_n = \hat{O}(\hat{O}\psi_n) = \hat{O}(O_n\psi_n) = O_n(\hat{O}\psi_n) = O_n^2\psi_n \quad ...(2)$$

and so $$<O^2> = \int_{-\infty}^{+\infty} \psi_n^* \hat{O}^2 \psi_n d\tau = O_n^2 \quad ...(3)$$

From equations (1) and (3), we see that the variance of measurements gives

$$\sigma_O^2 = <O^2> - (<O>)^2 = O_n^2 - O_n^2 = 0$$

This it confirms the postulate III, the only value we measure is the value O_n.

***Exercise* 2.24:** Calculate the average value of the linear momentum of a particle moving in one dimensional box of length a and wave function $\psi = A \sin \frac{n\pi}{a} x$.

The quantum mechanical operator for linear momentum is $\frac{h}{2\pi i} \cdot \frac{d}{dx}$. So

$$\bar{p}_x = \frac{\int_0^a A \sin \frac{n\pi}{a} x \left(\frac{h}{2\pi i} \frac{d}{dx} \right) A \sin \frac{n\pi}{a} x dx}{\int_0^a A \sin \frac{n\pi}{a} x . A \sin \frac{n\pi}{a} x \, dx}$$

$$= \frac{\frac{h}{2\pi i} \int_0^a \sin \frac{n\pi}{a} x . \frac{n\pi}{a} \cos \frac{n\pi}{a} x dx}{\int_0^a \sin^2 \frac{n\pi}{a} x \, dx}$$

$$= \frac{\frac{hn}{2\,ia} \int_0^a \sin \frac{n\pi}{a} x \cos \frac{n\pi}{a} x \, dx}{\frac{1}{2} \int_0^a \left(1 - \cos \frac{2n\pi x}{a} \right) dx}$$

$$= \frac{\frac{nh}{2\,ia} \left[\frac{a}{2n\pi} . \sin^2 \frac{n\pi}{a} x \right]_0^a}{\left[\frac{x}{2} \right]_0^a + \frac{a}{4n\pi} \left[\sin \frac{2n\pi}{a} x \right]_0^a} = \frac{\frac{h}{4i\pi} [\sin n\pi - \sin 0]}{a + \frac{a}{4n\pi} [\sin 2n\pi - \sin 0]}$$

$$= \frac{O}{a} = 0$$

Physically the result obtained is correct and can be explained as follows. Translational energy of a particle of mass m and velocity v is

$$E_{\text{trans}} = \frac{1}{2} m\text{v}^2 = \frac{(m\text{v})^2}{2m} = \frac{p^2}{2m}$$

or

$$p = \pm \left[E_{\text{trans}} \bullet 2m\right]^{\frac{1}{2}}$$

This means that linear momentum can have + ve and negative value. This means that some particles or moving towards left and some towards right. Average or net value will be zero.

***Exercise* 2.25:** Find the average value of $\bar{x}$ for a harmonic oscillator in the nth state.

The wave function of harmonic oscillator in the nth state is

$$\psi_n(x) = N_n \exp(-\xi^2/2)\, H_n(\xi) \qquad \xi = \sqrt{\alpha}\, x$$

where N_n is the normalisation constant and $H_n(\xi)$ is the Hermite polynomial. Now

$$\bar{x} = \frac{\int_{-\infty}^{+\infty} \psi_n^*(x)\, \hat{x}\psi_n(x)\, dx}{\int_{-\infty}^{+\infty} \psi_n^*(x)\, \psi_n(x)\, dx}$$

$$= \frac{N_n^2 \int \exp(-\xi^2/2)\, H_n(\xi)\, \hat{x} H_n(\xi) \exp(-\xi^2/2)\, dx}{N_n^2 \int \exp(-\xi^2/2)\, H_n(\xi) \exp(-\xi^2/2)\, H_n(\xi)\, dx} \qquad \text{...(2.42)}$$

Consider the numerator and substitute the value of x, *i.e.*

$$x = \frac{\xi}{\sqrt{\alpha}} \text{ and } dx = \frac{d\xi}{\sqrt{\alpha}}$$

$$\bar{x} = \frac{1}{\alpha} \int_{-\infty}^{+\infty} \exp(-\xi^2)\, H_n(\xi)\, \xi H_n(\xi)\, d\xi$$

Using the recursion formula for the Hermite polynomial, i.e.

$$\xi H_n(\xi) = \frac{1}{2} H_{n+1}(\xi) + nH_{n-1}(\xi) \qquad \text{...(2.43)}$$

the above integral becomes

$$\bar{x} = \frac{1}{\alpha} \int_{-\infty}^{+\infty} \exp(-\xi^2)\, H_n(\xi) \left[\frac{1}{2} H_{n+1}(\xi) + nH_{n-1}(\xi)\right] d\xi$$

$$= \frac{1}{2\alpha} \int_{-\infty}^{+\infty} \exp(-\xi^2) H_n(\xi)\, H_{n+1}(\xi)\, d\xi + \frac{n}{a} \int_{-\infty}^{+\infty} \exp(-\xi^2) H_n(\xi) H_{n-1}(\xi)\, d\xi$$

Since the wave functions of harmonic oscillator for different stationary states are orthogonal, so the above two integrals become zero.

As the numerator is zero in equation (2.42), the average value of $\bar{x} = 0$ for the harmonic oscillator in the nth state.

***Exercise* 2.26:** Show that the quantum mechanical operators give real eigen values.

Let $\hat{O}$ be a Hermitian operator of function f_1 (x) and let O be its eigen value, *i.e.*

$$\hat{O} f_1 (x) = O f_1 (x) \quad ...(2.44)$$

Taking the complex conjugate of each quantity in the equation (2.44)

$$\hat{O} f^*_1 (x) = O^* f^*_1 (x) \quad ...(2.45)$$

Integrate after multiplying (2.44) by f_1^* (x) and (2.45) by f_1 (x) respectively.

$$\int f_1^*(x) [\hat{\alpha} f_1 (x)] \; dx = O \int f_1^*(x) \; f_1 (x)] \; dx \quad ...(2.46)$$

and $$\int f_1(x) [\hat{O}^* f_1^* (x)] \; dx = O^* \int f_1(x) f_1^* (x)] \; dx \quad ...(2.47)$$

Since $\hat{O}$ is Hermitian so

$$\int f_1^*(x) [\hat{O} f_1 (x)] \; dx = \int f_1(x) (\hat{O}^* f_1^*(x)] \; dx \quad ...(1)$$

Thus, $\lambda \int f_1^*(x) \; f_1 (x) \, dx = O^* \int f_1(x) f_1^*(x) \, dx$ ...(2)

This will be true provided $O = O^*$, *i.e.* O is a real quantity.

The other possible ways of writing a condition for Hermitian operators can be:

$$\int f_1^*(x) \; \hat{O} f_2(x) \, dx = [\int f_2^*(x) \; \hat{O} f_1(x)] \, dx \,]^* \quad ...(3)$$

or $$\int f_1^*(x) [\hat{O} f_2(x)] \, dx = \int [(\hat{O} f_1(x)]^* f_2(x) \, dx \quad ...(4)$$

where f_1 and f_2 are two well behaved functions.

***Exercise* 2.27:** Continuing with Exercise 2.26, consider $f(x) = \frac{1}{\pi^{\frac{1}{4}}} e^{-\frac{x^2}{2}}$ in the range from $-\infty$ to $+\infty$ and $g(x) = \frac{\sqrt{2}}{\pi^{\frac{1}{4}}} xe^{-\frac{x^2}{2}}$ in the range from $-\infty$ to $+\infty$. $\hat{O} = \frac{h}{2\pi i} \frac{\partial}{\partial x}$. Prove the (2.26).

The constants in *f*(x) and g(x) are normalisation constants. Further *f*(x) and g(x) are the solution of the simple harmonic oscillator. Proceeding on the lines of Exercise (2.26), *i.e.*:

$$\hat{O} g(x) = \frac{h}{2\pi i} \frac{\partial}{\partial x} \left(\frac{2^{\frac{1}{2}}}{\pi^{\frac{1}{4}}} xe^{-\frac{x^2}{2}} \right) = \frac{h}{2\pi i} \frac{2^{\frac{1}{2}}}{\pi^{\frac{1}{4}}} \left(e^{-\frac{x^2}{2}} - x^2 e^{-\frac{x^2}{2}} \right)$$

and $$\int_{\text{all space}} f^*(x) \hat{O} g(x) dx = \int_{-\infty}^{+\infty} \frac{1}{\pi^{\frac{1}{4}}} e^{-\frac{x^2}{2}} \frac{h}{2\pi i} \frac{\partial}{\partial x} \left(\frac{2^{\frac{1}{2}}}{\pi^{\frac{1}{4}}} xe^{-\frac{x^2}{2}} \right) dx = \frac{h}{2\pi i} \left(\frac{2}{\pi} \right)^{\frac{1}{2}} \int_{-\infty}^{+\infty} \left(e^{-x^2} - x^2 e^{-x^2} \right) dx$$

$$= \left(\frac{h}{2\pi i}\right)\left(\frac{2}{\pi}\right)^{\frac{1}{2}}\left(\pi^{\frac{1}{2}} - \frac{\pi^{\frac{1}{2}}}{2}\right) = \left(\frac{h}{2\pi i}\right)\frac{1}{\sqrt{2}} \qquad ...(1)$$

Similarly $$\hat{O}^* f^*(x) = -\frac{h}{2\pi i}\frac{\partial}{\partial x}\frac{1}{\pi^{\frac{1}{4}}}e^{-\frac{x^2}{2}} = \frac{h}{2\pi i}\left(2x\frac{1}{\pi^{\frac{1}{4}}}e^{-\frac{x^2}{2}}\right) = \frac{h}{\pi i}x\frac{1}{\pi^{\frac{1}{4}}}e^{-\frac{x^2}{2}}$$

$$\int_{\text{all space}} g(x)\hat{O}^* f^*(x)dx = \int_{-\infty}^{+\infty}\frac{\sqrt{2}}{\pi^{\frac{1}{4}}}xe^{-\frac{x^2}{2}}\frac{h}{2\pi i}\frac{\partial}{\partial x}\left(\frac{1}{\pi^{\frac{1}{4}}}e^{-\frac{x^2}{2}}\right)dx = \frac{h}{2\pi i}\left(\frac{2}{\pi}\right)^{\frac{1}{2}}\int_{-\infty}^{+\infty}x^2e^{-\frac{x^2}{2}}dx$$

$$= \left(\frac{h}{2\pi i}\right)\left(\frac{2}{\pi}\right)^{\frac{1}{2}}\frac{\pi^{\frac{1}{2}}}{2} = \left(\frac{h}{2\pi i}\right)\frac{1}{\sqrt{2}} \qquad ...(2)$$

Equality of equations (1) and (2) led to new definition of postulate III, *i.e. to every observation in classical mechanics there corresponds to linear, Hermitian operator in quantum mechanics.*

2.10. SCHRÖDINGER WAVE EQUATION

Schrödinger wave equation was derived from the basic equation of wave properties and combining the wave and particle nature of the matter. The Schrödinger wave equation, describing the behaviour of particle, independent of time, can be written as

$$\nabla^2\psi + \frac{8\pi^2 m}{h^2}(E - V)\psi = 0 \qquad ...(2.48)$$

where ψ is the wave function with the characteristics already defined, m is the mass of the particle, E the total energy and V is the potential energy. The ∇ is the Laplacian operator and in cartesian coordinates, its value is

$$\nabla^2 = \frac{\partial^2}{\partial x^2} + \frac{\partial^2}{\partial y^2} + \frac{\partial^2}{\partial z^2} \qquad ...(2.49)$$

and in spherical polar coordinates

$$\nabla^2 = \frac{1}{r^2}\frac{\partial}{\partial r}\left(r^2\frac{\partial}{\partial r}\right) + \frac{1}{r^2\sin\theta}\frac{\partial}{\partial\theta}\left(\sin\theta\frac{\partial}{\partial\theta}\right) + \frac{1}{r^2\sin^2\theta}\frac{\partial^2}{\partial\phi^2} \qquad ...(2.50)$$

The solution of second order differential equation gives the values of ψ with corresponding value of E.

The wave equation (2.48) can also be written with the help of operators and the corresponding equation is

$$\hat{H}\psi = E\psi \qquad ...(2.51)$$

where $\hat{H}$ is the Hamiltonian operator.

The general equation of the form given by (2.51) is

$$\hat{O}\,\psi = O\psi \qquad ...(2.52)$$

where $\hat{O}$ is an operator for any observable quantity and O is its eigen value. The equation (2.52) is called eigen value eigen function equation. "Eigen function" is that function when operated by an operator generates the same function multiplied by a constant. The constant is called the "eigen value".

Equations (2.48) and (2.49) can be written as:

$$\left[-\frac{h^2}{8\pi^2 m}\nabla^2 + V\right]\psi = E\psi \qquad ...(2.53a)$$

$$\left\{-\frac{h^2}{8\pi^2 m}\left[\frac{1}{r^2}\frac{\partial}{\partial r}\left(r^2\frac{\partial}{\partial r}\right)+\frac{1}{r^2\sin\theta}\frac{\partial}{\partial\theta}\left(\sin\theta\frac{\partial}{\partial\theta}\right)+\frac{1}{r^2\sin^2\theta}\frac{\partial^2}{\partial^2\varphi}\right]+V\right\}\psi = E\psi \qquad ...(2.54a)$$

Comparing equations (2.51) with (2.53*a*) and with (2.54*a*), we get

$$\hat{H} = -\frac{h^2}{8\pi^2 m}\nabla^2 + V \qquad ...(2.55)$$

$$\hat{H} = -\frac{h^2}{8\pi^2 m}\left[\frac{1}{r^2}\frac{\partial}{\partial r}\left(r^2\frac{\partial}{\partial r}\right)+\frac{1}{r^2\sin\theta}\frac{\partial}{\partial\theta}\left(\sin\theta\frac{\partial}{\partial\theta}\right)+\frac{1}{r^2\sin^2\theta}\frac{\partial^2}{\partial^2\varphi}\right]+V \qquad ...(2.56)$$

$\hat{H}$ is known as Hamiltonian operator (*i.e.* operator for total energy, sum of kinetic and potential energies). It may be mentioned here that different systems have different forms of potential energy functions, $V(x_1, y_1, z_1, ...x_n, y_n, z_n)$ and thus leads to different sets of allowed wave functions and their corresponding energies when equation (2.51) is solved for different systems.

***Exercise* 2.28:** Write a Hamiltonian operator for a freely moving particle and hence the corresponding Schrödinger equation.

If m is the mass of the particle and moving with a velocity v_x in the direction of x, then

$$\text{K.E.} = T = \frac{1}{2}\,mv_x^2 = \frac{p_x^2}{2m}$$

Potential energy is zero for freely moving particle. So the classical Hamiltonian is

$$H = \text{K.E.} + \text{P.E.} = \frac{p_x^2}{2m}$$

Since it is a freely moving particle, the potential energy is zero.

The operator for momentum is $\frac{h}{2\pi i}\frac{d}{dx}$, *i.e.*

$$\hat{H} = \frac{1}{2m}\left[\frac{h}{2\pi i}\frac{d}{dx}\right]\left[\frac{h}{2\pi i}\frac{d}{dx}\right]$$

$$= -\frac{h^2}{8\pi^2 m}\frac{d^2}{dx^2}$$

Substituting the value of $\hat{H}$ in equation (2.51)

$$-\frac{h^2}{8\pi^2 m}\frac{d^2\psi}{dx^2} = E\psi$$

or

$$\frac{d^2\psi}{dx^2} + \frac{8\pi^2 m}{h^2}\psi E = 0.$$

2.11. PARTICLE IN A BOX

The Schrödinger equation can be applied to a particle of mass m moving in box with infinite walls, *i.e.* potential energy is ∞ outside the box but a constant and a finite value in the box and this fixed value can be taken as zero. The Schrödinger equation is

$$\frac{d^2\psi}{dx^2} + \frac{8\pi^2 m}{h^2}E\psi = 0 \qquad ...(2.57)$$

The general solution of equation (2.57) is:

$$\psi = A\cos kx + B\sin kx \qquad ...(2.58)$$

where A, B and k are constants to be determined. These constants can be determined from (*i*) *boundary conditions* and (*ii*) *total probability of the particle in the box is unity*.

(*i*) Boundary Conditions

Boundary conditions are some constraints put on the motion of the particle so that it undergo well behave motion or provide well behave solution of the problem which agree with the experimental results.

If we do not apply the boundary conditions or some constraints to the motion of the particle, the constants A, B and k cannot be determined and we can say that the particle is free to move. Under such conditions the particle is known as free particle.

The value of potential energy is either zero or constant puts the boundary con ditions as $\psi = 0$ when $x = 0$ and $x = a$.

After applying the boundary conditions, the solution of equation (2.57) for one dimensional box gives the value of energy levels and wave function for each energy, *i.e.*

$$E_{n_x} = \frac{n_x^2 h^2}{8ma^2} \qquad ...(2.59)$$

where $n_x = 1, 2, 3, ...$, and a is the length of the box and

$$\psi_n = \sqrt{\frac{2}{a}}\sin\frac{n_x\pi}{a}.x \qquad ...(2.60)$$

Similarly, for a particle moving in three-dimensional box, the Schrödinger wave equation is

$$-\frac{h^2}{8\pi^2 m}\left[\frac{\partial\psi^2(x,y,z)}{\partial^2 x}+\frac{\partial\psi^2(x,y,z)}{\partial^2 y}+\frac{\partial\psi^2(x,y,z)}{\partial^2 z}\right]\psi(x,y,z) = E\psi(x,y,z) \quad ...(2.61)$$

As such this equation is not easy to solve. There is a very important theorem in quantum mechanics which help in solving many other problems in quantum mechanics.

2.11.1. *Separation of Variables*

If the total energy (in general a property) which can be separated into sum of the energies due to different variables (each motion depending on the variable is independent of each other) the total wave function can be written as the product of wave functions depending on the individual variable, *i.e.* mathematically the general form can be written as

$$\hat{O} = \hat{O}_1 + \hat{O}_2 + ...\hat{O}_n \quad ...(2.62)$$

where $\hat{O}_i$ is the operator only of coordinate say (q_i), *i.e.* each operator is the function of only one coordinate. In such condition,

$$f(q_1, q_2, ...q_n) = f_1(q_1)\, f_2\,(q_2).........f_n(q_n) \quad ...(2.63)$$

Considering particle in three dimensional box, kinetic energy along each coordinate x, y and z are independent of each other. Thus

$$\hat{H} = \hat{H}_x + \hat{H}_y + \hat{H}_z \quad ...(2.64)$$

and $$\Psi(x, y, z) = X(x)\; Y(y)\; Z(z) \quad ...(2.65)$$

Since

$$E(x, y, z) = E(x) + E(y) + E(z) \quad ...(2.66)$$

Using this theorem for a particle moving in three-dimensional box, the wave function and the energy levels are

$$\psi_{n_1 n_2 n_3}(x, y, z) = \left(\frac{8}{abc}\right)^{1/2} \sin\frac{n_1 \pi x}{a} \sin\frac{n_2 \pi}{b} y \sin\frac{n_3 \pi}{c} z \quad ...(2.67)$$

and

$$E = \frac{h^2}{8m}\left(\frac{n_1^2}{a^2}+\frac{n_2^2}{b^2}+\frac{n_3^2}{c^2}\right) \quad n_1 = 1, 2, 3,... \quad ...(2.68)$$

$$n_2 = 1, 2, 3, ..., n_3 = 1, 2, 3, ...$$

where *a, b, c* are the dimensions of the box. In case of cubical box

$$E = \frac{h^2}{8ma^2}(n_1^2 + n_2^2 + n_3^2) \quad ...(2.69)$$

Two very important results are obtained from the above equations:

(i) Quantisation of Energy

If there are no restraints (or boundary conditions) on the motion of the particle, it is free to move and thus can possess any amount of energy, *i.e.* not quantised. It is the boundary conditions that impose restrictions on the motions of particle and thus as a result energy is quantised.

In summary it is the boundary conditions that lead to quantisation of energy .

(ii) Degeneracy

New features appear in case of particle moving in a cubical box, *i.e.*, for different wave functions the corresponding energies are same. Degeneracy is defined as the occurrence of more than one distinct wave function for the same eigen value. In quantum mechanics the terms state and energy level has different meanings.

A stationary state is specified by giving the wave function ψ. Each different ψ is a different state.

An energy level is specified by giving the value of energy. Each different value of E is a different energy group.

Comparing equations (2.67) and (2.68), it is clear that degeneracy appears as the rectangularoid structure becomes cubic, *i.e.*, less symmetric structure turns to more symmetric cubic structure. *In conclusion, higher the symmetry, higher the degeneracy will be.* For example, we will learn later (Chapter 3) that in hydrogen atom, electron moves under the highly symmetric field, *i.e.*, spherical and this leads to maximum degeneracy of energy levels. It depends only on the principal quantum number, *n, i.e.*

(*a*) *energy levels are degenerate till n is the same for one electron systems.*

(*b*) *energy levels are degenerate till n and l are the same for atoms containing two or more than two electrons.*

(*c*) *under the external magnetic field, all the levels are non-degenerate .*

Zero Point Energy

Minimum value of quantum number is 1 for one dimensional box and $n_x = n_y = n_z = 1$ for three dimensional box. If we substitute the value of quantum number in equations (2.59) and (2.69), the values of energy, we get are

$$E_{n_x} = \frac{h^2}{8ma^2} \quad \text{...(2.70)}$$

and

$$E = \frac{h}{8m}[\frac{1}{a^2}+\frac{1}{b^2}+\frac{1}{c^2}] \text{ in case of ractangularoid} \quad \text{...(2.71)}$$

and

$$E = \frac{3h}{8ma^2} \quad \text{...(2.72)}$$

It is clear that none of this value of energy is zero. Since this energy is obtained with the minimum value of quantum numbers, the value of this energy is called *zero point energy*.

***Exercise* 2.29:** A particle of mass 9×10^{-31} kg is moving in 0.05 nm one dimensional box. Suppose this particle comes down from $n = 4$ to 3, calculate the wavelength of this emission and also region in which wavelength lies.

Using equation (2.59):

$$E(3) = \frac{h^2(3)^2}{8ma^2} \text{ and } E(4) \frac{h^2(4)^2}{8ma^2}. \text{ So}$$

$$\Delta E = h\nu = \frac{h^2}{8ma^2}[16-9] = \frac{7h^2}{8ma^2}$$

$$\nu = \frac{7h^2}{8ma^2 h} = \frac{7h}{8ma^2} = \frac{(7)(6.626 \times 10^{-34} \text{ J s})}{(8)(9 \times 10^{-31} \text{ kg})(0.05 \times 10^{-9} \text{ m})^2} = 2.577 \times 10^{17} \text{ s}^{-1}$$

$$\lambda = \frac{c}{\nu} = \frac{3 \times 10^8 \text{ m s}^{-1}}{2.577 \times 10^{17} \text{ s}^{-1}} = 1.164 \times 10^{-9} \text{ m.}$$

***Exercise* 2.30:** If a particle of mass 9.1×10^{-31} kg in certain one-dimensional box, the $n = 2$ to $n = 1$ transition occurs at $\nu = 4 \times 10^{14}$ s^{-1}. Find the length of the box.

Using equation (2.58)

$$E(1) = \frac{h^2(1)^2}{8ma^2} \text{ and } E(2) = \frac{h^2(2)^2}{8ma^2}. \text{ So}$$

$$\Delta E = h\nu \; \frac{h^2}{8ma^2}[4-1] = \frac{3h^2}{8ma^2}$$

$$a^2 = \frac{3h}{8m\nu}$$

or $$a = \left[\frac{3h}{8m\nu}\right]^{\frac{1}{2}} = \left[\frac{(3)(6.626 \times 10^{-34} \text{ J s})}{(8)\,(9 \times 10^{-31} \text{ kg}) \times (4 \times 10^{14} \text{ s}^{-1})}\right]^{\frac{1}{2}} = 0.83 \times 10^{-9} \text{ m.}$$

***Exercise* 2.31:** Assume that prophyrine is a planar molecule. Number of π-electrons is 26. If the long wavelength absorption band occurs approximately at 640 nm, find the side length of this square.

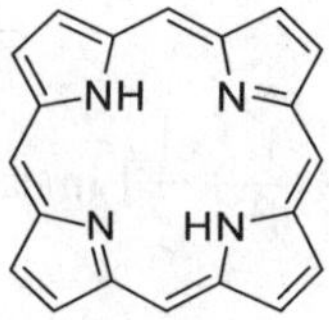

Porphyrine can be considered as two dimensional molecule and contains 26 π-electrons. The energy level equation is given as for two dimensional box

$$E = \frac{h^2}{8ma^2}[n_x^2 + n_y^2] = \frac{h^2 n^2}{8ma^2} \qquad ...(2.72a)$$

where $$n^2 = n_x^2 + n_y^2$$

The longest wavelength transition involves the electron from the highest occupied molecular orbital (HOMO) to the lowest unoccupied molecular orbital (LUMO). So the number of energy levels occupied will be 13. Energy level diagram and filling of electrons are shown in Fig. 2.2. Highest

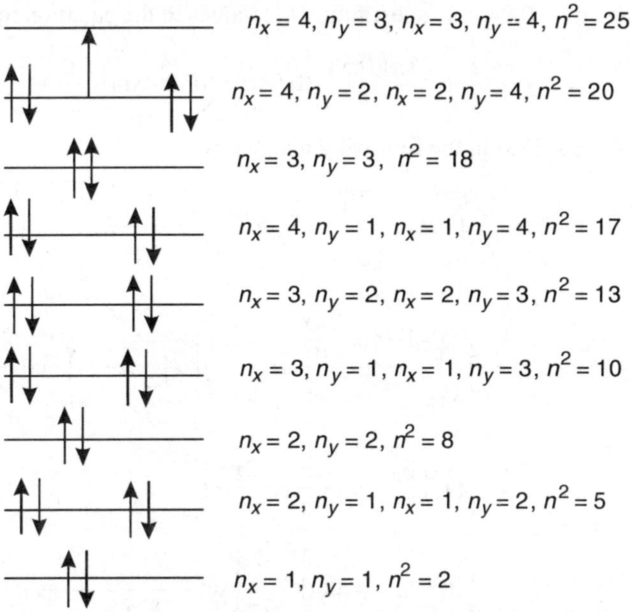

Fig. 2.2 *Energy level diagram and filling of 26-π electrons of porphyrine.*

occupied energy level is $n_x = 4$, $n_y = 2$ and lowest unoccupied energy level is $n_x = 4$, $n_y = 3$. The energies of these levels can be obtained using modified equation (2.72*a*), *i.e.*,

$$E(n_x = 4, n_y = 2) = \frac{h^2}{8ma^2}[n_x^2 + n_y^2] = \frac{h^2}{8ma^2}[4^2 + 2^2] = \frac{20h^2}{8ma^2}$$

$$E(n_x = 4, n_y = 3) = \frac{h^2}{8ma^2}[n_x^2 + n_y^2] = \frac{h^2}{8ma^2}[4^2 + 3^2] = \frac{25h^2}{8ma^2}$$

$$\text{So, } \Delta E = h\nu = E(n_x = 4, n_y = 3) - E(n_x = 4, n_y = 2)\ \frac{h^2}{8ma^2}[25 - 20] = \frac{5h^2}{8ma^2}$$

$$\text{or} \quad a = \left[\frac{5h}{8m\nu}\right]^{\frac{1}{2}} = \left[\frac{5h\lambda}{8mc}\right]^{\frac{1}{2}} = \left[\frac{(5)(6.626 \times 10^{-34}\ \text{J s})(640 \times 10^{-9}\ \text{m})}{(8)(9.1 \times 10^{-31}\text{kg})(3 \times 10^{8}\ \text{m s}^{-1})}\right]^{\frac{1}{2}} = 9.8 \times 10^{-10}\ \text{m}$$

The length of the square = 9.8×10^{-10} m = 9.8 Å.

***Exercise* 2.32:** Calculate the probability of finding the particle with in the range of 0.48*a* to 0.52*a*, where *a* is the length of the box in the ground state, *i.e.*, $n = 1$.

The ground state normalised wave function $\psi(1) = \sqrt{\frac{2}{a}} \sin\frac{\pi x}{a}$. The probability of finding the particle is given as

$$Pdx = \frac{2}{a}\sin^2\frac{\pi x}{a}dx$$

Here $x = 0.5\ a$ and $dx = 0.04a$. So substituting the values in the equation to depict P, we get

$$P = \frac{2}{a}\sin^2\frac{(\pi)(0.5a)}{a}(0.04a) = (0.08)\sin^2\frac{(\pi)}{2} = 0.08.$$

This problem can be solved in the following manner as:

$$P = \int_{0.48a}^{0.52a} Pdx = \int_{0.48a}^{0.52a}\sqrt{\frac{2}{a}}\sin\frac{\pi x}{a}\sqrt{\frac{2}{a}}\sin\frac{\pi x}{a}dx = \frac{2}{a}\int_{0.48a}^{0.52a}\sin^2\frac{\pi x}{a}dx$$

$$= \frac{2}{a}\int_{0.48a}^{0.52a}\frac{1-\cos\frac{2\pi x}{a}}{2}dx = \frac{2}{a}\frac{1}{2}\left[\int_{0.48a}^{0.52a}dx - \int_{0.48a}^{0.52a}\cos\frac{2\pi x}{a}dx\right]$$

$$= \frac{1}{a}[x]_{0.48a}^{0.52a} - \frac{1}{a}\left[\frac{a}{2\pi}\sin\frac{2\pi x}{a}\right]_{0.48a}^{0.52a}$$

$$= \frac{1}{a}[0.52a - 0.48a] - \frac{1}{2\pi}\left[\sin\frac{2\pi \times 0.52a}{a} - \sin\frac{2\pi \times 0.48a}{a}\right]$$

$$= 0.04 - \frac{1}{2 \times 3.14}[-0.125 - 0.125] = 0.08.$$

***Exercise* 2.33:** Wave functions and corresponding energy levels for the particle in one dimensional box of length a are given by equations (2.59) and (2.60) respectively. Plot these wave functions corresponding to their energy levels.

Let us consider 4 energy levels. Their corresponding wave functions are plotted in Fig. 2.3.

Following results can be elaborated:

(*i*) The wave function in the ground state attains a maximum value in the middle of the box.

$$\psi_4 = \sqrt{\frac{2}{a}}\sin\frac{4\pi}{a}x \qquad E_4 = \frac{16h^2}{8ma^2}$$

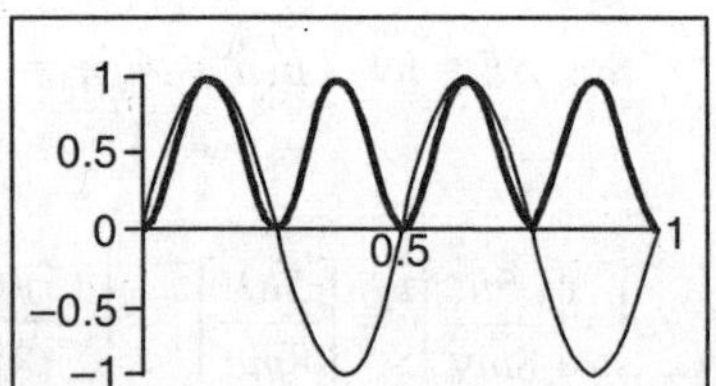

$$\psi_3 = \sqrt{\frac{2}{a}}\sin\frac{3\pi}{a}x \qquad E_3 = \frac{9h^2}{8ma^2}$$

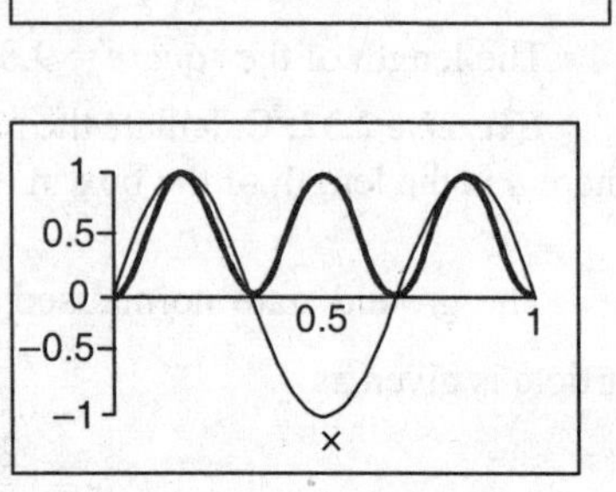

$$\psi_2 = \sqrt{\frac{2}{a}}\sin\frac{2\pi}{a}x \qquad E_2 = \frac{4h^2}{8ma^2}$$

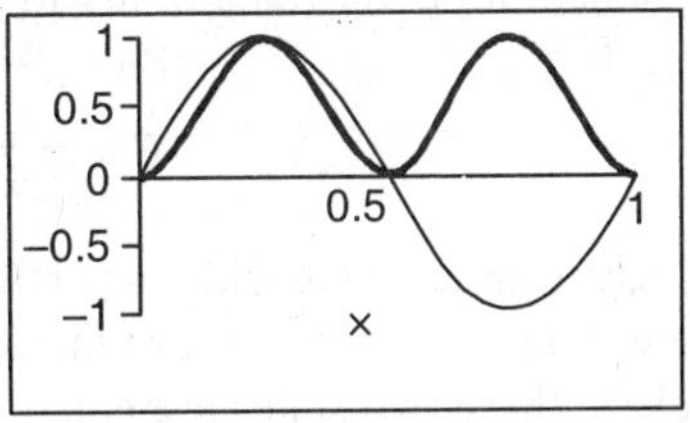

$$\psi_1 = \sqrt{\frac{2}{a}}\sin\frac{\pi}{a}x \qquad E_1 = \frac{h^2}{8ma^2}$$

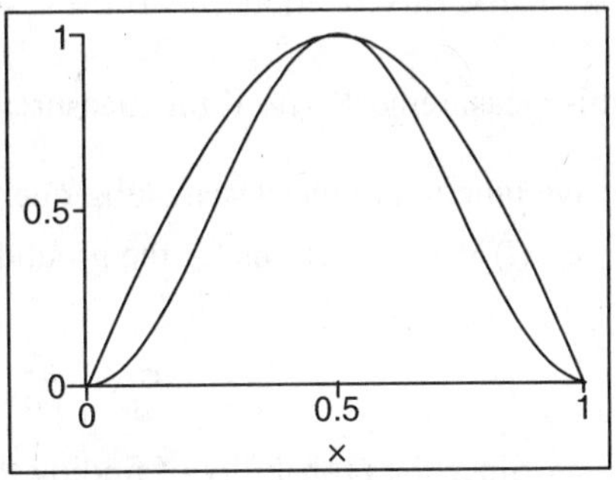

Fig. 2.3 *Plot of energies, wave functions (thin lines) and square of the wave functions (thick lines) for different quantum numbers of particle in one dimensional box. Energy (E) is in units of* $\left(\frac{h^2}{8\ ma^2}\right)$.

(*ii*) As the quantum number increases, wave functions behave differently from the ground state. The wave function becomes zero at certain point in the box. These points are called *nodes*, *i.e.,* where the wave function becomes zero. The number of nodes increases with the increase in quantum number and is given by $n - 1$.

(*iii*) Classically, all the locations for the particle in the box is equally probable. Quantum mechanically, the probability density is not uniform but shows oscillations.

(*iv*) When n is very large, oscillations in $|\psi|^2$ comes closer and closer together and ultimately become undetectable. This value of quantum number corresponds to the classical limits, *i.e.,* Bohr's correspondence principle (Section 2.12). *In conclusion, for large energy, i.e. large quantum number, the results obtained with the help of classical mechanics and quantum mechanics converge to the same point (i.e. they are equal) and happens for systems having macroscopic values of E, m and a.*

***Exercise* 2.34:** From Exercise 2.33, the length of the one dimensional box and wavelength are related by the relation, $\lambda = \frac{2a}{n}$, where n is the quantum number. Using de Broglie's relation show that the value of kinetic energy obtained is the same as obtained from Schrödinger equation.

Using de Broglie relation $\lambda = \frac{h}{mv}$. Comparing this relation and the given relation

$$\frac{h}{mv} = \frac{2a}{n} \Rightarrow \text{means } mv = \frac{nh}{2a}$$

Kinetic energy, $$E = \frac{mv^2}{2} = \frac{(mv)^2}{2m}$$

Substituting the value of mv in the above equation, we get

$$E = \frac{1}{2m}\left(\frac{nh}{2a}\right)^2 = \frac{h^2}{8ma^2}n^2$$

***Exercise* 2.35:** Consider a particle in one dimensional box of length a. Suppose we know that the particle is present either in ground state or in the first excited state and probabilities of finding the particle in these two states are equal. (*i*) Write down the appropriate expression of wave function. (*ii*) Normalise this function. (*iii*) If the energy of this system is measured, what are the possible results of this measurement? (*iv*) If the measurement shows that the energy of the system is $\frac{h^2}{2ma^2}$, what is the wave function of the system subsequent to the measurement?

(*i*) Wave functions for the ground and first excited states are

$$\phi_1 = \sqrt{\frac{2}{a}}\sin\frac{\pi}{a}x, \ \phi_2 = \sqrt{\frac{2}{a}}\sin\frac{2\pi}{a}x$$

Since the probability of finding the particle in these two states are equal, so

$$\psi = N[\phi_1 + \phi_2]$$

where N is the normalisation constant. We have taken the coefficients of the two wave functions equal as they have equal probabilities.

(*ii*) $\int_0^a \psi\psi dt = \int_0^a N(\phi_1+\phi_2)N(\phi_1+\phi_2)dt = N^2\{\int_0^a \phi_1\phi_1 dt + 2\int_0^a \phi_1\phi_2 dt + \int_0^a \phi_2\phi_2 dt\} = 1$

Since the wave functions are normalised, $\int_0^a \phi_1\phi_1 dt = \int_0^a \phi_2\phi_2 dt = 1$ and $\int_0^a \phi_1\phi_2 dt = 0$

So $\qquad N^2\{1+1\} = 1$

or $\qquad N = \frac{1}{\sqrt{2}}$

and the normalised wave function is $\psi = \frac{1}{\sqrt{2}}[\phi_1 + \phi_2]$.

(*iii*) Since there are only two possible energy levels and if we carry out the measurements, there will be only two possible values, *i.e.*, $\frac{h^2}{8ma^2}$ and $\frac{(2)^2 h^2}{8ma^2} = \frac{h^2}{2ma^2}$.

(*iv*) If the measured energy of the system is $\frac{h^2}{2ma^2}$, it is clear that the particle is present only in the first excited state and thus wave function of the system will be ϕ_2.

***Exercise* 2.36:** Calculate $<\Delta x^2> = <x^2> - <x>^2$ for the Particle in one dimensional box with infinite walls.

Using normalised wave function for the one dimensional box, *i.e.*, $\psi_n = \sqrt{\frac{2}{a}} \sin \frac{n\pi}{a} x$ and the average value of the position is given by the equation.

$$<x> = \int_0^a \psi^* \psi dx = \int_0^a \sqrt{\frac{2}{a}} \sin \frac{n\pi x}{a} x \sqrt{\frac{2}{a}} \sin \frac{n\pi x}{a} dx = \frac{2}{a}\int_0^a x \sin^2 \frac{n\pi x}{a} dx = \frac{2}{a}\int_0^a x \left[\frac{1-\cos\frac{2n\pi x}{a}}{2}\right] dx$$

$$= \frac{2}{a}\left[\int_0^a \frac{x}{2} dx - \int_0^a \frac{1}{2} x \cos\frac{2n\pi x}{a} dx\right] = \frac{2}{a}\left[\frac{x^2}{4}\right]_0^a - \frac{2}{a}\left[x\frac{a}{4n\pi}\sin\frac{2n\pi x}{a}\right]_0^a + \frac{2}{a}\int_0^a \frac{a}{4n\pi} \sin \frac{2npx}{a} dx$$

$$= \frac{2}{a}\left[\frac{x^2}{4}\right]_0^a - \frac{2}{a}\left[x\frac{a}{4n\pi}\sin\frac{2n\pi x}{a}\right]_0^a + \frac{2}{a}\left[-\frac{a^2}{8n^2\pi^2}\cos\frac{2n\pi x}{a}\right]_0^a$$

$$= \frac{2}{a}\left[\frac{a^2}{4}\right] - \frac{2}{a}\left[\frac{a}{4n\pi}\left(a \sin\frac{2n\pi a}{a} - 0 \sin\frac{2n\pi 0}{a}\right)\right] + \frac{2}{a}\left[-\frac{a^2}{8n^2\pi^2}\left(\cos\frac{2n\pi a}{a} - \cos\frac{2n\pi 0}{a}\right)\right]$$

$$= \frac{2}{a}\left[\frac{a^2}{4}\right] - \frac{2}{a}\left[\frac{a}{4n\pi}(0-0)\right] + \frac{2}{a}\left[-\frac{a^2}{8n^2\pi^2}(1-1)\right] = \frac{a}{2} \quad \text{... 273}(a)$$

Now calculating the value of $<x^2>$, we can write the equation as:

$$<x^2> = \int_0^a \psi^* x^2 \psi dx = \int_0^a \sqrt{\frac{2}{a}} \sin \frac{n\pi x}{a} x^2 \sqrt{\frac{2}{a}} \sin \frac{n\pi x}{a} dx = \frac{2}{a}\int_0^a x^2 \sin^2 \frac{n\pi x}{a} dx$$

$$= \frac{2}{a}\int_0^a x^2 \left[\frac{1-\cos\frac{2n\pi x}{a}}{2}\right] dx$$

Integrating by parts twice or using the standard

$$= \frac{2}{a}\left[\int_0^a \frac{x^2}{2} dx - \int_0^a \frac{x^2}{2}\cos\frac{2n\pi x}{a} dx\right] = \frac{2}{a}\left[\frac{x^3}{6}\right]_0^a - \frac{2}{a}\left[x^2\frac{a}{4n\pi}\sin\frac{2n\pi x}{a}\right]_0^a$$

$$+ \frac{2}{a}\left[-\frac{2xa^2}{8n^2\pi^2}\cos\frac{2n\pi x}{a}\right]_0^a - \frac{2}{a}\left[\frac{2a^3}{16n^3\pi^3}\sin\frac{2n\pi x}{a}\right]_0^a$$

$$= \frac{2}{a}\left[\frac{a^3}{6}\right] - \frac{2}{a}\left[a^2\frac{a}{4n\pi}\left(\sin\frac{2n\pi a}{a} - \sin\frac{2n\pi 0}{a}\right)\right] - \frac{2}{a}\left[\frac{2xa^2}{8n^2\pi^2}\left(\cos\frac{2n\pi a}{a} - \cos\frac{2n\pi 0}{a}\right)\right]_0^a$$

$$- \frac{2}{a}\left[\frac{2a^3}{16n^3\pi^3}\sin\frac{2n\pi a}{a} - \sin\frac{2n\pi 0}{a}\right]$$

$$= \frac{a^2}{3} - \frac{2}{a}\left[a^2 \frac{a}{4n\pi}(0-0)\right] - \frac{a^2}{4n^2\pi^2}(-1-1)\Bigg] - \left[\frac{2a^2}{8n^2\pi^3}(0-0)\right] = \frac{a^2}{3} - \frac{a^2}{2n^2\pi^2}$$

$$= a^2\left[\frac{1}{3} - \frac{1}{2n^2\pi^2}\right] \qquad \text{... (273(b))}$$

Combining equations (273 *a*) and (273 *b*) with the given equation, we get

$$= a^2\left[\frac{1}{3} - \frac{1}{2n^2\pi^2}\right]$$

$$<\Delta x^2> = <x^2> - <x>^2 = a^2\left[\frac{1}{3} - \frac{1}{2n^2\pi^2}\right] - \left(\frac{a}{2}\right)^2 = a^2\left[\frac{1}{12} - \frac{1}{2n^2\pi^2}\right]$$

2.12. BOHR'S CORRESPONDENCE PRINCIPLE

The correspondence principle states that the quantum mechanical results must go over to classical mechanics when the quantum number describing the system becomes very large.

***Exercise* 2.37:** Calculate the value of n if it has energy equal to k_bT at room temperature (27°C) for a ball of 0.5 g ball-bearing in 1.00 m long box.

Using equation (2.59), *i.e.* $E = k_bT = (1.38 \times 10^{-23} \text{ J K}^{-1})(300 \text{ K}) = (4.14 \times 10^{-21} \text{ J})$

$$\eta = \left[\frac{E_{nx} \cdot 8ma^2}{h^2}\right]^{\frac{1}{2}}$$

$$n = \left[\frac{(4.14\times10^{-21} \text{ J})(8)(0.5\times10^{-3} \text{ kg})(1 \text{ m})^2}{(6.63\times10^{-34} \text{ J s})^2}\right]^{1/2}$$

$$= 6.14 \times 10^{21}$$

i.e. there are 6.14×10^{21} maxima in 1.0 m long box. So these maxima are separated by a distance of $\frac{1 \text{ m}}{6.14\times10^{21}} = 1.6 \times 10^{-22}$ m, *i.e.* more or less a continuous position. This indicates that the results obtained for large particles by classical and quantum mechanics are same.

***Exercise* 2.38:** Show that the value of energy obtained for a particle of mass m moving in a one dimensional box can also be obtained with the help of relation used to find expectation value.

The normalised wave function for the particle in one dimensional box is

$$\psi_n(x) = \sqrt{2/a} \sin \frac{n\pi}{a} x$$

The Hamiltonian operator is

$$\hat{H} = \frac{1}{2}mv_x^2 = \frac{p_x^2}{2m}$$

$$= -\frac{h^2}{8\pi^2 m}\frac{d^2}{dx^2}$$

$$\bar{E} = \langle E \rangle = \int_0^a \psi_n^*(x)\hat{H}\psi_n(x)\,dx$$

$$= -\frac{h^2}{8\pi^2 m}\int_0^a \sqrt{2/a}\,\sin\frac{n\pi}{a}x.\frac{d^2}{dx^2}\left(\sqrt{2/a}.\sin\frac{n\pi}{a}x\right)dx$$

$$= -\frac{2h^2}{8\pi^2 am}\int_0^a \sin\frac{n\pi}{a}x.\left(-\frac{n^2\pi^2}{a^2}\sin\frac{n\pi}{a}x\right)dx$$

$$= \frac{2h^2}{8\pi^2 ma}\times\frac{n^2\pi^2}{a^2}\int_0^a \sin^2\frac{n\pi}{a}x.\,dx$$

$$= \frac{n^2h^2}{4ma^3}\left[\frac{1}{2}\int_0^a\left(1-\cos\frac{2n\pi}{a}x\right)dx\right]$$

$$= \frac{n^2h^2}{8ma^3}\left\{[x]_0^a - \frac{a}{2n\pi}\left[\sin\frac{2n}{a}x\right]_0^a\right\}$$

$$= \frac{n^2h^2}{8ma^3}(a-0) = \frac{n^2h^2}{8ma^2}$$

This value is the same as obtained by solving Schrödinger equation.

Note: For more problem, see Chapter 4.

2.13. HARMONIC OSCILLATOR

Harmonic oscillator is defined as oscillator in which restoring force is defined as:

$$F = -kx \qquad ...(2.74)$$

where k is the force constant and x is the displacement from the equilibrium position. The potential energy of the oscillator is given by

$$V(x) = \frac{1}{2}kx^2 \qquad ...(2.75)$$

The Schrödinger wave equation for the one dimensional harmonic oscillator is given as

$$\frac{d^2\psi(x)}{dx^2} + \frac{8\pi^2\mu}{h^2}(E - \frac{1}{2}x^2)\psi(x) = 0 \qquad ...(2.76)$$

The solution of Schrödinger wave equation gives wave function for the harmonic oscillator as

$$\psi_v(x) = \frac{1}{(2^v\,v!)^{\frac{1}{2}}}\left(\frac{\alpha}{\pi}\right)^{\frac{1}{4}} H_v\left(\alpha^{\frac{1}{2}}x\right)e^{-\frac{\alpha x2}{2}} \qquad ...(2.77)$$

where $$\alpha = (\frac{4\pi^2 k\mu}{h^2})^{\frac{1}{2}} \quad ...(2.78)$$

and $H_v\left(\alpha^{\frac{1}{2}}x\right)$ is known as Hermite polynomials and is defined by equation

$$H_v(x) = (-1)^v e^{x^2} \frac{d^v(e^{-x^2})}{dx^v} \quad ...(2.79)$$

where $v = 0, 1, 2, 3,\ldots\ldots\ldots$ The first five members of the series are given below.

$$H_0(x) = (-1)^0 e^{x^2} \frac{d^0(e^{-x^2})}{dx^0} = 1$$

$$H_1(x) = (-1)^1 e^{x^2} \frac{d^1(e^{-x^2})}{dx^1} = -e^{x^2}(-2x)e^{-x^2} = 2x$$

$$H_2(x) = (-1)^2 e^{x^2} \frac{d^2(e^{-x^2})}{dx^2} = e^{x^2}\frac{d}{dx}[-2x e^{-x^2}] = e^{x^2}[-2e^{-x^2} + 4x^2 e^{-x^2}]$$

$$= (4x^2 - 2)$$

$$H_3(x) = (-1)^3 e^{x^2} \frac{d^3(e^{-x^2})}{dx^3} = 8x^3 + 12x$$

$$H_4(x) = (-1)^4 e^{x^2} \frac{d^4(e^{-x^2})}{dx^4} = 16x^4 - 48x^2 + 12$$

The energies of the corresponding wave functions are given by the equation (2.80)

$$E_v = hí\,(v + 1/2) \quad ...(2.80)$$

where v is the vibrational quantum number.

***Exercise* 2.39:** Find the wave function for the Harmonic oscillator for v = 5, using Hermite polynomial.

Using equation (2.79) the Hermite polynomial for v = 5, is

$$H_5(x) = (-1)^5 e^{x^2} \frac{d^5(e^{-x^2})}{dx^5} = -e^{x^2}\frac{d^4}{dx^4}\left[-2x e^{-x^2}\right] = -e^{x^2}\frac{d^3}{dx^3}\left[-2e^{-x^2} + 4x^2e^{-x^2}\right]$$

$$= -e^{x^2}\frac{d^2}{dx^2}\left[4xe^{-x^2} + 8xe^{-x^2} - 8x^3e^{-x^2}\right] = -e^{x^2}\frac{d^2}{dx^2}\left[12xe^{-x^2} - 8x^3e^{-x^2}\right]$$

$$= -e^{x^2}\frac{d}{dx}\left[12e^{-x^2} - (24x^2 + 24x^2)e^{-x^2} + 16x^4e^{-x^2}\right]$$

$$= -e^{x^2}\frac{d}{dx}\left[12\,e^{-x^2} - 48x^2e^{-x^2} + 16\,x^4e^{-x^2}\right]$$

$$= -e^{x^2}\left[-24\,x\,e^{-x^2} - 96\,x\,e^{-x^2} + 96\,x^3e^{-x^2} + 64\,x^3e^{-x^2} - 32x^5e^{-x^2}\right]$$

$$= 32x^5 - 160x^3 + 120x = 16[2x^5 - 10x^3 + 15/2\,x]$$

Substituting the values of $H_5(x)$ in equation (2.77), the normalized wave function for v = 5 is:

$$\psi_5 = \left(\frac{\alpha}{\pi}\right)^{\frac{1}{4}}\left(\frac{1}{2^5\;5!}\right)^{\frac{1}{2}} e^{-\frac{\alpha x^2}{2}}\left(32\,x^5 - 160x^3 + 120x\right)$$

$$= \left(\frac{\alpha}{\pi}\right)^{\frac{1}{4}}\left(\frac{1}{2^5\;5!}\right)^{\frac{1}{2}} e^{-\frac{\alpha x^2}{2}}\left(2x^5 - 10x^3 + \frac{15}{2}x\right)$$

Note: The wave functions rotational motion can be seen from the angular motion of the hydrogen atom and corresponding energy levels can be seen from the rotational spectrum.

***Exercise* 2.40:** The Hermite polynomials for v = 1 and v = 2 are $H_1(x) = 2x$ and $H_2(x) = 4x^2 - 2$ respectively. Show that the wave functions of a Harmonic oscillator in the stationary states 1 and 2 are orthogonal. The total wave function for states 1 and 2 for harmonic oscillator are:

$$\psi_1(x) = 2\,N_1\,x\,e^{-\frac{x^2}{2}} \quad \text{and} \quad \psi_2(x) = N_2\,e^{-\frac{x^2}{2}}(4x^2 - 2)$$

Wave functions are orthogonal if $\int_{-\infty}^{+\infty}\psi_1\,\psi_2\,dq = 0$. Substituting the values of the wave functions and x:

$$\int_{-\infty}^{+\infty}\psi_1\,\psi_2\,dx = \frac{2\,N_1\,N_2}{\alpha^{1/2}}\int_{-\infty}^{+\infty} x\,e^{-x^2/2}(4x^2 - 2)\,e^{-x^2/2}dx$$

$$= \frac{2\,N_1\,N_2}{\alpha^{1/2}}\int_{-\infty}^{+\infty} 4\,x^3\,e^{-x^2}\,dx - \frac{4\,N_1\,N_2}{\beta^{1/2}}\int_{-\infty}^{+\infty} q\,e^{-q^2}\,dq$$

Both the integrals are odd functions, so th e values of both the integrals are zero. Thus ψ_1 and ψ_2 are orthogonal functions.

Some of the general relations between the Hermite polynomials and their derivatives are:

$$H_{v+1}(\zeta) - 2\zeta H_v(\zeta) + 2vH_{v-1}(\zeta) = 0 \qquad ...(2.81)$$

$$\frac{dH_v(\zeta)}{d\zeta} = 2\zeta H_v(\zeta) - H_{v+1}(\zeta) = 2_v H_{v-1}(\zeta) \qquad ...(2.82)$$

Note: More solved exercise on the spectrum will be given in Chapter 5.

2.14. APPROXIMATE METHODS

2.14.1. *First Order Perturbation Thoery for Non-degenerate Energy Levels*

Except for very few systems it becomes difficult to get the exact solutions of Schrödinger equation. Under these situations an approximate solution of the wave equation is obtained. This approximate solution differs from the exact one by omitting certain terms whose effects are very small. Such problems can be treated with the help of perturbation theory; First calculating the approximate solutions and then adding the additional terms of corrections.

If $\hat{H}_0$ is the unperturbed Hamiltonian, the total Hamiltonian $\hat{H}$, can be written as

$$\hat{H} = \hat{H}^0 + \lambda \hat{H} \qquad ...(2.83)$$

where λ, is the parameter such that as $\lambda \to 0$, $\hat{H} \to \hat{H}^0$. It is a measure of deviation from the unperturbed system. Further, for first order perturbation the value of λ is unity if the perturbation is fully used. In general the value of λ is small. Similarly the wave functions and their corresponding energies can be expressed as

$$\psi_k = \psi_k^0 + \lambda\psi_k' + \lambda^2 \psi_k'' + ... \qquad ...(2.84)$$

$$E_k = E_k^0 + \lambda E_k' + \lambda^2 E_k'' + ... \qquad ...(2.85)$$

For first-order perturbation theory the equations (2.84) and (2.85) become

$$\psi_k = \psi_k^0 + \lambda \sum_{\substack{j=0 \\ j\neq k}}^{\infty} \frac{H_{jk}\psi_j^0}{E_j^0 - E_k^0} \qquad ...(2.86)$$

$$E_k = E_k^0 + \lambda H_k' \qquad ...(2.87)$$

where

$$H'_{jk} = \int \psi_j^0 \hat{H} \psi_k^0 \, d\tau \qquad ...(2.88)$$

and

$$E_k^0 = \int \psi_k^0 \hat{H} \psi_k^0 \, d\tau \qquad ...(2.89)$$

and the $\sum_j$ means the sum over all the values of j except $j = k$.

2.14.2. *Variation Method*

There are many problems in quantum mechanics which cannot be solved accurately even with first-order perturbation theory. The amount of work required to achieve better results using higher order perturbation theory is too much. The approximate but better results can be obtained with the help of variation method and this is useful specially when one wants to find the energy of the ground state.

The variation method involves the following steps:

(*i*) Construct the wave function in terms of certain variable parameters C_i and independent functions χ_i, *i.e.*

$$\phi_n = \Sigma_n C_n \chi_n \qquad ...(2.90)$$

The C_n are the parameters which are to be determined in order to get the lowest value of E.

(*ii*) The energy of the system is given by the expression

$$\bar{E} = \frac{\int \phi^* \hat{H} \phi \, d\tau}{\int \phi^* \phi \, d\tau} = \frac{\sum_{n=1}^{i} \sum_{n'=1}^{i} C_n C_{n'} H_{nn'}}{\sum_{n=1}^{i} \sum_{n'=1}^{i} C_n C_{n'} S_{nn'}} \quad ...(2.91)$$

where

$$\Delta_{nn'} = 1 \text{ for } n = n',\ S_{nn'} = \int \chi_{n'}^* \chi_n \, d\tau$$

$$= 0 \text{ for } n \neq n',\ H_{nn'} = \int \chi_{n'}^* \hat{H} \chi_n \, d\tau$$

The equation (2.91) can also be written as

$$\sum_{n=1}^{i} \sum_{n'=1}^{i} C_n C_{n'} S_{nn'} = \sum_{n=1}^{i} \sum_{n'=1}^{i} C_n C_{n'} H_{nn'} \quad ...(2.92)$$

(*iii*) Differentiate equation (2.92) with respect to C_n and then equate $\frac{\partial E}{\partial C_n} = 0$ for $n = 1, 2, 3, ..., i$ one gets the following set of simultaneous equations:

$$\begin{aligned} C_1 (H_{11} - ES_{11}) + C_2 (H_{12} - ES_{12}) + \ ... + C_n (H_{1n} - ES_{1n}) &= 0 \\ C_1 (H_{21} - ES_{21}) + C_2 (H_{22} - ES_{22}) + \ ... + C_n (H_{2n} - ES_{2n}) &= 0 \\ \cdots\cdots\cdots\cdots\cdots \quad & \cdots\cdots \\ \cdots\cdots\cdots\cdots\cdots \quad & \cdots\cdots \\ C_1 (H_{n1} - ES_{n1}) + C_2 (H_{n2} - ES_{n2}) + \ ... + C_n (H_{nn} - ES_{nn}) &= 0 \end{aligned} \quad ...(2.93)$$

The above set of simultaneous equations can have non-trivial solution only if the following determinant is zero,

$$\begin{vmatrix} H_{11} - ES_{11} & H_{12} - ES_{12} & ... & H_{1n} - ES_{1n} \\ H_{21} - ES_{21} & H_{22} - ES_{22} & ... & H_{2n} - ES_{2n} \\ \cdots & \cdots & \cdots & \cdots \\ \cdots & \cdots & \cdots & \cdots \\ H_{n1} - ES_{n1} & H_{n2} - ES_{n2} & ... & H_{nn} - ES_{nn} \end{vmatrix} = 0 \quad ...(2.94)$$

The above determinant is called "secular determinant". The values of E determined from secular equation helps in calculating the values of C_i from the set of differential equations along with the condition that the wave function should be normalised.

(*iv*) One thing must be kept in mind is that the value of E obtained with help of the equation (2.94) is always equal to or greater than the actual value of the energy of the state.

***Exercise* 2.41:** Let the approximate wave function for the particle in a box with infinite walls be $\psi = x(a - x)$. Using the boundary conditions that $\psi = 0$ when $x = 0$ and $x = a$, calculate the energy of the particle.

First thing to see is whether the trial wave function satisfies the boundary conditions. Substituting the values of $x = 0$ and $x = a$, leads the value of $\psi = 0$. Using equation (2.37) and substituting the value of ψ and Hamiltonian operator, we get

$$\bar{E} = \frac{\int_0^a [x(a-x)]\left(-\frac{h^2}{8\pi^2 m}\frac{d^2}{dx^2}\right)[x(a-x)dx}{\int_0^a [x(a-x)][x(a-x)dx} \quad ...(A)$$

Solving for the numerator

$$\int_0^a [x(a-x)]\left(-\frac{h^2}{8\pi^2 m}\frac{d^2}{dx^2}\right)[x(a-x)dx = \int_0^a [x(a-x)]\left(-\frac{h^2}{8\pi^2 m}\right)(-2)dx$$

$$= \frac{h^2}{4\pi^2 \text{m}}\int_0^a (xa-x^2)dx = \frac{h^2}{4\pi^2 \text{m}}\left\{\left[\frac{ax^2}{2}\right]_0^a - \left[\frac{x^3}{3}\right]_0^a\right\} = \frac{h^2}{4\pi^2 \text{m}}\left[\frac{a^3}{2} - \frac{a^3}{3}\right] = \frac{a^3 h^2}{24\pi^2 \text{m}}$$

Solving denominator, we get

$$\int_0^a [x(a-x)][x(a-x)dx = \int_0^a [x(a-x)]^2 dx = \int_0^a (a^2x^2 - 2ax^3 + x^4)dx$$

$$= \left[\frac{a^2x^3}{3}\right]_0^a - \left[\frac{2ax^4}{4}\right]_0^a + \left[\frac{x^5}{5}\right]_0^a = \frac{a^5}{3} - \frac{2a^5}{4} + \frac{a^5}{5} = \frac{1}{60}(20a^5 - 30a^5 + 12a^5) = \frac{a^5}{30}$$

Substituting the values of numerator and denominator in equation (A), we get

$$\bar{E} = \frac{\frac{a^3h^2}{24\pi^2 m}}{\frac{a^5}{30}} = \frac{a^3h^2}{24\pi^2 m} \times \frac{30}{a^5} = \frac{5h^2}{4\pi^2 ma^2} = 0.12655\frac{h^2}{ma^2}$$

Actual value of ground state energy $= \bar{E} = \frac{h^2}{8ma^2} = 0.125\frac{h^2}{ma^2}$

$$\%\ \text{error} = \frac{0.12655\frac{h^2}{ma^2} - 0.125\frac{h^2}{ma^2}}{0.125\frac{h^2}{ma^2}} \times 100 = \frac{0.00105}{0.125} \times 100 = 1.24$$

PROBLEMS

Planck's Relation

2.1 Calculate the energy of the photon corresponding to the frequency: (*a*) 1 nm (X-rays), (*b*) 150 nm, (*c*) 330 nm, (*d*) 600 nm, (*e*) 10^5 Å (IR), (*f*) 100 cm (microwave), (*g*) 10 m (radiowave). Convert these into the units of Joules, calories and electron volt.

2.2 A mol of photons are associated with energy equal to 120 kcal. Calculate the wavelength and frequency of the photon.

2.3 The energy required to dissociate oxygen molecule is 118 kcal mol^{-1}. Find out the frequency of the photon corresponding to this energy.

2.4 Calculate the average energy of an oscillator if it is oscillating with the wavelength: (*a*) 150 nm, (*b*) 200 nm, (*c*) 330 nm, (*d*) 100 cm.

2.5 A cubical cavity of volume 16×10^{-6} m^3 is heated up to 1000 K. Calculate the amount of energy emitted corresponding to the wavelengths in the range (*a*) 300–330 nm, (*b*) 600–630 nm and (*c*) 1500–1530 nm. Comment on the amount of energy emitted in each range.

2.6 Calculate the number of photons of $\lambda = 235.7$ nm emitted per second at 1000 K by (*i*) 200 W, (*ii*) 50 W mercury lamp.

2.7 If it is assumed that heater is behaving as a black body and maximum in its energy occurs at 480 nm, calculate its surface temperature.

Photoelectric Effect

2.8 Calculate the maximum velocity of the electrons emitted from the surface of the sodium metal if the photons of wavelength 200 nm strikes it. The work function for metal is 6.95×10^{-12} erg.

2.9 In Problem 2.8, (*a*) calculate the voltage required such that most energetic electrons should not reach the collector, (*b*) calculate the threshold wavelength for the sodium metal.

2.10 If a photon of wavelength 100 pm strikes an atom and one of the inner bound electron is ejected out with a velocity 2×10^7 m s^{-1}, calculate the energy with which it is bound to the nucleus.

2.11 The following data is observed when lithium metal is shone with following wavelength of light. Calculate, (*a*) threshold frequency and (*b*) Planck's constant:

λ(nm)	500	480	450	400
v(m s^{-1}) $\times 10^{-5}$	1.83	2.64	3.61	5.02

2.12 The following data is observed in the photoelectric effect experiment when ceasium was used as metal. Calculate, (*a*) work function and (*b*) Planck's constant:

λ(nm)	500	540	570	600
v(m s^{-1}) $\times 10^{-5}$	4.87	4.15	3.6	3.02

Bohr's Theory of Atomic Structure and Spectra

2.13 Balmer's original formula to explain the emission spectrum of hydrogen atom is

$$\lambda = c\left(\frac{n^2}{n^2 - 4}\right)$$

where *c* is constant and $n > 2$. Show that this formula is agreeable with $n_1 = 2$ and $n_2 > 2$ in equation 2.11. Calculate the value of constant *c* in terms of Rydberg constant.

2.14 If the mass of the nucleus is *M* and that of electron is *m*, derive the expression for the energy of *n*th level with the help of Bohr's theory.

2.15 If mass of the nucleus is 2 a.u., what will be the shift in frequency observed in comparison to hydrogen atom for line $n_2 = 3$ in Balmer series?

2.16 Considering Bohr's model of hydrogen atom, calculate the velocity as well as the kinetic energy of the electron when it is present in $n = 4, 5, 6$ stationary states.

2.17 Considering the Problem 2.16, calculate the radii of the orbits when $n = 4$, 5 and 6.

2.18 Calculate the frequency of absorption when the electron in hydrogen atom is excited from (*a*) $n = 4$ to $n = 5$, (*b*) $n = 5$ to $n = 6$ and (*c*) $n = 4$ to $n = 6$.

2.19 If the Bohr's theory is applied to tritium atom, calculate the values of frequencies required to excite the electron from (*a*) $n = 4$ to $n = 5$, (*b*) $n = 5$ to $n = 6$ and (*c*) $n = 4$ to $n = 6$.

2.20 Repeat the Problems 2.16 and 2.17 for tritium atom.

2.21 Repeat the Problems 2.16, 2.17 and 2.18 for Li^{2+}.

2.22 In the spectrum of hydrogen atom, the Brackett series starts with $n_1 = 4$. Calculate the wavelengths of first three lines of this series.

2.23 Suppose that a particle is made up of positron (mass of positron = m_e) and electron (called positronium) and if the Bohr's theory is applicable to it, repeat the Problems 2.16, 2.17 and 2.18.

2.24 A particle mesonium consists of a proton and a negatively charged particle meson (μ) whose mass is 207 times that of an electron. Repeat the Problems 2.16, 2.17 and 2.18.

2.25 If the ionisation energy of hydrogen atom is 314.4 kcal (13.595 eV), find the ionisation energy for He^+, Li^{2+} and Be^{3+}.

2.26 Using the results of Problems 2.16, 2.17, calculate the number of times the electron will revolve per second around the nucleus if the electron is present in $n = 4$ and $n = 5$. Compare these frequencies with the frequency required to excite the electron from $n = 4$ to $n = 5$.

2.27 (*a*) Calculate the temperature at which the average kinetic energy of the molecule will be the same as required for the transitions $n = 4$ to $n = 5$, $n = 5$ to $n = 6$ and $n = 4$ to $n = 6$.

(*b*) Calculate the velocity of the electron if the energy of each of the above transition is supplied to it.

de Broglie Relation

2.28 An electron is accelerated by applying potential difference of 1000 eV. What is the de Broglie's wavelength associated with it?

2.29 A proton is moving with a kinetic energy of 10^4 eV. Calculate the de Broglie wavelength associated with it.

2.30 A car weighing 2.5 tons and going with the speed of 30 mile per hr never undergoes wave motion. Explain.

2.31 Derive Bohr's condition of quantisation of angular momentum from de Broglie's relation of wavelength and momentum.

2.32 The wavelength associated with an electron moving in a potential difference of 2000 V is 0.05 nm. How fast (or rather how slow) must 0.01 kg hockey ball travel to have the same de Broglie wavelength as a 2000 V electron.

2.33 Calculate the voltage required to accelerate an electron to have a velocity of 1.87×10^9 m s^{-1}. What will be the de Broglie's wavelength?

2.34 How much potential difference be applied between the plates to accelerate the protons such that they have the wavelength of 200 nm.

2.35 Calculate the de Broglie's wavelength for an electron in Bohr's orbits corresponding to $n = 2$, $n = 20$, $n = 200$ for hydrogen atom.

2.36 Diffraction pattern is observed when the wavelength of the radiation that strikes the crystal is same as the internuclear distance between the atoms. Beam of electrons were accelerated with 200 V potential difference. If these electrons strikes the crystal, will the diffraction be observed or not?

Heisenberg Uncertainty Principle

2.37 A tennis ball of mass 0.2 kg is moving with a velocity of 3×10^3 m s^{-1}. If the ball can be located within an error equal to wavelength of light used, *i.e.* 600 nm, calculate the uncertainty in momentum of the ball compared with the momentum of the ball itself.

2.38 In an experiment, the position of an electron can be measured with an accuracy of ± 0.005 nm. What will be the accuracy in measuring the momentum of an electron? Suppose if the momentum of the electron is $h/4\pi \times 0.05$ nm, would you feel any problem in describing its magnitude?

2.39 Do you think that Heisenberg's uncertainty principle is violated in the following observation? The speed of a car weighing 2.5 ton can be measured with an accuracy of ± 0.1 mile.

2.40 If the natural linewidth of 0.02 nm in the spectral line of 500 nm is considered as the uncertainty in energy of any particular state, calculate the uncertainty in the lifetime of this state.

2.41 In a cyclotron, the electron is accelerated by having a potential difference across the two plates. If the magnitude of the potential difference is 2.0 ± 0.025 kV, calculate the uncertainty of the position of the electron along its line of flight.

Quantum Mechanics

2.42 (*a*) If ψ is an eigen function of operator $\hat{Q}$ corresponding to eigen value of q_i, show that

$$\bar{q} = q_i$$

(*b*) If ψ is not an eigen function to $\hat{Q}$ but can be expressed as $\psi = \Sigma\, C_n \phi_n$ where

$\delta_{nn'} = \int \phi_n^* \, \phi \, d\tau$ and ϕ_n is an eigen function of $\hat{Q}$ then

$$\bar{q} = \Sigma \mid C_n \mid^2 q_n$$

2.43 If q is the eigen value of $\hat{Q}$ then q^n is the eigen value of $\hat{Q}^n$.

2.44 In the integral $\int \psi^* \, \psi \, d\tau$, what are the dimensions of ψ?

2.45 (*a*) Is e^{-ax} an eigen function of $\hat{p}_x^2$ (p_x is linear momentum along x axis)? What will be the eigen value if a is a constant?

(*b*) Is the function e^{-ax} an acceptable wave function when x varies from $-\infty$ to $+\infty$? Give reason. What conditions should be imposed on the constant a that it is an acceptable wave function in the range $x = 0$ to $x = \infty$?

(*c*) Which one of the followings are acceptable wave functions in the range given?

(*i*) $\frac{1}{x}(0, \infty)$, (*ii*) $e^{-2x} \sinh x\,(0, \infty)$, (*iii*) $e^{-x} \cos x\,(0, \infty)$, (*iv*) $e^{x}\,(-\infty, +\infty)$,

(*v*) $e^{-x^2/2}\,(-\infty, +\infty)$, (*vi*) $xe^{-x}\,(0, \infty)$

2.46 In case of a particle in a box (with infinite high walls) problem, if the potential energy (assumed to be constant but generally taken zero) is not assumed to be zero but a constant value is added to total energy, how will the state function change?

2.47 What is the reason that the value of $n \neq 0$ for a particle in box with infinite walls?

2.48 Consider a particle of mass m moving in a circle of radius a in a constant potential energy zero, (*a*) show that the solutions of Schrödinger equation for the particle are

$$\psi_n = N \sin n\theta,\ \psi_{n'} = N' \cos n\theta$$

with
$$E_n = E_{n'} = \frac{n^2 h^2}{8\pi^2 m a^2},\ n = 0, 1, 2, \ldots$$

where θ is angle of the radius vector from any chosen origin. (*b*) Does the lowest state violate the uncertainty principle?

2.49 Show that any two wave functions corresponding to different energy levels are orthogonal in case of particle in a box problem.

2.50 Solve the circular oscillator problem. Also solve the circular oscillator in cartesian coordinates. Account for the degeneracy of eigen function in r and θ space relative to the degeneracy of the energy eigen functions in x and y space.

2.51 Normalise the following wave functions

(*a*) $\sin (n\pi x/a)$ within the range $-a$ to $+a$,

(*b*) e^{-r/a_o} in three-dimensional space and

(*c*) $x^2 e^{-r/3a_o}$ in three-dimensional space.

2.52 Find out which one of the following functions are eigen functions to the operator $\frac{d}{dx}$, (*i*) e^{-ikx}, (*ii*) $\sin kx$, (*iii*) $e^{-\alpha x^2}$, (*iv*) kx^2.

2.53 The vibrational motion of diatomic molecule can be considered as two particles of masses m_1 and m_2, separated by a distance r and executing vibration around their mean position. If the restoring force acting on the particle is proportional to the distance from mean position, write the Hamiltonian. (This approximation is known as simple harmonic motion.)

2.54 The wave functions obtained by solving the Schrödinger equation for the simple harmonic motion are

$$\psi_v(x) = \left(\frac{1}{2^v v!}\right)^{1/2} \left(\frac{a}{\pi}\right)^{1/4} \exp(-\xi^2/2)\, H_v(\xi)$$

where $\xi = \sqrt{a}\, x$ and $v = 0, 1, 2, \ldots$ are vibrational quantum numbers. $H_v(\xi)$ is the Hermite polynomials and are defined as

$$H_v(\xi) = (-1)^v e^{\xi^2} \left(\frac{d^v e^{-\xi^2}}{d\xi^v}\right)$$

Calculate the first four wave functions, sketch them and sketch $\psi_v^* \psi_v$.

2.55 The vibrational energy levels of simple harmonic motion are given as

$$E_v = h\nu(v + \tfrac{1}{2}),\ v = 0, 1, 2, \ldots$$

Sketch the first four energy levels and find the difference between them. v is the classical frequency and given by

$$v = \frac{1}{2}\sqrt{k/\mu}$$

where k is the force constant and μ is the reduced mass.

2.56 Show that the wave function $\psi_2(x)$ obtained in the Problem 2.54 is normalised.

2.57 Plot the wave functions of the first four states and their probability distribution for the particle in box with infinitely high walls, problem.

2.58 (*a*) Find the expectation value of distance for the ground state wave function of a particle in a box with infinite high walls. Also find $\bar{x}$ for first and second excited state wave function and compare them.

(*b*) Repeat the above problem for x^2.

(*c*) Show that as the value of n increases, $\bar{x}$ values approaches the classical value.

2.59 Find the average value of $\overline{p^2}$ in case of particle in one dimensional box. Compare the results with that obtained in Exercise 2.23.

2.60 Calculate $\bar{p}$ for the function defined as

$$\psi = ke^{ikx} + k'e^{-ikx}$$

where k is constant.

2.61 A wave function ψ can be written as

$$\psi = \Sigma a_n \phi_n(x)$$

where a_n is a constant and $\phi_n(x)$ is a normalised wave function for harmonic oscillator. Show that $\bar{\mu}$ (μ is a dipole moment and $\mu = ex$, where e is the charge on the electron and x is the distance between two poles) is not zero. Given $a_0 = 1, a_1 = \frac{1}{\sqrt{2}}$ and $a_n = 0$ for $n > 1$ and recursion relation

$$xH_n = nH_{n-1} + \tfrac{1}{2} H_{n+1}$$

where H_n is the Hermite polynomial and $\phi_n(x)$ is given in Problem 2.54.

2.62 For a particle in one dimensional box, if the length of the box is 0.2 nm, calculate the probability of finding the particle within a distance of 0.002 nm on both sides of the point $x = 0.033$ nm for the state $n = 1$ and $n = 2$.

2.63 If a particle is moving in a three-dimensional cubical box of length 0.3 nm, calculate the probability of finding the electron moving in a volume space defined by $\Delta x = \Delta y = \Delta z = 0.001$ nm and whose centre is at $x = 0.01$ nm, $y = 0.02$ nm and $z = 0.03$ nm for the state $n_x = n_y = n_z = 1$, $n_x = n_y = 2$, $n_z = 1$ and $n_x = 1$, $n_y = 2$, $n_z = 2$.

2.64 If the uncertainty in distance and momentum is defined as

$$(\Delta x)^2 = [\langle(x - \langle x\rangle)^2\rangle] = \langle x^2\rangle - \langle x\rangle^2$$

$$(\Delta p)^2 = [\langle(p - \langle p\rangle)^2\rangle] = \langle p^2\rangle - \langle p\rangle^2$$

Find the product of Δx and Δp for a particle in one dimensional box for state $n = 1$.

2.65 Calculate $\langle p\rangle$ and $\langle p^2\rangle$, $\langle x\rangle$ and $\langle x^2\rangle$ for the ground state of a harmonic oscillator. If Δx and Δp are defined as in Problem 2.64, calculate $\Delta x \times \Delta p$.

2.66 Draw the energy level diagram for the first four states of a particle moving in one dimensional box. Calculate the frequency of radiation required to promote an electron from (*i*) $E_1 \rightarrow E_2$

(*ii*) $E_2 \rightarrow E_3$, and (*iii*) $E_2 \rightarrow E_4$ if the length of the box is 1 nm and mass of the electron is 9.11×10^{-31} kg.

2.67 In case of a particle moving in a cubical box calculate the degeneracy corresponding to energy (*E*) equal to

$$(1)\ 18\frac{h^2}{8ma^2},\ (2)\ 14\frac{h^2}{8ma^2},\ (3)\ 11\frac{h^2}{8ma^2}$$

2.68 If the energy absorbed between two excited states of a particle in three-dimensional cubical box is numerically equal to the absolute energy of the ground state, calculate the degeneracy of the lower excited state.

2.69 π-electrons in the linear chain conjugated polyene hydrocarbons can be regarded as the particle moving in one dimensional box under the influence of fixed potential energy. Consider the butadiene system which contain four π-electrons.

(*a*) Arrange these four electrons according to Pauli's principle.

(*b*) If the wavelength required for promoting the electron from the highest occupied orbital to lowest unfilled orbital is 210 nm, calculate the length of the chain.

(*c*) Compare the length of chain calculated in part (*b*) with the actual, *i.e.* 0.42 nm.

2.70 β-Carotene is another linearly conjugated polyene system containing 11 double bonds. If the actual length of the conjugated system is 3.0 nm, calculate the wavelength for the longest wavelength band. Compare this λ with observed one at 460 nm.

2.71 Show that $\int \psi^* \hat{Q}\phi d\tau = \int \phi(\hat{Q}\psi)^* d\tau$ if $\hat{Q}$ is a Hermitian operator.

2.72 Which of the following functions are eigen-functions of L_z?

$$(i)\ \phi_m = \left(\frac{1}{2m}\right)^{1/2} e^{im\phi}\quad m = 0, \pm 1, \pm 2, \ldots$$

$$(ii)\ \phi_{\cos m\phi} = \left(\frac{1}{\pi}\right)^{1/2} \cos m\phi,\ m = 1, 2, 3, \ldots$$

2.73 Show that

$$(a)\ \left[\hat{L}_y, \hat{L}_z\right] = i\frac{h}{2\pi}\hat{L}_x$$

$$(b)\ \left[\hat{L}_z, \hat{L}_x\right] = i\frac{h}{2\pi}\hat{L}_y$$

2.74 Starting with the cartesian coordinates show that

$$(i)\ \hat{L}_z = \frac{h}{2\pi i}\frac{\partial}{\partial\phi}$$

$$(ii)\ \hat{L}_x = -\frac{h}{2\pi i}\left[\cot\theta\cos\phi\frac{\partial}{\partial\phi} + \sin\phi\frac{\partial}{\partial\theta}\right]$$

(iii) $\hat{L}_y = -\frac{h}{2\pi i}\left[\cot\theta \sin\phi \frac{\partial}{\partial\phi} \cos\phi \frac{\partial}{\partial\theta}\right]$

2.75 Prove that the commutator of x and $\hat{p}$ is $i\, h/2\pi$.

2.76 If two functions are eigen functions of the same operator with different eigenvalues, show that they are orthogonal.

2.77 Prove that the eigenvalues of Hermitian operators are real.

2.78 Calculate $\Delta E = E_2 - E_1$ for (*a*) an electron moving in a box of length 5 Å and (*b*) an oxygen molecule moving in a box of 10 cm. Compare the ΔE in both the cases with thermal energy at 27°C.

2.79 Calculate the value of n when CO_2 molecule possessing an energy equal to k_bT and moving in a box of 0.1 m at 27°C. Show that the correspondence principle is valid here.

Addition Problems

2.80 The colour of chlorophyll is a consequence of the absorption of radiation at ~ 450 nm and at ~ 680 nm and thus mostly green light is transmitted. Calculate the energy of these photons per mol at these wavelengths. [**Ans.** 275 kJ mol^{-1}, 176 kJ mol^{-1}]

2.81 The minimum energy required for ionising the electron from zinc metal surface is 8.54 × 10^{14} Hz. Calculate the minimum amount of energy required to remove an electron from the metal in kJ and Kcal.

2.82 Calculate the value of λ_{max} at: (*i*) 1000 K, (*ii*) 2000 K and (*iii*) 8000 K.

2.83 500 g of water sample of water absorbs infrared radiation at 1.06 × 10^4 nm from CO_2 laser. Assuming that all the radiation absorbed by water is converted into heating the water sample, calculate the number of photons of this wavelength required to raise the temperature of water by 10º C.

2.84 It takes 334 J to melt one gram of ice at 0º C. How many photons of 650 nm must be absorbed to melt 1 kg of ice? On average, how many water molecules does one photon convert ice to water?

2.85 It has been observed that the retina of human eye can detect radiation when radiant energy incident on it is at least 4×10^{-17} J. Calculate the number of photons that correspond to the light of wavelength 600 nm.

2.86 The ionisation energy of certain element is 42.774 eV. When the atoms of the same elements are present in its first excited state, the ionisation energy reduces to 13.08 eV. Calculate the wavelength of light emitted when the atoms relaxes from the first excited state to the ground state.

(**Ans.** 418 nm)

2.87 Planck assumed that energy that emitted is quantised ($\varepsilon = nh\nu$) and $\Delta\varepsilon = h\nu$. If $\nu \to 0$, the energy gap between successive levels decreases and becomes continuous under this that conditions. Planck's distribution law should resemble the classical picture of energy. Show equation (2.5) reduces to Rayleigh-Jeans distribution when $\nu \to 0$.

2.88 The hottest star observed is known as Sirus. Calculate its temperature if the maximum blackbody radiation (λ_{max}) observed is found to be 260 nm.

2.89 In thermodynamic explosion, the temperature of the fireball can reach of the tune of 10^7 K. Calculate the value of λ_{max} corresponding to this temperature and assign the region of electromagnetic radiation.

2.90 Stefan-Boltzmann constant is given by the expression, $\sigma = \frac{2\pi^5 k_b^4}{15h^3c^2}$ and Wein displacement law, $\lambda_{max}\ T = \frac{hc}{4.965} = 2.90 \times 10^{-3}$ m K, where all the constants have the usual meanings. Taking $\sigma = 5.6697 \times 10^{-8}$ J m^{-2} K^{-4} s^{-1} and $c = 2.998 \times 10^{8}$ m s^{-1} (accurately known value in 1900). Calculate the value of h and k_b. Further using $R = N_{av}\ k_b$, where R is the gas constant = 8.314 J K^{-1}, calculate the value of Avogadro's number N_{av}.

2.91 Using the results of Exercise 2.8 and using $E = mc^2$, calculate the relativistic mass of the photons lost by radiation from the Sun per year. (*i.e.*, multiply the energy s^{-1} by 1 yr = 365 × 24 × 3600s)

(**Ans.** = 1.58×10^{18} kg)

2.92 200 W sodium vapour lamp emits light of wavelength 590 nm. Calculate the number of photons emitted per minute.

2.93 The fireball in a thermonuclear explosion can reach the temperature approximately up to 10^{7} K. Calculate the λ_{max} corresponding to this temperature and also region of electromagnetic spectrum.

2.94 A laser pulse has 4.0 mJ of power. If the wavelength of the lase r beam is (*i*) 1.06 μm and (*ii*) 337 nm, calculate the number of photons in this beam.

2.95 Just for information the power of the laser beam is measured in units of watts (W), *i.e.*, one joule per second (1 J s^{-1}). Calculate the number of photons emitted per second of radiation 537 nm.

2.96 Following data from the photoelectric effect in Na was observed.

$10^{12}\ E_{kin}$/erg	3.41	2.56	1.95	0.75
λ (nm)	312.5	365.0	404.7	546.1

where E_{kin} is the maximum kinetic energy of emitted electrons and λ is the wavelength of incident light. From the plot of E_{kin} versus ν, calculate the value of h and work function of Na.

2.97 The work function of Ni and that of K are 5.0 eV and 2.2 eV respectively. (*a*) Calculate the threshold frequency and wavelength for these metals. (*b*) Will the photoelectric effect be observed if we use 450 nm radiation? (*c*) If the photoelectric effect is observed in both cases, calculate the maximum kinetic energy of the ejected electrons.

2.98 The work function of some metals is listed below. Find out the number of metals which will show photoelectric effect when radiation of wavelength 300 nm falls on them. (**Ans.** 4)

Metal	Li	Na	K	Mg	Cu	Ag	Cr	Fe	Pt	W
Work function (eV)	2.4	2.3	2.2	3.7	4.8	4.3	4.4	4.7	6.3	4.75

2.99 Suppose that in the photoelectric effect experiment, two laser beams are used, *i.e.*, one of wave-length at 450 nm and the other at 560 nm. Assuming that (*i*) same number of photons is delivered to the metal surface by each laser and frequencies of laser light used are greater than the threshold frequencies of the metal surface, suggest (*i*) which light will generate more number of electrons and (*ii*) which radiation would generate greater kinetic energy?

2.100 In the photoelectric experiment, a student uses a certain radiation source whose frequency is greater than the threshold frequency of the metal surface to eject the electrons. After shining the same area of the metal for a long period of time with the same frequency of radiation, the student found that the maximum kinetic energy of the ejected electron begins to decrease, even though the frequency of the radiation is held constant. Explain this behaviour.

2.101 Following data are obtained in the photoelectric effect experiment.

V (volt)	1.475	1.268	1.207	0.886	0.667	0.381
λ (nm)	405	435.8	480	520	577.7	650

where V is the retarding voltage required such that the current due to electron read exactly zero. Rearrange equation (2.10*a*) to depict this data and calculate the values of h and work function.

2.102 Calculate (*i*) the first transition lines and (*ii*) series limit for Lyman, Balmer, Paschen, Bracket and Pfund series. Explain the meaning of each term obtained in (*i*) and (*ii*).

2.103 (*i*) Calculate magnitude of the centrifugal force of an electron in the Bohr orbits with quantum numbers, $n = 2$, 3 and 4 for a hydrogen atom. (*ii*) What is the centrifugal force for the Li^{2+} species?

2.104 Calculate the value of angular momentum of the electron in the orbits mentioned in Problem 2.103. Carry out the calculations for the Li^{2+} ion. Compare the values for the two species and explain.

2.105 Calculate the ionisation energy of the hydrogen atom predicted by Bohr theory. The experimentally measured result is 2.17866×10^{-18} J. Calculate the percent error in the Bohr result and explain.

2.106 From the data of Table 2.1, it is clear that Lyman and Balmer series do not overlap. Verify this statement by calculating the longest wavelength associated with the Lyman series and the shortest wavelength associated with Balmer series.

2.107 An electron in hydrogen atom from its excited state can come to the ground state by two paths. First path involves the direct loss of energy by emission for the excited state to the ground state and wavelength of emission is λ_1. Second path involves two steps, *i.e.*, first involves the loss of energy by emission to the intermediate state and wavelength of emission is λ_2 and then losing energy further by emission of wavelength λ_3 to ground state. Find out the relation between λ_1 to λ_2 and λ_3.

2.108 It has been observed by the scientist that interstellar hydrogen atoms are present in the highly excited states. Calculate the wavelength of emission if initial and final states of hydrogen atom are 245 and 244. Indicate the region of electromagnetic radiation.

2.109 In mid-nineteenth century, physicists studying the emission spectrum (continuum) from Sun found that there are some set of dark lines which did not match with any of the emission lines (bright) on the earth. They found out that these lines originated from yet another unknown element. Later this element was identified as helium atom. (*i*) What is the origin of dark lines? How these lines correlated with the emission lines of helium? (*ii*) Why was helium so difficult to detect in earth's atmosphere? (*iii*) Where is the most likely place to detect helium on Earth?

2.110 Hydrogen atom absorbs a radiation of 97.2 nm from its ground state and then emits radiation of wavelength 1875.6 nm. Calculate the principle quantum number of the final state.

2.111 From Table 2.1, calculate the series limit for each series in the units of Joules.

2.112 Neutron diffraction is used to study the structure of the molecules. Collimating beam of neutron at a particular temperature T is generated from a high energy neutron source and is accomplished at several accelerator facilities. The speed of the neutron is given by the relation, $v_n = (\frac{3k_bT}{m_n})^{\frac{1}{2}}$, where m_n is the mass of the neutron (1.67×10^{-27} kg). If the de Broglie's wavelength is 50 pm, calculate the temperature of source.

2.113 Calculate the de Broglie's wavelength and energy associated with α-particle when it is accelerated through a potential difference of 5.0 V. Mass of the α-particle is 6.64×10^{-27} kg.

2.114 Show that a small change in the speed of a particle, Δv, causes a change in the de Broglie's wavelength $\Delta\lambda$ of $|\Delta\lambda| = \frac{|\Delta v|\lambda_0}{v_0}$, where λ_0 and v_0 are the initial wavelength and initial velocity.

2.115 If the uncertainty in determining the position of the electron circling an atom in an orbit is 0.025 nm, calculate the uncertainty in its velocity.

2.116 A person weighing 80 kg runs at speed of 2 m s^{-1}. (*i*) Calculate the momentum and the wavelength associated with the jogger. (*ii*) If the uncertainty in measuring his speed is 5 %, calculate the uncertainty in measuring his position. (*iii*) If we take the value of Planck constant as 2 J s, predict the changes that would take place in this problem.

2.117 The diffraction phenomenon is observed whenever the wavelength of the radiation is nearly equal in magnitude to the size of the slit width. Now if the same phenomenon is to be observed in macro-system, *i.e.*, to be diffracted, how fast a man weighing 80 kg move through the door of 1 m wide?

2.118 Alveoli are tiny sacs of air in the lungs. Their average diameter is 5.0×10^{-5} m. Calculate the uncertainty in the velocity of an oxygen molecule trapped within sacs. (Hint; the maximum uncertainty in the position of the molecule is given by the diameter of the sac.)

2.119 Using Heisenberg uncertainty principle, show that electron cannot be confined to within nucleus. Repeat the calculations for a proton. Take the diameter of the nucleus as 2×10^{-15} m.

2.120 Calculate the uncertainty in the excited energy if the lifetime in this state is 100 ps.

2.121 When the excited nucleus decays it emits a γ-ray. If the lifetime of the excited state is of the order of 10^{-12} s, find the uncertainty in energy of the excited state.

2.122 Spectroscopy has shown the presence of atomic hydrogen atom in the surface of sun. Is the hydrogen present in the surface of Sun is atomic hydrogen or H^+ ion? The temperature of the Sun has been found to be 6000 K. If the ionisation energy of hydrogen is 2.179×10^{-18} J. Calculate the temperature of blackbody so that the maximum wavelength emitted by it can ionise the atomic hydrogen from its ground state.

2.123 A complex function is given as: $f(x) = 3e^{-4i} + 4i\,(4 - 5i)^2$. Show that the absolute value is real.

2.124 (*i*) Consider the function: $f(x) = A(e^{cx} + e^{-cx})$, where A and c are constants. Show that this function is an eigen function of $\frac{d^2}{dx^2}$ but not of $\frac{d}{dx}$. (*ii*) Calculate the eigen value of the operator $\frac{d}{dx}$.

2.125 Three possible solutions of the particle in one dimensional box Schrödinger wave equation are proposed. These are:

(*i*) $\Psi = N \sin(ax)$, where N and a are constants.

(*ii*) $\Psi = N x^{1/2} e^{-ax}$ for $0 \le x \le \infty$, where N and a are again constants.

(*iii*) $\Psi = N \tan(ax)$, for $0 \le x \le \infty$, where N and a are constants.

Find out which one is correct?

2.126 We have two functions $f(x)$ and $g(x)$ which are eigen functions to the operator $\hat{O}$ and give the same eigen value to both functions. Prove that the linear combination of these two functions also give the same eigen value.

2.127 Let the functions $f(x)$ and $g(x)$ be eigen function to the operator $\hat{O}$ giving the eigen values as o and p. Show that these two functions are orthogonal to each other.

2.128 Why product $\psi^*\psi$ have to be real everywhere, non-negative, finite and of definite value?

2.129 Indicate which of the following wave functions are normalised? If these are not normalised them. (*i*) $e^{i\theta}$ $(0, 2\pi)$, (*ii*) xe^{-x} $(0,\infty)$, (*iii*) $e^{-\frac{x^2}{2}}$ $(-\infty,+\infty)$ e^x $(0, \infty)$.

2.130 Following functions are two dimensional wave functions. Indicate which one are normalised. Otherwise normalise them if they are not.

(*i*) $e^{-\frac{x^2+y^2}{2}}$ $\begin{matrix}(0 \le x \le +\infty)\\(0 \le y \le +\infty)\end{matrix}$, (*ii*) $e^{-\frac{x+y}{2}}$ $\begin{matrix}(0 \le x \le +\infty)\\(0 \le y \le +\infty)\end{matrix}$,

(*iii*) $\left(\dfrac{4}{ab}\right)^{\frac{1}{2}} \sin\dfrac{n_x\pi}{a}x \sin\dfrac{n_y\pi}{b}y,\ 0 \le x \le a;\ 0 \le y \le b$

2.131 If the operators $\hat{A} = x^2$, $\hat{B} = \dfrac{d}{dx}$, find (*i*) $(\hat{A}+\hat{B})(x^3 + \sin x)$, (*b*) Find $\hat{A}\hat{B}(x^3 + \sin x)$ and $\hat{B}\hat{A}(x^3 + \sin x)$. Find out whether $\hat{A}\hat{B} = \hat{B}\hat{A}$. If the products of the two operators are not equal, find out the commutator.

2.132 Let the operators $\hat{A} = x^3$, $\hat{B} = \dfrac{d^2}{dx^2}$, find (*i*) $(\hat{A}+\hat{B})(x^5 + 1/x)$, (*b*) Find $\hat{A}\hat{B}(x^5 + 1/x)$ and $\hat{B}\hat{A}(x^3 + \sin x)$. Find out whether $\hat{A}\hat{B} = \hat{B}\hat{A}$. If the products of the two operators are not equal, find out the commutator.

2.133 Consider the operators. $\hat{A} = x^2$, $\hat{B} = \dfrac{d^2}{dx^2}$. Find $\hat{A}\hat{B}f(x)$ and $\hat{B}\hat{A}f(x)$. (c) Find out whether these operators commute with each other and if no find out the commutator.

2.134 It is well known that operator algebra differs from ordinary algebra. From the relation $\hat{H}\psi = E\psi$, we cannot conclude that $\hat{H} = E$, because $\hat{H}$ is an operator and E is a number. Prove this statement by taking example as $\dfrac{d}{dx}(e^{4x})$.

2.135 Consider the operators, $\hat{A} = \dfrac{d^2}{dx^2}$ and $\hat{B} = x^2$. (*a*) Find $\hat{A}\hat{B}f(x) - \hat{B}\hat{A}f(x)$ and (*b*) $(\hat{A}+\hat{B})[\sin 2x + e^{x^2}]$

2.136 Find out whether each of the following entities is an operator or function. (*a*) $\hat{A}\hat{B}f(x)$. (*b*) $\hat{A}\hat{B} + \hat{B}\hat{A}$, (*iii*), (*iv*) $f(x)\hat{A}$, (*v*) $g(x)\hat{A}f(x)$.

[**Ans.**(*a*) function, (*b*) operator]

2.137 Indicate which of the following operators are linear or non-linear? (*i*) $\frac{\partial^2}{\partial x^2}$, (*ii*) $\frac{\partial}{\partial y}$, (*iii*) $3\hat{z}^2$ (*iv*) $(\,)^2$, (*v*) $(\,)^*$.

2.138 Calculate the quantum mechanical operator for (*i*) p_x^3, (*ii*) p_x^4.

2.139 Tick mark which of the quantities is true or false? (*i*) $(\hat{A}+\hat{B})f(x)$ is always equal to $\hat{A}f(x)+\hat{B}f(x)$, (*ii*) $\hat{A}[f(x)+g(x)]$ is always equal to $\hat{A}f(x)+\hat{A}g(x)$, (*iii*) $\hat{A}\hat{B}f(x)$ is always equal to $\hat{B}\hat{A}f(x)$, (*iv*) $\frac{\hat{A}F(x)}{f(x)}$ is always equal to $\hat{A}$, provided $f(x) \neq 0$, (*v*) $3x$ is eigen value of $\hat{x}$, (*vi*) $3x$ is an eigen function of $\hat{x}$, (*vii*) $e^{\frac{2\pi ikx}{h}}$ where k is a constant, is an eigen function of $\hat{p}_x$.

[**Ans.** T, F, F, F, F, F, T]

2.140 Suppose a particle of mass 9.1×10^{-31} kg moves in a one dimensional box of length 5 Å. Calculate the frequencies emitted when electron jumps from $n = 5$ to $n = 4$ and $n = 4$ to $n = 3$. Is there exist relation between these frequencies and corresponding quantum numbers involved?

2.141 Suppose the particle is moving in a one dimensional box of length a and it is in $n = 3$. Calculate the probability of finding this particle (*a*) within $\pm$ 0.05a of $x = 0.25a$ (b) within $x = 0.1a$ and $x = 0.4a$.

2.142 Calculate the probability of finding the particle in three dimensional cubical box in a state defined by $n_x = 2$, $n_y = 1$ and $n_z = 2$, when $x = 0.2a$, $y = 0.5a$ and $z = 0.3a$ and $dx = 0.02a$, $dy = 0.03a$ and $dz = 0.04a$.

2.143 An electron is excited from the ground state ($n = 0$) to the $n = 5$ state. Based on this few statements are made below. Suggest whether these are right or wrong.
(*i*) This is first excited state, (*ii*) Energy required to ionise the electron from $n = 5$ state is less than that required to ionise the electron from the $n = 1$ state. (*iii*) The electron on the average is closer to the nucleus in $n = 5$ state as compared to $n = 1$ state. (*iv*) Wavelength of radiation emitted from $n = 5$ to $n = 1$ is shorter than that emitted when electrons falls from $n = 5$ to $n = 2$, (*v*) The wavelength of radiation emitted from the $n = 5$ state to $n = 1$ state is the same as the light absorbed in going from $n = 1$ to $n = 5$.

2.144 The variance in the measurements of observable O is defined as $\sigma_O^2 = \langle O^2 \rangle - (\langle O \rangle)^2$. Consider a two dimensional particle with the wave function and $\hat{E} = -\frac{h^2}{8\pi^2 m}[\frac{\partial^2}{\partial x^2} + \frac{\partial^2}{\partial y^2}]$,

$\psi_{n_x,n_y}(x, y) = \left(\frac{4}{ab}\right)^{\frac{1}{2}} \sin\frac{n_x\pi}{a}x \sin\frac{n_y\pi}{b}y$, where $n_x = 1, 2, 3, \ldots\ldots$; $n_y = 1, 2, 3, \ldots\ldots$ Show that the variance in energy is given as $\sigma_E^2 = \langle E^2 \rangle - (\langle E \rangle)^2 = 0$.

2.145 If the observable is momentum defined as $\hat{P} = -\frac{ih}{2\pi}(i\frac{\partial}{\partial x} + j\frac{\partial}{\partial y})$. Using the wave function as given in Problem 2.144, calculate the variance in the measurement of momentum.

2.146 In 1996, physicists created an anti-atom of hydrogen. In such an atom, which is an anti-matter equivalent of an ordinary atom, the electrical charges of all the components are reversed. Thus, the nucleus of anti-atom is made anti-proton, which has the same mass as that of proton but bears a negative charge, and the electron is replaced by an anti-electron, *i.e.*, positron with the same mass but carries positive charge. Would you expect the energy levels, emission spectra, and atomic orbitals of anti-atom to be different from those of a hydrogen atom? What would happen if anti-atom of hydrogen would collide with normal hydrogen atom?

2.147 Consider two non-interacting particles of masses m_1 and m_2 in a one dimensional box of length a. Derive and find out the formulas for the stationary-state wave functions and energies.

2.148 Alkali metals like sodium or potassium when dissolved in liquid ammonia gives blue colour solution which is due to absorption of solvated electrons. Considering a crude model, we can treat solvated electron as a particle in three dimensional box of length a. Assuming that the absorbed radiation ($\lambda = 600$ nm) excites the electron from its ground state to the first excited state, calculate a. (Hint, consider three quantum numbers n_x, n_y and n_z).

2.149 Consider a particle in one dimensional box of length 0.05 nm. Assume that the particle is present in the excited state which is having 16 nodes in it s plot between ψ and distance x. (*i*) Calculate the wavelength, (*ii*) the energy of the state and the wavelength of the transition to one lower state.

2.150 Consider a particle in one dimensional box of length a and also consider the first two states, *i.e.* n_1 and n_2. Suppose the particle is present in both states and weight-age of state n_2 is three times more than state n_1. (*i*) Write down the normalised wave function, (*ii*) Normalise the wave function, (*iii*) Further if we have 10^6 particles and distributed in these two wave functions, calculate the average energy of the system. [**Ans.** $\frac{1}{\sqrt{3}}(\psi_1 + \sqrt{2}\,\psi_2)$, $E = \frac{3h^2}{8ma^2}$]

2.151 Consider a particle in one dimensional infinite potential well of width 0.2 nm. (*i*) Calculate the maximum possible uncertainty in the position of electron. (*ii*) Calculate the minimum possible value of the linear momentum. (*iii*) Calculate the minimum kinetic energy value the electron can have, both in cm^{-1} and J.

2.152 (*a*) Few, trial functions are given for the particle in a one dimensional box. Indicate which one is valid? (*i*) $x^2(a-x)^2$, (*ii*) x, (*iii*) x^2, (*iv*) x^3, (*v*) $\cos\frac{\pi n}{a}x$

(*b*) Using the right trial function calculate the per cent error in the ground state energy.

2.153 Consider the trail wave function e^{-cx^2} for the one simple harmonic oscillator. (*a*) Calculate the value of $\overline{E}$, (*ii*) Find the value of c for the minimum energy and (*iii*) calculate the value of minimise of the energy.

2.154 Consider the trail wave function given for the particle in one dimensional box in Exercise 2.36 , *i.e.*, $\psi = x(a-x)$. (*i*) Normalise ψ, (*ii*) Calculate the average value of x^2 in the ground state taking this normalised wave function and compare it with the actual wave function.

2.155 Continuing with the Problem 2.154, consider the trial wave function as: $\psi = x^2 (a-x)^2$. (*i*) Normalise ψ, (*ii*) Calculate the average value of x^2 in the ground state taking this normalised wave function and compare it with the actual wave function, (*iii*) calculate the variance of the measurements of energy.

2.156 Extending the trial wave function for the particle in two dimensional box as: $\psi(x, y) = N x(a-x)(b-x)$. (*i*) Normalise $\psi(x, y)$ and (*ii*) average energy and calculate the error if the correct wave function is used.

2.157 Rather than taking the trail wave function for the particle in one dimensional box as $x\ (a-x)$ (taken in Exercise 2.41) take the trail wave function as $N x^k (a-x)^k$. Find the value of k that minimise the value of energy. Repeat Exercise 2.35 and Problem 2.138 for rest of problem.

2.158 Consider Exercise 2.31 and consider a M^{2+}-porphyrin complex that contains 26 π electrons, where M^{2+} is Mg^{2+} in chlorophyll or Fe^{2+} as in heme group. (*a*) Sketch the energy level diagram for this system. Note that number of energy levels is degenerate in energy as it is a two dimensional square box, (*b*) Take the box length as 0.1 nm. Calculate the wavelengths for the first two lowest energy electronic transitions in nm.

2.159 Consider a particle in a box of length a. Choose a trail wave function which is zero at $x = 0$ and $x = a$. Further, we know that ground state wave function do not possess nodes. Let these wave functions be: $\psi_1 = x\,(a^2 - x^2)$ and $\psi_2 = x^5\,(a^2 - x^4)$. Calculate the wave function that is the linear combination of these two wave functions and also the ground state energy by using variation method. Plot the wave functions obtained as a function of x and compare them with the true wave function, *i.e.*, $\psi = \sqrt{\dfrac{2}{a}}\,\sin\dfrac{n\pi}{a}x$. [**Ans.** $E_0 = 0.12656\dfrac{h^2}{ma^2}$

$$\psi = (3.06675a)^{-\frac{7}{2}}\,[(a^2x - x^3) - \frac{7.2981122}{a^4}(a^2x^5 - x^7)]$$

2.160 Consider the variation function, $\psi = Ne^{-ar^2}$, where N normalisation constant and a is the variation parameter. Using variation theorem, calculate the energy of hydrogen atom and determine the value of a that minimises the energy. Compare this minimum value of energy with the true hydrogen atom ground state energy.

2.161 Consider the variation function, $\psi = Ne^{-ax^2}$, where N normalisation is constant and a is the variation parameter to obtain the best possible value of energy of ground state of a harmonic oscillator. Using variation theorem, calculate the variation energy of parameter a for harmonic oscillator and determine the value of a that minimise the energy.

2.162 Consider a particle in one dimensional box of an infinite well of width a. Use the linear variational theory with the trail wave function $\Psi = C_1\,x(a-x) + C_2\,x^3\,(a-x)$, where C_1 and C_2 are variational parameters. Calculate the values of these parameters for the minimum value of energy.

CHAPTER 3

ATOMIC STRUCTURE

3.1. SYSTEM DESCRIPTION

In quantum mechanics, the problem of hydrogen atom or hydrogenlike atoms (*i.e.* atom with one electron) can be looked up as follows. An electron of negative charge e is moving around a nucleus of positive charge Ze (Z is atomic number). The hamiltonian can be written as a sum of kinetic energy and potential energy (see Problem 3.1*a*). The problem becomes simpler if the spherical polar coordinates (r, θ, ϕ) and centre of mass coordinates are used. The relation between the cartesian and polar coordinates

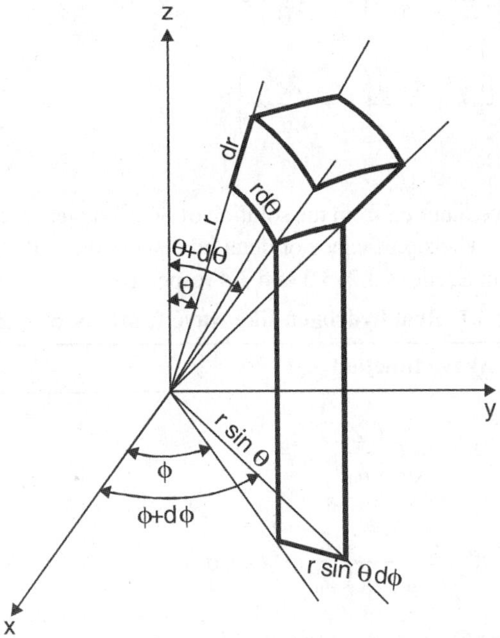

Fig. 3.1

is shown in Fig. 3.1. After separating the terms which describe the translational motion of atom as a whole (*i.e.* nucleus and electron), the Schrödinger equation for the relative motion of electron with respect to nucleus in terms of spherical polar coordinates becomes

$$\frac{1}{r^2 \partial r}\left[r^2 \frac{\partial \psi}{\partial r}(r,\theta,\phi)\right]+\frac{1}{r^2 \sin\theta}\frac{\partial}{\partial \theta}\left[\sin\theta \frac{\partial \psi}{\partial \theta}(r,\theta,\phi)\right]$$

$$+\frac{1}{r^2 \sin^2\theta}\frac{\partial \psi^2}{\partial \phi^2}(r,\theta,\phi)+\frac{8\pi^2\mu}{h^2}\left(E+\frac{Ze^2}{4\pi\varepsilon_0 r}\right)\psi(r,\theta,\phi)=0 \qquad ...(3.1)$$

where μ is the reduced mass of the electron and the nucleus $\mu = \frac{m_e m_n}{m_n + m_e}$ and $\left(\frac{-Ze^2}{4\pi\varepsilon_0 r}\right)$ is the potential energy of the system, due to interaction of the nucleus and electron in a medium of dielectric constant ε_0. If we assume that potential energy is only a function of r alone, the wave function $\psi(r, \theta, \phi)$ can be further separated as

$$\psi(r,\theta,\phi) = R(r)\Theta(\theta)\Phi(\phi) \qquad ...(3.2)$$

where $R(r)$ is a function of r only, $\Theta(\theta)$ the function of θ and Φ, the function of ϕ. Substituting this into equation (3.1), equation (3.1) can be separated into three equations as

$$\frac{d^2\Phi}{d\phi^2} = -m^2\Phi \qquad ...(3.3)$$

$$\frac{1}{\sin\theta\, d\theta}\left[\left(\sin\theta \frac{d\Theta}{d\theta}\right)\right]-\frac{m^2\Theta}{\sin^2\theta}+\beta\Theta = 0 \qquad ...(3.4)$$

$$\frac{1}{r^2}\frac{d}{dr}\left[\left(r^2\frac{dR}{dr}\right)\right]-\frac{\beta}{r^2}R+\frac{8\pi^2\mu}{h^2}\left(E+\frac{Ze^2}{4\pi\varepsilon_0 r}\right)R = 0 \qquad ...(3.5)$$

where m and β are constants.

The complete real wave function from the solution of Schrödinger wave equation (3.1) up to $n = 3$ is complied in the Table 3.1. The significance or detailed discussion of the functions depending on the angles φ, θ and r are given in Sections 3.2, 3.3 and 3.4 respectively.

TABLE 3.1. Real hydrogen-like wave funtions $\sigma = (Z\,r/a_0)$

n	l	m_l	Wave function	E_n	
1	0	0	$\Psi_{1s} = \frac{1}{\sqrt{\pi}}\left(\frac{Z}{a_0}\right)^{\frac{3}{2}} e^{-\sigma}$	$-\frac{Z^2\mu e^4}{2\hbar^2}$	1s
2	0	0	$\Psi_{2s} = \frac{1}{4\sqrt{2\pi}}\left(\frac{Z}{a_0}\right)^{3/2}(2-\sigma)\,e^{-\sigma/2}$	$-\frac{Z^2\mu e^4}{8\hbar^2}$	2s
2	1	0	$\Psi_{2p_z} = \frac{1}{4\sqrt{2\pi}}\left(\frac{Z}{a_0}\right)^{3/2}\sigma\, e^{-\sigma/2}\cos\theta$	$-\frac{Z^2\mu e^4}{8\hbar^2}$	2p_z

2	1	±1	$\Psi_{2px} = \frac{1}{4\sqrt{2\pi}}\left(\frac{Z}{a_0}\right)^{3/2} \sigma e^{-\sigma/2} \sin\theta\cos\varphi$	$-\frac{Z^2\mu e^4}{8\hbar^2}$	$2p_x$
2	1	±1	$\Psi_{2p_y} = \frac{1}{4\sqrt{2\pi}}\left(\frac{Z}{a_0}\right)^{3/2} \sigma e^{-\sigma/2} \sin\theta\sin\varphi$	$-\frac{Z^2\mu e^4}{8\hbar^2}$	$2p_y$
3	0	0	$\Psi_{3s} = \frac{1}{81\sqrt{3\pi}}\left(\frac{Z}{a_0}\right)^{3/2} \left(27-18\sigma+\sigma^2\right)e^{-\sigma/3}$	$-\frac{Z^2\mu e^4}{18\hbar^2}$	$3s$
3	1	0	$\Psi_{3p_z} = \frac{\sqrt{2}}{81\sqrt{\pi}}\left(\frac{Z}{a_0}\right)^{3/2} (6-\sigma)\sigma e^{-\sigma/3} \cos\theta$	$-\frac{Z^2\mu e^4}{18\hbar^2}$	$3p_z$
3	1	±1	$\Psi_{3px} = \frac{\sqrt{2}}{81\sqrt{\pi}}\left(\frac{Z}{a_0}\right)^{3/2} (6-\sigma)\sigma e^{-\sigma/3} \sin\theta\cos\varphi$	$-\frac{Z^2\mu e^4}{18\hbar^2}$	$3p_x$
3	2	±1	$\Psi_{3py} = \frac{\sqrt{2}}{81\sqrt{\pi}}\left(\frac{Z}{a_0}\right)^{3/2} (6-\sigma)\sigma e^{-\sigma/3} \sin\theta\sin\varphi$	$-\frac{Z^2\mu e^4}{18\hbar^2}$	$3p_x$
3	2	0	$\Psi_{3d_{z^2}} = \frac{1}{81\sqrt{6\pi}}\left(\frac{Z}{a_0}\right)^{3/2} \sigma^2 e^{-\sigma/3} \left(3\cos^2\theta-1\right)$	$-\frac{Z^2\mu e^4}{18\hbar^2}$	$3d_{z^2}$
3	2	±1	$\Psi_{3d_{zx}} = \frac{\sqrt{2}}{81\sqrt{\pi}}\left(\frac{Z}{a_0}\right)^{3/2} \sigma^2 e^{-\sigma/3} \sin\theta\cos\theta\cos\varphi$	$-\frac{Z^2\mu e^4}{18\hbar^2}$	$3d_{xz}$
3	2	±1	$\Psi_{3d_{yz}} = \frac{\sqrt{2}}{81\sqrt{\pi}}\left(\frac{Z}{a_0}\right)^{3/2} \sigma^2 e^{-\sigma/3} \sin\theta\cos\theta\sin\varphi$	$-\frac{Z^2\mu e^4}{18\hbar^2}$	$3d_{yz}$
3	2	±2	$\Psi_{3d_{x^2-y^2}} = \frac{1}{81\sqrt{2\pi}}\left(\frac{Z}{a_0}\right)^{3/2} \sigma^2 e^{-\sigma/3} \sin^2\theta\cos 2\varphi$	$-\frac{Z^2\mu e^4}{18\hbar^2}$	$3d_{x^2-y^2}$
3	2	± 2	$\Psi_{3d_{xy}} = \frac{1}{81\sqrt{2\pi}}\left(\frac{Z}{a_0}\right)^{3/2} \sigma^2 e^{-\sigma/3} \sin^2\theta \sin 2\varphi$	$-\frac{Z^2\mu e^4}{18\hbar^2}$	$3d_{xy}$

3.2. ANGULAR FUNCTIONS

The above equations (3.3, 3.4 and 3.5) can be solved. The solution for equation (3.3) is

$$\Phi(\phi) = \frac{1}{\sqrt{2\pi}} e^{im\phi} \quad ...(3.6)$$

where $1/\sqrt{2\pi}$ is a normalisation constant and $m = 0, \pm 1, \pm 2, ...$ are known as magnetic quantum numbers. This name is given to m because in the presence of external magnetic field, the states with different values of m have different energies. This effect is known as Zeeman effect.

The normalised solution of equation (3.4) is given as

$$\Theta(\theta)=\left[\frac{(2l+1)(l-|m|)!}{2(l+|m|)!}\right]^{1/2}P_l^{|m|}(\cos\theta) \quad ...(3.7)$$

with $\beta = l(l+1)$ and $P_l^{|m|}$ (cos θ) is an associated legendre polynomial given by

$$P_l^{|m|}(\cos\theta)=\frac{1}{2^l l!}\left(1-\cos^2\theta\right)^{|m|/2}\frac{d^{l+|m|}}{d\cos\theta^{l+|m|}}\left(\cos^2\theta-1\right)^l \quad ...(3.8)$$

and $l = 0, 1, 2, ..., l$ is known as azimuthal quantum number and describes the general shape of atomic subshells. The values of magnetic quantum number (m) is restricted by the value of l. If Θ(θ) is defined by the equation (3.7), the possible values of m can be

$$m = 0, \pm1, \pm2, \ldots , \pm l \quad ...(3.8a)$$

The product of functions Θ(θ) Φ(ϕ) is called angular functions (also known as spherical harmonics) and is denoted as $Y(\theta, \phi)$. It can be seen from equation (3.6) that for $m \neq 0$, the wave functions are imaginary and thus becomes difficult to visualise. It is possible and becomes convenient if we take the linear combination of them. This generates real functions of the same energy and generates the atomic orbitals. The symbols given for different atomic orbitals are connected with the values of l. For example if $l = 0$, orbital is s, for $l = 1$, it is p orbital, $l = 2$, d orbital and so on.

***Exercise* 3.1:** Prove that the values of m are: 0, ± 1, ± 2, ± 3, ...

The solution of equation (3.3) can be

$$\Phi(\varphi) = A_m e^{im\varphi} \text{ and } \Phi(\varphi) = A_{-m} e^{-im\varphi}. \quad ...(1)$$

where A_m and A_{-m} are normalisation constants. The boundary condition or Φ(φ) to be a single valued function of φ is that

$$\Phi(\varphi) = \Phi(2\pi + \varphi) \quad ...(2)$$

Substitute these values in equation (1),

$$A_m e^{im(2\pi+\varphi)} = A_m e^{2\pi im} e^{im\varphi} = A_m e^{im\varphi} \quad ...(3)$$

and

$$A_{-m} e^{-im(2\pi+\varphi)} = A_{-m} e^{-2im\pi} e^{-im\varphi} = A_{-m} e^{-im\varphi} \quad ...(4)$$

Equations (3 and (4) imply that

$$A_m e^{\pm 2\pi im} = 1 \quad ...(5)$$

In terms of real functions equation (5) can be written as

$$\cos(2\pi m) \pm i\sin(2\pi m) = 1 \quad ...(6)$$

Equation (6) can be only satisfied if the values of $m = 0, \pm1, \pm2, \pm3$, ..., because $\cos 2\pi m = 1$ for $m = 0, \pm1, \pm2, \pm3$, ... and $\sin 2\pi m = 0$ for $m = 0, \pm1, \pm2, \pm3$, ...The values of normalisation constants are obtained in Exercise 3.2.

Note: Problems regarding whether the angular or radial wave functions are normalised or orthogonal are given in Chapter 2. Further to generate the angular or radial wave functions from the given formula are also given in Chapter 2.

The solution of equation will show that the angular function will become zero l times, *i.e.* This function has l nodes.

***Exercise* 3.2:** Form the real functions from the functions defined by equation (3.6) for $m = \pm 2$.

Substitute the values of m in equation (3.6),

$$\Phi(\phi)_{m=2} = \frac{1}{\sqrt{2\pi}} e^{2i\phi} \qquad ...(3.9)$$

$$\Phi(\phi)_{m=-2} = \frac{1}{\sqrt{2\pi}} e^{-2i\phi} \qquad ...(3.10)$$

Take the linear combinations of equations (3.9) and (3.10).

$$\Phi_+ = \frac{1}{\sqrt{2}}\frac{1}{\sqrt{2\pi}}\left[e^{2i\phi} + e^{-2i\phi}\right] = \frac{2}{\sqrt{2}\sqrt{2\pi}}\cos 2\phi = \sqrt{1/\pi}\cos 2\phi$$

$$\Phi_- = \frac{1}{\sqrt{2}}\frac{1}{\sqrt{2\pi}}\left[\frac{e^{2i\phi} - e^{-2i\phi}}{i}\right] = \frac{2}{\sqrt{2}\sqrt{2\pi}}\sin 2\phi = \sqrt{1/\pi}\sin 2\phi$$

The additional $\frac{1}{\sqrt{2}}$ is an extra normalisation factor.

***Exercise* 3.3:** Show that the wave functions obtained in exercise 3.2 are normalised. The wave function is normalised if

$$\int_0^{2\pi} \Phi^* \Phi d\phi = 1$$

Substituting the values of Φ and Φ^*, we get

$$\int_0^{2\pi}\left(\frac{1}{2\pi}\right)^{1/2} e^{2i\phi}\left(\frac{1}{2\pi}\right)^{1/2} e^{-2i\phi} d\phi = \frac{1}{2\pi}\int_0^{2\pi} d\phi = 1$$

or otherwise

$$\int_0^{2\pi}\left(\frac{1}{\pi}\right)^{1/2}\cos 2\phi\left(\frac{1}{\pi}\right)^{1/2}\cos 2\phi d\phi = \frac{1}{\pi}\int_0^{2\pi}\cos^2 2\phi d\phi$$

$$= \frac{1}{2\pi}\int_0^{2\pi}(1+\cos 4\varphi)d\phi = \frac{1}{2\pi}\int_0^{2\pi}d\phi + \frac{1}{2\pi}\int_0^{2\pi}\cos 4\varphi d\phi$$

$$= \frac{1}{2\pi}[2\pi - 0] + \frac{1}{2\pi}\left[\frac{\sin 4\varphi}{4}\right]_0^{2\pi} = 1 + 0 = 1$$

Similarly, it can be shown that $\frac{1}{\sqrt{\pi}}\sin 2\phi$ is also normalised.

***Exercise* 3.4:** Calculate the normalised wave function $\Theta(\theta)$ for $l = 2$, $m = 1$. The wave function $\Theta(\theta)$ can be obtained by substituting the values in equation (3.7).

$$\Theta(\theta) = \left[\frac{(4+1)(2-1)!}{2(2+1)!}\right]^{1/2} P_2^1(\cos\theta) = \left(\frac{5}{12}\right)^{1/2} P_2^1(\cos\theta) \qquad ...(3.11)$$

Now $P_2^1(\cos\theta)$ can be calculated from (3.8),

$$P_2^1(\cos\theta) = \frac{1}{(2^2)(2!)}\left[1-\cos^2\theta\right]^{1/2} \frac{d^{2+1}}{d(\cos\theta)^{2+1}}\left[\left(\cos^2\theta-1\right)^2\right]$$

$$= \frac{1}{8}\left[\sin^2\theta\right]^{1/2} \frac{d^3}{d(\cos\theta)^3}\left(\cos^2\theta-1\right)^2$$

$$= \frac{1}{8}\sin\theta\ 24\cos\theta = 3\sin\theta\cos\theta$$

substituting the value of P_2^1 (cos θ) in (3.11)

$$\Theta(\theta) = \left(\frac{5}{12}\right)^{1/2} 3\sin\theta\ \cos\theta = \left(\frac{45}{12}\right)^{1/2} \sin\theta\cos\theta$$

$$= \frac{\sqrt{15}}{2}\sin\theta\cos\theta$$

***Exercise* 3.5:** The angular wave function for hydrogen like atoms for $l = 2$, $m = \pm 2$ is

$$\Psi_{3dx^2-y^2} = NR(r)\sin^2\theta\cos 2\phi$$

Prepare the plot for $\Psi_{3dx^2-y^2}$ in *xy* plane.

For fixed value of *r* and $\theta = \pi/2$, the angular wave function in *xy* plane will vary as shown in the Table 3.2 ($N.R(r)\sin^2\theta = 1.0$ arbitrary unit).

Plot cos 2ϕ vs ϕ, as shown in Fig. 3.2.

TABLE 3.2

φ	cos 2φ	φ	cos 2φ	φ	cos 2φ	φ	cos 2φ
0	1.0	90	– 1.0	180	1.0	270	–1.0
10	0.94	100	– 0.94	190	0.94	280	– 0.94
20	0.766	110	– 0.766	200	0.766	290	– 0.766
30	0.5	120	– 0.5	210	0.5	300	– 0.5
40	0.174	130	–0.174	220	0.174	310	– 0.174
50	–0.174	140	0.174	230	–0.174	320	0.174
60	–0.5	150	0.5	240	–0.5	330	0.5
70	–0.766	160	0.766	250	–0.766	340	0.766
80	–0.94	170	0.94	260	– 0.94	350	0.94
						360	1.0

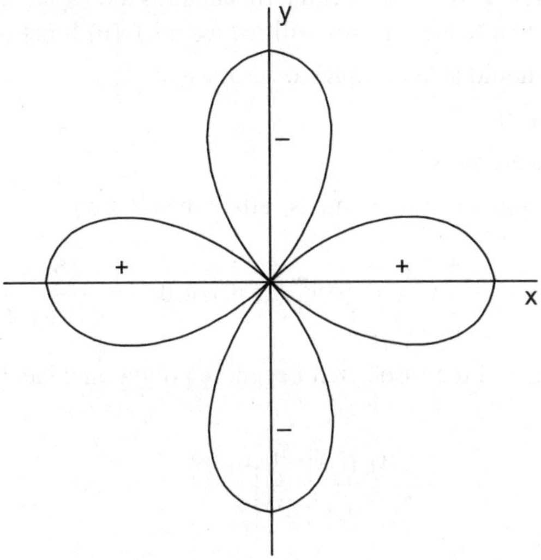

Fig. 3.2 *Angular plot of $dx^2 - y^2$ orbital in xy plane.*

***Exercise* 3.6:** Show that $\Theta(\theta) = \left(\frac{3}{2}\right)^{\frac{1}{2}} \cos\theta$ is a solution of equation (3.4) provided $m = 0$ and $(-2 + \beta)\sin^2\theta = 0$.

Equation (3.4) can be written as

$$\sin\theta \frac{d}{d\theta}\left[\sin\theta \frac{d\Theta(\theta)}{d\theta}\right] + \beta\sin^2\theta\Theta(\theta) - m^2\Theta(\theta) = 0 \qquad \text{...(A)}$$

Substituting the value of the given angular function, without using the normalisation constant, in equation (A) gives us

$$\sin\theta \frac{d}{d\theta}\left[\sin\theta \frac{d\cos\theta}{d\theta}\right] + \beta\sin^2\theta\cos\theta - m^2\cos\theta = 0 \qquad \text{...(B)}$$

The first term after applying differentiation, reduces to

$$\sin\theta \frac{d}{d\theta}\left[\sin\theta \frac{d\cos\theta}{d\theta}\right] = \sin\theta \frac{d}{d\theta}[\sin\theta(-\sin\theta] = -\sin\theta \frac{d}{d\theta}[\sin^2\theta]$$

$$= -\sin\theta[2\sin\theta\cos\theta] = -2\sin^2\theta\cos\theta \qquad \text{...(C)}$$

Substituting the first term and rearranging it, we get

$$[(-2 + \beta)\sin^2\theta]\cos\theta - m^2\cos\theta = 0 \qquad \text{...(D)}$$

For the given function, $\Theta(\theta) = \left(\frac{3}{2}\right)^{\frac{1}{2}} \cos\theta$ to be the solution of equation (A), each term in the equation (D) should be equal to zero. The second term contains the variable m and this should be equal to zero as cos θ cannot be equal to zero as this will reduce the Θ(θ) function to be zero.

Further the first term should also be equal to zero, *i.e.*,

$$(-2 + \beta) \sin^2 \theta = 0$$

The desired conditions are satisfied.

Note: The associated Legendre polynomials satisfy the relation

$$\int_{-1}^{+1} p_l^{|m|}(x) p_{l'}^{|m|}(x) dx = \int_0^{\pi} p_l^{|m|}(\cos\theta) p_{l'}^{|m|}(\cos\theta) \sin\theta d\theta = \frac{2}{(2l+1)} \frac{(l+|m|!}{(l-|m|)!} \delta_{ll'} \quad ...(3.12)$$

and the normalisation constant of the associated Legendre polynomial can be written as

$$N_{ll'} = \left[\frac{(2l+1)}{2} \frac{(l-|m|!}{(l+|m|)!} \right]^{\frac{1}{2}} \quad ...(3.13)$$

Couple of things need to be discussed. (*i*) The spherical harmonics or angular functions are not eigen functions of $\hat{L}_x$ or $\hat{L}_y$. (*ii*) Is there specific reason for choosing z direction in general? The answer is that there is nothing at all special for choosing z direction. We could choose either x or y direction as the unique direction. All the results obtained with z direction would have been obtained with x or y. For example, if we take x direction as the particular direction, the $\hat{L}^2$ and $\hat{L}_x$ precisely, in which case $\hat{L}_y$ and $\hat{L}_z$ do not have the precise values. It is customary to choose z direction because the expression for $\hat{L}_z$ in spherical harmonics coordinates is so much simpler than those for $\hat{L}_x$ or $\hat{L}_y$.

3.3. THE RADIAL FUNCTION

The equation (3.5) for the radial part can be solved to give a solution

$$R_{n,l}(r) = -\left[\frac{(2Z/na_0)^3 (n-l-1)!}{2n[(n+l)!]^3} \right]^{1/2} e^{-\rho/2} \rho^l L_{n+l}^{2l-1}(\rho) \quad ...(3.14)$$

where

$$\rho = \left(\frac{2Z}{na_0} \right) r \quad ...(3.15)$$

$$a_0 = \frac{(4\pi\varepsilon_0).h^2}{4\pi^2 \mu e^2} \quad ...(3.16)$$

L_{n+l}^{2l+1} (ρ) is the associated Laguerre polynomial and defined as

$$L_r^s(\rho) = \frac{d^s}{d\rho^s} \left[e^{\rho} \frac{d^r}{d\rho^r} \left(\rho^r e^{-\rho} \right) \right] \quad ...(3.17)$$

where $s = 2l + 1$ and $r = n + 1$ and $n = 1, 2, 3, 4, ...$ This is called principal quantum number and describes the distance of the electron from the nucleus as well as the energies of the different orbitals. If $R(r)$ is defined by the equation (3.12), the values of l are restricted and are

$$l = 0, 1, 2, \ldots (n-l) \qquad \ldots(3.18)$$

Further the probability of finding the electron between the distance r and $r + dr$ is $R^*(r)\, R(r)\, r^2\, dr$. The factor r^2 comes from the change of cartesian coordinates to spherical coordinates when volume element is calculated.

Similar to angular functions, radial wave functions also possesses points in its plot as a function of r, where the radial function is zero, called radial nodes. These are given as:

$$\begin{aligned} \text{Radial nodes} &= \text{Total number of nodes} - \text{Angular nodes} \\ &= (n-1) - l = n - l - 1 \end{aligned} \qquad \ldots(3.19)$$

***Exercise* 3.7:** Determine the radial part of the wave function for $2p$ orbital.

In $2p$ orbital, $n = 2$, $l = 1$, so associated Laguerre polynomial becomes

$$L_3^3(\rho) = \frac{d^3}{d\rho^3}\left[e^{\rho}\frac{d^3}{d\rho^3}\left(\rho^3 e^{-\rho}\right)\right]$$

$$\frac{d^3}{d\rho^3}\left[\rho^3 e^{-\rho}\right] = \frac{d^2}{d\rho^2}\left[\left(3\rho^2 - \rho^3\right)e^{-\rho}\right] = (6 - 18\rho + 9\rho^2 - \rho^3)e^{-\rho}$$

$$L_3^3(\rho) = \frac{d^3}{d\rho^3}\left[e^{\rho}\left(6 - 18\rho + 9\rho^2 - \rho^3\right)e^{-\rho}\right]$$

$$= \frac{d^3}{d\rho^3}\left[6 - 18\rho + 9\rho^2 - \rho^3\right] = 6$$

Substituting the respective values in equation (3.14)

$$R_{2,1}(r) = \left[\frac{(2Z/2a_0)^3(2-1-1)!}{2\times 2(2+1)!^3}\right]^{1/2} e^{-\rho/2}\rho.6 = \left(\frac{Z}{a_0}\right)^{3/2}\frac{1}{2\sqrt{6}}\rho e^{-\rho/2}$$

***Exercise* 3.8:** The radial part of the wave function of $2p$ ($n = 2$, $l = 1$) orbital of hydrogen atom is given by

$$R_{2,1}(r) = \left(\frac{Z}{a_0}\right)^{3/2}\frac{1}{2\sqrt{6}}\rho e^{-\rho/2} = \left(\frac{Z}{a_0}\right)^{3/2}\frac{1}{\sqrt{24}}\cdot\left(\frac{Z}{a_0}\right) r e^{-Zr/2a_0}$$

Plot $R_{2,1}(r)$ as a function of r and also $4\pi r^2 R_{2,1}^2(r)$ vs r.

Now, here $Z = 1$, so

$$R_{2,1}(r) = \left(\frac{1}{a_0}\right)^{5/2}\left(\frac{1}{24}\right)^{1/2} r e^{-r/2a_0}$$

and

$$4\pi r^2 R_{2,1}^2(r) = \frac{4\pi}{24a_0^5}.r^4 e^{-r/a_0}$$

The values of $R_{2,1}(r)$ and $4\pi r^2 R_{2,1}^2(r)$ as a function of r are given in Table 3.3 and are plotted as a function of r in Fig. 3.3.

TABLE 3.3

Sl. No.	$r(m)$	$R_{2,1}(r)\times 10^{-14}$ $(m^{-3/2})$	$4\pi r^2 R_{2,1}^2(r)\times 10^{10}$ (m^{-1})
0	0.5	3.134	
1	1.0	3.897	2.116
2	1.5	3.644	4.162
3	2.0	3.029	5.112
4	2.5	2.360	4. 850
5	3.0	1.766	3.909
6	3.5	1.284	2.814
7	4.0	0.915	1.866
8	4.5	0.642	1.161
9	5.0	0.444	0.688
10	5.5	0.305	0.391
11	6.0	0.207	0.215
12	7.0	0.094	0.060
13	8.0	0.042	0.016
14	9.0	0.018	0.0038
15	10.0	0.008	0.0009

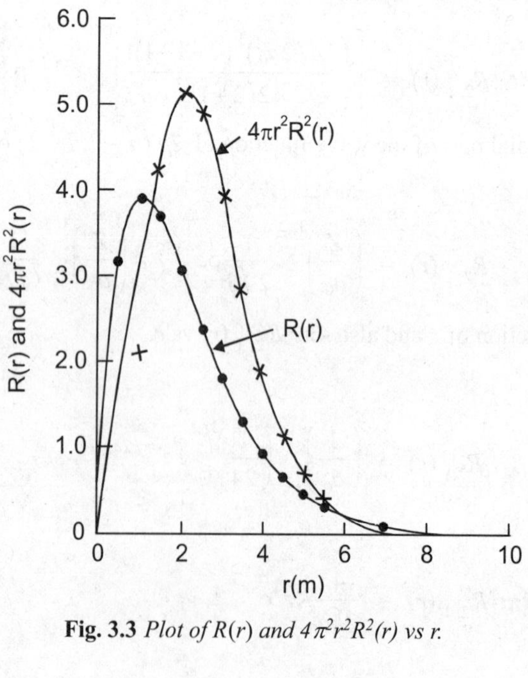

Fig. 3.3 *Plot of R(r) and $4\pi^2 r^2 R^2(r)$ vs r.*

***Exercise* 3.9:** Show that $R(r) = Ne^{-\frac{Zr}{a_0}}$, N is a normalisation constant, is the solution of the radial equation of the Schrödinger wave equation, provided $l = 0$ and $E_n = -\frac{Z^2e^2}{2a_0} = -\frac{2\pi^2\mu e^4Z^2}{(4\pi\varepsilon_0)^2h^2}$, where a_0 is Bohr radius $= \frac{(4\pi\varepsilon_0)h^2}{4\pi^2\mu e^2}$

The radial wave equation as given by equation (3.5) after multiplying by r^2 is

$$\frac{d}{dr}\left[r^2\frac{dR(r)}{dr}\right] - l(l+1)R(r) + \frac{8\pi^2\mu r^2}{h^2}\left(E + \frac{Ze^2}{4\pi\varepsilon_0 r}\right)R(r) = 0 \qquad \text{...(A)}$$

The given solution will be valid if equation (A) will hold good for all the values of radial coordinates. Substituting the values $R(r)$ in equation (A) and dividing by N, we get

$$\frac{d}{dr}\left[r^2\frac{d\left(e^{-\frac{Zr}{a_0}}\right)}{dr}\right] - l(l+1)\left(e^{-\frac{Zr}{a_0}}\right) + \frac{8\pi^2\mu r^2}{h^2}\left(E + \frac{Ze^2}{4\pi\varepsilon_0 r}\right)\left(e^{-\frac{Zr}{a_0}}\right) = 0 \qquad \text{...(B)}$$

Expanding the first term of equation (B), we get

$$\frac{d}{dr}\left[r^2\frac{d\left(e^{-\frac{Zr}{a_0}}\right)}{dr}\right] = \frac{d}{dr}\left[-r^2\frac{Z}{a_0}e^{-\frac{Zr}{a_0}}\right] = -\frac{Z}{a_0}\frac{d}{dr}\left[r^2e^{-\frac{Zr}{a_0}}\right] = -\frac{Z}{a_0}\left[2r - \frac{Zr^2}{a_0}\right]e^{-\frac{Zr}{a_0}} \qquad \text{...(C)}$$

Substituting equation (C) in (B), we get

$$\left[\frac{Z^2r^2}{a_0^2} - \frac{2Zr}{a_0}\right]\left(e^{-\frac{Zr}{a_0}}\right) - l(l+1)\left(e^{-\frac{Zr}{a_0}}\right) + \frac{8\pi^2\mu r^2}{h^2}\left(E + \frac{Ze^2}{4\pi\varepsilon_0 r}\right)\left(e^{-\frac{Zr}{a_0}}\right) = 0 \qquad \text{...(D)}$$

Divide equation (D) by the exponential term, we get

$$\left[\frac{Z^2r^2}{a_0^2} - \frac{2Zr}{a_0}\right] - l(l+1) + \frac{8\pi^2\mu r^2}{h^2}\left(E + \frac{Ze^2}{4\pi\varepsilon_0 r}\right) = 0 \qquad \text{...(E)}$$

Arrange the equation in terms of r^2 and r, we get

$$\left[\frac{Z^2}{a_0^2} + \frac{8\pi^2\mu E}{h^2}\right]r^2 + \left(-\frac{2Z}{a_0} + \frac{8\pi^2\mu Ze^2}{4\pi\varepsilon_0 h^2}\right)r - l(l+1) = 0 \qquad \text{...(F)}$$

A given solution is a solution of equation (F) if each term in equation (F) is equal to zero. So considering taking the third term will show that it is only possible if $l = 0$.

Now considering the first term and equating it to zero, we get

$$\frac{Z^2}{a_0^2} + \frac{8\pi^2\mu E}{h^2} = 0$$

i.e.,
$$E = -\frac{Z^2h^2}{8\pi^2\mu a_0^2} \qquad \text{...(G)}$$

Now substituting the value of a_0, *i.e.,* $a_0 = \dfrac{(4\pi\varepsilon_0)h^2}{4\pi^2\mu e^2}$ in equation (G) gives us

$$E = -\frac{Z^2h^2}{8\pi^2\mu a_0^2} \times \left(\frac{4\pi^2\mu e^2}{(4\pi\varepsilon_0)h^2}\right)^2 = -\frac{2\pi^2\mu e^4 Z^2}{(4\pi\varepsilon_0)^2 h^2} \qquad \text{...(H)}$$

The value of energy is the same as required by the problem.

Similarly the second term when equated to zero will also give the same value of energy as given by the equation (H). Proof of this is left for the students.

3.4. ENERGY VALUES

The energies of eigen functions of the hydrogen like atoms can be obtained by substituting equations (3.3), (3.4) and (3.5) into (3.2) and is given by

$$E_n = \frac{-2\pi^2\mu Z^2 e^4}{n^2h^2(4\pi\varepsilon_0)^2} \qquad \text{...(3.19}a\text{)}$$

in SI units.

In electrostatic units, the value of E_n is given as

$$E_n = -\frac{2\pi^2\mu e^4 Z^2}{h^2n^2} \qquad \text{...(3.19}b\text{)}$$

and in atomic units, it is

$$E_n = -\frac{Z^2}{2n^2}\text{(hartree)} \qquad \text{...(3.19}c\text{)}$$

where $n = 1, 2, 3,$

The values of energies for different orbitals obtained from this equation is the same as those obtained from Bohr theory. So the spectral lines obtained will have the same frequencies, as described in Section 2.4.

***Exercise* 3.10:** Arrange energies of orbitals of hydrogen-like atoms in the increasing order.

From equation (3.19), it is clear that energy of any orbital depends only on the principal quantum number (n). So orbitals with same n value have the same energy. Thus $\varepsilon_{1s} < \varepsilon_{2s} = \varepsilon_{2p} < \varepsilon_{3s} = \varepsilon_{3p} = \varepsilon_{3d} < \varepsilon_{4s} = \varepsilon_{4p} = \varepsilon_{4d} = \varepsilon_{4f}$ and so on.

3.5. QUANTUM NUMBERS

The solutions of equations (3.3), (3.4) and (3.5) give rise to three quantum numbers, n, l and m_l. Thus a wave function or state is characterised by three spatial quantum numbers (n, l, m_l). The characteristics of these quantum numbers are given below:

3.5.1. *Principle Quantum Number*

The label n is called the *principal quantum number*. It determines the size of the wave function and the total energy of the electron. It is usually defined as the sum of:

$$n = n_r + l + 1 \qquad \text{...(3.20)}$$

where n_r determines the number of zeros (nodes) of $R_{n,l}(r)$. The label n_r is called the *radial quantum number* and l the *azimuthal quantum number*. The minimum values of n_r and l are zero. Thus the minimum value of n will be 1. Since n_r is non-negative, the principal quantum number is greater than the azimuthal quantum number at least by one. Thus, *the principal quantum number determines the total numbers of nodes (i.e., n – 1), including radial and angular nodes*. The possible values of n can be

$$n = 1, 2, 3, \ldots\ldots\ldots\ldots \qquad \text{...(3.21)}$$

3.5.2. *Azimuthal Quantum Number or the Angular Momentum Quantum Number*

For a given value of n, the azimuthal quantum number l can take on values:

$$l = 0, 1, 2, 3 \ldots\ldots\ldots (n-1) \qquad \text{...(3.22)}$$

and determines the shape of the wave function or orbital. It also implies the following useful information:

(1) This quantum number specifies the number of angular nodes. The total number of radial nodes is: $n - l - 1$. Since the total number of nodes including radial and angular are:

$n - 1$, the total number of angular nodes = $n - 1 - n + l + 1 = l$

(2) It provides information about the total orbital angular momentum (M^2), which can be expressed as:

$$M^2 = \frac{h^2}{4\pi^2} l(l+1) \qquad \text{...(3.23)}$$

or

$$M = \frac{h}{2\pi}\sqrt{l(l+1)}\,,\ l = 0, 1, 2, \ldots\ldots\ldots\ldots (n-1) \qquad \text{...(3.24)}$$

For large values of l, equation (3.24) reduces to:

$$M \cong \frac{h}{2\pi} l \qquad \text{...(3.25)}$$

which is similar to Bohr's postulate

$$M = \frac{h}{2\pi} l \qquad \text{...(3.26)}$$

The only difference is that the value of $n = 0$ is not allowed in Bohr's treatment, while $l = 0$ is permitted in Schrödinger wave equation.

***Exercise* 3.11:** Show that the angular wave function $Y_1^1 = \left(\frac{3}{8\pi}\right)^{\frac{1}{2}} \sin\theta e^{i\varphi}$ is eigen function of $\hat{L}^2$.

The operator for $\hat{L}^2$ is

$$\hat{L}^2 = -\frac{h^2}{4\pi^2}\left[\frac{1}{\sin\theta}\frac{\partial}{\partial\theta}\left(\sin\theta\frac{\partial}{\partial\theta}\right)+\frac{1}{\sin^2\theta}\frac{\partial^2}{\partial^2\varphi}\right]$$

$$\hat{L}^2 Y_1^1 = -\frac{h^2}{4\pi^2}\left[\frac{1}{\sin\theta}\frac{\partial}{\partial\theta}\left(\sin\theta\frac{\partial}{\partial\theta}\right)+\frac{1}{\sin^2\theta}\frac{\partial^2}{\partial^2\varphi}\right]Y_1^1$$

$$= -\frac{h^2}{4\pi^2}\left[\frac{1}{\sin\theta}\frac{\partial}{\partial\theta}\left(\sin\theta\frac{\partial}{\partial\theta}\right)+\frac{1}{\sin^2\theta}\frac{\partial^2}{\partial^2\varphi}\right]\left(\frac{3}{8\pi}\right)^{\frac{1}{2}}\sin\theta e^{i\varphi}$$

$$= -\frac{h^2}{4\pi^2}\left(\frac{3}{8\pi}\right)^{\frac{1}{2}}\left\{\left[\frac{1}{\sin\theta}\frac{\partial}{\partial\theta}\left(\sin\theta\frac{\partial\sin\theta}{\partial\theta}\right)e^{i\varphi}\right]+\left[\frac{1}{\sin\theta}\frac{\partial^2 e^{i\varphi}}{\partial^2\varphi}\right]\right\}$$

$$= -\frac{h^2}{4\pi^2}\left(\frac{3}{8\pi}\right)^{\frac{1}{2}}\left\{\left[\frac{1}{\sin\theta}\frac{\partial}{\partial\theta}(\sin\theta\cos\theta)e^{i\varphi}\right]+\left[\frac{1}{\sin\theta}i^2 e^{i\varphi}\right]\right\}$$

$$= -\frac{h^2}{4\pi^2}\left(\frac{3}{8\pi}\right)^{\frac{1}{2}}\left\{\left[\frac{1}{\sin\theta}(\cos\theta\cos\theta - \sin\theta\sin\theta)e^{i\varphi}\right]+\left[\frac{1}{\sin\theta}i^2 e^{i\varphi}\right]\right\}$$

$$= -\frac{h^2}{4\pi^2}\left(\frac{3}{8\pi}\right)^{\frac{1}{2}}\left\{\left[\frac{1}{\sin\theta}(\cos^2\theta - \sin^2\theta - 1)e^{i\varphi}\right]\right\}$$

$$= -\frac{h^2}{4\pi^2}\left(\frac{3}{8\pi}\right)^{\frac{1}{2}}\left[-\frac{2\sin^2\theta}{\sin\theta}e^{i\varphi}\right]$$

$$= \frac{h^2}{4\pi^2}\left(\frac{3}{8\pi}\right)^{\frac{1}{2}}[2\sin\theta e^{i\varphi}]$$

It clearly shows that the given function is eigen function of $\hat{L}^2$.

3.5.3 *Magnetic Quantum Number*

It is designated as m_l and determines the orientation of the wave function or orbital in three dimensional space. The possible values can be

$$m_l = 0, \pm1, \pm1, \pm1, \ldots\ldots\ldots\ldots\ldots, \pm l \qquad \ldots(3.27)$$

The solution of radial equation (3.5) reveals that the energy levels $E_{n,\,l}$ do not depend on m_l as expected, since $m_l \frac{h}{2\pi}$ is the value of angular momentum component along Z-axis, as direction of the axis is well defined. But if the magnetic field is applied to the system (say in the Z-direction), the energy levels will depend on m_l as well on l and n. For this reason m_l is called *magnetic quantum number*. The value of Z-component of orbital angular momentum is given by the expression:

$$M_z = m_l \frac{h}{2\pi}, m_l = 0. \pm 1, \pm 2, \pm 3 \ldots\ldots\ldots\ldots\ldots\ldots \pm l \quad \ldots(3.28)$$

***Exercise* 3.12:** Show that angular wave function $Y_1^1 = \left(\frac{3}{8\pi}\right)^{\frac{1}{2}} \sin\theta e^{i\varphi}$ is eigen function of $\hat{L}_z$.

Angular part of the operator $\hat{L}_z$ is $-i\frac{h}{2\pi}\frac{\partial}{\partial\varphi}$. The eigen value and eigen function equations are

$$\hat{L}_z Y_1^1 = -i\frac{h}{2\pi}\frac{\partial}{\partial\varphi}\left[\left(\frac{3}{8\pi}\right)^{\frac{1}{2}} \sin\theta e^{i\varphi}\right] = -i\frac{h}{2\pi}\left(\frac{3}{8\pi}\right)^{\frac{1}{2}} \sin\theta\left[\frac{\partial}{\partial\varphi}e^{i\varphi}\right] = -i^2\frac{h}{2\pi}\left(\frac{3}{8\pi}\right)^{\frac{1}{2}} \sin\theta e^{i\varphi}$$

$$= \frac{h}{2\pi}\left[\left(\frac{3}{8\pi}\right)^{\frac{1}{2}} \sin\theta e^{i\varphi}\right]$$

The given function is an eigen function to the $\hat{L}_z$ operator and the eigen value is $\frac{h}{2\pi}$

***Exercise* 3.13:** Show that $l \geq |m|$.

Consider the angular wave function $Y_l^m(\theta,\varphi)$ and the operators $\hat{L}_z$ and $\hat{L}^2$. Using the eigen value and eigen function equations, we get

$$\hat{L}_z^2 Y_l^m(\theta,\varphi) = \hat{L}_z[\hat{L}_z Y_l^m(\theta,\varphi)] = \hat{L}_z[mY_l^m(\theta,\varphi)] = m[\hat{L}_z Y_l^m(\theta,\varphi)] = m^2 Y_l^m(\theta,\varphi) \quad \ldots(i)$$

Similarly

$$\hat{L}^2 Y_l^m(\theta,\varphi) = [l(l+1)\frac{h^2}{4\pi^2}]Y_l^m(\theta,\varphi) \quad \ldots(ii)$$

Subtracting (*i*) from (*ii*), we get

$$[\hat{L}^2 - L_z^2]Y_l^m(\theta,\varphi) = [l(l+1) - m^2]\frac{h^2}{4\pi^2}Y_l^m(\theta,\varphi) \quad \ldots(iii)$$

Knowing $\hat{L}^2 = L_x^2 + L_y^2 + L_z^2 \quad \ldots(iv)$

Combining equations (*iii*) and (*iv*)

$$[\hat{L}^2 - L_z^2]Y_l^m(\theta,\varphi) = [L_x^2 + L_y^2 + L_z^2 - L_z^2]Y_l^m(\theta,\varphi)$$

$$= [L_x^2 + L_y^2]Y_l^m(\theta,\varphi) = [l(l+1) - m^2]\frac{h^2}{4\pi^2}Y_l^m(\theta,\varphi) \quad \text{...}(v)$$

It is clear from equation (*v*), the angular function is also eigen function to the operator $L_x^2 + L_y^2$. Since the operator is the sum of two squared terms, it cannot be negative. Thus the right hand side of equation (5) will also be positive, *i.e.*,

$$[l(l+1) - m^2]\frac{h^2}{4\pi^2} \geq 0$$

or

$$l(l+1) \geq m^2 \quad \text{...}(vi)$$

Since l and m are integers equation (*vi*) indicates that

$$l \geq |m|$$

This proves the desired results.

Accordingly the total energy of hydrogen and hydrogen like atoms in a magnetic field is:

$$E_{n,\ell,m_\ell} = -\frac{2\pi^2\mu Z^2e^4}{(4\pi\varepsilon_0)^2h^2}\frac{1}{n^2} + \mu_B m_\ell H_{ext} \quad \text{...}(3.29)$$

The contribution of the first term is due to Coulomb interaction between electron and nucleus and of the second term is due to interaction between external magnetic field (H_{ext}) and magnetic moment (μ_B) of the circulating electron. Recall that for a given n and a given l, there are $2l + 1$ values possible for m_l ($0, \pm 1, \pm 2, \pm 3, \ldots\ldots\ldots \pm l$). Each permitted value of m_l produces a different value for E_{n,ℓ,m_ℓ}, thus splitting the original energy level into $2l + 1$ levels.

All the orbitals that have a given value of n form a single shell of the atom. The shells are referred to by letters:

n	1	2	3	4.....
	K	L	M	N.....

The orbitals with the same value of n but different values of l form the sub-shells of a given shell. These sub-shells are generally designated by the letters $s, p, d,$ .... as follows:

L	0	1	2	3	4	5
Name of sub-shell	s	p	d	f	g	h

***Exercise* 3.14:** (*i*) Calculate the energy of an excited state of hydrogen atom when it is in $3d$ orbital in SI units and in hartree unit, (*ii*) determine the magnitude of the Z-component of the angular momentum of the electron, (*iii*) calculate the energy of the quantum states evolved when this state is subjected to an external magnetic field of strength 10^6 G.

(*i*) Equations (3.18) and (3.18*b*) are used to calculate the energy of $3d$ state. Here $Z = 1$ and $n = 3$.

$$E(3d) = -\frac{2\pi^2 m_e e^4}{(4\pi\varepsilon_0)^2 h^2 n^2}$$

$$= -\frac{(2)(3.14)^2(9.109 \times 10^{-31}\ \text{kg})(1.6 \times 10^{-19}\text{C})^4}{[(4)(3.14)(8.85419 \times 10^{-12}\ \text{A}^{-2}S^4\ \text{kg}^{-1}\ \text{m}^{-3})]^2(6.626 \times 10^{-34}\text{J s})^2(3)^2} = 2.409 \times 10^{-19}\ \text{J}$$

Energy in hartree units is obtained from equation (3.19 *c*), *i.e.*,

$$E(3d) = -\frac{Z^2}{2n^2} = -\frac{(1)^2}{2(9)^2} = 6.1728 \times 10^{-3}\ \text{hartree}$$

(*ii*) The *Z*-component of ***L*** depends on the magnetic quantum number. For *d* state the value of $l = 2$, so there will be five m_l values, *i.e.*, 2, 1, 0, –1, –2. Thus there will be five possible directions of *L* and five possible values of *Z*-components. These are

$$\boldsymbol{M_l} = \frac{2h}{2\pi}, \frac{h}{2\pi}, 0, -\frac{h}{2\pi}, -\frac{2h}{2\pi}$$

(*iii*) The change in energy produced by the external magnetic field is given by equation (3.29). Because of this five *d* orbitals are no longer degenerate of energy 6.1728×10^{-3} hartree. They will split into five *d* orbitals. The energy will change as

$$[\Delta E_{mag}]_{ml=2} = 2\beta_e B = (2)(2.1272 \times 10^{-10}\ \text{hartree G}^{-1})(10^6\ \text{G}) = 4.2544 \times 10^{-4}\ \text{hartree}$$
$$[\Delta E_{mag}]_{m_1=1} = \beta_e B = (1)(2.1272 \times 10^{-10}\ \text{hartree G}^{-1})(10^6\ \text{G}) = 2.1272 \times 10^{-4}\ \text{hartree}$$
$$[\Delta E_{mag}]_{ml=0} = 0$$
$$[\Delta E_{mag}]_{m=-1} = -\beta_e = -(1)(2.1272 \times 10^{-10}\ \text{hartree G}^{-1})(10^6\ \text{G}) = -2.1272 \times 10^{-4}\ \text{hartree}$$
$$[\Delta E_{mag}]_{ml=1} = -2\beta_e B = -(1)(2.1272 \times 10^{-10}\ \text{hartree G}^{-1})(10^6\ \text{G}) = -4.2544 \times 10^{-4}\ \text{hartree}$$

Thus the energy of each state will be

$$[\Delta E_{mag}]_{ml=2} = [-6.1728 \times 10^{-3} + 4.2544 \times 10^{-4}]\ \text{hartree} = -57.4736 \times 10^{-4}\ \text{hartree}$$
$$[\Delta E_{mag}]_{ml=1} = [-6.1728 \times 10^{-3} + 2.1272 \times 10^{-4}]\ \text{hartree} = -59.6008 \times 10^{-4}\ \text{hartree}$$
$$[\Delta E_{mag}]_{ml=0} = [-6.1728 \times 10^{-3} + 0]\ \text{hartree} = -61.728 \times 10^{-4}\ \text{hartree}$$
$$[\Delta E_{mag}]_{ml=-1} = [-6.1728 \times 10^{-3} - 2.1272 \times 10^{-4}]\ \text{hartree} = -63.8552 \times 10^{-4}\ \text{hartree}$$
$$[\Delta E_{mag}]_{ml=-2} = [-6.1728 \times 10^{-3} - 4.2544 \times 10^{-4}]\ \text{hartree} = -65.9824 \times 10^{-4}\ \text{hartree}$$

3.6. ELECTRON POSITION

The probability of finding an electron within the volume *dV* is given as $\psi * \psi\, dV$ and over whole space is

$$p = \int_0^{+\infty}\int_0^{\pi}\int_0^{2\pi} \psi * \psi r^2 \sin\theta\ dr \cdot d\theta\ d\varphi \qquad ...(3.30)$$

The probability obtained by equation (3.30) is equal to unity if the wave functions used are normalised.

Similarly the average value of any function*F* of *r* or average value of function*F*(*r*) of any observable quantity.

$$F(r) = \int_0^\infty [R_n^l(r)]^* \hat{F}(r)[R_n^l(r)]r^2 dr \qquad ...(3.31)$$

If we wish to obtain the average value of a function that depends on r, θ and φ, equation (3.31) reduces to

$$G(r, \theta, \varphi) = \int_0^\infty \int_0^\pi \int_0^{2\pi} \psi^*(r,\theta,\varphi)\hat{G}(r,\theta,\varphi)\psi(r,\theta,\varphi)r^2 \sin\theta dr d\theta d\varphi \qquad ...(3.32)$$

where $\hat{G}(r,\theta,\varphi)$ is the operator of the quantity whose average value is to be determined.

***Exercise* 3.15:** (*a*) Calculate the average value of the potential energy between the electron and nucleus of the hydrogen and hydrogen like ions in its ground state. (*b*) Prove that potential energy is twice the kinetic energy.

(*a*) The potential energy in case of hydrogen or hydrogen like ions is given as: $V(r) = -\dfrac{Ze^2}{r}$. As we know that the potential energy depends only on the distance between the electron and proton, so we can only use the radial wave functions. Using this as an operator, the average potential energy is given as

$$R_{1,0} = 2\left(\frac{Z}{a_0}\right)^{\frac{3}{2}} e^{-\sigma}$$

(*b*) $$<V(r)> = \int_{r=0}^{r=\infty} R_1^0(r)\left|-\frac{Ze^2}{r}\right| R_1^0(r) r^2 dr = -\int_{r=0}^{r=\infty} 2\left[\frac{Z^3}{a_0^3}\right]^{\frac{1}{2}} e^{-\frac{Zr}{a_0}} \frac{Ze^2}{r} 2\left[\frac{Z^3}{a_0^3}\right]^{\frac{1}{2}} e^{-\frac{Zr}{a_0}} r^2 dr$$

$$<V(r)> = -\left[\frac{4Z^4e^2}{a_0^3}\right]\int_{r=0}^{r=\infty} e^{-2\frac{Zr}{a_0}} r dr \qquad ...(A)$$

Using the definite integral in its general form, *i.e.*, equation (A) can be written as

$$<V(r)> = -\left[\frac{4Z^4e^2}{a_0^3}\right]\left[\frac{a_0}{2Z}\right]^2 = -\frac{Z^2e^2}{a_0} \qquad ...(B)$$

Substituting the value of a_0 $\left[a_0 = \dfrac{(4\pi\varepsilon_0)h^2}{4\pi^2\mu e^2}\right]$ in (B), we get

$$<V(r)> = -Z^2e^2\frac{4\pi^2\mu e^2}{(4\pi\varepsilon_0)h^2} = -\frac{4\pi^2\mu e^4 Z^2}{(4\pi\varepsilon_0)h^2} \qquad ...(C)$$

Knowing that the value of total energy for the ground state is :

$$E = -\frac{2\pi^2\mu Z^2 e^4}{(4\pi\varepsilon_0)^2 h^2} \qquad ...(D)$$

Since $E = T + V$, so $T = E - V$.

$$T = E - V = -\frac{2\pi^2\mu Z^2 e^4}{(4\pi\varepsilon_0)^2 h^2} + \frac{4\pi^2\mu e^4 Z^2}{(4\pi\varepsilon_0)h^2} = \frac{2\pi^2\mu Z^2 e^4}{(4\pi\varepsilon_0)^2 h^2} \qquad ...(E)$$

Comparing the potential energy (equation C) and kinetic energy (equation E), we see that $\langle V\rangle = -2T$.

***Exercise* 3.16:** If the wave function for $1s$ orbital of hydrogen atom is $\psi_{100} = \frac{1}{\sqrt{\pi a_0^3}} e^{-r/a_0}$, find the average distance of $1s$ electron from the nucleus.

The average distance of electron in $1s$ orbital is given as

$$\langle r\rangle = \int_{\substack{\text{all}\\\text{space}}} \psi^* \hat{r}\psi d\tau = \int_0^\infty \int_0^\pi \int_0^{2\pi} \left(\frac{1}{\pi a_0^3}\right)^{1/2} e^{-r/a_0}\, r \left(\frac{1}{\pi a_0^3}\right)^{1/2} e^{-r/a_0} r^2 \sin\theta\, dr\, d\theta\, d\varphi$$

$$= \frac{1}{\pi a_0^3} \int_0^\infty r^3 e^{-2r/a_0} dr \int_0^\pi \sin\theta\, d\theta \int_0^{2\pi} d\varphi.$$

The first integral can be solved by seeing the footnote.

$$= \frac{1}{\pi a_0^3}\left[\frac{3!}{(2/a_0)^{3+1}}\right]\left[-\cos\theta\right]_0^\pi \left[\phi\right]_0^{2\pi} = \frac{3}{2}a_0$$

***Exercise* 3.17:** The normalised wave function for $1s$ electron is given as $\psi_{1s} = \left(\frac{1}{\pi a_0^3}\right)^{1/2} e^{-r/a_0}$.

Calculate the probability of this electron in a sphere of radius $r = a_0$.

The probability is given as

$$P = \int \psi^* \psi d\tau = \frac{1}{\pi a_0^3}\int_0^{a_0} e^{-2r/a_0} r^2 dr \int_0^\pi \sin\theta\, d\theta \int_0^{2\pi} d\varphi$$

$$= \frac{4}{a_0^3}\int_0^{a_0} e^{-2r/a_0} r^2\, dr$$

After integrating by parts, the above expression becomes

$$P = \frac{4}{a_0^3}\left\{\left[\frac{a_0}{2} r^2 e^{-2r/a_0}\right]_0^{a_0} + \left[-\frac{a_0^2}{2} r e^{-2r/a_0}\right]_0^{a_0} + \left[-\frac{a_0^3}{4} e^{-2r/a_0}\right]_0^{a_0}\right\}$$

$$P = -5e^{-2} + 1 = 0.323$$

***Exercise* 3.18:** Calculate the most probable distance at which the $1s$ electron of hydrogen-like atom with atomic number Z is to be found. Show that as Z increases the most probable distance decreases.

The normalised wave function of $1s$ orbital is $\left(\frac{Z^3}{\pi a_0^3}\right)^{1/2} e^{-Zr/a_0}$

$$\int_0^\infty r^n e^{-\beta r} dr = \frac{n!}{\beta^{n+1}}$$

and the radial distribution function is $4\pi r^2 \left(\dfrac{Z^3}{\pi a_0^3}\right) e^{-2Zr/a_0}$, *i.e.* $4\pi r^2\psi^2$.

The most probable value of r is obtained from the relation

$$\frac{d}{dr}\left[4\pi r^2\psi^2\right] = 0$$

i.e.

$$\frac{d}{dr}\left[4\pi r^2\left(\frac{Z^3}{\pi a_0^3}\right)e^{-2Zr/a_0}\right] = 0$$

$$4\pi\left(\frac{Z^3}{\pi a_0^3}\right)\left[2re^{-2Zr/a_0} - \frac{2Z}{a_0}r^2e^{-2Zr/a_0}\right] = 0$$

i.e.

$$r_{mp} = \frac{a_0}{Z}$$

where $a_0 = 53$ pm. The above relation shows that as Z increases, the r_{mp} will decrease.

3.7. MULTI-ELECTRONIC ATOMS

3.7.1. *Electron Spin Wave Functions*

Each electron in an atom can spin around its axis either clockwise or anticlockwise. The common notations for the spins are α and β. The former is having the quantum number + 1/2 and the latter −1/2. The corresponding spin angular momentum is a multiple of $h/2\pi$.

3.7.2. *Hamiltonian Operator and Wave Function*

The Hamiltonian operator for n electrons, present in an atom can be written as

$$\hat{H} = -\frac{h^2}{8\pi^2 m_e}\sum_{i=1}^{n}\nabla_i^2 - \sum_{i=1}^{n}\frac{Ze^2}{(4\pi\varepsilon_0)r_i} + \sum_{\substack{j \\ j>i}}^{n}\frac{e^2}{(4\pi\varepsilon_0)r_{ij}} \qquad ...(3.33)$$

r_{ij} is the distance between the electrons i and j. The other terms have the same meanings as explained earlier. The summation is to be carried over all the electrons.

The Schrödinger equation for n electron system cannot be separated into the equations containing only one variable (like hydrogen atom) due to the presence of r_{ij} terms. The approximate methods, *e.g.* neglect of the r_{ij} terms, self-consistent field theory, Hartree-Fock procedures, etc. are used to solve the Schrödinger equation.

Equation (3.33) consists of three terms. First one gives the kinetic energy of i electrons, second term describes the summation of attractive potential between each electron and nucleus, whereas the third term gives the repulsive potential among the electrons i and j summed over each electron-electron combination. The value of $j > i$ avoids the counting of repulsive terms twice. If there are N electrons in an atom, there will be (*i*) $N + 1$ kinetic energy terms, N from the electrons and one from the nucleus and (*ii*) N terms for the attractive potential. The number of terms in third summation involves pairwise repulsive interactions and the number of combinations of N electrons taken two at a time is given by

$$C(N, 2) = \frac{N!}{(N-2)!2!} = \frac{N(N-1)}{2} \qquad ...(3.34)$$

***Exercise* 3.19:** Write down the Hamiltonian operator for the oxygen atom.

In oxygen the number of electrons is eight. So there will be nine kinetic energy terms, eight attractive potential energy terms and the repulsive terms will be $\frac{8 \times 7}{2} = 28$. The Hamiltonian operator can be written as

$$-\frac{h^2}{8\pi^2 m_{nuc}}\nabla_n^2 - \frac{h^2}{8\pi^2 m_e}\sum_{i=1}^{8}\nabla_i^2 - \sum_{i=1}^{8}\frac{Ze^2}{r_i} + \sum_{i=1}^{8}\sum_{j=i+1}^{8}\frac{e^2}{r_{ij}}$$

The fourth term can be written as

$$\frac{e^2}{r_{12}}+\frac{e^2}{r_{13}}+\frac{e^2}{r_{14}}+\frac{e^2}{r_{15}}+\frac{e^2}{r_{16}}+\frac{e^2}{r_{17}}+\frac{e^2}{r_{18}}+\frac{e^2}{r_{23}}+\frac{e^2}{r_{24}}+\frac{e^2}{r_{25}}+\frac{e^2}{r_{26}}+\frac{e^2}{r_{27}}+\frac{e^2}{r_{28}}+\frac{e^2}{r_{34}}+\frac{e^2}{r_{35}}+\frac{e^2}{r_{36}}+$$

$$\frac{e^2}{r_{37}}+\frac{e^2}{r_{38}}+\frac{e^2}{r_{45}}+\frac{e^2}{r_{46}}+\frac{e^2}{r_{47}}+\frac{e^2}{r_{48}}+\frac{e^2}{r_{56}}+\frac{e^2}{r_{57}}+\frac{e^2}{r_{58}}+\frac{e^2}{r_{67}}+\frac{e^2}{r_{68}}+\frac{e^2}{r_{78}}$$

There are twenty-eight repulsive terms, agrees with the number of terms obtained from equation (3.34).

3.7.3. *One Electron Wave Functions*

From equation (3.33), we have seen that it is not possible to separate out the variables and it represents the wave functions of each electron as a function of its coordinates. Using crude approximation, *i.e.*, neglecting kinetic energy of nucleus and the electron repulsion terms (or in other words the Hamiltonian involves only the kinetic energy and attractive potential terms), equation (3.33) reduces to

$$\hat{H} = -\frac{h^2}{8\pi^2 m_e}\sum_{i=1}^{n}\nabla_i^2 - \sum_{i=1}^{n}\frac{Ze^2}{(4\pi\varepsilon_0)r_i} \qquad ..(3.35)$$

Equation (3.35) can also be written as sum of the terms belonging to each electron, just like hydrogen atom or hydrogen like ions

$$\hat{H} = \left(-\frac{h^2}{8\pi^2 m_e}\nabla_i^2 - \frac{Ze^2}{(4\pi\varepsilon_0)r_i}\right) \qquad ...(3.36)$$

This also suggests that the motion of each electron is independent of each other and each electron is observing the complete nuclear charge, *i.e.*, *Ze*. Under these situations the total wave function can be written as the product of the wave functions of individual electron, *i.e.*,

$$\Psi(r_1,\theta_1,\varphi_1, r_2,\theta_2,\varphi_2,......,r_n,\theta_n,\varphi_n) = \psi(r_1,\theta_1,\varphi_1)\psi(r_2,\theta_2,\varphi_2)............\psi(r_n,\theta_n,\varphi_n) \qquad ...(3.37)$$

and total energy will be given by

$$E_n(\text{J}) = -\sum_{i=1}^{n} \frac{2\pi^2 \mu e^4 Z^2}{(4\pi\varepsilon_0)^2 h^2 n_i^2} \qquad \text{...(3.38}a\text{)}$$

$$E_n(\text{hartree}) = -\sum_{i=1}^{n} \frac{Z^2}{2n_i^2} \qquad \text{...(3.38}b\text{)}$$

***Exercise* 3.20:** (*a*) Calculate the ground state energy of helium atom and (*b*) calculate the ionisation energy of helium atom.

(*a*) Helium atom has two electrons. Using equation (3.38 *b*), the total energy of the helium atom in its ground state will be

$$E_n(\text{hartree}) = -\frac{2^2}{2(1)^2} - \frac{2^2}{2(1)^2} = -4 \text{ hartrees} = -(4 \text{ hartree})(27.21 \text{ eV hartree}^{-1})$$

$$= -108.84 \text{ eV}$$

which is very large negative as compared to experimentally determined quantity.

(*b*) The first ionisation energy for helium atom is the energy required to remove an electron from its neutral atom, *i.e.*,

$$\text{He} \rightarrow \text{He}^+ + \text{e}^-$$

This energy is the energy difference between He and He^+, *i.e.*,

$$I(1) = E(\text{He}^+) - E(\text{He})$$

Now $E(He^+)$ can be calculated using the one electron hydrogen atom like model, *i.e.*,

$$\text{E}_{\text{He}^+}(\text{hartree}) = -\frac{2^2}{2(1)^2} = -2 \text{ hartrees} = -(2 \text{ hartree})(27.21 \text{ eV hartree}^{-1}) = -54.42 \text{ eV}$$

$$I(1) = E(\text{He}^+) - E(\text{He}) = -54.42 \text{ eV} - (-108.84 \text{ eV}) = 54.42 \text{ eV}.$$

which is very large as compared to the experimentally determined quantity, *i.e.*, 24.58 eV.

This clearly suggests that one electron model is not a correct model to depict the structure of multi-electron model of atoms.

3.7.4. *Electron Shielding*

In we take electron repulsion interactions, electrons cannot be treated as independent entity and repulsion terms between the electrons have to be included. Further when electrons are filled in different orbitals, each electron shields the nuclear charge coming before the first one. For example in helium atom there are two electrons. If we bring the first electron from infinity, it will experience complete nuclear charge of 2 units of positive charge and will lead to He^+ species. Now when we bring the second electron from infinity and if the first electron has completely shielded the nucleus the second electron will experience only one unit of positive charge. In reality when both the electrons are present in its equilibrium positions, *i.e.*, in its ground state, the net positive charge seen by the second electron will be much closer to +2e since not all of the electron density of electron 1 will lie between the second electron and the nucleus. In other words, first electron will not be completely shielding one proton from the second electron, *i.e.*, charge seen by each electron will be defined as effective nuclear charge, Z_{eff}. Using perturbation theory, the variation for the effective nuclear charge felt by the second electron in case of helium atom is given by

$$Z_{eff} = 1 + (2r_2 + 1)e^{-4r_2} \quad ...(3.39)$$

where r_2 is the distance of electron 2 from the nucleus in terms of Bohr's radius. The energy eigen value is given by

$$E_n(\text{hartree}) = -\frac{Z_{eff}^2}{2n^2} \quad ...(3.40)$$

where n is the principal quantum number and Z_{eff} is the effective nuclear charge felt by the electro. So one can calculate the Z_{eff} if we know the E_n.

***Exercise* 3.21:** Calculate the ground state energy of He atom as well as the Z_{eff} if the ionisation energy of He atom is 24.59 eV or 0.9037 hartree. Compare the value of Z_{eff} with that complied in Table 3.4.

Using equation (3.40) and since there are two electrons in He atom and both are present in 1 s orbital, $n = 1$, we get

$$E(\text{He}) = 2\left(-\frac{Z_{eff}^2}{2n^2}\right) = 2\left(-\frac{Z_{eff}^2}{2(1)^2}\right) = -Z_{eff}^2 \text{ (hartree)} \quad ...(A)$$

The ionisation process involves

$$\text{He} \rightarrow \text{He}^+ + e^- \quad ...(B)$$

Thus $$I = E(\text{He}^+) - E(\text{He}) \quad ...(C)$$

Using equation (3.40) the energy of the $E(\text{He}^+)$ will be

$$E(\text{He}^+) = \left(-\frac{2^2}{2(1)^2}\right) = -2 \text{ (hartree)} \quad ...(D)$$

Combining equations (A), (B), (C) and (D), we get

$$0.937 = -2 - (-Z_{eff}^2)$$

$$Z_{eff} = \sqrt{2.9037} = 1.704$$

The other term commonly used is the shielding constant (σ), which is defined as

$$Z_{eff} = Z - \sigma \quad ...(3.41)$$

The shielding constant in case of He atom is

$$\sigma = 2 - 1.704 = 0.296.$$

From Table 3.4, the value of parameter $\varsigma = 1.6875 = \frac{Z_{eff}}{n}$. Since $n = 1$ for He, $Z_{eff} = 1 \times 1.6875 = 1.6875$, which is very close to 1.704.

Slater has given the empirical rules to calculate the shielding constant, which are given in Section 3.7.5, whereas Clementi and Raimodni have reported the results obtained by variation calculations. These values in terms ζ and up to krypton have been complied in Table 3.4. The orbitals with the same value of n but different values of l form the sub-shells of a given shell. These sub-shells are generally designated by the letters s, p, d, …….. as

The orbitals with the same value of n but different values of l form the sub-shells of a given shell. These sub-shells are generally designated by the letters s, p, d, …….. as

TABLE 3.4. Effective nuclear in charges in terms of $\varsigma = \frac{Z_{\text{eff}}}{n}$ for ground state neutral atoms obtained by variational calculations

Atom	Z	1s	2s	2p	3s	3p	4s	3d	4p
He	2	1.6875							
Li	3	2.6906	0.6396						
Be	4	3.6848	0.9560						
B	5	4.695	1.2881	1.2107					
C	6	5.6727	1.6083	1.5679					
N	7	6.6651	1.9237	1.9170					
O	8	7.6579	2.2458	2.2266					
F	9	8.6501	2.5638	2.5506					
Ne	10	9.6421	2.8792	2.8792					
Na	11	10.6259	3.2857	3.4009	0.8358				
Mg	12	11.6089	3.6960	3.9129	1.1025				
Al	13	12.5910	4.1068	4.4817	1.3724	1.3552			
Si	14	13.5578	4.5100	4.9725	1.6344	1.4284			
P	15	14.5409	4.9125	5.4806	1.8806	1.6288			
S	16	15.5409	5.3144	5.9885	2.1223	1.8273			
Cl	17	16.5239	5.7152	6.4966	2.3561	2.0387			
Ar	18	17.5075	6.1152	7.0041	2.5856	2.2547			
K	19	18.4895	6.5031	7.5136	2.8933	2.5752	0.8738		
Ca	20	10.4730	6.8882	8.0207	3.2005	2.8861	1.0995		
Sc	21	20.4566	7.2868	8.5273	3.4466	3.1354	1.1581	2.2733	
Ti	22	21.4409	7.6883	9.0324	3.6777	3.3679	1.2042	2.7138	
V	23	22.4256	8.0907	9.564	3.9031	3.5950	1.2453	2.9943	
Cr	24	23.4138	8.4919	10.0376	4.1226	3.8220	1.2833	3.2522	
Mn	25	24.3957	8.8969	10.5420	4.3393	4.0364	1.3208	3.5094	
Fe	26	25.3810	9.2995	11.0444	4.5587	4.2593	1.3585	3.7266	

Contd...

Co	27	26.3668	9.7025	11.5462	4.7741	4.4782	1.3941	3.9518	
Ni	28	27.3526	10.1063	12.0476	4.9870	4.6950	1.4277	4.1765	
Cu	29	28.3386	10.5099	12.5485	5.1981	4.9102	1.4606	4.4002	
Zn	30	29.3245	10.9140	13.0490	5.4064	5.1231	1.4913	4.261	
Ga	31	30.3094	11.2995	13.5454	5.6654	5.4012	1.7667	5.0311	1.5554
Ge	32	31.2937	11.6824	14.0411	5.9299	5.6712	2.0109	5.4171	1.6951
As	33	32.2783	12.0635	14.5368	6.1985	5.9499	2.2360	5.7928	1.8623
Se	34	33.2622	12.4442	15.0326	6.4678	6.2350	2.4394	6.1590	2.0718
Br	35	34.2471	12.8217	15.5282	6.7395	6.5236	2.6382	6.5187	2.2570
Ar	36	35.2316	13.1990	16.0235	7.0109	6.8114	2.8289	6.8753	2.4423

3.7.5. *Slater's Rules to Calculate Shielding Constant*

Slater has given certain rules to calculate the shielding constant (σ) of the nucleus. The effective nuclear charge is given by equation (3.41*a*):

$$Z_{\text{eff}} = Z - \sigma \qquad \text{...(3.41}a\text{)}$$

where Z is the actual atomic number. These rules are:

(*a*) Divide atomic orbitals into the following groups:

(1*s*); (2*s*, 2*p*); (3*s*, 3*p*); (3*d*); (4*s*, 4*p*); (4*d*); (4*f*); (5*s*, 5*p*); (5*d*)

(*b*) The contribution to the shielding constant σ, from different electrons present in the atom, is obtained from the following set of instructions:

(*i*) From an orbital in a group, orbital outside the one (*i.e.*, coming after) considered is zero, *i.e.*, consider an orbital 3*s*, the contribution to the shielding constant from 3*d*, 4*s*, 4*p* etc. will be zero.

(*ii*) 0.35 from each other electron in the group considered, except that 0.3 is used in the 1*s* group.

(*iii*) For an *s* and *p* orbital, 0.85 from each electron with *n* one less than the orbital considered and 1.0 from each electron still further in.

(*iv*) For a *d* orbital, 1.0 for each electron inside the group.

***Exercise* 3.22:** Plot the potential energy curve in case of helium atom under the conditions (*i*) when the second electron is at infinity, (*ii*) when the second electron is close to nucleus and (*iii*) when the electron experience the effective nuclear charge given by equation (3.39).

(*i*) When the second electron is at infinity, it means that, $r_2 = \infty$, Z_e obtained from equation (3.39) will be equal to unity. The potential energy is given by $V(r_2) = -e^2/r_2$.

(*ii*) When the second electron is closed to nucleus, $r = 0$ and this will lead to Ze equal to two. The potential energy is given by $V(r_2) = -2e^2/r_2$.

Thus potential energies obtained under the conditions (*i*), (*ii*) and (*iii*) are plotted in Fig. (3.4).

(*iii*) In this case the Z_e varies as given by equation (3.39) and thus $V(r_2) = Z_e\, e^2/r_2$.

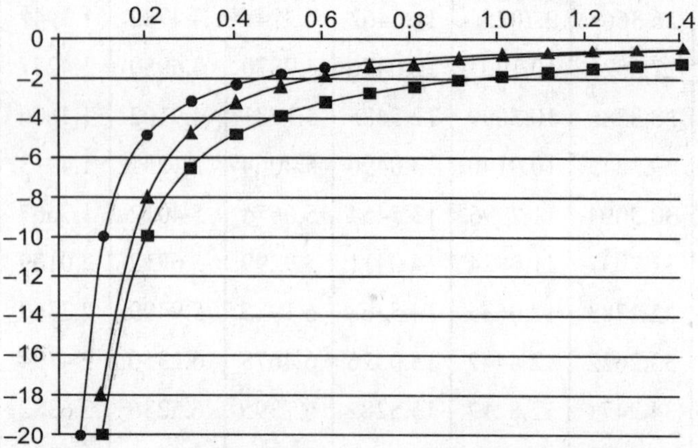

Figure 3.4: *The effect of shielding in the helium atom, i.e. potential experienced by the second 1s electron is plotted as a function of the distance of the electron from the nucleus. 1; $V(r_2) = -1/r_2$, 3; $V(r_2) = -2/r$, 2; $V(r_2)$*

$$V(r_2) = -\frac{1+[2r_2+1]e^{-4r_2}}{r_2}$$

***Exercise* 3.23:** Calculate the effective nuclear charge for: (*a*) N, (*b*) Na, (*c*) S, (*d*) Zn.

(*a*) For N, $Z = 7$

$$Z_{eff}(1s) = 7 - 0.3 = 6.7.$$

$$Z_{eff}(2s) = 7 - (2 \text{ electrons from } 1s) - (1 \text{ electron from } 2s)$$

$$= 7 - 2 \times 0.85 - 1 \times 0.35 = 7 - 1.7 - 0.35 = 4.95$$

$$Z_{eff}(2p) = 7 - 2 \times 0.85 - 4 \times 0.35 = 3.90$$

(*b*) For Na, $Z = 11$

$$Z_{eff}(3s) = 11 - (2 \text{ electrons from } 1s) - (8 \text{ electrons from } 2s \text{ and } 2p)$$

$$= 11 - 2 \times 1.0 - 8 \times 0.85 = 11 - 2.0 - 6.8 = 2.2$$

(*iii*) For S, $Z = 16$

$$Z_{eff}(3s) = 16 - (2 \text{ electrons from } 1s) - (8 \text{ electrons from } 2s \text{ and } 2p) - (5 \text{ electrons from } 2s \text{ and } 2p)$$

$$= 16 - 2 \times 1 - 8 \times 0.85 - 5 \times 0.35 = 16 - 2 - 6.8 - 1.75 = 5.45$$

(*iv*) For Zn, $Z = 30$

$$Z_{eff}(4s) = 30 - (2 \text{ electrons from } 1s) - (8 \text{ electrons from } 2s \text{ and } 2p) - (8 \text{ electrons from } 3s \text{ and } 3p) - (10 \text{ electrons from } 3d) - (\text{one electron from } 4s)$$

$$= 30 - 2 \times 1 - 8 \times 1 - 8 \times 0.85 - 10 \times 0.85 - 1 \times 0.35$$

$$= 30 - 2 - 6.8 - 8.5 - 0.35$$

$$= 4.35$$

$$Z_{eff}(3d) = 30 - (2 \text{ electrons from } 1s) - (8 \text{ electrons from } 2s \text{ and } 2p) - (8 \text{ electrons from } 3s \text{ and } 3p) - (9 \text{ electrons from } 3d)$$

$$= 30 - 2 \times 1 - 8 \times 1 - 8 \times 1.0 - 9 \times 0.35 = 30 - 2 - 8 - 8 - 3.15 = 8.85.$$

***Exercise* 3.24:** First ionisation energy of Li is 5.39 eV. (*i*) Calculate the effective nuclear charge using Slater's rules, (*ii*) energy of the 2*s* level and effective nuclear charge using ionisation energy. (*iii*)

Also compare Z_{eff} obtained from Table 3.4 and (*iv*) calculate the second ionisation energy of Li atom.

(*i*) There are two electrons in the 1*s* orbital of lithium. Using Slater's rules the shielding constant is

$$\sigma = 2 \times 0.85 = 1.7$$

Thus $$Z_{eff} = 3 - 1.7 = 1.3$$

(*ii*) The energy of one electron system in multi-electron systems is given by

$$E_n(\text{hartree}) = -\frac{Z_{eff}^2}{2n^2} \qquad ...(3.40)$$

Since ionisation energy is the energy supplied to ionise the electron it will have positive value,

$$I\,(\text{hartree}) = \frac{Z_{eff}^2}{2n^2} \qquad ...(3.42)$$

The principal quantum number of all the atoms in second period is equal to two so the effective nuclear charge is given as

$$Z_{eff} = \sqrt{I \times 2n^2} = \sqrt{2I}\ n \qquad ...(3.43)$$

$$Z_{eff} = \sqrt{2I}\ n = \left(\sqrt{2\left(\frac{5.39}{27.42}\right)}\right)(2) = 1.254$$

(*iii*) From Table 3.4, $\zeta = 0.6936 = Z_{eff}/n$, so $Z_{eff} = 2 \times 0.6396 = 1.2792$

The value of Z_{eff} calculated by three methods are quite close to each.

Note: For simple purposes Slater's rules are still preferred.

(*iv*) The chemical equation for the second ionisation energy is written as

$$Li^+ \rightarrow Li^{2+}$$

Thus $$I_2 = E(Li^{2+}) - E(Li^+) \qquad ...(i)$$

Energy of $E(Li^{2+})$ can easily be calculated as Li^{2+} is a one electron system, *i.e.*, using equation (3.40):

$$E_{1s}\,(Li^{2+}) = -\frac{Z_{eff}^2}{2n^2} - \frac{3^2}{2(1)^2} = -4.5 \text{ hartree} \qquad ...(ii)$$

From Table 2.4, ζ for 1*s* orbital of Li is 2.6906, so $Z_{eff} = 2.6906 \times 1 = 2.6906$

$$E_{1s}\,(Li^{2+}) = 2\left(-\frac{Z_{eff}^2}{2n^2}\right) = -2\left(\frac{2.6906^2}{2(1)^2}\right) = -7.2393 \text{ hartree} \qquad ...(iii)$$

Substituting (*ii*) and (*iii*) in (*i*) we get

$I_2 = -4.5 - (-7.2393) = 2.7393$ hartree $= (2.7393$ hartree$)(27.211$ eV hartree$^{-1}) = 74.534$ eV

$$\text{Per cent error} = \frac{75.64 - 74.534}{75.64} \times 100 = 14.6\%$$

3.7.6. *Pauli Exclusion Principle and Representation of Wave Function*

According to Pauli's exclusion principle, no two electrons in multi-electron atoms can have all the four quantum numbers (principal, azimuthal, magnetic and spin quantum numbers) same.

This can also be stated in other manner as "only states whose wave functions are antisymmetric with respect to interchange of any pair of electrons are allowed." It can also be stated that a wave function must be totally antisymmetric with respect to interchange of any other particles with half integral spin and totally symmetric with respect to the interchange of particles with integral spin. This type of wave function can be expressed by Slater's determinant.

$$\psi = \frac{1}{\sqrt{n!}}\begin{vmatrix} \phi_{1s}(1)\alpha(1) & \phi_{1s}(2)\alpha(2)\ldots & \phi_{1s}(n)\alpha(n) \\ \phi_{1s}(1)\beta(1) & \phi_{1s}(2)\beta(2)\ldots & \phi_{1s}(n)\beta(n) \\ \phi_{2s}(1)\alpha(1) & \phi_{2s}(2)\alpha(2)\ldots & \phi_{2s}(n)\alpha(n) \\ \ldots & \ldots & \ldots \\ \ldots & \ldots & \ldots \\ \ldots & \ldots & \ldots \end{vmatrix} \quad \ldots(3.44)$$

where $\phi_{1s}(1)\alpha(1)$ represents the product of the hydrogen-like wave function of 1 *s* orbital containing electron labelled 1 and the spin wave function α for the electron labelled 1 and so on. The number of columns correspond to the number of electrons and the number of rows to the lowest quantum states. The above determinant satisfies the antisymmetric nature of wave function (*i.e.* change of coordinates means the change of columns and this means the change in sign).

***Exercise* 3.25:** Write the Hamiltonian for He atom and also the wave function for the ground state.

Helium atom contains two electrons and let their position be represented as shown in Fig. 3.5. The Hamiltonian for helium atom can be written as

$$\hat{H} = -\frac{h^2}{8\pi^2 m_e}\left(\nabla_1^2 + \nabla_2^2\right) - \frac{2e^2}{r_1} - \frac{2e^2}{r_2} + \frac{e^2}{r_{1,2}}$$

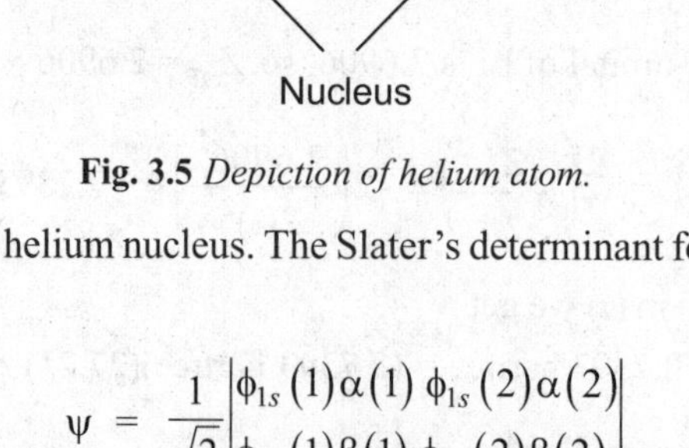

Fig. 3.5 *Depiction of helium atom.*

where + 2*e* is the charge on the helium nucleus. The Slater's determinant for helium atom can be written as

$$\psi = \frac{1}{\sqrt{2}}\begin{vmatrix} \phi_{1s}(1)\alpha(1) & \phi_{1s}(2)\alpha(2) \\ \phi_{1s}(1)\beta(1) & \phi_{1s}(2)\beta(2) \end{vmatrix}$$

The above determinant can be expanded as

$$\psi = \frac{1}{\sqrt{2}}\left[\phi_{1s}(1)\phi_{1s}(2)\alpha(1)\beta(2) - \phi_{1s}(1)\phi_{1s}(2)\alpha(2)\beta(1)\right]$$

3.7.7. *Energy Levels and Auf-Bau Principle*

Unlike hydrogen or hydrogen-like atoms energies of the orbitals do not depend only on the principal quantum number but also indirectly on the azimuthal quantum number. This is due to the presence of other electrons in the orbitals and their interaction with each other. The general order of the energies of the orbitals are $1s\,(1) < 2s(1) < 2p(3) < 3s(1) < 3p(3) < 4s(1) < 3d(5) < 4p(3) < 5s(1) < 4d(5) < 5p(3) < 6s(1) < 4f(7) < 5d\,(5) < 6p(3) < 7s(1) < 5f(7) = 6d(5) < 7p(3)$ etc. The numbers in the parentheses give number of orbitals in that subshells.The number of electrons those can occupy any orbital is given by Pauli's principle, *i.e.* two with opposite spin. The number of electrons present in any orbital is represented as nl^i where n is the principal quantum number, l is the azimuthal quantum number of that orbital and i is the number of electrons present in that orbital, *i.e.* $4p^2$ means that there are 2 electrons in $4p$ orbital. According to Pauli's principle, two electrons must have two different combinations of $m = \pm 1$, 0 and $s = \pm 1/2$.

The electrons in various orbitals are filled according to Hund's rules, *i.e.*

(*i*) Maximum two electrons can be placed in any orbital.

(*ii*) Electrons are placed in the orbitals according to their energies. Lowest is filled first and rest according to the scheme mentioned above.

(*iii*) In the degenerate orbitals, the electrons with parallel spin are placed singly until it is necessary to place a second electron.

(*iv*) As said later (Section 3.9), the orbitals with the most positive values of m and of s are lowest in energy and used first, *e.g.* $4p^2$ configuration means $m = +1$, $s = +1/2$, $m = 0$, $s = +1/2$.

(*v*) Further, half-filled and filled d and f subshells are more stable than other configurations. So slight rearrangement of electrons takes place from the normal arrangement, e.g. the electronic configuration of Cu (at. no. 29) is $1s^2\,2s^2\,2p^6\,3s^2\,3p^6\,3d^{10}\,4s^1$ rather than $1s^2\,2s^2\,2p^6\,3s^2\,3p^6\,3d^9\,4s^2$.

***Exercise* 3.26:** Write the electronic configuration of (*a*) Cr(24), (*b*) N(7), (*c*) Xe(54), (*d*) Hg(80), (*e*) Fe^{3+}(23), (*f*) Hg^{2+}(78):

(*a*) $1s^2\,2s^2\,2p^6\,3s^2\,3p^6\,3d^5\,4s^1$

(*b*) $1s^2\,2s^2\,2p^3$

(*c*) $1s^2\,2s^2\,2p^6\,3s^2\,3p^6\,3d^{10}\,4s^2\,4p^6\,4d^{10}\,5s^2\,5p^6$

(*d*) Xe $6s^2\,4f^{14}\,5d^{10}$ or Xe $5d^{10}\,4f^{14}\,6s^2$

(*e*) $1s^2\,2s^2\,2p^6\,3s^2\,3p^6\,3d^5$

(*f*) Xe $5d^{10}\,4f^{14}$

Note: In case of ions, the electrons are placed in different orbitals according to the increasing order of n and not as mentioned earlier.

Students do counter difficulty in remembering the above order of energy levels. The most common method, commonly called the *uncle Wiggling path*, is shown in Fig. 3.6 (*a*). We should follow the path as indicated by arrows in Fig. 3.6 (*a*). Recently a better method has also been suggested and procedure is as follows. (*i*). Draw a grid of 7 × 7 squares. (*ii*) Label the diagonal with known *s* orbital as shown in Fig. 3.6 (*b*), (*iii*) Fill in each row with all possible orbitals allowed for that energy level (or for that *n*). (*iv*) Read the orbital sequence by following the columns down beginning at the left. If all the orbitals in each column are written before advancing to the next column, the proper filling order is determined.

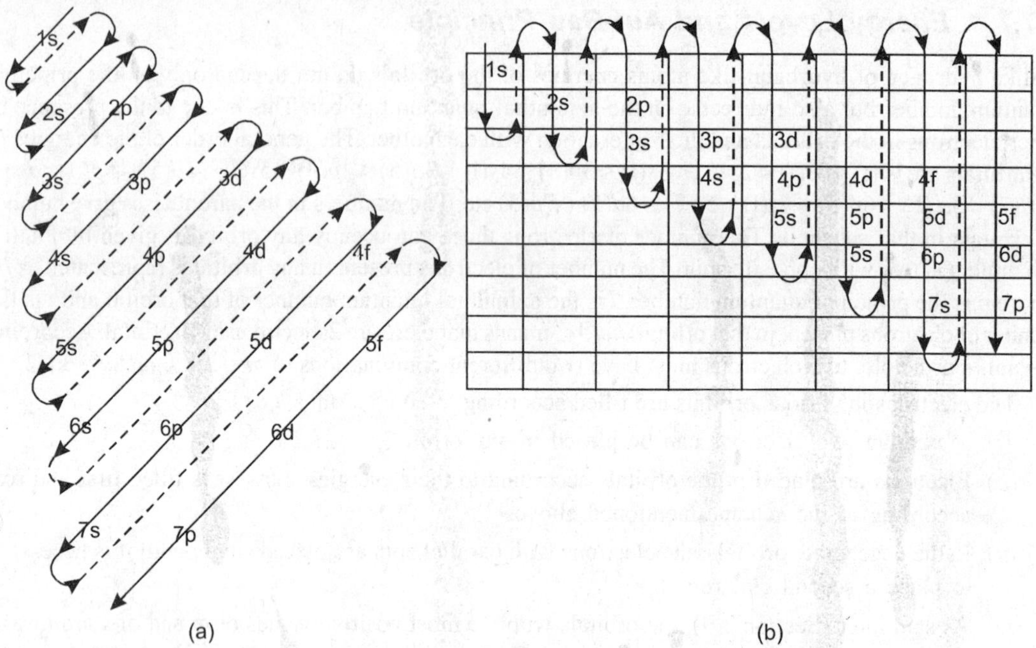

Fig. 3.6 *(a) Uncle Wiggly path for filling order or orbitals, (b) The orbital sequence by following the columns down beginning at the left.*

***Exercise* 3.27:** The electronic configuration for the ground state of Be can be(*i*)$1s^2 2s^2$, (*ii*) $1s^2 2s^1 2p^1$ or (*iii*) $1s^2 2p^2$. Indicate which one is more acceptable configuration based on the ground state energy.

Using Slater's rules one can calculate the effective nuclear charge attached to each orbital. Further shielding constant is same for $1s$ orbital in each configuration. The atomic number of Be is 4. The effective nuclear charge (Z_{eff}) on $1s$, $2s$ and $2p$ are:

$$Z_{eff}(1s) = 4 - (1)(0.3) = 3.7$$

$$Z_{eff}(2s) = 4 - (2 \text{ electrons from } 1s) - 1 \text{ electron from } 2s$$
$$= 4 - 2 \times 0.85 - 1 \times 0.35 = 4 - 1.7 - 0.35 = 1.95$$

$$Z_{eff}(2p) = 4 - 2 \times 0.85 - 3 \times 0.35 = 1.25$$

$$E(2s) = -\frac{Z_{eff}^2}{2n^2} = -\frac{(1.95)^2}{2(2)^2} = -0.475 \text{ hartree} = (-0.475 \text{ hartree})(27.42 \text{ eV hartree}^{-1})$$
$$= -13.0245 \text{ eV}$$

$$E(2p) = -\frac{Z_{eff}^2}{2n^2} = -\frac{(1.25)^2}{2(2)^2} = -0.195 \text{ hartree} = (-0.195 \text{ hartree})(27.42 \text{ eV hartree}^{-1})$$
$$= -5.356 \text{ eV}$$

Total energy for each configuration is

$$E(1s^2 2s^2) = E(1s^2) + 2 \times (-13.0245 \text{ eV}) = E(1s^2) - 26.049 \text{ eV}$$

$$E(1s^2 2s^1 2p^1) = E(1s^2) + 1 \times (-13.0245 \text{ eV}) + 1 \times (-5.356 \text{ eV}) = E(1s^2) - 18.3805 \text{ eV}$$

$$E(1s^2 2p^2) = E(1s^2) + 2 \times (-5.356 \text{ eV}) = E(1s^2) - 10.712 \text{ eV}$$

From the above data it is clear that the most stable configuration, *i.e.*, the configuration with most negative value is $1s^2 2s^2$.

***Exercise* 3.28:** Predict qualitatively the stable configuration of carbon atom, *i.e.*, $1s^2 2s^2 2p_{+1}^2$ or $1s^2 2s^2 2p_{+1}^1 2p_0^1$.

The subscript gives the magnetic quantum numbers. Since $2p$ atomic orbitals with values of magnetic quantum number are all degenerate, we could have used any of the three orbitals. All are equivalent. The answer is that the electrons tend to occupy different degenerate spatial orbitals with same spin function if permitted by the Pauli exclusion principle. The reason is that this configuration minimises the destabilisation repulsive interaction between negatively charged electrons. If the electrons occupy different orbitals, the average distance between them will be greater and their mutual repulsion is minimised. Since the one electron approximation ignores the electron-electron repulsions, there is no way for us to predict this result quantitatively as one requires more sophisticated theories and it is beyond the scope of this book.

The electronic configuration and some of the important properties of the elements are compiled in Table 3.5.

TABLE 3.5. Some atomic properties of the elements

Atomic number	Element	Configuration	Ground state term symbol	First ionization energy(eV)	Covalent radius (pm)	Atomic radius (pm)	Electron affinity kJmol^{-1}
1	H	1s	$^2S_{1/2}$	13.595	32	79	72.8
2	He	$1s^2$	$^1S_0$	24.580	93	49	– 21
3	Li	$[He]2s^1$	$^2S_{1/2}$	5.390	123	205	59.8
4	Be	$[He]2s^2$	$^1S_0$	9.320	90	140	– 18
5	B	$[He]2s^2 2p^1$	$^2P_{1/2}$	8.296	82	117	23
6	C	$[He]2s^2 2p^2$	$^3P_0$	11.264	77	91	122.5
7	N	$[He]2s^2 2p^3$	$^4S_{3/2}$	14.54	73	75	– 7
8	O	$[He]2s^2 2p^4$	$^3P_2$	13.614	73	65	141
9	F	$[He]2s^2 2p^5$	$^2P_{3/2}$	17.420	72	57	322
10	Ne	$[He]2s^2 2p^6$	$^1S_0$	21.559	71	51	– 29
11	Na	$[Ne]3s^1$	$^2S_{1/2}$	5.138	154	233	52.9
12	Mg	$[Ne]3s^2$	$^1S_0$	7.644	136	172	– 21

Contd...

13	Al	$[Ne]3s^23p^1$	$^2P_{1/2}$	5.984	118	182	44
14	Si	$[Ne]3s^23p^2$	$^3P_0$	8.149	111	146	133.6
15	P	$[Ne]3s^23p^3$	$^4S_{3/2}$	11.000	106	123	71.7
16	S	$[Ne]3s^23p^4$	$^3P_2$	10.357	102	109	200.4
17	Cl	$[Ne]3s^23p^5$	$^2P_{3/2}$	13.010	99	97	348.7
18	Ar	$[Ne]3s^23p^6$	$^1S_0$	15.755	98	88	– 35
19	K	$[Ar]4s^1$	$^2S_{1/2}$	4.339	203	277	48.3
20	Ca	$[Ar]4s^2$	$^1S_0$	6.111	174	223	– 188
21	Sc	$[Ar]4s^23d^1$	$^2D_{3/2}$	6.560	144	209	
22	TI	$[Ar]4s^23d^2$	$^3F_2$	6.830	132	200	
23	V	$[Ar]4s^23d^3$	$^4F_{3/2}$	6.740	122	192	
24	Cr	$[Ar]4s^13d^5$	$^7S_3$	6.736	118	185	
25	Mn	$[Ar]4s^23d^5$	$^6S_{5/2}$	7.432	117	179	
26	Fe	$[Ar]4s^23d^6$	$^5D_4$	7.900	117	172	
27	Co	$[Ar]4s^23d^7$	$^4F_{9/2}$	7.860	116	167	
28	Ni	$[Ar]4s^23d^8$	$^3F_4$	7.633	115	162	
29	Cu	$[Ar]4s^13d^{10}$	$^2S_{1/2}$	7.724	117	157	
30	Zn	$[Ar]4s^23d^{10}$	$^1S_0$	9.391	125	153	
31	Ga	$[Ar]4s^23d^{10}4p^1$	$^2P_{1/2}$	6.000	126	181	36
32	Ge	$[Ar]4s^23d^{10}4p^2$	$^3P_0$	7.880	122	152	116
33	As	$[Ar]4s^23d^{10}4p^3$	$^4S_{3/2}$	9.810	120	133	77
34	Se	$[Ar]4s^23d^{10}4p^4$	$^3P_2$	9.750	116	122	195
35	Br	$[Ar]4s^23d^{10}4p^5$	$^2P_{3/2}$	11.840	114	112	324.5
36	Kr	$[Ar]4s^23d^{10}4p^6$	$^1S_0$	13.996	112	103	– 39
37	Rb	$[Kr]5s^1$	$^2S_{1/2}$	4.176	216	298	46.9
38	Sr	$[Kr]5s^2$	$^1S_0$	5.692	191	245	– 146
39	Y	$[Kr]5s^24d^1$	$^2D_{3/2}$	6.500	162	227	
40	Zr	$[Kr]5s^24d^2$	$^3F_2$	6.950	145	216	
41	Nb	$[Kr]5s^24d^3$	$^4F_{3/2}$	6.770	134	208	
42	Mo	$[Kr]5s^14d^5$	$^7S_3$	7.100	130	201	
43	Tc	$[Kr]5s^24d^5$	$^6S_{5/2}$	7.280	127	195	
44	Ru	$[Kr]5s^24d^6$	$^5D_4$	7.364	125	189	

Contd...

45	Rh	[Kr]$5s^24d^7$	$^4F_{9/2}$	7.460	125	183	
46	Pd	[Kr]$5s^24d^8$	$^3F_4$	8.330	128	179	
47	Ag	[Kr]$5s^14d^{10}$	$^2S_{1/2}$	7.574	134	175	
48	Cd	[Kr]$5s^24d^{10}$	$^1S_0$	8.991	148	171	
49	In	[Kr]$4d^65s^25p^1$	$^2P_{1/2}$	5.785	144	205	34
50	Sn	[Kr]$4d^65s^25p^2$	$^3P_0$	7.342	141	172	121
51	Sb	[Kr]$4d^65s^25p^3$	$^4S_{3/2}$	8.639	140	153	101
52	Te	[Kr]$4d^65s^25p^4$	$^3P_2$	9.010	136	142	190.2
53	I	[Kr]$4d^65s^25p^5$	$^2P_{3/2}$	10.454	133	132	295.3
54	Xe	[Kr]$4d^65s^25p^6$	$^1S_0$	12.127	131	124	– 41
55	Cs	[Xe]$6s^1$	$^2S_{1/2}$	3.893	235	324	45.5
56	Ba	[Xe]$6s^2$	$^1S_0$	5.210	198	278	– 46
57	La	[Xe]$5d^16s^2$	$^2D_{3/2}$	5.610	169	274	
58	Ce	[Xe]$4f^15d^16s^2$	$^3H_4$	6.910	165	270	
59	Pr	[Xe]$4f^36s^2$	$^4I_{9/2}$	5.760	165	267	
60	Nd	[Xe]$4f^46s^2$	$^5I_4$	6.310	164	264	
61	Pm	[Xe]$4f^56s^2$	$^6H_{5/2}$	5.554	163	262	
62	Sm	[Xe]$4f^66s^2$	$^7F_0$	5.600	162	259	
63	Eu	[Xe]$4f^76s^2$	$^8S_{7/2}$	5.670	185	256	
64	Gd	[Xe]$4f^75d^16s^2$	$^9D_2$	6.160	161	254	
65	Tb	[Xe]$4f^96s^2$	$^6H_{15/2}$	6.740	159	251	
66	Dy	[Xe]$4f^{10}6s^2$	$^5I_8$	6.820	159	249	
67	Ho	[Xe]$4f^{11}6s^2$	$^4I_{15/2}$	6.016	158	247	
68	Er	[Xe]$4f^{12}6s^2$	$^3H_6$	6.080	157	245	
69	Tm	[Xe]$4f^{13}6s^2$	$^2F_{5/2}$	5.810	156	242	
70	Yb	[Xe]$4f^{14}6s^2$	$^1S_0$	6.200	174	240	
71	Lu	[Xe]$4f^{14}5d^16s^2$	$^2D_{1/2}$	5.000	156	225	
72	Hf	[Xe]$4f^{14}5d^26s^2$	$^3F_4$	6.650	144	216	
73	Ta	[Xe]$4f^{14}5d^36s^2$	$^4F_{3/2}$	7.880	134	209	
74	W	[Xe]$4f^{14}5d^46s^2$	$^5D_0$	7.980	130	202	
75	Re	[Xe]$4f^{14}5d^56s^2$	$^6S_{5/2}$	7.870	128	197	
76	Os	[Xe]$4f^{14}5d^66s^2$	$^5D_4$	8.700	126	192	

Contd...

77	Ir	[Xe]$4f^{14}5d^{7}6s^{2}$	$^{4}F_{9/2}$	9.000	127	187	
78	Pt	[Xe]$4f^{14}5d^{8}6s^{2}$	$^{3}D_{3}$	9.000	130	183	
79	Au	[Xe]$4f^{14}5d^{10}6s^{1}$	$^{2}S_{1/2}$	9.220	134	179	
80	Hg	[Xe]$4f^{14}5d^{10}6s^{2}$	$^{1}S_{0}$	10.430	149	176	
81	Tl	[Xe]$4f^{14}5d^{10}6s^{2}6p^{1}$	$^{2}P_{1/2}$	6.106	148	208	30
82	Pb	[Xe]$4f^{14}5d^{10}6s^{2}6p^{2}$	$^{3}P_{0}$	7.415	147	181	35.2
83	Bi	[Xe]$4f^{14}5d^{10}6s^{2}6p^{3}$	$^{4}S_{3/2}$	7.287	146	163	101
84	Po	[Xe]$4f^{14}5d^{10}6s^{2}6p^{4}$	$^{3}P_{2}$	8.430	146	153	186
85	At	[Xe]$4f^{14}5d^{10}6s^{2}6p^{5}$	$^{2}P_{3/2}$	-	145	143	270
86	Rn	[Xe]$4f^{14}5d^{10}6s^{2}6p^{6}$	$^{1}S_{0}$	10.740	-	134	–41
87	Fr	[Rn]$6s^{1}$	$^{2}S_{1/2}$	-	-	-	
88	Ra	[Rn]$6s^{2}$	$^{1}S_{0}$	5.277	-	-	
89	Ac	[Rn]$6d^{1}6s^{2}$	$^{2}D_{1/2}$	5.170	-	0	
90	Th	[Rn]$6d^{2}7s^{2}$	$^{3}F_{3/2}$	6.950	165	-	
91	Pa	[Rn]$5f^{2}6d^{1}7s^{2}$	$^{4}K_{13/2}$	5.890	-	-	
92	U	[Rn]$5f^{3}6d^{1}7s^{2}$	$^{5}L_{7}$	6.100	142	0	
93	Np	[Rn]$5f4^{2}6d^{1}7s^{2}$	$^{6}L_{13/2}$	6.190	-	-	
94	Pu	[Rn]$5f^{6}7s^{2}$	$^{7}F_{0}$	5.100	-	-	
95	Am	[Rn]$5f^{7}7s^{2}$	$^{8}S_{7/2}$	6.000	-	-	
96	Cm	[Rn]$5f^{7}6d^{1}7s^{2}$	$^{9}D_{6}$	6.02	-	-	
97	Bk	[Rn]$5f^{9}7s^{2}$	$^{6}H_{15/2}$	-	-	-	
98	Cf	[Rn]$5f^{10}7s^{2}$	$^{5}I_{8}$	-	-	-	
99	Es	[Rn]$5f^{11}7s^{2}$	$^{11}H_{5/2}$	-			
100	Fm	[Rn]$5f^{12}7s^{2}$	$^{3}H_{6}$	-			
101	Md	[Rn]$5f^{13}7s^{2}$	$^{2}F_{7/2}$	-			
102	No	[Rn]$5f^{14}7s^{2}$	$^{1}S_{0}$	-			
103	Lw	[Rn]$5f^{14}6d^{1}7s^{2}$	$^{2}D_{3/2}$	-			

3.8. SPIN-ORBIT COUPLING

Before we learn about the term symbols or spectroscopic terms let us try to learn little bit about the *orbital angular momentum* and *spin angular momentum* and their interactions. Spin-orbit coupling deals with interaction of spin angular momentum and orbital angular momentum. The resultant is known as the *total angular momentum*. The terms are denoted as ***l***, ***s*** and ***j*** respectively. The values of these momenta, obtained by quantum mechanics, are

$$\boldsymbol{l} = \frac{h}{2\pi}\sqrt{l(l+1)} \qquad ...(3.45)$$

$$\boldsymbol{s} = \frac{h}{2\pi}\sqrt{s(s+1)} \qquad ...(3.46)$$

$$\boldsymbol{j} = \frac{h}{2\pi}\sqrt{j(j+1)} \qquad ...(3.47)$$

where l, s and j are the orbital angular momentum quantum number, spin angular momentum quantum number and total angular momentum quantum number respectively.

Detailed calculations show that the magnitude of the spin-orbit splitting is proportional to the dot product of L and S. This can be obtained as follows

$$j^2 = \boldsymbol{j.j} = [\boldsymbol{l} + \boldsymbol{s}].[\boldsymbol{l} + \boldsymbol{s}] = l^2 + s^2 + 2\,\boldsymbol{l.s}$$

Solving for $\boldsymbol{l.s}$ yields

$$\boldsymbol{l.s} = \frac{1}{2}[j^2 - l^2 - s^2]^{\frac{1}{2}} \qquad ...(3.48)$$

Now substituting the values of the respective angular momenta from the equations (3.45), (3.46) and (3.47) in equation (3.48), we get

$$\boldsymbol{l.s} = \frac{1}{2}[j(j+1) - l(l+1) - s(s+1)]^{\frac{1}{2}}\frac{h^2}{4\pi^2} \qquad ...(3.49)$$

The change in the energy associated with spin-orbit splitting is therefore,

$$\Delta E_{\text{spin-orbit}} = C[j(j+1) - l(l+1) - s(s+1)]^{\frac{1}{2}}\frac{h^2}{8\pi^2} \qquad ...(3.50)$$

The spin-orbit coupling constant in equation (3.50) is approximately $C = 1.67 \times 10^{-6}$ hartrees for hydrogen-like atoms or ion with nuclear charge Z. Since C increases rapidly with increasing atomic number, spin-orbit coupling interactions become much more important as Z increases. When $l = 0$, $J = S$, it is clear from equation (3.50) that $\Delta E_{\text{spin-orbit}}$ will be zero. This is the expected result, since with no orbital angular momentum there will be no magnetic field with which the spin magnetic moment can interact.

***Exercise* 3.29:** Calculate the spin-orbit coupling splitting between the $^3P_1$ and $^3P_2$ states of the carbon atom.

For both the angular momentum states, $s = 1$, $l = 1$. Using equation (3.50) for the $^3P_1$ state, we get

$$\Delta E_{\text{spin-orbit}} = C[1(1+1) - 1(1+1) - 1(1+1)]\frac{h^2}{8\pi^2} = C[2-2-2]\frac{h^2}{8\pi^2} = -2C\frac{h^2}{8\pi^2} \qquad ...(A)$$

Similarly for $^3P_2$ state we obtain

$$\Delta E_{\text{spin-orbit}} = C[2(2+1) - 1(1+1) - 1(1+1)]\frac{h^2}{8\pi^2} = C[6-2-2]\frac{h^2}{8\pi^2} = 2C\frac{h^2}{8\pi^2} \qquad ...(B)$$

Therefore, the energy splitting (spacing) between these two states is

$$\text{Splitting} = \Delta E_{\text{spin-orbit}}(^3P_2) - \Delta E_{\text{spin-orbit}}(^3P_1) = 2C\frac{h^2}{8\pi^2} - (-2C\frac{h^2}{8\pi^2}) = 4C\frac{h^2}{8\pi^2}$$

Knowing the values of C, we can calculate the splitting between the states of angular momentum. For hydrogen and hydrogen like ions the value of $C = 1.67 \times 10^{-6} Z^4$ hartrees, where Z is the nuclear charge and the value of C increases with increase of Z.

3.8.1. *Magnitude of Various Momenta*

Similar to the various angular momenta for the one electron systems, the total angular momenta for multi-electron atom are given on the same lines, *i.e.*,

$$|\boldsymbol{L}| = \frac{h}{2\pi}[L(L-1)]^{\frac{1}{2}} \quad ...(3.51)$$

$$|\boldsymbol{S}| = \frac{h}{2\pi}[S(S-1)]^{\frac{1}{2}} \quad ...(3.52)$$

$$|\boldsymbol{J}| = \frac{h}{2\pi}[J(J-1)]^{\frac{1}{2}} \quad ...(3.53)$$

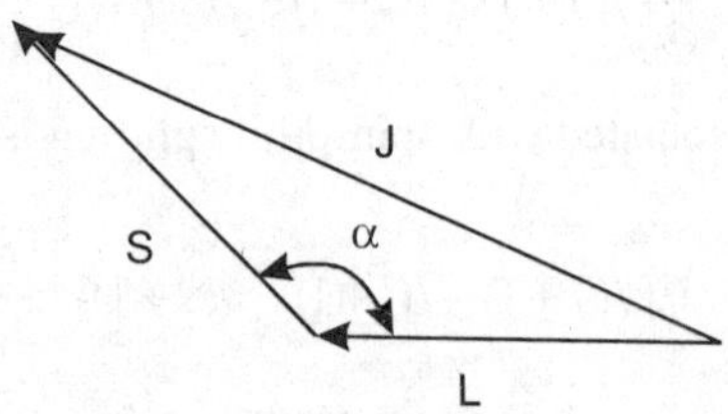

Fig. 3.7 *The vector addition of L and S.*

If we know the values of $\boldsymbol{L}$ and $\boldsymbol{S}$, the value of total angular momentum to be having specific values, the vectors $\boldsymbol{L}$ and $\boldsymbol{S}$ must add vectorially in such a manner to give the specific values of $\boldsymbol{J}$. In other words, the angle between $\boldsymbol{L}$ and $\boldsymbol{S}$ can have only some specific values. These can be calculated by applying the vector addition as shown in Fig. (3.7) and given by equation (3.54).

$$J^2 = L^2 + S^2 - 2\,LS\cos\alpha \quad ...(3.54)$$

$$\text{Cos}\,\alpha = \left[\frac{L^2 - S^2 - J^2}{2LS}\right]$$

or

$$\alpha = \cos^{-1}\left[\frac{L^2 - S^2 - J^2}{2LS}\right] \quad ...(3.55)$$

***Exercise* 3.30:** Consider the term symbol $^4F_{9/2}$. Calculate the angle between the vectors L and S to give the value of J as 9/2.

For the term symbol $^4F_{9/2}$, $S = 3/2$, $L = 3$ and $J = 9/2$. Using equations (3.51) to (3.53) the values of different angular momenta are

$$|L| = \frac{h}{2\pi}[3(3+1)]^{\frac{1}{2}} = \frac{h}{2\pi}[12]^{\frac{1}{2}} \quad \text{...(A)}$$

$$|S| = \frac{h}{2\pi}\left[\frac{3}{2}\left(\frac{3}{2}+1\right)\right]^{\frac{1}{2}} = \frac{h}{2\pi}\left[\frac{15}{4}\right]^{\frac{1}{2}} \quad \text{...(B)}$$

$$|J| = \frac{h}{2\pi}\left[\frac{9}{2}\left(\frac{9}{2}+1\right)\right]^{\frac{1}{2}} = \frac{h}{2\pi}\left[\frac{99}{4}\right]^{\frac{1}{2}} \quad \text{...(C)}$$

Substituting the values of angular momenta in equation (3.55) we get

$$\alpha = \cos^{-1}\left[\frac{12+\frac{15}{4}-\frac{99}{4}}{2\sqrt{(12)\left(\frac{15}{4}\right)}}\right] = \cos^{-1}\left[\frac{-9}{2\sqrt{45}}\right] = \cos^{-1}[-0.67] = 132.13^{\circ}.$$

3.9. ATOMIC TERM SYMBOLS AND RUSSELL-SAUNDERS COUPLING FOR TWO ELECTRONS

The spectroscopic term symbol for the atomic system is denoted by $^{2s+1}L_J$ where left hand side superscript is the spin multiplicity (S is the total spin quantum number), L is total orbital angular momentum quantum number and J is the total angular momentum quantum number. These three quantities, when there are two electrons, can be calculated as:

(*a*) The two electrons can interact with each other and total spin quantum number will be

$$S = 0 \text{ or } 1 \quad \text{...(3.56)}$$

This coupling or interaction is called spin-spin (s-s) coupling.

(*b*) L value is due to the coupling of the orbital angular momentum of two electrons and the resultant values of total orbital angular momentum quantum number (L) will be

$$L = l_1 + l_2, l_1 + l_2 - 1, \ldots |l_1 - l_2| \quad \text{...(3.57)}$$

(*c*) The coupling of orbital angular momentum and spin angular momentum is known as Russell-Saunders coupling, also called L-S coupling. This coupling gives rise to total angular momentum and quantum numbers for this is obtained as

$$J = L + S, L + S - 1, \ldots |L - S| \quad \text{...(3.58)}$$

The Russell-Saunders coupling is only applicable for light atomic weight atoms or where the coupling between orbital and spin angular momentum is weak. Similar to the s, p, d, f symbols, the symbols given for different values of total orbital angular momentum quantum number is when $L = 0$, it is S, for P, $L = 1$; for D, $L = 2$; for F, $L = 3$; for G, $L = 4$, etc.

***Exercise* 3.31:** Calculate the L, S and J values for the coupling between an electron belonging to p orbital and to f orbital.

The total S for two electrons can be

$$S = 1/2 + 1/2 = 1$$
$$= 1/2 - 1/2 = 0, \text{ i.e. } S = 1 \text{ and } S = 0$$

The values of l for p orbital is 1 and for f orbital 3.

$$L = 1 + 3 = 4$$
$$1 + 3 - 1 = 3$$
$$1 + 3 - 2 = 2, \quad \text{i.e. } |l_1 - l_2|$$

So this corresponds to letters D, F and G, the states can be ^3D_J, ^3F_J, ^3G_J, ^1D_J, ^1F_J, and ^1G_J. The values of J for each value of L can be obtained as:

(*i*) When $L = 2, S = 1$

$$J = 2 + 1 = 3$$
$$J = 2 + 1 - 1 = 2$$
$$J = 2 - 1 = 1, \text{ i.e. } |L - S|$$

Similarly (*ii*) for $L = 2, S = 0, J = 2$

(*iii*) for $L = 3, S = 1, J = 4, 3, 2$

(*iv*) for $L = 3, S = 0, J = 3$

(*v*) for $L = 4, S = 1, J = 5, 4, 3$

(*vi*) for $L = 4, S = 0, J = 4$.

The spectroscopic term symbols are

$$^3D_3, {}^3D_2, {}^3D_1; {}^1D_2; {}^3F_4, {}^3F_3, {}^3F_2; {}^1F_3; {}^3G_5, {}^3G_4, {}^3G_3; {}^1G_4.$$

3.10. TERM SYMBOLS FOR POLYELECTRONIC ATOMS

There are two kind of coupling between the orbital angular momentum and spin angular momentum: (*i*) *L-S coupling or Russell-Saunders coupling* and (*ii*) *j-j coupling*. In the former case coupling between spin angular momentum and orbital angular momentum is weak and thus applicable to the light elements or elements with smaller atomic number. In this case electrostatic interactions dominate. In the latter case the magnetic interactions and spin angular momentum and orbital angular momentum coupling is strong. Since the spin-orbit coupling is proportional to Z^4, magnetic interactions and *j-j* coupling become much more important for the heavier atoms. For lighter elements, the spin-orbit interactions are very small, and *L-S* coupling is the major appropriate method for representing the angular momentum states. Since a large majority of molecules that are of interest to chemists involve the lighter elements with atomic numbers less than 18,*L-S* coupling is, therefore, the scheme with which we are primarily concerned and most of the examples discussed are regarding these elements.

When Pauli exclusion principle does not have to be considered, there are few simple rules that permit use to write all the terms states.

(*i*) The maximum spin state is obtained when all the Z-components of the spin are aligned in the same direction, *i.e.*,

$$S_{max} = \frac{k}{2} \qquad ...(3.59)$$

where k is the number of electrons being considered. Recollect that we omit electrons filled in sub-shells. The minimum value of spin (S_{min}) is zero if the value of k is even and 1/2 is the value of k is odd. The possible total spin quantum numbers are the values from S_{max} to S_{min} by integer steps.

***Exercise* 3.32:** Calculate the possible values of total spin for the electronic configuration of the excited state of nitrogen atom, *i.e.*, $1s^22s^12p^13p^14p^15p^1$.

This electronic configuration consists of five unpaired electrons and two paired electrons. Since the paired electrons in any orbital leads to zero value of spin, thus we do not have to consider them. Since the number of electrons is odd, the minimum value of spin (S_{min}) will be 1/2 and S_{max} will be 5/2. The other values will be 3/2. Thus the possible total spin quantum numbers will be 5/2, 3/2 and 1/2.

(*ii*) The maximum orbital angular momentum state is obtained from

$$L_{max} = l_1 + l_2 + l_3 \ldots\ldots\ldots\ldots\ldots\ldots + l_k \quad \text{...(3.60)}$$

The L_{min} will be zero if all the l are equal. If one l_1 is larger than all the others, the minimum L is given by

$$L_{min} = L_{max} - \text{sum of all other } l_i \quad \text{...(3.61)}$$

subject to the constraint that $L_{min} > 0$.

***Exercise* 3.33:** Again consider the excited state of nitrogen atom with the electronic configuration as $1s^22s^22p^13p^13d^1$ and calculate the possible values of total orbital angular momentum quantum numbers.

The values of l_i's are: $l_1 = 0, l_2 = 0, l_3 = 1, l_4 = 1, l_5 = 2$.

So $L_{max} = 0 + 0 + 1 + 1 + 2 = 4$.

$L_{min} = 2 - 0 - 0 - 1 - 1 = 0$

The possible total orbital angular momentum quantum numbers are the values from L_{max} to L_{min} by integer steps, *i.e.*, 4, 3, 2, 1, 0.

(*iii*) For given values of the L and S quantum numbers, the possible total angular momentum quantum numbers J are the values from $L + S$ to $|L - S|$, by integer steps. *It should be kept in mind that the absolute-value bars about L – S are kept as J is never negative .*

***Exercise* 3.34:** Considering the excited state configuration of nitrogen atom ($1s^22s^22p^13p^13d^1$) as given in Exercise 3.32. Calculate the terms symbols for all the states.

The possible total orbital angular momentum quantum numbers are 4, 3, 2, 1 and 0. Thus the possible states can be written as *G, F, D, P* and *S*.

Since there are three unpaired electrons, $S_{max} = 3/2$ and $S_{min} = 1/2$. The value of spin multiplicity $(2S + 1)$ for each combination is $2 \times 3/2 + 1 = 4$ and $2 \times 1/2 + 1 = 2$. So this immediately gives the terms as $^4G, ^4F, ^4D, ^4P, ^4S$ and $^2G, ^2F, ^2D, ^2P, ^2S$.

Application of J rule gives the possible values of J. Here we will take one combination and rest we will leave for the students to carry out.

Consider ^4G term. Here $L = 4$, $S = 3/2$, the possible values of J are

$$J_{max} = 4 + 3/2 = 11/2$$

and

$$J_{min} = 4 - 3/2 = 5/2$$

The rest of the J values in this limit are 9/2, 7/2. So the terms are

$^4G_{11/2}, ^4G_{9/2}, ^4G_{7/2}, ^4G_{5/2},$

For ^4F, the terms are $^4F_{9/2}, ^4F_{7/2}, ^4F_{5/2}, ^4F_{3/2}$

For ^4D, the terms are $^4D_{7/2}, ^4D_{5/2}, ^4D_{3/2}, ^4D_{1/2}$

For ^4P, the terms are $^4P_{5/2}, ^4P_{3/2}, ^4P_{1/2}$

For ^4S, the terms are $^4S_{3/2}$

Similarly for ^2G, the terms are: $^2G_{9/2}$, $^2G_{7/2}$, ^2f, the terms are: $^2f_{7/2}$, $^2f_{5/2}$, ^2D, the terms are: $^2D_{5/2}$, $^2D_{3/2}$, ^2P, the terms are: $^2p_{3/2}$, $^2p_{1/2}$, ^2s, the terms are: $^2S_{1/2}$.

The total orbital angular momentum quantum number and total spin quantum number can be evaluated as

$$M_L = \Sigma m_l \qquad M_L = L, L-1, \ldots -L \qquad \ldots(3.62)$$

$$M_S = \Sigma m_s \qquad M_S = S, S-1, \ldots -S \qquad \ldots(3.63)$$

Further the contributions to the total orbital angular momentum is only from the orbitals which are partially filled.

The spectroscopic states or term symbols can be obtained as follows:

(*a*) Write all possible combinations of different wave functions (orbital angular momentum and spin angular momentum) which are allowed by Pauli's exculsion principle.

(*b*) Calculate the values of M_L and M_S for each combination, with the help of equations (3.62) and (3.63).

(*c*) Calculate the term symbols for each combination, having the largest value of M_S and M_L, *i.e.* $S = M_S$ and $L = M_L$.

(*d*) To this highest value term, assign all the other substates, whose M_L and M_s values are related by the equations (3.62) and (3.63).

(*e*) Repeat the steps (*c*) and (*d*) for next highest $M_L - M_S$ combinations in such a manner that all the combinations obtained in steps (*a*) are taken care, *i.e.* do not repeat.

Although the above rules are applicable to all systems, they can be further simplified by dividing the electrons into two classes.

3.10.1. *Non-Equivalent Electrons*

Non-equivalent electrons are those electrons which belong to different orbitals, *i.e.* at least with either different *n* or *l*. For example, each electron occupying *p*, *d*...orbitals or $n_1p^1n_2p^1$, *i.e.* at least one of *n* and *l* values are different. In this case, Pauli's principle is always satisfied and all the possible combinations are allowed.

If there are more than two electrons the term symbols can be obtained as: (*a*) calculate the M_L and M_S values by combining two electrons; (*b*) then combine each M_L and M_S value calculated in point (*a*) with the m_l and m_s value of third electron and so on (see Exercise 3.35).

***Exercise* 3.35:** Calculate the spectroscopic term symbol for the electronic configuration $2p^13p^13d^1$ system.

In the above system first take $2p^13p^1$, l_1 and l_2 for these electrons are 1 each, so

$$L = 1 + 1 = 2$$

$$1 + 1 - 1 = 1$$

$$1 - 1 = 0$$

$$S = \frac{1}{2} + \frac{1}{2} = 1, \quad 2S + 1 = 3$$

$$\frac{1}{2} - \frac{1}{2} = 0, \quad 2S + 1 = 0$$

The term symbols due to coupling of $2p^13p^1$ electrons can be ^3D, ^3P, ^3S and ^1D, ^1P and ^1S.

Now each of six states can couple with $3d^1$ electron to give the resultant L and S. For example ^3D $3d^1$ gives,

$$L = 2+2=4$$
$$2+2-1=3$$
$$2+2-2=2$$
$$2+2-3=1$$
$$2-2=0$$

$$S = 1+\frac{1}{2} = \frac{3}{2},\ 2S+1=4$$

$$1-\frac{1}{2} = \frac{1}{2},\ 2S+1=2$$

The term symbols for the coupling of $^3D3D^1$ configuration are

$$^4G\,^4F\,^4D\,^4P\,^4S \text{ and } ^2G\,^2F\,^2D\,^2P\,^2S$$

For $^3P\ 3d^1$ the terms are

$$^4F\,^4D\,^4P \text{ and } ^2F\,^2D\,^2P$$

For $^3S\ 3d^1$ the terms are ^4D and ^2D.

Similarly for combinations

$^1D\ 3d^1$ the terms are $^2G\,^2F\,^2D\,^2P\,^2S$

For $^1P\ 3d^1$, the terms are $^2F\,^2D\,^2P$

and for $^1S\ 3d^1$, the term is ^2D.

It can be seen that some of the term symbols occurs more than once in different combinations and they can be combined to give as

$$^4G\,^4F(2)\,^4D(3)\,^4P(2)\,^4S$$
$$^2G(2)\,^2F(4)\,^2D(6)\,^2P(4)\,^2S(2)$$

The number in the parentheses tells how many times the particular state is formed by different combinations.

3.10.2. *Equivalent Electrons*

Equivalent electrons are those electrons which belong to the same orbital, *i.e.* same n and l values; for example, np^2, nd^3, nd^5 etc. Due to this electronic configuration, some of the states which are obtained when the electrons are non-equivalent vanish due to Pauli's exclusion principle. The procedure is exactly same as described above to calculate the term symbols for non-equivalent electrons.

The total degeneracy (TD) of the microstates is given by the relation

$$T.D. = \frac{(4l+2)!}{(4l+2-N)!(N!)} \qquad ...(3.64)$$

where N is the number of electrons and l is the azimuthal quantum number.

***Exercise* 3.36:** Find the term symbols for the electronic configuration np^2.

The total degeneracy of microstates in this configuration will be

$$T.D. = \frac{(4+2)!}{(4+2-2)!\ 2!} = 15$$

***Method I*:** The different microstates which can be obtained by different combinations are:

$$\left(1,\frac{1}{2};1,-\frac{1}{2}\right), \left(1,\frac{1}{2};0,-\frac{1}{2}\right), \left(1,\frac{1}{2};-1,-\frac{1}{2}\right), \left(1,\frac{1}{2};0,\frac{1}{2}\right), \left(1,\frac{1}{2};-1,+\frac{1}{2}\right), \left(1,-\frac{1}{2};0,-\frac{1}{2}\right),$$

$$\left(1,-\frac{1}{2};-1,-\frac{1}{2}\right), \left(1,-\frac{1}{2};0,+\frac{1}{2}\right), \left(1,-\frac{1}{2};-1,+\frac{1}{2}\right), \left(0,\frac{1}{2};0,-\frac{1}{2}\right), \left(0,\frac{1}{2};-1,-\frac{1}{2}\right), \left(0,+\frac{1}{2};1,\frac{1}{2}\right),$$

$$\left(0,-\frac{1}{2};-1,\frac{1}{2}\right), \left(0,-\frac{1}{2};-1,-\frac{1}{2}\right), \left(-1,\frac{1}{2};-1,-\frac{1}{2}\right).$$

This gives rise to

(2, 0), (1, 0), (0, 0), (1, 1), (0, 1), (1, – 1), (0, – 1), (1, 0), (0, 0), (0, 0), (– 1, 0), (– 1, 1), (– 1, 0), (– 1, – 1), (– 2, 0).

Now starting with highest values of $M_L = 2$ and $M_S = 0$, the term symbol is ^1D and the microstates which originate with the help of equation (3.23) are (2, 0), (1, 0), (0, 0), (– 1, 0) and (– 2, 0).

Now starting with the next highest M_L, *i.e.* $M_L = 1, M_S = 1$, the possible microstates due to equations (3.62) and (3.63) are (1, 1), (0, 1), (–1, 1), (1, 0), (0, 0), (– 1, 0), (1, – 1), (0, – 1) and (– 1, – 1). These are nine. The only microstate left is (0, 0) and it is due to $M_L = 0, M_S = 0$.

Thus, the total degeneracy agrees with that obtained from equation (3.64).

Method II: The term symbols for the equivalent electrons if they are two in number, *i.e.* np^2, nd^2 etc. can be obtained as follows:

(*a*) Make a table of the m_l values of one of the electrons as row and the m_l values of the second electron as column as shown below in Exercise (3.37).

(*b*) Calculate each element of the square table by the addition of two m_l values of the elements.

(*c*) Elements lying on the diagonal of the table will be singlet and occurring off diagonal as triplet.

(*d*) Make use only the M_L values from one side of the diagonal as those on the other side merely correspond to a mirror image.

***Exercise* 3.37:** Calculate the term symbols for the electronic configuration np^2.

m_l values for each electron is 1, 0, – 1 and table can be constructed as

m_{l_2} \ m_{l_1}	1	0	–1
–1	0	–1	–2
0	1	0	–1
1	2	1	0

The absolute value of the diagonal elements are 2 and 0. Thus states *D* and *S* will be singlet states (rule *C*), *i.e.* ^1D and ^1S. If you draw the dotted line as shown in Table, the elements enclosed in it, *i.e.* $M_l = 2, 1, 0, -1, -2$ and 0 form the microstates for $L = 2$ and $L = 0$.

The off diagonal elements $L = 1$ will be triplet state, *i.e.* ^3P and elements $M_L = 1, 0, -1$ forms the microstates.

The total angular momentum quantum number (J) can be calculated for each states as:

for $^1D, L = 2, S = 0$, so

$$J = 2 + 0 = 2 = 2 - 0$$

for $^1S, L = 0, S = 0, J = 0$

for $^3P, L = 1, S = 1, J = 2, 1, 0$

The term symbols are

$$^1D_2, {}^1S_0, {}^3P_2, {}^3P_1, {}^3P_0$$

Method III: The above two methods are very good if . . number of electrons are two. But if the number of electrons are more than two, the procedure becomes cumbersome and confusion can arise. In order to avoid frustration the following procedure is very good.

(*a*) Calculate the M_S values for the possible arrangements obtained by grouping the electrons, taking care of Pauli principle. The possible arrangement for each M_S value and the total number of possible arrangements for unpaired electrons from zero to 7 are given in Table 3.6.

(*b*) For each values of M_S, make a table to calculate the M_L values of different possible arrangements of electrons, with Pauli exclusion principle in mind.

(*c*) Finally construct the major array that contains all the arrangements. This can be done by realizing that each M_L value that occurs in arrangement with one unpaired electron has two possible M_S values whereas each M_L value that occurs in arrangement with three unpaired electrons has eight M_S values and so on. It should be kept in mind that total number of microstates in the major array should be same as that obtained by the formula (3.64).

TABLE 3.6. Possible M_S values for groupings of arrangements with zero to 7 unpaired electrons

No. of unpaired electrons	M_S Val	Total arrangements
0	0	1
1	+1/2, –1/2	2
2	1, 0 (2), –1	4
3	+3/2, +1/2(3), –1/2(3), –3/2	8
4	+2, +1(4), 0(6), –1(4), –2	16
5	+5/2, +3/2 (5), +1/2(10), –1/2 (10), –3/2(5), –5/2	32
6	+3, +2 (6), +1 (15), 0 (20), –1 (15), –2(6), –3	64
7	+7/2, +5/2 (7), + 3/2 (21), 1/2 (35), –1/2 (–35), –3/2 (21), –5/2 (7), –7/2	128

The term symbols and the number of times the particular state is occurring can be obtained by solving this major array table as shown in the Exercise 3.38.

***Exercise* 3.38:** Calculate the term symbols for electronic configuration p^3.

There are three electrons, the M_S values can be 3/2 and 1/2. The total M_L values can be calculated for each M_S values as in the Table 3.7.

TABLE 3.7. Grouping of arrangement for $M_s = 1/2$

$m_l = 1$	0	–1	$M_L = \Sigma m_l$	$m_l = 1$	0	–1	$M_L = \Sigma m_l$
×	/		2		×	/	–1
×		/	1	/		×	–1
/	×		1		/	×	–2

The cross means two electrons with opposite spin and oblique means only one electron in any particular orbital.

There will be only one group for $M_S = 3/2$ because there are three p orbitals with $m_l = 1, 0, -1$, and total $M_L = 0$.

Now the major array can be constructed as

M_L \ M_S	3/2	1/2	–1/2	–3/2
2		1	1	
1		2	2	
0	1	3	3	1
–1		2	2	
–2		1	1	

i.e. by adding up all the arrangements for each M_L and M_S values from the two subtables (3.7) and the total number of microstates with the number obtained from equation (3.64.)

For calculation of term symbols remove the each term one by one from major array as shown below:

i.e.

M_L \ M_S	3/2	1/2	–1/2	–3/2
2		1	1	
1		2	2	
0	1	3	3	1
–1		2	2	
–2		1	1	

$L = 2, S = \frac{1}{2}$ → Subarray removed (^2D)

M_L \ M_S	3/2	1/2	–1/2	–3/2
1		1	1	
0	1	2	2	1
–1		1	1	

↓ $L = 1, S = 1/2$ Subarray removed (^2P) ↓

M_L	3/2	1/2	–1/2	–3/2
0	1	1	1	1

Array Vanishes ← $L = 0, S = 3/2$ Subarray removed (^4S)

Therefore, the term symbols are ^4S, ^2D, ^2P.

3.11. HUND'S RULES

Hund's rules are used to arrange the different spectroscopic states according to the increasing energy levels. These are:

1. The state which has the highest spin multiplicity has the lowest energy. When the spatial orbital occupancy is different. The average distance between the electrons is maximised and their mutual repulsions are minimised.
2. If the two states have the same spin multiplicity, then the state with greater L value will be lower in energy than that with lower L value, because larger L states require the electrons to be spread-over a wider range of ml values than states with smaller L values.
3. If the orbital is less than half filled, the state with smallest J value will be lower in energy, whereas if the orbital is more than half filled, the state with largest J value will be lower in energy. When L and S are identical for two states the energy difference are due to spin-orbit coupling, which is small for the light atoms. The approximate energy changs due to spin-orbit coupling for a one electron system is given by equation (3.48), whereas for many electron systems, the analogous equation is

$$\Delta E_{\text{spin-orbit}} = C' [J(J+1) - L(L+1) - S(S+1)] \frac{h^2}{8\pi^2} \quad ...(3.65)$$

where C is different from C.

Exercise 3.39: Arrange the different spectroscopic states evolved from p^3 electronic configuration according to energy scale.

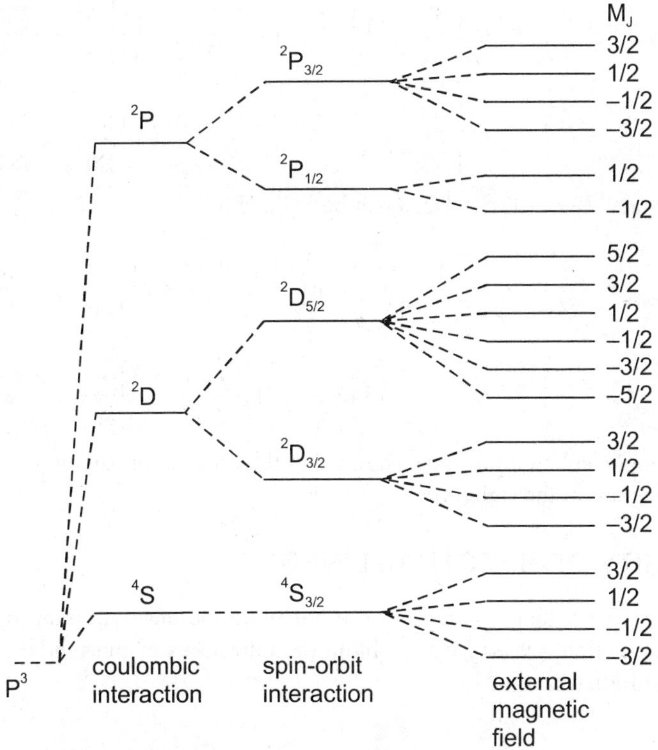

Fig. 3.8 *Term splitting in the p^3 system.*

The different term symbols obtained from p^3 configuration are ^4S, ^2D, ^2P (see Exercise 3.38). The different J values for each state are:

for			
^4S,	$L = 0$,	$S = 3/2$	$J = 3/2$
^2D,	$L = 2$,	$S = 1/2$	$J = 2 + 1/2 = 5/2$
			$2 - 1/2 = 3/2$
^2P,	$L = 1$,	$S = 1/2$	$J = 1 + 1/2 = 3/2$
			$1 - 1/2 = 1/2$

Since ^4S is the state with highest spin will be ground state and the diagram is shown in Fig. 3.8.

***Exercise* 3.40:** Calculate the total number of microstates for p^3 system, using the equation (3.64) and compare it with that obtained by other method.

Here $l = 1$, $N = 3$. Substituting the values in equation (3.64) one gets

$$\text{T.D.} \quad = \frac{[(4)(1)+2]!}{[(4)(1)+2-3]!\,3!} = \frac{6!}{(3!)(3!)} = 20$$

This agrees with that obtained in part (*a*).

***Exercise* 3.41:** Calculate the spin-orbit coupling energy change for each term arisen in Exercise (3.38).

For state $^4S_{3/2}$, $L = 0$ and $S = 3/2$. So there will not be any spin-orbit coupling, so $\Delta E_{\text{spin-orbit}} = 0$

For state ^2D, $L = 2$, $S = 1/2$, J can be 3/2 and 5/2. So

$$\Delta E_{\text{spin-orbit}}\left(J = \frac{3}{2}\right) = C'\left[\frac{3}{2}\left(\frac{3}{2}+1\right) - 2(2+1) - \frac{1}{2}\left(\frac{1}{2}+1\right)\right]\frac{h^2}{8\pi^2} = -\frac{3h^2C'}{8\pi^2}$$

$$\Delta E_{\text{spin-orbit}}\left(J = \frac{5}{2}\right) = C'\left[\frac{5}{2}\left(\frac{5}{2}+1\right) - 2(2+1) - \frac{1}{2}\left(\frac{1}{2}+1\right)\right]\frac{h^2}{8\pi^2} = \frac{2h^2C'}{8\pi^2}$$

Similarly for state ^2P, $L = 1$, $S = 1/2$, J can be 1/2 and 3/2.

$$\Delta E_{\text{spin-orbit}}\left(J = \frac{1}{2}\right) = C'\left[\frac{1}{2}\left(\frac{1}{2}+1\right) - 1(1+1) - \frac{1}{2}\left(\frac{1}{2}+1\right)\right]\frac{h^2}{8\pi^2} = -\frac{2h^2C'}{8\pi^2}$$

$$\Delta E_{\text{spin-orbit}}\left(J = \frac{3}{2}\right) = C'\left[\frac{3}{2}\left(\frac{3}{2}+1\right) - 1(1+1) - \frac{1}{2}\left(\frac{1}{2}+1\right)\right]\frac{h^2}{8\pi^2} = \frac{h^2C'}{8\pi^2}$$

These approximate calculations have shown that the order of arranging the states arisen because of spin-orbit coupling are in the right order.

3.12. INTENSITY OF SPECTRAL LINES

We will learn in Chapter 5 that it is not easy to calculate the absolute intensity of absorption lines theoretically. In this section, we will try to obtain the intensities of lines either experimentally or by means of semi-quantitative method.

3.12.1. *Selection Rules*

Selection rules can be derived by using the transition moment integral (equation 3.66), *i.e.*, the transition is allowed when the transition moment is non-zero and forbidden when it zero.

$$Z = e\iiint \Psi^*_{n'',l'',m''_l}\hat{r}\Psi^*_{n',l',m'_l} r^2 dr \sin\theta\cos\theta d\theta d\varphi \qquad ...(3.66)$$

where $\Psi_{n'',l'',m''l}$ and $\Psi_{n'',l',m'l}$ are the wave functions of the ground state and excited state respectively, $e\hat{r}$ is the dipole moment operator.

Selection rules obtained using equation (3.66) is:

$$\Delta S = 0 \qquad ...(3.67)$$

$$\Delta L = 0, \pm1,\ L = 0 \neq L = 0 \qquad ...(3.68)$$

$$\Delta J = 0, \pm1,\ J = 0 \neq J = 0 \qquad ...(3.69)$$

$$\Delta l = \pm1 \qquad ...(3.70)$$

where each symbol has the usual meaning. Sign $\neq$ indicates a forbidden transition.

As stated earlier, $\Delta S = 0$ selection rule is followed rigidly if the atoms are light and of small atomic number but fails down in case of heavy atoms because l and s are not well defined quantum numbers or the spin-orbit interaction is very strong between the individual electrons. So j the total angular momentum quantum number of individual electron is more important and *j*-*j* coupling becomes more predominant rather than *L*-*S* coupling. Further, the total wave function of heavy the atoms cannot be written as the product of the wave function of the space part and the wave function of the spin part.

Laporte rule: *According to this rule the only possible combination is between even and odd terms. A term is called even if the sum of azimuthal quantum numbers of all the electrons i in the atom is even otherwise the term is odd.*

3.12.2. *Intensity Rules*

These rules are in general stated for doublet spectra but they are valid for all spectral lines in general. These rules are stated qualitatively as well as quantitatively.

3.12.2.1. Qualitative rules

(*i*) The strongest line in a doublet is one which results due to transition for which the quantum numbers j and l change in the same way. If there are more such lines then the line corresponding to the largest value of j will be strongest.

(*ii*) Line due to the transition for which j and l do not change in the same way will be faint.

(*iii*) Transition for which l decreases will give rise to stronger line as compared to the transition for which l increases.

(*iv*) Transition for which l increases and j decreases or vice versa is forbidden since this would violate the selection principle viz.

If $\quad \Delta l = -1$ and $\Delta j = +1 \quad$ Forbidden

or $\quad \Delta l = +1$ and $\Delta j = -1 \quad$ Forbidden

***Exercise* 3.42:** Consider that principal series has a doublet. Indicate the strongest line.

The lines of doublet are

$$^2S_{1/2} \leftarrow {}^2P_{1/2};\ {}^2S_{1/2} \leftarrow {}^2P_{3/2}$$

In this case the second line is the strongest because both for l and j change by equal amount (–1), whereas in the first case change in l is by –1 and in change in j is zero.

***Exercise* 3.43:** Consider a diffuse series and suggest which will be the strongest line.

In case of diffuse series, there are three spectral lines, one being not permissible by the selection rules, *i.e.*, ${}^2P_{1/2} \leftarrow {}^2D_{5/2}$ as $\Delta J = -2$

Three allowed lines are

$$ {}^2P_{3/2} \leftarrow {}^2D_{5/2};\ {}^2P_{3/2} \leftarrow {}^2D_{3/2} \text{ and } {}^2P_{1/2} \leftarrow {}^2D_{3/2} $$

l and j change in the first and third lines are same and the amount of change is (–1). Thus they are strongest than the second line where $\Delta l = -1$ and $\Delta J = 0$ is relatively weak line. Out of two strongest lines, the first line is strongest because it has the largest j value.

3.12.2.2. Quantitative rules

(*i*) Sum of the intensities of all lines corresponding to transitions from a given level is proportional to the statistical weights or quantum weights of the level.

(*ii*) Sum of the intensities of all lines which end on a common level is proportional to the quantum weights of that level. Quantum weight of a particular level is given by $(2j + 1)$.

***Exercise* 3.44:** Consider the doublet of principal series as discussed in Exercise 3.42. Find out the intensity ratio of the lines.

The doublet results is due to the transitions

$$ {}^2S_{1/2} \leftarrow {}^2P_{1/2};\ {}^2S_{1/2} \leftarrow {}^2P_{3/2} $$

Both the transitions end on a common level, *i.e.*, ${}^2S_{1/2}$. The statistical weights of the lines corresponding to these two transitions are (2 × 1/2 +1) = 2 and (2 × 3/2 + 1) = 4. Hence the intensity ratio of second line to the first line will be 4 : 2, *i.e.*, 2 : 1.

***Exercise* 3.45:** Calculate the intensity ratios of the diffuse series.

The transitions in a diffuse series are shown in the tabular form as follows:

Level	Quantum weight $(2j + 1)$	${}^2P_{3/2}(4)$	${}^2P_{1/2}$ (2)
${}^2D_{5/2}$	6	I_1	I_4
${}^2D_{3/2}$	4	I_2	I_3

I_is represents the intensity of the spectral lines. $I_4 = 0$ since the transition ${}^2P_{1/2} \leftarrow {}^2D_{5/2}$ is forbidden since $\Delta J = -2$.

Applying the intensity of spectral lines, we get

$$ \frac{I_1}{I_2 + I_3} = \frac{6}{4} = \frac{3}{2} \qquad \text{...(A)} $$

and

$$ \frac{I_1 + I_2}{I_3} = \frac{4}{2} = \frac{2}{1} \qquad \text{...(B)} $$

From (A) and (B) we write

$$ 2I_1 = 3I_2 + 3I_3 $$

$$2I_1 + 2I_2 = 4I_3$$
$$3I_2 + 2I_2 + 3I_3 = 4I_3$$
$$I_3 = 5\,I_2$$
$$I_1 = 9I_2$$

Hence, the intensity ratio is

$$I_1 : I_2 : I_3 :: 9 : 1 : 5$$

***Exercise* 3.46:** Calculate the intensity ratios of the three lines observed in the fundamental series.

Different transitions in fundamental series can be written in the following tabular form as

Level	Quantum weight (2*j* + 1)	$^2d_{5/2}$ (6)	$^2D_{3/2}$ (4)
$^2F_{7/2}$	8	I_1	I_4
$^2F_{5/2}$	6	I_2	I_3

Here $I_4 = 0$, since the transition $^2D_{3/2} \leftarrow {}^2F_{7/2}$ is not allowed because $\Delta J = -2$. Thus we write

$$\frac{I_1}{I_2 + I_3} = \frac{8}{6} = \frac{4}{3} \qquad \text{...(A)}$$

and

$$\frac{I_1 + I_2}{I_3} = \frac{6}{4} = \frac{3}{2} \qquad \text{...(B)}$$

Solving (A) and (B) for the intensity ratio, we get

$$3I_1 = 4I_2 + 4I_3$$
$$2I_1 + 2I_2 = 3I_3$$
$$I_3 = 14\,I_2$$
$$I_1 = 20I_2$$

Hence the intensity ratio is

$$I_1 : I_2 : I_3 :: 20 : 1 : 14$$

3.13. ATOMIC SPECTRA OF ONE VALENCE ELECTRON (THEORY)

3.13.1 *Hydrogen Atom Spectra*

The spectral lines of hydrogen atom are discussed earlier and thus do not want to repeat here again. The energy of different levels can be expressed in terms joules as

$$E_n(J) = -\frac{2\pi^2\mu e^4 Z^2}{(4\pi\varepsilon_0)^2 h^2 n_i^2} = -\frac{RhcZ^2}{n^2} \qquad \text{...(3.18)}$$

where

$$R(\text{cm}^{-1}) = \frac{2\pi^2\mu e^4}{(4\pi\varepsilon_0)^2 h^3 c}, \text{ Rydberg constant} \qquad \text{...(3.71)}$$

whereas in spectroscopy energy levels are commonly expressed by *term possessing the units* cm^{-1}. Terms can be obtained by dividing equation (3.18) by *hc*, denoted by the letters *T*, *i.e.*,

$$T_n(\text{cm}^{-1}) = -\frac{2\pi^2 \mu e^4 Z^2}{(4\pi\varepsilon_0)^2 h^3 c n_i^2} = -\frac{RZ^2}{n^2} \quad ...(3.72)$$

Coarse spectra of hydrogen atom are discussed in Chapter 2 in terms of Bohr's model.

3.13.2. *Spectra of Alkali Metals*

The outer most orbital in alkali atoms is *ns* and contains only one electron. Thus series regularities like those of hydrogen atom spectrum are also observed in the spectra of alkali elements. On the basis of analysis of the optical spectra of alkali metals, it can be grouped into the following four main series.

Sharp series: $$\bar{\nu} = \frac{R}{(2+p)^2} - \frac{R}{(m+s)^2}; \quad m = 2, 3, ... \quad ...(3.73)$$

This series arises due to transitions from the *s*-levels (exclusive of the lowest) and to the lowest of *p*–level.

Principal series: $$\bar{\nu} = \frac{R}{(1+s)^2} - \frac{R}{(m+p)^2}; \quad m = 2, 3, ... \quad ...(3.74)$$

This series is due to transitions between various *p*-levels and the lower *s*-level.

Diffuse series: $$\bar{\nu} = \frac{R}{(2+p)^2} - \frac{R}{(m+d)^2}; \quad m = 3, 4, ... \quad ...(3.75)$$

This series is due to transitions between various *d*-levels and lowest *p*-level.

Fundamental (or Bergmann) series: $$\bar{\nu} = \frac{R}{(3+d)^2} - \frac{R}{(m+f)^2}; \quad m = 4, 5, ... \quad ...(3.76)$$

This series arises from transitions from various *f*-levels to the lowest *d*-levels.

Here *s*, *p*, *d* and *f* (for hydrogen like atoms) or *S*, *P*, *D* and *F* (for multi-electronic atoms) are the characteristic constants of sharp, principal, diffuse and fundamental series respectively. The following series formulas in the abbreviated forms may be expressed as

$$\bar{\nu} = 1s \leftarrow mp, m \geq 2 \quad \text{Principal series}$$
$$\bar{\nu} = 2p \leftarrow ms, m \geq 2 \quad \text{Sharp series}$$
$$\bar{\nu} = 2p \leftarrow md, m \geq 3 \quad \text{Diffuse series} \quad ...(3.77)$$
$$\bar{\nu} = 3d \leftarrow mf, m \geq 4 \quad \text{Fundamental series}$$

Denoting the parameters *s*, *p*, *d*, *f*, by the letters "*a*" we can represent any term of alkali element by the general formula, called Rydberg formula.

$$T = \frac{R}{(m+a)^2} = \frac{R}{n_{\text{eff}}^2} \quad ...(3.78)$$

Comparing equation (3.78) with Balmer series, *i.e.*,

$$T = \frac{R}{n^2} \quad ...(3.79)$$

$$n = n_{\text{eff}} = m + a = n + \delta$$

where n_{eff} is known as the *effective quantum number* and parameter δ, independent of the principal quantum number n was proposed by Rydberg.

From experiment it is found that δ is negative and called *Rydberg correction*. Detailed form of δ was proposed by Ritz and is known as *Ritz correction*. The quantity $\Delta = |\delta|$ is known as the *quantum defect.* It is defined as the deviation of *effective quantum number* (n_{eff}) from the closest principal quantum number n. The sequence of the value of $n : n, n + 1, n + 2,$ .... is determined by the sequence of the values of the number $m: m, m + 1, m + 2,$ .... When a and δ are small, the numbers m and n coincide.

The values of n_{eff}, the effective quantum number for the lower *s*-, *p*-, *d*- and *f*- terms of alkali metals are listed in Table 3.8.

TABLE 3.8. Values of the effective quantum numbers of the lower *s*-, *p*-, *d*- and *f*-terms of atoms *f* the alkali metals

Elements terms	*s*	*P*	*d*	*F*
^{1}H	1.000	2.000	3.000	4.000
^{3}Li	1.588	1.959	2.998	4.000
^{11}Na	1.627	2.117	2.990	3.999
^{19}K	1.770	2.234	2.854	3.993
^{37}Rb	1.805	2.289	2.767	3.988
^{55}Cs	1.869	2.351	2.552	3.978

Further if the hydrogen atom term, characterised by the quantum number n, is expressed as

$$T_H = \frac{R}{n^2} \quad ...(3.80)$$

then on the basis of equations (3.78) and (3.80), we write

$$\frac{T_H}{T} = \frac{R}{n^2} \times \frac{(m+a)^2}{R} = \frac{(m+a)^2}{n^2} = \frac{(n+\delta)^2}{n^2} = \left(1 + \frac{\delta}{n}\right)^2$$

As mentioned earlier that δ has negative value, so

$$\sqrt{\frac{T_H}{T}} = 1 - \frac{\delta}{n}; \quad ...(3.81)$$

Thus, as n increases, T approaches T_H. For multi-electron ions, with one outer electron, the energy levels are expressed as

$$E = -\frac{hcRZ_a^2}{n_{\text{eff}}^2} \quad ...(3.82)$$

which is similar to formula (3.18). Here Z_a represents the charge of the atomic core of the ion. Parallel to the terms of atoms of alkali elements, *H* atom and terms of ions with one electron (He^+, Li^{2+}, etc.) we can write

$$T = \frac{|E|}{hc} = \frac{RZ_a^2}{n_{eff}^2} \qquad ...(3.83)$$

In terms of *effective nuclear charge* or *effective atomic number*, *i.e.*, Z – σ, equation (3.83) can be written as

$$T = \frac{R}{n^2} Z^{*2} = R\frac{(Z - \sigma)^2}{n^2}$$

or

$$\sqrt{\frac{T}{R}} = \frac{Z - \sigma}{n} \qquad ...(3.84)$$

Accordingly, the quantity $\sqrt{\frac{T}{R}}$ for various atoms and ions with one outer electron is a linear function of the nuclear charge Z. The formula in terms of frequency provided the screening constant σ for two combining levels is approximately the same, can be written as

$$\nu_{nm} = \frac{R(Z - z)^2}{n^2} - \frac{R(Z - z)^2}{m^2} = R\frac{m^2 - n^2}{m^2 n^2}(Z - z)^2$$

or

$$\sqrt{\frac{\nu_{nm}}{R}} = \frac{\sqrt{m^2 - n^2}}{n\, m}(Z - z) \qquad ...(3.85)$$

Here the quantity $\sqrt{\frac{\nu_{nm}}{R}}$ for various atoms and ions with one outer electron also varies linearly with Z. This relationship (3.85) corresponds to Moseley law first established for *x*-ray spectroscopy.

***Exercise* 3.47:** The ionisation energy of sodium atom from its ground state configuration is 5.138 eV. Calculate the (*a*) effective quantum number of this orbital and (*b*) quantum defect.

$$R_{Na} = 109734 \text{ cm}^{-1}.$$

The term value of the ground state of sodium atom, using equation (3.83) can be written as

$$T = \frac{|E|}{hc} = \frac{R_H Z_a^2}{n_{eff}^2} = \frac{R_{Na}}{n_{eff}^2}$$

$$n_{eff} = \left[\frac{RZ_a^2}{T}\right]^{\frac{1}{2}} = \left[\frac{(109734 \text{ cm}^{-1})(6.626 \times 10^{-34} \text{ J s})(3 \times 10^{10} \text{ cm s}^{-1})}{(5.138 \text{ eV})(1.6 \times 10^{-19} \text{ J eV}^{-1})}\right]$$

$$= 1.63$$

(*b*) Quantum defect = Z – n_{eff} = 11 – 1.63 = 9.37.

***Exercise* 3.48:** The principal and sharp series for the Li atom converge at 43487 and 28583 cm^{-1} respectively. If R_{Li} = 109729 cm^{-1}, calculate the quantum defect for the common term in each series.

The common term for sharp and principal series is ^2P term. The wave number position of this term is 28535 cm^{-1}. Using equation (3.83)

$$T = \frac{|E|}{hc} = \frac{R_H Z_a^2}{n_{\text{eff}}^2} = \frac{R_{Li}}{n_{\text{eff}}^2}$$

$$n_{\text{eff}} = \left[\frac{RZ_a^2}{T}\right]^{\frac{1}{2}} = \left[\frac{(109729 \text{ cm}^{-1})}{28583 \text{ cm}^{-1})}\right]^{\frac{1}{2}} = 1.9593$$

(*b*) Quantum defect = $Z - n_{\text{eff}} = 3 - 1.9593 = 1.0407$.

***Exercise* 3.49:** The convergence limits of principal and sharp series of sodium atom are at 41450 and 24477 cm^{-1} respectively. Calculate the ionisation energy of the sodium atom.

The ground state electronic configuration of sodium atom is [Ne]3 s^1, ^2S, *i.e.*, $n_i = 3$.

Thus principal series will give the convergence limit in reality. So the ionisation energy of sodium atom is 41450 cm^{-1}.

or $\quad = (41450 \text{ cm}^{-1})(3 \times 10^{10} \text{ cm s}^{-1})(6.626 \times 10^{-34} \text{ J s})/1.6 \times 10^{-19} \text{ J eV}^{-1} = 5.1496 \text{ eV}$.

3.13.3. *Fine (Multiplet) Structure of Atomic Terms (States)*

If we do not take into account the spin-orbit coupling, the spectrum obtained is called the *coarse structure*. If the interaction of spin-orbit coupling is taken into account, splitting in energy levels takes place. Spectrum so obtained is called the *fine structure* or *multiplet structure*. The terms which arise because of these interactions have already been discussed in Section 3.10. The values of *J* obtained because of spin-orbit coupling for one electron and two electrons are compiled in Tables 3.9, 3.10 and 3.11. The vector addition of these interactions has been shown in Figs. 3.9 to 3.10. where as splitted terms of singlet and triplat states of two electrons are compiled in table 3.12 and 3.13.

TABLE 3.9. Terms (states) of atoms and ions with one outer electron described by means of total angular momentum quantum number (j)

Terms	L	J			
S	0	1/2			
P	1	1/2	3/2		
D	2		3/2	5/2	
F	3			5/2	7/2

TABLE 3.10. Singlet terms of an atom with two electrons

Term	*L*	*J*				Split terms
^1S	0	0				$^1S_0$
^1P	1		1			$^1P_1$
^1D	2			2		$^1D_2$
^1F	3				3	$^1F_3$

TABLE 3.11. Triplet terms of an atom with two electrons

Term	L	*J*				
^3S	0		1			$^3S_1$
^3P	1	0	1	2		$^3P_0, {}^3P_1, {}^3P_2$
^3D	2		1	2	3	$^3D_1, {}^3D_2, {}^3D_3$
^3F	3		1	2	3 4	$^3F_2, {}^3F_3, {}^3F_4$

Note that the quantum conditions imposed on this type of couplings is that J can have only non-negative values.

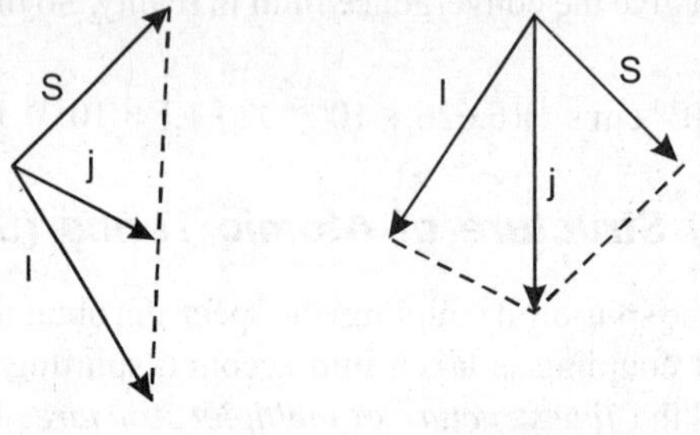

Fig. 3.9 *Interaction of orbital angular momentum l and the spin s.*

***Exercise* 3.50:** Consider two electrons with $l_1 = 2$ and $l_2 = 3$. Show pictorially and calculate the total values of orbital angular momentum and total angular momentum.

The mutual orientation of orbital angular momentum vectors, l_1 and l_2 (for $l_1 = 2$ and $l_2 = 3$) in the vector model of an atom with two electrons are shown in Fig. 3.10. The resultant L values can be 5, 4, 3, 2 and 1. Since this is a case of non-equivalent electrons, $S = 1/2 + 1/2 = 1$ and $S = 1/2 - 1/2 = 0$. The L and S vectors then combine to form a resultant J vector which represents the total orbital angular momentum of the whole electron system of the atom. Pictorial diagrams to show the vector addition of different L and S values are left for students.

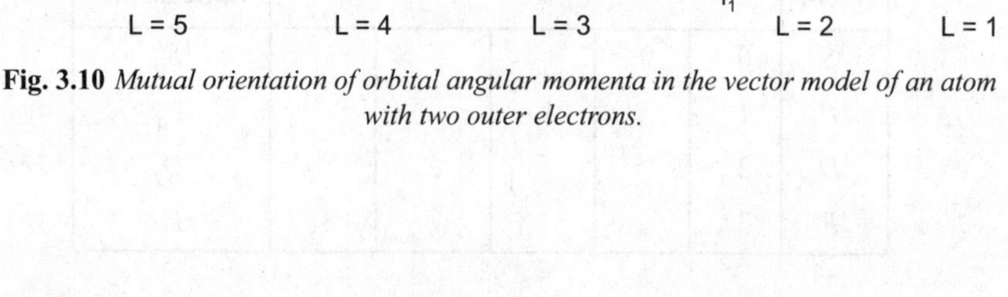

Fig. 3.10 *Mutual orientation of orbital angular momenta in the vector model of an atom with two outer electrons.*

TABLE 3.12. Singlet terms of an atom with two electrons

Term	L	J				Splitted terms
^1S	0	0				$^1S_0$
^1P	1		1			$^1P_1$
^1D	2			2		$^1D_2$
^1F	3				3	$^1F_3$

TABLE 3.13. Triplet terms of an atom with two electrons

Term	L	J					*Splitted terms*
^3S	0		1				$^3S_1$
^3P	1	0	1	2			$^3P_0, {}^3P_1, {}^3P_2$
^3D	2		1	2	3		$^3D_1, {}^3D_2, {}^3D_3$
^3F	3			2	3	4	$^3F_2, {}^3F_3, {}^3F_4$

Note that the quantum conditions imposed on this type of couplings is that J can have only non-negative values.

3.13.4. *Case (ii) j-j Coupling*

This type of interaction occurs in atoms of heavy elements or of higher atomic number. This type of interaction is between the orbital and spin angular momenta of one and the same electron. Consequently the interaction of vectors l_i with vector s_i gives rise to the resultant vector j_i such that

$$j_i = l_i + s_i \quad (i = 1, 2,)$$

or

$$j_1 = l_1 + s_1$$

$$j_2 = l_2 + s_2$$

The interaction of the vectors j_1 and j_2 which is weaker than the interaction of the vectors l_i and s_i gives rise to the total angular momentum vector of the atom

$$J = j_1 + j_2 \qquad ...(3.86)$$

The total angular momentum vector J is characterised by the quantum number J. In case of (j, j) coupling, the quantum numbers L and S have no meaning (or not specified nicely).

***Exercise* 3.51:** Consider *sp* configuration of a heavy metal. Calculate the resulting terms.

For *sp* configuration, we have

$$l_1 = 0 \text{ and } l_2 = 1$$

So

$$L = 1 \text{ and } S = 1 \text{ and } 0$$

Thus in the *R-S* scheme, the terms will be $^3P_0, {}^3P_1, {}^3P_2$ and $^1P_1$.

For (j, j) coupling we have

$$l_1 = 0, l_2 = 1$$

$$s_1 = 1/2, s_2 = 1/2$$

Hence

$$j_1 = l_1 + s_1 \text{ to } |l_1 - s_1|, j_1 = 1/2$$

and $$j_2 = l_2 + s_2 \text{ to } |l_2 - s_2|, j_2 = 1 + 1/2 \text{ to } 1 - 1/2 = 3/2 \text{ and } 1/2$$

Therefore (*i*) $$J = j_1 + j_2 \text{ to } |l_2 - j_2| = \frac{1}{2} + \frac{1}{2} \text{ to } \left|\frac{1}{2} - \frac{1}{2}\right| = 1 \text{ and } 0$$

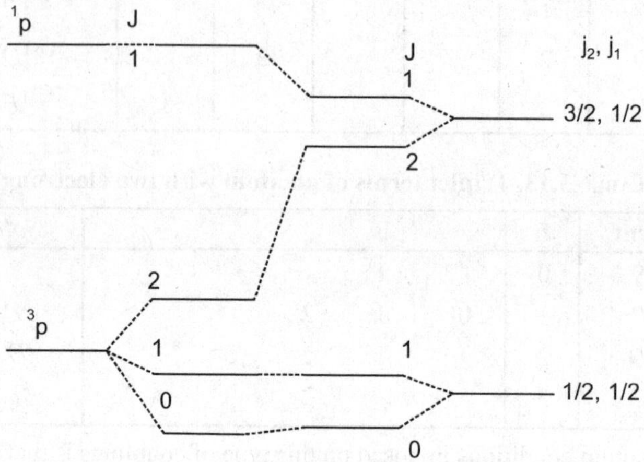

Fig. 3.11 *Relative positions of the terms of sp configuration in L-S coupling (left) and in (j, j) coupling (right).*

(*ii*) $$J = j_1 + j_2 \text{ to } |l_1 - j_2| = \frac{3}{2} + \frac{1}{2} \text{ to } \left|\frac{3}{2} - \frac{1}{2}\right| = 2 \text{ and } 1.$$

The relative positions of the atomic terms corresponding to this electron configuration (*sp*) are shown in Fig. 3.11. Terms obtained in *L-S* coupling are shown to the left while those obtained in (*j, j*) coupling is shown to the right. Since (*j, j*) coupling are not very common we will be considering only atomic terms based on *R-S* coupling. Term symbols arsing from different electronic configurations in case of non-equivalent and equivalent electrons are compiled in Tables 3.14 and 3.15.

TABLE 3.14. Term symbols of non-equivalent electrons for different n

Electronic configuration	Terms
s s	$^1S, ^3S$
s p	$^1P, ^3P$
s d	$^1D, ^3D$
p p	$^1S, ^1P, ^1D, ^3S, ^3P, ^3D$
p d	$^1P, ^1D, ^1F, ^3P, ^3D, ^3F$
d d	$^1S, ^1P, ^1D, ^1F, ^1G, ^3S, ^3P, ^3D, ^3F, ^3G$
s s s	^2S (2), ^4S
s s p	^2P (2), ^4P
s p p	^2S (2), ^2P (2), ^2D (2), $^4S, ^4P, ^4D$

$s\,p\,d$	^2P (2), ^2D (2), ^2F (2), ^4P, ^4D, ^4F
$p\,p\,p$	^2S (2), ^2P (5), ^2D (6), ^2F (2), ^4S, ^4P (3), ^4D (2), ^4F
$p\,p\,d$	^2S (2), ^2P (4), ^2D (6), ^2F (4), ^2G (2), ^4S , ^4P (2), ^4D (3), ^4F (2), ^4G
$p\,d\,f$	^2S (2), ^2P (4), ^2D (6), ^2F (6), ^2G (6), ^2H (4), ^2I (2), ^4S, ^4P (2), ^4D (3), ^4F (3), ^4G (3), ^4H (2), ^4I

TABLE 3.15. Term symbols for equivalent electrons (n and l are the same)

Electronic configuration	Terms
s^2	^1S
p^2 or p^4	^1S, ^1D, ^3P
p^3	^2P, ^2D, ^4S
p^1 or p^5	^2P
p^6	^1S
d^2	^1S, ^1D, ^1G, ^3P, ^3F
d^3	^2P, ^2D (2), ^2F, ^2G, ^2H, ^4P, ^4F
d^4	^1S (2), ^1D (2), ^1F, ^1G (2), ^1I, ^3P (2), ^3D, ^3F (2), ^3G, ^3H, ^5D
d^5	^2S, ^2P, ^2D(3), ^2F (2), ^2G (2), ^2H, ^2I, ^4P, ^4D, ^4F, ^4G, ^6S

3.13.5. *Atomic Spectra (EXPERIMENTAL)*

Atomic spectra of one electron and two electron systems will be discussed in the form of solved exercises.

3.13.5.1. Atomic spectra of hydrogen atom

***Exercise* 3.52:** Transitions observed in the hydrogen spectrum up to $n = 4$ are shown in Fig. 3.12. Left hand side depicts the principal quantum number, *n*, whereas the right hand side depicts *J* values. Spectroscopic term symbols are given at the top. Indicate the doublet transition observed in sharp series.

We know that hydrogen atom contains one electron. Energy level diagram shown in Fig. 3.12, depicts the effect of *l-s* coupling. There will be only two value of *j*. due to *l-s* coupling. The *l* (*L*) values is shown at the top of the energy level diagram, *i.e.*, *S*, *P*, *D*,.... . The higher states will also follow the same trend. *It may be mentioned here that the gap between states arisen by the l-s coupling is much smaller as compared to the gap between the two states based on the l values only. Lastly the gap between the j states decreases as the n and l values increases* .

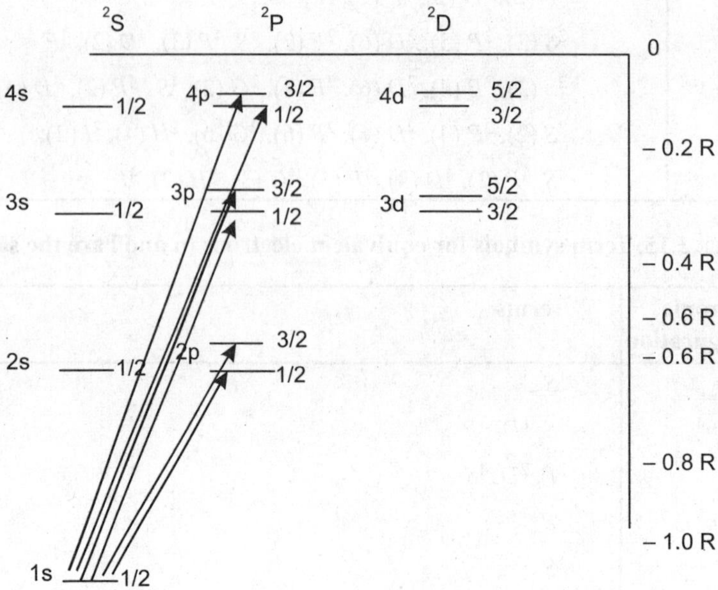

Fig. 3.12 *Energy level diagram of lower states of hydrogen atom and also depicting the l-s splitting. The splitting shown is not to the scale.*

Different transitions observed will be based on the following selection rules, *i.e.*

$$\Delta n = \text{any combination}, \Delta l = \pm 1 \text{ only}, \Delta j = 0, \pm 1$$

The transitions starting from the ground state will be

$$^2S_{1/2} \rightarrow {}^2P_{1/2} \quad (\Delta j = 0)$$

$$^2S_{1/2} \rightarrow {}^2P_{3/2} \quad (\Delta j = 1)$$

These transitions are identical to Lyman series with the difference that each Lyman line will be split into two lines, *i.e.*, *every line will be a doublet*. This doublet is only observed under high resolution spectrometer as the gap between these lines is quite small.

***Exercise* 3.53:** Now consider the transition between $^2P \rightarrow {}^2D$ in the hydrogen atom spectrum. Suggest the fine structure observed in this line.

In the initial state, $l = 1$ and $s = 1/2$. So the j values can be 1/2 and 3/2. Similarly for the final state, $l = 2$, $s = 1/2$. So j values can be 3/2 and 5/2. Each state is split into two states due to different j values and energies of the states have been shown in Fig. 3.13.

The allowed transitions are

(*i*) $^2P_{3/2} \rightarrow {}^2D_{3/2}$ $(\Delta j = 0)$ (*ii*) $^2P_{3/2} \rightarrow {}^2D_{5/2}$ $(\Delta j = 1)$ (*iii*) $^2P_{1/2} \rightarrow {}^2D_{3/2}$ $(\Delta j = 1)$

and the forbidden one is

$$^2P_{1/2} \rightarrow {}^2D_{5/2} \ (\Delta j = 2)$$

This is quite clear that this spectrum consists of three lines and is called *compound doublet spectrum*.

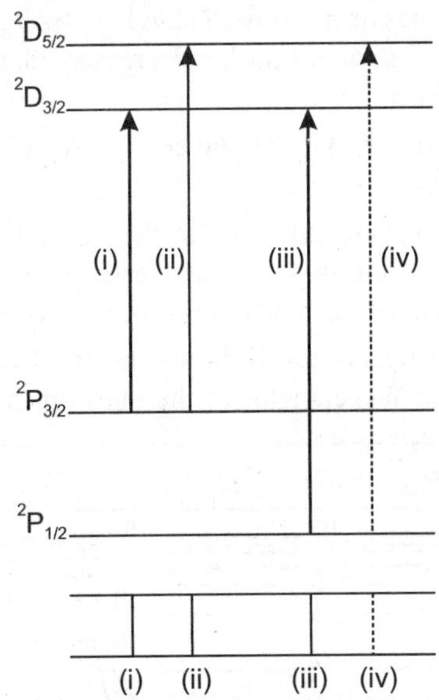

Fig. 3.13 *The fine structure observed in the transition between $^2P \rightarrow {}^2D$ of the hydrogen atom, also called compound doublet. The splitting between the states is not to the scale.*

In conclusion, it may be mentioned here that if in the transitions one of the state is s-orbital, the lines observed will be doublet and if the transitions are between other states, the spectral lines will be compound doublet.

3.13.5.2. Atomic spectra of alkali elements

In the electronic structure of alkali elements, the core of an inert gas is surrounded by an *s*-electron as shown in Table 3.16. Electron configuration of alkali elements is identical to hydrogen atom as core

TABLE 3.16. Electronic configuration of alkali elements

Element	Electron configuration
Li	$[He]2s^1$
Na	$[Ne]3s^1$
K	$[Ar]4s^1$
Rb	$[Kr]5s^1$
Cs	$[Xe]6s^1$

electrons do not play any role in the optical spectra but the extra electron called *valence* or *optical electrons* give rise to optical atomic spectra. Thus the spectra of alkali metals are very similar to each other.

The magnitude of doublet splitting of a term is measured by the values of Z, n and l (or n_{eff}). The doublet splitting of a term of alkali elements increases with Z. For example, the doublet splitting ΔT of lower ^2P terms falls in the range ^{3}Li (0.34) – ^{55}Cs (554.0) cm^{-1}. That is why the doublet splitting of *D*- and *F*-terms are resolved only in heavy elements. Further ΔT decreases sharply as the principal quantum number n (particularly for small values of n) grows, *i.e.*, $\Delta T_{3p} >> \Delta T_{4p} >> \Delta T_{5p}$ .... in case of sodium and for Rb; $\Delta T_{5p} >> \Delta T_{6p} >> \Delta T_{7p}$........ A still further decrease with the increase of azimuthal quantum number is noticed for fixed value of n.

i.e., $\Delta_{np} > \Delta_{nd} > \Delta_{nf}$.............For example, Δ_{5p} (237.60 cm^{-1}) > Δ_{5d} (2.96 cm^{-1}) in case of Rb and in case Cs, Δ_{6p} (554.0 cm^{-1}) > Δ_{6d} (42.8 cm^{-1}).

***Exercise* 3.54:** The longest wavelength lines in the series (n, l) → (4, 0) in potassium have wavelengths 344.77, 344.67, 404.72 and 404.1 nm respectively. Construct the relevant energy level diagram and compute in the splitting between the levels with same value of n and l but with different J.

The energy level diagram of potassium is shown in Fig. 3.14 *a*, given below and the relevant transitions are shown in Fig. 3.14 *b*. Wavelengths of the lines given in the example correspond to the

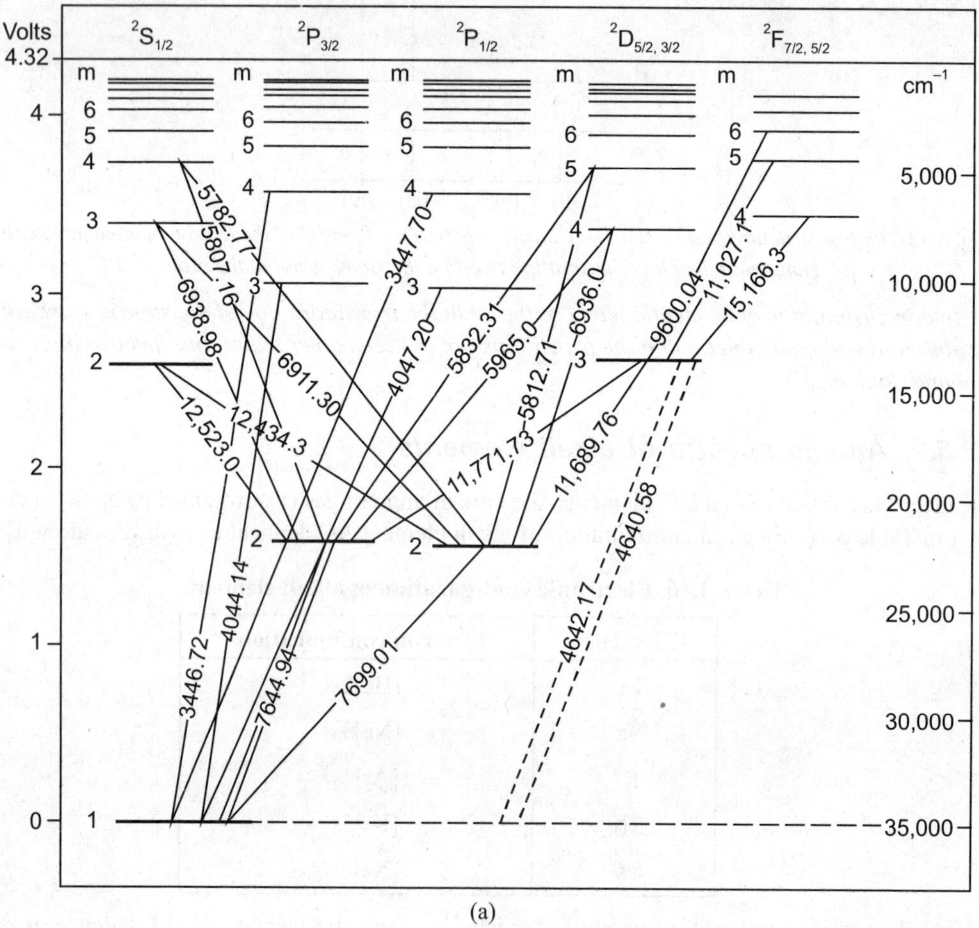

(a)

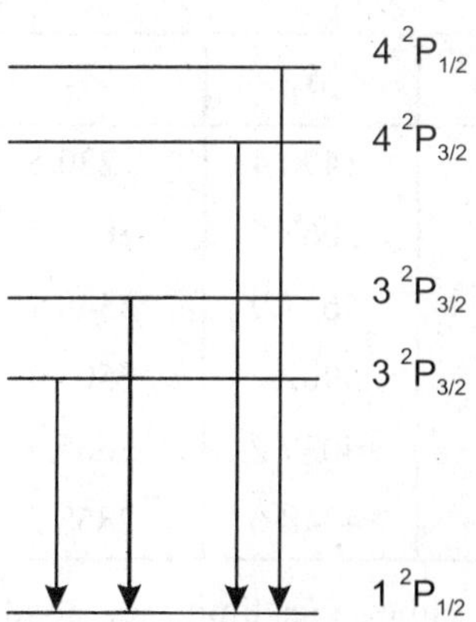

(b)

Fig. 3.14 *(a) The energy level diagram of potassium atom and (b) relevant transitions corresponding to the wavelengths give.*

following transitions:

$$4\,^2P_{1/2} \rightarrow 1^2S_{1/2}\,(344.77 \text{ nm})$$
$$4\,^2P_{3/2} \rightarrow 1^2S_{1/2}\,(344.67 \text{ nm})$$
$$3\,^2P_{1/2} \rightarrow 1^2S_{1/2}\,(404.72 \text{ nm})$$
$$3\,^2P_{3/2} \rightarrow 1^2S_{1/2}\,(404.41 \text{ nm})$$

The energy level splitting for $4P$ level = (344.77 – 344.67) = 0.10 nm and for $3P$ level = (404.72 – 404.41) = 0.31 nm. Since

$$\nu = \frac{c}{\lambda}$$

$$d\nu = \frac{c}{\lambda^2}|d\lambda| \quad \text{or} \quad \frac{d\nu}{c} = \frac{d\lambda}{\lambda_1\lambda_2} \quad \text{or} \quad \frac{\Delta\nu}{c} = \frac{\Delta\lambda}{\lambda_1\lambda_2}$$

Thus

$$\left(\frac{\Delta\nu}{c}\right)_{4P} = \frac{0.1 \times 10^{-9}}{344.77 \times 10^{-9} \times 344.67 \times 10^{-9}} m^{-1} = 841.52 \text{ m}^{-1}(8.415 \text{ cm}^{-1})$$

$$\left(\frac{\Delta\nu}{c}\right)_{3P} = \frac{0.31 \times 10^{9}}{404.72 \times 404.41} m^{-1} = 1894.02 \text{ m}^{-1}(18.9 \text{ cm}^{-1})$$

Note: Figure 3.14*a* is given so that one can construct many problems from the data given in it.

***Exercise* 3.55:** The wave numbers in mm^{-1} obtained from the spectrum of lithium have been divided into spectral series. The series limit of S_1 is 4348.6 mm^{-1} while that of both S_2 and S_3 is 2858.2 cm^{-1}. Construct to scale (approximately) an energy level diagram and identify each series. Determine the series principal quantum number of the lowest term of each sequence. The Rydberg constant R_{Li} = 10973.7 cm^{-1}.

Variation number	S_1	S_2	S_3
1	1490.4	1230.5	1638.4
2	3063.5	2011.3	2172.5
3	3647.9	2340.1	2420.0
4	3902.2	2508.9	2554.3
5	4039.2	2605.4	2685.3
Limit	4348.6	2858.2	2858.2

Since S_2 and S_3 series have the same series limit, *i.e.*, 2858.2 cm^{-1} so they must be based on the same energy level. Accordingly the series S_2 and S_3 most probably corresponds to sharp and diffuse spectral series, but this evidence is not sufficient to distinguish between the two spectral series. So in order to distinguish between the two spectral series, we find the other energy levels by the subtraction process which are listed below in Fig. 3.15.

$A = (2858.2 - {}^2S_{5,4,3,2,1})$, $B = (4348.6 - {}^1S_{5,4,3,2,1})$ $C = (2858.2 - {}^3S_{5,4,3,2,1})$

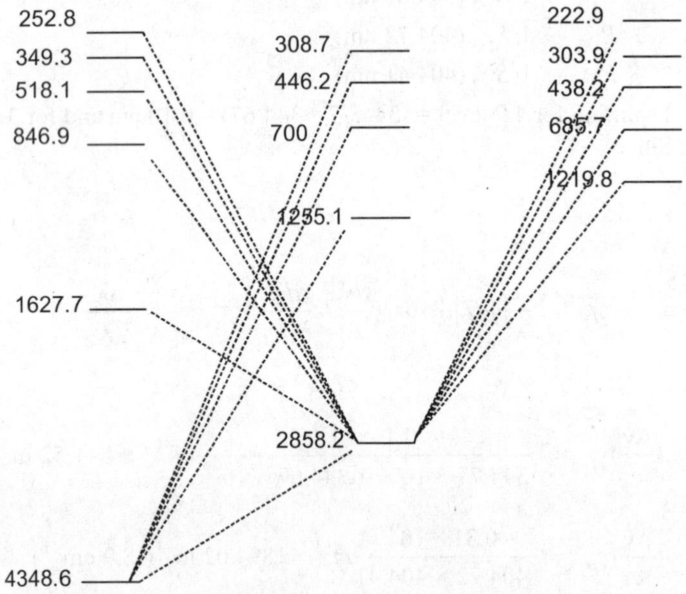

Fig. 3.15 *Energy levels obtained by the subtraction process. Lines drawn between the various figures indicate the given series S_1, S_2 and S_3.*

The principal equantum number is given by the formula

$$\bar{\nu} = \frac{R}{n^2} \quad \text{or} \quad n = \left[\frac{R}{\bar{\nu}}\right]^{\frac{1}{2}}$$

$$n_A = \left[\frac{10.937 \times 10^6}{4348.6 \times 10^3}\right]^{\frac{1}{2}} = 1.59$$

$$n_B = \left[\frac{10.937 \times 10^6}{42858.2 \times 10^3}\right]^{\frac{1}{2}} = 1.96$$

$$n_C = \left[\frac{10.937 \times 10^6}{1219.8 \times 10^3}\right]^{\frac{1}{2}} = 3.00$$

The term under C is free from quantum defect, *i.e.*, deviation from the closet integer, the principal quantum number so it belongs to D-terms. The term under number A have the largest quantum defect so it belongs to S-terms while those under B belong to P-terms. This is evident from the fact that the S-terms of alkali element are the lowest. Accordingly the three spectral series S_1, S_2 and S_3 correspond to principal, sharp and diffuse respectively.

***Exercise* 3.56:** (*a*) Using the limiting values of different series of lithium atom, calculate the ionisation energy of lithium atom. (*b*) If $R = 10972.9 \text{ mm}^{-1}$, calculate the quantum defect in the principal quantum numbers of these states.

As mentioned in Exercise (3.55) two series start from the same level and one from different level. Thus there will be two ionisation energies; one is arising most probably from 2 *s* level and other one from 2*p* level. So

$$\text{I.P.} = hc\,\bar{\nu}_{\text{lim}} = (6.626 \times 10^{-34} \text{ J s})(3 \times 10^{10} \text{ cm s}^{-1})(10 \text{ mm cm}^{-1})(4348.6 \text{ mm}^{-1})$$

$$= 8.644 \times 10^{-19} \text{ J} = \frac{8.644 \times 10^{-19} \text{ J}}{1.6 \times 10^{-19} \text{ J eV}^{-1}} = 5.40 \text{ eV}$$

Similarly from the IInd limiting term

$$\text{I.P.} = hc = (6.626 \times 10^{-34} \text{ J s})(3 \times 10^{10} \text{ cm s}^{-1})(10 \text{ mm cm}^{-1})(2858.2 \text{ mm}^{-1})$$

$$= 5.6815 \times 10^{-19} \text{ J} = \frac{5.6815 \times 10^{-19} \text{ J}}{1.6 \times 10^{-19} \text{ J eV}^{-1}} = 3.55 \text{ eV}$$

(*b*) Using equation (3.83), we get

$$T = \frac{R}{n_{\text{eff}}^2}$$

$$n_{\text{eff}} = \left[\frac{R_{Li}}{T}\right]^{\frac{1}{2}} = \left[\frac{10972.9}{4348.7}\right]^{\frac{1}{2}} = 1.59$$

So quantum defect = 2 – 1.59 = 0.41 for the ground state of lithium atom.

Similarly for the second limiting series

$$n_{\text{eff}} = \left[\frac{R_{Li}}{T}\right]^{\frac{1}{2}} = \left[\frac{10972.9}{2858.2}\right]^{\frac{1}{2}} = 1.96$$

and quantum defect for the next level of lithium, *i.e.*, 2*p*. = 2 – 1.96 = 0.04

3.13.5.3. Spectrum of helium atom (or two electron systems)

Helim atom contains two valence electrons and total binding energy of He-atom is 78.98 eV. The energy levels and the observed emission spectrum of He-atom are shown in Fig. 3.16. Using the Fig. 3.16 one

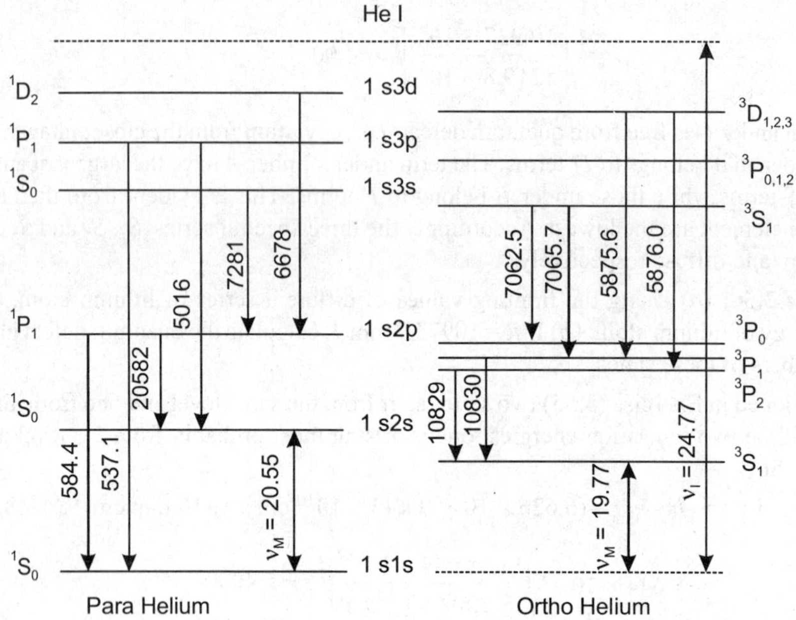

Fig. 3.16 *Energy levels and emission spectrum of helium atom.*

can form many problems as lot of data is given. The transitions on left hand side of the diagram are all between helium states in which two helium electrons are paired (anti-parallel) called the *singlet states*. So for singlet states $S = 0$ and ∞ and multiplicity $= 2S + 1 = 1$. While the transitions on the right hand side in Fig. 3.16 are between states in which electrons are unpaired, (*i.e.* parallel) called the *triplet sates*. Thus for triplet states $S = 1$ and multiplicity is $2S + 1 = 3$. The data for different series are given in Table 3.17.

TABLE 3.17. Different series observed in case of helium atom

Para-helium		Ortho-helium	
Series	**Transition**	**Series**	**Transition**
D	$1s3d\ ^1D_2 \rightarrow 1s2p\ ^1P_1$ (667.8 nm)	D	$1s3d\ ^3D_{1,2,3} \rightarrow 1s2p\ ^3P_2$ (587.60 nm) $1s3d\ ^3D_{1,2,3} \rightarrow 1s2p\ ^3P_1$ (587.56 nm)
P	$1s3p\ ^1P_1 \rightarrow 1s2s\ ^1S_0$ (501.6 nm) $1s3p\ ^1P_1 \rightarrow 1s1s\ ^1S_0$ (537.1 nm) $n = 3$ $m = 1$(normal state) $1s2p\ ^1P_1 \rightarrow 1s2s\ ^1S_0$ (2058.2 nm)	P	$1s2p\ ^3P_0 \rightarrow 1s2s\ ^3S_1$ (1083.0 nm) $1s2p\ ^3P_2 \rightarrow 1s2s\ ^3S_1$ (1082.9 nm)

Contd...

	$1s2p\ ^1P_1 \rightarrow 1s1s\ ^1S_0$ (584.4 nm) $n = 2 \quad m = 1$ (normal state)		
S	$1s3s\ ^1S_0 \rightarrow 1s2p\ ^1P_1$ (728.1 nm)	S	$1s3s\ ^3S_1 \rightarrow 1s2p\ ^3P_0$ (706.25 nm) $1s3s\ ^3S_1 \rightarrow 1s2p\ ^3P_2$ (706.57 nm)

The spectrum of Li^+ and Be^{++} etc. is similar to that of He-atom.

***Exercise* 3.57:** Six lines lying close to each other in the spectrum of strontium (Sr) are observed. The respective wave numbers are 20109, 20124, 20147, 20503, 20518 and 20690 cm $^{-1}$. Assign these transitions corresponding to the energy levels.

Sr atom has two valence electrons. Thus the ground state electronic configuration will be ns^2. The term symbol will be $^1S_0$. Similar to helium atom, it can have two set of lines, one arising from the singlet terms and other from triplet terms. The triplet states will have the electronic configurations as ns^1np^1 ^3P_J or ns^1nd^1 ^3D_J... . If we assume that spin-orbit coupling is taking place, the terms arising from the singlet states will be only one each as in this case $J = L + 0 = L$, whereas in case of triplet $J = L + S, L + S - 1 .....|L - S|$. Since the lines observed are quite close to each other (*i.e.*, gap or frequency

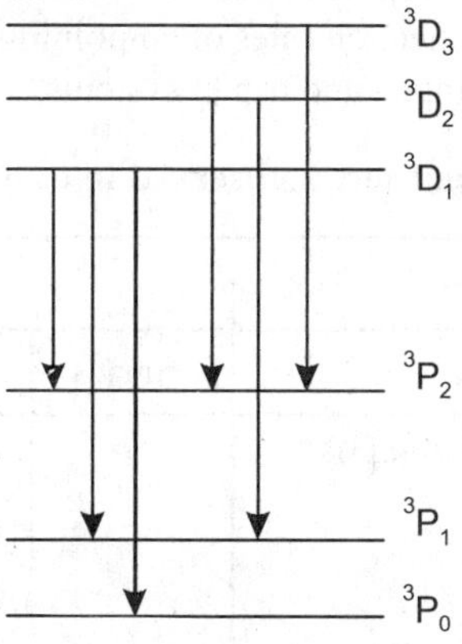

Fig. 3.17 *Energy level diagram for strontium atom.*

difference are 0, 15, 38, 394, 409 and 581 cm^{-1}), the possible six lines can arise from the $^3P \rightarrow {^3D}$ as shown in Fig. (3.17). Taking into account *L-S* coupling, the possible states are $^3P_2$, $^3P_1$ and $^3P_0$ for ^3P and $^3D_3$, $^3D_2$ and $^3D_1$ for ^3D terms. The energy level diagram is shown in Fig. 3.17. Above assignment can be looked from the interval rule, which will show up as the ratio of small integral numbers between some of the frequency differences. Using the above given frequency differences, it is found that

$$\frac{38-15}{15-0} = \frac{23}{15} = \frac{3}{2}$$

and
$$\frac{394-0}{581-394} = \frac{394}{187} = \frac{2}{1}$$

***Exercise* 3.58.** Calculate the effective quantum number ($n - \mu$) and effective nuclear charge ($Z - \sigma$) for the electron configuration $4s4p$ of Ca. The term value is 33989 cm^{-1} and R = 109737 cm^{-1}.

Using equation (3.83)

$$n_{\text{eff}} = \left[\frac{RZ_a^2}{T}\right]^{\frac{1}{2}} = \left[\frac{R_{Ca}}{T}\right]^{\frac{1}{2}} = \left[\frac{109737}{33989}\right]^{\frac{1}{2}} = 1.7968$$

The quantum defect, $\mu = 4 - 17968 = 2.2032$

Using equation (3.83) and substituting the desired values, we get

$$Z - \sigma = \left[\frac{T\, n^2}{R}\right]^{\frac{1}{2}} = \left[\frac{33989 \times 4^2}{109737}\right]^{\frac{1}{2}} = 2.226\,.$$

3.13.5.4. Spectrum of mercury atom

The atomic number of mercury is 80 and its electronic configuration is $1s^2 2s^2 2p^6 3s^2 3p^6 3d^{10} 4\ s^2$ $4p^6 4d^{10} 5s^2 5p^6 5d^{10} 6s^2$. Its energy levels and emission spectrum shown in Fig. 3.18 are due to the excitation of one of the two $6s$ optical electrons. The state of second $6s$ electron is held constant as that of second $1s$ electron in the He-atom. The spectrum is similar to that of He-atom except some inter-combinations which otherwise are forbidden in normal spectrum of He-atom are allowed in case of mercury. It follows from Fig. 3.18 that as per the normal selection rules of combining terms, the following first set of lines in the main series is observed in the singlet and triplet systems. These are compiled in Table 3.18.

TABLE 3.18. Different series observed in case of mercury atom

Singlet system		Triplet system	
Series	**Transition**	**Series**	**Transition**
P	$6s6p\ ^1P_1 \rightarrow (6s)^2\ ^1S_0$ (184.957)	P	$6s6p\ ^3P_0 \rightarrow 6s7s\ ^3S_1$
			$6s6p\ ^3P_1 \rightarrow 6s7s\ ^3S_1$
			$6s6p\ ^3P_2 \rightarrow 6s7s\ ^3S_1$
S	$6s7s\ ^1S_0 \rightarrow 6s6p\ ^1P_1$ (1014.0)	S	$6s7s\ ^3S_1 \rightarrow 6s6p\ ^3P_0$ (404.66 nm)
			$6s7s\ ^3S_1 \rightarrow 6s6p\ ^3P_1$ (435.83 nm)
			$6s7s\ ^3S_1 \rightarrow 6s6p\ ^3P_2$ (546.07nm)
D	$6s6d\ ^1D_2 \rightarrow (6s6p\ ^1P_1$ (579.07)	D	$6s6d\ ^3D_3 \rightarrow 6s6p\ ^3P_2$ (365.02 nm)
			$6s6d\ ^3D_2 \rightarrow 6s6p\ ^3P_1$ (312.57 nm)
			$6s6d\ ^3D_2 \rightarrow 6s6p\ ^3P_2$ (365.48 nm)
			$6s6d\ ^3D_1 \rightarrow 6s6p\ ^3P_0$ (296.73 nm)
			$6s6d\ ^3D_1 \rightarrow 6s6p\ ^3P_1$ (313.16 nm)
			$6s6d\ ^3D_1 \rightarrow 6s6p\ ^3P_2$ (366.292 nm)

TABLE 3.19. Transitions observed in mercury atom involved due to intercombination

Intercombination system	Wavelength in nm
$6s6d\,^3D_2 \rightarrow 6s6p\,^1P_1$	576.96
$6s6d\,^3D_1 \rightarrow 6s6p\,^1P_1$	578.97
$6s6d\,^1D_2 \rightarrow 6s6p\,^3P_1$	313.18
$6s6d\,^1D_2 \rightarrow 6s6p\,^3P_2$	386.33
$6s7s\,^1S_0 \rightarrow 6s6p\,^3P_1$	407.78
$6s6p\,\,^3P_1 \rightarrow 6s6p\,^1S_0$	263.652, this is one of the most intense line Hg spectrum and represent the resonance line of principal series of the inter-combination system.

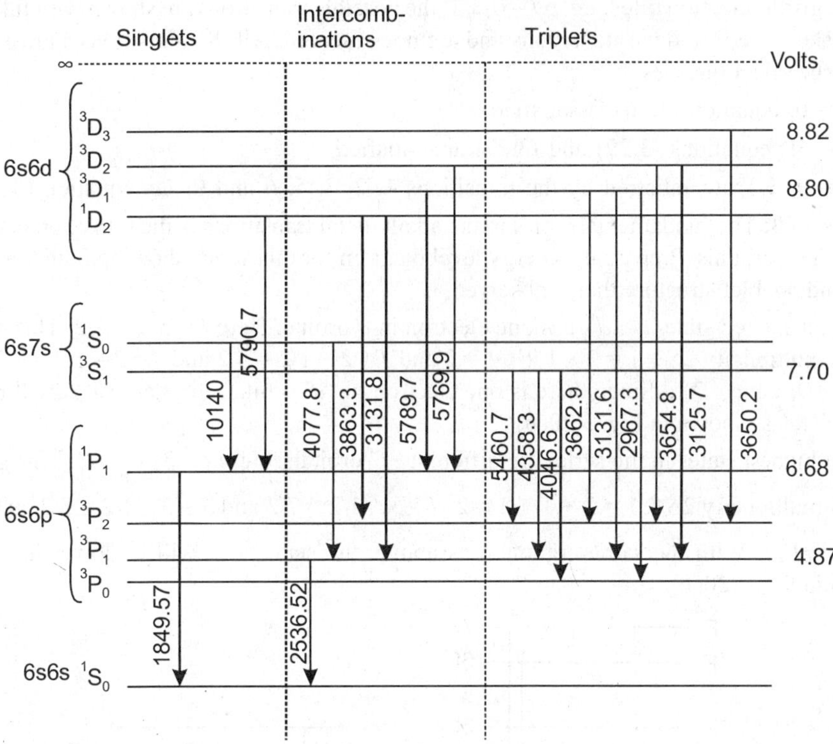

Fig. 3.18 *Energy levels and emission spectrum of mercury atom.*

***Exercise* 3.59:** Draw the energy level diagram for ^3D and ^3p states, taking into account the $L - S$ coupling and predict the possible transitions.

Considering the $L - S$ coupling the possible J values for each state are

$$^3D, L = 2, S = 1, J = 3, 2, 1$$

$$^3P, L = 1, S = 1, J = 2, 1, 0$$

If these states are formed from less than half filled shell, then these can be arranged as

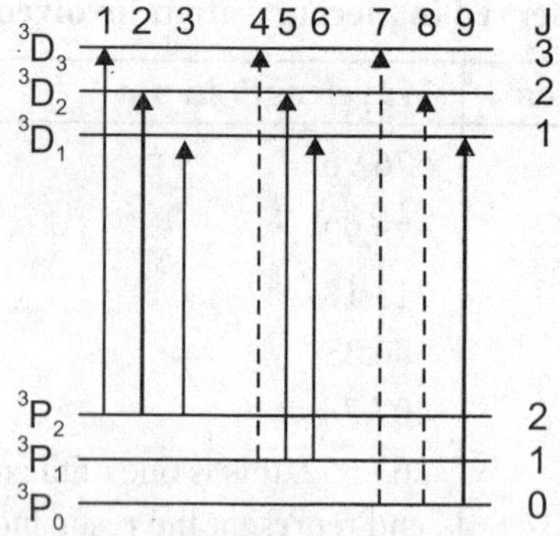

Fig. 3.19

With the given selection rules, *i.e.* $\Delta J = 0, \pm 1$, the possible transitions are shown with full lines. The dotted lines show the forbidden transitions and are not observed. All the transitions shown with solid lines satisfy the selection rules

1. $\Delta S = 0$, equation (3.26) is satisfied.
2. $\Delta l = +1$, equations (3.29) and (3.27) are satisfied.
3. $\Delta J = 0, \pm 1$ are satisfied by the transitions 1, 2, 3, 5, 6 and 9, *i.e.* equation (3.28).

***Exercise* 3.60:** The fundamental series in the alkali metal transitions is the transition between the *f* orbitals and the *d* orbitals. Prepare the energy level diagram for this series showing both the transitions and compound doublet structure that is observed.

The lowest excited state, *i.e.* d^1, has one electron in *d* orbital. Here $l = 2$, $s = 1/2$. This gives $L = 2$, $S = 1/2$, spin multiplicity $2S + 1 = 2 \times 1/2 + 1 = 2$ and $J = 2 + 1/2 = 5/2$ and $J = 2 - 1/2 = 3/2$. The term symbols are $^2D_{5/2}$ and $^2D_{3/2}$. Since there is one electron in '*d*' orbital, the state $^2D_{3/2}$ will be of lower energy than $^2D_{5/2}$ as shown in Fig. 3.20(*b*).

The next highest states in the series are different '*f*' orbitals. Here $l = 3$, $s = 1/2$. This gives $L = 3$, $S = 1/2$, spin multiplicity $2S + 1 = 2 \times \frac{1}{2} + 1 = 2$, $J = 3 + 1/2 = 7/2$ and $J = 3 - 1/2 = 5/2$, and the terms are $^2F_{7/2}$ and $^2F_{5/2}$. With the same argument as above, the state $^2F_{5/2}$ will be lower in energy than $^2F_{7/2}$, shown in Fig. 3.20(*b*).

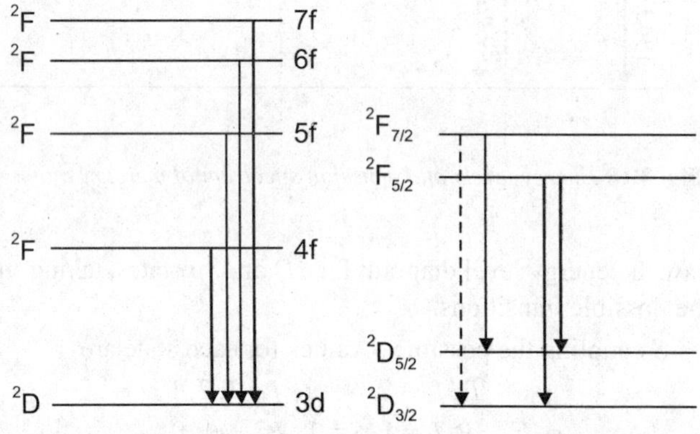

Fig. 3.20 *Fundamental transitions in alkali metals: (a) coarse structure (b) fine structure*

The spectrum consists of transitions between ^2F and the ^2D state. The selection rule (3.29) is satisfied, *i.e.* $\Delta l = 1$, (3.26) is obeyed because $\Delta S = 0$, *i.e.* doublet to doublet, (3.27) is satisfied because $\Delta L = 1$, *i.e.* $F \to D$, (3.28) is satisfied because $7/2 \leftrightarrow 5/2$, $5/2 \leftrightarrow 3/2$ satisfy $\Delta J = \pm 1$ and $5/2 \leftrightarrow 5/2$ satisfies $\Delta J = 0$. The transition shown as dotted line is forbidden because this involves $7/2 \leftrightarrow 3/2$, *i.e.* $\Delta J = 2$. Figure 3.20(*a*) shows the overall fundamental series of transitions and Fig. 3.20(*b*) shows the compound doublet.

3.14. AN ATOM IN MAGNETIC FIELD

When an atom is subjected to an external magnetic, as mentioned in Section (3.5.3) the degeneracy of atomic levels having the same value of azimuthal quantum number is lifted. Consequently splitting of the observed spectral lines takes place. The splitting of the atomic line under the influence of magnetic field is called *Zeeman effect*. For a given value of J, there are $2J + 1$ values of m_J, having the same energy in the absence of external magnetic field. But this degeneracy is removed in the presence of external magnetic field and split into $2J + 1$ energy levels.

3.14.1. *Normal Zeeman Effect*

In a strong magnetic field, the energy of interaction between the magnetic moment of the total angular momentum and external magnetic field varies by the magnitude

$$\Delta E = M_J \mu_B H \qquad \text{...(3.87)}$$

where μ_B is the Bohr magnetone and H is the intensity of the magnetic field. Thus the total energy of the atom in the upper and lower states in a magnetic field is equal to

$$E' = E_0' + M_J' \mu_B H \qquad \text{...(3.88)}$$

$$E'' = E_0'' + M_J'' \mu_B H \qquad \text{...(3.89)}$$

Hence it follows

$$\nu = \frac{E' - E''}{h} = \frac{E_0' - E_0''}{h} + \frac{(M_J' - M_J'')\mu_B H}{h} = \nu_0 + \frac{\Delta M_J \mu_B H}{h} \qquad \text{...(3.90)}$$

Applying the selection rules $\Delta M_J = 0, \pm 1$, we obtain

$$\begin{aligned} \nu_{+1} &= \nu_0 + \frac{\mu_B H}{h} \\ \nu_0 &= \nu_0 \\ \nu_{-1} &= \nu_0 - \frac{\mu_B H}{h} \end{aligned} \qquad \text{...(3.91)}$$

The original spectral line has thus been split into three components with frequencies.

$$\nu_0 - \frac{\mu_B H}{h} \qquad \nu_0 \qquad \nu_0 + \frac{\mu_B H}{h}$$

This set of three lines is also known as *Lorentz triplets*. The change in frequency $\Delta\nu = \frac{\mu_B H}{h}$ is proportional to the strength of the applied magnetic field. The pictorial view of normal Zeeman effect is shown in Fig. 3.21.

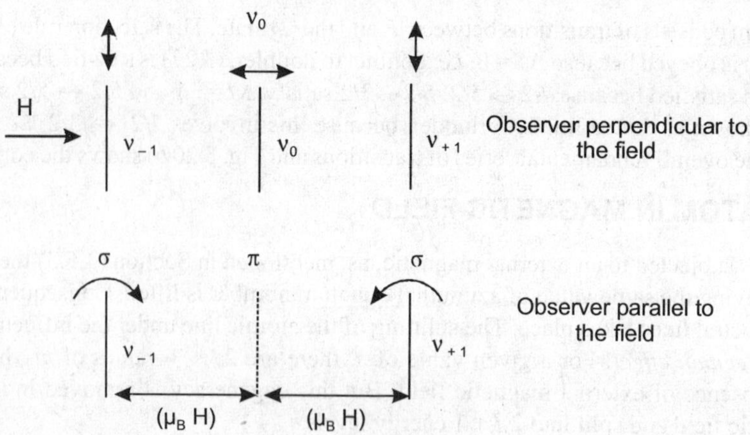

Fig. 3.21 *Normal Zeeman effect.*

***Exercise* 3.61:** A spectrometer can resolve spectral lines in the visible region ($\lambda = 600$ nm) when they are separated by $\Delta\lambda = 0.01$ nm. What will be the magnitude of the applied magnetic field required to confirm experimentally the normal Zeeman effect?

The energy separation between two successive spectral lines is $\Delta E = \mu_B H$. The corresponding frequency separation is

$$\Delta\nu = \frac{\mu_B H}{h} \qquad ...(3.92)$$

or in terms of wavelength

$$|\Delta\nu| = \left|\Delta\left(\frac{c}{\lambda}\right)\right| = \frac{c\Delta\lambda}{\lambda^2} = \frac{\mu_B H}{h} \qquad ...(3.92a)$$

$$\text{Thus } H = \frac{hc\Delta\lambda}{\mu_B\lambda^2} = \frac{(3\times10^8 \text{ m s}^{-1})(6.626\times10^{-34} \text{ J s})(0.01\times10^{-9}\text{ m})}{(9.3\times10^{-24} \text{ J Wb}^{-1} \text{ m}^{-2})(6.0\times10^{-7} \text{ m})^2} = 0.059 \text{ T Wb m}^2$$

***Exercise* 3.62:** The calcium line of wavelength $\lambda = 422.673$ nm ($P \to S$) shows normal Zeeman effect when placed in a uniform magnetic field of 4W m^{-2}. Calculate the wavelength of three components of normal Zeeman pattern and the separation between them.

We know that

$$\nu = \nu_0 \pm \frac{eH}{4\pi m_e}$$

$$\Delta\nu = \nu - \nu_0 = \left(\frac{e}{m_e}\right)\left(\frac{H}{4\pi}\right) = \frac{1.6\times10^{-19}\text{C}}{9.31\times10^{-31}\text{kg}}\frac{1}{4\pi}\times 4\ (\text{Wb m}^{-2}) = 0.56\times10^{11}\ \text{s}^{-1}$$

Since $$\nu = \frac{c}{\lambda} \Rightarrow d\nu = -\frac{c}{\lambda^2}d\lambda$$

$$|d\lambda| = \frac{\lambda^2}{c}d\nu = \frac{(4226.70 \times 10^{-10}\ \text{m})^2}{3 \times 10^8\,\text{m s}^{-1}} \times 0.56 \times 10^{11}\ s^{-1} = 0.33 \times 10^{-10}\ \text{m} = 0.33\ \text{Å}$$

Thus the wavelengths of three Zeeman components are: (4226.73 – 0.33) Å = 4226.40 Å; 4226.73 Å and (4226.73 + 0.33) Å = 4227.06 Å.

***Exercise* 3.63:** Determine the strength of the magnetic field H needed to observe the normal Zeeman effect if a spectrometer can resolve lines by 0.25 Å at 5000 Å.

We know

$$|d\lambda| = \frac{\lambda^2 |dE|}{c\,h}$$

or

$$\frac{|d\lambda|}{\lambda} = \frac{|dE|}{\frac{c\,h}{\lambda}} = \frac{\left[c\left(\frac{h}{2\pi}\right)\cdot\frac{1}{2m_e}\right]H}{\frac{c\,h}{\lambda}}$$

$$H = \left(\frac{|d\lambda|}{\lambda}\right)\left(\frac{c\,h}{\lambda}\right)\left(\frac{2m_e \times 2\pi}{c\,h}\right)$$

$$= \left(\frac{0.25}{5000}\right)\left(\frac{12.4 \times 10^3\ \text{eV} \times 10^{-8}\ \text{cm}}{5000 \times 10^{-8}\ \text{cm}}\right)\left(\frac{1}{5.79 \times 10^{-5}\ \text{eV T}^{-1}}\right) = 2.14\ \text{T}$$

The lines observed in the external magnetic field are also specified as σ lines and π lines. When the observer is at right angle to the direction of the applied magnetic field, components observed is always plane polarised parallel to the magnetic field and is called π component, while the other components are plane polarised perpendicular to the field and are called σ-component.

Components when observed perpendicular to the field, $\Delta M_J = \pm 1$, Lines observed are plane polarised perpendicular to the magnetic field direction (σ-component).

$\Delta M_J = 0$, lines are plane polarised parallel to the magnetic field direction (π-component).

Observer parallel to the field.

3.14.2. *Anomalous Zeeman Effect*

The *Zeeman effect that exhibits a complex splitting of spectral lines is known as anomalous* . This effect has been observed in atoms which have finite spin, *i.e.*, $S \neq 0$. For example in alkali metals, *D*-lines of sodium λλ 589.593 nm/588.996 nm ($^2P - {}^2S$ transition) exhibit a compound doublet in the absence of applied magnetic field (*i.e.* normal or natural splitting). However under the influence of external magnetic field one of the lines of *D*-doublet ($^2P_{1/2} \rightarrow {}^2S_{1/2}$ line) split into 4 components, while the second ($^2P_{3/2} \rightarrow {}^2S_{1/2}$ line) split into 6 lines as shown in Fig. 3.22.

Anomalous Zeeman effect arises due to the interaction of electron orbital angular momentum vector *L* with the spin angular momentum *S*. They precess around their resultant total angular momentum

Fig. 3.22 *Zeeman splitting of D-lines of sodium λλ 589.593 nm/588.996 nm ($^2P \rightarrow {}^2S$ doublet). Top is without applied magnetic field and bottom is with applied magnetic field.*

vector J which in turn precesses around the direction of the applied magnetic field. The observed complexity of splitting is due to the fact that the change of energy of the atom in a magnetic field (which is assumed to be weak compared to the internal magnetic field due to spin and orbital angular motions of valence electrons) in this case is expressed by the formula

$$\Delta E = g\, M_J\, \mu_B\, H \qquad ...(3.93)$$

rather than

$$\Delta E = M_J\, \mu_B\, H \qquad ...(3.94)$$

where g is the Lande splitting factor equal to

$$g = 1 + \frac{J(J+1) + S(S+1) - L(L+1)}{2J(J+1)} \qquad ...(3.95)$$

and is due to L-S interaction. For a single term where $S = 0$ and hence $J = L$, so $g = 1$. Hence equation (3.93) reduces to $\Delta E = M_J \mu_B H$, equation (3.94). Thus equation (3.94) due to normal Zeeman effect may be regarded as a special case of more general equation for anomalous Zeeman effect (3.93).

The Lande factor is not equal to unity in the case of multiplet terms (or $S \neq 0$) and is the cause of producing complex pattern of splitting of these terms in the magnetic field. The nonunity value of g is due to the "*anomalous*" value of the relation of the magnetic moment (μ_s) of the electron to its intrinsic spin angular momentum (p_s). The relation is

$$\frac{\mu_s}{\mathbf{p}_s} = \frac{e}{m} \qquad ...(3.96)$$

unlike the electron cloud (orbital), where ratio of magnetic moment to angular momentum is

$$\frac{\mu_l}{\mathbf{p}_l} = \frac{e}{2\,m} \qquad ...(3.97)$$

i.e., magnetic moment of the electron due to spin motion is twice the spin angular momentum, whereas magnetic moment of the electron due to orbital momentum to the orbital angular momentum.

***Exercise* 3.64:** Derive the expression for the Lande g Formula

The Lande formula (equation 3.95) may be obtained from the vector model of the atom as follows. The vector diagram of the L-S coupling along with the directions of the respective magnetic moments, also shown as dotted lines, is shown in Fig. 3.23. Suppose an atom has the values of L, S and J as shown in the Fig. 3.23. Based on quantum mechanical results, the length of these vectors is proportional to

$\sqrt{L(L+1)}, \sqrt{S(S+1)}$ and $\sqrt{J(J+1)}$ respectively. The magnetic moment vectors μ_L and μ_S associated with L and S are also shown in Fig. 3.23 *b* and their directions are opposite to the L and S vectors. The resultant total angular momentum J is also shown in the Fig. 3.23 *b*. Now the magnitude of the resulting magnetic moment would lie in the opposite direction of J if the magnitude of magnetic moment μ_S is related to p_S in a normal manner, *i.e.*, $\frac{\mu_S}{p_S} = \frac{\mu_L}{p_L}$. But we know from equations (3.96) and (3.97)

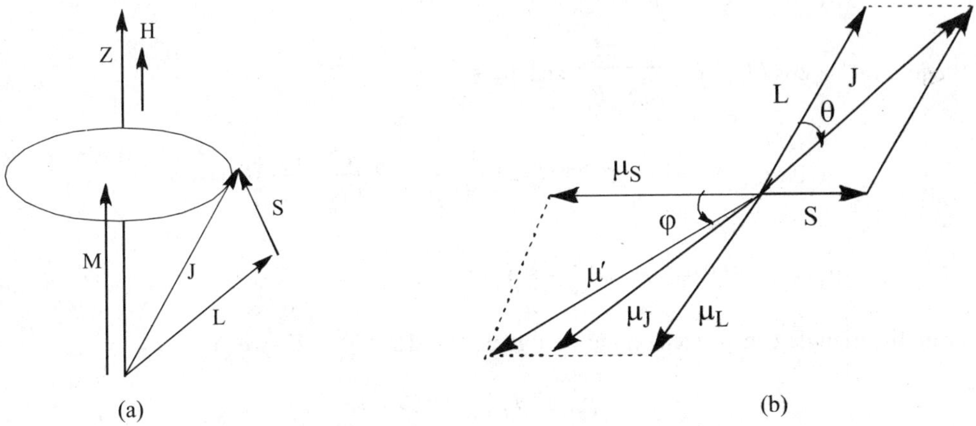

Fig. 3.23 *(a) Precession of J about the magnetic field direction Z in a weak magnetic field H. (b) Addition of Magnetic moments in an atom (in explanation of Lande g formula). The length of the vector μ_L is taken equal to L. Therefore μ_S is double the length of S. It should be noted that the direction of the angular momentum vectors is opposite to that of the corresponding magnetic moments. φ is the angle between the vectors L and J; θ is the angle between the magnetic moment μ_S and μ_J.*

$$\frac{\mu_S}{p_S} = \frac{2\mu_L}{p_L} \quad ...(3.98)$$

Thus the resulting μ' will not lie along the direction of J but it will lie in the direction as shown in the Fig. (3.23) which is different from J and precesses with L and S with the directions of the total angular momentum. Mathematically we can arrive at these relations as follows.

The projections of the magnetic moment (μ') and angular momentum (p') of the atom on the direction of the magnetic field is expressed as

$$\mu' = \mu_L \cos(L, H) + \mu_S \cos(S, H) \quad ...(3.99)$$

$$\mathbf{p}' = \mathbf{p}_L \cos(L, H) + \mathbf{p}_S \cos(S, H) \quad ...(3.100)$$

where μ_L, μ_S, $\mathbf{p}_L$, and $\mathbf{p}_S$ are the orbital and spin magnetic moments and orbital angular momentum and spin angular momentum respectively.

Substituting $\mathbf{p}' = \frac{h}{2\pi} M_J$, $\mathbf{p}_L = \frac{2m_e}{e}\mu_L$ and $\mathbf{p}_s = \frac{m_e}{e}\mu_s$ into equation (3.100), we get

$$\frac{h}{2\pi} M_J = \frac{2m_e}{e}[\mu_L \cos(L,H) + \frac{1}{2}\mu_S \cos(S,H)] \quad ...(3.101)$$

Lumping together equations (3.99) and (3.101), we obtain

$$\mu' = M\mu_B + \frac{1}{2}\mu_S \cos(S, H) \qquad \left(\mu_B = \frac{e\,h}{4\pi\,m}\right) \qquad ...(3.102)$$

Because of the rapid precession of the vector *S* (like *L*) around the resultant vector *J*, which in turn precess about the direction of the magnetic field, the mean values of the components of these vectors on the direction normal to the axis of precession becomes zero, so as a consequence there of

$$\mu_S \cos(S, H) = \mu_S \cos(S, J)\cos(J, H). \qquad ...(3.103)$$

Since $\cos(J, H) = \dfrac{h}{2\pi}\dfrac{M_J}{p}$ and $\mu_s = \dfrac{e}{m} p_s$ ...(3.104)

$$\mu_s \cos(S, H) = \frac{e}{m_e} p_S \cos(S,J)\frac{h}{2\pi}\frac{M_J}{p} = 2\frac{eh}{4\pi m_e}\frac{p_s}{p} M_J \cos(S,J)$$

$$\therefore \quad \frac{1}{2}\mu_s \cos(S,H) = M\mu_B \frac{p_s}{p}\cos(S,J) \qquad ...(3.105)$$

From the triangle constructed on the vectors *S*, *L* and *J*, Fig. 3.23, we get

$$p_L^2 = p^2 + p_s^2 - 2p_J\, p_S \cos(S,J)$$

or

$$\cos(S, J) = \frac{p_J^2 + p_s^2 - p_L^2}{2p\, p_S} \qquad ...(3.105a)$$

Combining equations (3.105) and (3.105 *a*), we get

$$\frac{1}{2}\mu_S \cos(S,H) = M\mu_B \frac{p_J^2 + p_S^2 - p_L^2}{2p_J^2} \qquad ...(3.106)$$

The quantum mechanical expressions of moment are

$$p_J = \frac{h}{2\pi}\sqrt{J(J+1)},\ p_L = \frac{h}{2\pi}\sqrt{L(L+1)},\ \text{and } p_S = \frac{h}{2\pi}\sqrt{S(S+1)}$$

Thus the factor

$$\frac{p_J^2 + p_s^2 - p_L^2}{2p_J^2} = \frac{\left(\frac{h}{2\pi}\right)^2 [J(J+1) + S(S+1) - L)(L+1)]}{2\left(\frac{h}{2\pi}\right)^2 J(J+1)}$$

$$= \frac{[J(J+1) + S(S+1) - L)(L+1)]}{2J(J+1)} \qquad ...(3.107)$$

Combining equations (3.102), (3.105) and (3.107), we get

$$\mu' = M\mu_B + M\,\mu_B \frac{J(J+1)+S(S+1)-L(L+1)}{2J(J+1)} \qquad ...(3.108)$$

Since $\mu' = g\,M\,\mu_B$

So we get the Lande formula (3.95)

$$g = 1 + \frac{J(J+1)+S(S+1)-L(L+1)}{2J(J+1)} \qquad ...(3.95)$$

The Lande formula enables us to compute g and hence the change in energy of most terms of many atoms in a magnetic field. The computed values of the splitting factor g from various terms of alkali metals are listed in Table 3.20.

TABLE 3.20. Lande splitting factor values for doublet terms

Term	L	$J=$	1/2	3/2	5/2	7/2
^2S	0	2				
^2P	1		2/3	4/3		
^2D	2			4/5	6/5	
^2F	3				6/7	8/7

The total energy of the atom in anomalous Zeeman effect is given by the expression

$$E = E_0 + g\,M_J\,\mu_B\,H \qquad ...(3.109)$$

Thus for upper (′) and lower (″) states, we write

$$E' = E'_0 + g'\,M'_J\,\mu_B\,H$$

$$E'' = E''_0 + g''\,M''_J\,\mu_B\,H$$

$$\frac{\Delta E}{h} = \frac{E'-E''}{h} = \frac{E_0' - E_0''}{h} + \frac{(g'M_J' - g''M_J'')\mu_B H}{h}$$

$$\nu = \nu_0 + (g'M_J' - g'M_J')\frac{\mu_B H}{h} \qquad ...(3.110)$$

Substituting $\Delta\nu_0 = \frac{\mu_B H}{h}$ in equation (3.110) we obtain

$$\nu = \nu_0 + (g'M_J' - g'M_J')\Delta\nu_0 \qquad ...(3.111)$$

Let us now consider the magnetic splitting of doublet terms. The quantum numbers necessary for the determination of magnitude of splitting of doublet terms $M_J g$ are given in Table 3.21.

TABLE 3.21. Magnitude of splitting of doublet terms in a magnetic field

Term	L	s	j/J	$M_J = 2J+1$	M_J	g	$M_J g$
$^2S_{1/2}$	0	$\frac{1}{2}$	$\frac{1}{2}$	2	$\frac{1}{2}, -\frac{1}{2}$	2	$1, -1$
$^2P_{1/2}$	1	$\frac{1}{2}$	$\frac{1}{2}$	2	$\frac{1}{2}, -\frac{1}{2}$	$\frac{2}{3}$	$\frac{1}{3}, -\frac{1}{3}$
$^2P_{3/2}$	1	$\frac{1}{2}$	$\frac{3}{2}$	4	$\frac{3}{2}, \frac{1}{2}, -\frac{1}{2}, -\frac{3}{2}$	$\frac{4}{3}$	$\frac{6}{3}, \frac{2}{3}, -\frac{2}{3}, -\frac{6}{3}$
$^2D_{3/2}$	2	$\frac{1}{2}$	$\frac{3}{2}$	4	$\frac{3}{2}, \frac{1}{2}, -\frac{1}{2}, -\frac{3}{2}$	$\frac{4}{5}$	$\frac{6}{5}, \frac{2}{5}, -\frac{2}{5}, -\frac{6}{5}$
$^2D_{5/2}$	2	$\frac{1}{2}$	$\frac{5}{2}$	6	$\frac{5}{2}, \frac{3}{2}, \frac{1}{2}, -\frac{1}{2}, -\frac{3}{2}, -\frac{5}{2}$	$\frac{6}{5}$	$\frac{15}{5}, \frac{9}{5}, \frac{3}{5}, -\frac{3}{5}, -\frac{9}{5}, -\frac{15}{5}$

Now by making use of the data of Table 3.21 and equation (3.111), the components of splitting of the line

$$\nu_0^{(1)} = {}^2P_{1/2} - {}^2S_{1/2} \text{ are}$$

$$\nu = \nu_0^{(1)} \pm \frac{2}{3}\Delta\nu_0,\ \nu_0^{(1)} \pm \frac{4}{3}\Delta\nu_0 \qquad \text{...(3.112)}$$

and for the line

$$\nu_0^{(2)} = {}^2P_{3/2} - {}^2S_{1/2} \text{ are}$$

$$\nu = \nu_0^{(2)} \pm \frac{1}{3}\Delta\nu_0,\ \nu_0^{(2)} \pm \Delta\nu_0,\ \nu_0^{(2)} \pm \frac{5}{3}\Delta\nu_0 \qquad \text{...(3.113)}$$

A comparison of theoretical data and the experimentally observed pattern of magnetic splitting of the *D*-line of sodium lines are in complete agreement as regards both the number of splitting components and the magnitude of splitting (Fig. 3.24). The polarisation of splitting components expected on the

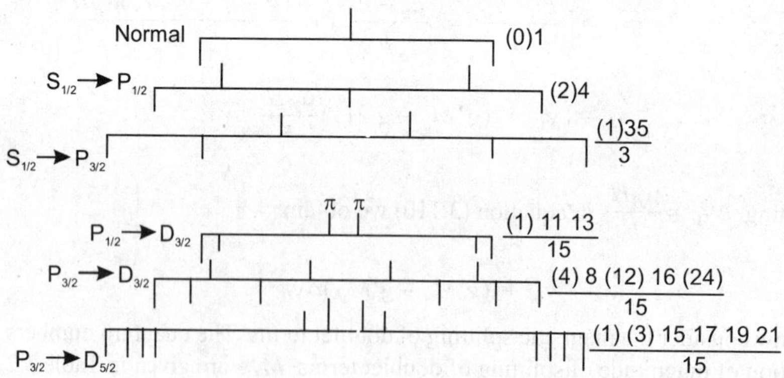

Fig. 3.24 *Magnetic splitting of lines in the normal (upper) and anomalous Zeeman effect. The dashes above the lines designate the π-components while under the line designate the σ-components. The numbers to right indicate the observed distance* $\left(\text{in units of normal splitting } \Delta\nu_0 = \frac{\mu_B H}{h}\right)$ *of each component of splitting from the initial line* μ_0. *The numbers as within brackets correspond to π-components while others to σ-components.*

basis of selection rules is also in complete agreement with experiment. The splitting of the D-doublet of sodium and terms in magnetic field as well as the polarisation of the separ ate components (π- and σ-components) is given in Fig. 3.24.

***Exercise* 3.65:** Derive the general expression for Lande g factor of a single electron.

The general expression for g value is

$$g = 1 + \frac{j(j+1)+s(s+1)-l(l+1)}{2j(j+1)} \qquad ...(3.95)$$

For a single electron, $s = 1/2$, and $j = l \pm 1/2$. Substituting the value of $s = 1/2$ and $l = j \mp 1/2$ in the equation (3.95), we get

$$g = 1 + \frac{j(j+1)+\frac{1}{2}\left(\frac{1}{2}+1\right)-\left(j-\frac{1}{2}\right)\left(j-\frac{1}{2}+1\right)}{2j(j+1)} = 1 + \frac{j(j+1)+\frac{3}{4}-\left(j-\frac{1}{2}\right)\left(j+\frac{1}{2}\right)}{2j(j+1)}$$

$$g = 1 + \frac{j(j+1)+\frac{3}{4}-\left(j^2-\frac{1}{4}\right)}{2j(j+1)} = 1 + \frac{j^2+j+\frac{3}{4}-j^2+\frac{1}{4}}{2j(j+1)}$$

$$g = 1+\frac{1}{2j} = \frac{2j+1}{2j}$$

Similarly substitute $j = l - 1/2$ and $s = 1/2$ in the above equation we obtain

$$g = 1 + \frac{j(j+1)+\frac{1}{2}\left(\frac{1}{2}+1\right)-\left(j+\frac{1}{2}\right)\left(j+\frac{1}{2}+1\right)}{2j(j+1)} = 1 + \frac{j(j+1)+\frac{3}{4}-\left(j+\frac{1}{2}\right)\left(j+\frac{3}{2}\right)}{2j(j+1)}$$

$$g = 1 + \frac{j^2+j+\frac{3}{4}-\left(j^2+\frac{1}{2}j+\frac{3}{2}j+\frac{3}{4}\right)}{2j(j+1)} = 1 + \frac{j^2+j+\frac{3}{4}-\left(j^2-2j-\frac{3}{4}\right)}{2j(j+1)}$$

$$g = 1 - \frac{j}{2j(j+1)} = \frac{2j+1-1}{2(j+1)} = \frac{j}{j+1}.$$

***Exercise* 3.66:** The ground state term symbol for iron is $^5D_4$. Calculate the g factor.

Using equation (3.95) and substituting the desired values we get

$$g = 1+\frac{4(4+1)+2(2+1)--2(2+1)}{2\times 4(4+1)} = 1+\frac{20+6-6}{40} = \frac{3}{2}$$

Exercise 3.67: The doublet observed in the sodium line spectrum is from $^2P_{1/2} \rightarrow {}^2S_{1/2}\,(D_1)$ and $^2P_{3/2} \rightarrow {}^2S_{1/2}\,(D_2)$ transitions. Predict the number of lines observed in the D_1 and D_2 lines in the external magnetic field.

Since in this case $s \neq 0$, the $g \neq 1$. Thus, the splitting in the J values into m_J components will not be equally spaced. First we should calculate g value for each state.

Using equation (3.95) and $S = 1/2$, $L = 0$ and $J = 1/2$ for $^2S_{1/2}$

$$g = 1 + \frac{1/2(1/2+1) + 1/2(1/2+1) - 0(0+1)}{2 \times 1/2(1/2+1)} = 1 + \frac{3/4 + 3/4}{3/2} = 2$$

For $\quad ^2P_{1/2}$, $S = 1/2$, $L = 1$, $J = 1/2$,

$$g = 1 + \frac{1/2(1/2+1) + 1/2(1/2+1) - 1(1+1)}{2 \times 1/2(1/2+1)} = 1 + \frac{3/4 + 3/4 - 2}{3/2} = -2/3$$

For $\quad ^2P_{3/2}$, $S = 1/2$, $L = 1$, $J = 3/2$,

$$g = 1 + \frac{3/2(3/2+1) + 1/2(1/2+1) - 1(1+1)}{2 \times 3/2(3/2+1)} = 1 + \frac{15/4 + 3/4 - 2}{15/2} = 4/3$$

The transitions observed in D_1 and D_2 lines in the external magnetic field are shown in Fig. (3.25).

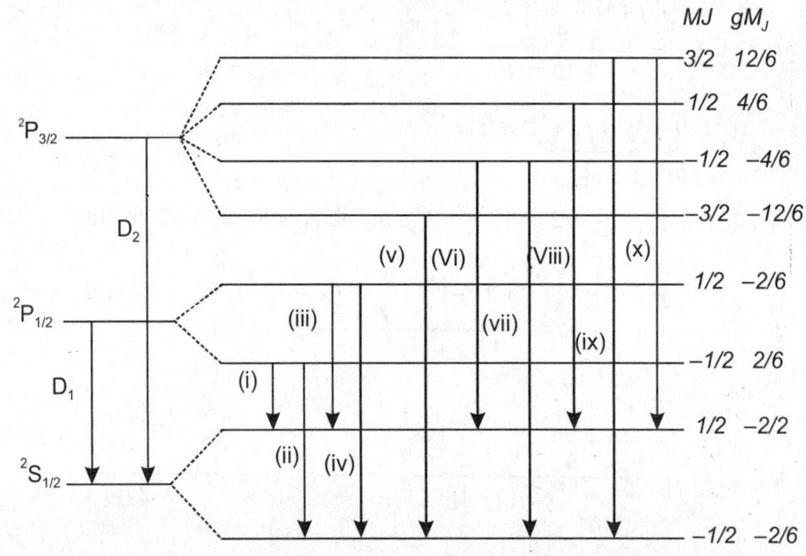

Fig. 3.25 *Anomalous Zeeman splitting for D_1 and D_2 lines of sodium atom.*

The transitions observed are

(i) $^2P_{-1/2} \rightarrow {}^2S_{1/2}$ $(\Delta M_J = +1)$ (ii) $^2P_{-1/2} \rightarrow {}^2S_{-1/2}$ $(\Delta M_J = 0)$

(iii) $^2P_{1/2} \rightarrow {}^2S_{1/2}$ $(\Delta M_J = 0)$ (iv) $^2P_{1/2} \rightarrow {}^2S_{-1/2}$ $(\Delta M_J = -1)$

(v) $^2P_{-3/2} \rightarrow {}^2S_{-1/2}$ $(\Delta M_J = +1)$ (vi) $^2P_{-1/2} \rightarrow {}^2S_{1/2}$ $(\Delta M_J = +1)$

(vii) $^2P_{-1/2} \rightarrow {}^2S_{-1/2}$ $(\Delta M_J = 0)$ (viii) $^2P_{1/2} \rightarrow {}^2S_{1/2}$ $(\Delta M_J = 0)$

(ix) $^2P_{1/2} \rightarrow {}^2S_{-1/2}$ $(\Delta M_J = -1)$ (x) $^2P_{3/2} \rightarrow {}^2S_{1/2}$ $(\Delta M_J = -1)$

TABLE 3.22. Multiplet splitting (in cm^{-1}) and transition probabilities (s^{-1}) of $^3P_1 \rightarrow {}^1S_0$ terms $\tau = \frac{1}{A}$ in Zn, Cd and Hg

Element	Zn	Cd	Hg
$^3P_2 \rightarrow {}^3P_1$	389	1172	4261
$^3P_1 \rightarrow {}^3P_0$	190	542	1767
A	1×10^5	5×10^5	15×10^6

The data in the Table 3.22 reveals that as the probability of the forbidden ($\Delta S \neq 0$) transition increases the magnitude of the multiplet splitting also increases.

***Exercise* 3.68:** Predict the transitions that would occur between $^1F_3$ and $^1D_2$ in the absence and in presence of external magnetic field.

In the transition $^1F_3 \rightarrow {}^1D_2$, the total angular momentum quantum numbers involved are $J = 3 \rightarrow J = 2$. In the presence of external magnetic field, these levels will split as shown in the Fig. 3.26. The different possible transitions are shown by the arrows and the transitions $\Delta M_L = 0$, $\Delta M_L = 1$ and $\Delta M_L = -1$ are grouped together. Since the spacing between each energy level is same, so the transitions grouped will have the same length and, therefore, will give rise to one and the same lines in the spectrum. The three lines observed are shown in the bottom of the Fig. 3.26.

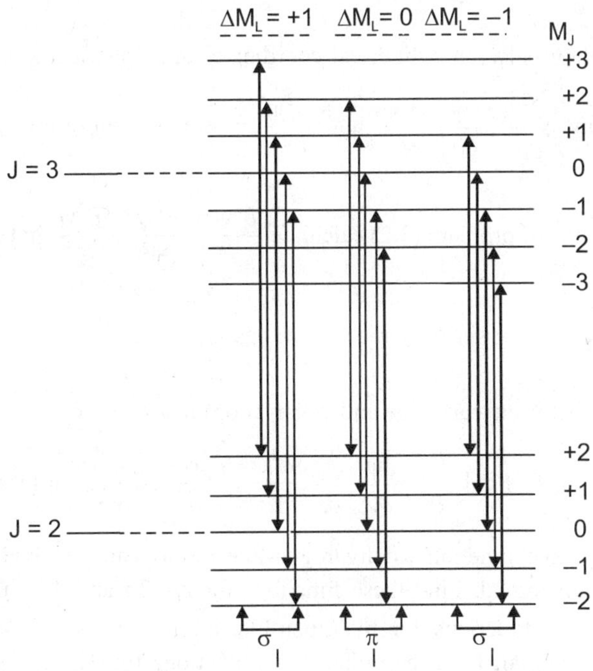

Fig. 3.26 *Normal Zeeman effect*

PROBLEMS

3.1 (*a*) If the electron with mass (m_e) and nucleus with mass M and charge e have the coordinates ($x_1 y_1 z_1$) and ($x_2 y_2 z_2$), write the Schrödinger equation for this two particle system where wave function ψ can be expressed as $\psi(x_1y_1z_1, x_2y_2z_2)$.

(*b*) Derive the relation between the centre of mass coordinates (x_c, y_c, z_c), relative coordinates (x, y, z) and the original coordinates (x_1, y_1, z_1) and (x_2, y_2, z_2).

(*c*) Convert the equation derived in part (*a*) to the coordinates mentioned in part (*b*).

3.2 The relation between the cartesian and cylindrical coordinates are

$$x = \rho \cos \phi,\ y = \rho \sin \phi,\ z = z$$

By using chain rule as mentioned in Appendix 2, find the expression for $\frac{\partial \psi}{\partial x}, \frac{\partial \psi}{\partial y}, \frac{\partial \psi}{\partial z}$ when ψ is a function of ρ, ϕ, z. Also derive expression for $\frac{\partial^2 \psi}{\partial x^2} + \frac{\partial^2 \psi}{\partial y^2} + \frac{\partial^2 \psi}{\partial z^2}$ in terms of $\frac{\partial^2 \psi}{\partial \rho^2}, \frac{\partial \psi}{\partial \rho}, \frac{\partial^2 \psi}{\partial \phi^2}$ and $\frac{\partial^2 \psi}{\partial z^2}$.

3.3 Find the relation between cylindrical coordinates and spherical polar coordinates. Using the chain rule, find $\frac{\partial^2 \psi}{\partial \rho^2}, \frac{\partial \psi}{\partial \rho}, \frac{\partial^2 \psi}{\partial \phi^2}$ and $\frac{\partial^2 \psi}{\partial z^2}$ where ψ is an arbitrary function of r, θ and ϕ.

Using the results of problem (3.2) calculate $\frac{\partial^2 \psi}{\partial x^2} + \frac{\partial^2 \psi}{\partial y^2} + \frac{\partial^2 \psi}{\partial z^2}$ in terms of $\frac{\partial^2 \psi}{\partial r^2}, \frac{\partial \psi}{\partial r}, \frac{\partial^2 \psi}{\partial \theta^2}$, and $\frac{\partial \psi}{\partial \phi} \frac{\partial^2 \psi}{\partial \phi^2}$. $\rho = r \sin \theta$, $\phi = \phi$, $Z = r \cos \theta$.

3.4 The relation between cartesian and polar coordinates are $x = r \sin \theta \cos \phi$, $y = r \sin \theta \sin \phi$, $z = r \cos \theta$. Find $\frac{\partial \psi}{\partial x}, \frac{\partial \psi}{\partial y}, \frac{\partial \psi}{\partial z}, \frac{\partial^2 \psi}{\partial x^2}, \frac{\partial^2 \psi}{\partial y^2}$ and $\frac{\partial^2 \psi}{\partial z^2}$ in terms of r, θ and ϕ.

3.5 The angular wave function for hydrogen-like atoms can be obtained by using associated Legendre's polynomial. Find these functions for 2*p*, 3*p* and 4*f* with $m = 1$.

3.6 The radial wave functions for hydrogen-like atoms can be obtained by using associated Laguerre polynomial. Find the radial wave functions for 2*s*, 3*p*, 3*d* and 4*f* orbitals.

3.7 Plot the following functions with fixed value of θ:

(*i*) $2p_x$, (*ii*) $3d_{xy}$, (*iii*) $3d_{yz}$ and (*iv*) $3d_{xz}$ for fixed value of ϕ.

3.8 Plot angular wave function of: (*i*) ψ_{2p_y} in *yz* plane, (*ii*) $\psi(d_{z^2})$ in *xz* and *yz* plane, (*iii*) $\psi(d_{xz})$ in *xz* and *xy* plane, (*iv*) $\psi(d_{xy})$ in *xy* and *xy* plane at angle of 45° with the *x* axis, (*v*) $\psi(d_{x^2-x^2})$ in *xz* and *xy* plane and (*vi*) $\psi(d_{yz})$ in *yz* and *xy* plane.

3.9 What will be the section of (*a*) 3*s* orbital by *xy* plane, (*b*) $3d_{z^2}$ by *xy* plane?

3.10 Plot the radial density and radial distribution curves for 2*s*, 3*s*, 4*d*, 3*p* and 3*d* up to 1.0 nm.

3.11 Plot qualitatively the radial wave function, *i.e.* $R(r)$ vs. r for H, He^+ and Li^{2+} on the same plot. Which wave function will approach zero first?

3.12 The radial wave function for 3*s* orbital can be written as $R(r) = N(a + br + cr^2)\,e^{-dr}$. Calculate the values of the constants *a, b, c* and *d.* Also the energy E_3.

3.13 Prove that the real angular functions for $l = 2$ and $m_l = \pm 2$ are orthogonal.

3.14 Prove that any two radial wave functions for hydrogen-like atoms are orthogonal.

3.15 Show that all functions of the form $R(r)\,Y(\theta, \varphi)$ give probability distribution independent of φ and therefore symmetrical about *z* axis.

3.16 The maximum radial probability distribution for 2*p* calculated with the help of quantum mechanics occurs at the same point as given by Bohr's relation. Prove it.

3.17 Under normal circumstances Bohr theory says that 1*s* electron will not be found beyond $r = a_0$. Calculate the quantum mechanical probability of finding the above electron within $r = 2a_0$.

3.18 Calculate the probability of finding 1*s* electron for hydrogen atom within a sphere of radius $r = 10a_0$. Compare this value with that obtained in Problem 3.17.

3.19 Calculate the probability of finding 1*s* electron for Li^{2+} ion. Compare the probability of finding the electron in case of Li^{2+} and that of hydrogen within a sphere of radius $r = 2a_0$ and $r = 10a_0$.

3.20 Calculate the probability of finding the electron present in 1*s* orbital within volume defined as $r = 2.0a_0$ and $r = 2.5a_0$, $\theta = \pi/3$ and $\theta = 0.4\pi$, $\phi = 0.5\pi$ and $\phi = 0.55\pi$.

3.21 Calculate the probability of finding $2p_x$ electron present in the volume defined as in problems 3.17, 3.18, 3.19 and 3.20. ($\phi = 0.6\pi$, $\phi = 0.65\pi$).

3.22 Find the radius of the sphere around the nucleus such that probability of finding 1*s* electron is 95%.

3.23 Calculate the average value of r and r^2 for (*a*) 2*s* (*b*) $2p_x$ and (*c*) $3d_{z^2}$ for hydrogen-like atoms

Hint: $\int_\theta^\infty x^n e^{-x} dx = nl$

3.24 Calculate the most probable distance from the nucleus for the electron occupying 2*s* and 2*p* orbitals. Is the distance same?

3.25 Calculate the most probable point at which an electron occupying $2p_x$ orbital will be found.

3.26 Calculate the value of angular momentum of the electron occupying: (*a*) 2*s*, (*b*) 2*p*, (*c*) 3*d*, (*d*) 4*d*, (*e*) 4*f*, and (*f*) 5*f*.

3.27 Calculate the number of radial and angular nodes present in each case of the orbital, mentioned in Problem 3.26.

3.28 If the energy of the level of hydrogen atom is: (*a*) $-R_H$, (*b*) $-R_H/4$, (*c*) $-R_H/16$, (*d*) $-R_H/36$, calculate the degeneracy of the level.

Multi-Electronic Atoms

3.29 Write the Hamiltonian for Li atom and also the Slater's wave function.

3.30 Write the Hamiltonian for boron atom and also the Slater's wave function.

3.31 Write the electronic configuration for (*i*) V, (*ii*) Cr, (*iii*) Mn, (*iv*) Ni, (*v*) Cu, (*vi*) Ge, (*vii*) Ce.

3.32 Write the electronic configurations for (*i*) Ni^{2+}, (*ii*) Cr^{2+}, (*iii*) Cr^{3+}, (*iv*) Ce^{4+}, (*v*) Co^{2+}, (*vi*) Fe^{2+}, (*vii*) Fe^{3+}.

3.33 The first ionisation potential for *N* is 14.54 eV. Using Bohr's relation (2.19) calculate Z_{eff}. Also calculate the value of orbital radius *r* with Z_{eff} calculated.

3.34 Calculate Z_{eff} using Bohr's relation and I.P. or electron affinity for the following species:

(*i*) O^-(EA = 1.465 eV), (*ii*) F (IP = 17.42 eV), (*iii*) F^- (EA = 3.4 eV),
(*iv*) K (IP = 4.339 eV), (*v*) S (IP = 10.357 eV), (*vi*) S^-(EA = 2.07 eV).

Also calculate the orbital radius for each case.

3.35 Explain briefly why the first IP of nitrogen atom is greater than that of oxygen atom whereas the second IP of oxygen is greater than that of nitrogen.

3.36 Which of the following wave functions are correct according to quantum mechanical description rules?

(*i*) 2, 1, 0, $-\frac{1}{2}$; (*ii*) 2, 2, 1, $\frac{1}{2}$; (*iii*) 4, 3, 0, $-\frac{1}{2}$; (*iv*) 3, 0, 3, $-\frac{1}{2}$;

(*v*) 3, 1, –1, $-\frac{1}{2}$; (*vi*) 3, 2, –1, $-\frac{1}{2}$; (*vii*) 4, 2, 3, $-\frac{1}{2}$; (*viii*) 4, 1, 0, $-\frac{1}{2}$;

(*ix*) 4, + 3, – 3, $-\frac{1}{2}$; (*x*) 4, 5, –1, $\frac{1}{2}$.

Atomic Term Symbols

3.37 What is the term symbol for the ground state electronic configuration for the atoms:

(*i*) V, (*ii*) Cr, (*iii*) Cl, (*iv*) N, (*v*) O,
(*vi*) F, (*vii*) Cu, (*viii*) Mn, (*ix*) Mg, (*x*) Fe.

3.38 Calculate the spectroscopic term symbols for (*i*) nd^4, (*ii*) nd^5, (*iii*) nf^2, (*iv*) nf^3, (*v*) nd^3 using all the three methods. Arrange the spectroscopic states according to increasing energy.

3.39 Calculate the spectroscopic term symbols for

(*i*) $1s^1\ 2s^1$, (*ii*) $2p^1\ 3p^1$, (*iii*) $2p^1\ 3d^1$, (*iv*) $2p^2\ 3d^1$, (*v*) $2p^3\ 3d^2$.

3.40 The ground state spectroscopic term symbols for the atoms Nb, Mo, Rh and W are $^6D_{1/2}$, $^7S_3$, $^4F_{9/2}$ and $^5D_0$ respectively. Write the electronic configurations of each atom compatible with the spectroscopic term. Which of them are not following normal auf-bau principle?

Atomic Spectrum

3.41 Which of the following transitions are allowed and which one are forbidden:

(*a*) 2*p* → 3*p*, (*b*) 2*p* → 3*s*, (*c*) 2*p* → 5*s*, (*d*) 3*d* → 4*f*, (*e*) 3*d* → 4*p*,
(*f*) 3*p* → 4*f*, (*g*) 4*p* → 4*f*, (*h*) 3*d* → 4*s*.

3.42 (*a*) The principal series of sodium involves the excitation of electron from ground state to various *p* excited levels. The smallest wavelength line consists of doublet, observed at 539 and 539.6 nm. Justify these two transitions.

(*b*) The sharp series of sodium involves the excitation of electron from *np* level to the excited *n′s* levels ($n' > n$). Under high resolution each line is a doublet. Justify it and find the separation between doublet.

3.43 The term values for Li atoms for different set of quantum numbers are as follows: 7017.0 cm^{-1} for $n = 4$, $l = 1$, 12559.9 cm^{-1} for $n = 3$, $l = 1$, 6855.5 cm^{-1} for $n = 4$, $l = 3$, 3031.0 cm^{-1} for $n = 6$, $l = 3$. If the value of Rydberg constant for Li is 109729 cm^{-1}, calculate the effective quantum number and hence the quantum defect for these states.

3.44 The term values (cm^{-1}) for Na atoms for different set of quantum numbers are as follows: 24492 for $n = 3$, $l = 1$, 11181.9 for $n = 4$, $l = 1$, 6858.6 for $n = 4$, $l = 3$, 3039.7 for $n = 6$, $l = 3$. If the value of Rydberg constant for sodium atom is 109734 cm^{-1}, calculate the effective quantum number and hence the quantum defect for these states. Compare the values obtained with those of Li, obtained in problem 3.43.

3.45 The spectrum of potassium atom studied under high resolution gave the following results. The first member of principal series doublet occurs at 766.7 nm and 770.1 nm. The first member of the fundamental series are at 151.67 nm and 151.72 nm. If the multiplet splitting in ^2F terms are small, calculate the relative separations of the first member of the diffuse series.

3.46 The ground state of oxygen atom is triplet (^3P). By means of labelled energy level diagram show three qualitative pattern of the fine structure of $^3p_j - {}^3D_j$ transition.

Addition Problems

3.47 One of the solutions of radial equation for the hydrogen atom can be expressed as

$$R(r) \quad = \quad N[1 - \frac{br}{2}]e^{-\frac{br}{2}}$$

Show that it is a solution of the general radial wave equation for hydrogen and hydrogen-like atoms provided $b = \frac{Z}{a_0}$, $l = 0$, $a_0 = \frac{(4\pi\varepsilon_0)h^2}{4\pi^2 m_e e^2}$ and $E_n = -\frac{2\pi^2 \mu Z e^4}{(4\pi\varepsilon 0)^2 h^2 n^2}$ in SI units.

3.48 Problem 3.47 can be written in different manner *i.e.*, show that the radial function $R(r) = Nre^{-\frac{br}{2}}$ is a solution of the radial wave equation for the hydrogen or hydrogen like atoms provided

$$b = \frac{Z}{a_0}, \; l = 1 \text{ and } E_n = -\frac{2\pi^2 \mu Z e^4}{(4\pi\varepsilon 0)^2 h^2 n^2}.$$

3.49 Considering that in hydrogen atom nucleus is stationary and electron is revolving around the nucleus, the Bohr radius is given as $a_0 = \frac{(4\pi\varepsilon_0)h^2}{4\pi^2 m_e e^2}$. If we consider that both nucleus and electron are revolving around the centre of mass coordinate, Bohr's radius is given $a_0 = \frac{(4\pi\varepsilon_0)h^2}{4\pi^2 \mu e^2}$ where μ is the reduced mass of the nucleus and electron. Calculate the percentage error in the value of a_0 for (*i*) a hydrogen atom, (*ii*) a deuterium and (*iii*) a tritium atom.

3.50 Considering the Problem 3.49 calculate the energies of $\left(E_n = -\frac{2\pi^2 m_e Z e^4}{(4\pi\varepsilon 0)^2 h^2 n^2} \simeq -\frac{2\pi^2 \mu Z e^4}{(4\pi\varepsilon 0)^2 h^2 n^2} \right)$ for the ground state of (*i*) a hydrogen atom, (*ii*) a deuterium atom and (*iii*) a tritium atom, as well as, the percentage error in each case.

3.51 Write the Hamiltonian operator of hydrogen atom or hydrogen like ions in terms of Cartesian coordinates and spherical polar coordinates.

3.52 Show that the angular function $\Theta(\theta) = \left[\frac{3}{2}\right]^{\frac{1}{2}} \cos\theta$ is a solution of the equation

$$\sin\theta \frac{d}{d\theta}\left[\sin\theta \frac{d\Theta(\theta)}{d\theta}\right] + \frac{8\pi^2 I E_R \sin^2\theta}{h^2}\Theta(\theta) - m^2\Theta(\theta) = 0 \text{, provided } m = 0 \text{ and } E_R = \frac{h^2}{8\pi^2 I}$$

3.53 Transform equation (3.4) into

$$(1 - x^2)\frac{d^2 P(x)}{dx^2} - 2x\frac{dP(x)}{dx} + \left[\beta - \frac{m^2}{1 - x^2}\right] P(x) = 0$$

3.54 Prove the general relation for orthogonality integral for the Legendre polynomials, *i.e.*,

$$\int_0^\pi p_l(\cos\theta) p_n(\cos\theta) \sin\theta d\theta \neq 0 \text{ for } l \neq n$$

3.55 Show that $Y_1^{-1}(\theta, \varphi)$ is not an eigen function to $\hat{L}_x$ or $\hat{L}_y$.

3.56 Show that $\hat{L}^2$ commute with $\hat{L}_x$ or $\hat{L}_y$ or $\hat{L}_z$.

3.57 The nitrogen atom has one electron in each of the three $2p$ orbitals. Show that the total electron density is spherically symmetric, *i.e.*, it is independent of angle θ and φ.

(**Hint:** Take the square of the angular wave functions).

3.58 Find the average value of r in the $2s$ state. **[Ans.** $6a_0$**]**

3.59 Using the general properties of associated Laguerre polynomials show that $<r> = \frac{3}{2} a_0 n^2$ for an electron in *ns* electron.

3.60 Calculate the probability of $1s$ electron of the hydrogen atom that can be found up to $r = 2a_0$.

3.61 Calculate the average value of r for the $2s$ state for a hydrogenlike atom.

3.62 The general expression for the average value of r in the states $2s$, $2p$ and $3s$ states of the hydrogen atom is given as

$$<r> = \frac{a_0}{2}[3n^2 - l(l+1)]$$

Calculate the values of $<r>$ by using this equation and from the general from of the average value of any observable.

3.63 Using the equation given in equation (3.62) calculate $<r>$ for the 2*s* and 2*p* electrons. Compare and explain.

3.64 Calculate the value of *r* for the 2*s* and 3*s* orbitals at which nodes exist.

3.65 Using the energy equation for the particle in one dimensional box, account for the difference in the standard molar entropies of He, Ne and Ar.

3.66 Show that for a given value of *n*, the number of orbitals is equal to n^2.

3.67 Consider a state with quantum numbers as: $n = 4$, $l = 2$ and $m_l = -2$ for hydrogen atom or hydrogen like atom. (*i*) Calculate the orbital angular momentum of the electron in this state, (*ii*) what is the value of the *Z*-component of the orbital angular momentum, (*iii*) find out the number of radial and angular nodes? Which angular coordinates exhibit the nodes? (*iv*) What is the degeneracy of this energy state? (*v*) Calculate the energy of this state in SI units.

3.68 Show that the virial theorem is satisfied for a hydrogen atom or hydrogen like atoms in the 2 *p* excited state.

3.69 How many radial nodes are in $R^0{}_4$ (*r*) radial eigen function for hydrogen like atoms? At what values of the radial coordinate will the radial nodes appear for the hydrogen atom?

3.70 Consider a carbon atom and write down the Hamiltonian operator. (*i*) If Born-Oppenheimer approximation is applied, determine the term or terms omitted, (*ii*) Identify the one electron terms, (*iii*) How many two electrons terms are present? How many electron-electron repulsion terms are present?

3.71 Consider an iron atom (Fe). (*i*) Using equation similar to (3.33) write down the Hamiltonian operator for Fe atom. (*ii*) Find out the number of electron-nuclear attraction terms, (*iii*) Calculate the number of electron-electron repulsion terms.

3.72 Calculate the spin-orbit coupling splitting between the (*i*) $2^2P_{1/2}$ and $2^2P_{3/2}$ states of the hydrogen atom.

3.73 The angular momentum states for a hydrogen atom in the 3 *d* excited states are 3 $^2D_{5/2}$ and 3 $^2D_{3/2}$. Calculate the spin-orbit splitting in (*i*) SI units and (*ii*) in hartree.

3.74 Using the Slater's rules for screening constants and then obtaining the effective nuclear charge felt by the electron, calculate the ground state energy of (*i*) nitrogen atom, (*ii*) neon atom, (*iii*) Fe atom.

3.75 Calculate the angle between the orbital and spin angular momentum to give the desired value of total angular momentum: (*i*) $^3D_3$, (*ii*) $^3D_2$, (*iii*) $^3D_1$, (*iv*) $^3P_2$, (*v*) $^3P_1$ and (*vi*) $^3P_0$.

[**Ans.** 125.26°, 73.22°, 30°, 120°, 60°, 0°]

3.76 Determine the possible term symbols for an excited nitrogen atom whose electronic configuration is $1s^22s^22p^23d^1$.

[**Ans.** $^2F_{7/2}$, $^2F_{5/2}$, $^2D_{5/2}$, $^2D_{3/2}$, $^2D_{5/2}$, $^2P_{3/2}$, $^2P_{1/2}$, $^2S_{1/2}$, $^4D_{7/2}$, $^4D_{5/2}$, $^4D_{3/2}$, $^4D_{1/2}$, $^4P_{5/2}$, $^4P_{3/2}$, $^4P_{1/2}$ and $^4S_{3/2}$].

3.77 Consider another excited state of nitrogen atom as $1s^22s^22p^23p^1$. Calculate the possible term symbols for this configuration and arrange them in order of increasing order of energy.

3.78 The electronic configuration for the sp^3 hybridised carbon atom is $1s^22s^12p^3$. Calculate the term symbols.

3.79 The excited state electronic configuration for oxygen is (*i*) $1s^22s^22p^33p^1$, (*ii*) $1s^22s^22p^23p^2$, and (*iii*) $1s^22s^22p^33d^1$. Calculate the term symbols.

3.80 Figure below represents the emission spectrum of a hydrogen-like ion in a gas phase. All these lines results from the electronic transitions from the excited state to the $n = 2$ state.

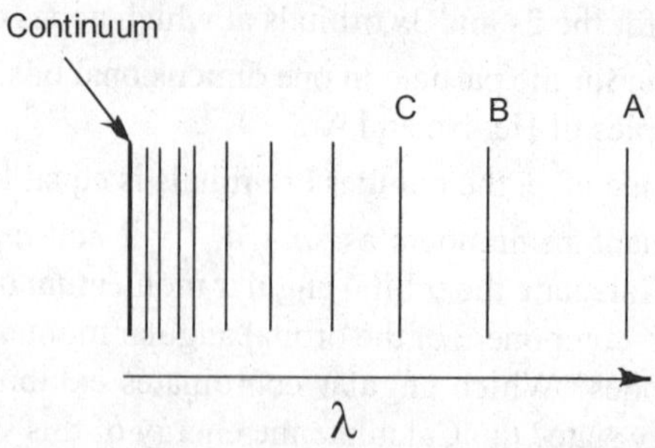

(*i*). What transitions corresponds to line B and C, (*ii*) if the wavelength of line *C* is 27.1 nm, calculate the wavelengths of lines *A* and *B*, (*iii*) calculate the energy required to move electron from the ion in the $n = 5$ state, (*iv*) what is the physical significance of this emission?

3.81 The first ionisation energy of Na, K and Cs are 5.318 eV, 4.339 eV and 3.893 eV respectively. Using equation for one electron hydrogen like ions, *i.e.*,

$$E_n = -\frac{(2.18 \times 10^{-18}\ \text{J})Z_{\text{eff}}^2}{n^2}$$

where Z_{eff} is defined as $Z - \sigma$ and σ is the shielding constant. Calculate the values of shielding constants for Na, K and Cs atoms. Compare them with those calculated from Slatr's rules.

3.82 Calculate the difference between the first line of Lyman series of hydrogen and lithium atoms. $R_\infty = 109737\ \text{cm}^{-1}$.

3.83 Hydrogen atom is excited to higher energy levels and it comes back to the ground state, *i.e.* $n = 1$. Let the emitting wavelength is 102.55 nm. First derive the relation that gives the value of the starting wavelength in this emission and then calculate the value of the quantum number of this state.

3.84 Calculate the spectroscopic term symbols for the electronic configuration $2p^1 3p^1 3d^1$ system.

[**Ans.** ${}^4G$, ${}^4F$ (2), ${}^4D$ (3), ${}^4P$ (2), ${}^4S$ and ${}^2G$ (2), ${}^2F$ (4), ${}^2D$ (6), ${}^2P$ (4), ${}^2S$ (2)]

3.85 Determine the relative intensities of the lines of compound doublet originating from transitions ${}^2D \leftarrow {}^2F$ and ${}^2G \leftarrow {}^2F$.

3.86 The term symbol for the state is: (*a*) ${}^3G_5$, (*b*) ${}^2D_{5/2}$. Calculate the values of orbital, spin and total angular momentum for each state.

3.87 When single line of cadmium 643.846 nm (${}^1D_2 \leftarrow {}^1P_1$) is subjected to external magnetic field, it splits into multiplet. Draw the energy level diagram in absence and presence of external magnetic field and depict the transitions observed.

3.88 Determine the normal Zeeman effect of the cadmium red line 6438 Å when the atoms are placed in a magnetic field of 0.009T. [**Ans.** $d\lambda = 1.74 \times 10^{-11}$ cm $= 1.74 \times 10^{-3}$ Å]

3.89 (*a*) The Zeeman pattern of a line is made up of 6 equidistant components. The upper term state term is ${}^3P_{3/2}$. What would be the lower state term? (*b*) The Zeeman pattern of a line consists of nine equidistant components. The upper state term is ${}^3P_2$ (*L-S* coupling). What would be the lowest state term?

3.90 Construct the energy level diagram when the transition $^1P_1 \leftarrow {}^1D_2$ are subjected to external magnetic field. Draw the transitions observed.

3.91 The first two lines of the principal series ($^2P - {}^2S$) observed in case of caesium atom, consist of doublets and arise due to the transitions: $6p\ {}^2P_{1/2} \rightarrow 6s\ {}^2S_{1/2}$ (894.346 nm), $6p\ {}^2P_{3/2} \rightarrow 6s\ {}^2S_{1/2}$ (852.112 nm); $7p\ {}^2P_{1/2} \rightarrow 6s\ {}^2S_{1/2}$ (459.3116 nm), $7p\ {}^2P_{1/2} \rightarrow 6s\ {}^2S_{1/2}$ (455.526 nm). Construct the approximate energy level diagram depicting the transitions. Calculate the intensities ratio of the above transitions, effective nuclear charge and effective quantum number.

3.92 The first line of the sharp series ($^2S - {}^2P$) of caesium atom also consists of a doublet and arises due to the following transitions: $7s\ {}^2S_{1/2} \rightarrow 6p\ {}^2P_{3/2}$ (1469.48 nm), $7s\ {}^2S_{1/2} \rightarrow 6p\ {}^2P_{1/2}$ (1358.81 nm). Calculate the effective quantum number and quantum defect. Calculate the intensities ratio of the above transitions, effective nuclear charge and effective quantum number.

3.93 The first line of the diffuse series ($^2D - {}^2P$) of caesium atom also consists of a triplet arising from the following transitions: $5d\,{}^2D_{3/2} \rightarrow 6p\ {}^2P_{3/2}$ (3612.7 nm), $5d\,{}^2D_{5/2} \rightarrow 6p\ {}^2P_{3/2}$ (3489.2 nm), $5d\ {}^2D_{3/2} \rightarrow 6p\ {}^2P_{1/2}$ (3010.0 nm), whereas the second line of diffuse series are at: $6\,d\,{}^2D_{3/2} \rightarrow 6p\ {}^2P_{3/2}$ (920.840 nm), $6d\,{}^2D_{5/2} \rightarrow 6p\ {}^2P_{3/2}$ (917.223 nm), $6d\,{}^2D_{3/2} \rightarrow 6p\ {}^2P_{1/2}$ (876.135 nm). Calculate the intensities ratio of the above transitions, effective nuclear charge and effective quantum number.

3.94 Just like the diffuse series of caesium atom, the selection rule $\Delta J = 0, \pm 1$, permits triplets owing to the combinations namely $4f\,{}^2F_{5/2} \rightarrow 5d\ {}^2D_{3/2}$, (1002.51 nm), $4f\,{}^2F_{7/2} \rightarrow 5d\ {}^2D_{5/2}$ (1012.41 nm). The apparent doublet nature of the lines of the fundamental *D*-series is due to the small width of the doublet splitting of the ^2F-terms. Calculate the relative intensity ratio of these lines, effective nuclear charge and effective quantum number.

3.95 Let us now consider the first doublet of the principal series. The doublet results due to the transitions $^2P_{1/2} \rightarrow {}^2S_{1/2}$, $^2P_{3/2} \rightarrow {}^2S_{1/2}$. Both the transitions end on a common level $^2S_{1/2}$. The statistical weights of the lines corresponding to these two transitions are (2 × 1/2 + 1) = 2 and (2 × 3/2 + 1) = 4. Thus the intensity ratio of the second line to the first line of the doublet is 2:1 and conforms to the experimentally observed intensity ratio of this doublet also.

3.96 Using the trail eigen function, $\psi = Ne^{-br}$, obtain the variational energy for a ground state hydrogen atom, *i.e.*, minimise the result with respect to *b*. **[Ans.** $b = 1$**]**

CHAPTER 4

MOLECULAR STRUCTURE

4.1. INTRODUCTION

Molecules are formed by the combination of different atoms. These atoms in the molecules are held together by different kinds of interactions. There are mainly two kinds of chemical bondings which are responsible to keep the atoms intact in molecules.

(*i*) *Ionic Bond*: In ionic bonds, the ions are formed by the actual transference of electrons from one atom to another and these ions are held together by electrostatic attractions.

(*ii*) *Covalent Bond*: In covalent bonds, the atoms are held together by sharing a pair of electrons between atoms. The pair of electrons held together by the atoms are considered to be responsible for achieving the stable configurations of the atoms.

The ionic bonding is relatively simple to consider than covalent and will be discussed first.

4.2. IONIC BOND

As said earlier, in ionic bonds, the electrons are first transferred from one atom to another and then are held together by the electrostatic forces. The indication about the presence of ionic character is obtained from the dipole moment of the molecules. The potential energy of the ionic molecules can be obtained from the relation

$$U(R) = \frac{-Z_A Z_B e^2}{4\pi\varepsilon_0 R} + ae^{-bR} + \Delta U(\infty) \qquad ...(4.1)$$

where Z_A and Z_B are the valence of ions, R is the distance between ions, $\Delta U(\infty)$ is energy possessed by the ions in gas phase when apart by infinite distance or the sum of the energies of the individual ion when separated, a and b are constants. These constants can be determined from the relations

$$\left(\frac{dU}{dR}\right)_{R=R_e} = 0 \qquad ...(4.2)$$

$$a = \frac{Z_A Z_B e^2}{4\pi\varepsilon_0 b\, R_e}\, e^{bRe} \qquad \text{...(4.3)}$$

and

$$\left(\frac{d^2U}{dR^2}\right)_{R=R_e} = k \qquad \text{...(4.4)}$$

$$b = \frac{4\pi\varepsilon_0 k\, R_e^2}{e^2} + 2/R_e \qquad \text{...(4.5)}$$

where k is the force constant of the bond and can be obtained if one knows the fundamental frequency of oscillation ($\overline{\nu}$). The relation is

$$\overline{\nu} = \frac{10^{-2}}{2\pi c}\sqrt{k/\mu} \qquad \text{...(4.6)}$$

where $\overline{\nu}$ is expressed in cm^{-1}, c in m s^{-1}, k in N m^{-1} and μ in kg.

The value of $\Delta U(\infty)$ is obtained, for example for a diatomic molecule A^+B^-, by the relation

$$\Delta U(\infty) = \text{IP (A)} - \text{EA (B)} \qquad \text{...(4.7)}$$

***Exercise* 4.1:** Show that in gas phase, the presence of $Na^+$and Cl^- ions are unstable as compared to Na and Cl atoms when NaCl molecule dissociates, IP (Na) = 5.14 eV and EA (C1) = 3.65 eV.

When NaCl molecule dissociates into neutral atoms, let the energy required for the process be E

$$\text{NaCl }(g) \rightarrow \text{Na }(g) + \text{Cl }(g) \qquad \text{(A)}$$

If the NaCl molecule dissociates into ions, let the energy required for the process

$$\text{NaCl }(g) \rightarrow \text{Na}^+ (g) + \text{Cl}^- (g) \qquad \text{(B)}$$

be E'. The difference of energies for the two pro cesses.

$\Delta E\,(\infty) = E' - E$ = Energy required to ionise gaseous sodium atom – the energy gained by chlorine atom for capturing an electron

= IP (Na) – EA (Cl)

= 5.14 – 3.65 = 1.49 eV

It means at infinite distance process A is more favourable than process B.

***Exercise* 4.2:** Prove the relation (4.5).

Differentiate equation (4.1) with respect to R and put it equal to zero.

$$\frac{dU}{dR} = \frac{Z_A Z_B e^2}{4\pi\varepsilon_0 R^2} - abe^{-bR} = 0 \text{ or } ab = \frac{Z_A Z_B e^2}{4\pi\varepsilon_0 R_e^2}\, e^{bR_e}$$

and also

$$\frac{d^2U}{dR^2} = -\frac{2Z_A Z_B e^2}{4\pi\varepsilon_0 R^3} + ab^2 e^{-bR_e}$$

Substitute $R = R_e$, then second relation reduces to (using the result of 1st equation) and using equation (4.4)

$$-2\frac{Z_A Z_B e^2}{4\pi\varepsilon_0 R_e^3}+\frac{Z_A Z_B e^2}{4\pi\varepsilon_0 R_e^2}\,b.\,e^{-bR}.\,e^{bR}=k$$

or

$$b=\frac{k.\,4\pi\varepsilon_0 R_e^2}{Z_A Z_B e^2}+2/R_e$$

where k is the force constant of the ionic bond, and can be obtained from the vibrational frequency of the ionic bond, *i.e,*

$$k = 4\pi^2 c^2 v^{-2}\mu_{AB}/10^{-4} \qquad \text{...(4.7a)}$$

where v is the vibrational frequency and μ_{AB} is the reduced mass of AB.

***Exercise* 4.3**

(*a*) Calculate the values of constants a and b of the equation (4.1) in case of NaCl molecule. $R_e = 0.236$ nm; $v = 1.093 \times 10^{13}\ s^{-1}$.

(*b*) Calculate the dissociation energy.

The reduced mass of NaCl is

$$\mu_{\text{NaCl}}=\frac{(23\text{ g mol}^{-1})\,(35.5\text{ g mol}^{-1})\,(10^{-3}\text{ kg g}^{-1})}{(23+35.5)\text{ g mol}^{-1}\,(6.02\times10^{23}\text{ mol}^{-1})}$$

$$= 2.303 \times 10^{-26}\text{ kg}$$

Using equation (4.6)

$$k = (4\pi^2)\,(v^2)\,\mu_{\text{NaCl}}$$
$$= (4\pi^2)\,(1.093 \times 10^{13}\text{ s}^{-1})^2\,(2.303 \times 10^{-26}\text{ kg})$$
$$= 1.09 \times 10^2\text{ N m}^{-1}$$

Substituting the respective values in equation (4.5)

$$b=\frac{(1.112\times10^{-10}\text{C}^2\text{ N}^{-1}\text{ m}^{-2})(1.09\times10^2\text{ N m}^{-1})\,(2.36\times10^{-10}\text{ m})^2}{(1)(1)(1.6\times10^{-19}\text{C})^2}+\frac{2}{2.36\times10^{-10}\text{ m}}$$

$$= 3.48 \times 10^{10}\text{ m}^{-1}$$

Substituting the respective values in equation (4.3)

$$a=\frac{(1)\,(1)\,(1.6\times10^{-19}\text{C})^2}{(1.112\times10^{-10}\text{C}^2\text{N}^{-1}\text{m}^{-2})(3.48\times10^{-10}\text{ m}^{-1})(2.36\times10^{-10}\text{m})^2}\exp\,[(3.48\times10^{10}\text{ m}^{-1})$$
$$(2.36\times10^{-10}\text{ m})]$$
$$= 0.438 \times 10^{-15}\text{ J}$$

The value of ΔU calculated at $R = R_e$ gives the dissociation energy.

Substituting the values of a, b and R_e in equation (4.1)

$$D_e = -\Delta U(R_e) = -(0.438 \times 10^{-15}\text{ J})\exp\,[-(3.48 \times 10^{10}\text{ m}^{-1})(2.36 \times 10^{-10}\text{ m})]$$

$$+\frac{(1.6\times10^{-19}\text{C})^2}{(1.112\times10^{-10}\text{C}^2\text{N}^{-1}\text{m}^{-2})\,(2.36\times10^{-10}\text{m})}-(1.49\text{ eV})\,(1.602\times10^{-19}\text{ J eV}^{-1})$$

$$= -1.188 \times 10^{-19}\text{ J} + 9.75 \times 10^{-19}\text{ J} - 2.387 \times 10^{9}\text{ J}$$
$$= 6.175 \times 10^{-19}\text{ J}$$

$$\bar{D}_e = (6.175 \times 10^{-19}\ \text{J})\ (6.02 \times 10^{-23}\ \text{mol}^{-1}) = 371.74\ \text{J mol}^{-1}$$

The actual value is 407.10 kJ mol^{-1}. The simple ionic model gives only 8.7% less than actual value.

4.3. COVALENT BOND

4.3.1. *Born-Oppenheimer Approximation*

Born-Oppenheimer approximation states that the motion of the electrons and the nuclei can be treated independently. The reason for this is: mass of the nucleus is nearly 2000 times greater than that of electron and thus during the motion of electrons, the positions of nuclei can be considered unchanged. The consequence of this is that (*i*) the total energy can be considered as the sum of the energies due to rotational, vibrational and electronic motion, *i.e.*

$$E = E_{\text{ele}} + E_{\text{vib}} + E_{\text{rot}} \qquad \text{...(4.8)}$$

(see Chapter 5)

(*ii*) The motion of electrons can be studied by keeping the positions of the nuclei fixed.

4.3.2. *Hamiltonian Operator*

The Hamiltonian operator for the polyatomic molecule can be written as

$$\hat{H} = -\frac{h^2}{8\pi^2}\sum_{j}^{\text{all nuclei}}\frac{1}{m_j}\nabla_j^2 - \frac{h^2}{8\pi^2 m_e}\sum_{i}^{\text{all electrons}}\nabla_i^2$$

$$+\sum_{ii'}\frac{e^2}{4\pi\varepsilon_0 r_{ii'}} - \sum\frac{Z_j e^2}{4\pi\varepsilon_0 r_{ij}} + \sum_{jj'}\frac{Z_j Z_{j'} e^2}{4\pi\varepsilon_0 R_{jj'}} \qquad \text{...(4.9)}$$

The first and second terms on the right hand side are the kinetic energy terms due to nuclei and electrons, third term is the repulsion between the electrons, 4th term attraction between the nuclei and electrons and fifth is the repulsion between the nuclei. The summation is being carried over all the nuclei and the electrons.

If the Born-Oppenheimer approximation is considered, the kinetic energy term due to nuclei will be neglected and the repulsion term between nuclei will be constant for a given positions of the nuclei.

If the diatomic molecule is considered the repulsion between the two nuclei will be

$$\frac{Z_j Z_{j'} e^2}{4\pi\varepsilon_0 R_{jj'}}$$

where $R_{jj'}$ is the distance between two nuclei. The electronic energy of each level (E_{ele}) can be calculated as a function of the bond distance $R_{jj'}$, and total energy E of molecule will be

$$E = E' + \frac{Z_j Z_{j'} e^2}{4\pi\varepsilon_0 R_{jj'}} \qquad \text{...(4.10)}$$

where E' is the energy of the system for the particular positions of the nuclei.

***Exercise* 4.4:** Write the Hamiltonian operator for H_2^+ and H_2 molecules.

In H_2^+, there is only one electron and using Born-Oppenheimer approximation, the equation (4.9) becomes

$$\hat{H} = -\frac{h^2}{8\pi^2 m_e}\nabla^2 + \frac{e^2}{4\pi\varepsilon_0 R_{12}} - \frac{e^2}{4\pi\varepsilon_0 r_1} - \frac{e^2}{4\pi\varepsilon_0 r_2}$$

where the distances are defined as shown in Fig. 4.1*a*. Similarly, using Fig. 4.1*b*, the Hamiltonian for H_2 can be written as

$$\hat{H} = -\frac{h^2}{8\pi^2 m_e}\nabla_a^2 - \frac{h^2}{8\pi m_e}\nabla_b^2 + \frac{e^2}{4\pi\varepsilon_0 r_{ab}} + \frac{e^2}{4\pi\varepsilon_0 R_{12}} - \frac{e^2}{4\pi\varepsilon_0}\left(\frac{1}{r_{a1}} + \frac{1}{r_{a2}} + \frac{1}{r_{b1}} + \frac{1}{r_{b2}}\right)$$

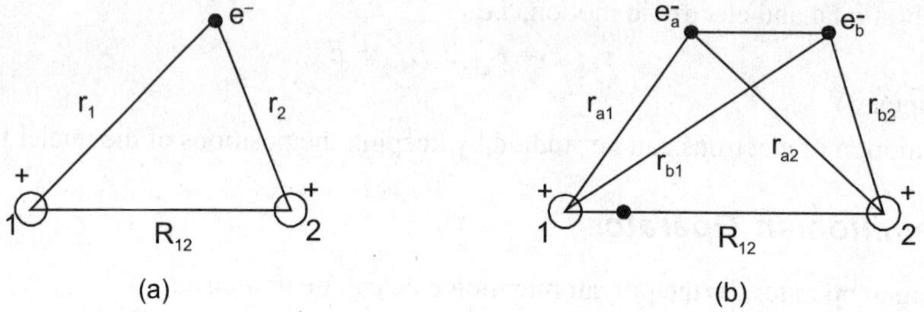

Fig. 4.1 (*a*) *Depiction of* H_2^+ *ion* (*b*) *Depiction of* H_2 *molecule*

4.3.3. *Wave Functions*

4.3.3.1. Molecular orbital theory

Similar to polyelectronic atoms, the total wave function of polyatomic molecules can be written as the product of the one electron molecular orbital wave functions, *i.e.*

$$\psi = \psi_1\psi_2\psi_3, \ldots \qquad \ldots(4.11a)$$

where ψ is the total molecular orbital wave function and ψ_i the one electron molecular orbital wave function. The molecular orbital wave functions can be obtained from the linear combination of atomic orbitals, the method is commonly known as LCAO (linear combination of atomic orbitals).

$$\psi_i = a_{i1}\phi_1(i) + a_{i2}\phi_2(i) + a_{i3}\phi_3(i) + \ldots \qquad \ldots(4.11b)$$

where $\phi_1(i)$ is the atomic orbital wave function and describes the position of electrons with respect to the nucleus 1 and so on. a_{i1}, a_{i2}, ... are the constants. Substituting the values of ψ_i in equation (4.11 *a*) and expanding them, *e.g.* in case of diatomic molecule, will lead to terms like $\varphi_1(1)\ \varphi_1(2)$ and $\phi_1(1)\ \phi_2(2)$, etc. The first and the similar ones are called the ionic terms as both the electrons are on the same atomic orbital, second and the similar ones covalent terms as the electrons are present on both the nuclei and shared. The contribution of each term to the molecular orbital depends on the nature of the molecule. The coefficient of each term gives the weighted average of the particular term. Further the coefficients a_{ik} are adjusted in such a way that the value of the energy is minimum.

***Exercise* 4.5:** Write the trial wave function for H_2^+ and H_2 molecule. In H_2 molecule, point out, which one are ionic and which one are covalent terms?

Both H_2^+ molecule ion and H_2 molecule are made up of two $1s$ nuclei. Let the nuclei be a and b. So trial wave function for H_2^+ molecule ion is

$$\psi = c_1\phi_a(1) + c_2\phi_b(1)$$

Similarly for H_2

$$\psi = \psi_1\psi_2$$

$$= [c_1\phi_a(1) + c_2\phi_b(1)]\,[c_1\phi_a(2) + c_2\phi_b(2)] \qquad ...(4.12)$$

where ϕ_i are the atomic wave functions. Expanding equation (4.12)

$$\psi = c_1^2\,\phi_a(1)\,\phi_a(2) + c_2^2\phi_b(1)\,\phi_b(2) + c_1c_2\,[\phi_a(1)\phi_b(2) + \phi_b(1)\,\phi_b(2)]$$

The first and second terms are ionic terms whereas third and fourth terms are covalent ones.

***Exercise* 4.6:** If $\hat{H}$ is the Hamiltonian and using the trial function for H_2^+ molecule ion from exercise 4.5, calculate the energies of bonding and antibonding orbitals. Find out also the corresponding wave functions.

The average energy of H_2^+ ion using the trial wave function

$$\psi = c_1\phi_a(1) + c_2\phi_b(1)$$

can be written as

$$E = \frac{\int[c_1\phi_a(1) + c_2\phi_b(1)]\,\hat{H}\,[c_1\phi_a(1) + c_2\phi_b(1)]\,d\tau}{\int[c_1\phi_a(1) + c_2\phi_b(1)][c_1\phi_a(1) + c_2\phi_b(1)]\,d\tau}$$

$$E = \frac{c_1^2\int\phi_a(1)\,\hat{H}\phi_a(1)\,d\tau + c_2^2\int\phi_b(1)\,\hat{H}\phi_b(1)\,d\tau + c_1c_2\int\phi_a(1)\hat{H}\phi_b(1)\,d\tau + c_1c_2\int\phi_b(1)\hat{H}\phi_a(1)\,d\tau}{c_1^2\int\phi_a(1)\phi_a(1)d\tau + c_2^2\int\phi_b(1)\phi_b(1)\,d\tau + c_1c_2\int\phi_a(1)\phi_b(1)\,d\tau + c_1c_2\int\phi_b(1)\phi_a(1)\,d\tau} \qquad ...(4.13a)$$

Let

$$\int\phi_a(1)\hat{H}\phi_a(1)\,d\tau = H_{aa},\quad \int\phi_b(1)\hat{H}\phi_b(1)\,d\tau = H_{bb}$$

$$\int\phi_a(1)\hat{H}\phi_b(1)\,d\tau = H_{ab},\quad \int\phi_b(1)\hat{H}\phi_a(1)\,d\tau = H_{ba}$$

$$\int\phi_a(1)\phi_a(1)\,d\tau = S_{aa},\quad \int\phi_b(1)\phi_b(1)\,d\tau = S_{bb}$$

$$\int\phi_a(1)\phi_b(1)\,d\tau = S_{ab},\quad \int\phi_b(1)\phi_a(1)\,d\tau = S_{ba}$$

Now, both ϕ_a and ϕ_b are normalised $1s$ atomic orbital wave functions, $H_{aa} = H_{bb}$, $S_{aa} = S_{bb} = 1$. It can also be shown that $H_{ab} = H_{ba}$. The equation (4.13a) becomes

$$E = \frac{c_1^2\,H_{aa} + c_2^2\,H_{aa} + 2c_1c_2H_{ab}}{c_1^2 + c_2^2 + 2c_1c_2S_{ab}} \qquad ...(4.13b)$$

Using the variation theorem and minimising the energy with respect to c_1 and c_2. The equations obtained after rearranging and setting

$$\frac{dE}{dc_1} = 0 \text{ and } \frac{dE}{dc_2} = 0, \text{ we get}$$

$$(H_{aa}-E)\,c_1+(H_{ab}-ES_{ab})\,c_2 = 0 \qquad ...(4.13c)$$

$$(H_{ab}-ES_{ab})\,c_1+(H_{aa}-E)\,c_2 = 0 \qquad ...(4.13d)$$

The solution of these equations other than $c_1 = c_2 = 0$, is if the following determinant is zero, *i.e.*

$$\begin{vmatrix} H_{aa}-E & H_{ab}-E\,S_{ab} \\ H_{ab}-E\,S_{ab} & H_{aa}-E \end{vmatrix} = 0 \qquad ...(4.13e)$$

or $\quad (H_{aa}-E)^2-(H_{ab}-ES_{ab})^2 = 0$

This is called "secular determinant". The solutions of the quadratic equation (4.13 *e*) are

$$E_1 = \frac{H_{aa}+H_{ab}}{1+S_{ab}} \qquad ...(4.13f)$$

$$E_2 = \frac{H_{aa}-H_{ab}}{1-S_{ab}} \qquad ...(4.13g)$$

Substituting the value of E_1 in (4.13*c*)

$$\left(H_{aa}-\frac{H_{aa}+H_{ab}}{1+S_{ab}}\right)c_1+\left(H_{ab}-\frac{H_{aa}S_{ab}+H_{ab}S_{ab}}{1+S_{ab}}\right)c_2=0$$

or

$$\left(\frac{H_{aa}S_{ab}-H_{ab}}{1+S_{ab}}\right)c_1+\left(\frac{H_{ab}-H_{aa}S_{ab}}{1+S_{ab}}\right)c_2=0 \qquad ...(4.13h)$$

or

$$c_1 = c_2 = c \qquad ...(4.13i)$$

Similarly for $\qquad E = E_2,\ c_1 = -\,c_2 = c$

Since, the total probability is unity, with $c_1 = c_2$ and ψ, we get

$$1 = \int_{\substack{\text{all}\\\text{space}}}\psi^*\psi d\tau = \int c^2[\phi_a(1)+\phi_b(1)][\phi_a(1)+\phi_b(1)]\,d\tau$$

$$= c^2\left[\int\phi_a(1)\phi_a(1)d\tau+\int\phi_b(1)\phi_b(1)d\tau+2\int\phi_a(1)\phi_b(1)d\tau\right]$$

$$= c^2\,[2+2\,S]$$

or

$$c = \frac{1}{\sqrt{2+2S}} \qquad ...(4.13j)$$

Similarly for $c_1 = -\,c_2 = c$, the value of c will be

$$c = \frac{1}{\sqrt{2-2S}} \qquad ...(4.13k)$$

So the wave functions and energies for the bonding and anti-bonding orbitals are

$$\psi_{\text{bonding}} = \frac{1}{\sqrt{2+2S}}\,(\phi_a(1)+\phi_b(1))\ E_1 = \frac{H_{aa}+H_{ab}}{1+S_{ab}}$$

$$\psi_{\text{anti-bonding}} = \frac{1}{\sqrt{2-2S}}\,(\phi_a(1) - \phi_b(1))\; E_2 = \frac{H_{aa} - H_{ab}}{1 - S_{ab}}$$

The integrals H_{aa}, H_{ab} and S_{ab} can be evaluated in closed analytic form by converting to elliptical coordinates. Without giving the details of getting them the results in atomic units are:

$$H_{aa} = H_{bb} = \frac{Z_e^2}{2} - Ze + \frac{(Z_e R + 1)e^{-2Z_e R}}{R} \quad ...(4.14)$$

$$H_{ab} = H_{ba} = -\frac{S_{12} Z_e^2}{2} + (Z_e - 2)(Z_e^2 \mathrm{R} + \mathrm{Z_e})\, e^{-Z_e R} + \frac{S_{12}}{R} \quad ...(4.15)$$

$$S_{ab} = [Z_e \mathrm{R} + \frac{Z_e^2 R^2}{3} + 1]e^{-Z_e R} \quad ...(4.16)$$

***Exercise* 4.7:** Using equation (4.16) construct the plot between S_{ab} and R for H_2 molecule. For hydrogen molecule, the Z value is 1, so equation (4.16) reduces to

$$S(H_2) = [1 + \mathrm{R} + \frac{R^2}{3}]e^{-R} \quad ...(4.16a)$$

Values of $S(H_2)$ were calculated and are compiled in the Table 4.1 and are plotted in Fig. 4.2.

TABLE 4.1. Values of $S(H_2)$ for hydrogen molecule using equation (4.16a)

R/a_0	$S(H_2)$	R/a_0	$S(H_2)$	R/a_0	$S(H_2)$	R/a_0	$S(H_2)$
0.0	1	1.0	0.858385	2.0	0.586453	3.0	0.348509
0.1	0.998337	1.1	0.833287	2.1	0.559626	3.1	0.329009
0.2	0.993393	1.2	0.8072	2.2	0.533333	3.2	0.310336
0.3	0.985288	1.3	0.780349	2.3	0.507644	3.3	0.292484
0.4	0.974198	1.4	0.752943	2.4	0.48262	3.4	0.275441
0.5	0.96034	1.5	0.725173	2.5	0.458308	3.5	0.259194
0.6	0.943956	1.6	0.697216	2.6	0.434748	3.6	0.243728
0.7	0.925304	1.7	0.669231	2.7	0.41197	3.7	0.229022
0.8	0.904649	1.8	0.64136	2.8	0.389995	3.8	0.215058
0.9	0.882256	1.9	0.61373	2.9	0.368839	3.9	0.201812
1.0						4.0	0.189262

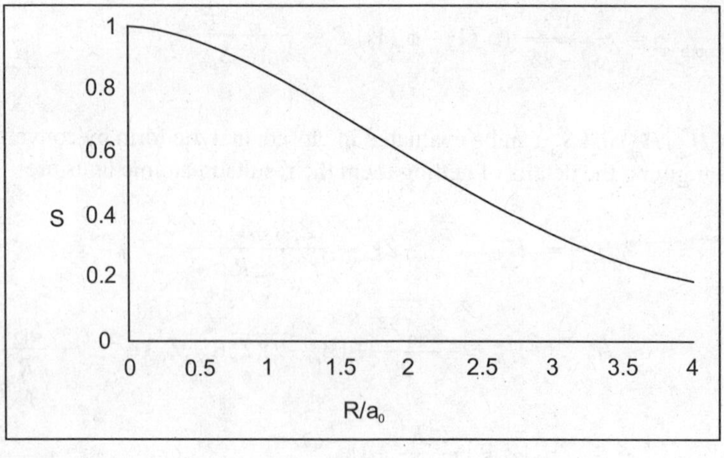

Fig. 4.2 *Plot of overlap integral (S) as a function of internuclear distance (R_{12}).*

It is evident from Exercise 4.7 that the overlap integral decreases as the distance between the incoming atomic orbitals increases.

***Exercise* 4.8**: Evaluate the integrals H_{aa} and H_{ab}. Explain the physical meaning briefly.

H_{aa} is defined as:

$$E_1 = \int[\varphi_a^*(1)+\varphi_b^*(1)]\left[-\frac{h^2}{8\pi^2 m_e}\nabla^2-\frac{e^2}{4\pi\varepsilon_0 r_1}-\frac{e^2}{4\pi\varepsilon_0 r_2}+\frac{e^2}{4\pi\varepsilon_0 R_{12}}\right][\varphi_a(1)+\varphi_b(1)]dr \quad ...(1)$$

$$= \int[\varphi_a^*(1)+\varphi_b^*(1)]\left[-\frac{h^2}{8\pi^2 m_e}\nabla^2-\frac{e^2}{4\pi\varepsilon_0 r_1}-\frac{e^2}{4\pi\varepsilon_0 r_2}+\frac{e^2}{4\pi\varepsilon_0 R_{12}}\right]\varphi_a(1)dr +$$

$$\int[\varphi_a^*(1)+\varphi_b^*(1)\left[-\frac{h^2}{8\pi^2 m_e}\nabla^2-\frac{e^2}{4\pi\varepsilon_0 r_1}-\frac{e^2}{4\pi\varepsilon_0 r_2}+\frac{e^2}{4\pi\varepsilon_0 R_{12}}\right]\varphi_b(1)dr \quad ...(2)$$

Recalling the one electron atomic system, *i.e.*, hydrogen atom

$$\int\varphi_a^*(1)\left[-\frac{h^2}{8\pi^2 m_e}\nabla^2-\frac{e^2}{4\pi\varepsilon_0 r_1}\right]\varphi_a(1)dr = E_{1s}\varphi_a \quad ...(3)$$

and $$\int\varphi_b^*(1)\left[-\frac{h^2}{8\pi^2 m_e}\nabla^2-\frac{e^2}{4\pi\varepsilon_0 r_2}\right]\varphi_b(1)dr = E_{1s}\varphi_b \quad ...(4)$$

$$E_1 = \int[\varphi_a^*(1)+\varphi_b^*(1)]\left[E_{1s}-\frac{e^2}{4\pi\varepsilon_0 r_2}+\frac{e^2}{4\pi\varepsilon_0 R_{12}}\right]\varphi_a(1)dr +$$

$$\int[\varphi_a^*(1)+\varphi_b^*(1)]\left[E_{1s}-\frac{e^2}{4\pi\varepsilon_0 r_1}+\frac{e^2}{4\pi\varepsilon_0 R_{12}}\right]\varphi_b(1)dr \quad ...(5)$$

Expanding the above integral we get:

$$= \int \varphi_a^*(1)E_{1s}\varphi_a(1)dr + \int[\varphi_b^*(1)E_{1s}\varphi_a(1)dr + \int \varphi_a^*(1)[-\frac{e^2}{4\pi\varepsilon_0 r_2} + \frac{e^2}{4\pi\varepsilon_0 R_{12}}]\varphi_a(1)dr$$

$$+ \int \varphi_b^*(1)\left[-\frac{e^2}{4\pi\varepsilon_0 r_2} + \frac{e^2}{4\pi\varepsilon_0 R_{12}}\right]\varphi_a(1)dr + \int[\varphi_a^*(1)E_{1s}\varphi_b(1)dr + \int[\varphi_b^*(1)E_{1s}\varphi_b(1)dr$$

$$+ \int \varphi_a^*(1)\left[-\frac{e^2}{4\pi\varepsilon_0 r_1} + \frac{e^2}{4\pi\varepsilon_0 R_{12}}\right]\varphi_b(1)dr + \int \varphi_b^*(1)[-\frac{e^2}{4\pi\varepsilon_0 r_2} + \frac{e^2}{4\pi\varepsilon_0 R_{12}}]\varphi_b(1)dr \quad ...(6)$$

This can be simplified to as:

$$= E_{1s} + E_{1s}S + \int \varphi_a^*(1)\left[-\frac{e^2}{4\pi\varepsilon_0 r_2} + \frac{e^2}{4\pi\varepsilon_0 R_{12}}\right]\varphi_a(1)dr + \int \varphi_b^*(1)\left[-\frac{e^2}{4\pi\varepsilon_0 r_2} + \frac{e^2}{4\pi\varepsilon_0 R_{12}}\right]\varphi_a(1)dr$$

$$+E_{1s} + E_{1s}S + \int \varphi_a^*(1)\left[-\frac{e^2}{4\pi\varepsilon_0 r_1} + \frac{e^2}{4\pi\varepsilon_0 R_{12}}\right]\varphi_b(1)dr + \int \varphi_b^*(1)\left[-\frac{e^2}{4\pi\varepsilon_0 r_1} + \frac{e^2}{4\pi\varepsilon_0 R_{12}}\right]\varphi_b(1)dr$$

$$= 2E_{1s}(1+S) + \int \varphi_a^*(1)\left[-\frac{e^2}{4\pi\varepsilon_0 r_2} + \frac{e^2}{4\pi\varepsilon_0 R_{12}}\right]\varphi_a(1)dr + \int \varphi_b^*(1)\left[-\frac{e^2}{4\pi\varepsilon_0 r_2} + \frac{e^2}{4\pi\varepsilon_0 R_{12}}\right]\varphi_a(1)dr$$

$$+ \int \varphi_a^*(1)\left[-\frac{e^2}{4\pi\varepsilon_0 r_1} + \frac{e^2}{4\pi\varepsilon_0 R_{12}}\right]\varphi_b(1)dr + \int \varphi_b^*(1)\left[-\frac{e^2}{4\pi\varepsilon_0 r_1} + \frac{e^2}{4\pi\varepsilon_0 R_{12}}\right]\varphi_b(1)dr \quad ...(7)$$

Physically the first integral on the right hand side of equation (7) reflects (*i*) the charge density of the electron around the nucleus A interacting with nucleus B via Coulomb potential and*i*() the internuclear repulsion. This integral is denoted by *J* and is called *Coulomb integral*.

$$J = \int \varphi_a^*(1)\left[-\frac{e^2}{4\pi\varepsilon_0 r_b} + \frac{e^2}{4\pi\varepsilon_0 R_{a12}}\right]\varphi_a(1)dr = -\frac{e^2}{4\pi\varepsilon_0}\int \varphi_a^*(1)\left[\frac{1}{r_b}\right]\varphi_a(1)dr + \frac{e^2}{4\pi\varepsilon_0 R_{a12}} \quad ...(8)$$

In evaluating this integral we have assumed that this calculation has been carried out at fixed positions of the nuclei. Further the charge over a volume element dr is $\varphi_a^*\varphi_a dr$ and the Coulomb energy of interaction with a proton separated by a distance r_b is $\frac{\varphi_a^*\varphi_a dr}{r_b}$. The second term reflects that if all the values are added or integrated, we get the Coulomb repulsion energy between two protons separated by a distance *R*.

The second integral is given the symbol *K* and is called exchange integral, *i.e.*,

$$K = \int \varphi_b^*(1)\left[-\frac{e^2}{4\pi\varepsilon_0 r_b} + \frac{e^2}{4\pi\varepsilon_0 R_{a12}}\right]\varphi_a(1)dr = -\frac{e^2}{4\pi\varepsilon_0}\int \varphi_b^*(1)\left[-\frac{1}{r_b}\right]\varphi_a(1)dr + \frac{Se^2}{4\pi\varepsilon_0 R_{a12}} \quad ...(9)$$

S defines the overlap of two orbitals $\varphi_a\varphi_b$ and defines the terms for a fixed internuclear separation *R*. It may be mentioned here that there is no counterpart for this term in classical mechanics and is purely a quantum mechanical term. It is direct result of the approximation that the molecular orbital is a linear combination of atomic orbitals centred on different atoms. Third and fourth integrals are the same as the first two integrals except that the indices are exchanged and thus the mathematical expressions are the same. We can write equation (1)

$$\int[\varphi_a^*(1)+\varphi_b^*(1)]\hat{H}[\varphi_a(1)+\varphi_b(1)]dr = E_+ = 2E_{1s}(H)[1+S]+2J+2K \quad ...(10)$$

Introducing the normalisation constants to the orbitals, equation 10 can be written as

$$E_1 = E_{1s}(H)+\frac{J+K}{1+S}$$

or

$$\Delta E_1 = E_1 - E_{1s}(H) = \frac{J+K}{1+S} \quad ...(11)$$

On the same lines the value of the E_2 and ΔE_2 can be obtained and these are:

$$E_2 = E_{1s}(H)-\frac{J-K}{1-S} \quad ...(12)$$

or

$$\Delta E_1 = E_2 - E_{1s}(H) = \frac{J-K}{1-S} \quad ...(13)$$

Equation (11) represents the energy of H_2^+ relative to the completely dissociated species H and H^+. Similarly equation (12) represents the energy of H_2^+ relative to the completely dissociated species H and H^+. Integrals J and K can be obtained analytically and the values are:

$$J = \left(1+\frac{1}{R_{12}}\right)e^{-2R_{12}} \quad ...(14)$$

$$K = \frac{S}{R_{12}} - (1+R_{12})e^{-R_{12}} \quad ...(15)$$

***Exercise* 4.9:** Calculate the values of E_1 and E_2 and plot them against different values of R_{12} and explain the salient features.

The values of J and K are calculated for different values of R_{12} and are compiled in Table 4.2. These values are further substituted in equations (11) and 13 to get the values of E_1 and E_2 respectively. The data are plotted in Fig. 4.3.

TABLE 4.2. Values of E_1 and E_2 at different internuclear distances

R_{12}	E_1	E_2	R_{12}	E_1	E_2
0			3.2	– 0.054	0.112
0.2	4.01	6.83	3.4	– 0.050	0.090
0.4	1.55	2.88	3.6	– 0.045	0.078
0.6	0.78	2.24	3.8	– 0.042	0.064
0.8	0.4	1.35	4.0	– 0.037	0.053
1.0	0.211	1.04	4.2	– 0.033	0.046

1.2	0.01	0.81	4.4	– 0.029	0.038
1.4	0.002	0.63	4.6	– 0.025	0.032
1.6	– 0.01	0.51	4.8	– 0.023	0.028
1.8	– 0.04	0.41	5.0	– 0.019	0.023
2.0	– 0.054	0.336	5.5	– 0.014	0.016
2.2	– 0.062	0.278	6.0	– 0.008	0.011
2.4	– 0.064	0.230	6.5	– 0.006	0.063
2.6	– 0.0637	0.189	7.0	– 0.004	0.004
2.8	– 0.0628	0.160	7.5	– 0.0026	0.0027
3.0	– 0.06	0.133	8.0	– 0.0017	0.002

The results of Table 4.2 and Figs. (4.3 *a*) and (4.3*b*) bring forth certain very useful informations which are given below:

(*i*) Values of E_2 are always positive, *i.e.*, greater than E_H and keeps on decreasing as the value of R_{AB} increases. On the other hand, the values of E_1 are always greater than E_H at $R_{AB} \leq 1.5$ a_0, becomes less than E_H at ≥ 1.5 a_0.

(*ii*) There is no minimum in the value of E_2 and asymptotically approach zero at $R_{AB} = \infty$. In case of E_1, it deceases with increase of R_{AB}, reaches minimum and then start increasing, reaching zero at $R_{AB} = \infty$.

(*iii*) At a given value of R_{AB}, the value of E_1 is always less than that of E_2, indicating that *m. o.* corresponding to E_1 is always stable than the m. o. corresponding to E_2.

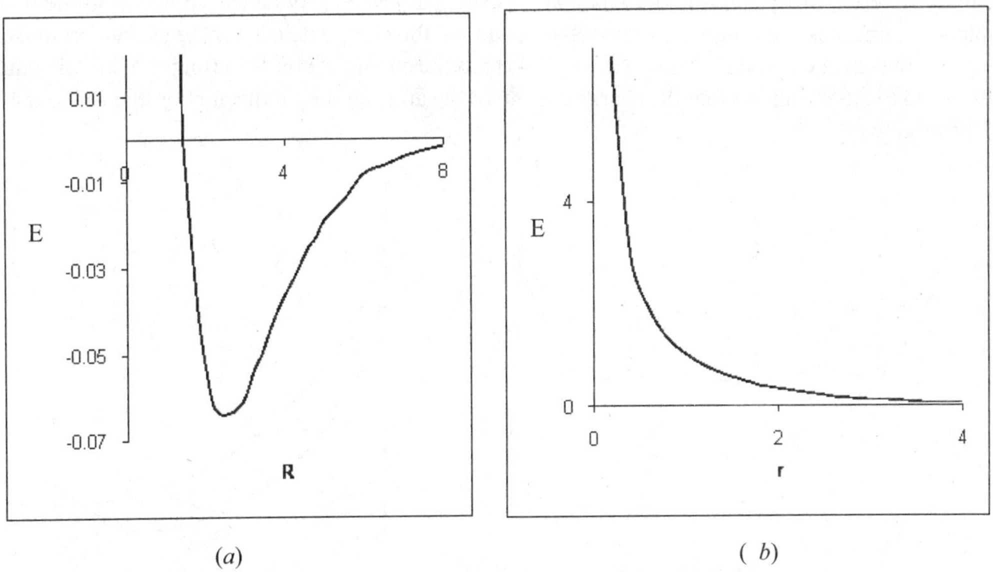

(*a*) (*b*)

Fig. 4.3 *Plot of (a) E_1 versus R_{AB} and (b) E_2 versus R_{AB}. R_{AB} is in terms of Bohr's radius.*

(*iv*) The value of E_1 and R_{AB} at the minimum value of energy are: 1.77 eV (170.8 kJ mol^{-1}) and 2.5 a_0 = 0.132 nm. It is clear from the above discussion that H_2^+ is a stable species with respect to H and H^+ system, the dissociation energy is 1.77 eV (170.8 kJ mol^{-1}) and the bond length of at the equilibrium position is 0.132 nm. The experimental value of E_1 (dissociation energy) and the equilibrium distance are found to be 2.8 eV (270.2 kJ mol^{-1}) and 0.106 nm respectively. Although the agreement between theoretical results and experimental observations is not good but one thing is clear that the above approach is leading in the right direction. The success of the variation method depends on the choice of the proper and correct wave function. In other words, we need better wave function to describe the H_2^+ molecule.

***Exercise* 4.10:** Using the wave functions for the bonding and anti-bonding molecular orbitals, show that the charge densities in case of bonding molecular orbital is greater than that of the two atomic orbitals and smaller in case of anti-bonding orbital.

$$\psi_{\text{bonding}} = \frac{1}{\sqrt{2+2S}}[\varphi_a(1)+\varphi_b(1)]$$

and

$$\psi^2_{\text{bonding}} = \frac{1}{2+2S}[\varphi_a(1)+\varphi_b(1)]^2 = \frac{1}{2+2S}[\varphi_a^2(1)+\varphi_b^2(1)+2\varphi_a(1)\varphi_b(1)] \quad ...(4.17)$$

Similarly for anti-bonding molecular orbital,

$$\psi^2_{\text{anti-bonding}} = \frac{1}{2+2S}[\varphi_a(1)+\varphi_b(1)]^2 = \frac{1}{2+2S}[\varphi_a^2(1)+\varphi_b^2(1)-2\varphi_a(1)\varphi_b(1)] \quad ...(4.18)$$

Interpreting the results of the equation (4.17) and knowing the fact that $\varphi_a(1)\varphi_b(1)$ is a positive quantity (overlap of two atomic orbitals), we can see that probability density function of the bonding molecular orbital is more in the centre of the bond than the charge density of the atomic orbitals in the centre of the chemical bond as shown in Fig. 4.4. Thus the chemical bond is stronger. Similarly equation (4.18) and Fig. (4.4) shows that the charge density of the anti-bonding molecular orbital is less than that of the atomic orbitals.

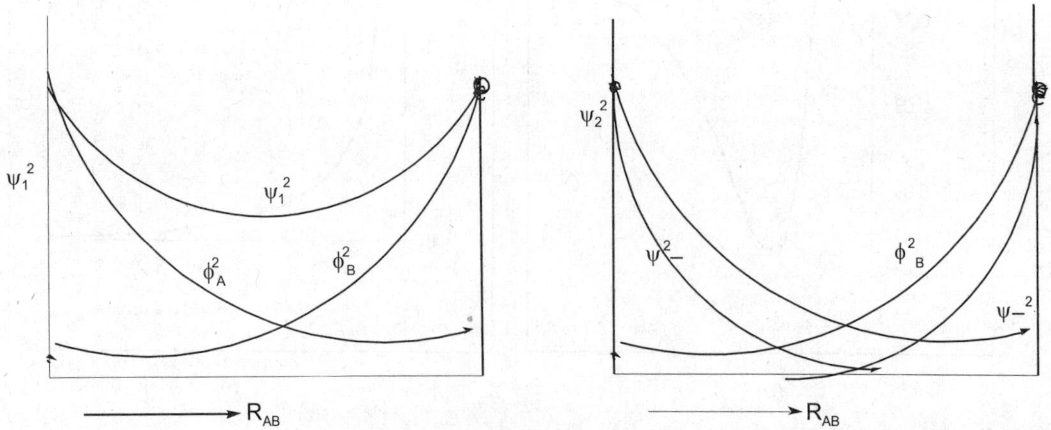

Fig. 4.4 *(a) Plot of* ψ_1^2 *versus* R_{12} *and (b) plot of versus* R_{12}.

***Exercise* 4.11:** Using the wave function obtained in case of H_2^+ write the ground state wave function of H_2 using Slater's determinant and show that the spin and space parts are separate parts.

$\psi_{Bonding}$ describes the molecular orbital for the ground state of H_2^+. Thus the wave function for the H_2 molecule can be obtained by filling the bonding molecular orbital with two electrons, keeping in mind the Pauli's exclusion principle. The Slater's form of wave function is

$$\psi_{bonding} = \frac{1}{\sqrt{2!}}\begin{vmatrix} \varphi_b(1)\alpha(1) & \varphi_b(1)\beta(1) \\ \varphi_b(2)\alpha(2) & \varphi_b(2)\beta(2) \end{vmatrix}$$

Expanding the determinant, we can write

$$\psi_{bonding} = \varphi_b(1)\varphi_b(2)\left[\frac{1}{\sqrt{2}}\right]\{\alpha(1)\beta(2) - \alpha(1)\beta(2)\}$$

One can see that once again the space and spin part are separately described.

4.3.4. *Valence Bond Theory*

The valence bond theory developed by Heitler and London, considers that molecules are formed of atoms (which to some extent preserve their distinct character) and these are held togther by chemical bonds. To be more exact, the atoms contribute unshared outer or valence electrons to pair up with the electrons of other atoms (to form bonds with other atoms) such that molecules consist of atomic cores and bonds between these cores. Generally the pairing of only those valence electrons takes place which are not paired. Sometimes some of the paired electrons are excited to other vacant orbitals to make them unpaired thus forming hybrid atomic orbitals (see Section 4.14).

The wave function of a molecule in valence bond theory, is the contributions of wave functions of different structures, arising from different possible arrangements of the electrons in the molecule, *i.e.*

$$\psi = c_1\psi_1 + c_2\psi_2 + c_3\psi_3 + ... \qquad ...(4.19)$$

where c_1, c_2, c_3 are the weighted average contribution of the structures, ψ_1, ψ_2, ψ_3.

The total wave function of molecules is written as

$$\psi = \psi_{cov} + \lambda\psi_{ion} \qquad ...(4.20)$$

where λ is the fraction of the contribution of ionic structure in the total wave function. The vaule of λ can be obtained from the relation (4.21) if one knows the percentage ionic character of the bond, *i.e.*

$$\%\text{ ionic character} = \frac{\lambda^2}{1+\lambda^2} \times 100 \qquad ...(4.21)$$

***Exercise* 4.12:** Write the wave function of H_2 molecule according to valence bond theory.

Consider H_2 molecule as hydrogen atom *a* and hydrogen atom *b*. Thus each electron of hydrogen atoms *a* and *b* can be paired. Since, the two electrons cannot be distinguished, the wave function can be written as

$$\psi = N\,[1s_a(1)\ 1s_b(2) + 1s_b(1)\ 1s_a(2)]$$

where *N* is normalisation constant.

The other possible arrangement of placing two electrons are: both the electrons can be either on atom *a* or on atom *b*, *i.e.* the wave functions of these two arrangements are: $1s_a(1)\,1s_a(2)$ and $1s_b(1)\,1s_b(2)$. The last two arrangements show the presence of ionic structure. But the first arrangement indicates the covalent nature. Thus total wave function will be

$$\psi = c_1\, 1s_a(1)\, 1s_b(2) + c_2\, 1s_a(2)\, 1s_b(1) + c_3\, 1s_a(1)\, 1s_a(2) + c_4\, 1s_b(1) 1s_b(2)$$

where c_1, c_2, c_3 and c_4 are the coefficients and their values indicate the contributions to the total wave function.

***Exercise* 4.13:** Calculate the fraction of the ionic wave function in HBr molecule if the percentage ionic character is 11%.

Using the equation (4.21)

$$11 = \frac{\lambda^2}{1+\lambda^2} \times 100$$

or $$11 + 11\,\lambda^2 = 100\,\lambda^2$$

or $$\lambda^2 = 11/89$$

$$\lambda = \sqrt{11/89} = 0.35$$

4.4. MOLECULAR ORBITALS

As clear from Section (4.3.1), that use of variation theorem results in the construction of molecular orbitals from the atomic orbitals and the number of molecular orbitals obtained are the same as the number of atomic orbitals taking part in the formation of molecule. Half of the molecular orbitals have the energy lower than those of atomic orbitals and the other half have higher energy than those of atomic ones. The former are called, "the bonding molecular orbitals" and the latter are called "antibonding molecular orbitals". The electrons not taking part in the bonding are called "non-bonding" electrons and have the same energy as possessed by the separated atoms. The relative energies of each kind of orbitals are shown in the Fig. 4.5. The atomic orbitals taking part in the formation of molecular orbitals should fulfil the following requirements.

(*a*) The atomic orbitals forming molecular orbitals should have comparable energy.

(*b*) The wave functions of the respective atomic orbitals should overlap in space as much as possible.

(*c*) The wave functions of the atomic orbitals should have the same symmetry relative to the molecular axis.

Table 4.3 lists the various combinations of atomic orbitals that give the molecular orbitals, taking *x*-axis as the molecular axis.

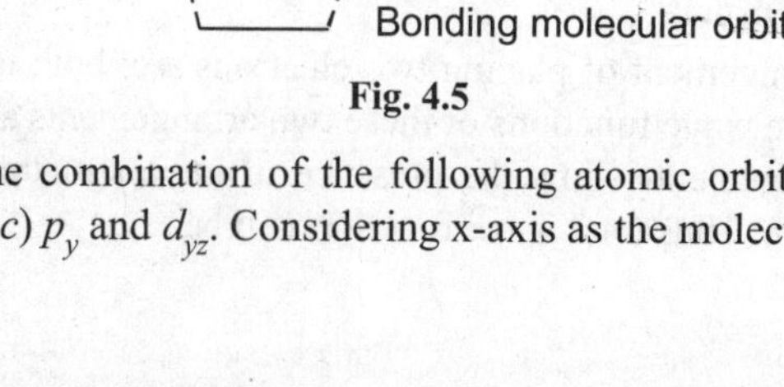

Fig. 4.5

***Exercise* 4.14:** Will the combination of the following atomic orbitals give molecular orbitals? (*a*) s and p_z, (*b*) p_x and p_x, (*c*) p_y and d_{yz}. Considering x-axis as the molecular axis.

Only p_x and p_x atomic orbitals will give molecular orbitals as the overlap of these will be finite. The overlap of s and p_z and p_y and d_{yz} will be zero, considering x axis as the molecular axis.

***Exercise* 4.15:** Write the trial wave function for (*a*) HCl molecule; (*b*) NH_3 molecule.

(*a*) $\psi_T = \psi_1\psi_2 = [c_1\ 1s_H(1) + c_2\ 2p_x(1)]\ [c_1 1s_H(2) + c_2 2p_x(2)]$

$= c_1^2\ 1s_H(1)1s_H(2) + c_2^2\ 2p_x(1)2p_x(2) + c_1c_2\ [1s_H(1)\ 2p_x(2) + 1s_H(2)\ 2p_x(1)]$

The first two are ionic terms and the last two are covalent terms.

(*b*) There are three unpaired electrons present in $2p$ orbitals of N. If these are listed as such in forming molecular orbitals, the wave function is

$$\psi_T = \psi_1\psi_2\psi_3 = K(N)\ 2s^2\ (N)\ [1s\ (H_A) + p_x(N)]^2\ [1s\ (H_B) + p_y(N)]^2\ [1s\ (H_C) + p_z(N)]^2$$

If sp^3 hybrid orbitals of nitrogen are used, then m.o. involves hybrid orbitals.

TABLE 4.3. Allowed Combinations of Atomic Orbitals in Linear Combination of Atomic Orbitals (LCAO)

A	Allowed	Forbidden
s	s, p_x, d_{z^2}	$p_y, p_z, d_{xy}, d_{yz}, d_{zx}, d_{x^2-y^2}$
p_x	s, p_x, d_{z^2}	$p_y, p_z, d_{xy}, d_{yz}, d_{xz}, d_{x^2-y^2}$
p_y	p_y, d_{xy}	$s, p_x, p_z, d_{yz}, d_{xz}, d_{z^2}, d_{x^2-y^2}$
d_{z^2}	s, p_x, d_{z^2}	$p_y, p_z, d_{xy}, d_{yz}\ d_{xz}, d_{x^2-y^2}$
$d_{x^2-y^2}$	$d_{x^2-y^2}$	$s, p_x, p_y, p_z, d_{xy}, d_{yz}, d_{xz}, d_{z^2}$
d_{xy}	p_y, d_{xy}	$s, p_x, p_{z'}, p_{yz'}, d_{zx'}, d_{z^2}, d_{x^2-y^2}$

From equation (4.13 *a* and *b*) it is clear that molecular wave function or molecular orbital is best represented if we take into account the linear combination of larger number of atomic orbitals. For example let us consider the atomic orbitals of the first 10 atoms and their atomic orbitals, *i.e.*, $1s$, $2s$ and $2p_x$, $2p_y$ and $2p_z$ atomic orbitals. Using variation method, we can generate 10×10 secular determinants. It is not easy to solve, although fast computers are available. Based on actual calculations certain principles have been formulated. By applying these principles, we can speculate qualitatively that many of the coefficients turn out to be zero and some of them can be of very small value. These principles are:

1. The lower energy atomic orbitals will tend to produce the lower energy molecular orbitals.

2. As said above that molecular orbitals are formed from the atomic orbitals with comparable energies, it may be mentioned that high energy atomic orbitals do not usually mix significantly with much lower energy atomic orbitals to form molecular orbitals.

3. A molecular orbital obtained from a particular pair of atomic orbitals that contains nodes will have a higher energy than one without nodes, *i.e.*, combination of s atomic orbitals will be of lower energy than obtained from the combination of p atomic orbitals and thus similarly the combination of d atomic orbitals.

4. σ, π and δ molecular orbitals do not mix to form diatomic molecular orbitals.

The molecular orbitals obtained from the linear combinations, their notation, the value of the m_l of the atomic orbitals are compiled in Table 4.4. The molecular axis has been assumed to be x-axis, although this is very arbitrary and one can choose any other axis as molecular axis.

TABLE 4.4. Nomenclature of molecular orbitals considering *X*-axis as the molecular axis

Atomic orbitals	m_l	Nodal plane	Designation
S	0	0	Σ
p_x	±1	0	Σ
p_y or p_z	±1, 0	1	Π
d_{xy} or d_{zx}	±1	1	Π
$d_{x^2-y^2}$	±2	2	Δ
d_{z^2}	±2	2	Δ

For the shapes of these orbitals, their geometrical representations and other characteristics are discussed in Section 4.5.

Taking into account these principles, we can qualitatively determine the respective coefficients of the 10 × 10 secular determinant and have been compiled in Table 4.5.

TABLE 4.5. Qualitative form of molecular orbitals for H_2^+ obtained from taking linear combination of 10 atomic orbitals. The notation + and – means variational coefficients of equal and large magnitude, but different sign. The notation s stands for the coefficients whose magnitude is small as compared to the ones designated as + and –. A zero entry means that the variational coefficient is zero.

a_1	a_2	a_3	a_4	a_5	a_6	a_7	a_8	a_9	a_{10}	Orbital
+	+	s	s	s	s	0	0	0	0	σ_{1s}
+	–	s	s	s	s	0	0	0	0	σ^*_{1s}
s	s	+	+	s	s	0	0	0	0	σ_{2s}
s	s	+	–	s	s	0	0	0	0	σ^*_{2s}
s	s	s	s	+	–	0	0	0	0	σ_{2px}
0	0	0	0	0	0	+	+	0	0	π_{2py}
0	0	0	0	0	0	0	0	+	+	π_{2pz}
0	0	0	0	0	0	+	–	0	0	π^*_{2py}
0	0	0	0	0	0	0	0	+	–	π^*_{2pz}
s	s	s	s	+	+	0	0	0	0	σ^*_{2px}

We will take up couple of cases. It is clear from Table 4.5 that 1 *s* orbital is of lowest energy in atomic system and thus will form lowest molecular orbital σ_{1s}. Thus the coefficient a_1 and a_2 will be quite large and thus positive, whereas those of 2*s* and $2p_x$ will be small, *i.e.*, a_3, a_4, a_5 and a_6 will be small. On the other hand, the coefficient a_7, a_8, a_9 and a_{10} will be zero as these atomic orbitals form π molecular orbitals. Similarly the next higher atomic orbital will be 2*s*. Thus in the formation of σ_{2s} molecular orbital a_1 and a_2 will be small: a_3 and a_4 will be large and a_5 and a_6 will be small and rest of them will be zero. Lastly $2p_z$ or $2p_y$ form π molecular orbitals. In case of $2p_y$, a_7 and a_8 will be positive and in the latter case a_9 and a_{10} will positive and rest all are zero. This is because a_1 to a_6 form σ orbital whereas $2p_y$ and $2p_z$ are perpendicular to each other. Similarly the other coefficients can be explained.

Electronic configurations, bond order, bond dissociation energy and bond lengths of few diatomic molecules are compiled in Table 4.6.

TABLE 4.6. Electron configuration and bond properties of homonuclear diatomic molecules of the second period elements

Molecules	Electron configuration	Bond order	Bond Enthalpy (kJ mol^{-1}	Bond Length (nm)
H_2	$(\sigma_{1s})^2$	1	436.4	0.074
Li_2	$KK(\sigma_{2s})^2$	1	104.6	0.267
B_2	$KK(\sigma_{2s})^2(\sigma^*_{2s})^2(\pi_y)^1\ (\pi_z)^1$	1	288.7	0.159
C_2	$KK(\sigma_{2s})^2(\sigma^*_{2s})^2(\pi_y)^2(\pi_z)^2$	2	627.6	0.131
N_2	$KK(\sigma_{2s})^2(\sigma^*_{2s})^2(\pi_y)^2(\pi_z)^2(\sigma_{2px})^2$	3	941.4	0.110
O_2	$KK(\sigma_{2s})^2(\sigma^*_{2s})^2(\sigma_{2px})^2(\pi_y)^2(\pi_z)^2(\pi^*_y)^1(\pi^*_z)^1$	2	498.8	0.121
F_2	$KK(\sigma_{2s})^2(\sigma^*_{2s})^2(\sigma_{2px})^2(\pi_y)^2(\pi_z)^2(\pi^*_y)^2(\pi^*_z)^2$	1	150.6	0.142

***Exercise* 4.16:** In the contrast of the coefficients of atomic orbitals mentioned in Table 4.6, construct the molecular orbitals in HF molecule.

HF molecule contains hydrogen atom, containing only one electron, *i.e.,* 1*s*, whereas fluorine atom consists of $1s^2 2s^2 2p^5$ atomic orbitals. We have assumed that molecular axis is *X*-axis. Thus based on linear combination of atomic orbitals and equation (4.11 *b*), the wave function for the molecular orbital can be written as

$$\Psi_i = a_{i1}\varphi_{1s}(\text{H}) + a_{i2}\varphi_{1s}(\text{F}) + a_{i3}\varphi_{2s}(\text{F}) + a_{i4}\varphi_{2px}(\text{F}) + a_{i5}\varphi_{2py}(\text{F}) + a_{i6}\varphi_{2pz}(\text{F}) \quad ...(1)$$

We know that the 1*s* atomic orbital of fluorine is of very low energy as compared to that of 1 *s* of hydrogen atom. This is because of very large positive charge on the fluorine nucleus. Thus we would expect very small overlap between these two orbitals, *i.e.,* the coefficient a_{i2} will be nearly zero. Before we proceed further we can also predict that coefficients a_{i5} and a_{i6} will be zero as these orbitals are π-forming orbitals. So we are left with coefficients a_{i3} and a_{i4} to be decided. Now the energies of 1*s*(H) and 2*s*(F) and $2p_x$(F) are nearly equal and thus these orbitals will overlap and thus the coefficients a_{i3} and a_{i4} will be non-zero. It suggests that 1 *s*(H) orbital of hydrogen will overlap with 2 *s* (F) and $2p_x$(F) orbitals of the fluorine atom. Although we will learn in detail in Section 4.13.1 about hybridisation, it may be mentioned here that combination of 2 *s*(F) and $2p_x$(F) is called hybridisation and atomic hybrid obtained from this combination is called hybrid orbitals.

From energy point of view, equation (1) suggests that there will be six energy levels obtained by the linear combination of one atomic orbital of hydrogen and five atomic orbitals from fluorine. We have found that a_{i2} will be nearly negligible when we consider the overlap. In other words, it can also be stated that 1*s*(F) atomic orbital will be confined mostly to fluorine atom and thus the value of a_{i2} can be very large from the point that it is localised on fluorine and will be of lowest energy , can be denoted as 1σ. The MO of next higher energy will be primarily a combination of 2 *s* and $2p_x$ atomic orbitals, a_{i3} and

a_{i4} will be very large and a_{i1} and a_{i2} will be small and $a_{i5} = a_{i6} = 0$. The next MO of higher energy will be the overlap of the $1s$(H) and hybrid orbital of fluorine. In this case a_{i1}, a_{i3} and a_{i4} will be very large and $a_{i5} = a_{i6} = 0$ (because σ and π orbitals do not overlap). $2p_y$ and $2p_z$ atomic orbitals of fluorine would form two degenerate π-orbitals, with a_{i5} and a_{i6} with large values and rest of the coefficients will be zero. The energies of these π-orbitals will be nearly equal to the energies of $2p_y$ and $2p_z$ orbitals. The energy level diagram is shown in Fig. 4.6. Such a depiction of energy level diagram is called as the *correlation diagram*. Thus the electronic configuration for HF can be written as: $(1\sigma)^2 (2\sigma)^2 (3\sigma)^2 (1\pi_y)^2 (1\pi_z)^2$. The electronic configuration can also be written as K(F) $(1\sigma)^2 (2\sigma)^2 (1\pi_y)^2 (1\pi_z)^2$.

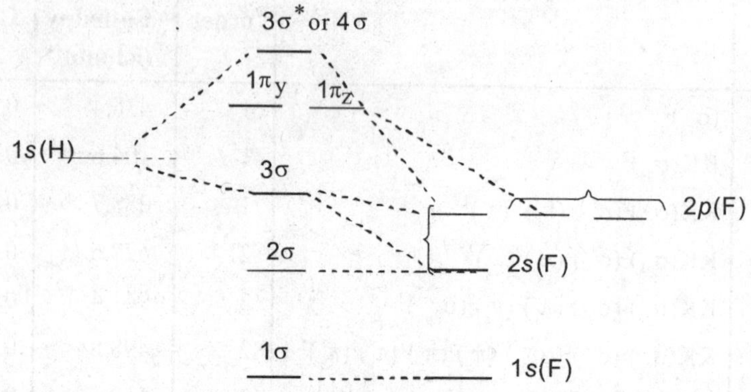

Fig. 4.6 *Qualitative energy level diagram of HF. Energies of different atomic or molecular orbitals are not drawn to the actual scale.*

4.5. SHAPES OF MOLECULAR ORBITALS

Most commonly occurring molecular orbitals are σ (sigma) and π (pi) orbitals. Sigma orbitals are symmetrical about the internuclear axis. The cross-section perpendicular to the internuclear axis (usually *x* axis) gives a ellipse. These are formed either from *s* orbitals or *p* and *d* orbitals having the lobes along the internuclear axis.

π-orbitals are formed when *p* orbitals on each atom point perpendicular to the internuclear axis. The overlap region is above and below the bond axis. The charge densities are above and below the nodal plane of *p* orbitals (see Fig. 4.7).

The other kinds of molecular orbitals called as δ and so on, are not commonly found. The asterisk indicates the anti-bonding orbitals and the symbol at subscript indicates the atomic orbitals used in forming the molecular orbitals. The subscript *g* and *u* refers to the symmetry of the molecular orbital, σ and π* are symmetric and σ* and π are anti-symmetric.

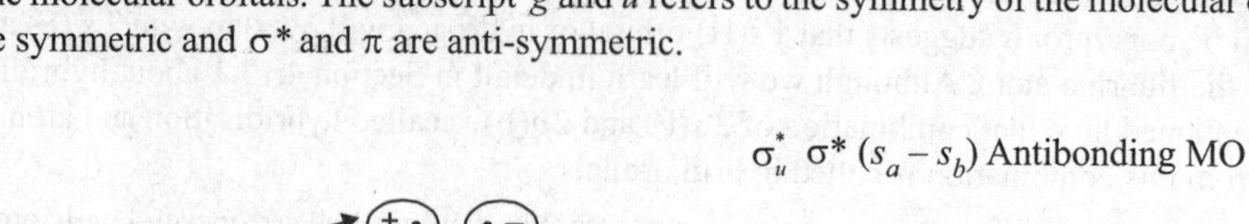

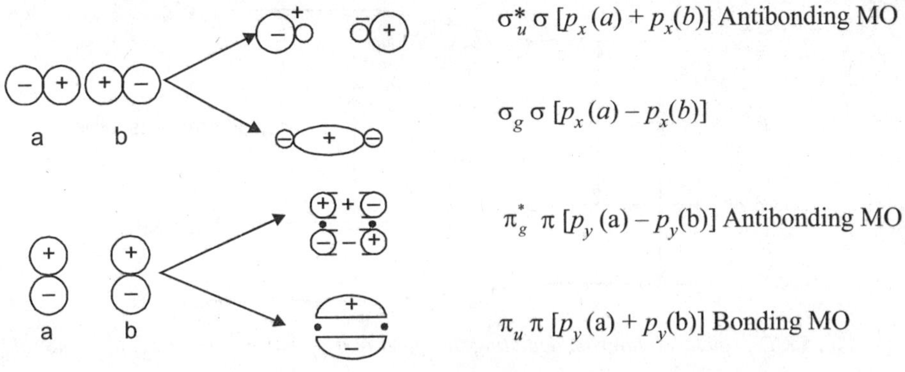

Fig. 4.7 *Shapes of molecular orbitals formed from atomic orbitals*

4.6. RELATIVE ENERGY LEVELS OF THE MOLECULAR ORBITALS

The relative energies of the molecular orbitals formed from the atomic orbitals of the second period elements, containing 7 or fewer electrons for homonuclear molecules are given in Fig. 4.8 *a*. The Fig. 4.8*b* gives the molecular orbitals formed from the atoms containing 8 to 10 electrons. The homonuclear molecules with total number of 36 electrons involves the participation of 3 *s* and 3*p* atomic orbitals and their energies are placed above the 2*p* orbitals. The notation and the relative energy of each molecular orbital is $\sigma_{1s} < \sigma^*_{1s} < \sigma_{2s} < \sigma^*_{2s} < \sigma_{2p_x} < \pi_{2p_y} = \pi_{2p_z} < \pi^*_{2p_y} = \pi^*_{2p_z} < \sigma^*_{2p_x}$

Lot of uncertainty existed about the position of σ_{2p_x} bonding orbital. It is only through experimental observations, *i.e.*, molecular spectroscopy, magnetic properties etc. that the position of σ_{2p_x} m. o. is ascertained in the energy level diagram. If the gap between 2*s* and 2*p* a. o.'s is very large, σ_{2p_x} bonding is more stable than π_{2p}, as shown in Fig. 4.8*b*. This energy level scheme holds good for oxygen, fluorine and neon atoms in the second period. In case of other elements of second period, is less stable than π_{2p} and is placed above π_{2p}, as depicted in Fig. 4.8*a*. This instability of σ_{2p_x} in comparison to π_{2p} has nothing to do with the poor overlap of σ_{2p_x} a. o. to form σ m. o., because if this would have been so then $\sigma^*_{2p_x}$ should have been less unstable than π^*_{2p}, which is not. This can be explained as follows. For 2*s* and 2*p* a. o.'s where the gap between them is small, it is better if we consider the formation of m. o.'s from the linear combination of two 2*s* and σ_{2p_x} together. The most stable m. o. from LCAO will be:

$$\psi = \sqrt{2+2x^2} \ [2s\,(A) + x\ 2p_x\,(A) + 2s\,(B) + x\ 2p_x\,(B)] \qquad ...(4.22)$$

where the coefficient *x* is less than unity and represents the proportion of $2p_x$ a. o. included in the σ_{2s} m. o. In other words, *sp* hybridization concept is used in forming the σ_{2s} and σ^*_{2s} m .o. and leads to their stability. But on the other hand, it will lead to the destabilization of σ_{2p_x} m. o. This and $\sigma^*_{2p_x}$ m. o., suggesting that some 2*s* character will be present in the σ_{2p_x} and $\sigma^*_{2p_x}$ effect is shown in Fig. 4.9. From the above explanation it can be concluded that any finite contribution from *s-p* hybridization on the LCAO m. o. would lead to the destabilisation of $\sigma^*_{2p_x}$ and in turn it may happen that the energy of $\sigma^*_{2p_x}$ may be larger than those of π_{2p}, as π_{2p} are still formed by the linear combination of only either $2p_z$ or $2p_y$ present on each atom. Thus in most of the cases, if not for all atoms, the energies of π_{2p} m. o. are less than that of $\sigma^*_{2p_x}$. The final energy level scheme of M.O.'s formed from the atoms up to nitrogen is shown in Fig. 4.9*a* and for rest of the atoms Fig. 4.9*b* represents the energies of the M.O.'s. It has been

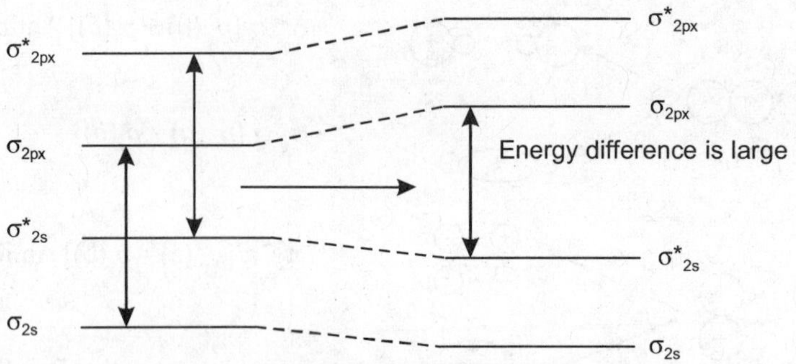

Fig. 4.8 *Energy level diagram depicting the effect of σ_{2s}– interaction on σ_{2s}, σ^*_{2s}, and M.O.*

established from the emission spectroscopy, *i.e.*, by observing transition from $2p \rightarrow 2s$, that $2s$ a. o. is more stable that $2p$ a. o. and this gap keeps on increasing with increase of atomic number along the second period. For example, the gap between 2 s and $2p$ orbitals is 1.85 eV (178.5 kJ mol^{-1}) in lithium and 20 eV (1929.7 kJ mol^{-1}) in fluorine.

In case of heterodiatomic molecules, the energies of the molecular orbitals are not symmetrical with respect to atomic orbitals. This is because the energies of atomic orbitals are different for different atoms. The atomic energy levels as well as the molecular orbitals are drawn in an approximate scale (Fig. 4.10). The designation of the molecular orbitals, formed from valence electrons only, as suggested by Mulliken is $\sigma_z < \sigma^*_y < \sigma_x < \pi_w = \pi_w < \pi^*_v = \pi^*_v\ \sigma^*_u$

The core electrons are not considered.

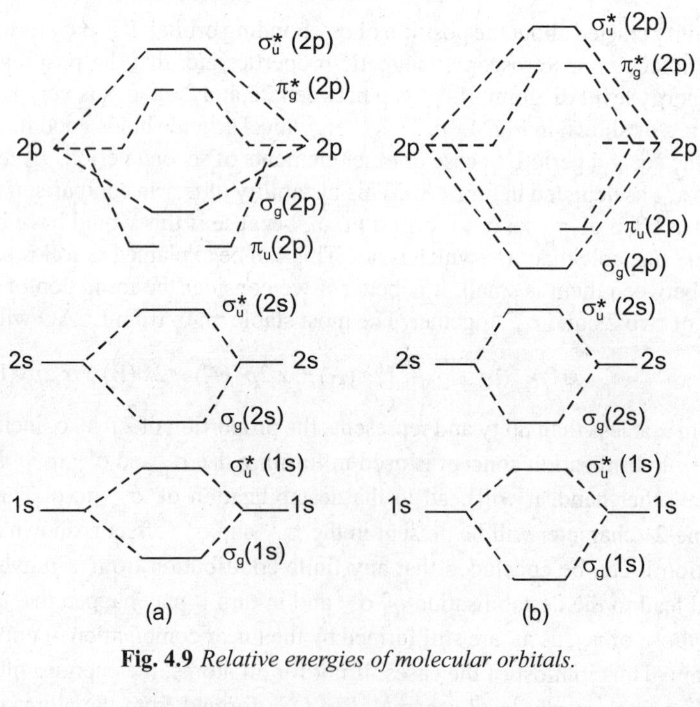

Fig. 4.9 *Relative energies of molecular orbitals.*

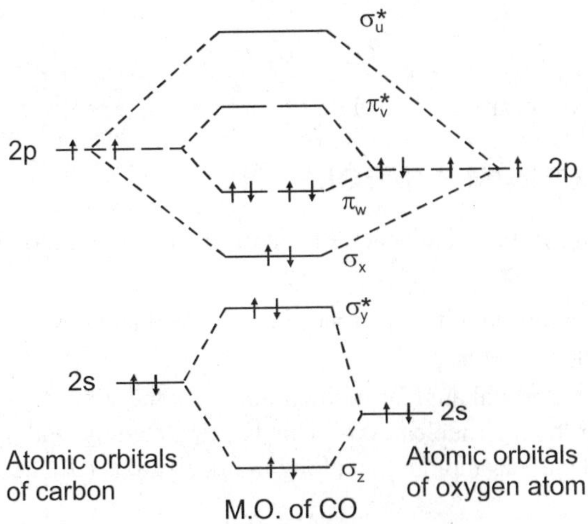

Fig. 4.10

4.7. MOLECULAR ELECTRONIC CONFIGURATIONS

The electrons in the molecular orbitals, shown in Fig. 4.9 *a* and 4.9*b* can be filled as follows:

(*i*) Each orbital can have maximum two electrons with the opposite spin.

(*ii*) In the degenerate orbitals, each molecular orbital can have one electron till all are occupied.

***Exercise* 4.17:** Write the electronic configuration of C_2.

The electronic configuration for carbon atom is $1s^2\ 2s^2\ 2p^2$, *i.e.* total number of electrons in C_2 are 12. So according to energy level scheme of molecular orbitals shown in Fig. 4.9 *a*, the electronic configuration is

$$\sigma_{1s}^2\ \sigma_{1s}^{*2}\ \sigma_{2s}^2\ \sigma_{2s}^{*2}\ \pi_{2py}^2\ \pi_{2pz}^2$$

***Exercise* 4.18:** Write the electronic configuration for N_2, N_2^- and N_2^+.

The number of electrons in N_2, N_2^- and N_2^+ are 14, 15 and 13 respectively. So according to the rules mentioned in Section (4.7), the electronic configuration for

(*i*) N_2 is $\sigma_{1s}^2\ \sigma_{1s}^{*2}\ \sigma_{2s}^2\ \sigma_{2s}^{*2}\ \pi_{2p}^4\ \sigma_{2p}^2$

(*ii*) N_2^- is $\sigma_{1s}^2\ \sigma_{1s}^{*2}\ \sigma_{2s}^2\ \sigma_{2s}^{*2}\ \pi_{2p}^4\ \sigma_{2p}^2 \pi_{2p}^{*1}$

(*iii*) N_2^+ is $\sigma_{1s}^2\ \sigma_{1s}^{*2}\ \sigma_{2s}^2\ \sigma_{2s}^{*2}\ \pi_{2p}^4\ \sigma_{2p}^1$

***Exercise* 4.19(*a*):** (*a*) Write the electronic configuration of CO molecule. (*b*) Draw the atomic as well as m.o. formed in CO.

(*a*) The electronic configuration of carbon and oxygen atoms are K $2s^2\ 2p^2$ and K $2s^2\ 2p^4$ respectively, *i.e.* there are 6 valence electrons in oxygen atom and 4 valence electrons in carbon atom. Total 4 + 6 = 10 electrons. So the electronic configuration of CO molecule is

$$KK\ \sigma_z^2\ \sigma_y^{*2}\ \sigma_x^2\ \pi_w^4$$

(*b*) The relative energies of molecular orbitals* formed in CO molecule are shown in Fig. 4.8.

4.8. MOLECULAR TERM SYMBOLS

Similar to the atomic term symbols, the general form of the molecular term symbols is

$$^{2S+1}\Lambda_{\Omega} \quad ...(4.23a)$$

where (*a*) $2S+1$ is the spin multiplicity. The total S can be evaluated by $s-s$ coupling technique in the same manner as in the case of atoms.

(*b*) Contrary to the total orbital angular momentum, Λ is the sum of the components of the orbital angular momentum along the internuclear axis. This is calculated by finding the algebraic sum of the components of the orbital angular momentum along molecular axis (λ), *i.e.*

$$\Lambda = \sum_i \lambda_i \quad ...(4.23b)$$

because it can only have positive or negative values. Except for $\Lambda = 0$, all the other states are doubly degenerate. Similar to the labelling of atomic orbitals, the molecular orbitals are also labelled as

$$\sigma \text{ for } \lambda = 0;\ \ \pi \text{ for } \lambda = 1,\ \delta \text{ for } \lambda = 2 \text{ and } \phi \text{ for } \lambda = 3$$

and spectroscopic states as

$$\Sigma \text{ for } \Lambda = 0;\ \ \prod \text{ for } \Lambda = 1;\ \Delta \text{ for } \Lambda = 2;\ \phi \text{ for } \Lambda = 3.$$

(*c*) Ω is the total angular momentum, *i.e.*

$$\Omega = |\Lambda + S|, |\Lambda + S - 1| ... |\Lambda - S| \quad ...(4.23c)$$

Lastly, the symmetry for the homonuclear diatomic molecules can be obtained by multiplying the symmetries of the orbitals being used, *i.e.*

$$(g)\ (g) = (u)\ (u) = (g) \text{ and } (g)\ (u) = (u)\ (g) = (u) \quad ...(4.24)$$

The term symbol for a heteronuclear diatomic molecule does not contain the subscript *g* or *u*, where *g* (gerade) symmetric and *u* (ungerade) is anti-symmetric with respect to centre of inversion. Table 4.7 gives the molecular term symbols for various electronic configurations.

TABLE 4.7

Configuration	Term symbols
σ	$^2\Sigma^+$
σ^2	$^1\Sigma^+$
π or π^3	$^2\Pi$
π^2	$^1\Delta,\ ^1\Sigma^+\ ^3\Sigma^-$
$\pi^2\sigma$	$^2\Sigma^+,\ ^2\Sigma^-,\ ^2\Delta,\ ^4\Sigma$
$\pi^2\pi$	$^2\Pi\ (3),\ ^2\Phi,\ ^4\Pi$
$\pi^2\rho$	$^2\Sigma^+,\ ^2\Sigma^-,\ ^2\Delta(2),\ ^2T,\ ^4\Delta$
$\pi^3\sigma$	$^1\pi,\ ^3\pi$
$\pi^3\pi$ or $\pi\pi$	$^1\Sigma^+,\ ^1\Sigma^-,\ ^1\Delta,\ ^3\Sigma^+,\ ^3\Sigma^-,\ ^3\Delta$
$\pi^3\delta$ or $\pi\delta$	$^1\Pi, ^1\Phi, ^3\Pi, ^3\Phi$
π^4	$^1\Sigma^+$

*One should be careful in drawing the energies of atomic orbitals in heterodiatomic molecules since their energies are not equal.

δ or δ^3	$^2\Delta$
δ^2	$^1\Sigma^+$, $^3\Sigma^-$, ^1T
δ^4	$^1\Sigma^+$

Two points concerning molecular symmetry are worth consideration from the molecular term symbols.

(*i*) The first of these is symmetry with respect to centre of inversion of the orbitals through the origin. As discussed earlier such inversion converts Cartesian coordinates (x, y, z) to ($-x, -y -z$) and gives information whether the given orbital is symmetrical (gerade, g) or asymmetric (ungerade, u). Just to revive, σ and π^* are gerade and $\sigma^*$and π are ungerade (u).

(*a*) If the molecular electronic configuration contains odd number of electrons in MOs with u symmetry, the overall wave function will have u symmetry.

(*b*) If the number of electrons in ungerade orbitals is even, the total symmetry will be g. This is a consequence of the fact when we invert the wave function through the origin, each electron in an ungerade orbital produces a sign change.

An even number of such changes results in no sign change on the total wave function; an odd number will change the sign on the total wave function .

(*ii*) The second symmetry operation involves the reflection through the plane and is conveyed by the term symbol involving whether the total wave function changes sign (–) or does not change sign (+) on reflection in a plane that contains both nuclei. Since such a reflection causes the angel φ to be replaced with $-\varphi$. Replacing φ with $-\varphi$ means replacing m_l with $-m_l$. This means that the MOs formed from the even value of m_l of atomic orbitals will not change signs, whereas those MOs formed from the odd values of m_l of atomic orbitals will change sign. For examples, MOs formed from $m_l = 0$ will not change sign, whereas those formed from $m_l = 1$ will change sign. That is π_{+1} and π_{-1} orbitals, which comprise atomic orbital with $m_l = 1$ or $m_l = -1$ respectively, will undergo the transformation: $\pi_{+1} \leftrightarrow \pi_{-1}$.

***Exercise* 4.20:** If the electronic configuration is $(\pi_{+1})^1(\pi_{-1})^1$, what kind of term symbols can be obtained?

We know that the λ (m_l) values for these MOs can be +1 and –1 respectively. Further in the present electronic configuration, reflection through a plane containing$x$-axis is equivalent to exchanging electrons in these two π-orbitals. We know that the total wave function must be anti-symmetric with respect to electron exchange. Therefore if the antisymmetry is due to an antisymmetry in the spatial part, exchanging the electrons in the $(\pi_{+1})^1$ and $(\pi_{-1})^1$ orbitals will produce an overall sign change on the wave function. If however, the antisymmetry is produced by the spin part of the wave function, no change will occur on reflection. We know that singlet spin functions are antisymmetric with respect to electron exchange, thus space part will have to be symmetric. On the other hand-triplet spin functions are symmetric with respect to electron exchange, the space part have to be antisymmetric.

So we will two kinds of wave functions. Antisymmetric spin functions will have symmetric space part and symmetric spin functions will have antisymmetric space part .

***Exercise* 4.21:** Calculate the molecular states for electronic configuration $\pi^1 \pi^1$.

Since these two electrons belong to two different molecular orbitals the Pauli's exclusion principle is satisfied. Here $\lambda_1 = 1$ or -1, $\lambda_2 = 1$ or -1. The values of Λ can be obtained by the following arrangements of the component of orbital angular momentum, *i.e.*

$$\begin{matrix}\rightarrow\\\rightarrow\end{matrix} \text{ or } \begin{matrix}\leftarrow\\\leftarrow\end{matrix} \text{ and } \begin{matrix}\rightarrow\\\leftarrow\end{matrix} \text{ or } \begin{matrix}\leftarrow\\\rightarrow\end{matrix}$$

$$\Lambda = 2, -2 \qquad \Lambda = 0$$

Similarly the $S = \frac{1}{2} + \frac{1}{2} = 1$ and $S = \frac{1}{2} - \frac{1}{2} = 0$.

Further for $\Lambda = 0$, there are two ways of arranging whereas this state is non-degenerate. So one of them will be Σ^+ and the other Σ^- (meaning explained already); so the possible terms can be

$$^3\Delta,\ ^3\Sigma^+,\ ^3\Sigma^-,\ ^1\Delta,\ ^1\Sigma^+,\ ^1\Sigma^-$$

***Exercise* 4.22:** Calculate the molecular term symbols for the electronic configuration π^2.

Here the electrons are equivalent so some of the molecular states obtained in Exercise 4.21 will not occur, due to Pauli's exclusion principle.

Δ state has been obtained when both the λ_i have the same value so electrons must have opposite spins, *i.e.*

$$S = 0 \text{ and } 2S + 1 = 1,\ ^1\Delta$$

According to Pauli's exclusion principle, total wave function ($\psi_{space} \times \psi_{spin}$) must be antisymmetric. Now in Σ^+, ψ_{space} is symmetric, so ψ_{spin} should be antisymmetric, *i.e.* opposite spin ($S = 0$, $2S + 1 = 1$), $^1\Sigma^+$.

Similarly for Σ^- state because ψ space is antisymmetric so ψ_{spin} should be symmetric $2S + 1 = 2 \times 1 + 1 = 3$, *i.e.* $^3\Sigma^-$.

The possible term symbols for π^2 are $^1\Delta$, $^1\Sigma^+$ and $^3\Sigma^-$.

***Exercise* 4.23:** Write the electronic configuration for the ground state and first excited state of N_2^-. Calculate the molecular term symbols.

Electronic configuration for the ground state of N_2^- is

$$\sigma_{1s}^2\ \sigma_{1s}^{*2}\ \sigma_{2s}^2\ \sigma_{2s}^{*2}\ \pi_{2p}^4\ \sigma_{2p}^2\ \pi_{2p}^{*1}$$

and for first excited state is

$$\sigma_{1s}^2\ \sigma_{1s}^{*2}\ \sigma_{2s}^2\ \sigma_{2s}^{*2}\ \pi_{2p}^4\ \sigma_{2p}^1\ \pi_{2p}^{*2}$$

For ground state, the only unfilled molecular orbital is π_{2p}^{*1}, $\Lambda = \lambda = 1$ and $S = \frac{1}{2}$. So the molecular state is $^2\Pi$. For excited state, there are two unfilled orbitals, *i.e.* σ_{2p}^1 and σ_{2p}^{*2}.

For π_{2p}^2, the possible molecular states are (see Exercise 4.22):

$$^1\Delta,\ ^1\Sigma^+,\ ^3\Sigma^-$$

and for

$$\sigma_{2p}^1,\ ^2\Sigma$$

Now combining $^2\Sigma$ with each state of π^2 will lead to following molecular states:

(*a*) $^1\Delta$ and $^2\Sigma$ will give $\Lambda = 2 + 0 = 2$

i.e. $^2\Delta$, $\quad S = 0 + \dfrac{1}{2}$

(*b*) $^1\Sigma^+$ and $^2\Sigma \rightarrow {}^2\Sigma^+$

(*c*) $^2\Sigma$ and $^3\Sigma^-$ will have $\quad \Lambda = 0 + 0 = 0$

$$S = 1 + \frac{1}{2} = 3/2 \text{ or } 1/2$$

i.e. $2S + 1 = 2 \times \dfrac{3}{2} + 1 = 4$ or $2 \times \dfrac{1}{2} + 1 = 2$

States are $^4\Sigma^-$ and $^2\Sigma^-$

Total possible states are

$$^4\Sigma^-, {}^2\Sigma^-, {}^2\Sigma^+ \text{ and } {}^2\Delta$$

4.9. BOND ORDER AND BOND ENERGY

The bond order (B.O.) in any molecule can be calculated by the relation

$$\text{B.O.} = \frac{1}{2} \text{ (Number of bonding electrons – number of antibonding electrons)} \qquad ...(4.25)$$

The bond order is roughly a measure of bond energy. Higher the bond order, greater will be the bond energy. But if the bond order is same, this does not mean same bond energy for the molecules.

***Exercise* 4.24:** Calculate the bond order of (*a*) N_2^- molecule ion and (*b*) CO molecule.

(*a*) From the Exercise 4.18 and looking at electronic configuration and using equation (4.25)

$$\text{B.O.} = \frac{1}{2}\ (10 - 5) = 2.5$$

(*b*) From Exercise 4.19 and using equation (4.25)

$$\text{B.O.} = \frac{1}{2}\ (8 - 2) = 3$$

***Exercise* 4.25:** Which of the following species has higher bond order and higher bond energy N_2, N_2^+, N_2^-.

The electronic configuration for N_2 and N_2^+ are

$$\sigma_{1s}^2\ \sigma_{1s}^{*2}\ \sigma_{2s}^2\ \sigma_{2s}^{*2}\ \pi_{2p_{y=2pz}}^4\ \sigma_{2px}^2 \text{ and}$$

$$\sigma_{1s}^2\ \sigma_{1s}^{*2}\ \sigma_{2s}^2\ \sigma_{2s}^{*2}\ \pi_{2p_{y=2p_z}}^4\ \sigma_{2px}^1$$

The bond orders for N_2, N_2^+ and N_2^- are

$$\text{B.O. }(N_2) = \frac{1}{2}(10 - 4) = 3,$$

$$\text{B.O. }(N_2^+) = \frac{1}{2}\ (9 - 4) = 5/2$$

$$\text{B.O. }(N_2^-) = \frac{1}{2}(10 - 5) = 5/2$$

Thus the bond energy of N_2 will be greater than N_2^+ and N_3^-.

4.10. ELECTRONEGATIVITY

Electronegativity is defined as the tendency of an atom in the molecule to pull the electrons towards itself. It can also be defined as the unequal attraction of electrons by the atoms in a heterodiatomic molecule. Two methods will be discussed to calculate it.

4.10.1. *Mulliken's Method*

The expression given by Mulliken to calculate the electronegativity (χ_i) of any element is

$$\chi_i = \frac{IP_i + EA_i}{5.6} \qquad ...(4.26)$$

where IP_i is the ionisation potential and EA_i is the electron affinity of atom *i.* Both the quantities are expressed in electron volts.

4.10.2. *Pauling's Method*

Pauling's method gives the relative values of the electronegativities and the expression is

$$\chi_y - \chi_x = 0.102\,\Delta^{1/2} \qquad \text{...(4.27)}$$

where the constant 0.102 converts $\Delta^{1/2}$ from $(\text{kJ mol}^{-1})^{1/2}$ to eV and

$$\Delta = BE_{xy} - (BE_{x_2} \times BE_{y_2})^{1/2} \qquad \text{...(4.28)}$$

where BE_x is the bond dissociation energy of a particular bond *i.*

***Exercise* 4.26:** Calculate the electronegativity of iodine atom if ionisation potential and electron affinity for it is 10.5 eV and 3.43 eV respectively.

$$IP = 10.5 \text{ eV} \quad EA = 3.43 \text{ eV}$$

$$\chi_I = \frac{10.5 + 3.43}{5.6} = \frac{12.93}{5.6} = 2.49$$

***Exercise* 4.27:** Calculate the electronegativity of iodine atom from the following data:

$$BE_{H_2} = 435.14 \text{ kJ mol}^{-1}$$
$$BE_{I_2} = 151.14 \text{ kJ mol}^{-1}$$
$$BE_{HI} = 297.1 \text{ kJ mol}^{-1}$$

and

$$\chi_H = 2.1$$

Using equation (4.28)

$$\Delta = BE_{HI} - (BE_{H_2} \times BE_{I_2})^{1/2}$$
$$= 297.1 - (435.14 \times 151.14)^{1/2}$$
$$= 297.1 - 256.4 = 40.7$$
$$\chi_I = 0.102 \times (40.7)^{1/2} + \chi_H$$
$$= 0.65 + 2.1 = 2.75$$

***Exercise* 4.28:** Calculate the dissociation energy of Li_2 from the following data. Bond dissociation energies of H_2 and LiH are 435.56 kJ mol^{-1} and 244.53 kJ mol^{-1} respectively. $\chi_{Li} = 1.0$, $\chi_H = 2.1$.

Substituting the values of χ_{Li} and χ_H in equation (4.27) one get

$$2.1 - 1.0 = 0.102\,\Delta^{1/2}$$

or

$$\Delta = \left(\frac{1.1}{0.102}\right)^2 = 116.30 \text{ kJ}$$

Using equation (4.28)

$$BE_{Li_2} = \frac{(BE_{LiH} - \Delta)^2}{BE_{H_2}}$$

$$= \frac{(244.50 - 116.3)^2}{435.6}$$

$$= 37.73 \text{ kJ mol}^{-1}$$

In poly atomic molecules, each bond is assigned a bond moment and resultant or overall dipole moment is determined by the vector addition of the bond moments. In general charge separation is difficult to estimate. Because of this although dipole moment is a product of two quantities (charge and distance), still dipole moment is not used generally to measure bond length and bond angles. Still the measurements of dipole moments are very useful in determining the symmetry or structure of the molecules. This has been explained in Exercise 4.29.

1 1A																	18 8A
H 2.1	2 2A											13 3A	14 4A	15 5A	16 6A	17 7A	He
Li 1.0	Be 1.5											B 2.0	C 2.5	N 3.0	O 3.5	F 4.0	Ne
Na 0.9	Mg 1.2	3 3B	4 4B	5 5B	6 6B	7 7B	8	9 8B	10	11 1B	12 2B	Al 1.5	Si 1.8	P 2.1	S 2.5	Cl 3.0	Ar
K 0.8	Ca 1.0	Sc 1.3	Ti 1.5	V 1.6	Cr 1.6	Mn 1.5	Fe 1.8	Co 1.9	Ni 1.9	Cu 1.9	Zn 1.6	Ga 1.6	Ge 1.8	As 2.0	Se 2.4	Br 2.8	Kr 3.0
Rb 0.8	Sr 1.0	Y 1.2	Zr 1.4	Nb 1.6	Mo 1.8	Tc 1.9	Ru 2.2	Rh 2.2	Pd 2.2	Ag 1.9	Cd 1.7	In 1.7	Sn 1.8	Sb 1.9	Te 2.1	I 2.5	Xe 2.6
Cs 0.7	Ba 0.9	La- Lu	Hf 1.3	Ta 1.5	W 1.7	Re 1.9	Os 2.2	Ir 2.2	Pt 2.2	Au 2.4	Hg 1.9	Tl 1.8	Pb 1.9	Bi 1.9	Po 2.0	At 2.2	Rn
Fr 0.7	Ra 0.9	Ac 1.1	Rf	Db	Sg	Bh	Hs	Mt	Ds								

Fig. 4.11 *Electronegativity scale of the elements. The electronegativity of the lanthanides (La-Lu) varies between 1.0 to 1.2, whereas those actinides (Th-Lr) vary between 1.1 to 1.5. This scale was developed by Pauling by arbitrarily defining* $\chi_F = 4.0$.

4.11. DIPOLE MOMENT

Unequal sharing of electrons by different atoms, creates a small positive charge on the less electronegative atom and small negative charge on the more electronegative atom. Results of this is a formation of dipole and dipole moment μ is defined as

$$\mu = qr \qquad ...(4.29)$$

where q is small positive or negative charge on the atom and r is the distance between two poles. The dipole moment is often expressed in unit of Debyes where 1D = 3.335641 × 10^{-30} C m. Since dipole moment is vector, the direction of the vector is from positive end to the negative end.

***Exercise* 4.29:** The electronegativity of carbon and oxygen atoms are 2.55 and 3.44 respectively. But the dipole moment of CO is only 0.122 D. Explain briefly.

The electronegativity of oxygen is 3.44 and for carbon it is 2.55 with a difference of 0.89. That is not a trivial difference. It translates to a bond that has 32% ionic character. The low dipole moment is more a result of the bonding electron distribution rather than electronegativity difference. That is based on Lewis dot structure carbon and oxygen can form two bonds only. This will complete the octet of oxygen atom but not of carbon atom. Thus oxygen atom donates one of its lone pair to carbon atom to complete its octet. Thus CO has a triple bond (Exercise 4.24 *b*). Since so many bonding electrons are found in the pi (π) and sigma (σ) bonds, there are fewer lone pairs on oxygen to give it a more negative partial charge.

Since there are no unpaired electrons on any of the atoms in CO or in molecular orbitas, so CO is diamagnetic.

4.12. IONIC CHARACTER

The percentage ionic character (IC) of a chemical bond can be calculated from the relation

$$\% \text{ IC} = \frac{\mu_{\text{actual}}}{\mu_{\text{cal}}} \times 100 \qquad \text{...(4.30)}$$

where μ_{actual} is the dipole moment measured experimentally and μ_{cal} is the dipole moment obtained using equation (4.29), with the assumption that $q = e$ and $r = r_{ab}$, *i.e.* bond length. The other relation which can also be used to calculate the % ionic character is

$$\% \text{ IC} = 16\,(\chi_a - \chi_b) + 3.5\,(\chi_a - \chi_b)^2 \qquad \text{...(4.31)}$$

No compound is 100% ionic. The most ionic compound is CsF, *i.e.* approximately 92%.

***Exercise* 4.30:** Calculate the dipole moment of HCl molecule if the equilibrium bond length is 1.2746 Å.

Using the equation (4.27), the dipole moment μ is

$$\begin{aligned}\mu &= (1.6022 \times 10^{-19}\text{ C})\,(1.2476 \times 10^{-10}\text{ m})\\ &= 1.999 \times 10^{-29}\text{ C m}\\ &= (1.999 \times 10^{-29}\text{ C m})\,(3.3356 \times 10^{-30}\text{ C m D}^{-1})^{-1}\\ &= 5.99\text{ D}\end{aligned}$$

***Exercise* 4.31:** Calculate the percentage ionic character in HCl molecule if the observed dipole moment is 1.08 D.

Using the calculated μ from Exercise (4.28) and the equation (4.30)

$$\begin{aligned}\% \text{ ionic character} &= \frac{(1.08\text{ D})\,(3.34\times10^{-30}\text{ C m D}^{-1})}{1.999\times10^{-29}\text{ C m}} \times 100\\ &= 18.08\end{aligned}$$

***Exercise* 4.32:** The electronegativities of hydrogen and chlorine atoms are 2.1 and 3.0. Using equation (4.29), calculate the percentage ionic character.

$$\begin{aligned}\% \text{ IC} &= 16(3.0 - 2.1) + 3.5\,(3.0 - 2.1)^2\\ &= 16 \times 0.9 + 3.5 \times 0.81 = 17.24\end{aligned}$$

The agreement with the value obtained in Exercise 4.24 is quite good.

***Exercise* 4.33:** Explain the following results of the dipole moments of polyatomic molecules.

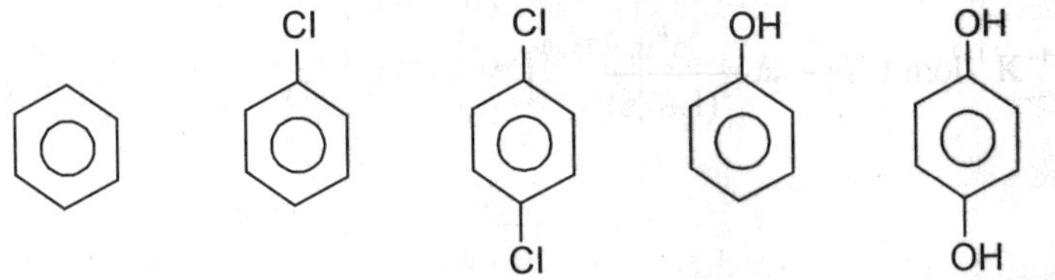

$\mu = 0$ D $\mu = 1.7$ D $\mu = 0$ D $\mu = 1.40$ D $\mu = 1.64$ D

(*i*) (*ii*) (*iii*) (*iv*) (*v*)

The molecule (*i*) is benzene and is highly symmetrical. All the bonds are symmetrical so the resultant dipole moment of all the 6 C – H bond moments is zero.

Molecules (*ii*) and (*iv*) are monosubstituted benzene and thus symmetry of benzene molecule is destroyed, so the resultant is not zero.

Molecule (*iii*) is *p*-dichlorobenzene. The direction of the bond moment vectors are shown in Fig. 4.12. So the resultant of all bond moments is zero. This confirms that the molecule is planar.

Although molecule (*v*), *i.e. p*-dihydroxyl benzene is similar to molecule (*iii*) but still has $\mu = 1.64$ D. This shows that the two hydroxy groups are not in the same plane as benzene molecule. So provide resultant dipole moment.

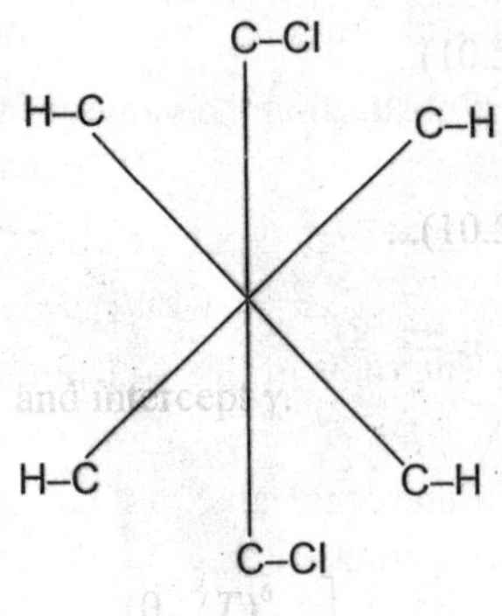

Fig. 4.12

***Exercise* 4.34:** The dipole moments of chloro- and nitrobenzene are 1.55 D and 3.80 D. Calculate the dipole moment of *m*-dichlorobenzene and *o*-dinitrobenzene.

The structure for *m*-dichlorobenzene is shown in Fig. 4.13(*a*). If no other effects are present and the given dipole moments are considered as bond moments the resultant dipole moment of *m*-dichlorobenzene will be vector addition of vectors of magnitude 1.55 D at an angle of 120°. The resultant of two vectors or dipole moment of *m*-dichlorobenzene will be

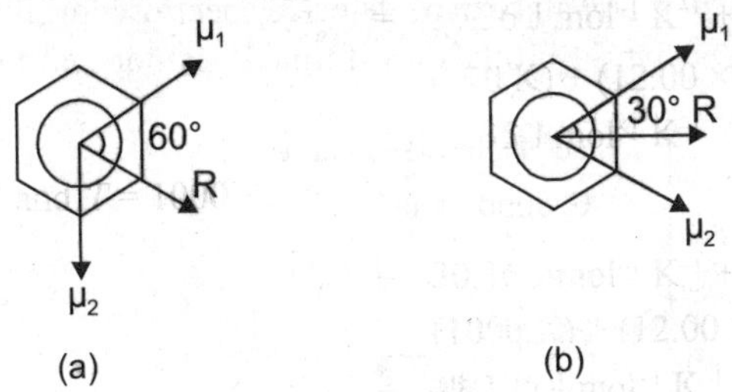

Fig. 4.13

$$\mu = \mu_{Cl} \cos 60 + \mu_{Cl} \cos 60 = 1.55 \times \frac{1}{2} + 1.55 \times \frac{1}{2} = 1.55 \text{ D}$$

Similarly for *o*-dinitrobenzene, the vector diagram is shown in Fig. 4.13(*b*)

$$\begin{aligned} \mu &= \mu_1 \cos 30° + \mu_1 \cos 30° \\ &= (3.80)\,(0.866) + (3.80)\,(0.866) \\ &= 6.58 \text{ D} \end{aligned}$$

***Exercise* 4.35:** If the dipole moment of LiF is 6.32 D and the $\chi_F = 4.0$, calculate χ_{Li} atom; R_e(LiF) = 0.156 nm.

The theoretical dipole moment of LiF can be calculated as

$$\mu_{cal} = (1.6 \times 10^{-19}\,\text{C})\,(0.156 \times 10^{-9}\,\text{m})$$
$$= 0.2496 \times 10^{-28}\,\text{C m} = 2.496 \times 10^{-29}\,\text{C m}$$

$$\%\ \text{Ionic character} = \frac{\mu_{exp}}{\mu_{cal}} \times 100$$

$$= \frac{(6.32\,\text{D})\,(3.34 \times 10^{-30}\,\text{C m D}^{-1})}{2.496 \times 10^{-29}\,\text{C m}} \times 100$$

$$= 84.33$$

Substituting the values of the known terms in equation (4.31)

$$84.33 = 16\,(4 - \chi_{Li}) + 3.5\,(4 - \chi_{Li})^2$$

Rearranging one gets

$$3.5^2\chi_{Li} - 44\chi_{Li} + 35.67 = 0$$

The roots of this equation are

$$\chi_{Li} = 0.9 \text{ and } \chi_{Li} = 13.5$$

Since, the value of χ cannot be greater than 4, so $\chi_{Li} = 0.9$, which is quite close to 1, calculated by other methods.

4.13. POLYATOMIC MOLECULES

4.13.1. *Hybridisation*

The study of bonding in polyatomic molecules requires that we able to account for their geometry as well. One such approach to this problem is *hybridisation* of atomic orbitals, which is based on valence bond theory. The most common atom used to explain hybridisation is carbon and that too involving sp^3 hybridisation.The energy level diagram depicting the formation of hybrid orbitals and bonding is depicted in Fig. 4.14. State labelled $2s^1p^3$ is real and can be detected spectroscopically. The valence state (*i.e.*, the state in which four equivalent hybrid orbitals are formed), is not real in the sense that it does not

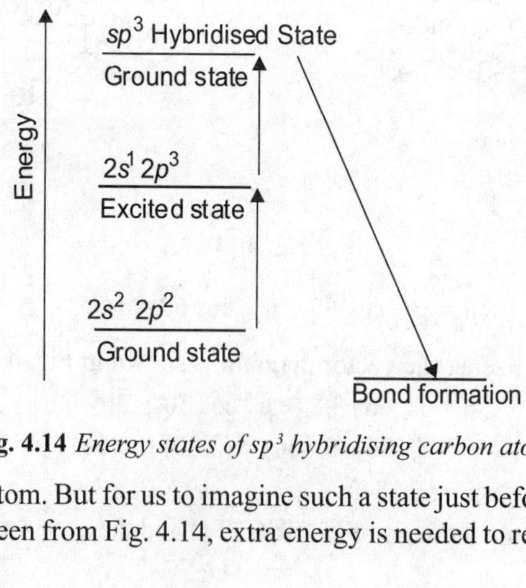

Fig. 4.14 *Energy states of sp^3 hybridising carbon atom.*

exist for an isolated carbon atom. But for us to imagine such a state just before the formation of methane molecule is convenient. As seen from Fig. 4.14, extra energy is needed to reach this state or to hybridise the atomic orbitals, but the investment is more than compensated for by the release of energy that results from four bond formation. The other kinds of hybridisation used in case of carbon atom can be *sp* and sp^2.

As mentioned above, hybridisation involves the linear combination of different atomic orbitals present on the same atom and as well as possess the directional properties. The shapes of different hybrid orbitals obtained by the linear combination of different atomic orbitals are depicted in Fig. 4.15 and discussed in Exercise 4.36. Mathematical expression or wave function for the hybrid orbitals obtained from the linear combination of atomic orbitals are given in Section 4.13.1. *The important point to be kept in mind is that the number of hybrid orbitals is exactly the same as the total number of atomic orbitals taking part in the formation of hybrid orbitals*. Further it must be kept in mind that idea of hybridisation is involved to explain the proper geometry of the molecule, i.e. we can predict hybridisation if we know the characteristics of the molecules. This can be best understood from the solved exercises.

***Exercise* 4.36:** Construct the hybrid orbitals obtained by the linear combination of (*i*) *sp* and (*ii*) sp^2.

The *s* and *p* (let us say p_x) orbitals is depicted in Fig. 4.15 *a*. Since there are two atomic orbitals, there will be two hybrid orbitals. These can be obtained by taking the linear combination of *s* and *p* orbitals, *i.e*., one hybrid orbital is obtained by taking plus combination and the other by taking the negative combination as shown in Fig. 4.15 *b*. In case of positive combination, positive lobe of the p_x orbital and positive part of the *s*-orbital towards the right of centre will increase the amplitude or size and in case of negative combination the negative lobe of the p_x orbital and positive part of the *s*-orbital

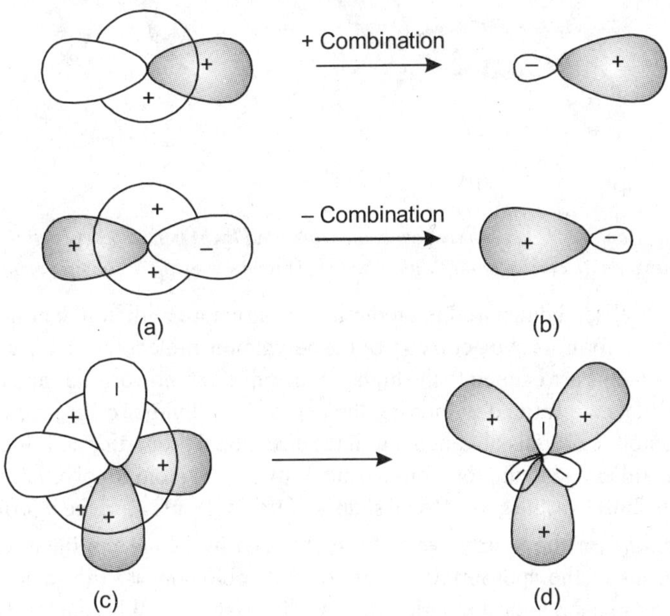

Fig. 4.15 *Formation of sp hybrid orbitals from the linear combination of s and p atomic orbitals.*

will decrease the size of the newly formed orbital. Thus one side of the hybrid orbital will have large size and the other side will have smaller size. In other words, the bonding will occur from one side and thus these hybrid orbitals are known to possess the directional properties. Since these orbitals possess electrons, the negatively charged particles, the best way of arranging them in space is to keep them apart, *i.e.* in the opposite side.

(*ii*) In case of sp^2 hybridisation, there are three atomic orbitals, one s and two p-orbitals. The two p orbitals can be any of three p-orbitals. For simplicity we take p_x and p_y orbitals. The manner in which three orbitals can combine is shown in Fig. 4.15 *c*. Similar to sp hybridisation, the only way to arrange three sp^2 hybrid orbitals in the space, with minimum repulsion between the negatively charged electrons or to have the maximum stability, can be to place them in the plane with each hybrid orbital at 120° from each other. This is shown in Fig. 4.15*d*. The plane will be XY-plane.

***Exercise* 4.37:** Find out the hybridisation involved in the structure of ethylene and ethyne molecules.

From the experimental results it is found that ethylene is a planar molecule and HCH angle is about 120°. Further each carbon atom is bonded to only three atoms (*i.e.*, two hydrogen atoms each and carbon atom). Thus both the geometry and bonding can be best understood if we assume that each carbon atom is sp^2 hybridised. As learnt in Exercise 4.36, the hybrid orbitals are arranged in XY-plane and each carbon atom is having one electron in p_z orbital which is perpendicular to the XY-plane. These three hybrid orbitals of each carbon atom are used to form three σ bonds, two with the $1s$-orbital of hydrogen atoms and third one form σ-bond with other carbon atom. This is shown in Fig. 4.16*a*. The $2p_z$ orbitals on the two carbon atoms can also overlap sideways to form a π-bond as shown in Fig. 4.16*b*.

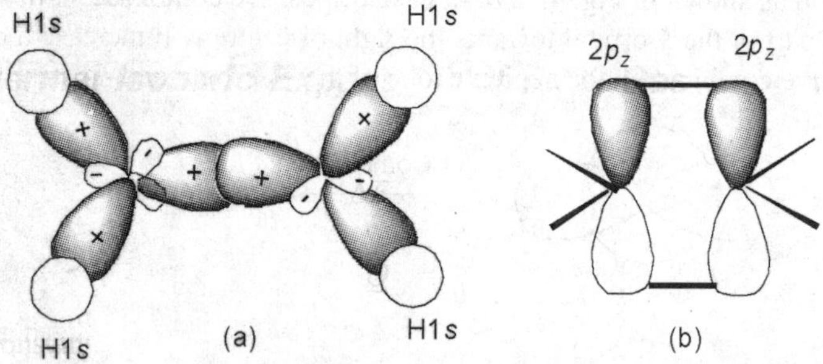

Fig. 4.16 *(a) The depiction of three σ-bonds between carbon and hydrogen and between the two carbon atoms in ethylene, (b) the π bond formed by the sideways overlap of the $2p_z$ orbitals.*

The concept of localised bonds used in predicting the structure of diatomic molecules often results in not getting the correct formula and geometry of the polyatomic molecules. In many cases, the electrons in the ground state of atoms are excited to the higher unoccupied atomic orbitals and the desired number of hybrid atomic orbitals are formed by mixing the appropriate number of ground and excited atomic orbitals. The hybrid atomic orbitals obtained thus have directional properties and shapes of the molecules are decided by the number and nature of atomic orbitals of central atom involved in forming the hybrid orbitals. Table 4.8 (p. 260) lists the geometrical shapes of the molecules and the hybrid orbitals involved.

The wave functions for the hybrid orbitals are obtained by taking the linear combination of the angular wave functions of the appropriate atomic orbitals belonging to the same atom. Most of the organic molecules involve the hybrid atomic orbitals which are formed by the linear combination of s and p orbitals. The number of hybrid atomic orbitals obtained are the same as the number of atomic orbitals. The wave function for the ith hybrid orbital formed from s and p atomic orbitals is

$$\phi_i = a_i\psi_s + b_i\psi_{p_x} - c_i\psi_{p_y} + d_i\psi_{p_z} \qquad ...(4.32)$$

The coefficients a_i, b_i, c_i and d_i can be calculated from the three relations mentioned below:

(1) Each wave function is normalised, *i.e.*

$$a_i^2 + b_i^2 + c_i^2 + d_i^2 = 1 \quad ...(4.33)$$

(2) Each hybrid wave function (orbital) in the set of hybrid orbitals must be orthogonal to other hybrids, *i.e.*

$$a_i a_j + b_i b_j + c_i c_j + d_i d_j = 0 \quad ...(4.34)$$

(3) The squares of the coefficients of any component atomic function summed overall the hybrids in which it participates must be equal to unity, *i.e.*

$$\Sigma a_i^2 = 1 \quad ...(4.35)$$

With the above rules, the following procedure can be used to generate the equivalent hybrid orbitals.

(4) Specify a cartesian coordinate system, and place one of the *i*th hybrid orbitals along one of the coordinate axis and as many of the other orbitals as possible in the plane containing this axis and other coordinate axis.

(5) Since the *s*-orbital has a spherical symmetry, each equivalent hybrid orbital of a set contains $1/n$ of the total of the unit *s* orbital distributed in *n* hybrid orbitals. Therefore coefficient in each hybrid orbital is $1/\sqrt{n}$.

***Exercise* 4.38:** Find the coefficients of the sp^3 hybrid orbitals.

There are four hybrid orbitals, *i.e.*

$$\phi_1 = a_1 s + b_1 p_x + c_1 p_y + d_1 p_z$$
$$\phi_2 = a_2 s + b_2 p_x + c_2 p_y + d_2 p_z$$
$$\phi_3 = a_3 s + b_3 p_x + c_3 p_y + d_3 p_z$$
$$\phi_4 = a_4 s + b_4 p_x + c_4 p_y + d_4 p_z$$

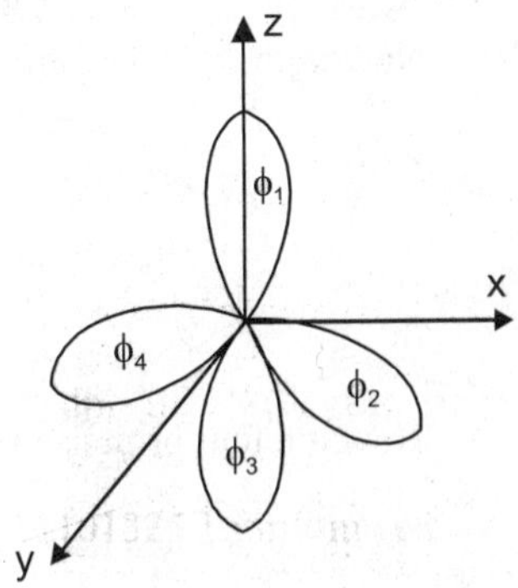

Fig. 4.17

Since there are four orbitals, so according to step 5, the values of $a_i = \sqrt{1/4} = 1/2$. Orient ϕ_1 along positive directions of *z*-axis as shown in Fig. 4.17. Then ϕ_2 has its plane of symmetry in the *xz*-plane and has its large positive lobe below *xy* plane as required by the angle 109°28′, the bond angle of tetrahedral molecule.

Now ϕ_1 will be only generated by *s* and p_z orbitals because p_x and p_y orbitals are perpendicular to ϕ_1.

$$a_1^2 + d_1^2 = 1 \text{ or } d_1 = \sqrt{1 - \frac{1}{4}} = \frac{\sqrt{3}}{2}$$

so

$$\phi_1 = \frac{1}{2} s + \frac{\sqrt{3}}{2} p_z$$

The other three sp^3 hybrid orbitals are equivalent with respect to *z* axis so $d_2 = d_3 = d_4$, using the step 3 and $d_1 = \sqrt{3/2}$

$$\frac{3}{4} + 3d^2 = 1 \text{ or } d_2^2 = \frac{1}{12} \text{ or } d = \frac{1}{\sqrt{12}}$$

Further, all the three sp^3 hybrid orbitals lie on the negative side of *z* axis so $d_2 = d_3 = d_4 = -\frac{1}{\sqrt{12}}$.

Now consider ϕ_2 hybrid orbital. According to the figure ϕ_2 lies in *xz* plane. Since this is a nodal plane for p_y atomic orbital, the contribution of this orbital will be zero in ϕ_2 so using the step 1,

$$a_2^2 + b_2^2 + c_2^2 + d_2^2 = \frac{1}{4} + b_2^2 + 0 + \frac{1}{12} = 1$$

$$b_2 = \sqrt{\frac{2}{3}}$$

So $$\phi_2 = \frac{1}{2}s + \sqrt{\frac{2}{3}}\,p_x - \frac{1}{\sqrt{12}}\,p_z$$

According to the orientation as mentioned in Fig. 4.17, the coefficients on p_x in the hybrid orbitals (b_3 and b_4) must be equal but negative. Using the step 3,

$$b_1^2 + b_2^2 + b_3^2 + b_4^2 = \frac{2}{3} + 2b^2 = 1,\ b = \sqrt{1/6}$$

So $$b_3 = b_4 = -\sqrt{\frac{1}{6}}$$

Now both ϕ_3 and ϕ_4 are equivalent with respect to *y* axis but with opposite signs. Using step 3,

$$c_3^2 + c_4^2 = 2c^2 = 1 \text{ or } c = \frac{1}{\sqrt{2}}$$

So $$c_3 = \frac{1}{\sqrt{2}},\ c_4 = -\frac{1}{\sqrt{2}}$$

Now the four orbitals are

$$\phi_1 = \frac{1}{2}s + \frac{\sqrt{3}}{2}p_z$$

$$\phi_2 = \frac{1}{2}s + \sqrt{\frac{2}{3}}\,p_x - \frac{1}{\sqrt{12}}\,p_z$$

$$\phi_3 = \frac{1}{2}s - \sqrt{\frac{1}{6}}\,p_x + \sqrt{\frac{1}{2}}\,p_y - \sqrt{\frac{1}{12}}\,p_z$$

$$\phi_4 = \frac{1}{2}s - \sqrt{\frac{1}{6}}\,p_x - \sqrt{\frac{1}{2}}\,p_y - \sqrt{\frac{1}{12}}\,p_z$$

Note: In this geometrical method, the coefficients of the atomic orbitals depend upon the way of tacklling the problems. But all the procedures give the equivalent hybrid atomic orbitals. The relation between the two sets of hybrid orbitals can be derived using the transformation matrix.

Alternate Method: The more familiar form of sp^3 hybrid orbitals can be drawn in the cubic symmetry as shown in Fig. 4.18. According to this figure all the four sp^3 hybrid orbitals are symmetrically placed with respect to three axes. So all have the equal contribution of each atomic orbitals, *i.e.* $\sqrt{\frac{1}{4}} = \frac{1}{2}$. The signs can be decided as follows. According to the diagram, ϕ_1 and ϕ_2 lie on the positive side of z-axis and ϕ_3 and ϕ_4 on negative side of z-axis. In ϕ_1 both p_x and p_y are also positive. In ϕ_2 p_x and p_y are negative, ϕ_3, p_x is positive and p_y negative and ϕ_4, p_x is negative and p_y is positive, so

$$\phi_1 = \frac{1}{2}(s + p_x + p_y + p_z)$$

$$\phi_2 = \frac{1}{2}(s - p_x - p_y + p_z)$$

$$\phi_3 = \frac{1}{2}(s + p_x - p_y - p_z)$$

$$\phi_4 = \frac{1}{2}(s - p_x + p_y - p_z)$$

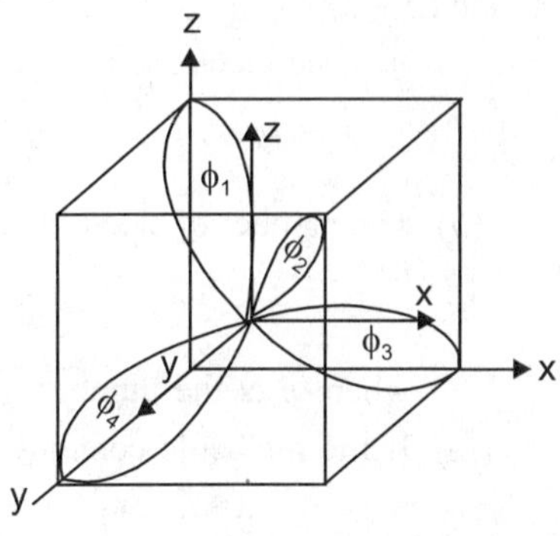

Fig. 4.18

Both the set of hybrid orbitals are same and equally acceptable.

4.14. BOND ANGLES IN HYBRID ORBITALS

The angles between the sp^x hybrid orbitals can be calculated by the relation

$$\cos\theta = \frac{S}{S-1} = \frac{P-1}{P} \quad \text{...(4.36)}$$

where θ is the angle between the two bonds, S and P are the share of the s and p orbitals in the hybrid orbitals. Knowing the angles between two hybrid orbitals one can calculate the share of s and p orbitals.

***Exercise* 4.39:** The bond angle between two O – F bonds in OF_2 molecule is 103.2°. Calculate the share of p orbitals in each hybrid orbitals.

The hybridisation on oxygen atom is sp^3, two hybrid atomic orbitals contains lone pairs and two hybrid atomic orbitals form bonds. Using equation (4.36):

$$\frac{S}{S-1} = \cos 103.2° = -0.228, \text{ so } S = 0.186$$

and

$$\frac{P-1}{P} = \cos 103.2° = -0.228, \text{ so } P = 0.814$$

i.e. total S and P character in two bonding hybrid orbitals is 2 × 0.186 = 0.372 and 2 × 0.814 = 1.628 respectively. Total P character in all the four hybrid orbitals is 3. Since two lone pairs are equivalent, so each hybrid orbital containing lone pair has $\frac{1}{2}(3 - 1.628) = 0.686$ and that of S character $\frac{1}{2}(1 - 0.372)$ = 0.314.

4.15. GEOMETRICAL SHAPES OF THE MOLECULES

Once the molecular formula of any compound is obtained, the next question comes into mind is how these atoms in the molecules are arranged in space and to write canonical forms if there are any. The most easy way is to first draw the Lewis dot structure for the molecule or molecular ion under consideration. But this is also not simple because confusion often arises about the correct number of multiple bonds to use, the correct number of valence electrons to use and the placing of formal charges. The following procedure* makes it very easy to write Lewis structures if the atoms belong up to second period ($n = 2$).

(*i*) Valence electrons (V) can be calculated by relation

$$V = (\sum_i \text{group number}) - \text{Charge} \qquad ...(4.37)$$

(*ii*) The number of electrons which must be shared through π bonding (P) in non-cyclic molecules is

$$P = 8n - 2(n-1) - V = 6n + 2 - V \qquad ...(4.38)$$

where n is the number of atoms in molecule.

(*iii*) If the molecule contains hydrogen atoms

$$P = 8(n-q) + 2q - 2(n-1) - V = 6n - 6q + 2 - V \qquad ...(4.39)$$

$$= 6X + 2 - V \qquad ...(4.40)$$

where q is the number of hydrogen atoms and X the number of atoms exclusive of hydrogen.

(*iv*) For cyclic molecules $2(n-1)\sigma$ electrons are to be replaced by $2n$.

(*v*) The formal charge (FC) on any atom is given as

$$\text{FC} = \text{Group number} - \text{Total number of electrons own by the atom concerned} \qquad ...(4.41)$$

(*vi*) If the molecule is of AX_n type and none of them hydrogen atom the number of σ bonds is given by the relation

$$\sigma \text{ bonds} = \frac{\text{Total number of valence electrons}}{8} \qquad ...(4.42)$$

If the division is not exact, the quotient is the number of σ bonds and the rest of the electrons is divided by two to give number of lone pairs. Thus total number of orbitals used around central atom = number of bonds + number of lone pairs.

The hybridization on the central atom will be the number of orbitals used. If the hydrogen atoms are involved the σ bonds are obtained dividing total number of electrons by *two*.

(*vii*) The shape of the molecule depends upon the number of bond pairs and number of lone pairs and this can be looked up in Table 4.8.

(*viii*) Further help about the geometry of the molecules can be obtained from the Valence Shell Electron Pair Repulsion (VSEPR) model. This states qualitatively that lone pair-lone pair interactions > lone pair-bond pair interactions > bond pair-bond pair interactions.

(*ix*) Arloth** has given quantitative features to the interactions mentioned in (*viii*) from empirical relations and has suggested the following procedure to find the shape of the molecule:

(*a*) When two or more alternative structures are present, consider all the interactions in each structure at 90° or less than 90°.

(*b*) Assign relative value of 3.5 repulsive units (RU) to all lone pair-lone pair interaction, 2.0 RU to lone pair-bond pair and 1.0 RU to bond pair-bond pair interactions.

---

*A.B.P. Lever, *J. Chem. Educ.*, **49**, 819 (1972).

R.J. Arloth *J. Chem. Educ.*, **54, 306 (1977).

(*c*) Count up the number of interactions of each type and multiply by their respective ratings.

(*d*) Add up the individual ratings for each form to arrive at a repulsive rating for each structure. The structure with lowest rating is preferred.

(*x*) The angles between the bonds depend upon the (*a*) electronegativities and (*b*) size of the central and other atoms present in the molecules.

***Exercise* 4.40:** (*a*) Draw the electron dot structure of (*i*) F_2O, (*ii*) N_2F_2, (*iii*) ClO_2^-, (*iv*) NO_3^-, (*b*) Compute the hybridisation on the central atom and (c) write the total wave function of each structure.

(*i*) Using equation (4.37) $V = 7 + 7 + 6 = 20$.

Using equation (4.38) $P = 6 \times 3 + 2 - 20 = 0$, *i.e.* there is no π bond. The dot structure can be

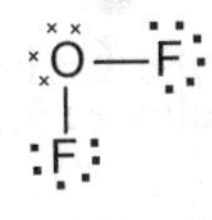

hybridisation on each central atom (*c*) write the total wave functions in terms of resonance wave functions and guess on the values of coefficients.

Using equation (4.41), $$\text{FC(O)} = 6 - (4 + \frac{1}{2} \times 4) = 0$$

$$\text{FC(F)} = 7 - (6 + \frac{1}{2} \times 2) = 0$$

i.e. there is no formal charge on any atom.

(*ii*) Using equation (4.41) $V = 5 + 5 + 7 + 7 = 24$. Using equation (4.38), $P = 4 \times 6 + 2 - 24 = 2$, *i.e.* there is one π bond. The dot structure can be

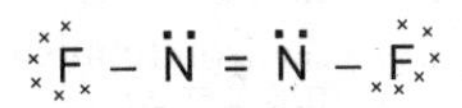

Using equation (4.41), the formal charges on each atom is zero.

(*iii*) Using equation (4.37), $V = 7 + 6 + 6 - (-1) = 20$.

Using equation (4.38), $P = 3 \times 6 + 2 - 20 = 0$. There should be zero π bond

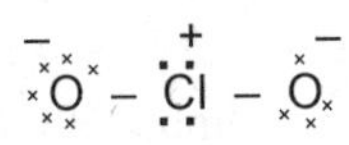

The formal charge on $$\text{Cl} = 7 - (4 + \frac{1}{2} \times 4) = 1$$

$$\text{FC}(\text{O}_1) = 6 - (6 + \frac{1}{2} \times 2) = -1$$

$$\text{FC}(\text{O}_2) = 6 - (6 + \frac{1}{2} \times 2) = -1$$

(*iv*) Using equation (4.37), $V = 5 + 6 \times 3 - (-1) = 24$

Using equation (4.38), $P = 4 \times 6 + 2 - 24 = 2$

So there is one π orbital between one of the three N – O bonds. The structures are

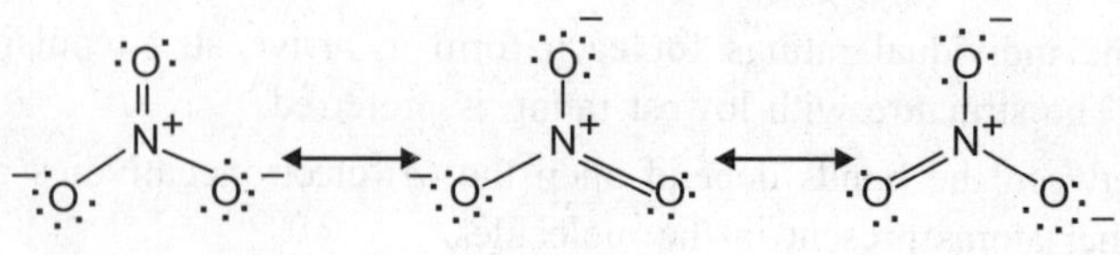

The formal charges on N and O are:

$$FC(N) = 5 - (\frac{1}{2} \times 8) = +1$$

$$FC(O_1) = 6 - (4 + \frac{1}{2} \times 4) = 0$$

$$FC(O_2) = 6 - (6 + \frac{1}{2} \times 2) = -1$$

$$FC(O_3) = 6 - (6 + \frac{1}{2} \times 2) = -1$$

(*b*) (*i*) In OF_2, the central atom is oxygen. It contains two lone pairs and two bond pairs. Thus the central atom oxygen is sp^3 hybridised.

(*ii*) There two central atoms in N_2F_2, two nitrogen atoms. Each nitrogen atom is connected to another nitrogen atom by a double bond, one hydrogen atom and one lone pair. Thus each nitrogen atom is sp^2 hybridised. N_2F_2 exists in two forms, *i.e.* cis form and trans form as shown below. There are two logical reasons why trans form is more stable than cis form: (1). The nonbonding lone pair orbitals in the cis form will be destabilised due to electron-electron repulsion, (2) The individual C – F dipoles are mutually repulsive (pointing in the same direction) in cis isomer.

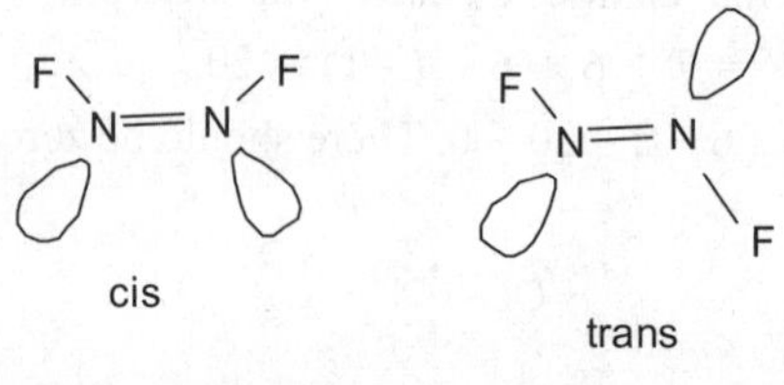

(*iii*) In ClO_2^-, the central atom chlorine has two lone pairs and two bond pairs. Thus chlorine atom is *sp*3 hybridised.

(*iv*) In NO_3^-, the central atom nitrogen atom is connected to three oxygen atoms and there is one π-bond between nitrogen atom and one of the oxygen atom. Thus nitrogen atom is sp^2 hybridised, further confirmed by the planar structure of NO_3^-.

(*c*) OF_2 and ClO_2^- have only one structure each. Thus the similar to equation (4.6), there is only one wave function. On the other hand, N_2F_2 have two structures, cis and trans, which have definite identity and can be prepared at low temperatures. So there is only one structure at low temperatures but at high temperature each form is interconvertible due to small activation barrier.

NO_3^- ion can be represented by three resonance structures and thus according to equation (4.43), the total wave function can be written as:

$$\Psi = C_a \psi_a + C_b \psi_b + C_c \psi_c \qquad ...(4.43)$$

where C_a, C_b and C_c are the coefficients of each structure and gives the weightage of each resonance structure. Since all the three structures are Equivalent, all the coefficients are equal.

***Exercise* 4.41:** What will be the shapes of (*i*) ICl_2^- and (*ii*) ClF_3 molecules?

(*i*) Since all the atoms belong to $n > 2$ periods, the octet rule is not strictly followed. Using equation (4.42)

$$\text{Number of orbitals} = \frac{7 \times 3 + 1}{8} = 2 + 6$$

i.e. there are two σ bonds and three lone pairs. So iodine is using dsp^3 hybridization. The shapes of the molecule can be as shown below:

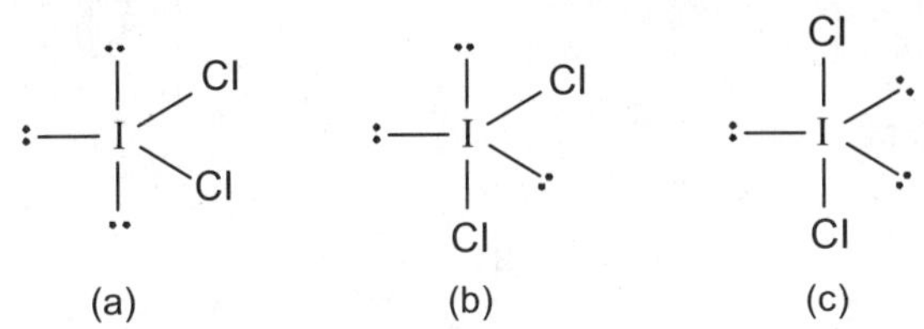

Now using step (*ix*), in structure (*a*) there are two lone pair-lone pair and four lone pair-bond pair interactions.

Total repulsive units = 3.5 × 2 + 4 × 2.0 = 15.0

In structure (*b*), two lone pair-lone pair, three lone pair-bond pair and one bond pair-bond pair interactions.

Total repulsive units = 2 × 3.5 + 3 × 2.0 + 1 × 1.0 = 14.0

In structure (*c*), there are six lone pair-bond pair interactions only, therefore,

Total repulsive units = 6 × 2.0 = 12

The structure (*c*) has the minimum repulsion which is most stable, *i.e.* linear structure.

(*ii*) In ClF_3, the number of σ bonds $= \frac{7 \times 4}{8} = 3 + 4$. Total number of orbitals involved are five, *i.e.* two lone pairs and three bonding. So the hybridisation on chlorine is dsp^3. On seeing the Table 4.6, the structure should be T shaped. This can also be tested with step (*ix*):

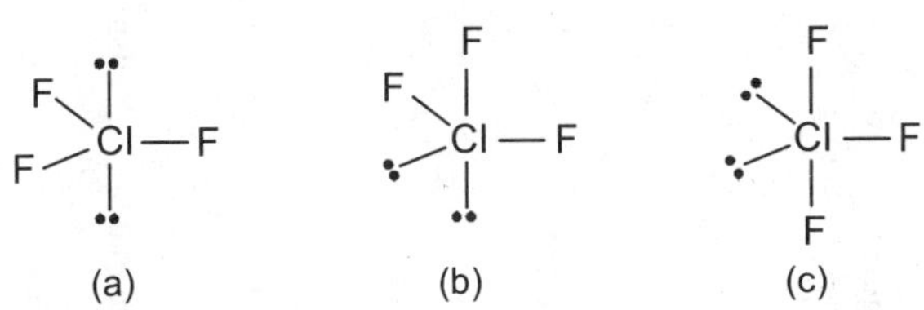

In structure (*a*) there are six lone pair-bond pair interactions, R.U. = 6 × 2 = 12 R.U. In structure (*b*) there are one lone pair-lone pair, three bond pair-lone pair and two bond pair-bond pair interactions, R.U. = 1 × 3.5 + 3 × 2.0 + 2 × 1.0 = 11.5 R.U. In structure (*c*), there are four lone pair-bond pair and two bond pair-bond pair interactions, R.U. = 4 × 2.0 + 2 × 1.0 = 10 R.U. The structure (*c*) has the minimum repulsive forces, so the most feasible.

TABLE 4.8

Structure SN_x	Geometrical shape	Hybridization	Molecular shape possibilities	Sketch of shape
1.	Spherical	–	–	
2.	Linear	sp	LP = 0, BP = 2, linear LP = 1, BP = 1, linear	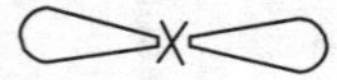
3.	Triangular	sp^2	LP = 0, BP = 3, triangular LP = 1, BP = 2, bent (∠ is 120°) LP = 2, BP = 1, linear	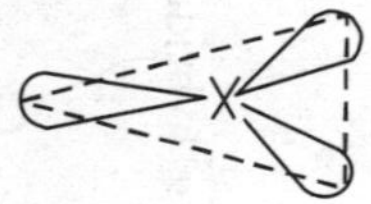
4.	Tetrahedron	sp^3	LP = 0, BP = 4, tetrahedral LP = 1, BP = 3, trigonal pyramidal LP = 2, BP = 2, bent (∠ is 109°) LP = 3, BP = 1, linear	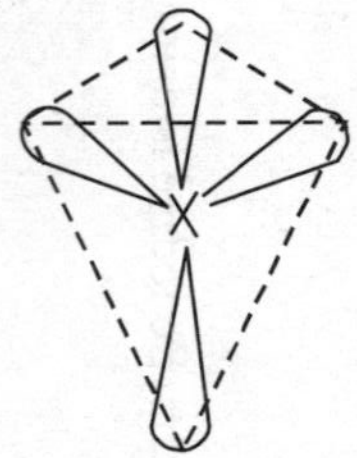
5.	Square planar	dsp^2 or sp^2d	LP = 0, BP = 4, square planar	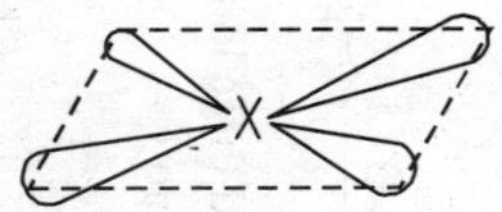
6.	Trigonal bipyramidal	dsp^3 or sp^3d	LP = 0, BP = 5, trigonal bipyramidal LP = 1, BP = 4, distorted square planar LP = 2, BP = 3, T-shaped LP = 3, BP = 2, linear LP = 4, BP = 1, linear	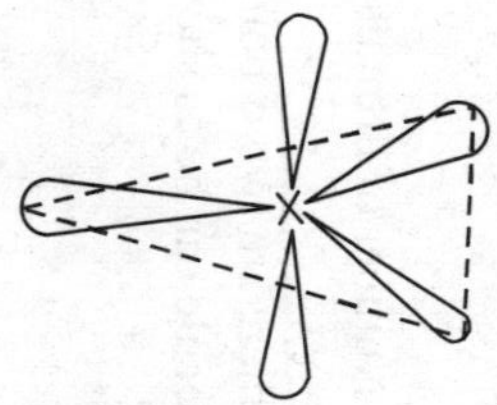

contd ...

7.	Octahedral	sp^3d^2 or d^2sp^3	LP = 0, BP = 6, octahedral LP = 1, BP = 5, square pyramidal LP = 2, BP = 4, square planar	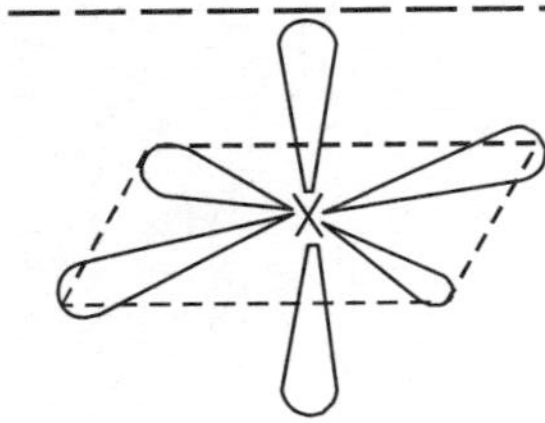
8.	Pentagonal pyramidal	–	–	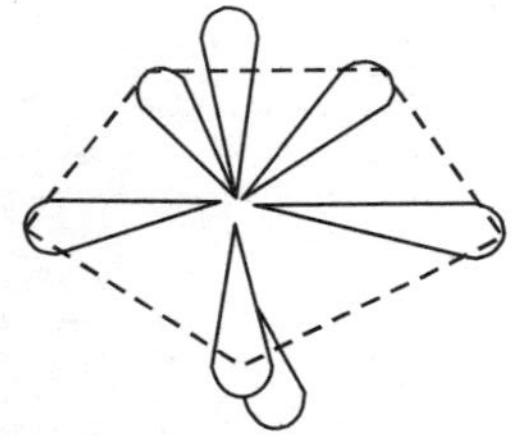
9.	Cubic	–	–	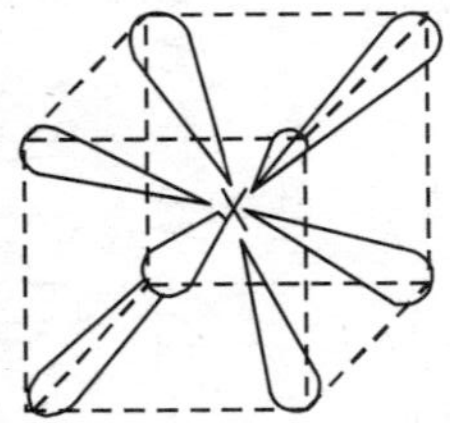
10.	Square antiprismal	–	–	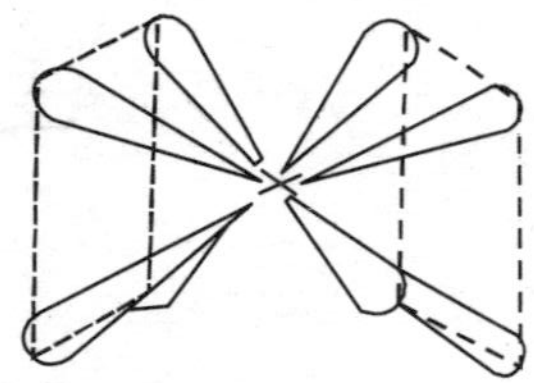

***Exercise* 4.42:** Consider the φ_2 and φ_3 hybrid orbitals involved on the carbon atom in methane obtained in Exercise 4.38. Show that they are orthogonal.

The wave functions for the φ_2 and φ_3 are:

$$\varphi_2 = \frac{1}{2}s + \sqrt{\frac{2}{3}}p_x - \frac{1}{\sqrt{12}}p_z \quad ...(1)$$

$$\varphi_3 = \frac{1}{2}s - \sqrt{\frac{1}{6}}p_x + \sqrt{\frac{1}{2}}p_y - \frac{1}{\sqrt{12}}p_z \quad ...(2)$$

$$\int \varphi_2\varphi_3 d\tau = \int \left[\frac{1}{2}s + \sqrt{\frac{2}{3}}p_x - \frac{1}{\sqrt{12}}p_z\right]\left[\frac{1}{2}s - \frac{1}{6}p_x + \frac{1}{\sqrt{2}}p_y - \frac{1}{\sqrt{12}}p_z\right]d\tau$$

$$\int \varphi_2\varphi_3 d\tau = \int \frac{1}{2}s\frac{1}{2}s d\tau - \int \frac{1}{2}s\frac{1}{\sqrt{6}}p_x d\tau + \int \frac{1}{2}s\frac{1}{\sqrt{2}}p_y d\tau - \int \frac{1}{2}s\frac{1}{\sqrt{12}}p_z d\tau$$
$$+ \int \sqrt{\frac{2}{3}}p_x\frac{1}{2}s d\tau - \int \sqrt{\frac{2}{3}}p_x\frac{1}{\sqrt{6}}p_x d\tau - \int \sqrt{\frac{2}{3}}p_x\frac{1}{\sqrt{2}}p_y d\tau - \int \sqrt{\frac{2}{3}}p_x\frac{1}{\sqrt{12}}p_z d\tau$$
$$- \int \frac{1}{\sqrt{12}}p_z\frac{1}{2}s d\tau + \int \frac{1}{\sqrt{12}}p_z\frac{1}{\sqrt{6}}p_x d\tau - \int \frac{1}{\sqrt{12}}p_z\frac{1}{\sqrt{2}}p_y d\tau + \int \frac{1}{\sqrt{12}}p_z\frac{1}{\sqrt{12}}p_z d\tau \quad ...(3)$$

Since each atomic orbitals, *i.e.*, *s* and *p* are orthogonal to each other, equation (3) reduces to:

$$\int \varphi_2\varphi_3 d\tau = \int \frac{1}{2}s\frac{1}{2}s d\tau - \int \sqrt{\frac{2}{3}}p_x\frac{1}{\sqrt{6}}p_x d\tau + \int \frac{1}{\sqrt{12}}p_z\frac{1}{\sqrt{12}}p_z d\tau$$

$$\int \varphi_2\varphi_3 d\tau = \frac{1}{4} - \frac{1}{3} + \frac{1}{12} = \frac{3-4+1}{12} = 0.$$

4.15.1. *Resonance and Electron Delocalisation*

The idea of orbital hybridisation not only provides us the explanation of the geometry of molecules, it enables us to continue thinking of a chemical bond as pairing of electrons. It has observed that the characteristics of a molecule, however, cannot always be completely represented by a single structure. A well known example is that of benzene. Structure proposed by Kekule and as shown below suggest that the molecule (*i*) will possess alternate single and double bonds, (*ii*) bond length of single bond is 0.154 nm and that of double bond is 0.133 nm, (*iii*) mono-substituted derivatives of benzene will dependent on the position where it is substituted. Although two structures of benzene are drawn but

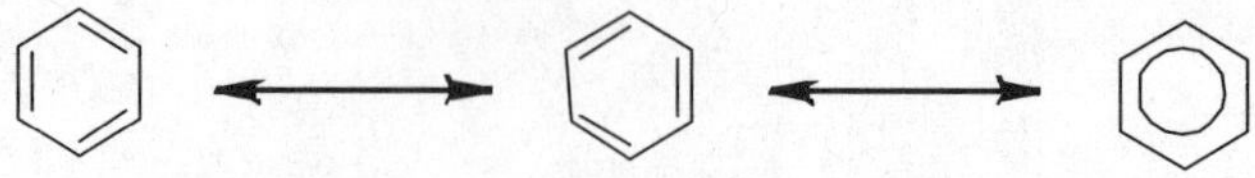

(a) (b) (c)

experimentally observed characteristics tells us that all the bonds have equal bond strength and bond length, (0.140 nm). In other words, structures drawn above cannot depict the properties of the benzene. This dilemma is solved by introducing the concept of *resonance, i.e., the term resonance means the use of two or more Lewis structures for a molecule (or ion) that cannot be described fully with a single Lewis structure*. The point is that none of the individual resonance structures accurately represents the benzene. Thus the character of each carbon-to-carbon bond is somewhat between a single and double bond, in agreement with experimental observations. No evidence whatsoever indicates that these two (actually not depicted but there are three more Dewar's structures also there) structures actually oscillate back and forth. In fact each of the two resonance structures is non-existent. It may be mentioned that model above merely enables us to solve the dilemma of trying to explain characteristics such as bond length, bond strength, some chemical properties etc. More commonly the structure of benzene is represented by (*c*), where each bond is equivalent and gives rise to a single mono-substituted derivative.

In valence bond theory, we can write the wave function for the benzene molecule as:

$$\Psi = C_a \psi_a + C_b \psi_b \qquad \text{...(4.44)}$$

where ψ_a and ψ_b are the wave functions for the Kekule's two individual resonance structures and C_a and C_b are the coefficients that determine the weight or importance of the resonance structures. In the present case both the structures are of equal importance and thus have the equal values.

***Exercise* 4.43:** (*i*) What is a peptide bond, (*ii*) Draw the possible resonance structures, (*iii*) predict the hybridisation on the carbon and nitrogen atom, (*iv*) based on molecular orbital theory write the wave functions of the π-orbitals involved and (*v*) assuming that peptide bond can be treated as a particle in one dimension box, compute the coefficients qualitatively.

(*i*) Peptide bond is a bond between the carbon and nitrogen atoms in the amide group, —CO—NH and is obtained by the reaction of carboxylic acid group and amino group as shown below.

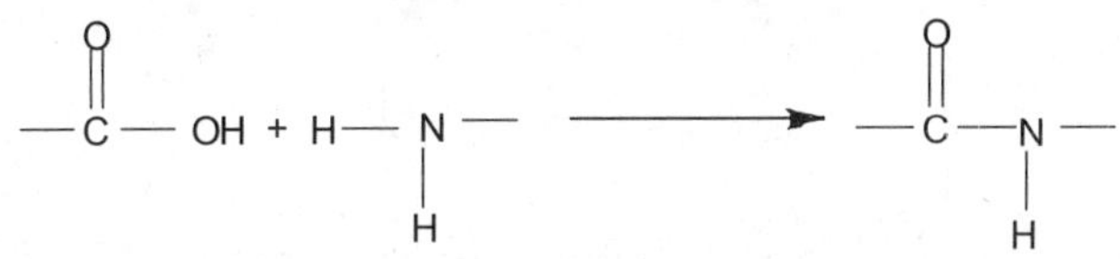

X-ray diffraction and other experimental techniques have shown that all the four atoms (O, C, N and H) lie in one plane, indicating a restricted rotation about the C-N bond. This suggests that bond between C and N is little more than a single bond. This is confirmed from the bond length of C-N which is 0.133 nm. This value lies in between the C and N single bond (0.149 nm) and C=N double bond (0.127 nm). The partial double bond character between the C and N restricts the rotation about the carbon-to-nitrogen bond and locks the four atoms in the same plane. Based on the above data and planarity of the molecule, the resonance structures of the amide group are

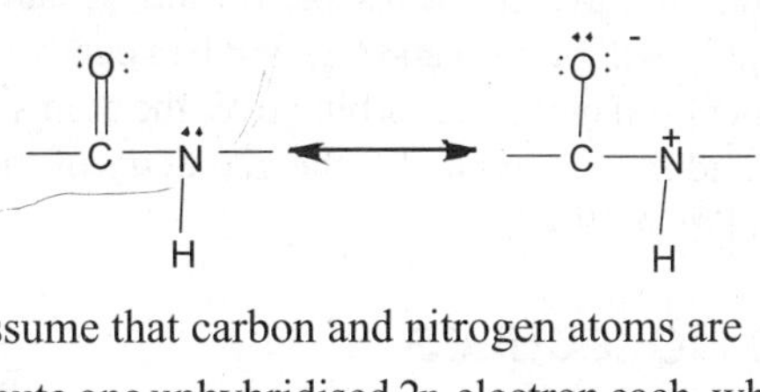

(*iii*) In MO theory we assume that carbon and nitrogen atoms are sp^2 hybridised.

(*iv*) Thus C and O contribute one unhybridised $2p_z$ electron each, whereas nitrogen atom contributes two $2p_z$ unhybridised electrons to the system. These four unhybridised $2p_z$ electrons combined linearly to form three π MOs. The wave functions can be written as

$$\Psi_1 = C_{11}\psi(O) + C_{12}\psi(C) + C_{13}\psi(N) \quad ...(1)$$
$$\Psi_2 = C_{21}\psi(O) + C_{22}\psi(C) + C_{23}\psi(N) \quad ...(2)$$
$$\Psi_3 = C_{31}\psi(O) + C_{32}\psi(C) + C_{33}\psi(N) \quad ...(3)$$

(*v*) Based on particle in one dimensional box problem, each state in this model is non-degenerate. Thus, one of them is the ground state, whereas the other two are excited states.

Further, the ground state wave function does not have any node, whereas the number of nodes

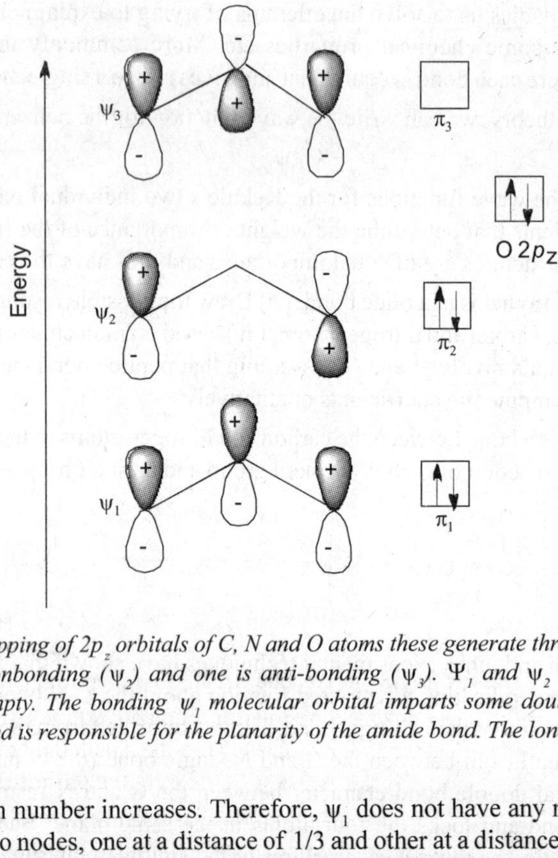

Fig. 4.19 *The overlapping of $2p_z$ orbitals of C, N and O atoms these generate three π-orbitals of which one is bonding (ψ_1), one is nonbonding (ψ_2) and one is anti-bonding (ψ_3). Ψ_1 and ψ_2 orbitals each contain two electrons and the ψ_3 is empty. The bonding ψ_1 molecular orbital imparts some double bond character to the carbon-nitrogen linkage and is responsible for the planarity of the amide bond. The lone pairs on O are placed in the 2s and $2p_y$ orbitals.*

increases as the quantum number increases. Therefore, ψ_1 does not have any node, ψ_2 has one node in the middle and ψ_3 hrs two nodes, one at a distance of 1/3 and other at a distance of 2/3 from the one end. This suggests that coefficients of wave functions ψ_1, ψ_2 and ψ_3 can be: C_{11}, C_{12} and C_{13} will all be positive; C_{21} will be positive, C_{22} will be zero and C_{23} will be negative; C_{31} and C_{33} will be positive and C_{32} will be negative. The delocalised molecular orbital over the atoms imparts double bond character to the carbon to nitrogen bond and thus accounts for the geometry of the amide bond. The shapes of the three π-orbitals are shown in Fig. 4.19.

4.16. CONJUGATED MOLECULES

4.16.1. *Linear Chain*

In the linear conjugated chain molecules, the electrons in the *p*-orbitals, which form the π bonds, are assumed to move freely in a constant potential along the chain of the molecule. The treatment is similar and can be approximated to the motion of a particle along one dimensional box (also known as free electron model, FEMO). The energy levels are given by the expression

$$E_n = \frac{n^2h^2}{8m_e a^2},\ n = 1, 2, 3, \ldots \qquad \text{...(4.45)}$$

where a is the length of the conjugated chain, m_e is the mass of electron and other terms have the usual meaning. Electrons in the molecular orbital (π) are filled according to Pauli's exclusion principle and Hund's rule. Thus only $\frac{N}{2}$ π orbitals are filled where N is the number of π orbitals (*i.e.* Np atomic orbitals).

The longest wavelength transition involves the excitation of electron from $N/2$ (highest occupied Molecular orbital HOMO) to $N/2 + 1$ (lowest unoccupied molecular orbital, LUMO). The frequency of excitation is

$$\nu = \frac{E_{(N/2)+1} - E_{N/2}}{h} = \frac{h}{8m_e a^2}\left[\left(\frac{N}{2}+1\right)^2 - \left(\frac{N}{2}\right)^2\right]$$

$$= \frac{h(N+1)}{8m_e a^2} \qquad \text{...(4.46)}$$

or

$$\bar{\nu} = \frac{\nu}{c} = \frac{h(N+1)}{8m_e a^2 c} \qquad \text{...(4.47)}$$

The only discrepancy in the relation (4.47) is about the calculation of length of the chain a. It is generally considered as:

$$a = (\text{Number of actual bonds between } N \text{ carbon atoms} + 2)l$$
$$= (N - 1 + 2)l$$
$$a = (N + 1)l$$

where l is the bond length between each carbon atom of the benzene molecule (most delocalised system, *i.e.* $l = 1.39 \times 10^{-10}$ m). In the above relation, allowance is made for molecular orbitals to extend one bond length l past each end carbon atoms. The relation (4.47) becomes

$$\bar{\nu} = \frac{h(N+1)}{8m_e c\,(N+1)^2 l^2} = \frac{h}{8m_e c\,(N+1)\,l^2} \qquad \text{...(4.48a)}$$

$$\bar{\nu} = \frac{148,137.6}{N+1}\ \text{cm}^{-1} \qquad \text{...(4.48b)}$$

A better agreement with the experimental value can be obtained if the sinusoidal form of potential energy function with a minimum at the centre of the double bond is considered. The equation (4.48 *b*) becomes

$$\bar{\nu} = \frac{153,000\ \text{cm}^{-1}}{N+1} + (16000\ \text{cm}^{-1})\left(1 - \frac{1}{N}\right) \qquad \text{...(4.49)}$$

***Exercise* 4.44:** Show that the frequency (cm^{-1}) for the longest wavelength band in polymethine dyes is given by the relation

$$\bar{\nu} = \frac{h(n_c + 4)}{8m_e c\,(n_c + 3)^2\, l^2} \qquad ...(4.50)$$

where n_c is the number of carbon atoms having $p\pi$-electron, l is the length of the bond between two atoms in the chain.

The general formula for polymethine dyes is shown here.

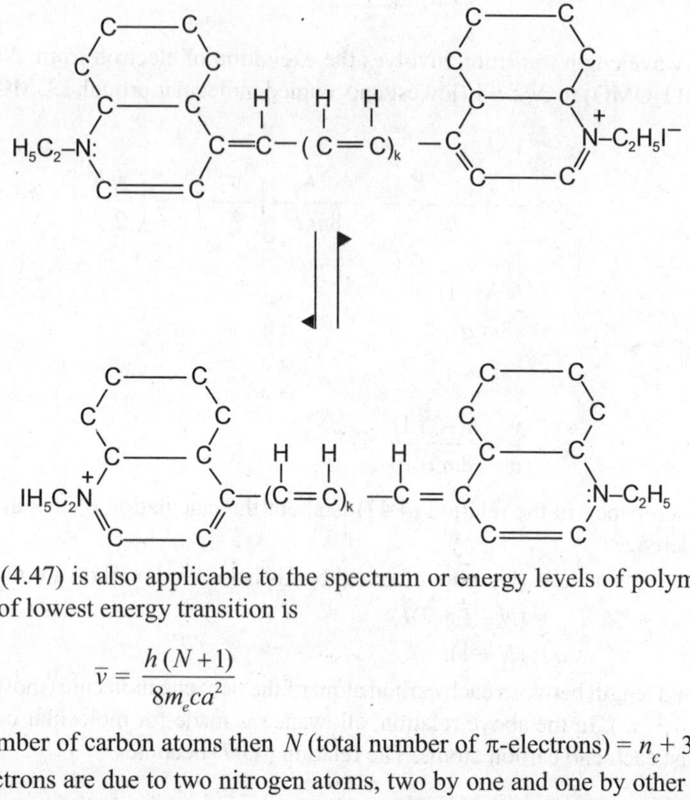

The equation (4.47) is also applicable to the spectrum or energy levels of polymethine dyes. The frequency (cm^{-1}) of lowest energy transition is

$$\bar{\nu} = \frac{h\,(N + 1)}{8m_e ca^2}$$

If n_c is the number of carbon atoms then N (total number of π-electrons) = n_c + 3.

The three electrons are due to two nitrogen atoms, two by one and one by other nitrogen. As said earlier, if it is supposed that the electrons extend their domain one bond length on either side and all the bonds are equal

$$a = (n_c + 2 - 1 + 2)\, l = (n_c + 3)\, l$$

Substituting the values of a and N in the equation (4.47)

$$\bar{\nu} = \frac{h\,(n_c + 3 + 1)}{8m_e c\,(n_c + 3)^2\, l^2} = 1.48 \times 10^5\ \frac{n_c + 4}{(n_c + 3)^2}\ \text{cm}^{-1} \qquad ...(4.51)$$

***Exercise* 4.45:** If the value of $k = 1$ in polymethine formula (1,1′-diethyl-4, 4′-carbocyanine, kryptocyanine), calculate the wavelength of lowest energy transition and compare it with the experimental value*

---

*H. Kunhn, *J. Chem. Phys.*, 17, 1198 (1949); *Rev. Mod. Phys.*, **14**, 294 (1942), K.F. Herzfeld and A.L. Sklai.

If $k = 1$ in the above formula then $n_c = 9$ and the equation (4.51) gives

$$\bar{v} = 1.48 \times 10^5 \frac{(9+4)}{(9+3)^2} = \frac{1.48 \times 10^5 \times 13}{144}$$

$$= 1.3361 \times 10^4 \text{ cm}^{-1}$$

$$\lambda = 748.4 \text{ nm.}$$

***Exercise* 4.46:** Apply the free electron molecular orbital theory to hexatriene molecule to calculate the frequency of lowest energy transition. Also apply the modified free electron model.

In hexatriene molecule, $N = 6$ so substituting the value in equation (4.48 *b*)

$$\bar{v} = \frac{148,137.6}{6+1} = 21162.5 \text{ cm}^{-1}$$

Similarly substituting the values of $N = 6$ in equation (4.49)

$$\bar{v} = \frac{153,000}{6+1} + (16,000)\,(1 - 1/6)$$

$$= 35,190.47 \text{ cm}^{-1}$$

The observed value of the frequency is 37313.4 cm^{-1}.

4.16.2. *Hückel Molecular Orbital Method*

The Hückel molecular orbital (HMO) approximation assumes that (*a*) all $S_{ij} = 0$ and $S_{ii} = 1$, *i.e.* overlap integral $\left(\int \phi_i \phi_j d\tau\right)$ between different atoms are zero; (*b*) all $H_{ij} = 0$ $\left(\int \phi_i \hat{H} \phi_j d\tau\right)$ unless the *i*th and *j*th orbitals are on adjacent atoms. The $H_{ii} = \alpha$ is the coulomb integral and $H_{ij} = \beta$ is the exchange integral. The secular determinant to calculate the energy of the π molecular orbitals for *n* atomic *p* orbitals can be written as

$$\begin{vmatrix} \alpha - E & \beta & 0 & 0 & \cdots & 0 \\ \beta & \alpha - E & \beta & 0 & \cdots & 0 \\ \cdot & \cdot & \cdot & \cdot & \cdots & \cdot \\ \cdot & \cdot & \cdot & \cdot & \cdots & \cdot \\ 0 & 0 & 0 & 0 & \beta & \alpha - E \end{vmatrix} = 0 \qquad \text{...(4.52)}$$

The secular determinant can also be written as

$$\begin{vmatrix} x & 1 & 0 & 0 & \cdots & 0 & 0 \\ 1 & x & 1 & 0 & \cdots & 0 & 0 \\ 0 & 1 & x & 1 & \cdots & 0 & 0 \\ \cdot & \cdot & \cdot & \cdot & \cdots & \cdot & \cdot \\ \cdot & \cdot & \cdot & \cdot & \cdots & \cdot & \cdot \\ 0 & 0 & 0 & & & 1 & x \end{vmatrix} = 0 \qquad \text{...(4.53)}$$

where $x = \dfrac{\alpha - E}{\beta}$

The secular determinant has N rows and N columns. The solution to (4.53) can be written as

$$\frac{\alpha - E}{\beta} = x = -2\cos\left(\frac{i\pi}{N+1}\right) \qquad ...(4.54)$$

where $i = 1, 2, 3, \ldots$. It can also be written as

$$E_i = \alpha + 2\,\beta\cos\left(\frac{i\pi}{N+1}\right) \qquad ...(4.55)$$

The values of α and β are negative, so the lowest energy level will be for $i = 1$.

***Exercise* 4.47:** Apply the HMO theory to hexatriene molecule to calculate energy levels and also the frequency of lowest electronic transition.

The secular determinant for hexatriene molecule is

$$\begin{vmatrix} x & 1 & 0 & 0 & 0 & 0 \\ 1 & x & 1 & 0 & 0 & 0 \\ 0 & 1 & x & 1 & 0 & 0 \\ 0 & 0 & 1 & x & 1 & 0 \\ 0 & 0 & 0 & 1 & x & 1 \\ 0 & 0 & 0 & 0 & 1 & x \end{vmatrix} = 0$$

The energies of the electronic levels can be calculated from the equation (4.55)

$$E_1 = \alpha + 2\beta\cos\left(\frac{\pi}{7}\right)$$

$$= \alpha + 2\beta\cos(25°\,42'\,52'') = \alpha + (2\beta)(0.\,90108)$$

$$= \alpha + 1.802\beta$$

$$E_2 = \alpha + 2\beta\cos\left(\frac{2\pi}{7}\right) = \alpha + 1.247\beta$$

$$E_3 = \alpha + 2\beta\cos\left(\frac{3\pi}{7}\right) = \alpha + 0.4448\beta$$

$$E_4 = \alpha - 0.4448\beta,\ E_5 = \alpha - 1.247\beta,\ E_6 = \alpha - 1.802\beta$$

The energy levels are drawn in the Fig. 4.20 and the electrons are filled according to Pauli's exclusion principle. The frequency required to excite the electron from E_3 to E_4 is

$$v = \frac{E_4 - E_3}{h} = \frac{\alpha - 0.4448\beta - \alpha - 0.4448\beta}{h}$$

$$= -\frac{0.8896\beta}{h}$$

or $$\bar{\nu} = -\frac{0.8896\beta}{hc}$$

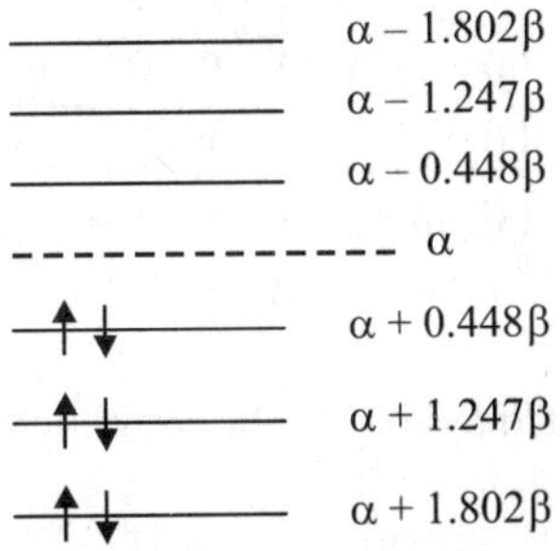

Fig. 4.20 *Energy levels of hexatriene.*

4.16.3. *Cyclic Molecules*

The free electron molecular orbital (FEMO) in case of cyclic molecules assumes that the *p*-electrons are moving freely on the circumference of circle, formed by drawing through the conjugated nuclei. The energies of the molecular orbitals are given by

$$E_n = \frac{h^2 n^2}{8\pi^2 m_e a^2} \qquad \text{...(4.56)}$$

where a is the radius of the circle and $n = 0, \pm 1, \pm 2, \ldots$. One can see that except for $n = 0$, all the other levels are doubly degenerate.

The secular determinant obtained with the help of HMO theory for conjugated cyclic molecule is

$$\begin{vmatrix} x & 1 & 0 & 0 & . & . & 1 \\ 1 & x & 1 & 0 & . & . & 0 \\ 0 & 1 & x & 1 & . & . & 0 \\ . & . & . & . & . & . & . \\ . & . & . & . & . & . & . \\ 1 & 0 & 0 & 0 & 0 & 1 & x \end{vmatrix} = 0 \qquad \text{...(4.57)}$$

and solutions are

$$E_k = \alpha + 2\beta \cos\left(\frac{2\pi k}{N}\right) \qquad \text{...(4.58)}$$

where $k = 0, 1, 2, \ldots N-1$. Since cosine is a cyclic function, the values of E_k can be degenerate. It can be seen that the minimum value of $E_k = \alpha + 2\beta$ and maximum value $\alpha - 2\beta$. The energies of all the other molecular orbitals lie in between the two extreme. The aromatic character of the carbocyclic systems like $(CH)_n$ is given by $N = 4m + 2$ rule where $m = 0, 1, 2, \ldots$. If $N = 4m + 1$, the compound is free radical and the ground state is doublet. If $N = 4m$, the compound is biradical and ground state is triplet. The ground state for $N = 4m + 2$ is singlet.

***Exercise* 4.48:** Calculate the energy levels for benzene molecule using HMO theory and also the frequency required to have the lowest energy transition.

Benzene is a cyclic molecule and has six atoms. The secular determinant will be

$$\begin{vmatrix} x & 1 & 0 & 0 & 0 & 1 \\ 1 & x & 1 & 0 & 0 & 0 \\ 0 & 1 & x & 1 & 0 & 0 \\ 0 & 0 & 1 & x & 1 & 0 \\ 0 & 0 & 0 & 1 & x & 1 \\ 1 & 0 & 0 & 0 & 1 & x \end{vmatrix} = 0$$

The energy levels can either be calculated by solving the above determinant or by using the equation (4.58)

$$E_0 = \alpha + 2\beta \cos\left(\frac{0\pi}{6}\right) = \alpha + 2\beta$$

$$E_1 = \alpha + 2\beta \cos\left(\frac{2\pi}{6}\right) = \alpha + \beta$$

$$E_2 = \alpha + 2\beta \cos\left(\frac{4\pi}{6}\right) = \alpha - \beta$$

$$E_3 = \alpha + 2\beta \cos\left(\frac{6\pi}{6}\right) = \alpha - 2\beta$$

$$E_4 = \alpha + 2\beta \cos\left(\frac{8\pi}{6}\right) = \alpha + 2\beta \cos\left(\pi + \frac{2\pi}{6}\right) = \alpha - \beta$$

$$E_5 = \alpha + 2\beta \cos\left(\frac{10\pi}{6}\right) = \alpha + 2\beta \cos\left(\pi + \frac{4\pi}{6}\right) = \alpha + \beta$$

Thus one can see that E_1 and E_5 and E_2 and E_4 are degenerate. These are drawn in Fig. 4.21. The energy required to excite the electron from E_1 to E_2, *i.e.*

$$v = \frac{\alpha - \beta - \alpha - \beta}{h} = -\frac{2\beta}{h}$$

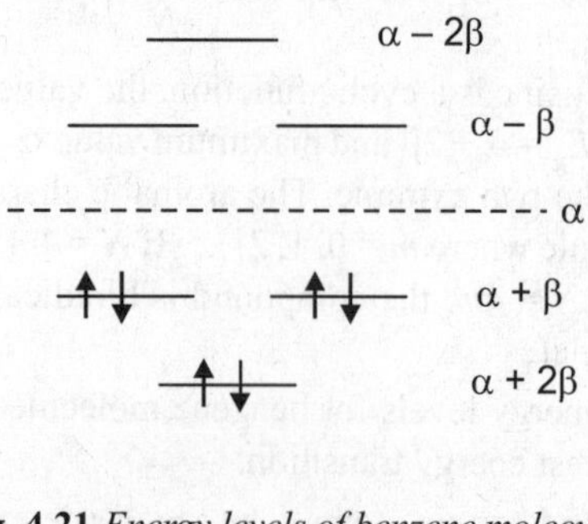

Fig. 4.21 *Energy levels of benzene molecule.*

4.17. BOND ORDER AND BOND LENGTH

The procedure given below is only to determine the extent of bond order for π bonds. Since there is already a σ bond present between the two atoms, the total bond order between any bond in the molecule is

$$P_{lm} = 1.0 + p_{lm} \qquad ...(4.59)$$

where p_{lm} is the bond order for π bond between atoms l and m. This can be determined as

$$p_{lm} = \sum_k \frac{n_k}{2} (C^*_{kl}\, C_{km} + C^*_{km}\, C_{kl}) \qquad ...(4.60)$$

where n_k is the number of electrons in the kth orbital and the sum is over all the π molecular orbital. C_{kl}'s are the coefficients of the different p atomic orbitals which form the π molecular orbitals by LCAO. These coefficients for linear conjugated molecules can be calculated by equation (4.61) and for cyclic molecules by (4.62)

$$C_{kl} = \left(\frac{2}{n+1}\right)^{1/2} \sin\left(\frac{kl\pi}{n+1}\right) \qquad ...(4.61$$

and

$$C_{kl} = \left(\frac{1}{n}\right)^{1/2} e^{2\pi i l(k-1)/n} \qquad ...(4.62)$$

w here n is the number of carbon atoms in the chain molecules and $k = 1, 2, 3, ...$

The length of the bond P_{lm} in Å for linear congugated systems and cyclic systems are given by equations (4.63) and (4.63a)

$$r_{lm} = 01.521 - 0.186\, P_{lm} \qquad ...(4.63)$$
$$r_{lm} = 1.707 - 0.186\, P_{lm} \qquad ...(4.63a)$$

***Exercise* 4.49:** Calculate the bond orders and bond lengths for the bonds 1–2, 2–3, 3–4, 4–5 and 5–6 for hexatriene molecule.

To calculate bond orders for the bond one needs the coefficient of wave functions for different molecular levels. These can be calculated by using equation (4.61).

For level $l = 1$

$$C_{11} = \left(\frac{2}{7}\right)^{1/2} \sin\left(\frac{\pi}{7}\right) = (0.5345)\,(0.4339) = 0.232$$

$$C_{12} = \left(\frac{2}{7}\right)^{1/2} \sin\left(\frac{2\pi}{7}\right) = (0.5345)\,(0.7818) = 0.418$$

$$C_{13} = \left(\frac{2}{7}\right)^{1/2} \sin\left(\frac{3\pi}{7}\right) = (0.5345)\,(0.9749) = 0.521$$

Similarly $C_{14} = 0.521$, $C_{15} = 0.418$, $C_{16} = 0.232$.

For level $l = 2$

$$C_{21} = \left(\frac{2}{7}\right)^{1/2} \sin\left(\frac{2\pi}{7}\right) = (0.5345)\,(0.7818) = 0.418$$

$$C_{22} = \left(\frac{2}{7}\right)^{1/2} \sin\left(\frac{4\pi}{7}\right) = (0.5345)\,(0.9749) = 0.521$$

$$C_{23} = \left(\frac{2}{7}\right)^{1/2} \sin\left(\frac{6\pi}{7}\right) = (0.5345)\,(0.4339) = 0.232$$

$$C_{24} = \left(\frac{2}{7}\right)^{1/2} \sin\left(\frac{8\pi}{7}\right) = (0.5345) \sin\left(\pi + \frac{\pi}{7}\right) = -\,0.232$$

$$C_{25} = \left(\frac{2}{7}\right)^{1/2} \sin\left(\frac{10\pi}{7}\right) = (0.5345)\,(-\,0.9749) = -\,0.521$$

$$C_{26} = \left(\frac{2}{7}\right)^{1/2} \sin\left(\frac{12\pi}{7}\right) = (0.\,5345)\,(-\,0.7818) = -\,0.418$$

For level $l = 3$

$C_{31} = 0.521$, $C_{32} = -\,0.232$, $C_{33} = -\,0.418$

$C_{34} = 0.418$; $C_{35} = 0.232$, $C_{36} = -\,0.521$.

For level $l = 4$

$C_{41} = 0.521$, $C_{42} = -\,0.232$, $C_{43} = -\,0.418$

$C_{44} = 0.418$, $C_{45} = 0.232$, $C_{46} = -\,0.521$

For level $l = 5$

$C_{51} = 0.418$, $C_{52} = -\,0.521$, $C_{53} = 0.232$

$C_{54} = 0.232$, $C_{55} = -\,0.521$, $C_{56} = 0.418$

For level $l = 6$

$C_{61} = 0.232$, $C_{62} = -\,0.418$, $C_{63} = 0.521$

$C_{64} = -\,0.521$, $C_{65} = 0.418$, $C_{66} = -\,0.232$.

The bond order p_{12} can be found using equation (4.60):

$$p_{12} = \frac{2}{2}\,[(0.418)\,(0.232) + (0.418)\,(0.232)] + \frac{2}{2}\,[0.418)\,(0.521) + (0.521)\,(0.418)] + \frac{2}{2}\,[(0.521)$$

$$(0.232) + (0.232)\,(0.521)] + \frac{0}{2}\,[(0.521)\,(-\,0.232) + (-\,0.232)\,(0.521)] + \frac{0}{2}\,[(0.418)$$

$$(-\,0.521) + (-0.521)\,(0.418)] + \frac{0}{2}\,[(0.232)\,(-\,0.418) + (-\,0.418)\,(0.232)]$$

$$= 0.871$$

and equation (4.59) gives

$$p_{12} = 1.0 + 0.871 = 1.871$$

$$p_{23} = \frac{2}{2}\,[(0.418)\,(0.521) + (0.521)\,(0.418)] + \frac{2}{2}\,[(0.521)\,(0.232) + (0.232)\,(0.521)] + \frac{2}{2}$$

$$[(0.232)(-0.418)+(-0.418)(0.232)]+\frac{0}{2}[(-0.232)(-0.418)+(-0.418)(-0.232)]+\frac{0}{2}$$

$$[(-0.521)(0.232)+(0.232)(-0.521)]+\frac{0}{2}[(-0.418)(0.521)+(0.521)(-0.418)]$$

$$= 0.483$$

So $P_{23} = 1 + 0.483 = 1.483$

$$p_{34} = \frac{2}{2}[0.521)(0.521)+(0.521)(0.521)]+\frac{2}{2}[(0.232)(-0.232)+(-0.232)(0.232)]+\frac{2}{2}$$

$$[(-0.418)(-0.418)+(-0.418)(-0.418)]+\frac{0}{2}[(0.418)(-0.418)+(-0.418)(0.418)]+$$

$$[(0.232)(-0.232)+(0.232)(-0.232)]+\frac{0}{2}[(0.521)(-0.521)+(-0.521)(0.521)]$$

$$= 0.892$$

So $P_{34} = 1.0 + 0.892 = 1.892$

Since hexatriene is symmetrical molecule

$$P_{45} = P_{23} \text{ and } P_{56} = P_{12}$$

So the bond length for the above values of bond orders can be calculated from equation (4.63).

$r_{12} = 1.707 - (0.186)(1.871) = 1.359$ Å

$r_{23} = 1.707 - (0.186)(1.483) = 1.43$ Å

$r_{34} = 1.707 - (0.186)(1.892) = 1.355$ Å

Similarly, $r_{45} = r_{23} = 1.43$ Å

and $r_{56} = r_{12} = 1.359$ Å

***Exercise* 4.50:** Calculate the bond orders and bond lengths of all the six bonds in benzene molecule.

Since benzene is a cyclic molecules, equation (4.59) will be used by taking different values of k, we can get

For first level $l = 0$,

$$C_{11} = \frac{1}{\sqrt{6}}e^{\frac{2\pi i.0\,(1-1)}{6}} = \frac{1}{\sqrt{6}}e^0 = \frac{1}{\sqrt{6}};\ C_{12} = \frac{1}{\sqrt{6}}e^{\frac{2\pi i.0\,(2-1)}{6}} = \frac{1}{\sqrt{6}}e^0 = \frac{1}{\sqrt{6}};$$

$$C_{13} = \frac{1}{\sqrt{6}}e^{\frac{2\pi i.0\,(3-1)}{6}} = \frac{1}{\sqrt{6}}e^0 = \frac{1}{\sqrt{6}};\ C_{14} = \frac{1}{\sqrt{6}}e^{\frac{2\pi i.0\,(4-1)}{6}} = \frac{1}{\sqrt{6}}e^0 = \frac{1}{\sqrt{6}}$$

$$C_{15} = \frac{1}{\sqrt{6}}e^{\frac{2\pi i.0\,(5-1)}{6}} = \frac{1}{\sqrt{6}} = e^0 = \frac{1}{\sqrt{6}};\ C_{16} = \frac{1}{\sqrt{6}}e^{\frac{2\pi i.0\,(6-1)}{6}} = \frac{1}{\sqrt{6}}e^0 = \frac{1}{\sqrt{6}}$$

For second level $l = 1$

$$C_{21} = \frac{1}{\sqrt{6}}e^{\frac{2\pi i.1\,(1-1)}{6}} = \frac{1}{\sqrt{6}}e^0 = \frac{1}{\sqrt{6}};\ C_{22} = \frac{1}{\sqrt{6}}e^{\frac{2\pi i.1\,(2-1)}{6}} = \frac{1}{\sqrt{6}}e^{\frac{\pi i}{3}}$$

$$C_{23} = \frac{1}{\sqrt{6}} e^{\frac{2\pi i.1\,(3-1)}{6}} = \frac{1}{\sqrt{6}} e^{\frac{2\pi i}{3}} ; C_{24} = \frac{1}{\sqrt{6}} e^{\frac{2\pi i.1\,(4-1)}{6}} = \frac{1}{\sqrt{6}} e^{\frac{6\pi i}{6}} = -\frac{1}{\sqrt{6}}$$

$$C_{25} = \frac{1}{\sqrt{6}} e^{\frac{2\pi i.1\,(5-1)}{6}} = \frac{1}{\sqrt{6}} e^{\frac{4\pi i}{3}} ; C_{26} = \frac{1}{\sqrt{6}} e^{\frac{2\pi i.1\,(6-1)}{6}} = \frac{1}{\sqrt{6}} e^{\frac{5\pi i}{3}}$$

For third level $l = 5$, but the coefficients can be written as:

$$C_{31} = \frac{1}{\sqrt{6}} e^{\frac{2\pi i.5\,(1-1)}{6}} = \frac{1}{\sqrt{6}} e^{0} = \frac{1}{\sqrt{6}};$$

$$C_{32} = \frac{1}{\sqrt{6}} e^{\frac{2\pi i.5\,(2-1)}{6}} = \frac{1}{\sqrt{6}} e^{\frac{5\pi i}{3}} = \frac{1}{\sqrt{6}} e^{2\pi i} e^{-\frac{\pi i}{3}} = \frac{1}{\sqrt{6}} e^{-\frac{\pi i}{3}}$$

$$C_{33} = \frac{1}{\sqrt{6}} e^{\frac{2\pi i.5\,(3-1)}{6}} = \frac{1}{\sqrt{6}} e^{\frac{10\pi i}{3}} = \frac{1}{\sqrt{6}} e^{4\pi i} e^{-\frac{2\pi i}{3}} = \frac{1}{\sqrt{6}} e^{-\frac{2\pi i}{3}}$$

$$C_{34} = \frac{1}{\sqrt{6}} e^{\frac{2\pi i.5\,(4-1)}{6}} = \frac{1}{\sqrt{6}} e^{5\pi i} = -\frac{1}{\sqrt{6}}$$

$$C_{35} = \frac{1}{\sqrt{6}} e^{\frac{2\pi i.5\,(5-1)}{6}} = \frac{1}{\sqrt{6}} e^{\frac{20\pi i}{3}} = \frac{1}{\sqrt{6}} e^{8\pi i} e^{-\frac{4\pi i}{3}} = \frac{1}{\sqrt{6}} e^{-\frac{4\pi i}{3}}$$

$$C_{36} = \frac{1}{\sqrt{6}} e^{\frac{2\pi i.5\,(6-1)}{6}} = \frac{1}{\sqrt{6}} e^{\frac{25\pi i}{3}} = \frac{1}{\sqrt{6}} e^{10\pi i} e^{-\frac{5\pi i}{3}} = \frac{1}{\sqrt{6}} e^{-\frac{5\pi i}{3}}$$

For fourth level $l = 2$ but the coefficients can be written as:

$$C_{41} = \frac{1}{\sqrt{6}} e^{\frac{2\pi i.2\,(1-1)}{6}} = \frac{1}{\sqrt{6}} e^{0} = \frac{1}{\sqrt{6}}; C_{42} = \frac{1}{\sqrt{6}} e^{\frac{2\pi i.2\,(2-1)}{6}} = \frac{1}{\sqrt{6}} e^{\frac{2\pi i}{3}};$$

$$C_{43} = \frac{1}{\sqrt{6}} e^{\frac{2\pi i.2\,(3-1)}{6}} = \frac{1}{\sqrt{6}} e^{\frac{4\pi i}{3}}; C_{44} = \frac{1}{\sqrt{6}} e^{\frac{2\pi i.2\,(4-1)}{6}} = \frac{1}{\sqrt{6}} e^{2\pi i} = \frac{1}{\sqrt{6}}$$

$$C_{45} = \frac{1}{\sqrt{6}} e^{\frac{2\pi i.2\,(5-1)}{6}} = \frac{1}{\sqrt{6}} e^{\frac{8\pi i}{3}} = \frac{1}{\sqrt{6}} e^{\frac{2\pi i.}{3}}; C_{46} = \frac{1}{\sqrt{6}} e^{\frac{2\pi i.2\,(6-1)}{6}} = \frac{1}{\sqrt{6}} e^{\frac{4\pi i}{3}}$$

For fifth level $l = 4$ and coefficients can be written as:

$$C_{51} = \frac{1}{\sqrt{6}} e^{\frac{2\pi i.4\,(1-1)}{6}} = \frac{1}{\sqrt{6}}, C_{52} = \frac{1}{\sqrt{6}} e^{\frac{2\pi i.4\,(2-1)}{6}} = \frac{1}{\sqrt{6}} e^{2\pi i} e^{-\frac{2\pi i}{3}} = \frac{1}{\sqrt{6}} e^{-\frac{2\pi i}{3}}$$

$$C_{53} = \frac{1}{\sqrt{6}} e^{\frac{2\pi i.4\,(3-1)}{6}} = \frac{1}{\sqrt{6}} e^{4\pi i} e^{-\frac{4\pi i}{3}} = \frac{1}{\sqrt{6}} e^{-\frac{4\pi i}{3}}$$

$$C_{54} = \frac{1}{\sqrt{6}} e^{\frac{2\pi i.4\,(4-1)}{6}} = \frac{1}{\sqrt{6}} e^{4\pi i} = \frac{1}{\sqrt{6}},$$

$$C_{55} = \frac{1}{\sqrt{6}} e^{\frac{2\pi i.4\,(5-1)}{6}} = \frac{1}{\sqrt{6}} e^{6\pi i} e^{-\frac{2\pi i}{3}} = \frac{1}{\sqrt{6}} e^{-\frac{2\pi i}{3}}$$

$$C_{56} = \frac{1}{\sqrt{6}} e^{\frac{2\pi i.4\,(6-1)}{6}} = \frac{1}{\sqrt{6}} e^{8\pi i} e^{-\frac{4\pi i}{3}} = \frac{1}{\sqrt{6}} e^{-\frac{4\pi i}{3}}$$

For sixth level, $l = 3$

$$C_{61} = \frac{1}{\sqrt{6}} e^{\frac{2\pi i.3\,(1-1)}{6}} = \frac{1}{\sqrt{6}} e^{0} = \frac{1}{\sqrt{6}};\; C_{62} = \frac{1}{\sqrt{6}} e^{\frac{2\pi i.3\,(2-1)}{6}} = \frac{1}{\sqrt{6}} e^{i\pi} = -\frac{1}{\sqrt{6}}$$

$$C_{63} = \frac{1}{\sqrt{6}} e^{\frac{2\pi i.3\,(3-1)}{6}} = \frac{1}{\sqrt{6}} e^{2\pi i} = \frac{1}{\sqrt{6}};\; C_{64} = \frac{1}{\sqrt{6}} e^{\frac{2\pi i.3\,(4-1)}{6}} = \frac{1}{\sqrt{6}} e^{3\pi i} = -\frac{1}{\sqrt{6}}$$

$$C_{65} = \frac{1}{\sqrt{6}} e^{\frac{2\pi i.3\,(5-1)}{6}} = \frac{1}{\sqrt{6}} e^{4\pi i} = \frac{1}{\sqrt{6}};\; C_{66} = \frac{1}{\sqrt{6}} e^{\frac{2\pi i.3\,(6-1)}{6}} = \frac{1}{\sqrt{6}} e^{5\pi i} = -\frac{1}{\sqrt{6}}$$

Thus the Hückel M.O.'s can be written as:

$$\varphi_1 = \frac{1}{\sqrt{6}} [2p_1 + 2p_2 + 2p_3 + 2p_4 + 2p_5 + 2p_6]$$

$$\varphi_2 = \frac{1}{\sqrt{6}} \left[2p_1 + e^{\frac{\pi i}{3}} 2p_2 + e^{\frac{2\pi i}{3}} 2p_3 - 2p_4 + e^{\frac{4\pi i}{3}} 2p_5 + e^{\frac{5\pi i}{3}} 2p_6\right]$$

$$\varphi_3 = \frac{1}{\sqrt{6}} \left[2p_1 + e^{-\frac{\pi i}{3}} 2p_2 + e^{-\frac{2\pi i}{3}} 2p_3 - 2p_4 + e^{-\frac{4\pi i}{3}} 2p_5 + e^{-\frac{5\pi i}{3}} 2p_6\right] \quad \text{...(4.63}b\text{)}$$

$$\varphi_4 = \frac{1}{\sqrt{6}} \left[2p_1 + e^{\frac{2\pi i}{3}} 2p_2 + e^{\frac{4\pi i}{3}} 2p_3 + 2p_4 + e^{\frac{2\pi i}{3}} 2p_5 + e^{\frac{4\pi i}{3}} 2p_6\right]$$

$$\varphi_5 = \frac{1}{\sqrt{6}} \left[2p_1 + e^{-\frac{2\pi i}{3}} 2p_2 + e^{-\frac{4\pi i}{3}} 2p_3 + 2p_4 + e^{-\frac{2\pi i}{3}} 2p_5 + e^{-\frac{4\pi i}{3}} 2p_6\right]$$

$$\varphi_6 = \frac{1}{\sqrt{6}} \left[2p_1 - 2p_2 + 2p_3 - 2p_4 + 2p_5 - 2p_6\right]$$

The degenerate Hückel M.O.'s depicted in equation (4.63 *b*) are of complex in nature. Since the two M.O.'s of each degenerate levels are complex conjugate of each other, the real forms of these m.o.'s can be obtained by taking the plus and minus combinations of the complex ones. The real form of these M.O.'s are:

$$\varphi_{2,\,\text{real}} = \frac{1}{\sqrt{2}}[\varphi_2 + \varphi_3],\ \varphi_{3,\,\text{real}} = -i\,\frac{1}{\sqrt{2}}[\varphi_2 - \varphi_3]$$

$$\varphi_{2,\,\text{real}} = \frac{1}{\sqrt{12}}[2\,2p_1 + 2p_2 - 2p_3 - 2\,2p_4 - 2p_5 + 2p_6]$$

$$\varphi_{3,\,\text{real}} = \frac{1}{2}[2p_2 + 2p_3 - 2p_5 - 2p_6]$$

Similarly taking the linear combination of the 4th and 5th orbitals, we can form the real form of these m.o.'s and are given below:

$$\varphi_{4,\,\text{real}} = \frac{1}{\sqrt{12}}[2\,2p_1 - 2p_2 - 2p_3 + 2\,2p_4 - 2p_5 - 2p_6]$$

$$\varphi_{5,\,\text{real}} = \frac{1}{2}[2p_2 - 2p_3 + 2p_5 - 2p_6]$$

Bond order can be calculated using Equation (4.63) and real coefficients of each m.o. obtained earlier.

$$p_{12} = \frac{2}{2}[C_{11}C_{12} + C_{12}C_{11}] + \frac{2}{2}[C_{21}C_{22} + C_{22}C_{21}] + \frac{2}{2}[C_{31}C_{32} + C_{32}C_{31}]$$

In the summation the last three terms have been neglected as there will not be any electron in orbitals 4, 5 and 6. Substituting the values of the coefficients, we get:

$$p_{12} = \left[\frac{1}{\sqrt{6}}\frac{1}{\sqrt{6}} + \frac{1}{\sqrt{6}}\frac{1}{\sqrt{6}}\right] + \left[\frac{2}{\sqrt{12}}\frac{1}{\sqrt{12}} + \frac{1}{\sqrt{12}}\frac{2}{\sqrt{12}}\right] + \left[0.\frac{1}{\sqrt{6}} + \frac{1}{\sqrt{6}}0\right]$$

$$= \frac{1}{6} + \frac{1}{6} + \frac{2}{12} + \frac{2}{12} = \frac{8}{12} = 0.66$$

$$p_{23} = \frac{2}{2}[C_{12}C_{13} + C_{13}C_{12}] + \frac{2}{2}[C_{22}C_{23} + C_{23}C_{22}] + \frac{2}{2}[C_{32}C_{33} + C_{33}C_{32}]$$

$$= \left[\frac{1}{\sqrt{6}}\frac{1}{\sqrt{6}} + \frac{1}{\sqrt{6}}\frac{1}{\sqrt{6}}\right] + \left[\frac{1}{\sqrt{12}}\left(-\frac{1}{\sqrt{12}}\right) + \left(-\frac{1}{\sqrt{12}}\right)\frac{1}{\sqrt{12}}\right] + \left[\frac{1}{2}\frac{1}{2} + \frac{1}{2}\frac{1}{2}\right]$$

$$= \frac{1}{6} + \frac{1}{6} - \frac{1}{12} - \frac{1}{12} + \frac{1}{4} + \frac{1}{4} = \frac{8}{12} = 0.66$$

$$p_{34} = \frac{2}{2}\left[C_{13}C_{14} + C_{14}C_{13}\right] + \frac{2}{2}\left[C_{23}C_{24} + C_{24}C_{23}\right] + \frac{2}{2}\left[C_{33}C_{34} + C_{34}C_{33}\right]$$

$$= \left[\frac{1}{\sqrt{6}}\,\frac{1}{\sqrt{6}} + \frac{1}{\sqrt{6}}\,\frac{1}{\sqrt{6}}\right] + \left[\left(-\frac{2}{\sqrt{12}}\right)\left(-\frac{1}{\sqrt{12}}\right) + \left(-\frac{1}{\sqrt{12}}\right)\left(-\frac{2}{\sqrt{12}}\right)\right] + \left[0.\,\frac{1}{2} + \frac{1}{2}\,0\right]$$

$$= \frac{1}{6} + \frac{1}{6} + \frac{2}{12} + \frac{2}{12} = \frac{8}{12} = 0.66.$$

Similarly we also get the value of p_{45}, p_{56} and p_{61} and all are equal to 0.666. In other words, all the values of p_{lm} are equal. Using equation (4.63), we can calculate the total bond order:

$$P_{lm} = 1 + p_{lm} = 1.0 + 0.66 = 1.666$$

It shows that bond order of each bond in benzene is the same. The bond length of each bond will also be equal and can be calculated by using Equation (4.63 *b*), *i.e.*

$$R_{kl} = 1.521 - 0.186\, P_{kl} = 1.521 - 0.186 \times 1.66 = 1.521 - 0.309 = 1.21\ \text{Å}. = 0.121\ \text{nm}.$$

4.18. COORDINATION COMPOUNDS

4.18.1. *Valence Bond Theory*

The valence bond theory, developed by Pauling, involves the donation of pair of electrons by the ligand to the empty orbitals of the central metal ion. Thus the bonding between the ligands and central metal ion is of coordinate covalent nature. The '*d*' orbitals of metal ion can be either inner '*d*' orbitals or outer '*d*' orbitals and this can be decided from the known magnetic moment of the compounds. Generally, if the ligands, *e.g.* CN^- ion, can donate the electrons easily, the '*d*' orbitals involved are the inner one and involves covalent bonding. These complexes are called inner-orbital complexes. These are relatively, inert, *i.e.* these compounds undergo slow ligand displacement. On the other hand, if the ligands, like F^- ions, do not donate the pair of electrons easily, the outer '*d*' orbitals of the metal ions are used and bending or interaction between the metal ion and ligands are of ionic nature. These complexes are called outer-orbital complexes and the ligand substitution is quite rapid in these complexes. The shape of the complex is determined from the orbitals used: *sp* is linear, sp^3 or sd^3 is tetrahedral, dsp^2 is square planar, dsp^3 or spd^3 is trigonal bipyramidal, sp^2d^2 or sd^4 is square pyramidal and sp^3d^2 is octahedral.

The magnetic moment of the coordinate complexes can be calculated from the relation (4.64)

$$\mu = \sqrt{n(n+2)}\ \text{BM} \qquad \text{...(4.64)}$$

where *n* is the number of unpaired electrons and BM is Bohr magneton. The major disadvantage of *VB* Theory is that it does not satisfactorily account for crystal field splitting and therefore cannot readily explain colour and magnetism of complex ions. Nevertheless, the use of hybridisation to account for geometry is helpful.

4.18.2. *Crystal Field Theory*

The crystal field theory assumes that the ligands can be considered as point negative charges and are surrounding the central metal ion. Due to this, the degeneracy of the '*d*' orbitals are removed and splitting of the *d* orbitals takes place depending upon the geometrical arrangement of the ligands around the metal ion. Figure 4.22 shows the splitting of the *d* orbitals in different environments of ligands. The separation between each energy levels, drawn in Fig. 4.22 is relative to that when the *d* orbitals are surrounded by the spherical field. Further if the charge density of the ligands are very high, the separation between the orbitals will be large compared to the pairing energy of electrons and the electrons filled

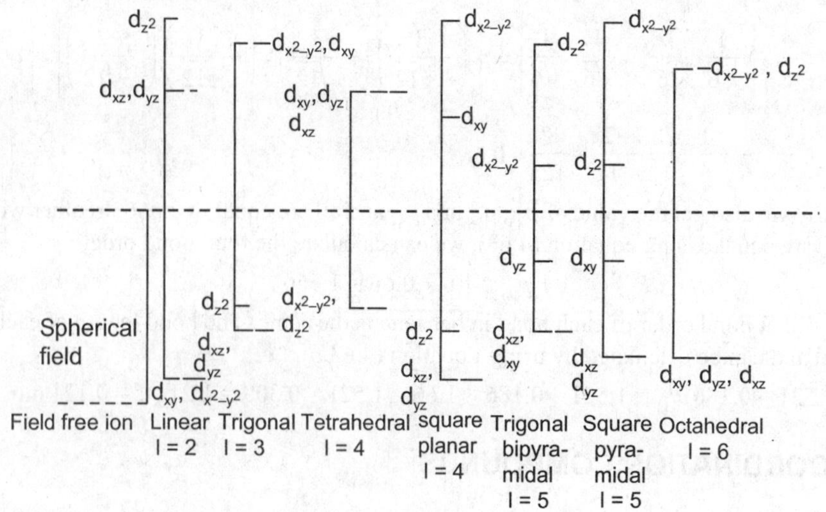

Fig. 4.22 *Splitting of d orbitals under different geometric arrangements of ligands.*

according to Hund's rule, will occupy the lower orbitals. The complexes obtained are called "low spin complexes." If the separation between the orbitals are small compared to the pairing energy of the electrons, high spin complexes are obtained. This is because all the orbitals will be filled with the parallel spins before a second electron is placed in any orbital.

The differences between the energy levels generally lie in the visible region of the spectrum and thus most of the transition element complexes are coloured. The simplest transition in the octahedral complexes is when the metal ion has d^2 electronic configuration. The Russell-Saunders or spectroscopic term for free ion is ^2D. The crystal field splitting generates T_{2g} (triply degenerate state) and E_g (doubly degenerate state) states. The transition can be depicted as $^2T_{2g} \rightarrow {}^2E_g$.

***Exercise* 4.51:** F^- ion is a weak ligand compared to CN^- ion. If the pairing energy is larger than the octahedral field splitting of F^- ion and vice versa for CN^- ion, (*a*) calculate the number of unpaired electrons in $[FeF_6]^{3-}$ and $[Fe(CN)_6]^{3-}$, (*b*) also mention which one is high spin and which one is low spin?, and (*c*) calculate the magnetic moment of each complex.

The electronic configuration of Fe^{3+} is $1s^22s^22p^63s^23p^63d^5$, *i.e.* Fe^{3+} is a d^5 system. The *d* orbitals in octahedral crystal field splits as shown in Fig. 4.23.

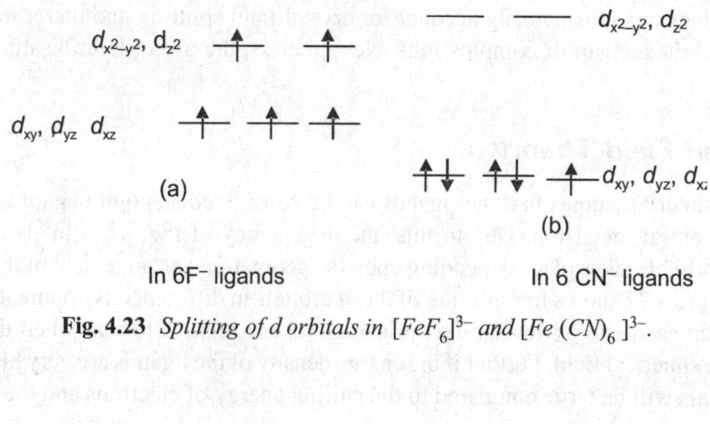

Fig. 4.23 *Splitting of d orbitals in* $[FeF_6]^{3-}$ *and* $[Fe(CN)_6]^{3-}$.

As mentioned, Δ in F^- octahedral field is smaller than pairing energy of the electrons, all five electrons are singly occupied in 5*d* orbitals, *i.e.* unpaired electron in $[(FeF_6)]^{3-} = 5$, whereas in $[Fe(CN)_6]^{3-}$ unpaired electron is only 1 (Fig. 4.23). $[FeF_6]^{3-}$ is a high spin with the electronic configuration $t_{2g}^3\ e_g^2$ and $[Fe(CN)_6]^{3-}$ is a low spin complex with electronic configuration $t_{2g}^5\ e_g^0$.

$$\text{Magnetic moment } \mu = \sqrt{n(n+2)} \text{ BM}$$

i.e. $$\mu\ ([Fe(CN)_6]^{3-}) = \sqrt{1 \times 3} = 1.732 \text{ BM}$$

and $$\mu\ ([FeF_6]^{3-}) = \sqrt{5 \times 7} = 5.92 \text{ BM}$$

***Exercise* 4.52:** Explain the results of Exercise 4.51 with the help of valence bond picture. Indicate which one is outer orbital complex and which one is inner orbital complex?

The magnitude of magnetic moments indicate that $[Fe(CN)_6]^{3-}$ has only one unpaired electron and $[FeF_6]^{3-}$ 5, unpaired electrons. Further six ligands form octahedral structure so the hybridisation should be d^2sp^3 or sp^3d^2. The electronic configurations of Fe^{3+}, $[FeF_6]^{3-}$, and $[Fe(CN)_6]^{3-}$ is shown below:

	3*d*	4*s*	4*p*	4*d*
Fe^{3+}	↑ | ↑ | ↑ | ↑ | ↑	(empty)	(empty) (empty) (empty)	(empty) (empty) (empty) (empty) (empty)
$[Fe(CN)_6]^{3-}$	↑↓ | ↑↓ | ↑ | xx | xx	xx	xx | xx | xx	(empty) (empty) (empty) (empty) (empty)
$[FeF_6]^{3-}$	↑ | ↑ | ↑ | ↑ | ↑	xx	xx | xx | xx	xx | xx | (empty) (empty) (empty)

Electronic configurations mentioned above show that $[Fe(CN)_6]^{3-}$ is an inner orbital complex and $[FeF_6]^{3-}$ is an outer orbital complex.

***Exercise* 4.53:** $[Fe(H_2O)_6]^{3+}$ absorbs at 10,400 cm^{-1} and $[Fe(CN)_6]^{3-}$ absorbs at 33,000 cm^{-1}. Predict the strength of the ligands.

Since the frequency of light absorbed by transition metal complex is proportional to crystal field splitting, the ligand CN^- will be stronger than H_2O because the $\bar{\nu}$ = 33,000 cm^{-1} is greater than $\bar{\nu}$ = 10,400 cm^{-1}.

Another approach to the bonding of ligands to central metal atom or metal ion was proposed by Ralph Gottfrid Pearson. According him the central metal ion or acid in a complex can be designated either hard or soft, according to its ability to attach various ligands. Hard acids are small, compact and less easily polarised metal ions (electron density is not easily distorted). Soft acids, on the other hand, are large, fairly polarisable metal ions. Similarly, we can classify ligand as hard or soft according to how tightly the lone pair of electron is held by the donor atom. The more electronegative the atom the harder the base will be several hard acids and hard bases are compiled in Table 4.9. *The significance of this designation is that generally hard acids preferentially bind to hard bases and soft acids to soft bases*. The correlation helps us to understand and predict many chemical and biological properties. Another term which is commonly used is the coordination number, which is defined as the number of donor atoms (in ligands) surrounding the central metal atom in a complex. Different geometries are

TABLE 4.9. List of hard and soft acids and bases in ligand-metal reactions

	Acids	Bases[a]
Hard	H^+, Li^+, Na^+, K^+, Mg^{2+}, Ca^{2+}, Mn^{2+}, Al^{3+}, Ga^{3+}, Co^{3+}, Cr^{3+}, Fe^{3+}	H_2O, NH_3, RNH_2, OH^-, F^-, Cl^-, NO_3^-, CO_3^{2-}, CH_3COO-, PO_4^{2-}
Intermediate	Fe^{2+}, CO^{2+}, Ni^{2+}, Cu^{2+}, Zn^{2+}, Sn^{2+}	N_2, Br^-, NO_2^-, SO_3^{2-}, N_3^-, imidazole
Soft	Cu^+, Au^+, Pt^{2+}, Ag^+, Cd^{2+}, Hg^{2+}, Pb^{2+}	H^-, CN^-, SCN^-, I^-, RSH, RS^-, R_3P, CO

$a = R$ is an alky & ary & group.

give in Fig. 4.24. These demonstrate the different number of atoms or lignads surrounding the central metal or metal ion.

The splitting of *d*-orbitals is simple in case of octahedral and tetrahedral environment. The energy difference between the two sets of *d*-orbitals is called the crystal field splitting, denoted by Δ. The crystal field stabilisation energy (CFSE) is the net decrease in energy of the *d*-orbital in a complex ion relative to the *d*-orbital energy in a spherically symmetric crystal field. The value of Δ depends on the nature of ligands and the number of electrons present. It can be calculated as follows:

$$\text{CSFE} = n(e_g)(0.6\Delta) - n(t_{2g})(0.4\Delta) \qquad ...(4.65)$$

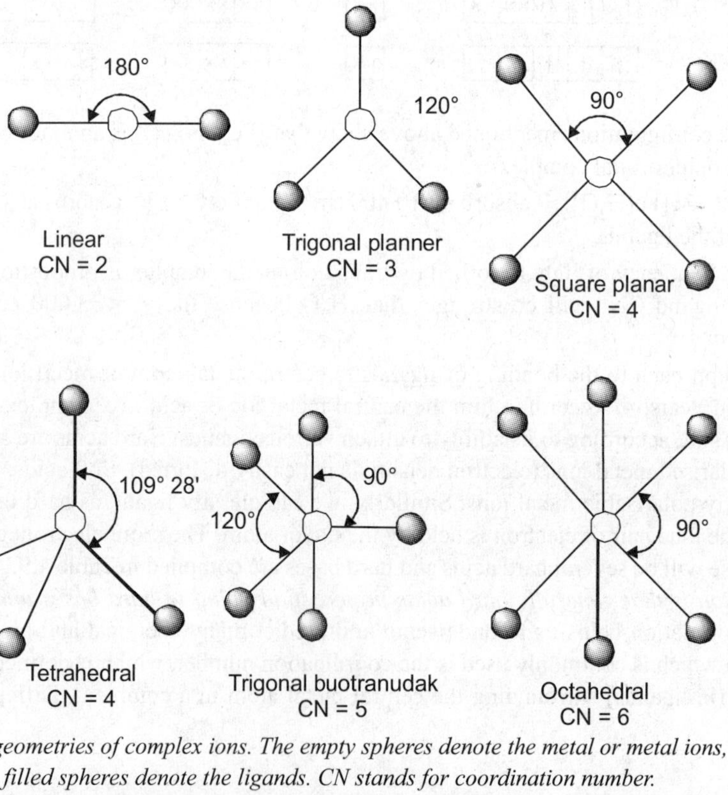

Fig. 4.24 *Common geometries of complex ions. The empty spheres denote the metal or metal ions, whereas the filled spheres denote the ligands. CN stands for coordination number.*

where $n(e_g)$ and $n(t_{2g})$ are the numbers of electrons in e_g and t_{2g} orbitals respectively. Further $\Delta_{tet} = (4/9)\Delta_{oct}$. Because of this reason most of the tetrahedral complexes are high-spin complexes.

The splitting of d-orbitals in other geometries is complicated. Thus pattern of splitting does not permit us to define crystal field splitting in the same way as in the octahedral and tetrahedral geometries.

***Exercise* 4.54:** Compute the CFSE in terms of Δ and predict the magnetic properties of (*i*) $[Fe(CN)_6]^{4-}$ and (*ii*) $[FeF_6]^{4-}$.

The electronic configuration of Fe^{2+} is $[Ar]3d^6$.

(*i*) Since CN^- is a strong field ligand, all the six electrons will be present in t_{2g} orbitals. Thus using Equation (4.65) the CFSE is given as

$$\text{CFSE} = (0)(0.6\Delta) - (6)(0.4\Delta) = -2.4\Delta.$$

All the six electrons are paired, giving us a diamagnetic and zero spin.

(*ii*) Since F^- is a weak field ligand, first five electrons will be present in t_{2g} and e_g orbitals. Sixth electron will be present in t_{2g} orbital. Out of six electrons, four will be unpaired. Thus using equation(4.63) the CFSE is given as

$$\text{CFSE} = (2)(0.6\Delta) - (4)(0.4\Delta) = 1.2\Delta - 1.6\Delta = -0.4\Delta$$

Only two electrons are paired and four are unpaired. So this complex is high spin and the magnetic will be non-zero. This complux will be paramagnetic.

4.18.3. *Molecular Orbital Theory*

Although crystal field theory is conceptually easy to understand and provides an explanation for the spectral and magnetic properties of some complexes, it has a serious defect. By considering only electrostatic interaction, it completely neglects the covalent character of the metal-ligand bonds. A more satisfactory treatment that takes this property into consideration is MO theory.

As explained earlier in Section (4.3.3), the molecular orbital theory involves the combination of the atomic orbitals of the central atom and the bonding orbitals of the ligands. The relative splittings of the molecular orbitals depend upon (*a*) the strength of the ligands, (*b*) nature of the symmetry group, *i.e.* number of ligands and their arrangements around the central atom, (*c*) nature of bonding, *i.e.* formation of σ or π bonds.

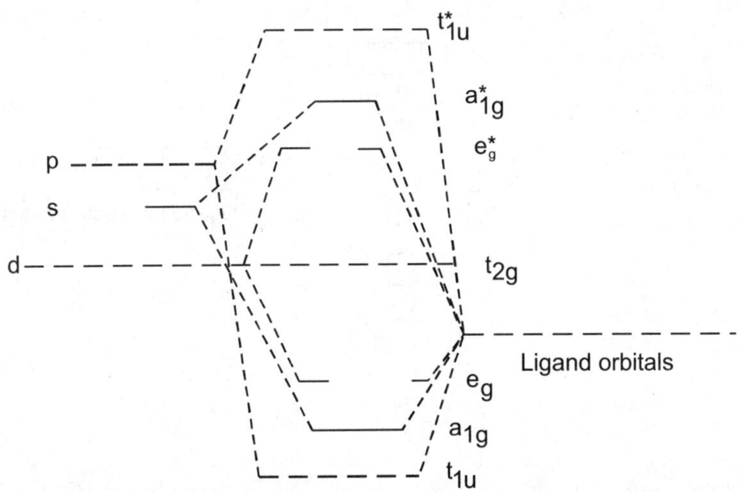

Fig. 4.25 *Energy level diagram of the m.o. in octahedral field.*

Figure 4.25 shows the energy level diagram of the molecular orbitals obtained in the octahedral symmetry and forming only σ bonds. Since only particular kinds of orbitals lead to octahedral symmetry, *i.e.* $d_{x^2-y^2}$, d_{z^2}, s, p_x, p_y, p_z so the d_{xy}, d_{yz} and d_{xz} will be unaffected by the bonding as these orbitals are not involved in the bond formation.

This is because the d_{z^2} and $d_{x^2-y^2}$ orbitals point at the ligands. Therefore they will form σ molecular orbitals with the ligands lone pair orbitals. The 4*s* orbital has spherical symmetry and will overlap with all the ligands lone pair orbitals. The three *p* orbitals will overlap with ligand lobe pair orbitals on the *x*, *y* and *z* axes. On the other hand, the d_{xy}, d_{yz} and d_{zx} orbitals point between the ligands and thus will not be involved in the σ-bonding with the ligands. Thus ,d_{z^2}, $d_{x^2-y^2}$, s, p_x, p_y and p_z orbitals will be involved in the σ molecular orbital formation and the d_{xy}, d_{yz} and d_{zx} orbitals will be the non-bonding molecular orbitals.

***Exercise* 4.55:** Draw the molecular orbital energy level diagram for $[FeF_6]^{3-}$.

Since F^- ion is a weak ligand, the electrostatic field splitting Δ will be small and will lead to high spin complex. This is because the e_g^* orbitals are primarily composed of d_{z^2} and $d_{x^2-y^2}$ atomic orbitals, with relatively little contribution from the ligand orbitals. This lack of mixing is due to the large energy difference between the ligand orbitals and the metal 3 *d* orbitals. In other words, MO theory provides the same type of *d* orbital splitting as does the crystal field theory. In addition, MO theory provides a sounder basis for bonding and offers a clearer picture of crystal field splitting. For example, a ligand with very high electronegative donor atom, say *F*, will have lone pair orbitals with very low energy,

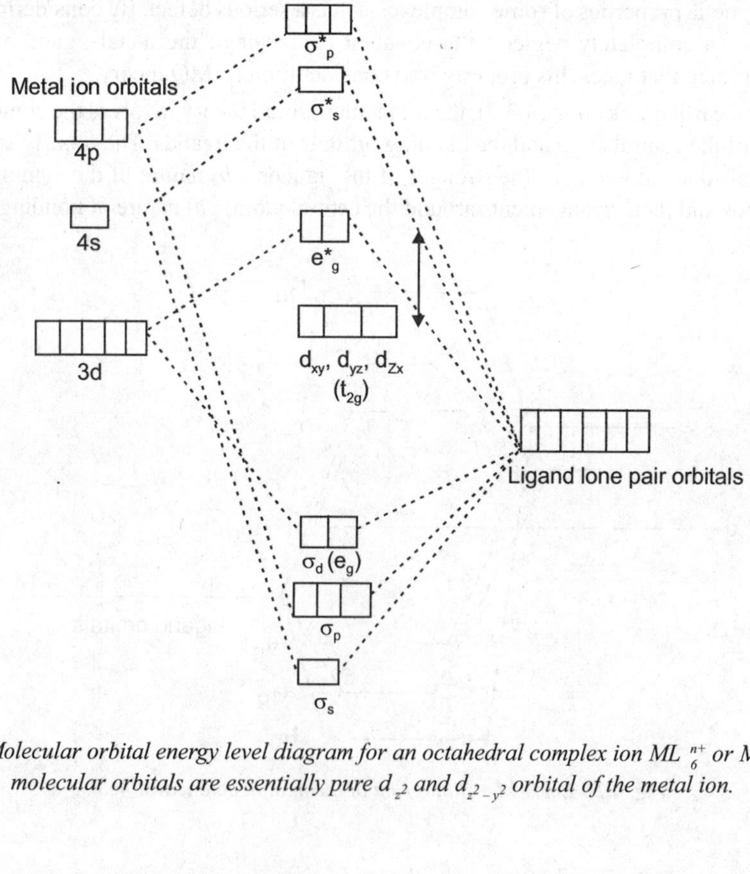

Fig. 4.26 *Molecular orbital energy level diagram for an octahedral complex ion* ML_6^{n+} *or* ML_6^{n-}. *The* e_g *molecular orbitals are essentially pure* d_{z^2} *and* $d_{z^2-y^2}$ *orbital of the metal ion.*

because electrons are strongly held by the atom. Consequently, these orbitals do not mix well with the metal orbitals, and crystal field splitting is minimal. The molecular orbital energy level diagram is drawn in Fig. 4.26.

As we know that most of the transition metal complexes are coloured and this is due to the fact that part of the visible radiation is absorbed and rest is transmitted. It is the transmitted light that makes them coloured. Table 4.10. gives the absorption transition of the transition metal complexes and contemporary colour emitted.

TABLE 4.10. Relationship between the absorption transition and colour of the complex

Absorption wavelength (nm)	Colour of the complex observed
400 (violet)	Greenish yellow
450 (blue)	Yellow
490 (blue-green)	Red
570 (yellow-green)	Violet
580 (yellow)	Dark blue
600 (orange)	Blue
650 (red)	Green

The Δ, the splitting between T_{2g} and e_g^* depends upon the strength of the ligands and this will also determine whether one is high spin or low spin complex. The energy level diagram for π bonding will be different.

4.19. INTERMOLECULAR FORCES (INTERACTIONS)

Before we discuss the intermolecular interactions it is useful to distinguish between force and potential energy. In mechanics, work done is force (F) multiplied by distance. The work done (dw) in moving two interacting molecules apart by an infinitesimal distance (dr) is given by

$$dw = -Fdr \qquad ...(4.66)$$

The sign convention for work is that if the molecules attract each other (*i.e.*, F is negative) and dr is positive (*i.e.*, molecules are moved farther apart), then dw is positive. Expression for potential energy when the molecules are separated by a distance r is defined as work done in bringing another molecule from infinite separation to a distance r from this molecule. This work is the potential energy acquired by the system, given by

$$V = \int_{\infty}^{r} dw \qquad ...(4.67)$$

Combining equations (4.66) and (4.67), we get

$$V = -\int_{\infty}^{r} Fdr \qquad ...(4.68)$$

Equation (4.68) relates the potential energy to the force between the two molecules. Similarly we can define the force as the negative slope of the curve that describes the dependence of V or r, *i.e.*, equation (4.66) shows the relationship between the potential energy and force. We are now ready to discuss the different types of intermolecular forces. The species which we will consider will be molecules-molecules and ion-molecules.

4.19.1. Dipole-Dipole Interactions

Dipole-dipole interaction involves interactions between polar molecules. Consider two molecules A and B with dipole moments μ_A and μ_B, separated by a distance r. We will define the dipole-dipole interactions for two extreme situations. The first one is when the molecules are oriented linearly (Fig. 4.27*a*) and the other case is when orientation is parallel (Fig. 4.27 *b*). The expression for the linear orientation is given as

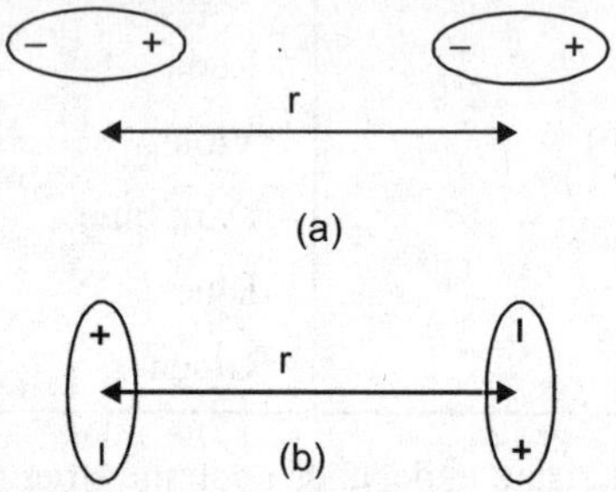

Fig. 4.27 *Schematic representation of the extreme orientation of two permanent dipoles for attraction interaction: (a) aligned linearly and (b) aligned perpendicularly.*

$$V = -\frac{2\mu_A\mu_B}{4\pi\varepsilon_0 r^3} \qquad ...(4.69)$$

and for parallel orientation the expression is

$$V = -\frac{\mu_A\mu_B}{4\pi\varepsilon_0 r^3} \qquad ...(4.70)$$

where the negative sign indicates that the interaction is attractive, *i.e.*, energy is released when these two molecules interact. Reversing the sign of one of the dipoles makes V a positive quantity, *i.e.*, the interaction between the two molecules will be repulsive. In bulk system the dipoles are oriented in all possible directions. Thus we may expect a mean value of V to be zero because there would be as many repulsions as attractions. But even under conditions of free rotation in a liquid or gaseous state, orientations giving rise to a lower potential energy are favoured over those resulting in a higher potential energy. A detailed calculations has shown that the average or net energy of interaction of permanent dipoles is given by

$$V = -\frac{2}{3}\frac{\mu_A^2\mu_B^2}{(4\pi\varepsilon_0)^2 r^6 k_b T} \qquad ...(4.71)$$

where k_b is the Boltzmann's constant and T is the temperature in absolute. Here V is inversely proportional to the sixth power of r, so that the energy of interaction falls off rapidly with distance. Also V is inversely proportional to T, because at higher temperatures the average kinetic energy of the molecules will be greater. This will not allow the favourable condition to aligning dipoles for attractive interaction. In other words, *the dipole-dipole interaction will gradually average out to be zero with increasing temperature*.

***Exercise* 4.56:** Calculate the dipole-dipole interactions when the HF molecules are oriented (*i*) linearly, (*ii*) in a parallel manner and (*iii*) in a random manner. The dipole moment of HF is 1.92 D and two HF molecules are separated by 0.4 nm.

$$\mu(\text{HF}) = 1.92\text{ D} = (1.92\text{ D})(3.3356 \times 10^{-30}\text{ C m D}^{-1}) = 6.4 \times 10^{-30}\text{ C m.}$$

The potential energy for linear orientation is given by equation (4.71) and

$$V = -\frac{2\mu_A\mu_B}{4\pi\varepsilon_0 r^3} = -\frac{(2)(6.4 \times 10^{-30}\text{ C m})(6.4 \times 10^{-30}\text{ Cm})}{(1.113 \times 10^{-10}\text{ C}^2\text{ N}^{-1}\text{m}^2)(0.4 \times 10^{-9}\text{m})^3} = -1.15 \times 10^{-20}\text{ N m}$$

$$= -1.15 \times 10^{-20}\text{ J}$$

The potential on per mol basis can be obtained by multiplying the above number by Avogadro's number, *i.e.*,

$$V = -(1.15 \times 10^{-20}\text{ J})(6.02 \times 10^{23}\text{ mol}^{-1}) = -6.9\text{ kJ mol}^{-1}$$

The potential energy for parallel orientation will be half of linear orientation, *i.e.*, –3.45 kJ mol^{-1}, whereas that of average value can be obtained by using equation (4.72),

$$V = -\frac{2}{3}\frac{\mu_A^2\mu_B^2}{(4\pi\varepsilon_0)^2 r^6 k_b T}$$

$$= -\frac{2}{3}\frac{(6.4 \times 10^{-30}\text{ C m})^4}{(1.113 \times 10^{-10}\text{ C}^2\text{N}^{-1}\text{ m}^2)^2(0.4 \times 10^{-9}\text{m})^6(1.38 \times 10^{-23}\text{J K}^{-1})(300\text{ K})}$$

$$= -532.46 \times 10^{-23}\text{ J}$$

and
$$V = -(532.46 \times 10^{-23}\text{ J})(6.02 \times 10^{23}\text{ mol}^{-1}) = 3.205\text{ kJ mol}^{-1}$$

4.19.2. Ion-Dipole Interaction

The potential energy of interaction between ion of charge q at a distance r from a molecule having dipole moment μ is given by the expression (4.72)

$$V = -\frac{q\mu}{4\pi\varepsilon_0 r^2} \quad \text{...(4.72)}$$

Equation (4.72) holds good only when the ion and the dipole lies along the same axis. This attractive interaction is mainly responsible for the dissolution of ionic compounds in polar solvents.

In the presence of ion with charge q, charge distribution in a neutral nonpolar species is distorted as shown in Fig. 4.28 and the magnitude of the induced dipole is given by the equation (4.73)

$$\mu_{\text{ind}} \propto \text{E}$$

$$\mu_{\text{ind}} = \alpha' E \quad \text{...(4.73)}$$

where α' is the proportionality constant and is called the *polarisability*. The potential energy of interaction is given by the work done in bringing the neutral particle from infinity ($E = 0$) to distance r ($E = E$) is given as

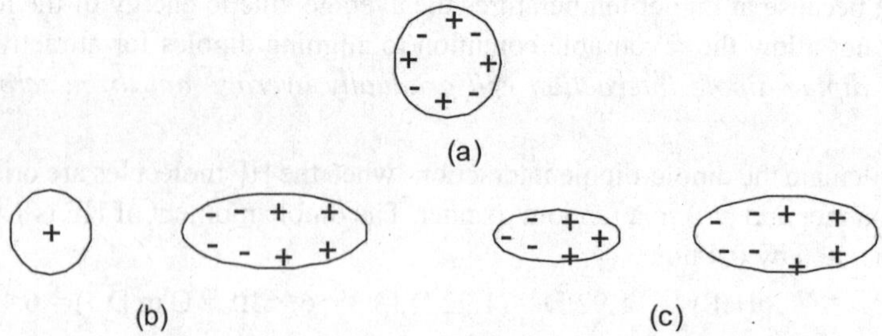

Fig. 4.28: *(a) An isolated neutral particle, having spherically symmetrical electron density, (b) induced dipole moment in neutral species by a cation and (c) induced dipole moment in a neutral species caused by a species with permanent dipole moment. The plus and minus signs depict the charges on the species.*

$$V = -\int_0^E \mu_{ind} dE = -\int_0^E \alpha' E dE = -\frac{1}{2}\alpha' E^2 \qquad ...(4.74)$$

The electric field by the ion of charge q on the atom is

$$E = \frac{q}{4\pi\varepsilon r^2}$$

So equation (4.74) reduces to

$$V = -\frac{1}{2}\frac{\alpha' q^2}{(4\pi\varepsilon_0)^2 r^4} \qquad ...(4.75)$$

Qualitatively polarisability measure how easily the electron density in an atom or in a molecule can be distorted by an external electrical field or permanent dipole moment. For example, unsaturated double bonds such as C = C, C = N, the nitro group ($-NO_2$), the phenyl group ($-C_6H_5$), the base pairs in DNA and negative charged species are highly polarisable. Generally, the larger the number of electrons and more diffuse the electron charge cloud in the molecule, the greater is the polarisability. From equation (4.73) the units of polarisability are little awkward, *i.e.*, C $m^2 V^{-1}$. For this reason, a new units for polarisbility α were derived as m^3, obtained dividing α' by the permittivity factor $(4\pi\varepsilon_0)$, *i.e.*,

$$\alpha = \frac{\alpha'}{4\pi\varepsilon_0} \qquad ...(4.76)$$

Substituting this value of α' in equation (4.75), we get

$$V = -\frac{1}{2}\frac{\alpha q^2}{(4\pi\varepsilon_0) r^4} \qquad ...(4.77)$$

Molecules with permanent dipole moment can also induce a dipole in nonpolar molecule (Fig. 4.29c). The potential energy of interaction for the dipole-induced dipole interaction is given by

$$V = -\frac{\alpha' \mu^2}{(4\pi\varepsilon_0)^2 r^6} = -\frac{\alpha \mu^2}{4\pi\varepsilon_0 r^6} \qquad ...(4.78)$$

where α is the polarisability of the nonpolar molecule and μ is the permanent dipole moment of polar molecule. It may be noted that potential energies given by equations (4.77) and (4.78) are independent of temperature. This is because the dipole moment can be induced instantaneously so the value of V is unaffected by the thermal motion of the molecules. The polarisability values of some molecules are given in Table 4.11 (this can help in the construction of more problems for the teachers).

TABLE 4.11. Polarisabilities of some molecules

Atom/Molecule	α (10^{-30} m^3)	Molecule	α (10^{-30} m^3)
He	0.22	H_2	0.911
Ne	0.40	N_2	1.97
Ar	1.66	CO_2	2.93
Kr	2.54	NH_3	2.47
Xe	4.15	CH_4	2.89
I	4.96	C_6H_6	11.6
Cs	42	CCl_4	11.7
HF	0.57	H_2O	1.65
CO	2.20	CH_3OH	3.59
HCl	2.93	CH_3Cl	5.04
HBr	4.01	CH_2Cl_2	9.46
HI	6.06		

It may be mentioned that the ion-induced dipole or dipole-induced dipole interactions are weaker than the ion-dipole or dipole-dipole interactions. This is the reason that ionic compounds like KCl and polar molecules like alcohols do not dissolve in nonpolar solvents such hydrocarbons, carbon tetrachloride or benzene etc.

***Exercise* 4.57:** Suppose Ca^{2+} ions are present in air at a distance of 0.5 nm from HF molecules with dipole moment 1.92 D. Calculate the potential energy of this interaction.

Make use of equation (4.71) and using the data

$$V = -\frac{q\mu}{4\pi\varepsilon_0 r^2} = -\frac{(2)(1.6 \times 10^{-19}\text{C})(1.92\text{D})(3.3356 \times 10^{-30}\text{ Cm D}^{-1})}{(1.113 \times 10^{-10}\text{ C}^2\text{N}^{-1}\text{m}^{-2})(0.5 \times 10^{-9}\text{ m})^2}$$

$$= -7.365 \times 10^{-20}\text{ N m} = -7.365 \times 10^{-20}\text{ J} = 44.34\text{ kJ mol}^{-1}.$$

***Exercise* 4.58:** In the Exercise (4.57) replace HF molecules by CO_2 molecules and calculate the potential energy of the ion-induced dipole interactions in kJ mol^{-1}. α (CO_2) = 2.91×10^{-30} m^3.

Using equation (4.77) and substituting the values of the constants, we get

$$V = -\frac{1}{2}\frac{\alpha q^2}{(4\pi\varepsilon_0)r^4} V = -\frac{(2 \times 1.6 \times 10^{-19}\text{ C})^2(2.91\times10^{-30}\text{m}^3)}{(2)(1.113 \times 10^{-10}\text{ C}^2\text{N}^{-1}\text{m}^{-2})(0.5 \times 10^{-9}\text{m})^4}$$

$$= -2.14 \times 10^{-20}\text{ Nm} = -2.14 \times 10^{-20}\text{ J} = 12.88\text{ kJ mol}^{-1}.$$

***Exercise* 4.59:** In Exercise 4.58, replace Ca^{2+} by HF molecules. Calculate the potential energy of the dipole-induced dipole interaction energy per mole.

Potential energy of the dipole-induced dipole interaction is obtained by using equation (4.77), *i.e.*,

$$V = -\frac{\alpha\mu^2}{4\pi\varepsilon_0 r^6} = -\frac{(2.91 \times 10^{-30}\ \text{m}^3)(1.92\ \text{D} \times 3.3356 \times 10^{-30}\ \text{C m D}^{-1})^2}{(1.113 \times 10^{-10}\ \text{C}^2\ \text{m}^{-2}\ \text{N}^{-1})(0.5 \times 10^{-9}\ \text{m})^6}$$

$$= 6.863 \times 10^{-23}\ \text{Nm} = 6.863 \times 10^{-23}\ \text{J} = 41.31\ \text{kJ mol}^{-1}.$$

4.19.3. Induced Dipole-Induced Dipole Interactions

Induced dipole-induced dipole interactions involve the interactions between nonpolar molecules. As we know that all the neutral molecules consists of equal number of positive and negative charges and on the average over a certain period of time (for example, a time long enough for us to carry out measurement on the system), the electron density at a certain fixed distance away from the nucleus is the same in every direction. Now if we take snapshots of instantaneous configuration of each individual nonpolar species we would most likely find varying degrees of deviation from spherical symmetry, owing to interactions between the particles. This temporary dipole created at every instant can induce a dipole in its neighbouring atom(s) so an attractive interaction will result. Thus it is expected that these interactions will be weak, and are confirmed by the low boiling points of the nonpolar molecules. However, for molecules with large polarisabilities, this interaction can be comparable to or even greater than dipole-dipole or dipole-induced dipole interactions. For example, carbon tetrachloride (CCl_4), nonpolar molecule with quite large polarisability, possesses a considerable higher boiling point (76.5°C) than that of methyl fluoride (CH_3F), a polar molecule (–141.8 °C).

The expression for this interaction was first derived by German physicist Fritz London, using quantum mechanical treatment and potential energy arising from the interaction of two identical atoms or nonpolar molecules is given by

$$V = -\frac{3}{4}\frac{\alpha'^2 I}{(4\pi\varepsilon_0)^2 r^6} = -\frac{3}{4}\frac{\alpha^2 I}{r^6} \qquad \text{...(4.79)}$$

where I is the first ionisation energy of the atom or molecule. For unlike atoms or molecules A and B equation (4.79) reduces to

$$V = -\frac{3}{2}\frac{I_A I_B}{I_A + I_B}\frac{\alpha'_A \alpha'_B}{(4\pi\varepsilon_0)^2 r^6} = -\frac{3}{2}\frac{I_A I_B}{I_A + I_B}\frac{\alpha_A \alpha_B}{r^6} \qquad \text{...(4.80)}$$

Forces which arise from this kind of interaction are also called *dispersion*, or *London forces*.

Dipole-dipole interactions, dipole-induced dipole interactions and induced dipole-induced dipole interactions are collectively called *van der Waals forces*.

***Exercise* 4.60:** Calculate the potential energy of interaction between two helium atoms separated by a distance of 0.5 nm. Given α (He) = 0.2×10^{-30} m³, I = 2373 kJ mol⁻¹

Using equation (4.78)

$$V = -\frac{3}{4}\frac{\alpha'^2 I}{(4\pi\varepsilon_0)^2 r^6} = -\frac{3}{4}\frac{(0.2 \times 10^{-30}\text{m}^3)^2(2373\ \text{kJ mol}^{-1})}{(0.5 \times 10^{-9}\text{m})^6} = 4.7 \times 10^{-3}\ \text{kJ mol}^{-1}$$

This value is very small. This is the reason that the boiling point of helium is only 4 K.

***Exercise* 4.61:** Calculate the induced dipole-induced dipole interactions between C_6H_6 and Xe when the species are separated by 0.5 nm. α (Xe) 4.15×10^{-30} m³, α (C_6H_6) = 10.4×10^{-30} m³, I(Xe) 1170.4 kJ mol⁻¹, I(C_6H_6) = 890.4 kJ mol⁻¹.

Using equation (4.79)

$$V = -\frac{3}{2}\frac{I_A I_B}{I_A + I_B}\frac{\alpha_A \alpha_B}{r^6}$$

$$= -\frac{3}{2}\frac{(1170.4 \text{ kJ mol}^{-1})(890.4 \text{ kJ mol}^{-1})}{(1170.4 + 890.4) \text{ kJ mol}^{-1}}\frac{(4.15 \times 10^{-30} \text{m}^3)(10.4 \times 10^{-30} \text{m}^3)}{(0.5 \times 10^{-9} \text{m})^6}$$

$$= 2.096 \text{ kJ mol}^{-1}.$$

4.19.4. Repulsive and Total Interactions

In addition to the attractive forces, atoms and molecules also repel each other, otherwise they would eventually fuse. Fusion is prevented by the strong repulsive forces between electron clouds and between nuclei. The repulsion between atoms or molecules is a direct consequence of the Pauli exclusion principle, which prevents electrons from sharing the same region in space. *It is proportional to $1/r^n$, where n is between 8 and 12*. The expression for the total interactions to represent the attractive and repulsive interaction between non-ionic system was given by British physicist Sir John Edward Lennard-Jones, *i.e.*,

$$V = -\frac{A}{r^6} + \frac{B}{r^{12}} \quad \text{...(4.81)}$$

where A and B are constants for two interacting atoms or molecules. First term represents the attractive potential and second term represents the repulsive potential, *i.e.*, repulsive potential is effective over much shorter distance in comparison to that of attractive potential. A more common form of equation (4.81) is Lennard-Jones 6-12 potential, given by

$$V = 4\varepsilon\left[\left(\frac{\sigma}{r}\right)^{12} - \left(\frac{\sigma}{r}\right)^{6}\right] \quad \text{...(4.82)}$$

where ε is the depth of the potential well and σ is the separation at which $V = 0$. Values of σ and ε of some molecules are compiled in Table 4.12.

TABLE 4.12. Lennard-Jones parameters for few atoms and molecules

Species	σ (nm)	ε (kJ mol^{-1})
Ar	0.340	0.997
Xe	0.410	1.77
H_2	0.293	0.307
N_2	0.392	0.765
O_2	0.365	0.943
CO_2	0.433	1.65
CH_4	0.382	1.23
C_6H_6	0.860	2.02

The value of σ gives a measure of how closely two non-bonded atoms can approach each other. One half of internuclear distance between two non-bonded atoms in crystal is called *van der Waals radius*. Van der Waals radii of some atoms and few molecular species are compiled in Table 4.13. Lastly it may be mentioned here that magnitude of the intermolecular potential energy (van der Waals forces) is much smaller than the magnitude of the intramolecular potential energy (bond energy), for example, the depth of the intermolecular potential well is 0.3 kJ mol^{-1}, whereas intramolecular depth of the potential well is 432 kJ mol.$^{-1}$ Further different kind of interactions between molecules, their order of magnitude is summarised in Table 4.13.

TABLE 4.13. van der Waals radii of some atoms and methyl group

Atom	H	C	N	O	P	S	F	Cl	Br	I	$-CH_3$
Radius (nm)	0.12	0.15	0.15	0.14	0.19	0.185	0.135	0.18	0.195	0.22	0.20

TABLE 4.14. Interaction between molecules and their range of magnitudes, actual values depend on the distance between the molecules

Type of Interaction	Distance dependence	Example	Order of magnitude (kJ mol^{-1})
Covalent	No simple expression	H – H	200-800
Ion-ion	$\frac{q_A q_B}{4\pi\varepsilon_0 r}$	K^+	40-400
Ion-dipole	$\frac{q\mu}{4\pi\varepsilon_0 r^2}$	$K^+(H_2O)_n$	5-60
Dipole-dipole	$\frac{2}{3}\frac{\mu_A^2\mu_B^2}{(4\pi\varepsilon_0)^2 r^6 k_b T}$	SO_2 SO_2	0.5-15
Ion-induced dipole	$\frac{1}{2}\frac{\alpha q^2}{4\pi\varepsilon_0 r^4}$	K^+ C_6H_6	0.4-4
Dipole-induced dipole	$\frac{\alpha\mu^2}{4\pi\varepsilon_0 r^6}$	HCl C_6H_6	0.4-4
Dispersion	$\frac{3}{4}\frac{\alpha^2 I}{r^6}$	CH_4 CH_4	4-40
Hydrogen bond	No simple expression	H_2O.....H_2O	4-40

4.19.5. Hydrogen Bonding

The hydrogen bonding is a special kind of interaction between molecules. It forms whenever a polar bond containing the hydrogen atom (*e.g.*, O – H, N – H) interacts with an electronegative atom such as nitrogen, oxygen or fluorine. This interaction is represented as $A - H \cdots B$, where A and B are the electronegative atoms and the dotted line denotes the hydrogen bond. Figure 4.29 depicts several examples of hydrogen bonding. Although bonding are relatively weak (about 40 kJ mol $^{-1}$), they play a central role in determining the properties of many compounds. Few characteristics which are observed due to hydrogen bonding are:

(*i*) In general the boiling points increases with increase in molar mass, but in case of binary hydrogen compounds of group 15, 16 and 17 elements do not follow this trend. In each of these series, the lightest compounds (NH_3, H_2O and HF) have the highest boiling points, because there is extensive hydrogen bonding between molecules of these compounds.

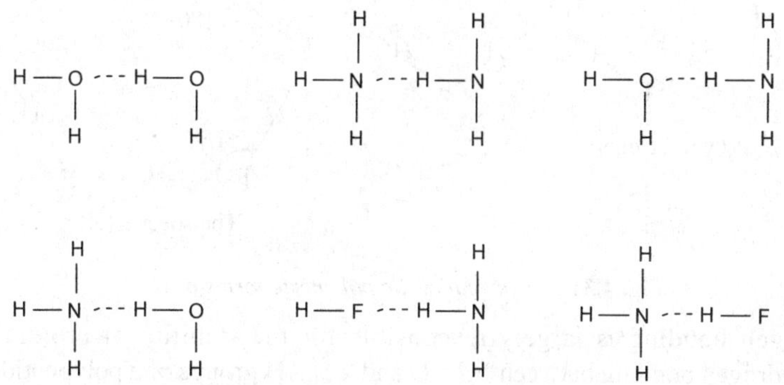

Fig. 4.29 *Some examples of hydrogen bonding. The dotted lines represent the hydrogen bond.*

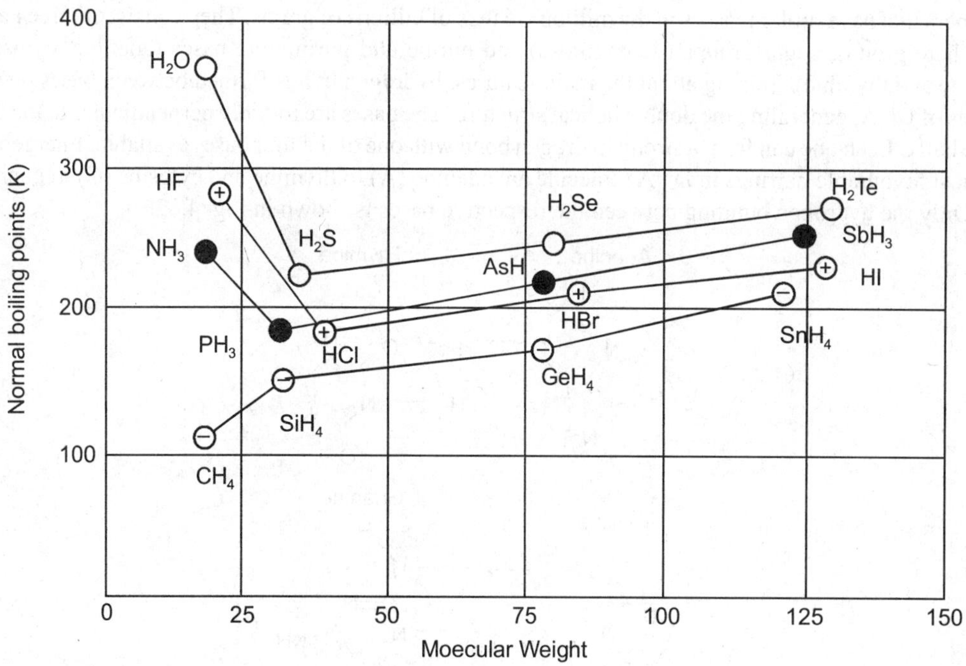

Fig. 4.30 *Correlation between the molecular weight and boiling points of the group 14, 15, 16 and 17 hydrides.*

(*ii*) In molecules containing electronegative atom, the only electron on hydrogen atom is pulled towards the electronegative element, thus leaving the hydrogen atom partially unshielded. Consequently, its proton can interact directly with another electronegative atom on a different molecule. Depending on its strength, hydrogen bonding can exist in gas phase, besides in the solid and liquid phase. In solids and liquids, HF forms a polymeric as well as cyclic chain of few molecules, as shown in Fig. 4.31. For maximum stability the donor pair (AH) and the acceptor (B) are usually collinear (*i.e.*, $\angle AHB = 180^o$), but deviations up to 30° are known.

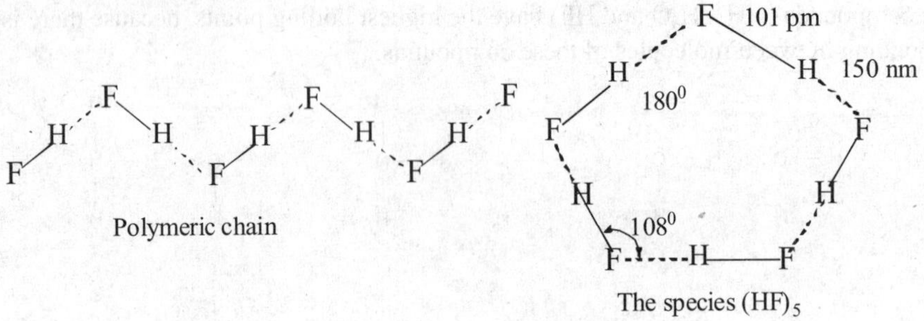

Fig. 4.31 *Linear and cyclic polymeric forms of HF.*

(*iii*) Hydrogen bonding is largely responsible for the stability of protein conformations. Intramolecular hydrogen bonding between >C = O and > N – H groups of a polypeptide chain results in the α-helix. On the other hand, intermolecular hydrogen bonding between two polypeptide chains accounts for β-pleated-sheet structures. Hydrogen bonding in DNA is of great importance. DNA molecules are polymers that have molar masses in the millions to tens of billions of grams. They consist of three parts: phosphate groups, sugar groups (deoxyribose) and purine and pyrimidine bases (adenine, cytosine, guanine and thymine). Talking about the main features, hydrogen bonds formed between bases on two strands of DNA, generating the double helical structure. The bases are roughly perpendicular to the axis of the helix. Each one can form a strong hydrogen bond with one of the four bases available. Energetically, the most favourable pairings in DNA molecule are adenine (A) to thymine and cytosine (C) to guanine (G). Only the hydrogen bonding between the respective bases is shown in Fig. 4.32.

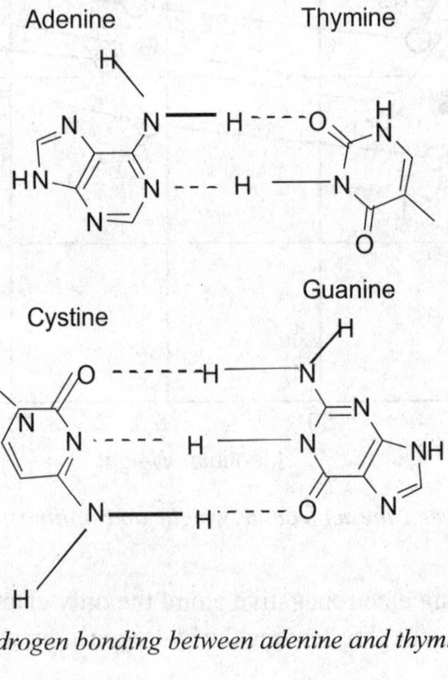

Fig. 4.32 *Intermolecular hydrogen bonding between adenine and thymine: cystine and guanine.*

So far we have talked the hydrogen bonding taking place between the strong electronegative atoms, like N, O and F. Ample evidence suggests that hydrogen bonds also exist in compounds showing weak hydrogen bonds as shown in Fig. 4.33. For example electron rich triple bond in acetylene can form hydrogen bond with HF. Similarly the delocalisation in benzene can form weak hydrogen bonds.

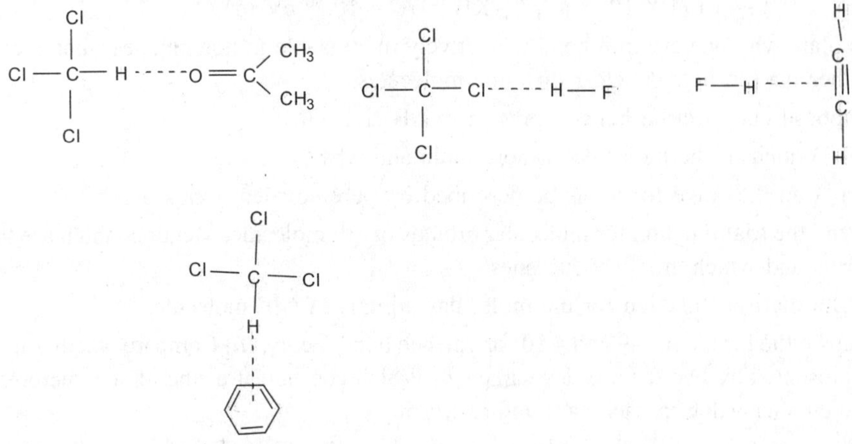

Fig. 4.33 *Molecules showing weak hydrogen bonding.*

PROBLEMS

Ionic Bond

4.1 Show that the dissociation of RbF into rubedium and fluorine atoms is more favourable than into Rb^+ and F^- ions. IP(Rb) = 4.176 eV, EA(F) = 3.47 eV.

4.2 Calculate the values of the constants *a* and *b* of equation (4.1) for RbF molecule from the following data: $\bar{\nu}$ = 373.3 cm^{-1}, R_e (RbF) = 2.270 Å, IP (Rb) = 4.176 eV, EA(F) = 3.47 eV.

4.3 Calculate the dissociation energy of RbF from the data given in Problem 4.2.

4.4 (*a*) Calculate the values of the constants *a* and *b* of equation (4.1) for KI molecule from the following data: $\bar{\nu}$ = 186.53 cm^{-1}, R_e (KI) = 3.0478×10^{-10} m; IP(K) = 4.339 eV, EA(I) = 3.063 eV.

(*b*) Calculate the dissociation energy of KI.

4.5 Equation (4.1) can be modified by including terms like

(*a*) van der Waals forces ($-C_6/R^6$).

(*b*) Polarizabilities terms for both + ve and – ve ions $\left(-\frac{1}{2}e^2\frac{(\alpha_+ + \alpha_-)}{(4\pi\varepsilon_0)R^4} - \frac{2e^2\,\alpha_+\alpha_-}{(4\pi\varepsilon_0)^2\,R^7}\right)$, *i.e.*

equation (4.1) becomes

$$U(R) = -\frac{Z_A Z_B e^2}{(4\pi\varepsilon_0)R} + a\,e^{-br} - \frac{C_6}{R^6} - \frac{e^2(\alpha_+ + \alpha_-)}{2(4\pi\varepsilon_0)R^4} - \frac{2e^2\alpha_+ + \alpha_-}{R^7(4\pi\varepsilon_0)^2} + \Delta\,U(\infty).$$

Calculate the dissociation energy of RbF given $C_6(\text{RbF}) = 19 \times 10^{-60}$ eV m^6, $\alpha(\text{Rb}^+) = 1.797 \times 10^{-30}$ m^3, $\alpha(\text{F}^-) = 0.759 \times 10^{-30}$ m^3. Rest of the data is in Problems 4.1 and 4.2. Compare the values of the D_e in both cases.

4.6 Repeat the Problem (4.5) for KI with part of data from Problem (4.4). $\alpha(\text{K}^+) = 1.201 \times 10^{-30}$ m^3, $\alpha(\text{I}^-) = 6.194 \times 10^{-30}$ m^3, $C_6(\text{KI}) = 76 \times 10^{-60}$ eV m^6.

4.7 Explain why one electron bond is relatively more stable in homonuclear diatomic molecules rather than in heteronuclear diatomic molecules.

4.8 Suppose one electron bond is present in AB and AA;

(*a*) Which of the molecule is more ionic and why?

(*b*) Can the wave functions be described by pure covalent terms only?

4.9 Write the trial function for molecular orbitals of HF molecule. Mention which are the covalent terms and which are the ionic ones.

4.10 Write the trial function for the molecular orbitals of CIF molecule.

4.11 Repeat the Problems 4.9 and 4.10 for valence bond theory, (*a*) Compare whether the functions represented by two theories are same, (*b*) Which coefficient of the ionic structures will have the smaller value in each case and justify it.

4.12 What are the factors those determine the importance of the ionic configurations in the heterodiatomic molecules?

4.13 HF molecule is more ionic than HCl despite the electron affinity of fluorine is less than that of chlorine.

4.14 Among HF, HCl, HBr and HI, which will have the highest coulombic interaction?

4.15 If the molecular axis in diatomic molecules coincides with the *x*-axis, which of the following combinations of the atomic orbitals will lead to the formation of molecular orbitals?

(*i*) s and p_x, (*ii*) s and p_y, (*iii*) p_x and p_z or p_y, (*iv*) p_z and p_z,
(*v*) d_{z^2} and p_x, (*vi*) d_{z^2} and $d_{z^2-y^2}$, (*vii*) $d_{x^2-y^2}$ and s, (*viii*) d_{xy} and d_{yz},
(*ix*) d_{xy} and p_y, (*x*) p_x and d_{xy}.

4.16 Draw the shapes of the antibonding molecular orbitals formed by the overlap of

(*i*) s and p_x atomic orbitals (*ii*) p_x and p_x atomic orbitals.

4.17 Write the electronic configuration for the following molecules:

(*i*) N_2, (*ii*) He_2, (*iii*) Cl_2, (*iv*) LiH, (*v*) CN,
(*vi*) SO, (*vii*) CIF, (*viii*) CIO, (*ix*) NO.

4.18 Find the bond order in each of the molecule mentioned in Problem 4.17.

4.19 Draw the molecular orbital energy level diagram for each of the molecule mentioned in Problem 4.17, clearly mentioning the relative energies of the combining atoms before forming molecular orbitals.

4.20 The electronic configuration of atoms, *A*, *B*, *C* and *D* are:

A: $1s^2\,2s^2\,2p^5$ *B*: $1s^2\,2s^2\,2p^6\,3s^2\,3p^6\,4s^1$ *C*: $1s^2\,2s^2\,2p^3$ *D*: $1s^2\,2s^2\,2p^6\,3s^2\,3p^6$

(*a*) Write the molecular formula of the molecules formed by the combination of atoms (*i*) *A* and *B*, (*ii*) *A* and *C*, (*iii*) *C* and *C* and (*iv*) *D* and *D*.

(*b*) Write the electronic configuration of each of the molecule and thus calculate the bond order.

4.21 The ground state of molecular oxygen is paramagnetic (*i.e.* presence of unpaired electrons.) Show that it is so on the basis of molecular orbital theory (one needs more detailed valence bond wave function than mentioned in this book).

4.22 In the following molecules and molecular ions, (*a*) which species has the highest bond energy, (*b*) which species has the highest bond order and (*c*) draw the moleculor orbital energy level diagram.

(*i*) OF, OF^-, OF^+ (*ii*) CN^+, CN, CN^- (*iii*) O_2, O_2^-, O_2^+, O_2^{2+}.

4.23 Find the molecular term symbols for the molecules mentioned in Problem 4.17.

4.24 Find the molecular term symbols for the molecules and molecular ions mentioned in the Problem 4.22.

4.25 The spectroscopic term symbol for the ground state of B_2 and C_2 are $^3\Sigma^-$ and $^1\Sigma$ respectively. Write the electronic configurations compatible with the term symbols.

Electronegativity

4.26 Calculate the electronegativity of fluorine atom with Pauling's method from the following data. Bond dissociation energies of H_2, F_2 and HF are 435.14 kJ mol^{-1}, 158.99 kJ mol^{-1} and 568.19 kJ mol^{-1} respectively, and $\chi_H = 2.1$.

4.27 The ionization energy and electron affinity for fluorine atom are 1680.29 kJ mol^{-1} and 349.36 kJ mol^{-1} respectively. Calculate the electronegativity of fluorine atom using Mulliken's method and compare this value with that obtained in Problem 4.26.

4.28 Using the data of the Problems 4.26 and 4.27, calculate the percentage ionic character in HF molecule. Compare your results with that obtained in Problem 4.33.

4.29 The bond dissociation energies of Cl_2, Br_2 and I_2 are 238.91 kJ mol^{-1}, 192.46 kJ mol^{-1} and 151.04 kJ mol^{-1} respectively. The bond dissociation energies of HCl, HBr and HI are 431.79 kJ mol^{-1}, 366.1 kJ mol^{-1} and 298.74 kJ mol^{-1}. If bond dissociation energy of H_2 is 435.14 kJ mol^{-1} and $\chi_H = 2.1$, calculate the electronegativities of chlorine, bromine and iodine atoms. Also find the percentage ionic character in HCl, HBr and HI molecules.

4.30 The electronegativities of chlorine and bromine atoms are 3.0 and 2.8 respectively. If the bond dissociation energies of Cl_2 and Br_2 are 238.91 kJ mol^{-1} and 192.46 kJ mol^{-1}, calculate the dissociation energy of BrCl molecule.

4.31 The electronegativities of sodium and hydrogen atoms are 0.9 and 2.1 respectively. If the bond dissociation energies of H_2 and NaH are 435.14 kJ mol^{-1} and 201.67 kJ mol^{-1}, calculate the bond dissociation energies of disodium molecule.

Dipole Moment and Ionic Character

4.32 NO and CO molecules have very small dipole moment. Account for this with the help of different resonating structures.

4.33 The equilibrium bond distances in the hydrogen halides, *i.e.* HF, HCl, HBr and HI are 0.092 nm, 0.128 nm, 0.143 nm and 0.162 nm respectively. If the actual dipole moment of HF, HCl, HBr and HI are 1.98D, 1.03D, 0.79D and 0.38D, calculate the percentage ionic character.

4.34 In the valence bond theory, the wave function of molecule can be written as $\psi = \psi_{cov} + \lambda\psi_{ion}$. Calculate the value of λ for the hydrogen halides from the percentage ionic character obtained in Problems 4.28 and 4.29.

4.35 The linear molecule HCN has a dipole moment 2.95D. Calculate the charges on H, C and N atoms compatible with this dipole moment if δ^+ (charge) on hydrogen atom is 10^{-19} C. Given $\chi_H = 2.1$, $\chi_C = 2.5$ and $\chi_N = 3.0$. $r_{H\text{–}C} = 0.12$ nm, $r_{CN} = 0.15$ nm.

4.36 The dipole moment of NH_3 molecule is 1.46D. If the angle ∠HNH = 108°, calculate the dipole moment of N–H bond.

4.37 The dipole moment of bonds C–H and C–Cl are 0.4 D and 1.5 D respectively. Calculate the dipole moments of CH_3C1, CH_2C1_2 and $CHC1_3$ molecules. Compare the calculated values with experimental values, *i.e.* $\mu_{CH_3Cl} = 1.85D$, $\mu_{CH_2Cl_2} = 1.59$ D and $\mu_{CHCl_3} = 1.15$ D.

4.38 The measured dipole moment of CsF is 7.884 D. If the equilibrium bond distance is 0.235 nm, calculate the percentage ionic character of the CsF bond, the value of λ (the weighted factor of ionic function), the value of $\chi_F - \chi_{Cs}$ and hence the electronegativity of Cs atom provided $\chi_F = 4.0$.

4.39 Which of the molecules mentioned below possess permanent dipole moment?

(*i*) SO_4^{2-}, (*ii*) $C1O_2$, (*iii*) C_2H_2, (*iv*) 1, 3, 5-trinitro-benzene, (*v*) SO_2.

Hybridisation

4.40 In isolated atoms, do the hybrid atomic orbitals exist? Explain briefly.

4.41 (*a*) What kinds of orbitals are the followings?

$$\psi = \frac{1}{\sqrt{2}}(\psi_{2s} + \psi_{2p_x}),\ \psi' = \frac{1}{\sqrt{2}}(\psi_{2s} - \psi_{2p_x})$$

(*b*) How many bond/bonds can an each orbital form?

4.42 Calculate the coefficients of the *sp*, sp^2, dsp^2, dsp^3 and d^2sp^3 hybrid atomic orbitals by the procedure mentioned in Section 4.12.

4.43 Draw the relative energy level diagram for the atomic orbitals, hybrid orbitals and the bonded hybrid orbitals.

4.44 What are the major factors those control the angle (XAX) in the polyatomic molecules?

4.45 Explain the trend of angles observed in the following compounds

(*a*) H_2O, ∠HOH = 104.5°, $r_{O\text{–}H} = 0.096$ nm;

(*b*) OF_2, ∠FOF = 103.2, $r_{OF} = 0.142$ nm;

(*c*) OCl_2, ∠ClOCl = 111°, $r_{OCl} = 0.17$ nm;

(*d*) O $(CH_3)_2$, $\angle CH_3OCH_3 = 111°$, $r_{O\text{–}CH_3} = 0.142$ nm.

4.46 Explain the trend of angles observed in the following compounds even though the central atom possesses sp^3 hybridisation:

(*i*) H_2O, ∠HOH = 104.5°; (*ii*) NH_3, ∠HNH = 107.3° and

(*iii*) CH_4, ∠HCH = 109.5°.

4.47 The angles HXH in NH_3 and PH_3, are 107.3°, 93° respectively. Explain briefly. Does the importance of hybridisation decrease as one goes down the group?

4.48 In the compounds NH_3, NF_3, $N(CH_3)_3$, NH_4^+ and $ON(CH_3)_3$, the angles ∠XNX are 107.3°, 102.1°, 108°, 109.5° and 114° respectively. Explain briefly.

4.49 BX_3 compounds can easily form coordination complexes or act as a powerful lewis acids. Explain briefly.

4.50 Write the M.O. description of the following compounds:
(*a*) H_2O, (*b*) C_2H_2, (*c*) N_2F_2, (*d*) CH_4,
(*e*) C_2H_4, (*f*) C_2H_6.

4.51 Write the resonance structures of the following molecules:
(*i*) CO, (*ii*) NO, (*iii*) NO_2, (*iv*) NO_3^-, (*v*) CO_3^{2-},
(*vi*) (cyanogen azide) NCNNN, (*vii*) CH_3COO^-, (*viii*) N_2^-, (*ix*) ClO_2^-,
(*x*) PO_4^{3-}, (*xi*) O_3, (*xii*) SO_4^{2-}.

4.52 Calculate the percentage of *s* and *p* atomic orbitals used in the formation of hybrid orbitals which are used in bond formation and lone pairs, in the compounds mentioned in Problems 4.45, 4.46 and 4.48.

4.53 Explain briefly why N_2F_2 is planar whereas H_2O_2 is not?

4.54 Is there any similarity between $H_2N–NH_2$ and $CH_3–CH_3$? Which one of the following ions $N_2H_5^+$ and $N_2H_6^{2+}$ are closely related to C_2H_6?

4.55 What will be the shapes of the following molecules on the basis of valence shell electron pair repulsion (VSEPR) model? Also mention the hybridisation on the central atom,
(*i*) IF_2^-, (*ii*) ICl_2^-, (*iii*) XeF_2, (*iv*) ClF_3, (*v*) $TeCl_4$,
(*vi*) XeF_4, (*vii*) ICl_4^-, (*viii*) IF_4^-, (*ix*) PCl_5, (*x*) IF_5,
(*xi*) $SbCl_5\ (CH_3)_3$, (*xii*) SF_6, (*xiii*) XeF_6, (*xiv*) IF_7,
(*xv*) N_3^-, (*xvi*) ClO_3^-, (*xvii*) PO_4^{3-}, (*xviii*) POF_3, (*xix*) SeF_4,
(*xx*) CH_3COO^-,(*xxi*) XeO_3, (*xxii*) $XeOF_3$, (*xxiii*) XeO_4, (*xxiv*) XeO_4^{3-}.

4.56 From the following set of molecules, which are the linear one and which are the bent? Give reasons:
(*a*) CS_2, (*b*) ClO_2, (*c*) C_3, (*d*) CSTe, (*e*) OCl_2,
(*f*) $CdCl_2$, (*g*) $SnCl_2$, (*h*) ONCl, (*i*) CN_2^{2-}, (*j*) S_3^{2-}.

4.57 What will be the shapes of the following tetratomic molecules or ions?
(*i*) BO_3^{3-}, (*ii*) PCl_3, (*iii*) BrO_3^-, (*iv*) AsH_3, (*v*) OH_3^+,
(*vi*) SeO_3^{2-}, (*vii*) SbF_3.

4.58 What will be the shape of the following molecules?
(*i*) $IBrCl^-$, (*ii*) $Tl(CH_3)_2^+$, (*iii*) $Pb(CH_3)_4$, (*iv*) $SbCl_5$, (*v*) $[TlCl_5H_2O]^{2-}$,
(*vi*) $BiCl_6^{3-}$, (*vii*) $F_6S_2F_6$, (*viii*) TeF_6.

4.59 Calculate the formal charges if any on the central atoms present in the molecules mentioned in the Problem 4.60.

4.60 Draw the resonating structures of the following molecules or ions:
(*i*) N_3^-, (*ii*) CO_3^{2-}, (*iii*) IO_3^-, (*iv*) HCN, (*v*) NO_2Cl,
(*vi*) ClO_3^-, (*vii*) SO_2, (*viii*) $(CH_2)_3NO$, (*ix*) O_3 and (*x*) FNO.

Hückel Molecular Orbital Theory

4.61 Using Hückel molecular orbital theory, calculate the energy levels, wave functions, electron density and bond order for the molecule $CH_2 = CH–CH = CH_2$. Compare the energy of the longest wavelength transition with that obtained in the free electron and modified free electron model.

4.62 Apply the Hückel molecular orbital theory to cyclobutadiene to calculate the parameters mentioned in Problem 4.61. Show with the help of resonance energy that this is more stable than butadiene.

4.63 Calculate the total electronic energy, wave functions for the molecular orbitals, electron density and bond order for each of the following:

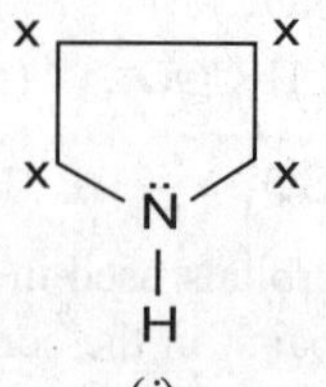

(*i*)

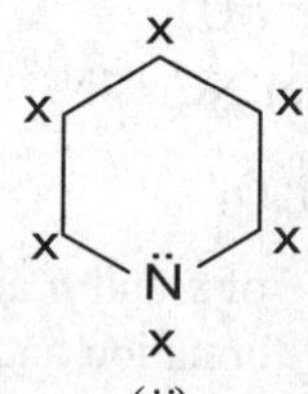

(*ii*)

The cross at each carbon atom shows the p orbital to form π bond. In the Problem 4.63 (*i*) and (*ii*) one can take $\alpha_N = \alpha_C + \frac{3}{2}\beta_C$ and $\alpha_N = \alpha_C + \frac{1}{2}\beta_C$ respectively. $\beta_N = \beta_C$.

4.64 In a chemical reaction

$$A + B_2 \rightleftharpoons \underset{\text{complex}}{\text{activated}} \rightarrow AB + B$$

The activated complex can be a linear or a cyclic triatomic molecule. Consider that each electron on atoms A and B is a part of $p\pi$ electron. Find out which of the complex is more stable?

4.65 Which of the following molecules will have aromatic character?

(*i*) $C_3H_3^+$, (*ii*) C_3H_2, (*iii*) C_4H_4, (*iv*) C_5H_5, (*v*) $C_5H_5^-$, (*vi*) $C_{10}H_8$, (*vii*) $C_7H_7^+$, (*viii*) $C_{14}H_{10}$.

4.66 In the monosubstituted ethylene, the substituted group contains a lone-pair and the structure can be

$$\ddot{X} - \dot{C} - \dot{C}$$

α_x can be written as $\alpha_x = \alpha_c + \delta\beta$ and $\beta_x = \beta$ (α_c is the Coulomb integral for carbon and β is the exchange integral, α_x and β_x are the corresponding terms for x). Solve the above problem by Hückel molecular orbital theory for two extreme cases: (*i*) $\delta = 0$, and (*ii*) $\delta = \infty$. Calculate the frequency of the longest wavelength transition for each case.

Conjugated Dienes

4.67 Apply the free electron model to butadiene molecule to calculate the wavelength of longest wavelength transition. Also apply the modified free electron model. Compare this with the transition at 210 nm obtained experimentally. The distance between the carbon—carbon atoms in benzene is 1.39×10^{-10} m.

4.68 Calculate $\bar{\nu}$ absorption for the lowest energy band of the molecules of type: H– $(CH = CH)_x$ – H when $x = 1, 8$; using free electron model and modified free electron model. Compare with the experimentally observed $\bar{\nu}$, *i.e.* for $x = 1$, $\bar{\nu} = 61{,}500\ cm^{-1}$, $x = 8$, $\bar{\nu} = 3289.47\ cm^{-1}$.

4.69 In the trans β-carotene, there are 22 carbon atoms containing $p\pi$ electrons. If the λ for longest wavelength band is 451 nm, calculate the length of the box as well as the bond length of each bond.

4.70 If one applies free electron model to benzene molecule (*i.e.* moving in circular ring), calculate $\bar{v}$ of the lowest energy transition if radius = 1.39×10^{-10} m. Compare it with the 256 nm band.

Coordinate Complexes

4.71 According to crystal field theory, arrange the 'd' electrons of the transition metal ions and indicate whether they are high spin or low spin complexes:

(*i*) $K_3[Mn(CN)_6]$, (*ii*) $K_3[Mn(OX)_3]$. $3H_2O$, (*iii*) Mn $(Py)_6Br_2$,
(*iv*) $[Cr(dipy)_3]$ $Br_2.4H_{11}O$, (*v*) $(H_4N)_3$ $[FeF_6]$, (*vi*) $[Fe(NH_3)_6]Cl_2$,
(*vii*) $[Co(NH_3)_6]Cl_3$, (*viii*) $K_3[CoF_6]$, (*ix*) $[Co(NH_3)_6]$ $(C10_4)_2$,
(*x*) $[Ni(NH_3)_6]Cl_2$, (*xi*) $[Ni(Br_3)(C_2H_5)_3P]$, (*xii*) K_3CuF_6.

4.72 Calculate the magnetic moment associated with each compound in Problem 4.71.

4.73 Explain the structure of each compound listed in Problem 4.72 on the basis of valence bond picture, indicating whether they are innerfield or outerfield complexes. Make sure that your results are consistent with the magnetic moments obtained in the Problem 4.72.

4.74 It has been found that following coordinate complexes are diamagnetic. Explain the structure and arrange the electrons both by crystal field theory and valence bond picture, (*i*) $K_2[Ni(CN)_4]$, (*ii*) $[Cu(H_2O)_4]^{2+}$, (*iii*) $[Au(CN)_4]^-$, (*iv*) [Ni $(dmg)_2$], (*v*) $[Cu(acac)_2]$.

4.75 Pd (II) and Pt (II) transition metal ions can form either tetrahedral or square planar complexes with four ligands. If for certain ligands the splittings among *d* orbitals is always greater than the pairing energies of the electrons, which complexes will have greater magnetic moment?

4.76 $[Ni(CN)_4]^{2-}$ complex ion is diamagnetic whereas $[Ni(NH_3)_4]^{2+}$ SO_4 is a paramagnetic. Find the structure of each complex if the *d* orbital splitting by the crystal field is always larger than the pairing energy of electrons.

4.77 $Pt(NH_3)_2Cl_2$ is a square planar complex. Arrange the ligands in such a manner that (*a*) complex is non-polar, (*b*) complex is polar.

4.78 Tetrahedral complexes are generally high spin as compared to octahedral complexes. Explain briefly.

Addition Problems

4.79 Consider an electron around the nucleus in the spherically symmetric potential well of radius R_0 where the potential inside the well ($r \leq R_0$) is zero and outside the well is infinite. (*a*) Show that electron cannot reside in the nucleus. Plot the total energy as a function of R_0, (*b*) Calculate the minimum value of R_0 as well as the total minimum energy.

4.80 Derive the normalisation constant for the $\psi_{\text{anti-bonding}}$ wave function.

4.81 Similar to equation (11) of Exercise (4.8) show that the energy of anti-bonding orbital of H_2^+ can be written as: $\Delta E_- = E_- - E_{1s} = \dfrac{J-K}{1+S}$.

4.82 Consider H_2 molecule. Calculate the average value of (*a*) kinetic energy, (*b*) potential energy and (*c*) total energy. Plot these energies as a function of distance between the nuclei.

4.83 Using the values of integrals, H_{aa}, H_{ab} and S_{ab} from equations (4.14), (4.15) and (4.16), derive the expression for the energy levels (4.13 *e*) and (4.13*f*).

4.84 Based on the arguments given or using Table 4.5 data, construct qualitatively the molecular wave function of (*i*) NO, (*ii*) CO, (*ii*) FO molecules.

4.85 Disulphide bond plays an important role in the three dimensional structure of protein molecules. Discuss the nature of the –S–S– bond.

4.86 Consider HF molecule and let ψ_{1s} and ψ_{2p} are the wave functions of the 1*s* orbital of hydrogen atom and 2*p* orbital of fluorine atom respectively. Based on the valence bond theory, write the ground state wave function of (*i*) structure which involves a complete transfer of electron from the hydrogen atom to fluorine atom, (*ii*) structure which involves pure covalent bond and (*iii*) structure in which bond is polar.

4.87 Carbene or methylene (CH_2) molecule, isolated as an intermediate in some chemical reactions, has been shown to be unstable by spectroscopic techniques. Suggest the type of bonding present in this molecule and how would you determine which type of bonding is present in CH_2.

4.88 Write the Hamiltonian operator for (*i*) H_2^+ and (*ii*) H_2 molecule.

4.89 Write down the Hamiltonian operator for H_2O molecule. How many of these are the (*i*) one electron terms, (*ii*) two electron terms and (*iii*) zero electron terms?

4.90 Considering the Exercise 4.5, suggest under what condition molecular orbital wave function (equation 4.12) and wave function given by valence bond theory are identical?

4.91 By calculating the bond order He_2^+ and He_2, indicate which of the species is more stable and can exist.

4.92 The bond length and bond dissociation energies of the species H_2^+, H_2, He_2^+ and He_2 are 106, 76, 108 and ~ 6000 pm and 268, 457, 241 and < < 1 kJ mol^{-1}. Explain these results by calculating the bond orders.

4.93 Consider the set of molecules. (*i*) F_2 and F_2^+, (*ii*) He and He_2^+ and (*iii*) NO^+, NO and NO^-. Speculate which of the species will have the longest bond length and highest bond energy?

4.94 Acetylene (C_2H_2) molecule is known to form ionic compounds by losing two protons and giving the structure as C_2^{2-} species. This species can form compounds known as MgC_2 and CaC_2. Discuss the bonding scheme in the C_2^{2-} species in terms of molecular orbital theory and compare the bond length and bond strength of C_2^{2-} and C_2 species.

4.95 Consider the OH radical. Write the energy level diagram and write the ground state electronic configuration and predict the highest occupied molecular orbital.

4.96 In HF molecule the highest occupied orbital is nonbonding orbital on fluorine atom. The ionisation energy for an electron from this nonbonding molecular orbital of HF is 1550 kJ mol^{-1}. On the other hand, the ionisation energy of a 2*p* electron from a fluorine atom is 1795 J mol^{-1}. Why is the ionisation energy of an electron from 2*p* atomic electron from 2*p* atomic orbital on fluorine atom greater for fluorine atom than for the HF molecule?

4.97 In continuation to the Problem 4.96, the photoelectron spectroscopy has shown that the ionisation energy of 1*s* electron of HF molecule is 66.98 MJ mol^{-1} and that fluorine molecule is 67.22 MJ mol^{-1} respectively. Explain this difference.

4.98 The values of energies of different $2s$ and $2p$ orbitals of carbon and oxygen atoms are:

Atom	Valence orbital	Ionisation energy kJ mol^{-1}
Carbon atom	$2s$	180.31
	$2p$	98.54
Oxygen atom	$2s$	360.13
	$2p$	146.79

Draw the energy levels of both atoms on the two extreme sides of the energy level diagram and construct the molecular energy level diagram of the CO molecule. Write the ground state electronic configuration of CO molecule, determine the bond order and indicate whether the CO molecule is paramagnetic or diamagnetic?

4.99 Consider a molecule N_2F_2. (*a*) Draw the different possible structures of this molecule (*b*) predict the hybridisation on nitrogen atom and (*c*) dipole moment of each structure.

4.100 Taking cubic geometry, construct the structure of NH_3 molecule, depicting the three σ bonds and one lone pair, considering that orbitals on nitrogen atom are sp^3 hybridised.

4.101 Single bond between the two atoms is generally represented by the σ-orbital and double bond is represented by including one σ-bond and one π-bond. Can you identify the exceptions in the homonuclear diatomic molecules of the second period?

4.102 Between the pairs of species which has lower first ionisation energy, (*i*) N or N_2 and (*ii*) O or O_2? Explain briefly.

4.103 Hydroxyl radical (OH) is found to be present in trace amount in the atmosphere and it plays an important role in the atmospheric chemistry because it is an anti-oxidant and reacts with many pollutants. Assume that molecular orbitals results from the overlap of 1 *s* orbital of hydrogen atom and $2p_z$ orbital of the oxygen atom, considering *Z*-axis as the bond axis (or molecular axis). (*i*) Draw the pictures of the σ and σ* molecular orbitals in OH, (*ii*) Which of the two molecular orbitals has more hydrogen $1s$ character? (*iii*) Draw a molecular orbital energy-level diagram and write the electronic configuration for OH radical, (*iv*) Note that the non-bonding orbitals on oxygen atom have π character and should be assigned as such, (*v*) calculate the bond order of OH radical. Compare the bond order of OH with OH^+.

4.104 Determine the symmetry (*g* and *u*) for the term symbol of the ground state electron configuration of B_2.

4.105 Consider the electronic configuration of the excited state of H_2 molecule, *i.e.*, $\sigma_{1s}^1(g)\sigma_{1s}^1(u)$. Calculate the term symbols.

4.106 The ground state electronic configuration of oxygen molecule is: $KK\sigma_{2s}^2(g)\sigma^{*2}_{2s}(u)\sigma_{2p}^2(g)\pi_{2p}^2(u)\,\pi_{2p}^2(u)\pi_{2p}^{1*}(g)\sigma_{2p}^{*1}(u)$ and the electronic configuration of the first excited electronic state is $KK\sigma_{2s}^2(g)\sigma^{*2}_{2s}(u)\sigma_{2p}^2(g)\pi_{2p}^2(u)\;\pi_{2p}^2(u)\pi_{2p}^{1*}(g)\sigma_{2p}^{*1}(u)$. Determine the term symbols of the ground state and excited states.

4.107 Show that $H_{aa} = H_{bb} = -\dfrac{1}{2} + J$ and $H_{aa} = H_{bb} = -\dfrac{S}{2} + K$ in terms of hartrees for H_2^+.

4.108 Consider the formation of molecular bond by the overlap of s and p_z atomic orbitals when the molecular axis is X-axis. Using the atomic wave functions for these orbitals, show that this overlap will be zero.

4.109 Consider the wave function for the ground state of H_2, show that this function is an eigen function of $\hat{S}_z = \hat{S}_{z1} + \hat{S}_{z2}$ with $S = 0$.

4.110 The dissociation energy of N_2 is greater than that of O_2, whereas the dissociation energy of O_2^+ is greater than that of N_2^+.

4.111 Consider the wave function for anti-bonding m.o. of H_2^+ ion. Calculate the normalisation and plot the overlap integral versus inter-nuclear distance.

4.112 Which of the following molecules has the shortest nitrogen-to-nitrogen bond? Explain briefly. N_2, N_2O_4, N_2O, N_2H_4.

4.113 Discuss the kind of bonding present in CO_2 and C_3H_4 (allene) in terms of hybridisation. Depict the σ and π bonds present in the formation of these molecules.

4.114 The chlorine nitrate molecule ($ClONO_2$) is believed to be involved in the removal of ozone in the Antarctic atmosphere. Draw a possible Lewis dot structure and calculate the formal charges on each atom.

4.115 BF_3 is a triangular planar molecule. Formation of three single bonds does not complete the octet of boron atom. Explain the formation on the boron atom by using the resonance structures.

4.116 Ozone is O_3. Describe the state of hybridisation of the central oxygen atom in O_3? Describe the bonding in O_3 in terms of delocalised molecular orbitals.

4.117 Molecule borazine ($B_3N_3H_6$) is iso-electronic with benzene molecule. Discuss qualitatively the bonding in borazine in terms of (*i*) molecular orbital theory and (*ii*) resonance.

4.118 The bond moment of HF is 1.92 D and bond length of HF molecule is 0.92 Å(92 pm). Calculate the percent ionic character of the HF bond. [**Ans.** 43.2%, No compound is 100% ionic)

4.119 In tetrahedral AX_4 molecules, the resultant of bond moments of the three bonds (A – X) forming a triangular base acts in the opposite direction to the bond moment of the fourth A – X bond, giving rise to zero dipole moment. Show that all the bond angles are equal to 109.5 ^{0}.

4.120 Consider the derivative of allene molecule as HXC = C = CHX. Comment whether this molecule possesses dipole moment?

4.121 Which of the two molecules has the greater degree of π-electron delocalisation? Naphthalene or biphenyl.

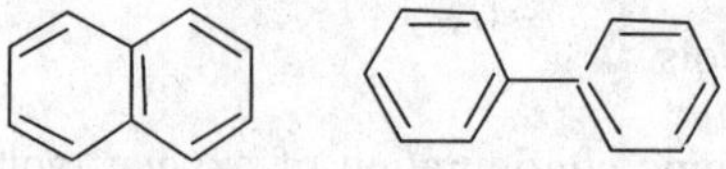

4.122 The dipole moment of trans-dichloroethylene is zero at 27 °C. On heating the dipole moment of this molecule starts increasing. Explain briefly.

4.123 How many geometric isomers can the square planar complex platinum Pt(abcd), where all the letters are mono-dentate ligands, can have?

4.124 Draw the energy level diagrams to show the low- and high-spin octahedral complexes of the transition-metal ions that have the electron configuration d^4, d^5, d^6, and d^7.

4.125 Draw qualitative diagrams for the crystal field splitting in (*i*) a linear complex ion ML_2, (*ii*) trigonal complex ML_3 and (*iii*) a square planar complex ML_4 and (*iv*) tetrahedral complex ML_4.

4.126 Calculate the number of unpaired electrons in the following complex ions: (*i*) $[Cr(CN)_6]^{4-}$ and (*ii*) $[Cr(H_2O)_6]^{2+}$ as well as the magnetic moment.

4.127 Transition metal complexes containing CN^- ligands are often yellow in colour, whereas the complexes containing H_2O as the ligands tend to be green or blue. Explain briefly.

4.128 Hydrated Mn^{2+} ions are more or less colourless although they possess five 3*d* electrons. Explain qualitatively.

4.129 Oxyhemoglobin is bright red in colour, whereas deoxyhemoglobin is purple. Explain the difference in colour in terms of the electronic configurations of the iron in these two complexes.

Intermolecular forces

4.130 Mention the kind of intermolecular forces present in the following molecules: Kr, O_3, C_6H_5F and KCl.

4.131 Arrange the following molecules in order of decreasing melting points: CaO, H_2O, CO_2, LiF, Xe.

4.132 The compounds Br_2 and ICl have the same number of electrons, yet Br_2 melts at –7.2 °C, whereas ICl melts at 27.2 °C. Explain briefly.

4.133 Molecular formula of n-Pentane and neopentane is same. The boiling point of n-pentane is 36.1 °C and that neopentane is 9.5 °C.

4.134 Let the two water molecules in air are separated by a distance of 0.4 nm. Calculate the (*i*) dipole-dipole interactions, (*ii*) induced dipole-induced dipole interaction and (*iii*) induced dipole-dipole interaction. Dipole moment of water is 1.82 D, $\alpha = 1.45 \times 10^{-30}$ m³.

4.135 Give the kind of intermolecular forces present between the molecules given below: (*i*) benzene, (*ii*) $CHCl_3$, (*iii*) NH_3, (*iv*) CO_2, (*v*) KBr.

4.136 We have learnt in Section 4.19.1 that Coulombic forces are usually referred to as long range forces and depend on $1/r^2$, Whereas van der waals forces are called short range forces and depend on $1/r^7$. (*i*) Assuming that forces depend on distance only, plot *F* as a function of *r*, if *r* varies between 1 to 8 Å. (*ii*) Based on these plots explain that non-electrolytes behave as ideal solution over a large concentrations Whereas electrolytes over a small concentrations.

4.137 Calculate the induced dipole moment of benzene due to Ca^{2+} ion when Ca^{2+} ion is 0.5 nm away from the centre of benzene molecule. The polarisability of benzene is 10.4×10^{-30} m³.

4.138 Using the data of Table 4.11 determine the van der Waals radius of Xe. Using this radius to determine the fraction of the volume occupied by 1 mole of xenon at 27 °C.

4.139 Differentiate equation (4.82) with respect to *r* to obtain the expression for σ and ε. Express the equilibrium distance r_e in terms of σ and show that V = – ε.

4.140 Which of the following properties exhibit very strong intermolecular forces of liquid? (*i*) a very low surface tension, (*ii*) a very low critical temperature, (*iii*) a very low boiling point and (*iv*) a very low vapour pressure.

4.141 The potential energy of the helium dimer (He_2) is given by

$$V = -\frac{B}{r^6} + \frac{C}{r^{13}}$$

where B = 97.7 kJ A^6 and C = 9.29×10^4 kJÅ^{-13} (mol dimer)$^{-1}$. (*i*) Calculate the equilibrium distance between the He atoms, (*ii*) Calculate the binding energy of dimer, (*iii*) Would do you expect the dime rot be stable at room temperature (300 K)?

4.142 The internuclear distance between the two closest Ar atoms in solid argon is nearly 3.8 Å. The polarisability of argon is 1.66×10^{-30} m^3, and the first ionisation energy is 1521 kJ mol^{-1}. Compute the boiling point of argon.

4.143 Diethyl ether ($C_2H_5OC_2H_5$) and 1-butanol (C_4H_9OH) are isomer to each other. The former has the boiling point as 34.5 ºC and the latter one being 117 ºC. Explain the difference in their boiling points.

4.144 Suppose that water has a linear structure like CO_2. (*i*) Would water be polar, (*ii*) will water be able to form hydrogen bonding?

4.145 NH_3 is soluble in water, whereas nitrogen trichloride is not. Explain this briefly.

4.146 What type of chemical analysis is required to test the A –T and G – C pairing in DNA?

4.147 Among the following, molecules which, one will have higher melting point?

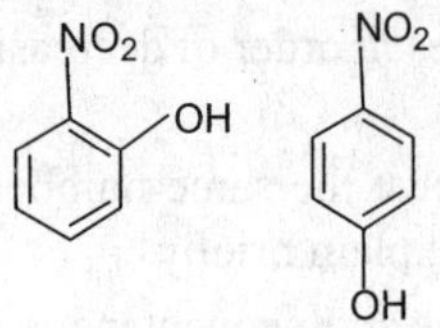

4.148 Acetic acid is soluble in water as well as carbon tetrachloride or benzene. Explain briefly.

CHAPTER 5

SPECTROSCOPY OF DIATOMIC AND POLYATOMIC MOLECULES

5.1. INTRODUCTION

Spectroscopy is the most powerful tool available to the scientists for probing the microscopic world of atoms and molecules. Molecular events occurring on widely different time scales, ranging from days to picoseconds and, most recently, even femto-seconds, can be monitored and studied in remarkable depth. Modern spectroscopy allows the probe and examination of the transition states of chemical reactions. *Put together it may be concluded that quantum mechanics and spectroscopy provide powerful tools those allow us to unlock the innermost secrets of molecular dynamics and structure*.

Spectroscopic measurements involves in general two parameters:

(*i*) Wavelength (λ) and in general it is denoted by *band maxima* (λ_{max}) where the intensity of the band is maximum.

(*ii*) Intensity and it is in general denoted by the *molecular extinction coefficient* at the band maxima (ε_{max}).

It may be mentioned here that no two compounds in general possess the same band maxima and molecular extinction coefficients We will discuss first the intensity followed by the band maximum.

In this chapter, we will solve and discuss the basic examples to illustrate the ideas given above. In Table 5.1 we have compiled the different branches of spectroscopy and their range in the electromagnetic radiation.

TABLE 5.1. Different types of spectroscopy and range of energies

Type of spectroscopy	Approximate energy spacing (kJ mol^{-1})	Approximate wavelength (cm)
Nuclear magnetic resonance (NMR)	10^{-6}	10^4
Electron spin resonance (ESR) Rotational	10^{-3} 10^{-1}	10 0.1 (microwave)
Vibrational	10	10^{-3} (infrared)
Electronic	10^8	10^{-5} to 10^{-6} (ultraviolet-visible)

5.1.1. *Born-Oppenheimer Approximation*

As mentioned in Section (4.1), each degree of freedom can be treated independently *i.e.*, the total energy of the system is due to the sum of rotational, vibrational and electronic energy, neglecting the other forms of energies:

$$E_T = E_{\text{rot}} + E_{\text{vib}} + E_{\text{ele}} \quad ...(5.1)$$

These individual energies can be calculated with the help of Schrödinger equation. It has been

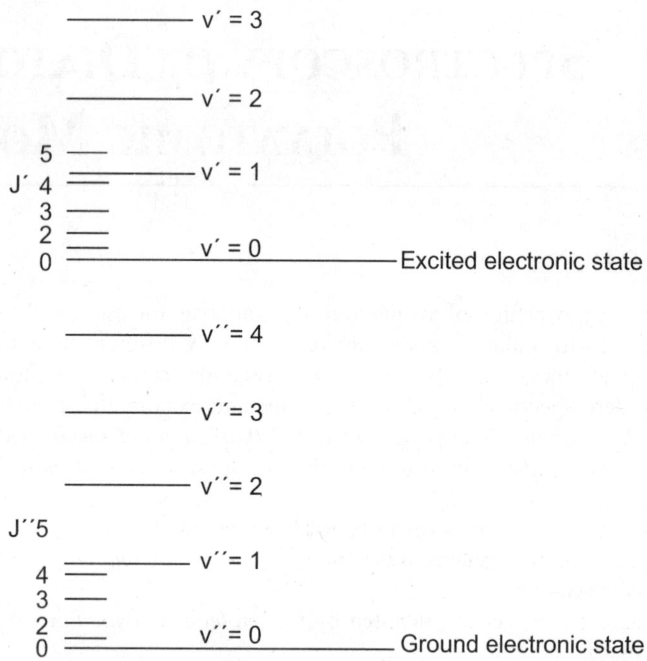

Fig. 5.1 *Relative energies of rotational, vibrational and electronic levels*

found that rotational quantum is much smaller than vibrational and vibrational quantum is much smaller than electronic. The relative energy levels are drawn in Fig. 5.1. It has also been found that during electronic excitation, vibrational as well as rotational excitations also take place. While during vibrational excitation, rotational levels are also excited. So instead of getting individual lines, like atomic spectra, bunch of lines or band spectra are observed, in case of molecules.

5.2. INTENSITY

Intensity of absorption or emission in molecular or atomic spectrum is as important as the frequency difference between the two energy states. The practical measure of intensity of absorption spectrum at any wavelength is the value of *molecular extinction coefficient*, ε, also called the *molar absorbance*, on *molar decadic coefficient*. In this section, we will discuss first the experimental features, *i.e.* Beer-Lambert's law and it will be followed by the theoretical treatment of the subject.

5.2.1. *Beer-Lambert Law*

Beer-Lambert law involves the change observed in electromagnetic radiation of intensity I_0 when passed through the sample of concentration C (mol L^{-1}) in a cell of path length l (cm), *i.e.* intensity of radiation absorbed by the sample is proportional to

(*i*) intensity of the electromagnetic radiation

(*ii*) concentration of the sample

(*iii*) length of the sample through which the electromagnetic radiation passes. Mathematically it can be written as:

$$-\log \frac{I}{I_0} = a\,C\,l \qquad \text{...(5.2)}$$

or

$$I = I_0\, e^{-aCl}$$

where I is the intensity of radiation transmitted by the sample and a is proportionality constant. Expression (5.2) in term of logarithm to the base 10 can be written as:

$$-\log \frac{I}{I_0} = \frac{a\,C\,l}{2.303} = \varepsilon\, C\, l \qquad \text{...(5.3)}$$

where ε is the *molar absorption coefficient* or *molar absorptivity* (molar extinction coefficient). In old literature quantity a is also known as molar extinction coefficient and the two are related by

$$a = 2.303\ \varepsilon \qquad \text{...(5.4)}$$

Two scales are used to depict the change in the intensity of the electromagnetic radiation.

(*i*) Transmittance (T) is defined as

$$T = \frac{I}{I_0} \qquad \text{...(5.5)}$$

(*ii*) Absorbance (A) also known as optical density (OD) and is defined as

$$A = \ln \frac{I_0}{I} \qquad \text{...(5.6)}$$

A chemist is more familiar with absorbance or optical density (A) and is related to equation (5.2) as

$$A = \log \frac{I_0}{I} = \varepsilon\, C\, l \qquad \text{...(5.7)}$$

The *optical density has no units*. The unit of ε is $dm^3\ mol^{-1}\ cm^{-1}$. In SI system, concentration is measured in mol m^{-3} and l in m, so the units for ε (expressed as κ) will be $m^3\ mol^{-1}\ m^{-1}$ and equation (5.7.) can be written as

$$A = \log \frac{I}{I_0} = \kappa\, C'\, l \qquad (5.7a)$$

In terms of molecular dimensions equation (5.2) can also be written as

$$I = I_0\, e^{-\sigma N x} \qquad \text{...(5.2}a\text{)}$$

where N is the number of molecules per m^3, x is the thickness in m, so the unit for σ is m^2, known as *absorption cross-section.* It may be mentioned here that σ, although has the unit of area it should not be interpreted as the area of the absorbing molecule. σ and ε are related to each other by the relation:

$$\sigma = \frac{2303\,\varepsilon}{N_{av}} \qquad ...(5.8)$$

Following points should be kept in mind as these are very important

(*i*) *It may be mentioned clearly that equations* (5.2*a*), (5.7) *and* (5.7*a*) *have been derived for a particular monochromatic radiation. Thus these expressions tell that molar absorption coefficient* ε *not only depends on the wavelength of excitation but also on the nature of the material.*

(*ii*) ε_{max} *also varies with the nature of solvent and temperature, because the absorption band maximum also depends on these two parameters.*

The most intense electronic spectrum or *fully allowed transition* of a compound has the value of ε_{max} of the order of 10^5 $dm^3\ mol^{-1}\ cm^{-1}$. Values of ε_{max} below the order of 10^3 $dm^3\ mol^{-1}\ cm^{-1}$ correspond to the *quasi-forbidden* or *probable transition* and ε_{max} below 100 $dm^3\ mol^{-1}\ cm^{-1}$ refers to the *forbidden transition.*

The other quantity, which is also commonly used in the absorption spectrum, is the *integrated molar absorption coefficient, i.e.*

$$A = \int_{\nu_1}^{\nu_2} \varepsilon\, d\nu \qquad ...(5.9)$$

where ν_1 and ν_2 (frequencies) are the limits of the absorption spectrum. As seen from this relation, absorption spectra are mostly spreadover a wide range of frequencies, and so monitoring the molar absorption coefficient for a single frequency may not depict a true indication of intensity of absorption of a molecular transition. Unit for integrated absorption coefficient is $M^{-1}\ cm^{-1}\ s^{-1}$ or $cm^2\ mol^{-1}\ s^{-1}$. If the shape of absorption band is of Gaussian type, the integrated absorption coefficient can be approximated to the relation (5.9*b*).

$$\int_{\nu_1}^{\nu_2} \varepsilon\, d\nu = 1.06\,\varepsilon\,\Delta\upsilon_{\frac{1}{2}} \qquad ...(5.9b)$$

where $\Delta\upsilon_{\frac{1}{2}}$ is the bandwidth at half the maximum height. In general, ν_{max} depends on temperature and thus value of ε_{max} will also be a function of temperature. This is because the population distribution of the energy levels is governed by the Maxwell-Boltzmann law, which changes with change of temperature. On the other hand, the value of integrated absorption coefficient does not depend on temperature. This is because the integrated absorption coefficient involves the transition from all the energy levels of the lower electronic state to all the energy levels of the upper electronic state. However bandwidth ($\Delta\upsilon_{\frac{1}{2}}$) does depend on temperature.

5.2.2. *Multi-component System*

When more than one kind of absorbing species is present in the solution, the absorbance is given as.

$$A = l \Sigma \varepsilon_i C_i \qquad ...(5.10)$$

where ε_i is the molar absorption coefficient and C_i is the concentration of the ith species. When there are more than one absorbing species (n) in the solution and their absorption spectra overlap, the concentration of each species can be calculated by generating as many equations as the number of species. This is done by measuring the absorbance at n wavelengths and independently calculating the molar absorption coefficients of each species at these wavelengths. For example, consider a solution containing two species A and B. Measure the molar absorption coefficients of A and B at two wavelengths λ_1 and λ_2. Thus the absorbance of this solution containing mixture of A and B for $l = 1$ cm is given as:

$$A(\lambda_1) = C_A \varepsilon_A(\lambda_1) + C_B \varepsilon_B(\lambda_1) \qquad ...(5.11a)$$

$$A(\lambda_2) = C_A \varepsilon_A(\lambda_2) + C_B \varepsilon_B(\lambda_2) \qquad ...(5.11b)$$

Solving these equations, one gets:

$$C_A = \frac{A(\lambda_1)\varepsilon_B(\lambda_2) - A(\lambda_2)\varepsilon_B(\lambda_1)}{\varepsilon_A(\lambda_1)\varepsilon_B(\lambda_2) - \varepsilon_B(\lambda_1)\varepsilon_A(\lambda_2)} \qquad ...(5.12)$$

$$C_B = \frac{A(\lambda_1)\varepsilon_A(\lambda_2) - A(\lambda_2)\varepsilon_A(\lambda_1)}{\varepsilon_B(\lambda_1)\varepsilon_A(\lambda_2) - \varepsilon_A(\lambda_1)\varepsilon_B(\lambda_2)} \qquad ...(5.13)$$

The additivity law is of great importance to chemist, because it allows the spectroscopists to carry out the analysis of solution containing several species.

5.2.3. *Deviations from Beer-Lambert Law*

Deviations from Beer-Lambert law may be real or apparent. These are:

(*i*) Although in the derivation of equation (5.8) radiation used is assumed to be a perfectly monochromatic, this law is valid over certain range of wavelengths, where molar absorption coefficients of the species should not vary too much.

(*ii*) Although this law is valid over large range of concentration of the species, still a non-linear relation is observed between absorbance and concentration. This happens when a molecule undergoes association or dissociation, *e.g.* aqueous solutions of dyes, which are prone to dimerize or polymerize, rarely follow Beer-Lambert law over an extended concentration range.

***Exercise* 5.1:** (*a*) Light of wavelength 400 nm is passed through a cell of 1 mm path length containing 10^{-3} mol dm^{-3} of compound A. If the absorbance of this solution is 0.25, calculate the molar absorption coefficient and the transmittance. (b) Derive the expression relating absorption cross-section (σ) and molar absorption coefficient (ε).

$$\varepsilon = \frac{0.25}{(10^{-3}\text{ mol dm}^{-3})(0.1\text{ cm})} = \frac{0.25}{10^{-4}\text{ mol dm}^{-3}\text{ cm}} = 2.5 \times 10^3 \text{ dm}^3\text{ mol}^{-1}\text{ cm}^{-1}$$

Now $\quad A = \log \frac{I_0}{I} = 0.25$

$$\frac{I_0}{I} = 1.778 \text{ and } \frac{I}{I_0}\ 0.56$$

Transmittance $$T = \frac{I}{I_0} = 0.56 \times 100 = 56\%$$

Note: The molar absorption coefficient is sometimes expressed in mol^{-1} cm^2 unit. This choice of unit suggests that ε can be interpreted as a molar cross-section for photon absorption. Thus the present exercise gives the value of ε as 2.5×10^6 mol^{-1} cm^2 as 1 dm^3 = 1000 cm^3.

(*b*) Equation (5.2*a*) can be written as:

$$\log \frac{I_0}{I} = \frac{\sigma N x}{2.303} \tag{B}$$

Comparing equations (B) and (5.2*a*), we get:

$$\frac{\sigma N x}{2.303} = \varepsilon C l \tag{C}$$

Now in the above equation, before we compare anything, we must express the quantities in proper units.

$$C\,(\text{mol dm}^{-3}) = C\,(\text{mol dm}^{-3})\,(10^3\text{ dm m}^{-3})$$
$$x\,(\text{m}) = l\,(\text{cm})\,(10^{-2}\text{ m cm}^{-1})$$
$$N = \text{Number of molecules per m}^3\text{ in the concentration } C$$
$$= C\,(\text{mol dm}^{-3})\,(10^3\text{ dm}^3\text{ m}^{-3})\,(N_{av}).$$

Now substituting the above quantities in equation (C),

$$\frac{\sigma\,(C\text{ mol dm}^{-3})\,(10^3\text{ dm}^3\text{ m}^{-3})\,(N_{av})\,(1\times10^{-2}\text{ m})}{2.303} = \varepsilon C l$$

$$\sigma = \frac{(2.303)\,(\varepsilon)\,(10^2\text{ cm m}^{-1})}{(N_{av})\,(10^3\text{ dm}^3\text{ m}^{-3})}$$

Substituting the values of the respective quantities, we get

$$\sigma = \frac{(2.303)\,(2.5\times10^3\text{ dm}^3\text{ mol}^{-1}\text{ cm}^{-1})\,(10^2\text{ m}^{-1}\text{cm})}{(6.022\times10^{23}\text{ mol}^{-1})\,(10^3\text{ dm}^3\text{ m}^{-3})} = 9.56\times10^{-22}\text{ m}^2$$

***Exercise* 5.2:** A solution contains a mixture of two substances A and B, whose absorption spectra overlap. The absorbance of the given solution at wavelength 300 and 350 nm is 0.24 and 0.66 respectively. Calculate the concentration of species A and B if ε_{300} (A) = 2×10^3 dm^3 mol^{-1} cm^{-1}, ε_{350} (A) = 10^2 dm^3 mol^{-1} cm^1, ε_{300} (B) = 10^5 dm^3 mol^{-1} cm^{-1}, ε_{350} (B) = 5×10^5 dm^3 mol^{-1} cm^{-1}. The path length is 1 cm.

Using equation (5.10)

$$A_{300} = l\,\varepsilon_{300}\,(\text{A})\,C_A + l\,\varepsilon_{300}\,(\text{B})\,C_B \quad ...(5.14)$$
$$A_{350} = l\,\varepsilon_{350}\,(\text{A})\,C_A + l\,\varepsilon_{350}\,(\text{B})\,C_B \quad ...(5.15)$$

where C_A and C_B are the concentrations of species *A* and *B*. Substituting the respective values and taking l = 1 cm, we get:

$$0.24 = (2\times10^3\text{ dm}^3\text{ mol}^{-1}\text{ cm}^{-1})\,(1\text{ cm})\,(C_A) + (10^5\text{ dm}^3\text{ mol}^{-1}\text{ cm}^{-1})\,(1\text{ cm})\,(C_B)$$
$$= 2\times10^3\,C_A + 10^5\,C_B \quad ...(5.16)$$

and $$0.64 = 10^3\, C_A + 5 \times 10^5\, C_B \qquad \text{...(5.17)}$$

Solving equations (5.16) and (5.17), one gets: $C_A = 5.45 \times 10^{-5}$ mol dm^{-3} and $C_B = 1.31 \times 10^{-6}$ mol dm^{-3}.

5.3. THEORIES OF ABSORPTION OR EMISSION SPECTRUM

In this section, we will briefly discuss the outlines of the results of classical and quantum mechanical theories of intensities of absorption of radiation and comparing them with the experimental results. In both theories, electrons are displaced from their mean position by the interaction of electrical field of the radiation, when the frequency of the radiation matches with the characteristic frequency of the electronic transition of the atoms or molecules. Due to this, the induced electrical dipole moment is created in the molecule. Thus intensity of absorption depends on the strength of this induced electrical dipole moment. Only difference between the two theories is the procedure of calculating the induced dipole moment and the interpretation of the characteristic frequencies of electrons in the atoms or molecules.

5.3.1. *Classical Theory*

The quantity, which is quite commonly used to estimate the intensity of the line or λ_{max}, is known as *oscillator strength (f)*. It is defined as the effective number of dispersion electrons, which are oscillating with a characteristic frequency v_i. The oscillating strength can be expressed in many ways and their relation to each other will be given in Exercise 5.4. The oscillator strength can be related to the experimentally known quantity ε, determined at each frequency by the expression (5.18) in cgs units:

$$f = \frac{2303\, m_e\, c^2}{\pi\, N_{av}\, e^2} \int \varepsilon(\bar{\nu})\, d\bar{\nu} = 4.39 \times 10^{-9} \int \varepsilon(\bar{\nu})\, d\bar{\nu} \qquad \text{...(5.18)}$$

where m_e is the mass of an electron in gram, e is the charge of the electron in electrostatic unit, c the velocity of light, N_{av} is the Avogadro number, $\bar{\nu}$ is the frequency expressed in wave numbers (cm^{-1}), factor 2303 involves the conversion from natural logarithms as well as the number of cubic centimetres in a litre required by the introduction of the molar absorption coefficient. The advantage of introducing the oscillator strength is that it can be interpreted theoretically and can be calculated with the help of quantum mechanics. The integral $\int \varepsilon(\bar{\nu})\, d\bar{\nu}$ gives the area of the absorption transition band under consideration. The approximate value of the area under the band can be determined from the relation:

$$\int \varepsilon\, (\nu) d\nu = 1.06 \times \varepsilon \times \text{width of the band at half the maximum height} \qquad \text{...(5.9}b\text{)}$$

The oscillator strength in SI units can be expressed as:

$$f = \left[\frac{2303\, (4\pi\varepsilon_0)\, m_e\, c}{\pi\, N_{av}\, e^2}\right] A \qquad \text{...(5.19}a\text{)}$$

where $4\pi\varepsilon_0$ is the permittivity factor. After substituting the values of the constants, the above relation can be written as:

$$f = 1.44 \times 10^{-19}\, A \text{ (}A\text{ is expressed as cm}^2\text{ mmol}^{-1}\text{ s}^{-1}\text{)} \qquad ...(5.19\ b)$$

A point to be noted here is that although the expression for f contains so many constants but the units of each quantity is such that the products of all the quantities give rise to a dimensionless quantity and for three dimensional oscillator its value is 1.

For example, spectral lines emitted by an atom can be associated with the corresponding electronic oscillator strength with a particular value of f. It turns out that all the intense transitions have oscillator strengths of *about unity*.

***Exercise* 5.3:** In the longest wavelength band of 5-aminoindazole in water as a solvent, molar absorption coefficient at the band maximum is 5.4×10^3 dm^3 mol^{-1} cm^{-1} and the bandwidth at half the maximum height is 4047 cm^{-1}. Calculate the oscillator strength of this band.

Using equation (5.9*b*):

$$\int \varepsilon(\bar{\nu})\, d\bar{\nu} = 1.06\ (5.4 \times 10^3 \text{ dm}^3 \text{ mol}^{-1} \text{ cm}^{-1})\ (4047 \text{ cm}^{-1}) = 2.317 \times 10^7 \text{ dm}^3 \text{ mol}^{-1} \text{ cm}^{-1}$$

Using equation (5.19)

$$f = 4.31 \times 10^{-9} \int \varepsilon\,(\nu)\, d\nu = 1.06 \times 4.31 \times 10^{-9} \times 2.317 \times 10^7 = 0.1$$

5.3.2. *Quantum Mechanical Theory*

Einstein was the first to explain the intensities of the transitions involved between two quantum states. His theory is based on the following facts:

(*i*) Energies of the states are quantized, as suggested by Bohr.

(*ii*) Molecules in the two states are present according to the Maxwell-Boltzmann's distribution.

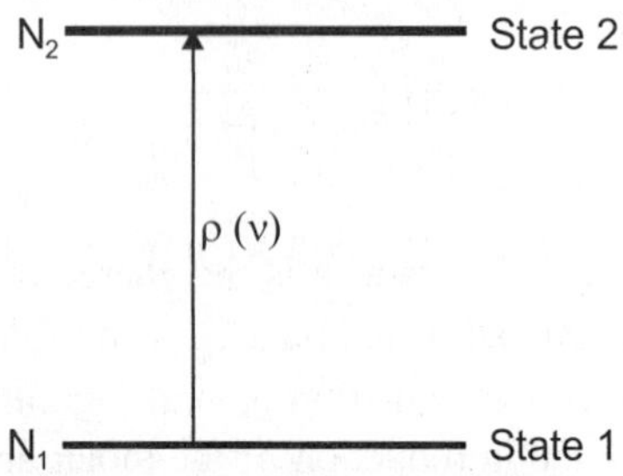

Fig. 5.2 *Absorption of radiation between two quantum states depicting Einstein's theory of absorption of radiation. N_1 and N_2 are the number of molecules per unit volume present in the state 1 and state 2 respectively, e(v) erg cm^{-3} per unit interval of frequency unit (s^{-1}) is the density of the radiation at the frequency v, which is capable of inducing transition between the states 1 and 2*

(*iii*) Equilibrium distribution of energy among the frequencies of radiation is given by Planck's equation for black body radiation.

Based on the above assumptions, Einstein defined three quantities. (*i*) Einstein's coefficient of induced absorption (B_{12}), (*ii*) Einstein's coefficient of induced emission (B_{21}), and (*iii*) Einstein's coefficient of spontaneous emission (A_{21}). It may be mentioned here that Einstein's coefficient of induced emission led to the discovery of LASERs after nearly 44 years of his prediction. The three coefficient are related to each other as

$$\frac{A_{21}}{B_{21}} = \frac{8\,\pi\,h\nu^3}{c^3} \quad ..(5.20)$$

and
$$B_{12}\,g_1 = B_{21}\,g_2 \quad ...(5.21)$$
where g_1 and g_2 are the degeneracies of the state 1 and state 2 respectively.

5.3.3. *Quantum Mechanical Formulation*

Time dependent Schrödinger wave equation was used to develop equations to relate the Einstein's coefficients with wave functions of the lower and upper states. These equations are:

$$A_{21} = \frac{64\,\pi^4\,\nu^3\,M_{12}^2}{3\,h\,c^3} \quad ...(5.22)$$

and
$$B_{12} = B_{21} = \frac{8\,\pi^3\,M_{12}^2}{3\,h^2} \quad ...(5.23)$$

where M_{12} and M_{21} are the quantum mechanical description of absorbance and emission processes, called *transition moment.* It can be defined as:

$$M_{12} = M_{21} = e\,r_{12} = e\int \psi_2{}^*\,\hat{r}\,\psi_1 dr \quad ...(5.24)$$

where ψ_1 and ψ_2 are the wave functions of the lower and upper states, e is the charge on the electron and r is the distance between the centre of gravities of the positive and negative charges. Here the integral gives the average distance between the electron and nucleus when it is oscillating between the lower and upper states. Thus it is nearly similar to dipole moment. Since dipole moment is a vector quantity, the transition moment is also a vector quantity and can be resolved into three different components, *i.e.* M_x, M_y and M_z. Three components are related to M_{12} or M_{21} = M by the relation (5.25).

$$M_{21}^2 = M_{12}^2 = M_x^2 + M_y^2 + M_z^2 \quad ...(5.25)$$

where

$$M_x = \int \psi_2{}^*\,\hat{x}\,\psi_1\,dx \quad ...(5.25a)$$

$$M_y = \int \psi_2{}^*\,\hat{y}\,\psi_1\,dy \quad ...(5.25b)$$

$$M_z = \int \psi_2{}^*\,\hat{z}\,\psi_1\,dz \quad ...(5.25c)$$

If the wave functions for the respective states are known, M_{12} or M_{21} can be calculated by evaluating the integral. The connection between the transition moment and the observed intensity is:

$$f = \frac{8\,\pi^2\,m_e\,\nu}{3\,h\,e^2}\,|M_{12}|^2 \quad ...(5.26)$$

where m_e and e are the mass and charge of the electron respectively and v is the frequency (s^{-1}) of absorption band and M is measured in electrostatic units (debyes). If the value of f is obtained from equation (5.18), one can calculate the value of M_{12}^2 by the relation:

$$M_{12}^2 = \left(\frac{3\,h\,e^2 f}{8\,\pi^2 m_e\,\nu}\right)^{\frac{1}{2}} = 0.0958\left(\frac{\int \varepsilon d\nu}{\nu}\right)^{\frac{1}{2}} \quad \text{...(5.26a)}$$

Knowing wave functions, the intensity of the line can be calculated and thus it is an important link between theory and experiment.

The other important quantities which can be obtained from these relations are: This lifetime

(*i*) *radiative lifetime*: The mean radiative lifetime in the excited state is defined as:

$$\tau_{FM} = \frac{1}{A_{21}} \quad \text{...(5.27)}$$

The value of A_{21} (without derivation), can be related to the experimental quantity oscillator strength by the relation (in cgs units) :

$$A_{21} = \frac{64\,\pi^4 \nu^3 M_{21}^2}{3\,h} = \frac{8\,\pi^2\,e^2\,\nu^2\,f_{12}}{m_e\,c} \quad \text{...(5.22a)}$$

$$\text{Thus: } \tau_{FM} = \frac{m_e\,c}{8\,\pi^2\,\nu^2\,e^2\,f_{12}} = \frac{1.5}{\nu^2 f_{12}} = \frac{N_{av}}{8\,\pi\,2303\,\nu^2 c\int \varepsilon\,d\nu} = \frac{3.47\times 10^8}{\nu^2\int \varepsilon\,d\nu} \quad \text{...(5.28)}$$

It is obvious from equation (5.28) that *relatively shorter mean lifetime of the excited state is associated with an intense absorption band and longer mean lifetime with weak absorption band.* If the molecule is deactivated to the lower state by only fluorescence quantum, the value of τ_{FM} from equations (5.21*a*) or (5.28) is the actual and maximum possible mean lifetime of the excited state. This is also called *intrinsic lifetime* of the excited state. In general the intrinsic lifetime of the molecule is of the order of 10^{-8} to 10^{-9}s for the stronger absorption band. More sophisticated theoretical methods are available but still the experimental procedure gives the actual value of mean lifetime.

***Exercise* 5.4:** (*a*) The oscillator strength of the electronic transition in 5-aminoindazole at 316.7 nm is 0.1. Calculate the transition moment as well as dipole length of this transition.

(*b*) Calculate the mean lifetime of this molecule in its first excited state using the data of exercise (5.3). ($\bar{\nu}$ = 4047 cm^{-1}, $\int \varepsilon\, d\bar{\nu}$ = 2.317 × 10^7.

(*a*) We can make use of the equation (5.26). Here

$$\bar{\nu} = \frac{1}{\lambda} = \frac{1}{(316.7\times 10^{-9}\text{m})\,(10^2\text{ cm m}^{-1})} = 31575.6\text{ cm}^{-1}$$

$$M = 0.0958\left(\frac{2.317\times 10^7}{31575.6}\right)^{1/2} = 2.595\ D$$

$$D_{12} = \text{Dipole length} = \frac{M}{e} = \frac{2.59\times 10^{-18}\text{ cgs cm}}{4.8\times 10^{-10}\text{ cgs}}\ 0.54\times 10^{-8}\text{ cm.}$$

(*b*) Using equation (5.28):

$$\tau_{FM} = \frac{3.47\times 10^8}{(4047)^2\,(2.317\times 10^7)} = 9.14\times 10^{-7}\text{ s}$$

5.3.4 *Selection Rules*

Equations (5.22) and (5.23) give the accurate values of the intensities. But the evaluation of these integrals is quite cumbersome job. If we are only interested in knowing whether these transitions will occur or not, certain rules (called *selection rules*) are formulated by which we can easily find out qualitatively whether the equations (5.25, *a*, *b*, *c*) are finite or zero. *If the transition moment integral is zero, the transition is called forbidden and if the dipole moment integral is non-zero or finite it is called 'allowed transition'*. In the following section, we will discuss certain rules.

(*i*) Spin selection rules: Since dipole moment operator works only on the space part of the wave function, equation (5.24) can be written as:

$$M = \int \Psi_2^{\text{space}} (e\hat{r}) \Psi_1^{\text{space}} d\tau^{\text{space}} \int \Psi_2^{\text{spin}} \Psi_1^{\text{spin}} d\tau^{\text{spin}} \qquad ...(5.29)$$

Since the second integral of equation (5.29) only involves the dot product of the spin wave functions of the initial and final states respectively. it is quite evident that this integral will be zero until the wave functions for the spin belong to the same state, *i.e.*

$$\Delta S = 0 \qquad ...(5.30)$$

This means that for allowed transitions, the change in the spin quantum number should be zero.

Equation (5.29) has been written with the assumption that orbital motion and spin motion of an electron are independent of each other. This is only true for atoms and molecules of light masses. For heavy particles, the total wave function cannot be written as the product of wave functions depending only on either space part of the function or spin part of the function, (*i.e.* one variable). Therefore for heavy atoms or molecules (or atoms of higher atomic number) the selection rule ($\Delta S = 0$) is not valid and the transitions are observed even for $\Delta S \neq 0$. For example the strongest transition in mercury atom corresponding to 253.7 nm is due to $6(^3P_1) \rightarrow 6(^1S_0)$.

***Exercise* 5.5:** Show that the transition between the singlet and triplet state is forbidden.

The wave functions for the spin part of the singlet and triplet states are:

$$\Psi_S = [\alpha(1)\,\beta(2) - \alpha(2)\,\beta(1)] \qquad ...(5.31)$$

and

$$\Psi_T = \alpha(1)\,\alpha(2) \qquad ...(5.31a)$$

$$= \beta(1)\,\beta(2) \qquad ...(5.31b)$$

or

$$= [\alpha(1)\,\beta(2) + \alpha(2)\,\beta(1)] \qquad ...(5.31c)$$

The normalization constants are omitted for sake of simplicity. Now considering one of the triplet state wave functions, equation (5.29) becomes:

$$\int \Psi_1^{\text{spin}} \Psi_2^{\text{spin}} d\tau^{\text{spin}} = \int [\alpha(1)\,\beta(2) - \alpha(2)\,\beta(1)]\,[\alpha(1)\,\beta(2) + \alpha(2)\,\beta(1)]\, d\tau^{\text{spin}}$$

$$= \int \alpha(1)\,\alpha(1)\, d\tau_1^{\text{spin}} \int \beta(2)\,\beta(2)\, d\tau^{\text{spin}}{}_2 + \int \alpha(1)\,\beta(1) d\tau_1^{\text{spin}} \int \beta(2)\,\alpha(2)\, d\tau_2^{\text{spin}}$$

$$- \int \alpha(2)\,\beta(2)\, d\tau_2^{\text{spin}} \int \alpha(1)\,\beta(1)\, d\tau_1^{\text{spin}} - \int \alpha(2)\,\alpha(2)\, d\tau_2^{\text{spin}} \int \beta(1)\,\beta(1)\, d\tau_1^{\text{spin}}$$

Since the wave functions are normalized, the first and fourth integral will be unity, whereas the second and third integrals will be zero. Therefore

$$\int \Psi_1^{\text{spin}} \Psi_2^{\text{spin}} d\tau^{\text{spin}} = 1 + 0 + 0 - 1 = 0$$

Thus, this transition is forbidden. It can also be shown similarly that the integrals

$$\int [\alpha(1)\,\beta(2) - \alpha(2)\,\beta(1)]\,\alpha(1) \langle (2)\, d\tau^{\text{spin}}$$

and $$\int[\alpha(1)\,\beta(2) - \alpha(2)\,\beta(1)]\,\beta(1)\,\beta(2)\,d\tau^{spin}$$

are also zero, indicating that the singlet to triplet transition is not allowed.

***(ii)* Laporte's selection rules:** In molecules, which have centre of symmetry, the selection rules

$$u \leftrightarrow g \text{ and } g \leftrightarrow u \qquad ...(5.32)$$

give allowed transitions. The selection rules

$$g \leftrightarrow g \text{ and } u \leftrightarrow u \qquad ...(5.33)$$

give forbidden transitions and are sometimes called as *parity forbidden.*

***(iii)* Symmetry Selection Rules:** Laporte's selection rules are based on only one of the symmetry elements, *i.e.* centre of symmetry (or centre of inversion). Based on the symmetry point group to which the given molecule belongs, the transition moment integral must give rise to totally symmetric irreducible representation or the dot product of $\int \psi_1 \hat{M} \psi_2 d\tau$ if reducible representation must contain at least one totally symmetric irreducible representation. Since M is anti-symmetric and if ψ_1 is symmetric, the transition will be allowed only if ψ_2 should be anti-symmetric. Because the three components of the dipole moment operator transform differently under the covering operations of the various point groups, a transition may be allowed with polarization along one or two axis, but forbidden with polarization along the other axis.

There are also many selection rules but these are very specific and will depend on the branch of spectroscopy. These will be discussed at the proper places. Further use of symmetry and group theory to find the allowedness and forbiddenness of the transitions have already been discussed in Chapter 0.

If we consider that $\pi\ \pi^*$ is the fully allowed transition of oscillator strength (F_A) of the order of unity, then any other transition will have its oscillator strength F as to order of magnitude given by:

$$F = f_s f_0 f_m f_p F_A \qquad ...(5.34)$$

where f factors differ from unity as follows for the different kinds of forbiddenness which have been described so far:

f_s = Spin forbidden transition involving change of multiplicity, 10^{-5} for second row elements.

f_o = Overlap forbidden involving change of position of electronic charge, 10^{-2} for $n \rightarrow \pi^*$ transition of second row heteroatom.

f_m = Momentum forbidden or orbitally forbidden involving large change of linear or angular momentum, 10^{-1} to 10^{-3} for condensed ring system.

f_p = Parity forbidden, initial and final states both are either even or odd, 10^{-1} for condensed systems

***Exercise* 5.6:** In octahedral complexes of transition elements, the five degenerate orbitals split into a set of two, t_{2g} (three orbitals) and e_g (two orbitals). Will the transition from d to d orbital be allowed or forbidden?

According to Lapport's rule, the transition is allowed between the orbitals of parity g and u. On this basis the *d-d* transition is forbidden.

***Exercise* 5.7:** The ground state and the first three excited states of oxygen molecule have the following symbols: $^3\Sigma_g$, $^1\Sigma_g$, $^1\Delta_g$, and $^3\Sigma_u$. Based on the above selection rules, indicate the allowed and forbidden transitions starting from the ground state.

Above selection rules indicate that the allowed transitions starting from the ground state are:

(*i*) $^3\Sigma_g$ to $^1\Sigma_g$ transition is forbidden, based on spin and parity selection rules.

(*ii*) $^3\Sigma_g$ to $^1\Delta_g$ transition is forbidden, based on parity, spin and component of orbital angular momentum along the molecular axis ($\Delta\lambda = 0, \pm 1$). Here $\lambda = 0$ and 2 for Σ and Δ states respectively.

(*iii*) $^3\Sigma_g$ to $^3\Sigma_u$, is an allowed transition, based on all counts. $\Delta S = 0$, $u \leftrightarrow g$, and $\Delta\lambda = 0$.

5.4. ROTATIONAL SPECTRA OF DIATOMIC MOLECULES

To the first approximation, the diatomic molecules can be considered as rigid rotator and energy levels for the rotational motion are given by equation (5.35):

$$E_J = \frac{h^2}{8\pi^2 I} J(J+1), J = 0, 1, 2, ... \qquad ...(5.35)$$

where J is the rotational quantum number. The degeneracy of each level is $2J + 1$, I is the moment of inertia, given by the expression:

$$I = \mu r^2 \qquad ...(5.36)$$

Again r is the bond length and μ is the reduced mass of the system *i.e.,*

$$\mu = \frac{m_a m_b}{m_a + m_b} \qquad ...(5.37)$$

Generally, the energy in spectroscopy is expressed in cm^{-1}. This is obtained by dividing the equation (5.35) by *hc* (*c* is velocity of light) and is represented by $F(J)$, is called rotational term value:

$$F(J) = \frac{h}{8\pi^2 Ic} J(J+1) \qquad ...(5.38)$$

The constant term $h/8\pi^2 Ic$ for each molecule is called *rotational constant* (B). The unit for B is cm^{-1} whereas μ and r are expressed in kg and metre respectively, so the expression for B is

$$B = \frac{h \times 10^{-2}}{8\pi^2 Ic} \qquad ...(5.39)$$

Equation (5.38) can then be written as

$$F(J) = B\,J(J+1) \qquad ...(5.40)$$

The energy level diagram is depicted in Fig. 5.3. Energy levels are not equally spaced and gap between them keeps on increasing.

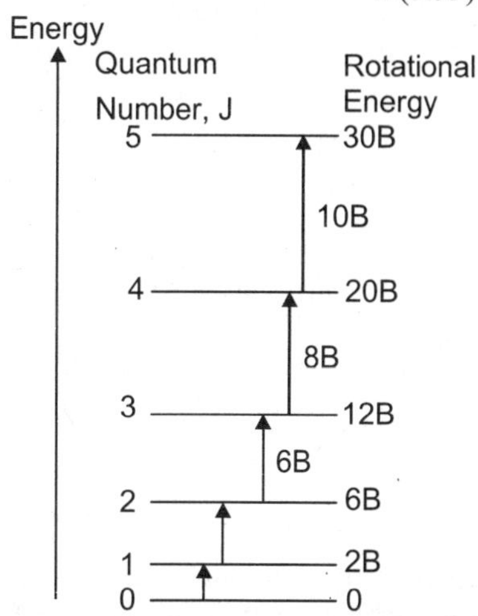

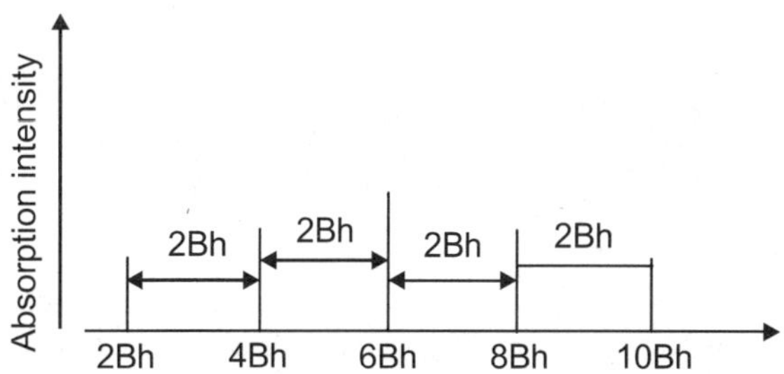

Fig. 5.3 *Allowed rotational energy values of diatomic molecules as expressed by Eqn. (5.40). Arrows between two states represent the absorption transitions. At the bottom of the figure are shown the transitions which are equally spaced and separated by 2Bh.*

The selection rules for the rotational spectrum are:

(*a*) The molecule must possess permanent dipole moment.

(*b*) The transitions are only observed between the levels of $\Delta J = \pm 1$.

These selection rules are obtained using equation (5.25), *i.e.*

$$M_{12} = M_{21} = \int Y_{J'}^{M'}(\theta,\varphi)[M_x + M_y + M_z]Y_J^M(\theta,\varphi)r^2 \sin\theta dr d\theta d\varphi \quad (5.41)$$

Derivation of these selection rules can be looked in books on spectroscopy [S. K. Dogra and H. S. Randhawa, *Atom, Molecule and Spectrum*, New Age International (P) limited Publishers, New Delhi, 2011]. It is also given as Appendix A-5.1

For the rotational transitions the wave numbers observed are

$$\begin{aligned}\bar{\nu} &= F(J+1) - F(J)\\ &= B(J+1)(J+2) - BJ(J+1)\\ &= 2B(J+1) \qquad ...(5.42)\end{aligned}$$

where $J = 0, 1, 2, 3, ...$

Since the energy difference between the rotational levels is small, many higher levels are also populated at room temperature. Due to this number of intense rotational lines are observed. It can be seen that the spacing between each rotational line is $2B$ *i.e.,* they are equally spaced. Knowing B, the bond distance between the two atoms can be calculated from the expression:

$$r = \left[\frac{h \times 10^{-2}}{8\pi^2 \mu c B}\right]^{1/2} \text{ m} \qquad ...(5.43)$$

***Exercise* 5.8:** The rotational spectrum of HC1 molecules shows that the rotational lines are equally separated by 20.70 cm^{-1}. Calculate the internuclear bond length.

$$2B = 20.70 \text{ cm}^{-1} \text{ or } B = 10.35 \text{ cm}^{-1}$$

Using equation (5.37)

$$\begin{aligned}\mu &= \frac{(1.0 \text{ g mol}^{-1})(35.5 \text{ g mol}^{-1})}{(1.0+35.5)\text{g mol}^{-1}(6.02\times10^{23}\text{ mol}^{-1})}\\ &= 1.627 \times 10^{-24}\text{g}\\ &= 1.627 \times 10^{-27} \text{ kg}\end{aligned}$$

Using equation (5.43)

$$\begin{aligned}r &= \left[\frac{(6.626\times10^{-34}\text{J s})(10^{-2}\text{m cm}^{-1})}{(8\pi^2)(1.627\times10^{-27}\text{ kg})(3.0\times10^8\text{ m s}^{-1})(10.35\text{ cm}^{-1})}\right]^{1/2}\\ &= 0.129 \text{ nm}\end{aligned}$$

***Exercise* 5.9:** The bond length of NO molecule is 1.151×10^{-10} m. Calculate the frequencies in cm^{-1} for the pure rotational lines in the spectrum of NO corresponding to the following changes in the rotational quantum numbers: $0 \rightarrow 1$, $1 \rightarrow 2$, $2 \rightarrow 3$, $3 \rightarrow 4$.

Before equation (5.42) can be used, μ and I must be calculated.

$$\mu_{NO} = \frac{m_N m_o}{m_N + m_o} = \frac{(14 gmol^{-1})(16 gmol^{-1})}{(14+16)\, gmol^{-1}} \times \frac{1}{(6.023 \times 10^{23}\, mol^{-1})} g$$

$$= 1.24 \times 10^{-23}\, g = 1.24 \times 10^{-26}\, \text{kg}$$

The moment of inertia can be obtained

$$I = \mu_{NO} r_{NO}^2$$

$$= (1.24 \times 10^{-26}\, \text{kg})\,(1.151 \times 10^{-10}\, \text{m})^2$$

$$= 1.428 \times 10^{-46}\, \text{kg m}^2$$

Substituting the values of I, h and c in equation (5.39) B can be calculated

$$B = \frac{h \times 10^{-2}}{8\pi^2 Ic} = \frac{(6.62 \times 10^{-34}\, \text{J s})(10^{-2}\, \text{m cm}^{-1})}{(8)(3.143)^2 (1.428 \times 10^{-46}\, \text{kg m}^2)(3 \times 10^8\, \text{m s}^{-1})}$$

$$= 1.956\, \text{cm}^{-1}$$

The frequency in cm^{-1} required to change the quantum number from $0 \rightarrow 1$ is obtained by substituting the quantum number of lower state in equation (5.42)

$$\bar{v}_{0-1} = (2)\,(1.956)\,(0 + 1) = 3.912\, \text{cm}^{-1}$$

Similarly,

$$\bar{v}_{1-2} = (2)\,(1.956)\,(1 + 1) = 7.824\, \text{cm}^{-1}$$

$$\bar{v}_{2-3} = (2)\,(1.956)\,(2 + 1) = 11.736\, \text{cm}^{-1}$$

$$\bar{v}_{3-4} = (2)\,(1.956)\,(3 + 1) = 15.648\, \text{cm}^{-1}$$

The rotational wave functions up to $J = 2$ are compiled in Table 5.2.

TABLE 5.2. Spherical harmonic wave functions up to $J = 2$

Orbital	Function
$J = 0,\ M = 0$	$\frac{1}{\sqrt{4\pi}}$
$J = 1,\ M = +1$	$\frac{1}{2}\sqrt{\frac{3}{\pi}}\sin\theta\cos\varphi$
$J = 1,\ M = -1$	$\frac{1}{2}\sqrt{\frac{3}{\pi}}\sin\theta\sin\varphi$
$J = 1,\ M = 0$	$\frac{1}{2}\sqrt{\frac{3}{\pi}}\cos\theta$
$J = 2,\ M = 0$	$\frac{1}{4}\sqrt{\frac{5}{\pi}}(3\cos^2\theta - 1)$
$J = 2,\ M = +1$	$\frac{1}{2}\sqrt{\frac{15}{\pi}}\sin\theta\cos\theta\cos\varphi$

$J = 2, M = -1$	$\frac{1}{2}\sqrt{\frac{15}{\pi}}$ sin θ cos θ sin φ
$J = 2, M = -2$	$\frac{1}{4}\sqrt{\frac{15}{\pi}}$ $\sin^2\theta\ \sin^2\varphi$
$J = 2, M = +2$	$\frac{1}{4}\sqrt{\frac{15}{\pi}}$ $\sin^2\theta\ \cos^2\varphi$

***Exercise* 5.10:** Show that the rotational transition is observed from the state $J = 0, M = 0$ to the state $J = 1, M = 0$.

The wave functions (which are also called spherical harmonics) for the two states are:

$$\psi_0^0(\theta,\varphi) = \frac{1}{\sqrt{4\pi}} \text{ and } \psi_1^0(\theta,\varphi) = \sqrt{\frac{3}{4\pi}}\cos\theta \qquad \text{...(5.44)}$$

The values of M_x, M_y and M_z in spherical polar coordinates, without using the charge (e) which is constant, are given as:

$$\therefore \quad M_x = r\sin\theta\cos\varphi.$$

$$M_y = r\sin\theta\sin\varphi$$

$$M_z = r\cos\theta$$

Substituting these values in equation (5.41), we get:

$$M_{12} = M_{21} = \int \frac{1}{\sqrt{4\pi}}[r\sin\theta\cos\varphi + r\sin\theta\sin\varphi + r\cos\theta)\sqrt{\frac{3}{4\pi}}\cos\theta\sin\theta d\theta d\varphi$$

$$= \frac{\sqrt{3}}{4\pi}r[\int_0^\pi \sin^2\theta\cos\theta\, d\theta \int_0^{2\pi}\cos\varphi\, d\varphi + \int_0^\pi \sin^2\theta\cos\theta\, d\theta\int_0^{2\pi}\sin\varphi\, d\varphi + \int_0^\pi \cos^2\theta\sin\theta\, d\theta\int_0^{2\pi} d\varphi]$$

The first and second integrals are zero because:

$$\int_0^{2\pi}\cos\varphi d\varphi = [\sin\varphi]_0^{2\pi} = [\sin 2\pi - \sin 0] = 0 - 0 = 0$$

$$\int_0^{2\pi}\sin\varphi d\varphi = [-\cos\varphi]_0^{2\pi} = [-\cos 2\pi - (-\cos 0)] = -1 + 1 = 0$$

whereas the third integral is:

$$M_{12} = \frac{\sqrt{3}}{4\pi}r\int_0^\pi \cos^2\theta\sin\theta d\theta\int_0^{2\pi} d\varphi] = \frac{\sqrt{3}}{4\pi}r[-\frac{1}{3}\cos^3\theta]_0^\pi[\varphi]_0^{2\pi}]$$

$$M_{12} = \frac{\sqrt{3}}{4\pi}(-\frac{1}{3})r(\cos^3\pi - \cos^3 0]_0^\pi(2\pi - 0) = \frac{\sqrt{3}}{4\pi}(-\frac{1}{3})r[(-1)^3 - (1)^3](2\pi) = \frac{r}{\sqrt{3}}$$

which is not equal to zero. Thus the transition from state $J = 0, M = 0$ to $J = 1, M = 0$ is an allowed transition.

***Exercise* 5.11:** The rotational transition for NO molecule from $J = 0$ to $J = 1$ is observed at 3.912 cm^{-1}. Assume that the energy density per unit frequency of the radiation field at the location of molecule is that of black body radiation at 1500 K. Indicate whether spontaneous emission is more important or stimulated?

The transition frequency for $J = 0$ to $J = 1$ rotational transition of NO given as wave number can be converted to frequency by the relation as

$$\nu = c\bar{\nu} = (3 \times 10^{10}\,\text{cm s}^{-1})(3.912\,\text{cm}^{-1}) = 1.1736 \times 10^{11}\,\text{s}^{-1}$$

Energy density per unit wavelength for Planck's expression for the black body radiation is given as:

$$U(T,\lambda) = \frac{8\pi hc}{\lambda^5}\frac{1}{e^{\frac{hc}{\lambda\kappa T}} - 1}d\lambda \tag{A}$$

Equation (A) can be converted into the energy density per unit frequency by using the relation

$$\nu = \frac{c}{\lambda} \tag{B}$$

and

$$\left|\frac{d\nu}{d\lambda}\right| = \frac{c}{\lambda^2} \tag{C}$$

Combining equations (B), (C) and (A), we get

$$U(T,\nu)d\nu = \frac{8\pi h\nu^3}{c^3}\frac{1}{e^{\frac{h\nu}{\kappa T}} - 1}d\nu = \rho d\nu \tag{D}$$

Therefore

$$\rho = \frac{8\pi h\nu^3}{c^3}\frac{1}{e^{\frac{h\nu}{\kappa T}} - 1} \tag{E}$$

It is known that the important of spontaneous emission relative to stimulated emission depends on the ratio

$$\frac{A_{21}\text{ (spontaneous)}}{B_{21}\text{ (stimulated)}} = \frac{\frac{8\pi h\nu^3}{c^3}}{\rho} \tag{F}$$

Substituting the value of ρ in equation (F), we get

$$\frac{A_{21}\text{ (spontaneous)}}{B_{21}\text{ (stimulated)}} = \frac{\frac{8\pi h\nu^3}{c^3}}{\rho} = e^{\frac{h\nu}{\kappa T}} - 1 \tag{G}$$

$$\frac{h\nu}{\kappa T} = \frac{(6.626 \times 10^{-34}\,\text{J s})(1.1736 \times 10^{11}\,\text{s}^{-1})}{(1.38 \times 10^{-23}\,\text{J K}^{-1})(1500\,\text{K})} = 3.7567 \times 10^{-3}$$

Equation (G) can be written as

$$\frac{A_{21}\text{ (spontaneous)}}{B_{21}\text{ (stimulated)}} = \frac{\frac{8\pi h\nu^3}{c^3}}{\rho} = e^{\frac{h\nu}{\kappa T}} - 1 = e^{3.7567 \times 10^{-3}} - 1 = 3.7738 \times 10^{-3}$$

Thus the spontaneous emission is only about 0.37567 % of the total emission.

Table 5.3 lists the rotational constants and bond lengths of the few selected molecules.

TABLE 5.3. Rotational constants and bond lengths of few selected molecules

Molecule	B_e (cm^{-1})	r_e(nm)	Molecule	B_e (cm^{-1})	r_e (nm)
HBr	8.473	0.1413	^{79}Br^{19}F	0.357	0.1756
HCl	10.591	0.1275	^{2}H^{35}Cl	5.445	0.1274
HF	20.939	0.0917	^{2}H^{19}F	11.007	0.0917
HI	6.551	0.1604	^{127}I^{35}Cl	0.114	0.2301
CO	1.931	0.1128	^{16}O^{1}H	18.872	0.09706
NO	1.705	0.1151	^{16}O^{2}H	10.01	0.0977

5.4.1. Limitations of Simple Model

It may be stated here that it is not so simple even for diatomic molecules to get the exact information about the structure of the molecules from the simple model considered above. Here are few points which will be discussed without going into the detailed discussion.

5.4.1.1. *Effect of Isotopes on the Rotational Spectrum*

The bond length in case of isotopic substituted molecules remained unchanged, so the only change to be observed will be in the reduced mass and thus in the rotational constant. Thus the ratio of the rotational constants are given as

$$\frac{B}{B'} = \frac{h \times 10^{-2}}{8\pi^2 \mu r^2 c} \times \frac{8\pi^2 \mu' r^2 c}{h \times 10^{-2}} = \frac{\mu'}{\mu} \tag{5.45}$$

where B' and μ' are the respective parameters for the rotational motion of the isotopic systems. Thus the value of the rotational constant in case of substitution by a heavy isotope will decrease.

***Exercise* 5.12:** Consider molecule DF. Assuming that, the bond distance does not change with change of isotope, calculate: (*a*) reduced mass, (*b*) moment of inertia, (*c*) value of *B* and (*d*) $\bar{\nu}$ for the transition $J = 4 \rightarrow J = 5$. Compare this value with that obtained in case of HF molecule. The bond length of both DF and HF is 0.093 nm.

(*a*) $$\mu = \frac{m_D\, m_F}{m_D + m_F} = \frac{(2\text{ g mol}^{-1}) \times (19\text{ g mol}^{-1})}{(2+19)(\text{g mol}^{-1}) \times 6.02\times10^{23}\text{mol}^{-1}} = 3.0 \times 10^{-24}\text{g} = 3.0 \times 10^{-27}\text{kg}$$

(*b*) $$I = \mu r^2 = (3.0 \times 10^{-27}\text{ kg})\,(0.093 \times 10^{-9}\text{ m})^2 = 2.6 \times 10^{-47}\text{ kg m}^2$$

(*c*) $B = \dfrac{h}{8\,\pi^2\, I\, c} = \dfrac{(6.62\times10^{-34}\text{ J s})\,(10^{-2}\text{ m cm}^{-1})}{(8\,\pi^2)\,(2.6\times10^{-47}\text{kg m}^2)\,(3\times10^8\text{m s}^{-1})} = 10.74\text{ cm}^{-1}$

(*d*) $\bar{\nu}_{4\to5} = 2\,B\,(J+1) = (2)\,(10.74)\,(4+1) = 107.4\text{ cm}^{-1}$

Now carrying out the same calculation for HF molecule

(*a*) $\mu = \dfrac{m_H\, m_F}{m_H + m_F} = \dfrac{(1\text{ g mol}^{-1})\times(19\text{ g mol}^{-1})}{(1+19)(\text{g mol}^{-1})\times6.02\times10^{23}\text{mol}^{-1}} = 1.578\times10^{-24}\text{ g}$

$= 1.578\times10^{-27}\text{ kg}$

(*b*) $I = \mu\, r^2 = (1.578\times10^{-27}\text{ kg})\,(0.093\times10^{-9}\text{ m})^2 = 1.365\times10^{-47}\text{ kg m}^2$

(*c*) $B = \dfrac{h}{8\,\pi^2\, Ic} = \dfrac{(6.62\times10^{-34}\text{ J s})\,(10^{-2}\text{ m cm}^{-1})}{(8\,\pi^2)\,(1.365\times10^{-47}\text{kg m}^2)\,(3\times10^8\text{m s}^{-1})} = 20.46\text{ cm}^{-1}$

(*d*) $\bar{\nu}_{4\to5} = 2\,B\,(\text{J}+1) = (2)\,(20.46)\,(4+1) = 204.6\text{ cm}^{-1}$

Utilizing B and reduced masses μ and μ' one can calculate B' directly from B,

$$B' = \frac{B\,\mu}{\mu'} = \frac{(20.46\text{ cm}^{-1})\,(1.578\times10^{-27}\text{ kg})}{3.0\times10^{-27}\text{ kg}} = 10.76\text{ cm}^{-1}$$

One arrives at the *conclusion that rotational constant is inversely proportional to the reduced mass*.

Calculation of Mass of the Atom from the Rotational Spectrum

Equation (5.45) can be modified to calculate the atomic mass of the isotope as follows:

$$\frac{\mu'}{\mu} = \frac{\dfrac{m_{A'}\, m_B}{m_A' + m_B}}{\dfrac{m_A\, m_B}{m_A + m_B}} = \frac{m_{A'}(m_A + m_B)}{m_A(m_{A'} + m_B)} \qquad \text{...(5.46)}$$

or it can be arranged as the ratio of the isotopic masses

$$\frac{m_{A'}}{m_A} = \left(\frac{\mu'}{\mu}\right)\left(\frac{m_B}{m_A}\right)\left[\frac{1}{1+\dfrac{m_B}{m_A} - \dfrac{\mu'}{\mu}}\right] \qquad \text{...(5.47)}$$

or

$$\frac{m_{A'}}{m_A} = \left(\frac{B}{B'}\right)\left(\frac{m_B}{m_A}\right)\left[\frac{1}{1+\dfrac{m_B}{m_A} - \dfrac{B}{B'}}\right] \qquad \text{...(5.48)}$$

Knowing the values of μ, μ', m_B and m_A, one can calculate the mass $m_{A'}$.

***Exercise* 5.13:** The average value of spacing between the adjacent lines in the rotational spectrum of diatomic molecule CX is 3.84235 cm^{-1}. The equilibrium distance is 0.1131 nm. Find the atomic mass of X.

Given $2B = 3.84235\ \text{cm}^{-1}$. Thus, $B = 1.921175\ \text{cm}^{-1}$ and $r = 0.1131$ nm

Using the equation

$$I = \frac{h}{8\pi^2 Bc} = \frac{(6.626 \times 10^{-34}\ \text{J s})(10^{-2}\ \text{m cm}^{-1})}{(8)(3.14)^2(1.921175\ \text{cm}^{-1})(3 \times 10^8\ \text{m s}^{-1})} = 1.458 \times 10^{-46}\ \text{kg m}^2$$

and $$\mu = \frac{1.458 \times 10^{-46}\ \text{kg m}^2}{\left(0.1131 \times 10^{-9}\ \text{m}\right)^2} = 113.98 \times 10^{-28}\ \text{kg}$$

Let the mass of atom X in CX is m_x. Then applying equation (5.37), we obtain

$$113.98 \times 10^{-28}\ \text{kg} = \left(\frac{12m_x}{12 + m_x}\right)\left(\frac{1}{6.02 \times 10^{23}}\right) \text{ or } m_x = 16.023\ \text{g}$$

Hence the CX molecule is actually CO.

***Exercise* 5.14:** The rotational constants for $^{12}C^{16}O$ and $^{13}C^{16}O$ are found to be 1.92118 and 1.83669 cm^{1} respectively. Calculate the ratio of ^{13}C and ^{12}C.

Using Equation. (5.48), we get

$$\frac{m_{^{13}C}}{m_{^{12}C}} = \left(\frac{1.92118}{1.83669}\right)\left(\frac{15.9994}{12}\right)\left[\frac{1}{\left(1 + \frac{15.9994}{12}\right) - \left(\frac{1.92118}{1.83669}\right)}\right]$$

$$\frac{m_{^{13}C}}{m_{^{12}C}} = (1.04600)\ (1.33328)\left[\frac{1}{(1 + 1.33328 - 1.04600)}\right] = 1.08338$$

$$m_{^{13}C} = 12 \times 1.08338 = 13.0006$$

This is within an accuracy of 0.02% of the best value obtained by other methods.

5.4.1.2. *Non-rigid Rotator*

In real sense the atoms in a molecule are not connected to each other by rigid bars, but can be compared to the ball and spring assemblies. Thus, as the molecules acquire rotational energy, the chemical bonds get extended due to *centrifugal distortion*. This will increase the bond length (inter-nuclear distance), moment of inertia and change the energy of the rotational levels. The values of the energy levels are given as

$$E_J = \frac{h^2}{8\pi^2 I} J(J+1) - \frac{h^4}{32\pi^4 I^2 r^2 k} J^2 (J+1)^2 \qquad \text{...(5.49)}$$

or $$F(J) = \frac{h}{8\pi^2 I c} J(J+1) - \frac{h^3}{32\pi^4 I^2 r^2 k c} J^2 (J+1)^2$$

$$F(J) = B\,J(J+1) - D\,J^2 (J+1)^2 \qquad \text{...(5.50)}$$

where B has been defined earlier and the centrifugal distortion constant (D, cm^{-1}) is defined as

$$D = \frac{h^3 \times 10^{-2}}{32\,\pi^4\, I^2\, r^2\, k\, c} \qquad ...(5.51)$$

D is a positive quantity. The value of D is much smaller than the value of B ($D << B$) and thus the centrifugal distortion will only affect the energy of rotational levels for higher value of J. This can be seen from the following equation relating B, D and ω_e.

$$D = \frac{h^3}{32\,\pi^4\, I^2\, r^2\, k\, c} = \frac{16\, B^3\, \pi^2\, \mu\, c^3}{k} \qquad ...(5.52)$$

$$= \frac{4\, B^3}{\omega_e^2} \qquad ...(5.53)$$

The value of B is generally of the order of 10 cm^{-1} and that of ω_e is 10^3 cm^{-1} (see Section 5.4). Thus the value of D is of the order of ~ 10^{-3} of B cm^{-1}. The frequency of rotational lines can be calculated by using the selection rules, $\Delta J = \pm 1$. For absorption spectrum, $\Delta J = +1$.

$$\overline{\nu} = 2\,B\,(J' + 1) - 4\,D\,(J' + 1)^3 \qquad ...(5.54)$$

or general expression will be

$$\overline{\nu} = 2\,B\,(J + 1) - 4\,D\,(J + 1)^3 \qquad ...(5.55)$$

where J = 0, 1, 2, ...

In case of non-rigid rotator, rotational lines in the rotational spectrum are no longer equally spaced but the gap between the successive lines keeps on decreasing with increase of J. Besides calculating the bond length, one can also calculate the approximate value of vibrational frequency knowing B and D from equation (5.55).

***Exercise* 5.15:** The equilibrium value of bond distance (r) in HF is 0.0927 nm (for simplicity assume, it to be 0.09 nm). Suppose that in vibrational level v =1 it is oscillating between 0.08 and 0.1 nm. Calculate the average value of r and $1/r^2$.

The average value of r can be obtained by

$$<r>_{av} = \frac{r_1 + r_2}{2} = \frac{0.08 + 0.1}{2} = 0.09 \text{ nm}$$

Similarly

$$<\frac{1}{r2}>_{av} = \frac{1}{2}\left[\left(\frac{1}{0.08}\right)^2 + \left(\frac{1}{0.1}\right)^2\right] = \frac{1}{2}[156.25+100] = 128.125$$

$$<r>_{av} = \left(\frac{1}{128.125}\right)^{\frac{1}{2}} = 0.0884 \text{ nm}$$

Although the difference between average value of r and average value of r obtained from $1/r^2$ is small but it is there. The percentage difference is 1.78 per cent. The percentage difference increases with increase of v.

***Exercise* 5.16:** Calculate the value of centrifugal distortion constant (D) from the given data. Force constant of HF molecule (k) = 966 N m^{-1}, equilibrium bond distance = 0.0927 nm, B = 20.46 cm^{-1}, ω_e = 4138.5 cm^{-1}.

Using equation (5.52), *i.e.*

$$D = \frac{h^3}{32\,\pi^4\,I^2\,r^2\,k\,c} = \frac{16\,B^3\,\pi^2\,\mu\,c^3}{k}$$

Using the data from exercise (5.12), $I = 1.365 \times 10^{47}$ kg m^2 and substituting the values of the respective constants, we get

$$D = \frac{(6.626 \times 10^{-34}\,\text{J s})^3}{(32)(3.143)^4(1.365 \times 10^{-47}\ \text{kg m}^2)^2(0.093 \times 10^{-9}\,\text{m})^2(966\text{N m}^{-1})(3.0 \times 10^8\,\text{m s}^{-1})}$$

$$D = \frac{290.901 \times 10^{-102}\ \text{m}^{-1}}{(32)(97.584)\ (1.863 \times 10^{-94})\ (8.649 \times 10^{-21})\ (966)(3.0 \times 10^8)}$$

$$= 1.995 \times 10^{-1}\ \text{m}^{-1}$$

$$= (1.995 \times 10^{-1}\ \text{m}^{-1})(10^2\ \text{cm m}^{-1})^{-1} = 1.995 \times 10^{-3}\ \text{cm}^{-1}.$$

Value of D can also be obtained from equation (5.53),

$$D = \frac{4\,B^3}{\omega_e^2} = \frac{(4)(20.46\ \text{cm}^{-1})^3}{4138.5\ \text{cm}^{-1})^2} = 2.00 \times 10^{-3}\ \text{cm}^{-1}$$

which is very close to the value obtained from above relation. It is left for the students to use the third relation to obtain the value of D.

***Exercise* 5.17:** The wave numbers for fourth and fifth rotational lines of HF molecule are at 164.0 and 204.62 cm^{-1} respectively. Calculate B, D, ω_e and r.

The fourth and fifth rotational lines will be observed for $J = 3$ and $J = 4$ respectively. Substituting the respective values in the equation (5.55), we get

$$164 = 2\,B\,(3 + 1) - 4\,D\,(3 + 1)^3 = 8\,B - 256\,D \qquad (i)$$

$$204.62 = 2\,B\,(4 + 1) - 4\,D\,(4 + 1)^3 = 10\,B - 500\,D \qquad (ii)$$

Multiply first equation (*i*) by 10 and second (*ii*) by 8 and subtracting them as we get

$$1640 - 1636.96 = (-2560 + 4000)\ D$$

$$D = \frac{3.04}{1440} = 2.11 \times 10^{-3}\ \text{cm}^{-1}$$

Substituting D in the first equation (*i*), we get

$$B = \frac{1}{8}\ (164 + 256 \times 2.11 \times 10^{-3}) = 20.568\ \text{cm}^{-1}$$

Equation (5.53) gives ω_e

$$\omega_e = \left(\frac{4\ (20.568\ \text{cm}^{-1})^3}{2.11 \times 10^{-3}\ \text{cm}^{-1}}\right)^{\frac{1}{2}} = 4061.4\ \text{cm}^{-2}$$

Using equation (5.43)

$$r = \left[\frac{h \times 10^{-2}}{8\pi^2\ \mu\ B\,c}\right]^{\frac{1}{2}} = \left[\frac{(6.62 \times 10^{-34}\,\text{J s})\ (10^{-2}\,\text{m cm}^{-1})}{(8\pi^2)\ (1.58 \times 10^{-27}\,\text{kg})\ (20.568\ \text{cm}^{-1})\ (3 \times 10^8\ \text{m s}^{-1})}\right]^{\frac{1}{2}}$$

$$= 0.926 \times 10^{-10}\ \text{m} = 0.0926\ \text{nm}.$$

5.5. VIBRATIONAL SPECTRA OF DIATOMIC MOLECULES

5.5.1. *Simple Harmonic Oscillator Problem*

Simplest case for the vibrational motion is to consider that the diatomic molecule is undergoing simple harmonic motion. With this approximation, the vibrational energy levels can be obtained by the equation (5.56):

$$E_{\text{v}} = h\nu\left(\text{v}+\frac{1}{2}\right) \quad \text{v} = 0, 1, 2,... \qquad ...(5.56)$$

where v is the vibrational quantum number, ν is the classical vibrational frequency given as

$$\nu = \frac{1}{2\pi}\sqrt{\frac{k}{\mu}} \qquad ...(5.57)$$

where k is the force constant expressed as N m^{-1}. The vibrational energy levels can be expressed in cm^{-1}, dividing equation (5.56) by hc

$$G(v) = \frac{E_v}{hc} = \frac{v}{c}\left(\text{v}+\frac{1}{2}\right) = w_e\left(\text{v}+\frac{1}{2}\right) \qquad ...(5.58)$$

where

$$w_e = \frac{v}{c} = \frac{10^{-2}}{2\pi c}\sqrt{\frac{k}{\mu}} \qquad ...(5.59)$$

and is expressed in cm^{-1}.

The vibrational transitions can occur only if (*a*) molecule possesses oscillating dipole moment, (*b*) $\Delta\text{v} = \pm 1$. The derivation of these selection rules quantum mechanically are described in Appendix A-5.2. Under these conditions

$$\bar{v} = G(\text{v}+1) - G(\text{v})$$

$$= w_e\left(\text{v}+\frac{1}{2}+1\right) - w_e\left(\text{v}+\frac{1}{2}\right) = w_e \qquad ...(5.60)$$

This means that pure vibrational spectrum consists of only one line at wave number $\bar{v}$. Further as the spacing between the vibrational levels are quite large, the vibrational levels with v > 1 are scarcely populated at normal temperature. Thus, the main contribution to the spectrum will be the transition between v = 0 to v = 1 levels. The vibrational frequency is found in infrared region.

The homonuclear molecules do not possess the vibrational spectra as there is no oscillating dipole moment. Vibrational data for homonuclear molecules are calculated from Raman spectra or from electronic spectrum.

***Exercise* 5.18:** The fundamental vibrational frequency of CO molecule is 2170.2 cm^{-1}. Calculate the force constant of CO molecule.

Equation (5.57) can be used if one knows the reduced mass of CO molecule

$$\mu_{CO} = \frac{m_C m_O}{m_C + m_O} = \frac{(12)(16)}{(12+16)} \times \frac{1}{6.023\times10^{23}}\, g$$

$= 1.138 \times 10^{-23}$ g $= 1.138 \times 10^{-26}$ kg

Equation (5.59) can be rearranged as

$$k = \frac{w_e^2 4\pi^2 c^2 \mu}{(10^{-2}\ \text{m cm}^{-1})^2}$$

$$k = \frac{(2170.2\,\text{cm}^{-1})^2 (4)(3.143)^2 (3.0 \times 10^8\ \text{m s}^{-1})^2 (1.138 \times 10^{-26}\ \text{kg})}{(10^{-2}\ \text{m cm}^{-1})^2}$$

$= 1.907 \times 10^3$ kg s^{-2} $= 1.906 \times 10^3$ N m^{-1}

i.e., the force constant of CO molecule is 2.78×10^2 N m^{-1}.

***Exercise* 5.19:** The force constant for HF molecule is 9.7×10^2 N m^{-1}. Calculate the frequency of radiation to excite the molecules from v = 0 to v = 1.

Solution : $\mu_{HF} = \dfrac{m_H \times m_F}{m_H + m_F} = \dfrac{1 \times 19}{1 + 19} \times \dfrac{1}{6.023 \times 10^{23}}\ g$

$= 1.577 \times 10^{-24}$ g $= 1.577 \times 10^{-27}$ kg

Substituting values of respective terms in equation (5.59)

$$\bar{v} = w_e = \frac{10^{-2}\,\text{m cm}^{-1}}{(2)(3.143)(3.0 \times 10^8\,\text{m s}^{-1})} \left(\frac{9.7 \times 10^2\ \text{Nm}^{-1}}{1.577 \times 10^{-27}\ \text{kg}} \right)^{1/2}$$

$= 4158.9$ cm^{-1}.

5.5.2. *Anharmonic Oscillator*

The potential energy curve representing the simple harmonic motion is shown in Fig. 5.4 by a dotted line whereas the actual one by the solid line. It can be seen that two curves do not match with each other except for small values of v. In case of anharmonic oscillator, the vibrational term can be written as

$$G(\text{v}) = w_e\left(\text{v} + \frac{1}{2}\right) - w_e x_e \left(\text{v} + \frac{1}{2}\right)^2 + w_e y_e \left(\text{v} + \frac{1}{2}\right)^3 + \ldots \quad \ldots(5.61)$$

where $w_e x_e$, $w_e y_e$, ... are the anharmonic constants. In some books the anharmonic constants are expressed simply as x_e, y_e ... . The former are expressed as cm^{-1}, whereas the later are unitless numbers. The value of $w_e x_e << w_e$ and that of $w_e y_e << w_e x_e$. Due to this, the difference between the vibrational energy levels are not constant and the transitions like Δv $= \pm 2$, Δv $= \pm 3$ are also observed. But their intensities are very weak. These transitions are called overtones.

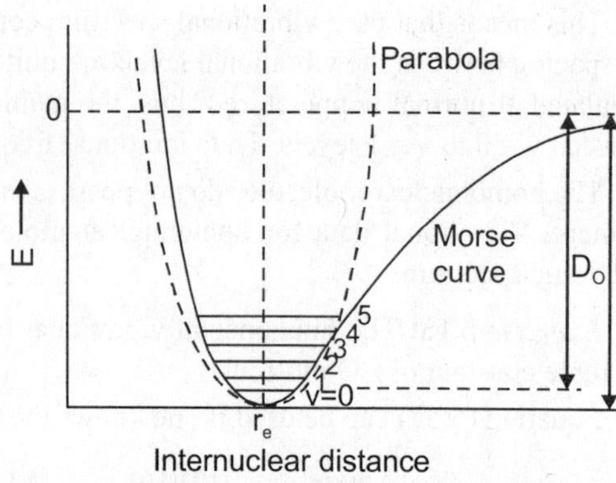

Fig. 5.4 *Shape of potential energy curves*

The zero point energy of the anharmonic oscillator is

$$G(0) = \frac{1}{2}w_e - \frac{1}{4}w_e x_e + \frac{1}{8}w_e y_e + ... \quad ...(5.62)$$

So the energy levels referred to zero point-energy will be

$$G_0(\text{v}) = w_0\text{v} - w_0 x_0 \text{v}^2 + w_0 y_0 \text{v}^3 + ... \quad ...(5.63)$$

where

$$w_0 = w_e - w_e x_e + \frac{3}{4} w_e y_e + ... \quad ...(5.64)$$

$$w_0 x_0 = w_e x_e - \frac{3}{2} w_e y_e + ... \quad ...(5.65)$$

$$w_0 y_0 = w_e y_e + ... \quad ...(5.66)$$

Neglecting the cubic terms, the absorption frequency

$$\bar{v}_{abs} = G_0(\text{v}') - G_0(0)$$
$$= w_0\text{v}' - w_0 x_0 \text{v}'^2 \quad ...(5.67)$$

Separation of the successive absorption bands (in cm^{-1}) which is equal to the separation of successive levels is

$$\Delta G_{\text{v}+1/2} = G(\text{v}+1) - G(\text{v}) = G_0(\text{v}+1) - G_0(\text{v})$$
$$= w_e - 2w_e x_e - 2w_e x_e \text{v} = w_0 - 2w_0 x_0 - 2w_0 x_0 \text{v} \quad ...(5.68)$$

The second difference, neglecting cubic terms is given by

$$\Delta^2 G_{\text{v}+1} = G_{\text{v}+3/2} - G_{\text{v}+1/2}$$
$$= -2w_e x_e = -2w_0 x_0 \quad ...(5.69)$$

The values of anharmonic constants $w_0 x_0$ and $w_e x_e$ can be calculated by the use of equations (5.68) and (5.69).

***Exercise* 5.20:** Calculate the value of $w_e x_e$ and $w_0 x_0$ from the following vibrational spectrum of HCl:

v	0	1	2	3	4	5
$\bar{v}_{obs}$ (cm–1)	0	2885.9	5668.0	8346.9	10923.1	13396.5

The above data can be arranged in the following table:

v	$\bar{v}_{obs}(cm^{-1})$	ΔG_{obs}	$\Delta^2 G_{obs}$
0	0		
1	2885.9	2885.9	–103.7
2	5668.0	2782.1	– 103.2
3	8349.6	2678.9	– 102.8
4	10923.1	2576.1	– 102.6
5	13396.5	2473.4	

The second difference is nearly constant within approximation limit.

The average of $\Delta^2 G_{obs}$ = – 103.2 cm^{-1}. So

$$w_0x_0 = \frac{103.2}{2} = 51.6 \text{ cm}^{-1} = w_ex_e$$

The value of w_0 can be obtained either from the plot of v *vs* $\Delta G_{v+1/2}$ (*i.e.* intercept) or analytically by taking $\bar{v}_{obs}$ from different bands and using equation (5.67) *e.g.*, using the data of 0–1 band:

$$2885.9 = w_0 - 51.6(1)^2$$

$$w_0 = 2885.9 + 51.6 = 2937.5 \text{ cm}^{-1}$$

and w_e from equation (5.58) *i.e.*,

$$2937.5 = w_e - 51.6$$

$$w_e = 2937.5 + 51.6 = 2989.1 \text{ cm}^{-1}$$

The best values of w_e and w_0 are 2988.90 and 2937.30 cm^{-1} respectively.

***Exercise* 5.21:** Plot the energies of the first four vibrational levels of CO molecule. Given $w_e = 2214.24$ cm^{-1}, $w_ex_e = 15.164$ cm^{-1} and $w_ey_e = -0.0007$ cm^{-1}. Note whether the difference between the successive vibrational levels is same or not.

Substituting the values of the respective terms in Equation (5.61) and v = 0.

$$G(0) = (2214.24 \text{ cm}^{-1})(1/2) - (15.164 \text{ cm}^{-1})(1/4) + (-0.0007 \text{ cm}^{-1})(1/8)$$

$$= 1103.33 \text{ cm}^{-1}.$$

Similarly,

$$G(1) = (2214\ 24)(3/2) - (15.164)(9/4) + (-0.0007)(27/8) = 3287.24 \text{ cm}^{-1}$$

and $G(2) = 5440.814$ cm^{-1}, $G(3) = 7564.05$ cm^{-1}, $G(4) = 9656.948$ cm^{-1}

These are plotted in the Fig. 5.5. The difference between successive levels *i.e.*,

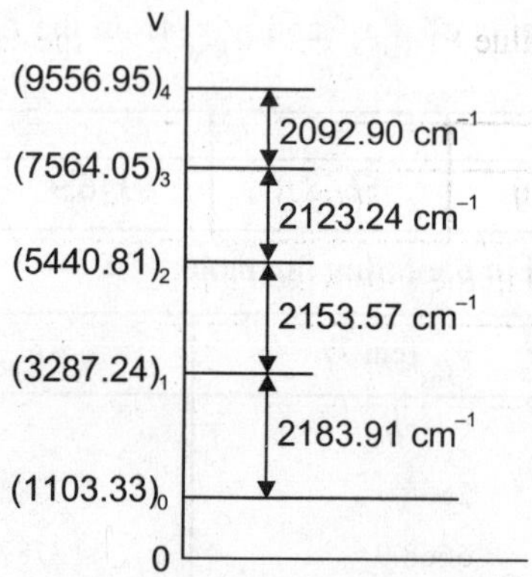

Fig. 5.5 *Plot of vibrational energy levels*

$$G(1) - G(0) = 3287.24 - 1103.33 = 2183.91 \text{ cm}^{-1}$$

$$G(2) - G(1) = 5440.84 - 3287.24 = 2153.57 \text{ cm}^{-1}$$

$$G(3) - G(2) = 7564.05 - 5440.81 = 2123.24 \text{ cm}^{-1}$$

$$G(4) - G(3) = 9656.95 - 7564.05 = 2092.90 \text{ cm}^{-1}$$

The differences between the successive vibrational levels are not equal and this is due to the anharmonicity in the molecule.

***Exercise* 5.22**: The wave functions for the ground state and first excited state of simple harmonic oscillator are $\psi_0(x) = \left(\frac{\alpha}{\pi}\right)^{\frac{1}{4}} e^{-\frac{\alpha x^2}{2}}$ and $\psi_1(x) = \left(\frac{4\alpha^3}{\pi}\right)^{\frac{1}{4}} x e^{-\frac{\alpha x^2}{2}}$ respectively. Show that the transition between these vibrational levels, *i.e.* from v = 0 to v = 1 is allowed.

Transition will be allowed if the transition moment integral is non-zero, *i.e.*

$$M_{0-1} = e \int_{-\infty}^{+\infty} \psi_0(x)\hat{x}\psi_1(x)dx = e \int_{-\infty}^{+\infty} \left(\frac{\alpha}{\pi}\right)^{\frac{1}{4}} e^{-\frac{\alpha x^2}{2}} x \left(\frac{\alpha}{\pi}\right)^{\frac{1}{4}} x e^{-\frac{\alpha x^2}{2}} dx$$

$$= e\left(\frac{4\alpha^4}{\pi^2}\right)^{\frac{1}{4}} \int_{-\infty}^{+\infty} x^2 e^{-\alpha x^2} dx$$

Since the value of integrand is always positive for all the values of x in the range of $-\infty \le x \le +\infty$, it follows that M_{0-1} is non-zero and thus the vibrational transition is allowed.

Table 5.4 lists the vibrational wave numbers, anharmonicity factors and dissociation energies (D_e) for some selected molecules in their electronic ground state.

TABLE 5.4. Vibrational wave numbers, anharmonicity factors and dissociation energies of few molecules in their ground electronic state

Molecule	α_e(cm^{-1})	ω_e(cm^{-1})	$x_e\omega_e$ (cm^{-1})	$y_e\omega_e$(cm^{-1})	K(N m^{-1})	D_e(eV)
$^1H-{}^1H$	3.0622	4,395.2	117.99	0.29		4.476
$^7Li-{}^7Li$		351.43	2.592	–0.0058		1.03
$^{14}N-{}^{14}N$	0.01732	2,359.61	14.456	0.00751		9.756
$^{16}O-{}^{16}O$	0.0159	1,580.36	12.073	0.0546		5.080
$^{19}F^{19}F$	0.13847	916.64	11.236			154.6
$^{35}Cl-{}^{35}Cl$	0.00149	564.9	4.0	–		2.475
$^{79}Br^{79}Br$	0.0003187	325.321	1.0774			190.1
$^{129}I-{}^{129}I$	0.0001138	214,57	0.617	–0.0058		1.03
$H^{19}F$	0.798	4,138.52	90.069	0.980	960.4	≤ 6.40
$H^{35}Cl$	0.3072	2,989.74	52.04	0.056	513	4.430
$H^{79}Br$	0.2333	2648.975	45.218			362.6
$H^{127}I$	0.1689	2309.014	39.644			294.7
$^{35}Cl-{}^{19}F$	0.004358	793.2	9.9	–		2.616
^7LiH		1,405.1	13.228	0.1633		2.5
^{12}CH		2,561.6	64.3	–		3.47
^{16}OH		3,735.21	82.81	–	775	4.35
$^{12}C^{16}O$	0.0175	2,170.21	13.4611	0.0308	1906	11.108
$^{14}N^{16}O$	0.0171	1,904.03	13.97	–0.0012	1598	6.49

***Exercise* 5.23:** The vibrational transitions observed in case of certain molecule, starting from the ground vibrational state are at 4,161.14 cm^{-1}, 8087.11 cm^{-1} and 11,782.35 cm^{-1}. Calculate the values of ω_e, $x_e\,\omega_e$ and $y_e\,\omega_e$ for this molecule.

Using equations (5.61) and (5.62), the transition frequencies of fundamental, and overtones are written as

$$G(0) = \omega_e\,(1/2) - x_e\omega_e(1/2)^2 + y_e\omega_e(1/2)^3 = \frac{1}{2}\omega_e - \frac{1}{4}x_e\omega_e + \frac{1}{8}y_e\omega_e \qquad (A)$$

$$G(1) = \omega_e\,(1+1/2) - x_e\omega_e\,(1+1/2)^2 + y_e\omega_e\,(1+1/2)^3 = \frac{3}{2}\omega_e - \frac{9}{4}x_e\omega_e + \frac{27}{8}y_e\omega_e \qquad (B)$$

$$G(2) = \omega_e\,(2+1/2) - x_e\omega_e\,(2+1/2)^2 + y_e\omega_e(2+1/2)^3 = \frac{5}{2}\omega_e - \frac{25}{4}x_e\omega_e + \frac{125}{8}y_e\omega_e \qquad (C)$$

$$G(3) = \omega_e\,(3+1/2) - x_e\omega_e\,(3+1/2)^2 + y_e\omega_e\,(3+1/2)^3 = \frac{7}{2}\omega_e - \frac{49}{4}x_e\omega_e + \frac{343}{8}y_e\omega_e \qquad (D)$$

Fundamental transition is obtained by taking the difference of (A) and (B), *i.e.*

$$\omega_e\,(0-1) = \omega_e - 2x_e\omega_e + 3.25\,y_e\omega_e = 4161.14 \qquad (E)$$

$$\omega_e\,(0-2) = 2\omega_e - 6x_e\omega_e + 15.5\,y_e\omega_e = 8087.11 \qquad (F)$$

$$\omega_e\,(0-3) = 3\omega_e - 12x_e\omega_e + 42.75\,y_e\omega_e = 11782.35 \qquad (G)$$

There are three unknowns and three linear equations. One can solve these equations by generating simultaneous equations of two variables by eliminating one variable or using Cramer's rule to write the solutions as ratios of the determinants. We will solve here by the first method and second method is left for students. Writing $(F) - 2(E)$ and $(G) - 3(E)$, we get

$$(F) - 2(E) = -2x_e\omega_e + 9y_e\omega_e = -235.17 \qquad (H)$$

$$(G) - 3(E) = -6x_e\omega_e + 33y_e\omega_e = -701.07 \qquad (I)$$

Now multiply (H) by 3 and subtract from (I) and we get

$$6\,y_e\omega_e = 4.44 \quad \text{or} \quad y_e\omega_e = 0.74\ \text{cm}^{-1}$$

Substituting the value $y_e\omega_e$ in (H), we get

$$-2x_e\omega_e + 9\times 0.74 = -235.17$$

or $$-2x_e\omega_e = -235.17 - 6.66 = -241.83 = -235.17$$

or $$x_e\omega_e = 120.92\ \text{cm}^{-1}$$

Now substituting the values of $x_e\omega_e$ and $y_e\omega_e$ in (E) we get

$$\omega_e - 2\times 120.92 + 3.25\times 0.74 = 4161.14$$

or $$\omega_e = 4400.58\ \text{cm}^{-1}$$

So the values of ω_e , $x_e\omega_e$ and $y_e\omega_e$ are 4400.58 cm^{-1}, 120.92 cm^{-1} and 0.74 cm^{-1} respectively. Energy of the $v = 4$ can be obtained by substituting the respective quantities.

$$G(4) = \omega_e\,(4+1/2) - x_e\omega_e\,(4+1/2)^2 + y_e\omega_e(4+1/2)^3 = \frac{9}{2}\omega_e - \frac{81}{4}x_e\omega_e + \frac{729}{8}y_e\omega_e$$

The value of the third overtone is obtained as follows:

i.e. $$G(4) - G(0) = \frac{9}{2}\omega_e - \frac{81}{4}x_e\omega_e + \frac{729}{8}y_e\omega_e - \left(\frac{1}{2}\omega_e - \frac{1}{4}x_e\omega_e + \frac{1}{8}y_e\omega_e\right)$$

$$= 4\omega_e - 20x_e\omega_e + 91y_e\omega_e = 4(4400.58) - 20(120.92) + 91(0.74) = 15{,}251.26 \text{ cm}^{-1}$$

5.6. ROTATIONAL VIBRATIONAL SPECTRA

The assumption that molecule is rotating and vibrating independently is not valid. But in reality molecule is undergoing rotation and vibration simultaneously. The energy of this vibrating rotator is given by

$$\begin{aligned} T(\text{v}, J) &= G(\text{v}) + F(J) \\ &= w_e(\text{v} + 1/2) - w_e x_e(\text{v} + 1/2)^2 + \ldots \\ &\quad + B_{\text{v}'} J(J+1) - D_{\text{v}'} J^2(J+1)^2 + \ldots \end{aligned} \qquad \ldots(5.70)$$

where first term represents the contribution due to harmonic oscillation, the second term and so on, represents the anharmonic vibrations, the third term represents rotational motion and the fourth term the contribution due to the centrifugal distortion or centrifugal stretching of the bonds. The subscript v on B and D indicates the rotational constants of the vibrational level v (*i.e.*, for v = 0 the rotational constants will be B_0, D_0 etc.). This means that the rotational constants are not same for all the vibrational levels. The values of different rotational constants can be calculated from the relation

$$B_{\text{v}} = B_e - \alpha_e(\text{v} + 1/2) + \gamma_e(\text{v} + 1/2)^2 \qquad \ldots(5.71)$$

where B_e is the rotational constant when the equilibrium bond distance is used to calculate moment of inertia, α_e and γ_e are the rotational-vibrational coupling constants. In general, the terms D_{v} and α_e, γ_e are very small.

The selection rules for the rotational-vibrational spectra of hetero-diatomic molecules with $\Lambda = 0$ are $\Delta\text{v} = \pm 1$, $\Delta J = \pm 1$, where Λ is the component of angular momentum of molecule along molecular axis. At room temperature, since many rotational levels are populated, one gets many transitions involving rotational levels. Thus instead of one line one gets a band spectrum corresponding to different rotational levels. Under high resolution, the band consists of number of lines.

For a particular vibrational transition between v′ and v″, the wave numbers for the resulting lines (neglecting the rotational constants D_{v}) is given by

$$\bar{\nu} = \bar{\nu}_0 + B_{\text{v}'} J'(J'+1) - B_{\text{v}''} J''(J''+1) \qquad \ldots(5.72)$$

where $\nu_0 = G(v') - G(v'')$, is the frequency of pure vibrational line without taking account of rotation ($J' = J'' = 0$). With $\Delta J = +1$, one gets a series of lines, called R branch and with $\Delta J = -1$, one gets another series of lines, called P branch. The position of these lines are given by equations (5.73) and (5.74).

$$\bar{\nu}_R = \bar{\nu}_0 + 2B_{\text{v}'} + (3B_{\text{v}'} - B_{\text{v}''})J + (B_{\text{v}'} - B_{\text{v}''})J^2 \qquad \ldots(5.73)$$

where $J = 0, 1, 2, \ldots$

$$\bar{\nu}_P = \bar{\nu}_0 - (B_{\text{v}'} + B_{\text{v}''})J + (B_{\text{v}'} - B_{\text{v}''})J^2$$

where $J = 1, 2, \ldots$ $\qquad \ldots(5.74)$

If we assume that $B_{\text{v}'} = B_{\text{v}''}$ *i.e.*, no interaction between vibrational and rotational motions, the equations (5.73) and (5.74) becomes

$$\bar{\nu}_R = \bar{\nu}_0 + 2B + 2BJ \qquad \ldots(5.75)$$

$$\bar{\nu}_P = \bar{\nu}_0 - 2BJ \qquad \ldots(5.76)$$

i.e., one gets series of equally spaced lines on both sides of $\bar{\nu}_0$ and separation between each line is $2B$. On longer wavelength side it is called P branch and R branch is towards shorter wavelength side of v_0. Due to restriction on the value of J, the line at $\bar{\nu}_0$ is missing and is called the 'zero gap'. Figure 5.6

represents the rotational-vibration spectrum of HCl. Further, due to interaction between rotational and vibrational motion, the values of $B_{v'}$ and $B_{v''}$ are different. If $B_{v'} < B_{v''}$, the fourth term in equation (5.73) will be negative, so the lines in R branch draw closer to each other and for higher values of J, the fourth term may become greater than third. Due to this reversal of lines takes place and the point where the reversal of lines takes place, is called *'band head'*. In equation (5.74), if $B_{v'} < B_{v''}$, The third term will also be megative and increase is more as it involves J^2. Thus the gap between the successive lines increases with increase of J.

The two equations (5.73) and (5.74) can be represented by one equation (5.77):

$$\bar{v} = \bar{v}_0 + (B_{v'} + B_{v''})m + (B_{v'} - B_{v''})m^2 \qquad ...(5.77)$$

$$m = 1, 2, 3, ...$$

where $m = J + 1$ for R branch and $m = -J$ for ρ branch. The line at $\bar{v} = \bar{v}_0$ is also called *'zero line'* or *'null line'* or *'band origin'*.

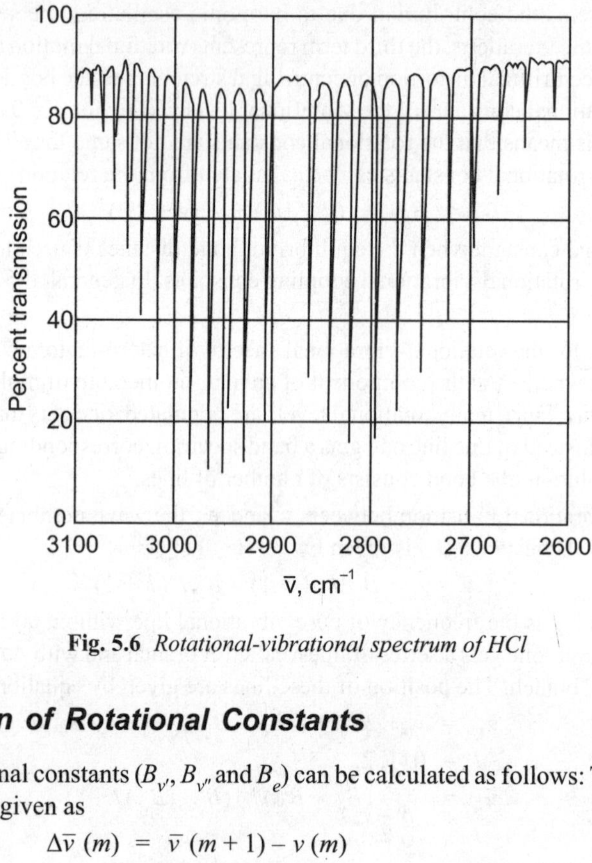

Fig. 5.6 *Rotational-vibrational spectrum of HCl*

5.6.1. *Calculation of Rotational Constants*

METHOD I: The rotational constants ($B_{v'}$, $B_{v''}$ and B_e) can be calculated as follows: The separation between the successive lines is given as

$$\Delta\bar{v}\,(m) = \bar{v}\,(m+1) - v\,(m)$$

$$= 2B_{v'} + 2(B_{v'} - B_{v''})m \qquad ...(5.78)$$

(*i*) Plot between $\Delta\bar{v}\,(m)$ and m will give a straight line. From the slope and intercept, one can calculate $B_{v'}$ and $B_{v''}$.

(*ii*) Second method involves the calculation of second difference*i.e.,* $D^2v(m)$

$$\Delta^2 v(m) = \Delta v\,(m+1) - \Delta v(m)$$

$$= 2(B_{v'} - B_{v''}) \qquad ...(5.79)$$

Thus using the data from equations (5.78) and (5.79) one can calculate $B_{v'}$ and $B_{v''}$.

Further, using the above data and equation (5.71), one can calculate B_e and α_e.

METHOD II: The values of the rotational constants can also be obtained in the following manner. Looking at the equations (5.73) and (5.74), the combinations of $\bar{\nu}_R$ and $\bar{\nu}_P$ are chosen in such a manner that either the transitions in the upper state end on the same J value or it starts from the same lower J values. See for Example (5.25).

(*a*) *Calculation of* $B_{v''}$: If the transition in R and P branch starts from $J-1$ and $J+1$ rotational level, these will end on the same J value of the upper state. The difference of $\bar{\nu}_R\,(J-1)$ and $\bar{\nu}_P(J+1)$ should give the separation of the rotational energy of the lower state, *i.e.*

$$\begin{aligned}\bar{\nu}_R(J-1) - \bar{\nu}_P(J+1) &= \bar{\nu}_0 + 2B_{v'} + (3B_{v'} - B_{v''})(J-1) + (B_{v'} - B_{v''})(J-1)^2 \\ &\quad - \bar{\nu}_0 + (B_{v'} - B_{v''})(J+1) - (B_{v'} - B_{v''})(J+1)^2 \\ &= 2B_{v'} + 4B_{v''}\,J \qquad \text{...(5.80)}\end{aligned}$$

i.e., knowing the value of J one can calculate $B_{v''}$.

(*b*) *Calculation of* $B_{v'}$: Similarly $B_{v'}$ can be calculated by the combination of $\bar{\nu}_R(J)$ and $\bar{\nu}_P(J)$ line *i.e.*, using equations (5.73) and (5.74),

$$\begin{aligned}\bar{\nu}_R(J) - \bar{\nu}_P(J) &= \bar{\nu}_0 + 2B_{v'} + (3B_{v'} - B_{v''})\,J + (B_{v'} - B_{v''})J^2 \\ &\quad - \bar{\nu}_0 + (B_{v'} + B_{v''})\,J - (B_{v'} - B_{v''})J^2 \\ &= 2B_{v'} + 4B_{v''}\,J \qquad \text{...(5.81)}\end{aligned}$$

For heterodiatomic molecules with $\Lambda \neq 0$, the selection rules are $\Delta_v = \pm 1$ and $\Delta J = 0, \pm 1$. Thus for given values of v′ and v″, $\Delta J = 0$ gives a series of lines, called Q branch. The frequencies of these lines are given by the equation.

***Exercise* 5.24:** Predict the rotational-vibrational spectrum of HCl molecule from the given data at 298 K. $\nu = 8.9630 \times 10^{13}\ s^{-1}$, $B = 10.35\ \text{cm}^{-1}$.

Since we deal the transitions in cm^{-1}, so $\omega_e = \dfrac{\nu}{c} = \dfrac{8.9630 \times 10^{13}\ \text{s}^{-1}}{3 \times 10^{10}\ \text{cm s}^{-1}} = 2987.7\ \text{cm}^{-1}$.

The selection rules for the rotational-vibrational absorption spectrum for R-branch are: $\Delta v = +1$, $\Delta J = +1$, whereas for P-branch are: $\Delta v = +1$, $\Delta J = -1$. Assume that the rotational constants in the ground vibrational and first excited vibrational states are the same. We will calculate the transition wavelength for the first line in each branch and rest of the transition lines are compiled in Table 5.5. Using equation (5.75) the first line in R-branch will be:

$$\bar{\nu}_R(\text{v} = 0, J = 0 \rightarrow\ = \text{v} = 1, J = 1) = \omega + 2B = 2989.7\ \text{cm}^{-1} + 2 \times 10.35\ \text{cm}^{-1} = 3010.4\ \text{cm}^{-1}$$

Similarly the first line for the P-branch can be obtained by using equation (5.76), *i.e.*

$$\bar{\nu}_P(\text{v} = 0, J = 1 \rightarrow\ = \text{v} = 1, J = 0) = \omega - 2B = 2989.7\ \text{cm}^{-1} - 2 \times 10.35\ \text{cm}^{-1} = 2969\ \text{cm}^{-1}$$

$$\begin{aligned}\bar{\nu}_Q &= \bar{\nu}_0 + (B_{v''} - B_{v'})\,\Lambda^2 + (B_{v'} - B_{v''})\,J + (B_{v'} - B_{v''})\,J^2 \\ J &= 0, 1, 2, \ldots \qquad \text{...(5.82)}\end{aligned}$$

TABLE 5.5. Transitions predicted for harmonic oscillator-rigid rotator for the R- and P-branch

Initial State v	J	Excited State v	J	Initial Energy (cm^{-1})	Final Energy (cm^{-1})	ΔE (cm^{-1})	ΔE (cm^{-1})
				R-Branch	$\Delta v = 1, \Delta J = +1$		
0	0	1	1	$0.5\ \omega_e$	$1.5\ \omega_e + 2B$	$\omega_e + 2B$	3010.4
0	1	1	2	$0.5\ \omega_e + 2B$	$1.5\ \omega_e + 6B$	$\omega_e + 4B$	3031.1
0	2	1	3	$0.5\ \omega_e + 6B$	$1.5\ \omega_e + 12B$	$\omega_e + 6B$	3051.8
0	3	1	4	$0.5\ \omega_e + 12B$	$1.5\ \omega_e + 20B$	$\omega_e + 8B$	3072.5
0	4	1	5	$0.5\ \omega_e + 20B$	$1.5\ \omega_e + 30B$	$\omega_e + 10B$	3093.2
0	5	1	6	$0.5\ \omega_e + 30B$	$1.5\ \omega_e + 42B$	$\omega_e + 12B$	3113.9
0	6	1	7	$0.5\ \omega_e + 42B$	$1.5\ \omega_e + 56B$	$\omega_e + 14B$	3134.6
0	7		8	$0.5\ \omega_e + 56B$	$1.5\ \omega_e + 72B$	$\omega_e + 16B$	3155.3
				P-Branch	$\Delta v = 1, \Delta J = -1$		
0	1	1	0	$0.5\ \omega_e + 2B$	$1.5\ \omega_e$	$\omega_e - 2B$	2969
0	2	1	1	$0.5\ \omega_e + 6B$	$1.5\ \omega_e + 4B$	$\omega_e - 4B$	2948.3
0	3	1	2	$0.5\ \omega_e + 12B$	$1.5\ \omega_e + 6B$	$\omega_e - 6B$	2927.6
0	4	1	3	$0.5\ \omega_e + 20B$	$1.5\ \omega_e + 12B$	$\omega_e - 8B$	2906.9
0	5	1	4	$0.5\ \omega_e + 30B$	$1.5\ \omega_e + 20B$	$\omega_e - 10B$	2886.2
0	6	1	5	$0.5\ \omega_e + 12B$	$1.5\ \omega_e + 30B$	$\omega_e - 12B$	2865.5
0	7	1	6	$0.5\ \omega_e + 56B$	$1.5\ \omega_e + 42B$	$\omega_e - 14B$	2844.8
0	7		7	$0.5\ \omega_e + 72B$	$1.5\ \omega_e + 56B$	$\omega_e - 16B$	2824.1

Intensity of each rotational line relative to $J = 0$ state can be calculated using Maxwell-Boltzmann relation. It should be kept in mind that intensity of rotational lines in P- and R-branch depends on the initial value of J. So in P-branch the intensity of $J = 2$ to $J = 1$ will be the same as that of R-branch with $J = 2$ to $J = 3$ line as the starting rotational level is same. We will do just one calculation just to explain.

$$\frac{N_J}{N_0} = \frac{g_J}{g_0} e^{-\frac{E_J - E_0}{\kappa T}} = \frac{2J+1}{1} e^{-\frac{BhcJ(J+1)}{\kappa T}}$$

The common factor

$$\frac{Bhc}{\kappa T} = \frac{(10.35\ \text{cm}^{-1})(6.626 \times 10^{-34}\ \text{J s})(3 \times 10^{10}\ \text{cm s}^{-1})}{(1.38 \times 10^{-23}\ \text{J K}^{-1})(298\ \text{K})} = 0.05$$

$$\frac{N_1}{N_0} = \frac{g_1}{g_0} e^{-\frac{E_1 - E_0}{\kappa T}} = \frac{2+1}{1} e^{-0.05[1(1+1)]} = 2.71$$

TABLE 5.6. Relative intensities of the rotational lines in comparison to $J = 0$ line

Initial rotational state (J)	0	1	2	3	4	5	6	7	8	9
Relative intensity	1	2.71	3.70	3.84	3.32	2.45	1.59	0.91	0.466	0.21

Band Head Formation

As mentioned above in case of R-branch, due to reversal of rotational lines ($B_{v'} < B_{v''}$), the band hand formation takes place. The value of J for the band head formation can be calculated from the relation when

$$(B_v' < B_v'')\, J^2 > -(3\,B_v' < B_v'')\, J, \; i.e.$$

$$J \geq -\frac{3\,B_V' - B_V''}{B_V' - B_V''} \qquad ...(5.83)$$

Since the difference between B_v' and B_v'' is not very large for the vibrational levels involved in the rotational-vibrational spectrum, the value of J in general for Band head formation is quite large. Due to this reason, band head formation is often not observed in the infrared spectrum. This is because the relative population of molecules up to such large rotational levels in a given vibrational level is quite small (See Exercise 5.27).

***Exercise* 5.25:** Calculate the frequencies of the lines of P-and R-branch for $J = 4$ in the infrared spectrum of HF. Given $\omega_e = 4138.52$ cm^{-1}, $B_e = 20.94$ cm^{-1}, $\alpha_e = 0.77$ cm^{-1} and the change in the vibrational quantum number is v = 0 to v = 1.

The equations (5.73) and (5.74) can be used to calculate $\bar{v}_R$ and $\bar{v}_P$ for $J = 4$ but one needs B_0, B_1. This can be calculated by using equation (5.71), neglecting γ_e,

$$B_0 = 20.94 - 0.77\,(0 + 1/2) = 20.555 \text{ cm}^{-1}$$

$$B_1 = 20.94 - 0.77(1 + 1/2) = 19.785 \text{ cm}^{-1}$$

$\bar{v}_R$ can be obtained by substituting the respective values in equation (5.73).

$$\bar{v}_R = 4138.52 + (2)(19.785) + (3 \times 19.785 - 20.555)\,(4) + (19.785 - 20.555)\,(4)^2$$

$$= 4332.06 \text{ cm}^{-1}$$

Similarly,

$$\bar{v}_P = 4138.52 - (19.785 + 20.555)\,(4) + (19.785 - 20.555)\,(16)$$

$$= 3975.93 \text{ cm}^{-1}$$

***Exercise* 5.26:** The following data is obtained in IR spectrum of ClO. Calculate, (i) w_0, (ii) B_e, B_0, B_1.

J	*R(J)*(cm^{-1})	*P(J)*	*J*	*R(J)* (cm^{-1})	*P(J)*
0	2906.25		4	2980.90	2798.78
1	2925.78	2865.09	5	2997.78	2775.79
2	2944.89	2843.56	6	3014.29	2752.03
3	2963.24	2821.49	7	3029.96	2727.75

The above problem can be tackled in two ways:

(*i*) Arrange the above data according to the equation (5.77) in the following table.

m	$\bar{\nu}_{obs}$(cm^{-1})	$\Delta\bar{\nu}_m$	$\Delta^2\bar{\nu}_m$
8	3029.96		
		15.67	
7	3014.29		–0.84
		16.51	
6	2997.78		–0.37
		16.88	
5	2980.90		–0.78
		17.66	
4	2963.24		–0.77
		18.43	
3	2944.89		–0.60
		19.03	
2	2925.78		–0.50
		19.53	
1	2906.25		
0			
–1	2865.09		
		21.53	
–2	2843.56		–0.54
		22.07	
–3	2821.49		–0.28
		22.71	
–4	2798.78		–0.77
		22.99	
–5	2775.79		–0.41
		23.76	
–6	2752.03		–0.64
		24.24	
–7	2727.75		

(*ii*) The values of the rotational constants can be found out as:

(*a*) Plot $\Delta\bar{\nu}_m$ vs m (shown in Fig. 5.7) and according to equation (5.78) it should be a straight line. The slope gives $2(B_{v'} - B_{v''})$ and intercept at $m = 0$ gives $2B_{v'}$.

$$\text{From Fig. 5.7, } 2(B_1 - B_0) = \frac{23.85 - 18.35}{-6 - 3} = -0.61$$

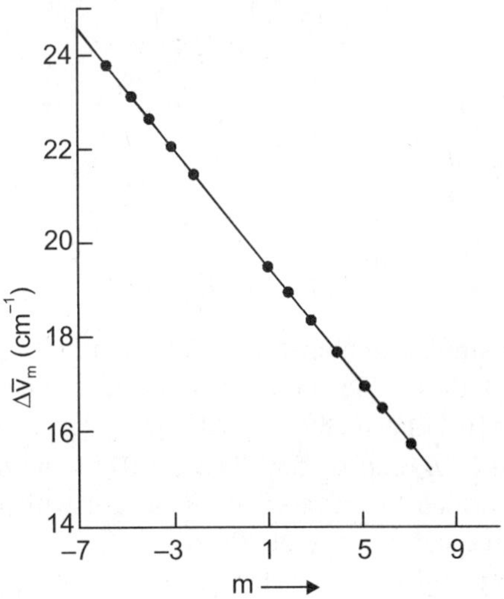

Fig. 5.7 *Plot of* $\Delta\bar{\nu}_m$ vs m

or $\quad B_1 - B_0 = -0.305 \text{ cm}^{-1}; \quad 2B_1 = 20.2 \text{ cm}^{-1}$

Hence $\quad B_1 = 10.1 \text{ cm}^{-1}; \quad B_0 = 10.1 + 0.305 = 10.405 \text{ cm}^{-1}$

Using equation (5.71) and neglecting γ_e

$$10.405 = B_e - \alpha_e/2 \qquad \alpha_e = 0.305$$
$$10.10 = B_e - 3\alpha_e/2 \qquad B_e = 10.557 \text{ cm}^{-1}$$

(*b*) The data in the fourth column is pretty well constant according to the limits of measurements of lines. The average is – 0.591 cm^{-1}, so according to equation (5.79):

$$2(B_1 - B_0) = -0.591 \quad \text{or} \quad B_1 - B_0 = -0.296 \text{ cm}^{-1}$$

Solving equations (5.78) and (5.79) for $m = 7$

$$\Delta\bar{\nu}\,(7) = 2B_1 + 2\,(-0.296)\,(7) = 15.67$$

i.e., $\quad B_1 = 9.907$

and $\quad B_0 = 9.907 + 0.296 = 10.203$

Using equation (5.71), $\quad \alpha_e = +0.296 \text{ cm}^{-1}$, so

$$B_e = 10.203 + 0.296 = 10.499 \text{ cm}^{-1}$$

Of course, the best value can be obtained by taking the average of all the calculations made like this.

METHOD III: Using second method one can form lots of combinations, large number of values of B_1 and B_0 can be obtained. Thus average can be taken.

Here only one such combination will be taken, e.g.,

consider $\bar{\nu}_R\,(3)$ and $\bar{\nu}_p\,(5)$

i.e., $\quad \bar{\nu}_R\,(3) - \bar{\nu}_p\,(5) = 2963.24 - 2775.79 = 187.45 \text{ cm}^{-1}$

Using equation (5.80) and $\quad J = \dfrac{5+3}{2} = 4$

$$2\,B_0 + 4\,B_0\,4 = 187.45$$
$$18\,B_0 = 187.45$$

$$B_0 = \frac{187.45}{18} = 10.414 \text{ cm}^{-1}$$

Similarly, $\bar{\nu}_R(3) - \bar{\nu}_P(3) = 2963.24 - 2821.49 = 141.75 \text{ cm}^{-1}$

Using equation (5.81) and $J = 3$

$$2B_1 + 4B_1.3 = 141.75$$

$$B_1 = \frac{141.75}{14} = 10.125 \text{ cm}^{-1}$$

Similarly by taking other similar combinations other values of B_0 and B_1 can be calculated. The best value will be average of all these. The values of B_e and α_e can be calculated using equation (5.71) *i.e.*, $\alpha_e = 0.289 \text{ cm}^{-1}$ and $B_e = 10.414 + 0.289 = 10.703 \text{ cm}^{-1}$.

***Exercise* 5.27:** The value of B_e and α_e for CO are 1.9313 and 0.0.01747 cm^{-1} respectively. (*a*) Calculate the value of J for which the reversal of R branch will occur. (*b*) Calculate the relative population of this level with respect to $J = 0$ at 300 K.

Using equation (5.71) and neglecting the third term

$$B_0 = 1.9313 - (0.01747)(1/2) = 1.9226 \text{ cm}^{-1}$$

$$B_1 = 1.9313 - (0.01747)(3/2) = 1.9051 \text{ cm}^{-1}$$

Now using equation (5.83)

$$J_{max} \geq -\frac{3B_1 - B_0}{B_1 - B_0} = -\frac{(3)(1.95051) - (1.9226)}{1.9051 - 1.9226} = 217$$

(*b*)
$$\frac{N_{217}}{N_0} = (2J+1)\exp(-B\,h\,c\,J(J+1)/k_bT]$$

$$\frac{B\,h\,c\,J(J+1)}{k_bT} = \frac{(1.9226 \text{ cm}^{-1})(6.62\times10^{-34} \text{ J s})(3\times10^{10} \text{cm s}^{-1})(217)(218)}{(1.38\times10^{-23} \text{ J K}^{-1})(300 \text{ K})} = 436$$

$$\frac{N_{217}}{N_0} = (2\times217+1)\,e^{-436} = 0$$

i.e. band head will not be observed in case of rotational-vibrational spectrum of CO at 300 K.

5.6.2. *Vibrational Spectra of Polyatomic Molecules*

In polyatomic molecules, if there are N atoms then there will be $3N - 6$ vibrational degrees of freedom in non-linear molecules and $3N - 5$ if the molecule is linear. These vibrations can be represented nicely either in terms of internal coordinates (*i.e.*, changes in the bond lengths or bond angles) or in terms of normal modes of vibrations which are obtained by taking linear combinations of vibrations represented in internal coordinates or generalised coordinates in such a manner that the cross terms do not appear in the expression for potential energy.

The vibrational spectrum is observed if either any of the vibrational mode is excited or any combination of the vibrations are involved. The gross-selection rule is that either the molecule must have permanent dipole moment or must possess oscillating dipole moment. Fundamental transitions are those in which all the v_i are zero except one which is equal to unity e.g., v (0, 0, 0, ...) $\rightarrow$ v (1, 0, 0,...).

Overtone bands are those in which all the v_i are zero except one where it is greater than unity, *i.e.*, v (0, 0, 0,...) → v (0, 2, 0,...) etc. Combination bands are those in which various combinations of v_i exist v (0, 0, 0....) → v (0, 1, 1...) etc.

Rotational transitions are also present in vibrational spectra for polyatomic molecules but the analysis of the data is complicated. This is generally observed under very high resolution, since the separation between rotational energy levels is small.

***Exercise* 5.28:** In polyatomic molecules, the vibrational transitions are generally represented in shorthand notation as

(*i*) (0, 0, 0) → (0, 2, 0)	(*ii*) (0, 0, 2) → (0, 1, 1)
(*iii*) (0, 1, 1, 1) → (0, 1, 2, 2)	(*iv*) (0, 0, 0, 0) → (0, 1, 1, 2).

Explain the meanings of the shorthand notations.

Solution: (*i*) The transition (0, 0, 0) → (0, 2, 0) means that the vibration v_2 is excited from $v_2 = 0$ to $v_2 = 2$ and this is first overtone. In all there are three discrete modes of vibrations.

(*ii*) The transition (0, 0, 2) → (0, 1, 1) means v_2 is excited from $v_2 = 0$ to $v_2 = 1$ and v_3 is dexcited from $v_3 = 2$ to $v_3 = 1$, *i.e.* it is a combination band with frequency $v_2 + v_3$. Overall there are three modes of vibrations.

(*iii*) Similarly the transition (0, 1, 1, 1) → (0, 1, 2, 2) is a combination band with frequency $v_3 + v_4$. There are four degrees of vibrations.

(*iv*) The transition (0, 0, 0, 0) → (0, 1, 1, 2) is a combination band with frequency $v_2 + v_3 + 2v_4$ and four degrees of vibrational freedom.

***Exercise* 5.29:** The molecular formula of a compound is XY_2. If this molecule shows three infrared bands, speculate the structure of the molecule.

Solution: If the molecule YXY is linear, it can have four vibrational degrees of freedom. The symmetric stretching will be inactive, asymmetric stretching active and two bending modes which are degenerate, are also active. In all two infrared bands can be observed.

Since, the molecule shows three infrared active bands the molecule can be either YXY bent or YYX linear.

***Exercise* 5.30:** The fundamental vibrations in H_2O have the frequencies: $\bar{v}_1$ (symmetric stretching) 3652 cm^{-1}, $\bar{v}_2$ (bending) 1595 cm^{-1} and $\bar{v}_3$ (asymmetric stretching) 3756 cm^{-1}. Besides these the following transition at 3151.4 cm^{-1}, 5332.0 cm^{-1} and 6874 cm^{-1} are also observed. These are either weak or medium. Assign these transitions and also calculate the energy in Joules for each transition.

Solution: Since, the last three transitions are either weak or medium in intensities, they must be either overtones or combination bands.

(*i*) 3151.4 cm^{-1} band seems to be the overtone of bending mode because 2 × 1595 = 3190 cm^{-1} is close to 3151.4 cm^{-1}. The difference of 48.6 cm^{-1} could be due to anharmonicity in the molecular bond.

(*ii*) 5332.0 cm^{-1} band could be the combination band due to $\bar{v}_2 + \bar{v}_3 = 1595 + 3755.8 = 5350.8$ cm^{-1}. The first overtones of $\bar{v}_1$ or $\bar{v}_3$ are having frequencies which are greater than 5332.0 cm^{-1} band.

(*iii*) 6874.0 cm^{-1} band could be assigned as the combination band $2\bar{v}_2 + \bar{v}_3 = 2 \times 1595 + 3755.8 =$ 6945.8 cm^{-1}.

The energy in Joules for corresponding transition is

(*i*) (3151.4 cm^{-1}) (3.0 × 10^{10} cm s^{-1}) (6.62 × 10^{-34} J s) = 6.26 × 10^{-20} J (*ii*) $(5332.0 \text{ cm}^{-1})(3.0 \times 10^{-10} \text{ cm s}^{-1})(6.62 \times 10^{-34} \text{ J s}) = 1.059 \times 10^{-19} \text{ J}$

(*iii*) $(6874.0 \text{ cm}^{-1})(3.0 \times 10^{10} \text{ cm s}^{-1})(6.62 \times 10^{-34} \text{J s}) = 1.366 \times 10^{-19} \text{ J}$

***Exercise* 5.31**: (*i*) Benzene is a symmetric molecule and thus permanent dipole moment is zero. The number of vibrational modes in benzene is $3 \times 12 - 6 = 30$. Overall dipole moment of benzene molecule is zero as the bond moment of the opposite bonds are acting in opposite directions. The vibrational mode shown in (*i*) is also showing the same symmetry, *i.e.* even the temporary oscillatory motion of each bond is in phase and thus will lead to oscillating dipole moment zero. Normal mode is inactive.

(*ii*) Ethene molecule is again a symmetric molecule and thus the dipole moment is zero. The vibrational motions of the two bonds shown in (*ii*) are acting in opposite direction and will lead to zero dipole moment, whereas the bond moments arising from the oscillatory motion of the other two bonds are acting in the same direction. Thus leading to finite oscillating dipole moment, leading to allowed vibration transition.

(*iii*) NO_2 is a bent molecule and possess the resultant dipole moment. Bond oscillations shown are in phase and thus the temporary oscillation motions will also lead to finite oscillatory dipole moment, leading to allowed vibration transition.

(*iv*) HCN is a linear hetero-triatomic molecule possessing permanent dipole moment and oscillation also lead to finite oscillatory dipole moment. Vibrational transition is allowed.

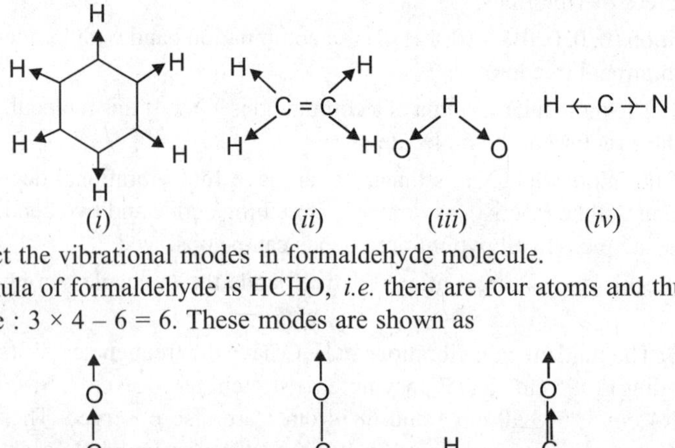

(*i*) (*ii*) (*iii*) (*iv*)

Exercise 5.32: Depict the vibrational modes in formaldehyde molecule.

Molecular formula of formaldehyde is HCHO, *i.e.* there are four atoms and thus the number of vibrational modes are : $3 \times 4 - 6 = 6$. These modes are shown as

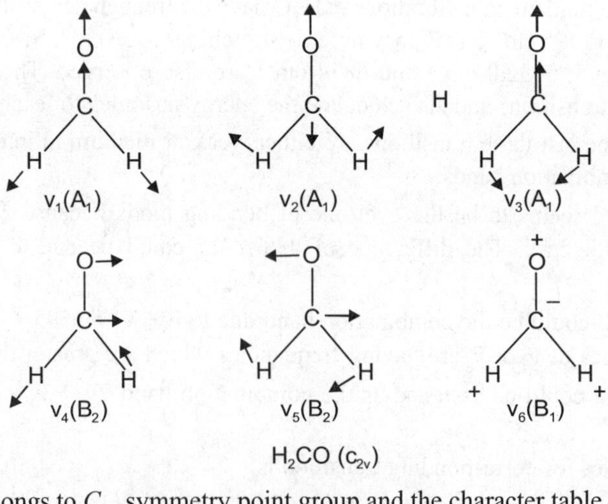

H_2CO (c_{2v})

This molecule belongs to C_{2v} symmetry point group and the character table of C_{2v} symmetry point group is

C_{2v}	E	C_2	σ_v	σ'_v		
A_1	1	1	1	1	T_z	T_{x2}, T_{y2}, T_{z2}
A_2	1	1	–1	–1	R_z	xy
B_1	1	–1	1	–1	T_x, R_y	zx
B_2	1	–1	–1	1	T_y, R_x	yz
Tr	12	–2	2	4	$4A_1 + A_2 + 3B_1 + 4B_2$	

Using the procedure describe in Chapter 0, the trace obtained will be 12, – 2, 2, 4. The reducible representation can be grouped into irreducible ones as

$$a(A_1) = \frac{1}{4}(1 \times 1 \times 12 + 1 \times 1 \times (-2) + 1 \times 1 \times 2 + 1 \times 1 \times 4 = \frac{1}{4}(12 - 2 + 2 + 4) = 4$$

$$a(A_2) = \frac{1}{4}(1 \times 1 \times 12 + 1 \times 1 \times (-2) + 1 \times (-1) \times 2 + 1 \times (-1) \times 4 = \frac{1}{4}(12 - 2 - 2 - 4) = 1$$

$$a(B_1) = \frac{1}{4}(1 \times 1 \times 12 + 1 \times (-1) \times (-2) + 1 \times 1 \times 2 + 1 \times (-1) \times 4 = \frac{1}{4}(12 + 2 + 2 - 4) = 3$$

$$a(B_2) = \frac{1}{4}(1 \times 1 \times 12 + 1 \times (-1) \times (-2) + 1 \times (-1) \times 2 + 1 \times 1 \times 4 = \frac{1}{4}(12 + 2 - 2 + 4) = 4$$

Out of these 12 one dimensional irreducible representations, three are translational (*i.e.* $A_1 + B_1 + B_2$) and three are rotational (*i.e.* $A_2 + B_1 + B_2$). So the irreducible representations belonging to vibrational modes are: $4A_1 + A_2 + 3B_1 + 4B_2 - (A_1 + B_1 + B_2) - (A_2 + B_1 + B_2) = 3A_1 + B_1 + 2B_2$. That is three vibrational modes correspond to A_1, one B_1 and two to B_2 irreducible representations of C_{2v} symmetry point group. To prove these assignments, we have shown below by carrying out the operations as shown below:

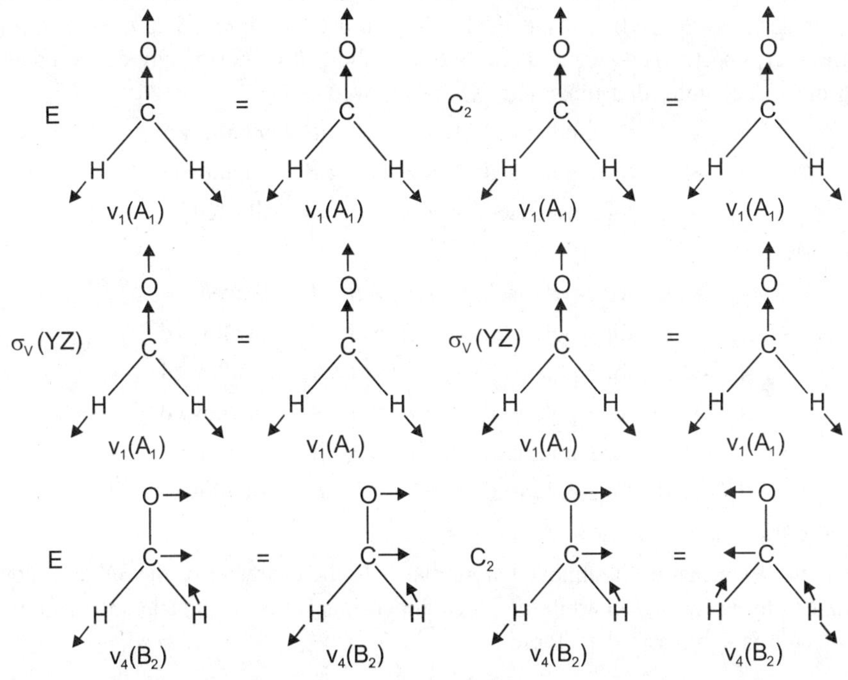

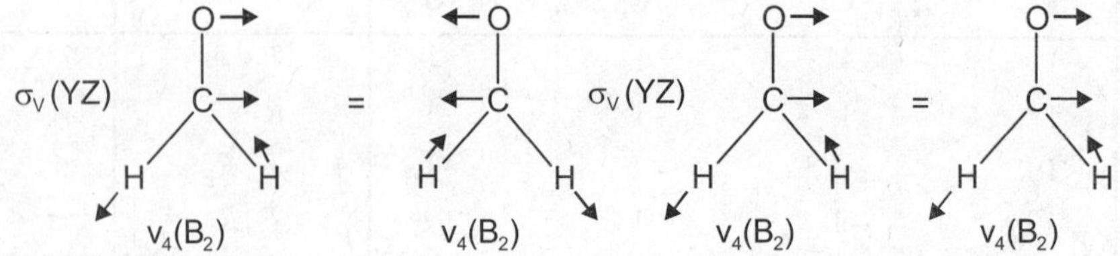

We have carried out two of symmetry operations of C_{2v} symmetry point group to two different types of vibrational modes. The similar treatment to rest of the four vibrational modes is left for student to carry out. The characters obtained are summarised in the Table 5.7 form and desired result is confirmed.

TABLE 5.7. The characters of the symmetry operations of C_{2v} symmetry point group applied to all the vibrational modes

C_{2v}	E	C_{2v}	$\sigma_v(YZ)$	$\sigma_v(XZ)$
$\nu_1\ (A_1)$	1	1	1	1
$\nu_2\ (A_1)$	1	1	1	1
$\nu_3\ (A_1)$	1	1	1	1
$\nu_4\ (B_2)$	1	–1	–1	1
$\nu_5\ (B_2)$	1	–1	–1	1
$\nu_6\ (B_1)$	1	–1	1	–1

***Exercise* 5.33**: Carry out the Exercise 5.32 and treat by group theory to prove that these transitions are allowed.

We know that the ground state of the vibrational motion is totally symmetric and thus belongs to A_1 irreducible representation of C_{2v} symmetric group. The *x*, *y* and *z* vectors of the magnetic moment belong to B_1, B_2 and A_1 respectively. We can use equations (5.25*a*, *b*, *c*) to prove whether the transitions $A_1 \to A_1$, $A_1 \to B_1$ and $A_1 \to B_2$ are allowed or not. If any of the equations (5.25*a*, *b*, *c*) turn out to be finite, that transition will be allowed.

$A_1 \to A_1$: for *x*-polarised: $A_1 \times B_1 \times A_1 : B_1$: not allowed
$A_1 \to A_1$: for *y*-polarised: $A_1 \times B_2 \times A_1 : B_2$: not allowed
$A_1 \to A_1$: for *z*-polarised: $A_1 \times A_1 \times A_1 : A_1$: allowed

Similarly

$A_1 \to B_1$: for *x*-polarised: $A_1 \times B_1 \times B_1 : A_1$: allowed
$A_1 \to B_1$: for *y*-polarised: $A_1 \times B_2 \times B_1 : A_2$: not allowed
$A_1 \to B_1$: for *z*-polarised: $A_1 \times A_1 \times B_1 : B_1$: not allowed
$A_1 \to B_2$: for *x*-polarised: $A_1 \times B_1 \times B_2 : A_2$: not allowed
$A_1 \to B_2$: for *y*-polarised: $A_1 \times B_2 \times B_2 : A_1$: allowed
$A_1 \to B_2$: for *z*-polarised: $A_1 \times A_1 \times B_2 : B_2$: not allowed

All these transitions are allowed

The contribution that each unmoved atom makes to the character of the 3*N*-dimesional representation obtained by operating on arbitrary (three dimensional) vectors attached to each of the *N* atoms of the molecule is summarised in Table 5.8.

TABLE 5.8. The contribution that each unmoved atom makes to the character of the 3N-dimesional representation obtained by operating on arbitrary (three dimensional) vectors attached to each of the N atoms of the molecule

Operation, $\hat{R}$	Contribution to $\chi(\hat{R})$ per unmoved atom
E	3
σ	1
i	–3
C_2	–1
C3, C_3^2	0
C_4^1, C_4^3	1
C_6^1, C_6^5	2
S_2	–3
S_3^1, S_3^2	–2
S_4^1, S_4^3	–1
S_6^1, S_6^5	0

5.7. RAMAN EFFECT

When a light beam is passed through the sample, the light beam is scattered in all directions. The frequency of the scattered light may or may not be different from the original one. If the scattered beam has the same frequency, it is known as Rayleigh scattering. The lines observed at higher frequencies, called 'anti-Stokes' lines are due to the energy absorbed by the photons from the rotationally and vibrationally excited molecules. The lines observed at lower frequencies, called Stokes lines, are due to the energy lost by the photons to the molecules to excite them into different rotational and vibrational levels. This effect of scattering is called Raman effect. The Raman frequencies in cm^{-1} can be calculated with the help of the relation (5.84):

$$\bar{\nu} = \frac{\lambda_{scat\ t} - \lambda_{inc}}{\lambda_{scatt} \cdot \lambda_{inc}} \qquad ...(5.84)$$

Raman effect can be used to find the vibrational frequencies as well as the rotational constants for homonuclear and heteronuclear molecules.

The gross selection rule for both rotational and vibrational spectrum is polarisability or oscillating polarisability in the molecule should be non-zero. The selection rules for pure rotational lines are $\Delta J = 0, \pm 2$ and anti-Stokes lines are generally weaker in intensity than Stokes lines. Since the spacing between rotational energy levels are small, the rotational lines are observed closed to the incident light. The selection rules for vibrational spectra are the same as for infrared spectroscopy, *i.e.* $\Delta v = \pm 1$. $\Delta v = +1$ will give a line at lower frequency than ν_{inc} ($\nu_{inc} - \nu_0$, called Stokes line) and $\Delta v = -1$ will give a line at higher frequency than ν_{inc} ($\nu_{inc} + \nu_0$ anti-Stokes line). Since higher vibrational levels are not populated at room temperature, anti-Stokes lines are generally much weaker than the Stokes lines.

Similar to the rotational-vibrational spectra in infrared, each vibrational band when viewed under high resolution consists of many lines. These lines correspond to the rotational transitions. The selection rules for rotational lines are $\Delta J = 0, \pm 2$.

Position of the lines for particular combination of v′ and v″, the position of the rotational lines for $\Delta J = 0$ (*Q* branch) is given as

$$\bar{\nu} = \bar{\nu}_0 + (B_{v'} - B_{v''})J + (B_{v'} - B_{v''})J^2, \; J = 0, 1, 2, 3, ... \qquad ...(5.85)$$

Position of rotational lines for $\Delta J = +2$ (*S* branch) and for $\Delta J = -2$ (*O* branch are given by the equations (5.86) and (5.87):

$$\bar{\nu} = \bar{\nu}_0 + 6B_{v'} + (5B_{v'} - B_{v''})J + (B_{v'} - B_{v''})J^2 \; J = 0, 1, 2, ... \qquad ...(5.86)$$

and

$$\bar{\nu} = \bar{\nu}_0 + 2B_{v'} - (3B_{v'} - B_{v''})J + (B_{v'} - B_{v''})J^2 \; J = 2, 3, 4, ... \qquad ...(5.87)$$

The equations (5.85), (5.86) and (5.87) have been derived by neglecting the centrifugal correction. All the other terms have the same meaning as explained in Section (5.5). The *S*-and *O*-branches observed are similar to the *P*-and *R*-branches except that line separation is twice as large.

***Exercise* 5.34:** Show that the separation between each rotational line, in case of Rotational Raman Spectrum is $4B$.

The expression for the energies (cm^{-1}) of rotational levels is

$$F(J) = BJ(J+1)$$

The selection rules for the rotational transitions are $\Delta J = \pm 2$. So frequency (cm^{-1}) of absorption is

$$\bar{\nu} = F(J') - F(J) = B[J'(J'+1) - J(J+1)]$$

Now $J' = J + 2$, so

$$\bar{\nu} = B[(J+2)(J+3) - J(J+1)]$$
$$= B(J^2 + 5J + 6 - J^2 - J) = B(4J + 6)$$
$$= 4B(J + 3/2)$$

Since $J = 0, 1, 2, 3, ...$ *i.e.,* it increases by 1, so the difference between each line in rotational Raman Spectrum is $4B$.

***Exercise* 5.35:** If $H^{35}Cl$ is irradiated with 435.8 nm mercury line, calculate the Raman line in nm if the fundamental vibrational frequency of $H^{35}Cl$ is 8.667×10^{13} s^{-1}.

The frequency of mercury line

$$\nu = \frac{c}{\lambda} = \frac{3.0 \times 10^8 \text{ m s}^{-1}}{435.8 \times 10^{-9} \text{ m}} = 68.839 \times 10^{13} \text{ s}^{-1}$$

The Stokes Raman line is observed at

$$\nu_{\text{Raman}} = \nu_{\text{inc}} - \nu_{\text{vib}}$$
$$= 68.839 \times 10^{13} - 8.667 \times 10^{13}$$
$$= 60.172 \times 10^{13} \text{ s}^{-1}$$

The Stokes Raman line in metre, can be obtained.

$$\lambda_{\text{Raman}} = \frac{3.0 \times 10^8 \text{ m s}^{-1}}{60.172 \times 10^{13} \text{ s}^{-1}}$$
$$= 498.57 \times 10^{-9} \text{ m}$$

The above value can also be calculated using equation (5.84).

***Exercise* 5.36:** If the $H^{35}Cl$ molecule is irradiated with 404.7 nm Hg line, calculate the first two Stokes as well as anti-Stokes lines observed in the rotational Raman spectrum. The bond distance of $H^{35}Cl$ molecule is 0.1275 nm

$$\mu = \frac{(1\,\text{g mol}^{-1})\,(35\,\text{g mol}^{-1})}{(36\,\text{g mol}^{-1})\,(6.02\times10^{23}\,\text{mol}^{-1})} = 1.615 \times 10^{-24}\ \text{g}$$

$$= 1.615 \times 10^{-27}\ \text{kg}$$

$$I = (1.615 \times 10^{-27}\ \text{kg})\,(1.275 \times 10^{-10}\ \text{m})^2 = 2.615 \times 10^{-47}\ \text{kg m}^2$$

Using equation (5.39)

$$B = \frac{(6.63\times10^{-34}\ \text{J s})\,(10^{-2}\ \text{m cm}^{-1})}{(8)(\pi^2)(2.615\times10^{-47}\ \text{kg m}^2)(3.0\times10^{8}\ \text{m s}^{-1})} = 10.65\ \text{cm}^{-1}$$

$$v_1\,(J = 0, J = 2) = (4)\,(10.65\ \text{cm}^{-1})\,(0 + 3/2) = 63.9\ \text{cm}^{-1}$$

$$\bar{v}_2\,(J = 1, J = 3) = (4)\,(10.65\ \text{cm}^{-1})\,(1 + 3/2) = 106.5\ \text{cm}^{-1}$$

Using equation (5.84)

$$(10^2\ \text{cm m}^{-1})\,(63.9\ \text{cm}^{-1}) = \frac{\lambda - 404.7\times10^{-9}\text{m}}{(\lambda)(404.7\times10^{-9}\text{m})}$$

$$2.585 \times 10^{-3}\lambda = \lambda - 404.7 \times 10^{-9}\ \text{m}$$

$$\lambda_1 = 405.85 \times 10^{-9}\ \text{m} = 405.85\ \text{nm}$$

Similarly $\lambda_2 = 406.45$ nm

***Exercise* 5.37:** Using the data of Exercise 5.24, predict the rotational-vibrational Raman spectrum of HCl at 298 K. Also calculate the relative intensities of the rotational lines of the *P*- and *R*-branches assuming that these intensities are proportional to the population of the originating states. The wave number of the incident radiation is 23,004 cm^{-1}.

Proceeding on the same lines as carried out in Exercise 5.24

$$\bar{v}_S\,(v = 0, J = 0 \rightarrow = v = 1, J = 2) \quad = \omega + 6B = 2989.7\ \text{cm}^{-1} + 6 \times 10.35\ \text{cm}^{-1} = 3049.8\ \text{cm}^{-1}$$

So the first Raman line of the *S*-branch of vibrational-rotational spectrum of HCl will be at:

$$23004 - 3049.8 = 19954.2\ \text{cm}^{-1}.$$

$$\bar{v}_O\,(v = 0, J = 2 \rightarrow = v = 1, J = 0) \quad = \omega - 6B = 2989.7\ \text{cm}^{-1} - 6 \times 10.35\ \text{cm}^{-1} = 2863.5\ \text{cm}^{-1}$$

So the first Raman line of the *O*-branch of vibrational-rotational spectrum of HCl will be at:

$$23004 - 3049.8 = 19954.2\ \text{cm}^{-1}.$$

Successive lines will be at 4*B* apart and are compiled in Table 5.9.

Similarly the first Raman line of the *O*-branch of rotational Raman line can be calculated.

TABLE 5.9. Transitions predicted for harmonic oscillator-rigid rotator for the S- and O-branches

Initial state v	J	Excited state v	J	Initial energy (cm^{-1})	Final Energy (cm^{-1})	ΔE (cm^{-1})	$\Delta E_R = DE_{inc} - \Delta E$ (cm^{-1})
				S-Branch	$\Delta v = 1, \Delta J = +2$		
0	0	1	2	$0.5\ \omega_e$	$1.5\ \omega_e + 6B$	$\omega_e + 6B$	19954.2
0	1	1	3	$0.5\ \omega_e + 2B$	$1.5\ \omega_e + 12B$	$\omega_e + 10B$	19912.8
0	2	1	4	$0.5\ \omega_e + 6B$	$1.5\ \omega_e + 20B$	$\omega_e + 14B$	19871.4

0	3	1	5	$0.5\ \omega_e + 12B$	$1.5\ \omega_e + 30B$	$\omega_e + 18B$	19830.0
0	4	1	6	$0.5\ \omega_e + 10B$	$1.5\ \omega_e + 42B$	$\omega_e + 22B$	19786.6
0	5	1	7	$0.5\ \omega_e + 30B$	$1.5\ \omega_e + 56B$	$\omega_e + 26B$	19747.2
0	6	1	8	$0.5\ \omega_e + 42B$	$1.5\ \omega_e + 72B$	$\omega_e + 30B$	19705.8
0	7		9	$0.5\ \omega_e + 56B$	$1.5\ \omega_e + 90B$	$\omega_e + 34B$	19664.4
				O-Branch	$\Delta v = 1, \Delta J = -2$		
0	2	1	0	$0.5\ \omega_e + 6B$	$1.5\ \omega_e$	$\omega_e - 6B$	20140.5
0	3	1	1	$0.5\ \omega_e + 12B$	$1.5\ \omega_e + 2B$	$\omega_e - 10B$	20181.9
0	4	1	2	$0.5\ \omega_e + 20B$	$1.5\ \omega_e + 6B$	$\omega_e - 14B$	20223.3
0	5	1	3	$0.5\ \omega_e + 30B$	$1.5\ \omega_e + 12B$	$\omega_e - 18B$	20264.7
0	6	1	4	$0.5\ \omega_e + 42B$	$1.5\ \omega_e + 20B$	$\omega_e - 22B$	20306.1
0	7	1	5	$0.5\ \omega_e + 56B$	$1.5\ \omega_e + 30B$	$\omega_e - 26B$	20347.5
0	8	1	6	$0.5\ \omega_e + 72B$	$1.5\ \omega_e + 42B$	$\omega_e - 30B$	20385.9
0	9		7	$0.5\ \omega_e + 90B$	$1.5\ \omega_e + 56B$	$\omega_e - 34B$	20430.3

5.8. EFFECT OF TEMPERATURE ON THE INTENSITIES OF ROTATIONAL VIBRATIONAL SPECTRA (THERMAL DISTRIBUTION OF QUANTUM STATES)

The intensity of any spectral line depends not only on the transition probability and the frequency (ν) but also on the number of molecules present in different states. Thus, it is necessary to calculate the population of each quantum state to find the theoretical intensities of each line.

The population of each quantum state can be obtained with Maxwell-Boltzmann distribution, *i.e.*,

$$n_i = \frac{N g_i \exp(-\varepsilon_i / k_b T)}{q} \qquad \text{...(5.88)}$$

where n_i is the number of molecules in the *i*th state, N is the total number of molecules, g_i is the degeneracy and ε_i is the energy of the *i*h state, k_b is the Boltzmann constant and q is the partition function (see Chapter 9). The relative population with respect to ground state level is given by

$$\frac{n_i}{n_0} = \frac{g_i}{g_0} \exp\left\{-(\varepsilon_i - \varepsilon_0)/k_b T\right\} \qquad \text{...(5.89)}$$

***Exercise* 5.38:** The difference between the successive rotational lines in case of HCl, is found to be 20.7 cm^{-1}. Calculate the rotational level which will have the highest population at 27°C.

Solution: The population of molecules in any rotational level is given as

$$N_J = \frac{N}{q_{\text{rot}}}(2J+1)\exp\left\{-\varepsilon_{\text{rot}}/k_b T\right\} \qquad \text{...(5.89a)}$$

where q_{rot} is the rotational partition function and independent of rotational quantum number, $(2J + 1)$ is degeneracy of rotational levels and

$$\varepsilon_{\text{rot}} = \frac{h^2}{8\pi^2 I} J(J+1)$$

$$= Bhc\, J\,(J+1)$$

The expression (5.89*a*) can be written as

$$N_J = \frac{N}{q_{\text{rot}}}(2J+1)\exp\left\{-\frac{Bhc(J^2+J)}{k_bT}\right\} \qquad \text{...(5.89}b\text{)}$$

The maximum population of any rotational level at any fixed temperature is obtained by setting of $dN_J/dJ = 0$ and solving the respective equation. So differentiating (5.89*b*) with respect to *J*

$$0 = \frac{dN_J}{dJ} = \frac{N}{q_{\text{rot}}}(2)\exp\left\{-\frac{Bhc(J^2+J)}{k_bT}\right\}$$

$$-\frac{N}{q_{\text{rot}}}(2J+1)\frac{(2J+1)Bhc}{k_bT}\exp\left\{-\frac{Bhc(J^2+J)}{k_bT}\right\}$$

or

$$\frac{N}{q_{\text{rot}}}\exp\left\{-\frac{Bhc(J^2+J)}{k_bT}\right\}\left(2-\frac{(2J+1)^2Bhc}{k_bT}\right) = 0$$

$$J = -\frac{1}{2}+\frac{1}{2}\left(\frac{2k_bT}{Bhc}\right)^{1/2} \qquad \text{...(5.90)}$$

Substituting the values of k_b, *B*, *h* and *c* the expression (5.90) becomes

$$J = -\frac{1}{2}+\frac{1}{2}\left[\frac{(2)(1.38\times10^{-23}\,\text{J K}^{-1})(300\text{ K})}{(10.35\text{ cm}^{-1})(6.62\times10^{-34}\,J\,s)(3.0\times10^{8}\,\text{m s}^{-1})(10^{2}\,\text{cm m}^{-1})}\right]^{1/2}$$

$$= 2.67 = 3$$

At 27°C, $J = 3$ rotational level of HCl will be populated maximum.

***Exercise* 5.39:** Justify that vibrational transitions in HCl are generally observed from v = 0 to v = 1 at room temperature (27°C) *i.e.*, v = 0 is generally populated more than higher level. Given the fundamental vibrational frequency is $8.66 \times 10^{13}\ \text{s}^{-1}$.

In this problem, one has to show that N_1/N_0 or N_2/N_0 is much less than unity.

According to Maxwell-Boltzmann distribution, ratio of molecules in v = 1 to v = 0

$$\frac{N_1}{N_0} = \frac{\exp\left\{-hv\left(1+\frac{1}{2}\right)/k_bT\right\}}{\exp\left\{-hv\left(0+\frac{1}{2}\right)/k_bT\right\}} = \exp\,(-\,hv/k_bT)$$

Now

$$\frac{hv}{k_bT} = \frac{(6.623\times10^{-34}\,\text{J s})(8.667\times10^{13}\ \text{s}^{-1})}{(1.38\times10^{-23}\,\text{JK}^{-1})(300\text{ K})}$$

$$= 13.86$$

$$\frac{N_1}{N_0} = e^{-13.86} = 9.59 \times 10^{-7}$$

This clearly shows that v = 1 is very less populated than v = 0 and higher levels (v > 1) will be even less populated than v = 0.

***Exercise* 5.40**: Using the data of the exercise 5.38, *i.e.* $2B = 20.7\ \text{cm}^{-1}$ for HCl molecule. Calculate the intensity ratio of the rotational line arising from $J = 5$ to $J = 3$ at 500 K.

Since $2B = 20.7\ \text{cm}^{-1}$, $B = 10.35\ \text{cm}^{-1}$.

Using equation (5.88), ratio of population of $J = 5$ to $J = 3$ is given as:

$$\frac{N_5}{N_3} = \frac{g_5}{g_3} e^{-\frac{E_5-E_3}{\kappa T}} = \frac{2 \times 5+1}{2 \times 3 + 1} e^{-\frac{E_5-E_3}{\kappa T}} = \frac{11}{7} e^{-\frac{Bhc}{\kappa T}[5(5+1) - 3(3+1)]} = \frac{11}{7} e^{-\frac{18Bhc}{\kappa T}}$$

$$\frac{18Bhc}{\kappa T} = \frac{(18)(10.35\ \text{cm}^{-1})(6.6.26 \times 10^{-34}\ \text{J s})(3 \times 10^{10}\ \text{cm s}^{-1})}{(1.38 \times 10^{-23}\ \text{J K}^{-1})(500\ \text{K})} = 0.536$$

$$\frac{N_5}{N_3} = \frac{11}{7} e^{-0.536} = \frac{11}{7} \times 0.585 = 0.919.$$

***Exercise* 5.41**: The oscillation frequency of $^{129}I^{35}Cl$ is $394.2\ \text{cm}^{-1}$. Assuming that the molecule follows simple harmonic motion, plot $\frac{N_v}{N_0}$ at 300 K and 1000 K.

Since $^{129}I^{35}Cl$ follows a simple harmonic motion

$$E_v = h\,\omega_e\,(v + \frac{1}{2})$$

and Maxwell-Boltzmann relation gives

$$\frac{N_v}{N_0} = \frac{e^{-\frac{E_v}{k_b T}}}{e^{-\frac{E_0}{k_b T}}} = e^{-\frac{h\,\omega_e\,c\,v}{k_b T}}$$

As the degeneracy of vibrational levels is unity.

At 300 K

$$\frac{h\,\omega_e\,c}{k_b\,T} = \frac{(6.62\times10^{-34}\ \text{J s})\,(354.2\ \text{cm}^{-1})\,(3\times10^{8}\ \text{m s}^{-1})}{(1.38\times10^{-23}\text{J K}^{-1})\,(300\ \text{K})\,(10^{-2}\text{m cm}^{-1})} = 1.843$$

At 1000 K,

$$\frac{h\,\omega_e\,c}{k_b\,T} = \frac{(6.62\times10^{-34}\ \text{J s})\,(354.2\ \text{cm}^{-1})\,(3\times10^{8}\ \text{m s}^{-1})}{(1.38\times10^{-23}\text{J K}^{-1})\,(1000\ \text{K})\,(10^{-2}\text{m cm}^{-1})} = 0.553$$

Values of $\frac{N_v}{N_0}$ at 300 K and 1000 K as a function of v are compiled in Table 5.10.

TABLE 5.10. Population ratios with respect to $v = 0$ in different vibrational levels at 300 and 1000 K

V	0	1	2	3	4	5	6	7	8
$\frac{N_v}{N_0}$ 300 K	1	0.16	0.025	0.004	0.00063	0.0001	–	–	–
$\frac{N_v}{N_0}$ 1000K	1	0.58	0.33	0.14	0.095	0.063	0.04	0.02	0.01

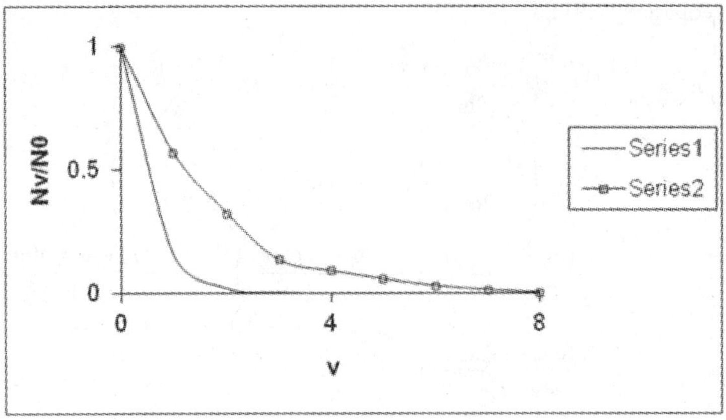

Fig. 5.8 *Plot of N_v/N_0 for $^{129}I\,^{35}Cl$ at 300 K and 1000 K versus* v. *Top line is at 1000 K and bottom line at 300 K*

The data compiled in Table 5.8 are plotted in Fig. 5.8 (1). Plot in the Fig. 5.8 (1) suggests that higher the vibrational levels are more populated than that at lower temperature.

5.9. DISSOCIATING ENERGY OF DIATOMIC MOLECULES

Before we derive the expression for calculating the dissociation energy of diatomic molecules, we would like to explain few terms.

5.9.1. *Maximum Value of Vibrational Quantum Number and Energy*

Neglecting the higher anharmonic constants, equation (5.61) in terms of Joules units can be written as:

$$E_v = hc\omega_e(v + \frac{1}{2}) - hc\omega_e x_e(v + \frac{1}{2})^2 \qquad ...(5.91)$$

$$E_{v+1} = hc\omega_e\left(v + \frac{3}{2}\right) - hc\omega_e x_e\left(v + \frac{3}{2}\right)^2 \qquad ...(5.92)$$

The spacing (ΔE) between the two successive vibrational levels can be obtained by subtracting (5.91) from (5.92) and it is

$$\Delta E = E_{v+1} - E_v = \left[hc\omega_e\left(v+\frac{3}{2}\right) - hc\omega_e x_e\left(v+\frac{3}{2}\right)^2\right] - \left[hc\omega_e\left(v+\frac{1}{2}\right) - hc\omega_e x_e\left(v+\frac{1}{2}\right)^2\right]$$

$$= hc\omega_e - hc\omega_e x_e\,(2v+2) \qquad \text{...(5.93)}$$

For the maximum value of quantum number, *i.e.* $v = v_{max}$, ΔE the spacing between consecutive levels will tend to zero. Consequently equation (5.93) reduces to

$$v_{max} = \frac{\omega_e - 2\omega_e x_e}{2\omega_e x_e} \qquad \text{...(5.94)}$$

The vibrational energy of the maximal vibrational quantum level is determined by substituting the value of v_{max} in equation (5.91), i.e.

$$E_{v_{max}} = hc\omega_e\left(\frac{\omega_e - 2\omega_e x_e}{2\omega_e x_e} + \frac{1}{2}\right) - hc\omega_e x_e\left(\frac{\omega_e - 2\omega_e x_e}{2\omega_e x_e} + \frac{1}{2}\right)^2$$

$$= hc\omega_e\left(\frac{\omega_e - \omega_e x_e}{2\omega_e x_e}\right) - hc\omega_e x_e\left(\frac{\omega_e - \omega_e x_e}{2\omega_e x_e}\right)^2$$

$$= \frac{hc(\omega_e - \omega_e x_e)}{2\omega_e x_e}\left(\omega_e - \frac{\omega_e - \omega_e x_e}{2}\right) = \frac{hc(\omega_e - \omega_e x_e)(\omega_e + \omega_e x_e)}{4\omega_e x_e}$$

$$= \frac{hc(\omega_e^2 - \omega_e^2 x_e^2)}{4\omega_e x_e} \qquad \text{...(5.95)}$$

Since $\omega_e^2 x_e^2 << \omega_e^2$, therefore

$$E_{v_{max}} = \frac{hc\omega_e^2}{4\omega_e x_e} = D_e \qquad \text{...(5.96)}$$

5.9.2. *Bond Energy of Diatomic Molecule*

The bond energy (BE) is the product of the difference between the vibrational energies on the maximal and zeroth vibrational levels and the Avogadro number, N_{av}, *i.e.*

$$\text{BE} = E_{v = v_{max}} - E_{v=0} = \left[\frac{hc\omega_e^2}{4\omega_e x_e} - \frac{hc\omega_e}{2} + \frac{hc\omega_e x_e}{4}\right]N_{av} = \frac{hc}{4}\left[\frac{\omega_e^2}{\omega_e x_e} - 2\omega_e + \omega_e x_e\right]N_{av} \qquad (5.97)$$

***Exercise* 5.41(*a*):** (*a*) Calculate the maximal vibrational quantum number for ClF, if $\omega_e = 78615$ m^{-1}. $\omega_e x_e = 616.3\ m^{-1}$ and $r_e = 1.63 \times 10^{-10}$ m. (*b*) Using the data from part (*a*) determine the vibrational energy of molecule ClF on the zeroth and maximal vibrational quantum levels. Determine also the bond energy, (*c*) with the aid of data from part (*b*) find the vibrational energies of the molecule ClF on the vibrational levels v = 0, 1, 2, 3, 5, 8, 12, 20, 30 and 63.

(*a*) According to equation (5.94)

$$v_{max} = \frac{(78615\ cm^{-1}) - 2 \times (616.3\ cm^{-1})}{2 \times (616.3\ cm^{-1})} = 63$$

(*b*) The maximum value of vibrational energy corresponding to v_{max} = 63 can be determined from equation (5.96).

$$E_{v = v_{max}} = \frac{hc\omega_e^2}{4\omega_e x_e} = \frac{(6.62\times10^{-34}\text{ J s})(3\times10^8\text{ m s}^{-1})(78615\text{ m}^{-1})^2}{4\times616.3\text{ m}^{-1}} = 49.786\times10^{-20}\text{ J}$$

The zero point vibrational energy is determined as follows:

$$E_{v=0} = \frac{hc}{2}\left[\omega_e - \frac{\omega_e x_e}{2}\right]$$

$$= \frac{(6.62\times10^{-34}\text{ J s})(3\times10^8\text{ m s}^{-1})}{2}\left[78615 - \frac{616.3}{2}\right](\text{m}^{-1}) = 0.778\times10^{-20}\text{ J}$$

Therefore, by definition from equation (5.97)

$$\text{BE} = (49.786 - 0.778)\times10^{-20}\times6.023\times10^{23} = 295.18\times10^3\text{ J mol}^{-1} = 295.18\text{ kJ mol}^{-1}$$

(*c*) Here

$$\omega_e = 78615\text{ m}^{-1},\ \omega_e x_e = 616.3\text{ m}^{-1},\ c = 3\times10^8\text{ m s}^{-1},\ h = 6.62\times10^{-34}\text{ J s}^{-1}$$

In order to determine the vibrational energies on the different vibrational levels, determine the terms $hc\,\omega_e$ and $hc\,\omega_e x_e$.

$$hc\,\omega_e = (78615\text{ m}^{-1})(3\times10^8\text{ m s}^{-1})(6.62\times10^{-34}\text{ J s}) = 156.12\times10^{-22}\text{ J}$$

$$hc\,\omega_e x_e = (6.62\times10^{-34}\text{ J s})(3\times10^8\text{ m s}^{-1})(616.3\text{ m}^{-1}) = 1.22\times10^{-22}\text{ J}$$

Now, using equation (5.94), *i.e.*,

$$E_v = \left(v + \frac{1}{2}\right)hc\omega_e - \left(v + \frac{1}{2}\right)^2 hcw_e x_e$$

E_v for different values of v has been determined and is listed in Table 5.11.

TABLE 5.11. Vibrational energy of ClF vibrator on different vibrational levels

V	$\left(v+\frac{1}{2}\right)hcw_e\times10^{+22}\,J$	$\left(v+\frac{1}{2}\right)^2 hc\omega_e x_e\times10^{+22}\,J$	$E_v\times10^{+22}$ J
0	78.06	0.3060	77.76
1	234.19	2.75	233.44
2	390.32	7.65	392.67
3	546.45	14.99	531.46
5	858.71	37.02	821.69
8	1327.09	88.43	1231.66
12	1951.61	191.25	1760.36
20	3200.65	514.40	2686.25
30	4761.94	1138.67	3623.27
63	9913.62	4719.34	4994.27

Now we see from Table 5.11 that as the vibrational quantum number increases the contribution of anharmonicity also increases. Also, we see that for the vibrational quantum numbers 0 and 1, the vibrations are almost harmonic.

5.9.3. *Determination of Dissociation Energy of Diatomic Molecules H_2*

The dissociation energy of diatomic molecules can be determined from the spectroscopic data by the following three methods:

(*i*) Approximate values of D_0 and D_e in terms of cm^{-1} can be determined from the vibration constants of the diatomic molecule, *i.e.* (equation 5.98)

$$D_e = \frac{\omega_e^2}{4\omega_e x_e};\ D_0 = \frac{\omega_0^2}{4\omega_0 x_0} \qquad ...(5.98)$$

***Exercise* 5.42:** Determine the values of D_e and D_0 for FO molecule if $\omega_e = 1060\ cm^{-1}$ and $\omega_e x_e = 15.476\ cm^{-1}$

From equation (5.98)

$$D_e = \frac{(1060\ cm^{-1})^2}{4 \times 15.476\ cm^{-1}} = 18.15 \times 10^3\ cm^{-1}$$

$$= (18.15 \times 10^3\ cm^{-1}) \times (3\times10^{10}\ cm\ s^{-1}) \times (6.62 \times 10^{-34}\ J\ s) \times (6.023 \times 10^{23}\ mol^{-1})$$

$$= 2171.12 \times 10^2\ J\ mol^{-1} = 217.112\ kJ\ mol–1$$

Using equation (5.64)

$$\omega_0 = \omega_e - \omega_e x_e = 1060 - 15.476 = 1044.524\ cm^{-1}$$

Since $\omega_e x_e = \omega_0 x_0$ (By equation 5.65)

Therefore

$$D_0 = \frac{(1045\ cm^{-1})^2}{4\ (1045\times14.6 \times 10^{-3} cm^{-1})} = 17.64 \times 10^3 cm^{-1} = 211.018\ kJ\ mol^{-1}$$

(*ii*) *Band convergence method* is the most accurate method since the discrete spectrum is joined by continuous spectrum. The position of convergence limit corresponds to D_e and D_0. The method fails if convergence limit is not achieved. In such case Birge-Sponer method is used.

We have already proved that the maximum value of vibrational quantum number is given by

$$v_{max} = \frac{\omega_e - 2\omega_e x_e}{2\omega_e x_e} \qquad ...(5.94)$$

The vibrational energy corresponding to v_{max} will then correspond to D_e as shown earlier.

$$E_{v = v_{max}} = \frac{hc}{4\omega_e x_e}\left(\omega_e - \omega_e^2 x_e^2\right) \qquad ...(5.95)$$

$$= \frac{hc\omega_e}{4\omega_e x_e} = D_e \qquad ...(5.95a)$$

i.e. neglecting $\omega_e^2 x_e^2$ with respect to ω_e

It is evident from this expression that vibrational energy of a diatomic molecule cannot exceed its dissociation energy.

(*iii*) *Birge-Sponer method* gives accurate values of D_0 and D_e if large number of vibrational bands is observed. This method is based on the transition energy between two successive quantum vibration levels.

$$\Delta G = -2\omega_e x_e (\text{v} + 1) \qquad ...(5.99)$$

A graph is plotted between ΔG (cm^{-1}) and v. If the molecule is a Morse oscillator, such a plot will be a straight line. Otherwise it is always curved as shown in Fig. 5.9.

The linear extrapolation to x'_m always overestimates v_{max} and hence dissociation energy. The

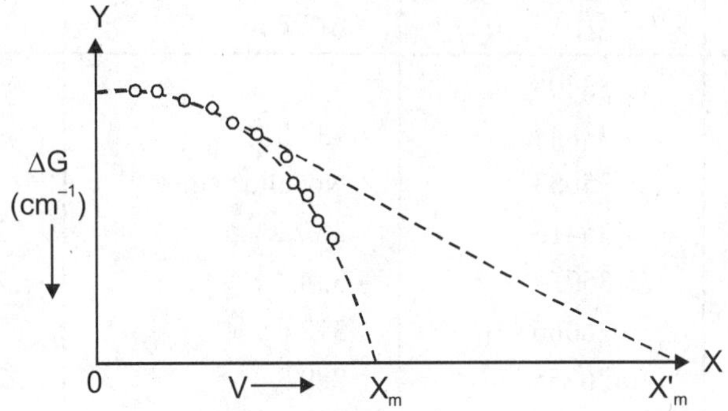

Fig. 5.9 *Plot of DG versus v.*

curve is extrapolated to meet the *x*-axis at x_m. This point corresponds to maximum value of vibrational quantum number, *i.e.* v_{max} since at this point $\Delta G = 0$. To determine D_e. substitute the value of v_{max} into the energy expression (5.91) for the vibrational level.

***Exercise* 5.43**: (*a*) The following data are obtained from the electronic spectrum of ClO radical (Table 5.12).

TABLE 5.12. Rotational-vibrational spectrum of ClO radical

Band, v″, v″	$\bar{\nu}$ cm^{-1}	Band v″, v″	$\bar{\nu}$, cm^{-1}
5	33406.6	17	37264.8
6	33844.4	18	37425.1
9	35053.6	19	37561.9
10	35415.8	20	37678.7
11	35754.5	21	37772.9
12	36066.5	22	37849.5
13	36355.8	23	37917.8
14	36619.5	24	37960.0
15	36864.3	25	37997.4
16	37072.8	—	—

(*i*) Compute the values of the constants ω_e, $\omega_e x_e$ for the upper level and determine the dissociation energy from these constants, and (*ii*) calculate the value of dissociation energy from Birge-Sponer plot and compare it with the one determined in part (*i*).

(*b*) The average value of $\Delta^2 G$ as deduced from the electronic spectrum of BrO radical is –6.40 cm^{-1}. ΔG for the first band is 477 cm^{-1}. The value of v_{max} found from Birge-Sponer plot is 24. Calculate the energy of the vibrational state for v_{max} = 24, ω_0 , ω_e, D_0 and D_e.

(*a*) In order to determine the constants, ω_e, $\omega_e x_e$ and the dissociation energy, the values of ΔG and $\Delta^2 G$ mentioned in equations (5.68) and (5.69) respectively have been calculated and recorded in the Table 5.13.

TABLE 5.13. Data of ClO radical from Table 5.12 are arranged in terms of equations (5.68) and (5.69)

Bond	$\bar{\nu}$,	ΔG, cm^{-1}	$-\Delta^2 G$, cm^{-1}
5	33406.		
6	33844.	437.8	
9	35053.	Not allowed	
10	35415.	362.2	23.5
11	35754.	338.7	26.7
12	36066.	312.0	22.7
13	36355.	289.3	25.6
14	36619.	263.7	18.9
15	36864.	244.8	36.3
16	37072.	208.5	16.5
17	37264.	192.0	31.7
18	37425.	160.3	23.5
19	37561.	136.8	20.0
20	37678.	116.8	22.6
21	37772.	94.2	17.6
22	37849.	76.6	8.3
23	37917.	68.3	26.1
24	37960.	42.2	4.8
25	37997.	37.4	

The average value of $-\Delta^2 G$ has been found to be 21.65 cm^{-1}, (*i*) Proceeding as in Example (5.40), we get

$$\omega_e x_e = \omega_0 x_0 = \frac{21.65}{2} = 10.83 \text{ cm}^{-1}$$

With the aid of equation (5.68), for v = 5, we obtain

$$\Delta G_{(v+1/2)} = \omega_0 - \omega_0 x_0 - 2\omega_0 x_0 \times 5$$

or $$437.8 = \omega_0 - 10.83 - 5 \times 21.66$$

Hence, $$\omega_0 = 546.88 \text{ cm}^{-1}$$

From equation (5.65), we get

$$546.88 = \omega_e - 10.83; \ \omega_e = 567.71 \text{ cm}^{-1}$$

Since $$\omega_e x_e = \omega_0 x_0, \text{ therefore}$$

$$\omega_e x_e = 19.07 \times 10^{-3}$$

and $$D_e = \frac{\omega_e^2}{4\omega_e x_e} = \frac{(567.71 \times 10^3 \text{ cm}^{-1})^2}{4 \times 19.07} = 7442.45 \text{ cm}^{-1} = 88.98 \text{ kJ mol}^{-1}$$

(*ii*) In order to determine dissociation energy from Birge-Sponer method, the maximum value of vibrational quantum number, *i.e.* v_{max} is determined from the plot of ΔG against v shown in Fig. 5.10.

Here $$v_{max} = 28$$

Using equation (5.61), we get

$$G(28) = 567.71\left(\frac{57}{2}\right) - 10.83\left(\frac{57}{2}\right)^2 = 7083.6 \text{ cm}^{-1} = 88.27 \text{ kJ mol}^{-1}.$$

The error with respect to $$G(28) = \frac{|88.27 - 88.98|}{88.27} \times 100 = 0.79 \text{ percent}$$

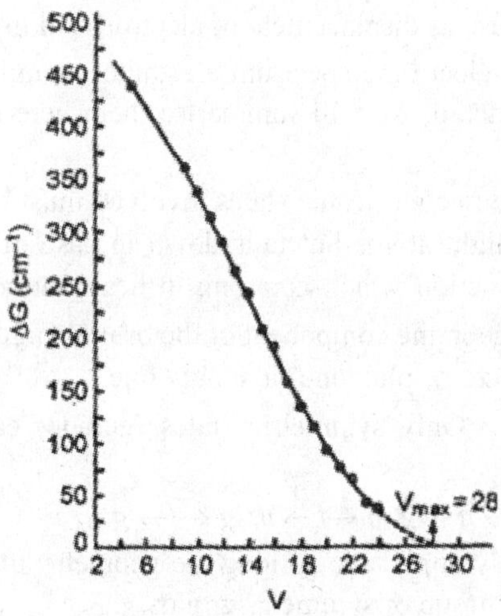

Fig. 5.10 *Plot of ΔG (cm^{-1}) versus v*

(*b*) Proceeding as in Exercise 5.26

$$\omega_e x_e = \omega_0 x_0 \frac{16.4}{2} = 8.20 \text{ cm}^{-1}$$

From equation (5.68), for v = 1, we get

$$\Delta G_{(v+1/2)} = G_{(1+1)} - G_{(1)}$$

$$477 = \omega_e - \omega_e x_e - 2\,\omega_e x_e$$

$$= \omega_0 - 8.20 - 16.4$$

$$\omega_0 = 501.6 \text{ cm}^{-1}$$

So from equation (5.64)

$$501 = \omega_e - 8.2, \omega_e = 509.8 \text{ cm}^{-1}$$

The energy corresponding to the maximum value of vibrational quantum number, *i.e.* 24 is determined from equation (5.63).

$$G(24) = 509.8\left(\frac{49}{2}\right) - 8.2\left(\frac{49}{2}\right)^2 = 7568 \text{ cm}^{-1}$$

$$D_0 = \frac{\omega_0^2}{4\omega_0 x_0} = \frac{(501.6)^2}{4 \times 8.2} = 7670.80 \text{ cm}^{-1} = 91.70 \text{ kJ mol}^{-1}$$

$$D_0 = \frac{\omega_e^2}{4\omega_e x_e} = \frac{(509.8)^2}{4 \times 8.2} = 7923.65 \text{ cm}^{-1} = 94.73 \text{ kJ mol}^{-1}$$

5.10. ELECTRONIC TRANSITIONS

5.10.1. *Diatomic Molecules*

Unlike the rotational and vibrational spectra, the electronic transitions (intense or weak) are possible in all kinds of molecules because as displacement of electron will involve a change of dipole moment.

The selection rules for electronic spectrum are quite complicated. Although we have discussed them in Section (5.3.4) in detail, we will summarise them here in general following rules are quite helpful:

(*i*) $\Delta S = 0$, the spin of the two electronic states involved must be same. This rule is rigidly followed by the molecules of light atoms but fails down in case of heavy atom molecules. This is due to the spin-orbit interaction which is strong in heavy atoms (as explained in Section 5.3.4).

(*ii*) $\Delta\Lambda = 0, \pm 1$ *i.e.*, change in the component of the orbital angular momentum along the internuclear axis should be either zero, plus one or minus one.

(*iii*) *Parity selection rules:* Only symmetric states (gerade) can combine with asymmetric states (ungerade)

$$u \leftrightarrow g, \; u \not\leftrightarrow u, \; g \not\leftrightarrow g \qquad ...(5.100)$$

These rules (*iii*) are only applicable so long the geometry of the molecule is conserved as these rules are based on the point group or symmetry groups.

As vibrational transitions involve the participation of rotational one, the electronical transitions do involve the presence of vibrational and rotational lines. The selection rules for rotational lines are same *i.e.*, $\Delta J = 0, \pm 1$. The participation of vibrational levels of the lower and upper electronic state is governed by "Franck-Condon Principle." Classically, according to Franck-Condon Principle, the transition between those vibrational levels will be intense which do not involve the change in the position or velocity of the nuclei *i.e.*, only verticle transitions (see Fig. 5.11) will be very intense and others will be weak. Quantum mechanically, the intensity of the vibrational bands in electronic transition depends upon the value of M *i.e.*, dipole moment integral, given by the expression:

$$M = \int \psi'_e \hat{M}_e \psi'_e \, d\tau_e \int \psi'_{v'} \, \psi'_{v''} \, d\tau_v \qquad ...(5.101)$$

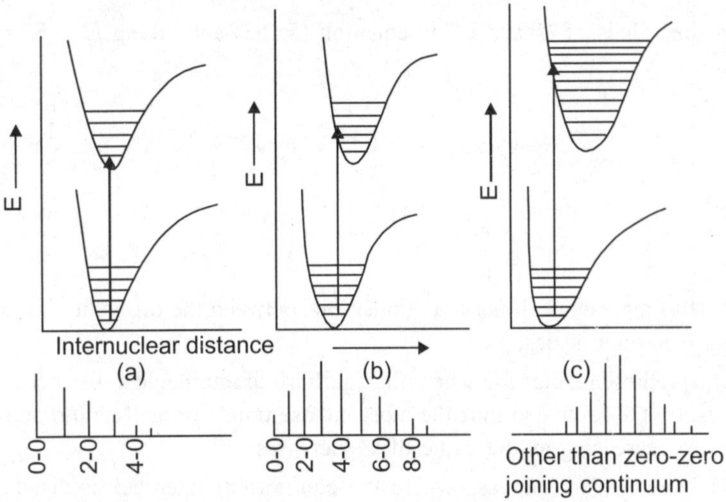

Fig. 5.11 *Franck-Condon Principle*

where ψ_e' and ψ_e'' are the electronic wave functions ψ_v' and ψ_v'' are the vibrational wave functions of the vibrational levels of excited and ground states and $\hat{M}$ is the electric dipole moment operator.. Figure 5.11 shows three different arrangement of the potential energy curves and the resulting intensities of the different vibrational bands.

The transitions between the two electronic states (whether in absorption or emission) in wave numbers of the spectral lines can be obtained by making use of the relation (5.102), *i.e.*

$$\bar{\nu} = T' - T'' = (T'_e - T''_e) + (G' - G'') + (F' - F'') \qquad ...(5.102)$$

where single prime refers to the final state and double prime refers to the initial state. Equation (5.102) indicates that the electronic frequency $\bar{\nu}$ can be expressed as the sum of the three frequencies, corresponding to the electronic, rotational and vibrational, *i.e.*

$$\bar{\nu} = \bar{\nu}_{elec} + \bar{\nu}_{vib} + \bar{\nu}_{rot} \qquad (5.103)$$

5.10.1.1. Vibrational Coarse Structure

It is not necessary that all of the terms in equation (5.103) are positive. In equation (5.103), for a given electronic transition, $\bar{\upsilon}_{elec} = (T'_e - T''_e)$ is constant and $\bar{\upsilon}_{\text{vib}} + \bar{\upsilon}_{\text{rot}}$ is a variable part. The variable part has an expression which is similar to that of rotational-vibrational spectrum. The main differences between the rotational-vibrational spectrum and the electronic transition is that in latter case the values of G' and G'' belong to different electronic states with different values of ω_e and $\omega_e x_e$. It can also be possible that G' may be smaller than G''. Similar is the behaviour of F' and F'' with quite distinct values of B and α_e. As mentioned above the values of F' and F'' are quite small as compared to T' and T'' and for time being rotational terms can be neglected. The spectrum so observed is called *coarse structure* or it is also called as the *vibrational structure* because the value of v is variable.

Substituting the values of G' and G'' in equation (5.103) and taking $F' - F'' = 0$, we obtain the following relation.

$$\bar{\nu} = \bar{\nu}_e + \omega_e'\left(v' + \frac{1}{2}\right) - \omega_e' x_e'(v' + \frac{1}{2})^2 + \omega_e' y_e'(v' + \frac{1}{2})^3 + \ldots\ldots\ldots$$

$$- \omega_e''\left(v'' + \frac{1}{2}\right) - \omega_e'' x_e'\left(v'' + \frac{1}{2}\right)^2 + \omega_e'' y_e'\left(v'' + \frac{1}{2}\right)^3 \ldots. \qquad \ldots(5.104)$$

Equation (5.104) represents all possible transitions between the different vibrational levels of the two participating electronic states.

Figure (5.11,*a*) represents the case where the equilibrium internuclear distance between the ground and excited states are the same and thus the most intense transition will be 0-0 and intensity of other transitions keep on decreasing as the value of v′ increases.

Figure (5.11,*b*) represents the case where the equilibrium internuclear distance for the excited state is *greater* than that of the ground state. In this case the intensity of the transitions increases with increase of v′, reaches a maximum value for some value of v′ and then starts decreasing.

Figure (5.11,*c*) describes the case where the equilibrium internuclear distance of the upper state is very large as compared to that of lower state. In this case, it may happen that the 0-0 transition may be missing, band starts appearing for some value of v′, reaches a maximum value and then starts decreasing with increase of v′.

***Exercise* 5.44:** The mixed orbital wave function of the triplet state is given by $\Psi_T = \Psi_T^0 + \lambda_{TS}\Psi_S^0$, where λ_{TS} is the mixing coefficient of singlet and triplet states. Show that singlet to triplet transition is allowed partially using the mixed wave function.

Using only the spin part of the integral given by equation (5.25), we get

$$\int \Psi''_{\text{spin}}\ \Psi'_{\text{spin}}\ d\tau_{\text{spin}} = \int[\alpha(1)\beta(2) - \alpha(2)\beta(1)] \bullet \begin{bmatrix} \alpha(1)\beta(2) + \alpha(2)\beta(1) \\ +\lambda_{TS}[\alpha(1)\beta(2) - \alpha(2)\beta(1)] \end{bmatrix} d\tau_{\text{spin}}$$

$$= \int[\alpha(1)\beta(2) - \alpha(2)\beta(1)] \bullet [\alpha(1)\beta(2) + \alpha(2)\beta(1)]\, d\tau_{\text{spin}}$$

$$+ \int[\alpha(1)\beta(2) - \alpha(2)\beta(1)] \bullet \lambda_{TS}[\alpha(1)\beta(2) - \alpha(2)\beta(1)]\, d\tau_{\text{spin}}$$

The first integral will be zero as shown in Exercise 5.5, but the second integral will be finite as both the functions belong to singlet states. The magnitude of the second term depends on the value of λ_{TS}, the mixing coefficient of singlet-triplet state. Larger the value of mixing coefficient λ_{TS} greater will be intensity of the singlet-triplet transition.

5.10.1.2. *Some Terms used in the Electronic Spectrum*

Sequence: Consider the lower energy electronic state Ψ''_{elec}. Accordingly allowed transitions are

$$\Psi''_{\text{elec}} : v''(0, 1, 2, 3, \ldots) \rightarrow \Psi'_{\text{elec}} : v'(0, 1, 2, 3, \ldots)$$

$$0 \to 0, 0 \to 1, 0 \to 2, 0 \to 3,...,$$
$$1 \to 0, 1 \to 1, 1 \to 2, 1 \to 3,....;$$
$$2 \to 0, 2 \to 1, 2 \to 2, 2 \to 3,...;$$
$$3 \to 0, 3 \to 1, 3 \to 2, 3 \to 3,...; \text{etc.}$$

Such a set of transitions is termed as band. A series of bands having a constant value of (v′ – v″) *is known as sequence, e.g.* the following sets of transitions constitute sequences:

$$(v' - v''): v' - v'' = 0: 0 \to 0\ 1 \to 1, 2 \to 2,....$$
$$= +1: 0 \to 1, 1 \to 2, 2 \to 3,....$$
$$= -1: 1 \to 0, 2 \to 1, 3 \to 2,....$$
$$= +2: 0 \to 2, 1 \to 3, 2 \to 4,....$$
$$= -2: 2 \to 0, 3 \to 1, 4 \to 2.....$$

....

transitions $1 \to 0$, $2 \to 0$, $3 \to 0$ do not form a sequence since $v' - v'' \neq$ constant.

Progression: *A set of bands having a definite value of* v″ *but varying values of* v′ *and vice versa is called a progression*. For example the following set of transition in bands consitute v″ *progressions, i.e.*

$v''(= 0) = \to v'(0, 1, 2, 3....)$,

$v''(= 1) \to v'(0, 1, 2, 3....)$,

$v''(= 2) \to v'(0, 1, 2, 3....)$, etc.

while for v′ progressions the value of v′ is fixed and values of v″ vary. Both v′ and v″ are

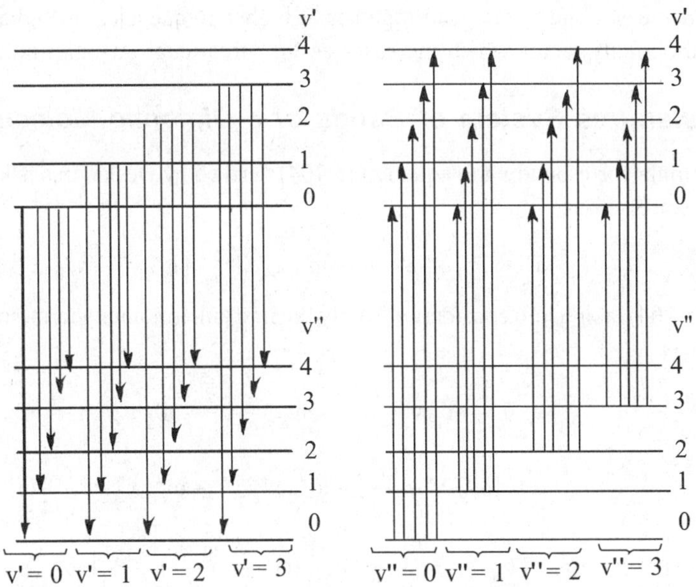

(a) (b)

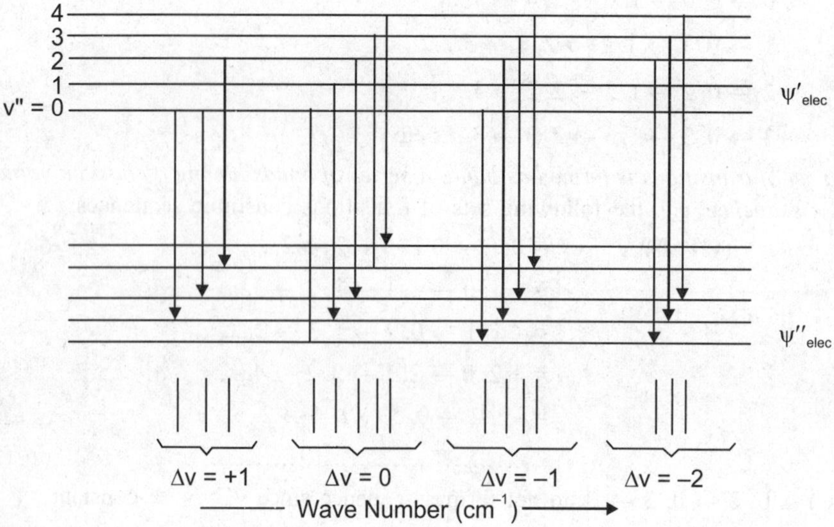

(c)

Fig. 5.12 *Energy level diagrams depicting progressions of (a)* v′*-progression and (b)* v″ *progression are shown in Fig. 5.12a and b. (c) Progressions in the emission spectrum of a pre-excited diatomic molecule.*

depicted in Fig. (5.12*a* and *b*). The vibrational coarse structure of a band v′ (= 0) → v″ (= 1, 2, 3, 4, 5, 6, 7) is presented in Fig. 5.12*c*. Coarse structure reveals that the spacing between the adjacent lines in a band decreases and is crowded together at higher frequencies or higher v′ values. This is because of anharmonicity factor which increases as the vibrational quantum number increases.

5.10.1.3. *Deslandres System of Bands of a Diatomic Molecule*

An empirical formula corresponding to equation (5.104) derived by Deslandres is known as Deslandres formula and is expressed as

$$\bar{\nu} = \bar{\nu}_0 + [\omega'_0 v' - \omega'_0 x_0' v'^2] - [\omega''_{00} v'' - \omega''_{00} x''_0 v''^2] \qquad ...(5.105)$$

Equation (5.104) taking into consideration only the first anharmonicity term, in order of increasing v, can be written as

$$\bar{v} = \bar{v}_e + \omega'_e v' + \frac{\omega'_e}{2} - \omega'_e x'_e v'^2 - \omega'_e x'_e v' - \frac{1}{4}\omega'_e x'_e ....$$

$$- [\omega''_e v'' + \frac{\omega''_e}{2} - \omega''_e x''_e v''^2 - \omega''_e x''_e v'' - \frac{1}{4}\omega''_e x''_e]$$

$$\bar{v} = \bar{v}_{\text{elec-vib}} = [\bar{v}_{\text{elec}} + \frac{1}{2}(\omega'_e - \frac{\omega' x'_e}{2}) - \frac{1}{2}\left(\omega''_e - \frac{\omega'' x''_e}{2}\right) +$$

$$[(\omega'_e - \omega'_e x'_e)v' - \omega'_e x'_e v'^2] - [(\omega''_e - \omega''_e x''_e)v'' - \omega''_e x''_e v''^2] \qquad ...(5.106)$$

Compare equations (5.104) and (5.105) in order to obtain the relations between the constants involved in them,

$$\bar{\nu}_0 = \bar{\nu}_{\text{elec}} + \left(\frac{\omega_e'}{2} - \frac{\omega_e' x_e'}{4}\right) - \left(\frac{\omega_e''}{2} - \frac{\omega_e'' x_e''}{4}\right)$$

$$\bar{\omega}_0'' = \omega_e'' - \omega_e'' x_e''$$

$$\omega_0'' x_0'' = \omega_e'' x_e''$$

$$\omega_0'' = \omega_e' - \omega_e' x_e'$$

$$\omega_0' x_0' = \omega_e' x_e' \qquad \text{...(5.107)}$$

Further, if each frequency of the band is designated as $\bar{\nu}_{v''v'}$ then equation (5.105) can be rewritten as

$$\bar{\nu}_{v''v'} = \bar{\nu}_0 + (\omega_0' v' - \omega_0' x_0' v'^2) - (\omega_0'' v'' - \omega_0'' x_0'' v''^2) \qquad \text{...(5.108)}$$

From this expression, the entire system of bands expressed by the Deslandres formula can be deduced.

The v′ progressions (correspond to v″ = Constant, *i.e.* 0, 1...) as deduced from equation (5.107) can be expressed as

$$\bar{\nu} = \bar{\nu}_{0\,v'} + \omega_0' v' - \omega_0' x_0' v'^2 \qquad \text{...(5.109)}$$

They are due to transitions from a definite vibrational state (v″ = 0, 1 ...) of the ground electronic state Ψ_{elec}'' to all possible vibrational levels (v′ = 0, 1, 2, 3...) of the upper electronic state Ψ_{elec}' (or vice versa). The v′ progressions are a group of bands that converge to a certain limit at high frequency because of anharmonicity parameters and are characteristic of the absorption spectra of molecules such as I_2, Br_2, N_2, O_2, NO, O_3, SO_3, H_2CO, C_6H_6, etc. CO (in the Schumann region, < 2000Å).

The v″ progressions (correspond to v′ = Constant, *i.e.* 0, 1, 2 ....) are expressed as

$$\bar{\nu} = \bar{\nu}_{0\,v''} - \omega_0'' v'' + \omega_0'' x_0'' v''^2 \qquad \text{...(5.110)}$$

The v″ progressions are due to the transitions from a definite vibrational state (v′ = 0, 1, ...) of an upper electronic state Ψ_{elec}' to all possible vibrational levels (v″ = 0, 1, 2, 3, 4...) in the ground electronic state Ψ_{elec}'' of the molecule (or vice versa), *i.e.* 0 ↔ 0, 0 ↔ 1, 0 ↔ 2, 0 ↔ 3 ... and 1 ↔ 0, 1 ↔ 1, 1 ↔ 2, 1 ↔3.... *These progressions are characteristics of the emission spectrum particularly in the fluorescence spectrum, e.g.* fluorescence spectra of I_2 (excited by green line of Hg 546.1 nm), O_2, S_2, NO, SO_3, etc.

The characteristic constant of each system of bands is the band, $\bar{\nu}''_{00}$, *i.e.* the zero line of the band, which corresponds to the transition (v′ = 0 ↔ v″ = 0) and is known as the *zero band.* The frequency of the zero line of the zero band does not coincide with the frequency $\bar{\nu}_{\text{elec}}$ because of zero vibrational (or point) energy of the molecule, *i.e.* $\nu_0 \neq \nu_{\text{elec}}$. *In other words, we can say that the pure electronic transition is strictly forbidden.* This is often called the 0—0 band and is devoid of vibrational quanta, both in the ground and excited states. The band corresponding to a transition from a vibrationally excited state of a ground electronic state (v'', Ψ''_{elec}) to the lower vibrational state of the upper electronic state (v', Ψ'_{elec}) is called the *hot band.*

5.10.1.4. *Energy Difference Between Two Electronic States*

The ground (Ψ''_{elec}) and excited (Ψ'_{elec}) electronic states of a molecule are usually represented, by the different Morse potential curves. The energy difference, $\bar{\nu}_{elec}$, between two electronic states will thus correspond to the energy difference between the minima of two Morse curves. The energy difference between the two minima as deduced from equation (5.105) is given by

$$\bar{\nu}_{elec\text{-}vib}(v''=0 \rightarrow v'=0) = \bar{\nu}_{elec} + \left(\frac{\omega'_e}{2} - \frac{\omega'_e x'_e}{4}\right) - \left(\frac{\omega'_e}{2} - \frac{\omega'_e x'_e}{4}\right) \text{ cm}^{-1}$$

or
$$\bar{\nu}_{elec} = \bar{\nu}_{elec\text{-}vib}(v''=0 \rightarrow v'=0) - \left(\frac{\omega'_e}{2} - \frac{\omega'_e x'_e}{4}\right) + \left(\frac{\omega''_e}{2} - \frac{\omega''_e x''_e}{4}\right) \text{ cm}^{-1}$$

or
$$\bar{\nu}_{elec} = \bar{\nu}_{elec\text{-}vib}(v''=0 \rightarrow v'=0) - \text{zero point energy of the upper state}$$
$$+ \text{ zero point energy of the ground state} \qquad ...(5.111)$$

$[1 \text{ cm}^{-1} \equiv 11.958 \text{ J mol}^{-1}]$

***Exercise* 5.45:** The zero point energy of the ground state of C_2 is 817.8 cm^{-1} and that of its lowest excited state is 890 cm^{-1}. The energy difference between the two minima of the electronic potential energy curves is 3.832 × 10$^{"19}$ J. Determine the energy of v' = 0 → v″ = 0 transition in cm^{-1}. What is the corresponding wavelength?

$$\bar{\nu}_{elec} = \frac{\Delta E}{hc} = \frac{(3.832 \times 10^{-19} \text{ J})}{(6.62 \times 10^{-34} \text{ J s})(3 \times 10^{10} \text{cm s}^{-1})} = 19295.8 \text{ cm}^{-1}$$

Substituting the values of zero point energies of the ground and excited states together with the value of $\bar{\nu}_{elec}$ into equation (5.108), we get

$$\bar{\nu}_{elec\text{-}vib(v''=0 \rightarrow v'=0)} = (19295.8 + 890 - 817.8)\text{cm}^{-1} = 19368.0 \text{ cm}^{-1}$$

The corresponding wavelength is $\dfrac{1}{19{,}368 \text{ cm}^{-1}} = 516.31 \text{ nm}$

***Exercise* 5.46:** The analysis of the vibrational structure of a band in the ultraviolet spectrum of CO leads to the following data:

Ground state $(^1\Sigma^+)$			Excited state $(^1\pi)$
$\omega''_e(\text{cm}^{-1})$	$\omega''_e x''_e$ (cm^{-1})	ω'_e (cm^{-1})	$\omega'_e x'_e$ (cm^{-1})
2170.21	13.46	1515.61	17.24

The $\bar{\nu}_{00}$ transition is observed at 64746.55 cm^{-1}. Calculate (*a*) the zero point energies for the upper and lower states, and (*b*) the energy difference between the two electronic states. What is the corresponding wavelength?

(*a*) From equation (5.112), we write the expression for the zero point energies of the upper and lower states as

$$\text{Zero point energy of the upper state} = \left(\frac{\omega_e'}{2} - \frac{\omega_e' x_e'}{4}\right) \text{cm}^{-1}$$

$$\text{Zero point energy of the lower state} = \left(\frac{\omega_e''}{2} - \frac{\omega_e'' x_e''}{4}\right) \text{cm}^{-1}$$

Substituting the values of the respective constants from the example, we get the zero point energy of the upper state

$$= \left(\frac{1515.61}{2} - \frac{17.24}{4}\right) \text{cm}^{-1} = (757.80 - 4.31)\ \text{cm}^{-1} = 753.49\ \text{cm}^{-1}$$

and zero point energy of the ground state

$$= \left(\frac{2170.21}{2} - \frac{13.46}{4}\right) \text{cm}^{-1} = (1085.1 - 3.36)\ \text{cm}^{-1} = 1081.74\ \text{cm}^{-1}$$

(*b*) Substituting these values of energies together with the value of ν_{00} from the problem into equation (5.110), we get the energy difference between the two electronic states:

$$\begin{aligned} \overline{\nu}_{\text{elec}} &= (64746.55 - 753.49 + 1081.74)\ \text{cm}–1 = 65074.8\ \text{cm}^{-1} \\ &= (65074\ \text{cm}^{-1})\ (3 \times 10^{10}\ \text{cm s}^{-1})(6.626 \times 10^{-34}\ \text{J s}) = 1.29 \times 10^{-18}\ \text{J} \\ &= (1.29 \times 10^{-18}\ \text{J})\ (6.02 \times 10^{23}\ \text{mol}^{-1}) = 778.1\ \text{kJ mol}^{-1} \end{aligned}$$

or $$\overline{\nu}_{\text{elec}} = (65074.8\ \text{cm}^{-1})(11.958\ \text{J cm}^{-1}) = 778.1\ \text{kJ mol}^{-1}$$

$$[1\ \text{cm}^{-1} = 11.958\ \text{J mol}^{-1}]$$

$$\text{Corresponding wavelength} = \frac{1}{65074.8\ \text{cm}^{-1}} = 1536.69 \times 10^{-8}\ \text{cm} = 153.669\ \text{nm}.$$

***Exercise* 5.47**: The 0 → 0 transition in the electronic vibrational spectrum of C_2 arising from the ground $B_{V'}$ and $B_{V''}$ and excited $\left({}^3\pi_g\right)$ electronic states of C_2 is observed at 19378 cm^{-1}. The values of ω_e and for ground and excited electronic states of C_2 are 1641.4 cm^{-1}, 11.67 cm^{-1} and 1788.2 cm^{-1}, 16.43 cm^{-1} respectively. Calculate (*a*) the energy difference between the two minima of the Morse curves corresponding to the electronic states of the molecule, and (*b*) the force constants for the normal modes of vibration of C_2 in the ground and excited electronic states of the molecule.

(*a*) In order to determine the energy difference between the two electronic states of C_2, we are in need of the quantities $\omega_e'' x_e''$ and $\omega_e' x_e'$, from the data reported in the example, we get Substituting the values of $\omega_e'' x_e''$ and $\omega_e' x_e'$, and also the wave number of 0–0 band into equation (5.111), we obtain

$$\overline{\nu}_{\text{elec}} = \left[\overline{\nu}_{\text{elec-vib (V}'=0\rightarrow V''=0)} - \left(\frac{\omega_e'}{2} - \frac{\omega_e' x_e'}{4}\right) + \left(\frac{\omega_e''}{2} - \frac{\omega_e'' x_e''}{4}\right)\right] \text{cm}^{-1}$$

$$= \left[19378 - \left(\frac{1788.2}{2} - \frac{16.43}{4} \right) + \left(\frac{1641.4}{2} - \frac{11.67}{4} \right) \right] \text{cm}^{-1}$$

$$= (19378 - 889.99 + 817.78) \text{ cm}^{-1} = 19306.79 \text{ cm}^{-1}$$

$$= (19306.79 \text{ cm}^{-1}) (3 \times 10^{10} \text{ cm s}^{-1}) (6.62 \times 10^{-34} \text{ J s}) (6.02 \times 10^{23} \text{ mol}^{-1}) = 230.85 \text{ kJ mol}^{-1}.$$

(*b*) The fundamental mode of vibration of C_2 in the ground and excited states is determined from equation (5.112), *i.e.*

$$\bar{\nu} = \omega_e - \omega_e x_e \text{ (cm}^{-1}) \qquad ...(5.112)$$

Thus

$$\bar{\nu} \text{ for ground state} = (1641.4 - 11.67) \text{ cm}^{-1} = 1629.7 \text{ cm}^{-1}$$

$$\bar{\nu} \text{ for excited state} = (1788.2 - 16.43) \text{ cm}^{-1} = 1771.7 \text{ cm}^{-1}$$

$$\mu = \left(\frac{12 \times 12}{12 + 12} \text{g mol}^{-1} \right) \frac{1}{6.02 \times 10^{23} \text{ mol}^{-1}} = 9.967 \times 10^{-24} g = 9.967 \times 10^{-27} \text{ kg}$$

From equation (5.112), we now obtain the C—C force constant in the ground and excited states of the molecule.

$$k \text{ for the ground state} \quad = = 4\pi^2 \omega_e''^2 c^2 \mu$$

$$= (4) (3.14)^2 (1629.7 \text{ cm}^{-1})^2 (3 \times 10^{10} \text{ cm s}^{-1})^2 (9.967 \times 10^{-24} \text{ g})$$

$$= 9.38 \times 10^5 \text{ dynes cm}^{-1} = 9.38 \text{ millidynes (Å)}^{-1}$$

***Exercise* 5.48 (a):** The frequencies assigned to the vibrational bands in the absorption spectrum of AgI near 3200 Å are listed here.

***Frequencies* (cm^{-1})**	**v″ → v″**
31642	0 → 5
31393	0 → 2
30956	1 → 0
31577	0 → 4
31281	0 → 1
30753	2 → 0
31491	0 → 3
31163	0 → 0
30549	3 → 0
31666	0 → 6

(*i*) Suggest assignment for bands at 31256, 31086, 30779, 30868, 30667, 31167, 30963 cm^{-1}.

(*ii*) Assign frequencies corresponding to

2 → 2, 2 → 3, 2 → 1, 1 → 3 bands.

(b) The frequencies assigned to the vibrational bands in the electronic absorption spectrum of nitrogen are listed below:

$v'' \rightarrow v'$	*Frequencies*(cm^{-1})
$0 \rightarrow 0$	29662
$0 \rightarrow 1$	31652
$1 \rightarrow 0$	27957
$0 \rightarrow 3$	35463
$2 \rightarrow 0$	26281
$3 \rightarrow 0$	24634
$0 \rightarrow 2$	33593
$4 \rightarrow 0$	23022
$5 \rightarrow 0$	21425

(*i*) Suggest assignment for the bands at 28271, 32082, 26624, 27226 cm^{-1} and (*ii*) assign frequencies corresponding to the $2 \rightarrow 2$, $3 \rightarrow 2$, $1 \rightarrow 3$ and $4 \rightarrow 2$ bands.

(a) Based on the data in the example, construct the Deslandres table. The missing elements in the table are generated on the basis of vibrational intervals between the adjacent elements in the first row and first column of the table. The numbers in parenthesis represent the vibrational intervals.

v'/v"	0		1		2		3
0	31163	(207)	30956	(203)	30753	(204)	30549
	(118)						
1	31281		31074		30871		30667
	(112)						
2	31393		31186		30983		30779
	(98)						
3	31491		31284		31086		30868
	(86)						
4	31577		31370		31167		30963
	(65)						
5	31642		31435		31232		31028
	(24)						
6	31666		31459		31256		31052

(*i*) From the table the following bands are assigned to the frequencies in question: $2 \rightarrow 6$, $2 \rightarrow 3$, $3 \rightarrow 2$, $3 \rightarrow 3$, $3 \rightarrow 1$, $2 \rightarrow 4$, $3 \rightarrow 4$.

(*ii*) From the table, the frequencies assigned to the bands in question are 30983, 31086, 30871, 31284.

(*b*) Proceeding as in the part (*a*) of this problem, we arrive at the following assignments:

(*i*) $2 \rightarrow 1$, $2 \rightarrow 3$, $3 \rightarrow 1$, $5 \rightarrow 3$

(*ii*) 30212, 28565, 33758, 26953 cm^{-1}

5.10.2. *Transition between Rotational Levels Associated with a Particular Vibrational Level in the Electronic Spectrum of a Diatomic Molecule*

Frequencies of separate lines in the electronic spectrum of a diatomic molecule are represented by

$$\bar{\nu} = \bar{\nu}_{\text{elec}} + \bar{\nu}_{\text{vib}} + \bar{\nu}_{\text{rot}} \qquad ...(5.101)$$

The frequencies of the individual rotational lines of a particular vibrational band characterised by vibrational quantum numbers v′ and v″, thus, can be represented by the formula

$$\bar{\nu}_{\text{rot-elec}} = \bar{\nu}_{\text{elec}} + \bar{\nu}_{\text{vib}} + B_{v'}J'(J'+1) - B_{v''}J''(J''+1)$$

$$\bar{\nu}_{\text{rot-elec}} = \bar{\nu}_{\text{elec-vib(v'v'')}} + B_{v'}J'(J'+1) - B_{v''}J''(J''+1) \qquad ...(5.113)$$

Here, $B_{v''}$ and J'' refer to the lower electronic state and $B_{v'}$ and J', to the upper electronic state of the diatomic molecule contrary to infrared spectroscopy where they belong to the same electronic state. Very often, when the electron is excited from the bonding orbital of the molecule in the ground state, the bond in the higher electronic state becomes weaker and longer. Further, since the quantity B, the rotational constant, is inversely proportional to the equilibrium moment of inertia and hence the square of the equilibrium bond distance of the molecule, it decreases during the electronic transition. In such case, $B_{v'} < B_{v''}$. Sometimes, when the electron excitation is from the anti-bonding orbital of the molecule, the reverse happens, *i.e.* $B_{v'} > B_{v''}$ $\bar{\nu}_{\text{elec-vib(v'v'')}}$ is the wave number of an electronic vibrational transition. This corresponds to any of the transitions represented by v″ (= 0 1, 2 ...) → v′ (0, 1, 2, 3 ...), *i.e.* 0 → 0, 0 → 1...., 1 → 0, 1 → 2, etc. Thus, it represents the origin of a particular band. Applying the selection rules $\Delta J = \pm 1$, $\Delta J = 0, \pm 1$; J' (= 0) ↔ J'' (= 0) (not allowed) to equation (5.113) for the spectral lines, we arrive at the expressions for the frequencies of P–, Q = and R = branches of the electronic spectrum of the diatomic molecule

$$P: \Delta J = -1,\ J' - J'' = -1 \text{ or } J' = J'' + 1$$

$$\bar{\nu}_P = \bar{\nu}_{\text{elec-vib}(v'v'')} + B_{v'}(J''-1)(J'') - B_{v''}J''(J''+1)$$

$$= \bar{\nu}_{\text{elec-vib}(v'v'')} + (J'')\left[B_{v'}(J''-1) - B_{v''}(J''+1)\right]$$

$$\bar{\nu}_P = \bar{\nu}_{\text{elec-vib}(v'v'')} - (B_{v''} + B_{v'})J'' + (B_{v'} - B_{v''})J''^2 \qquad ...(5.114)$$

where J'' = 1, 2, 3... $J'' \neq 0$ otherwise $J' = -1$ which has no meaning.

Equation (5.114) reveals that no rotational line of the P-branch appears at the band origin

$$\bar{\nu}_P = \bar{\nu}_{\text{elec-vib}(v'v'')}$$

$$R: \Delta J = +1,\ J' - J'' = +1,\ J' = J'' + 1$$

$$\bar{\nu}_{\text{rot-elec}} = \bar{\nu}_{\text{elec-vib}(v'v'')} + B_{v'}(J''+1)(J''+2)) - B_{v''}J''(J''+1)$$

$$\bar{\nu}_R = \bar{\nu}_{\text{elec-vib}(v'v'')} + 2B_{v'} + (3B_{v'} - B_{v''})J'' + (B_{v'} - B_{v''})J''^2 \text{ where } J'' = 0, 1, 2 \qquad ...(5.115)$$

Here too no rotational line of the R–branch will appear at the band origin.

$$Q{:}\Delta J = 0,\ J' - J'' = 0,\ J'' = J',\ J' = 0 \neq J'' = 0$$

$$\bar{\nu}_Q = \bar{\nu}_{\text{elec-vib}(v'v'')} + (B_{v'} - B_{v''})J'' + (B_{v'} - B_{v''})J''^2, \text{ where } J'' = 1, 2, 3, \qquad \text{...(5.116)}$$

since $J' = 0 \neq J'' = 0$, no line from the Q = branch will appear at the origin. However, for $\Lambda \neq 0$, the rule $J' = 0 \leftrightarrow J'' = 0$ is also allowed and thus J" = 0, 1, 2, 3,..., the lines of *Q*-branch will start from the origin.

Equations (5.114) and (5.115) are clubbed together into

$$\bar{\nu}_{PR} = \bar{\nu}_{\text{elec-vib}(v'v'')} + (B_{v'} + B_{v''})m + (B_{v'} - B_{v''})m^2 \text{ where } m = \pm 1, \pm 2, \pm 3 \qquad \text{...(5.117)}$$

Here, *m* is negative for *P*–branch ($\Delta J = -1$) and positive for *R* branch ($\Delta J = +1$). *m* cannot be zero otherwise for *P*-branch $J'' = -1$ which is not possible.

The wave numbers of the rotational fine lines in the *P*, *R* – and *Q*-branches, as worked out from equations (5.117) and (5.116) respectively for $B_{v'}(= 10\ \text{cm}^{-1}) < B_{v''}(= 12\ \text{cm}^{-1})$, 20 per cent difference between the two in the electronic vibrational spectrum of the diatomic molecule have been listed in Table 5.10, and the corresponding spectrum is sketched in Fig. 5.13. We already know that even the

TABLE 5.14. Position of rotational fine lines relative to $\bar{\nu}_{\text{elec-vib}}$ for $B_{v'} < B_{v''}$ and 20 per cent difference between the two in the electronic vibrational structure of diatomic molecule, $B_{v'} = 10\ \text{cm}^{-1}, B_{v''} = 12\ \text{cm}^{-1}, \bar{\nu}_{PR} = \bar{\nu}_{\text{elec-vib}} + 22\text{ m} - 2\text{ m}^2$, *m* is positive for *R*-branch and negative for *P*-branch. $\bar{\nu}_Q = \bar{\nu}_{\text{elec-vib}} - 2J'(J+1)$.

m/J"	$\bar{\nu}_R$ (cm^{-1})	$-\bar{\nu}_P$ (cm^{-1})	$-\bar{\nu}_{(Q)}$ (cm^{-1})
1	20	24	4
2	36	52	12
3	48	84	24
4	56	120	40
5	60	160	60
6	60	204	84
7	56	252	112
8	48	304	144
9	36	360	180
10	20	420	220

higher rotational levels are populated even at room temperature, consequently a large number of rotational lines appear in the *P*–and *R*–branches and wherever possible in the *Q*–branch also in the spectrum. The lines in the *R*–branch become closely spaced on the high wave number side and crowd together and eventually a maximum value of $\bar{\nu}_R$ is obtained at a particular value of *m*. *This limiting position of the band corresponding to maximum value of $\bar{\nu}_R$ is called band head, and the lines become widely spaced on the low wave number side of the band head called the tail of the band.*

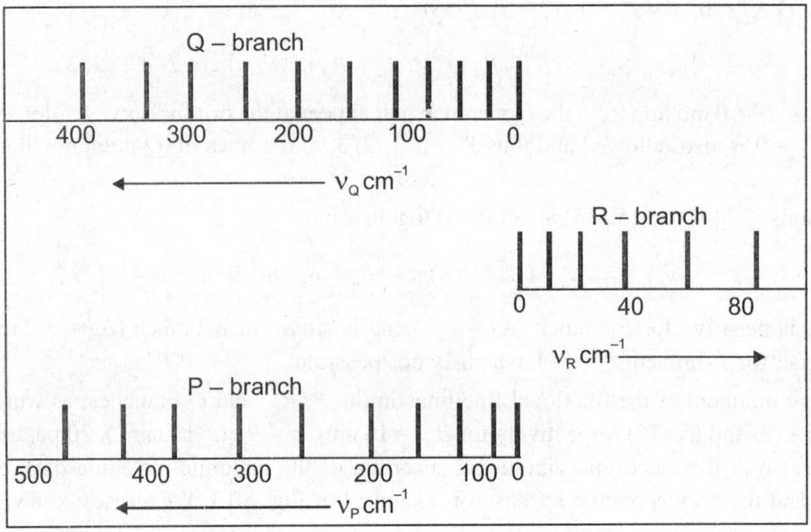

Fig. 5.13 *Electronic vibrational spectrum of a diatomic molecule*

This decline in total intensity away from the band head forms a line distribution in the band called *shading* or *degrading of band*. The shading of a band towards longer wavelengths or low frequencies is termed *red shading* and the band is said to *red shaded* as in the present case and *violet-shaded* if the shading is towards the shorter wavelengths. Further looking at the curve, the lines spread out in the *P*–branch as *m* increases. The *Q*-branch also follows the direction of the *P*-branch.

The characteristic band constant whose value corresponds to wave number of the band origin also called the zero line, i.e. $\overline{\nu}_{\text{elec-vib(00)}}$ is absent from the spectra of diatomic molecules. The position of zero line in a band can be determined from the characteristic gap that disrupts the line sequence. The exact value of $\overline{\nu}_{\text{elec-vib(00)}}$ band can be determined from a detailed analysis of the rotational structure of the bands. The set of values of $\overline{\nu}_{\text{elec-vib}}$ so obtained aids in establishing the Deslandres formula for a given system of bands, *i.e.* determining the constant ν_{elec} and also the vibrational constants w_e and x_e for both the combining electronic terms.

In case $B_{v'} > B_{v''}$, all the discussion is reversed and (*i*) the band head appears in the *P*-branch towards lower wave numbers or higher wavelengths away from the band origin $\overline{\nu}_{\text{elec-vib(00)}}$ and is shaded towards longer wave numbers, *i.e.* lower wavelengths away from the band head and is said to be violet shaded, (*ii*) the lines in the *R*-branch still appears on the higher wave number side of the band origin and show an intense separation with *m,* and (*iii*) the lines of the *Q*-branch spreads out to lower wavelengths or higher wave number side of the band origin and the separation between adjacent lines increases with *J*″.

***Exercise* 5.49:** The fine structure line of C_2 band at 19378 cm^{-1} fits well into the equation

$$\overline{\nu}_{\text{rot-elec}} = 19378 + 3.3853m + 0.1201m^2) \text{ cm}^{-1}$$

(*a*) Find the value of the band head and position of the band head. (*b*) Calculate the separation between zero line and band head and state the direction of shading of the band. (*c*) Determine moments of inertia and internuclear distances in the two electronic states of the molecule.

(*a*) The rotational fine structure of lines of a band is expressed by the relation

$$\bar{\nu}_{\text{rot-elec}} = \bar{\nu}_{\text{rot-elec}} + (B_{v'} + B_{v''})m + (B_{v'} - B_{v''})m^2, \text{cm}^{-1} \qquad \text{(A)}$$

Comparing equation (5.117) with the given relation (A), we obtain

$$\bar{\nu}_{\text{elec-vib}(v''=0 \to v'=0)} \text{ or } \bar{\nu}_{00} = 19378 \text{cm}^{-1}$$

$$B_{v'} + B_{v''} = 3.3853 \text{ cm}^{-1}$$

$$B_{v'}\ B_{v''} = 0.1201 \text{ cm}^{-1}$$

Solving these for $B_{v'}$ and $B_{v''}$, we get

$$Bv' = 1.7527 \text{ cm}^{-1}$$

$$Bv'' = 1.6321 \text{ cm}^{-1}$$

The value of *m* at band head is given by

$$m = -\frac{(B_{v'} + B_{v''})}{2(B_{v'} - B_{v''})} = -\frac{3.3853(\text{cm}^{-1})}{2 \times 0.1201(\text{cm}^{-1})} = -14$$

The negative value of *m* reveals that the band head lies in the low wave number side of the band origin or null line, *i.e.* *P*-branch of the spectrum. The band is shaded towards the high wave number side or towards violet. The position of the head, *i.e.* maximum value of $\bar{\nu}_{\text{rot-elec}}$ corresponding to $m = -14$ is determined from the equation

$$\begin{aligned}\bar{\nu}_{\text{head}} &= [19378 + 3.3853\,(-14) + 0.1201(-14)^2] \text{ cm}^{-1} \\ &= (19378 - 47.39 + 23.53) \text{ cm}^{-1} \\ &= (19378 - 23.86) \text{ cm}^{-1} \\ &= 19354.14 \text{ cm}^{-1}\end{aligned}$$

(*b*) The separation between band head and the null line is (19354.14 – 19378) cm^{-1} = –23.86 cm^{-1}. Thus the head is shaded towards the violet.

(*c*) The moments of inertia and bond distances in both the states of the molecule are determined from $B_{v'}$ and $B_{v''}$.

For the ground state,

$$I'' = \frac{h}{8\pi^2 B_{v''} c} = \frac{6.626 \times 10^{-34} \text{ J s}}{(8)(3.14)^2 (1.6321 \text{ cm}^{-1})(3 \times 10^{10} \text{cm s}^{-1})} = 1.716 \times 10^{-46} \text{ kg m}^2$$

$$\text{Since } \mu_m \text{ for } C_2 = \frac{12 \times 12}{12 + 12} \times \frac{1}{6.02 \times 10^{23} \text{ mol}^{-1}} \text{ g mol}^{-1}$$

$$= 9.967 \times 10^{-24}\ g = 9.967 \times 10^{-27} \text{ kg}$$

the bond length for C_2 in the ground state is

$$r'' = \sqrt{\frac{I''}{\mu_m}} = \sqrt{\frac{1.716 \times 10^{-46} \text{ kg m}^2}{9.967 \times 10^{-27} \text{ kg}}} = \sqrt{1.7213 \times 10^{-20} \text{ m}^2} = 1.311 \times 10^{-10} \text{ m Å}$$

Similarly, for the excited state,

$$I' = \frac{h}{8\pi^2 B_{v'} c} = \frac{6.626 \times 10^{-34} \text{ J s}}{(8)(3.14)^2(1.7527 \text{ cm}^{-1})(3 \times 10^{10} \text{cm s}^{-1})} = 1.5979 \times 10^{-46} \text{ kg m}^2$$

and $$r' = \sqrt{\frac{I'}{\mu_m}} = \sqrt{\frac{1.5979 \times 10^{-46} \text{ kg m}^2}{9.967 \times 10^{-27} \text{kg}}} = \sqrt{1.603 \times 10^{-20} \text{m}^2} = 1.266 \times 10^{-10} \text{ m}$$

The designation of the different electronic transitions are mentioned in Table 5.15.

TABLE 5.15. Designation of the transitions

Transition	**Bonding→Anti-bonding**	**Non-bonding to Anti-bonding**	**Rydberg transition**
Mulliken Notation	N→V	N→Q	N→R
M.O. Notation	σ–σ*,	n→π*	
	π–π*	n→σ*	

5.10.3. *Polyatomic Molecules*

Mainly, we encounter three kinds of molecular orbitals *i.e.*, σ and σ*, π and π* and *n* (non-bonding) orbitals. The asterisk means antibonding. They are arranged on a relative energy scale as shown in the Fig. 5.14. Since an electronic transition involves an excitation of the bonding or non-bonding electron to an antiboinding level, the possible transitions are (*i*) *n* → π*, (*ii*) *n* → σ*, (*iii*) π → π* and (*iv*) σ → σ*.

The first two are relatively weak and the last two are strong. Further the first two lie at longer wavelength compared to the last two.

The relative energy level spacing of $\rangle$C = O group in ketone $\nu_{n-\pi^*} < \nu_{n\to\sigma^*} < \nu_{\pi\to\pi^*} < \nu_{\sigma\to\sigma^*}$.

Generally in polyatomic molecules, specially the organic molecules, the bonding or non-bonding orbitals, are filled, so the longest wavelength electronic transition involved the promotion of an electron from highest filled orbital to the lowest unoccupied molecular orbital.

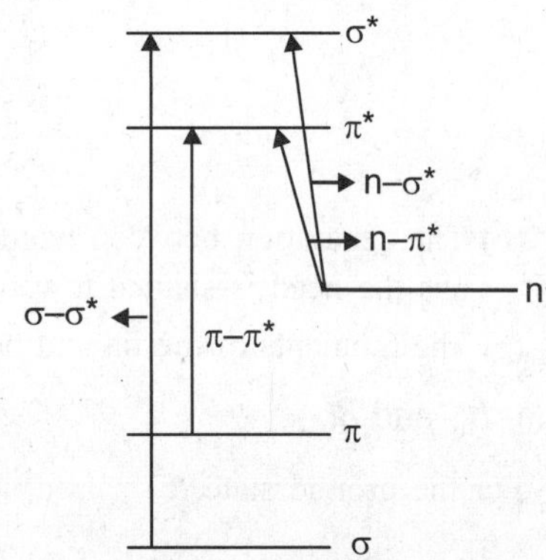

Fig. 5.14 *Electronic transitions between molecular orbitals of polyatomic molecules*

***Exercise* 5.50:** Which of the following transitions will be allowed and which one will be forbidden?

(*i*) $^1\Delta \to {}^1\Delta$ (*ii*) $^2\Sigma \to {}^2\Sigma$ (*iii*) $^1\Pi \to {}^1\Pi$ (*iv*) $^2\Sigma \to {}^1\Pi$ (*v*) $^3\Pi \to {}^1\Delta$.

All the three transitions (*i*), (*ii*) and (*iii*) are allowed because $\Delta S = 0$ and $\Delta \Lambda = 0$ or ± 1 for all. In the transitions (*iv*) and (*v*) $\Delta S \neq 0$ although $\Delta\Lambda = +1$ so these are forbidden.

***Exercise* 5.51:** Predict the kind of pattern observed in the electronic transitions depicted in the Fig. 5.15*a* and Fig. 5.15*c*.

(*i*) In Fig. 5.15*a*, since upper state has no minima in its potential energy curve *i.e.,* unstable, the spectrum will be continuous *i.e.,* no discrete bands are observed.

(*ii*) Since, the minima of the upper and lower states is at different distance, the 0 – 0 band will not be the most intense. As v′ increases the intensity of bands will increase till the arrow meet the vibrational level and then it will decrease.

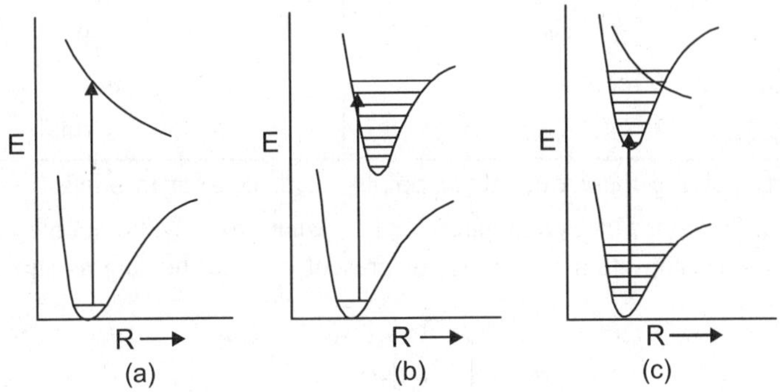

Fig. 5.15 *Various kinds of potential energy curves*

In upper electronic state, another unstable curve of some other electronic state crosses it. Due to this the bands arising from v″ = 0 to v′ of upper state near the crossing of curve will have diffuse band structure. After this again the discrete band structure will appear.

***Exercise* 5.52:** Predict the kind of transitions present in (*a*) Cl_2 molecule, (*b*) C = O group. Which one will be allowed and which one will be forbidden?

(*a*) In Cl_2 molecule, the molecular orbitals present are σ, σ* and non-bonding ones. So the allowed transition will be σ → σ*, intense and at shorter wavelength relative to *n* → σ*, forbidden, weak transition and at longer wavelength.

(*b*) In carbonyl group, the molecular orbitals are σ, π, *n*, σ* and π*. The transitions present will be

(*i*) *n* → π*, *n* → σ* both weak

(*ii*) *n* → π*, σ → σ* both strong.

5.10.3.1. *Conjugated Dienes*

In the case of conjugated dienes, the long wavelength band maxima decreases with the increase in the conjugated double bonds. Fieser and Fieser formulated rules to predict the long wavelength band maxima in case of conjugated double bonds and are complied in Table 5.16.

TABLE 5.16. Fieser's rules for predicting the absorption band maxima (nm) of dienes

Base value for heteroannular diene	214
Base value for homoannular diene	253
Increments for	
Double bond extending conjugation	+30
Alkyl substituent or ring residue	+5

Exocyclic double bond	+5
Polar groupings: OAc	+0
OAlk	+6
SAlk	+30
Cl, Br	+5
$N(Alk)_2$	+60
Solvent Correction	0
λ_{max}	= Total

Before we start applying these rules three points are to be kept in mind:

(*i*) *Exocyclic double bond*: In a conjugated cyclic system, excocyclic double bond is such that it starts from one of the carbon atom of a ring but present outside the ring as shown below:

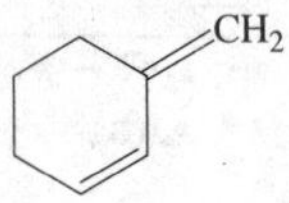

(*ii*) *Hetero- and Homoannular Systems*: If the conjugated double bonds are present in the same ring, it is called homoannular (I) and if these are present in different rings, it is called heteroannular (II).

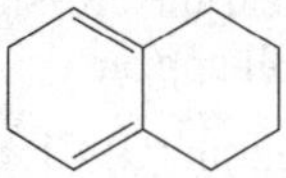
I

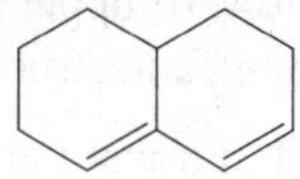
II

(*iii*) Rules mentioned in Table 5.16 do not give good results if applied to cross conjugated polyene system (III) such as

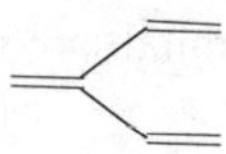
III

TABLE 5.17. Predicted and experimental spectral data of some conjusted dienes

Compound	Formula	Calculated (nm)	Observed (nm)	ε_{max}	Solvent
Butadiene		–	217	20,900	Hexane
Piperylene		217 + 1 × 5 = 222	223.5	23,000	Ethanol
Isoprene	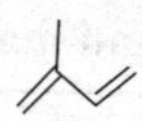	217 + 1 × 5 = 222	220	23,900	Hexane
β, γ-dimethylbutadiene	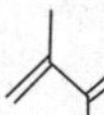	217 + 2 × 5 = 227	226		Hexane
2, 4-Δ-Hexadiene	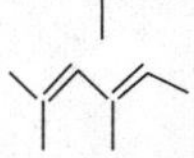	217 + 2 × 5 + 1 × 5 = 232	227	25,000	Hexane
3-methylcyclohexene	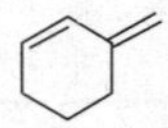	217 + 2 × 5 + 1 × 5	231	21,000	Cyclohexene

		= 232			
Allyidenecyclohexene	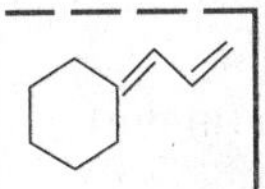	217 + 2 × 5 + 1 × 5 = 232	236.5	–	Ethanol
β-phelandiene	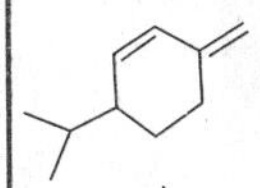	217 + 2 × 5 + 1 × 5 = 232	232	8,100	–
3, 5Δ-Choestadiene	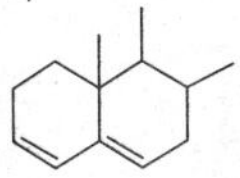	217 + 3 × 5 + 1 × 5 = 237	235	–	–

Empirical rules given by Woodward or Fieser and coworkers, compiled in Table 5.16, predict the absorption band maximum of conjugated dienes very accurately if the number of double bonds is four or less. Some of the example are compiled in table 5.17 along with the observed and calculated results.

Fieser and Kuhn Relation: Fieser and Kuhn have given another empirical relation which holds good even for conjugated polyenes containing more than four double bonds. This relation also predicts ε_{max} and takes care of some other structural aspects. The equation is

$$\lambda_{max} = 114 + 5M + n(48 - 1.7\,n) - 16.5\,R_{endo} - 10R_{exo} \qquad ...(5.118)$$

$$\epsilon_{max} = (1.74 \times 10^4)\,n \qquad ...(5.118a)$$

where n = Number of conjugated double bonds

M = Number of alkyl groups or alkyl like substituents on the conjugated system

R_{endo} = Number of rings with endocyclic double bonds in the conjugated system

R_{exo} = Number of rings with exocyclic double bonds

***Exercise* 5.53:** Calculate the absorption band maximum and molar extinction coefficient for β-carotene.

The structure of β-carotene is

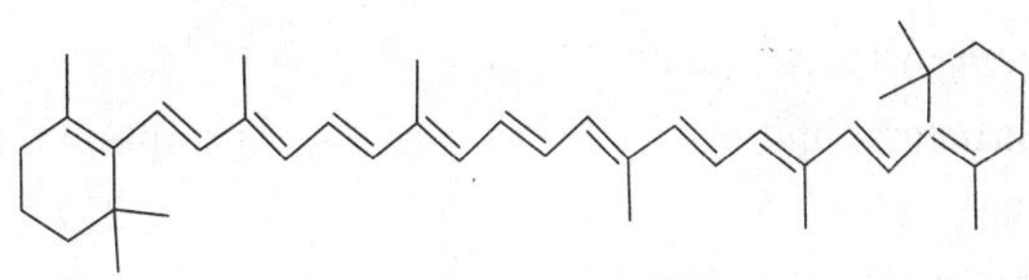

Number of conjugated double bonds is 11, number of methyl groups attached 10 and exo double bonds are 2. Using equations (5.118 and 5.119), we get

$$\lambda_{max} \text{ (calculated)} = 114 + 5 \times 10 + 11\,(48 - 1.7 \times 11) - 16.5 \times 2 - 10 \times 0 = 453.3 \text{ nm}$$

$$\epsilon_{max} = (1.74 \times 10^4) \times 10 = 19.14 \times 10^4$$

λ_{max} (observed) = 452 nm, ϵ_{max} (observed) = 15.2 × 10^4

The agreement in absorption band maximum and molar extinction coefficient is not bad.

***Exercise* 5.54:** Calculate the absorption maximum for cholesta-3, 5,-diene in ethanol.

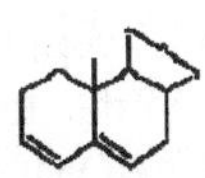

It is a heteroannular diene and its spectroscopic conjugated skeleton is

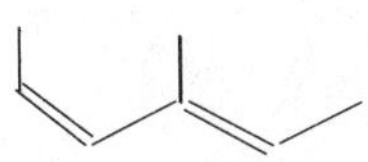

Thus, from Table 5.16, we write

Structural unit	Absorption (in nm)
Base value	214
Ring residues (= 3)	15
Exocyclic double bond (= 1)	5

$$\lambda_{max} = 234 \text{ nm}$$

$$\lambda_{max} \text{ (obs)} = 235 \text{ nm } (\varepsilon_{max} = 19000)$$

$$\text{Obs } \lambda_{max} \text{ for cholesta-2, 4,-diene} = 275 \text{ nm } (\varepsilon_{max} = 10000)$$

***Exercise* 5.55:** Calculate the absorption maximum for 4-cyclohexenyl-2-pentene.

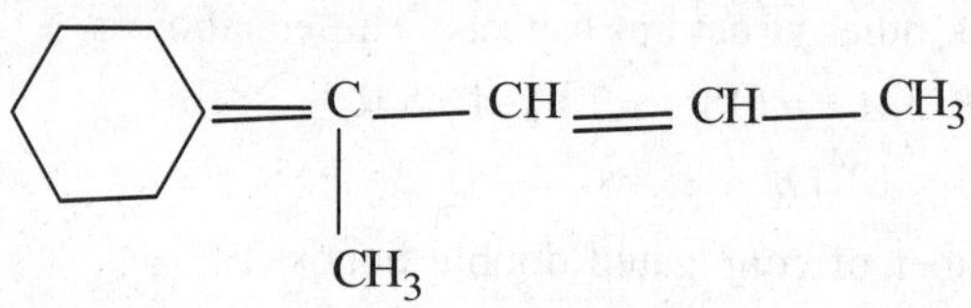

It belongs to a butadiene system and its spectroscopic conjugated skeleton is

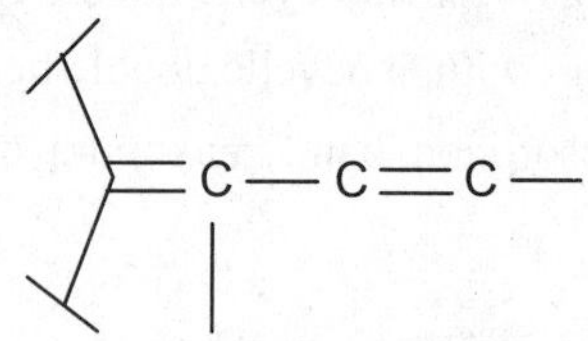

Thus, from Table 5.16, we write

Structural unit	Absorption (in nm)
Base value	217
Alkyl substituent (= 2)	10
Ring residues (= 2)	10
Exocyclic double bond (=1)	5

$$\lambda_{max} = 242 \text{ nm}$$

$$\lambda_{max} \text{ (obs)} = 242 \text{ nm}$$

5.10.8.3.2. α,β-*Unsaturated Carbonyl Compounds*

In this class of compounds, the carbonyl group is conjugated to an ethylenic double bond or to polyenes, thus making the theoretical treatment similar to conjugated double bonds, except with a presence of nonbonding energy state in between the HOMO and LUMO (π orbitals) of the conjugated polyenes. Two kinds of transitions are observed; one a weak transition above 280 nm and the other intense one above 200 nm. Both these bands are red shifted with the extension of conjugated double bond, increase being greater in the short wavelength band than that in the long wavelength one. In the compounds with five olefinic bonds, the tail part of the $\pi^* \leftarrow \pi$ transition in α, β-unsaturated carbonyl compounds appear in the visible region with the weak band appearing as a shoulder at the red side. Because of this reason the relative feable $\pi^* \leftarrow n$ band becomes difficult to identify when the chain length become long. Based on the following observations, the long wavelength band is assigned to $\pi^* \leftarrow n$ transition and short wavelength band to $\pi^* \leftarrow \pi$ transition.

(*i*) The long wavelength band maximum is blue shifted and the short wavelength band maximum is red shifted with the increase in the polarity and specifically hydrogen bond forming capacity of the solvents.

(*ii*) Similar to polyenes, extension by one further double bond decreases the gap between the π and π* molecular orbital. Energy level diagram of α, β-unsaturated carbonyl compounds is shown in Fig. 5.16, indicating clearly that introduction of double bond decreases the gap both between

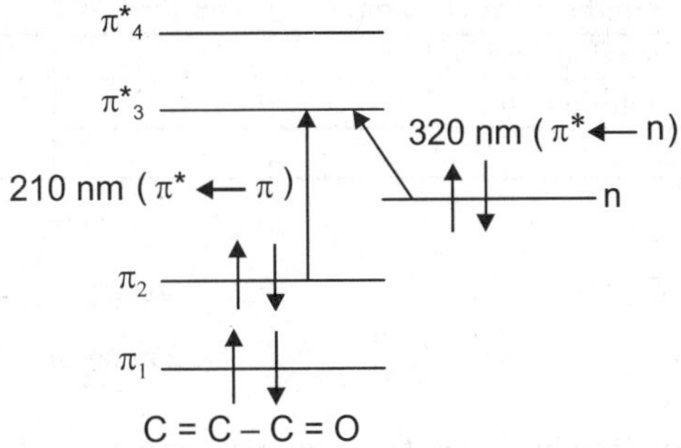

Fig. 5.16 *Energy level diagram for acrolein*

$n\pi^*$ and $\pi\pi^*$ energy levels, thus causing a red shift in the absorption spectra.

(*iii*) Effect of basic or alkyl groups on these bands are opposite as expected, the long wavelength band is blue shifted and the short wavelength band is red shifted, for example, $\pi^* \leftarrow n$ band in acrolein ($CH_2 = CH - CHO$) is at 328 nm, whereas the same band in methyl vinyl ketone [$CH_2 = CH - C(CH_3)O$] is at 324 nm.

Absorption spectra of these molecules are also sensitive to the structural arrangement of the carbonyl group and the double bond. For example, due to the steric hindrance of methyl group, the coplanarity of the double bond and carbonyl group is prevented. This decreases the molecular extinction coefficient of $\pi^* \leftarrow \pi$ transition to less than 10^4 in compound I. This kind of behaviour is observed specifically in cyclic ketones.

Similar to conjugated polyenes, Woodward and Feiser have developed empirical rules to predict the absorption maximum of $\pi^* \leftarrow \pi$ transition in α, β-unsaturated aldehydes, ketones, both acyclic and cyclic ones. The rules have been summarised in Tables 5.18 and 5.19. The effect of solvents on the $\pi^* \leftarrow \pi$ transition is summarised in Table 5.20.

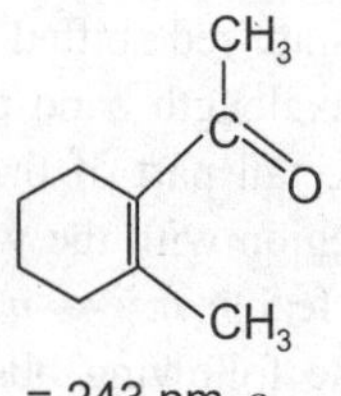

λ_{max} = 243 nm, ε_{max} = 1400

TABLE 5.18. Rules of eneones and dienones absorption spectra ($\pi^* \leftarrow \pi$)

Base value	(nm)
Acyclic α,β-unsaturated ketones	215
Six membered cyclic α,β-unsaturated ketones	215
Five membered cyclic α,β-unsaturated ketones	202
α,β-unsaturated aldehydes	210
α,β-unsaturated carboxylic acids and esters	195
Increment for double bond	+30
Exocyclic double bond	+5
Homodiene	+39

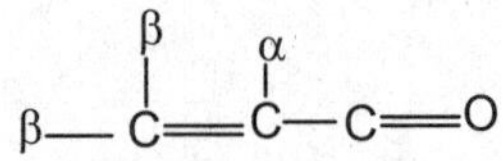

Enone

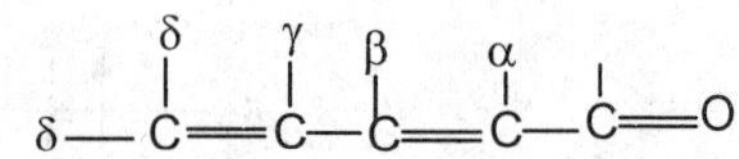

Dieneone

TABLE 5.19. Rules for increments for substituents at α, β, γ, δ....(in nm)

Basic Group	α	β	γ	δ
R-	10	12	18	18
OH	35	30	-	50
OR	35	30	17	31
OAc	6	6	6	6
Cl	15	12	-	-
Br	25	30	-	-
NR_2	-	95	-	-
SR	-	85	-	-

TABLE 5.20. Solvent corrections

Solvent	Corrections (nm)
Cyclohexane	+11
Ether	+7
Dioxane	+5
Chloroform	+1
Water	–8
CH_3OH, C_2H_5OH	0

***Exercise* 5.56**: Using Woodward's rule, predict and compare the band maximum of tiglicaldehyde. The formula of tiglicaldehyde is

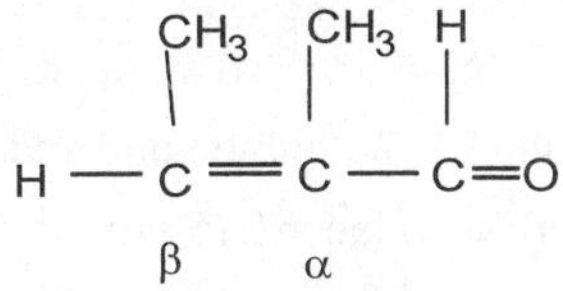

Base value = 210 nm
Methyl group at α = 10 nm
Methyl group of β = 12 nm

Total = 232 nm λ_{max}^{obs} (ethanol) = 228 nm

***Exercise* 5.57**: The structure of 2-cyclohexylidene cyclohexane-1-one is as given. Calculate the absorption band maximum, using Woodward's rules.

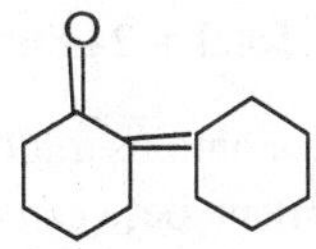

Base value = 218 nm
2-exo double bonds = 2 × 5
= 10 nm
α-ring residue = 10 nm
β-ring residue = 2 × 12
= 24 nm

Total = 259 nm, λ_{max}^{obs} (ethanol) = 258 nm

***Exercise* 5.58**: Calculate the absorption band maximum of diosphenol. The structural formula of diosphenol is

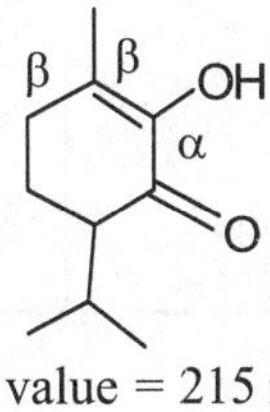

Base value = 215 nm
α-OH group = 35 nm
2 β-methyl groups = 2 × 12
= 24 nm

Total = 274 nm λ_{max}^{obs} (ethanol) = 270 nm

***Exercise* 5.59:** The structural formula of Cholesta-1,4-dien-3-one is as given below. Calculate the absorption band maximum by using Woodward's rules. This molecule can be treated in two ways.

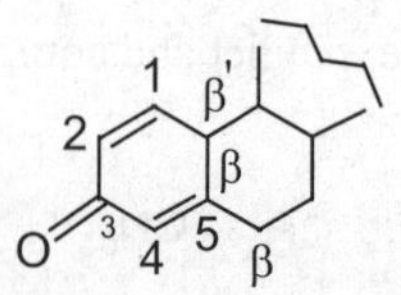

(*i*) α, β-unsaturated ketone by having 1, 2-double bond and β′-ring residue

Base value = 215 nm

β-ring residue = 12 nm

Total = 227 nm

(*ii*) α, β-unsaturated ketone by having 3,4-double bond.

Base value = 215 nm

1-exo double bond = 5 nm

2- β-ring residue = 2 × 12

= 24 nm

Total = 244nm λ_{max}^{obs} (ethanol) = 245 nm

This example suggests if the molecule contains more than one choice to consider it as α, β-unsaturated carbonyl compound, the agreement between the calculated and the observed values is better for the arrangement where the molecule is more highly substituted. The absorption spectrum of *p*-benzoqionone can be treated in the above manner. The weak n and the strong π* ← π bands are observed at 435 and 245 nm respectively.

Similar rules for the absorption spectrum of α, β-unsaturated carboxylic acids are given in Table 5.21.

TABLE 5.21. Absorption band maximum of α, β-unsaturated carboxylic acids and esters

	λ_{max} (±5 nm, in ethyl alcohol)
α or β monosubstituted	208
αβ or ββ disubstituted	217
αββ trisubstituted	225
Exocyclic αβ double bond	+5
Endocyclic αβ double bond in five or seven membered ring	

A.T. Neilson, J. Org. Chem., 22,1539 (1957).

***Exercise* 5.60:** Rhodospin is the protein opsin linked to retinene (or retinal) which is the vitamin A_1 aldehyde. The chain of retinal has an all *trans*-configuration. Rhodospin absorbs light and retinal bound by rhodospin isomerises to 11-cis retinene. The structural formula of retinal is

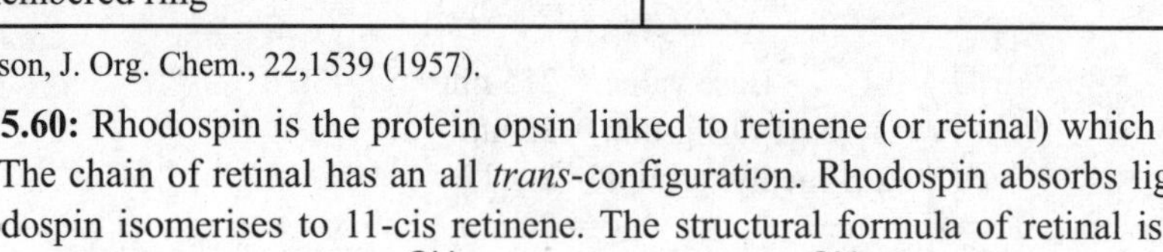

Calculate the absorption maximum for retinal.

It is a α, β-unsaturated aldehyde.

Thus, from Table 5.20, we write

Retinene (structural units)	*Absorption (in nm)*
Parent value	= 210
α-substituent (=0)	= 0
β-substituent (=1)	= 12
γ and higher substituents (= 4)	= 72 (= 4 × 18)
Double bonds extending conjugation (=4)	= 120 (= 4 × 30)
λ_{max}(calc)	= 414

***Exercise* 5.61:** Calculate the values of absorption maximum for

(*a*) 4-aminobenzoic acid

(*b*) 3-aminobenzoic acid

(*c*) 2-hydroxy-4-methoxy acetophenone

(*d*) 2-hydroxy-5-methoxy acetophenone in ethanol.

(*e*) 6-methoxytetralone

(*f*) 4-dimethylamino benzaldehyde.

From Table 5.20, we get

(*a*) λ_{max} = 230 nm (Parent chromophore) + 58 nm (*p*-NH_2)
= 288 nm (obs: 289 nm; log ε = 4.27)

(*b*) λ_{max} = 230 nm (Base value) + 13 nm (*m*-NH_2)
= 243 nm (obs: 241 nm, log ε = 3.85)

(*c*) λ_{max} = 246 nm (Base chromophore) + 7 nm (*o*-OH) + 25 nm (*p*-OCH_3)
= 278 nm (obs: 273.5 nm; log ε = 4.2)

(*d*) λ_{max} = 246 nm (Parent chromophore) + 7 nm (*o*-OH) + 7 nm (*m*-OCH_3) = 260 nm
(obs: 255.5 nm, log ε = 3.95)

(*e*) The principal spectroscopic skeleton in the basic molecular structure is

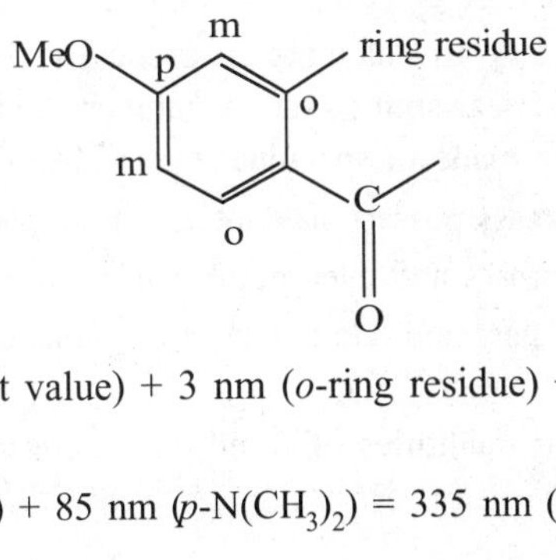

Thus, λ_{max} = 246 nm (parent value) + 3 nm (*o*-ring residue) + 25 nm (*p*-OMe) = 274 nm (obs: 276 nm, log ε = 4.21)

(*f*) λ_{max} = 250 nm (Base value) + 85 nm (*p*-$N(CH_3)_2$) = 335 nm (obs: 340nm, log ε = 4.5)

5.11. NUCLEAR MAGNETIC RESONANCE

Similar to the electron spin motion, each proton and neutron can have spinning motion around some axis. Both protons and neutrons have spin quantum number 1/2 each and the resultant spin quantum number of the nuclei will be the resultant of the spins of the protons and neutrons given as $I = \Sigma i$. It is designated as I. The resultant spin quantum number of the nucleus with odd mass has the multiple of half-integer. If the mass is even, there can be two cases: (*i*) for even mass and even atomic number, $I = 0$, (*ii*) for even mass and odd atomic number, I is an integer. The spin angular momentum of nucleus is given by $\sqrt{I(I+1)}\ h/2\pi$ and the component of the spin angular momentum of the nucleus along z axis by $m_I h/2\pi$ where $m_I = I, I-1, \ldots 0, \ldots -I$. The energy levels corresponding to I are in general degenerate but splits into $2I + 1$ components when external magnetic field is applied.

The magnetic moment (μ_N) of the nucleus is given by

$$\mu_N = \gamma_N \frac{h}{2\pi} I \qquad \text{...(5.119)}$$

where γ_N is the gyromagnetic ratio of the nucleus. The unit for γ_N is $G^{-1}\ s^{-1}$, for h, is J s, so the unit for μ_N is J G^{-1}. The magnetic moment (μ_N) can also be expressed as the product of a dimensionless quantity 'g_N' the nuclear 'g' factor and the magnitude of the nuclear magneton (β_N). Thus the magnetic moment can be written as

$$\mu_N = g_N\ \beta_N I \qquad \text{...(5.120)}$$

where

$$\beta_N = \frac{eh}{4\pi m_p} \qquad \text{...(5.121)}$$

e, the charge of electron expressed in coulombs and m_p mass of proton in kg.

In the presence of magnetic field of flux density H, the energy levels of the isolated nucleus are given by

$$E = -g_N\ \beta_N\ H\ m_I \qquad \text{...(5.122)}$$

The selection rules for given transitions are $\Delta m_I = \pm 1$. The frequency of the transition for the proton is

$$\Delta E = h\nu = g_N\ \beta_N\ H$$

or

$$\nu = g_N\ \beta_N\ H/h \qquad \text{...(5.123)}$$

Unlike electrons, it is not easy to calculate theoretically the value of the total nuclear spin quantum number. The value of nuclear spin quantum number can be determined experimentally only and following empirical rules are made to speculate the value of nuclear spin. These are:

(1) Nuclei having odd nuclear mass possess half integral multiple, *i.e.* 1/2, 3/2, ...

(2) Nuclei having even nuclear mass and even number of protons possess zero nuclear spin.

(3) Nuclei having even nuclear mass and odd number of protons possess integral nuclear spin *i.e.* 1, 2, 3......

The characteristics of nuclear quantities of some nuclei are compiled in Table 5.22.

TABLE 5.22. Nuclear spin quantum numbers, relative abudance, nuclear g_N factors and resonance frequencies for magnetic field of 1T

Nucleus	*I* (spin)	Relative abundance	g_N	ν (MHz)
^{1}H	1/2	99.985	5.5856948	42.5764
^{2}H	1	0.015	0.8574382	6.53573
^{11}B	3/2	80.1	1.7924327	13.6626
^{13}C	1/2	1.10	1.4048236	10.7081
^{14}N	1	99.634	0.4037610	3.0776
^{15}N	1/2	0.366	–0.5663776	4.3172
^{19}F	1/2	100	5.257736	40.0765
^{29}Si	1/2	4.67	–1.11058	8.4653
^{31}P	1/2	100	2.26320	17.2510
^{35}Cl	3/2	75.77	0.5479162	4.1764
^{79}Br	3/2	50.69	2.106400	10.7039
^{77}Se	1/2	7.63	1.070084	8.1566
^{89}Y	1/2	100	–0.2748308	2.0949
^{103}Rh	1/2	100	–0.17680	1.3476
^{107}Ag	1/2	51.839	–0.2273592	1.7330
^{109}Ag	1/2	48.161	–0.2613812	1.9924
^{111}Cd	1/2	12.80	–1.1897722	9.0689
^{113}Cd	1/2	12.22	–1.2446018	9.4868
^{117}Sn	1/2	7.68	–2.00208	15.2606

In the coming section only the proton magnetic resonance will be considered although many other nuclei do possess the finite nuclear spin quantum number and can have NMR spectrum.

***Exercise* 5.62:** Calculate the value of 'Bohr magneton' β_N for a proton.

Substituting the values of each term in equation (5.121)

$$\beta_N = \frac{(1.60219\times10^{-19}\,\text{C})(6.6262\times10^{-34}\,\text{J s})}{(4)(\pi)(1.67265\times10^{-27}\,\text{kg})}$$

$$= 5.0508 \times 10^{-27}\ \text{J T}^{-1}$$

Now $1\text{T} = 10{,}000$ G, so $1\text{T}^{-1} = 10^{-4}\ \text{G}^{-1}$ and

$$\beta_N = (5.0508 \times 10^{-27}\ \text{J T}^{-1}\ (10^{-4}\ \text{T G}^{-1})$$

$$= 5.0508 \times 10^{-31}\ \text{J G}^{-1}$$

***Exercise* 5.63:** Calculate the frequency *ν*, required to excite proton from $m_I = \frac{1}{2}$ to $m = -\frac{1}{2}$, given $g_N = 5.585$ and magnetic field applied $H = 14{,}000$ G.

Substituting the values of each quantity in (5.123)

$$\nu = \frac{(5.585)(5.0508\times10^{-31}\,\text{J G}^{-1})(14000\ \text{G})}{(6.6262\times10^{-34}\ \text{J s})}$$

$$= \frac{5.585 \times 5.0508 \times 14}{6.6262} \times 10^6\ s^{-1} = 59.66 \times 10^6\ s^{-1}$$

***Exercise* 5.64:** A free electron is placed in the magnetic field of strength 14000 G. Calculate the frequency required for the transition if $g_e = 2.0023$.

The Bohr magneton for the electron can be obtained as

$$\beta_e = \frac{eh}{4\pi m_e} = \frac{(1.60219 \times 10^{-19}\ C)(6.6262 \times 10^{-34}\ J\,s)}{(4\pi)(9.1095 \times 10^{-31}\ kg)}$$

$$= 9.2742 \times 10^{-24}\ J\,T^{-1}$$

$$= (9.2742 \times 10^{-24}\ J\,T^{-1})(10^{-4}\ T\,G^{-1})$$

$$= 9.2742 \times 10^{-28}\ J\,G^{-1}$$

$$\nu = \frac{g_e\,\beta_e\,H}{h}$$

$$= \frac{(2.0023)(9.2742 \times 10^{-28}\ J\,G^{-1})(14000\ G)}{(6.6262 \times 10^{-34}\ J\,s)}$$

$$= 32.27 \times 10^9\ s^{-1}$$

***Exercise* 5.65:** What is the ratio of number of proton spins in the lower state to the number in the higher state in a magnetic field of 14,000 G at temperature 25°C.

Using equation (5.123) and the value of β_N from exercise 5.21,

$$\Delta E = -\,g_N \beta H \left[\tfrac{1}{2} - \left(-\tfrac{1}{2}\right)\right] = -\,gN\,\beta H$$

$$= -\,(5.585)\,(5.051 \times 10^{-31}\ J\,G^{-1})\,(14{,}000\ G)$$

$$= -\,3.949 \times 10^{-26}\ J$$

The Boltzmann distribution law gives

$$\frac{N_+}{N_-} = \exp\,(-\,\Delta E/k_b T) = \exp\,\{-\,(-\,3.949 \times 10^{-26}\ J)/(1.3806 \times 10^{-23} J\,K^{-1})(298\ K)\}$$

$$= \exp\,(+\,9.596 \times 10^{-6}) = 1.000096$$

Thus the lower level is very slightly over populated than the upper one as the energy gap (ΔE) is quite small. Due to this very sensitive equipment and very low temperature are needed to obtain strong transition.

***Exercise* 5.66**: Calculate the value of Z-component of nuclear spin magnetic moment for a bare ^2H nucleus classically and quantum mechanically. Compare the two values.

Classical value of μ_Z can be obtained from, *i.e.*

$$\mu_Z = \frac{eh}{4\pi m_N} m_I \qquad ...(5.124)$$

The ^2H nucleus contains one unit of positive charge is 1.602177×10^{-19} C. Mass of deuterium atom is

$$m_N = (2.01335\ a.u.)(1.67265 \times 10^{-27}\ kg\ a.u.^{-1}) = 3.3679644 \times 10^{-27}\ kg$$

Substituting the values of desired parameters in equation (5.124), we get

$$\mu_Z = \frac{(1.602177 \times 10^{-19}\ \text{C})(6.626 \text{ x } 10^{-34}\ \text{J s})}{(4)(3.14)(3.3679644 \times 10^{-27}\ \text{kg})} m_I = 0.25096 m_I \times 10^{-26}\ \text{J T}^{-1}$$

Whereas quantum mechanical value of μ_Z can be obtained from

$$\mu_Z = g_N \frac{eh}{4\pi m_N} m_I = g_N \beta_N m_I$$

Substituting the values of g_N and β_N, we get:

$$\mu_Z = (0.8574382)(5.0508 \times 10^{-27}\ \text{J T}^{-1})\ m_I = 4.3307489 \times 10^{-27}\ \text{J T}^{-1}\ m_I$$

$$\frac{\mu_Z(\text{clas})}{\mu_Z(\text{quant})} = \frac{2.5096 \times 10^{-27} m_I\ \text{J T}^{-1}}{4.3307489 \times 10^{-27} m_I\ \text{J T}^{-1}} = 0.5795.$$

***Exercise* 5.67:** Suppose in proton NMR we are using 600 MHz as the radiation frequency. What magnetic field will be required to satisfy the resonance condition for the observation of this transition?

Using equation (5.123), we get

$$H = \frac{h\nu}{g_N \mu_N} = \frac{(6.626 \times 10^{-34}\ \text{J s})(600 \times 10^6\ \text{s}^{-1})}{(5.58569)(5.0508 \times 10^{-27}\ \text{J T}^{-1})} = 14.09\ \text{T}.$$

5.11.1. *Chemical Shift*

Each proton nucleus is surrounded differently by the electrons in different chemical-environments. Due to this each proton nuclei is shielded to different extent and thus experience different external magnetic field and is given by the expression

$$H_{\text{eff}} = H(1 - \sigma) \qquad ...(5.125)$$

where σ is called screening or shielding constant. The frequencies of NMR lines are given by

$$\nu = \frac{\beta_N (g_N)_j (1-\sigma) H}{h} \qquad ...(5.126)$$

Since g_N has different values for each nucleus j, well separated lines will undergo small shifts as a result of the respective chemical environments.

In organic chemistry, $Si(CH_3)_4$ is considered as a reference compound, occurrence of the NMR lines are measured with respect to the line of the protons of $Si(CH_3)_4$. The relative displacement is called chemical shift and is expressed as

$$\delta = \frac{H_{\text{sample}} - H_{\text{ref}}}{H_{\text{ref}}} \times 10^6 \qquad ...(5.127)$$

or it can be expressed as

$$\tau = 10.00 - \delta \qquad ...(5.127a)$$

If we are using fixed field instruments, the chemical shift is defined as

$$\delta = \frac{\text{Resonance frequency of nuclaus H relative to TMS}}{\text{Spectrometer frequency}} \times 10^6$$

$$= \frac{{}^{\nu}H - {}^{\nu}TMS}{{}^{\nu}\text{Spectrometer}} \times 10^6 \qquad ...(5.127b)$$

It may be mentioned here that chemical shifts expressed in terms of δ-scale or τ-scale is independent of the NMR spectrometer. The difference between the chemical shifts of two nuclei can be related to their frequencies by the equation (5.129)

$$v_i - v_j = \frac{gN\mu NH}{h}(\delta_i - \delta_j) \times 10^{-6} \tag{5.129}$$

The peak height or more exactly, area under each absorption peak is directly proportional to the number of protons having that type of environments.

As mentioned above that the protons at different places in a given molecule experience different effective magnetic field and thus different resonance frequency is observed for different protons. The resonance frequencies for different protons in comparison to tetramethylsilane are compiled in Table 5.23. The ranges of other nuclei in comparison to proton are compiled in Table 5.24 and the chemicals shifts of ^{13}C nuclei under different environments are compiled in Table 5.25.

TABLE 5.23. δ-values of some typical protons

Types of proton	δ (ppm)	Types of protons	δ (ppm)
$Si(CH_3)_4$	0	CH_3CH_2Cl	1.4
$C\ H_4$	0.2	$CH_3CH_2CH_2Cl$	1.1
$R–CH_3$	0.9	$F–CH_3$	4.3
$R_2–CH_2$	1.3	$Cl–CH_3$	3.0
$(R)_3-CH$	1.5	$Br–CH_3$	2.7
$CH_3-C\langle^{O}_{OH}$	2.1	$I–CH_3$	2.2
CH_3NO_2	4.3	$R–CH_2–OH$	3.6
CH_3NH_2	2.1	$R_2–CH–OH$	3.6
CH_3Ar	2.3	$(R)_2CHOH$	3.9
$R–C\equiv C–H$	2.3	$R_2C=CH_2$	4.6
$O=\underset{R}{\underset{\vert}{C}}-CH_3$	2.3	$(R)_2C=CHR$	5.2
$R-O-CH_3$	3.3	$Ar-H$	7.3
ArSH	3 – 4	$R-C\langle^{O}_{H}$	9.9
ROH	0.5 – 4.5	RSH	1 – 2
RNH_2, RNHR	1 – 5	ArOH	4.5 –6.5
$\rangle C=N-OH$	9 – 12	ArNH2, ArNHR	3 – 6
$\rangle C=O\ ....\ H-O-C\langle$	7 – 13	$RCONH_2$, RCONHR	5 – 12
CH_2Cl_2	5.3	RCOOH	9 – 13
$CHCl_3$	7.3		

TABLE 5.24. The range of chemical shifts for other nuclei when compared with H-nucleus

Nucleus	Standard	δ (ppm)
^{1}H	TMS	1 – 20
^{13}C	TMS	0 – 220, may extend to 400 ppm or more
^{19}F	CF_4	0 – 800
^{31}P	H_3PO_4 (85%)	0 – 300

TABLE 5.25. ^{13}C NMR chemical shifts in comparison to $(CH_3)_4Si$ as a reference

Carbon group	δ (ppm)	Carbon group	δ
– C = O, ketones	206 – 212	– C = C, alkenes	123.2 – 136.2
– C = O, aldehydes	199 – 200	– C ≡ N, nitriles	117.7
– C = O,		– C ≡ C, alkynes	71.9 – 73.9
α, β unsaturated carbonyls			
$CH_3CH = CHCHO$	192.4	– C – O, esters	57.6 – 67.9
$CH_2 = CHCOCH_3$	169.9	– C – O, alcohols	49.0 – 57.0
– C = O, carboxylic acid	166.0 – 178.1	– C – NH_2,	26.9 – 35.9
– C = O, amides	165 – 172.7	– S – CH_3, sulphides	15.6
– C = O, esters	165.5 – 170.3	– C – H, alkanes, cycloalkanes	–2.3 – 25.2
– C = C, aromatic	128.5		

It is clear from Tables 5.24 and 5.25 that chemical shifts are larger for ^{13}C than for ^{1}H. This is due to the fact that there are many more electrons about a carbon atom than about a proton. These additional electrons are perturbed by the applied magnetic field and thereby produce small internal magnetic field that give rise to larger chemical shifts. Thus, the chemical shifts of the two active nuclei can be related to their respective screening constants by the equation (5.130*a*)

$$\delta_2 - \delta_1 = \frac{\nu_2 - \nu_1}{\nu_{\text{spectrometer}}} \times 10^6 = \frac{\gamma H_0}{2\pi\nu_{\text{spectrometer}}}(\sigma_2 - \sigma_1) \times 10^6 \qquad \text{...(5.130}a\text{)}$$

Realising that typical values of σ are the order of 10^{-5} compared to 1 in equation (5.125) and replace $\nu_{\text{spectrometer}}$ with $\frac{\gamma B_0}{2\pi}$ in equation (5.130*a*) to get

$$\delta_2 - \delta_1 = (\sigma_2 - \sigma_1) \times 10^6 \qquad \text{...(5.130}b\text{)}$$

It is clear from equation (5.130*b*) that the separation between the two absorption lines for the chemical shift scale is independent of the applied magnetic field. Further the chemical shifts depend on many factors. These are: (*i*) inductive effect, (*ii*) mesomeric or resonance effect, (*iii*) influence of induced magnetic moment of the neighbouring atoms or molecules, (*iv*) ring current in aromatic molecules or molecules showing aromatic character, hydrogen bonding etc.

***Exercise* 5.68:** In the allyl cation or anion, the chemical shifts of the terminal and the middle protons are different from each other. The NMR spectra of these species are recorded with a 400 MHZ NMR spectrometer that uses a 9.395 T. Calculate the difference in the resonant frequencies of the

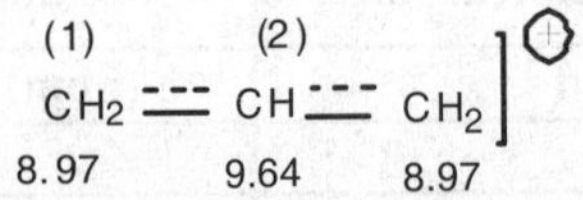

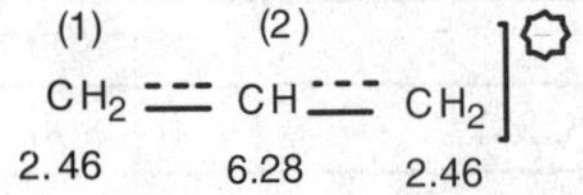

spectrum. (*ii*) Repeat the calculations if the NMR spectra are recorded at 700 MHz spectrometer that uses 16.4410 T.

Substituting the values of the desired quantities in equation (5.130*b*), the differences in the chemical shifts of two different kinds of protons and the frequency difference can be obtained

$$\nu_2 - \nu_1 = \frac{(5.58569)(5.0508\times 10^{-27}\ \mathrm{J\,T^{-1}})(9.395\ \mathrm{T})}{6.626\times 10^{-34}\ \mathrm{J}}[9.64 - 8.97]\times 10^{-6} = 268\ \mathrm{s^{-1}}$$

Similarly using 700 MHz frequency we get

$$\nu_2 - \nu_1 = \frac{(5.58569)(5.0508\times 10^{-27}\ \mathrm{J\,T^{-1}})(16.04410\ \mathrm{T})}{6.626\times 10^{-34}\ \mathrm{J}}[9.64 - 8.97]\times 10^{-6} = 457.7\ \mathrm{s^{-1}}$$

That is the frequency required for resonance for the H_2 atom is 268 s^{-1} greater than that for the H_1 proton in the case of allyl cation using 400 MHz spectrometer and 457.7 s^{-1} using 700 MHz spectrometer.

Similar exercise for the allyl anion is left for the students.

***Exercise* 5.69:** Let the two different kind of proton signals are observed at 540 Hz and 120 Hz in comparison to the signals for TMS in case of 60 MHz NMR spectrometer. Estimate the chemical shifts for these two protons. If we are using 400 MHz NMR spectrometer, calculate the separation of these two signals.

Using equation (5.127*b*) the chemical shifts of these two protons are

$$\delta_2 = \frac{\nu_A - \nu_{TMS}}{\nu_{\text{spectrometer}}}\times 10^6 = \frac{540}{60\times 10^6}\times 10^6 = 9\ \text{ppm}$$

Similarly δ_1 corresponding to 120 HZ will be 2 ppm. So using equation (5.130*b*) the difference in the screening effects can be obtained as

$$(\sigma_2 - \sigma_1)\times 10^{-6} = (9.0 - 2.0) = 7.0\times 10^{-6} \qquad (1)$$

If the frequency of the NMR spectrometer is 400 MHz, the separation in the frequency between the two signals would be

$$\nu_2 - \nu_1 = \nu_{\text{spectrometer}}\times(\delta_2 - \delta_1)\times 10^6 = 400\times 10^6\times 7\times 10^{-6}\ \text{Hz} = 2800\ \text{Hz}$$

5.11.2. *Spin-Spin Splitting*

It has been observed that under high resolution, the NMR lines (chemical shift peak) for one type of protons often split into number of lines. This is due to the extra magnetic fields created by the immediate neighbouring nuclei. If there are two kinds of sites *i.e.*, *A* and *B* having n_A and n_B protons, the number of lines observed at the peak of *A* are $(n_B + 1)$ centred around that of *A* and $(n_A + 1)$ number of lines centred around that of *B*. If the neighbouring nuclei around *A*, contain n_B, n_c, ... number of protons at site *B*,

C, ..., the number of lines observed are $(n_B + 1)(n_C + 1)$... . The intensity of each line is given by the coefficients of the binomial expansion of $(1 + x)^{n_A}$ where n_A is the total number of protons at the site A. This can be easily obtained if Pascal's triangle of binomial coefficients is used. It is where the coefficients in the additional rows can be determined by adding the coefficients to the left and right of the desired coefficient in the previous row.

n	*Relative Intensities* (Binomial coefficient)
0	1
1	1 1
2	1 2 1
3	1 3 3 1
4	1 4 6 4 1
5	1 5 10 10 5 1
6	1 6 15 20 15 6 1

This splitting is due to the interaction which causes the splitting of a single NMR absorption line into various multiplets arising from the interaction of nuclear spin magnetic moments of the nuclei in close proximity. For this reason this effect is called *spin-spin coupling*. The spin-spin coupling term produces an energy change, whose form is $\Delta E_{\text{spin-spin}} \propto m_{I_A} m_{I_B}$. Writing this proportionality in form of equality, we get

$$\Delta E_{\text{spin-spin}} = J_{AB}\, h\, m_{I_A}\, m_{I_B} \qquad ...(5.131)$$

where J_{AB} is called coupling constant between nuclei A and B and possesses the units of frequency, *i.e.* Hz or s^{-1}. Combining equations (5.125) and (5.131), the energy changes occurring at the nucleus A can be written as

$$\Delta E_A = -g_N \mu_N H(1-\sigma) m_{I_A} + J_{AB}\, h m_{I_A}\, m_{I_B} \qquad ...(5.132)$$

***Exercise* 5.70:** (*a*) Write the energy terms for the nucleus A in presence of nucleus B when both the nucleus are having $I = 1/2$, (*b*) predict the number of lines observed using the selection rules $\Delta m_I = \pm 1$ (*c*) calculate the magnitude of the splitting and (*d*) construct the energy level diagram in the absence of external magnetic field, in presence of external magnetic field without and with shielding and then in presence of spin-spin coupling.

Since both the nuclei have $I = 1/2$, the m_I values will be $\pm 1/2$. Considering $m_{I_A} = +\frac{1}{2}$ and taking $m_{I_B} = -\frac{1}{2}$ and $m_{I_B} = +\frac{1}{2}$, equation (5.131) can be written as

$$E_1 = -\frac{g_A \mu_N H(1-\sigma)}{2} - \frac{J_{AB} h}{4} \qquad ...(5.133)$$

$$E_2 = -\frac{g_A \mu_N H(1-\sigma)}{2} + \frac{J_{AB} h}{4} \qquad ...(5.134)$$

$$E_3 = \frac{g_A \mu_N H(1-\sigma)}{2} - \frac{J_{AB} h}{4} \qquad ...(5.135)$$

$$E_4 = \frac{g_A \mu_N H(1-\sigma)}{2} + \frac{J_{AB} h}{4} \qquad ...(5.136)$$

Similarly considering $m_{I_A} = -\frac{1}{2}$ and taking $m_{I_B} = +\frac{1}{2}$ and $m_{I_B} = -\frac{1}{2}$, equation (5.132) can be written as equations (5.135) and (5.136) respectively.

(*b*) Applying the selection rules $\Delta m_I = \pm 1$, the possible combinations of four energy levels lead to two transitions, *i.e.* E_1 and E_4 as well as, E_2 and E_3. The energies of the allowed transitions are:

$$\Delta E_{41} = E_4 - E_1 = \frac{g_A\mu_N H(1-\sigma)}{2} + \frac{J_{AB}h}{4} - \left(-\frac{g_A\mu_N H(1-\sigma)}{2} - \frac{J_{AB}h}{4}\right)$$

$$= g_A\mu_N H(1-\sigma) + \frac{J_{AB}h}{2} \qquad ...(5.137)$$

and

$$\Delta E_{32} = E_3 - E_2 = \frac{g_A\mu_N H(1-\sigma)}{2} - \frac{J_{AB}h}{4} - \left(-\frac{g_A\mu_N H(1-\sigma)}{2} + \frac{J_{AB}h}{4}\right)$$

$$= g_A\mu_N H(1-\sigma) - \frac{J_{AB}h}{2} \qquad ...(5.138)$$

Thus the NMR resonance line of nucleus *A* will appear as a doublet in the NMR spectrum at the energies given by the equations (5.137) and (5.138) respectively. The resonance frequencies of these lines can be obtained dividing the equations (5.137) and (5.138) by *h*, *i.e.*

$$v' = \frac{\Delta E_{41}}{h} = \frac{g_A\mu_N H(1-\sigma)}{h} + \frac{J_{AB}}{2} \qquad ...(5.139)$$

$$v'' = \frac{\Delta E_{32}}{h} = \frac{g_A\mu_N H(1-\sigma)}{h} - \frac{J_{AB}}{2} \qquad ...(5.140)$$

(*c*) Thus the frequency spacing between the doublet peaks produced by the spin-spin coupling is

$$\Delta v = v' - v'' = J_{AB} \qquad ...(5.141)$$

(*d*) The energy level diagram showing the spiliting of energy levels and transition involved are given Fig. 5.17

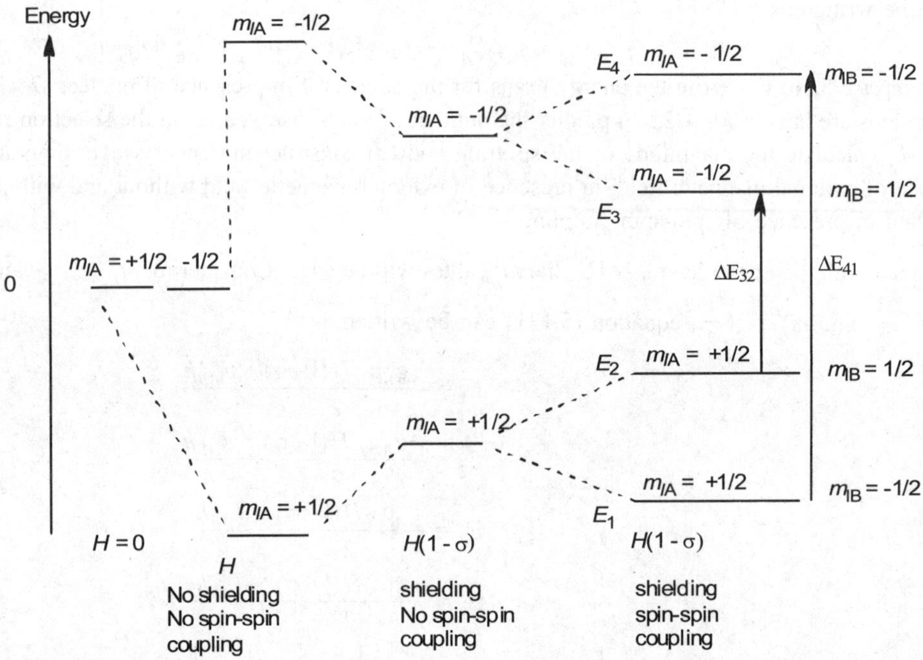

Fig. 5.17 *Energy level diagram showing the nucleus A in the absence and presence of external magnetic field, in presence of shielding and in presence of spin-spin coupling with nucleus B and showing two transitions that are occurring.*

5.11.3. Interpretation of NMR Spectra

After learning the simple basics, we are in a position to interpret most of the NMR spectra provided the absorption bands do not overlap to form complex patterns. The NMR spectra of this type are called *first order spectra*. The characteristics which are followed by this kind of NMR spectra are:

(*i*) The resonant frequency separation between magnetically non-equivalent nuclei is large in comparison to the spin-spin coupling constants between them, $(\delta/J) > 10$.

(*ii*) The absorption bands are narrow. The broad bands which result in the NMR spectra of nuclei $I > 1/2$ that have nuclear quadrupole moments make absorption band overlap more likely to occur.

(*iii*) There is a unique spin-spin coupling constant between any two sets of chemically equivalent $I = 1/2$ nuclei. We have explicitly invoked this assumption in all the preceding examples.

***Exercise* 5.71:** Suppose we have a compound whose molecular formula is $C_3H_6O_2$ and is an ester. Two of the signals observed are having chemical shifts of 2.0 and 3.5, having the same area, *i.e.* (1:1). Determine the structure of the compound.

Since the given compound is an ester, it could be either $HCOOCH_2CH_3$ (ethyl formate) or CH_3COOCH_3 (methyl acetate). Looking at the first compound, it is clear that there are three kinds of protons and thus would observe the three signals at different chemical shifts of areas in the ratio of 1 : 2 : 3. On the other hand, the second compound contains only two kinds of protons belonging to methyl groups. One of the methyl groups is connected to Oxygen atom and the other methyl group is connected to carbon atom. Thus, these two methyl groups are not equivalent and will give rise to two different signals. Table of chemical shifts also confirms the presence of methyl group. Thus the structure of the molecule will be methyl acetate.

***Exercise* 5.72:** What will be the pattern of proton NMR of ethyl alcohol, both under low resolution and under high resolution? Also indicate the relative field strength where the signal will appear.

(*a*) Ethyl alcohol (CH_3CH_2OH) molecule contains three kinds of protons *i.e.*, (*i*) alcoholic proton, (*ii*) two methylene protons and (*iii*) three methyl protons. Thus under low resolution, three NMR signals will appear, the relative intensities of three lines being in the ratio of 1: 2: 3 respectively.

(*b*) *Under high resolution:* (*i*) the methyl protons are under the influence of the magnetic field produced by the different orientation of two methylene protons. Since two methylene protons can arrange themselves in three ways *i.e.*, m_I can have the value 1, 0, – 1 and statistical weight

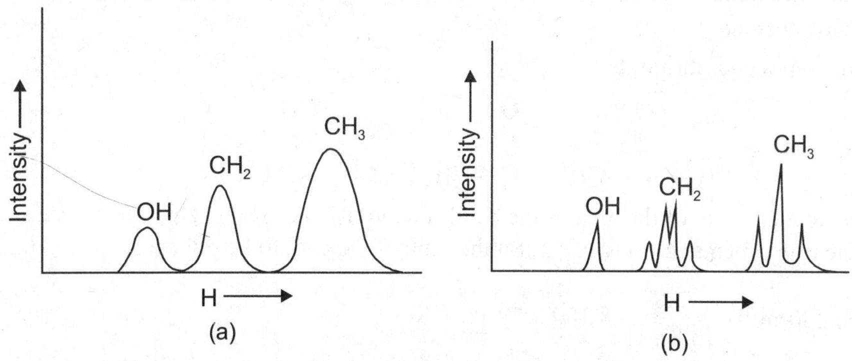

Fig. 5.18 *NMR spectrum of CH_3CH_2OH: (a) low resolution (b) high resolution*

of each arrangement is in the ratio of 1: 2: 1. So the methyl signal will split into three lines, whose heights are in the ratio of 1: 2: 1. (*ii*) Similarly the methylene protons are having three methyl neighbouring protons such that m_I can have the values 3/2, 1/2, – 1/2 and – 3/2 with statistical weight as 1: 3: 3: 1, *i.e.,* methylene protons will split into four peaks, with heights in the ratio of 1: 3 : 3 : 1. (*iii*) The alcoholic proton will have one line. This is because the alcoholic proton is always exchanging its position with other protons. Thus the average signal is obtained.

(*c*) Lastly the alcoholic proton **NMR** signal will appear at low magnetic field followed by methylene and then methyl. This is because, the alcoholic proton is attached to the oxygen atom whose electronegativity is higher than that of proton. Due to this the pair of electrons between the O–H bond will be close to oxygen atom rather than hydrogen. In other words the alcoholic proton nucleus is less shielded and have direct interaction with external magnetic field.

The inductive effect of oxygen atom decreases towards methylene and methyl protons. Thus next signal will be due to methylene protons followed by methyl protons.

The low and high resolution NMR of ethyl alcohol is shown in Fig. 5.18.

***Exercise* 5.73:** The proton NMR spectrum of a compound with molecular formula C_2H_6O was obtained. Under low resolution three signals with relative intensities of 1: 2: 3 were recorded. Under high resolution, the high field signal split into three lines of relative intensities 1: 2 : 1, medium field signal into four lines of intensities in the ratio of 1: 3 : 3 : 1. What is structural formula?

There can be two structural formula for this molecular formula *i.e.,* (*i*) CH_3OCH_3 dimethyl ether, (*ii*) CH_3CH_2OH ethyl alcohol. The low resolution NMR spectrum of dimethyl ether should have only one signal due to six equivalent methyl protons. So it cannot be dimethyl ether.

The three signals under low resolution NMR with relative intensities 1: 2: 3 is compatible with the formula (*ii*) because there are three different kinds of protons. The high resolution spectrum is also compatible with the ethyl alcohol as explained in the Exercise 5.72.

The structural formula is ethyl alcohol *i.e.,* CH_3CH_2OH.

***Exercise* 5.74:** The NMR spectrum of acetyl acetone was taken at 43°C. The band observed at δ 5.52 is due to olefinic hydrogen of enol form and band observed at δ 3.66 is attributed to keto methylene group. If the areas under the two curves are in the ratio of 37 :19.5, calculate the percentage of enol form in acetyl acetone.

The keto-enol equilibrium is

$$CH_3—\overset{\overset{\large O}{\|}}{C}—CH_2—\overset{\overset{\large O}{\|}}{C}—CH_3 \rightleftharpoons CH_3—\overset{\overset{\large OH}{|}}{C}{=}CH—\overset{\overset{\large O}{\|}}{C}—CH_3$$

The theoretical ratio of the area of the bands due to the enol H and keto methylene should be 1: 2 whereas the experimental ratio is 1.9: 1. So the ratio of the enol to keto form in the mixture is 3.8:1.

$$\% \text{ enol form} = \frac{3.8}{(3.8+1)} \times 100 = 79.1\%$$

$$\% \text{ keto form} = 100 - 79.1 = 20.9\%$$

5.12. ELECTRON SPIN RESONANCE

Similar to the NMR, the electron spin wave function for an odd electron also splits into two states when magnetic field is applied. By convention, $s = -1/2$ is the lower energy than the state with $s = +\frac{1}{2}$. The energies of the two states are obtained from the expression (5. 142) similar to that obtained for interaction of nuclear spin angular momentum and external magnetic field *i.e.*,

$$E = g_e\beta_e H m_s \quad ...(5.142)$$

where g_e is the electronic g factor, β_e is the Bohr magneton for electron. The transitions will be observed between the states if $\Delta m_s = \pm 1$ *i.e.*, the frequency of absorption for electron spin resonance spectrum is given as

$$\Delta E = h\nu = g_e\beta H\left[\tfrac{1}{2}-\left(-\tfrac{1}{2}\right)\right]$$

$$= g_e\beta_e H \quad ...(5.143)$$

$$\nu = \frac{g_e\beta_e H}{h} \quad ...(5.144)$$

Each of these states is further split into $(n + 1)$ lines for n equivalent nuclei with finite nuclear spin quantum number, present in the molecule. Again by convention, the energies of the positive nuclear spins are lower than those of negative nuclear spin in the $s = -\frac{1}{2}$ and vice versa for $s = +\frac{1}{2}$. The selection rule for allowed transitions is $\Delta M_I = 0$ where $M_I = I, I - 1, ... 0, ... -I + 1, -I$ and $I = n\frac{1}{2}$.

The intensities of lines can be calculated by the use of Pascal's triangle binomial coefficients as explained in Section (5.9.2).

***Exercise* 5.75:** Calculate the value of Bohr's magneton for the electron.

Solution: The expression for the Bohr's magneton for the electron is

$$\beta_e = -\frac{eh}{4\pi m_e} = -\frac{(1.6022177 \times 10^{-19}\ \text{C})(6.626 \times 10^{-34}\ \text{J s})}{(4)(3.14)(9.10939 \times 10^{-31}\ \text{kg})} = -9.2742 \times 10^{-24}\ \text{J T}^{-1}$$

5.12.1. Hyperfine Splitting

Proceeding on the same lines as in spin-spin coupling, interactions are also observed between the spin angular momentum of the electron and spin angular momentum of the nucleus. This leads to the spin-spin splitting of the ESR bands. This effect is called *hyperfine splitting*. The mathematical treatment is similar to the spin-spin splitting of the NMR band.

The simplest system is bare hydrogen atom with ground state electronic configuration as ^2S state. Since there is no electronic orbital angular momentum, the magnetic moment will be only due to the spin motion of the electron. The interaction of the magnetic moment with the external magnetic field will lead to energy change, given by the equation (5.142). The spin magnetic moment of the nucleus also produces small magnetic field and couple with the electronic spin magnetic moment. This interaction affects the electron spin energy state. The magnitude of the effect is given by equation (5.145)

$$\Delta E_{\text{spin-spin}} = a\,h\,m_s\,m_{I_N} \qquad ...(5.145)$$

where a is *hyperfine splitting constant*, m_s and m_{I_N} are the spin magnetic quantum numbers of the electron and nucleus (proton). The total energy change is similar to that observed for the spin-spin interaction for NMR and is given as

$$\Delta E = g_e\beta_e m_s H + ahm_s m_{I_N} \qquad ...(5.146)$$

Since the value of m_s and m_{I_N} can be ±1/2, the magnetic interactions split the degenerate electronic spin state into four energy levels, given by

$$E_1 = -\frac{g_e\beta_e H}{2} + \frac{ah}{4} \qquad \left(m_s = +\frac{1}{2}, m_{I_N} = -\frac{1}{2}\right) \qquad ...(5.147)$$

$$E_2 = -\frac{g_e\beta_e H}{2} - \frac{ah}{4} \qquad \left(m_s = -\frac{1}{2}, m_{I_N} = +\frac{1}{2}\right) \qquad ...(5.148)$$

$$E_3 = \frac{g_e\beta_e H}{2} + \frac{ah}{4} \qquad \left(m_s = +\frac{1}{2}, m_{I_N} = +\frac{1}{2}\right) \qquad ...(5.149)$$

$$E_4 = \frac{g_e\beta_e H}{2} - \frac{ah}{4} \qquad \left(m_s = +\frac{1}{2}, m_{I_N} = -\frac{1}{2}\right) \qquad ...(5.150)$$

The splitting is shown in Fig. 5.19. The selection rules for the ESR spectrum is $\Delta m_s = \pm 1$ and $m_{I_N} = 0$. The number of allowed transitions is only two as shown in Fig. 5.19.

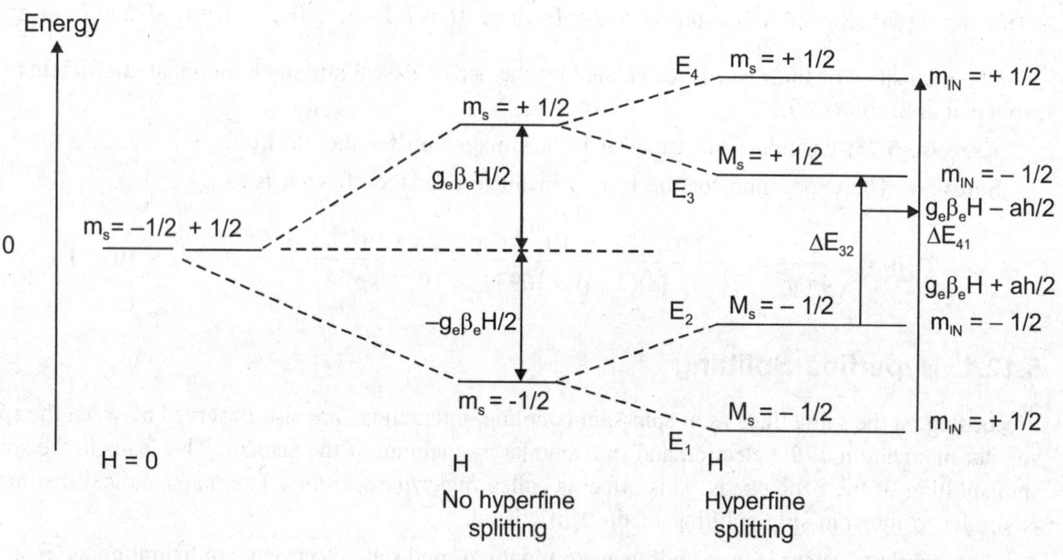

Fig. 5.19 *Energy level diagram showing the electron in the ground state in the presence of external magnetic field, in presence of hyperfine with nucleus (proton) and showing two transitions that are occurring.*

$$\Delta E_{41} = E_4 - E_1 = \frac{g_e\beta_e H}{2} + \frac{ah}{4} - \left(-\frac{g_e\beta_e H}{2} - \frac{ah}{4}\right) = g_e\beta_e H + \frac{ah}{2} \qquad ...(5.151)$$

and $$\Delta E_{32} = E_3 - E_2 = \frac{g_e\beta_e H}{2} - \frac{ah}{4} - \left(-\frac{g_e\beta_e H}{2} + \frac{ah}{4}\right) = g_e\beta_e H - \frac{ah}{2} \quad ...(5.152)$$

Thus the ESR resonance line of electron will appear as a doublet in the ESR spectrum at the energies given by the equations (5.151) and (5.152) respectively. The resonance frequencies of these lines can be obtained dividing the equations (5.151) and (5.152) by h, *i.e.*

$$\nu' = \frac{\Delta E_{41}}{h} = \frac{g_e\beta_e H}{h} + \frac{a}{2} \quad ...(5.153)$$

$$\nu'' = \frac{\Delta E_{32}}{h} = \frac{g_e\beta_e H}{h} - \frac{a}{2} \quad ...(5.154)$$

Thus the frequency spacing between the doublet peaks produced by the spin-spin coupling is

$$\Delta\nu = \nu' - \nu'' = a \quad ...(5.155)$$

The ESR band separation $\nu' - \nu''$ is therefore equal to the hyperfine splitting constant.

Equation (5.146) reveals that the magnitude of the hyperfine coupling constant depends on the magnitude of the magnetic moments of the nuclei and electron. The other factor which also affects its magnitude is the proximity of the nucleus to the electron. This quantity is measured by the electron density at the nucleus, which is proportional to the absolute square of the electronic wave function evaluated at the nucleus, $|\psi(0)|^2$. Equation relating the hyperfine coupling constant with charge density of the electron near the nucleus is known as *Fermi contact interaction* and is given by the equation (5.156).

$$a = \frac{8\pi}{3h} g_e\beta_e g_N\beta_N |\psi(0)|^2 \quad ...(5.156)$$

Thus ESR spectrum helps us in inferring the electron spin density at the various nuclei.

***Exercise* 5.76:** How many ESR lines are observed if an unpaired electron, when closed to two non-equivalent protons, is placed in magnetic field?

In the presence of magnetic field the unpaired electron will split into two energy levels. Each energy state will split further into two states due to the first proton ($I = \frac{1}{2}$) and each line further into two lines due to second proton because each proton has different splitting constants. Thus each state due to electron spin, will split into four states due to two non-equivalent protons as shown in Fig. 5.20. With selection rules $\Delta M_I = 0$, four lines will be observed in EPR spectrum.

***Exercise* 5.77:** If the two equivalent protons are considered in the Exercise (5.76), predict the number of lines.

Since the two protons are equivalent, the hyperfine splitting constants are same *i.e.*, $a_1 = a_2 = a$. Second proton will split the first two lines due to first proton but the central one will be at the same place. Three slates are observed with the intensity of central line is twice that of the outer ones(Fig. 5.21). So only three lines are observed.

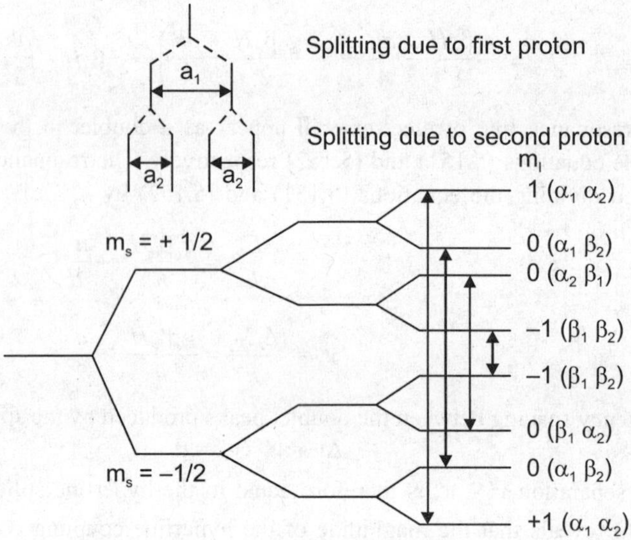

Fig. 5.20 *ESR spectrum of an electron in presence of non-equivalent protons*

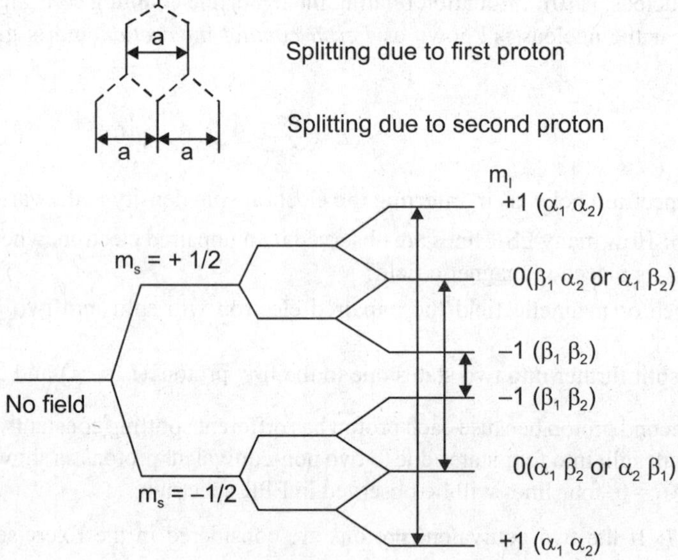

Fig. 5.21 *ESR spectrum of an electron in presence of two equivalent protons*

PROBLEMS

Rotational Spectra

5.1 Calculate the reduced mass and the moment of inertia of $^{79}Br^{35}Cl$ molecule. The bond length of Br—Cl is 0.214 nm.

5.2 Calculate the energies of rotational levels for $J = 0, 1, 2, 3, 4$ and 5 for BrCl molecule using the data from problem 5.1. Also calculate the frequencies required for the transitions $0 \rightarrow 1, 1 \rightarrow 2, 2 \rightarrow 3, 3 \rightarrow 4$ and $4 \rightarrow 5$.

5.3 Repeat the problems 5.1 and 5.2 for $H^{35}Cl$ molecule. The bond length in HCl is 0.128 nm. Compare the separations of the energies of the levels in BrCl and HCl. (It will be clear that as the mass of the molecule increases, the separation between the levels decreases).

5.4 Repeat the problem 5.3 for DC1, $D^{35}C1$ and $H^{37}Cl$. Compare the separation of energies of the rotational levels with that of $H^{35}Cl$.

5.5 The difference between the successive lines in the rotational spectrum of CO molecule is 2.98 cm^{-1} (or 3.86 cm^{-1}). Calculate the bond length.

5.6 The equilibrium bond length of $^{208}Pb^{32}S$ molecule is 2.394×10^{-10} m. Calculate the value of B_e and compare with that of HCl obtained in problem 5.3.

Vibrational Spectra

5.7 The force constant of HF molecule is 9.7×10^2 N m^{-1}. Calculate its fundamental frequency as well as the zero point energy.

5.8 If the force constants of HF and DF molecules are same *i.e.*, (9.7×10^2 N m^{-1}), calculate the zero point energy of DF molecule and find the difference between the zero point energies of HF and DF molecules.

5.9 In the vibrational spectrum of CO, the vibrational band origins ($\bar{v}_0$) are observed at $1-0 = 2143.38$ cm^{-1}; $2-0 = 4260\ 12$ cm^{-1}; $3-0 = 6350.39$ cm^{-1}; $4-0 = 8414.45$; $5-0 = 9564$ cm^{-1}. Calculate the anharmonicity constants.

5.10 The force constants for N – N, N = N and N ≡ N bonds are 4.5×10^2 N m^{-1}, 13.25×10^2 N m^{-1} and 22.9×10^2 N m^{-1} respectively. Which of three bonds will have higher frequency of vibration and why? Calculate the value of frequency in each case.

5.11 The fundamental vibrational frequencies observed for molecules H_2, HD and D_2 are 4395 cm^{-1}, 3817 cm^{-1} and 3118 cm^{-1} respectively. Calculate the force constant for each molecule. Do the results indicate that the values of force constants are independent of isotopic change?

5.12 The anharmonicity constant for $^{35}C1\ ^{19}F$ is $w_e x_e = 9.9$ cm^{-1}. If the fundamental frequency is 793.2 cm^{-1}, calculate the energies of the first four levels.

5.13 Using the data of the problem 5.12 find the difference between v = 24 and v = 25 vibrational levels, considering (*a*) harmonic oscillator model and (*b*) anharmonic oscillator model. Find the difference between the frequencies observed if any.

5.14 If the vibrations in a diatomic molecule (say HF) obeys Hook's law *i.e.*, $U = \frac{1}{2}kx^2 = \frac{1}{2}k(r - r_e)^2$, construct the potential energy curve for HF molecule. Given

$$k = 9.7 \times 10^2 \text{ N m}^{-1}, \quad r_e = 0.0917 \text{ nm}$$

5.15 The Morse potential is given by

$$U(r) = D_e[1 - e^{-a(r - re)}]^2$$

where D_e is the equilibrium bond dissociation energy, r_e is the equilibrium bond length and a is a constant. If $r_e = 0.0917$ nm, $D_e = 568.35$ kJ mol^{-1}, $w_e = 4138.52$ cm^{-1}, $w_e x_e = 90.07$ cm^{-1} for HF molecule, plot $U(r) - D_e$ versus r, showing first 10 vibrational levels. Compare the curve obtained with that observed in problem 5.14.

5.16 The vibrational constants of one of the upper states of Br_2 molecule are $w_e = 151.2$ cm^{-1}, $w_e x_e = 0.33$ cm^{-1}. Calculate the dissociation energy of Br_2 molecule in this state.

5.17 The following data is obtained for electronic spectrum of ClO radical by Durie and Ramsay (*Can. J. Phys.*, **36**, 35 (1958)):

(*a*) Calculate w_e, $w_e x_e$ for the upper state.

(*b*) From Birge-Sponer plot, calculate the dissociation energy of this state.

(*c*) Using equation (5.98), also calculate the dissociation energy.

Band (v′, v″)	$\bar{\nu}$ (cm^{-1})	Band (v′, v″)	$\bar{\nu}$ (cm^{-1})	Band (v′, v″)	$\bar{\nu}$ (cm^{-1})
5, 0	33406.6	13, 0	36355.8	19, 0	37561.9
6, 0	33844.4	14, 0	36619.5	20, 0	37678.7
9, 0	35053.6	15, 0	36864.3	21, 0	37772.9
10, 0	35415.8	16, 0	37072.8	22, 0	37849.5
11, 0	35754.5	17, 0	37264.8	23, 0	37917.8
12, 0	36066.5	18, 0	37425.1	24, 0	37960.7
				25, 0	37997.4

5.18 Briggs and Norrish observed the electronic transient spectrum of Cl_2 (*Proc. Roy. Soc.*, London, **A 276**, 51 (1963)). The following data is given:

Band (v′, v″)	$\bar{\nu}_{vac}$ (cm^{-1})	Band (v′, v″)	$\bar{\nu}_{vac}$ (cm^{-1})	Band (v′, v″)	$\bar{\nu}_{vac}$ (cm^{-1})
1, 0	40140	10, 0	42292	21, 0	44769
2, 0	40385	11, 0	42536	22, 0	44980
3, 0	40630	12, 0	42771	23, 0	45192
4, 0	40871	13, 0	43001	24, 0	45402

5, 0	41112	14, 0	43227	25, 0	45609
6, 0	41351	15, 0	43451	26, 0	45815
7, 0	41589	16, 0	43675	27, 0	46017
8, 0	41825	17, 0	43896	28, 0	46222
9, 0	42059	18, 0	44116	29, 0	46425
		19, 0	44335	30, 0	46623
		20, 0	44553	31, 0	46824
				32, 0	47824

Using the above data calculate (*a*) w_e, $w_e x_e$ for the upper state of Cl_2 molecule; (*b*) use Birge-Sponer plot to calculate the value of D_e; (*c*) use relation (5.98) to calculate the dissociation energy. Compare with that obtained in part (*b*).

Rotational Vibrational Spectra

5.19 (*a*) Find the energies of first three vibrational levels of HF and DF molecules. Force constants for HF and DF is 9.7×10^2 N m^{-1}.

(*b*) Find the frequency of absorption of light when molecules go from v = 0 to v = 1 in each case. Also calculate the difference in frequencies v_{HF} and v_{DF}.

(*c*) Calculate the energies of the first four rotational levels in case of HF and DF molecules. Bond length of HF and DF molecules are same and the value is 0.0917 nm.

(*d*) Assuming that the rotational constants for HF are equal for $v = 0$ and v = 1, draw the first four rotational levels in each vibrational level. Repeat the calculations for DF.

(*e*) Calculate the frequencies required for the transition v = 0, J = 3 to v = 1, J = 4, both for HF and DF. Note the difference in frequencies if any.

5.20 Following is the data available for $H^{35}Cl$ molecule: w_e = 2989.7 cm^{-1}, $w_e x_e$ = 52.05 cm^{-1}, B_e = 10.59 cm^{-1} and α_e = 0.302 cm^{-1}. Calculate the frequencies of P and R branch for J = 8 for v = 0 to v = 1 band.

5.21 In the ground electronic state of a diatomic molecule, the J = 35 rotational level of v = 0 vibrational state has the same total energy (electronic + vibrational + rotational) as the lowest rotational level of v = 1 vibrational state. If the pure vibrational frequency of absorption between v = 0 and v = 1 is 6.0×10^{13} s^{-1}, find the equilibrium distance for v = 0 state in Å if

$$h^2/8\pi^2\mu = 4.2 \times 10^8 \text{ erg Å}^2\text{/mol}$$

5.22 In the electronic spectrum of BrO, the following data is obtained for the rotational analysis of 0–0 band (*i.e.*, zero vibrational level of ground electronic state to zero vibrational level of the upper electronic state). Calculate rotational constants for the upper and lower state and thus the internuclear distance by using method I. Repeat the above process by using method II.

J	$R(J)$ (cm^{-1})	$P(J)$ (cm^{-1})	J	$R(J)$ (cm^{-1})	$P(J)$ (cm^{-1})
0	211199.81		6	211215.58	211174.82
1	211202.88	211193.25	7	211217.71	211170.65

2	211205.74	211189.97	8	211219.65	211166.93
3	211208.52	211186.41	9	211221.42	211161.93
4	211211.12	211182.66	10	211223.12	211157.28
5	211213.58	211178.88	11	211224.62	211152.57

5.23 In the infrared spectrum of CO, the fine structure of 0–1 band is as follows. Calculate the rotational constants from the following data:

J	*R(J)* (cm^{-1})	*P(J)* (cm^{-1})	*J*	*R(J)* (cm^{-1})	*P(J)* (cm^{-1})	*J*	*R(J)* (cm^{-1})	*P(J)* (cm^{-1})
0	2147.19		1	2150.97	2139.53	2	2154.71	2135.64
3	2158.41	2131.74	4	2162.08	2127.81	5	2165.71	2123.81
6	2169.31	2119.99	7	2172.87	2115.74	8	2176.39	2111.61

Vibrational Spectra of Polyatomic Molecules

5.24 Calculate the vibrational degrees of freedom in the following molecules:

(*i*) O_2, (*ii*) CH_2 radical, (*iii*) N_2O, (*iv*) CH_2O, (*v*) CH_4,
(*vi*) C_6H_6 (benzene), (*vii*) CH_3Cl, (*viii*) $CHC1_3$, (*ix*) CH_3CHO, (*x*) NO_2.

5.25 Infrared spectrum of H_2O shows three bands with maxima at 3756 cm^{-1}, 3652 cm^{-1} and 1595 cm^{-1}. Indicate whether the molecule is linear or bent.

5.26 In triatomic molecule of type MX_2, two infrared and one Raman frequencies are observed. Indicate the structure of the molecule.

5.27 Explain the following vibrational transitions mentioned in short hand notations:

(*i*) (0, 0, 0) → (2, 0, 0); (*ii*) (0, 0, 0) → (0, 2, 1);
(*iii*) (0, 2, 0) → (1, 1, 0); (*iv*) (0, 0, 0, 2) → (1, 1, 0, 1);
(*v*) (0, 0, 1, 1) → (2, 0, 1, 2).

5.28 Three fundamental vibrational frequencies of NO_2 molecule are observed at 750 cm^{-1}, 1323 cm^{-1} and 1616 cm^{-1}. If it is assumed that molecular vibrations follow simple harmonic motion, calculate the energies of the absorption bands for

(*i*) (0, 0, 0) → (0, 2, 0); (*ii*) (0, 0, 0) → (0, 1, 1);
(*iii*) (0, 0, 0) → (2, 1,1); (*iv*) (0, 0, 0) → (0, 2, 1);
(*v*) (0, 0, 0) → (1, 1, 1); (*vi*) (0, 0, 0) → (2, 2, 1).

5.29 Which of the following vibrations will be IR active and which are inactive?

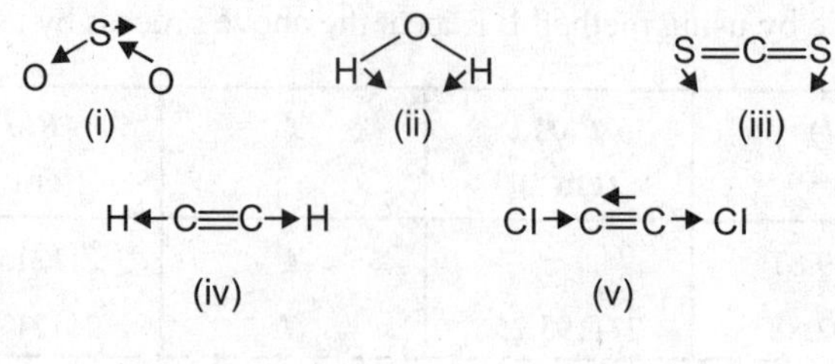

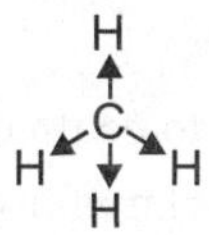

(vi)

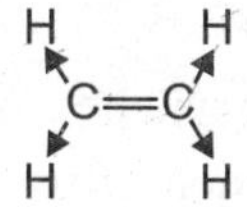

(vii)

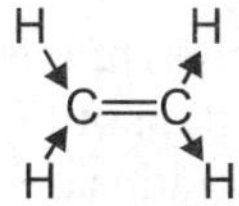

(viii)

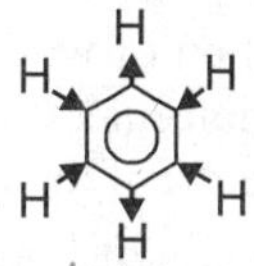

(ix)

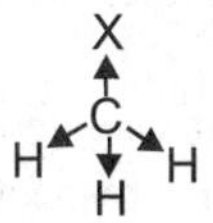

(x)

The vibrations (*iv*) to (*x*) are all stretching vibrations.

Temperature Dependence

5.30 The successive difference between the rotational lines, in case of $^{35}Cl^{19}F$, is found to be 0.516 cm^{-1}. Calculate the rotational level which will have the highest population at 27°C and 227°C.

5.31 Calculate the temperature at which the relative population of $J = 7$ level is maximum. Given, $\frac{h^2}{\pi^2 Ik} = 100$ K and degeneracy of the rotational levels is $2J + 1$.

5.32 The difference between the successive rotational levels in case of HCl is 20.70 cm^{-1}. Plot the populations of molecules in different rotational levels vs rotational quantum numbers at 27°C and 127°C. Compare the maximum obtained at both temperatures.

5.33 The fundamental vibrational frequency of HCl molecule is 2988.90 cm^{-1}. Plot the population of molecules in various vibrational levels vs vibrational quantum numbers at 27°C and 127°C. Compare this plot with that obtained in problem 5.32?

5.34 The fundamental vibrational frequency of $^{35}Cl^{19}F$ molecule is 793.20 cm^{-1}. Plot the population of molecules in various vibrational levels vs v at 4000 K.

5.35 Justify that vibrational absorption transitions are generally observed in case of HF from v = 0. $\bar{\nu}_{HF} = 4138.52$ cm^{-1}.

5.36 If the internuclear distance between HF is 0.0917 nm, calculate the relative population of the molecules in the state, $J = 1$, 2, 3 and 4 with respect to $J = 0$ at 27°C and at liquid nitrogen temperature *i.e.,* 70 K. Represent the relative population of each level as a height of a bar. What effect will be cooling has on the intensities of IR (rotational vibrational) spectrum of HF molecule?

5.37 In the rotational spectrum, many lines are observed at room temperature. If the rotational spectrum for same molecule is observed at lower temperature, will the number of lines decrease or increase and also the intensity of each line will be different or same. Plot the intensity vs rotational quantum number.

Raman Spectrum

5.38 What kind of light source will be more economical to be used for Raman Spectrum?

5.39 A mono-chromatic radiation of wavelength 438.5 nm is used to study the Raman rotational spectrum. Find the positions of Stokes and anti-Stokes lines in case of rotational spectrum of HCl molecule taking place from $J = 2$ level. Given $2B = 20.70\ \text{cm}^{-1}$.

5.40 A mono-chromatic radiation of wavelength 253.6 nm Hg line is used to study the vibrational Raman spectrum of O_2 molecule. Calculate the position of first Stokes and anti-Stokes Raman line arising from O_2 molecule if the fundamental vibrational frequency is of 1554.7 cm^{-1}.

5.41 In Raman spectrum, two sets of lines are observed. One set of lines on both sides of exciting light are only roughly 20 cm^{-1} apart and other set of lines are observed at longer displacements. Explain this behaviour.

5.42 The fluorine molecules were exposed to Hg lines of wavelength 404.7 nm and 435.8 nm. The Stokes lines are observed at 419.9 nm and 453.4 nm respectively. Calculate the ground state vibrational frequency of F_2.

Electronic Transitions (Electronic Spectrum)

5.43 Which of the following transitions will be observed and which one will be forbidden according to the selection rules?

(*i*) $^3\Sigma \rightarrow {}^3\Sigma$ (*ii*) $^3\Sigma \rightarrow {}^8\Pi$ (*iii*) $^3\Sigma \rightarrow {}^1\Delta$ (*iv*) $^2\Pi \rightarrow {}^2\Pi$

(*v*) $^2\Pi \rightarrow {}^1\Delta$ (*vi*) $^3\Sigma \rightarrow {}^3\Delta$ (*vii*) $^3\Sigma \rightarrow {}^1\Sigma$ (*viii*) $^2\Pi \rightarrow {}^2\Sigma$

(*ix*) $^4\Sigma \rightarrow {}^2\Pi$ (*x*) $^2\Pi \rightarrow {}^1\Sigma$

5.44 The potential energy diagrams are drawn (Figs. 5.11 and Fig. 5.15) and the transitions are depicted by arrows:

(*a*) The transition mentioned in Fig. 5.11 (*a*) will be observed or not.

(*b*) The transitions mentioned in Fig. 5.15(*b*) are from v″ = 0 to v′ = 0, 1 and 2. Arrange them in order of decreasing intensities.

(*c*) The transition mentioned in Fig. 5.15(*b*) will be discrete one or continuous one.

5.45 The spectroscopic bond dissociation energy (D_0) of $^{35}\text{Cl}^{16}\text{O}$ radical is 1.9 eV. Calculate the equilibrium bond dissociation energy of ClO if the fundamental vibrational frequency is 780 cm^{-1}.

5.46 Calculate the equilibrium dissociation energy (D_e) of the following molecules with the given data.

Specie	w_e (cm^{-1})	$w_e x_e$ (cm^{-1})	D_0 (eV)
CH	2862. 6	64. 3	3. 47
CO	2170. 21	13. 46	11. 108
$H^{35}Cl$	2989. 79	52. 05	4. 43
$^{14}N^{16}O$	1904. 03	13. 97	6. 49

5.47 If the following frequencies of Schumann-Runge absorption band system of O_2 corresponds to $J'' = 0$, v″ = 0 to $J' = 0$ and v′ = v, calculate the value of w_e and $w_e x_e$ for the upper electronic state:

v	$\bar{\nu}$ (cm^{-1})	v	$\bar{\nu}$ (cm^{-1})	v	$\bar{\nu}$ (cm^{-1})
0	49357.6	4	51969.8	8	54158.9
1	50045.6	5	52561.6	9	54624.4
2	50710.7	6	53122.8	10	55053.3
3	51352.2	7	53656.8	11	55441.5

5.48 In the electronic spectrum of ClO radical, the band structure joins the continuum at 38020 cm^{-1}. The ClO radical dissociates in the excited state as:

$$\text{ClO} \rightarrow \text{Cl}(^2P_{3/2}) + \text{O}(^1D)$$

and in the ground electronic state as

$$\text{ClO} \rightarrow \text{Cl}(^2P_{3/2}) + \text{O}(^3P_2)$$

If the energy difference between O($^3P_2$) and O(^1D) state is 15868 cm^{-1}, calculate the dissociation energy of the ClO radical in its ground state. Express the answer in Joules.

5.49 Predict the number of electronic transitions taking place in the molecules,
(*a*) CH_3Cl, (*b*) $CH_2 = CH_2$, (*c*) NH_2 group, (*d*) $-NO_2$ group,
(*e*) SO group.

Nuclear Magnetic Resonance

5.50 The nuclear spin quantum number of nitrogen atom is unity.
(*a*) Will there be any NMR spectrum due to nitrogen atom?
(*b*) How many lines will be observed?
(*c*) If the g_N value for nitrogen atom is 0.403, calculate the frequency required for NMR line. The strength of the external field is 14,000 guass.

5.51 In the proton NMR spectrum if a proton is replaced by deuterium, the NMR signal of proton disappears from its position. Explain this from the given data. Spin quantum numbers of ^{1}H and ^{2}H are 1/2 and 1 respectively, the frequency employed is 100 MHz.

5.52 Calculate the strength of the magnetic field to give a precessional motion: The frequency is 100 MHz.
(*a*) ^{17}O nuclei, $g_N = -0.757$, I = 5/2, $\gamma_N = -3628$ G^{-1}s^{-1}
(*b*) ^{19}F nuclei, I = 1/2, $g_N = 5.257$, $\gamma_N = 25179$ G^{-1}s^{-1}
(*c*) ^{63}Na nuclei, I = 3/2, $g_N = 1.478$, $\gamma_N = 7081$ G^{-1}s^{-1}.

5.53 The molecular formula for certain compound is C_2H_4O. In low resolution NMR spectrum two peaks are observed. The intensity of one of the peaks is thrice the other. Under high resolution, the peak with small intensity or at low field split into four components of relative intensity 1:3:3:1 and peak with high intensity into two components with relative intensity 1:1. Give the structural formula of the compound.

5.54 The molecular formula of an unknown compound is $C_2H_3Cl_3$. Under low resolution of NMR, only two peaks are observed, with the intensity of one peak twice the other. Under high resolution, the peak at low field splits into three components and peak at high field into two components with relative intensities 1:2:1 and 1:1 respectively. Find the structural formula.

5.55 Another compound with molecular formula $C_2H_3Cl_3$ studied under proton NMR gives only one peak. Find the structural formula.

5.56 What will be the pattern of NMR spectrum of the following compounds, both under low resolution and high resolution? Also label the NMR lines compatible with functional groups and the field strength. Also indicate the relative intensity ratio of each line:

(*i*) H_2O_2, (*ii*) Isopropyl alcohol, (*iii*) ter-butyl alcohol,
(*iv*) benzene, (*v*) 1:1 mixture of H_2O and $(CH_3)_4Si$ assuming no reaction.

5.57 Compare the NMR spectrum of HD molecules under high resolution.

5.58 (*a*) Which of the following species will possess the ESR spectrum?

(*i*) CH_2, (*ii*) CH_3, (*iii*) H_2, (*iv*) O_2,
(*v*) O_2+, (*vi*) C_6H_6, (*vii*) $CH_3CH_2CH_2$, (*viii*) CH_3CHCH_3,
(*ix*) $(CH_3)_3C$.

(*b*) Give the relative intensity of each line.

(*c*) In each case where there is an ESR spectrum draw the spectrum indicating also the hyperfine splitting.

5.59 In the low resolution NMR spectrum of a mixture containing cycloheptane and toluene, three bands were observed. The band at $\delta - 1.5$ is due to cyclic methylene protons, band at $\delta - 2.0$ is due to methyl protons and the band at $\delta - 8.0$ is due to ring protons of benzene. If the ratio of the bands are 90:30:115. Calculate the percentage of cycloheptane in the mixture.

Additional problems

5.60 Convert the following per cent transmittance to absorbance: (*a*) 90%, (*b*) 60% and (*c*) 30%.

5.61 Convert the following per cent absorbance to transmittance: (*a*) 0.18, (*b*) 0.54 and (*c*) 0.75.

5.62 We have learnt that absorbance of radiation by the molecules results in the formation of molecules in the excited state. After long interval, *i.e.* given enough time, it will appear that all of the molecules in a sample would have been excited and no more absorption would occur. Yet in practice we find that the absorbance of the sample at any give wavelength remains unchanged with time. Explain.

5.63 The microwave spectrum of HF is recorded in certain microwave spectrometer with initial intensity I_0 and radiation transmitted at certain wavelength $I' = 80\%$. If the path length of cell is increased by 50% and concentration of HF is doubled while the intensity I_0 remains unchanged, calculate ratio of the new transmitted intensity I'' to I' as well as I'' to I_0 at the same wavelength.

5.64 A student while recording a spectrum found that if he increases the path length of the electromagnetic radiation beam through the sample cell by 60%, the ratio of the transmitted radiation intensity to the incident intensity drops by a factor of 2.5 provided the measure ments are carried out at the same wavelength and concentration. Is the data given above enough to determine the fraction of radiation transmitted in the original experiment? If so calculate the fraction. If it is not, state what additional information is required to do the calculation?

5.65 The path length of the electromagnetic radiation beam through the sample cell can be increased by increasing the length of the sample cell or by placing the mirrors so that the electromagnetic radiation gets reflected and thus path length is increased and it is shown below. Here the electromagnetic radiation passes five times the cell. It is equivalent to a cell of length 5*l*. (*i*)

Consider a similar cell in which the radiation traverses the cell n times before reaching the detector. Derive a general equation relating the fraction of the incident intensity transmitted in such a cell to the fraction transmitted for a cell using only a single pass of path length l, (*ii*) If the fraction of the incident intensity transmitted in a cell using a single pass is f, derive an equation in terms of f that gives the value of n must be to ensure the absorption of at least 70% of the incident light, (*iii*) If $f = 0.85$, calculate the value of n for the conditions given in (*ii*).

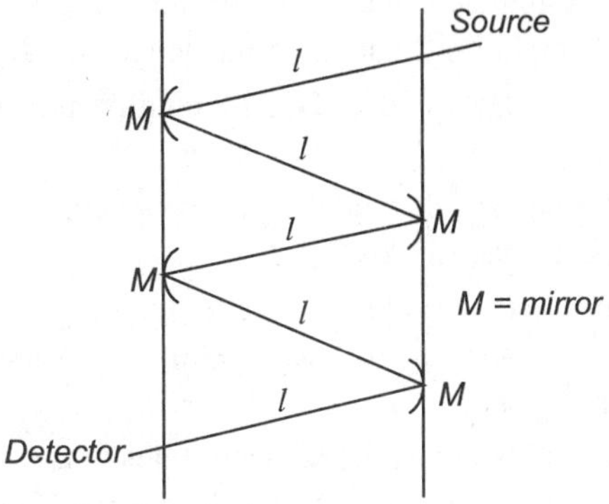

5.66 The familiar yellow *D*-lines of sodium is actually a doublet at 589.0 nm and 589.6 nm. Calculate the difference in energy in *J* between the two lines.

5.67 The resolution of visible and UV spectra can usually be improved by recording the spectra at low temperature. Why does this procedure work?

5.68 The molar absorptivity of a hexane solution of an organic compound is 4.2×10^4 L mol^{-1} cm^{-1} at 450 nm. If path length of cell is 1 cm calculate the percentage transmission of radiation if the concentration is 4.5×10^{-3} M.

5.69 The mean lifetime of an electronically excited molecule is 3×10^{-9} s. If the emission of the radiation occurs at 550 nm, calculate the uncertainties in frequency ($\Delta\nu$) and ($\Delta\lambda$).

5.70 The line width observed of the spectral line is found to be (*i*) 0.5 cm and (*ii*) 1.0 Hz. Assuming that the line width is only because of the uncertainty in the lifetime, calculate the lifetime of the excited state.

5.71 Singal to noise (S/N) ratio observed in one scan of NMR is 2.0. If each scan takes 8.0 minutes, calculate the minimum time required to generate a spectrum with S/N ratio to be 30.

5.72 Out of the following molecules which one will show microwave spectrum?

(*i*) NH_3, (*ii*) C_2H_2, (*iii*) CH_4, (*iv*) CH_3Cl, (*v*) $CHCl_3$, (*vi*) C_6H_6

5.73 The first rotational transition in HCl, initiating from $J = 0$ to $J = 1$ is observed at 20.70 cm^{-1}. (*i*) Calculate the ratio of spontaneous emission to stimulated emission for the above transition assuming that the incident radiation density per unit frequency is that for a black body radiator at a temperature of 500 K, (*ii*) calculate the temperature that would give spontaneous emission to be 30% that of stimulated emission.

5.74 Using the spherical harmonic wave functions given in Table 5.2, prove the followings.

(*i*) rotational transition from $J = 1$, $M = 0$ to $J = 2$, $M = 0$ is allowed

(*ii*) rotational transition from $J = 0, M = 0$ to $J = 2, M = 0$ is forbidden
(*iii*) rotational transition from $J = 1, M = 0$ to $J = 2, M = +1$ is forbidden
(*iv*) rotational transition from $J = 1, M = 0$ to $J = 3, M = 0$ is forbidden

Spherical harmonics for $J = 3, M = 0$ is $\psi_3^0 = \left[\frac{7}{16\pi}\right]^{\frac{1}{2}} [2\cos\theta - 5\sin^2\theta\cos\theta]$

5.75 If the equilibrium bond length in BH is 0.12325 nm, calculate (*a*) moment of inertia, (*b*) energy spacing between the rotational absorption bands, (*c*) energy of the absorption band arising from $J = 5$ to $J = 6$, (*d*) wave number of the transition observed in (*c*), (*e*) repeat the steps (*a*) to (*d*) for the molecule BD, assuming that the bond length in BH and BD are the same.

5.76 Repeat the Problem 5.75 for BeH and BeD. The equilibrium bond length is 2.538 bohr (*i.e.* 0.1343 nm).

5.77 Derive equation (5.55). Suggest whether the gap between successive rotational line will increase or decrease with increase of *J*?

5.78 The average value of difference between the wave numbers of adjacent lines in the rotational spectra of $^{12}C^{16}O$ and $^{12}CX^*$ are 3.8436 and 3.6606 cm^{-1} respectively. Determine the atomic mass of isotope of atom X, *i.e.* X^*.

5.79 The average value of spacing between the adjacent lines in the rotational spectrum of diatomic molecule CX is 3.84235 cm^{-1}. The equilibrium distance is 0.1131 nm. Find the atomic mass of X.

5.80 The rotational constants for $^{12}C^{16}O$ and $^{13}C^{16}O$ are found to be 1.92118 and 1.83669 cm^{-1} respectively. Calculate the ratio of ^{13}C and ^{12}C.

5.81 Following lines have been identified in the high resolution rotational spectrum of CO.

Line No.	1	2	3	4	5
J	0	1	2	3	4
$\bar{\nu}_{rot}$, cm–1	3.845	7.690	11.534	15.379	19.222

Determine analytically and graphically the values of r_o and D_0 for the molecule if the vibrational frequency of CO, is at 2300 cm^{-1}.

5.82 Determine the energy of rotation and the rate of rotation for $J = 10$ for the following molecules from their microwave spectra:

Molecule	$^2H^{37}Cl$	$^2H^{79}Br$	$^2H^{19}F$	$^3H^{19}F$	$^{35}Cl^{19}F$	$^{37}Cl^{81}Br$
2B(cm^{-1})	10.94	8.64	42.17	15.48	1.03	0.187

5.83 For $^1H^{19}F$, the rotational transitions corresponding to $J = 4 \rightarrow J = 5$ is observed at 204.62 cm^{-1} and $J = 5 \rightarrow J = 6$ is observed at 244.93 cm^{-1}. Using this data calculate the rotational as well as centrifugal distortion constant. [**Ans.** $B = 20.578$ cm^{-1}, $D = 2.325 \times 10^{-3}$ cm^{-1}]

5.84 Calculate the moment of inertia for the $^1H^{12}C^{14}N$ and $^2H^{12}C^{14}N$ molecules. $R_{HC} = 106.8$ pm, $R_{CN} = 115.6$ pm.

5.85 Using the value of rotational constant of $^1H^{19}F$ from Problem 5.82, calculate the frequency of rotation and period of rotation.

5.86 Estimate the values of centrifugal distortion constants from vibrational frequency and rotational constant data for the following non-rigid diatomic rotators. Also comment upon the values so obtained.

Molecule	H_2	I_2	IO	BrCl
$B(cm^{-1})$	60.86	0.037	0.340	0.152
$\omega_e(cm^{-1})$	4400.4	214.55	681.47	444.27

5.87 The data observed in infrared spectrum of HCl under the higher resolution is given in the tabular form as follows. Calculate the value of α_e, B_e and r_e.

v	0	1	2	3	4	5
B_v (cm^{-1})	10.4400	10.1366	9.8329	9.5343	9.232	8.933
ΔB_v (cm^{-1})		–0.3034	–0.3037	–0.2986	–0.302	–0.299

5.88 (*a*) Amongst the following molecules, which one will be IR active?

(*i*) NH_3, (*ii*) C_2H_2, (*iii*) CH_4, (*iv*) CH_3Cl, (*v*) $CHCl_3$, (*vi*) C_6H_6, (*vii*) Ne, (*viii*) H_2O_2, (*ix*) NO (*x*) O_3.

(*b*) Calculate the number of normal modes present in each molecule.

5.89 The fundamental vibrational frequency observed for $H^{35}Cl$ and $D^{35}Cl$ are 2886 and 2081 cm^{-1} respectively. Calculate the force constant in each case. Compare the values and explain.

5.90 The force constant for the $^{12}C^{14}N$ is 1630 N m^{-1} and the equilibrium bond length is 117 pm. Calculate the first five P-and R-branch of the rotational vibrational lines.

5.91 Calculate the maximum value of rotational quantum number corresponding to the vibrational transition observed in case of CO. Given ω_e = 2170 cm^{-1}, $\omega_e x_e$ = 13.26 cm^{-1}, B = 1.931 cm^{-1}.

5.92 The laser Raman and infrared spectra of tribromoacetyl chloride reveal that the position of the C—O fundamental vibration is at 1772 cm^{-1} and that of its first overtone is at 3540 cm^{-1}, calculate the anharmonicity constants for the C—O stretching frequency.

5.93 The equilibrium vibrational frequency and the first anharmonic constant of HI molecule are 2309.5 cm^{-1} and 38.72 cm^{-1}. Calculate the fundamental and first overtone transitions.

5.94 Calculate the frequencies of the fundamental, first, second and third overtones, observed in case of co. Given ω_e = 2170 cm^{-1}, $\omega_e x_e$ = 13.26 cm^{-1}, B = 1.931 cm^{-1}.

5.95 Calculate the frequencies of first, second and third overtone transition for $^{23}Na^{19}F$ (*g*). Given = 536.10 cm^{-1} and $\omega_e x_e$ = 3.4 cm^{-1}.

5.96 First two lines of *R*-and *P*-branch of the rotational vibrationcal spectrum of $^1H^{127}I$ are observed at 2242.087, 2254.257 and 2216.723, 2203.541. Calculate the values of R_e (v = 0) and R_e (v = 1).

5.97 Determine the wave numbers of the first five lines in the *P*- and *R*-branches of the rotation vibration absorption spectrum of ClF if ω_e = 786.15 cm^{-1}; $\omega_e x_e$ = 6.16 cm^{-1}, B_e = 0.516 cm^{-1}.

5.98 Show that the transition from v = 0 to v = 2 is forbidden.

5.99 The vibrational spectrum of $H^{127}I$ exhibits absorption bands at the following wave numbers: 2,257.1, 2,270.2, 2,283.3, 2,296.4, 2,322.6, 2,335.7 and 2,348.8 cm^{-1}. Calculate (*i*) vibrational frequency, (*ii*) force constant, (*iii*) moment of inertia, (*iv*) equilibrium bond length. Assume that the rotational constant is independent of the vibrational state.

5.100 The high resolution *IR* spectrum of some unknown molecule shows the following lines: 2932.229, 2954.236, 2976.243, 3020.257, 3042.264, 3064.271 and 3,086.278. Further it is also observed that the intensity of the band at 3,086.278 cm^{-1} is a factor of 1.9752 greater than the intensity of the band at 3,042.264 cm^{-1}. (*i*) Calculate the band origin in wave numbers, (*ii*) what transition produces the band at 3,042.264 cm^{-1}, (*iii*) Calculate the temperature at which this spectrum is recorded. Again assume that rotational and vibrational motions are independent of each other.

5.101 The vibrational constants of $^{12}C^{16}O$ are: ω_e = 2,170.21 cm^{-1}, $\omega_e x_e$ = 13.461 cm^{-1}, $\omega_e y_e$ = 0.0308 cm^{-1}. Calculate the wave numbers for the first two vibrational hot bands arising from v = 2.

5.102 The following data are observed in the microwave spectrum of $^{1}H^{127}I$ and $^{2}H^{127}I$ between 60 cm^{-1}, and 90 cm^{-1}. Calculate *B*, *I* and R_e.

		$\bar{\nu}$ (cm^{-1})		
$^{1}H^{127}I$	64.275	77.130	89.985	
$^{2}H^{127}I$	65.070	71.577	78.084	84.591

5.103 Taking data from Table 5.4, repeat the Problem 5.99 for $^{14}N^{16}O$ but with the following information, *i.e.* v = 2, *J* = 4 to v = 4, *J* = 5 and v = 5, *J* = 3. Rotational constant is independent of vibrational interaction.

5.104 Write shorthand notation for the following bands: (*i*) $2\nu_2$; $3n - 6 = 3$, (*ii*) $\nu_2 + \nu_3 + \nu_6$; $3n - 6 = 6$ and (*iii*) $2\nu_2 + \nu_3' + 2\nu_4$; $3n - 5 = 4$.

5.105 State the meaning of the following vibrational transitions in a polyatomic molecule: (*i*) (0, 0, 0) → (0, 0, 2); (*ii*) (0, 2, 0, 0) → (0, 1, 1, 1); (*iii*) (0, 0, 0, 0) → (0, 2, 1, 1); and (*iv*) (0, 1, 1, 1, 2) → (0, 2, 1, 2, 0).

5.106 The frequencies of three fundamental modes of vibration of CO_2 have been observed at 667, 1337 and 2349 cm^{-1}. Calculate the frequencies of the absorption bands for the following arbitrary transitions: (*i*) *v*(0 0, 0) → *v*(2, 0, 0); (*ii*) *v* → (0, 0, 0) → *v*(0, 1, 2); (*iii*) *v*(0, 0, 2) → *v*(l, 1, 0); and (*iv*) *v*(0, 0, 2) → *v*(2, 2, 0).

5.107 C – H stretching frequency of cyclopentane is smaller than that of cylcopropane (3010 cm^{-1}). Explain briefly.

5.108 Low frequencies vibrations of cycloparaffins, as shown below, are given in the tabular form. Explain briefly.

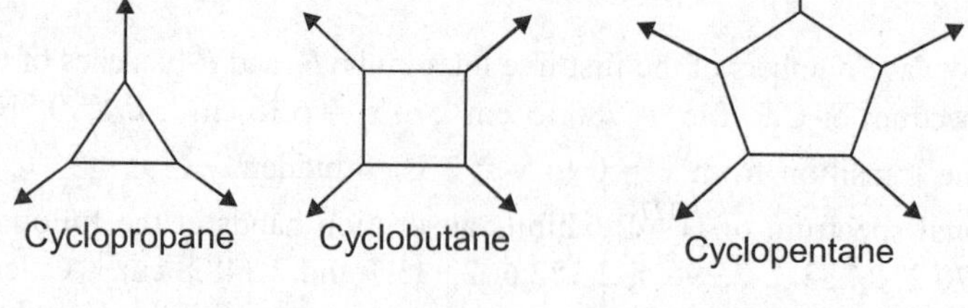

Cyclopropane Cyclobutane Cyclopentane

Molecule	$\bar{\nu}(cm^{-1})$
Cycloproane	1180
Cyclobutane	970
Cyclopentane	890
Cyclohexane (boat form)	822
Cyclohexane (chair form)	802
Cycloheptane	733
Cycloctane	703

5.109 The four C – H stretching vibrations in ethylene molecule is shown below. Suggest which one is IR active and which one is Raman active.

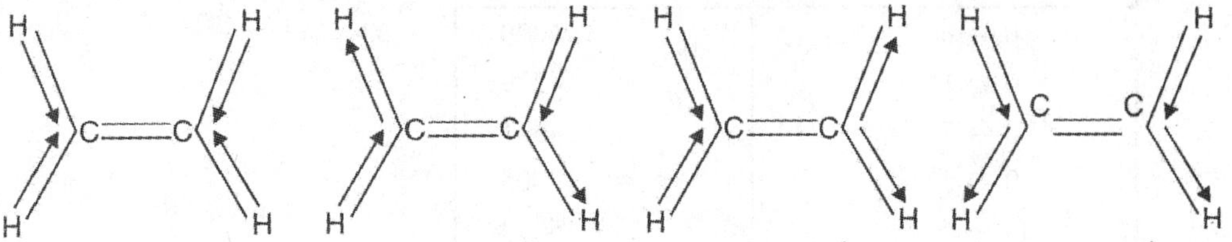

3019 cm^{-1} (I) 3019 cm^{-1} (II) 3019 cm^{-1} (III) 3019 cm^{-1} (IV)

5.110 Find out whether the following vibrational mode in benzene is Raman active or IR active.

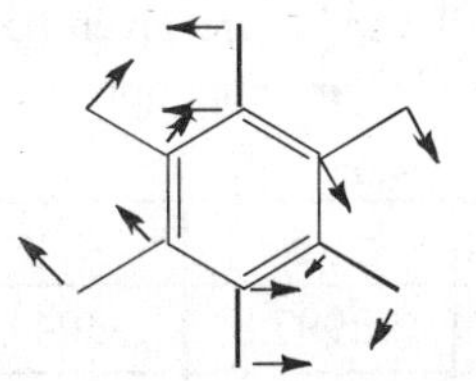

5.111 In the molecules of type AX_4, AX_3Y and AX_2Y_2, having tetrahedral structures, the number of fundamental transitions observed are 4, 6 and 9. Explain briefly.

5.112 Taking dialkyl ketone as the reference C = O stretching vibrational frequency (1715 cm^{-1}), explain the same stretching frequency observed in (*a*) carbonyl difluoride as 1942 cm^{-1}, (*b*) ester as 1735 cm^{-1}, (*c*) amide as 1681 cm^{-1} and (*d*) carboxylate as between 1610 to 1550 cm^{-1}.

5.113 In a triangular planar AX_3 molecule, (*i*) calculate the number of normal modes of vibration, (*ii*) draw the vibrational modes which are IR inactive and (*iii*) which are IR active?

5.114 Without doing any calculations, explain qualitatively which of the following molecules has the highest fundamental vibrational frequency: H_2, HD and D_2.

5.115 Show all the fundamentals normal modes of CS_2 and OCS. Indicate which one are IR active and which one IR inactive?

5.116 Suppose one wants to use the intensity ratio of rotational line from $J = 5$ and rotational line starting from $J = 3$ to make a calibration curve of temperature scale. Consider the molecule $^1H^{19}F$ with rotational constant = 20.578 cm^{-1}, molecule is rigid rotator and the intensity is only proportional to the population of HF in the rotational levels. Start from $T = 50$ K.

5.117 Based on symmetry and group theory assign the vibrational modes to the following molecules. (*i*) planar AX_3, (*ii*) cis-N_2F_2, (*iii*) trans-N_2F_2, (*iv*) C_2H_4, (*v*) CH_4.

5.118 Compute the maximal displacement above equilibrium distance as a percentage of r_e for the following systems from their vibrational frequencies:

Molecule	Br_2	CO	FO	K_2	O_2	N_2
w_e (cm^{-1})	325.3	2169.81	1060	92.96	1580.19	2358.03
r_e(Å)	2.28	1.128	1.32	3.92	1.207	1.098

Hint: Total potential energy = Zero point vibrational energy, i.e. $\frac{1}{2}k(r-r_e)2 = \frac{1}{2}h\nu_v$. [**Ans.** 24 per cent for O_2]

5.119 The frequencies of the first few vibronic transitions to an excited state $^7Li_2$ are given below.

Vibronic transitions	$\bar{\nu}_{obs}$ **(cm^{-1})**
$0 \rightarrow 0$	14020
$0 \rightarrow 1$	14279
$0 \rightarrow 2$	14541
$0 \rightarrow 3$	14805
$0 \rightarrow 4$	15074
$0 \rightarrow 5$	15345

From these data calculate the values of ω_e and $\omega_e x_e$ for the excited state of $^7Li_2$.

5.120 The prominent maxima for different rotational levels in the*R*-branch of the rotation vibration band in the absorption spectrum of HCI vapour are given below. Determine the number of molecules on each rotational quantum level if their extinction coefficients are independent of the rotational energy.

J	0	1	5	8	10	12
w_e(cm^{-1})	2906.3	2925.8	2997.8	3044.9	3072.8	3098.4
D	0.155	0.418	0.362	0.065	0.011	0.001

Ans. $J = 0$, 3.07×10^{22} molecules; $J = 12$, 18.06×10^{19} molecules]

5.121 Determine (*a*) the maximal vibration quantum number, (*b*) the vibrational energy on the zeroth level, (*c*) maximal vibrational energy, and (*d*) the bond energy for mole of the following systems.

Molecule	DI	FO	H_2	K_2	N_2	OH	HF
w_e(cm^{-1})	1639.6	1060	4400.4	92.96	2358.03	3737.9	4138.7
$x \times 10^3$	12.1	14.6	27.4	3.82	6.00	29.7	21.7

[**Ans.** For HF, (*a*) 22 (*b*) 4.0677 × 10–25kJ

(*c*) 94.154 × 10–25kJ (*d*) 542.54 kJ mole–1]

5.122 Calculate the (*a*) v^{max}, (*b*) $E_{v_{max}}$ and (*c*) bond energies of (*i*) $^{35}Cl^{35}Cl$, (*ii*) $^7Li^1H$, (*iii*) $^{16}O^{16}O$, using the data from Table 5.4.

5.123 The frequencies of the first few vibronic transitions to an excited electronic state of PN are: $0 \rightarrow 0$, 39699.10 cm^{-1}, $0 \rightarrow 1$, 40786.80 cm^{-1}, $0 \rightarrow 2$, 41858.90 cm^{-1}. Use this data to calculate ω_e' and $\omega_e' x_e'$ for the excited state of electronic state of PN.

[**Ans.** $\omega_e' = 1103.3$ cm^{-1} and $\omega_e' x_e' = 7.80$ cm^{-1}].

5.124 Morse potential is a simple function that describes the potential energy of anharmonic oscillator is $V(r) = D_e(1 - e^{-\beta x})^2$, where D_e is the equilibrium dissociation energy, β is a constant and x is the displacement of the particle from equilibrium position. Show that the force constant $k = 2D_e\beta^2$. Calculate the value of k if $D_e = 4.5688$ eV, $\beta = 1.83 \times 10^{10}$ m^{-1} for HCl.

5.125 The overtone spectra of is H_2^+ compiled in the table given below:

v	$\Delta\omega_e$ (cm^{-1})	v	$\Delta\omega_e$ (cm^{-1})
0	2,191	8	1,257
1	2,064	9	1,145
2	1,941	10	1,033
3	1,821	11	915
4	1,705	12	800
5	1,591	13	677
6	1,479	14	548
7	1,368		

Use the Birge-Spooner plot extrapolation method to obtain the dissocition energy of H_2^+.
The measured dissociation energy is 2.648 eV. Calculate the percentage error.

5.126 Use the following data for H_2 molecule, calculate dissociation energy.

v	G(v) (cm^{-1})	v	G(v) (cm^{-1})
0	4161.12	7	26830.97
1	8087.11	8	29123.93
2	22782.35	9	31150.19
3	15250.36	10	32886.85
4	18497.92	11	34301.83
5	21505.65	12	35351.01
6	24287.83	13	35972.97

5.127 The band origin and overtone spectra of HgH show that the successive vibrational spacings are 1,203.7, 956.6, 632.4 and 172 cm^{-1}. Determine the dissociation energy as accurately as possible from the given data. The measured dissociation energy is 0.376 eV. Calculate the percent error.

5.128 Raman spectrum of a mixture of two diatomic gaseous gave a series of vibrational-rotational line whose Raman shifts are: 3,300.64, 3,543.88, 3,761.62, 3,787.11, 3,845.38, 3,929.13, 4,012.89, 4,030.35, 4,138.52, 4,264.15, 4,347.91, 4,395.20, 4,431.67, 4,515.42, 4,760.05,

5,003.29, 5,246.53 and 5,489.76 cm $^{-1}$. (*i*) Using the information given in Table 5.4, identify the two molecules. (*ii*) Identify the O-and S-branch lines for each molecule and (*iii*) which Raman line for each molecule corresponds to Q-branch?

5.129 Consider the ground state of H_2 molecule and find out the term symbol of this state. (*b*) Suppose H_2 molecule is excited into a state with electronic configuration $(1s\sigma_g)^1(1s\sigma_u^*)^1$. What are the appropriate spectroscopic designations for the term state arising from this configuration?

5.130 Anthracene is colourless, whereas tetracene is light orange. Explain qualitatively.

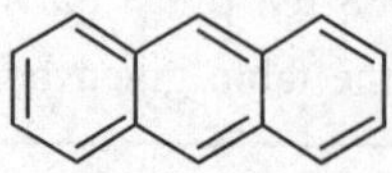
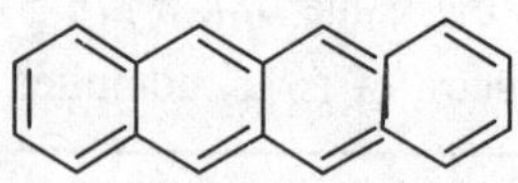

5.131 Many of the aromatic hydrocarbons are colourless but their cation and anion radicals are often strongly coloured. Explain qualitatively (Hint only use π-electrons).

5.132 A sample of nonferrous alloy weighing 0.325 g is dissolved and treated with ammonia and 250 ml of a coloured solution is obtained. The absorbance of this coloured solution in a cell of 2 cm path length is 0.254. Determine the percentage content of copper in the alloy given that the molar extinction coefficient of copper ammoniate is 423 L mol^{-1} cm^{-1}.

5.133 Calculate the molar extinction coefficient of copper if the absorbance of a solution containing 0.48 mg of copper in 500 mL is 0.14 in a cell that is 2 cm in length.

5.134 Both coffee and tea contain an alkaloid called caffeine with molecular formula $C_8H_{10}O_2N_4H_2O$. This alkaloid is responsible for the mildly stimulating effect of beverages made from coffee beans, tea leaves, mate leaves or cola nuts. Caffeine is also present in several common carbonated beverages (soft drinks), *e.g.* Coca-Cola. Caffeine shows an absorbance of 0.560 for a concentration of 5.17×10^{-5} mole L^{-1} at 272 nm. A 2.00 g commercial sample of an instant soluble coffee is mixed with water to a volume of 500 ml, 20 mL of this solution is treated with 25 mL of 0.1N sulphuric acid in a 500 mL flask for clarification by the prescribed procedure and the solution is then made up to 500 mL. This solution shows an absorbance of 0.402 at 272 nm. (*a*) Determine the molar decadic coefficient. (*b*) What is the percentage of caffeine in the given sample of instant coffee? Assume l = 1 cm.

5.135 100 mL of water from an industry is treated with o-phenanthroline and evaporated to 25 mL. The absorbance of this coloured solution has been found to be 0.460. Determine the iron content of industrial water (in mg L^{-1}) if the molar extinction coefficient of this coloured compound is 1100 L mol^{-1} cm^{-1} for a path length of 1 cm.

5.136 To determine phosphorous in steel by dark-blue colour of a reduced phosphomolybdic complexes on absorbance scale, the following data are observed for standard samples of steel:

Amount of phosphorous in steel (per cent)	0.15	0.024	0.031	0.040	0.055
Absorbance	0.318	0.538	0.722	0.921	1.302

(*a*) Determine the percentage content of phosphorous in the unknown sample whose reading on the absorbance scale is 0.619.

(*b*) Determine also the percentage content of phosphorous from the straight-line expression derived from the standard data.

5.137 A steel sample weighing 0.0532 g is dissolved in acid, the solution is treated with diethyl dithiocarbaminate and its volume is made up to 100 mL. The solution is analysed spectr photometrically at 328 and 368 nm in cell with a path length of 2 cm. The results obtained are A_{328} = 0.67 and A_{368} = 0.45. Determine the percentage contents of nickel and cobalt in the steel if ε_{Ni} = 35210 L mol^{-1} cm^{-1} and ε_{Co} = 3910 L mol^{-1} cm^{-1} at 328 nm and ε_{Ni} = 21820 L mol^{-1} cm^{-1} and ε_{Co} = 14340 L mol^{-1} cm^{-1} at 368 nm.

5.138 The solution in a dying vat after a hundred fold dilution yields the following absorbance data in a cell with path length 5 of cm at two wavelengths λ_1 and λ_2 : before dying $A_{\lambda 1}$ = 2.2 and $A_{\lambda 2}$ = 1.6; after dying, $A_{\lambda 1}$ = 0.40 and $A_{\lambda 2}$ = 0.25. Determine the degree of extraction of dyes from the vat if $\varepsilon_{\lambda 1}$ = 3400 L mol^{-1} cm^{-1} and $\varepsilon_{\lambda 2}$ = 1500 L mol^{-1} cm^{-1} for the dye *A* and $\varepsilon_{\lambda 1}$ = 1300 L mol^{-1} cm^{-1} and $\varepsilon_{\lambda 2}$ = 2500 L mol^{-1} cm^{-1} for the dye *B*.

5.139 When incident light is passed through an empty cell placed in a single beam photoelectric colourimeter, the galvanometer scale reading is observed at 100 divisions while it is at 80 divisions when the cell contains a solution having concentration *C* moles lit^{-1}. What will be the reading in the absorbance scale?

5.140 Calculate the absorption maximum for

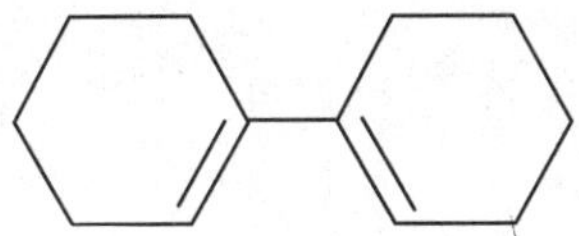

5.141 Calculate the absorption maximum for 2, 3-dimethylene bicyclo [2, 2, 1] heptanes.

5.142 Yellow plant pigments α-, β- and γ-carotene (*e.g.* carrots contain 15 per cent α, 85 per cent β, 0.1 per cent γ) and the red pigment of tomatoes, *i.e.* lycopene, are converted into vitamin A in the liver, the lack of which in the diet causes hardening of the conjunctiva, corneal softening and to complete blindness, weight loss and poor general physical condition; and a deficiency of vitamin A causes night blindness. All the four pigments have the same molecular formula (C_{40} H_{56}) but different structures given below:

(*a*) *α-carotene*

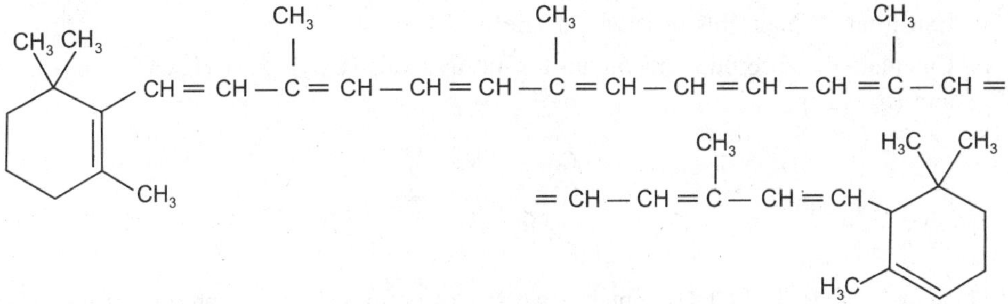

(*b*) β-carotene: The ring on the righthand side is

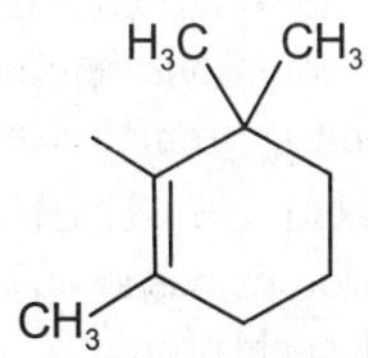

(*c*) y-carotene:

(d) Lycopene: 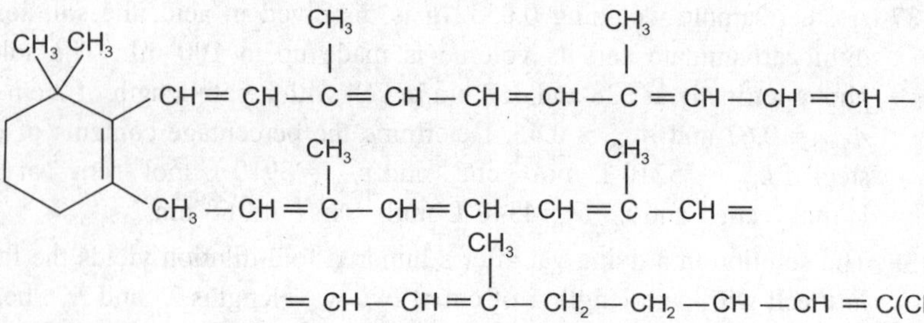

5.143 Calculate the absorption maximum for (*a*) Vitamin A_1, and (*b*) Vitamin A_2.

5.144 Calculate the absorption maximum for the following compounds:

(a) 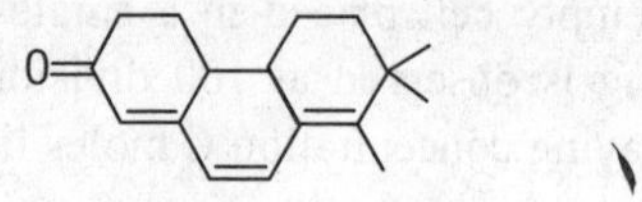

(b) 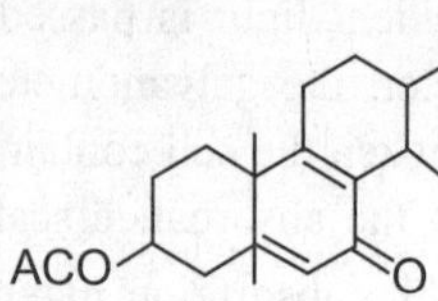

(c) 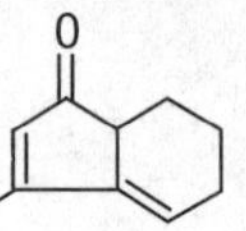

(d) 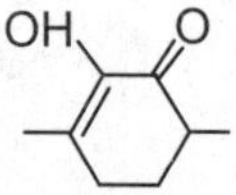

(e) 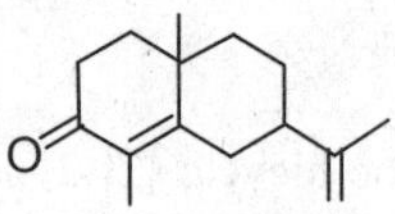

5.145 Three absorption band systems observed in compound I, are at 190 nm (ε_{max} ~ 10,000),

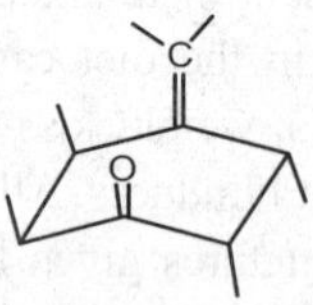

a moderately intense band in the region of 210-240 nm and third band belongs to $\pi^* \neg n$ transition. Explain this observation briefly.

5.146 Calculate the absorption maximum for mesityloxide $(CH_3)_2C{=}CHCOCH_3$ in (*a*) chloroform, and (*b*) water.

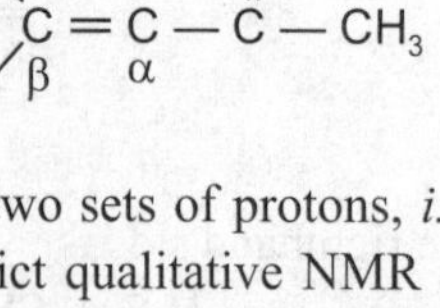

5.147 Acetaldehyde (CH_3CHO) contains two sets of protons, *i.e.* three protons of methyl group and one proton of CHO group. (*i*) Predict qualitative NMR spectrum with intensity ratio of each multiplet line, (*ii*) write down the equations representing energy levels, (*iii*) number of lines observed in the NMR spectrum and (*iv*) energy level diagram.

5.148 Consider the molecule propanoic acid, i.e. CH_3CH_2COOH. (*i*) Arrange the NMR spectrum of $-CH_3$, $-CH_2$ and carboxylic acid protons based on the chemical shifts, (*ii*) predict qualitatively the spin-spin coupling splitting of each proton, (*iii*) what would be approximate ratio of the total absorption intensities for each of these groups and (*iv*) what are the approximate intensity ratios of the spin-spin bands with each individual proton group?

5.149 Equation (5.154) defines hyperfine constant with the ESR separation in terms of frequency, $\Delta v = v' - v''$. One can also define the hyperfine constant if we fix the frequency and record the ESR spectrum by scanning over magnetic field. Prove that this relation is

$$a = \frac{g_e \beta_e}{h} \Delta H,$$

where $\Delta H = H' - H''$ corresponds to the magnetic field for the transition frequencies v' and v'' repectively.

5.150 Predict the ESR spectrum of CH_4^- radical anion.

[**Hint:** Consider carbon atom is to be ^{12}C, thus $I = 0$ and will not contribute to hyperfine splitting. Only nuclei are four protons.]

5.151 The NMR signal of a compound is found to be at 450 HZ downfield from the TMS signal using a NMR spectrometer operating at 100 MHz. Calculate its chemical shift in ppm relative to TMS.

5.152 Suppose the NMR spectrum of CH_3CHO is recorded at 200 MHz and 500 MHz. State whether each of the following quantities remains unchanged or is different from 200 MHz to 500 MHz: (*a*) sensitivity of detection, (*b*) $|\delta_{CH_3} - \delta_H|$, (*c*) $|v_{CH_3} - v_H|$, (*d*) J.

5.153 Sketch the NMR spectrum of isobutyl alcohol, given the following chemical shift data: —CH_3 = 0.89 ppm, —C —H = 1.67 ppm, —C —H_2 = 3.27 ppm and —O—H = 4.50 ppm.

5.154 The methyl radical can be considered as planar species. Predict the number of lines observed in its ESR spectrum.

5.155 Account for the number of lines observed in the ESR spectrum of naphthalene anion radical. How would you use the isotopic substitution to assign the two hyperfine couplings in naphathalene?

5.156 NMR spectrum of toluene when recorded at 60 MHz and field strength of 1.41 T gave two peaks, one due to aromatic protons and the other to the methyl group. (*i*) Calculate the field strength when the frequency is 400 MHz, (*ii*) at 60 MHz the resonance frequencies for the methyl group and aromatic protons are at 140 MHz and 430 MHz. Calculate the frequencies when the NMR spectrum is recorded at 400 MHz, (*iii*) calculate the chemical shifts of the signals when recorded at 60 MHz and 400 MHz.

5.157 The NMR spectrum of N,N′-dimethylformamide shows two methyl peaks at 25 ^{0}C. There is only one peak due to methyl groups when NMR spectrum is recorded at 140 ^{0}C. Explain.

5.158 (*a*) Calculate the value of Z-component of the spin magnetic moment of (*i*) ^{13}C, (*ii*) ^2H, (*iii*) ^{14}N, (*iv*) ^{11}B and (*v*) ^{19}F, (*b*) using the data from Table 5.4, calculate the measured values of spin magnetic moment (*c*) calculate the ratio of the values obtained in (*a*) and (*b*).

5.159 Commercially available NMR uses the fixed magnetic fields such that ^1H resonance occurs at a frequency of 60, 100, 300, 400, 500, 600 and 750 MHz. Progress is on to develop the NMR instruments of higher frequency. (*i*) Calculate the value of magnetic field required to produce proton resonance at the given frequencies, (*ii*) determine the frequencies at which we would observe resonance frequencies

5.160 Two unknown tubes containing O_2 and N_2 are lying on the table. How would you identify them using ESR spectrometer?

5.161 NMR spectrum of an unknown nucleus is recorded at a magnetic field of 2.3486 and resonance frequency observed is at 9.809 MHz. Identify the nuclear particle whose transitions are being observed under these conditions.

Appendix A-5.1

Evaluation of Transition Moment Integral for Rotational Spectrum:

For rotational spectrum, the transition moment integral **M** should be non-zero. For simplicity we will only consider the *z*-component $\hat{M}_z$ in detail. $\hat{M}_x$ and $\hat{M}_y$ components can also be solved in a similar manner. The *z*-component M_z (J', m'; J'', m'', *i.e* m' and m'' are the quantum numbers representing the non-degenerate rotational levels in the presence of external magnetic field) can be written as:

$$\mathbf{M_z}\,(J'.\, m'\,;\, J''\, m'') = \int \psi'_{\text{rot}}(J', m')\, \hat{M}_z\, \psi''_{\text{rot}}(J'', m'')\, d\tau \qquad \text{(A 5.1)}$$

Substituting the values ψ'_{rot} and ψ''_{rot}, $\hat{M}_z = \hat{M}_0 \cos\theta$ and $dt = \sin\theta\, d\theta$

$$M_z\,(J', m'\,;\, J', m') =$$

$$N'_r\, N''_r\, M_0 \int_0^{\pi} P_{J'}^{|m'|}(\cos\theta)\, \cos\theta\, P_{J''}^{|m''|}(\cos\theta)\, \sin\theta\, d\theta \int_0^{2\pi} e^{im'\varphi}\, e^{-im''\varphi}\, d\varphi \qquad \text{(A 5.2)}$$

where N'_r and N''_r are the normalizing constants and M_0 is the dipole moment vector. The second integral of the equation (A 5.2) will be zero except for $m' = m''$ and its value is 2π. For the evaluation of first integral, certain mathematical relations existing for spherical harmonics have to be used, *i.e.*

$$\cos\theta\, P_J^{|m|}(\cos\theta) = \frac{J + |m|}{2J + 1} P_{J-1}^{|m|}(\cos\theta) + \frac{J - |m| + 1}{2J + 1} P_{J+1}^{|m|}(\cos\theta) \qquad \text{(A 5.3)}$$

$$\sin\theta\, P_J^{|m|}(\cos\theta) = \frac{J + |m|}{2J + 1} P_{J-1}^{|m|}(\cos\theta) + \frac{J - |m| + 1}{2J + 1} P_{J+1}^{|m|}(\cos\theta) \qquad \text{(A 5.4)}$$

Substituting the first recursion relation in equation (A 5.2) and rearranging for $m' = m''$,

$$\hat{M}_z\,(J', m'\,;\, J'', m'') = 2\pi N'_r\, N''_r\, \mathrm{M}_0\, \frac{\mathrm{J} + |m|}{2\mathrm{J} + 1} \int_0^{\pi} P_{J'}^{|m|}(\cos\theta)\, P_{J''-1}^{|m|}(\cos\theta)\, \sin\theta\, d\theta$$

$$+\; 2\pi N'_r\, N''_r\, \mathrm{M}_0\, \frac{\mathrm{J} + |m| + 1}{2\mathrm{J} + 1} \int_0^{\pi} P_{J'}^{|m|}(\cos\theta)\, P_{J''+1}^{|m|}(\cos\theta)\, \sin\theta\, d\theta \qquad \text{(A 5.5)}$$

Other useful relations about spherical harmonics are :

$$\int_0^{\pi} P_{J'}^{|m|}(\cos\theta)\, P_{J''-1}^{|m|}(\cos\theta)\, \sin\theta\, d\theta = 0 \text{ If } J_1 \neq J_2 \qquad \text{(A 5.6)}$$

$$\int_0^{\pi} P_{J'}^{|m|}(\cos\theta)\, P_{J''-1}^{|m|}(\cos\theta)\, \sin\theta\, d\theta = \frac{2}{2\mathrm{J} + 1}\, \frac{(J + |m|)}{(J - |m|)} \text{ if } J_1 = J_2 \qquad \text{(A 5.7)}$$

If the above relations are used to evaluate the integrals of the equations (A 5.5), it can be seen that ($\hat{M}_z$ J', m' ; J'', m'') will be non-zero for $J' = J'' - 1$ or $J' = J'' + 1$ and also $m' = m''$. Similarly it can also be shown that M_x (J', m' ; J'', m'') and M_y (J', m' ; J'', m''), i.e. x and y components of the transition moment integral will be non-zero for $J' = J'' - 1$ or $J' = J'' + 1$ and at the same time $m' = m'' \pm 1$. This proves that the selection rules $\Delta J = \pm 1$. The selection rule $\Delta m = 0, \pm 1$ are valid when external magnetic field is applied.

Appendix 5.2

Evaluation of Transition Moment Integral for the Vibrational Spectrum

The expression for the transition moment integral in case of vibrational transitions is:

$$\mathbf{M}(v', v'') = \int \psi_{v'} \, \hat{M} \, \psi_{v''} \, d\tau \qquad \text{(A 5.8)}$$

where $\psi_{v'}$ and $\psi_{v''}$ are the wave functions of the final and initial states. $\hat{M}$ is the oscillating dipole moment operator. For diatomic molecule and considering that molecular axis is the x-axis, the oscillating dipole moment μ is given as:

$$\mu = \mu_0 + \left(\frac{\partial \mu}{\partial x}\right)_{x=0} x + \ldots\ldots \qquad \text{(A 5.9)}$$

where μ_0 is the permanent dipole moment and other terms depict the change of dipole moment with the displacement of the nuclei (x) from their mean positions. For simple harmonic oscillator, the higher terms in equation (A 5.9) can be neglected. Substituting the value of M in equation (A5.8), we get

$$M_x(v', v'') = \int \psi_{v'} \, \mu_0 \, \psi_{v''} \, dx + \left(\frac{\partial \mu}{\partial x}\right)_{x=0} \int \psi_{v'} \, x \, \psi_{v''} \, dx \qquad \text{(A 5.10)}$$

The first term is zero because μ_0 is constant and $\psi_{v'}$ and $\psi_{v''}$ are orthogonal to each other. For finite value of first integral $\Delta v = 0$ ($v' = v''$). It will only mean pure rotational spectrum. Further $\psi_{v'}$ and $\psi_{v''}$ are as Hermit polynomial and following recursion relation exist for Hermite polynomial.

$$x\, H_v(x) = 1/2\, H_{v+1}(x) + v\, H_{v-1}(x) \qquad \text{(A 5.11)}$$

$$= N_v \, e^{-\frac{\alpha x^2}{2}} \, H_v(\sqrt{\alpha}, x) \qquad \text{(A 5.12)}$$

where $\alpha = \text{Constant} = \dfrac{4\pi^2 \mu \, v}{h}$

$$M_x(v' v'') = \left(\frac{\partial \mu}{\partial x}\right)_{x=0} N_{v'} \, N_{v''} \int e^{-\frac{\alpha x^2}{2}} H_{v'}(\sqrt{\alpha}.x)\, x\, e^{-\frac{\alpha x^2}{2}} H_{v''}(\sqrt{\alpha}, x)\, dx \qquad \text{(A 5.13)}$$

$$= \left(\frac{\partial \mu}{\partial x}\right)_{x=0} N_{v}', N_{v}'' \left[\frac{1}{2} \int e^{-\alpha x^2} H_{v'}(\sqrt{\alpha}, x)\, H_{v''+1}(\sqrt{\alpha}, x)\, dx\right.$$

$$+ \int v e^{-\alpha x^2} \; Hv' (\sqrt{\alpha}, x) \, H_{v''-1}(\sqrt{\alpha}, x) \, dx] \qquad \text{(A 5.14)}$$

Both these integrals will be zero unless $v' = v'' + 1$ or $= v' = v'' - 1$. So the transition will be observed between the vibrational levels if $\Delta v = \pm 1$. This is called fundamental transition.

For an-harmonic oscillator, the higher terms in the expression (A 5.10) will also be included *i.e.*

$$\mu = \mu_0 + \left(\frac{\partial \mu}{\partial x}\right)_{x=0} x + \left(\frac{\partial^2}{\partial x^2}\right)_{x=0} x^2 + \left(\frac{\partial^3}{\partial x^3}\right)_{x=0} x^3 + \ldots\ldots \qquad \text{(A 5.15)}$$

Due to this the transition moment integral will be non-zero for $\Delta v = \pm 2$, $\Delta v = \pm 3$ ... besides $\Delta v = \pm 1$. The $\Delta v = \pm 2$ is called first overtone, $\Delta v = \pm 3$, the second overtone and so on.

CHAPTER 6

SOLIDS

6.1. DIFFERENT TYPES OF SOLIDS

Solids form one of the highly ordered arrangement of its constituent units (may be atoms or molecules) and these units are held together by different forces. Depending upon the nature of forces, the solids can be divided into four broad categories, though there may be some more. The characteristics of these categories are mentioned below and will be discussed briefly.

6.1.1. *Covalent Solids*

In covalent solids, the atoms are connected to one another by covalent linkages forming a three dimensional structure. The atoms present may be of same element or of different elements having nearly the same electronegativities. Non-metallic elements form the main source of examples e.g., carbon, silicon and silicon carbide (SiC). In diamond, each carbon atom is connected to other four carbon atoms through covalent linkages forming tetrahedral structure (sp^3 hybridization) and thus form a large three dimensional network. The graphite, although consists of pure carbon, has different arrangement of carbon atoms. In one plane, forming a sheet, each carbon atom is connected to three other by sp^2 hybridization and each sheet is connected to other by van der Waals forces.

Depending upon the arrangement of atoms, the solids can be 'isotropic' if all the bonds are equivalent *i.e.,* its properties are the same in all directions. The solids are anisotropic if the bonding varies within the substance or it is not equivalent in three dimensions. Lower melting points and high boiling points are expected of anisotropic solids.

6.1.2. *Ionic Solids*

In ionic solids, the constituents are ions, both positive and negative. These ions are held together by electrostatic forces maintaining the overall electro neutrality. There is a difference in the arrangement of ions in solids than in ion-pair formed in gas phase. In solids, a particular positive ion may not be associated to negative ion but a particular kind of ion is surrounded by a set of ions of opposite charges. This particular number depends upon the formula of a compound and sizes of the two kinds of ions *i.e.*

by the radius ratio R_+/R_-, as shown in Table 6.1. The other quantities, one has to find are (as mentioned in later sections) different values for cell content (Z_+, Z_-), different values of R_+ and R_-, different coordinates for different kind of ions and the coordination numbers (CN_+, CN_-) for the positive and negative ions. In ionic solids, the ratio of CN_+/CN_- is in the inverse ratio of the subscripts in the empirical formula and unit cell length will be related to the radii R_+ and R_-.

TABLE 6.1

Empirical formula	Radius ratio	CN	Structure of cubic
MX	$\frac{R_+}{R_-} = 0.225$ to 0.414	4	Sphalerite
	$= 0.414$ to 0.732	6	Halite
	$= 0.732$ to 1.0	8	CsCl
MX_2	$\frac{R_+}{R_-} = 0.225$ to 0.414	4 & 2	SiO_2 and Cu_2O
or	$= 0.414$ to 0.732	6 & 3	TiO_2, CdI_2, NiS_2
M_2X	$= 0.732$ to 1.00	8 & 4	CaF_2

Further, molecular formula of ionic compound depends upon (*a*) ionization energy of the metal atom to its various oxidation states, (*b*) electron affinity of the electronegative atom, and (*c*) the electrostatic attraction of the oppositely charged ions, indirectly depends upon the distance between the oppositely charged ions, determined by the geometrical arrangement of the ions.

Ionic solids have high melting and boiling points because of very strong bonds between the ions throughout the crystal and have poor electrical conductivity as the electrons are tightly bound to each ion whether positive or negative.

In ionic solids, the coulombic forces existing between the ions and responsible for the formation of solids are given by

$$F = \frac{q_1 q_2}{4\pi\varepsilon_0 r^2} \qquad ...(6.1)$$

and potential energy existing between the ions as

$$V(r) = \frac{q_1 q_2}{4\pi\varepsilon_0 r} \qquad ...(6.2)$$

where q_1 and q_2 are the charges on the two ions, r is the distance between these two ions and ε_0 is the permittivity in vacuum, has a value of 8.854×10^{-12} J^{-1} C^2 m^{-1} (N^{-1} C^2 m^{-2}) if the charges are expressed in coulombs and distance is measured in meters.

***Exercise* 6.1:** From the following data show that lithium fluoride has a molecular formula as LiF

1st IP(Li) = 520.07 kJ mol^{-1}; EA(F) = – 337.23 kJ mol^{-1}; D(F – F) = 161.46 kJ mol^{-1}; sublimation energy of Li = 160.67 kJ mol^{-1}

$$U(\text{LiF}) = -1020.9 \text{ kJ mol}^{-1}$$

According to Born-Haber cycle, ΔH(heat of reaction) for the reaction: Li(s) + $\frac{1}{2}$ F_2(g) → LiF(s)

$$\Delta H = \Delta H_{sub}(\text{Li}) + \text{IP (Li)} + \frac{1}{2}[D\,(F_2) + \text{RT}] - \text{EA (F)} + U\,(\text{LiF}) \qquad ...(6.3)$$

$$= 160.67 + 520.07 + 80.73 - 337.23 - 1020.9$$

$$= -596.66 \text{ kJ mol}^{-1}$$

Negative ΔH indicates the spontaneity of the reaction. Any other stoichiometry will give ΔH + ve.

***Exercise* 6.2:** Calculate the molar coulombic force and coulombic potential energy between Na^+ and Cl^{-1} ions when separated by 0.1 nm.

The charge on Na^+ and Cl^{-1} ions are $+1.6 \times 10^{-19}$ C and -1.6×10^{-19} C, r = 0.1 nm. Using equation (6.1), the coulombic force can be

$$F = \frac{q_1 q_2}{4\pi\varepsilon_0 r^2} = \frac{(1.6 \times 10^{-19}\text{ C})(-1.6 \times 10^{-19}\text{ C})}{(1.113 \times 10^{-10}\text{ J}^{-1}\text{ C}^2\text{ m}^{-1})(1.0 \times 10^{-10}\text{ m})^2}$$

$$= -2.31 \times 10^{-8}\text{ J m}^{-1} = -2.31 \times 10^{-8}\text{ J m}^{-1}$$

Similarly the potential energy is given by equation (6.2) as

$$V(r) = \frac{q_1 q_2}{4\pi\varepsilon_0 r} = \frac{(1.6 \times 10^{-19}\text{ C})(-1.6 \times 10^{-19}\text{ C})}{(1.113 \times 10^{-10}\text{ J}^{-1}\text{ C}^2\text{ m}^{-1})(1.0 \times 10^{-10}\text{ m})} = -2.31 \times 10^{-18}\text{ J}$$

$$= -2.31 \times 10^{-18}\text{ J} \times 6.02 \times 10^{23}\text{ mol}^{-1} = -1.3906 \times 10^{6}\text{ J mol}^{-1}$$

In other words, we need 1,391 kJ mol^{-1} to dissociate one mole of NaCl, which is nearly required by 4 moles of C – C single bond.

6.1.3. *Molecular Solids*

The main constituents of the molecular solids are molecules but it can be atoms of rare gases. Since, mostly, the bonding capacity of the molecules and the rare gases is saturated, the molecules are held together by weak forces called van der Waals forces (also called induced dipole-induced dipole interactions). Due to these weak attractive forces, one needs a very low temperature to solidify. This means that molecular solids have very low melting and boiling points, high vapour pressure and soft. The molecules retain their identity in solids because the intermolecular forces are weaker than the intramolecular ones.

6.1.4. *Metallic Solids*

Most of the elements in the periodic table are metals and crystals of these metals consist of either cubic closed-packed or hexagonal closedpacked unit cells. These unit cells are mainly face-centred cubic packing or hexagonal body centred packing. The coordination number in both the packing is 12. Due to this, metals are densely packed and are stable. Some of the metals like alkali metals have body centred cubic close packing and have coordination number eight. The common properties of the metals are; good conductor of heat and electricity, strong but malleable and ductile and possess shining luster.

6.1.5. *Melting and Boiling Points*

Melting and boiling points reflect the change of solid phase to liquid phase and liquid phase to vapour phase. We have learnt earlier that intermolecular forces qualitatively set the trend regarding these characteristics. Stronger are these forces, higher will be the melting and boiling points.

(*i*) *Ionic Solids*: Intermolecular forces present in ionic solids comprise of coulombic in nature, equation (6.2), *i.e.* potential energy depends on the *charges of cations and anions and the distance between them, basically ionic radii*. For example, the melting points of ionic solids having $q_1 = -q_2 = 2$ will be having higher melting point than that ionic solids with $q_1 = -q_2 = 1$ and ionic solids with $q_1 = 2$, $-q_2 = 1$ will be intermediate of the two extreme cases discussed above.

Charges on the cations are obtained by removing the electrons from the neutral or ionic species and thus are endothermic process, whereas the potential energy because of coulombic interaction is exothermic process. So the net effect reflects the trend in these properties. Table 6.2 compiles the melting and boiling points of few ionic solids, Table 6.3 gives the first and second ionisation energies and the ionic radii and Table 6.4 lists the boiling points of few carboxylic acids and ketones.

TABLE 6.2. Melting points of some ionic and covalently bonded solids

Solids	Cation charge	Anion charge	Melting point (K)
NaCl	+1	–1	1,074
KCl	+1	–1	1,049
RbCl	+1	–1	988
NaBr	+1	–1	1,028
KBr	+1	–1	1,003
RbBr	+1	–1	955
Na_2O	+1	–2	1548
K_2O	+1	–2	Decomposes before melting
CaO	+2	–2	2887
BaO	+2	–2	2191
$CaCl_2$	+2	–1	1,045
$SrCl_2$	+2	–1	1,146
$BaCl_2$	+2	–1	1,235
BaSO4	+2	–2	1,853
Diamond	0	0	>3,773
Silicon	0	0	1,693

TABLE 6.3. First and second ionisation energies and ionic radii of some group A elements

Element	Group	Ionisation energy (eV)		Ionic radii (nm)
		First	Second	
H	IA	13.60	–	–
Li		5.363	75.26	0.059(+1)[2]
Na		5.12	47.06	0.102
K		4.318	31.66	0.138
Rb		4.159	27.36	0.149
Cs		3.87	23.4	0.012(+2)
Be	IIA	9.28	18.12	0.072
Mg		7.61	14.96	0.100
Ca		6.09	11.82	0.116
Sr		5.667	10.98	0.136
Ba		5.19	9.95	0.012(+3)
B	IIIA	8.257	25.00	0.053
Al		5.96	18.74	0.062
Ga		5.97	20.43	0.079
In		5.76	18.79	0.088
Tl		6.07	20.32	0.171(–3)
N	VA	14.48	29.47	0.212
P		10.9	19.56	0.140(–2)
As		10.5	20.1	0.184
O	VIA	13.55	34.93	0.198(–2)
S		10.30	23.3	0.221
Se		9.7	21.3	0.133
Te		8.96	–	0.181(–1)
F	VIIA	17.34	34.81	0.196
Cl		12.952	23.67	0.220
Br		11.80	19.1	
I		10.6	19.4	

TABLE 6.4. Boiling points of some aliphatic carboxylic acids and aliphatic 2-ketones

Aliphatic carboxylic acids	Boiling point (K)	Aliphatic 2-ketones	Boiling point (K)
Formic acid	373.8	Acetone	329.4
Acetic acid	391.8	2-Butanone	352.8
Propanoic acid	414.1	2-Pentanone	375.2
Pentanoic acid	459.2	2-Hexanone	401.2
Hexanoic acid	478.2	2-Heptanone	424.6
Heptanoic acid	496.2	2-Nonanone	468.4
Nananoic acid	528.2	2-Decanone	483.2
Decanoic acid	543.2		

***Exercise* 6.3:** Taking the data from the Tables 6.2 and 6.3, explain briefly that the melting point of NaCl is higher than that of RbCl.

Two points should be considered to account for different melting points.

(i) As the ionisation potential of the element increases, (as this process is endothermic), the transfer of negative charge from cation to the anion will correspond to less than a full electronic charge. Because of this the charge on the cation will not be complete one unit or two units but slightly less than these values. Thus the coulombic potential energy will be small for the cations of higher ionisation energy as compared to the smaller ionisation energy.

(ii) Smaller the ionic radii, more will be potential coulombic energy. In other words the two effects are of competing in nature.

Now taking these factors, ionisation energy of Na is higher (0.961 eV) than that of Rb and thus will have less attractive potential in NaCl as compared to RbCl. On the other hand, the ionic radius of Rb is 0.047 nm larger than that of Na and thus NaCl will have higher coulombic potential. It is latter part which plays the major role in having higher melting point of NaCl than that of RbCl.

It is this reason that melting points of alkali metal halides decreases in going down the group .

***Exercise* 6.4:** Taking the data from the Tables 6.2 and 6.3, explain briefly that the melting point of $BaCl_2$ is higher than that of $CaCl_2$.

It is clear from the data of Table 6.3, the second ionisation energy of Ba is 1.87 eV (180.42 kJ mol^{-1}) smaller than that of Ca. On the other hand, the ionic radius of Ba is 0.036 nm larger than of Ca. Following the same explanation as discussed in exercise 6.3, it is clear that the ionisation energy is playing the major role regarding the higher melting point of $BaCl_2$ than that of $CaCl_2$.

***(ii) Molecular solids*:** Organic molecules in crystals are held together by dipole-dipole, induced dipole-dipole or induced dipole-induced dipole interactions. We have learnt that these interactions are much weaker than ionic interactions. Further also we have learnt that dipole-dipole interactions > dipole –induced dipole interactions > induced dipole-dipole interactions, *e.g.* the expression for dipole-dipole interactions are given as *i.e.*, interactions are proportional to square of the dipole and inversely as the cube of distance.

$$V(r) = -\frac{2\mu^2}{4\pi\varepsilon_0 r^3} \qquad \text{...(6.4)}$$

In other words, higher the dipole moment higher will be melting point and smaller the distance between the molecules in the crystals higher the melting and boiling points.

The other factor which is important is that how easily the induced dipole moment can be induced depends on the polarisability of the molecule and magnitude of the external charge. *So qualitatively it may be concluded that as the molecular weight increases the melting point or the boiling point also increases*.

Lastly for the molecules of same molecular weight, melting or boiling points also depends on the structure of molecules, *e.g.*, linear molecules will have higher melting or boiling points than those of branched chain molecules. Although qualitatively it is true that melting or boiling points increase with increase in molecular weights, but the correlation is not that good if we plot molar mass versus molecules containing different functional groups. On the other hand, if we restrict the data to compounds in a homologous series, such as primary alcohols, alkyl chlorides etc., the correlation becomes much stronger. The following correlation relations explain the boiling points (T_b^0) of a homologous series of *n*-alkanes, primary alcohols and aromatics are given by equations (6.5) to (6.7).

$$T_b^0\ (n\text{-alkanes}) = 81.518 + 3.741\ \text{M} - 0.00831\ M^2\ \text{K} \quad \text{...(6.5)}$$

$$T_b^0\ (\text{primary alcohols}) = 287.68 + 1.377\ \text{M K} \quad \text{...(6.6)}$$

$$T_b^0\ (\text{aromatics}) = 139.83 + 2.73\ \text{M K} \quad \text{...(6.7)}$$

where M is the molar mass in units of g mol^{-1} and the normal boiling points are given in degrees Kelvin.

***Exercise* 6.5:** Using the correlation relation given by (6.5), calculate the boiling point of pentane (C_5H_{12}). Calculate the percent error.

***Solution*:** The molar mass of pentane is $5 \times 12 + 12 \times 1 = 72$ g mol^{-1}. Substitute the value of M in equation (6.4), we get

$$T_b^0\ (n\text{-alkanes}) = 81.518 + 3.741\ (72) - 0.00831\ (72)^2 = 81.518 + 269.35 - 43.08 = 307.96\ \text{K}$$

The boiling points of different forms of pentanes are 282.5 K, 300.9 K and 309.1 K. The three isomers can be

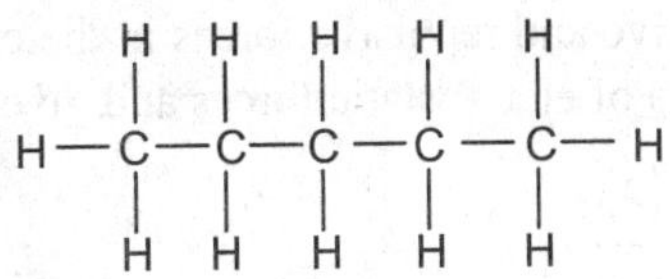

n-Pentane

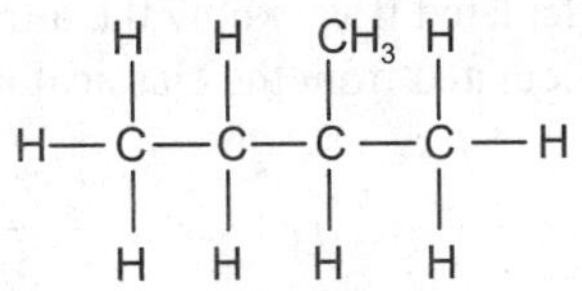

2-Methyl butane

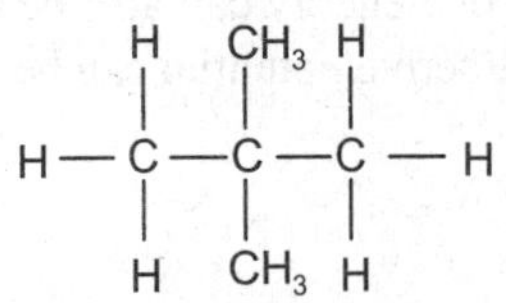

Neopentane

The boiling point 309.1 K can be assigned to *n*-pentane, 300.9 K to 2-methyl butane and 282.5 K to 2,3-dimethyl propane respectively. *n*-Pentane will be having the maximum polarisability and 2,3-dimethyl the least polarisability.

$$\% \text{ error} = \frac{309.1 - 307.96}{309.1} \times 100 = 0.37$$

6.2. LATTICE ENERGY

Lattice energy is the net energy with which an Avogadro number of molecules are held together with in the lattice. Alternatively, it is the energy required to carry out the following process.

$$A_mB_n(\text{crystal}) \rightarrow mA^{n+}(\text{g}) + nB^{m-}(\text{g}) \quad \text{...(6.8)}$$

Since, it is difficult to find the energy of the above process by direct experiment, lattice energy has been calculated by indirect methods. The most common one is the 'Born-Haber' cycle, which lists the following steps, the example used is the reaction

$$NaCl\ (crystal) \rightarrow Na^+(g) + Cl^-(g)$$

$$Na(s) \rightarrow Na(g) \qquad \Delta H_{sub}\ \text{(sublimation)}$$

$$Na(g) \rightarrow Na^+(g) + e^- \qquad \text{I.P. (Ionization)}$$

$$\frac{1}{2}Cl_2(g) \rightarrow Cl(g) \qquad D\ \text{(Bond energy)}$$

$$Cl(g) + e^- \rightarrow Cl^-(g) \qquad -\text{E. A. (Electron affinity)}$$

$$Na^+(g) + Cl^-(g) \rightarrow NaCl(s) \qquad U\ \text{(Lattice energy)}$$

$$Na(s) + \frac{1}{2}Cl_2(g) \rightarrow NaCl(s)\ \Delta H \qquad \text{(Heat of reaction)}$$

$$\Delta H = \Delta H_{sub} + \text{I.P.} + \text{D} + \text{E.A.} + U \qquad \text{...(6.9)}$$

So knowing all other quantities U can be calculated.

***Exercise* 6.6:** Calculate the lattice energy for the reaction

$$Li^+(g) + Cl^-(g) \rightarrow LiCl(s)$$

from the data, $\Delta H_{sub}(Li) = 160.67$ kJ mol^{-1}; $\frac{1}{2}(D + RT)_{Cl_2} = 122.17$ kJ mol^{-1}; I.P.(Li) = 520.07 kJ mol^{-1}; EA(Cl) = – 365.26 kJ mol^{-1} and ΔH_f^0 (LiCl) = – 401.66 kJ mol^{-1}.

Using the equation (6.9) and substituting the respective values:

$$-401.66 = 160.67 + 520.07 + 122.17 - 365.26 + U$$

$$U = -839.31 \text{ kJ mol}^{-1}$$

6.2.1. *Lattice Energy*

The lattice energy can also be calculated if we know the attractive and repulsive forces in the crystal. The attractive potential can be calculated from the classical laws of electrostatic forces and given by

$$U_{att} = -\frac{N_{av}Z_+Z_-e^2M}{4\pi\varepsilon_0 r} \qquad \text{...(6.10)}$$

where N_{av} is the Avogadro number, e is the charge on the electron, Z_+, Z_- are the positive integer, M is the unitless Madlung constant, r is the distance between the ions and $(4\pi\varepsilon_0)$ is the permittivity constant ($4\pi\varepsilon_0 = 1.113 \times 10^{-10}$ C^2 N^{-1} m^{-2}). The values of Madlung constants for some crystal structures are compiled in Table 6.5. The repulsive potential is given by empirical relation

$$U_{rep} = B\,e^{-r/C} \qquad \text{...(6.11)}$$

where B and C are constants, the total potential energy or lattice energy is

$$U = U_{att} + U_{rep} = -\frac{N_{av}Z_+Z_-e^2M}{(4\pi\varepsilon_0)r} + Be^{-r/c} \qquad \text{...(6.12)}$$

TABLE 6.5. Values of Madlung constants

Type of crystal	Madlung constant	Example
Face-centered cubic	1.74756	NaCl
Body-centered cubic	1.76267	CsCl
Hexagonal	1.4985	ZnO
Hexagonal	1.64132	ZnS

The value of B in terms of C and other known quantities can be calculated by using the relation

$$\left(\frac{dU}{dr}\right)_{r=r_e} = 0 \qquad ...(6.13)$$

i.e., first differential of U with respect to r at r_e is zero. This is because at $r = r_e$, the total potential energy must exhibit minimum. The expression for B is

$$B = \frac{Z_+Z_-e^2Me^{r_e/C}CN_{av}}{(4\pi\varepsilon_0)r_e^2} \qquad ...(6.14)$$

Substituting B in equation (6.12), the total potential energy

$$U = -\frac{Z_+Z_-e^2N_{av}M}{(4\pi\varepsilon_0)r_e}\left[\frac{r_e}{r} - \frac{C}{r_e}e^{-(r-r_e)/C}\right] \qquad ...(6.15)$$

If the equilibrium value of internuclear distance is substituted in equation (6.15), the energy evaluated is called the lattice energy of crystal energy

$$\Delta H \text{ (lattice energy)} = -\frac{N_{av}Z_+Z_-e^2M}{(4\pi\varepsilon_0)r_e}\left[1 - \frac{C}{r_e}\right] + 2\,RT \qquad ...(6.16)$$

where $2\,RT$ is term added to convert ΔU to ΔH.

For a series of alkali halides the variation in C is small and average value of 0.311×10^{-10} m can be used. The equation (6.16) is called Born-Mayer equation. The value of C can be calculated by the method described in exercise (6.6).

***Exercise* 6.6:** Calculate the lattice energy of LiCl by Born-Mayer method from the following data: $C = 0.311 \times 10^{-10}$ m, $r_e = 2.57 \times 10^{-10}$ m, $M = 1.74756$. Compare with the value obtained in Exercise 6.2.

Substituting the values of the respective terms in equation (6.16).

$$\Delta H = \frac{-(6.022 + 10^{23}\text{ mol}^{-1})\,(1)\,(1)\,(1.6022 \times 10^{-19}C)^2(1.74756)}{(1.113 \times 10^{-10}\text{C}^2\text{ N}^{-1}\text{ m}^{-2})(2.57 \times 10^{-10}\text{ m})}\left[1 - \frac{0.311 \times 10^{-10m}}{2.57 \times 10^{-10}\text{m}}\right]$$

$$+ (2)\,(8.314\text{ J mol}^{-1}\text{ K}^{-1})\,(298\text{ K})$$

$$= -8.30 \times 10^5\text{ J mol}^{-1} + 4.955\text{ J mol}^{-1} = -8.3 \times 10^2\text{ kJ mol}^{-1}$$

The value is quite agreeable with that obtained by Born-Haber cycle *i.e.*, 8.38×10^2 kJ mol^{-1}.

***Exercise* 6.7:** Calculate the value of constant 'C' for NaBr in equation (6.9) from the given data: $r_e = 2.97 \times 10^{-10}$ m; $M = 1.74756$; $U = -736384$ J mol^{-1}.

Substituting the respective values in equation (6.16)

$$-736384 = -\frac{(6.02\times10^{23}\ \text{mol}^{-1})\,(1)\,(1)\,(1.6\times10^{-19}C)^2(1.74756)}{(1.113\times10^{-10}C^2\ \text{N}^{-1}\ \text{m}^{-2})(2.97\times10^{-10}\ \text{m})}\left[1-\frac{C}{2.97\times10^{-10}\text{m}}\right]$$

$$8.147\times10^5\left[1-\frac{C}{2.97\times10^{-10}}\right] = 736384$$

or

$$1-\frac{C}{2.97\times10^{-10}} = \frac{736384}{8.147\times10^5} = -0.904$$

$$C = (1-0.904)\,(2.97\times10^{-10})\ \text{m} = 0.286\times10^{-10}\ \text{m}$$

i.e., nearly same (0.311×10^{-10} m) accepted for general purposes.

6.3. BONDING IN METALS

The bondings in the metals can be explained by the following two theories:

6.3.1. *The free electron model*

The free electron model assumes that kernals of atoms of metals are placed according to the geometrical arrangements as mentioned earlier and the valence electrons are freely moving particles in a three dimensional box with uniform potential energy. The bonding between these positively charged kernals and negatively charged electrons are quite strong and thus metals possess medium to high values of melting and boiling points, heats of sublimation and vaporisation.

The packing of the electrons in the various quantum states are done in the same manner as explained in Section (4.7) and the energy of the highest filled orbital at absolute zero is

$$E_F = \frac{h^2}{8m_e}\left(\frac{3N}{\pi V}\right)^{2/3} \qquad ...(6.17)$$

where E_F is the Fermi energy, N is the number of valence electrons per mol of metal, V is the molar volume of the metal. The kinetic energy of these electrons, at any temperature other than O K, is given by the expression

$$E = \frac{3}{5}N_{\text{av}}E_F\left[1+\frac{5\pi^2}{12}\left(\frac{k_bT}{E_F}\right)^2+...\right] \qquad ...(6.18)$$

***Exercise* 6.8:** (*a*) Calculate the Fermi energy for Al metal. The density of Al is 2.702×10^3 kg m^{-3}. (*b*) Also calculate the E_{th} at 27°C for Al.

$$V = \frac{(27\,\text{g mol}^{-1})\,(10^{-3}\ \text{kg g}^{-1})}{2.7\times10^3\ \text{kg m}^{-3}} = 10^{-5}\ \text{m}^3\ \text{mol}^{-1}$$

Substituting the values in (6.17)

$$E_F = \frac{(6.625\times10^{-34}\ \text{J s})^2}{(8)(9.11\times10^{-31}\ \text{kg})}\left[\frac{3\times6.02\times10^{23}\ \text{mol}^{-1}}{\pi\times10^{-5}\ \text{m}^3\ \text{mol}^{-1}}\right]^{2/3} = 8.97\times10^{-19}\ J$$

(*b*) According to relation (9.18), the thermal energy is

$$\overline{E}_{th} = \overline{E} - \overline{E}_0 = \tfrac{3}{5} N_{av} E_F \left[1 + \frac{5\pi^2}{12}\left(\frac{k_b T}{E_F}\right)^2\right] - \frac{3}{5} N_{av} E_F$$

$$= \frac{\pi^2 N_{av} k_b^2 T^2}{4E_F} \qquad \text{...(6.19)}$$

Substituting the respective value in above relation

$$\overline{E}_{th} = \frac{(\pi^2)(6.02\times10^{23}\,\text{mol}^{-1})(1.38\times10^{-23}\,\text{J K}^{-1})(300\,\text{K})^2}{(4)(8.97\times10^{-19}\,\text{J})}$$

$$= 28.41\ \text{J mol}^{-1}$$

***Exercise* 6.9:** (*a*) What is the contribution of Fermi energy to the heat capacity at constant volume; (*b*) Use the above relation taking potassium as metal. Density of potassium is 8.6×10^{2} kg m^{-3}; $T = 27$°C. Compare this value with that obtained in Exercise 10.18.

(*a*) Differentiating equation (6.19) with respect to temperature

$$\overline{C}_v = \frac{\pi^2 N_{av} k_b^2}{4E_F} \times 2T = \frac{\pi^2 N_{av} k_b^2 T}{2E_F} \qquad \text{...(6.20)}$$

Using the equation (6.17) and

$$V = \frac{\text{At.Wt.}}{d} = \frac{(39\,\text{g mol}^{-1})(10^{-3}\,\text{kg g}^{-1})}{8.60\times10^{2}\,\text{kg m}^{-3}} = 4.53\times10^{-5}\,\text{m}^3\,\text{mol}^{-1}$$

$$E_F = \frac{(6.63\times10^{-34}\,\text{J s})^2}{(8)(9.11\times10^{-31}\,\text{kg})}\left[\frac{(3)(6.02\times10^{23}\,\text{mol}^{-1})}{(\pi)(4.53\times10^{-5}\,\text{m}^3\,\text{mol}^{-1})}\right]^{2/3}$$

$$= 3.28\times10^{-19}\,\text{J}$$

Substituting the respective values in (6.20)

$$\overline{C}_V = \frac{(\pi)^2(6.02\times10^{23}\,\text{mol}^{-1})(1.38\times10^{-23}\,\text{J K}^{-1})^2(300\,\text{K})}{(2)(3.28\times10^{-19}\,\text{J})}$$

$$= 0.518\ \text{J K}^{-1}\,\text{mol}^{-1}$$

$$\gamma = \frac{\overline{C}_V}{T} = \frac{0.518\,\text{J mol}^{-1}\,\text{K}^{-1}}{300\,\text{K}} = 1.73\times10^{-3}\,\text{J K}^{-2}\,\text{mol}^{-1}$$

The value of γ obtained in Exercise 10.18 is 2×10^{-3} J K^{-2} mol^{-1}. The agreement is not bad considering the free electron model.

6.3.2. Molecular orbital theory

Molecular orbital theory is commonly known as *band model* and the concept is same as explained earlier in Section (4.5), *i.e.*, number of molecular orbitals formed are equal to the number of atomic orbitals. Half of them are bonding and the other half antibonding molecular orbitals. Since, the number of atoms in the smallest perceptible piece of metals are so large that the energies of the resulting M.O.'s are very close to each other and thus can be regarded as continuous. These groups of energy levels are called the band. Each kind of atomic orbitals gives different bands *i.e.*, 1s band, 2s band and so on. Generally each band is separated from each other but in some cases they do overlap.

Conduction in metals is due to the movement of electrons. This occurs if these bands are: (*a*) partially filled or (*b*) filled bands overlap the empty ones or (*c*) the barrier to cross over to the empty bands from filled one is not large. 'Insulators' are those substances where the gap between the conduction band and completely filled valence band is very large. The relative spacing between the different bands in different kinds are shown in Fig. 6.1.

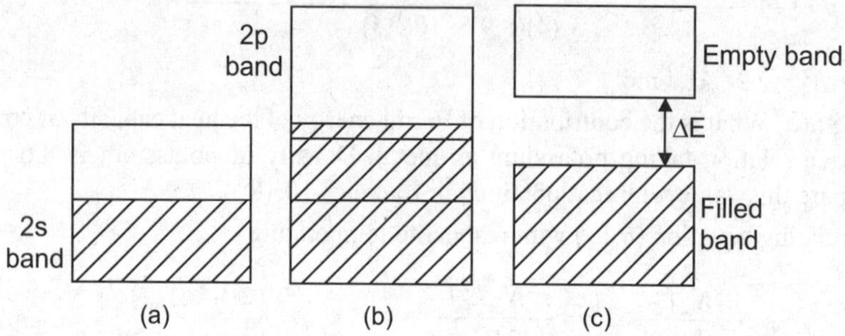

Fig. 6.1 *Band model of solids*

***Exercise* 6.10:** Explain the conduction of electricity in Be and Li metals with the help of m.o. theory.

The electronic configuration for Li is $1s^2\ 2s^1$ so there will be 1*s* and 2*s* bands as shown in Fig. 6.1*a*. Since only half of 2*s* orbitals are occupied by the electrons, the conduction will take place.

The electronic configuration for Be is $1s^2\ 2s^2$ and the bands are formed due to 1*s* and 2*s* electrons. Both the bands are completely filled but since 2*p* band which is completely empty, overlaps the 2*s* band (Fig. 6.1*b*) there is sufficient space for electrons to move and explains the conduction part.

***Exercise* 6.11:** Explain the conduction phenomenon in semiconductors.

In semiconductors, the gap between the filled and unfilled band is reduced by the presence of electrons due to impurities and the electrons of the impurities are promoted to the empty conduction band either by increasing the voltage or by shining light on the semiconductor.

6.4. SPACE LATTICE

A space lattice is defined as an extension of regular distribution of points in space indefinitely *i.e.*, any point selected in the array has the similar surroundings everywhere.

A space lattice can be described by the distance between each point along three axis (*a, b, c*) and three angles α, β, γ. The angles α, β and γ are defined as the angles between the arms *b* and *c*; *c* and *a*; *a* and *b* respectively (see Fig. 6.2, Table 6.6.).

TABLE 6.6 The Crystal Systems

System	Lattice Symmetry	Description of Co-ordinate System	Lattice Type
Triclinic	None	$a \neq b \neq c$ $\alpha \neq \beta \neq \gamma$	P
Monoclinic	2-fold axis and/or mirror planes	$a \neq b \neq c$ $\alpha = \gamma = 90^\circ \neq \beta$	P, I
Orthorhombic	3, 2-fold axis and/or 3 mirror planes	$a \neq b \neq c$ $\alpha = \beta = \gamma = 90^\circ$	P, I, C, F
Tetragonal	4-fold axis	$a = b \neq c$ $\alpha = \beta = \gamma = 90^\circ$	*P, I*
Hexagonal	3-fold axis 6-fold axis	$a = b \neq c$ $\alpha = \beta = 90^\circ$ $\gamma = 120^\circ$	P, R P
Cubic	4, 3-fold axis	$a = b = c$ $\alpha = \beta = \gamma = 90^\circ$	P, I, F

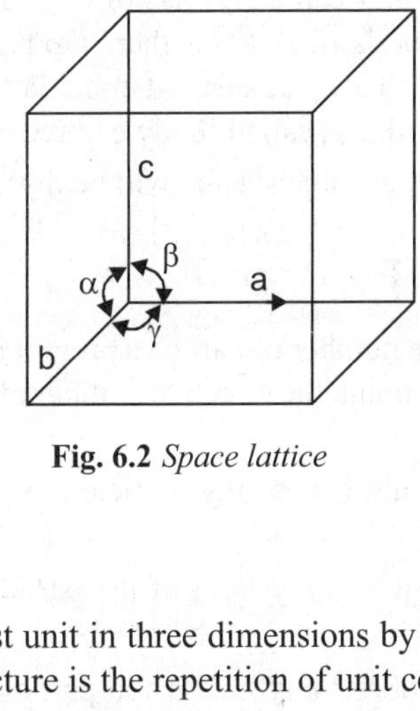

Fig. 6.2 *Space lattice*

6.5. ▮ THE UNIT CELL

The unit cell is defined as the smallest unit in three dimensions by which the crystal structure can be completely defined. Thus crystal structure is the repetition of unit cells touching each other.

The whole crystal structure can be defined completely by seven crystal formed by different combinations of *a, b, c,* and α, β and γ. Further all the seven crystal structures can be further divided into "Fourteen lattices" and these are called "Bravais lattices" as shown in Fig. 6.3.

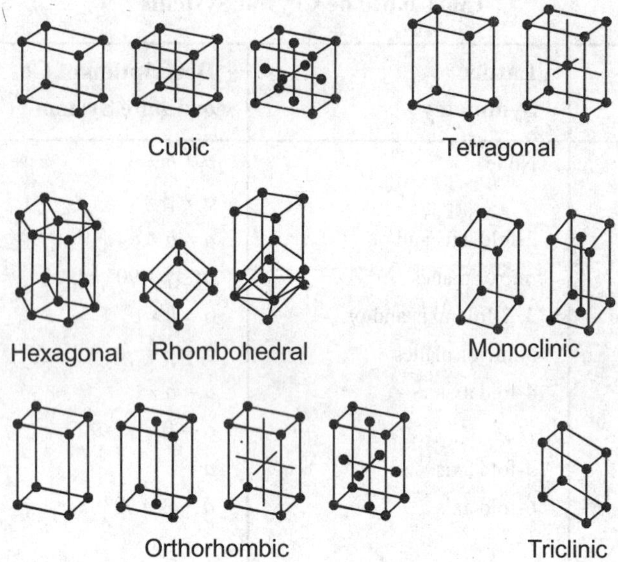

Fig. 6.3 *Crystal structure*

The most simplest is the cubic unit cell and it has three Bravais lattices: (1) If each corner has a particle this is called 'primitive cubic lattice'; (2) If there are particles at the eight corners and at the centre of each face of cube this is called 'face-centred cubic lattice'; (3) If there are particles at each corner and at the centre of the cube, this is called 'body centred cubic lattice'.

In all the coming sections only the cubic system will be discussed.

6.6. UNIT CELL CONTENT

Unit cell content means the effective number of particles present in the unit cell. Since in a lattice, each point is further connected to another point, the effective number of particles in a unit cell can be calculated as follows:

(*i*) As a corner is shared by eight cubes, any particle at a corner gives 1/8th of its share to each cube;

(*ii*) A side is shared by four cubes, only 1/4th of the particle effectively belongs to a unit cubic cell.

(*iii*) A face is shared by two cubes, only 1/2 of the particle effectively belongs to cube.

(*iv*) Any particle inside the unit cubic cell completely gives its share to the unit cell.

***Exercise* 6.12:** Calculate the effective number of particles in a face centred cubic system.

In a face-centred cubic system, there are eight particles at the corners and six particles at the centre of each face. The effective number of particles are: $(8 \times \frac{1}{8}) + (6 \times \frac{1}{2}) = 1 + 3 = 4.$

6.7. UNIT CELL COORDINATES

The position of each point in the three cubic cell units can be described with respect to unit cell axis. Any particle at the corner of the cube can be considered as the origin of the cubical system and be assigned the coordinates (0, 0, 0). The length of the cube in any direction can be considered as unit length and be equal to 1. With these two points in mind, the coordinates to the particles present in three cubical unit cells can be assigned and are given in the Table 6.7.

TABLE 6.7. Unit cell coordinates for different cubic lattices

	Coordinates	Number of unique Coordinates
Simple cubic ←	0, 0, 0	1
Face-centred	(0, 0, 0), (1/2, 1/2, 0), ← (1/2, 0, 1/2), (0, 1/2, 1/2)	4
Body centred	(0, 0, 0), (1/2, 1/2, 1/2)	2

***Exercise* 6.13:** In NaCl crystal Cl^- ions occupy the face centred arrangement, whereas Na^+ are present in between the two Cl^- ions, as shown in the Fig. 6.4. Label each ion in the system.

Each chloride ion is either present at the corners or at the centre of each face of a cube. So the discrete coordinates describing any Cl^- ion are (0, 0, 0), $(\frac{1}{2}, \frac{1}{2}, 0)$, $(\frac{1}{2}, 0, \frac{1}{2})$, $(0, \frac{1}{2}, \frac{1}{2})$. Each Na^+ is present either at the centre of the edge or in the centre of the cube, so discrete set of coordinates are $(\frac{1}{2}, 0, 0)$, $(0, \frac{1}{2}, 0)$, $(0, 0, \frac{1}{2})$ and $(\frac{1}{2}, \frac{1}{2}, \frac{1}{2})$.

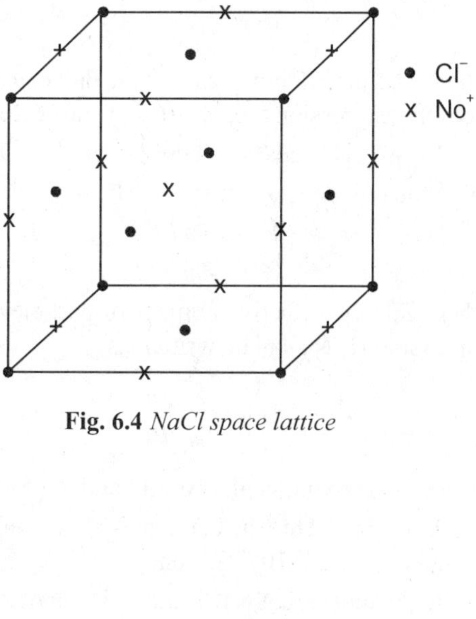

Fig. 6.4 *NaCl space lattice*

6.8. COORDINATION NUMBER

The number of particles touching the particular particle is defined as the coordination number. It may also be defined as the nearest number of neighbours. In the Bravais lattices all the atoms or molecules have the same coordination number. The coordination number can be obtained if one knows the cation to anion radius ratio and this has been mentioned in Table 6.1.

***Exercise* 6.14:** The ionic radii of Na^+ and Cl^- ions are 0.98×10^{-10} m and 1.81×10^{-10} m. Find the coordination number to each ion.

$$\text{Radius ratio} = \frac{\text{radius of Na}^+}{\text{radius of Cl}^-} = \frac{0.98\times10^{-10}\,\text{m}}{1.81\times10^{-10}\,\text{m}} = 0.54$$

Looking at the Table 6.1, it is clear that the CN of Na^+ and Cl^- ions are 6.

***Exercise* 6.15:** Find the radius of the smallest atom that can fit in the octahedral hole.

The cross-section of octahedral arrangement is described in Fig. 6.5. From the geometry of the diagram

$$ac^2 = ab^2 + bc^2; (2r + 2R)^2 = (2R)^2 + (2R)^2$$

i.e., $$r + R = \sqrt{2}R$$

or $$\frac{r}{R} = 0.414$$

i.e., radius of the smallest atom that can fit in the octahedral hole should be 0.414 R.

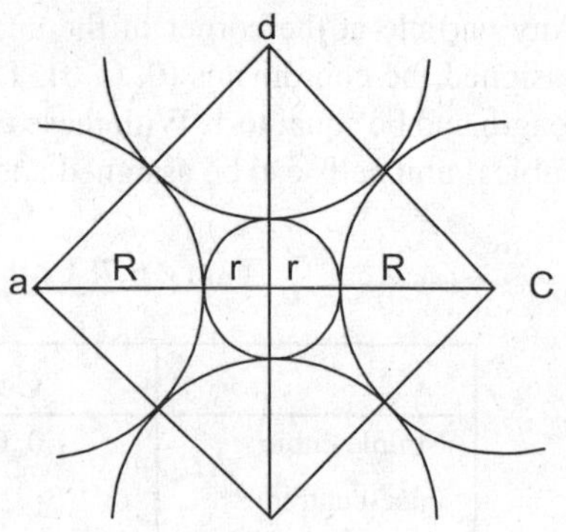

Fig. 6.5 *Section of octahedral space lattice by a plane*

6.9. IDEAL DENSITY OF CRYSTAL

By the ideal density, we mean here the theoretical density of the unit cell of the crystal and is defined as

$$d = \frac{M}{V} \quad ...(6.21)$$

where M is the effective mass of the unit cell and V is the volume of the unit cell. These can be calculated from the expressions (6.24), (6.22) and (6.23) respectively.

$$V = abc\,[1 - \cos^2\alpha - \cos^2\beta - \cos^2\gamma + 2\cos\alpha\cos\beta\cos\gamma]^{1/2} \quad ...(6.22)$$

for cubic unit cell, $\alpha = \beta = \gamma = 90°$, so

$$V = abc \quad ...(6.23)$$

$$M = Zm \quad ...(6.24)$$

where Z is the effective number of particles in the unit cell and m is the mass of each particle. Thus expression (6.21) can be written as

$$d = \frac{Z\ M}{N_{av} V} \quad ...(6.25)$$

where M is the molecular weight and N_{av} is the Avogadro's number.

***Exercise* 6.16:** NaCl crystal is a face centred cubic lattice. Calculate its ideal density if the edge of the cube is 5.62×10^{-10} m long.

In the unit cell of NaCl, each Cl^- ion occupy the corners and centre of faces of the cube. So effective number of Cl^- ions $= 8 \times \frac{1}{8} + 6 \times \frac{1}{2} = 4$. Na^+ ions occupy the centre of the edges and the centre of the cube, so $Z_{Na+} = 1 + 12 \times \frac{1}{4} = 4$. Effective number of units of NaCl = 4. Molecular weight of NaCl = 58.5 g mol^{-1}

$$d = \frac{(4)\,(58.5\ \text{g mol}^{-1})\,(10^{-3}\ \text{kg g}^{-1})}{(6.22 \times 10^{23}\ \text{mol}^{-1})\,(5.62 \times 10^{-10}\ \text{m})^3}$$

$$= 2.19 \times 10^3\ \text{kg m}^{-3}$$

Theoretical density obtained above is with the assumption that each lattice point is occupied by the species. Generally this is not so and the experimental density is always less than the ideal one. This points out that not all the lattice points are occupied by the atoms and some are left unoccupied. These points are called "vacancies". Percentage of occupancy of lattice sites can be obtained by the ratio of experimental to the theoretical density *i.e.*,

$$\% \text{ occupancy} = \frac{d_{exp}}{d_{cal}} \times 100 \quad \text{...(6.26)}$$

***Exercise* 6.17:** Li forms a body centred cubic lattice. If the lattice constant is 3.50×10^{-10} m and the experimental density is 5.30×10^2 kg m^{-3}, calculate the percentage occupancy of Li metal.

Since, the Li forms a body centred cubic lattice, the effective number of atoms per unit cell $= 8 \times \frac{1}{8} + 1 = 2.$

$$\text{Mass of unit cell} = \frac{2 \times 7}{6.023 \times 10^{23}} = 2.324 \times 10^{-23} \text{ g}$$

$$= 2.324 \times 10^{-26} \text{ kg}$$

$$\text{Volume of unit cell} = (3.50 \times 3.50 \times 3.50)\,(10^{-30})\text{ m}^3$$

$$= 42.875 \times 10^{-30} \text{ m}$$

$$\text{Ideal density} = \frac{2.324 \times 10^{-26}}{42.875 \times 10^{-30}} = 0.054 \times 10^4 \text{ kg m}^3$$

$$= 5.4 \times 10^2 \text{ kg m}^3$$

$$\text{Experimental density} = 5.3 \times 10^2 \text{ kg m}^3$$

$$\% \text{ occupancy} = \frac{5.3 \times 10^2 \text{ kg m}^3}{5.4 \times 10^2 \text{ kg m}^3} \times 100 = 98\%$$

6.10. CRYSTAL RADII

Depending upon the unit cell, the atoms, if assumed to be spheres, can touch along the edges, face diagonal, body diagonal etc. The crystal radii *i.e.,* radius of the atom or molecule can be calculated from the dimensions of the unit cell.

***Exercise* 6.18:** Calculate the radius of the atom which are occupying the body centred cubic unit cell. Length of the cube $= 4 \times 10^{-10}$ m.

In the body diagonal cubic lattice, the atoms along the body diagonal touch each other, *i.e.* (see Fig. 6.6):

$$hb = R + 2R + R = 4R$$

Now in right angle triangle hdb,

$$hb^2 = (4R)^2 = dh^2 + db^2$$

$$= dh^2 + (ad^2 + ab^2)$$

$$= a^2 + 2a^2 = 3a^2$$

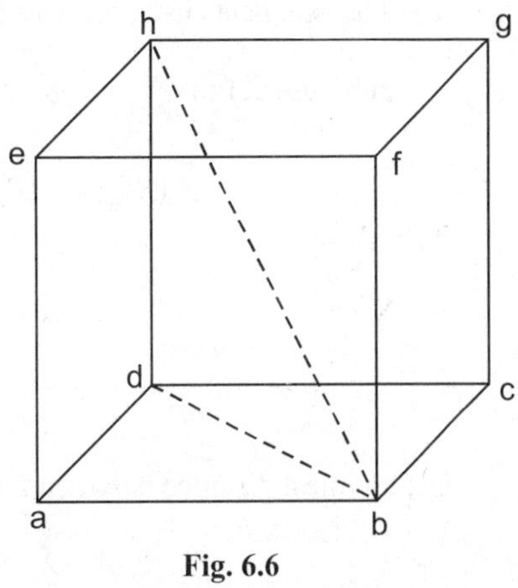

Fig. 6.6

or $$R = \frac{\sqrt{3}}{2}a = 0.866\,a$$

Substituting the value of *a*,

$$R = 0.866 \times 4 \times 10^{-10}\text{ m}$$
$$= 3.464 \times 10^{-10}\text{ m}$$

***Exercise* 6.19:** Zinc sulphide forms a cubic unit cell of length 6×10^{-10} m. Zinc ions form a face-centred cubic lattice and sulphide ions occupy the centre of the alternate small cubes: (*a*) Find the position of zinc and sulphide ions; (*b*) assign the coordinates; (*c*) effective number of ZnS molecules; (*d*) density of zinc sulphide if the atomic weight of zinc and sulphur are 65.4 and 32 g mol^{-1}; (*e*) formula of zinc sulphide; (*f*) smallest distance between zinc and sulphide ions and (*g*) coordination number of each ion.

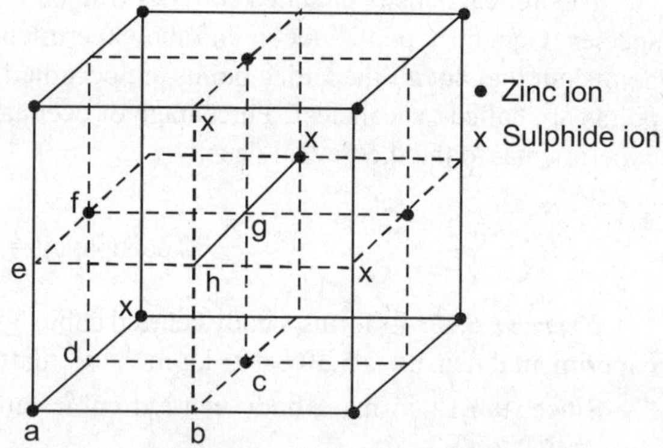

Fig. 6.7

(*a*) The • represents zinc ions and × represents sulphide ions.

(*b*) The coordinates for zinc ions are (0, 0, 0), $(\frac{1}{2}, \frac{1}{2}, 0)$, $(\frac{1}{2}, 0, \frac{1}{2})$, and $(0, \frac{1}{2}, \frac{1}{2})$.

For sulphide ions are $(\frac{1}{4}, \frac{1}{4}, \frac{1}{4})$, $(\frac{1}{4}, \frac{3}{4}, \frac{3}{4})$, $(\frac{3}{4}, \frac{1}{4}, \frac{3}{4})$, $(\frac{3}{4}, \frac{3}{4}, \frac{1}{4})$.

(*c*) Effective number of zinc ions in unit cell are

$$8 \times \frac{1}{8} + 6 \times \frac{1}{2} = 4$$

Effective number of sulphide ions are $4 \times 1 = 4$, so molecular formula is ZnS.

(*d*) $$\text{Density} = \frac{(4)(65.4+32)\times 10^{-3}\text{ kg mol}^{-1}}{(6.023\times 10^{23}\text{ mole mol}^{-1})\,(6\times 10^{-10})^3\text{ m}^3}$$
$$= 2.995 \times 10^3\text{ kg m}^3$$

(*e*) The smallest distance between zinc and sulphide ions is $\frac{1}{2}$ of the body diagonal of small cube abcdefgh

$$(fb)^2 = (db)^2 + (df)^2$$
$$(R_{Zn^{2+}} + 2R_{s^{2-}} + R_{Zn^{2+}})^2 = (3 \times 10^{-10})^2 + [(3 \times 10^{-10})^2 + (3 \times 10^{-10})^2]$$
$$= 27 \times 10^{-20}\text{ m}^2$$
$$2\,(R_{Zn^{2+}} + R_{s^{2-}}) = 3\sqrt{3} \times 10^{-10} = 3 \times 1.732 \times 10^{-10}\text{ m}$$
$$R_{Zn^{2+}} + R_{s^{2-}} = \frac{5.196\times 10^{-10}}{2} = 2.598\times 10^{-10}\text{ m}$$

i.e., smallest distance between zinc and sulphide ions is 2.598×10^{-10} m.

(*f*) Each sulphide ion has four zinc as the nearest neighbours arranged tetrahedrally and each zinc ion is surrounded by four sulphide ions arranged tetrahedrally.

PROBLEMS

6.1 From the following data calculate the lattice energy of $MgCl_2$. ΔH_{sub} (Mg) = 147.6 kJ mol^{-1}, IP(Mg^{2+}) = 736.8 kJ mol^{-1}, $\frac{1}{2}(D + RT)Cl_2$ = 122.17 kJ mol^{-1}, EA(Cl) = – 365.26 kJ mol^{-1}; ΔH_f^0 ($MgCl_2$) = – 641.83 kJ mol^{-1}.

6.2 The experimental crystal energy of SnO_2 is –11371.11 kJ mol^{-1}, ΔH_{sub}(Sn) = 292.88 kJ mol^{-1}, IP(Sn^{4+}) = 3929.19 kJ mol^{-1}. $\frac{1}{2}(D + RT)O_2$ = 247.7 kJ mol^{-1}; EA(0) = – 191.04 kJ mol^{-1}.

Calculate the heat of formation of SnO_2. Compare this value with 580.74 kJ mol^{-1}.

6.3 Calculate the lattice energy of RbBr from the following data; ΔH_{sub}(Rb) = 75.73 kJ mol^{-1}, IP(Rb^+) = 401.66 kJ mol^{-1}; $\frac{1}{2}(D + RT)Br_2$ = 111.7 kJ mol^{-1}, EA (Br) = – 338.66 kJ mol^{-1}; ΔH_f^0(RbBr) = – 391.2 kJ mol^{-1}.

6.4 Find the lattice energy of CaO if, ΔH_{sub} (Ca) = 153.72 kJ mol^{-1}, IP(Ca^{++}) = 1156 kJ mol^{-1}; $\frac{1}{2}(D + RT)O_2$ = 247.7 kJ mol^{-1}. EA (O) = – 191.04 kJ mol^{-1}, and ΔH_f^0 = – 635.55 kJ mol^{-1}.

6.5 ΔH_{sub} (Tl) = 163.34 kJ mol^{-1}, IP ($T1^+$) = 589.94 kJ mol^{-1}; $\frac{1}{2}(D + RT)_{Br_2}$ = 111.7 kJ mol^{-1}; EA (Br) = – 338.66 kJ mol^{-1}; ΔH_f(T1Br) = – 175.31 kJ mol^{-1}. Calculate the lattice energy of T1Br crystal.

6.6 Rubidium bromide structure can have either sodium chloride structure or cesium chloride structure, depending upon the external conditions. If the lattice constant is 4.06 × 10^{-10} m, calculate (*a*) the internuclear distance, (*b*) density, (*c*) the lattice energy for each form. Compare this lattice energy with that calculated in problem 6.3. Given

$$C = .311 \times 10^{-10} \text{ m};$$
$$M(\text{NaCl}) = 1.74756; \ M(\text{CsCl}) = 1.76267.$$

6.7 Calcium oxide exists as a sodium chloride structure with lattice constant 4.80 × 10^{-10} m. If the value of C = 0.311 × 10^{-10} m, calculate the lattice energy and enthalpy of the crystal by using Born-Mayer equation, M(NaCl) = 1.74756.

6.8 Calculate the crystal energies for (*i*) T1Br (cesium chloride structure), (*ii*) $SrCl_2$ (flouride structure) from the following data;

$$M(\text{CsCl}) = 1.76767;$$
$$M(\text{fluoride}) = 2.51939;$$
$$r_{T1+} = 1.45 \times 10^{-10} \text{ m}$$
$$r_{Br-} = 1.96 \times 10^{-10} \text{ m};$$

$$r_{Sr^{2+}} = 1.10 \times 10^{-10} \text{ m};$$
$$r_{Cl^-} = 1.81 \times 10^{-10} \text{ m}.$$

Assume that the value of $C = 0.311 \times 10^{-10}$ m can be used in above cases.

6.9 Born and Lande gave another expression for the calculation of the repulsive energy; $U_{rep} = \frac{B}{r^n}$ where B and n are constants. The value of constant n varies according to the inert gas configuration of the ion involved, e.g. 5 for the configuration He, 7 for Ne, 9 for Ar, 10 for Kr and 12 for Xe, or the arithmetic mean when the electronic configuration are not the same. Derive the expression for crystal energy similar to Born-Mayer expression.

6.10 Repeat the problems 6.6, 6.7 and 6.8 and calculate the lattice energies using the expression derived in problem 6.9.

6.11 The conductivity of Mg can be explained only if we assume that bands formed by 2s and 2p atomic levels overlap. Draw the energy level diagram.

6.12 Explain by drawing the energy level diagram that graphite is a good conductor whereas diamond is a bad conductor of electricity.

6.13 Explain briefly with the help of band model that conductivity increase in case of insulator while decreases in case of metals when temperature is increased.

6.14 Certain insulators become conductors when exposed to radiation. Explain.

6.15 Calculate C_v (electronic) for lithium metal. The density of Li metal is 0.53 kg dm^{-3} at 27°C.

6.16 If the density of Cu metal is 8.96×10^3 kg m^{-3}, calculate the value of E_F and η (C_v).

6.17 Rubedium forms a body centred cubic lattice. The atomic radius of rubedium atom is 2.48×10^{-10} m. Calculate the length of cubic lattice or lattice constant.

6.18 Strontium exists as a face centred cubic lattice. The atomic radius is 2.15×10^{-10} m. Calculate the ideal density of the metal.

6.19 If the atomic volumes of rubedium and strontium are 55.9×10^{-3} and 33.7×10^{-3} dm^3 mol^{-1}, calculate the percentage voids in the structures of Rb and Sr from the data of problems 6.17 and 6.18.

6.20 Uranium forms a simple cubic lattice and its atomic radius is 1.38×10^{-10} m. If the atomic volume of uranium is 12.5×10^{-3} dm^3 mol^{-1} calculate the ratio of actual volume to the theoretical volume.

6.21 Calculate the coordination number of the atoms in problems 6.17 and 6.20.

6.22 Label the coordinates of the atoms in the problems 6.17 and 6.18.

6.23 The edge of cubic lattice is 5×10^{-10} m. Calculate the radius of an atom which (*a*) forms a body centred cubic lattice, (*b*) face centred cubic lattice and (*c*) simple cubic lattice.

6.24 Find the size of the atom that can fit in the tetrahedral void created by the packing in crystals.

6.25 Show that space is more efficiently used in case of face centred cubic lattice rather than in primitive cubic or body centred cubic lattice. Calculate the value of each.

6.26 Show that the percent space occupied in case of face-centred lattice and hexagonal close packing is the same.

6.27 In a cubic lattice, if the corners are occupied by the atoms of kind A and centre of the face by the kind B, calculate the formula of the compound.

6.28 In a cubic lattice if the corners and centres of the faces are occupied by the cations X^{2+} and the centres of the each octant is occupied by the anions Y^-. Find (*a*) the molecular formula of this compound, (*b*) if the length of the cube is 6×10^{-10} m calculate the smallest distance between X^{2+} and Y^-.

6.29 RbI forms a cesium chloride structure. Calculate the (*a*) coordinates of Rb^+ and I^- ions: (*b*) If the ionic radii of Rb^+ and I^- ions are 0.148 nm and 0.216 nm, calculate the coordination number of each ion and show it is compatible with CsCl structure. (*c*) Calculate the density of RbI.

6.30 MgS forms NaCl structure. (*a*) Calculate the coordinates of Mg^{2+} and S^{2-} ions, (*b*) If the radii of Mg^{2+} and S^{2-} are 0.065 nm and 0.184 nm, calculate the coordination number of each ion. (*c*) Calculate Z^{2+}, Z^{2-} and Z for each species. (*d*) Calculate the density of the crystal.

6.31 Calculate the crystal radii of the structure mentioned in problems 6.29 and 6.30.

6.32 A compound AB forms NaCl çrystal a structure and compound XY forms CsCl structure. The mass of XY is three time, that of AB but the cubic edge of AB is twice that of XY. Calculate the ratio of the densities of AB and XY.

6.33 The density of Fe is 7.86×10^3 kg m^{-3}. It is found that Fe can form either body centred cubic lattice with a side length 2.9×10^{-10} m or face-centred cubic lattice with side length 3.68×10^{-10} m. Calculate the per cent voids in each case. $M_{Fe} = 56$ g mol^{-1}.

6.34 A compound AB_2 has a cubic unit cell. The coordinates of atoms A are (0, 0, 0), $(0, \frac{1}{2}, \frac{1}{2})$, $(\frac{1}{2}, 0, \frac{1}{2})$, $(\frac{1}{2}, \frac{1}{2}, 0)$ and those of B $(\frac{3}{4}, \frac{1}{4}, \frac{1}{4})$, $(\frac{1}{4}, \frac{1}{4}, \frac{1}{4})$, $(\frac{1}{4}, \frac{1}{4}, \frac{3}{4})$, $(\frac{1}{4}, \frac{3}{4}, \frac{1}{4})$, $(\frac{3}{4}, \frac{3}{4}, \frac{1}{4})$, $(\frac{3}{4}, \frac{1}{4}, \frac{3}{4})$, $(\frac{1}{4}, \frac{3}{4}, \frac{3}{4})$ and $(\frac{3}{4}, \frac{3}{4}, \frac{3}{4})$.

(*a*) Find the positions of these atoms in a cube.

(*b*) If length of the cube is 6×10^{-10} m, $M_A = 20$; $M_B = 40$ g mol^{-1} respectively; calculate the density of AB_2.

(*c*) Calculate the closest distance between atoms A and B.

(*d*) Find the coordination number of each atom.

6.35 (*a*) Na_2O has an antifluoride structure *i.e.*, cations forms a face centred cubic structure and the anions are at the centres of each octant. Draw the structure and show that the formula derived from the coordination number does agree with the above formula.

(*b*) Drive the general expression for the internuclear distance in this structure in terms of lattice constant $a = 0.555$ nm.

(*c*) Calculate the coordinates of each ion.

6.36 The SrO has a NaCl structure. If the lattice constant $a = 0.514$ nm and the ionic radius of $Sr^{2+} = 0.113$ nm, calculate the radius of O^{2-} ion.

6.37 In LiI structure, the two successive iodide ions touch each other and Li^+ ions are present in the interstatial positions. If LiI has a NaCl structure, calculate the radius of I^- ion. Given the internuclear distance between Li^+ and I^- ions is 0.3 nm and the lattice constant 0.6 nm.

6.38 If the effective nuclear charges on Na^+ and F^- are 6.48 and 4.48 respectively, the lattice constant is 0.628 nm, calculate the ionic radii of Na^+ and F^- ions. (The semi-empirical relation is $r = C_n/Z_{eff}$, C_n is a constant.)

6.39 Calculate the effective nuclear charge on Na^+ ion if the $Z_{ett}(F^-)$ is 4.48 and the ionic radii of F^- and Na^+ ions are 0.136 nm and 0.095 nm respectively.

6.40 The ionic radii (10^{-10} m) of alkali and alkaline metals, of sixth and seven group anions are given below. With the help of radius ratio rule, find the structures of alkali halides and alkaline earth chalcogenides. Note: that basic crystal lattice is cubic:

Li^+, 0.68	O_2^{2-}, 1.45; F^-, 1.33
Na^+, 0.98; Mg^{2+}, 0.65	S_2^{2-}, 1.90; Cl^-, 1.81
K^+, 1.33; Ca^{2+}, 0.94	Se_2^{2-}, 2.02; Br^-, 1.96
Rb^+, 1.48; Sr^{2+}, 1.10	I^-, 2.19
Cs^+, 1.67; Ba^{2+}, 1.29	

Addition Problems

6.41 From the data of Table 6.2, predict which one will have the higher melting point, (*i*) RbCl or RbBr, (*ii*) CsCl or CsBr?

6.42 Looking at the data of Table 6.2, answer qualitatively the following quarries:

(*i*) Is the normal melting point of Li_2O is higher or lower than that of Na_2O (1548 K)?

(*ii*) Data of Table 6.2 gives the melting point of CaO and BaO. Suggest whether the melting point of SrO will be in between those of CaO and BaO or higher or lower?

(*iii*) Data of Table 6.2 did not give the melting points of Li_2O and MgO. Make qualitative guess regarding their melting points.

6.43 Using the data of Table 6.4 regarding the boiling points of aliphatic carboxylic acids, plot the normal boiling points versus molar mass using the least square fit. Formulate the equation similar to equation (6.5). Using this equation predict the boiling point of butanoic and octanoic carboxylic acids and percent error.

6.44 Repeat the Problem (6.43) for the 2-aliphatic ketones and predict the boiling point of 2-octanone

6.45 Halogen elements form ionic crystals with alkali elements having valence electron as ns^1, whereas although hydrogen atom is placed in IA does not form ionic crystal. Explain briefly.

CHAPTER 7

LIQUIDS

7.1. HEAT OF VAPORISATION AND INTERMOLECULAR FORCES

Heat of vaporisation of any liquid can either be calculated experimentally (see Chapter 11) or theoretically (Chapter 9). It has also been found that heat of vaporisation is temperature dependent and decreases as the temperature is increased. It tends to zero as the critical temperature is approached. The critical temperature is a temperature where the properties of liquid and equilibrium vapour merge with each other. Thus

$$\frac{d(\Delta H)}{dT} = \Delta C_p = (C_p)_{\text{vap}} - (C_p)_{\text{liq}} \qquad ...(7.1)$$

$\left(\frac{d(\Delta H_{\text{vap}})}{dT} \neq 0\right)$ means that heat capacity of vapour and that of liquid is different.

The theoretical basis for heat of vaporisation implies some sort of interactions between the molecules *i.e.*, repulsive or attractive and in calculating these interactions between molecules in liquids one needs the intermolecular distance which is considered as some average value between a molecule and its nearest neighbours.

The repulsive potential energy is generally given by the expression

$$U_{\text{rep}} = be^{-ar} \qquad ...(7.2)$$

where a and b are empirical constants or by

$$U_{\text{rep}} = \frac{B}{r^9} \qquad ...(7.3)$$

again 9 and B are repulsive constants.

The attractive forces can be divided into four categories. Although we have discussed these attractive forces in detail in chapter 4, here we will summarise briefly.

Dipole-Dipole Interaction

Dipole-dipole interaction energy is given by

$$U_{D-D} = -\frac{2\mu_1^2\mu_2^2}{3r^6 k_b T(4\pi\varepsilon_0)^2} \quad ...(7.4)$$

where μ_1 and μ_2 are the dipole moments of the molecules, r is the intermolecular distance, greater than as compared to the distance between the two poles and k_b is the Boltzmann constant.This interaction is only present if the molecules have permanent dipole moments.

Dipole Moment-Induced Dipole Moment Attraction

A polar molecule can induce a polar or non-polar molecule to have such an orientation that attraction between the two molecules is experienced. The attractive energy, can be calculated as

$$U_{\text{ind}} = -\frac{2\alpha\mu^2}{r^6(4\pi\varepsilon_0)^2} \quad ...(7.5)$$

μ is the dipole moment, α is the polarisability. The polarisability depends upon the direction of the dipole moment but here it is assumed that it is same towards all direction. This is also called induction effect.

Induced Dipole-Induced Dipole Attraction

This effect was first considered by London and is also known as London's or dispersive forces. This effect can be treated only quantum mechanically and dependent mainly on the detailed motion of electrons in the neighbouring molecules. Though a time average would indicate that each atom has a zero dipole moment, instantaneous dipole moment is present and this can induce the neighbouring molecules to have such positions such that attractive forces are present. These are given by

$$U_{\text{London}} \,\alpha\, \frac{\alpha^2}{(4\pi\varepsilon_0)^2 r^6} = C\frac{\alpha_1\alpha_2}{r^6(4\pi\varepsilon_0)^2} \quad ...(7.6)$$

The proportionality constant C has same value for many molecules and it is -1.8×10^{-11} erg or -1.8×10^{-18} J.

Hydrogen Bonding

The hydrogen bonds are present generally between acidic hydrogen atom having small partial positive charge (*e.g.*, in N–H, O–H, F–H) and electronegative atom, having high concentration of negative charge (*e.g.* nitrogen atom in amines, oxygen atom in alcohols, ethers, water). The hydrogen bonding is not due to dipole-moment of the molecule but due to the sharing of proton between two electronegative atoms such as F, O or N, which are at right distance apart and can be best explained due to the localised electron attraction. The minimum value of energy due to hydrogen bonding between a pair of molecules is 6 kcal mole^{-1} or 400×10^{-22} J/molecule pair.

***Exercise* 7.1:** (*a*) Calculate the total attractive energy between a pair of HCl molecules if $\mu = 1.03$ D, $\alpha = 4.7 \times 10^{-40}$ C m^{2}V^{-1} and $r = 5 \times 10^{-10}$ m.

(*b*) If it is assumed that these are the only forces responsible for heat of vaporisation, calculate $\Delta\overline{H}_{vap}$.

(*a*) Using equations (7.4), (7.5) and (7.6)

$$U_{D-D} = -\frac{(2/3)(1.03\,\text{D})^4(3.3356\times10^{-30}\ \text{C m D}^{-1})^4}{(5\times10^{-10}\ \text{m})^6(1.36\times10^{-23}\ \text{J K}^{-1})(300\,\text{K})}$$

$$(1.1127\times10^{-10}\ \text{C}^2\ \text{N}^{-1}\ \text{m}^{-2})^2$$

$$= -1.16\times10^{-22}\ \text{J}$$

$$U_{ind} = -\frac{(2)(4.7\times10^{-40}\ \text{C m}^2\ \text{V}^{-1})(1.03\times3.3356\times10^{-30}\ \text{C m})^2}{(5\times10^{-10}\ \text{m})^6(1.1127\times10^{-10}\ \text{C}^2\ \text{N}^{-1}\ \text{m}^{-2})^2}$$

$$= -0.574\times10^{-22}\ \text{J}$$

$$U_L = \frac{(-1.8\times10^{-18}\ \text{J})(4.7\times10^{-40}\ \text{C m}^2\ \text{V}^{-1})^2}{(5\times10^{-10}\ \text{m})^6(1.1127\times10^{-10}\ \text{C}^2\ \text{N}^{-1}\ \text{m}^{-2})^2} = -20.55\times10^{-22}\ \text{J}$$

$$U_T = -(1.16+0.574+20.55)\times10^{-22}\ \text{J} = -22.28\times10^{-22}\ \text{J/molecule}$$

(*b*) $$\Delta\overline{H}_{vap} = -U_T\times N_{av} = (22.28\times10^{-22}\ \text{J molecule}^{-1})$$

$$(6.023\times10^{23}\ \text{molecule mol}^{-1})$$

$$= 1.34\ \text{kJ mol}^{-1}$$

The actual value of $\Delta\overline{H}_{vap}$ for HCl = 16.15 kJ mol^{-1}.

***Exercise* 7.2:** The heat of vaporisation of CCl_4 is 29.874 kJ mol^{-1} and $\mu = 0$. If the closest distance between CCl_4 molecule is 5×10^{-10} m and assuming that $\Delta\overline{H}_{vap}$ is only due to attractive forces, calculate α.

Since $\mu = 0$, the only attractive forces are the London's forces. Using equation (7.6)

$$\Delta\overline{H}_{vap} = \frac{(-1.8\times10^{-18}\ \text{J})(\alpha^2)N_{av}}{(4\pi\varepsilon_0)^2 r^6}$$

$$\alpha = \left[\frac{(29.874\times10^3\ \text{J mol}^{-1})(1.1127\times10^{-10}\ \text{C}^2\ \text{N}^{-1}\ \text{m}^{-2})^2(5\times10^{-10}\ \text{m})^6}{(-1.8\times10^{-18}\ \text{J})(6.023\times10^{23}\ \text{molecules mol}^{-1})}\right]^{1/2}$$

$$= 2.31\times10^{-39}\ \text{C m}^2\ \text{V}^{-1}$$

7.2. VISCOSITY

The flow of liquid can be classified into two types. One is 'laminar' or viscous flow, which generally describes the small flow rate, through a pipe of small diameter. The other is 'turbulent' flow, which describes the large flow rate through a pipe of larger diameter. These are further classified by Reynold number defined as

$$RN = \frac{d\bar{v}R}{\eta} \qquad ...(7.7)$$

where R is the radius of the tube, d is the density of the liquid, $\overline{v}$ is the average velocity of the fluid along the tube and η is the coefficient of viscosity. If RN is greater than 4000 the flow is turbulent and if it is less than 2100, the flow is laminar.

The volume of fluid passed through a fixed cross-section per second is

$$V = \frac{\pi(\Delta p)R^4}{8\eta l} \qquad ...(7.8)$$

where Δp is the pressure drop over a length l. The velocity profile in a laminar flow is:

$$\overline{v} = \frac{\Delta p}{4\eta l}(R^2 - r^2) \qquad ...(7.9)$$

where r is the distance from the axis of the pipe.

***Exercise* 7.3:** Find out whether flow is turbulent or laminar from the following data. $l = 1$ m, $R = 2 \times 10^{-3}$ m, $\overline{v} = 2 \times 10^2$ m s^{-1}, $d = 1.2$ kg dm^{-3} and $\eta = 1.001 \times 10^{-3}$ Pl.

Using equation (7.7)

$$RN = \frac{(1.2\times10^3 \text{ kg m}^{-3})(2\times10^2 \text{ m s}^{-1})(2\times10^{-3} \text{ m})}{1.001\times10^{-3} \text{ Pl or N s m}^{-2}}$$

$$= 4.795 \times 10^5$$

i.e., flow is turbulent.

***Exercise* 7.4:** (*a*) Find the velocity contour when $n - C_4C_9OH$ is flowing in a tube of radius 2 cm. η at 20°C is 2.948×10^{-3} Pl, the pressure drop is 2×10^5 N m^{-2} and $l = 1$ m.

(*b*) Calculate the volume of liquid flown in one second.

(*a*) Equation (7.9) can be used to calculate $\overline{v}$ for different values of r. Let $r = 0.2$ cm $= 2 \times 10^{-3}$ m, then $\overline{v}$ at this point will be

$$\overline{v} = \frac{(2\times10^5 \text{ N m}^{-2})(20^2 - 2^2)(10^{-6} \text{ m}^2)}{(4)(2.948\times10^{-3} \text{ N s m}^{-2})(1 \text{ m})}$$

$$= 67.162 \times 10^2 \text{ m s}^{-1}$$

Similarly for other values of r, $\overline{v}$ is calculated and are listed below:

$r \times 10^3$ (m)	1	2	4	8	12	16	18
$\overline{v}\times10^{-2}$ (m s^{-1})	67.67	67.16	65.01	56.99	43.42	24.42	12.89

(*b*) Using equation (7.8)

$$V = \frac{(22/7)(2\times10^5 \text{ N m}^{-2})(2\times10^{-2} \text{ m})^4}{(8)(2.948)\times10^{-3} \text{ N s m}^{-2}(1 \text{ m})}$$

$$= 42.6 \text{ m}^3 \text{ s}^{-1}$$

$$= 42.6 \times 10^3 \text{ dm}^3 \text{ s}^{-1}$$

7.3. MEASUREMENT OF VISCOSITY

The coefficient of viscosity is generally measured by two methods:

Ostwald Viscometer: Time required for the flow of fixed amount of fluid is noted and η is calculated with the relation

$$\eta = \frac{\pi(\Delta p)R^4 t}{8Vl} \quad ...(7.10)$$

Generally, the coefficient of viscosity is calculated by comparing the rate of flow of the liquid with the rate of flow of the fluid whose coefficient of viscosity is known. The relation is

$$\frac{\eta_1}{\eta_2} = \frac{d_1 t_1}{d_2 t_2} \quad ...(7.11)$$

Falling Sphere Method: The falling sphere method involves the balancing of the gravitational force against the viscous drag and relation is

$$\eta = \frac{2r_b^2(d_b - d)g}{9\mathrm{v}} \quad ...(7.12)$$

where b refers to the dropping sphere or bead and g is the gravitational constant. In case the comparison method is used we get

$$\frac{\eta_1}{\eta_2} = \frac{(d_b - d_1)t_1}{(d_b - d_2)t_2} \quad ...(7.13)$$

***Exercise* 7.5:** In a Ostwald Viscometer, water takes 25 s to flow between lower and upper marks while the liquid A takes 38 s. The densities of water and liquid at 20°C are 0.9982 and 0.78945 kg dm^{-3} respectively. If η of water is 1.005 cP, calculate η of liquid A.

Using equation (7.11)

$$\eta_A = (1.005\text{ cp})\frac{(0.7894\text{ kg dm}^{-3})(38\text{s})}{(0.9982\text{ kg dm}^{-3})(25\text{s})} = 1.208\text{ cp}$$

7.4. TEMPERATURE DEPENDENCE

The coefficient of viscosity varies with temperature and the relation is

$$\log \eta = A + B/T \quad ...(7.14)$$

where A and B are constants depending upon the liquid. The above equation can be written as

$$\eta = A' \exp(-\Delta E_{vis}/RT) \quad ...(7.15)$$

where A' is the undetermined constant. Comparing with the equation (7.14) $B = \frac{\Delta E_{vis}}{-2.303R}$. For many substances

$$\Delta E_{vis} = 0.3\,\Delta E_{(VAP)} \quad ...(7.16)$$

***Exercise* 7.6:** The viscosities of CCl_4 at various temperatures are given as

T(K)	273	293	313	333	353	373
$10^{-3} \times \eta$ (Pl)	1.329	0.964	0.739	0.585	0.468	0.384
log η	– 2.8765	– 3.02	–3.13	–3.23	–3.33	–3.42
$\frac{1}{T} \times 10^3$	3.663	3.423	3.195	3.00	2.83	2.68

(*a*) Calculate ΔE_{visc}; (*b*) ΔE_{vap} = 27.614 kJ mol^{-1}. See whether relation (7.16) is followed or not.

(*a*) Logarithms of the given values of η and $1/T$ are calculated and given in the above table. log η are plotted against $1/T$ in Fig. 7.1. The slope is 0.55×10^3 K^{-1}.

$$\Delta E_{visc} = (2.303)\,(8.314 \text{ J K}^{-1} \text{ mol}^{-1})\,(0.550 \times 10^3 \text{ K}^{-1})$$
$$= 10.53 \text{ kJ mol}^{-1}$$

(*b*) Using equation (7.16)

$$\frac{10.53 \text{ kJ mol}^{-1}}{27.14 \text{ kJ mol}^{-1}} = 0.388$$

i.e., ratio of $\Delta E_{visc} / \Delta E_{vap}$ is close to 0.3.

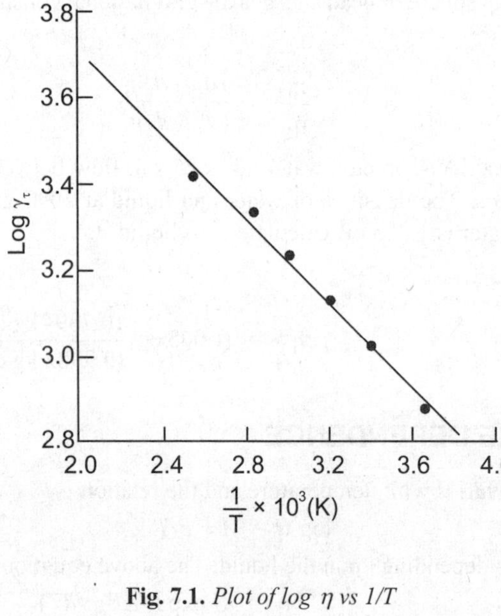

Fig. 7.1. *Plot of log η vs 1/T*

7.5. SURFACE TENSION

It is defined as the work done in expanding the surface of a liquid by unit area. The unit for surface tension (γ) is J m^{-2} or dyn cm^{-1} or N m^{-1}. The most common method to measure surface tension is the rise or depression of the liquid in the capillary tube *i.e.*,

$$\gamma = \frac{drgl}{2} \qquad ...(7.17)$$

where d is the density of the liquid, r is the radius of the capillary, l is the length of liquid being depressed or raised and g is gravitational constant. In the comparison method, surface tension of unknown liquid can be calculated from the relation (7.18) provided

$$\frac{\gamma_1}{\gamma_2} = \frac{d_1 l_2}{d_2 l_2} \qquad ...(7.18)$$

γ_2 of the reference liquid is known.

***Exercise* 7.7:** (*a*) The γ ($CHCl_3$) at 20°C is 27.4 dyne cm^{-1} = 27.4 × 10^{-3} N m^{-1}. The d_{20} (H_2O) = 0.9982 kg dm^{-3} and d_{20} ($CHCl_3$) = 1.595 kg dm^{-3}. If the rise of $CHCl_3$ and H_2O in a capillary of certain diameter is 2.33 cm and 9.9 cm, calculate the γ for H_2O. (*b*) Also calculate the radius of the tube.

(*a*) Using equation (7.18)

$$\gamma = \frac{(27.4\times10^{-3}\text{ N m}^{-1})(0.9982\text{ kg dm}^{-3})(9.9\times10^{-2}\text{ m})}{(1.595\text{ kg dm}^{-3})(2.33\times10^{-2}\text{ m})}$$

$$= 72.86 \times 10^{-3}\text{ N m}^{-1}$$

(*b*) Using equation (7.17)

$$r = \frac{2\gamma}{gld} = \frac{(2)(27.4\times10^{-3}\text{ N m}^{-1})}{(9.80\text{ m s}^{-2})(1.595\times10^{3}\text{ kg m}^{-3})(2.33\times10^{-2}\text{ m})}$$

$$= 1.505 \times 10^{-4}\text{ m} = 0.1505\text{ mm}$$

7.6. SURFACE TENSION AND VAPOUR PRESSURE OF SMALL DROPLETS

The surface tension (γ) and the vapour pressure of very small drops of liquids are related by the expression

$$\ln\frac{p}{p_0} = \frac{2\gamma M}{drRT} \qquad ...(7.19)$$

where p_0 is the normal vapour pressure of the liquid and p that of small drop.

The vapour pressure (p_0) for any liquid can be calculated with the Clausius-Clapeyron equation (see Chapter 14.2):

$$\ln p_0 = -\frac{\Delta\bar{H}_{vap}}{RT} + \text{constant} \qquad ...(7.20)$$

***Exercise* 7.8:** Calculate the vapour pressure of water bubble of radius 10^{-8} m at 20°C if the vapour pressure of water at this temperature is 23.76 mm Hg. $\gamma_{20}(H_2O)$ = 72.75 dyne cm^{-1} = 72.75 × 10^{-3} N m^{-1} and $d_{20}(H_2O)$ = 0.9982 kg dm^{-3}.

Using equation (7.19)

$$\ln\frac{p}{p_0} = \frac{(2)(72.75\times10^{-3}\text{ N m}^{-1})(18\text{ g mol}^{-1})(10^{-3}\text{ kg g}^{-1})}{(0.9982\times10^{3}\text{ kg m}^{-3})(10^{-8}\text{ m})(8.314\text{ J mol}^{-1}\text{ K}^{-1})(293\text{K})}$$

$$= 0.1077$$

$$\frac{p}{p_0} = e^{0.1077} = 1.1137$$

$$\cdot \quad p = (23.76 \text{ mm})(1.1137) = 26.46 \text{ mm}$$

7.7. CONTACT ANGLE

The contact angle (θ) is defined as the angle that the surface of the drop makes with the solid surface. The contact angle is calculated by the relation

$$\gamma_s = \gamma_{s.liq} + \gamma_{liq} \cos\theta \qquad ...(7.21)$$

where γ_s and γ_{liq} are the surface tension of solid and liquid respectively, $\gamma_{s.liq}$ is the surface tension (also called interfacial tension) between the solid and liquid phases. Similarly the interfacial tension can also be defined between the two liquids if they are insoluble in each other. If $\theta < 90°$, the liquid wets the surface of solid and this can occur if $\gamma_s > \gamma_{s.liq}$.

***Exercise* 7.9:** The interfacial tension of Hg-benzene at 20°C is 357×10^{-3} N m^{-1} whereas the γ_{Hg} and $\gamma_{benzene}$ are 484×10^{-3} N m^{-1} and 28.85×10^{-3} N m^{-1}. Calculate the angle of contact.

Using equation (7.21)

$$\cos\theta = \frac{\gamma_s - \gamma_{s-liq}}{\gamma_{liq}} = \frac{(484 - 375)\times 10^{-3} \text{ N m}^{-1}}{484 \times 10^{-3} \text{ N m}^{-1}}$$

$$= 0.225$$

$$\theta = 76.98°.$$

PROBLEMS

7.1 (*a*) Calculate the total attractive energy between the pair of He–He and Ar–Ar atoms: $\alpha_{He} = 3.41 \times 10^{-41}$ C m^2 V^{-1}, $\alpha_{Ar} = 1.95 \times 10^{-40}$ C m^2 V^{-1}, $r = 5 \times 10^{-10}$ m.

(*b*) Compare the U_{att} between He-He and Ar–Ar and suggest which will have higher heat of vaporisation, if this energy is equal to $\Delta \overline{H}_{vap}$.

7.2 If total attractive forces between the molecules are the cause of heat of vaporisation then calculate this value for $CHC1_3$, $\mu(CHCl_3) = 1.16$ D, α $(CHCl_3) = 7.50 \times 10^{-40}$ C m^2 V^{-1} and say $r = 5 \times 10^{-10}$ m.

7.3 Compare U_D, U_{ind} and U_L for CO molecule, $\mu = 0.12$ D and $\alpha = 2.39 \times 10^{-40}$ C m^2 V^{-1}, $r = 5 \times 10^{-10}$ m.

7.4 $\Delta \overline{H}_{vap}(H_2O) = 39.41$ kJ mol^{-1}. If $\Delta \overline{H}_{vap}$ is only due to the forces of attraction (U_D, U_{ind} and U_L), calculate the value of α for H_2O. $\mu_{H_2O} = 1.84$ D. Actual value of α is 1.6×10^{-40} C m^2 V^{-2}. Explain the discrepancy.

7.5 The molar polarisability P is related to α by the expression

$$P = \frac{4\pi N_{av}}{3}\left[\alpha + \frac{\mu^2}{3k_b T}\right]$$

If P for HBr is 2.79 × 10–15 C m^2 V^{-1} mol^{-1} and $\mu = 0.78$ D, calculate α.

Viscosity

7.6 What is the relation between the CGS unit and SI unit of viscosity?

7.7 (*a*) The water is flowing through a pipe of diameter 0.1 m at an average velocity of 0.04 m s^{-1}. Find whether the flow is viscous or turbulent. Given at 20°C = 1.005 × 10^{-3} *Pl*, $d = 1$ kg dm^{-3}.

(*b*) If this flow rate is to be maintained, calculate the pressure head to be applied $l = 1$ m.

7.8 Calculate the velocity of water at r = 0.002 m, 0.004 m, 0.016 m, 0.020 m, 0.03 m and 0.04 m. Use the other data from problem 7.7.

7.9 Calculate the rate of flow (volume s^{-1}) of a liquid through a tube of diameter 0.2 × 10^{-3} m and length 1 m. $\eta = 3Pl$ and $\Delta p = 10$ atm.

7.10 The flow rate of benzene at 20°C is 1.5 mL s^{-1} through a capillary of diameter 5 × 10^{-3} m and length 0.5 m. Calculate the value of pressure head to maintain this flow rate $\eta = 0.652 \times 10^{-2}$ P.

7.11 Compare the time taken by the same volume of benzene and ethanol to flow through acapillary of certain diameter and length. η (benzene) = 0.652 × 10^{-2} P, η (ethanol) = 1.2 × 10^{-2} P, *d*(ethanol) = 0.7893 × 10^3 kg m^{-3}, *d*(benzene) = 0.8785 × 10^3 kg m^{-3}.

7.12 The flow rate of liquid (η = 30 P) through a capillary of diameter 0.2 mm and 1.0 m long is 1.31 × 10^{-3} dm^3 s^{-1}, for a pressure of 10 atm. Calculate the flow rate of Hg under the same conditions if η = 1.547 cP.

7.13 The times taken by *V* dm^3 of water and $CHCl_3$ to flow through a capillary of diameter 0.2 mm and length 0.5 m are 10 s and 36 s. Calculate the viscosity of $CHCl_3$ if η_{H_2O} = 1.005 cP, d_{H_2O} = 0.9982 × 10^3 and d_{CHCl_3} = 1. 595 × 10^3 kg m^{-3}.

7.14 The time taken by a metal ball to drop through a liquid *A* of height *h* is 5 s, whereas that in liquid *B* is 7.5 s. If the densities of metal ball, liquids *A* and *B* are 7.8 × 10^3 kg m^{-3}, 1.5 × 10^3 kg m^{-3} and 4.6 × 10^3 kg m^{-3} respectively, calculate the viscosity of liquids, *A*, η (*B*) = 2.5 cP.

7.15 Calculate the activation energy for the viscous flow of ethyl ether from the following data:

t(°C)	0	20	40	60	80	100
$10^2 \times \eta$(P)	0.284	0.233	0.197	0.166	0.140	0.118

(*b*) Also calculate the viscosity at 30°C.

Surface Tension

7.16 Calculate the height of ethyl alcohol that must have risen in a capillary tube of diameter 0.2 mm, γ(EtOH) = 22.75 × 10^{-3} N m^{-1} at 20°C, d = 0.7893 × 10^3 kg m^{-3}.

7.17 The surface tension of Hg at 25°C is 484 × 10^3 N m^{-1}. If the depression caused by Hg is 3.63 cm and density is 13.6 × 10^3 kg m^{-3}, calculate the radius of the tube.

7.18 If the levels of water and benzene that rose in the same capillary are 9.9 cm and 4.5 cm respectively, calculate the surface tension of benzene. γ_{H_2O} = 72.75 dynes cm^{-1} at 20°C. Density of H_2O and benzene at 20°C are 0.9982 × 10^3 kg m^{-3} and 0.8785 × 10^3 kg m^{-3} respectively. Calculate the radius of the tube.

7.19 Calculate the contact angle between water and hexane at 20°C if interfacial tension is 51.1 × 10^{-3} N m^{-1} . $\gamma_{(H_2O)}$ = 72.75 × 10^{-3} N m^{-1} and $\gamma_{(hexane)}$ = 18.43 × 10^{-3} N m^{-1}.

7.20 Calculate the ratio of p/p_0 for ethanol at 20°C if radius of the drop is 10^{-8} m. Take rest of the data from problem 7.16.

CHAPTER 8

GASES

8.1. EQUATIONS OF STATE

The state of a gas is described by four variables, pressure, volume, temperature and quantity of material. These variables are denoted by p, V, T and n, where n measures the amount of gas in terms of number of moles present. We will discuss these quantities as follows.

8.1.1. *Mass and Chemical Amount*

The important factor in chemical reactions is the relative number of atoms or molecules involved. This number is obtained by dividing the mass of the gas by their atomic or molecular mass, which is a number that measures a mass relative to a standard. The standard chosen in SI system is carbon-12. The base unit of chemical amount, the *mole*, is defined as the amount of substance that contains the same number of chemically defined particles as exactly 12 g of carbon-12. This number is called *Avogadro's number* or *constant*. This number has the value of 6.02214×10^{23}. Thus for any substance, the amount of mass of Avogadro's number of particles is called mole.

8.1.2. *Pressure*

Force per unit area is called pressure. The common unit of pressure is atmosphere or torr, one torr is pressure exerted by a 1 mm column of mercury. The SI unit for pressure is 'Pascal' *i.e.*, Newton metre^{-2} (N m^{-2}), 1 atm = 1.0135×10^5 N m^{-2} and 1 torr = 133.3224 N m^{-2}. Other unit of pressure is pound per square inch (6894.7572 N m^{-2}) and 1 bar (1×10^5 N m^{-2} = 10^5 P$_a$ = 10^2 kPa). In laboratory pressure is commonly measured by either (*a*) closed end manometer or by (*b*) open end manometer. Pressure exerted by a column of any fluid = $\dfrac{\text{force exerted by that fluid}}{\text{area}}$. If h is height of the fluid in m, d the density kg m^{-3}, r(m) the radius of the column then

$$p = \frac{(\pi r^2 \text{ m}^2)(h\text{m})(d \text{ kg m}^{-3})(\text{g m s}^{-2})}{\pi r^2 \text{m}^2}$$

$$= h\,dg \text{ (kg m s}^{-2}\text{ m}^{-2})$$

$$= h\,dg\ (\text{N m}^{-2}) \qquad ...(8.1)$$

Differential form of barometric pressure is

$$d \ln p = -\frac{Mg}{RT}\,dh \qquad ...(8.2)$$

where M is the molecular weight and g is the gravitational constant. It can be integrated to

$$\ln p_2/p_0 = -\frac{Mg}{RT}h \qquad ...(8.3)$$

where p_0 is the pressure at $h = h_0 = 0$.

***Exercise* 8.1:** What will be the pressure in system as shown by the closed end (Fig. 8.1*a*) and open end manometer (Fig. 8.1*b*)? Calculate in SI unit. The liquid used is Hg. The pressure in the system depicted by Fig. 8.1*a* is

$$\begin{aligned} &= 50 \text{ mm of Hg} \\ &= (50 \text{ mm of Hg})\,(133.32 \text{ N m}^{-2} \text{ mm}^{-1} \text{ of Hg}) \\ &= 6.666 \times 10^3 \text{ N m}^{-2} \end{aligned}$$

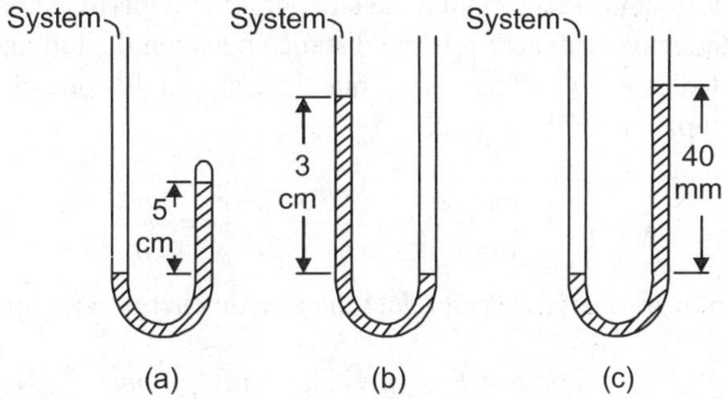

Fig. 8.1 *Various shapes of manometers*

The pressure depicted by the Fig. 8.1*b* in the system

$$\begin{aligned} &= 760 - 30 = 730 \text{ mm of Hg} \\ &= (730 \text{ mm})\,(133.32 \text{ N m}^{-2} \text{ mm}^{-1}) \\ &= 9.733 \times 10^4 \text{ N m}^{-2} \end{aligned}$$

Similarly in Fig. 8.1*c* = 760 + 40 = 800 mm of Hg

$$\begin{aligned} &= (800 \text{ mm})\,(133.32 \text{ N m}^{-2} \text{ mm}^{-1}) \\ &= 1.067 \times 10^5 \text{ Nm}^{-2} \end{aligned}$$

***Exercise* 8.2:** Dibutylphthalate is generally used as a fluid in the manometer where gases are reactive to mercury. If the density of this fluid is 1.047×10^3 kg m^{-3}, find the relation between 1 torr of each fluid.

The heights of various liquids corresponding to the same pressure at the same temperature are inversely proportional to the densities of the liquids *i.e.*,

$$\frac{h_f}{h_{\text{Hg}}} = \frac{d_{\text{Hg}}}{d_f} \quad \text{or} \quad h_f = h_{\text{Hg}}\frac{d_{\text{Hg}}}{d_f}$$

i.e., $$h_f = (1.0 \text{ mm})\frac{13.6\times10^3 \text{ kg m}^{-3}}{1.047\times10^3\times\text{kg m}^{-3}}$$

$$= 12.99 \text{ mm}.$$

***Exercise* 8.3:** Calculate the pressure exerted by NH_3 gas at the height of 5000 m if the pressure exerted at $h = 0$ is 400 torr at 27°C.

Using equation (8.3) and $M = (17 \text{ g mol}^{-1})(10^{-3} \text{ kg g}^{-1})$

$$p_2 = (400 \text{ torr}) \exp\left\{-\frac{(17\times10^{-3}\text{ kg mol}^{-1})(9.80 \text{ m s}^{-2})(5000 \text{ m})}{(8.314 \text{ JK}^{-1}\text{mol}^{-1})(300\text{K})}\right\}$$

$$= 286.4 \text{ torr}$$

The other relations which, can be obtained are

$$\boldsymbol{F} = \frac{dp}{dt} \qquad ...(8.4)$$

as we know the bold face is a vector quantity. Now if we assume that the gas particle is moving with a constant velocity v from left hand side to right hand side, before the particle hits the left hand side wall again, it will cover a distance of $2a$ where a is the distance between the left and right hand walls. The time taken to cover this distance $2a$ will be $2a$/v. Thus the force in this one direction, *i.e.*, from left to right can be obtained by replacing $\boldsymbol{F}$ by F_x, *i.e.*,

$$\boldsymbol{F} = F_x = \frac{dp}{dt} = \frac{dp_x}{dt} = \frac{\Delta p}{\Delta t} = \frac{2m\text{v}}{2a/\text{v}} = \frac{m\text{v}^2}{a} \qquad ...(8.5)$$

Using the definition of pressure, we can relate the kinetic energy with pressure as

$$p = \frac{\text{Force}}{\text{Area}} = \frac{F}{A} = \frac{(m\text{v}^2/a)}{a^2} = \frac{m\text{v}^2}{a^3} = \frac{m\text{v}^2}{V} = 2\frac{\text{Kinetic Energy}}{V} \qquad ...(8.6)$$

That is $p\,V$ = Two times the kinetic energy. ...(8.7)

Similarly force is also defined as change of momentum, obtained as

$$\boldsymbol{F} = \frac{d\boldsymbol{p}}{dt} = \frac{d(m\boldsymbol{v})}{dt} = m\frac{d(\boldsymbol{v})}{dt} \qquad ...(8.8)$$

***Exercise* 8.4:** Consider a vassal of volume 10 L containing one mole of Xe gas and having a pressure of 0.75 atm. Calculate the total kinetic energy of xenon atoms. Also calculate the average kinetic energy of each xenon atom and units of the answers.

Before we use equation (8.7), let us convert the quantities into SI units.

$V = 10 \text{ L} = (10 \text{ L}) \times (1{,}000 \text{ cm}^3 \text{ L}^{-1}) \times (10^{-2} \text{ m cm}^{-1})^3 = 10^{-2} \text{ m}^3$

$p = (0.75 \text{ atm}) \times (1.10325 \times 10^5 \text{ Pa atm}^{-1}) \times (1 \text{ N m}^{-2} \text{ Pa}^{-1}) = 0.76 \times 10^5 \text{ N m}^{-2}$

$p\,V = (0.76 \times 10^5 \text{ N m}^{-2})(10^{-2} \text{ m}^3) = 760 \text{ N m} = 760 \text{ J}$

Now using equation (8.7) we get the kinetic energy of one mole of xenon as

$$\frac{m\text{v}^2}{2} = \frac{pV}{2} = \frac{760 \text{ N m}}{2} = 380 \text{ J mol}^{-1}$$

Since the number of xenon atoms in one mole of xenon is Avogadro's number, so the average kinetic energy of per atom will be

$$< \text{Kinetic energy} > = \frac{380 \text{ J}}{6.02 \times 10^{23} \text{ atoms mol}^{-1}} = 63.13 \times 10^{-23} \text{ J} = 6.313 \times 10^{-22} \text{ J atom}^{-1}$$

The pointed brackets < > have been used to denote the average quantity and thus will be used in future also. The unit of energy is N m and also called Joule (J).

8.2. TEMPERATURE SCALE

The construction of temperature scale was constructed how the length of a rod, made up of particular material, was used and defined as

$$T = aL + b \qquad \text{...(8.9)}$$

where are a and b are constants and L is the length of the rod at temperature T °C and 1 atm. pressure. If at 0 °C the length of the rod be L_0, then $b = -aL_0$. The substitution of b in equation (8.9) gives us

$$T = a(L - L_0) \qquad \text{...(8.10)}$$

The size of the degree is determined by the temperature scale we use. If we use Celsius scale, let the length of the rod at 100 °C be L_{100}, *i.e.*, length of the rod when immersed in boiling water at one atmosphere. Substitution of these values in equation (8.10) we get the value of a.

$$a = \frac{100}{L_{100} - L_0}$$

and

$$T = \frac{100}{L_{100} - L_0}(L - L_0) \qquad \text{...(8.11)}$$

This allows us to determine the temperature by measuring the length of the rod. It has been observed that the increase in the length of metals is different for different metals so it is not a good choice. The other choice used is the volume of the rare gas, *i.e.*, helium or neon. With the same choice of reference point and degree size, equation (8.11) can be written as

$$T = \frac{100}{V_{100} - V_0}(V - V_0) \qquad \text{...(8.12)}$$

provided the pressure is kept constant. It has been observed that the temperature scale obtained by the use of equation (8.12) is independent of the gas used. Let us define the relative coefficient of thermal expansion as

$$\alpha = \frac{V_0}{V_{100} - V_0} \qquad \text{...(8.13)}$$

Equation (8.12) can be written as

$$T = \frac{100}{(V_{100} - V_0)/V_0}\left(\frac{V}{V_0} - 1\right) = \frac{100}{1/\alpha}\left(\frac{V}{V_0} - 1\right) = 100\,\alpha\left(\frac{V}{V_0} - 1\right) \qquad \text{...(8.14)}$$

Equation (8.14) was used to establish the absolute scale of temperature, *i.e.*, the lowest temperature is that when the value of $V = 0$. Substituting the value of $V = 0$ in equation (8.14), the lowest value of T we get is: $T = -100\ \alpha$ or -273.15 °C. Changing our reference point from the normal freezing point of water to this lowest temperature, which we can take to be zero, the temperature scale can be written by the equation (8.15).

$$T(\text{K}) = T(°\text{C}) + 273.15 \quad \text{...(8.15)}$$

Temperature is an intensive variable that is independent of the quantity of material present and is a measure of the quanttly of thermal energy contained in a materical.

The commonly used temperature scales are centigrade scale or Celsius and Fahrenheit scale. The former is based on the freezing point of water as 0°C and boiling point of water as 100°C. The same points on the Fahrenheit scale are 32°F and 212°F. The zero degree on Fahrenheit scale is the freezing point of saturated solution of NaCl in water. The relation between the two scales is

$$\frac{C}{5} = \frac{F-32}{9} \quad \text{...(8.16)}$$

where C and F are the temperatures on two scales. The absolute scale of temperature or commonly known as Kelvin scale (K) is based on the properties of gases and defined as

$$\frac{T_f}{T_i} = \lim_{p_i \to 0} \frac{p_f}{p_i} = 1.366 \text{ and } T_f - T_i = 100 \quad \text{...(8.17)}$$

where f stands for final and i for initial.

***Exercise* 8.5:** The highest temperature recorded in 1978 at Kanpur was 46.2°C. What is its value in Fahrenheit scale?

Using equation (8.16)

i.e.,
$$\frac{F-32}{9} = \frac{46.2}{5} = 9.24$$
$$F = 115.16°\text{F}$$

***Exercise* 8.6:** The normal body temperature for healthy human being is 98.4°F. What is its value in Celsius scale?

Using equation (8.16)

$$\frac{C}{5} = \frac{98.4-32}{9}$$
$$C = (5)\frac{(66.4)}{9} = 36.89°\text{C}$$

8.3. LAWS OF IDEAL GASES

8.3.1. *Boyle's Law*

Pressure (p) is inversely proportional to volume (V) at a constant temperature and for a fixed amount of gas, *i.e.*,

$$p \propto \frac{1}{V} \quad \text{...(8.18)}$$

or $$pV = \text{constant} \qquad ...(8.19)$$

or $$p_1V_1 = p_2V_2 \qquad ...(8.19a)$$

***Exercise* 8.7:** Calculate the pressure required to compress 4.24 dm ^{3} of gas at 412 torr to 1.56 dm ^{3} under isothermal conditions. Convert the pressure into SI unit.

Under isothermal (at constant temperature) conditions, Boyle's law is valid, *i.e.*,

$$p_2 = \frac{p_1V_1}{V_2} = \frac{(412 \text{ torr}) \times (4.24 \text{ dm}^3)}{(1.56 \text{ dm}^3)}$$

$$= 1119.795 \text{ torr}$$

$$= (1119.795 \text{ torr})\ (133.32 \text{ N m}^{-1} \text{ torr}^{-1})$$

$$= 1.493 \times 10^5 \text{ N m}^{-2}.$$

***Exercise* 8.8:** A fixed amount of gas is expanded from a pressure of 760 torr to 250 torr of mercury at constant temperature. If the initial volume is 10 dm ^{3}, calculate the final volume:

Using equation (8.19*a*)

$$V_2 = \frac{p_1V_1}{p_2} = \frac{(760 \text{ torr})\ (10 \text{ dm}^3)}{(250 \text{ torr})}$$

$$= \frac{760}{25} = 30.4 \text{ dm}^3.$$

8.3.2. *Charles' Law or Gay-Lussac Law*

For fixed amount of gas at constant pressure (under isobaric conditions) the volume (V) is directly proportional to temperature (T). The relation is

$$V \propto T \qquad ...(8.20)$$

or

$$V/T = \text{Constant} \qquad ...(8.21)$$

where T is the temperature in absolute scale;

or $$\frac{V_1}{T_1} = \frac{V_2}{T_2} \qquad ...(8.21a)$$

***Exercise* 8.9:** Gay-Lussac found that the volume of the gas under isobaric condition can be expressed as $V = V_0\ (1 + \alpha_0 t)$ where V_0 is the volume at 0°C, t is the temperature on centigrade scale and α_0 is constant. Using equation (8.21) calculate the value of α_0.

According to equation (8.21 *a*) the volume at any temperature T is given by

$$V = \frac{V_0}{T_0} \times T$$

Now $$T = 273 + t \text{ and } T_0 = 273 + 0$$

Combining the above relations

$$V = V_0 \frac{(273° + t)}{273°} = V_0\left(1 + \frac{1}{273}t\right)$$

Comparing the above equation with

$$V = V_0 (1 + \alpha_0 t).$$

$$\alpha_0 = \frac{1}{273} = 3.66 \times 10^{-3}\ °C^{-1}$$

***Exercise* 8.10:** 22.4 dm^3 of a gas at 50°C was allowed to expand to 40.8 dm^3 at fixed pressure. Calculate the new temperature. According to Charles' law (8.21 *a*)

$$T_2 = V_2 \cdot \frac{T_1}{V_1} = (323\ K)\frac{40.8\ dm^3}{22.4\ dm^3}$$

$$= 740.9\ K$$

or $$t_2 = 740.9 - 273 = 467.9°C$$

8.3.3. *The Ideal Gas Law*

Boyle's law and Charles' law or Gay-Lussac's law can be combined together, *i.e.* for fixed mass of gas

$$\frac{pV}{T} = \text{Constant} \qquad ...(8.22)$$

The condition of fixed mass can be removed with the help of Avogadro's hypothesis, which states that under the same conditions of temperature and pressure, equal volumes of gases contain same number of molecules. Thus equation (8.22) becomes

$$\frac{pV}{nT} = R \qquad ...(8.23)$$

where n is the number of mol and R is a gas constant. The *number of mol* is defined as the ratio of mass (w) of gas to its molecular weight (M), *i.e.*

$$n = \frac{w}{M} \qquad ...(8.24)$$

For 1 mol of gas, the ideal gas equation is

$$p\bar{V} = RT$$

where $\bar{V}$ is the volume of 1 mol of gas or molar volume.

***Exercise* 8.11:** If the volume is measured in dm^3, pressure in atmosphere, temperature in degree Kelvin, what will be the unit for R and calculate its numerical value? Ideal gas law equation is

$$pV = nRT$$

or $$R = \frac{pV}{nT}$$

Units of R will be atm dm^3 K^{-1} mol^{-1}. Now according to Avogadro's hypothesis, 1 mol of every gas at 0°C and 760 mm of pressure occupies 22.4 dm^3 of volume. Substituting this in equation (8.23)

$$R = \frac{p_0 V_0}{nT_0}$$

$$= \frac{(1\ \text{atm.})(22.414\ dm^3)}{(1\ \text{mol})(273.15\ K)}$$

$$= 0.08205\ dm^3\ atm\ mol^{-1}\ K^{-1}$$

R can be expressed in other units and its values are listed in Table 8.1.

TABLE 8.1. Values of *R* in other units

Type of units	Value	Units
Mechanical	0.082054	dm^3 atm mol^{-1} K^{-1}
Mechanical	82.054	mL atm mol^{-1} K^{-1}
cgs	8.3144×10^7	ergs mol^{-1} K^{-1}
Electrical	8.3144	Joules mol^{-1} K^{-1}
Thermal	1.9872	cal mol^{-1} K^{-1}

***Exercise* 8.12:** Prove the relation $\frac{p_1V_1}{T_1} = \frac{p_2V_2}{T_2}$ using Boyle's law and Charles' law.

Divide the process $(p_1V_1T_1) \rightarrow (p_2V_2T_2)$ as $(p_1V_1T_1) \rightarrow (p_2V'T_1) \rightarrow (p_2V_2\,T_2)$

State 1 to state 2 is Boyle's law *i.e.*,

$$p_1V_1 = p_2V'$$

or

$$V' = \frac{p_1V_1}{p_2}$$

State 2 to state 3 is Charles' law, *i.e.*

$$\frac{V'}{T_1} = \frac{V_2}{T_2}$$

Substituting the value of V', *i.e.*

$$\frac{p_1V_1}{p_2T_1} = \frac{V_2}{T_2} \text{ or } \frac{p_1V_1}{T_1} = \frac{p_2V_2}{T_2} \quad ...(8.25)$$

***Exercise* 8.13:** A cylinder contains 100 g of an ideal gas (mol wt. 40 g mol^{-1}) at 27°C and 2 atm pressure. In transference process, the cylinder fell and a dent was created *i.e.*, there was a decrease in the volume of cylinder. But valve attached to the cylinder cannot keep the pressure greater than 2 atm, so 10 g of gas leaked out. (*a*) Calculate volume of the cylinder before and after dent; (*b*) Had the valve been pretty strong, what would have been the pressure after dent? Temperature remained constant during this process.

(*a*) Initial volume can be calculated from the relation;

$$pV = nRT$$

$$V_1 = \left(\frac{100\text{ g}}{40\text{ g mol}^{-1}}\right)\frac{(0.0821\text{ dm}^3\text{ atm deg}^{-1}\text{ mol}^{-1})\,(300\text{ K})}{(2\text{ atm})}$$

$$= 30.79\text{ dm}^3$$

Mass of gas after dent = 100 – 10 = 90 g

so

$$V_2 = \left(\frac{90\,\text{g}}{40\,\text{g mol}^{-1}}\right)\left(\frac{(0.0821\,\text{dm}^3\ \text{atm K}^{-1}\,\text{mol}^{-1})(300\,\text{K})}{(2\,\text{atm})}\right)$$

$$= 27.71\ \text{dm}^3$$

(*b*) If there would have been no leak, then

$$p_1 = 2\ \text{atm},\ p_2 = ?$$

$$V_1 = 30.79\ \text{dm}^3,\ V_2 = 27.71\text{dm}^3$$

Using equation (8. 19*a*)

$$p_2 = \frac{(2\,\text{atm})\,(30.79\,\text{dm}^3)}{27.71\,\text{dm}^3} = 2.22\ \text{atm.}$$

***Exercise* 8.14:** A certain amount of ideal gas occupies a volume of 2.56 dm ^{3} at a pressure of 200 torr and temperature 40°C. The gas is compressed to a volume of 1.6 dm ^{3} by applying pressure 400 torr. Calculate the final temperature.

Now $\quad p_1 = 200\ \text{torr},\ V_1 = 2.56\ \text{dm}^3,\ T_1 = 273 + 40$

$$= 313\ \text{K}$$

$$p_2 = 400\ \text{torr},\ V_2 = 1.6\ \text{dm}^3,\ T_2 = ?$$

Using equation (8.25)

$$T_2 = p_2 V_2 \cdot \frac{T_1}{p_1 V_1}$$

$$= \frac{(400\ \text{torr})(1.6\ \text{dm}^3)(313\ \text{K})}{(200\ \text{torr})(2.56\ \text{dm}^3)}$$

$$= 391.25\ \text{K}$$

$$t_2 = 391.25 - 273 = 118.25°\text{C.}$$

***Exercise* 8.15:** 4 g of an ideal gas was introduced into a bulb of volume 10 dm ^{3} at certain pressure p and temperature T. The above bulb was placed in a thermostat maintained at temperature (T + 125) K. 0.8 g of the gas was let off to keep the original pressure. Calculate p and T if the molecular weight of the gas 40 g mol^{-1}.

Number of mol originally $\quad = \dfrac{4\ \text{g}}{40\,\text{g mol}^{-1}} = 0.1\,\text{mol}$

Number of mol in the end $\quad = \dfrac{3.2\ \text{g}}{40\,\text{g mol}} = 0.08\,\text{mol}$

i.e., $\quad n_1 = 0.1\ \text{mol}\ p_1 = p,\ V_1 = 10\ \text{dm}^3,\ T_1 = T_1 K$

$$n_2 = 0.08\ \text{mol}\ p_2 = p, V_1 = 10\ \text{dm}^3,\ T_2 = (T_1 + 125)\ \text{K}$$

Since the initial and final pressure and volume are same the ideal gas equation (8.23) becomes

$$n_1 T_1 = n_2 T_2$$

i.e., $$0.1\,T_1 = 0.08\,(T_1 + 125)$$
$$(0.1 - 0.08)\,T_1 = 0.08 \times 125$$
$$T_1 = \frac{0.08 \times 125}{0.02} = 500 \text{ K}$$

The ideal gas equation can be written as

$$p = \frac{nRT}{V} = \frac{(0.1 \text{ mol})\,(0.0821 \text{ atm dm}^3 \text{ K}^{-1} \text{ mol}^{-1})\,(500 \text{ K})}{10 \text{ dm}^3}$$
$$= 0.411 \text{ atm.}$$

***Exercise* 8.16:** An unknown ideal gas of volume 0.202 dm^3 at 625 torr and 40°C weighs 0.2058×10^{-3} kg. Calculate its molecular weight.

Molecular weight can be calculated in two ways:

METHOD I: It is known that at 760 torr and at 0°C (273 K), 22.4 dm^3 gas weighs to its molecular weight. So first calculate the volume of the unknown gas at these conditions:

$$p_1 = 625 \text{ torr},\ V_1 = 0.202 \text{ dm}^3,\quad T_1 = 273 + 40 = 313 \text{ K}$$
$$p_2 = 760 \text{ torr},\ V_2 = ?,\quad T_2 = 273 \text{ K}$$

Using equation (8.25)

$$V_2 = \frac{p_1 V_1}{T_1} \cdot \frac{T_2}{p_2} = \frac{(625 \text{ torr})\,(0.202 \text{ dm}^3)\,(273 \text{K})}{(313 \text{K})(760 \text{ torr})}$$
$$= 0.1457 \text{ dm}^3$$

0.1457 dm^3 weighs 0.2058×10^{-3} kg

$$22.40 \text{ dm}^3 \text{ weighs } \frac{(0.2058 \times 10^{-3} \text{ kg})\,(22.4 \text{ dm}^3)}{(0.1457 \text{ dm}^3)}$$
$$= 31.58 \times 10^{-3} \text{ kg}$$

i.e., molecular weight = 32 g mol^{-1}

METHOD II: See Section 8.3.6.

8.3.4. *Dalton's Law of Partial Pressure*

The total pressure of a mixture of ideal gases is the sum of the partial pressure of each gas *i.e.*,

$$P = p_1 + p_2 + p_3 + \ldots + p_i + \ldots \qquad \ldots(8.26)$$
$$= \Sigma p_i$$

where p_i is called partial pressure of *i*th gas in the mixture. The partial pressure of any component in a mixture of ideal gases can be related to the total pressure by the relation:

$$p_i = x_i P \qquad \ldots(8.27)$$

where x_i is the mol fraction of *i*th gas and defined as

$$x_i = \frac{\text{Number of mol of } i\text{th gas}}{\text{Total number of mol of the system}} = \frac{n_i}{\sum_i n_i} \qquad \ldots(8.28)$$

***Exercise* 8.17:** A flask has a capacity of 4 dm^3. If 4 g of O_2 and 10 g of N_2 at 27°C are added to it, calculate the (*a*) mole fraction of each gas, (*b*) calculate the partial pressure, (*c*) total pressure due to this mixture.

$$\text{Number of Mol of } O_2 = \frac{4\ g}{32\ g\ mol^{-1}} = 0.125\ \text{mol}$$

$$\text{Number of Mol of } N_2 = \frac{10\ g}{28\ g\ mol^{-1}} = 0.357\ \text{mol}$$

$$x_{O_2} = \frac{0.125\ \text{mol}}{(0.125+0.357)\ \text{mol}} = \frac{0.125}{0.482} = 0.259$$

$$x_{N_2} = \frac{0.357\ \text{mol}}{(0.125\ \text{mol}+0.357)\ \text{mol}} = 0.741$$

According to ideal gas law equation

$PV = nRT$ (where n is total number of mol).

$$P = \frac{nRT}{V} = \frac{(0.482\ \text{mol})\ (0.082\ \text{dm}^3\ \text{atm mol}^{-1}\ \text{K}^{-1})\ (300\ \text{K})}{4\ \text{dm}^3}$$

$$= 2.966\ \text{atm} = 3.005 \times 10^5\ \text{N m}^{-2}$$

$$p_{O_2} = x_{O_2} . P$$

$$= 0.259 \times 2.966\ \text{atm}$$

$$= 0.768\ \text{atm} = 0.778 \times 10^5\ \text{N m}^{-2}$$

$$p_{N_2} = x_{N_2} . P = 0.741 \times 2.966\ \text{atm}$$

$$= 2.198\ \text{atm} = 2.227 \times 10^5\ \text{N m}^{-2}$$

This can also be calculated from the relation

$$p_i V = n_i RT$$

where p_i and n_i are the partial pressure and number of mol of ith component.

8.3.5. *Graham's Law of Effusion*

According to Graham's law of effusion, the rates of effusion (or leaking through an orifice) of gases at the same temperature and pressure is inversely proportional to the square root of density *i.e.*, if v_1 and v_2 are rates of effusion of two gases, d_1 and d_2 are densities, then

$$\frac{v_1}{v_2} = \sqrt{d_2/d_1} \qquad \text{...(8.29)}$$

and also

$$\frac{d_2}{d_1} = \frac{M_2}{M_1} \qquad \text{...(8.30)}$$

The common relation is derived with the help of equation (8.30) *i.e.*, at the same temperature and pressure, the density of each gas is proportional to its molecular weight. Thus,

$$\frac{v_1}{v_2} = \sqrt{M_2 / M_1} \qquad ...(8.31)$$

General equation for effusive rate, if the two gases are at different pressures, is

$$\frac{v_1}{v_2} = \frac{p_1}{p_2}\sqrt{M_2 / M_1} \qquad ...(8.32)$$

***Exercise* 8.18:** Oxygen (one dm^3 at 1 atm) takes two minutes to effuse through a orifice. How much time will be taken by N_2, He and SF_6 vapour to effuse under the same conditions?

$$\text{Rate of effusion} = \frac{\text{Volume effused}}{\text{Time taken}}$$

$$v_{O_2} = \frac{1\,dm^3}{2\,min} = 0.5\ dm^3\ min^{-1}$$

Molecular weight of

$$O_2 = 32\ g\ mol^{-1}$$
$$SF_6 = 146\ g\ mol^{-1}$$
$$He = 4\ g\ mol^{-1}$$
$$N_2 = 28\ g\ mol^{-1}$$

Using equation (8.31)

$$v_{He} = (0.5\ dm^3\ min^{-1})\left(\frac{32\,g\ mol^{-1}}{4\,g\ mol^{-1}}\right)^{1/2}$$

$$= 1.4140\ dm^3\ min^{-1}$$

Time taken by 1 dm^3 of He will be

$$= \frac{V}{\text{rate}} = \frac{1\ dm^3}{1.414\ dm^3\ min^{-1}} = 0.7\ min$$

Similarly time taken by N_2 and SF_6 will be 1.87 min and 4.27 min respectively.

8.3.6. *Molecular Weight of an Ideal Gas*

Ideal gas equation for *n* mol of gas is

$$pV = nRT \qquad ...(8.23)$$

$$\text{Number of mol } (n) = \frac{\text{Weight}}{\text{Molecular weight}} = \frac{w}{M} \qquad ...(8.24)$$

Combining equations (8.23) and (8.24)

$$pV = \frac{w}{M}RT \qquad ...(8.33)$$

or

$$M = \frac{RTw}{pV}$$

Rewriting equation (8.33) as

$$p = \left(\frac{w}{V}\right).\frac{RT}{M} = d\frac{RT}{M}$$

or
$$\frac{d}{p} = \frac{M}{RT}$$

where d is the density of the gas and (d/p) for ideal gas is independent of pressure.

***Exercise* 8.19:** A 1 dm³ sample of air at a pressure of 1 atm and 27°C weighs 0.0012 kg. Calculate the effective molecular weight of air if air behaves as an ideal gas.

Using equation (8.33)

$$M = \frac{(0.0012\,\text{kg})\,(10^3\,\text{g kg}^{-1})\,(0.0821\,\text{atm dm}^3\,\text{K}^{-1}\,\text{mol}^{-1})\,(300\,\text{K})}{(1\,\text{atm})\,(1\,\text{dm}^3)}$$

$$= 29.6\ \text{g mol}^{-1}$$

8.4. REAL GASES

The behaviour of real gases can be expressed by many equations of states, valid only under certain conditions of temperature and pressure. Few of them are mentioned below:

8.4.1. *van der Waals Equation*

van der Waals equation for 1 mol and for n mol respectively are

$$\left(p + \frac{a}{\overline{V}^2}\right)(\overline{V} - b) = RT \qquad \text{...(8.34)}$$

$$\left(p + \frac{n^2 a}{V^2}\right)(V - nb) = nRT \qquad \text{...(8.35)}$$

where $\overline{V}$ is the volume of 1 mol, a and b are constants. The constant a is a correction for the attractive forces between molecules and b for the effective size of the molecules, also called 'excluded volume'.

The physical meanings of the van der Waals constants can be easily obtained as follows:

Equation (8.34) can be rewritten as

$$\overline{V} = \frac{RT}{p + \dfrac{a}{\overline{V}^2}} + b \qquad \text{...(8.36)}$$

If we now take the limit $p \to \infty$, *i.e.* $\overline{V} \to 0$, the denominator of the first term on right hand side becomes ∞ and $\dfrac{RT}{\infty} = 0$, equation (8.36) reduces to

$$\lim_{p\to\infty} \overline{V} = b \qquad \text{...(8.37)}$$

i.e., when the gas is compressed with a very large pressure the molar volume cannot decrease below the value of b. Equation (8.34) can also be rewritten as

$$p = \frac{RT}{(\bar{V} - b)} - \frac{a}{\bar{V}^2} \qquad ...(8.38)$$

At low pressure, when $\bar{V}$ becomes large, $\bar{V} - b \rightarrow \bar{V}$ as the value b becomes very small as compared to $\bar{V}$. Equation (8.38) can be written as

$$\xrightarrow[\bar{V}\to\infty]{Lim} p = \left[\frac{RT}{\bar{V}} - \frac{a}{\bar{V}^2}\right] = p_{\text{ideal}} - \frac{a}{\bar{V}^2} \qquad ...(8.39)$$

i.e., the pressure exerted by the real gas will be less than the pressure exerted by the ideal gas. The second term in equation (8.39) relates to the attractive forces acting between the molecules. Value of van der Waals constants of some gases are compiled in Table 8.2.

TABLE 8.2. van der Waals constants for some of the gases

Gas	b(L mol^{-1})	a(L^2 bar mol^{-2})	Gas	b(L mol^{-1})	a(L^2 bar mol^{-2})
He	0.0238	0.0346	H_2O	0.03049	5.537
Ne	0.01672	0.208	CO_2	0.04286	3.658
Ar	0.03201	1.355	C_2H_2	0.0522	4.516
Kr	0.0396	2.325	C_2H_6	0.0638	5.489
N_2	0.0387	1.37			

It may be mentioned here that *correlation between the melting points of the gases and the values a, obtained from fitting pressure, volume and temperature data suggests that a is associated with the magnitude of the attractive interactions between the molecules*.

***Exercise* 8.20:** Calculate the value of the 'excluded volume'.

Consider the molecules A and B of equal radius r of any gas. Two molecules when touching from any direction will always have their centres separated by 2 r. This means that effective volume occupied by two molecules is $\frac{4}{3}\pi(2r)^3$ *i.e.*, $8.\frac{4}{3}\pi r^3$.

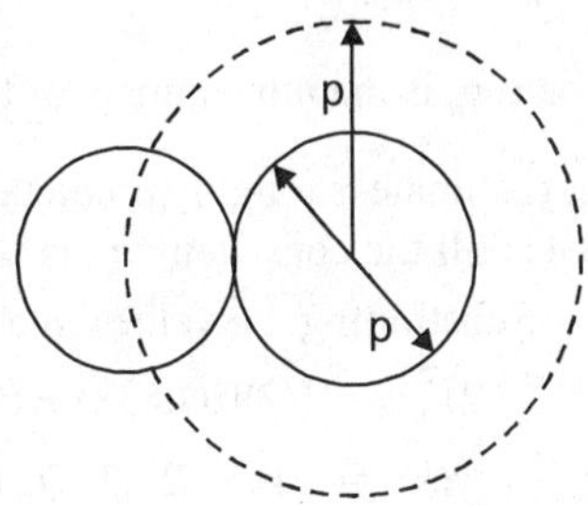

The effective volume occupied by each molecule

$$= \frac{1}{2}\, 8.\frac{4}{3}\pi r^3 = 4.\frac{4}{3}\pi r^3.$$

The volume of the individual isolated molecule,

$$= \frac{4}{3}\pi r^3$$

Thus the excluded volume is four times the actual volume of a molecule.

***Exercise* 8.21:** The molecular radius of O_2 is 3.81×10^{-10} m. Calculate the excluded volume per mol of O_2.

The excluded volume per molecule

$$= 4 \times \text{Actual volume of molecule} = 4.\frac{4}{3}\pi r^3$$

$$= (4)\left(\frac{4}{3}\right)\left(\frac{22}{7}\right)(3.6\times10^{-10}\ \text{m})^3 = 7.82\times10^{-25}\ \text{dm}^3$$

Excluded volume per mol = $(7.82 \times 10^{-25}\ \text{dm}^3)(6.023 \times 10^{23}\ \text{mol}^{-1})$
$= 0.48\ \text{dm}^3\ \text{mol}^{-1}$.

***Exercise* 8.22:** Calculate the pressure exerted by 2 dm^3 mol^{-1} of ethane at 27°C if it follows van der Waals equation. $a = 5.489$ dm^6 atm mol^{-2}, $b = 0.0638$ dm^3 mol^{-1}. Compare the value if the gas is ideal. The equation (8.34) can be rewritten as

$$p = \frac{RT}{\bar{V}-b} - \frac{a}{\bar{V}^2}$$

$$= \frac{(0.0821\ \text{dm}^3\ \text{atm mol}^{-1}\ \text{K}^{-1})(300\text{K})}{(2.0-0.0638)\ \text{dm}^3\ \text{mol}^{-1}} - \frac{5.489\ \text{dm}^6\ \text{atm mol}^{-2}}{(2\ \text{dm}^3\ \text{mol}^{-1})^2}$$

$$= 12.72 - 2.7445 = 9.976\ \text{atm}$$

From ideal gas equation

$$p = \frac{RT}{\bar{V}} = \frac{(0.0821\ \text{atm dm}^3\ \text{mol}^{-1}\ \text{K}^{-1})(300\text{K})}{2\ \text{dm}^3\ \text{mol}^{-1}}$$

$$= 12.315\ \text{atm.}$$

***Exercise* 8.23:** Calculate the volume of 1 mol of ethane at 2 atm pressure and at 27°C if it obeys van der Waals equation. Constants a and b are given in exercise 8.22.

Equation (8.34) can be rewritten as

$$p\bar{V}^3 - (pb + RT)\ \bar{V}_2 + a\bar{V} - ab = 0 \qquad \text{...(8.40)}$$

The above equation is a cubical equation. There are many ways to solve it but Newton-Raphson iterative process will be used here. For any equation $f(x) = 0$, the $(n + 1)$th estimate of the root is given by

$$x_{n+1} = x_n - \frac{f(x_n)}{f'(x_n)} \qquad \text{...(8.41)}$$

where x_n is the nth estimate of the root, $f(x_n)$ is the function $f(x) = 0$ at x_n and $f'(x_n) = \frac{df}{dx}$ at $x = x_n$. The first estimate is roughly predicted and the further estimates are calculated by making use of the equation (8.41) till the consistent value is obtained.

Substituting the values in equation (8.40)

$$2\bar{V}^3 - [(2)(0.0638) + (0.0821)(300)]\ \bar{V}^2 + 5.489\ \bar{V} - (0.0638)(5.489) = 0$$

$$f(x) = 2\bar{V}^3 - 24.7576\ \bar{V}^2 + 5.489\ \bar{V} - 0.35 = 0 \qquad \text{...(8.40}a\text{)}$$

$$f'(x) = 6\ \bar{V}^2 - 49.5152\ \bar{V} + 5.489 = 0 \qquad \text{...(8.40}b\text{)}$$

The first estimate can be predicted by using ideal gas equation

$$\bar{V} = \frac{(1\ \text{mol})(0.0821\ \text{dm}^3\ \text{atm K}^{-1}\ \text{mol}^{-1})(300\text{K})}{(2\ \text{atm})} = 12.315\ \text{dm}^3$$

So using equation (8.41)

$$V_{n+1} = V_n - \frac{2\overline{V}^3 - 24.7576\,\overline{V}^2 + 5.489\,\overline{V} - 0.35}{6\overline{V}^2 - 49.515\,\overline{V} + 5.489}$$

Now taking $V_1 = 12.315$ dm^3 mol^{-1}

$$V_2 = 12.315 - \frac{(2)(12.315)^3 - (24.7576)(12.315)^2 + (5.489)(12.315) - 0.35}{(6)(12.315)^2 - (49.5152)(12.315) + 5.489}$$

$$= 12.315 - 0.155 = 12.16 \text{ dm}^3 \text{ mol}^{-1}$$

Similarly,

$$V_3 = 12.16 - \frac{(2)(12.16)^3 - (24.7576)(12.16)^2 + (5.489)(12.16) - 0.35}{(6)(12.16)^2 - (49.5152)(12.16) + 5.489}$$

$$= 12.16 - 0.0058 = 12.154 \text{ dm}^3 \text{ mol}^{-1}$$

The difference between V_3 and V_2 is so small that it can be considered accurate enough.

8.4.2. *Virial Equation*

Virial equation can be expressed as a function of pressure or as a function of volume *i.e.*,

$$p\overline{V} = RT\left(1 + \frac{B}{\overline{V}} + \frac{C}{\overline{V}^2} + \frac{D}{\overline{V}^3} + ...\right) \qquad ...(8.42)$$

$$p\overline{V} = RT\left(1 + \frac{B}{RT}p + \frac{C}{(RT)^2}p^2 + \frac{D}{(RT)^3}p^3 + ...\right) \qquad ...(8.43)$$

where *B*, *C*, *D*,... are second, third, fourth virial coefficients and can be determined with the help of statistical mechanics, experimental result or from van der Waals equation. Of course these have different values in two equations.

8.4.3. *Critical Points*

The plot between pressure and volume at constant temperature is known as an isotherm and for ideal gases this plot is a smooth curve (Fig. 8.2). The same plot for real gases exhibits smoothness at higher temperature but as the gases are cooled, these nicely shaped isotherms become distorted. Generally at one particular temperature, known as critical temperature, a point of inflection is obtained *i.e.*, $\left(\frac{\partial p}{\partial V}\right)_{T_c} = \left(\frac{\partial^2 p}{\partial V^2}\right)_{T_c} = 0$. The values of volume and pressure at this temperature are known as critical volume and critical pressure. These critical variables can be calculated from the equations of state. Thus, the highest temperature at which liquefaction is still possible, is called the critical temperature. At this temperature, the discontinuity in the isotherm at the point of condensation will have been reduced to a single point, which will be an inflection point in the curve.

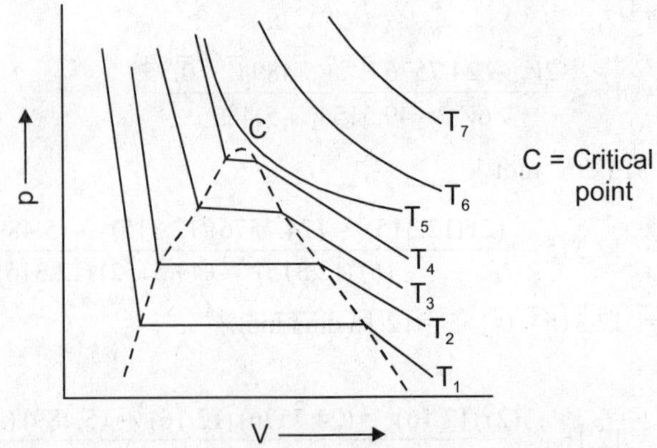

Fig. 8.2 *Plot of pressure vs volume at different temperature*

***Exercise* 8.24:** Calculate the values of van der Waals constant a and b by using the values of critical parameters. van der Waals equation for one mol of real gas can be rewritten as

$$p = \frac{RT}{\overline{V}-b} - \frac{a}{\overline{V}^2}$$

$$\left(\frac{\partial p}{\partial V}\right)_T = -\frac{RT}{(\overline{V}-b)^2} + \frac{2a}{\overline{V}^3} = 0$$

$$\left(\frac{\partial^2 p}{\partial V^2}\right)_T = \frac{2RT}{(\overline{V}-b)^3} - \frac{6a}{\overline{V}^4} = 0$$

Under these conditions, the values of volume ($\overline{V}$), temperature (T) and pressure (p) are critical values and can be denoted as $\overline{V}_c$, p_c and T_c. Solving the three equations simultaneously

$$b = \frac{\overline{V}_c}{3},\ a = 3p_c\overline{V}_c^2,\ R = \frac{8p_c\overline{V}_c}{3T_c}$$

The value of R can be calculated from other experiments, thus elimination $\overline{V}_c$ we get

$$a = \frac{27R^2T_c^2}{64p_c},\ b = \frac{RT_c}{8p_c} \quad \text{...(8.44)}$$

or

$$V_c = 3b,\ p_c = \frac{a}{27b^2},\ T_c = \frac{8a}{27bR} \quad \text{...(8.45)}$$

The equation of state in terms of critical constants can be written as

$$p_c V_c = 3b.\frac{a}{27b} = \frac{a}{9b}$$

$$= \frac{1}{9}.\frac{27R^2T_c^2}{64p_c}.\frac{8p_c}{RT_c}$$

$$= \frac{3}{8}RT_c \quad \text{...(8.46)}$$

8.4.4. *Compressibility Factor*

The van der Waals and other equatio.is of states are quite complicated. So to keep the equation of state simple (to describe the behaviour of real gases and finally resembling the equation of state for ideal gas), a factor called compressibility factor (Z) is introduced. The equation of state becomes:

$$pV = ZnRT \qquad ...(8.47)$$

For ideal gases,

$$\frac{pV}{nRT} = Z = 1 \qquad ...(8.48)$$

and for real gases

$$\underset{p \to 0}{\frac{pV}{nRT}} = Z = 1 \qquad ...(8.49)$$

Equation (8.49) can be written as

$$Z(T, p) = \frac{\overline{V}}{\overline{V}_{\text{ideal}}} = \frac{p}{p_{\text{ideal}}} \qquad ...(8.49a)$$

compression factor can be either greater or less than unity. When $Z(T, p) < 1$, the actual molar volume of the real gas is lessthan that of ideal gas under similar conditions. Compression factors less than unity are usually observed in gas molecules with a relatively large intermolecular forces that contract the molecules, thereby producing a volume less than that which would be observed for an ideal gas. Compression factors greater than unity are the result of the non-zero volume of the molecules. When this factor becomes important compression factors larger than unity are observed.

8.4.5. *The Law of Corresponding States*

Using the values of a and b, calculated above, the van der Waals equation can be written as

$$p = \frac{8p_c\overline{V}_cT}{3T_c(\overline{V} - \overline{V}_c/3)} - \frac{3p_c\overline{V}_c^{\,2}}{\overline{V}^2} \qquad ...(8.50)$$

and this equation can also be written as

$$\frac{p}{p_c} = \frac{8(T/T_c)}{3(\overline{V}/\overline{V}_c) - 1} - \frac{3}{(\overline{V}/\overline{V}_c)^2} \qquad ...(8.51)$$

or

$$\pi = \frac{8\tau}{3\phi - 1} - \frac{3}{\phi^2} \qquad ...(8.52)$$

where

$$\pi = \frac{p}{p_c}, \ \tau = \frac{T}{T_c} \text{ and } \phi = \frac{\overline{V}}{\overline{V}_c} \qquad ...(8.53)$$

are called reduced variables. The equation (8.50) is a general equation of state since it does not contain any variable which depends on the properties of any gas. The beauty of law of corresponding states is that gases at the same reduced pressure and temperature must have same reduced volume.

The plots of compressibility factor versus reduced pressure at various reduced temperatures are plotted in Fig. 8.3. This helps in calculating the various data regarding the real gases.

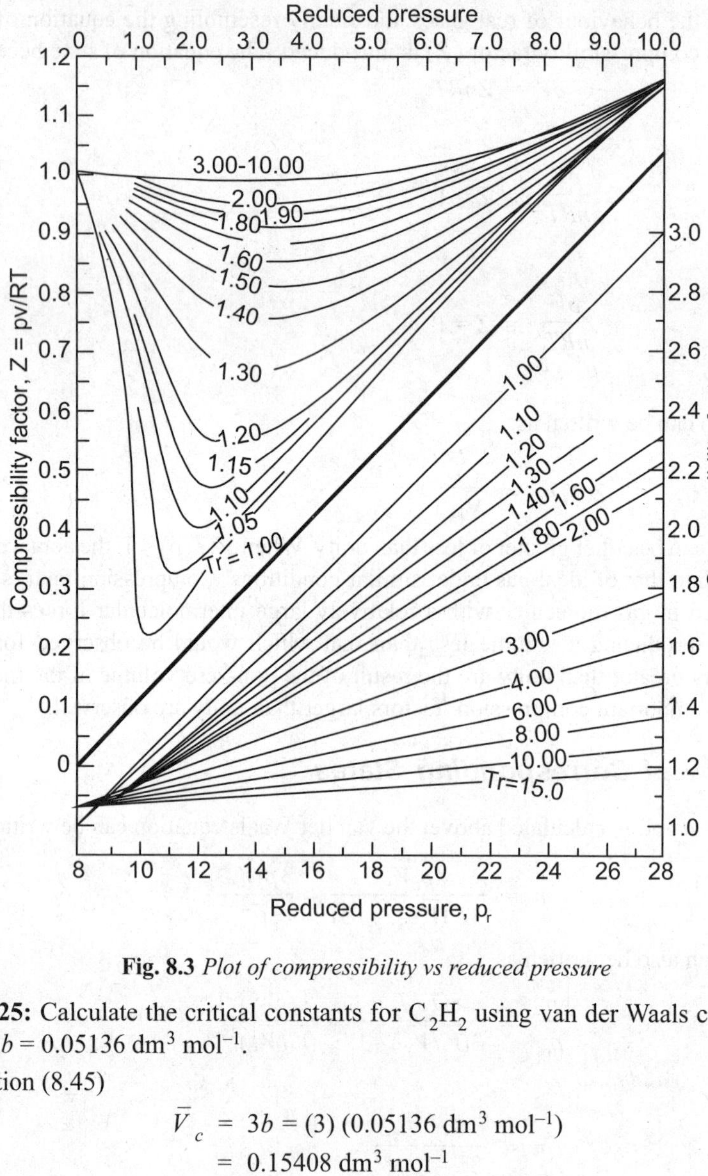

Fig. 8.3 *Plot of compressibility vs reduced pressure*

***Exercise* 8.25:** Calculate the critical constants for C_2H_2 using van der Waals constants $a = 4.390$ dm^6 atm mol^{-2}, $b = 0.05136$ dm^3 mol^{-1}.

Using equation (8.45)

$$\bar{V}_c = 3b = (3)\,(0.05136 \text{ dm}^3 \text{ mol}^{-1})$$

$$= 0.15408 \text{ dm}^3 \text{ mol}^{-1}$$

$$p_c = \frac{a}{27b^2} = \frac{4.390 \text{ dm}^6 \text{ atm mol}^{-2}}{(27)\,(0.05136 \text{ dm}^2 \text{ mol}^{-1})^2}$$

$$= 61.638 \text{ atm.}$$

$$T_c = \frac{8a}{27bR} = \frac{(8)(4.39 \text{ dm}^6 \text{ atm mol}^{-2})}{(27)(0.05136 \text{ dm}^3 \text{ mol}^{-1})(0.0821 \text{ atm dm}^3 \text{ mol}^{-1} \text{ K}^{-1})}$$

$$= 308.48 \text{ K}$$

***Exercise* 8.26:** The critical constants for benzene are $T_c = 561.6$ K and $p_c = 47.89$ atm. Calculate the van der Waals constants.

Using equation (8.44)

$$a = \frac{27R^2T_c^2}{64p_c}$$

$$= \frac{(27)(0.0821 \text{ atm dm}^3 \text{ mol}^{-1} \text{ K}^{-1})^2 (561.6 \text{ K})^2}{(64)(47.89 \text{ atm})} = 18.73 \text{ dm}^6 \text{ atm mol}^{-2}$$

$$b = \frac{RT_c}{8p_c} = \frac{(0.0821 \text{ dm}^3 \text{ atm K}^{-1} \text{ mol}^{-1})(561.6\text{K})}{(8)(47.89 \text{ atm})}$$

$$= 0.12 \text{ dm}^3 \text{ mol}^{-1}.$$

***Exercise* 8.27:** Calculate the second and third virial coefficients in terms of van der Waals constants. Use equation (8.42) as virial equation.

The equation (8.42) can be written as

$$Z = \frac{p\overline{V}}{RT} = 1 + \frac{B}{\overline{V}} + \frac{C}{\overline{V}^2} \qquad ...(8.42a)$$

leaving higher terms.

The van der Waals equation can be written as

$$Z = \frac{p\overline{V}}{RT} = \frac{1}{1-\left(\frac{b}{\overline{V}}\right)} - \frac{a}{RT\overline{V}} \qquad ...(8.54)$$

The first term on right hand side can be expanded as

$$Z = 1 + \frac{b}{\overline{V}} + \left(\frac{b}{\overline{V}}\right)^2 + ... - \frac{a}{RT\overline{V}}$$

$$= 1 + \left(b - \frac{a}{RT}\right)\frac{1}{\overline{V}} + \frac{b^2}{\overline{V}^2} \qquad ...(8.55)$$

leaving higher terms of $b/\overline{V}$ since $b/\overline{V}$ is small.

Comparing equations (8.42) and (8.55)

$$B = b - \frac{a}{RT}, C = b^2 \qquad ...(8.56)$$

***Exercise* 8.28:** Calculate the values of B and C for ethane gas at 27°C where $a = 5.489$ dm^6 atm mol^{-2}, $b = 0.0638$ dm^3 mol^{-1}.

Using the results of exercise 8.27

$$C = b^2 = (0.0638 \text{ dm}^3 \text{ mol}^{-1})^2$$

$$= 4.07 \times 10^{-3} \text{ dm}^6 \text{ mol}^{-2}$$

$$B = 0.0638\,\text{dm}^3\,\text{mol}^{-1} - \frac{5.489\,\text{dm}^6\,\text{atm mol}^{-2}}{(0.0821\,\text{atm dm}^3\,\text{mol}^{-1}\,\text{K}^{-1})(300\,\text{K})}$$

$$= -0.159\,\text{dm}^3\,\text{mol}^{-1}.$$

***Exercise* 8.29:** Calculate expression for the Boyle's temperature for a gas obeying van der Waals gas.

The *Boyle's temperature* is defined as the temperature at which the plot of Z vs p approached $Z = 1$ line asymptotically as p approaches zero, *i.e.*,

$$\left(\frac{\partial Z}{\partial p}\right)_{T,\,p\to 0} \to 0 \qquad ...(8.57)$$

***Note*:** A gas at Boyle's temperature behaves as ideal one over a very large variation of pressure. This is due to the fact that intermolecular forces of attraction and repulsion are compensated.

Now writing van der Waals equation as

$$Z = 1 + \frac{p}{RT}\left(b - \frac{a}{RT}\right) + \left(\frac{b}{RT}\right)^2 p^2 \qquad ...(8.57a)$$

The above equation is obtained by replacing $\bar{V}$ by RT/P in equation (8.55). This is valid only if $\bar{V}$ is large

$$\left(\frac{\partial Z}{\partial p}\right)_T = \frac{1}{RT}\left(b - \frac{a}{RT}\right) + 2\left(\frac{b}{RT}\right)^2 p$$

$$\left(\frac{\partial Z}{\partial p}\right)_{T,\,p\to 0} = 0 = \frac{1}{RT_B}\left(b - \frac{a}{RT_B}\right)$$

i.e.,

$$T_B = \frac{a}{Rb} \qquad ...(8.58)$$

***Exercise* 8.30:** Calculate the value of Boyle's temperature for ethane gas, taking the data from exercise (8.28):

$$T_B = \frac{5.489\,\text{dm}^6\,\text{atm mol}^{-2}}{(0.0821\,\text{atm dm}^3\,\text{K}^{-1}\,\text{mol}^{-1})\,(0.0638\,\text{dm}^3\,\text{mol}^{-1})}$$

$$= 1047.92\,\text{K}.$$

***Exercise* 8.31:** Calculate the molar volume of ethylene (C_2H_4) at 35.6°C and 100 atm using the chart (Fig. 8.3): $p_c = 61.659$ atm, $T_c = 308.6$ K. Calculate also the molar volume if ideal gas equation is obeyed.

Figure 8.3 is a plot of compressibility versus reduced pressure at various reduced temperature. One needs T_r and p_r *i.e.*,

$$T_r = \frac{T}{T_c} = \frac{(273 + 35.6)\,\text{K}}{308.6\,\text{K}} = 1.0$$

$$p_r = \frac{p}{p_c} = \frac{100 \text{ atm}}{61.65 \text{ atm}} = 1.622$$

From Fig. 8.3, the value of Z for $p_r = 1.622$ and $T_r = 1.00$ is 0.26.

Now
$$Z = \frac{p\overline{V}}{RT}$$

or
$$\overline{V} = \frac{(0.26)(0.0821 \text{ atm dm}^3 \text{ mol}^{-1} \text{K}^{-1})(308.6 \text{ K})}{(100 \text{ atm})}$$

$$= 6.59 \times 10^{-2} \text{ dm}^3 \text{ mol}^{-1}$$

Using ideal gas equation

$$\overline{V} = \frac{RT}{P} = \frac{(0.0821 \text{ atm dm}^3 \text{ mol}^{-1} \text{K}^{-1})(308.6 \text{ K})}{(100 \text{ atm})}$$

$$= 25.3 \times 10^{-2} \text{ dm}^3 \text{ mol}^{-1}.$$

***Exercise* 8.32:** The volume and pressure of neon gas at 88.23 K are 9.42 L and 0.6055 atm respectively. Calculate the volume of Ar gas at 300 K and 1.15 atm. The critical constants for Ne and Ar are: T_c(Ne) = 44.45 K, V_c(Ne) = 0.0417 L, p_c(Ne) = 25.9 atm.; T_c(Ar) = 151.15 K, V_c(Ar) = 0.0752 L and p_c(Ar) = 47.7 atm.

Using the data given let us calculate the reduced parameters of each given gas,

$$T_c(\text{Ne}) = \frac{88.23\text{K}}{44.45\text{K}} = 1.985\,;\; p_c(\text{Ne}) = \frac{0.6055 \text{ atm}}{25.9 \text{ atm}} = 0.0234\,;\; V_c(\text{Ne}) = \frac{9.42 \text{ L}}{0.417 \text{ L}} = 226$$

Similarly for Ar gas the reduced parameters are

$$T_c(\text{Ar}) = \frac{300 \text{ K}}{151.15 \text{ K}} = 1.985\,;\; p_c(\text{Ar}) = \frac{1.115 \text{ atm}}{47.7 \text{ atm}} = 0.0234$$

Since the reduced pressure and reduced temperature of Ne and Ar are same, the law of corresponding states requires that the reduced volume of Ar should also be equal to Ne, *i.e.*, 226. Thus the volume of Ar at the given conditions will be

$$V_c(\text{Ar}) = \frac{V(\text{Ar}) \text{ L}}{0.0752 \text{ L}} = 226$$

or
$$V(\text{Ar}) = (226)(0.0752 \text{ L}) = 17.0 \text{ L}.$$

8.4.6. *Molecular Weight and Density of Non-ideal Gases*

At low pressure, the real gases tend to behave like ideal gases. From the pV vs p diagram, it is clear that the plot is not linear. So d/p or $\left(\frac{W}{Vp}\right)$ will not be independent of p. One has to get the value of d/p at $p = 0$ to calculate the density and thus actual molecular weight. At low pressure, the van der Waals equation can be written as

$$p(V - nb) = nRT \qquad ...(8.59)$$

or

$$pV = n(RT + pb) = W/M\,(RT + pb)$$

$$p = W/V\left(\frac{RT}{M} + \frac{bp}{M}\right) = d\left(\frac{RT}{M} + \frac{bp}{M}\right)$$

$$\frac{d}{p} = \frac{M}{RT + bp} = \frac{M/RT}{1 + bp/RT} = \frac{M}{RT}\left(1 + \frac{bp}{RT}\right)^{-1} \qquad ...(8.60)$$

At low pressure $\left(1 + \frac{bp}{RT}\right)^{-1} = \left(1 - \frac{bp}{RT}\right)$

$$\frac{d}{p} = \frac{M}{RT}\left(1 - \frac{bp}{RT}\right) = \frac{M}{RT} - \frac{Mb}{(RT)^2}p \qquad ...(8.61)$$

i.e., plot of $\frac{d}{p}$ *vs p* is linear, slope = $\frac{-Mb}{(RT)^2}$ and intercept = $\frac{M}{RT}$. Molecular weight can be obtained only from intercept because only at $p = 0$, the gas will behave ideal and M is independent of p.

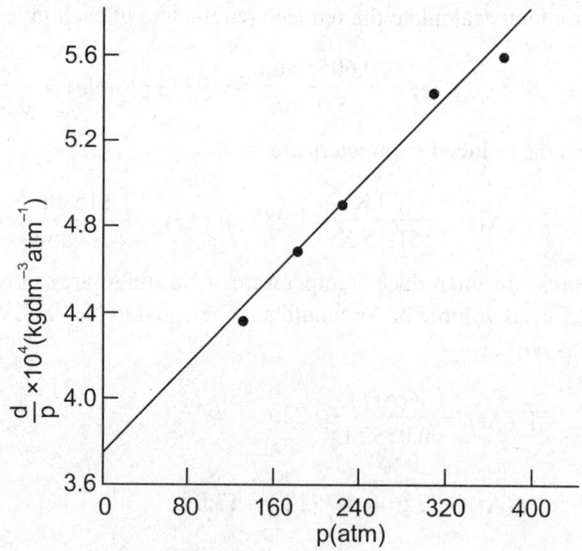

Fig. 8.4 *Plot of d/p vs pressure*

***Exercise* 8.33:** The densities of NH_3 observed at 275°C and at various pressures are mentioned below. Calculate the molecular weight of NH_3 by plotting $\frac{d}{p}$ vs p.

p(atm)	125.4	181.0	228.2	313.9	380
d(kg dm^{-3})	0.0548	0.085	0.1133	0.17	0.2125
$10^4 \times \frac{d}{p}$	4.37	4.696	4.965	5.416	5.59

$\frac{d}{p}$ is plotted vs p in Fig. 8.4. The intercept obtained

$$= 3.74 \times 10^{-4} \text{ kg dm}^{-3} \text{ atm}^{-1}$$

From equation (8.61), intercept $= \frac{M}{RT}$

$$M = (3.74 \times 10^{-4} \text{ kg dm}^{-3} \text{ atm}^{-1})\,(0.0821 \text{ atm dm}^{-3} \text{ K}^{-1} \text{ mol}^{-1})\,(548 \text{ K})$$
$$= 16.83 \times 10^{-3} \text{ kg mol}^{-1}$$

i.e., molecular weight of NH_3 = 17 g mol^{-1}.

8.4.7. *The Residual Volume*

We have learnt earlier that ratio of $\bar{V}$ to $\frac{RT}{p}$ is unity as the pressure approached zero, but the actual difference between these quantities is not zero for real gases. The non-zero difference between $\bar{V}$ and $\frac{RT}{p}$ in the limit of zero pressure or infinite volume is defined as the *residual volume*

$$\bar{V}_{\text{res}} = \xrightarrow[p\to 0]{Lim} \left[\bar{V} - \frac{RT}{p}\right] \qquad \text{...(8.62)}$$

***Exercise* 8.34:** Calculate the expression for residual volume of a van der Waals gas.

The van der Waals gas equation can be written as

$$p\bar{V} - pb + \frac{a}{\bar{V}} - \frac{ab}{\bar{V}^2} = RT \qquad \text{...(8.63)}$$

Divide equation (8.63) by p, we get

$$\bar{V} - b + \frac{a}{p\bar{V}} - \frac{ab}{p\bar{V}^2} = \frac{RT}{p}$$

which can be rewritten as

$$\bar{V} - \frac{RT}{p} = b - \frac{a}{p\bar{V}} + \frac{ab}{p\bar{V}^2}$$

Taking limits on both sides as $p \to 0$, we get

$$\bar{V}_{\text{res}} = \xrightarrow[p\to 0]{Lim} \left[\bar{V} - \frac{RT}{p}\right] = \xrightarrow[p\to 0]{Lim} \left[b - \frac{a}{p\bar{V}} + \frac{ab}{p\bar{V}^2}\right] \qquad \text{...(8.64)}$$

Knowing that $p\bar{V}$ tends to RT when the limit p $\to 0$ is applied (because at very low pressure the ideal gas behaviour is observed), so equation (8.64) can be written as

$$\bar{V}_{\text{res}} = \xrightarrow[p\to 0]{Lim} \left[b - \frac{a}{p\bar{V}} + \frac{ab}{p\bar{V}^2} \right] \qquad ...(8.65)$$

As p tends to zero, we know that the value $\bar{V}$ will tends to infinite, *i.e.*, very large and thus the third term can be neglected. So equation (8.65) reduces to

$$\bar{V}_{\text{res}} = \left[b - \frac{a}{p\bar{V}} \right] \qquad ...(8.66)$$

Just to give physical meaning to each term in equation (8.66), let us write equation (8.66) as

$$\bar{V}_{\text{res}} = \xrightarrow[p\to 0]{Lim} \left[\bar{V} - \frac{RT}{p} \right] = \xrightarrow[p\to 0]{Lim} [\bar{V}_{\text{real}} - \bar{V}_{\text{ideal}}] \qquad ...(8.67)$$

We see that the expression for residual volume (8.66) consists of two terms, b and $-\frac{a}{RT}$. The parameter b, *i.e.*, first term gives the volume occupied by gas and is equal to zero for ideal gas. It is equal to volume of the van der Waals gas, b L mol^{-1}. This parameter has a positive value and thus makes positive contribution to $\bar{V}_{\text{res}}$. The second term arises because of intermolecular attractive forces present in the van der Waals gas, but not in the ideal gas. These attractive forces whose, magnitude is determined by the parameter a, causes the molecules to be pulled closer together. The net effect is that the resulting volume is, therefore, less than would be observed by the ideal gas and contribution to the $\bar{V}_{\text{res}}$ will be negative.

From the above discussion it is evident that *the two terms are having opposite effects. Thus, the resulting value can be negative or positive. At low pressure, it is generally observed that the second term is larger in magnitude than the first term. As a result the* $\bar{V}_{res}$ *is usually negative*.

***Exercise* 8.35:** Using van der Waals data for the (*i*) Ar, (*ii*) CO_2 and (*iii*) of He, calculate the residual volume of these gases,

(*i*) For Ar, b = 0.03201 L mol^{-1}, a = 1.355 L^2 bar mol^{-1}. Using equation (8.66), *i.e.*,

$$\bar{V}_{\text{res}} = \left[b - \frac{a}{RT} \right] = (0.03021 \text{ L mol}^{-1}) - \frac{1.355 \text{ L}^2 \text{ bar mol}^{-2}}{(0.083145 \text{ L bar mol}^{-1} \text{ K}^{-1})(300 \text{ K})}$$

$$= 0.03021 - 0.05432 = -0.0241 \text{ L mol}^{-1}$$

That is value of residual volume is negative.

(*ii*) For CO_2 gas, b = 0.04286 L mol^{-1}; a = 3.658 L^2 bar mol^{-2}

$$\bar{V}_{\text{res}} = \left[b - \frac{a}{RT} \right] = (0.04286 \text{ L mol}^{-1}) - \frac{3.658 \text{ L}^2 \text{ bar mol}^{-2}}{(0.083145 \text{ L bar mol}^{-1} \text{ K}^{-1})(300 \text{ K})}$$

$$= -0.1038 \text{ L mol}^{-1}$$

(*iii*) For He, $b = 0.0238$ L mol^{-1}, $a = 0.0346$ L^2 bar mol^{-2}

$$\bar{V}_{\text{res}} = \left[b - \frac{a}{RT}\right] = (0.0238 \text{ L mol}^{-1}) - \frac{0.346 \text{ L}^2 \text{ bar mol}^{-2}}{(0.083145 \text{ L bar mol}^{-1} \text{ K}^{-1})(300 \text{ K})}$$

$$= 0.0238 - 0.01387 = 0.0093 \text{ L mol}^{-1}$$

It is clear that the molar volume play a major role in the case of He gas whereas in case of Ar and CO_2, the intermolecular attractive forces play the major role .

Since the $\bar{V}_{res}$ and second virial coefficient are similar for van der Waals gas, the temperaturee obtained when $\bar{V}_{res}$ is equal to zero, from the van der Waals gas is called Boyle's temperature. That is a gas at Boyle's temperature behaves as ideal over a very large variation of pressure. This is due to the fact that intermolecular forces of attraction and repulsion are compensated .

$$b = \frac{a}{RT} \qquad \text{...(8.68)}$$

***Exercise* 8.36:** Using the data from Exercise 8.35, calculate the Boyle's temperature. For Ar gas, substituting the values of parameters in equation (8.68), we get

or
$$T = \frac{a}{bR} = \frac{1.355 \text{ L}^2 \text{ bar mol}^{-2}}{(0.03201 \text{ L mol}^{-1})(0.083145 \text{ L bar mol}^{-1} \text{ K}^{-1})} = 509.1 \text{ K}$$

Similarly for CO_2,

$$T = \frac{a}{bR} = \frac{3.658 \text{ L}^2 \text{ bar mol}^{-2}}{(0.04286 \text{ L mol}^{-1})(0.083145 \text{ L bar mol}^{-1} \text{ K}^{-1})} = 1374.43 \text{ K}$$

and for He,
$$T = \frac{a}{bR} = \frac{0.346 \text{ L}^2 \text{ bar mol}^{-2}}{(0.0238 \text{ L mol}^{-1})(0.083145 \text{ L bar mol}^{-1} \text{ K}^{-1})} = 174.85 \text{ K}$$

***Exercise* 8.37:** Calculate the compressibility factor for van der Waals gas.

van der Waals equation (8.34) can be written as

$$(\bar{V} - b)\left(p + \frac{a}{\bar{V}^2}\right) = RT \qquad \text{...(8.34)}$$

or
$$p = \frac{RT}{(\bar{V} - b)} - \frac{a}{\bar{V}^2} \qquad \text{...(1)}$$

or
$$p\bar{V} = \frac{RT\bar{V}}{(\bar{V} - b)} - \frac{a}{\bar{V}} \qquad \text{...(2)}$$

Divide equation (2) by RT, the compressibility factor, *i.e.*,

$$Z(p,T) = \frac{p\bar{V}}{RT} = \frac{\bar{V}}{(\bar{V} - b)} - \frac{a}{RT\bar{V}} \qquad \text{...(8.69)}$$

The equation so obtained explains all the features discussed earlier or discussed in (8.37).

***Exercise* 8.38:** Calculate the compressibility factor for ethane gas with the following data. $\bar{V} = 12.16\ dm^3\ mol^{-1}$, $b = 0.0638\ L\ mol^{-1}$, $a = 5.489\ L^2\ atm\ mol^{-2}$.

Using equation (8.69) and substituting the values of van der Waals parameters, we get

$$Z(p,T) = \frac{12.16\ dm^3\ mol^{-1}}{(12.16 - 0.0638)\ dm^3\ mol^{-1}} - \frac{5.489\ dm^6\ atm\ mol^{-2}}{(0.0821\ dm^3\ atm\ mol^{-1}\ K^{-1})(300\ K)(12.16\ dm^3\ mol^{-1})}$$

$$= 1.0053 - 0.0018 = 1.0035$$

It may be mentioned that the agreement of the computed value of $Z(p, T)$ with van der Waals parameters and the experimental results are not that good. Better agreement can be obtained if we use virial equation of state or by using the law of corresponding states.

8.5. KINETIC THEORY OF GASES

8.5.1. *Assumptions*

The model used in the kinetic theory of ideal gases involves following assumptions:

(1) A gas consists of very large number of particles (atoms or molecules).

(2) The actual volume of the gas particles is negligible as compared to the actual volume of the container. This implies that particles can move freely.

(3) There is no interaction (*i.e.*, neither attraction nor repulsion) between the particles *i.e.*, these particles move in a straight lines (Newton's first law of motion).

(4) There are elastic collisions between particles and walls. This implies that total kinetic energy and linear momentum is conserved before and after the collision.

(5) The kinetic energy of the gas is directly proportional to the temperature.

8.5.2. *Pressure of a Gas*

The relation between the pressure (p) exerted by the gas particles moving with root mean square speed (C_{rms}) on the walls of box is

$$p = \frac{Nmc^2_{rms}}{3\,V} \qquad ...(8.70)$$

where V is the volume of container, m, the mass of each particle, N, the number of particles in the container. The equation (8.47) can be written as

$$p = \frac{Mc^2_{rms}}{3\,V} \text{ if } N = N_{av} \qquad ...(8.70a)$$

8.5.3. *Pressure-Volume and Kinetic Energy*

The relation between pressure, volume and kinetic energy of gas derived on the bases of kinetic theory is

$$pV = \frac{2}{3}N\bar{\varepsilon} \qquad ...(8.71)$$

where $\bar{\varepsilon}$ is average kinetic energy of a particle. Also

$$pV = nRT$$

so

$$\frac{2}{3}N\bar{\varepsilon} = nRT = \frac{N}{N_{av}}RT$$

$$\bar{E} = \frac{3}{2}RT = N_{av}\bar{\varepsilon}$$

$$= N_{av}\cdot\frac{1}{2}mc_{rms}^2 = \frac{1}{2}Mc_{rms}^2$$

where $\overline{E} = N_{av}\,\bar{\varepsilon}$, total energy of the 1 mol of particles. Combining

$$p\overline{V} = RT = \frac{1}{3}Mc_{rms}^2 = \frac{2}{3}\text{KE}$$

or

$$c_{rms} = \sqrt{\frac{3RT}{M}} \qquad \text{...(8.72)}$$

M is the molecular weight.

***Exercise* 8.39:** Calculate the root mean square velocity (c_{rms}) of N_2 at temperature 27°C.

Substitute the proper values in equation (8.72)

$$c_{rms} = \left[\frac{(3)\,(8.314\,\text{J mol}^{-1}\,\text{K}^{-1})\,(300\,\text{K})}{(28\,\text{g mol}^{-1})\,(10^{-3}\,\text{kg g}^{-1})}\right]^{1/2}$$

$$= 517\ \text{m s}^{-1}.$$

***Exercise* 8.40:** Using the data of exercise 8.39, calculate the pressure exerted by 1 mol of N_2 contained in 2 dm^3 container.

Using equation (8.70*a*)

$$p = \frac{(28\,\text{g mol}^{-1})\,(10^{-3}\,\text{kg g}^{-1})\,(517\,\text{ms}^{-1})\,(1\,\text{mol})}{(2\,\text{dm}^3)}$$

$$= 3.728 \times 10^3\ \text{J dm}^{-3}$$

This is not the proper unit of pressure so

$$p = \frac{(3.728\times10^3\,\text{J dm}^{-3})\,(0.0821\,\text{dm}^3\,\text{atm k}^{-1}\text{mol}^{-1})}{8.314\ \text{J k}^{-1}\text{mol}^{-1}}$$

$$= 36.81\ \text{atm}.$$

8.5.4. *Velocity Distribution*

The molecules in a gas at any temperature (except $T = 0$ K) had a wide spectrum of velocities. The number of molecules, in a mol possessing velocity between c_x and $c_x + dc_x$ along the component x is given by the Maxwell relation as

$$dN_{cx} = N\,A\,\exp\left(-\frac{1}{2}Mc_x^2/k_bT\right)dc_x \qquad \text{...(8.73)}$$

where N is total number of particles, c_x is the component of velocity along x axis, A is a constant and k_b the Boltzmann constant. In three-dimensions, the Maxwell relation for the fraction of molecules in a mol dN_c/N having a velocity between c and $c + dc$ is

$$\frac{dN_c}{N} = 4\pi\left(\frac{m}{2\pi k_b T}\right)^{3/2} c^2 \exp(-mc^2/2k_bT)dc \qquad ...(8.74)$$

***Exercise* 8.41:** Show that the root mean square (rms) velocity obtained with the help of equation (8.74) and $pV - KE$ relation is same.

The average value of any function using the velocity distribution equation (8.74) is given by

$$c_{\text{rms}}^2 = \int_0^\infty f(u)\frac{dN}{N}$$

Here $\qquad f(u) = c^2$ so

$$c_{\text{rms}}^2 = \int_0^\infty c^2\frac{dN}{N} = \int_0^\infty 4\pi\left(\frac{m}{2\pi k_b T}\right)^{3/2} c^4 \exp(-mc^2/2k_bT)dc \qquad ...(8.74a)$$

$$= 4\pi\left(\frac{m}{2\pi k_b T}\right)^{3/2} \int_0^\infty c^4 \exp(-mc^2/2k_bT)\,dc$$

Let

$$Z^2 = \frac{1}{2}\frac{mc^2}{k_bT} \text{ or } Z = \left(\frac{m}{2k_bT}\right)^{1/2} c$$

$$dZ = \left(\frac{m}{2k_bT}\right)^{1/2} dc \text{ or } dc = \left(\frac{2k_bT}{m}\right)^{1/2} dZ$$

The equation (8.74a) becomes

$$\bar{c}^2 = 4\pi\left(\frac{m}{2\pi k_b T}\right)^{3/2}\left(\frac{2k_bT}{m}\right)^{5/2}\int_0^\infty Z^4 e^{-Z^2}\,dZ$$

Using $\int_0^\infty Z^4 e^{-Z^2}\,dZ = \frac{3}{8}\pi^{1/2}$ from standard integral table

$$\bar{c}^2 = 4\pi\left(\frac{2k_bT}{m}\right)\cdot\frac{1}{(\pi)^{3/2}}\cdot\frac{3}{8}(\pi)^{1/2}$$

$$= \frac{3k_bT}{m}$$

$$(\bar{c}^2)^{1/2} = c_{\text{rms}} = \left(\frac{3k_bT}{m}\right)^{1/2} = \left(\frac{3k_bN_{\text{av}}T}{mN_{\text{av}}}\right)^{1/2}$$

$$= \left(\frac{3RT}{M}\right)^{1/2} \qquad \text{...(8.75)}$$

Same as obtained from $pV - KE$ relation.

***Exercise* 8.42:** Derive an expression for the most probable velocity and relate it to the c_{rms}.

The most probable velocity is obtained by differentiating dN/N with respect to c. Equate the result to zero and find the value of c.

The three dimensional distribution of velocities is given by the equation (8.74) as

$$\frac{dN/N}{dc} = 4\pi c^2 \left(\frac{m}{2\pi k_b T}\right)^{3/2} \exp\left(\frac{-mc^2}{2k_b T}\right)$$

Differentiating R.H.S. with respect to c, one get

$$4\pi\left(\frac{m}{2\pi k_b T}\right)^{3/2}\left[2c - \frac{mc^3}{k_b T}\right]\exp\left(-mc^2/2k_b T\right)$$

Setting the above equation = 0 and substituting c as c_{mp}, we get

$$2c_{\text{mp}} - \frac{mc_{\text{mp}}^3}{k_b T} = 0$$

or

$$c_{\text{mp}} = \left(\frac{2k_b T}{m}\right)^{1/2} = \left(\frac{2RT}{M}\right)^{1/2} \qquad \text{...(8.76)}$$

$$= \left(\frac{3RT}{M}\right)^{1/2}\left(\frac{2}{3}\right)^{1/2}$$

$$= 0.83\ c_{\text{rms}} \qquad \text{...(8.76}a\text{)}$$

***Exercise* 8.43:** Obtain an expression for the average velocity from the equation (8.74) and point the three *i.e.*, c_{mp}, $\bar{c}$ and c_{rms} on the Maxwell distribution of velocity.

Proceeding on the same lines as exercise (8.42)

$$\bar{u} = \int_0^\infty f(u)\, dN/N$$

$$\bar{c} = \int_0^\infty c\frac{dN}{N} = \int_0^\infty 4\pi\left(\frac{m}{2\pi k_b T}\right)^{3/2} c^3 \exp\left(-\frac{1}{2}\frac{mc^2}{k_b T}\right)dc$$

which gives

$$\bar{c} = \left(\frac{8k_b T}{\pi m}\right)^{1/2} = \left(\frac{8RT}{\pi M}\right)^{1/2} = 0.92\ c_{\text{rms}} \qquad \text{...(8.77)}$$

c_{mp}, $\bar{c}$ and c_{rms} has been shown in Fig. 8.5.

***Exercise* 8.44:** Calculate the value of $p(c)$, probability distribution $\left(\dfrac{dN}{dc.N}\right)$ for N_2 at 27°C and $c = 2 \times 10^3$ m s^{-1}.

Making use of equation (8.74),

$$\left(\frac{m}{2\pi k_b T}\right)^{3/2} = \left(\frac{M}{2\pi RT}\right)^{3/2} = \left[\frac{(28\text{ g mol}^{-1})(10^{-3}\text{ kg g}^{-1})}{(2)(\pi)(8.314\text{ J K}^{-1}\text{ mol}^{-1})(300\text{K})}\right]^{1.5}$$

$$= 2.39 \times 10^{-9}\text{ m}^{-3}\text{ s}^3$$

$$\frac{mc^2}{2k_bT} = \frac{Mc^2}{2RT} = \frac{(28\text{ g mol}^{-1})(10^{-3}\text{ kg g}^{-1})(2\times10^3\text{ m s}^{-1})^2}{(2)(8.314\text{ J K}^{-1}\text{ mol}^{-1})(300\text{K})}$$

$$= 22.45$$

$$e^{-22.45} = 1.775 \times 10^{-10}$$

Now

$$p(c) = (4\pi)(2.39 \times 10^{-9}\text{ m}^{-3}\text{ s}^3)(1.775 \times 10^{-10})(2 \times 10^3\text{ m s}^{-1})^2$$

$$= 2.13 \times 10^{-11}\ (\text{m}^{-1}\text{ s})$$

***Exercise* 8.45:** Plot $p(c)$ for N_2 at 27°C and 227°C. Calculate c_{rms}, $\bar{c}$ and c_{mp} at each temperature and show them on the figure.

The plot can be drawn by calculating $p(c)$ for different values of c. The procedure is exactly same as done in exercise 8.44.

At 27°C $c\times10^{-2}$	0	1	2	2.5	3	4	5	6	7	8	9	10	11	and	12
(m s^{-1}) $10^{-4} \times p(c)$	0	2.84	9.6	13.2	16.3	19.6	18.5	14.3	9.4	5.3	2.6	1.1	0.41	and	0.13

$p(c)$ vs c is plotted in Fig. 8.6.

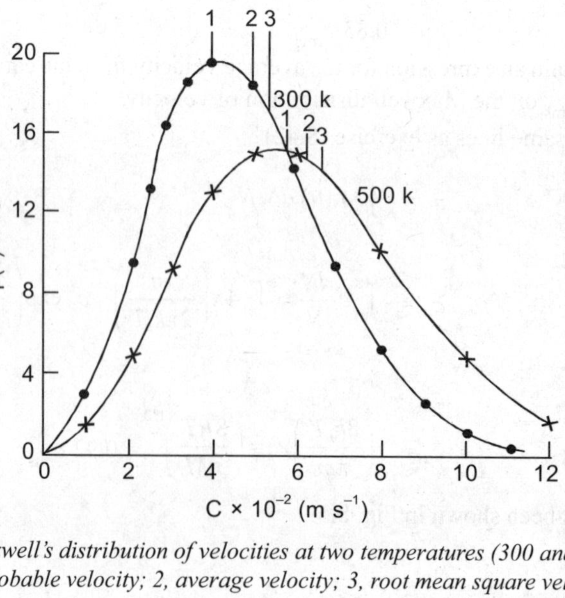

Fig. 8.6 *Maxwell's distribution of velocities at two temperatures (300 and 500 k). 1, most probable velocity; 2, average velocity; 3, root mean square velocity.*

At 227°C $c \times 10^{-2}$	0	1	2	3	4	5	6	8	10	12	14
$(m\ s^{-1})$ $10^4 \times p(c)$	0	1.35	4.88	9.27	13.02	15.02	14.94	10.34	4.81	1.57	0.37

Plot $p(c)$ vs c in Fig. 8.6.

Using equations (8.72), (8.76) and (8.77) at 27°C:

$$c_{rms}(300\text{ K}) = \left[\frac{3RT}{M}\right]^{1/2} = \left[\frac{(3)(8.314\text{ J mol}^{-1}\text{ K}^{-1})(300\text{ K})}{(28\text{ g mol}^{-1})(10^{-3}\text{ kg g}^{-1})}\right]^{1/2} = 516.95\text{ m s}^{-1}$$

$$c_{rms}(500\text{ K}) = \left[\frac{(3)(8.314\text{ J mol}^{-1}\text{ K}^{-1})(500\text{ K})}{(28\text{ g mol}^{-1})(10^{-3}\text{ kg g}^{-1})}\right]^{1/2} = 667.38\text{ m s}^{-1}$$

$$\bar{c}(300\text{ K}) = \left[\frac{(8)(8.314\text{ J K}^{-1}\text{ mol}^{-1})(300\text{ K})}{(\pi)(28\text{ g mol}^{-1})(10^{-3}\text{ kg g}^{-1})}\right]^{1/2} = 476.18\text{ m s}^{-1}$$

$$\bar{c}(500\text{ K}) = \left[\frac{(8)(8.314\text{ J K}^{-1}\text{ mol}^{-1})(500\text{ K})}{(\pi)(28\text{ g mol}^{-1})(10^{-3}\text{ kg g}^{-1})}\right]^{1/2} = 614.74\text{ m s}^{-1}$$

$$c_{mp}(300\text{ K}) = (516.95)\left(\frac{2}{3}\right)^{1/2} = 422.09\text{ m s}^{-1}$$

$$c_{mp}(500\text{ K}) = (667.38)\left(\frac{2}{3}\right)^{1/2} = 554.91\text{ m s}^{-1}$$

The values of c_{rms}, $\bar{c}$ and c_{mp} at 27° C and 227° C are plotted in Fig. 8.6.

***Exercise* 8.46:** Calculate the fraction of krypton atoms that are moving with speeds in the range of 457 to 463 m s^{-1} at 450 °C.

Using equation (8.74) the probability distribution of krypton atom can be obtained as

$$\frac{dN_c}{N}dc = 4\pi\left(\frac{m}{2\pi k_b T}\right)^{\frac{3}{2}} c^2 e^{-\frac{mc^2}{2k_b T}} dc$$

or

$$\frac{dN_c}{N}dc = 4\pi\left(\frac{M}{2\pi RT}\right)^{\frac{3}{2}} c^2 e^{-\frac{Mc^2}{2RT}} dc \qquad ...(1)$$

where M is the molar mass (0.084 kg mol^{-1}) of krypton and R is the gas constant. Substitute the values of constants in equation (1), we get

$$\frac{dN_c}{N}dc = (4)(3.143)\left[\frac{0.084\text{ kg mol}^{-1}}{(2)(3.143)(8.314\text{ J K}^{-1}\text{ mol}^{-1})(723\text{ K})}\right]^{\frac{3}{2}} e^{-\frac{(0.084\text{ kg mol}^{-1})c^2}{(2)(8.314\text{ J K}^{-1}\text{ mol}^{-1}(723\text{ K})}} c^2 dc$$

$$\frac{dN_c}{N}\, dc = 4.171 \times 10^{-8} e^{-6.987 \times 10^{-6} c^2} c^2 dc$$

or
$$\frac{dN_c}{N} = 4.171 \times 10^{-8} \int_{457}^{463} e^{-6.987 \times 10^{-6} c^2} c^2 dc \qquad ...(2)$$

It is not simple integral. One needs to solve by numerical methods. We will try another method that may not be as accurate but it is fine. Since the range of speed is small, *i.e.*, 460 ± 3 m s^{-1}, the fraction of molecules in this small range is nearly constant at 460 m s^{-1}. So the equation (2) can be written as

$$\frac{dN_c}{N} = 4.171 \times 10^{-8} e^{-6.987 \times 10^{-6} (450\ \text{m s}^{-1})^2} (450\ \text{m s}^{-1})^2 \int_{457}^{463} dc$$

$$\frac{dN_c}{N} = (8.446 \times 10^{-3})(0.243)[463 - 457] = 1.23 \times 10^{-2}.$$

8.5.5. *Collision Frequency on a Plane Surface*

Since, all the directions in space are equivalent, the collision frequency derived on any wall will be equally probable for other directions. So considering only x direction and using velocity component in that direction, one can calculate the number of collisions occurring per second on the wall. The expression is

$$Z = N\left(\frac{k_b T}{2\pi m}\right)^{1/2} = N\left(\frac{RT}{2\pi M}\right)^{1/2} \qquad ...(8.78)$$

where Z is the collision frequency and N is the number of molecules per unit volume. This can also be written as

$$Z = \frac{1}{4} N\bar{c} \qquad ...(8.79)$$

Alternatively, assuming that the vapours are behaving as an ideal gas, $N = p/k_b T$, so

$$Z = p\left(\frac{1}{2\pi m k_b T}\right)^{1/2} = pN_{av}\left(\frac{1}{2\pi MRT}\right)^{1/2} \qquad ...(8.80)$$

***Exercise* 8.47:** Consider the N_2 gas at 27°C and at 0.2 atm pressure. Calculate Z.
Using equation (8.80)

$$Z = \frac{(0.2\ \text{atm})(1.0133 \times 10^5\ \text{N m}^{-2}\ \text{atm}^{-1})(6.02 \times 10^{23}\ \text{mol}^{-1})}{\left[(2\pi)(28\ \text{g mol}^{-1})(10^{-3}\ \text{kg g}^{-1})(8.314\ \text{J K}^{-1}\ \text{mol}^{-1})(300\ \text{K})\right]^{1/2}}$$

$$= 5.823 \times 10^{26}\ \text{molecule m}^{-2}\ \text{s}^{-1}$$

8.5.6. *Vapour Pressure of Solids*

Knudsen method make use of equation (8.79) to calculate the vapour pressure of solids. Actually the vapours impinging on the walls are allowed to leak through a hole of unit area. The mass collected per unit time (w) is related to the equation $w = \frac{ZM}{N_{av}}$.

Using ideal gas behaviour and per unit volume $p = \frac{nRT}{N_{av}}$ and combining equation (8.78) with the above, *i.e.*

$$p = \frac{nRT}{N_{av}} = \frac{nRT}{ZM/\omega} = \frac{wRT}{MZ}$$

$$= nw\frac{RT}{M} \times \left[\frac{2\pi M}{RT}\right]^{\frac{1}{2}} = w\left[\frac{2\pi RT}{M}\right]^{\frac{1}{2}}$$

$$p = w\left(\frac{2\pi RT}{M}\right)^{1/2} \qquad ...(8.81)$$

or

$$p = \frac{W}{At}\left(\frac{2\pi RT}{M}\right)^{1/2} \qquad ...(8.82)$$

where A is the area of hole, W the amount collected in t sec.

***Exercise* 8.48:** The solid A is heated at 1000 K. The vapours are allowed to effuse through a pin hole of radius 4 mm and the amount collected is 1.70×10^{-4} kg in 40 min. Calculate the vapour pressure if $M_A = 24$ g mol^{-1}.

Using equation (8.82)

$$p = \frac{(1.7\times10^{-4}\text{ kg})}{(\pi)(4\times10^{-3}\text{ m})^2(40\times60\text{s})}\left[\frac{(2\pi)(8.314\text{ J K}^{-1}\text{ mol}^{-1})(1000\text{K})}{(24\text{g mol}^{-1})(10^{-3}\text{ kg g}^{-1})}\right]^{1/2}$$

$$= 2.0786\text{ N m}^{-2}$$

$$= \frac{2.0786\text{ N m}^{-2}}{101325\text{ N m}^{-2}\text{ atm}^{-1}} = 2.05 \times 10^{-5}\text{ atm.}$$

8.5.7. *Mean Free Path*

Mean free path is defined as the average distance travelled between the collisions. If σ_1 and σ_2 are the collision diameters of two kinds of molecules and N_{12} is the total number of molecules in 1 m^3, the mean free path is given by

$$L = \frac{1}{2^{1/2}\,\pi\,\sigma_{12}^2\,N_{12}} \qquad ...(8.83)$$

and N_{12} can be calculated by using ideal gas law expression

$$N_{12} = \frac{nN_{av}}{V} = \frac{N_{av}p}{RT} = \frac{p}{k_bT}$$

where V is the volume of n mol of gas at pressure p and temperature T. σ_{12} *i.e.*, average diameter is given by

$$\sigma_{12} = \frac{1}{2}(\sigma_1 + \sigma_2) \qquad ...(8.84)$$

or

$$L = \frac{k_bT}{\sqrt{2}\pi\sigma_{12}^2 p} \qquad ...(8.85)$$

For pure gas, $N_{12} = N$ and

$$\sigma_{12} = \frac{1}{2}[\sigma + \sigma] = \sigma$$

so equation becomes

$$L = \frac{k_bT}{2^{1/2}\pi\sigma^2 p} \qquad ...(8.85a)$$

***Exercise* 8.49:** Calculate the mean free path (L) for N_2 at 27°C and 1 atm pressure. The collision diameter for N_2 is 0.374 nm.

Using equation (8.85)

$$L = \frac{(1.38\times10^{-23}\text{ J K}^{-1})(300\text{K})}{(\sqrt{2})(\pi)(0.374\times10^{-9}\text{ m})^2(10150\text{ N m}^{-2})}$$

$$= 6.57 \times 10^{-8}\text{ m.}$$

8.5.8. *Collision Number*

Collision number denoted as Z_1 is defined as the number of collisions experienced per second by a single molecule in a container having N_1 molecules per unit volume. It is obtained dividing average velocity by mean free path *i.e.*, $\bar{c}/L$ and the expression is

$$Z_1 = 2^{1/2}\pi\sigma_1^2 \bar{c} N_1 \qquad ...(8.86)$$

$$= 4\sigma_1^2\left[\frac{\pi k_bT}{m}\right]^{1/2} N_1 = 4\sigma_1^2\left[\frac{\pi RT}{M}\right]^{1/2} N_1 \qquad ...(8.86a)$$

Collision number (Z_{12}) is also defined as the number of collisions between two kinds of molecules per unit volume per unit time and can be obtained by multiplying the expression (8.86) by N_2.

$$Z_{12} = \pi\sigma_{12}^2\left(\frac{k_bT}{\pi\mu}\right)^{1/2} N_1N_2 \qquad ...(8.87)$$

where μ is the reduced mass of molecules 1 and 2, the average relative speed $\bar{c}_r$ is defined as

$$\bar{c}_r = \left(\frac{8k_bT}{\pi\mu}\right)^{1/2} \qquad ...(8.88)$$

If the molecules are alike,

$$Z_{11} = \frac{1}{2}4\pi\sigma^2\left(\frac{k_bT}{\pi m}\right)^{1/2} N_1^2 = 2\sigma^2\left(\frac{\pi k_bT}{m}\right)^{1/2} N_1^2 \quad ...(8.89)$$

***Exercise* 8.50:** Calculate Z_1 and Z_{11} for N_2 at 27°C and pressure 1 atm.

$$N = nN_{av} = N_{av}\frac{pV}{RT}$$

$$= \frac{(6.02\times10^{23}\text{ mol}^{-1})\,(1\text{ atm})(1\text{ m}^3)}{(0.0821\times10^{-3}\text{ atm m}^3\text{ mol}^{-1}\text{ K}^{-1})\,(300\text{ K})}$$

$$= 2.44\times10^{25}\text{ molecule m}^{-3}$$

Using equation (8.86)

$$Z_1 = (4)\,(3.74\times10^{-10}\text{ m})^2\,(529.12\text{ m s}^{-1})\,(2.44\times10^{25}\text{ molecule m}^{-3})$$

$$= 7.22\times10^9\text{ collisions s}^{-1}$$

Using equation (8.87)

$$Z_{11} = (\pi)\left(\frac{1}{2}\right)^{1/2}(3.74\times10^{-10}\text{ m})^2\,(529.12\text{ m s}^{-1})\,(2.44\times10^{25}\text{ molecules m}^{-2})^2$$

$$= 7.79\times10^{34}\text{ collision m}^{-3}\text{ s}^{-1}$$

$$= \frac{9.79\times10^{34}\text{ molecules of collision m}^{-2}\text{s}^{-1}}{6.02\times10^{23}\text{ molecule mol}^{-1}}$$

$$= 1.626\times10^{11}\text{ mol of collision m}^{-3}\text{ s}^{-1}.$$

8.5.9. *Kinetic Theory of Gas Viscosity*

Viscosity is a measure of resistance that is present when adjacent layers of the fluid are moving at different speeds. If the flow is considered as the movement of one layer over the other layer, then the force required for the motion is

$$f = \frac{\eta Ac}{d} \quad ...(8.90a)$$

where f is the force, A is the area of the layers, c is the difference in velocity of the layers, d is the distance between the layers and η is the coefficient of viscosity of the fluid.

Equation (8.90) is not easy to apply. The alternative equation to calculate η is

$$\eta = \frac{\pi(p_1 - p_2)r^4t}{8Vl} \quad ...(8.90b)$$

where $(p_1 - p_2)$ is the pressure drop down the tube, generally capillary, of radius r and length l, V is the volume of fluid flowing in time t. In case of gases, V should be evaluated at $(p_1 + p_2)/2$ because volume is a strong function of pressure.

The coefficient of viscosity can be calculated on the basis of kinetic theory of gases by the expression:

$$\eta = \frac{\overline{c}m}{2^{3/2}\pi\sigma^2} = \frac{1}{\pi\sigma^2}\left[\frac{k_bTm}{\pi}\right]^{1/2} \quad ...(8.91)$$

where $\bar{c}$ is the average speed. The cgs unit for viscosity is the 'poise' *i.e.*, 1 g s^{-1} cm^{-1}. The SI unit of viscosity is the 'Pascal second' (Pas). The relation between the two is 1 poise = 0.1 Pas.

Generally the viscosity is obtained from experimental results and its value is used to calculate the collision diameter.

***Exercise* 8.51:** The viscosity of N_2 is 1.78 × 10^{-4} poise. Calculate the collision diameter if the temperature is 27°C.

Using equation (8.91) and rearranging it

$$\sigma = \left[\frac{1}{\pi\eta}\left(\frac{k_b T m}{\pi}\right)^{1/2}\right]^{1/2}$$

$$= \left[\frac{1}{(\pi)(1.78\times10^{-5}\ \text{Ns m}^{-2})}\left(\frac{(1.38\times10^{-23}\ \text{J K}^{-1})(300\text{K})(28\text{g mol}^{-1})(10^{-3}\ \text{kg g}^{-1})}{(\pi)(6.02\times10^{23}\ \text{mol}^{-1})}\right)^{1/2}\right]^{1/2}$$

$$= 3.74\times 10^{-10}\ \text{m}.$$

***Exercise* 8.52:** Calculate the coefficient of viscosity of methane at 27°C using Kinetic theory of gases, σ for methane is 4.14 × 10^{-10} m.

Using equation (8.91)

$$\eta = \left[\frac{1}{(\pi)(4.14\times10^{-10}\ \text{m})^2}\left(\frac{(1.38\times10^{-23}\ \text{J K}^{-1})(300\text{K})(16\text{g mol}^{-1})(10^{-3}\ \text{kg g}^{-1})}{(\pi)(6.02\times10^{23}\ \text{mol}^{-1})}\right)\right]^{1/2}$$

$$= 1.1\times10^{-3}\ \text{Ns m}^{-2} = 1.1\times10^{-4}\ \text{poise}$$

PROBLEMS

Pressure and Temperature

8.1 The 0 K corresponds to –273°C. What is its value in Fahrenheit scale?

8.2 The boiling point of O_2 is 90.1 K. What is its value in Celsius and Fahrenheit scale?

8.3 What is the temperature which has the same value both on Celsius and Fahrenheit scale?

8.4 Calculate the pressure exerted by 40 cm of Hg in cgs units, MKS units. Given d (Hg) =13.596 kg dm^3, g = 9.80 m s^{-2}.

8.5 Assume that air has the average molecular weight of 29 g mol^{-1}. Calculate the pressure of air at the height of 10,000 meter. At $h = 0$, $p = 760$ mm of Hg.

8.6 Assume that air at $h = 0$ contains 80% N_2 and 20% O_2. Calculate the composition of air at 80,000 meter.

8.7 Calculate the heights where the partial pressures of O_2 and N_2 will be half that at $h = 0$.

8.8 Calculate the height of ethyl alcohol which will have the same pressure as exerted by 2 mm of Hg, d (Hg) =13.6 kg dm^{-3} and $d(C_2H_5OH)$ = 0.7983 kg dm^{-3}.

8.9 When two thermometers (one reading Celsius scale and other Fahrenheit scale) dipped in a liquid, it is found that the reading given by Fahrenheit thermometer is numerically twice that of the reading given by Celsius scale. Calculate this temperature in absolute degree.

8.10 At 20°C if some quantity of N_2 occupies 200 cm^3 at 50 mm, what will be the pressure so that the same amount of gas can be contained in 60 cm^3 vessel at the same temperature?

8.11 5 g of gas gives a pressure of 200 mm in certain container at 25°C. If 12 g of the same gas is added to the same container, what change in temperature would have to be made to have the same pressure?

8.12 At 20°C if the pressure in the scooter tyre measured by pressure gauge is 15 psi, what will be the pressure in scooter tyre if it attains the temperature of 45°C after running? How much air has to be removed so that the pressure at the new temperature is 15 psi? (*Note:* pressure read by the gauge is always the excess over the pressure of the external atmosphere *i.e.,* 14.7 psi).

8.13 A cylinder containing compressed N_2 at 120 atm and 20°C when opened, occupies a volume of 200 dm^3 at 20°C and 1 atm. What is the volume of the cylinder?

8.14 Two flasks containing gases are connected by a stop-cock. Before opening the stop-cock, if the pressure in flask *A* of volume V_A is p_A and that flask *B* of volume V_B is p_B, what will be the pressure if the stop-cock is opened and gases are allowed to equilibrate?

8.15 Repeat the problem 8.14 with the following data: V_A = 100 dm^3; p_A = 5 atm; V_B = 150 dm^3; p_B = 1 atm.

8.16 A certain amount of gas present in a container of unknown volume shows a pressure of 750 mm. A small amount of gas is taken out and measured a volume of 2 dm^3 when expanded to 600 mm. The pressure of the gas remaining in the container was 700 mm. Calculate the volume of the container if all the measurements were done at the same temperature.

8.17 A bubble of diameter 1 cm is formed at a depth of 5 m. What will be the diameter of the same bubble of air when brought just below the surface? Assume that density of water 1 kg dm^{-3}.

8.18 A cylinder contains 100 g of an ideal gas (molecular wt. 64 g mol^{-1}) at 40°C and 3 atm. In transportation, the cylinder fell down and a dent was created *i.e.,* the effective volume of the cylinder decreases. But the valve attached to the cylinder cannot keep the pressure greater than 3 atm, so 15 g of gas leaked out.

(*a*) Calculate volume of the cylinder after and before the dent.

(*b*) Had the valve been pretty strong what would have been the pressure after dent? Temperature remained constant during this process.

8.19 A one dm^3 flask when evacuated weighs 65.05 g. The flask is filled with a gas of molecular weight 48 g mol^{-1} at pressure 760 mm and at temperature 27°C. Calculate the final weight of the flask.

8.20 Vapours of certain compounds are found to contain 12 g of carbon for every 1 g of hydrogen indicating that the formula of the compound is either CH or some multiple of CH. If 0.45 dm^3 of these vapours at 40°C and 1 atm weighs 1.365 g, calculate the molecular formula of the compound.

8.21 0.5 dm^3 of H_2 gas was collected over water when prepared by the action of dil. sulphuric acid on zinc. If the total pressure of the gas, saturated with water vapours, is 740 mm at 27°C, calculate the weight of the gas collected. The vapour pressure of water at 27°C is 26.74 mm. Also calculate the amount of zinc required to produce this amount of H_2 when added to excess of acid.

8.22 (*a*) 2.4 g of NH_4Cl was placed in 1.5 dm^3 flask and was completely vaporised at 375°C. Calculate the pressure observed.

(*b*) The chances are that NH_4Cl may be dissociated at this temperature according to

$$NH_4Cl(g) \rightleftharpoons NH_3(g) + HCl(g)$$

Calculate the partial pressure of each components under the average molecular weight of the system.

(*c*) Also calculate the average molecular weight of the system.

8.23 2.5 g of H_2 (molecular weight = 2 g mol^{-1}) is added to a container at 27°C and 1 atm pressure. How much CO_2 should be added to the container at 27°C such that the combined weight of the two gases should be the same as when the air is filled in the flask at 27°C and 1 atm. Average molecular weight of air is 29 g mol^{-1}. Calculate the average molecular weight of H_2 and CO_2 mixture.

8.24 Calculate the average molecular weight of air with the composition: N_2:O_2: H_2O:CO_2: : 73.44:15.13:6.32:5.11.

8.25 The degree of dissociation of Cl_2 at 1500 K is 0.45, according to the reaction

$$Cl_2 \rightarrow 2\ Cl$$

Assuming that both Cl_2 and chlorine atoms behave ideal gases calculate the density of the mixture if the pressure of the mixture is 1.5 atm.

8.26 4 g of gas *A* when added to a container at 25°C has the pressure 0.85 atm. To this gas 6 g of gas *B* was added and the pressure increased to 1.30 atm at 25°C. If the molecular weight of *A* is 40 g mol^{-1}, calculate the molecular weight of *B*.

8.27 Two bulbs of equal volume are connected by a stop-cock and contains O_2. When both the bulbs are at the same temperature and the stop-cock is open, the equilibrium pressure is 1.2 atm. Now one of the bulb is placed in a bath maintained at 150°C and the other is kept at the original temperature of 40°C.

(*a*) Calculate the equilibrium pressure in flask. (Stop-cock is open),

(*b*) If the volume of each bulb is 2 dm^3, calculate the number of mol of O_2 in each bulb after attaining equilibrium.

8.28 A bulb contains 2 mol of H_2 at a pressure of 0.8 atm and temperature *T* K. 0.6 mol of O_2 is added to this bulb and the temperature of the bulb has to be lowered by 15 K to keep the same pressure:

(*a*) Calculate the volume of the bulb and temperature *t:*

(*b*) Calculate the partial pressure of each gas.

Non-ideal Gases

8.29 Calculate the molar excluded volume of CH_4, CO_2, HI and He. Given $\sigma_{CH_4} = 3.82 \times 10^{-10}$ m, $\sigma_{HI} = 3.50 \times 10^{-10}$ m, $\sigma_{CO_2} = 4.49 \times 10^{-10}$ m and $\sigma_{He} = 2.18 \times 10^{-10}$ m.

8.30 (*i*) Calculate the pressure exerted by 2 mol of methane present in 2 dm^3 container at 25°C. Assume methane follows van der Waals equation: $a = 2.253$ dm^6 mol^{-2} atm; $b = 0.04278$ dm^3 mol^{-1}.

(*ii*) Repeat the above calculations for Cl_2, CO_2 and He. Given a (Cl_2) = 6.493 dm^6 atm mol^{-2}; b (Cl_2) = 0.05622 dm^3 mol^{-1}; a(CO_2) = 3.592 dm^6 atm mol^{-2}; b (CO_2) = 0 .04267 dm^3 mol^{-1}; a (He) = 0.3412 dm^6 atm mol^{-2}; b (He) = 0.0237 dm^3 mol^{-1}.

(*iii*) Calculate also the pressure exerted by each gas at the above conditions assuming that they obey ideal gas laws.

8.31 Using the data of the above problem (8.30) calculate the volume of 2 mol of each gas at 25°C and at a pressure of 4 atm.

8.32 Calculate the second and third virial coefficients for (*i*) methane, (*ii*) CO_2, (*iii*) Cl_2 and (*iv*) He, using van der Waals constants given in problem 8.30.

8.33 Calculate Boyle's temperature for CH_4, CO_2, Cl_2 and He using the data from problem (8.30).

8.34 The van der Waals equation can be written in terms of reduced parameters as

$$\left(p_r + \frac{3}{V_r^2}\right)\left(V_r - \frac{1}{3}\right) = \frac{8}{3}T_r$$

Find the expression such that T_r represents the Boyle's temperature. The Boyle's temperature in this is expressed as

$$\lim_{p_r \to 0} \frac{d(p_r V_r)}{dT_r} = 0$$

8.35 The coefficient of thermal expansion is defined as

$$\alpha = \frac{1}{V}\left(\frac{\partial V}{\partial T}\right)_p$$

Calculate its value (*a*) for ideal gas and (*b*) for van der Waals gas.

8.36 Draw the isotherm for C_2H_5OH at 123°C using the van der Waals equation. Given $a = 12.02$ dm^6 atm mol^{-2}; $b = 0.08401$ dm^3 mol^{-1}. The plot is extended up to 25 dm^3 mol^{-1}.

8.37 Calculate the critical constants for CH_4, CO_2, Cl_2 and He using van der Waals constants from problem 8.30.

8.38 Given the critical constants for CH_4, CO_2, Cl_2 and He. Calculate the van der Waals constants for each gas:
$T_c(\text{He}) = 5.2$ K, $p_c(\text{He}) = 2.25$ atm,
$T_c(CO_2) = 304.16$ K, $p_c(CO_2) = 72.83$ atm, $T_c(Cl_2) = 417.1$ K, $p_c(Cl_2) = 76.1$ atm, $T_c(CH_4) = 190.25$ K, $P_c(CH_4) = 45.6$ atm.
Compare the values of van der Waals constants given in the problem (8.30).

8.39 Calculate the volume of 1 mol each of CH_4, CO_2, Cl_2 and He at 150°C and 10 atm. (Use the Fig. 8.3). The values of critical constants are given in problem (8.38). Compare the values of the volumes calculated with the help of ideal gas law.

8.40 Using the data from the problem 8.31 and pressure calculated there, find the compressibility factor for each gas.

8.41 The densities of a gas at temperature 27°C and at various pressures are given

p(atm)	0.25	0.4	0.6	0.9	1.0
d(kg m^{-3})	0.75	1.082	1.302	1.3029	1.3033

Calculate the molecular weight of the gas by plotting $\frac{d}{p}$ vs p.

8.42 The densities of CO_2 at 47°C and at various pressures are given as

p(atm)	1	10	40	100
(kg m^{-3})	1.6794	17.46	81.48	275.6

Calculate the molecular weight of CO_2 by plotting $\frac{d}{p}$ vs p.

Kinetic Theory of Gases

8.43 Given below are the relations on the left between the variables for two gases. What can you conclude regarding the variables on the right:

Infer (>, <, =)

(*a*) equal $p, V, T, M_1 > M_2$	$\bar{c}_1\ \bar{c}_2$
(*b*) equal $p, V, T, M_1 > M_2$	$N_1\ N_2$
(*c*) equal $p, V, N_1 > N_2$	$T_1\ T_2$
(*d*) equal $T, N, p_1 > p_2, M_1 > M_2$	$V_1\ V_2$
(*e*) equal $V, N, \bar{c}, M_1 > M_2$	$p_1\ p_2$

8.44 O_2 gas, present in a vessel at 37°C and at 1 atm pressure, is replaced by N_2 under the same conditions: (*i*) Will there be any change in number of impacts made by each kind of molecules? If yes, calculate ratio of number of impacts made by N_2 and O_2. (*ii*) What will be the change in the average momentum of the molecules?

8.45 Calculate the average, root mean square and most probable speed of SO_2 at 25°C. Also calculate the ratio of each speed.

8.46 Repeat the above calculations for CO_2.

8.47 (*a*) Find the ratio of c_{rms} of SO_2 and CO_2 at 25°C.

(*b*) Calculate the temperature at which the c_{rms} of SO_2 is same as that of CO_2 at 27°C.

8.48 Plot the probability density $p(\bar{c})$ at different speed vs the speeds at 27°C for SO_2.

8.49 Calculate the $\bar{c}$ of (*i*) SO_2 and, (*ii*) CO_2 at 27°C and 120°C if one considers, (*a*) 20 molecules, (*b*) 10 molecules and (*c*) 100 molecules of each gas.

8.50 Air is blowing with 60 km hr^{-1} and a nail of diameter 0.1 cm is sticking on the wall. How many molecules of air will strike the head of the nail per sec if the average molecular weight of air is 29 g mol^{-1} and temperature is 40°C.

8.51 In the Knudsen tube, the diameter of the pinhole is 0.26 mm. 15.2 g of CsCl was placed in the tube at 1490°C. Calculate the vapour pressure of CsCl at 1490°C if amount of CsCl left after 50 min is 3.7 g.

8.52 The vapour pressure UF_6 at 291.2 K is 100 mm. Calculate the amount of compound left after 60 sec if 2.0 g is placed in Knudsen tube, having a pinhole of diameter 0.02 cm.

8.53 The proportion of O_2, SO_2 and SO_3 present in mixture of gases is 0.5:0.3:0.2. They are allowed to effuse through a pinhole at 27°C. Calculate the composition of the mixture leaving initially.

8.54 The mixture of rare gases (He, Ne, Ar, Kr) having composition 0.7:0.2:0.0.8:0.02 is placed in a vessel with a pinhole. If the temperature of the container is 27°C, calculate the composition of the mixture leaving initially.

8.55 1 mol of NO_2 is present in 5 dm^3 container at 25°C. Calculate (*a*) the number of molecules hitting a 1 cm^2 area per sec. (*b*) the number of g of molecules hitting the above area.

8.56 Repeat the problem (8.55) keeping the volume and temperature same but reducing the pressure to half of the original.

8.57 Calculate the ratio of the collisions of certain gas molecules with the wall if the pressure is increased to 2.5 but the density remains the same.

8.58 A container was having 20 g of He at 3 atm and 27°C. Calculate the time taken for 2.0% of helium to diffuse through a pinhole of diameter 0.02 cm.

8.59 Calculate the ratio of the collision frequencies hitting a wall if

(*a*) pressure is increased by three times keeping density same,

(*b*) pressure is increased by three times but temperature is same,

(*c*) pressure is kept constant but density is decreased to half,

(*d*) pressure is kept constant but temperature is increased to $1\frac{1}{2}$ time.

8.60 Two cubical containers are of equal volume, one contains SO_2 at 50°C and 0.8 atm. The other contains CO_2 at 27°C and 2 atm.

(*a*) Which gas will be colliding more per unit area per sec?

(*b*) Calculate the ratio of the number of moles colliding with walls per unit area per sec.

8.61 A one dm^3 flask is evacuated and the stop-cock is closed. After 2 hrs it is found that the pressure inside the flask is 0.04 atm. This shows that there is small pinhole through which air is entering. Calculate the diameter of this hole with the following data: $\overline{M_{air}} = 29$ g mol^{-1}; $p = 1$ atm; $T = 300$ K.

8.62 Composition of the air present in 2 dm^3 vessel is 80% N_2 and 20% O_2. If the pressure is 0.4 atm and temperature 50°C, calculate (*i*) mean free path of O_2, (*ii*) mean free path of N_2, (*iii*) the number of collisions made with each other per unit volume per unit time. $\sigma_{O_2} = 3.61 \times 10^{-10}$m; $\sigma_{N2} = 3.75 \times 10^{-10}$ m.

8.63 Methane is present in a container of volume of 2 dm^3 at a pressure of 200 mm and temperature 0°C. Calculate (*a*) mean free path, (*b*) the collision frequency in collision mol^2 cc^{-2}, collision mol^2 cc^{-2}, collision mol^2 dm^{-6}. $\sigma_{CH_4} = 4.14 \times 10^{-10}$ m.

8.64 Repeat the above problem for CO_2, $\sigma_{CO_2} = 3.63 \times 10^{-10}$ m.

8.65 Calculate the coefficient of viscosity of CO_2 at 27°C with the data of problem 8.64.

8.66 Repeat problem 8.65 for O_2, $\sigma_{O_2} = 3.61 \times 10^{-10}$ m.

8.67 The coefficient of viscosity for Ar is 0.951×10^{-4} P. Calculate the collision diameter for Ar and compare it with that obtained from self diffusion *i.e.*, 3.47×10^{-10} m.

8.68 Repeat problem 8.67 for CO_2, $\eta = 1.402 \times 10^{-4}$ P.

Additional Problems

8.69 One mole of neon gas is present in a container whose volume is 3,000 cm^3. If all the atoms are moving with the similar velocities of 500 cm s^{-1} in the *x*-, *y*- and *z*-directions, calculate the pressure exerted by this much gas in N m, atm. and bar.

8.70 (*i*) Two moles of oxygen gas is present in a container of volume 5 L at a pressure of 3 atm at 27 °C. Calculate the kinetic energy of the system and also the average kinetic energy of the oxygen molecule. (*ii*) Also calculate the kinetic energy using equi-partition principle. Compare the two values.

8.71 Repeat the Problem 8.70 for the system containing argon atoms present in the vessel of volume 2 L and at a pressure of 1 atm. at 227 °C.

8.72 The deepest part of the sea is nearly 35,000 ft deep. Assume that the density of water is 1 kg L^{-1} and water is incompressible. Calculate the pressure at this depth. We have neglected the effect of dissolved materials in water on its density. Also express this pressure in bar, atm. and psi (pounds per square inch). From this results calculate the height of water which can produce 1 atm pressure.
[**Ans.** 1.046 × 105 kPa; 1,046 bar; 1.33 atm.; 15,180 psi; 33.88 ft height]

8.73 Density of mercury is 13.596 kg L^{-1}. What is the height of the mercury which can produce 1 atm of pressure? [760 mm]

8.74 (*i*) The density of the gas *A* at 0 °C and 1 atm. pressure in a container *A* is 1.784 g L^{-1}. Calculate the molar mass of the gas. (*ii*) Suppose we have another container *B* containing another gas. The temperature and pressure of both the container were reduced to same values. If the ratio of the densities of the gases in container *A* and *B* is 1.248, (*i.e.*, $\frac{d_A}{d_B} = 1.248$), calculate the molar mass of the gas in container *B*.

8.75 A right circular cylinder whose radius is 5 cm contains a mixture of Hg ($d = 13.556$ kg L^{-1}) and H_2O ($d = 1$ kg L^{-1}) with mercury at the bottom of the cylinder and the water at the top of the cylinder. It is found that the pressure at the bottom of the cylinder is 1.4 atm. It is also determined that the mass of Hg in the cylinder is eight times the mass of the water present. Determine the length of the Hg column, h_{Hg} and the length of the water column h_{wa} in the cylinder.

8.76 Using the idea from the equation (8.10), consider the two metals rods denoted by *A* and *B*, whose lengths are linear functions of absolute temperature T, *i.e.*,

$$L_A = L_{A0} + a_1 T \text{ and } L_B = L_{B0} + b_1 T$$

where L_{A0} and L_A are the lengths of the rod at temperature $T = 0$ K and temperature T and a_1 is a positive constant. Similar terms are defined for the rod *B*. Similarly the lengths of these rods in terms of another temperature scale (t_A) are defined as

$$t_A = \frac{100}{L_{100} - L_{A0}}(L_A - L_{A0}) \text{ and } t_B = \frac{100}{L_{100} - L_{B0}}(L_B - L_{B0})$$

where L_{A0} and L_{B0} and L_{100} is the length of rods *A* and *B* at the freezing and boiling points of water respectively. (*i*) Determine the relationship between t_A and *T*. (*ii*) Show that $t_A = t_B$ at all values of *T* even if $L_{A0} \pm L_{B0}$ and $A_1 \pm B_1$.

8.77 Instead of ideal gas laws, assume that ideal gas somewhere not on this planet follows: (*i*) at constant temperature the volume is inversely proportional to cube of the pressure, (*ii*) at constant pressure, the volume is proportional to 3/2 power of temperature, (*iii*) at 273.15 K and 1 atm. the volume of 1 mole of gas is 22.414 L.

(*a*) Using these laws, construct the ideal gas equation, (*b*) The coefficient of thermal expansion is defined as: $\alpha = \frac{1}{V_0}\left(\frac{dV}{dT}\right)_p$. Calculate the value of α for the ideal gas with which we are familiar and the ideal gas which we derive above.

8.78 The coefficient of thermal expansion is defined as $\alpha = \dfrac{1}{V_0}\left(\dfrac{dV}{dT}\right)_p$. (*i*) Calculate the value of α for ideal gas, (*ii*) for van der waals gas and (*iii*) for a gas following Dieterici equation given as

$$p(\bar{V} - b)e^{\frac{a}{RT\bar{V}}} = RT \qquad \textbf{[Ans. } \alpha = \frac{RV + \frac{a}{T}}{pV2e^{\frac{a}{RTV}} - a}\textbf{]}$$

8.79 40 g of Xe and 15 g of CO_2 are mixed in a vessel of volume 15 dm^3 at 27 °C. Assuming the mixture to be ideal, determine the partial pressure of each gas and the total pressure inside the vessel.

8.80 A container is divided into two parts, *A* and *B* and both the compartments contain ideal gases. In compartment *A*, the gas is at a pressure of 4 atm and at a temperature of 127 °C. The gas in compartment *B* is at 10 atm and at a temperature of 127 °C. When the compartment is removed the gases is allowed to mix and is found that the temperature does not change. If the mole fraction of gas *A* in the mixture is = 0.48, calculate the volume of each compartment.

8.81 If the density of rare gas is 0.003733 g cm^{-3} at STP (standard temperature and pressure) calculate the molar mass of the rare gas and give the name of the gas.

8.82 A container contains a mixture of He and Ar. If the density of the mixture at temperature 300 K and pressure of 1 atm is 0.925 g dm^3, calculate the mole fraction of each gas.

8.83 Calculate the mass of ethane gas present in a container of volume 60 dm^3 and a pressure of 15 atm and 400 K if the gas is assumed to be (*i*) ideal and (*ii*) if it follows van der Waals equation.

8.84 Calculate the second and third virial coefficients of (*i*) C_2H_2 and (*ii*) water using the data given in Table 8.1.

8.85 Suppose a gas obeys the equation of state as $pV = RTe^{-\frac{a}{RTV}}$, where *a* is a constant. (*i*) Calculate the second and third virial coefficients for this gas as a function of *a*, *R* and *T*. (*ii*) Calculate the expression for the residual volume of this gas in terms of *a*, *R* and *T*.

8.86 A gas obeys the van der Waals equation of state. The critical parameters for this gas are $p_c = 47.7$ atm, $T_c = 151.15$ K and $V_c = 0.0752$ dm^3 mol^{-1}. Calculate the reduced pressure of the gas at 400 K and a volume of 10 dm^3.

8.87 A gas is represented by the equation of state as

$$p = RT\left[\frac{1}{V} + \frac{a}{V^2} + \frac{b}{V^3}\right]$$

where *a* and *b* are constants. Assuming that this equation of state follows the constraints at the critical point, calculate the values of *a*, *b* and *R* in terms of critical variables and express the equation of state in reduced form.

8.88 Express the value of residual volume in terms of virial coefficients if a gas follows a virial equation of state.

8.89 A non-ideal gas that follows an equation is $p\bar{V}(1 - \alpha p) = RT$, where α is a function of temperature only. Calculate the expression for the residual volume of this gas as a function *R*, *T* and α.

8.90 Similar to coefficient of expansion, the coefficient of compressibility is defined as

$$\beta = -\frac{1}{V}\left(\frac{dV}{dp}\right)_T$$

where subscript T relates the change in volume with pressure is at constant temperature. Calculate the value of β for (*i*) an ideal gas and (*ii*) for van der Waals gas and (*iii*) if the limits of infinite volume are considered, the value of β for van der Waals gas reduces to ideal gas.

8.91 In the molecular bean experiments, the magnitude of translational energy is controlled by allowing the beam of either atoms or molecules are allowed to pass through slotted disc mounted on the same axis and moving with the same speed but the slots out of phase in such a manner that by the time particles pass from one disc to another the whole has reached and thus can pass through. Determine the fraction of Xe atoms at 300 K with speed in the range of 397 to 403 m s^{-1}.

8.92 Let us define a reduced speed as $c_R = c/c_{mp}$, where c_{mp} is the most probable speed. Transform the molecular speed probability distribution of reduced speeds.

8.93 (*a*) Calculate $< c >$ for $^{235}UF_6(g)$ and $^{238}UF_6(g)$ at 298 K if the speed distribution is Maxwell-Boltzmann. (*b*) If an ensemble of $^{235}UF_6(g)$ and $^{238}UF_6(g)$ molecules are released into long diffusion tunnel, what distance would they have to travel before the average distance between the $^{235}UF_6(g)$ and $^{238}UF_6(g)$ molecules is 115 m.

CHAPTER 9

STATISTICAL MECHANICS

9.1. DEFINITIONS

Distribution: The procedure of arranging total number of systems among different energy states is known as distribution. For example, arranging the total number of molecules N into different energy states, $\varepsilon_1, \varepsilon_2, \varepsilon_3, \ldots$ with $n_1, n_2, \ldots$ system.

Distribution Numbers: $n_1, n_2, \ldots$ are called distribution numbers.

Complexion: Complexion can be defined as one physically distinct way of assigning molecules into different energy levels in a particular distribution.

Combination: In mathematics, in general it is defined as partition of a number of things into smaller groups and calculation of different combinations can be obtained by combinatory rule.

9.2. COMBINATORY RULES

9.2.1. *System with no Restriction*

The number of distinguishable complexions obtained by arranging N number of systems into $n_1, n_2, \ldots$ in different energy levels $\varepsilon_1, \varepsilon_2, \ldots$ without any restriction on the total energy is given as

$$\Omega = \frac{N!}{\prod n_i!} = \frac{N!}{n_1!\,n_2!\ldots n_i!} \qquad \ldots(9.1)$$

where ! is a factorial sign and $\prod$ is a sign of product.

***Exercise* 9.1:** Find the number of complexions for arranging four distinguishable systems such that two should be in energy level ε_1 and two in energy level ε_2. Compare the results using the equation (9.1) and calculating mechanically.

Let the two energy levels be ε_1 and ε_2. The systems can be arranged as shown in Table below and total number is six:

cd	*bd*	*bc*	*ad*	*ac*	*ab*
ab	*ac*	*ad*	*bc*	*bd*	*cd*

Using the equation (9.1), $N = 4$, $n_1 = n_2 = 2$, so

$$\Omega = \frac{4!}{2!2!} = 6$$

The two values agree but for large number of systems, mechanical procedure is very cumbersome and equation (9.1) is used.

9.2.2. *Systems with Definite Energy*

In real physical system, if a definite amount of energy E, is to be divided among $n_1, n_2, \ldots$ particles such that

$$E = n_1\varepsilon_1 + n_2\varepsilon_2 + \ldots + n_i\varepsilon_i = \Sigma n_i\varepsilon_i \qquad \ldots(9.2)$$

$$N = n_1 + n_2 + n_3 + \ldots + n_i = \Sigma n_i \qquad \ldots(9.3)$$

(where $n_1, n_2, \ldots$ are distribution numbers), there can be many distributions which have the same total energy. In such cases the total number of complexions, are

$$\Omega_T = \sum \Omega_i = \sum \frac{N!}{\prod n_i!} \qquad \ldots(9.4)$$

***Exercise* 9.2:** Find the total number of complexions in arranging four distinguishable systems among energy levels 0, 1 ε, 2ε, 3ε such that the total energy is 3 ε.

The possible ways of arranging the particles in the energy levels with total energy 3 ε are

(*i*) three particles in 0 and 1 in energy level 3ε

(*ii*) two particles in energy 0, 1 in energy level ε and 1 in 2ε

(*iii*) one particle in energy level 0 and three in energy level ε.

The number of complexion in each case is

$$\Omega_1 = \frac{4!}{3!\,1!} = \frac{4.3.2.1}{3.2.1} = 4$$

$$\Omega_2 = \frac{4!}{2!\,1!\,1!} = \frac{4.3.2.1}{2.1} = 12$$

$$\Omega_3 = \frac{4!}{1!\,3!} = \frac{4.3.2.1}{3.2.1} = 4$$

$$\Omega_T = \Sigma\Omega_i = 4 + 12 + 4 = 20$$

The above procedure can nicely be arranged in Table form as

No. of ways	0	E	$2E$	$3E$	Ω	$p = \frac{\Omega}{\Omega_T}$
1	3	0	0	1	4	0.2
2	2	1	1	0	12	0.6
3	1	3	0	0	4	0.2

$$\Omega_T = 4 + 12 + 4 = 20$$

9.2.3. *Systems with Definite Total Energy and Degenerate Energy Levels*

The total number of complexions obtained by arranging n_1 systems in energy level ε_1 with degeneracy g_1, n_2 in energy level ε_2 with degeneracy g_2, ... such that

$$n_1 + n_2 + n_3 + ... = N \qquad ...(9.3)$$

and

$$n_1\varepsilon_1 + n_2\varepsilon_2 + ... = E \qquad ...(9.2)$$

can be calculated by the expression

$$\Omega_T = \Sigma N! \prod \frac{g_i^{n_i}}{n_i\,!} \qquad ... (9.5)$$

***Exercise* 9.3:** Calculate the total number of complexions of the system in exercise 9.2, if $\varepsilon = 0$, $\varepsilon_1 = \varepsilon$ and $\varepsilon_2 = 2\varepsilon$ are 3-, 2-and 2-fold degenerate.

Continuing with the exercise (9.2) and using the relation (9.5)

$$\Omega_1 = \frac{(4!)\,(3^3)}{(3!)\,(1!)} = 4 \times 27 = 108$$

$$\Omega_2 = \frac{(4!)\,(3^2)\,(2^1)\,(2^1)}{(2!)\,(1!)\,(1!)} = 4 \times 3 \times 9 \times 2 \times 2 = 432$$

$$\Omega_3 = \frac{(4!)\,(3^1)\,(2^3)}{(3!)\,(1!)} = 4 \times 3 \times 8 = 96$$

$$\Omega_T = 108 + 432 + 96 = 636$$

9.3. PROBABILITY AND MOST PROBABLE DISTRIBUTION

Probability of the occurrence of any distribution or event is the ratio of number of complexions in that distribution to total number of complexions obtained by dividing the fixed amount of energy (E) in 'N number of particles, *i.e.*, if Ω_i is the number of complexions in the ith distribution and $\Sigma\Omega_i$ sum of the total number of complexions, then probability of distribution i is

$$p\,(i) = \frac{\Omega_i}{\Sigma\Omega_i} \qquad ...(9.6)$$

and the value of $p\,(i) \leq 1$.

Most probable distribution is that distribution which has the maximum value of Ω and is denoted as Ω_D.

***Exercise* 9.4:** In exercise 9.2, find the probability of distribution in which two particles are in energy level 0 and 1 each in energy level ε and 2ε such that total number of particles are four and total energy is 3ε. Also find the number of complexions in most probable distribution. From exercise 9.2 the total number of complexions are = 20;

Number of complexions in 2nd distribution = 12,

$$p\,(2) = \frac{12}{20} = 3/5$$

Most probable distribution has the number of complexions = 12.

9.3.1. *Average Values*

In the last section we have obtained the probability of occurrence of any distributions or events as given by equation (9.6). Other important quantity that we will encounter is the average value. It is given by the equation (9.7), *i.e.*,

$$<Q(X)> = \sum_{i=1}^{M} p(X_i)q(X_i) = \frac{\sum_{i=1}^{M} q(X_i)\Omega(X_i)}{\sum_{i=1}^{M} \Omega(X_i)} \quad ...(9.7)$$

where $p(i)$ is normalised probability of observing a value, X_i, the discrete variable. If $q(X_i) = X$, equation (9.7) gives $<Q(X_i)>$ which is also called the *first moment of the distribution*. If $q(X_i) = X^2$, the result is the average of square of X^2, $<X^2>$ which is termed as the *second moment of the distribution*.

***Exercise* 9.5:** Combining the exercises (9.2) and (9.3), calculate (*i*) $<\varepsilon>$ and (*ii*) $<\varepsilon^2>$.

In Exercises (9.2) and (9.3), we have observed that there are three distributions of arranging four distinguishable particles in four energy levels to have three units of energy (3ε).

Consider the Exercise (9.2)

$$<\varepsilon>(\Omega_1) = \frac{(3)(0)+(0)(\varepsilon)+(0)(2\varepsilon)+(1)(3\varepsilon)}{1+1+1+1} = 0.75\varepsilon \quad ...(1)$$

Similarly for Ω_2 and Ω_3 are

$$<\varepsilon>(\Omega_2) = \frac{(2)(0)+(1)(\varepsilon)+(1)(2\varepsilon)+(0)(3\varepsilon)}{1+1+1+1} = 0.75\varepsilon \quad ...(2)$$

$$<\varepsilon>(\Omega_3) = \frac{(1)(0)+(3)(\varepsilon)+(0)(2\varepsilon)+(0)(3\varepsilon)}{1+1+1+1} = 0.75\varepsilon \quad ...(3)$$

The average value of ε in each distribution is same, *i.e.*, 0.75. The average of ε for the total distribution can be obtained by using equation (9.7), *i.e.*,

$$<\varepsilon> = (0.2)(0.75) + (0.6)(0.75) + (0.2)(0.75) = 0.15 + 0.45 + 0.15 = 0.75\varepsilon \quad ...(4)$$

i.e., the average of the total distribution is 0.75ε

Now let us calculate $<\varepsilon^2>$ for the each distribution and total distribution. These are given as:

$$<\varepsilon^2>(\Omega_1) = \frac{(3)(0)^2+(0)(\varepsilon)^2+(0)(2\varepsilon)^2+(1)(3\varepsilon)^2}{1+1+1+1} = 2.25\varepsilon^2 \quad ...(5)$$

Similarly for Ω_2 and Ω_3 are

$$<\varepsilon^2>(\Omega_2) = \frac{(2)(0)^2+(1)(\varepsilon)^2+(1)(2\varepsilon)^2+(0)(3\varepsilon)^2}{1+1+1+1} = 1.25\varepsilon^2 \quad ...(6)$$

$$<\varepsilon^2>(\Omega_3) = \frac{(1)(0)^2 + (3)(\varepsilon)^2 + (0)(2\varepsilon)^2 + (0)(3\varepsilon)^2}{1+1+1+1} = 0.75\varepsilon^2 \quad ...(7)$$

The average value of ε^2 in each distribution is different. The average of ε^2 for the total distribution can be obtained by using equation (9.7), *i.e.*,

$$<\varepsilon^2> = (0.2)(2.25\varepsilon^2) + (0.6)(1.25\varepsilon^2) + (0.2)(0.75\varepsilon^2) = .45\ \varepsilon^2 + 0.75\ \varepsilon^2 + 0.15\ \varepsilon^2 = 1.35\varepsilon^2 \quad ...(8)$$

Unlike $<\varepsilon>$ the $<\varepsilon^2>$ for the complete system is close to the most probable distribution.

(*ii*) Now repeat this exercise for the Exercise (9.3). The distribution is the same and the only difference is the probability of occurrence of each event or complexions. Using second term of equation (9.7) and substituting the terms we get

$$<Q\ (\varepsilon)> = \frac{(108)(0.75\varepsilon) + (432)(0.75\varepsilon) + (96)(0.75\varepsilon)}{108 + 432 + 96}$$

$$= \frac{81\varepsilon + 324\varepsilon + 72\varepsilon}{636} = 0.764\varepsilon$$

$$<Q\ (\varepsilon^2)> = \frac{(108)(2.25\varepsilon^2) + (432)(1.25\varepsilon^2) + (96)(0.75\varepsilon^2)}{108 + 432 + 96}$$

$$= 1.344\varepsilon^2$$

Although the difference in $<\varepsilon^2>$ obtained in exercise (*i*) and (*ii*) are not very different but trend shows that it is tending toward the most probable distribution. *From the above results, although the number is small, it is clear that the average value of any quantity is always close to the property observed for the most probable distribution*. We will also show the same by taking most general case later in section (9.6.1).

***Exercise* 9.6:** In a fair, a stall has the following game. You have to throw a ball of radius 10 cm on a board with 25 holes evenly divided into the sizes of radii 10 cm, 20 cm, 30 cm, 40 cm and 50 cm and for every ball thrown you have to invest ₹ 10. For every ball passing through 50 cm hole one gets nothing, ₹ 10 for the hole 40 cm, ₹ 15 for hole 30 cm, ₹ 20 for hole 15 cm and ₹ 25 for 10 cm hole. (*i*) Suppose you invest ₹ 10 and you throw the ball only once. Calculate the average return in this act, (*ii*) what would the payoff for the 10 cm hole have to be to make the average return to the person equal to his or her investment? (*iii*) Suppose the player wants to earn ₹ 20 in only two throws. Calculate the probability of achieving this.

In this exercise the probability that the ball will enter the hole of radius r is proportional to the area of the hole, *i.e.*, r^2. So the probability factors $p_1 = 10^2 = 100$ for 10 cm hole, $p_2 = 20^2 = 400$ for 20 cm hole, $p_3 = 30^2 = 900$ for 30 cm hole, $p_4 = 40^2 = 1600$ for 40 cm hole and $p_5 = 50^2 = 2500$ for 50 cm hole. The return is q_1 = ₹ 25, q_2 = ₹ 20, q_3 = ₹ 15, q_4 = ₹ 10 and $q_5 = 0$. So the average return can be calculated using equation (9.7), *i.e.*,

$$<Q(X)> = \frac{\sum_{i=1}^{M} q(X_i)\Omega(X_i)}{\sum_{i=1}^{M} \Omega(X_i)}$$

$$= \frac{(₹\,25)(100) + (₹\,20)(400) + (₹\,15)(900) + (₹\,10)(1600) + (₹\,0)(2500)}{100 + 400 + 900 + 1600 + 2500}$$

$$= \frac{(₹\,2500) + (₹\,8000) + (₹\,13{,}500) + (₹\,16{,}000) + (₹\,0)}{100 + 400 + 900 + 1600 + 2500} = \frac{₹\,40{,}000}{5{,}500} = ₹\,7.27 \quad ...(1)$$

So the player will lose an average of ₹ (10 – 7.27) = ₹ 2.73 each time the player plays the game.

(*ii*) Our next job is to calculate the value of $q(X_1)$ must be charged in order to make X_1 = ₹ = 10. Using equation (1) with unknown value of $q(X_1)$, we can write

$$\frac{(q(X_1))(100) + (₹\,20)(400) + (₹\,15)(900) + (₹\,10)(1600) + (₹\,0)(2500)}{100 + 400 + 900 + 1600 + 2500} = ₹\,10$$

$$q(X_1) = \frac{₹\,[55{,}000 - 8{,}000 - 13{,}500 - 16{,}000]}{100} = ₹\,175$$

(*iii*) To earn ₹ 20 in two throws the player must earn ₹ 40 since for each throw he has to invest ₹ 10 each time. This can be possible if the player earns (*a*) ₹ 15 in one throw and ₹ 25 in second throw, (*b*) ₹ 25 in one throw and ₹ 15 in second throw and (*c*) ₹ 20 in each throw. The probability of each case is:

$$\text{Case } (a) = p(X_1)\,p(X_3) = \frac{100}{5500} \times \frac{900}{5500} = 2.98 \times 10^{-3}$$

$$\text{Case } (b) = p(X_3)\,p(X_1) = \frac{900}{5500} \times \frac{100}{5500} = 2.98 \times 10^{-3}$$

$$\text{Case } (c) = p(X_2)\,p(X_2) = \frac{400}{5500} \times \frac{400}{5500} = 5.29 \times 10^{-3}$$

So the total probability can be obtained by adding all the three cases, *i.e.*,

$$= 2.98 \times 10^{-3} + 2.98 \times 10^{-3} + 5.29 \times 10^{-3} = 11.25 \times 10^{-3}.$$

This exercise is similar to distribution of fixed amount of energy (say *E*) into *n* particles with the constraint that *x* number of molecules will be present in certain energy level.

***Exercise* 9.7:** Suppose we have electrons and each electron can have two values of spins, α and β. Calculate the probability that five electrons have α spins and five electrons have β spins.

Since each electron can have two spins and there are 10 electrons, the total number of possible arrangements can be = 2^{10} = 1024.

The number of arrangements or complexions that five electrons have α spin and five electrons β

$$\text{spins} = \Omega(5,5) = \frac{10!}{5!5!} = \frac{10 \times 9 \times 8 \times 7 \times 6}{5 \times 4 \times 3 \times 2 \times 1} = 252$$

$$p(5,5) = \frac{252}{1024} = 0.2461$$

***Exercise* 9.8:** Now let us apply the above results to the real system. Consider a bare proton present in the magnetic field of strength *H*. Derive the expression for the $< E >$.

Using the equation (5.123), *i.e.*,

$$E_{\text{mag}}\left(\pm\frac{1}{2}\right) = \mp\frac{1}{2}\, g_N\mu_z H_0$$

where each term has been explained in Chapter 5. So the system involves only two energy states and thus the partition function is given as:

$$Q(T, H) = e^{-\frac{g_N\mu_z H_0}{2k_bT}} + e^{\frac{g_N\mu_z H_0}{2k_bT}}$$

Using equation (9.42)

$$E = Nk_bT^2\left(\frac{\partial \ln Q}{\partial T}\right)_V = Nk_bT^2\frac{1}{Q}\left(\frac{\partial Q}{\partial T}\right)_V$$

$$E = Nk_bT^2\,\frac{(e^{-\frac{g_N\mu_Z H}{2k_bT}})\left(-\frac{g_N\mu_Z H}{2k_b}\right)\left(-\frac{1}{T^2}\right)+(e^{\frac{g_N\mu_Z H}{2k_bT}})\left(\frac{g_N\mu_Z H}{2k_b}\right)\left(-\frac{1}{T^2}\right)}{e^{-\frac{g_N\mu_Z H}{2k_bT}}+e^{\frac{g_N\mu_Z H}{2k_bT}}}$$

or

$$E = \left(\frac{Ng_N\mu_Z H}{2}\right)\frac{(e^{-\frac{g_N\mu_Z H}{2k_bT}})-(e^{\frac{g_N\mu_Z H}{2k_bT}})}{e^{-\frac{g_N\mu_Z H}{2k_bT}}+e^{\frac{g_N\mu_Z H}{2k_bT}}} \qquad ...(1)$$

The total energy can be obtained by using the actual data for different nuclei. But the limiting values can be obtained by taking $T = 0$ and $T = \infty$. For $T = \infty$, equation (1) reduces to zero. This means that the number of molecules present in both the states is equal and thus the contribution to the energy from each state is equal. Similarly for $T = 0$, equation (1) reduces to $E = \dfrac{Ng_N\mu_Z H}{2}$, *i.e.*, all the species are present in the ground state and thus the total energy corresponds to the energy of the ground state.

***Exercise* 9.9:** Prove the Sterling's theorem.

$n!$ can be written as:

$$n! = n\,(n-1)\,(n-2)\,(n-3)\,\ldots\ldots\ldots\ldots 3\;2\;1$$

Taking logarithm of both sides we get

$$\begin{aligned} lnn! &= ln[n\,(n-1)\,(n-2)\,(n-3)\,\ldots\ldots\ldots +3+2+1] \\ &= lnn + ln(n-1) + ln(n-2) + ln(n-3)\,\ldots\ldots\ldots + ln3 + ln2 + ln1] \\ &= \sum_{x=1}^{n} ln\, x \end{aligned}$$

Applying Euler-Maclaurin summation formula the above summation can be written as:

$$\sum_{x=1}^{n}\ln x \simeq \int_1^n \ln x = [x\ln x - x]_1^n = (n\ln - n) - (1\ln 1 - 1) = n\ln n - n + 1 = n\ln n - n$$

Neglecting 1 when n is very large. So

$$lnn! = \sum_{x=1}^{n} \ln x \simeq n \ln n - n.$$

9.4. DISTRIBUTION OF NON-DISTINGUISHABLE PARTICLES

The number of complexions in any distribution of dividing total energy (E) among total number of (N) similar particles occupying $\varepsilon_1, \varepsilon_2, \varepsilon_3, ..., \varepsilon_i$ energy levels, are given by the expression (9.8):

$$\varepsilon_T = \Sigma_i \prod \frac{g_i^{n_i}}{n_i\,!} \qquad ...(9.8)$$

i.e., number of complexions in this case can be obtained by dividing the equation (9.5) with $N\,!$

***Exercise* 9.10:** Repeat the exercise 9.3 for the case where all the four particles are non-distinguishable.

The number of complexions obtained can be calculated by expression (9.8), or obtained by dividing expression (9.5) by $N\,!$

$$N = 4$$

$$\Omega = \frac{1}{N\,!} \; \Sigma N! \; \pi_i \; \frac{g_i^{n_i}}{n_i\,!}$$

$$= \frac{636}{4!} = 26.5$$

The results of exercise 9.3 are used here *i.e.*, total number of complexions obtained for non-distinguishable molecules or systems are much less than that obtained for distinguishable ones, for the same number and total energy; or using equation (9.7)

$$\Omega_1 = \frac{3^3}{3!\;1!} = 4.5, \; \Omega_2 = \frac{3^2.\,2.\,2}{(2!)\,(1!)\,(1!)} = 18$$

$$\Omega_3 = \frac{(3^1)\,(2^3)}{3!\;1!} = 4$$

$$\Omega_T = \Omega_1 + \Omega_2 + \Omega_3 = 26.5$$

The same as that obtained above.

9.5. STIRLING'S THEOREM

Stirling's theorem or approximation gives

$$\ln n! = n \ln n - n \qquad ...(9.9)$$

provided n is very large number. Applying Stirling's theorem to equation (9.5) *i.e.*,

$$\Omega = N! \prod_i \frac{(g_i)^{n_i}}{n_i\,!} \qquad ...(9.5)$$

Taking logarithms of both sides

$$\ln \Omega = \ln N! + \ln (g_1)^{n_1} + (g_2)^{n_2} + ... - \ln (n_1!)\,(n_2!) ...$$

$$= \ln N! + \sum_i n_i \ln g_i - \sum_i \ln (n_i !).$$

Applying Stirling's theorem

$$\ln \Omega = N \ln N - N + \Sigma n_i \ln g_i - \Sigma (n_i \ln n_i - n_i)$$

$$= N \ln N + \Sigma n_i \ln g_i - \Sigma n_i \ln n_i \quad ...(9.10)$$

Since $\Sigma n_i = N$

***Exercise* 9.11:** Calculate the value of ln 6! with and without Stirling's theorem. Find the difference between the values if any.

The values of $6! = 6 \times 5 \times 4 \times 3 \times 2 \times 1 = 720$

$$\ln 720 = 6.579$$

By Stirling's theorem

$$\ln 6! = 6 \ln 6 - 6 = 4.750$$

$$\ln n! - (n \ln n - n) = 1.828$$

This shows that Stirling's theorem is only valid when n is large.

9.6. MAXWELL-BOLTZMANN STATISTICS

Maxwell-Boltzmann distribution helps in calculating the average vibrational energy of an ensemble of harmonic oscillators to obtain the heat capacities of solids, derived by Einstein and Debye (9.11.2.1 and 9.11.2.2); ionic activity coefficients, derived by Debye-Huckel, computing the intensities of absorption bands (Chapter 5), reaction rates (Chapter 19) etc.

Maxwell-Boltzmann statistics helps in calculating the values of $n_1, n_2, ...$ such that the most probable distribution which leads to maximum values of Ω_D can be calculated. Maxwell-Boltzmann statistics can be applied to a system with following characteristics:

1. Thermodynamic system is an assembly of N distinguishable or non-distinguishable particles in any state and is defined by the volume V and internal energy E.
2. Any individual particle can occupy any energy level. The distribution of the particles in different energy levels should be such that total energy of the system is E. The energy levels are the eigen values of the Schrödinger equation (See Section 2.11).
3. All the complexions corresponding to the given value of E and V are equally probable.

With these characteristics, the number of particles (n_i) occupying any energy level ε_i is

$$n_i = \frac{N g_i \exp(-\varepsilon_i / k_b T)}{\Sigma\, g_i \exp(-\varepsilon_i / k_b T)} = \frac{N g_i}{q} \exp(-\varepsilon_i / k_b T) \quad ...(9.11)$$

The summation is given the symbol 'q' called partition function or sum over all states. Sometimes also called particle partition function. The ratio of the number of molecules in different energy levels is given by

$$\frac{n_i}{n_j} = \frac{g_i \exp(-\varepsilon_i / k_b T)}{g_j \exp(-\varepsilon_j / k_b T)} \quad ...(9.12)$$

The equation (9.11) is applicable to both distinguishable and non-distinguishable systems. Equation (9.11) is obtained by considering only the most probable distribution. It has been found that any thermodynamic property calculated by considering the most probable distribution agrees nicely with that obtained experimentally.

***Exercise* 9.12:** Calculate the ratios of n_i/n_0, where n_0 is lowest energy level and n_i are higher energy levels at 300 K with $\varepsilon_i - \varepsilon_0$ having the values 10^{-21} J, 10^{-20} J, 10^{-19} J and 10^{-18} J and $g_i/g_0 = 1$.

According to equation (9.12)

$$\frac{n_i}{n_0} = \frac{g_i}{g_0} \exp\{-(\varepsilon_i - \varepsilon_0)/k_bT\} = \exp\{-(\varepsilon_i - \varepsilon_0)/k_bT\}$$

$$k_bT = (1.381 \times 10^{-23} \text{ J K}^{-1})(300 \text{ K})$$
$$= 4.143 \times 10^{-21} \text{ J}$$

Now

$$n_1/n_0 = \exp\{-(\varepsilon_1 - \varepsilon_0)/k_bT\}$$
$$= \exp\{-(10^{-21} \text{ J})/(4.143 \times 10^{-21} \text{ J}) = 0.7855$$

$$n_2/n_0 = \exp\{-(\varepsilon_2 - \varepsilon_0)/k_bT\} = \exp\left\{-\frac{10^{-20} \text{ J}}{4.143 \times 10^{-21} \text{ J}}\right\} = 0.0895$$

Similarly $\frac{n_3}{n_0}$ and $\frac{n_4}{n_0}$ are 3.29×10^{-11} and 0 respectively.

***Exercise* 9.13:** The value of the equilibrium rotational constant (B) for one of the electronic state of Br_2 molecule is 0.642 cm^{-1}. Assume that it has the same value for v = 0, vibrational state, plot $\frac{n_J}{n_0}$ vs J and find the value of J where $\frac{n_J}{n_0}$ is maximum at 27°C. Also calculate J_{max} by using equation (5.42). Compare the two values.

According to Maxwell-Boltzmann distribution

$$\frac{n_J}{n_0} = \frac{(2J+1)\exp(-\varepsilon_J / k_bT)}{\exp(-\varepsilon_0 / k_bT)} = (2J+1)\exp\{-(\varepsilon_J - \varepsilon_0)k_bT\}$$

The degeneracy for rotational levels is given as $g_J = 2J + 1$ and the expression for rotational energy is

$$\varepsilon_J = B \cdot hcJ(J+1) \text{ in Joules}$$

Now $\varepsilon_0 = 0$, so

$$\frac{n_J}{n_0} = (2J+1)\exp(\varepsilon_J/k_bT) = (2J+1)\exp\left\{-\frac{BhcJ(J+1)}{k_bT}\right\}$$

Now

$$\frac{Bhc}{k_bT} = \frac{(0.642 \text{ cm}^{-1})(6.62 \times 10^{-34} \text{ J s})(3.0 \times 10^{10} \text{ cm s}^{-1})}{(1.381 \times 10^{-23} \text{ J K}^{-1})(300 \text{ K})}$$

$$= 3.078 \times 10^{-3}$$

By taking different values of J, n_j/n_0 can be calculated and are listed in table below. As a sample of calculation

$$\frac{n_1}{n_0} = (2 \times 1 + 1)\exp(-2 \times 3.078 \times 10^{-3}) = 4.91$$

J	0	1	5	10	15	20	25	30	35	40	45	50
$\frac{n_J}{n_0}$	1.0	4.91	11.03	14.97	14.81	11.25	6.90	3.48	1.47	0.52	0.16	0.03

The $\frac{n_J}{n_0}$ is plotted in Fig. 9.1 and the value of J for n_J/n_0 to be maximum is $J = 13$. Also using equation (5.42)

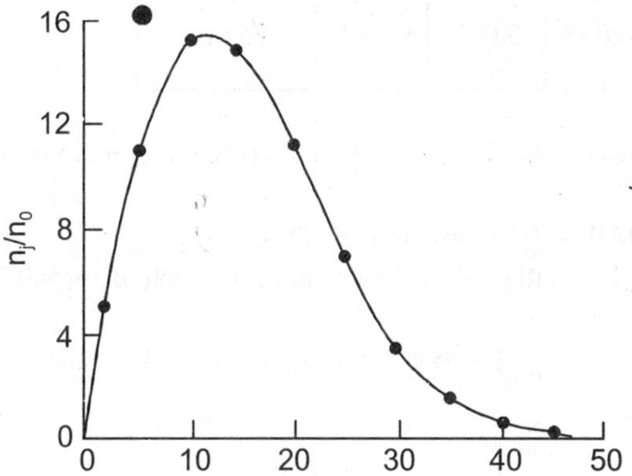

Fig. 9.1 *Plot of n_J/n_0 vs J*

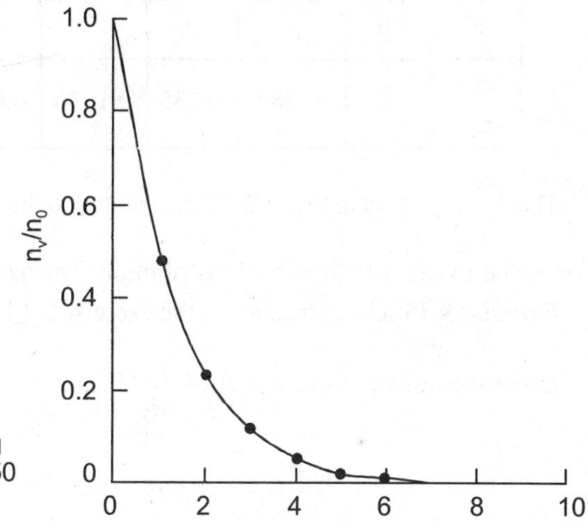

Fig. 9.2 *Plot of n_v/n_0 vs v*

$$J = -\frac{1}{2} + \frac{1}{2}\left[\frac{(2)(1.38\times10^{-23}\text{ J K}^{-1})(300\text{ K})}{(0.642\text{ cm}^{-1})(3.0\times10^{10}\text{ cm s}^{-1})(6.62\times10^{-34}\text{ J s})}\right]^{1/2}$$

$$= 12.24 \textit{ i.e., } 12$$

The agreement is very good and the error could be in reading from the plot.

***Exercise* 9.14:** The w_e value for one of the electronic state of molecule A is 151.2 cm^{-1}. Assuming that the energy (in cm^{-1}) of the vibrational levels is given as $\varepsilon_v = w_e(v + \frac{1}{2})$, and $\epsilon_o = \frac{1}{2}hc\omega_e$ Plot $\frac{n_v}{n_0}$ vs v at 27°C and point out the difference between the two curves (Figs. 9.1 and 9.2)

The equation (9.12) becomes

$$\frac{n_v}{n_0} = \frac{g_v \exp(-\varepsilon_v / k_bT)}{g_0 \exp(-\varepsilon_0 / k_bT)} = \exp(-hcw_e v/k_bT)$$

because the degeneracy of each vibrational level is one. Now

$$\frac{hcw_e}{k_bT} = \frac{(151.2\ \text{cm}^{-1})(6.62\times10^{-34}\ \text{J s})(3.0\times10^{10}\ \text{cm s}^{-1})}{(1.381\times10^{-23}\ \text{J K}^{-1})(300\ \text{K})}$$
$$= 0.725$$

By taking different values of v, n_v/n_0 can be calculated and is tabulated in table below. As a sample of calculation

$$\frac{n_1}{n_0} = e^{-(1)(0.725)} = e^{-0.725} = 0.484$$

v	0	1	2	3	4	5	6	7
$\frac{n_v}{n_0}$	1	0.484	0.235	0.1136	0.055	0.027	0.013	0.25×10^{-3}

The $\frac{n_v}{n_0}$ is plotted in Fig. 9.2. Comparing with the results of exercise 9.13, it is clear that no maxima is observed in case of vibrational distribution. This is due to the fact that $\varepsilon_v \gg k_bT$.

***Exercise* 9.15:** Using the data of the exercise 9.14, calculate the temperature for which $n_2/n_1 = 0.75$.

Using the results of exercise 9.14 $\frac{n_v}{n_0} = \exp(-w_e hcv/k_bT)$. From exercise 9.14 it is clear that

$$\frac{hcW_e}{(k_b)(300)} = 0.725$$

or

$$\frac{h}{k_b} = 0.725 \times 300 = 217.5$$

$$\text{Now } 0.75 = e^{-217.5/T}$$

Taking logarithms of both sides,

$$\ln 0.75 = -\frac{217.5}{T}$$

or

$$T = \frac{217.5}{\ln 0.75} = -\frac{217.5}{-0.288} = 756.0\ \text{K}$$

Note: One can see from the results of the Figs. 9.1 and 9.2, the distribution of molecules in various energy levels depends upon the successive difference between the respective energy levels.

9.6.1. *Average Value of Molecular Properties*

Maxwell-Boltzmann relation has been derived for the most probable distribution and average properties have been obtained using the most probable distribution. We also know that there are many other distributions but we neglect all and only use the most probable one. The agreement between the calculated average property and experimental one is quite good. This means that other distributions hardly make any contribution to the average property. We will show this mathematically.

Let D_0 be the most probable distribution and D_1 be any other distribution obtained from D_0 by rearranging the molecules among the energy levels in such a way that the total energy and total number of particles is unchanged. The ratio of the probabilities of D_1 and D_0 is given as

$$\frac{\Omega_{D1}}{\Omega_{D0}} \qquad ...(9.13)$$

and physical importance of D_1 is determined by this ratio. The number of molecules in D_0 is:

n_1, n_2, n_3 ...... and the corresponding values for D_1 can be written as

$$(n_1 + \Delta n_1), (n_2 + \Delta n_2), (n_3 + \Delta n_3), \ldots\ldots\ldots\ldots (n_i + \Delta n_i)\ldots\ldots \qquad ...(9.14)$$

where $\Delta n_1, \Delta n_2, \Delta n_3, \ldots\ldots\ldots\ldots \Delta n_i \ldots\ldots$ are finite numbers and can be negative or positive. Now

$$\ln \frac{\Omega_{D1}}{\Omega_{D0}} = \ln \Omega_{D1} - \ln \Omega_{D0} = \Delta \ln \Omega_D \qquad ...(9.15)$$

Taylor series expansion of Ω_D about Ω_{D0} yields

$$\Delta \ln \Omega_D = \sum_i \left(\frac{\partial \ln \Omega_D}{\partial n_i} \right) \Delta n_i + \frac{1}{2} \sum_i \left(\frac{\partial^2 \ln \Omega_D}{\partial n_i^2} \right) (\Delta n_i)^2 + \ldots\ldots \qquad ...(9.16)$$

We have not considered the mixed derivatives as these will be zero. Using equation (9.17), which have been used in solving the Maxwell-Boltzmann distribution (*i.e.*, method of undetermined multipliers),

$$\frac{\partial \ln \Omega_D}{\partial n_i} - \alpha - \beta \varepsilon_i = 0 \qquad ...(9.17)$$

or

$$\frac{\partial \ln \Omega_D}{\partial n_i} = \alpha + \beta \varepsilon_i \qquad ...(9.17a)$$

where α and β are undetermined constants. Thus

$$\sum_i \left(\frac{\partial \ln \Omega_D}{\partial n_i} \right) \Delta n_i = \alpha \sum_i \Delta n_i + \beta \sum_i \varepsilon_i \Delta n_i = 0 \qquad ...(9.18)$$

zero value arises because of the constancy of the total number of molecules ($\Sigma \Delta n_i = 0$) and total energy ($\Sigma \varepsilon_i \Delta n_i = 0$). We know that the value of Ω is given as

$$\Omega = N! \Pi \frac{(g_i)^{n_i}}{n_i!} \qquad ...(9.5)$$

Taking log of both sides of equation (9.5), we get

$$\ln \Omega_D = \ln N! + \sum_i \ln (g_i)^{n_i} - + \sum_i \ln(n_i!) \qquad ...(9.19)$$

Using Sterling's theorem, we get

$$\ln \Omega_D = N \ln N - N + \sum_i n_i \ln g_i - \sum_i n_i \ln n_i + \sum_i n_i$$

or $$\ln \Omega_D = N \ln N + \sum_i n_i \ln g_i - \sum_i n_i \ln n_i \quad \text{...(9.20)}$$

Because $N = \sum_i n_i$, the total number of molecules. Differentiation equation (9.20) twice, we get

$$\frac{\partial^2 \ln \Omega_D}{\partial n_i^2} = -\frac{1}{n_i} \quad \text{...(9.21)}$$

Using equations (9.17*a*) and (9.21) in equation (9.16), we get

$$\Delta \ln \Omega_D = \sum_i (\alpha + \beta\varepsilon_i)\Delta n_i + \frac{1}{2}\sum_i -\frac{1}{n_i}(\Delta n_i)^2 + ...$$

$$\Delta \ln \Omega_D = \alpha \sum_i \Delta n_i + \beta \sum_i \varepsilon_i \Delta n_i + \frac{1}{2}\sum_i -\frac{1}{n_i}(\Delta n_i)^2 + ... = -\frac{1}{2}\sum_i \frac{1}{n_i}(\Delta n_i)^2 \quad \text{...(9.22)}$$

Since as said above $\Sigma \Delta n_i = 0$ and $\Sigma \varepsilon_i \Delta n_i = 0$. From equations (9.15) and (9.22), we get

$$\ln \frac{\Omega_{D_1}}{\Omega_{D_0}} = \Delta \ln \Omega_D = -\frac{1}{2}\sum_i \frac{1}{n_i}(\Delta n_i)^2$$

or $$\frac{\Omega_{D_1}}{\Omega_{D_0}} = \exp\left(-\frac{1}{2}\sum_i \frac{1}{n_i}(\Delta n_i)^2\right) \quad \text{...(9.23)}$$

This equation can be used to show that the average properties will not be affected neglecting other distributions. Consider that the distribution number of D_1 differ from the corresponding numbers D_0 by approximately by 0.1% of the latter (we may assume that the signs and numerical values of Δn_i are chosen in such a manner so that $\Sigma \Delta n_i = 0$ and $\Sigma \varepsilon_i \Delta n_i = 0$), *i.e.*,

$$\frac{|\Delta n_i|}{n_i} \simeq \frac{1}{10^3}$$

So that $$\frac{|\Delta n_i|^2}{n_i} \simeq \frac{n_i}{10^6} \quad \text{...(9.24)}$$

Then $$-\frac{1}{2}\sum_i \frac{1}{n_i}(\Delta n_i)^2 \approx -\frac{1}{2}\sum_i \frac{n_i}{10^6} = \approx -\frac{1}{2 \times 10^6}\sum_i n_i = -\frac{N}{2 \times 10^6}$$

If we consider one mole of molecules, $N = 6.02 \times 10^{23}$, then

$$\frac{\Omega_{D_1}}{\Omega_{D_0}} = \exp\left(-\frac{N}{2 \times 10^6}\right) = \exp\left(-\frac{6.02 \times 10^{23}}{2 \times 10^6}\right) = \exp(-3.01 \times 10^{17}) \quad \text{...(9.25)}$$

which is an extremely small numbers. This justifies the neglect of distribution other than Ω_{D_0}.

9.7. PARTITION FUNCTIONS

Partition function is defined by the equation (9.26) as

$$q = \sum_i g_i \exp(-\varepsilon_i/k_bT) \qquad ...(9.26)$$

where g_i is the degeneracy of ith energy level with energy ε_i above the zero point energy and the summation is overall the energy levels. If the system is assumed to be ideal (*i.e.*, no interaction among the molecules) and Born-Oppenheimer approximation is valid, then energy of any level can be written as the sum of translational energy, rotational energy, vibrational energy, electronic energy and nuclear energy *i.e.*,

$$\varepsilon = \varepsilon_{\text{nuc}} + \varepsilon_{\text{ele}} + \varepsilon_{\text{vib}} + \varepsilon_{\text{rot}} + \varepsilon_{\text{tra}} + \varepsilon_0 \qquad ...(9.27)$$

Thus equation (9.26) becomes

$$q = \sum_i g_{i(\text{nuc})} \exp(-\varepsilon_{\text{nuc}}/k_bT) \sum_i g_{i(\text{ele})} \exp(-\varepsilon_{i(\text{ele})}/k_bT)$$

$$\sum_i g_{i(\text{vib})} \exp(-\varepsilon_{i(\text{vib})}/k_bT) \sum_i g_{i(\text{rot})} \exp(-\varepsilon_{i(\text{rot})}/k_bT)$$

$$\sum_i g_{i(\text{tr})} \exp(-\varepsilon_{i(\text{tr})}/k_bT) \exp(-\varepsilon_0/k_bT) \qquad ...(9.28)$$

where ε_0 is the zero point energy and the energy of the vibrational levels are measured from this level.

Thus,

$$q = q_{\text{nuc}}\, q_{\text{ele}}\, q_{\text{vib}}\, q_{\text{rot}}\, q_{\text{tr}}\, q_0 \qquad ...(9.29)$$

where q_i's are respective partition functions. It may also be mentioned here that it represents the effective number of states available to the system and thus in turn is very important to the development of statistical thermodynamics.

9.8. CALCULATION OF PARTITION FUNCTIONS

9.8.1. *Nuclear Partition Function*

The values of the energies of the excited nuclear states are so large that hardly any contribution to the sum is made by the higher terms *i.e.*,

$$q_{\text{nuc}} = g_{0(\text{nuc})} \exp(-\varepsilon_{0(\text{nuc})}/k_bT) + g_{1(\text{nuc})} \exp(-\varepsilon_{1(\text{nuc})}/k_bT) + \ldots \text{ becomes}$$

$$q_{\text{nuc}} = g_{0(\text{nuc})} \exp(-\varepsilon_{0(\text{nuc})}/k_bT) = g_{0(\text{nuc})} \qquad ...(9.30)$$

if the energy of the ground nuclear state is considered zero. It is normally observed that in almost every calculation g_{nuc} cancels out and detailed knowledge of $g_{0\,(\text{nuc})}$ is not necessary. Important exceptions to this are calculations for symmetric diatomic molecules such as H_2 and D_2. It will be shown in Section (9.13) that these systems can be regarded as a mixtures of ortho and para molecules. These differ in the symmetry properties of their nuclear spin wave functions.

The nuclear partition function of any molecule or atom is the degeneracy of the ground state nuclear energy level. The degeneracy of the ortho state is given as

$$g_{0(\text{nuc})}^{\text{ortho}} = (I+1)(2I+1) \qquad ...(9.31)$$

and for para molecules $g_{0(\text{nuc})}^{para} = I(2I+1)$ ...(9.32)

where I is the spin quantum number of the nucleus.

9.8.2. *Ortho and Para Nuclear States*

The group of states of higher nuclear spin degeneracy is called *ortho state*. The group of states of group of lower spin degeneracy is called *para state*. Nucleus of many atoms possesses a resultant spin angular momentum. The spin vectors of two nuclei yield a resultant spin quantum number S. If I is the spin of one of the nuclei, then for a homonuclear diatomic molecule, the resultant spin S will be

$$S = 2I, 2I - 1, 2I - 2, \ldots\ldots\ldots\ldots 0 \qquad \text{...(9.32a)}$$

and the degeneracy of the resultant states will be $2S + 1$. Out of these $2I, 2I - 2, 2I - 4, \ldots$ are even (symmetrical) states and $2I - 1, 2I - 3, 2I - 5, \ldots$, are odd (antisymmetric) states. In general the resultant nuclear spin is $2I - n$, where n is an integer and the value of S cannot exceed $2I$ (*i.e.*, $n = 0, 1, 2, 3, \ldots$). For ortho states, n must be either zero or even integer. For para states, n must be odd. So the multiplicity of nuclear spin state is:

$$\text{Multiplicity} = 2(2I - n) + 1 = 4I - 2n + 1 \qquad \text{...(9.33)}$$

***Exercise* 9.16:** Find the value of nuclear partition function.

Equation (9.31) shows that the nuclear partition function of any molecule or atom is the degeneracy of the ground state nuclear energy level. The degeneracy of the ortho is given as

$$g_0(nuc) = \sum_{n=0,2,4,6..} 2(2I - n) + 1 = (4I + 1) + (4I - 3) + (4I - 7) + \ldots\ldots \qquad \text{...(9.34)}$$

This is an arithmetic progression with difference of 4.

$$\text{The sum is given as} = \frac{\text{Number of terms}}{2}[\text{First term} + \text{Last term}] \qquad \text{...(9.35)}$$

$$\text{Last term} = \text{First term} + (\text{Number of terms} - 1)(\text{difference } d) \qquad \text{...(9.36)}$$

Substituting the values of different terms and assuming that N be the number of terms in equation (9.36), we get

$$1 = 4I + 1 + (N - 1)(-4) = 4I + 1 - 4N + 4$$

or
$$N = I + 1 \qquad \text{...(9.37)}$$

Substituting the values of terms in equation (9.35), we get

$$\text{Sum} = g_0(\text{nuc, ortho}) = \frac{I+1}{2}[4I + 1 + 1] = (I + 1)(2I + 1) \qquad \text{...(9.38)}$$

For odd number of para states, the number of states present can be

$$1 = 4I - 1 + (N - 1)(-4) = 4I - 1 - 4N + 4$$

or
$$N = \frac{4I+2}{4} = \frac{2I+1}{2}$$

$$\text{Sum} = g_0(\text{nuc, para}) = \frac{2I+1}{2 \times 2}[4I - 1 + 1] = I(2I + 1) \qquad \text{...(9.39)}$$

$$\text{Total number of states} = (I + 1)(2I + 1) + I(2I + 1) = (2I + 1)(I + 1 + I) = (2I + 1)^2 \qquad \text{...(9.40)}$$

From the above discussion it is clear that homonuclear diatomic molecule gas behaves like a mixture of two kinds of molecules and these molecules are present in a fixed ratio of numbers. One kind of molecules has only odd and the other with only even J values (rotational quantum number) in order to satisfy the symmetry requirements. *Therefore for homonuclear diatomic molecules, the product of nuclear and rotational partition function for different kind of particles should be*

(*a*) **Boson (even mass nuclei):** Since the complete wave function of Boson has to be symmetric, thus in taking the summation to write the product of rotational and nuclear partition functions, *i.e.*, q_{rot} q_{nuc} ($q_{rot.nuc}$), when J is even, the nuclear spin states must be symmetric and for odd J nuclear spin states should be antisymmetric

$$q_{nuc,rot} = q_{nuc}q_{rot}$$

$$= (I+1)(2I+1)\sum_{J=0,2,4..}(2J+1)e^{-\frac{h^2(J^2+J)}{8\pi^2 Ik_bT}} + I(2I+1)\sum_{J=1,2,5..}(2J+1)e^{-\frac{h^2(J^2+J)}{8\pi^2 Ik_bT}} \quad ...(9.41)$$

(*b*) **Fermion (odd mass number):** For such particles the total wave function should be antisymmetric. Thus $\psi_{nuc}\ \psi_{rot}$ should also be antisymmetric, *i.e.*, ψ_{nuc} and ψ_{rot} should have opposite symmetry, *i.e.*, if ψ_{rot} is symmetric (J = even), ψ_{nuc} should be antisymmetric (odd) and if ψ_{rot} is antisymmetric (J = odd), even nuclear state are accessible. The particle function

$$q_{nuc,rot} = q_{nuc}q_{rot} \quad ...(9.42)$$

$$= I(2I+1)\sum_{J=0,2,4..}(2J+1)e^{-\frac{h^2(J^2+J)}{8\pi^2 Ik_bT}} + (I+1)(2I+1)\sum_{J=1,3,5..}(2J+1)e^{-\frac{h^2(J^2+J)}{8\pi^2 Ik_bT}} \quad ...(9.43)$$

If $I = 0$ (which occurs only in nuclei having even number of protons and even number of neutrons) then the partition function (equation 9.41) becomes

$$q_{nuc,rot} = q_{nuc}q_{rot} = \sum_{J=0,2,4..}(2J+1)e^{-\frac{h^2(J^2+J)}{8\pi^2 Ik_bT}} \quad ...(9.44)$$

In this case, rotational states with odd J do not occur.

Heteronuclear Diatomic Molecules: For heteronuclear diatomic molecules XY, the rotational states are distinguishable and partition function is

$$q_{rot} = \sum_{J=0}^{\infty}(2J+1)e^{-\frac{h^2(J^2+J)}{8\pi^2 Ik_bT}} \quad ...(9.45)$$

whereas the nuclear spin partition function is

$$q_{nuc} = (2I_1 + 1)(2I_2 + 1) \quad ...(9.46)$$

where I_1 and I_2 are the nuclear spins of X and Y respectively.

$$q_{nuc}q_{rot} = (2I_1+1)(2I_2+1)\sum_{J=0}^{\infty}(2J+1)e^{-\frac{h^2(J^2+J)}{8\pi^2 Ik_bT}} \quad ...(9.47)$$

At high temperature (9.45) and (9.47) reduces to

$$= [I(2I+1)+(I+1)(2I+1)]\frac{8\pi^2\mu r^2 k_b T}{h^2} = (2I+1)^2\frac{8\pi^2\mu r^2 k_b T}{h^2} \quad ...(9.48)$$

TABLE 9.1. Molecular parameters of few characteristic diatomic molecules, data have been taken different sources

Molecule	Electronic state	Θ_{vib} K	Θ_{rot} K	D_0(kJ mol^{-1})	D_0(kJ mol^{-1})
H^2	$^1\Sigma_g^+$	6215	85.3	432.1	457.6
HD					
D_2	$^1\Sigma_g^+$	4394	42.7	435.6	453.9
Cl_2	$^1\Sigma_g^+$	805	0.352	239.2	242.3
Br_2	$^1\Sigma_g^+$	463	0.116	190.1	191.9
I_2	$^1\Sigma_g^+$	308	0.0537	148.8	150.3
O_2	$^3\Sigma_g^+$	2256	2.07	493.6	503.0
N_2	$^1\Sigma_g^+$	3374	2.88	941.6	953.0
CO	$^1\Sigma^+$	3103	2.77	1070	1085
NO	$^2\Pi_{1/2}$	2719	2.39	626.8	638.1
HCl	$^1\Sigma^+$	4227	15.02	427.8	445.2
HBr	$^1\Sigma^+$	3787	12.02	362.6	377.7
HI	$^1\Sigma^+$	3266	9.25	284.7	308.6
Na_2	$^1\Sigma_g^+$	229	0.221	71.1	72.1
K_2	$^1\Sigma_g^+$	133	0.081	53.5	54.1

TABLE 9.2. Values of the characteristic rotational temperatures, the vibrational temperatures, D_0, for the ground state, and the symmetry number, σ, for some polyatomic molecules. The values in the parenthesis indicate the degeneracy of that vibrational mode

Molecule	Θ_{rot} K	Θ_{vib} K	D_0 (kJ mol^{-1})	σ
CO_2	0.561	3360,954(2), 1890	1596	
H_2O	40.1,20.9, 13.4	5360, 5160, 2290	917.6	
NH_3	13.6, 13.6, 8.92	4800, 1360, 4880(2), 230(2)	1158	
ClO_2	2.50, 0.478, 0.400	1360, 640, 1600	378	
SO_2	2.92, 0.495, 0.422	1660, 750, 1960,	1063	
N_2O	0.603	3200, 850(2), 1840	1104	
NO_2	11.5, 0.624, 0.590	1900, 1080, 2330	928	

CH_4	7,54, 7.54, 7.54	4170,2180(2), 4320(3), 1870(3)	1642	2
CH_3Cl	7,32, 0.637, 0.637	4270,1950,1050,4380(2), 2140(2), 1460(2)	1551	
CCl_4	0.0823, 0.0823, 0.0823	660, 310(2), 1120(3), 450(3)	1292	2

***Exercise* 9.17:** Calculate the nuclear partition functions of ortho H_2 and D_2.

Using equation (9.31), we get

For hydrogen atom $I = \frac{1}{2}$.

$$g_{0(\text{nuc})}^{\text{ortho}} = \left(\frac{1}{2}+1\right)\left(2\times\frac{1}{2}+1\right) = 3/2 \times 2 = 3$$

$$q_{0(\text{nuc})}^{\text{ortho}} = g_{0(\text{nuc})}^{\text{ortho}} = 3$$

For deuterium atom $I = 1$

$$g_{0(\text{nuc})}^{\text{ortho}} = (1+1)(2\times 1+1) = 6$$

$$q_{0(\text{nuc})}^{\text{ortho}} = g_{0(\text{nuc})}^{\text{ortho}} = 6$$

9.8.3. *Electronic Partition Functions*

Similar to the energies of excited nuclear states, the differences between the successive electronic energy levels are also large so electronic partition function is given as

$$q_{\text{ele}} = g_{0(\text{ele})} \qquad ...(9.49)$$

if $\varepsilon_0 = 0$ is considered. But in some cases the difference in the energy levels of the excited states is small or if the temperature is very high then the contribution of individual term is also considered *i.e.*,

$$q_{\text{ele}} = g_{0(\text{ele})} + g_{1(\text{ele})} \exp\left(-\varepsilon_{1(\text{ele})}/k_bT\right) + ... \qquad ...(9.50)$$

In general $g_{0(\text{ele})} = 1$ for most of the molecules but exceptions are there, for example O_2, free radicals etc.

***Exercise* 9.18:** Calculate the electronic partition function of chlorine atom if it is assumed that energies of first and higher excited states are very large.

The ground state spectroscopic term for chlorine atom is $^2p_{3/2}$ where subscript is the value of total electronic angular momentum (J). The degeneracy of any state is $2J + 1$.

Thus $$g_{0(\text{ele})} = 2J+1 = 2\times\frac{3}{2}+1 = 4$$

$$q_{\text{ele}} = g_{0(\text{ele})} = 4$$

***Exercise* 9.19:** If energy of the first electronically excited state of chlorine atom is 0.11 eV, calculate the contribution to the partition function of this term at 1000 K.

g_0 obtained in exercise 9.18 is 4. The J value for 1st excited electronic state ($^2P_{1/2}$) is $\frac{1}{2}$, so

$$g_{1(ele)} = \frac{1}{2} \times 2 + 1 = 2$$

$$\varepsilon_{1(ele)} = 0.11 \text{ eV} = (0.11\text{eV})\,(1.6 \times 10^{-19} \text{ J eV}^{-1}) = 1.8 \times 10^{-20} \text{ J}$$

so using equation (9.50)

$$q_{ele} = 4 + 2 \exp\left[-\frac{1.8\times10^{-20} \text{ J}}{(1.381\times10^{-23} \text{ J K}^{-1})\,(1000 \text{ K})}\right]$$

$$= 4.56$$

$$\text{The contribution by the second term} = \frac{0.56}{4.56} \times 100$$

$$= 12.3\%$$

9.8.4. *Vibrational Partition Function*

9.8.4.1. Diatomic molecule

Considering vibrations as a simple harmonic motion, the energy levels of the vibrational states are given as

$$\varepsilon_{vib} = h\nu(\text{v} + \frac{1}{2}),\ \text{v} = 0, 1, 2 \qquad ...(9.51)$$

where each term has the usual meanings. The vibrational frequency can be expressed by ν (Hz), $\bar{\nu}$ or ω_e (cm^{-1}).

Since the lowest level in the vibrational motion has the energy $\frac{1}{2}h\nu$ and if the energies of excited levels are measured relative to this, equation (9.51) becomes

$$\varepsilon_{vib} = h\nu\text{v} \qquad ...(9.52a)$$

and partition function will be

$$q_{vib} = \sum_{\text{v}=0}^{\infty} g_{\text{v}} \exp(-h\nu\text{v} / k_b T) \qquad ...(9.53)$$

Since $g_{\text{v}} = 1$ for all vibrational levels

$$q_{vib} = \frac{1}{1-\exp(-h\nu / k_b T)} \qquad ...(9.54)$$

Substituting $h\nu/k_b = \theta_{vib}$, the characteristic temperature for vibration

$$q_{vib} = \frac{1}{1-\exp(-\theta_{vib} / T)} \qquad ...(9.55)$$

9.8.4.2. Polyatomic molecules

The number of vibrational degrees of freedom for linear diatomic molecules are 3 $N-5$ and for non-linear $3N-6$ (N is the number of atoms in the molecule). Expressing the vibrations in terms of normal modes of vibrations, the energy levels of each normal mode is independent of others so the total vibrational energy is the sum of the energy due to each normal mode. The total vibrational partition function of each normal mode, *i.e.*,

$$q_{\text{vib}} = q_{\text{vib}(1)}\, q_{\text{vib}(2)} \ldots q_{\text{vib}}\,(3N-5 \text{ or } 6) \quad \ldots(9.56)$$

$$= \prod_{i=1}^{3N-5 \text{ or } 3N-6} \frac{1}{1-\exp(-h\nu_i / k_b T)} = \prod_{i=1}^{3N-5 \text{ or } 3N-6} \frac{1}{1-\exp(-\theta_{\text{vib}(1)} / T)} \quad \ldots(9.57)$$

***Exercise* 9.20:** (*a*) Calculate the vibrational partition function for Br_2 at 300 K if the vibrational frequency (w_e) = 151.2 cm^{-1}. Also repeat the above calculation for 1000 K, (*b*) Calculate the vibrational partition by taking sum upto v = 7. Find the percentage by comparing the results in a

From the exercise 9.13

$$\frac{hcw_e}{k_b T} = \frac{hcw_e}{k_b \cdot 300} = 0.725$$

$$q_{\text{vib}} = \frac{1}{1-e^{-0.725}} = \frac{1}{1-0.484} = \frac{1}{0.516} = 1.938$$

$$\frac{hcw_e}{k_b T} = \frac{hcw_e}{k_b \times 1000} = 0.2175$$

$$q_{\text{vib}} = \frac{1}{1-e^{-0.2175}} = \frac{1}{0.1954} = 5.118$$

(*b*) By taking the data from exercise (9.13), the sum is

$$q'_{\text{v}} = 1.000 + 0.484 + 0.235 + 0.1136 + 0.055 + 0.027 + 0.00025$$
$$= 1.91485$$

$$\% \quad \text{Error: } \frac{1.938 - 1.91485}{1.938} \times 100 = 1.195\%$$

At 1000 K, the population ratio of $\frac{N_{\text{v}}}{N_o}$ upto v = 7 is

v	0	1	2	3	4	5	6	7
$\frac{N_{\text{v}}}{N_o}$	1	0.8045	0.6473	0.5207	0.4190	0.3371	0.2712	0.2712

$$q'_{\text{v}}\,(1000 \text{ K}) = 3.9998$$

$$\% \quad \text{Error : } \frac{5.118 - 3.9998}{5.118} \times 100 = 21.84\%$$

Results show that at 1000 k, higher vibrational levels (v > 7) are populated, which we are not considering and thus the population error increases if the temperature increases.

***Exercise* 9.21:** Calculate the vibrational partition function for H_2O at 300 K.The vibrational frequencies are 1654 cm^{-1}, 3825 cm^{-1} and 3935 cm^{-1}.

The $hc\bar{v}_i/k_bT$ terms for each vibrational frequencies are:

$$\frac{hc\bar{v}_1}{k_bT} = \frac{(1654\ \text{cm}^{-1})(6.63\times10^{-34}\ \text{J s})(3.0\times10^{8}\ \text{m s}^{-1})(10^{2}\ \text{cm m}^{-1})}{(1.381\times10^{-23}\ \text{J K}^{-1})(300\ \text{K})}$$

$$= 7.929$$

$$e^{-7.929} = 3.60\times10^{-4}$$

$$\frac{hc\bar{v}_2}{k_bT} = \frac{(3825\ \text{cm}^{-1})(6.62\times10^{-34}\ \text{J s})(3.0\times10^{8}\ \text{m s}^{-1})(10^{2}\ \text{cm m}^{-1})}{(1.381\times10^{-23}\ \text{J K}^{-1})(300\ \text{K})}$$

$$= 18.336$$

$$e^{-18.336} = 1.088\times10^{-8}$$

$$\frac{hc\bar{v}_3}{k_bT} = \frac{(3935\ \text{cm}^{-1})(6.62\times10^{-34}\ \text{J s})(3.0\times10^{8}\ \text{m s}^{-1})(10^{2}\ \text{cm m}^{-1})}{(1.381\times10^{-23}\ \text{J K}^{-1})(300\ \text{K})}$$

$$= 18.863$$

$$e^{-18.863} = 6.425\times10^{-9}$$

The total vibrational partition function

$$= \frac{1}{1-\exp(-hv_1/k_bT)}\times\frac{1}{1-\exp(-hv_2/k_bT)}\times\frac{1}{1-\exp(-hv_3/k_bT)}$$

$$= \frac{1}{1-3.60\times10^{-4}}\times\frac{1}{1-1.088\times10^{-8}}\times\frac{1}{1-6.425\times10^{-9}}\approx 1$$

Note: Generally for most cases the vibrational partition function at room temperature is unity.

9.8.5. *Rotational Partition Function*

The energies of the rotational levels of diatomic molecule are given by

$$\varepsilon_{\text{rot(J)}} = \frac{h^2}{8\pi^2 I}J(J+1),\ J=0,1,2,... \quad ...(9.58)$$

Each term is explained in Chapter 5 (Section 3).

Rotational energy levels are degenerate and degeneracy is expressed as

$$g_{\text{J(rot)}} = 2J+1 \quad ...(9.59)$$

Rotational partition function is

$$q_{\text{rot}} = \sum_{J=0}^{\infty}(2J+1)\exp\left(-\frac{h^2J(J+1)}{8\pi^2 Ik_bT}\right) \quad ...(9.60)$$

Except for hydrogen and hydrogen isotopes the separation between successive rotational levels is very small and can be considered as continuous. Replacing summation by integration

$$q_{rot} = \frac{8\pi^2 Ik_bT}{h^2} = \frac{T}{\theta_{rot}} \qquad ...(9.61)$$

where θ_{rot} is the rotational characteristic temperature and

$$\theta_{rot} = \frac{h^2}{8\pi^2 Ik_b} \qquad ...(9.62)$$

For homonuclear diatomic molecules only half the rotational states are populated so the rotational partition function will be

$$q_{rot} = \frac{T}{2\theta_{rot}} \qquad ...(9.63)$$

In general

$$q_{rot} = \frac{T}{\sigma\theta_{rot}} \qquad ...(9.64)$$

where σ is the symmetry number. It is 2 for homonuclear and 1 for heteronuclear diatomic molecules.

The rotational partition function for polyatomic molecules is

$$q_{rot} = \frac{1}{\sigma}\left(\frac{8\pi^2 k_bT}{h^2}\right)^{3/2} (\pi I_A I_B I_C)^{1/2} \qquad ...(9.65)$$

where I_A, I_B and I_C are the principal moments of inertia.

***Exercise* 9.22:** If equilibrium internuclear distance of $^{127}I_2$ is 2.666×10^{-10} m, calculate (1) moment of inertia; (2) characteristic rotation temperature; (3) rotational partition function at 300 K and 1000 K.

Moment of inertia is given as

$$I = \mu r^2 = \frac{m_A m_B}{m_A + m_B} \times r^2$$

$$= \frac{(127/N_{av})(127/N_{av})}{(127/N_{av}) + (127/N_{av})} (2.666 \times 10^{-10}\text{ m})^2$$

$$= \frac{(127\text{ g mol}^{-1})(10^{-3}\text{ kg g}^{-1})(2.666 \times 10^{-10}\text{ m})^2}{2(6.023 \times 10^{-23}\text{ mol}^{-1})}$$

$$= 7.49 \times 10^{-45}\text{ kg m}^2$$

$$\theta_{rot} = \frac{h^2}{8\pi^2 Ik_b}$$

$$= \frac{(6.625 \times 10^{-34}\text{ J s})^2}{(8\pi^2)(7.49 \times 10^{-45}\text{ kg m}^2)(1.38 \times 10^{-23}\text{ J K}^{-1})} = 0.05376\text{ K}$$

$$q_{rot}(300\text{ K}) = \frac{T}{\sigma\theta_{rot}} = \frac{300\text{ K}}{2\times 0.05376\text{ K}}$$

$$= 2790$$

$$q_{rot}(1000\text{ K}) = \frac{1000\text{ K}}{2\times 05376} = 9300$$

***Exercise* 9.23:** Calculate the rotational partition function for H_2 at 300 K using equations (9.60) and (9.63). $B = 60.80\text{ cm}^{-1}$ for H_2.

According to equation (9.60)

$$q_{rot} = 1 + 3\exp(-\varepsilon_1/k_bT) + 5\exp(-\varepsilon_2/k_bT) + 7\exp(-\varepsilon_3/k_bT)$$
$$+ 9\exp(-\varepsilon_4/k_bT) + 11\exp(-\varepsilon_5/k_bT) + ...$$

The common factor in the calculation of ε_i is

$$\frac{h^2}{8\pi^2 Ik_bT} = \frac{Bhc}{k_bT} = \frac{(60.80\text{ cm}^{-1})(6.62\times10^{-34}\text{ J s})(3.0\times10^{10}\text{ cm s}^{-1})}{(1.381\times10^{-23}\text{ J K}^{-1})(300\text{ K})}$$

$$= 291.45\times10^{-3}$$

$$\frac{\varepsilon_i}{k_bT} = \frac{Bhc}{k_bT}J(J+1) = 291.45\times10^{-3}\,J(J+1)$$

Calculating each term in the above expression

$$q_{rot} = 1 + 1.6748 + 0.870 + 0.212 + 0.026 + 0.002 = 3.785$$

Since only half the levels are populated so the partition function is $\frac{3.785}{2} = 1.8925$.

According to equation (9.63), the rotational partition function can be calculated as

$$q_{rot} = \frac{8\pi^2 Ik_bT}{\sigma h^2} = \frac{k_bT}{\sigma Bhc}$$

$$= \frac{1}{(2)(0.29145)} = 1.715$$

There is some difference and it is 1.8925 – 1.715 = 0.1775

$$\%\text{ Error}: \frac{1.8925 - 1.715}{1.8925}\times100 = \frac{0.1775}{1.8925}\times100 = 9.4.$$

***Exercise* 9.24:** Calculate the rotational partition function for NH_3 at 27°C. The three moments of inertia are 2.78×10^{-47}; 2.78×10^{-47} and 4.33×10^{-47} kg m^2 respectively, σ for $NH_3 = 3$.

Using equation (9.65),

$$q_{rot} = \frac{1}{(3)}\left[\frac{(8)(\pi^2)(1.38\times10^{-23}\text{ J K}^{-1})(300\text{ K})}{(6.62\times10^{-34}\text{ J s})^2}\right]^{3/2}$$

$$[(\pi)(2.78\times10^{-47}\text{ kg m}^2)^2(4.33\times10^{-47}\text{ kg m}^2)]^{1/2}$$

$$= 69.73.$$

9.8.6. *Translational Partition Function*

The quantum mechanical solution of a particle of mass m moving in a three dimensional cubical box of side a gives the energy, which is the sum of the energies in three directions *i.e.*, x, y, z. The expression is

$$\varepsilon_{tr} = \frac{h^2}{8ma^2}(n_x^2 + n_y^2 + n_z^2) \quad ...(9.66)$$

where n are quantum numbers having values 1, 2, 3, ... Since these components of energy are independent of each other, the total partition function will be a product of translational partition function in each direction,

$$q_{tr} = q_x q_y q_z = q_x^3 \quad ...(9.67)$$

in case of cube. Thus the translational partition function in x-axis is

$$q_x = \Sigma \exp\left(-\frac{h^2 n_x^2}{8ma^2 k_b T}\right) \quad ...(9.68)$$

Since the translational energy levels are very closely spaced and separation is much less than $k_b T$, the summation can be replaced by integration. Equation (9.68) reduces to

$$q_x = \frac{(2\pi m\, k_b T)^{1/2} a}{h} \quad ...(9.69)$$

So

$$q_{tr} = \frac{(2\pi m\, k_b T)^{3/2} a^3}{h^3} = \frac{(2\pi m\, k_b T)^{3/2} V}{h^3} \quad ...(9.69a)$$

where V is the volume.

***Exercise* 9.25:** Calculate the translational partition function of 1 mol of hydrogen at 300 K and at 1 atm (1.01325×10^5 N m^{-2}) pressure.

The problem for ideal gases can be first treated a general case as

(*i*) $\bar{V}^0$ *i.e.*, molar volume at 1 atm (1.01325×10^5 N m^{-2}) can be expressed as

$$\bar{V}^0 = \frac{RT}{p} = \frac{(8.21\times10^{-5}\text{ atm m}^3\text{ mol}^{-1}\text{ K}^{-1})(T\text{ K})(1\text{ mol})}{1\text{ atm}}$$

$$= 8.21 \times 10^{-5}\, T\text{ m}^3 \quad ...(9.70)$$

$$(ii)\quad q_{tr}^0 = \frac{(2\pi m\, k_b T)^{3/2}}{h^2}\bar{V}^0 = \frac{\left(2\pi \frac{M}{N_{av}}\cdot\frac{R}{N_{av}}T\right)^{3/2}\bar{V}^0}{h^3}$$

$$\text{Partition function per molecule} = \frac{q_{tr}^0}{N_{av}} = \frac{(2\pi\, MRT)^{3/2}\,\bar{V}^0}{h^3 N_{av}^4}$$

$$= \frac{[(2\pi)(M \times 10^{-3}\ \text{kg mol}^{-1})(8.314\ \text{J mol}^{-1}\ \text{K}^{-1})(T\ \text{K})]^{3/2}}{(6.63\times10^{-34}\ \text{J s})^3\ (6.023\times10^{23}\ \text{mol}^{-1})^4} \times 8.21 \times 10^{-5}\ \text{m}^3 T$$

$$= 0.02563\ T^{5/2}\ M^{3/2} \qquad ...(9.71)$$

where M is the molecular weight in g.

(*iii*) Now for H_2, $M = 2$, $T = 300$

$$\frac{q^0_{\text{tr}}}{N_{\text{av}}} = (0.02563)\,(2)^{3/2}\,(300)^{5/2} = 1.156 \times 10^5$$

$$q^0 = (1.156 \times 10^5)\,(6.02 \times 10^{23}) = 6.959 \times 10^{28}$$

9.8.7. *Number of Quantum States in Molecular Systems* (TRANSLATIONAL DEGENERACY)

We have derived expressions to calculate different kinds of partition functions and calculate their values. The magnitude of partition functions or sum over all the quantum states effectively give the number of quantum states available for filling these quantum states with molecules. At first instance it appears that the number of molecules available is larger than the quantum states available. For examples, we know that there are three translation degrees of freedom, three rotational degrees of freedom and 3 N – 6 degrees of vibrational degrees of freedom for non-linear molecules. Otherwise there are 3 N + 1 total number of quantum numbers. This number seems to be small considering the number of molecules and earlier calculations have further confirmed this. For example, magnitude of rotational partition function have shown that effective number of rotational quantum states available for occupation is of the order of 10^2, the effective number of vibrational states are even less than this and does not exceed 10, followed by the smaller number of effective electronic quantum states. This is because the gap between the energy levels increases as we proceed from translational to rotational to vibration to electronic ones. We will show in Exercise (9.26) that number of translational quantum states is very large as compared to number of molecules.

***Exercise* 9.26:** Calculate the value of translational degeneracy and show that number of translational quantum states is very large.

Translation energy for a particle in a three dimensional box is given by the expression

$$\varepsilon_{tr}\left(\frac{n_x}{a}, \frac{n_y}{b}, \frac{n_z}{c}\right) = \frac{h^2}{8m}\left[\frac{n_x^2}{a^2} + \frac{n_x^2}{b^2} + \frac{n_x^2}{c^2}\right] \qquad ...(9.72)$$

Let $$r^2 = \frac{n_x^2}{a^2} + \frac{n_x^2}{b^2} + \frac{n_x^2}{c^2} \qquad ...(9.73)$$

So the equation (9.72) becomes

$$\varepsilon_{tr} = \frac{h^2 r^2}{8m} \qquad ...(9.74)$$

Suppose that the possible values of $\frac{n_x}{a}, \frac{n_y}{b}$ and $\frac{n_z}{c}$ are represented as Cartesian coordinates

(Fig. 9.3). Each quantum state can then be represented as a point, for example a point P has the coordinates, n_x/a, n_y/b and n_z/c.

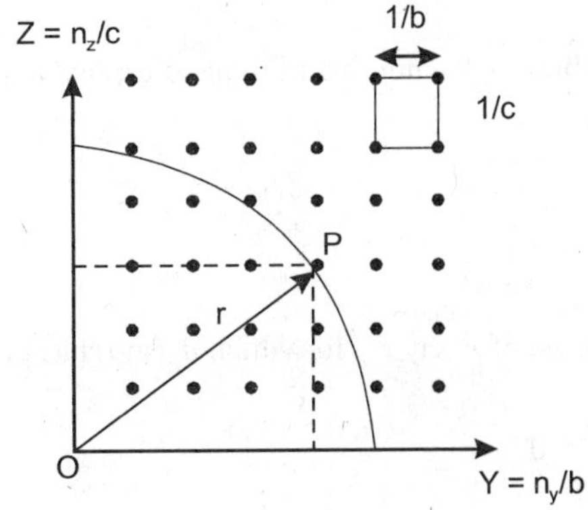

Fig. 9.3. *Representation of permitted* $\frac{n_y}{b}$ *and* $\frac{n_z}{c}$ *values as two dimensional array of points.*

If all such points are joined by straight lines to the neighbouring points, the Cartesian space is divided into rectangular boxes (units cells), the sides of which have the lengths $1/a$, $1/b$ and $1/c$. The volume of each cell

$$V = \frac{1}{abc} \quad \text{...(9.75)}$$

Since $$\text{OP} = \left(\frac{n_x^2}{a^2} + \frac{n_y^2}{b^2} + \frac{n_z^2}{c^2}\right)^{\frac{1}{2}} = r$$

all the quantum states having the same value of r must be represented by points which lie on a sphere of radius r. Further n_x, n_y and n_z are positive numbers, all the quantum states thus lie on the positive octant of this sphere. Therefore all the quantum states with smaller value of r than this must be represented by points which lie inside the positive octant on the sphere of radius r. The volume of this octant is:

$$\frac{1}{8}\left(\frac{4}{3}\pi r^3\right) = \frac{1}{6}\pi r^3 \quad \text{...(9.76)}$$

Now each unit cell has eight corners, each corner is shared with seven other unit cells. Therefore the number of unit cells within the octant (calculated by dividing the volume of the octant by the volume V of a unit cell) determines the number of quantum states. Thus the number of quantum states (g_{tr}) with energies less than $r^2h^2/8m$ is:

$$g_{tr} = \frac{1\pi r^3}{6V} = \frac{abc\pi r^3}{6} = \frac{V\pi r^3}{6} \quad \text{...(9.77)}$$

If r is increased by a small amount Δr, the number of quantum states is increased by:

$$\frac{dg_{tr}}{dr}\Delta r = \frac{1}{2}V\pi r^2 \Delta r = g_{tr}\,(\text{say}) \quad \text{...(9.78)}$$

If Δr is sufficiently small, the translational energy can be treated as constant over this range of r-values, and, in this sense $\frac{1}{2}V\pi r^2\Delta r$ can be regarded as the degeneracy of the energy $r^2h^2/8m$.

In the most probable distribution, the number of molecules, say n_{tr}(mp), with this energy is given, in the usual way, by

$$n_{tr}(mp) = \frac{N}{q_{tr}}\left(\frac{1}{2}\pi r^2\Delta r\right)e^{-\frac{r^2h^2}{8mk_bT}} \qquad ...(9.79)$$

Now let us calculate the value of r, say r_m, for which n_{tr}(mp) has the greatest value. For this value

$$\frac{dn_{tr}}{dr} = 0$$

i.e.,
$$\frac{d}{dr}\left(\frac{N\pi}{2q_{tr}}r^2e^{-r^2B}\right) = \frac{N\pi}{2q_{tr}}[2re^{-r^2B} - 2r^3Be^{-r^2B}] = \frac{N\pi}{2q_{tr}}2re^{-r^2B}[1-r^2B] = 0$$

where $B = \dfrac{h^2}{8mk_bT}$. It is thus found that

i.e.,
$$r_m = \left(\frac{1}{B}\right)^{\frac{1}{2}} = \left(\frac{8mk_bT}{h^2}\right)^{\frac{1}{2}} = \left(\frac{8MRT}{N_{av}^2h^2}\right)^{\frac{1}{2}} \qquad ...(9.80)$$

The value of g_{tr} thus is

$$g_{tr} = \frac{1}{2}V\pi\frac{8MRT}{N_{av}^2h^2}\Delta r \qquad ...(9.81)$$

(*b*) Using equation (9.80) and substituting the values for hydrogen gas, we get

$$r_m = \left(\frac{8MRT}{N_{av}^2h^2}\right)^{\frac{1}{2}} = \left(\frac{(8)(2\times10^{-3}\text{ kg})(8.314\text{ J K}^{-1}\text{ mol}^{-1})(300\text{ K})}{(6.02\times10^{23}\text{ mol}^{-1})^2(6.626\times10^{-34}\text{ J s})^2}\right)^{\frac{1}{2}} = 1.58\times10^{10}\text{ m}^{-1}$$

Volume of one mole of hydrogen gas at 1 atm. (101.325×10^5 N m^{-2}) can be obtained from ideal gas equation, *i.e.*,

$$\bar{V} = \frac{nRT}{p} = \frac{(1\text{ mol})(8.314\text{ J mol}^{-1}\text{ K}^{-1})(300\text{ K})}{101.325\times10^3\text{ N m}^{-2}} = 2.46\times10^{-2}\text{ m}^3$$

Substituting the values of $\bar{V}$ and r_m in equation (9.81), we get

$g_{tr} = (1/2)(2.46 \times 10^{-2}\text{ m}^3)(3.143)(1.58 \times 10^{10}\text{ m}^{-1})^2(1\text{ m}^{-1}) = 6.11 \times 10^{18}$

i.e., degeneracy is very large.

The number of molecules in any one translational state i for the most probable distribution is

$$\bar{m}_{tr}(i) = \frac{N_{av}}{q_{tr}} e^{-\varepsilon_{tr}(i)/k_bT}$$

Since $\frac{\varepsilon_{tr}}{k_bT}$ is a positive number, $\bar{m}_{tr}(i)$ cannot be greater than $\frac{N_{av}}{q_{tr}}$. From Exercise (9.25), we know that

$\frac{N_{av}}{q_{tr}} = 1.156 \times 10^{-5}$. It means physically, that, although the number of molecules N is very larger, the number of available quantum states is even larger so that the probability of any one quantum state being occupied by a molecule is very small.

***Exercise* 9.27:** Calculate the translational partition function when water molecules are present in a cubic container with $a = b = c = 0.3$ m at 298 K.

Using the equation (9.69a) and substituting the variables, we get

$$q_{tr} = (2\pi MRT)^{\frac{3}{2}} \frac{V}{N_{av}^3 h^3} = \frac{[(2)(3.143)(18 \times 10^{-3}\ \text{kg})(8.314\ \text{J K}^{-1}\ \text{mol}^{-1})(298\ \text{K})]^{\frac{3}{2}} (0.3\ \text{m})^3}{(6.02 \times 10^{23}\ \text{mol}^{-1})^3 (6.626 \times 10^{-34}\ \text{J s})^3}$$

$$= 2 \times 10^{30}$$

From the above result we have come with the same conclusion mentioned above that the number of states available is very large as compared to the number of molecules present in one mole of the gas. In other words many quantum states will not have any molecule present in it.

***Exercise* 9.28:** The criterion used to see that the number of available states is larger than the number of particles, given by equation especially in case of translational partition function is given by

$$\left(\frac{N}{V}\right)\left(\frac{h^2}{8mk_bT}\right)^{\frac{3}{2}} << 1 \qquad ...(9.82)$$

This criterion is generally obeyed by large particle mass, high temperature and low density.

Test this inequality (9.82) for He atoms at 4 K although it is light particle and temperature is low.

$$\left(\frac{N_{av}}{V}\right)\left(\frac{h^2}{8mk_bT}\right)^{\frac{3}{2}} = \left(\frac{p}{k_bT}\right)\left(\frac{h^2}{8mk_bT}\right)^{\frac{3}{2}} \qquad \left(\text{since } \frac{N_{av}}{V} = \frac{p}{k_bT}\right)$$

$$= \left(\frac{10^5\ Pa}{1.38 \times 10^{-23}\ \text{J K}^{-1} \times 4\ K}\right)\left[\frac{(6.626 \times 10^{-34})^2}{(8)(6.645 \times 10^{-27}\ \text{kg})(1.38 \times 10^{-23}\ \text{J K}^{-1})(4\ \text{K})}\right]^{\frac{3}{2}}$$

$$= (0.181 \times 10^{28}\ \text{m}^{-3})[0.1496 \times 10^{-18}]^{\frac{3}{2}} = (0.181 \times 10^{28}\ \text{m}^{-3})[0.05787 \times 10^{-27}\ \text{m}^3] = 0.105$$

Although the result is not much less than one but still it shows that the number of states are larger than the number of He atoms. If one increases the temperature the fraction will be much less than one.

***Exercise* 9.29:** Repeat Exercise 9.28 for krypton atoms at the 127 °C.

Using equation (9.82)

$$\left(\frac{N_{av}}{V}\right)\left(\frac{h^2}{8mk_bT}\right)^{\frac{3}{2}} = \left(\frac{p}{k_bT}\right)\left(\frac{h^2}{8mk_bT}\right)^{\frac{3}{2}}$$

$$= \left(\frac{10^5\,Pa}{1.38 \times 10^{-23}\ \text{J K}^{-1} \times 300\ \text{K}}\right)\left[\frac{(6.626 \times 10^{-34})^2}{(8)(15.282 \times 10^{-26}\ \text{kg})(1.38 \times 10^{-23}\ \text{J K}^{-1})(300\ \text{K})}\right]^{\frac{3}{2}}$$

$$= (2.416 \times 10^{25}\ \text{m}^{-3})[8.674 \times 10^{-23}]^{\frac{3}{2}} = (2.416 \times 10^{25}\ \text{m}^{-3})[807.85 \times 10^{-36}\ \text{m}^3]$$

$$= 1.95 \times 10^{-10}$$

Clearly the results agree with the criterion for high temperature and for high molecular mass.

9.9. THERMODYNAMIC PROPERTIES

The thermodynamical quantities can be calculated with the help of statistical mechanics and following relations can be derived:

$$E = Nk_bT^2\left(\frac{\partial \ln q}{\partial T}\right)_V \quad ...(9.83)$$

$$S = Nk_b \ln q + \frac{E}{T} \quad ...(9.84)$$

$$S = Nk_b\left[\ln q + T\left(\frac{\partial \ln q}{\partial T}\right)_V\right] \quad ...(9.85)$$

$$A = E - TS$$
$$= -Nk_bT \ln q \quad ...(9.86)$$

$$p = -\left(\frac{\partial A}{\partial V}\right)_T = Nk_bT\left(\frac{\partial \ln q}{\partial V}\right)_T \quad ...(9.87)$$

$$C_V = \left(\frac{\partial E}{\partial T}\right)_V$$

$$= 2Nk_bT\left(\frac{\partial \ln q}{\partial T}\right)_V + Nk_b\,T^2\left(\frac{\partial^2 \ln q}{\partial T^2}\right)_V \quad ...(9.88)$$

The above quantities are molar quantities if $N = N_{av}$.

9.10. THERMODYNAMIC PROPERTIES OF IDEAL GASES

The difference between the ideal gases and the condensed phase is that in the latter case the particles can be considered (a situation can be equated to) as independently localised (distinguishable particles), whereas in the former case the particles are independently non-localised. The thermodynamic properties will differ in two cases by the difference in the number of complexions in two cases *i.e.*,

$$\Omega_D \text{ (localised)} = N! \prod \frac{(g_i)^n}{n_i!} \quad \text{...(9.5)}$$

$$\Omega_D \text{ (non-localised)} = \prod \frac{(g_i)^{n_i}}{n_i} \quad \text{...(9.8)}$$

9.10.1. *Entropy*

The entropy for gases or non-localised particles is given as

$$S = k_b \ln \Omega_D$$

$$= k_b \ln \left(\prod_i \frac{(g_i)^{n_i}}{n_i!} \right)$$

$$= k_b \ln \left(N! \prod_i \frac{(g_i)^{n_i}}{n_i!} \right) - k_b \ln N!$$

Now $\quad k_b \ln \left(N! \prod \frac{(g_i)^{n_i}}{n_i!} \right) = Nk_b \ln q + \frac{E}{T}$ ...(9.85)

So

$$S = k_b N \ln q + \frac{E}{T} - k_b N \ln N + k_b N \ln e$$

$$= k_b N \ln \left(\frac{qe}{N} \right) + \frac{E}{T} \quad \text{...(9.89)}$$

The entropy of non-distinguishable particles can be obtained from the equation (9.85 *a*) *if q is multiplied by the factor e/N.*

9.10.2. *Energy*

For non-distinguishable particles, the energy can be obtained by replacing q by qe/N in equation (9.86)

$$E = Nk_b T^2 \left[\frac{\partial}{\partial T} (\ln qe/N)_V \right]$$

$$= Nk_b T^2 \left[\frac{\partial}{\partial T} \ln q + \frac{\partial}{\partial T} \ln \frac{e}{N} \right]_V$$

$$= Nk_b T^2 \left[\frac{\partial}{\partial T} \ln q\right]_V \quad ...(9.84c)$$

i.e., there is no difference between expression (9.84) and (9.84 *c*).

***Exercise* 9.30:** Obtain an ideal gas equation with the help of results obtained above.

For *n* mol of ideal gas the Gibb's free energy is related to work function by

$$G = n\left(\frac{\partial A}{\partial n}\right)_{T,V} \quad ...(9.90)$$

For localised system, *A* is given by Equation. (9.85), where as for gases or non-localised system

$$A = -Nk_b T \ln \frac{qe}{N} \quad ...(9.86a)$$

Since

$$N = nN_{av}$$

$$A = -nN_{av}k_b T \ln \frac{qe}{nN_{av}}$$

$$\left(\frac{\partial A}{\partial n}\right)_{T,V} = -N_{av}k_b T \ln \frac{qe}{nN_{av}} + N_{av}k_b T$$

$$= \frac{A}{n} + N_{av}k_b T \quad ...(9.91)$$

$$G = A + nN_{av}k_b T \quad ...(9.92)$$

For any system, classical thermodynamics defines G as

$$G = A + pV \quad ...(9.93)$$

Comparing equations (9.92) and (9.93)

$$pV = nN_{av}k_b T$$

$$= nRT$$

9.10.3. *Molar Thermal Energy*

Molar thermal energy of ideal gas is the difference of the energy content of ideal gas and the energy contributed by the molecular ground state: E_0^0. This is due to the occupation of higher energy levels when the temperature of the gas is increased.

$$E^0_{\text{ther}} = E^0 - E_0^0 \quad ...(9.94)$$

where superscript tells the standard state energy (at 1 atm pressure) as the translational partition function is pressure dependent.

***Note*:** $E_0^0 = H_0^0 = A_0^0 = G_0^0$ (9.95)

H_0^0 and E_0^0; A_0^0 and G_0^0

for ideal gases because the difference between H_0^0 and E_0^0, A_0^0 and G_0^0 is RT which is zero at absolute zero.

***Exercise* 9.31:** Show that molar translational and rotational energy calculated with the help of statistical mechanics resemble with that obtained by the principle of equipartition of energy, whereas that of vibrational only at higher temperature.

According to equation (9.84)

$$E^0 = RT^2 \left(\frac{\partial \ln q}{\partial T}\right)_V$$

$$q = q_{tr}\, q_{rot}\, q_{vib}\, q_{ele}\, q_{nuc}\, q_0$$

$$\bar{E}^0 = RT^2 \left(\frac{\partial \ln q_{tr}^0}{\partial T}\right)_V + RT^2 \left(\frac{\partial \ln q_{rot}}{\partial T}\right)_V + RT^2 \left(\frac{\partial \ln q_{vib}}{\partial T}\right)_V$$

$$+ RT^2 \left(\frac{\partial \ln q_{ele}}{\partial T}\right)_V + RT^2 \left(\frac{\partial \ln q_{nuc}}{\partial T}\right)_V$$

$$= \bar{E}^0_{tr} + \bar{E}^0_{rot} + \bar{E}^0_{vib} + \bar{E}^0_{ele} + \bar{E}^0_{nuc}$$

Substituting the values of respective partition function and differentiating with respect to temperature:

$$\bar{E}^0_{tr} = RT^2 \left(\frac{\partial \ln q_{tr}^0}{\partial T}\right)_V = 3/2\ RT$$

$$\bar{E}^0_{rot} = \begin{Bmatrix} RT \text{ for diatomic or linear molecule} \\ 3/2\ RT \text{ for non-linear molecule.} \end{Bmatrix}$$

$$\bar{E}^0_{vib} = \sum_{i=1}^{3N-5,\, \text{or}\, 3N-6} \frac{RT\,(h\nu / k_bT)}{\exp\,(hw / k_bT) - 1}$$

$$\bar{E}^0_{ele} = 0$$

$$\bar{E}^0_{nuc} = 0$$

If T is large, $h\nu/k_bT \ll 1$ and exponential term can be expanded as $e^{+h\nu/k_bT} = 1 + h\nu/k_bT$

$$\bar{E}^0_{vib} = \sum_{i=1}^{3N-5,\, 3N-6} \frac{RT\,(h\nu / k_bT)}{1 + \dfrac{h\nu}{k_bT} - 1} = \sum_{i=1}^{3N-5 \text{ or } 3N-6} RT$$

The thermal energy obtained from equipartition principle for translational, rotational and vibrational degrees of freedom are 3/2 RT, $RT \left(\frac{3}{2} RT\right)$ and $\sum_{i}^{3N-5, 3N-6} RT$.

***Exercise* 9.32:** Calculate the value $\bar{E}^0$ (thermal) for CO_2 at 200°C. The vibrational characteristic temperatures are $\theta_{1\,(vib)} = 1890$ K, $\theta_{2\,(vib)} = 3360$ K and $\theta_{3\,(vib)} = \theta_{4\,(vib)} = 954$ K.

$$\bar{E}^0_{(tr)} = \frac{3}{2}\,(8.314 \text{ J mol}^{-1} \text{ K}^{-1})\,(473.0 \text{ K}) = 5898 \text{ J mol}^{-1}$$

$$\bar{E}^0_{(rot)} = (8.314 \text{ J mol}^{-1} \text{ K}^{-1})\,(473.0 \text{ K}) = 3932 \text{ J mol}^{-1}$$

because CO_2 is a linear molecule:

$$\bar{E}^0_{(vib)} = RT\left[\frac{\theta_1/T}{\exp(\theta_1/T)-1}+\frac{\theta_2/T}{\exp(\theta_2/T)-1}+\frac{2\theta_3/T}{\exp(\theta_3/T)-1}\right]$$

$$= \left[\frac{3.996}{e^{3.996}-1}+\frac{7.103}{e^{7.103}-1}+\frac{2\times 2.017}{e^{2.017}-1}\right](8.314 \text{ J mol}^{-1}\text{ K}^{-1})(473 \text{ K})$$

$$= 3932\,[0.0749 + 0.0059 + 0.619] \text{ J mol}^{-1}$$

$$= 2751.6 \text{ J mol}^{-1}$$

$$\bar{E}_{ele} = \bar{E}^0_{nuc} = 0$$

Adding up the contributions

$$\bar{E}^0_{therm} = 12581.6 \text{ J mol}^{-1} \qquad \text{...(9.95)}$$

9.10.4. *Molar Thermal Enthalpy*

The molar thermal enthalpy H^0 is defined as

$$H^0_{ther} = \bar{H}^0 - \bar{H}^0_0 \qquad \text{...(9.96)}$$

$$= \bar{E}^0 + RT - \bar{E}^0_0 \qquad \text{...(9.96a)}$$

where $H^0_0 = E^0_0$ and $H^0_{ther} = E^0_{ther} + RT$ ...(9.97)

***Exercise* 9.33:** Calculate the $\bar{H}^0$ (thermal) for CO_2 at 200°C.

Substituting the results of exercise (9.32) in equation (9.97).

$$H^0_{ther} = 12581.6 + (8.314)(473)$$

$$= 12581.6 + 3932 = 16513.6 \text{ J mol}^{-1}$$

$$= 16.5 \text{ kJ mol}^{-1}$$

9.10.5. *Molar Entropy*

The standard molar entropy of gases is defined by equation

$$\bar{S}^0 = R\ln\frac{q^0 e}{N_{av}} + \frac{\bar{E}^0}{T} \qquad \text{...(9.98)}$$

i.e.,

$$\bar{S}^0 = R\ln\frac{q^0_{tr} e}{N_{av}} + \frac{E^0_{tr}}{T} + R\ln q_{rot} + \frac{E^0_{rot}}{T} +$$

$$R\ln q_{vib} + \frac{E^0_{vib}}{T} + R\ln q_{ele} + \frac{E^0_{ele}}{T}$$

$$= S^0_{tr} + \bar{S}^0_{rot} + \bar{S}^0_{vib} + \bar{S}^0_{ele} \qquad \text{...(9.98a)}$$

Using the values of each partition functions already calculated:

$$\bar{S}^0_{tr} = R\left[\frac{5}{2} + \ln\frac{(2\pi m k_b T)^{3/2}\bar{V}^0}{h^3 N_{av}}\right] \qquad \text{...(9.98b)}$$

$$\bar{S}^0_{rot} = R\left(1+\ln\frac{8\pi^2 Ik_bT}{\sigma h^2}\right) \text{ for diatomic or linear polyatomic} \quad ...(9.99a)$$

$$= R\left[\frac{3}{2}+\ln\left(\frac{(\pi I_A I_B I_C)^{1/2}(8\pi^2 k_b T)^{3/2}}{h^3}\right)\right] \quad ...(9.99b)$$

$$\bar{S}^0_{vib} = R\sum_{i=1}^{3N-6,\,3N-6}\left[\frac{h\nu_i/k_bT}{\exp(+h\nu_i/k_bT)-1} - \ln\{1-\exp(-h\nu_i/k_bT)\}\right] \quad ...(9.100)$$

$$\bar{S}^0_{ele} = R\ln g_{0(ele)} \quad ...(9.101)$$

$\bar{S}^0_{nuc}$ is generally neglected for most of chemical reactions.

This equation (9.98) is known as Sackur Tetrode equation. This is one of the most valuable equation in statistical mechanics because one can calculate the absolute entropy of ideal gas.

***Exercise* 9.34:** Calculate the molar absolute entropy of Kr at 27°C and 1 atm pressure.

Using equations (9.98) and (9.83)

$$\bar{S}^0_{tr} = \frac{5}{2}R + R\ln(0.02563\ T^{5/2}\ M^{3/2}) \quad ...(9.99a)$$

$$= R\ln T^{5/2}\ M^{3/2} - 9.68 \quad ...(9.99b)$$

is the general equation or transformed Sackur-Tetrode equation,T in absolute scale and$M$ in g. Substituting the respective values in (9.98 *b*):

$$\bar{S}^0_{tr} = (8.314\text{ J mol}^{-1}\text{ K}^{-1})\ln(300)^{5/2}(83.8)^{3/2} - 9.68$$

$$= 164.1\text{ J mol}^{-1}$$

Since Kr is a monatomic gas $\bar{S}^0_{rot} = \bar{S}^0_{vib} = 0$ and

Since $g^0_{ele} = 1,\ \bar{S}^0_{ele} = 0,$

$$\bar{S}^0_T = \bar{S}^0_{tr} = 164.1\text{ J mol}^{-1}$$

***Exercise* 9.35:** Calculate the entropy of $H^{35}Cl$ gas at 1 atm pressure and 27°C. $\theta_{rot} = 15.2$ K, $\theta_{vib} = 4152$ K.

Using equation (9.99*b*),

$$\bar{S}^0_{tr} = (8.314\text{ J K}^{-1}\text{ mol}^{-1})\ln(300)^{5/2}(36.5)^{3/2} - 9.68$$

$$= 153.61\text{ J K}^{-1}\text{ mol}^{-1}$$

Equation (9.99*a*) can be written as

$$\bar{S}^0_{rot} = R + R\ln\frac{T}{\theta_{rot}} \quad ...(9.99b)$$

Substituting respective values

$$\bar{S}^0_{rot} = 8.314\text{ J K}^{-1}\text{ mol}^{-1} + (8.314\text{ J K}^{-1}\text{ mol}^{-1})\ln\left(\frac{300\text{ K}}{15.2\text{ K}}\right)$$

$$= 33.11\text{ J K}^{-1}\text{ mol}^{-1}$$

Similarly equation (9.100) can be written as

$$\bar{S}^0_{vib} = R\left[\frac{\theta_{vib}/T}{\exp(\theta_{vib}/T)-1} - \ln\{1-\exp(-\theta_{vib}/T)\}\right] \quad ...(9.100a)$$

Substituting the respective value

$$\bar{S}^0_{vib} = (8.314\,\text{J K}^{-1}\,\text{mol}^{-1})\left[\frac{4152\,\text{K}/300\,\text{K}}{\exp(4152\,\text{K}/300\,\text{K})-1} - \ln(1-e^{-4152\,\text{K}/300\,\text{K}}\right]$$

$$= 10.41 \times 10^{-5}\,\text{J K}^{-1}\,\text{mol}^{-1}$$

Since $\bar{S}^0_{ele} = 0$

$$\bar{S}^0 = \bar{S}^0_{tr} + \bar{S}^0_{rot} + \bar{S}^0_{vib} = (153.61 + 33.11 + 1.041 \times 10^{-4})\,\text{J K}^{-1}\,\text{mol}^{-1}$$

$$= 186.72\,\text{J K}^{-1}\,\text{mol}^{-1}$$

It is clear from the data that contribution from the translational motion is maximum

9.11. MOLAR HEAT CAPACITY

9.11.1. *Molar Heat Capacity of Ideal Gases*

Heat capacity at constant volume is defined as

$$\bar{C}_V = \left(\frac{\partial \bar{E}}{\partial T}\right)_V$$

with the definition given by the above equation, the contribution to molar heat capacity from different degrees of freedom can be obtained as follows:

With the results already obtained in exercise 9.31 differentiate the energy with respect to temperature and the expressions are

$$\bar{C}^0_{V(tr)} = \frac{3}{2}R \quad ...(9.102)$$

$$\bar{C}^0_{V(rot)} = R \text{ for diatomic and linear polyatomic molecule} \quad ...(9.103a)$$

$$= \frac{3}{2}R \text{ for a nonlinear polyatomic molecule} \quad ...(9.103b)$$

$$\bar{C}^0_{V(vib)} = \sum_{i=1}^{3N-5,\,3N-6} \frac{R(h\nu/k_bT)^2\exp(h\nu/k_bT)}{\{\exp(h\nu/k_bT)-1\}^2} \quad ...(9.104)$$

$$\bar{C}^0_{V(ele)} = 0$$

$$\bar{C}^0_{V(nuc)} = 0$$

***Exercise* 9.36:** Calculate $\bar{C}^0_V$ and $\bar{C}^0_p$ for CO_2 at 200°C, using the data of exercise 9.32.

Combining the data of the exercise 9.32 and the equations (9.102–9.104):

$$\bar{C}^0_{V(tr)} = \left(\frac{3}{2}\right)(8.314\,\text{J mol}^{-1}\,\text{K}^{-1}) = 12.471\,\text{J mol}^{-1}\,\text{K}^{-1}$$

$$\bar{C}^0_{V(\text{rot})} = (1)\,(8.314 \text{ J mol}^{-1} \text{ K}^{-1}) = 8.314 \text{ J mol}^{-1} \text{ K}^{-1}$$

$$\bar{C}^0_{V(\text{vib})} = (8.314)\left[\frac{(3.996)^2 e^{3.996}}{(e^{+3.996}-1)^2} + \frac{(7.103)^2 e^{7.103}}{(e^{7.103}-1)^2} + \frac{2\times(2.017)^2 e^{2.017}}{(e^{2.017}-1)^2}\right]$$

$$= (8.314)\,[0.3046 + 0.0417 + 1.440] = 14.85 \text{ J mol}^{-1} \text{ K}^{-1}$$

$$\bar{C}^0_{V(\text{ele})} = \bar{C}^0_{V(\text{nuc})} = 0$$

Adding the three together

$$\bar{C}^0_V = 35.64 \text{ J K}^{-1} \text{ mol}^{-1}$$

$$\bar{C}^0_p = \bar{C}^0_V + R = 35.64 + 8.314 = 43.954 \text{ J K}^{-1} \text{ mol}^{-1}.$$

9.11.2. *Molar Heat Capacity of Solids*

9.11.2.1. Einstein's theory

Einstein's theory is based on the three assumptions:

(*a*) Only vibrational motion is possible and that too when atoms undergo small oscillations about their equilibrium positions.

(*b*) Each atom vibrates independently of others and each vibrations of an atom can be resolved into three normal coordinates (*i.e.*, vibrations of atoms along *x*, *y* and *z* directions).

(*c*) Each normal mode is a simple harmonic motion and frequency of each mode is same.

With the above three assumptions the molar heat capacity at constant volume is

$$\bar{C}_{V(\text{vib})} = \frac{3R\,(\theta_E/T)^2 \exp(\theta_E/T)}{\{\exp(\theta_E/T)-1\}^2} \qquad \text{...(9.105)}$$

where $\theta_E = \dfrac{hv}{k_b}$, Einstein characteristic temperature.

***Exercise* 9.37:** Using equation (9.105) show that the limiting value of heat capacity of metals agree with that obtained by Dulong and Petit.

If the temperature is very high $\left(\dfrac{\theta_E}{T} \ll 1\right)$ then exponential term can be expanded

$\exp\left(+\dfrac{\theta_E}{T}\right) = 1 + \dfrac{\theta_E}{T}$, neglecting higher terms. Substituting this in equation (9.105) gives

$$\bar{C}_{V(\text{vib})} = \frac{3R(\theta_E/T)^2(1+\theta_E/T)}{(1+\theta_E/T-1)^2} = 3R\left(1+\frac{\theta_E}{T}\right) \approx 3R \text{ as } \frac{\theta_E}{T} << 1$$

$$= 3R = 3 \times 8.314 = 24.942 \text{ J mol}^{-1} \text{ K}^{-1}$$

Dulong and Petit's value is at constant pressure *i.e.*,

$$\bar{C}_p = 6.4 \text{ cal K}^{-1} \text{ mol}^{-1}$$

$$= 26.765 \text{ J mol}^{-1} \text{ K}^{-1}$$

$\bar{C}_p - \bar{C}_V$, for solids can be calculated using coefficient of expansion and coefficient of compression, equation (10.29) and is

$$\bar{C}_p - \bar{C}_V \approx 0.4 \text{ cal mol}^{-1} \text{ K}^{-1}$$
$$= 1.673 \text{ J mol}^{-1} \text{ K}^{-1}$$
$$\bar{C}_V = 26.765 - 1.673 = 25.092 \text{ J mol}^{-1} \text{ K}^{-1}$$

Two values agree nicely.

9.11.2.2. Debye's theory of heat capacity of solids

Debye's theory assumes:

(*a*) The atoms in a crystal can be considered as a part of one large molecule.

(*b*) The number of vibrational modes in 1 mol is considered to be $3N_{av}$ rather than $3N_{av} - 6$ as six is very small compared to $3N_{av}$.

(*c*) The frequencies of vibrations can have values from 0 to v_i and since the value of v_i will be finite, let v_{max} is maximum of all the vibrational modes. Some of them can be degenerate.

With these assumptions the molar average value of heat capacity at constant volume is

$$\bar{C}_{V(\text{vib})} = 9R\left(\frac{T}{\theta_D}\right)^3 \int_0^{\theta_D/T} \frac{u^4 e^u}{(e^u - 1)^2}\, du \qquad ...(9.106)$$

where $u = \frac{hv}{k_b T} = \frac{\theta_D}{T}$, $\theta_D = \frac{hv}{k_b}$, Debye characteristic temperature.

***Exercise* 9.38:** Show that the limiting value of heat capacity of solids at constant volume and at high temperature, calculated by Debye's formula is 3 *R*.

At higher temperature $\theta_D/T \ll 1$ so

$$\frac{u^4 e^u}{(e^u - 1)^2} = \frac{u^4\,(1 + u + u^2/2! + ...)}{(1 + u + u^2/2! + ... - 1)^2} = \frac{u^4\,(1 + u + u^2/2! + ...)}{u^2\,(1 + u/2 + u^2/3! + ...)}$$
$$= u^2$$

So equation (9.106) becomes

$$\bar{C}_{V(\text{vib})} = 9R\left(\frac{T}{\theta_D}\right)^3 \int_0^{\theta_D/T} u^2\, du$$

$$= 9R\left(\frac{T}{\theta_D}\right)^3 \left[\frac{1}{3}\cdot u^3\right]_0^{\theta_D/T} = 9R\left(\frac{T}{\theta_D}\right)^3 \left(\frac{1}{3}\right)\left(\frac{\theta_D}{T}\right)^3 = 3R$$

***Exercise* 9.39:** Show that at low temperature, heat capacity of solids at constant volume is proportional to cube of absolute temperature.

At low temperature where θ_D/T is very large (*i.e.* ∞), the integral can be approximated as

$$\int_0^{\infty} \frac{u^4 e^4\, du}{(e^u - 1)^2} = \frac{4}{15}\pi^4 \qquad ...(9.107)$$

So the equation (9.106) becomes

$$\bar{C}_{V(\text{vib})} = 9R\left(\frac{T}{\theta_D}\right)^3 \times \frac{4}{15}\pi^4$$

$$= \frac{36}{15}\pi^4\left(\frac{T}{\theta_D}\right)^3 R = 239.3R\left(\frac{T}{\theta_D}\right)^3 \qquad \text{...(9.108)}$$

i.e. $\bar{C}_{V(\text{vib})} \; \alpha \; T^3$

Note: For more problems on Debye theory see Chapter 10: First law of thermodynamics (Section 10.9.3).

9.12. EQUILIBRIUM CONSTANT

The equilibrium constant is related to standard free energy as

$$RT \ln K_p = -\Delta G^0$$

where ΔG^0 is the difference in the standard free energies of the products and reactants. The equilibrium constant can be calculated by two methods. Both are similar in the sense that these are used to obtain the free energy value.

9.12.1. *Method I*

This involves the calculation of free energy functions for the reactants and products and then calculating ΔG^0. Since,

$$\bar{G}^0 = \bar{H}^0 - T\bar{S}^0$$

and for ideal gases

$$\bar{G}^0 = \bar{E}^0 + RT - T\bar{S}^0$$

or

$$\frac{\bar{G}^0 - \bar{E}_0^0}{T} = \frac{\bar{E}^0 - \bar{E}_0^0}{T} + R - \bar{S}^0 \qquad \text{...(9.109)}$$

where $\bar{E}^0$ and $\bar{S}^0$ are the standard molar energy and entropy respectively and $\bar{E}_0^0$ is standard molar zero point energy. Further

$$\bar{E}^0 - \bar{E}_0^0 = \bar{E}_{tr}^0 + \bar{E}_{\text{rot}}^0 + \bar{E}_{\text{vib}}^0 + \bar{E}_{\text{ele}}^0$$

and

$$\bar{S}^0 = \bar{S}_{\text{tr}}^0 + \bar{S}_{\text{rot}}^0 + \bar{S}_{\text{vib}}^0 + \bar{S}_{\text{ele}}^0$$

Both $\bar{E}^0 - \bar{E}_0^0$ and $\bar{S}^0$ can be calculated as described in Sections 9.10.3 and 9.10.5. In general the free energy functions in the form

$$\frac{\bar{G}^0 - \bar{H}_0^0}{T} \text{ or } \frac{\bar{G}^0 - \bar{E}_0^0}{T}$$

are tabulated for different temperatures and also the values of $\bar{S}^0$, $\bar{G}^0$ and hence ΔG^0 can be evaluated for a reaction.

***Exercise* 9.40:** Consider a reaction $H_2 + CO_2 \rightleftharpoons H_2O + CO$. The following data is available for different components and at 900 K.

Compound	H_2	CO_2	H_2O	CO
$\frac{\bar{G}-E^0}{T}$ (J K^{-1} mol^{-1})	134	222.1	193.0	201.3

Calculate K_p at 900 K of the above reaction if $\Delta E_0^0 = 40{,}330$ J.

The free energy change for the above reaction is

$$\frac{\Delta (G^0 - E_0^0)}{T} = (193 + 201.3) - (134 + 222.1)$$

$$= 38.2 \text{ J K}^{-1} \text{ mol}^{-1}$$

$$\frac{\Delta G^0}{T} - \frac{\Delta E_0^0}{T} = 38.2 \text{ J K}^{-1} \text{ mol}^{-1}$$

i.e.,

$$\frac{\Delta G^0}{T} = \frac{40{,}330}{900} + 38.2 \text{ J K}^{-1} \text{ mol}^{-1}$$

$$= 83.0 \text{ J K}^{-1} \text{ mol}^{-1}$$

Now

$$-\frac{\Delta G^0}{T} = -83.0 \text{ J K}^{-1} \text{ mol}^{-1} = (8.314 \text{ J K}^{-1} \text{ mol}^{-1}) \log K_p$$

$$\ln K_p = -\frac{83.0}{8.314} = -9.984$$

$$K_p = 462 \times 10^{-5}$$

9.12.2. *Method II*

This method involves the calculation of standard free energy of the components by calculating partition functions. Thus the equilibrium constant K_p for the reaction

$$aA + bB \rightleftharpoons cC + dD \qquad \text{...(9.110)}$$

is given as

$$K_p = \frac{(q'_C)^c (q'_D)^d}{(q'_A)^a (q'_B)^b} \left(\frac{1}{N_{av}}\right)^{\Delta n} \exp(-\Delta E^0/RT) \qquad \text{...(9.111)}$$

where

$$q'_i = q^0_{tr}\, q_{rot}\, q_{vib}\, q_{ele}\, q_{nuc}$$

and q^0_{tr} is the translational partition function at 1 atm. ΔE^0 is the difference in zero point energies of products and reactants *i.e.*,

$$\Delta E^0 = c\,\bar{E}^0_{(C)} + d\,\bar{E}^0_{(D)} + a\,\bar{E}^0_{(A)} + b\,\bar{E}^0_{(B)} \qquad \text{...(9.112)}$$

and

$$\Delta n = c + d - a - b \qquad \text{...(9.113)}$$

The equilibrium constant in concentration units can be written as

$$K_c = \frac{(q'_C)^c (q'_D)^d}{(q'_A)^a (q'_B)^d} \left(\frac{1}{V^0 N_{av}} \right)^{\Delta n} \exp(-\Delta E^0/RT) \quad ...(9.114)$$

where $\bar{V}^0$ is the molar volume at 1 atm.

***Exercise* 9.41:** Calculate the equilibrium constant for the following ionisation process at 2000 K.

$$Na \rightarrow Na^+ + e^-$$

Ionisation potential of sodium atom is 497.896 kJ mol^{-1}.

Using the equation (9.111) K_p can be written as

$$K_p = \frac{(q'_{Na+})(q'_{e^-})}{(q'_{Na})} \left(\frac{1}{N_{av}} \right) \exp(-\Delta E^0/RT)$$

In this problem, ΔE^0 is the minimum energy required to remove the electron from sodium atom in its ground state and this is ionisation potential.

The electron can have only translational degrees of freedom and two directions for spin, thus q'_e is

$$q_e^{0'} = 2 \frac{(2\pi m_e k_b T)^{3/2}}{h^3} \bar{V}^0$$

Using equation (9.83)

$$\frac{q_e^{0'}}{N_0} = 2 \times 0.02563 \times \left(\frac{1}{1837} \right)^{3/2} (2000)^{5/2}$$

$$= 1.16 \times 10^2$$

whereas the Na and Na$^+$ can have only translational and electronic partition functions. The mass of Na and Na$^+$ are same so the translational partition functions for both are equal.

Na has ground state degeneracy 2 as its

$$J = \frac{1}{2}, \text{ whereas } g_0 (Na^+) = 1$$

$$q_{ele}(Na) = 2 \text{ and } q_{ele}(Na^+) = 1$$

$$\frac{\Delta E^0}{RT} = \frac{497896 \text{ J mol}^{-1}}{(2000 \text{ K})(8.314 \text{ J mol}^{-1} \text{ K}^{-1})} = 29.94$$

$$e^{-29.94} = 9.90 \times 10^{-14}$$

Substituting the respective values

$$K_p = \frac{1.16 \times 10^2 N_{av}}{2 \times N_{av}} \times 9.90 \times 10^{-14}$$

$$= 5.742 \times 10^{-12}$$

***Exercise* 9.42:** Calculate the dissociation constant of $H_2 \rightarrow 2H$ at 2000 K from the following data.

$\left(\frac{\bar{G}^0 - \bar{H}^0}{T} \right)$ 2000 K for H_2 and H are 157.61 J deg^{-1} and 133.39 J deg^{-1} respectively. $\Delta \bar{H}_0^0$ for H = 215.98 kJ mol^{-1}. D (H_2) = 431.956 kJ mol^{-1}.

Method I:

$$\frac{\Delta(G^0 - E_0^0)}{T} = -2 \times 133.39 - 157.61 = -109.17 \text{ J deg}^{-1}$$

i.e.,

$$\frac{\Delta G^0}{T} - \frac{\Delta E_0^0}{T} = -109.17 \text{ J deg}^{-1}$$

$$-\frac{\Delta G^0}{T} = -\frac{2 \times 215978}{2000 \text{ K}} + 109.17 = -106.808 \text{ J deg}^{-1}$$

$$\ln K_p = -\frac{\Delta G^0}{RT}$$

or

$$\log K_p = -\frac{\Delta G^0}{2.303\,RT} = -\frac{106.808 \text{ J deg}^{-1}}{(2.303)(8.314 \text{ J deg}^{-1})}$$

$$= -5.578$$

$$K_p = 10^{-5.578} = 2.6 \times 10^{-6}$$

Method II:

According to equation (9.111)

$$K_p = \frac{(q'_H)^2}{q'_{H_2}} \cdot \left(\frac{1}{N_{av}}\right)^1 \exp(-\Delta E^0/RT)$$

Hydrogen atom will have only electronic and translation partition functions

$$q_{ele}(\text{H}) = g_{ele}^0 = 2$$

$$\frac{q_{tr}(\text{H})}{N_{av}} = (0.02562)\,(2000)^{5/2}\,(1)^{3/2}$$

$$= 4.58 \times 10^6$$

For H_2 molecule $\quad \dfrac{q_{tr}}{N_{av}} = (0.02562)\,(2000)^{5/2}\,(2)^{3/2} = 1.296 \times 10^7$

The translational partition functions for H and H_2 have been calculated using equation (9.83), using the results of exercise (9.23) the rotational partition function for H_2 is

$$q_{rot} = \frac{k_b T}{\sigma B h c} = \frac{(1.381 \times 10^{-23} \text{ J K}^{-1})(2000 \text{ K})}{(2)(60.80 \text{ cm}^{-1})(6.62 \times 10^{-34} \text{ J s})(3.0 \times 10^{10} \text{ cm s}^{-1})}$$

$$= 1.14 \times 10 = 11.4$$

The vibrational and electronic partition function for H_2 are unity.

Now $\exp(-\Delta E^0/RT) = \exp\left(\dfrac{-431956 \text{ J mol}^{-1}}{(8.314 \text{ J K}^{-1} \text{ mol}^{-1})(2000 \text{ K})}\right) = 5.225 \times 10^{-12}$

Substituting all the values

$$K_p = \frac{(2 \times 4.58 \times 10^6 \cdot N_{av})^2}{(1.296 \times 10^7 \times N_{av})(11.4)(N_{av})} \times 5.225 \times 10^{-12}$$

$$= 29.67 \times 10^{-7} = 2.967 \times 10^{-6}$$

The agreement between the values of K_p obtained by two methods is quite good.

9.13. INDEPENDENT AND DISTINGUISHABLE (LOCALISED) MOLECULES OR PARTICLES

As learnt earlier when the system consists of N independent particles, the total energy or Hamiltonian will be the sum of the individual particle energy or Hamiltonian and total wave functions will be the product of individual wave function, *i.e.*,

$$H = H_1 + H_2 + H_3 \ldots\ldots\ldots\ldots\ldots + H_N \qquad ...(9.115)$$

$$E = E_1 + E_2 + E_3 \ldots\ldots\ldots\ldots\ldots + E_N \qquad ...(9.116)$$

$$\Psi = \psi_1 \times \psi_2 \times \psi_3 \times \ldots\ldots\ldots\ldots\ldots \Psi_N \qquad ...(9.117)$$

Then the total partition function Q will be

$$Q = q^N \qquad ...(9.118)$$

if the particles are identical and if the particles are not identical, then Q will be

$$Q = q_1\, q_2\, q_3 \ldots\ldots\ldots\ldots\ldots q_N \qquad ...(9.119)$$

The model of independent and distinguishable particles in practice is a crystal in which each atoms, molecule or ion is confined to a particular lattice. Suppose we label these sites, as we can do in a given coordinate system, then in this sense and in this model we regard the atoms, molecules or ions of the system to be identical, independent and distinguishable.

9.13.1. *Independent and Indistinguishable (Non-localised) Molecules or Particles*

In an ideal gas at low pressure in which the intermolecular forces are negligible, the molecules can be treated to be independent of each other. However, we cannot locate a molecule at a given instant of time (we cannot label the molecule), as we can do in case of crystal, because of random kinetic motion of the molecules. The molecules we are handling are essentially indistinguishable. The total partition can be given in terms of molecular particle partition function as

$$Q = \frac{q^N}{N!} \qquad ...(9.120)$$

***Exercise* 9.43:** Consider two distinguishable and independent particles$a$, and b and non-degenerate energy levels available are ε_1, ε_2 and ε_3. Calculate the total partition function.

Let us first consider the occupation of energy levels one by one as the number is small.

$E_1 = \varepsilon_1(a) + \varepsilon_1(b)$	Both particles are in the same state ε_1
$E_2 = \varepsilon_2(a) + \varepsilon_2(b)$	Both particles are in the same state ε_2
$E_3 = \varepsilon_3(a) + \varepsilon_3(b)$	Both particles are in the same state ε_3
$E_4 = \varepsilon_1(a) + \varepsilon_2(b)$	Particle a in ε_1 and b in ε_2

$E_5 = \varepsilon_1(a) + \varepsilon_3(b)$ Particle a in ε_1 and b in ε_3

$E_6 = \varepsilon_2(a) + \varepsilon_3(b)$ Particle a in ε_2 and b in ε_3

$E_7 = \varepsilon_1(b) + \varepsilon_2(a)$ Particle b in ε_1 and a in ε_2

$E_8 = \varepsilon_1(b) + \varepsilon_3(a)$ Particle b in ε_1 and a in ε_3

$E_9 = \varepsilon_2(b) + \varepsilon_3(a)$ Particle a in ε_3 and b in ε_2

Hence, the total partition function of the system (Q) by the definition

$$Q = \sum_1^9 g_i e^{-\varepsilon_i / k_b T}$$

$$Q = e^{-E_1/k_bT} + e^{-E_2/k_bT} + e^{-E_3/k_bT} + e^{-E_4/k_bT} + e^{-E_5/k_bT} + e^{-E_6/k_bT} + e^{-E_7/k_bT} + e^{-E_8/k_bT} + e^{-E_9/k_bT}$$

or

$$Q = e^{-\varepsilon_1(a)/k_bT} e^{-\varepsilon_1(b)/k_bT} + e^{-\varepsilon_2(a)/k_bT} e^{-\varepsilon_2(b)/k_bT} + e^{-\varepsilon_3(a)/k_bT} e^{-\varepsilon_3(b)/k_bT}$$
$$+ e^{-\varepsilon_1(a)/k_bT} e^{-\varepsilon_2(b)/k_bT} + e^{-\varepsilon_1(a)/k_bT} e^{-\varepsilon_3(b)/k_bT} + e^{-\varepsilon_2(a)/k_bT} e^{-\varepsilon_3(b)/k_bT}$$
$$+ e^{-\varepsilon_1(b)/k_bT} e^{-\varepsilon_2(a)/k_bT} + e^{-\varepsilon_1(b)/k_bT} e^{-\varepsilon_3(a)/k_bT} + e^{-\varepsilon_2(b)/k_bT} e^{-\varepsilon_3(a)/k_bT}$$

or

$$Q = e^{-\varepsilon_1(a)/k_bT}[e^{-\varepsilon_1(b)/k_bT} + e^{-\varepsilon_2(b)/k_bT} + e^{-\varepsilon_3(b)/k_bT}]$$
$$+ e^{-\varepsilon_2(a)/k_bT}[e^{-\varepsilon_2(b)/k_bT} + e^{-\varepsilon_3(b)/k_bT} + e^{-\varepsilon_1(b)/k_bT}]$$
$$+ e^{-\varepsilon_3(a)/k_bT}[e^{-\varepsilon_1(b)/k_bT} + e^{-\varepsilon_2(b)/k_bT} + e^{-\varepsilon_3(b)/k_bT}]$$

$$Q = e^{-\varepsilon_1(a)/k_bT} q_b + e^{-\varepsilon_2(a)/k_bT} q_b + e^{-\varepsilon_3(a)/k_bT} q_b$$

$$= q_b[e^{-\varepsilon_1(a)/k_bT} + e^{-\varepsilon_2(a)/k_bT} + e^{-\varepsilon_3(a)/k_bT}]$$

$$= q_a q_b \qquad ...(1)$$

Now if the two particles are identical, $q_a = q_b = q$, equation (1) becomes

$$Q = q^2 \qquad ...(2)$$

Further if we have N distinguishable particles, equation (1) becomes

$$Q = q_a\, q_b\, q_c \ldots\ldots\ldots\ldots\ldots q_N \qquad ...(3)$$

and if we have N distinguishable and identical particles, equation (1) becomes

$$Q = q_a\, q_a\, q_a \ldots\ldots\ldots\ldots\ldots q_a = q^N \qquad ...(4)$$

***Exercise* 9.44:** Prove equation (9.120).

Taking again the same system as described in Exercise (9.43). Now the particles a and b are identical and indistinguishable. Here from quantum mechanical point of view the distribution $\varepsilon_1(a) + \varepsilon_2(b)$ and $\varepsilon_2(a) + \varepsilon_1(b)$ are indistinguishable and the two belong to a single quantum state. This should be counted only once instead of twice as in the sum in Exercise (9.43), *i.e.*,

$$Q = \sum_1^9 g_i e^{-\varepsilon_i / k_b T}$$

In general, the system partition function of N independent particles is of the form

$$Q = \sum_i e^{-\varepsilon_i(a)/k_bT} \cdot \sum_j e^{-\varepsilon_j(a)/k_bT} \cdot \sum_k e^{-\varepsilon_k(a)/k_bT}$$

or

$$Q = \sum_{i,j,k} e^{-[\varepsilon_i(a)+\varepsilon_j(b)+\varepsilon_k(c)+....]/k_bT} \quad ...(1)$$

Now $a, b, c...$ are indistinguishable and identical molecules each in different molecular quantum state $(i \neq j \neq k \neq l...)$. These will occur N times in the summation for Q if the number of states is very large compared to the number of molecules. This follows because N states $i, j, k, l...$ etc. can be permuted $N!$ ways among N molecules and all the $N!$ ways will appear in the product Q^N. Then to get the corrected partition function, we will simply divide Q^N by $N!$ so as not to weigh each quantum state of the system $N!$ times but count only once, and thus for a system of N identical, independent and indistinguishable molecules, satisfying the condition that the number of available states is larger than N.

$$Q = \frac{q^N}{N!} \quad ...(9.120)$$

9.14. QUANTUM EFFECTS

So far we have calculated the thermodynamic properties discussed above by assuming that energy levels increase continuously, except in calculating the contribution of vibrational degrees of freedom to heat capacity. There are few other properties in which quantum effects are important and should be considered. They fall into two classes: (*i*) effects modifying the distribution law and (*ii*) effects modifying the partition function.

9.14.1. *Bose-Einstein Statistics*

Unlike Maxwell-Boltzmann distribution law, Bose-Einstein distribution law in which the indistinguishability of the molecules, is taken into account in a rigorous way, the expression is given as where each term has its own meaning explained earlier.

$$n_i = \frac{g_i}{e^{\alpha} e^{-\beta\varepsilon_i} - 1} \quad ...(9.121)$$

9.14.2. *Fermi-Dirac Statistics*

The Fermi-Dirac distribution law is consistent with the hypothesis, the Pauli exclusion principle, *i.e.*, in any system of indistinguishable molecules(or other particles) for which number of electrons (ν) has an odd value, no molecular quantum state can be occupied by more than one molecule at a time. This is a generalisation of Pauli exclusion principle. Fermi-Dirac distribution law is given by

$$n_i = \frac{g_i}{e^{\alpha} e^{-\beta\varepsilon_i} + 1} \quad ...(9.122)$$

9.14.3. *Limiting Limits*

The three laws can be summarised here for convenience, *i.e.*,

Maxwell-Boltzmann distribution law

$$n_i = \frac{g_i}{e^{\alpha} e^{\beta\varepsilon_i}} \qquad ...(9.11)$$

If $e^{\alpha} e^{\beta\varepsilon_i}$ happens to be much greater than 1, then the expression differs negligibly from

$$n_i = \frac{g_i}{e^{\alpha} e^{\beta\varepsilon_i} + 1} \qquad ...(9.122)$$

and from

$$n_i = \frac{g_i}{e^{\alpha} e^{-\beta\varepsilon_i} - 1} \qquad ...(9.121)$$

Consequently, when $e^{\alpha}e^{\beta\varepsilon_i}$ is large enough, the three distribution laws become identical. It is said, therefore, that the Maxwell-Boltzmann distribution law is the classical limit of the Bose-Einstein and Fermi-Dirac distribution laws.

***Exercise* 9.45:** Show that Maxwell-Boltzmann distribution law is valid for the classical limit.

In Maxwell-Boltzmann distribution statistics, e^{α} is given by the equation

$$e^{\alpha} = \frac{q}{N}$$

Since we are concerned with the gaseous assemblies, the largest factor in q is the translational partition function:

$$q_{\text{trans}} = \frac{(2\pi m k_b T)^{\frac{3}{2}} V}{h^3} \qquad ...(9.69a)$$

Using equation (9.111), we have seen that for q_{trans} for H_2 at its boiling point, *i.e.*, 20.3 K, $e^{\alpha} = 1.4 \times 10^2$ and for N_2 at its boiling point, *i.e.*, 27.2 K, $e^{\alpha} = 9.3 \times 10^3$. The lowest value of $e^{\beta\varepsilon_1}$ is $e^0 = 1$. Hence, in both of the above cases:

$$e^{\alpha}e^{\beta\varepsilon_i} >> 1,$$

and the Maxwell-Boltzmann distribution law can be used without significant error. For gasses with larger values of *m*, *T*, and *V*, the error becomes even less. There are two systems of interest in physical chemistry for which e^{α} is not large and for which Maxwell-Boltzmann law cannot be used but it is beyond the scope of the book.

PROBLEMS

9.1 The letters 'CHEMISTRY' are written separately one on each card. The nine cards are shuffled. Calculate the probability of obtaining the word 'Chemistry'.

9.2 The energy levels for certain kind of molecule are 0, ε, 2ε, 3ε, 4ε and are non-degenerate. Let there be six molecules of this kind. Calculate the number of complexions for each distribution if the total energy of the system is 6 ε. Also find the total number of complexions, probability of each distribution and the most probable distribution.

9.3 If the energy levels ε and 3ε are degenerate and their degeneracy are 6 and 10 fold respectively, repeat the problem 9.2.

9.4 The most probable distribution for arranging N_A atoms and N_B atoms in a mixed crystal is

$$\Omega = \frac{(N_A + N_B)!}{N_A!\,N_B!}$$

Using Sterling's theorem show that

$$\Omega = \frac{(N_A + N_B)^{N_A+N_B}}{N_A^{N_A}\,N_B^{N_B}}$$

9.5 If in the problem 9.4, $N_A = 6$, and $N_B = 3$, calculate the number of complexions using the equations (9.5) and (9.9). Explain the discrepancy if any.

9.6 Repeat the problem 9.2 if all the six molecules are indistinguishable. Use the equation (9.5) to calculate the number of complexions. Explain if there is any difference.

9.7 Show that the maximum number of complexion obtained in any distribution is when $n_1 = n_2 = n_3 = \ldots = n_i = n$ and $n = \frac{N}{i}$ where i is the total number of energy levels. There is no other restriction of energy.

9.8 Calculate the ratio $\frac{\Omega}{\Omega_{\max}}$ if n' molecules from one level is transferred to another. Again assume that only restriction is $\sum n_i = N$.

9.9 Repeat the problem 9.2 such that, there should always be at least three particles in energy level equal to zero.

9.10 Repeat the problem 9.3 such that there is at least three particles in level with energy equal to zero.

9.11 Using the data of the problem 9.2, calculate the value of $\bar{\varepsilon}^2$ for each distribution and $\bar{\varepsilon}^2$ for the complete distributions.

9.12 Using the data of the problems 9.2 and 9 3, calculate the values of ε^2 for each distribution and for the complete ensemble.

9.13 A molecule has only three energy levels, spaced ΔE apart in energy. There is only one level for each value of energy. The population ratios are: $\frac{n_2}{n_1} = 10^{-5}$ and $\frac{n_3}{n_2} = 10^{-3}$, where n_1 n_2 and n_3 are the number of molecules in energy levels 1, 2 and 3 respectively. Is this system describable by a temperature ? Explain briefly.

9.14 In an experiment it is found that ratios of molecules present in excited levels with respect to ground level is as follows:

$\varepsilon_i - \varepsilon_0$	0	10^{-21}	10^{-20}	10^{-19}
n_i/n_0	1	0.78	0.08	4.32×10^{-11}

Calculate the temperature of the system.

9.15 If in the problem 9.14, the temperature of the system is raised to 1000 K, what will be the ratios n_i/n_0 for the above energy levels?

9.16 If in the problem 9.14, the degeneracies for the levels ε_0, ε_1, ε_2 and ε_3 are 2, 3, 6 and 4, find the ratios n_i/n_0 at (*a*) temperature 300 K and (*b*) 1000 K.

9.17 Repeat the exercise 9.13 when temperature is 1000 K. How does the distribution behave by plotting the two curves on the same graph?

9.18 Repeat the exercise 9.14 when the temperature is 2000 K. Compare the two distribution curves.

9.19 Derive the following expression relating internal energy (ε) with the number of complexions (Ω_D) and partition function *i.e.*,

$$\beta_\varepsilon = \ln \Omega_D - N \ln q.$$

9.20 Calculate the value Ω_D for one mol of helium at 300 K and 1 atm pressure. Take $q = 1.924 \times 10^{29}$ and $\beta = 3/2$ N.

9.21 HCl molecules were heated. The relative population in excited rotational states were obtained with the help of spectroscopy and data is

J	0	1	2	3	4	5
N_J/N_0	1	2.90	4.48	5.69	6.25	6.60

(*a*) Show that the gas is in thermodynamic equilibrium with respect to distribution of rotational energy,

(*b*) Calculate the temperature of the gas.

9.22 Similar to the problem 9.21, the relative population of HCl gas into different excited state vibrational levels were also calculated and the data

v	0	1	2	3
N_v/N_0	1	5.65×10^{-3}	3.199×10^{-5}	1.81×10^{-7}

(*a*) Show that gas is in thermodynamic equilibrium with respect to the distribution of vibrational energy.

(*b*) Calculate the temperature of the gas.

(*c*) Calculate the fraction of the energy of the gas present as vibrational one.

9.23 In the problem 9.13, if the molecules occupying the energy levels are 1000; 100; 10, show that the system can be expressed by a temperature. Calculate the valueΩ_D. Show that this distribution gives the maximum value of Ω_D and any departure from this distribution with same energy will have lower value of Ω_D.

9.24 (*a*) One mol of HBr gas is present in a container at 1 atm and 400 K. Calculate the translation partition function; (*b*) If the pressure of the gas is reduced to 0.5 atm, what would be the value of translational partition function?

9.25 Calculate the translational partition function for argon gas at 1 atm pressure and at temperatures (*a*) 300 K and (*b*) 1000 K, (*c*) Repeat the parts (*a*) and (*b*) when the pressure is 0.4 atm.

9.26 The internuclear distance between the HBr molecule is 0.142 nm. Calculate (*a*) reduced mass, (*b*) moment of inertia and (*c*) rotational partition function at 300 K and 1000 K.

9.27 Repeat the exercise 9.23 for $^1H^3H$ molecule. See if there is any difference in the partition functions $B_e = 40.58$ cm^{-1}.

9.28 The vibrational frequency of HBr is 264.67 cm^{-1}. Calculate (*a*) θ_v, (*b*) relative population in the first four vibrational levels at 300 K and plot them, (*c*) Calculate the vibrational partition function by using equations (9.53) and (9.54). Explain the difference if any.

9.29 The following data is available for D_2, equilibrium distance is 0.074 nm (B_e = 30.43 cm^{-1}), vibrational frequency 3118.4 cm^{-1}. Calculate the total as well as the individual partition functions if $^1\Sigma$ is the ground electronic state, temperature is 300 K and pressure 1 atm.

9.30 Repeat the problem 9.29 for HD molecule with the following data, B_e = 46.655 cm^{-1}, vibrational frequency 3817.09 cm^{-1}, pressure 1 atm and temperature 300 K and 1000 K.

9.31 The equilibrium bond distance in T_2 molecule is 0.074 nm, (B_e = 20.3243 cm^{-1}) and vibrational frequency, 2553.8 cm^{-1}. Calculate the total partition function of T_2 molecule at 1 atm pressure at temperature 300 K and 1000 K.

9.32 Calculate $\bar{E}^0$ (thermal) and $\bar{H}^0$ (thermal) for HBr molecule using the data of the problems 9.26 and 9.28 at 300 K.

9.33 Calculate $\bar{E}^0$ (thermal) and $\bar{H}^0$ (thermal) for D_2, HD and T_2 at 300 K using the data of problems 9.29, 9.30 and 9.31.

9.34 Using the data of the problems 9.26 and 9.28, calculate the entropy of 1 mol of HBr at 300 K.

9.35 Repeat the problem 9.34 for D_2, HD are T_2 molecules at 300 K, use the data of the problems 9.29, 9.30 and 9.31.

9.36 Calculate the molar heat capacity at constant volume and at constant pressure for HBr molecule using the data of the problems 9.26 and 9.27.

9.37 Calculate the molar heat capacity at constant volume and entropy at 300 K for HD molecule using the data of problems 9.29, 9.30 and 9.31.

9.38 Calculate $\bar{E}^0$, $\bar{H}_0$ and $\bar{C}_V^0$ for methane at 300 K if $\bar{\nu}_1$ = 2917 cm^{-1}, $\bar{\nu}_2 = \bar{\nu}_3 = \bar{\nu}_4$ = 3018.9 cm^{-1} (triply degenerate), $\bar{\nu}_5 = \bar{\nu}_6 = \nu_7$ = 1306.2 cm^{-1} and $\bar{\nu}_8 = \bar{\nu}_9$ = 1533.6 cm^{-1}.

9.39 Calculate the molar entropy for methane using the data of problem 9.38.

9.40 Calculate the ΔH^0 of the following reaction at 300 K and 1000 K, given

$$H(^2S_{1/2}) + Cl_2\ (^1\Sigma) \rightarrow HCl + Cl\ (^2P_{3/2})$$

$$\bar{\nu}_{Cl_2} = 561.1\ \text{cm}^{-1};\ r^0_{Cl_2} = 0.1986\ \text{nm},\ \sigma_{Cl_2} = 2$$

$$\bar{\nu}_{HCl} = 2989.7\ \text{cm}^{-1};\ r_{HCl} = 0.128\ \text{nm};\ \sigma_{HCl} = 1$$

9.41 Derive an expression for the equilibrium constant of the reaction

$$A_2 + B_2 \rightleftharpoons 2\,AB$$

where A_2 and B_2 are isotopes.

9.42 If $A_2 = H_2$, $B_2 = D_2$, $AB = HD$, calculate the value of equilibrium constant at 300 K. If necessary, use the data from the problems 9.29, 9.30 and exercise 9.23.

9.43 Calculate the equilibrium constant for the reaction mentioned in problem 9.40 and using the data mentioned there.

9.44 Calculate the equilibrium constant for the reaction mentioned in problem 9.40 from the following data at 500 K;

species	H	Cl_2	HCl	Cl
$-\left(\frac{(\bar{G}^0 - \bar{H}_0^0)}{T}\right)$ (J deg^{-1})	109.56	208.57	172.84	155.06
$\Delta \bar{H}_0^0 = \Delta E_0^0$ (kJ)	215.98	0	–91.74	119.41

Compare the results obtained with that of above problem 9.43.

9.45 Using the free energy functions, calculate the equilibrium constant for the reaction at 1000 K

$$C_6H_6(s) + \frac{15}{2}\,O_2(g) \rightleftharpoons 6CO_2(g) + 3H_2O(g)$$

species	C_6H_6	O_2	CO_2	H_2O
$-\left(\frac{\bar{G}^0 - \bar{H}_0^0}{T}\right)$ (J deg^{-1})	320.37	212.13	226.40	196.69
$\Delta \bar{H}^0 = \Delta E_0^0$ (kJ)	100.416	0.0	– 339.166	– 238.936

9.46 Calculate the equilibrium constant for reaction

$$Cl_2 \rightleftharpoons 2\,Cl$$

at 500 K by both the methods. Data can be obtained from problems 9.40 and 9.44.

Addition Problems

9.47 Calculate the summation given below:

$$(i)\ \sum_{i=1}^{3}\sum_{j=1}^{3} x^{i+j},\ (ii)\ \sum_{i=1}^{4}\sum_{j=1}^{5} x^{i+j},\ (iii)\ \sum_{i=1}^{3}\sum_{j=1}^{2}\sum_{k=1}^{3} x^i y^j z^k .$$

This problem helps in finding the answer to write the total partition function in terms of the product of the partition function of the individual partition functions belonging to each degree of freedom.

9.48 A dice contains six numbers from 1 to 6 in its face. There is equal probability that each number can come when one throws the dice. Calculate the probability that number 3 will arrive when a dice is thrown once.

9.49 Continuing with the Problem 9.48 suppose that when the dice is thrown, the probability of getting each number is equal. It has been observed that face with dot 2 lands 1.7 times when face with dot 1 lands, face with dot 3 lands 1.5 times when face with dot 1 lands, face with dot 4 lands 1.2 times when face with dot 1 lands, whereas the faces with dots 5 and 6 lands with the same probability as the face with 1 dot lands. Calculate the normalised probability distribution for the appearance of the various faces when this dice is thrown.

9.50 Take coins of 25 paisa, 50 paisa, one rupee and 5 rupee. (*i*) Calculate the probability that each coin when land on table is head and (*ii*) when all the coins are thrown simultaneously that head will land in each case.

9.51 Continuing with Problem 9.48, we have now two dices and both the dices have the similar kind of biases. (*i*) Calculate the probability that when two dices are thrown the one dice will have 5 and the other dice will have 3, (*ii*) in part (*i*) the total numbers add up to 8. Now calculate total probability if both the dices are thrown, a total of eight dots will show on the top faces of the two dices.

9.52 Repeat the Problem 9.51 if the sum of the dots on the top faces of the two dices is 10.

Note: The Problems 9.51 and 9.52 introduce the idea of degeneracy.

9.53 10 coins are randomly tossed onto a board. Calculate the probability that (*i*) four will be heads and 6 will be tails, (*ii*) 3 will be tails and 7 will be heads, (*iii*) five will be heads and five will be tails.

9.54 Consider that 16 coins are tossed onto a table. Calculate (*i*) the possible distributions of tails and heads, (*ii*) the probability of each distribution obtained in part (*i*) and which distribution is the most probable?

9.55 Consider that we have N coins and they are tossed onto a table. Calculate the (*i*) probability of most probable distribution (assume that the most probable distribution is when half of them are head and half of them tails), (*ii*) Calculate the probability of the distribution which is x% away from the most probable distribution and (*iii*) the ratio of the two distributions. Using Sterling's approximation to express the result obtained in (*ii*) as

$$R = \exp[2x\ln(x) - (x + c)\ln(x + c) - (x - c)\ln(x - c)]$$

where $x = N/2$ and $c = xp/100$, and (*iii*) the ratio of the two distributions.

9.56 In Problem 9.55, consider that we have (*i*) 1000 coins, (*ii*) 10^8 coins and (*iii*) 6.02×10^{23} coins. Repeat the calculations as done in Problem 9.55 and (*iii*) the ratio of the two distributions.

9.57 Consider first four rotational energy levels, $J = 0$, $J = 1$, $J = 2$ and $J = 3$. Suppose we have Avogadro's number of molecules. Calculate the ratio of number of molecules to be placed in each rotational energy level so that the total energy E is divided equally in each energy levels.

9.58 (*i*) Six dices are thrown randomly. What is the probability that the sum of the up faces of the dice will add up 10. (*ii*) Calculate the total number of arrangements and probability of occurrence of distribution.

9.59 Repeat the Problem 9.58 when the number of dice is 5 and sum of the up faces of the dice add up to 12.

9.60 (*i*) Consider nitrogen gas at 293 K and at pressure of 1 bar (*i.e.* 10^5 Pa). Show that the number of translational states available is much larger than the Avogadro's number nitrogen molecules. (*ii*) Repeat the calculations for again nitrogen molecules at its boiling point (–195.8 °C) and density of liquid nitrogen is 0.808 g mL^{-1}.

9.61 The energies of the first and second electronic states of hydrogen atom relative to ground state are 82503.3 cm^{-1} and 97595.33 cm^{-1} respectively. If the degeneracies of the ground, first and second electronic states are 1, 8 and 18 respectively, calculate percentage of hydrogen atoms in the first and second state relative to ground state at 10,000 and 15,000K.

9.62 The bond length of HF molecule is 0.0917 nm. Assuming HF to be rigid rotor, calculate (*i*) moment of inertia, (*ii*) B in cm^{-1}, (*iii*) plot n_J/n_0 versus J using Maxwell-Boltzmann relation at 400 K, (*iv*) calculate the maximum value of J from this plot and compare it with the result obtained using equation (5.89) and (*v*) calculate the rotational partition function and compare it with that obtained from the summation using equation (9.61) and compare of n_J/n_0.

9.63 Consider a system containing N ($N >> 6$) molecules with four energy levels with their energies and degeneracy's given below. (*i*) If the number of molecules are N, calculate the fraction of the molecules present, in the (*a*) ground state and first state at 300 K. (*ii*) Four molecules are picked out at random. Calculate the temperature at which the probability that the sum of the energies of the four molecules add up to 600 J mol^{-1} will be the maximum. What is the value of maximum probability?

Level (n)	1	2	3	4
E (J mol^{-1})	0	100	200	300
g_n	2	3	1	2

9.64 Consider the same system as given in 9.63 and out of the four molecules one is present in the level 1, second is present in level 2, third is present in level 3 and fourth is present in level 4. Calculate the probability of this distribution.

9.65 In case of F, Br and I atoms, the gap between the ground and first excited electronic states is 0.050 eV, 0.46 eV and 0.94 eV respectively. Calculate the contribution to the partition function of this state at 300, 500 K and 100K. The results show that the contribution to the electronic partition function of the higher states decrease as the gap between the electronic states increase.

9.66 The ground state electronic configuration of silicon is [Ne]3s^{2}3p^2. (*i*) Using *L-S* coupling determine the term symbols, (*ii*) the energies of the first five states are: 0.0; 76.40; 222.25; 6292.27 and 15383.47 cm^{-1}. Including the J (total angular momentum quantum number) values, arrange and assign them to each state, (*iii*) determine the g_i values and the electronic partition function at 298 K, 1000 K, 4000 K, and (*iv*) calculate the fraction of the molecules in each state.

9.67 The ground state term symbol of O_2 molecule is three fold degenerate, ($^3\Sigma^-$) and the next one is $^1\Delta$, which is doubly degenerate and lies 7997 cm^{-1} above the ground state. Calculate the percentage contribution to q_{el} from this level at (*i*) 1000 K, (*ii*) 2000 and (*iii*) 5000 K.

9.68 Consider the following data for the ground state and the first excited state of the fluorine atom as given below:

Electronic configuration	$1s^2 2s^2 2p^5$		$1s^2 2s^2 2p^4 3s^1$				
Term symbol	$^2P_{3/2}$	$^2P_{3/2}$	$^4P_{5/2}$	$^4P_{3/2}$	$^4P_{1/2}$	$^2P_{3/2}$	$^2P_{1/2}$
Energy (cm^{-1})	0.0	404.0	102406.5	102681.24	102841.2	104731.9	105057.1

Calculate the electronic partition function at 300 K, 1000 K and 2000 K. Calculate the number of fluorine atoms in the first excited state at these temperature.

9.69 In a sample containing atomic hydrogen, what is the proportion of atoms in the first excited state at (*i*) 300 K, (*ii*) 500 K and (*iii*) 1000 K. Rydberg constant is 109,737.32 cm^{-1}.

9.70 In CO, the vibrational frequency is 2170.2 cm^{-1}. Calculate the proportional of the population in the first excited state above the ground state at (*i*) 300 K, (*ii*) 500 K and (*iii*) 1000 K.

9.71 Based on Exercise 9.39, show that the translational partition function for a three dimensional ensemble containing N particles is

$$Q(N, V, T) = \frac{1}{N!}\left(\frac{2\pi m k_b T}{h^2}\right)^{\frac{3N}{2}} V^N,$$

whereas for a two dimensional ensemble the partition function is

$$Q(N, V, T) = \frac{1}{N!}\left(\frac{2\pi m k_b T}{h^2}\right)^{N} A^N$$

where A is the area of the two dimensional box.

9.72 Pressure is defined by the equation (9.86), *i.e.*, if the partition function is defined as:

$Q(N, V, T) = \frac{1}{N!}\left(\frac{2\pi k_b T m}{h^2}\right)^{\frac{3N}{2}} (V - Nb)^N e^{\frac{aN^2}{k_b TV}}$, determine the expression for the pressure.

9.73 Using the relation of partition function defined by Problem 9.72, calculate the contribution to the heat capacity.

9.74 Using the equation given in Problem 9.72, calculate the total heat capacity of nitrogen at 100 K, 300 K, 500 K and 1000 K. Given $a = 1.3661$ mol^{-2} dm^6 bar, $b = 0.038577$ dm mol^{-1}.

9.75 Repeat Problem 9.74 for ethene. $a = 4.6112$ mol^{-2} dm^6 bar, $b = 0.058199$ dm mol^{-1}.

9.76 The internuclear bond distance in CO molecule is 0.128 nm. Calculate (*i*) moment of inertia, (*ii*) rotational constant and (*iii*) proportion of the population in the first excited state at (*a*) 300 K, (*b*) 500 K and (*c*) 1000K.

9.77 Derive equations (9.86), (9.87) and (9.88).

9.78 According to equipartition principle the kinetic energy for three dimensional cubic system is equal to $\frac{3}{2}k_b T$. Taking this as the kinetic energy, calculate the value of quantum number corresponding to the energy at 300 K for the oxygen atom. $a = 1$ dm.

9.79 Calculate the molar entropy of helium atoms at 1bar pressure and 300 K. The experimental value at this temperature is 126.3 J mol^{-1} K^{-1}. Calculate the % error.

9.80 Calculate the rotational energy of HF molecules using the classical expression, *i.e.* assuming the rotational levels increase continuously and using the summation expression. Compare the two values. $r_e = 0.0917$ nm.

9.81 Consider a system consisting of three non-interacting non-distinguishable fermions, each of which has states with energies E_1, E_2, E_3 and E_4. Calculate the allowed total energies in the summation in equation $Q(N, V, T) = \sum_{i,j,k,l...} e^{-\frac{E_i + E_j + E_k + E_l .....}{k_b T}}$.

9.82 Using the data from 9.80 for HF, calculate the entropy of 1 mole of HF at 50 K and 300 K using both the concepts. And compare the results.

9.83 Expression (9.54) for the vibrational partition function has been derived assuming the molecules follow simple harmonic motion. Calculate the q_{vib} for hydrogen molecule using above equation and summation equation. Given $\omega_e = 4395.2$ cm^{-1}, $\omega_e x_e = 117.99$ cm^{-1}.

9.84 Repeat the Problem 9.83 for O_2 and calculate the vibrational partition function O_2. $\bar{v} = 1580.36\ \text{cm}^{-1}$.

9.85 Equation (9.54) has been derived by taking the sum of infinite vibrational terms. If we take the sum of only v = k terms instead of v = ∞, then equation (9.96) is replaced by

$$\frac{1-e^{-k\Theta vib/T}}{1-e^{-\Theta vib/T}}$$

9.86 If $N\left(\frac{1}{2}\right)$ is the number of protons in the ground and $N\left(-\frac{1}{2}\right)$ in the upper state in presence of magnetic field of strength 10 T. Show that

$$\frac{N\left(-\frac{1}{2}\right)}{N\left(\frac{1}{2}\right)} = e^{-\frac{\gamma N h}{2\pi}H}$$

where γ_N is gyromagnetic ratio and its value is 26.7522×10^7 rad T^{-1} s^{-1} for a proton. At what temperature is $N\left(\frac{1}{2}\right) = N\left(-\frac{1}{2}\right)$. Suggest any physical meaning.

9.87 Verify equation (9.84) for (*i*) hydrogen gas at 300 K and (*ii*) at its boiling point 20.3 K. The density of liquid hydrogen at its boiling point is 0.067 g mL^{-1}.

9.88 Repeat the Problem (9.87) for electrons in sodium metal at 300 K. Density of sodium metal is 0.97 g mL^{-1}.

9.89 Repeat the Problem (9.87) for neon at (*i*) 200 K and (*ii*) 300 K.

9.90 Use the data of from Table 9.1 for H_2, calculate (*i*) the vibrational partition function by taking the sum of v = 10 vibrational levels and (*ii*) the vibrational partition function using equation (9.96) at 300 K and 1000 K. Calculate the percentage error.

9.91 The partition function is defined as

$$q = \sum_i g_i e^{-\frac{\varepsilon i}{kbT}}$$

Show that the entropy at 0 K is $k_b \log g_0$, where g_0 is the degeneracy of the ground state.

9.92 Calculate the number of complexions (Ω) for hydrogen gas if the molar entropy of H_2 is 130.6 J K^{-1} mol^{-1}.

9.93 If p_i is defined as the fraction of molecules in the *i*th energy level or as the probability of n_i molecules in the *i*th state, prove that the entropy will be

$$S = -k_b \sum_i p_i \ln p_i$$

9.94 Repeat the Problem 9.92 for I_2 molecule (Take the data from Table 9.1). Compare the percentage error obtained in Problem (9.80) and Problem (9.92). Explain briefly.

9.95 The vibrational frequency of $P_2(g)$ is 780 cm^{-1} and follows simple harmonic motion. Assuming that rotational levels are very closed spaced and follows equipartition principle, calculate the heat capacity of $P_2(g)$ at 298 K. The measured value is 32.05 J mol^{-1} K^{-1}.

[**Ans.** 31.96 J K^{-1} mol^{-1}]

9.96 The vibrational frequencies of water molecule are: $\bar{\nu}_1 = 3752$ cm^{-1}; $\bar{\nu}_2 = 3648$ cm^{-1}; $\bar{\nu}_3$ 1780 cm^{-1}. Calculate the heat capacity of H_2O (g) at 298 K, 500 K and 1000 K. Compare the results and explain briefly.

9.97 Considering $Cl_2(g)$ and $I_2(g)$ following equipartition principle for translational and rotational degrees of freedom, calculate the heat capacities of Cl_2 and I_2 at 298 K, 500 K and 1000 K. Compare the values with each other and draw and correlation. $\bar{\nu}(Cl_2) = 564$ cm^{-1}; $\bar{\nu}(I_2) = 214$ cm^{-1}.

9.98 The experimental hear capacity of copper at 30 K is 1.693 J K^{-1} mol^{-1}. Using equation (9.108) *i.e.*, limiting value of heat capacity at low temperature, calculate the Debye's characteristic temperature, express the heat capacity equation and using this expression for heat capacity calculate the entropy at 30 K.

9.99 Calculate the heat capacity of (*i*) Au, (*ii*) Cr, (*iii*) Ni and (*iv*) Nb at 40 K. The characteristic temperatures of the given elements are: 165 K, 630 K, 450 K, 275 K.

9.100 Calculate the total entropy of HF molecule at 298 K and 1 atm., assuming that HF follows rigid rotor and simple harmonic approximation.

9.101 Show that the contribution from rotation to the molar entropy of a diatomic molecule is

$\bar{S}_{rot} = R\left[\ln\frac{T}{\sigma\Theta_{rot}} + 1\right]$. Calculate the $\bar{S}_{rot}$ and $\bar{A}_{rot}$ for N_2 and D_2 at 300 K and 500 K. $\Theta_{rot}(N_2)$ $= 2.863$ K and $\Theta_{rot}(N_2) = 43.03$ K.

9.102 Calculate the symmetry number (σ) of CH_4, CH_3Cl, CH_2Cl_2, benzene, toluene, cis- and trans-butadiene and SF_6 molecules.

9.103 For most of the molecules the electronic partition is equal to g_0(elc). If we assume that the second term is also contributing to the electronic partition function, *i.e.*,

$$q_{elc} = g_0(elc) + g_1(elc)e^{-x}$$

where $x = \varepsilon_{elc}(1)/k_bT$, show that the electronic contribution to the heat capacity is

$$\bar{C}_v(elc) = \frac{Rx^2}{\left(1+\frac{g_0}{g_1}e^{x}\right)\left(1+\frac{g_0}{g_1}e^{-x}\right)}$$

If $\frac{g_0}{g_1} = 1$, the above equation reduces to

$$\bar{C}_v(elc) = R\left(\frac{x}{e^{x/2}+e^{-x/2}}\right)^2$$

9.104 Which of the following particles are fermions or bosons? ^3He, α-particle, deuterium, hydrogen gas, methane, electron, deuterated methane (CH_3D), $D^{35}Cl$, HD, photon and $^{35}Cl_2$.

9.105 Distribute 6 units of energy among 3 particles. The particles are (*i*) Boltzmann particle, (*ii*) corrected Boltzmann particle, (*iii*) Fermion and (*iv*) Boson. [W_{total} = 28; 28/3!; 3; 7]

9.106 Find the ratio of iodine molecules in the ground, first and second excited vibrational states at 298 K. The vibrational energy levels are separated by 214.6 cm^{-1}. [**Ans.** $P_0 = 0.645$; $P_1 = 0.229$; $P_2 = 0.081$; $P_3 = 0.029$].

9.107 Consider that there are 10^4 molecules of iodine. Prepare the table showing the distribution for various energy levels assuming that iodine follows harmonic motion.

9.108 Consider two subsystems, one containing *N* independent, non-localised particles *A* occupying a volume of V_1 and the second of *M* independent, non-localised particles *B* occupying a volume V_2. These are separated by a diathermic wall. Show that number of microstates accessible in the system as a whole is given by

$$\ln W_{max} = N \ln \frac{q_a}{N} + N + N \ln \frac{q_a}{N} + M + \beta E$$

9.109 Find the term symbol of ground and excited states of NO molecule. It has doubly ground state and doubly excited state at 121.1 cm^{-1} above the ground state. Calculate the electronic partition function at 300 K.

9.110 Following spectroscopic data is available for tellurium atom

Energy level (cm^{-1})	0	4707	4751	10559
Degeneracy	5	1	3	5

Calculate the electronic partition function of tellurium atom at (*i*) 300 K and (*ii*) 5000 K.

[**Ans.** 5; 6.26]

9.111 The energies of the first five vibrational states of iodine molecule relative to ground state are

Vibrational quantum numbers	0	1	2	3	4	5
Vibrational Energy level (cm^{-1})	0	213.3	425.39	636.27	845.93	1054.38

Calculate the vibrational partition function by using the (*i*) summation method and (*ii*) by using the equation (9.54) at 100 K, (*ii*) 300 K and 1000 K.

9.112 Calculate the average vibrational energy of iodine molecules using the data of Problem 9.111 at (*i*) 100 K, (*ii*) 300 K and (*iii*) 1000 K.

9.113 The rotational constants of water molecule are: A = 27.878 cm^{-1}, B = 14.509 cm^{-1} and C = 9.287 cm^{-1}. Calculate the rotational partition function at 300 K, 500 K and 1000 K.

9.114 Calculate the equilibrium constant for the reaction

$$H_2(g) + I_2 \rightleftarrows 2HI(g)$$

TABLE 9.3. Debye's temperature for some elements in cryogenics

Elements	Θ_D (K)	Elements	Θ_D (K)
Al	428	Hg	71.9
Au	165	In	108
Cd	209	Nb	275
Cr	630	Ni	450
Cu	343	Pb	105
Fe	470	Sn	200
Ga	320	Ti	420
Hf	252	V	380
		Zn	329

CHAPTER 10

FIRST LAW OF THERMODYNAMICS

Thermodynamics is the science of heat and temperature and, in particular, of the laws governing the conversion of thermal energy into mechanical, electrical or other forms of energy. It is the central branch of science which has applications in chemistry, physics, biology and engineering. On the positive side, main feature of thermodynamics is that it is a macroscopic science concerning such properties such as pressure, temperature and volume. Unlike quantum mechanics, thermodynamics is not based on a specific model, and thus remains unaffected by changing the concepts of atomic or molecular structure. On the negative side, equations derived from laws of thermodynamics do not provide us molecular interpretation of complex phenomenon. Lastly although thermodynamics helps us in predicting the feasibility of any process (chemical reaction) but does not give information about the rate of the reaction, *i.e.* how long it will take to complete the process. It can be summed up as: *a NO in thermodynamics is emphatic (i.e. reaction or process will not occur) but YES in thermodynamics means the reaction may or may not take place.*

10.1. SOME DEFINITIONS

10.1.1. *Reference Points and Derivatives*

The most important point in science is that *all measurements in science are difference measurements*. For example, the body temperature with a clinical thermometer consists of measuring the difference in length of column of mercury at two different temperature, height of the building relative to the ground floor, all the thermodynamic measurements etc.

10.1.2. *Total Differential*

Let us start with a function of one variable, defined as $y = f(x)$. If we change x_1 to $x_1 + \Delta x$, the change in y will be from y_1 to y_2. If Δx is very small, then the ratio $\Delta y/\Delta x$ will be a good approximation to the derivative required to compute Δy. In the limit as $\Delta x \to 0$, the ratio will approach the derivative required for the calculation:

$$\frac{dy}{dx} = \lim_{\Delta x \to 0} \frac{\Delta y}{\Delta x} \quad ...(10.1)$$

The total differential of y is written as:

$$dy = \left(\frac{dy}{dx}\right)dx \quad ...(10.2)$$

Integrating both sides of equation (10.2) gives the change in the variable y, *i.e.*,

$$\Delta y = \int_{y_1}^{y_2} dy = \int_{x_1}^{x_1+\Delta x} \left(\frac{dy}{dx}\right)dx \quad ...(10.3)$$

That is to calculate the change in variable y, we require the knowledge of dy/dx, *i.e.*, differential of y with respect to x.

If we are using a function of two variables, $z = f(x, y)$, the total change in z can be written as:

$$dz = dz_1 + dz_2 = \left(\frac{\delta z}{\delta x}\right)_y dx + \left(\frac{\delta z}{\delta y}\right)_x dy \quad ...(10.4)$$

Here the first term defines the change produced in z by changing x to $x + \Delta x$ keeping y as constant, *i.e.*, partial change produced in z by the change in x and multiplying it by dx gives the actual change occurred in z by changing small amount of x (dx). Similarly the second term gives the changes occurred in z by changes taking place in y only. The sum defines the total changes obtained when x and y are changed by dx and dy. Similarly the changes occurred in the function of n variables is defined as

$$dz(q_1, q_2, ...q_n) = \sum_{i=1}^{n}\left(\frac{\delta z}{\delta q_i}\right)dq_i \quad ...(10.5)$$

Partial derivatives are taken with respect to one variable and all other variables are held constant. Equation (10.5) gives the general expression for the total differential of n variable function.

Now the desired change in z can be obtained by integrating the equation (10.5). Simple example is taken for the function of two variables as

$$\Delta z = \int_{z_1}^{z_1+dz} dz = \int_{x_1, y_1}^{x_1+dx, y_1+dy} \left(\frac{\delta z}{\delta x}\right)_y dx + \left(\frac{\delta z}{\delta y}\right)_x dy = \int_{x_1}^{x_1+dx} \left(\frac{\delta z}{\delta x}\right)_y dx + \int_{y_1}^{y_1+dy} \left(\frac{\delta z}{\delta y}\right)_x dy \quad ...(10.6)$$

During integration it must be remembered that the integral over x should be taken with y held constant at the value of y_1 and integral over y is taken with x held constant at the value of $x_1 + dx$.

10.1.3. *Exact and Inexact Differential*

In Section 10.1.2, knowing partial derivatives $\left(\frac{\delta z}{\delta x}\right)_y$ and $\left(\frac{\delta z}{\delta y}\right)_x$, we have obtained the changes occurring in dz when we follow a path that gives the changes in z due to variation occurring in x from x to $x + dx$ and in y from y to $y + dy$, provided the changes in x and y are executed during the integration of equation

(10.6), as mentioned in Section 10.1.2. Now question arises whether the similar changes in dz will be observed if we follow some different path. These kinds of questions are very important in understanding thermodynamics.

In Section 10.1.2, Δz is an exact differential if it does not depend on path how one gets this change. This means that if Δz is path independent the change in Δz can be determined if we know the values of initial and final states of the independent variables. On the other hand, if Δz is path dependent then we must have detailed knowledge of the precise pathway along which the independent variables are changed. To know whether a function is exact or inexact, the Euler criterion provide the answer, *i.e.*, for dz to be exact differential, we must have

$$\left(\frac{\partial f(x,y)}{\partial x}\right)_y = \left(\frac{\partial g(x,y)}{\partial y}\right)_x \quad ...(10.7)$$

***Exercise* 10.1:** (*i*) Is the differential $[(3x^2y - 6x)\,dx + (x^3 + 2y)dy]$ exact or inexact? (*ii*) If it is exact, find the function $z(x,y)$ of which it is exact differential.

(*i*) We can take the help of Euler's criterion to decide whether the given function is exact or inexact. We have:

$$f(x, y) = x^3 + 2y \quad ...(1)$$

and

$$g(x, y) = 3x^2y - 6x \quad ...(2)$$

According to equation (10.7) the two partial derivatives that are required are

$$\left(\frac{\partial f(x,y)}{\partial x}\right)_y = \left(\frac{\partial (x^3+2y)}{\partial x}\right)_y = 3x^2 \quad ...(3)$$

$$\left(\frac{\partial g(x,y)}{\partial y}\right)_x = \left(\frac{\partial (3x^2y-6x)}{\partial y}\right)_x = 3x^2 \quad ...(4)$$

The above result suggests that the given function is an exact differential as Euler's criterion is satisfied. In other words, it may be written as

$$dz = \left(\frac{\partial z}{\partial x}\right)_y dx + \left(\frac{\partial z}{\partial y}\right)_x dy = [(3x^2y - 6x)dx + (x^3 + 2y)dy] \quad ...(5)$$

(*ii*) Equation (5) shows that

$$\left(\frac{\partial z}{\partial x}\right)_y = 3x^2y - 6x$$

Integration with respect to x keeping y constant, we can write

$$\int\left(\frac{\partial z}{\partial x}\right)_y dx = \int (3x^2y - 6x)dx = x^3y - 3x^2 + F(y) = z(x,y) \quad ...(6)$$

where $F(y)$ is a constant of integration and is a constant function of integration because y is held constant during integration. This can be easily verified by differentiating (6) with respect to x. Now similarly

$$\left(\frac{\partial z}{\partial y}\right)_x = x^3 + 2y$$

Integration with respect to y keeping y constant, we can write

$$\int\left(\frac{\partial z}{\partial y}\right)_x dx = \int(x^3+2y)dy = x^3y+y^2+G(x) = z(x,y) \qquad ...(7)$$

where $G(x)$ is a constant of integration and is a constant function of integration because x is held constant during integration. The values of constants can be evaluated by equating equations (6) and (7) as both the functions are same, *i.e.*,

$$x^3y - 3x^2 + F(y) = x^3y + y^2 + G(x) \qquad ...(8)$$

This can be true only if $F(y) = y^2 + C$ and $G(x) = -3x^2 + C$, where C is a constant. Therefore the required function will be:

$$z(x, y) = x^3y + y^2 - 3x^2 + C \qquad ...(9)$$

10.2. WORK

In thermodynamics, work is generally defined as the force multiplied by the distance. If the displacement of body under the force F is ds, the work done will be

$$DW = F\,ds$$

The symbol DW stands for the small amount of work and also inexactness of the function *i.e.*, work done depends on the path of reaction. There can be different kinds of work, defined by the equations:

$$\begin{aligned} &\text{Mechanical work} && DW = F\,ds \\ &\text{Expansion work} && DW = p_{ext}\,dV \\ &\text{Gravitational work} && DW = mgdh \\ &\text{Surface work} && DW = \gamma dA \\ &\text{Electrical work} && DW = \varepsilon dq \end{aligned} \qquad ...(10.8)$$

where F is the force, p_{ext} is the pressure exerted on the system by the surroundings, m is the mass, g is the acceleration due to gravity, γ is the surface tension, ε is the potential difference. V is the volume of the system, h is the height, A is the surface area and q is charge (*i.e.*, current × time = IdT).

The sign convention followed in the coming sections is that (*a*) the work is + ve if the system is doing work on the surroundings; (*b*) the work is – ve if the work is done on the system by the surroundings.

Total work done by the system will be obtained by integrating the above quantities. For example, the work of expansion is given as

$$W = \int p_{ext}\,dV \qquad ...(10.9)$$

There is another sign convention which is based on the energy of the system. During expansion (*i.e.*, dV = positive), the system is doing work against the external pressure and thus transferring energy to the surroundings and the system losing energy. Thus work is said to be *negative*, *i.e.*,

$$DW = -p_{ext}\,dV \qquad ...(10.10)$$

On the other hand, during compression, there is decrease in the volume, (*i.e.*, dV = negative), DW will be positive, *i.e.*, work is done on the system by the surroundings and thus gains energy. In this book we will follow the former convention, *i.e.*, based on expansion (*i.e.*, ΔV = positive), the work done is positive and in case of compression (*i.e.*, ΔV = negative) the work will be negative. It must be kept in mind that the pressure used will be always external pressure and not the internal pressure of the gas or system.

If the system expands against vacuum, *i.e.*, $p_{ext} = 0$, the integral (10.10) reduces to:

$$W = \int p_{ext} dV = \int 0 \times dV = 0 \qquad ...(10.11)$$

i.e., work done will be zero.

Work done is also zero if the work is done against constant volume, $\Delta V = 0$, *i.e.*,

$$W = \int p_{ext} dV = p_{ext} \int dV = p_{ext} \, \Delta V = p_{ext} \times 0 = 0 \qquad ...(10.12)$$

If the work is done against a constant external pressure, *i.e.*, p_{ext} = constant, the integral (10.10) reduces to:

$$W = \int_{V_1}^{V_2} p_{ext} dV = p_{ext} \int_{V_1}^{V_2} dV = p_{ext}(V_2 - V_1) = p_{ext} \Delta V \qquad ...(10.13)$$

***Exercise* 10.2:** What is the work done on the system if a gas is compressed against a constant pressure of 5 atm and the gas is compressed from 5 dm ^{3} to 1 dm^3 at 300 K. Express W in term of kJ

$$DW = p_{ext} \, dV$$

Using equation (10.13), we get

$$W = \int_{V_1}^{V_2} p_{ext} \, dV = p\,(V_2 - V_1)$$

Since p is constant, therefore

$$W = (5 \text{ atm}) (1 - 5) \text{ dm}^3 = -20 \text{ atm dm}^3$$

$$= -(20 \text{ atm dm}^3) (1.0132 \times 10^2 \text{ J atm}^{-1} \text{ dm}^{-3}) (10^{-3} \text{ kJ J}^{-1})$$

$$= -2.026 \text{ kJ}$$

***Exercise* 10.3:** A piece of metal of mass M kg and volume V m^3 is immersed in liquid whose density is d kg m^{-3}. Calculate the work done in lifting the piece by a height h m. What is the change in the potential energy?

When any substance is immersed in liquid its weight decreases by the amount of liquid displaced. Here liquid displaced is V m^3. So weight of liquid displaced

$$= (V \text{ m}^3) (d \text{ kg m}^{-3})$$

The effective weight of metal piece

$$= (M - Vd) \text{kg}$$

The work done against the action of gravity

$$= (M - Vd)\, g \times h \text{ Joules}$$

Change in potential energy = Mgh Joules

***Exercise* 10.4:** One mol of the ideal gas at 3 atm and 300 K is expanded isothermally to double its initial volume against an external pressure of 1.5 atm. Calculate W.

$$V_1 = \frac{RT}{p} = \frac{(0.082 \text{ dm}^3 \text{ atm K}^{-1} \text{ mol}^{-1}) (300 \text{ K}) (1 \text{ mol})}{3 \text{ atm}}$$

$$= 8.2 \text{ dm}^3$$

$$V_2 = \frac{p_1V_1}{p_2} = \frac{(3\,\text{atm})\,(8.2\,\text{dm}^3)}{(1.5\,\text{atm})} = 16.4\ \text{dm}^3$$

$$W = p_{ex}\,\Delta V = (1.5\ \text{atm})\,(16.4 - 8.2)\ \text{dm}^3$$
$$= 12.3\ \text{atm dm}^3$$
$$= 1.246 \times 10^3\ \text{J}$$

Since work is done by the system so it is positive.

***Exercise* 10.5:** One mol of ideal gas at 3 atm and 300 K is expanded isothermally and reversibly to twice its original volume. Calculate the work done.

Since the expansion is reversible, the expression for W is

$$W = nRT \ln \frac{V_2}{V_1} \tag{10.14}$$

$$= (1\ \text{mol})\,(8.314\ \text{J K}^{-1}\ \text{mol}^{-1})\,(300\ \text{K})\left(\ln \frac{2V_1}{V_1}\right)$$

$$= 1.72 \times 10^3\ \text{J}$$

Comparing with the results of the exercise 10.4, the work done in a reversible expansion is greater.

The work done in the reversible expansion is the maximum. Comparison of the results of exercises (10.3) and (10.4) led to an important canclusion. If Dw were an exact differential , the magnitude of w obtained in exercises (10.3) and (10.4) would have been equal. The pathways, reversible or not, would have made no difference.

***Exercise* 10.6:** What is the work done on the system if one mol of ideal gas at 300 K is compressed isothermally and reversibly to one fifth of its original volume?

$$W = nRT \ln \frac{V_2}{V_1}$$

$$= (1\ \text{mol})\,(8.314\ \text{J K}^{-1}\ \text{mol}^{-1})\,(300\ \text{K})\,(\ln 1/5)$$

$$= -4.014 \times 10^3\ \text{J}$$

Comparing the results of the exercise 10.5, here the gas is compressed. So the work is done on the system and the sign is negative.

10.3. HEAT

Heat is also a form of energy and similar changes in the states can be brought with heat as can be achieved with work. Similar to gravitational pull, electrical potential, heat also flows from higher temperature to lower temperature unless some work is done on the system.

The convention adopted here is that Q (heat) is positive when heat is absorbed by the system from the surroundings and it is negative if the heat is transferred from the system to the surroundings. Further similar to work heat gained or lost also depends upon the path followed by the system.

10.4. ENERGY

A system has number of degrees of freedom or motions and the internal energy is sum of the contributions due to these modes. The broad division between the nature of energies is kinetic and potential energies but otherwise it can be sum of translational, rotational, vibrational, electronic, nuclear, position and gravitational. Since in thermodynamics it is difficult to find the absolute value of energy, it is always expressed as the difference in the final and the initial states of the system *i.e.*,

$$\Delta E = E_2 - E_1 \quad ...(10.15)$$

Unlike work or heat, internal energy is an exact quantity, does not depend on the path but only on the initial or final state of the system. It is called *exact quantity* or *state function.*

10.5. FIRST LAW OF THERMODYNAMICS

Mathematically first law can be written as

$$\Delta E = Q - W \quad ...(10.16)$$

where Q is the total heat absorbed, W is the work done by the system and ΔE is the change in the energy that has occurred. In the differential form,

$$dE = DQ - DW \quad ...(10.17)$$

Further equation (10.17) can be written under different conditions as

$$dE = DQ \quad \text{if} \quad DW = 0 \quad ...(10.18)$$

$$dE = -DW \quad \text{if} \quad DQ = 0 \quad ...(10.19)$$

The last process is called adiabatic process *i.e.*, neither heat is absorbed nor given out.

***Exercise* 10.7:** One mol of diatomic gas is heated. The system absorbs 100 J of heat but does work equivalent to 50 J on the surroundings. If the contribution from the vibrational degrees of freedom is negligible, calculate the change in temperature of the gas.

According to first law of thermodynamics

$$\Delta E = 100 - 50$$

$$= 50 \text{ J mol}^{-1}$$

The contributions to the internal energy is only due to translational and rotational motions *i.e.*,

$$\Delta E = \frac{3}{2} R\Delta T + R\Delta T = \frac{5}{2} R\Delta T$$

$$\frac{5}{2} R\Delta T = 50 \text{ J mol}^{-1}$$

$$\left(\frac{5}{2}\right)(8.314 \text{ J}k^{-1}\text{mol}^{-1})(\Delta T) = 50 \text{ J mol}^{-1}$$

$$\Delta T = \frac{50 J\, mol^{-1}}{20.785 J\, mol^{-1} k^{-1}}$$

$$= 2.4 \text{ deg}$$

***Exercise* 10.8:** One mol of gas absorbs 400 J of heat at constant volume and its temperature is raised from 20°C to 25°C. Calculate the values of W, Q, ΔE.

Since the gas is absorbing heat at constant volume, i.e. equation (10.12)

$$W = 0$$

From first law of thermodynamics

$$\Delta E = Q = 400 \text{ J}$$

***Exercise* 10.9:** Hydrogen and O_2 are enclosed in a bomb calorimeter immersed in a thermal bath at 27°C. Mixture is ignited electrically and after the reaction is complete the temperature of the mixture in bomb calorimeter is 35°C. Calculate the values of W, Q and ΔE of the bomb calorimeter and its contents, (*a*) just after the completion of the reaction (*b*) when the temperature of the bomb calorimeter and its contents are brought down to the thermal bath after losing 10 kJ of heat.

In 1st case (*a*), $W = 0$, constant volume process, $Q = 0$, since no heat has left the calorimeter. So by first law

$$\Delta E = Q - W = 0 - 0 = 0$$

(*b*) $W = 0$, constant volume process $Q = -10$ kJ, heat is given out, $\Delta E = -10$ kJ, the energy of the system decreases.

10.6. REVERSIBLE AND IRREVERSIBLE PROCESS

A process is called irreversible if it cannot be reversed and if it is reversed, it cannot do so without leaving the surroundings permanently changed. A process is called a reversible one if the system concerned and the surroundings are in equilibrium and remain so through out the process. In real sense, it is not true equilibrium but a quasi-equilibrium.

A reversible process should also follow the undermentioned points:

(*i*) All parts of the system are in complete thermodynamic equilibrium *i.e.*, all the parts are at the same temperature (thermal equilibrium), all parts are at the same pressure (the mechanical equilibrium) and all parts of the system are in chemical equilibrium and remain so throughout.

(*ii*) The system and surroundings are free to exchange energy in any form, but maintaining the complete equilibrium with each other throughout the process *i.e.*, if the system and surroundings exchange heat energy, they must remain in thermal equilibrium, if the free exchange is in the form of mechanical work, the pressure exerted by the system must equal that of the surroundings throughout the process and so on with any type of energy exchanged freely, with the surroundings.

***Exercise* 10.10:** Prove the equation (10.14).

Equation (10.14) is obtained for an isothermal reversible process. As mentioned in Section 10.6, a process is reversible if the pressure difference between the system and surroundings (p_{ext}) is infinitesimally small (dp), *i.e.*, $p_{ext} = p_{int} - dp$ for expansion process. Substituting the value of p_{ext} in equation (10.9), we get:

$$W = \int_{V_1}^{V_2} p_{ext} dV = \int_{V_1}^{V_2} (p_{int} - dp) dV = \int_{V_1}^{V_2} p_{int} dV - \int_{V_1}^{V_2} dp dV \quad ...(1)$$

Assuming that gas is ideal one and p_{int} is the pressure exerted by n moles of ideal gas at temperature T, using ideal gas, p_{int} can be expressed as

$$p_{int} = \frac{nRT}{V} \quad ...(2)$$

Substituting the value of p_{int} in equation (2), we get

$$W = \int_{V_1}^{V_2} \frac{nRT}{V} dV - \int_{V_1}^{V_2} dp dV \quad ...(3)$$

Since the process is carried out at constant temperature, equation (3) can be written as

$$W = nRT \int_{V_1}^{V_2} \frac{dV}{V} - \int_{V_1}^{V_2} dp dV = nRT \ln \frac{V_2}{V_1} - \int_{V_1}^{V_2} dp dV \quad ...(4)$$

Under the limits, $dp \to 0$, the second term will be negligibly small and thus can be neglected. So equation (4) reduces to:

$$W = nRT \ln \frac{V_2}{V_1} \qquad ...(10.14)$$

***Exercise* 10.11:** A gas initially at pressure p_1 and volume V_1 undergoes expansion in two paths, (*i*) against the p_{ext} equal to p_3 and volume undergoes change from V_1 to V_3, (*ii*) expansion takes place in two steps, (*a*) against $p_{ext} = p_2$ and volume change from V_1 to V_2 and (*b*) against $p_{ext} = p_3$ and volume increases from V_2 to V_3. Calculate the work done in both cases and compare the results.

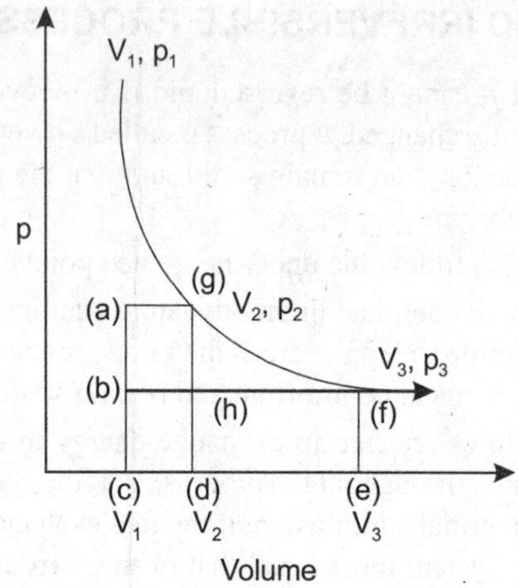

(*i*) Using equation (10.9),

$$W = p_3 (V_3 - V_1) \qquad ...(1)$$

and it is represented by the area of the rectangle *bcef*

(*ii*) In the path (*a*) the work done is:

$$W_1 = p_2 (V_2 - V_1) \qquad ...(2)$$

and it is represented by the area of rectangle *acdg*.

In the path (*b*) the work done is:

$$W_2 = p_3 (V_3 - V_2) \qquad ...(3)$$

and it is represented by the area of rectangle *defh*. The total work done is given as:

$$W' = W_1 + W_2 \qquad ...(4)$$

and it is represented by the area of rectangle *acefhg*. It is clear that the area in (*ii*) is larger than that of the area *bcef*. This suggests that if the expansion is carried out in more steps rather than in one step, the work done will be larger in the former case as compared to the latter step. The more the number of steps, the process will be more close to reversible process and work done in a reversible process is always the maximum as learnt earlier.

10.7. PROCESS AT CONSTANT VOLUME

If the system performs only the expansion work then

$$W = p_{ext} \Delta V \qquad ...(10.13)$$

Since $\Delta V = 0$, *i.e.*, isochoric process, the first law becomes

$$\Delta E_V = Q_V \qquad ...(10.20)$$

i.e., heat lost or gained at constant volume by the system is a direct measure of changes in the energy of the system. The examples of this process are the reactions carried out in bomb calorimeter. Q_V is a state function.

10.8. ENTHALPY AND PROCESS AT CONSTANT PRESSURE

Enthalpy is defined as

$$H = E + pV \qquad ...(10.21)$$

and changes in enthalpy

$$\Delta H = \Delta E + \Delta(pV) \qquad ...(10.22)$$

In differential form

$$dH = dE + d(pV) \qquad ...(10.23)$$

Now

$$dH = dE + pdV + Vdp \qquad ...(10.24)$$

If the process is carried at constant pressure *i.e.*, under isobaric conditions, $dp = \Delta p = 0$, so

$$dH = dE + pdV \qquad ...(10.25)$$

According to first law

$$dE = DQ - DW \qquad ...(10.26)$$

and if expansion is the only kind of work

$$dW = pdV$$

Combining equations

$$dH = DQ - pdV + pdV$$

$$= DQ_p = dQ_p \qquad ...(10.27)$$

Thus heat changes under isobaric conditions is a direct measure of the enthalpy change of the system. The examples are, heat of reaction carried out in laboratories in calorimeter, in reaction vessels and so on. Q_p is also a state function.

10.9. RELATION BETWEEN ENTHALPY AND ENERGY

For ideal gases the relation between the changes in enthalpy and energy can be derived as

$$\Delta H = \Delta E + \Delta(pV) \qquad ...(10.28)$$

But

$$\Delta(pV) = \Delta nRT \quad \text{at } T = \text{constant} \qquad ...(10.29)$$

where Δn is the change in the number of mol of reaction.

$$\Delta n = \sum_i n_i \text{ (products)} - \sum_j n_j \text{ (reactants)} \qquad ...(10.30)$$

$$\Delta H = \Delta E + \Delta nRT \qquad ...(10.31)$$

***Exercise* 10.12:** If ΔH for reaction at 298 K

$$CO(g) + \frac{1}{2}O_2(g) = CO_2(g)$$

is 282.85 kJ mol^{-1}, calculate ΔE of the reaction.

$$\Delta n = 1\,(CO_2) - 1\,(CO) - \frac{1}{2}\,(O_2)$$

$$= -\frac{1}{2}$$

$$\Delta E = \Delta H - \Delta nRT$$

$$= 282.85 \text{ kJ mol}^{-1} - (-\frac{1}{2} \text{ mol})\,(8.314 \text{ J mol}^{-1} \text{ K}^{-1})\,(298 \text{ K})$$

$$= 283.85 \text{ kJ mol}^{-1}$$

10.10. TOTAL DIFFERENTIALS OF *E* AND *H*

Consider an open system, that is both E and H are function of T, p, V and number of moles of each substance present in the system and let us denote them as $n_1, n_2, n_3, \ldots\ldots\ldots, n_i$. As usual our choice of variables is governed by the consideration of convenience and each of computation. We know that virtually in every chemical reaction, choice of temperature is a critical variable and thus necessary. Out of p and V it is preferable to choose one. This is because out of four variables, three can be defined independently and the fourth one will be dependent. So, we define E and H as

$$E = E(T, V, n_1, n_2, n_3, \ldots\ldots\ldots, n_i) \qquad \ldots(10.32)$$

$$H = H(T, p, n_1, n_2, n_3, \ldots\ldots\ldots, n_i) \qquad \ldots(10.33)$$

After defining the variables for E and H, the total differentials of both E and H are written as:

$$dE = \left(\frac{\partial E}{\partial T}\right)_{V,n} dT + \left(\frac{\partial E}{\partial V}\right)_{T,n} dV + \left(\frac{\partial E}{\partial n_1}\right)_{T,V,n_j} dn_1 + \left(\frac{\partial E}{\partial n_2}\right)_{T,V,n_j} dn_2 \ldots\ldots + \left(\frac{\partial E}{\partial n_i}\right)_{T,V,n_j} dn_i$$

or

$$dE = \left(\frac{\partial E}{\partial T}\right)_{V,n} dT + \left(\frac{\partial E}{\partial V}\right)_{T,n} dV + \sum_{i=1}^{K}\left(\frac{\partial E}{\partial n_i}\right)_{T,V,n_j} dn_i \qquad \ldots(10.34)$$

and

$$dH = \left(\frac{\partial H}{\partial T}\right)_{p,n} dT + \left(\frac{\partial H}{\partial p}\right)_{T,n} dp + \left(\frac{\partial H}{\partial n_1}\right)_{T,p,n_j} dn_1 + \left(\frac{\partial H}{\partial n_2}\right)_{T,p,n_j} dn_2 \ldots\ldots + \left(\frac{\partial H}{\partial n_i}\right)_{T,p,n_j} dn_i$$

or

$$dH = \left(\frac{\partial H}{\partial T}\right)_{p,n} dT + \left(\frac{\partial H}{\partial p}\right)_{T,n} dp + \sum_{i=1}^{K}\left(\frac{\partial H}{\partial n_i}\right)_{T,p,n_j} dn_i \qquad \ldots(10.35)$$

Now the quantity $\left(\frac{\partial E}{\partial n_i}\right)_{T,V,n_j}$ represents the change in the internal energy per mole of the substance i when T, V and all other components except i are held constant. Similarly $\left(\frac{\partial H}{\partial n_i}\right)_{T,p,n_j}$ represents change in the enthalpy per mole of the substance i when T, p and all other components other than i are held constant. These quantities are defined as *the partial molal quantities*, for example the first term defines the *partial molal energy* and the second term as the *partial molal enthalpy* and will be dealt in details in Chapter 18. If we are dealing with closed systems, *i.e.*, there is no change in mass, it is customary to delete the subscript n and equations (10.34) and (10.35) reduce to:

$$dE = \left(\frac{\partial E}{\partial T}\right)_V dT + \left(\frac{\partial E}{\partial V}\right)_T dV \qquad ...(10.36)$$

$$dH = \left(\frac{\partial H}{\partial T}\right)_p dT + \left(\frac{\partial H}{\partial p}\right)_T dp \qquad ...(10.37)$$

***Exercise* 10.13:** Consider the following processes: (*i*) 1 mol of ice at 273 K → 1 mol of H_2O (*l*) at 273 K and 1 atm. pressure, (*ii*) 1 mol of H_2O (*l*) at 273 K → 1 mol of H_2O (*g*) at 373 K and 1 atm pressure. Compare the values of $\Delta H - \Delta E$ for the two processes. Given, molar volume of ice and water at 273 K are 0.0196 L mol^{-1} and 0.018 L mol^{-1}respectively, whereas the molar volume of water and steam at 373 K are 0.0188 L mol^{-1} and 36.61 L mol^{-1} respectively.

Both processes are occurring at 1 atm., so these are constant pressure processes, *i.e.*,

$$\Delta H = \Delta E + \Delta(pV) = \Delta E + p\Delta V$$

or

$$\Delta H - \Delta E = p\Delta V \qquad ...(1)$$

(*i*) The change in molar volume of ice when it melts is:

$$\Delta V = 0.018 - 0.0196 = -0.0016 \text{ L mol}^{-1}$$

Substituting the values of p and ΔV in equation (1), we get

$$\Delta H - \Delta E = p\Delta V = (1 \text{ atm})(-0.0018 \text{ L mol}^{-1}) = -0.0018 \text{ atm L mol}^{-1}$$
$$= (-0.0018 \text{ atm L mol}^{-1})(1.0132 \times 10^2 \text{ J atm}^{-1} \text{ L}^{-1}) = -0.1986 \text{ J mol}^{-1}$$

(*ii*) The change in molar volume of water when it vaporises to steam is:

$$\Delta V = 30.61 - 0.0188 = 30.59 \text{ L mol}^{-1}$$

Substituting the values of p and ΔV in equation (1), we get

$$\Delta H - \Delta E = p\Delta V = (1 \text{ atm})(30.59 \text{ L mol}^{-1}) = 30.59 \text{ atm L mol}^{-1}$$
$$= (30.59 \text{ atm L mol}^{-1})(1.0132 \times 10^2 \text{ J atm}^{-1} \text{ L}^{-1}) = 3099.379 \text{ J mol}^{-1}$$

That is, based on the results of two exercises, it is clear that $\Delta H - \Delta E$ is very small for the condensed phase and very large for the changes in gas phase. In (*i*) $\Delta H < \Delta E$, the increase in internal energy is greater than the heat absorbed at constant pressure. This is because there is decrease in volume and thus work is done by the surroundings on the system. The vice versa is the case (*ii*), as steam is doing work on the system.

10.11. HEAT CAPACITY

10.11.1. *Definitions*

Heat capacity is defined as the heat required to raise the temperature of the system by one degree. The symbol used for heat capacity is C *i.e.*,

$$C = \frac{Q}{\Delta T} \text{ in differential form } \frac{DQ}{dT} \quad ...(10.38)$$

$$\Delta T \to 0$$

Molar heat capacity is the heat capacity of 1 mol of the substance and *specific heat* is the heat capacity per gram of substance, *i.e.*

$$C = n\bar{c} = wc' \quad ...(10.39)$$

where n is the number of mol, w is the weight of the substance, $\bar{c}$ is the molar heat capacity and c' is the specific heat capacity. Q is the heat required to raise the temperature of the substance by ΔT degrees. From sections (10.7) and (10.8) it is clear that heat may be absorbed at constant volume or at constant pressure. The two heat capacities may be defined as;

$$C_p = \frac{DQ_p}{dT} = \left(\frac{\partial H}{\partial T}\right)_p \quad ...(10.40)$$

$$C_V = \frac{DQ_V}{dT} = \left(\frac{\partial E}{\partial T}\right)_V \quad ...(10.41)$$

where C_p and C_V are the heat capacities at constant pressure and at constant volume respectively. The significance of heat capacity defined by equation (10.38) is as follows. It is the ratio of path dependent function heat by temperature and thus it gives a quantity which is path dependent and has no meaning unless the path of heating or cooling is defined. For example, consider melting of ice at 273 K. We know that we have to supply the heat till whole of ice has melted and during this process temperature will remain constant. In other words, the heat capacity will $DQ/0 = \infty$. Similarly during the cooling process, it is the reverse process and thus heat capacity will be $-\infty$. Using first law of thermodynamics and for closed system (*i.e.*, n_i are constant) DQ can be written as

$$DQ = dE - DW = \left(\frac{\partial E}{\partial T}\right)_V dT + \left(\frac{\partial E}{\partial V}\right)_T dV + p_{\text{ext}} dV$$

and substituting in equation (10.38), we get

$$C = \frac{DQ}{\Delta T} = \frac{DQ}{dT} = \left(\frac{\partial E}{\partial T}\right)_V + \left[\left(\frac{\partial E}{\partial V}\right)_T + p_{\text{ext}}\right]\frac{dV}{dT} \quad ...(10.42)$$

Now the change of volume with temperature only can be evaluated if we know the path followed by the system. For example consider the behaviour of ideal gas, $V = \dfrac{nRT}{p}$.

$\dfrac{dV}{dt} = \dfrac{nR}{p} - \dfrac{nR}{p^2}\dfrac{dp}{dt}$. Here also we cannot determine the value of dp/dt unless path is defined.

Similarly we can write equation (10.38) in terms of enthalpy

$$C = \frac{DQ}{\Delta T} = \frac{DQ}{dT} = \left(\frac{\partial H}{\partial T}\right)_p + \left[\left(\frac{\partial H}{\partial p}\right)_T - V\right]\frac{dp}{dT} \quad ...(10.43)$$

Now the four derivatives we require in equations (10.42) and (10.43) can be defined if we define the path. That is for equation (10.42) if we carry out the process at constant volume, ($dV = 0$) or if we carry out the process at constant pressure ($dp = 0$), equations (10.42) and (10.43) reduces to equations (10.41) and (10.40)

$$C = C_V = \left(\frac{\partial E}{\partial T}\right)_V \qquad ...(10.41)$$

and

$$C = C_p = \left(\frac{\partial H}{\partial T}\right)_p \qquad ...(10.40)$$

***Exercise* 10.14:** If a non-linear triatomic gas obeys the principle of equipartition of energies, calculate the molar heat capacity.

In a triatomic gas, there will be nine degrees of freedom, three translational, three rotational and three vibrational, contribution to the energy by each degree of translational and rotational motion is $\frac{1}{2}RT$ per mol, whereas that of vibrational mode it is RT per mol of gas. Thus,

$$\begin{aligned} \overline{E}_T &= 3 \times \frac{1}{2}RT + 3 \times \frac{1}{2}RT + 3RT \\ &= 6RT \\ \overline{C}_V &= \left(\frac{\partial E}{\partial T}\right)_V = 6R \\ &= (6 \times 8.314 \text{ J mol}^{-1} \text{ deg}^{-1}) \\ &= 49.884 \text{ J mol}^{-1} \text{ deg}^{-1} \end{aligned}$$

***Exercise* 10.15:** If the contribution to the internal energy due to vibrational mode is

$$\overline{E}_{(vib)} = \frac{Rh\nu / k_b}{\exp(h\nu / k_b T) - 1}$$

then find the value of $\overline{C}_V$ in the above exercise.

$$\overline{E}_{(vib)} = \frac{Rh\nu / k_b}{\exp\,(h\nu / k_b T) - 1}$$

$$\overline{C}_{V(vib)} = \left(\frac{\partial \overline{E}_{(vib)}}{\partial T}\right)_V = \frac{R(h\nu / k_b)^2 1/T^2 \exp\,(h\nu / k_b T)}{\{\exp\,(h\nu / k_b T) - 1\}^2}$$

$$= \frac{R(h\nu / k_b T)^2 \exp\,(h\nu / k_b T)}{\{\exp\,(h\nu / k_b T) - 1\}^2}$$

$$\overline{C}_V = 3R + \frac{R(h\nu / k_b T)^2 \exp\,(h\nu / k_b T)}{\{\exp\,(h\nu / k_b T) - 1\}^2}$$

***Exercise* 10.16:** The temperature of 2.0 kg of water dropped from 95°C to 40° *C* when 1.25 kg of a metal piece at 25°C was added to water. Calculate the heat capacity of metal if heat capacity of water is 4.184 J g^{-1} K^{-1}.

Since water and metal piece form isolated system, heat lost by water is gained by metal piece. Let c' be the specific heat of metal piece,

Heat gained by metal piece = $mc'\Delta T$

$$= (1.25 \times 10^3 \text{ g}) (40 - 25) \text{ deg } c'$$

Heat lost by water = – (2.0 × 10^3 g) (4.184 J K^{-1} g^{-1}) (– 55 deg). Equating the two

$$c' = \frac{2 \times 10^3 \times 4.184 \times 55}{1.25 \times 10^3 \times 15} \text{J K}^{-1} \text{g}^{-1}$$

$$= 24.55 \text{ J K}^{-1} \text{ g}^{-1}$$

10.11.2. *Relationship between C_p and C_V*

The following relationships between C_p and C_V can be derived;

$$C_p - C_V = \left[p + \left(\frac{\partial E}{\partial V}\right)_T\right]\left(\frac{\partial V}{\partial T}\right)_p \quad ...(10.44)$$

$$C_p - C_V = \left[V - \left(\frac{\partial H}{\partial p}\right)_T\right]\left(\frac{\partial p}{\partial T}\right)_V \quad ...(10.45)$$

Now for *ideal gases* $\left(\frac{\partial E}{\partial V}\right)_T = \left(\frac{\partial H}{\partial p}\right)_T = 0$ ...(10.46)

and $\left(\frac{\partial V}{\partial T}\right)_p = \frac{nR}{p}$ and $\left(\frac{\partial p}{\partial T}\right)_V = \frac{nR}{V}$, so the equations (10.44) and (10.45) become

$$C_p - C_V = nR \quad ...(10.47)$$

For solids and liquids, the more useful expression, can be written as

$$C_p - C_V = \frac{\alpha^2 VT}{\beta} \quad ...(10.48)$$

where α is the coefficient of thermal expansion and β is the coefficient of compression or compressibility factor. These are defined as

$$\alpha = \frac{1}{V}\left(\frac{\partial V}{\partial T}\right)_p \quad ...(10.49)$$

$$\beta = \frac{-1}{V}\left(\frac{\partial V}{\partial p}\right)_T \quad ...(10.50)$$

Heating at constant volume only increases the temperature whereas heating at constant pressure; (*i*) increases the temperature; (*ii*) work done in pushing back the environments and (*iii*) extension of the material against the intermolecular forces. The last two effects are responsible for the difference between C_p and C_V.

***Exercise* 10.17:** Derive the relation (10.44) *i.e.*,

$$C_p - C_V = \left[p + \left(\frac{\partial E}{\partial V}\right)_T\right]\left(\frac{\partial V}{\partial T}\right)_p$$

$$C_V = \left(\frac{\partial E}{\partial T}\right)_V$$

Since

$$E = E(T, V)$$

$$dE = \left(\frac{\partial E}{\partial T}\right)_V dT + \left(\frac{\partial E}{\partial V}\right)_p dV \qquad ...(10.36)$$

From 1st law

$$dE = DQ - pdV = \left(\frac{\partial E}{\partial T}\right)_V dT + \left(\frac{\partial E}{\partial V}\right)_T dV$$

$$DQ = \left(\frac{\partial E}{\partial T}\right)_V dT + \left[\left(\frac{\partial E}{\partial V}\right)_T + p\right] dV$$

Divide this equation by dT at constant pressure

$$\left(\frac{DQ}{\partial T}\right)_p = C_V + \left[\left(\frac{\partial E}{\partial V}\right)_T + p\right]\left(\frac{\partial V}{\partial T}\right)_p$$

$$C_p - C_V = \left[p + \left(\frac{\partial E}{\partial V}\right)_T\right]\left(\frac{\partial V}{\partial T}\right)_p \qquad ...(10.44)$$

***Exercise* 10.18:** Calculate the difference between the heat capacities at constant pressure and at constant volume for copper at 20°C. Given: $\alpha = 0.492 \times 10^{-4}\ K^{-1}$, $\beta = 0.70 \times 10^{-6}\ atm^{-1}$ and density of copper = 8.96 kg dm^{-3}.

If one mol of copper is taken, then the molar volume can be

$$\overline{V} = \frac{\text{wt.of 1 mol of copper}}{\text{density}}$$

$$= \frac{(63.54\ \text{g mol}^{-1})(10^{-3}\ \text{kg g}^{-1})}{8.96\ \text{kg dm}^{-3}} = 7.092 \times 10^{-3}\ \text{dm}^3\ \text{mol}^{-1}$$

Now

$$\overline{C}_P - \overline{C}_V = \frac{\alpha^2 \overline{V} T}{\beta} = \frac{(0.492 \times 10^{-4}\ \text{K}^{-1})^2 (7.092 \times 10^{-3}\ \text{dm}^3\ \text{mol}^{-1})(298\ \text{K})}{0.78 \times 10^{-6}\ \text{atm}^{-1}}$$

$$= 6.451 \times 10^{-6} \text{ atm m}^3 \text{ mol}^{-1} \text{ K}^{-1}$$
$$= (6.451 \times 10^{-6} \text{ atm m}^3 \text{ mol}^{-1} \text{ K}^{-1}) (101325 \text{ J m}^{-3} \text{ atm}^{-1})$$
$$= 0.654 \text{ J mol}^{-1} \text{ K}^{-1}$$

10.11.3. *Some Expressions for Heat Capacity*

Heat capacity cannot be predicted theoretically with the help of thermodynamics but can be calculated using statistical mechanics (Section 9.13). However, it is generally measured experimentally with the help of calorimeter. Some empirical expressions are:

$$\overline{C}_P = a + bT + cT^2 + dT^3 \quad ...(10.51)$$

and

$$\overline{C}_P = a' + b'T + c'/T^2 \quad ...(10.52)$$

where *a, b, c, d* and *a′, b′, c′* are constants depending upon the nature of the substance.

It was recognised by Dulong and Petit that molar heat capacity ($\overline{C}_V$) of solid elements is generally close to 25.9 J mol^{-1} K^{-1}. Thus knowing the specific heat, the approximate atomic weight of the element can be calculated.

More exact expression is due to Debye (see Section 9.13). The limiting values of $\overline{C}_V$ due to Debye theory are:

as $T \to \infty$

$$\overline{C}_V = 3R = 3\,(8.314 \text{ J mol}^{-1} \text{ K}^{-1}) \quad ...(10.53)$$
$$= 24.942 \text{ J mol}^{-1} \text{ K}^{-1}$$

and $T \to 0$

$$\overline{C}_V = \frac{12\pi^4}{5} R \frac{T^3}{\theta_D^3} \quad ...(10.54a)$$

$$= 1.947 \times 10^3 \frac{T^3}{\theta_D^3} \text{ J mol}^{-1} \text{ K}^{-1} \quad ...(10.54b)$$

where θ_D is called Debye's characteristic temperature.

The exact expression is

$$\overline{C}_V = \frac{9RT^3}{\theta_D^3} \int_0^{\theta_D/T} \frac{u^4 e^u}{(e^u - 1)^2} du \quad ...(10.55)$$

where

$$u = \frac{h\nu}{k_b T} \quad ...(10.56)$$

The contribution to $\overline{C}_V$ due to electronic motion in metals can be calculated with the help of Fermi-Dirac statistics and equation (10.56) becomes,

$$\overline{C}_V = \frac{9RT^3}{\theta_D^3} \int_0^{\theta_D/T} \frac{u^4 e^u}{(e^u - 1)^2} du + \frac{R\pi^2 k_b T}{2E_F} \quad ...(10.57)$$

where E_F is the Fermi energy; or

$$\overline{C}_V = \frac{9RT^3}{\theta_D^3}\int_0^{\theta_D/T}\frac{u^4e^u}{(e^u-1)^2}du + \gamma T \text{ J mol}^{-1}\text{ K}^{-1} \quad ...(10.58)$$

where $$\gamma = \frac{\pi^2 R k_b}{2E_F}$$

The electronic contribution to heat capacity is only significant at very low temperatures. The low temperature limit of equation (10.57) can be written as

$$\overline{C}_V = 1.947 \times 10^3 \frac{T^3}{\theta_D^3} + \gamma T \quad ...(10.54a)$$

or $$\frac{\overline{C}_V}{T} = \frac{1.947\times10^3}{\theta_D^3}T^2 + \gamma \quad ...(10.54b)$$

Plot of ($\overline{C}_V/T$) vs T^2 gives a straight line with slope = $\frac{1.947\times10^3}{\theta_D^3}$ and intercept γ.

At high temperature equation (10.55) becomes

$$\overline{C}_V = 3R\left[1-\frac{(\theta_D/T)^2}{20}+\frac{(\theta_D/T)^4}{560}-\frac{(\theta_D/T)^6}{18144}...\right] \quad ...(10.55b)$$

***Exercise* 10.19:** Calculate the molar heat capacity of H_2O (*g*) at 300 K and 1000 K if the values of constants *a, b, c* in equation (10.51) are

$$a = 30.36 \text{ J mol}^{-1}\text{ K}^{-1};\ b = 9.615 \times 10^{-3}\text{ J mol}^{-1}\text{ K}^{-2};$$
$$c = 12.00 \times 10^{-7}\text{ J mol}^{-1}\text{ K}^{-3}$$

Using equation (10.51) and T = 300 K

$$\overline{C}_p = 30.36 \text{ J mol}^{-1}\text{ K}^{-1} + (9.615 \times 10^{-3}\text{ J mol}^{-1}\text{ K}^{-2})(300\text{ K}) + (12.00 \times 10^{-7}\text{ J mol}^{-1}\text{ K}^{-3})(300\text{ K})^2$$
$$= 33.35 \text{ J mol}^{-1}\text{ K}^{-1}$$

Using equation (10.51) and T = 1000 K,

$$\overline{C}_p = 30.36 \text{ J mol}^{-1}\text{ K}^{-1} + (9.615 \times 10^{-3}\text{ J mol}^{-1}\text{ K}^{-2})(1000\text{ K}) + (12.00 \times 10^{-7}\text{ J mol}^{-1}\text{ K}^{-3})(1000\text{ K})^2$$
$$= 41.175 \text{ J mol}^{-1}\text{ K}^{-1}$$

***Exercise* 10.20:** Calculate the change in enthalpy of 1 mol of CO_2 (*g*) if it is heated from 300 to 1000 K. Given $a = 26.86$ J mol^{-1} K^{-1}; $b = 6.966 \times 10^{-3}$ J mol^{-1} K^{-2}; $c = 8.243 \times 10^{-7}$ J mol^{-1} K^{-3}.

From equation (10.40) at constant *p*

$$d\overline{H} = \overline{C}_p\, dT$$

$$\Delta\overline{H} = \int_{T_1}^{T_2}\overline{C}_p\, dT \quad ...(10.40a)$$

But $\overline{C}_p = a + bT + cT^2$, then equation (10.40$a$) becomes,

$$\Delta\overline{H} = \int_{T_1}^{T_2} (a + bT + cT^2)dT$$

$$= a(T_2 - T_1) + \frac{1}{2}b(T_2^2 - T_1^2) + \frac{1}{3}c(T_2^3 + T_1^3)$$

Substituting the values of a, b, c and T_2 and T_1

$$\Delta H = (26.86 \text{ J mol}^{-1} \text{ K}^{-1}) (1000 - 300) \text{ K}$$

$$+\frac{1}{2}(6.966 \times 10^{-3} \text{ J mol}^{-1} \text{ K}^{-2}) (1000^2 - 300^2) \text{ K}^2$$

$$+\frac{1}{2}(8.243 \times 10^{-7} \text{ mol}^{-1} \text{ K}^{-3}) (1000^3 - 300^3) \text{ K}^{-3}$$

$$= 22.24 \text{ kJ mol}$$

***Exercise* 10.21:** The characteristic Debye temperature for Fe metal is 420 K and the value of γ is 50.208×10^{-4} J g at^{-1} K^{-2}. Calculate the $\overline{C}_V$ for Fe at 6 K.

Substituting the respective values in equation (10.57 a)

$$\overline{C}_V = 1.947 \times 10^3 \left(\frac{6}{420}\right)^3 + (50.208 \times 10^{-4} \text{ J g at}^{-1} \text{ K}^{-2})$$

$$= 5.676 \times 10^{-3} + 30.125 \times 10^{-3} = 35.80 \text{ mJ g at}^{-1} \text{ K}^{-1}$$

***Exercise* 10.22:** The heat capacity of K metal has been measured at very low temperatures and following data is obtained:

T(K)	0.2604	0.6901	1.013	2.13	2.914	3.53	4.10
$\overline{C}$ (mJ mol^{-1}K^{-1})	0.5852	2.303	4.899	31.36	79.98	146.4	236.2

By plotting $\overline{C}/T$ vs T^2, calculate the value of γ and θ_D. Calculate $\overline{C}_V$ at 27°C by using the equation (10.55b).

Using the above data, construct the Table:

T^2	0.07	0.48	1.03	4.54	8.49	12.46	16.81
$\overline{C}/T$	2.247	3.337	4.84	14.72	27.45	41.47	57.61

Plot of $\overline{C}/T$ vs T^2 (Fig. 10.1) gives intercept

$$= 2.0 \text{ mJ mol}^{-1} \text{ K}^{-1} = \gamma$$

$$\text{Slope} = \frac{1.947 \times 10^3}{\theta_D^3} = 3.028$$

$$\theta_D = \left(\frac{1.947 \times 10^3}{3.028}\right)^{1/3} = 8.63 \text{ K}$$

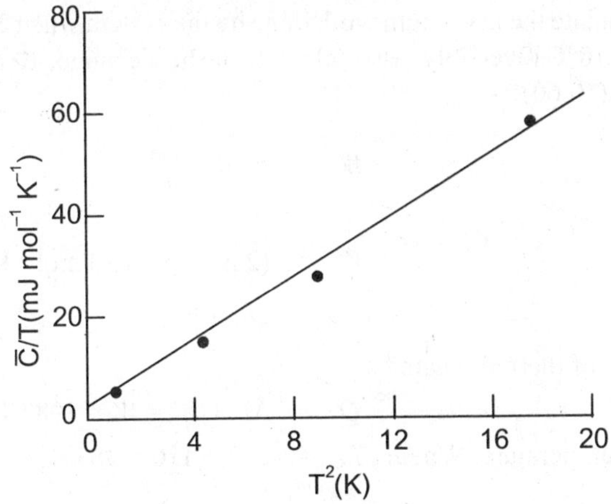

Fig. 10.1 *Plot of $\bar{C}/T$ vs T^2*

$$\frac{\theta_D}{T} = \frac{8.63\,\text{K}}{300\,\text{K}} = 2.877 \times 10^{-2}$$

Using equation (10.55*b*)

$$\bar{C}_V = (3)(8.314\,\text{J K}^{-1}\,\text{mol}^{-1})$$

$$\left[1 - \frac{(2.877\times10^{-2})^2}{20} + \frac{(2.877\times10^{-2})^4}{560} - \frac{(2.877\times10^{-2})^6}{18144}\right]$$

$= (24.94\ \text{J K}^{-1}\ \text{mol}^{-1})\ [1 - 4.1386 \times 10^{-5} + \text{very small terms}]$

$= 24.94\ \text{J mol}^{-1}\ \text{K}^{-1}$

10.12. APPLICATIONS OF FIRST LAW TO IDEAL GASES

10.12.1. *Definition of Ideal Gas*

A gas is ideal which obeys (*a*) $pV = nRT$ ...(10.59)

(*b*) $$\left(\frac{\partial E}{\partial V}\right)_T = \left(\frac{\partial E}{\partial p}\right)_T = \left(\frac{\partial H}{\partial p}\right)_T = \left(\frac{\partial H}{\partial V}\right)_T = 0 \quad \text{...(10.46)}$$

10.12.2. *Reversible Isothermal Expansion*

Since the internal energy of an ideal gas is only a function of temperature, at constant temperature. i.e. under isothermal conditions

$$\Delta E = \Delta H = 0$$

Combining equations (10.17) and (10.46)

$$Q = W = \int_{V_1}^{V_2} p\,dV = nRT \ln V_2/V_1 \quad \text{...(10.60)}$$

***Exercise* 10.23:** Calculate the maximum work done by the system when 2 mol of ideal gas expands from 2.4 dm³ to 5.6 dm³ at 10°C reversibly. Also calculate the heat change. Repeat the above calculations at 110°C. Using equation (10.60)

$$W = nRT \ln \frac{V_2}{V_1}$$

$$W = (2 \text{ mol})(8.314 \text{ J mol}^{-1} \text{ K}^{-1})(283 \text{ K}) \ln \left(\frac{5.6 \text{dm}^3}{2.4 \text{ dm}^3}\right)$$

$$= 3987 \text{ J}$$

According to first law of thermodynamics

$$Q = \Delta E + W = W = 3987 \text{ J}$$

since $\Delta E = 0$ at constant temperature. When $T_2 = 273 + 110 = 383$ K.

$$W = (2 \text{ mol})(8.314 \text{ J mol}^{-1} \text{ K}^{-1})(383 \text{ K}) \ln \frac{5.6}{2.4}$$

$$= 5396 \text{ J}$$

Similarly, $Q = W = 5396$ J

10.12.3. *Isothermal Isobaric Expansion of Ideal Gas*

For this process

$$\Delta E = \Delta H = 0$$

$$Q = W = \int_{V_1}^{V_2} pdV = p\Delta V \qquad ...(10.61)$$

since pressure is constant.

10.12.4. *Isothermal Isobaric Phase Change*

For this process, heat absorbed at constant plessure is enthalpy change.

$$Q = \Delta H \qquad ...(10.62)$$

$$W = \int_{V_1}^{V_2} pdV = p\Delta V \qquad ...(10.63)$$

and ΔE can be determined with the help of first law (10.17).

***Exercise* 10.24:** 1 mol of ideal gas at 300 K is compressed isothermally under the external pressure of 200 atm. Calculate the quantities Q, W, ΔH, ΔE if the initial and final pressure of the gas is 2 atm and 100 atm respectively.

Ideal gas equation is

$$V = \frac{nRT}{p}$$

$$V_1 = \frac{(1 \text{ mol})(0.0821 \text{ atm dm}^3 \text{ mol}^{-1} \text{ K}^{-1})(300 \text{ K})}{(2 \text{ atm})}$$

$$= 12.315 \text{ dm}^3$$

$$V_2 = \frac{(1 \text{ mol})(0.0821 \text{ atm dm}^3 \text{ mol}^{-1} \text{ K}^{-1})(300 \text{ K})}{100 \text{ atm}}$$

$$= 0.2463 \text{ dm}^3$$

$$W = p\Delta V = (200 \text{ atm})(0.2463 - 12.315) \text{ dm}^3$$

$$= -2413.74 \text{ atm dm}^3 = -2.414 \text{ atm m}^3$$

$$= -(2.414 \text{ atm m}^3)(101\cdot325 \text{ J m}^{-3} \text{ atm}^{-1})$$

$$= -246.6 \text{ kJ}$$

Since at constant T, $\Delta E = \Delta H = 0$, first law says

$$Q = W = -246.6 \text{ kJ}$$

***Exercise* 10.25:** Calculate ΔH, ΔE, Q and W for the following process at 100°C and 1 atm.

1 mol H_2O (l) $\rightarrow$ 1 mol $H_2O(g)$, $\Delta \overline{H} = 40668.48 \text{ J mol}^{-1}$

when the density of water (l) at 100°C is 0.9573 kg dm^{-3} and volume of one mol of H_2O (g) is 30.2 dm^3.

Under the above conditions

$$Q = \Delta H = 40668.48 \text{ J mol}^{-1}$$

$$W = p(\overline{V}_2 - \overline{V}_1)$$

Volume of 1 mol of H_2O at 1 atm and 100°C (373 K)

$$= \frac{(10^{-3} \text{ kg g}^{-1})(18 \text{ g mol}^{-1})}{0.9573 \text{ kg dm}^{-3}} = 0.0188 \text{ dm}^3 \text{ mol}^{-1}$$

$$W = (1 \text{ atm}) [(30.2 - 0.0188) \text{ dm}^3 \text{ mol}^{-1}] (1 \text{ mol})$$

$$= 30.18 \text{ atm dm}^3$$

$$= (30.18 \times 10^{-3} \text{ atm m}^3)(101325 \text{ J atm}^{-1} \text{ m}^{-3})$$

$$= 3060 \text{ J}$$

According to equation (10.17)

$$\Delta E = Q - W = 40668.48 - 3060 = 37608.48 \text{ J}$$

***Exercise* 10.26:** A balloon containing air at 27°C and 1 atm initially is filled with air further isothermally and reversibly till the pressure is 4 atm. If the initial diameter of the balloon is 1 m and the pressure at each stage is proportional to diameter of the balloon, calculate the work done.

$$p \propto D \text{ or } p_1 = kD_1$$

where k is constant

$$k = \frac{1 \text{ atm}}{1 \text{ m}} = 1 \text{ atm m}^{-1}$$

Now $p_2 = kD_2$

or

$$D_2 = \frac{4 \text{ atm}}{1 \text{ atm m}^{-1}} = 4 \text{ m}$$

$$V = \frac{4}{3}\pi\left(\frac{D}{2}\right)^3 = \frac{4}{3}\pi\left(\frac{D^3}{8}\right) = \frac{1}{6}\pi D^3$$

$$dV = \frac{1}{6}\pi\, 3D^2 dD = \frac{1}{6}\pi\, D^2\, dD$$

Equation (10.9) reduces to

$$W = \int_{D_1}^{D_2} p dV = \int_{D_1}^{D_2} kD.\frac{1}{2}\pi D^2 dD = \frac{1}{2}\pi k \int_{D_1}^{D_2} D^3 dD$$

$$= \frac{1}{8}\pi k \left(D_2^4 - D_1^4\right)$$

$$W = \left(\frac{1}{8}\right)\left(\frac{22}{7}\right)(1 \text{ atm m}^{-1})(4^4 - 1^4)\text{ m}^4 = 100.18 \text{ atm m}^3$$

$$= 1.015 \times 10^4 \text{ kJ}$$

10.12.5. *Reversible Adiabatic Expansion of Ideal Gas*

Under these conditions

$$Q = 0$$

So

$$\Delta E = -W = \int_{T_1}^{T_2} n\overline{C}_V\, dT \qquad ...(10.64)$$

$$\Delta H = \int_{T_1}^{T_2} n\overline{C}_p\, dT \qquad ...(10.65)$$

For the calculation of total change in the internal energy or enthalpy, one needs the initial and final temperature. This can be calculated from one of the following equations:

$$T_1 V_1^{R/C_V} = T_2 V_2^{R/C_V} \qquad ...(10.66)$$

$$p_1 V_1^{\gamma} = p_2 V_2^{\gamma} \qquad ...(10.67)$$

where

$$\gamma = C_p / C_V \qquad ...(10.68)$$

10.12.6. *Isobaric Adiabatic Expansion of Ideal Gas*

Under these conditions

$$Q = 0$$

$$\Delta E = \int_{T_1}^{T_2} n\overline{C}_V dT \qquad ...(10.64)$$

$$\Delta H = \int_{T_1}^{T_2} n\overline{C}_p dT \qquad ...(10.65)$$

$$W = p\Delta V \qquad ...(10.63)$$

The final temperature can be calculated with the help of first law *i.e.*,

$$-p\Delta V = \int_{T_1}^{T_2} n\overline{C}_V dT \qquad ...(10.69a)$$

$$= n\overline{C}_V (T_2 - T_1) \qquad ...(10.69b)$$

assuming that $\overline{C}_V$ is independent of temperature.

***Exercise* 10.27:** 1 mol of ideal gas is compressed reversibly and adiabatically. The initial volume is 6 dm³, final is 2 dm³ and the initial temperature is 27°C. Calculate the thermodynamic quantities like W, Q, ΔE, if the heat capacity at constant volume is 20.91 J mol^{-1} K^{-1}.

Substituting in the equation (10.66)

$$(300\,\text{K})(\sigma\,\text{dm}^3)^{\frac{8.314\,\text{J mol}^{-1}\text{K}^{-1}}{20.91\,\text{J mol}^{-1}\text{K}}} = (T_2)(2\,\text{dm}^3)^{\frac{8.314\,\text{J mol}^{-1}\text{K}^{-1}}{20.91\,\text{J mol}^{-1}\text{K}}}$$

$$T_2 = (300\text{ K})\left(\tfrac{6}{2}\right)^{\frac{8.314}{20.91}}$$

$$T_2 = 464.3\text{ K}$$

$$\Delta E = C_V(T_2 - T_1)$$

$$= (20.91\text{ J mol}^{-1}\text{ K}^{-1})(464.3 - 300)\text{K}$$

$$= 3435\text{ J mol}^{-1}$$

$$W = -\Delta E = -3435\text{ J mol}^{-1}$$

$$Q = 0$$

***Exercise* 10.28:** In the exercise 10.27, the gas is compressed from 6 dm³ to 3 dm³ against constant external pressure of 15 atm at 27°C. Calculate Q, W, ΔE, ΔH for this process.

Using equation (10.69*b*) the final temperature can be determined as

$$(1\text{ mol})(20.91\text{ J mol}^{-1}\text{ K}^{-1})(T_2 - 300)\text{K}$$

$$= -(15\text{ atm})(-6\text{ dm}^3 + 3\text{ dm}^3)(101325\text{ J m}^{-3}\text{ atm}^{-1})$$

$$T_2 = 300 + \frac{15 \times 3 \times 10^{-3} \times 101325}{20.91}\text{ K} = 518.1\text{K}$$

$$Q = 0$$

$$\Delta E = -W = (20.91\text{ J mol}^{-1}\text{ K}^{-1})(218.1\text{ K}) = 4.56\text{ kJ}$$

$$\Delta H = C_p\Delta T = (20.91 + 8.314)\text{ J mol}^{-1}\text{ K}^{-1}(218.1\text{ K})$$

$$= 6.374\text{ kJ}$$

Note: Although process mentioned in the exercise 10.28 is not reversible, but the laws of ideal gases are applicable to both initial and final states.

***Exercise* 10.29:** Calculate the value of ΔH in the exercise 10.28 by the relation

$$\Delta H = \Delta E + (p_2V_2 - p_1V_1)$$

Compare the value with that obtained above.

Now $p_1 = ?;\ V_1 = 6\text{ dm}^3;\ T_1 = 300\text{ K}\ n = 1\text{ mol}$

$p_2 = ?;\ V_2 = 3\text{ dm}^3;\ T_2 = 518.1\text{ K}$

Using ideal gas equation of state

$$pV = nRT$$

$$p_1 = \frac{(1\text{ mol})(0.082\text{ atm dm}^3\text{ K}^{-1}\text{ mol}^{-1})(300\text{ K})}{(6\text{ dm}^3)}$$

$$= 4.100\text{ atm}$$

$$p_2 = \frac{(1\text{ mol})(0.082\text{ atm dm}^3\text{ K}^{-1}\text{ mol}^{-1})(518.1\text{K})}{(3\text{ dm}^3)}$$

$= 14.18$ atm

Now

$$\Delta H = (14.18 \text{ atm})(3 \text{ dm}^3) - (4.1 \text{ atm})(6 \text{ dm}^3) + 4.56 \text{ kJ}$$
$$= 4.56 \text{ kJ} + 17.94 \text{ dm}^3 \text{ atm}$$
$$= 4.56 \text{ kJ} + 1.82 \text{ kJ} = 6.38 \text{ kJ}$$

The value of ΔH obtained in exercises (10.28) and (10.29) are same.

Note: In the above two exercises $Q = 0$ and also the process is carried out at constant pressure. One confusion can arise that $\Delta H = Q_p = 0$, but it is not *i.e.*, $\Delta H \neq Q_p \neq 0$ since $p_{\text{initial}} \neq p_{\text{final}} \neq p_{\text{ext}}$.

***Exercise* 10.30:** A following process is carried out:

$$1 \text{ mol } H_2O \text{ (s, 250 K)} \rightarrow 1 \text{ mol } H_2O \text{ (g at 400 K)}$$

Calculate the energy required to carry out this process if latent heat of vaporisation of H_2O is 2.257 kJ g^{-1} at 100°C and 1 atm, latent heat of fusion of H_2O is 333.51 J g^{-1} at 0°C and 1 atm. Specific heats of ice and steam can be considered as 1.925 J deg g^{-1} and 1.996 J g^{-1} deg^{-1}.

The above process can be broken into

$$1 \text{ mol } H_2O \text{ (s, 250 K)} \xrightarrow[\text{I}]{\text{at const. pressure}} 1 \text{ mol } H_2O \text{ (s, 273 K)}$$

$\downarrow$ II (at constant pressure and temperature)

$$1 \text{ mol } H_2O \text{ (}l\text{, 373 K)} \xleftarrow[\text{III}]{\text{at constant pressure}} 1 \text{ mol } H_2O \text{ (}l\text{, 273 K)}$$

$\downarrow$ IV (at constant temperature and pressure)

$$1 \text{ mol } H_2O \text{ (g, 373 K)} \xrightarrow{\text{V}} 1 \text{ mol } H_2O \text{ (g, 400 K)}$$

The heat required in this process is

$$\Delta H = \Delta H\text{(I)} + \Delta H\text{(II)} + \Delta H\text{(III)} + \Delta H\text{(IV)} + \Delta H\text{(V)} \quad \text{...(A)}$$

$$\Delta H\text{(I)} = \int_{T_1}^{T_2} C_p\text{(ice) } dT = C_p \text{(ice) (273 – 250) K} \quad \text{...(10.40}a\text{)}$$
$$= (18 \text{ g mol}^{-1})(1.925 \text{ J deg}^{-1} \text{ g}^{-1})(23 \text{ deg}) = 796.95 \text{ J mol}^{-1}$$
$$\Delta H\text{(II)} = (333.51 \text{ J g}^{-1})(18 \text{ g mol}^{-1}) = 6.003 \text{ kJ mol}^{-1}$$
$$\Delta H\text{(III)} = C_p(l)(T_2 - T_1) = C_p(l)(373 - 273) \text{ K}$$
$$= (18 \text{ g mol}^{-1})(4.184 \text{ J g}^{-1} \text{ K}^{-1})(100 \text{ K}) = 7.53 \text{ kJ mol}^{-1}$$
$$\Delta H\text{(IV)} = (2.257 \text{ kJ g}^{-1})(18 \text{ g mol}^{-1})$$
$$= 40.626 \text{ kJ mol}^{-1}$$
$$\Delta H\text{(V)} = C_p\text{(g)}(400 - 373) \text{ K}$$
$$= (1.996 \text{ J K}^{-1} \text{ g}^{-1})(18 \text{ g mol}^{-1})(27 \text{ K}) = 970 \text{ J mol}^{-1}$$

Substitute ΔH(I), ΔH(II)......in equation (A),

$$\Delta H = 0.797 + 6.003 + 7.53 + 40.63 + 0.970 = 55.93 \text{ kJ mol}^{-1}$$

***Exercise* 10.31:** Two mol of ideal gas at 1 atm and 27°C undergoes the following processes: (*i*) heat is absorbed at constant volume till the pressure is doubled; (*ii*) isothermal and reversible expansion till the pressure is 0.8 atm; (*iii*) adiabatic compression till the initial stage is reached. Calculate ΔE, Q and W for each step and for the total cycle if $\overline{C_V} = 40$ J mol^{-1} K^{-1}. Plot p vs V for the cycle.

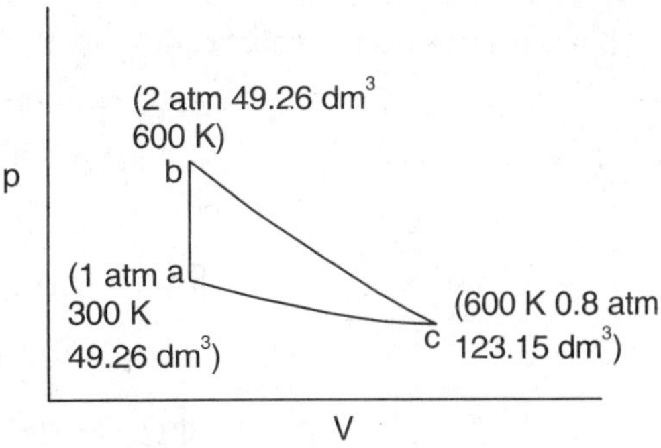

Step I

$$W = 0 \text{ since volume is constant.}$$

From first law,

$$Q = \Delta E = C_V\, dT$$
$$= C_V (T_2 - T_1)$$
$$p_1 = 1 \text{ atm}, V_1 = ?, T_1 = 27 + 273 = 300 \text{ K}$$
$$p_2 = 2 \text{ atm}, V_2 = V_1 = ?, T_2 = \frac{p_2}{p_1} T_1 = \frac{(2\,\text{atm})}{(1\,\text{atm})} \times (300 \text{ K})$$
$$= 600 \text{ K}$$
$$Q = \Delta E = (2 \text{ mol})\,(40 \text{ J mol}^{-1} \text{ K}^{-1})\,(600 - 300) \text{ K}$$
$$= 24.00 \text{ kJ}$$

Step II

$$V_2 = ?, p_2 = 2 \text{ atm}, T_2 = 600 \text{ K}$$
$$V_3 = ?, p_3 = 0.8 \text{ atm}, T_2 = 600 \text{ K}$$

Using ideal gas equation $pV = nRT$

$$V_2 = \frac{(2\,\text{mol})(0.0821\,\text{atm dm}^3\,\text{K}^{-1}\,\text{mol}^{-1})(600\,\text{K})}{(2\,\text{atm})}$$
$$= 49.26 \text{ dm}^3$$
$$\frac{p_2 V_2}{p_3} = V_3 = \frac{(2.0\,\text{atm})(49.26\,\text{dm}^3)}{(0.8\,\text{atm})} = 123.15 \text{ dm}^3$$

Since temperature is constant, $\Delta E = 0$

$$Q = W = \int_{V_2}^{V_3} p dV = nRT_2 \ln \frac{V_3}{V_2}$$
$$= (2 \text{ mol})\,(8.314 \text{ J mol}^{-1} \text{ K}^{-1})\,(600 \text{ K}) \ln \left(\frac{123.15 \text{ dm}^3}{49.26 \text{ dm}^3}\right)$$
$$= 9.142 \text{ kJ}$$

Step III

$$p_3 = 0.8 \text{ atm}, V_3 = 123.15 \text{ dm}^3, T_3 = 600 \text{ K}$$
$$p_1 = 1 \text{ atm}, V_1 = 49.26 \text{ dm}^3, T_1 = 300 \text{ K}$$

Since the process is adiabatic, $Q_{3-1} = 0$

$$\Delta E_{3-1} = -W_{3-1} = n\overline{C}_V(T_1 - T_3) = (2 \text{ mol})(40 \text{ J K}^{-1} \text{ mol}^{-1})(300 - 600) \text{ K}$$

$$= -24 \text{ kJ}$$

$$\oint \Delta E = \Delta E_{1-2} + \Delta E_{2-3} + \Delta E_{3-1}$$

$$= 24 + 0 - 24 = 0$$

$$\oint Q = 24.00 + 9.142 = 33.142 \text{ kJ}$$

$$\oint W = 0 + 9.142 + 24 = 33.142 \text{ kJ}$$

10.13. REAL GASES

10.13.1. *Equations of States*

(See Chapter 8)

10.13.2. *Joule-Thomson Experiment*

Under this experiment, a gas is allowed to expand through a porous plug and it measures the change of temperature with the variation of pressure under isoenthalpic nature *i.e.*, at constant enthalpy. The relation is

$$\mu_{JT} = \left(\frac{\partial T}{\partial p}\right)_H = -\frac{1}{C_p}\left(\frac{\partial H}{\partial p}\right)_T \qquad ...(10.70)$$

$$= -\frac{1}{C_p}\left[\left(\frac{\partial E}{\partial p}\right)_T + \left(\frac{\partial [pV]}{\partial p}\right)_T\right] \qquad ...(10.71)$$

where μ_{JT} is called Joule-Thomson coefficient. If μ_{JT} is positive and the gas is allowed to expand adiabatically, the temperature of the gas drops *i.e.*, cooling takes place. Physically it means that gas is doing work on the surroundings without getting heat energy, so the temperature drops. If μ_{JT} is negative and the gas is allowed to expand, the temperature of gas increases.

***Exercise* 10.32:** If a real gas obeys the van der Waals equation, calculate the changes in thermodynamic quantities if the gas expands isothermally and reversibly.

The work done against expansion *i.e.*,

$$W = \int_{V_1}^{V_2} p dV$$

and van der Waals equation for 1 mol of gas

$$\left(p + \frac{a}{\overline{V}^2}\right)(\overline{V} - b) = RT$$

Substituting the value of pressure in above equation

$$W = \int_{V_1}^{V_2}\left(\frac{RT}{\overline{V}-b}-\frac{a}{\overline{V}^2}\right)dv$$

$$= RT\ln\left(\frac{\overline{V}_2-b}{\overline{V}_1-b}\right)+\left(\frac{a}{\overline{V}_2}-\frac{a}{\overline{V}_1}\right) \quad ...(10.72)$$

The change in the internal energy cannot be calculated with the help of first law and needs second law of thermodynamics (see Chapter 12). It can be shown that

$$\left(\frac{\partial E}{\partial V}\right)_T = T\left(\frac{\partial p}{\partial T}\right)_V - p$$

The van der Waals equation can be written as

$$p = \frac{RT}{\overline{V}-b}-\frac{a}{\overline{V}^2}$$

$$\left(\frac{\partial p}{\partial T}\right)_V = \frac{R}{\overline{V}-b}$$

and

$$T\left(\frac{\partial p}{\partial T}\right)_V - p = \frac{RT}{\overline{V}-b}-\frac{RT}{\overline{V}-b}+\frac{a}{\overline{V}^2} = \frac{a}{\overline{V}^2}$$

$$\left(\frac{\partial E}{\partial V}\right)_T = \frac{a}{\overline{V}^2}$$

$$\Delta E = \int_{V_1}^{V_2}\frac{a}{\overline{V}^2}dV = \frac{a}{\overline{V}_1}-\frac{a}{\overline{V}_2} \quad ...(10.73a)$$

Applying first law heat change can be obtained as

$$Q = \Delta E + W$$

$$= a\left(\frac{1}{\overline{V}_1}-\frac{1}{\overline{V}_2}\right)+RT\ln\frac{\overline{V}_2-b}{\overline{V}_1-b}+a\left(\frac{1}{V_2}-\frac{1}{V_1}\right)$$

$$= RT\ln\frac{\overline{V}_2-b}{\overline{V}_1-b} \quad ...(10.73b)$$

***Exercise* 10.33:** 1 mol of O_2 at 27°C expands isothermally and reversibly from 5 dm³ to 20 dm³. Calculate the work done if $a = 1.36$ dm⁶ atm mol⁻² and $b = 0.0318$ dm³ mol⁻¹ *i.e.*,

Substituting the values in equation (10.72)

$$W = (8.314 \text{ J mol}^{-1}\text{ K}^{-1})(1 \text{ mol})(300 \text{ K})\ln\left(\frac{20-0.0318}{5.0-0.0318}\right)$$

$$+ (1.36 \text{ dm}^6 \text{ atm mol}^{-2})\left(\frac{1}{20}-\frac{1}{5}\right)$$

$$\text{dm}^{-3}\text{ mol } (101325 \text{ J m}^{-3}\text{ atm}^{-1}) (1 \text{ mol})$$
$$= 3477 - 20.67 = 3456.33 \text{ J}$$

Substituting the values in equation (10.73)

$$\Delta E = (1.36 \text{ dm}^6 \text{ atm mol}^{-2}) \left(\frac{1}{20} - \frac{1}{5}\right) \text{dm}^{-3}\text{ mol}$$
$$(101325 \text{ J m}^{-3}\text{ atm}^{-1}) (1 \text{ mol})$$
$$= 20.67 \text{ J}$$

Using first law

$$\Delta E = Q - W$$

or

$$Q = \Delta E + W$$
$$= 20.67 + 3456.33 = 3477 \text{ J}$$

***Exercise* 10.34:** The Joule-Thomson coefficient for van der Waals gas can be written as

$$\mu_{JT} = \frac{2a/RT - b}{C_p}$$

Using the data of exercise 10.33, calculate the value of ΔH for the isothermal ($T = 300$ K) compression of 1 mol of oxygen from the pressure of 1 to 200 atm.

Combining equation (10.70) and the given relation

$$\mu_{JT} = -\frac{1}{Cp}\left(\frac{\partial H}{\partial p}\right)_T = \frac{1}{C_p}\left(\frac{2a}{RT} - b\right)$$

$$\Delta H = -\int_{p1}^{p2}\left(\frac{2a}{RT} - b\right)dp$$

At constant temperature

$$\Delta H = \left(b - \frac{2a}{RT}\right)(p_2 - p_1) \quad \text{...(10.74)}$$

Substituting the values in (10.74)

$$\Delta H = \left(0.0318 \text{ dm}^3\text{ mol}^{-1} - \frac{2 \times 1.36 \text{ dm}^6 \text{ atm mol}^{-2}}{(0.0821 \text{ atm dm}^3\text{ mol}^{-1}\text{ K}^{-1})(300 \text{ K})}\right)$$
$$(200\text{–}1) \text{ atm}$$
$$= -15.46 \text{ atm dm}^3\text{ mol}^{-1}$$
$$= -(15.46 \text{ atm dm}^3\text{ mol}^{-1})(101.325 \text{ dm}^{-3}\text{ atm}^{-1}\text{ J})$$
$$= -1556.5 \text{ J mol}^{-1}$$

PROBLEMS

10.1 A body of mass 300 g has fallen a distance of 250 cm. Calculate the work that can be done during this process.

10.2 A constant current of 2.30 amp and 6 V was passed through heater dipping in CCl_4. If after 20 min 85 g of CCl_4 was collected, calculate the heat of vaporisation for one mol.

10.3 The work done by the surroundings in stretching the length of the spring by a distance dL is

$$DW = -\tau dL$$

where τ is the tension of the spring of length L. Calculate the work done for an isothermal increase in tension of spring using the following relations.

$$\left(\frac{\partial \tau}{\partial L}\right)_T = -\alpha \, AY, = \frac{L}{A}\left(\frac{\partial \tau}{\partial L}\right)_T, \; \alpha = \frac{1}{L}\left(\frac{\partial L}{\partial T}\right)_\tau$$

where α, A and Y are coefficient of linear expansion, area crosssection and isothermal Youngs modulus, respectively. Also assume that α, A and Y remain constant as the tension is increased.

10.4 Which of the fallowings are state functions?

(*i*) q, (*ii*) $q+W$, (*iii*) $q - W$, (*iv*) q_p, (*v*) q/W, (*vi*) $\Delta H/T$, (*vii*) $E + pV$, (*viii*) $H + TS$, (*ix*) $E - pV$.

10.5 Define extensive and intensive variables. Give examples of each.

10.6 CO_2 is produced by the action of dilute mineral acid on carbonate. If the gas collected is 35 dm ^{3} against the atmospheric pressure of 1 atm, calculate the work done by the gas in pushing back the atmosphere in (*a*) dm^3 atm, (*b*) Joules, (*c*) ergs, (*d*) calories.

10.7 A man weighing 70 kg climbs Qutub Minar which is nearly 200 ft high. How much work will be done in this case? Suppose now, that only 20% of the energy produced by the oxidation of food is used in this mechanical work, how many kcal should be supplied ?

10.8 A copper spring is compressed by a distance of 8 cm by a weight of 200 kg. (*a*) If the acceleration due to gravity is 9.80 m s^{-2}, what is the internal energy change (ΔE) of the spring when there is no heat lost to surroundings, (*b*) If the specific heat of copper is 0.385 J g^{-1} deg^{-1} and the spring weights 2 kg, what is the rise in temperature due to compression?

10.9 Heat of reactions (ΔH^0) for the following reactions are given at 25°C, calculate the values of ΔE^0 for each reaction:

(*a*) $H_2(g) + \frac{1}{2} O_2(g) \rightarrow H_2O(g)$ $\quad \Delta H^0 = -241.83$ kJ

(*b*) $H_2(g) + F_2(g) \rightarrow 2\,HF(g)$ $\quad \Delta H^0 = 268.49$ kJ

(*c*) $C_6H_6(g) + \frac{15}{2} O_2(g) \rightarrow 6\,CO_2(g) + 3H_2O(g)$ $\quad \Delta H^0 = -3301.51$ kJ

(*d*) $CH_3COOH(l) + 2O_2(g) \rightarrow 2\,CO_2(g) + 2H_2O(g)$ $\quad \Delta H^0 = -872.36$ kJ

(*e*) $C_2H_2(g) + \frac{5}{2} O_2(g) \rightarrow 2\,CO_2(g) + H_2O(g)$ $\quad \Delta H^0 = -1299.63$ kJ

10.10 Calculate the heat capacity ($\bar{C}_p$) of propane at 25°C and 2000°C if the value of a = 10.08 J

mol^{-1} deg^{-1}, $b = 239.30 \times 10^{-3}$ J mol^{-1} deg^{-2}, $c = -733.58 \times 10^{-7}$ J mol^{-1} deg^{-3}. Calculate $\overline{C}_V$ also assuming the gas is behaving ideally under these conditions.

10.11 Calculate the heat of reaction at 500°C for the reaction:

$$H_2S(g) + \frac{3}{2}O_2(g) \rightarrow H_2O(g) + CO_2(g)$$

if $\Delta H^0_{298} = -538.73$ kJ and $\overline{C}_p$ (J mol^{-1} deg^{-1}) of various components are:

$$\overline{C}^0_p(H_2S) = 26.715 + 23.866 \times 10^{-3}\,T - 50.626 \times 10^{-7}\,T^2$$

$$\overline{C}^0_p(O_2) = 25.503 + 13.612 \times 10^{-3}\,T - 42.555 \times 10^{-7}\,T^2$$

$$\overline{C}^0_p(H_2O) = 30.204 + 9.933 \times 10^{-3}\,T + 11.171 \times 10^{-7}\,T^2$$

$$\overline{C}^0_p(SO_2) = 25.719 + 57.923 \times 10^{-3}\,T - 380.87 \times 10^{-7}\,T^2$$

Also calculate the value ΔE^0.

10.12 Calculate ΔH^0 at 1000 K for the reaction

$$CH_4\,(g) + 2\,O_2\,(g) \rightarrow CO_2\,(g) + 2H_2O\,(g)$$

if $\Delta H^0_{298} = -890.36$ kJ mol^{-1} and $\overline{C}_p$ (J mol^{-1} deg^{-1}), are,

$$\overline{C}_p(CH_4) = 14.146 + 75.496 \times 10^{-3}\,T - 179.91 \times 10^{-7}\,T^2$$

$$\overline{C}_p(O_2) = 25.503 + 13.612 \times 10^{-3}\,T - 42.555 \times 10^{-7}\,T^2$$

$$\overline{C}_p(CO_2) = 26.648 + 42.258 \times 10^{-3}\,T - 142.465 \times 10^{-7}\,T^2$$

$$C_p(H_2O) = 30.204 + 9.933 \times 10^{-3}\,T + 11.171 \times 10^{-7}\,T^2$$

10.13 Calculate $\overline{C}_p - \overline{C}_V$ for a gas obeying van der Waals equation of state, *i.e.*, $\left(p + \frac{a}{\overline{V}^2}\right)(\overline{V} - b)$ $= RT$ and $E\,(V, T) = \overline{C}_p T - \frac{a}{\overline{V}}$.

10.14 Derive the expressions

(*i*) $\overline{C}_p - \overline{C}_V = \dfrac{\alpha^2 \overline{V} T}{\beta}$

(*ii*) $C_p - C_V = \left[V - \left(\dfrac{\partial H}{\partial p}\right)_T\right]\left(\dfrac{\partial p}{\partial T}\right)_V$

10.15 The Debye characteristic temperature for diamond is 1860 K. Calculate the heat capacity at constant volume at 6 K and 300 K using respective relations, *i.e.* (10.54 *a*) and (10.54 *b*). Does the value of $\overline{C}_V$ at 300 K agrees with that due to Dulong-Petit ?

10.16 The Debye characteristic temperatures for Cu and Ag are 315 K and 215 K respectively. Calculate the values of $\overline{C}_V$ at 10 K and 300 K if γ for Cu and Ag are 0.695 mJ mol^{-1} deg^{-2} and 0.645 mJ mol^{-1} deg^{-1} respectively.

10.17 The following data is obtained for $\overline{C}_V$ of Rb at low temperature (W.H. Lien and N. E. Phillips, *Phys. Rev.,* **133A,** 1370 (1964)).

T(K)	0.2002	0.5002	0.8708	1.11	2.161	2.5961	3.053	3.5784	4.0718
$\overline{C}_V$ (mJ mol^{-1} deg^{-1})	0.5701	2.681	10.02	14.93	149.7	269.1	445.1	711.6	1009.0

Calculate the value θ_D and γ for Rb. Calculate $\overline{C}_V$ at 300 K and see whether it agrees with Dulong-Petit value.

10.18 The following data is obtained for $\overline{C}_V$ of *Cs* at low temperature (W.H. Lien & N. E. Phillips, *Phys. Rev.,* **133A,** 1370 (1964)).

T(K)	0.208	0.4056	0.655	1.002	0.8072	1.6006	2.1302	3.5012	3.024	4.088
$\overline{C}_V$ (mJ mol^{-1}) deg^{-1})	0.9499	3.431	11.29	38.42	20.20	161.6	397.5	1566	1077	2217

Calculate the value of θ_D and γ for *Cs*.

10.19 If the low temperature limit of equation (10.57) is used, then find the relation at which the contributions to $\overline{C}_V$ due to electronic and vibrational motions are equal. Use the data given in problem 10.16.

10.20 Repeat the calculations for *Cs* after calculating the data from problem 10.18.

10.21 Calculate $\overline{C}_V$ of Al at 27°C from the following data: C_p = 0.92 J g^{-1} deg^{-1}; α = 71.4 × 10^{-6} deg^{-1}; β = 1.34 × 10^{-13} cm^2 dyne^{-1}; atomic wt. = 27.0 g mol^{-1}, density = 2.7 kg dm^{-3}.

10.22 If the value of α and β for Ag and Cu are 0.57 × 10^{-4} deg^{-1}; 0.48 × 10^{-4} deg^{-1} and 9.9 × 10^{-13} cm^2 dyne^{-1}; 7.2 × 10^{-13} cm^2 dyne^{-1} respectively, calculate the values of C_V for Ag and Cu at 27°C. Given: *V*(*Ag*) = 10.2 × 10^{-3} dm^3, $\overline{V}$ (Cu) = 7.1 × 10^{-3} dm^3, $\overline{C}_P$ (Cu) = 24.476 J deg^{-1} mol^{-1}. $\overline{C}_P$ (Ag) = 25.481 J deg^{-1} mol^{-1}. Compare these values with those obtained in problem 10.16.

10.23 Calculate the value $\overline{C}_V$ of H_2O at 25°C from the following data: α = 257.05 × 10^{-6} deg^{-1}, β = 43.85 × 10^{-6} atm^{-1}, *d* = 0.9971 kg dm^{-3} and C_p = 75.291 J mol^{-1} deg^{-1}.

10.24 Calculate the heat capacity $\left(\frac{Dq}{dT}\right)$ of an ideal monatomic gas along a reversible path, at each point of which $P/V^{1.4}$ = constant.

10.25 A cyclic process involving 2 mol of an ideal gas has W of 466.6 J/cycle. What is the value of Q per cycle?

10.26 Three mol of ideal gas was allowed to expand isothermally and reversibly from V_1 to $6V_1$. The initial pressure was 20 atm. The work done in the above expansion was 45.6 kJ. (*a*) Calculate V_1, (*b*) Calculate the initial temperature.

10.27 Two mol of ideal gas at 27°C and volume 20 dm^3 undergoes the following reversible processes: (*i*) heating at constant volume till the temperature was three times, (*ii*) isothermal expansion till the volume is three times, (*iii*) cooling at, constant pressure till the original state is reached. (*a*) Calculate Q, W, ΔE for each step and for the complete cycle; (*b*) Show that $\oint W = \oint Q$; (*c*) Draw p–V and V–T diagrams for the cycle. $C_p = 29.234$ J mol^{-1} deg^{-1}; $C_V = 20.92$ J mol^{-1} deg^{-1}.

10.28 One kg mol of an ideal gas initially at 300 K and 1 atm pressure is isothermally compressed till its pressure is doubled. It is then heated at constant volume to pressure p_3 and then expanded reversibly and adiabatically to reach the initial state. Calculate W, Q and ΔE for each process. Draw the $p - V$, $E - V$ diagram, for the cycle, $C_p = 29.239$ J mol^{-1} deg^{-1}, $C_V = 20.92$ J mol^{-1} deg^{-1}.

10.29 Five mol of monatomic ideal gas was initially at 0°C and 1 atm pressure and were carried through the following reversible steps to complete the cycle: (*i*) heated at constant volume till $\Delta E = 81.5 \times 10^3$ J; (*ii*) adiabatic reversible expansion till the temperature is 0°C and (*iii*) isothermal compression till the original state is reached, (*a*) Calculate W, Q and ΔE for each step, (*b*) Draw $p - T$, $p - V$ and E–V diagrams.

10.30 A 50 g sample of O_2 initially at 30 atm and 25°C is allowed to expand isothermally against a constant pressure of 1 atm. Calculate Q, W, ΔE and ΔH assuming the gas is ideal.

10.31 Heat capacities (C_p and C_V) of *n*-propane have been computed at 25°C and 200°C in problem 10.10. If it is assumed that the values of heat capacities remain constant with in this temperature range, gas is ideal and 35.2 g of propane at 1 atm was heated from 25°C to 200° C. Calculate (*i*) final volume, Q, W, ΔE and ΔH if the process is carried at constant pressure; (*ii*) final pressure, Q, W, ΔE and ΔH if the process is carried at constant volume.

10.32 If the empirical relation for C_p as a function of temperature is used in calculating Q, integrate and repeat the calculations for Q, W, ΔE and ΔH in case of problem 10.31.

10.33 Three mol of ideal gas at 27°C and 2 atm pressure is expanded isothermally and reversibly to a volume of 675 dm^3 and then heated at constant volume till the pressure is again 2 atm. Calculate Q, W, ΔE and ΔH for the total process.

10.34 The problem 10.33 was carried in a different manner *i.e.*, gas under initial conditions was expanded under constant external pressure of 0.65 atm and then cooled at constant pressure till the initial volume is attained. Calculate Q, W, ΔE and ΔH for the complete process.

10.35 If the following combustion reaction is complete CH_4 (g) + $2O_2$ (g) $\rightarrow$ $2H_2O$ (g) + CO_2 (g) and it is assumed that reagents are initially at 300 K. Calculate the maximum rise in temperature attained by the system. Consider it a closed system and $\Delta H = -890.36$ kJ.

10.36 An ideal gas is kept in a cylinder and piston whose radius is 0.5 m. A constant pressure is maintained during reversible compression and the piston is pushed inside by 0.25 m. If the work done on the system is 10.46 J, calculate the pressure in atmosphere.

10.37 The following process is carried out at constant pressure of 1 atm and at 78.5°C:

$$3 \text{ mol, } C_2H_6OH\ (l, 78.5°C) \rightarrow 3 \text{ mol, } C_2H_6OH\ (g, 78.5°C)$$

Calculate the values of V_1, V_2, W, Q and ΔE. Given: heat of vaporisation of EtOH = 7.14 J g^{-1}, density of EtOH at 78.5°C = 0.7893 kg dm^{-3} and it is assumed that vapours are behaving ideally.

10.38 20 g of water was evaporated at 100°C and 1 atm. d (H_2O, g) = 5.98 × 10^{-4} kg dm^{-3}, ΔH_{Vap} = 2.26 kJ g^{-1}. Calculate the work done on the surroundings and ΔE of the system.

10.39 (*a*) A waterfall falls through the height of 400 m. If there is no other loss of heat, calculate the rise in temperature of water when it reaches the ground.

(*b*) If the evaporation is also taken into account, will the height of the fall has to be increased or decreased to have the same rise in temperature ?

10.40 A man produces about 10.46 kJ of heat per day through metabolic activity, (*a*) If a man of mass 70 kg was a closed system with heat capacity of H_2O = 75.318 J mol^{-1} deg^{-1}, what will be his temperature rise per day?

(*b*) Man is an open system and main reason of heat lost is due to evaporation of water. How much water needs to be evaporated in a day to maintain constant temperature ?

10.41 Derive an expression for the work done under reversible and isothermal expansion in case of gas obeying

(*i*) an equation of state $p\,\overline{V} = RT + Bp + Cp^2$

(*ii*) Berthelot equation of state

$$p\overline{V} = RT\left[1 + \frac{9}{128}\frac{p}{p_c}\frac{T_c}{T}\left(1 - \frac{T_c^2}{T^2}\right)\right]$$

10.42 If a gas obeys the equation of state as $p\,\overline{V} = RT + Bp$ and the heat capacity $\overline{C}_V$ is assumed to be independent of temperature then (*a*) derive equation relating T and V in an reversible adiabatic expansion.

10.43 For an ideal gas, prove the following relations:

(*i*) $\left(\frac{\partial C_V}{\partial V}\right)_T = 0,$ (*ii*) $\left(\frac{\partial Cp}{\partial p}\right)_T = 0,$

(*iii*) $\left(\frac{\partial C_V}{\partial p}\right)_T = 0,$ (*iv*) $\left(\frac{\partial Cp}{\partial V}\right)_T = 0,$

(*v*) Calculate C_V and $\left(\frac{\partial C_V}{\partial V}\right)_T$ for a van der Waals

gas *i.e.*, $\left(p + \frac{n^2 a}{V^2}\right)(V - nb) = nRT$

10.44 A 50 g sample of O_2 initially at 30 atm and 25°C is allowed to expand adiabatically against a pressure of 1 atm. Considering the gas to be ideal and C_p = 29.50 J mol^{-1} deg^{-1}; $\overline{C}_V$ = 21.13 J K^{-1} mol^{-1}, calculate Q, W, ΔE and ΔH.

10.45 In the problem 10.44, if the gas is allowed to expand adiabatically and reversibly, calculate the final temperature, Q, W, ΔE and ΔH. See if there is any difference in thermodynamic quantities.

10.46 A 50 g sample of O_2 is compressed adiabatically with a constant opposing pressure of 30 atm. Initially the gas is at 25°C and 1 atm. Calculate the final temperature, Q, W, ΔE and ΔH.

10.47 Repeat the problem 10.46 if it is assumed that the compression is carried out reversibly to final pressure of 30 atm.

10.48 van der Waals constant for CO_2 are: $a = 3.61$ dm^6 atm mol^{-2}; $b = 0.0429$ dm^3 mol^{-1}. If 2 mol of CO_2 at 27°C expands isothermally and reversibly from 10 dm^3 to 40 dm^3, calculate the work done: (*a*) assuming CO_2 is behaving ideally, (*b*) following van der Waals equation of state, (*c*) calculate Q and ΔE for each step.

10.49 The equation of state describing the behaviour of a real gas is $p(\overline{V} - b) = RT$. One mol of gas compressed reversibly to 5 dm^3 from 35 dm^3 at 300 K: (*a*) Calculate the work done if $b = 0.06$ dm^3 mol^{-1}; (*b*) compare the results if the gas is behaving as an ideal gas.

10. 50 Work done in compressing the condensed system can be expressed in terms of compressibility factor β. Calculate the work done in compressing the liquid of volume V_1 to V_2 reversibly if p_0 is the initial pressure andβ is compressibility factor and independent of temperature and pressure.

10.51 Prove the following relation for van der Waals gas for expression $p(V - nb) = nRT$, when it expands under reversible and adiabatic condition, *i.e.*, $T^{C_V/R}(V - b)$ = constant. Assume C_V is independent of volume.

10.52 If $C_V = A + BT$ where A and B are constants, prove the following relation: $V^R T^A e^{BT}$ = constant, when a gas expands under adiabatic and reversible conditions.

10.53 Calculate the decrease in temperature when O_2 expands adiabatically and reversibly from 2 dm^3 to 10 dm^3. The initial temperature is 27°C and the value of A and B for O_2 are 25.503 J mol^{-1} deg^{-1} and 13.612 × 10^{-3} J mol^{-1} deg^{-2}.

10.54 The $\mu_{JT} = 0.179$ for N_2 at 50°C and $p = 10$ atm and $\mu_{JT} = 0.066$ at 50°C and 200 atm. If it is assumed that average value of μ_{JT} taken to be independent of pressure, calculate the drop in temperature when N_2 expands from 200 atm to 1 atm under Joule-Thomson experiment.

10.55 If the Joule-Thomson experiment is carried with a mol of gas and the system is kept strictly under adiabatic conditions:

(*a*) Show that $\left(\dfrac{\partial T}{\partial V}\right)_E = -\dfrac{1}{C_V}\left(\dfrac{\partial E}{\partial V}\right)_T$

(*b*) Combining the results of problem (10.55 *a*), *i.e.*,

$$= \left(\frac{\partial E}{\partial V}\right)_T = \frac{a}{V} \text{ for van der Waals gas, show that}$$

$$T = \frac{a}{C_V}\left(\frac{1}{V_2} - \frac{1}{V_1}\right)$$

(*c*) Find the change in temperature when O_2 at 27°C and 2 atm is expanded from 4 dm^3 to 16 dm^3, $a = 1.36$ atm dm^6 mol^{-2}.

10.56 Prove the following relations:

$$(a)\ \mu_{JT} = \frac{1}{C_p}\left[T\left(\frac{\partial V}{\partial T}\right)_p - V\right]$$

(b) $\mu_{JT} = \frac{RT^2}{C_p}\left[\frac{dB'}{dT} + \frac{dC'}{dT} + \ldots\right]$

For a gas obeying an Virial equation

$$V = \frac{RT}{p} + RT\,B'(T) + RTC'(T)p + \ldots$$

(c) $\mu_{JT} = \frac{1}{C_p}\left(\frac{2a}{RT} - b\right)$

For a van der Waals gas.

10.57 The inversion temperature is defined as the temperature when $\mu_{JT} = 0$ and Boyle's temperature when second virial coefficient for van der Waals equation is zero, *i.e.*,

$$B'(T) = \frac{bRT - a}{R^2T^2} = 0$$

Show that $T_{inv}/T_{Boyle} = 2$.

Note: The above relation is not followed rigorously as the van der Waals equation of state is not followed over the whole range.

10.58 The Boyle's temperatures of N_2 and H_2 are 323 K and 123 K, the inversion temperature 600 K and 193 K respectively. Show that the results arrived in problem 10.57 is only followed approximately.

10.59 Calculate the Joule-Thomson coefficient of N_2 at 27°C and 100 atm pressure and the value of C_p equal to 29.04 J K^{-1} mol^{-1}.

10.60 The N_2 gas at 0°C and 200 atm pressure is released. Calculate the drop in temperature if $T_{inv} = 323$ K and $\mu_{JT} = 0.87$ deg atm^{-1} at 0°C.

Additional Problems

10.61 Determine whether the expressions that follow are exact differentials. If so, determine the function F whose total differential is equal to the expression given.

(*i*) $[\cos(3y) + xe^{-y^2}]\,dx + [2xye^{-y^2} - 3x\sin(3y)]dy$, (*ii*) $[2x^3 - xy^2 - 2y + 3]\,dx - [x^2y + 2x]dy$

10.62 Consider the differential: $dZ = [2xy + \sin(y)]\,dx + [x^2 + x\cos(y)]\,dy$. (*i*) The state of the system is changed from ($x = 0, y = 0$) to ($x = 1, y = 1$) along the path $x = y$. Calculate the total change in Z along this path. (*ii*) The change in (*i*) is now made along the path $x = y^2$. Calculate the total change in Z along this path. (*iii*) Show that dZ is an exact differential.

10.63 With help of first law of thermodynamics, explain the following observations. (*i*) Artificial snow is formed by quickly releasing a mixture of compressed air and water vapour at about 20 atm. from a snow making machine to the surroundings, (*ii*) when a bicycle tire is inflated with a hand pump, the temperature inside rises.

10.64 Assuming that neon is an ideal gas, calculate the magnitude of internal energy of 5 moles of neon at 27 °C. Suggest two ways to increase its internal energy by 20 J.

10.65 An ideal gas compressed by a force of 100 newtons acting along the distance of 0.5 m. Calculate the values of ΔE and Q.

10.66 A thermos bottle containing some liquid is shaken vigorously. Assume that the liquid forms the system and thermos as surroundings. (*i*) Will the temperature of the liquid increase as a result of shaking? (*ii*) Has the work done on the system? (*iii*) Has the heat added to the system? (*iv*) Has the internal energy added to the system?

10.67 10 kg copper piece at a temperature of 40 °C is dropped from a height of 10 km. (*i*) Calculate the speed of copper piece when it hits the ground and its kinetic energy, (*ii*) suppose all its kinetic energy is converted into it internal energy when it hits the ground, if the specific heat of copper is 0.385 J g^{-1} K^{-1}.

10.68 2 moles of SO_2 is placed in a container of volume 3 L at 27 °C. If this gas is compressed isothermally against a constant external pressure of p_{ext} so that its final volume is 0.75 L, calculate the smallest value of p_{ext} can have assuming that SO_2 satisfies van der Waals equation of state under the given conditions. Calculate the work involved using this value of p_{ext}.

10.69 Consider the reaction between sodium and water as

$$2Na\,(s) + 2H_2O(l) \rightarrow 2NaOH(aq) + H_2(g)$$

Calculate the amount of Q, W, ΔE and ΔH when 46 g of sodium metal is treated with water.

10.70 Consider the vapourisation of water at 373 K and 1 atm. pressure as given

$$H_2O\,(l, 373\text{ K}, 1\text{ atm.}) \rightarrow H_2O\,(g, 373\text{ K}, 1\text{ atm}).$$

Calculate the value of ΔE and ΔH from the given data. Heat of vapourisation of water is 40.79 kJ mol^{-1}, molar volume of liquid water and that of steam are 0.0188 L mol^{-1} and 30.6 L mol^{-1}.

10.71 Two moles of an ideal gas undergoes isothermal expansion at 300 K from 5 atm. to a final pressure p atm. If the work performed during this process is 100 J, calculate the final pressure of the gas if the external pressure is 0.5 atm.

10.72 Use equation (10.43) and $H = E + pV$ and find out the expression for $C_p - C_V$.

10.73 Using equation (10.42) and $E = H - pV$, find out the expression for $C_p - C_V$.

10.74 Derive equation (10.45).

10.75 5 moles of an ideal gas at 8 atm. and $T = 310$ K are expanded reversibly and isothermally to a final volume of 65 L. (*i*) Calculate the work done in L atm. and J, (*ii*) 5 moles of an ideal gas at 8 atm. and $T = 310$ K are expanded against the constant pressure of 2 atm. until equilibrium is reached. Calculate the work done in this expansion, (*iii*) Compare the work done in (*i*) and (*ii*). (*iv*) Is the DW is exact or inexact variable.

10.76 5 moles of an ideal gas is reversibly heated at $dp = 0$ from $p_1 = 5$ atm. and $T_1 = 300$ K to $p_2 = 5$ atm. and $T_2 = 600$ K. If the molar heat capacity of the system is expressed by the equation (10.51). (*i*) Calculate the value of ΔH, ΔE, Q and W for this process in terms of the constants a, b and c. (*ii*) If the heating were carried out at $dV = 0$, calculate the heat associated with this process.

10.77 22 g of solid carbon dioxide (also known as the dry ice) is allowed to sublime (*i.e.*, evaporate) in a container with moveable piston. Calculate the expansion work done against a constant external pressure of 1 atm. and at a temperature of 295 K, assuming that volume of dry ice is negligible.

10.78 5 moles of an ideal gas at 300 K and 12 atm. pressure are expanded isothermally to a final pressure of 3 atm. Calculate the minimum and maximum work in expansion.

10.79 Calculate if possible (*i*) ΔH, ΔE, Q and W from the data given below. $T_1 = 300$ K, $p_1 = 2$ atm.; $T_2 = 600$ K, $p_2 = 5$ atm.; $\bar{C}p = 2.5$ R and $\mu_{JT} = 0.27$ K atm^{-1}.

10.80 Five moles of an ideal gas at 300 K and at 12 atm. pressure are expanded isothermally against a constant external pressure of 3 atm. until the internal pressure is reached the value of 5 atm. At this point, the expansion is halted. Calculate ΔE, ΔH, Q and W for this process.

10.81 Five moles of an ideal gas at 300 K at 12 atm. pressure are expanded isothermally in two manners, (*i*) expanded against a constant external pressure of 2 atm. till it reaches 2 atm. (*ii*) the expansion is carried in three stages, against a constant pressure of 8 atm. till the internal pressure reaches 8 atm.; against a constant external pressure of 4 atm. till the internal pressure reaches 4 atm. and in third stage external pressure of 2 atm. till the internal pressure is 2 atm. Calculate the work done in each path and compare it if the process is carried out reversibly.

10.82 One mole of an ideal gas with $\bar{C}p = 2.5\ R$ is heated reversibly along a path such that $V = Ae^{bT}$ where $A = 1.2257$ L and $b = 0.01000\ \text{K}^{-1}$. (*i*) How does the pressure vary with T for heating along the given path? (*ii*) Obtain the molar heat capacity of the gas as a function of T for heating along the given path? (*iii*) If 5 moles of gas are heated along this path reversibly from 300 K to 500 K, calculate ΔE, W and Q.

10.83 Assuming that O_2 molecular gas is following equipartition principle, calculate the values of Q, ΔH, W and ΔE if 100 g of O_2 is heated from 300 to 600 K under constant pressure and under constant volume. In which case the value of heat absorbed will be larger.

10.84 25 g of copper sheet at a temperature of 50 °C is laid on silver sheet weighing 30 g at a temperature of 100 °C. Calculate the common temperature when equilibrium is reached. Specific heat of copper and silver are 0.39 J g^{-1} K^{-1} and 0.23 J g^{-1} K^{-1}.

10.85 The molar heat capacity of water at 298 K is 44.01 kJ mol^{-1} and at 373 K is 40.79 kJ mol^{-1}. Explain the qualitatively difference between the two values.

10.86 Nitrogen gas (assumed to be an ideal) is present initially at 27 °C and its volume is 20 L. Calculate the final temperature if nitrogen gas is expanded to 50 L. Assume that nitrogen gas follows equi-partition principle.

10.87 Starting with the expression $p_1V_2^{\gamma} = p_1V_2^{\gamma}$, prove that $T_1V_1^{R/C_V} = T_2V_2^{R/C_V}$

10.88 Consider the figure (see page 560) where different paths are given. Ca lculate W, Q, ΔE and ΔH for the (*i*) path I, (*ii*) path II, (*iii*) path III, (*iv*) path IV and (*v*) path V. Path I is carried out at constant pressure, path II is carried out reversibly and isothermally, path III is carried out Adiabatically and reversibly, paths IV and V are carried out under constant volume.

(*b*) Total work done and heat changes if we follow a cyclic path A $\rightarrow$ B $\rightarrow$ C $\rightarrow$ A and A $\rightarrow$ B $\rightarrow$ C $\rightarrow$ D $\rightarrow$ A.

10.89 2 moles of a diatomic gas following equi-partition principle initially at a pressure of 2 atm and at a temperature of 27 °C is taken to a final pressure of 5 atm by the reversible path defined by p/V = constant. Calculate the value of ΔE, ΔH, Q and W.

10.90 Relation between the heat capacity of a gas and speed of sound (c) in the gas is given as $c = \left(\frac{\gamma RT}{M}\right)^{\frac{1}{2}}$, where γ is the ratio of heat capacities at constant pressure to constant volume and M is the molar mass of the gas. Calculate the velocity of sound in neon gas at 27 °C.

10.91 Dulong-Petit rule, *i.e.* (Molar mass of solid x specific heat $\approx$ 25 J C^{-1}) and thus was used to estimate the specific heats of the metals. Verify this law for Be (1.83 J g^{-1} C^{-1}), copper (0.39 J g^{-1} C^{-1}), silver (0.23 J g^{-1} C^{-1}).

10.92 Explain qualitatively the followings basics on thermodynamics. (*i*) Efficiency of refrigerator increases if its freezer compartments should be fully packed with food. (*ii*) Starting at the same temperature, tea and coffee remain hot longer in thermos flaks than soup.

10.93 The compressibility at constant temperature is defined as: $\beta = -\frac{1}{V}\left(\frac{\partial V}{\partial p}\right)_T$. Show that the value of β for ideal gas is given by $1/p$. If we assume that the value of β is nearly constant over a pressure range, show that

$$\frac{V}{V_0} = e^{-\beta(p-p_0)}$$

where V_0 is the volume at a pressure p_0. Use this result to show that the reversible isothermal work of compressing the liquid from a volume V_0 (at a pressure p_0) to a volume V (for a pressure p) is given as

$$W = -p_0(V - V_0) + \frac{V_0}{\beta}\left(\frac{V}{V_0}\ln\frac{V}{V_0} - \frac{V}{V_0} + 1\right)$$

or $$W = -p_0V_0[e^{-\beta(p-p_0)} - 1] + \frac{V_0}{\beta}\{1 - [1 + \beta(p - p_0)]e^{-\beta(p-p_0)}$$

We know that for liquids the value of β is very small and thus we can assume $\beta(p - p_0) << 1$ for moderate pressure. Show that

$$W = \beta p_0V_0[p - p_0] + \frac{\beta V_0(p - p_0)^2}{2} + O(\beta^2) = \frac{\beta V_0(p^2 - p_0^2)}{2} + O(\beta^2)$$

Calculate the work required to compress one mole of toluene reversibly and isothermally from 10 atm. to 100 atm. at 27 °C. Take the value of β to be 8.95×10^{-5} bar and molar volume of to be 0.106 mol L^{-1}.

10.94 Assuming that a non-ideal gas obeys the gas equation: $p(\bar{V} - b) = RT$, where b is a function of temperature. (*i*) Prove that the work done whenever gas is heated at constant pressure from T_1 and T_2 is given by

$$W = R(T_1 - T_2) + (b_1 - b_2)p$$

wherever b_1 and b_2 are the values of b at temperature T_1 and T_2 respectively. (*ii*) Show that if the gas is expanded isothermally from V_1 to V_2, the reversibly work will be

$$W = RT\ln\left[\frac{V_1 - b}{V_2 - b}\right]$$

10.95 Two moles of diatomic gas (assume that equi-partition principle is valid) initially at 127 °C and pressure of 12 atm. expands to a final pressure of 1.00 atm. The expansion can take place in the following manner. (*i*) isothermal and reversible manner, (*ii*) isothermal and irreversible, (*iii*) adiabatic and reversible, and (*iv*) adiabatic and irreversible. In all the irreversible processes, the expansion occurs against the external pressure of 1 atm. calculate for each case the values of Q, W, ΔE and ΔU.

10.96 Consider the following processes. (*i*) melting of ice at 0 °C and 1 atm., (*ii*) melting of solid cyclohexane at 1 atm. and the normal melting point, (*iii*) reversible isothermal expansion of ideal gas and (*iv*) reversible adiabatic expansion of an ideal gas. Predict whether the values of Q, W, ΔE and ΔH are positive, zero or negative.

10.97 A cyclic process is carried out in the following manner. 2.1 mole sample of crystalline acetic acid, initially at 17 °C, is allowed to melt at 17 °C and is then heated to 118.1 °C (its normal boiling point) at 1.00 atm. The sample is allowed to vapourise at 118.1 °C and is then rapidly quenched to 170 °C, so that it recrystallizes. Calculate the value of $\Delta_r H^0$ for the total process as described above.

10.98 State the conditions for each of the followings. (*i*) $W = -nRT \ln \frac{V_2}{V_1}$, (*ii*) $dH = Q$, (*iii*) $\frac{p_2}{p_1} = \left(\frac{V_2}{V_1}\right)^{-\gamma}$, (*iv*) $C_p - C_V = nR$, (*v*) $W = n\bar{C}_V (T_2 - T_1)$, (*vi*) $\Delta H = \Delta E + p\Delta V$, (*vii*) $W = -p\Delta V$ and (*viii*) $\gamma = \frac{5}{3}$.

10.99 Consider whether the following statements are true or false. (*i*) The internal energy of a real gas depends only on temperature, (*ii*) C_V is independent of temperature for gases, (*iii*) ΔE is a state function, (*iv*) $\Delta E \approx \Delta H$ except for gases or high pressure processes, (*v*) In gas compression, the reversible and isothermal process involves minimum work and (*vi*) $\Delta E = Q - W$ for open system.

10.100 Using thermodynamics explain (*i*) why hot, humid air is more uncomfortable that hot dry air and (*ii*) cold damp air is more uncomfortable than cold dry air.

10.101 A real gas at 300 K and 75 atm. is allowed to expand to an atmospheric pressure of 1 atm. Assuming, that expansion is adiabatic and conducted under constant enthalpy, estimate the final temperature after expansion.

10.102 Two moles of an ideal gas at a temperature of 500 K and a pressure of 6 atm. is subjected to the following changes:

(*i*) The gas is expanded isothermally and reversibly to a final pressure of 5 atm.

(*ii*) After completion of step (*i*) the gas is expanded adiabatically and reversibly until the pressure is 4 atm.

(*iii*) After step 2 is completed, the gas is compressed isothermally and reversibly to a pressure of 4.8 atm.

(*iv*) After step 3 the gas is compressed reversibly and adiabatic to a pressure of 6 atm. at which point the temperature is found to be 500 K.

Taking the molar heat capacity at constant volume is equal to 1.5 R. (*i*) Calculate W, Q and ΔE, at the end of each step, (*ii*) Calculate the total values of W, Q and ΔE at the end of step (*iv*).

10.103 Show that $\Delta H = \Delta E$ for any isothermal change of state for an ideal gas.

10.104 Two moles of an ideal gas is heated reversibly at a constant pressure of 2 atm. from 273 K to 373 K. (*i*) Calculate the work involved in the process, (*ii*) If the gas is expanded isothermally and reversibly at 273 K from the initial pressure of 2 atm., what would be the final pressure required if the work done in this step is equal to that obtained in step (*i*).

10.105 A real gas follows the gas equation $p(\bar{V}-b)=RT$, where $\bar{V}$ is the molar volume. 2 moles of this gas at a temperature T_1 is expanded adiabatically and reversibly from an initial pressure p_1 to a final pressure of p_2. If the molar heat capacity at constant volume is 1.5 R, (*i*) calculate the final temperature of gas as a function of T_1, p_1 and p_2, (*ii*) if $T_1 = 400$ K, $p_1 = 15$ atm. and $p_2 = 2$ atm., calculate the final temperature.

10.106 We can derive equation (10.44) in many ways. Follow the following procedure and derive this expression. (*i*) Consider *E* as a function of *E*(*p*,*T*), write *dE*, (*ii*) Consider *E* as a function of *E*(*V*,*T*) and write *dE*, (*iii*) *V* as a function of *V*(*p*,*T*) and write *dV* and (*iv*) $E = H - pV$ and write *dE*. Combine all these results in such a manner that we get

$$C_p - C_V\left[p+\left(\frac{\partial E}{\partial V}\right)_T\right]\left(\frac{\partial V}{\partial T}\right)_p$$

10.107 Also prove

$$C_p - C_V = \left[V - \left(\frac{\partial H}{\partial p}\right)_T\right]\left(\frac{\partial p}{\partial T}\right)_V$$

10.108 Starting with $H - E = pV$, show that

$$\left(\frac{\partial E}{\partial T}\right)_p = C_p - p\left(\frac{\partial V}{\partial T}\right)_p$$

10.109 Two mole of an ideal gas with molar heat capacity at constant volume 1.5 *R*, is heated along a reversible path such that $p = aT^{\frac{2}{3}}$, where a is constant all along the path. Determine the expression for the heat capacity of the gas when heated along this path.

10.110 A real gas, over a certain range of temperature follows the gas equation as

$$p\bar{V} = RT + aT^2$$

where a is a constant and $\bar{V}$ is a molar volume. Further we will learn in Chapter 12, the following equation which hold good is

$$\left(\frac{\partial E}{\partial V}\right)_T = \frac{aT^2}{V}$$

(*i*) Calculate the work associated with this path if the gas is expanded isothermally and reversibly from volume V_1 to V_2. (*ii*) Calculate the value of change in internal energy *E* for the process undergoing in step (*i*).

10.111 Show that the Joule-Thomson coefficient can be rewritten as

$$\mu_{JT} = \left(\frac{\partial T}{\partial p}\right)_H = -\frac{1}{C_p}\left[\left(\frac{\partial E}{\partial V}\right)_T\left(\frac{\partial V}{\partial p}\right)_T + \left(\frac{\partial (pV)}{\partial p}\right)_T\right]$$

Show that for ideal gas $\mu_{JT} = 0$.

CHAPTER 11

THERMOCHEMISTRY

11.1. HEAT OF REACTION

The heat of reaction can be expressed as the energy change of the products and the reactants at constant volume ($\Delta_r E$) or at constant pressure ($\Delta_r H$). For example, for the reaction

$$\text{Reactants } (T) \rightarrow \text{Products } (T)$$

$$\Delta_r E = E_{(\text{products})} - E_{(\text{reactants})} \quad ...(11.1)$$

at constant temperature and constant volume and

$$\Delta_r H = H_{(\text{products})} - H_{(\text{reactants})} \quad ...(11.2)$$

at constant temperature and constant pressure.

The SI units for E or H are Joules *i.e.,* unit of energy but other common unit is calories. Generally the value of E or H for each reactant or product is expressed as Joules mol $^{-1}$ (J mol^{-1}) or kJ mol^{-1} at a particular constant temperature, generally 298 K.

If $\Delta_r E$ or $\Delta_r H$ is positive, the reaction is said to be "endothermic" and if $\Delta_r E$ or $\Delta_r H$ is negative it is called "exothermic" reaction.

A chemical equation is only complete if besides writing the balanced equation and giving the energy values, one writes the states of the reactants and the products. The symbol for solids is '*s*' for liquids '*l*' and for gases it is '*g*'. If the solid state has more than one structure then the nature of the structure should also be mentioned. For example, carbon is present as diamond or as graphite. Similarly sulphur is present as "rhombic" or "monoclinic" sulphur.

A superscript at the right hand side of E or H indicates the values at the standard states. The standard state for solid substance is a specified crystalline state at 1 atm, at the temperature concerned (*e.g.*, standard form of carbon is graphite and that of sulphur is rhombic sulphur), for a liquid it is a pure form of the liquid at 1 atm, at a specified temperature and for a gas it is the ideal gas at 1 atm pressure at a specified temperature.

***Exercise* 11.1:** Calculate $\Delta_r H^0$ for the following reaction:

$$CaC_2\,(s) + 2H_2O\,(l) = Ca\,(OH)_2\,(s) + C_2H_2\,(g)$$

$$\Delta_f H^0\,(CaC_2, s) = -62.76 \text{ kJ mol}^{-1}, \Delta_f H^0\,(H_2O, l) = -285.84 \text{ kJ mol}^{-1},$$

$$\Delta_f H^0\,(Ca(OH)_2, s) = -986.59 \text{ kJ mol}^{-1}, \; \Delta_f H^0\,(C_2H_2, g) = 226.75 \text{ kJ mol}^{-1}$$

Using equation (11.2)

$$\Delta_r H^0 = (-986.59 + 226.75) - (-62.76 + 2 \times -285.84)$$

$$= -125.4 \text{ kJ}$$

***Exercise* 11.2:** Calculate ΔH^0 for the following reaction:

$$TiO_2\,(s) + 2Cl_2\,(g) \rightarrow TiCl_4\,(l) + O_2\,(g)$$

$$\Delta_f H^0\,(TiO_2, s) = -912.11 \text{ kJ mol}^{-1}, \; \Delta_f H^0\,(TiCl_4, l) = -750.19 \text{ kJ mol}^{-1}$$

Using equation (11.2)

$$\Delta_r H^0 = (-750.19 + 0) - (-912.11 + 0) = 161.92 \text{ kJ}$$

The standard state enthalpy of formation of elements at 298 K is taken as zero.

11.2 MEASUREMENTS OF HEAT OF REACTIONS

The heat of reaction is measured with the help of calorimeter. $\Delta_r E$ value is obtained when the reaction is carried out in bomb calorimeter *i.e.*, at constant volume and $\Delta_r H$ is the heat of the reaction measured, at constant pressure, in insulated beaker or flask, thermos bottle, Dewar's flask etc. Since the processes are well specified the heat evolved or absorbed are only the state functions, *i.e.*, $Q_p = \Delta_r H$ or $Q_V = \Delta_r E$ are state functions. These quantities can be measured by the equation:

$$Q = \Delta_r E \text{ or } \Delta_r H = \int_{T_1}^{T_2} \Delta C_i \text{(products, calorimeter)}\, dT \quad ...(11.3)$$

where C_i can be C_V for measurements of E and C_p for H. In many of the experiments the C_i for calorimeter is kept constant and maintained.

***Exercise* 11.3:** (*a*) Calculate the heat absorbed when 5 litres of water at 20°C is evaporated. Heat capacity of the container along with water is 200 J K^{-1}.

(*b*) Suppose a current of 5 A and 50 V is passed to evaporate the above quantity of water, calculate time taken.

(*a*) Part (*a*) consists of heating the contents from 20°C to 100°C and then vaporising. Assuming heat capacity and density to be constant during this temperature rise and taking $\Delta_{vap}H = 40.66$ kJ mol^{-1}.

$$Q = (200 \text{ J K}^{-1})(80 \text{ K}) + \frac{(5000 \text{ ml})(1 \text{ g ml}^{-1})(40.66 \text{ kJ mol}^{-1})}{(18 \text{ g mol}^{-1})}$$

$$= 11310.444 \text{ kJ}$$

(*b*)

$$Q\,(J) = I\,(A)\, t\,(s)\, V\,(V)$$

$$t = \frac{11310444 J}{(5A)(50V)} = 45241.8 \text{ s} = 12.57 \text{ hr}$$

***Exercise* 11.4:** A current of 4 A and 8 V was passed for 15 min through a liquid A and the amount of vapours collected is 25.6 g. If this much heat is only used to vaporise the liquid, calculate $\Delta_{vap}H$. Molecular weight of A is 84 g mol^{-1}.

$$Q\,(\text{J}) = I\,(\text{A})\,t\,(\text{s})\,V\,(\text{Volts})$$
$$= (4)\,(15 \times 60)\,(8) = 28.8 \text{ kJ}$$

$$\Delta_{vap}H = \frac{Q \times M}{W} = \frac{(28.8 \text{ kJ})\,(84 \text{ g mol}^{-1})}{25.6 \text{ g}} = 94.5 \text{ kJ mol}^{-1}$$

11.3. RELATIONSHIP BETWEEN $\Delta_r H$ AND $\Delta_r E$

In the equation (10.17)

$$\Delta_r H = \Delta_r E + \Delta(PV) \qquad ...(10.17)$$

the term $\Delta(PV)$ is very small for solids and liquids, and can be neglected. Thus $\Delta_r H$ and $\Delta_r E$ are same in case of solids and liquids. In case of ideal gases at constant temperature, the above equation reduces to

$$\Delta_r H = \Delta_r E + RT\,\Delta n \qquad ...(10.20)$$

where Δn is the number of mol of gaseous products minus the number of mol of gaseous reactants.

***Exercise* 11.5:** Consider the reaction mentioned in exercise 11.1 and calculate $\Delta_r E$

$$\Delta n = 1 - 0 = 1$$

Using equation (10.20) and data from exercise 11.1

$$\Delta_r E = -\,125.4 \text{ kJ} — (1 \text{ mol})\,(8.314 \text{ J K}^{-1} \text{ mol}^{-1})\,(298 \text{ K})\,(10^{-3} \text{ kJ J}^{-1})$$
$$= -\,127.88 \text{ kJ}$$

***Exercise* 11.6:** The $\Delta_r H^0$ for the following reaction:

$C_2H_5OH\,(l) + 3O_2\,(s) \rightarrow 2\,CO_2\,(g) + 3H_2O\,(l)$ is – 1366.95 kJ. Calculate ΔE^0.

Here $\Delta n = 2 - 3 = -\,1$. The number of mols of condensed phases are not counted.

Using equation (10.20)

$$\Delta_r E^0 = -\,1366.95 - (-\,1 \text{ mol})\,(8.314 \text{ J K}^{-1} \text{ mol}^{-1})\,(298 \text{ K})\,(10^{-3} \text{ kJ J}^{-1})$$
$$= 1364.47 \text{ kJ}$$

***Exercise* 11.7:** Consider the reaction given in exercise 11.6 and calculate the value of $\Delta_r E$ by calculating the pressure volume work.

The given reaction is carried out at 1 atm. and so the heat of the reaction given is $\Delta_r H = Q_p =$ –1366.95 kJ. To calculate $\Delta_r E$, we have to first calculate the ΔV. Neglecting the volume of the liquid states (*i.e.*, ethanol and water), initially we are having 3 moles of oxygen and finally we are having 2 moles of CO_2. If we assume that these gases behave as ideal gases, we can use the ideal gas equation to calculate ΔV, *i.e.*,

$$\Delta V = V_2 - V_1 = \frac{n_2 RT}{p} - \frac{n_1 RT}{p} = \frac{(0.0831 \text{ L atm. K}^{-1} \text{ mol}^{-1})(298 \text{ K})}{1 \text{ atm.}}[2-3] \text{ mol} = -24.764 \text{ L}$$

$$p\Delta V = (-24.764 \text{ L})(1 \text{ atm.}) = -\,(24.764 \text{ L atm.})$$
$$= -\,(24.764 \text{ L atm.})(1.0132 \times 10^2 \text{ J atm}^{-1} \text{ L}^{-1}) = -2509.07 \text{ J} = -2.5097 \text{ kJ}$$
$$\Delta E = \Delta H - \Delta(pV) = -1366.95 \text{ kJ} - (-2.5097 \text{ kJ}) = -1364.\,44 \text{ kJ}$$

The results obtained in Exercise 11.6 are the same as obtained in Exercise 11.7.

11.4.1. *Temperature dependence of heat of reaction (Kirchhoff Relation)*

From equation (10.23), the change of enthalpy with temperature is

$$dH = C_p\, dT \qquad ...(10.23)$$

Similarly enthalpy change for reaction with temperature is

$$d(\Delta_r H) = \Delta C_p\, dT \qquad ...(11.4)$$

where $\Delta_r H$ is the heat of reaction, given by (11.2) and ΔC_p is the heat capacity change and is

$$\Delta C_p = \sum_i n_i \overline{C}_{p\,(\text{products})} - \sum_j n_j \overline{C}_{p\,(\text{reactants})} \qquad ...(11.5)$$

If ΔC_p is independent of temperature then

$$\Delta_r H_{T_2} - \Delta_r H_{T_1} = \Delta C_p (T_2 - T_1) \qquad ...(11.6)$$

If heat capacity depends upon temperature and is given by equation (10.33), equation (11.6) becomes

$$\Delta_r H_T = \Delta_r H_{298} + \Delta a(T - 298) + \frac{1}{2}\Delta b(T^2 - 298^2)$$

$$+ \frac{1}{3}\Delta c(T^3 - 298^3) + \frac{1}{4}\Delta d(T^4 - 298^4) \qquad ...(11.7)$$

or

$$\Delta_r H_T = \Delta_r H_{298} + \Delta a'(T - 298) + \frac{1}{2}\Delta b'(T^2 - 298^2)$$

$$- \Delta c'(T^{-1} - 298^{-1}) \qquad ...(11.8)$$

If C_p is given by equation (10.34), equations (11.7) and (11.8) can be written as

$$\Delta_r H_T = A + \Delta a T + \frac{1}{2}\Delta b T^2 + \frac{1}{3}\Delta c T^3 + \frac{1}{4}\Delta d T^4 \qquad ...(11.9)$$

or $$\Delta_r H_T = A' + \Delta a' T + \frac{1}{2}\Delta b' T^2 - \frac{1}{3}\Delta c' T^{-1} \qquad ...(11.10)$$

where $$A = \Delta_r H_{298} - (\Delta a)\, 298 - \frac{1}{2}(298^2)\,\Delta b - \frac{1}{3}(298^3)\,\Delta c$$

$$- \frac{1}{4}\Delta d\,(298^4) \qquad ...(11.11)$$

$$A' = \Delta_r H_{298} - \Delta a'(298) - \frac{1}{2}\Delta b'(298^2) - \frac{1}{3}\Delta c'(298^{-1}) \qquad ...(11.12)$$

***Exercise* 11.8:** Consider the reaction

$$C\,(\text{graphite}) + O_2\,(g) = CO_2\,(g).$$

If $\Delta_r H^0 = -393.51$ kJ mol^{-1}, and C_p's are given as

$$C_p(C) = -5.293 + 58.61 \times 10^{-3}\, T - 432.25 \times 10^{-7}\, T^2$$

$$C_p(O_2) = 25.723 + 12.98 \times 10^{-3}\, T - 38.62 \times 10^{-7}\, T^2$$

$$C_p(CO_2) = 26.00 + 43.50 \times 10^{-3}\, T - 148.32 \times 10^{-7}\, T^2$$

Calculate $\Delta_r H^0$ at 1000 K.

Now

$$\Delta C_p = 5.57 - 28.09 \times 10^{-3}\, T + 322.54 \times 10^{-7}\, T^2$$

i.e.,

$$\Delta a = 5.57;\ \Delta b = -28.09 \times 10^{-3};\ \text{and } \Delta c = 322.54 \times 10^{-7}$$

Using equation (11.11)

$$\begin{aligned} A &= -393510 - (5.57)(298) - \frac{1}{2}(-28.09 \times 10^{-3})(298)^2 \\ &\quad - \frac{1}{3}(322.54 \times 10^{-7})(298)^3 \\ &= -394207.12 \text{ J} \end{aligned}$$

Using equation (11.9)

$$\begin{aligned} \Delta_r H^0{}_{1000} &= -394207.12 + (5.57)(1000) + \frac{1}{2}(-28.09 \times 10^{-3})(1000)^2 \\ &\quad + \frac{1}{3}(322.54 \times 10^{-7})(10^3)^3 \\ &= -391.93 \text{ kJ} \end{aligned}$$

11.4.2. *Effect of Pressure on $\Delta_r H$*

In real gases or solutions heat of reaction does change with pressure and it is given as

$$d\Delta_r H = \Delta C_p\, dT + d(-\mu C_p)\, dp \qquad \ldots(11.13)$$

where μ is the Joule-Thomson coefficient.

***Exercise* 11.9:** Prove the equation (11.13).

General form of the heat of the reaction can be written as

$$\Delta_r H = \sum_{j=1}^{l} \nu_j \bar{H}_j - \sum_{k=1}^{m} \nu_k \bar{H}_k \qquad \ldots(11.14)$$

where í$_j$ and í$_k$ are the coefficients of the products and reactants respectively. Differentiating with respect to pressure on both sides

$$\left(\frac{\partial \Delta_r H}{\partial p}\right)_T = \frac{\partial}{\partial p}\left[\sum_{j=1}^{l} \nu_j \bar{H}_j - \sum_{k=1}^{m} \nu_k \bar{H}_k\right]_T \qquad \ldots(11.15)$$

The summation and derivative operation commute, therefore we can write:

$$\left(\frac{\partial \Delta_r H}{\partial p}\right)_T = \sum_{j=1}^{l} \nu_j \left(\frac{\partial \bar{H}_j}{\partial p}\right)_T - \sum_{k=1}^{m} \nu_k \left(\frac{\partial \bar{H}_k}{\partial}\right)_T \qquad \ldots(11.16)$$

The derivative can be written as

$$\left(\frac{\partial \bar{H}_j}{\partial p}\right)_T = \left(\frac{\partial}{\partial p}\left\{\left(\frac{\partial H_j}{\partial n_j}\right)_{T,p,n_s}\right\}_T\right) = \left(\frac{\partial}{\partial n_j}\left\{\left(\frac{\partial H_j}{\partial p}\right)_T\right\}_{T,p,n_s}\right) \qquad \ldots(11.17)$$

Since enthalpy is a state function, order of differentiation does not matter. From Joule-Thomson relation, we know

$$\left(\frac{\partial H}{\partial p}\right)_T = -\mu C_p = -\mu\, n\, \bar{C}_p \qquad ...(11.18)$$

Substituting equation (11.18) in equation (11.17), we get

$$\left(\frac{\partial \bar{H}_j}{\partial p}\right)_T = \left(\frac{\partial}{\partial n_j}\{(-\mu_j n_j \bar{C}_p)_j\}\right)_{T,p,n_s} = -\mu_j[\bar{C}_p]_j \qquad ...(11.19)$$

Substituting equation (11.19) in equation (11.16), we get

$$\left(\frac{\partial \Delta_r H}{\partial p}\right)_T = \sum_{j=1}^{l} -\nu_j(\mu_j[\bar{C}p]_j - \sum_{k=1}^{m} -\nu_k(\mu_k[\bar{C}p]_k = \Delta[-\mu\bar{C}_p] \qquad ...(11.20)$$

Here $\Delta[-\mu\bar{C}_p]$ gives the rate of change of the reaction enthalpy with pressure. Since in many cases the Joule-Thomson coefficients are small, thus $\Delta_r H$ tends to highly insensitive to the changes in the pressure. If we combine the effect of temperature and pressure, the change in the heat reaction can be written as

$$d\Delta_r H = \Delta C_p\, dT + d(-\mu C_p)\, dp \qquad ...(11.13)$$

11.5. HESS'S LAW

A desired chemical reaction can be written as a sequence of many chemical reactions. If one knows heat of reaction of each of the above steps, one can calculate heat of the desired reaction by adding or subtracting the heat of reaction of each step. This principle in which the heat of reactions are added or subtracted algebraically is called Hess's law of constant heat summation. The basis for this law is that enthalpy or internal energy is path-independent quantities *i.e.,*

$$\Delta H_p = q_p \text{ and } \Delta E = q_V \text{ and hence} \qquad (11.22)$$

$$\Delta H = \Delta H_1 + \Delta H_2 + \Delta H_3 + \ldots \text{ or } q_p = q_p' + q_p'' + q_p'' + \ldots$$

***Exercise* 11.10:** Calculate $\Delta_r H$ for the following reaction:

$$\text{C (graph)} + H_2O\text{ (g)} = H_2\text{ (g)} + \text{CO (g)}$$

if the following data is available

$$\text{C (graph)} + \frac{1}{2} O_2\text{ (g)} = \text{CO (g) } \Delta H = -110.50 \text{ kJ} \qquad ...(A)$$

$$H_2\text{ (g)} + \frac{1}{2} O_2\text{ (g)} = H_2O\text{ (g) } \Delta H = -241.84 \text{ kJ} \qquad ...(B)$$

The desired chemical reaction is obtained if reaction (B) is subtracted from (A), *i.e.*

$$\text{C (graph)} + \frac{1}{2} O_2\text{ (g)} - H_2\text{(g)} - \frac{1}{2} O_2\text{ (g)} = \text{CO (g)} - H_2O\text{ (g)}$$

Rearranging the terms

$$C\ (graph) + H_2O\ (g) = CO\ (g) + H_2\ (g)$$
$$= -110.50 - (-241.84) = 131.34\ kJ$$

***Exercise* 11.11:** Calculate ΔH for the reaction

$$C\ (graph) = C\ (dia)$$

from the following results:

$$C\ (graph) + O_2(g) = CO_2(g)\ \Delta_r H = -393.514\ kJ$$
$$C\ (dia) + O_2(g) = CO_2(g)\ \Delta_r H = -395.408\ kJ$$

The desired reaction can be obtained by subtracting second reaction from first, *i.e.*,

$$C\ (graph) + O_2(g) - C\ (dia) - O_2(g) = CO_2(g) - CO_2(g)$$

or

$$C\ (graph) = C\ (dia)\ \Delta H = -393.54 - (3.95.498)$$
$$= 1.894\ kJ$$

11.6. HEAT OF FORMATION

Heat of formation of any compound is the enthalpy of the reaction which represents the formation of one mol of compound from its elements.

If all the species of the above chemical reactions are in their standard states, the heat of formation ($\Delta_f H^0$) is called standard heat of formation. The standard heat of formation of elements in their most stable states at 298 k are taken as zero. As usual the tables only mention $\Delta_f H^0$ per mol of the compound at 298 K and $\Delta_f H^0$ at any other temperature can be obtained with the help of equation (11.9) or (11.10).

If heat of formation of the reactants and the products of chemical reactions are known, heat of reaction can be calculated from the relation:

$$\Delta_f H^0_{298} = \sum_i n_i\ \Delta_f \bar{H}^0_f(\text{products}) - \sum_j n_j\ \Delta_f \bar{H}^0_f(\text{reactants}) \quad ...(11.22)$$

where n_i and n_j are number of mol of product species and reactant species respectively. $\Delta_f \bar{H}^0$ is the molar heat of formation.

***Exercise* 11.12:** Calculate $\Delta_r H^0$ for the reaction

$$CH_3OH(l) + \frac{3}{2}\ O_2(g) \rightarrow CO_2(g) + 2H_2O(g)$$

Given
$$\Delta_f H^0\ (CH_3OH,\ l) = -238.57\ kJ\ mol^{-1},$$
$$\Delta_f H^0\ (H_2O,\ g) = -241.84\ kJ\ mol^{-1},$$
$$\Delta_f H^0\ (CO_2,\ g) = -393.55\ kJ\ mol^{-1}$$

Using equation (11.14) and $\Delta_f \bar{H}^0\ (O_2) = 0$

$$\Delta_r H^0 = (-393.55\ kJ\ mol^{-1})\ (1\ mol) + (2\ mol)(-241.84\ kJ\ mol^{-1}) - (1\ mol)(-238.57\ kJ\ mol^{-1})$$
$$= -638.66\ kJ$$

***Exercise* 11.13:** Consider the reaction

$$C(graph) + H_2O(g) \rightarrow CO(g) + H_2(g)$$

If $\Delta_r H^0$ for the above reaction is 131.34 kJ and $\Delta_f H^0 (H_2O, g) = -241.84$ kJ mol^{-1}, calculate $\Delta_f H^0$ (CO, g).

Using equation (11.14) and $\Delta_f H^0 (H_2, g) = 0$ and $\Delta_f H^0$ (C, graph) = 0

$$131.34 \text{ kJ} = \Delta_f H^0 \text{ (1 mol)} - (-241.84 \text{ kJ mol}^{-1}) \text{ (1mol)}$$

$$\Delta_f H^0 \text{ (CO)} = -110.5 \text{ kJ mol}^{-1}$$

11.7. HEAT OF COMBUSTION

Heat of combustion is the heat of reaction in which 1 mol of a substance is oxidised completely.

If a compound contains C, H, O and N, the oxidised products are CO_2, $H_2O(l)$ and N_2, and equation can be balanced. On the other hand, for compounds containing halogens, sulphur, phosphorus etc., balancing the chemical equations becomes difficult as these elements form many oxides.

Relation (11.22) also holds for the data of heat of combustion. This equation (11.22) has been used quite often to calculate the heat of formation of compounds, specifically hydrocarbons.

***Exercise* 11.14:** Calculate heat of formation of methane, *i.e.*,

$$C(graphite) + 2H_2(g) \rightarrow CH_4(g)$$

from the following data:

$$CH_4(g) + 2O_2(g) = CO_2(g) + 2H_2O(l), \quad \Delta_{comb}H = -890.35 \text{ kJ mol}^{-1} \quad (1)$$

$$H_2(g) + \frac{1}{2} O_2(g) = H_2O(l) \quad \Delta_{comb}H = -285.84 \text{ kJ mol}^{-1} \quad (2)$$

$$C(graphite) + O_2(g) = CO_2(g) \quad \Delta_{comb}H = -393.51 \text{ kJ mol}^{-1} \quad (3)$$

The desired reaction can be written as

2 reaction (2) + reaction (3) – reaction (1)

i.e., $2H_2(g) + O_2(g) + C(graph) + O_2(g) - CH_4(g) - 2O_2(g)$

$$= 2H_2O(1) + CO_2(g) - 2H_2O(1) - CO_2(g)$$

or $$C(graph) + 2H_2(g) = CH_4(g)$$

$$\Delta_r H = (2 \text{ mol})(-285.84 \text{ kJ mol}^{-1}) + (-393.51 \text{ kJ mol}^{-1})(1 \text{ mol}) - (-890.35 \text{ kJ mol}^{-1})$$

$$= -74.84 \text{ kJ mol}^{-1}$$

***Exercise* 11.15:** 2 g of ethyl alcohol is burnt in excess of O_2 in bomb calorimeter and the products analysed are $H_2O(l)$ and $CO_2(g)$ and unreacted O_2. If the heat capacity of the system is 1600 J K^{-1} and the rise in temperature is 37.1 deg, calculate the heat evolved in this reaction.

The chemical equation is

$$C_2H_5OH(l) + 3O_2(g) \rightarrow 2CO_2(g) + 3H_2O(l)$$

so $$\Delta n = -1.$$

Heat evolved here is at constant volume, so

$$\Delta E = Q_V = c\Delta T = (1600 \text{ J K}^{-1})(37.1 \text{ K})$$

$$= 59360 \text{ J}$$

This heat is given by 2 g of C_2H_5OH, so 1 mol will give

$$= \frac{(59360 \text{ J})(46 \text{ g mol}^{-1})}{2 \text{ g}} = 1365.28 \text{ kJ mol}^{-1}$$

i.e., $\Delta E = -1365.28 \text{ kJ mol}^{-1}$

Using equation $\Delta H = \Delta E + \Delta nRT$

$$\Delta H = (-1365.28 \text{ kJ mol}^{-1}) + (-1 \text{ mol})(8.314 \text{ J K}^{-1} \text{ mol}^{-1})(298 \text{ K})(10^{-3} \text{ kJ J}^{-1})$$

$$= -1367.75 \text{ kJ mol}^{-1}$$

***Exercise* 11.16:** Calculate heat of combustion of $H_2S(g)$ from the following data:

$\Delta_f H^0 (H_2S, g) = -20.15 \text{ kJ mol}^{-1}$,
$\Delta_f H^0 (H_2O, l) = -285.84 \text{ kJ mol}^{-1}$
$\Delta_f H^0 (SO_2, g) = -296.9 \text{ kJ mol}^{-1}$

The balanced equation for one mol of H_2S is

$$H_2S(g) + \frac{3}{2} O_2(g) \rightarrow H_2O(l) + SO_2(g)$$

Using equation (11.22)

$$\Delta_{comb} H^0 = (-285.84 \text{ kJ mol}^{-1}) + (-296.9 \text{ kJ mol}^{-1}) - (-20.15 \text{ kJ mol}^{-1})$$
$$= -562.59 \text{ kJ mol}^{-1}$$

11.8. BOND ENERGY

Bond energy is defined as the heat of reaction associated with breaking of chemical bonds of gaseous molecules to the gaseous fragments. Sometimes it is also called as bond enthalpy, actually more correct name.

The bond dissociation energy (B.E.) can be used to calculate the heat of reaction by the relation

$$\Delta_r \bar{H}_0 = -\sum_{i}^{\text{products}} n_i BE_i + \sum_{j}^{\text{reactants}} n_j BE_j \qquad \text{...(11.23)}$$

where BE is the bond dissociation energy per mol of bonds, n_j and n_i are the number of mol of bonds broken or formed in the case of reactants and products.

Similarly heat of reaction data can also be used to calculate the bond dissociation energy of any particular bond provided other data in the equation (11.23) is known. One thing must be kept in mind that environments around an atom do affect the bond energy of a particular bond, therefore, the values obtained from equation (11.23) are the average ones or approximate ones.

Although the bond energies refer to the molecules in gas phase yet the approximate values of heat of reactions can be calculated for the condensed phase, *i.e.,* liquid or solid states. The heat of reaction in condensed phase can be corrected if the heat of vaporisation, heat of sublimation etc. can be included.

***Exercise* 11.17:** Calculate $\Delta_r H^0$ of the reaction

$$2H_2 (g) + \frac{1}{2} O_2 (g) + C \text{ (graph)} \rightarrow CH_3OH \text{ (g)}$$

Given: $D_{O_2} = 495.0 \text{ kJ mol}^{-1}$; $D_{H_2} = 436 \text{ kJ mol}^{-1}$; C(graph) = $712.96 \text{ kJ mol}^{-1}$; $D_{C-H} = 414.2 \text{ kJ mol}^{-1}$; $D_{O-H} = 464.4 \text{ kJ mol}^{-1}$; $D_{C-O} = 351.5 \text{ kJ mol}^{-1}$.

Using equation (11.23), *i.e.,*

$$\Delta_r H^0 = (2)(H-H) + (0.5)(0-0) + (C-C) - (3)(C-H) - (l)(O-H) - (l)(C-O)$$

$$\Delta_r H^0 = (2 \text{ mol})(436 \text{ kJ mol}^{-1}) + (0.5 \text{ mol})(495.0 \text{ kJ mol}^{-1}) + (712.95 \text{ kJ mol}^{-1})(1 \text{ mol}) - (3 \text{ mol})(414.2 \text{ kJ mol}^{-1}) - (1 \text{ mol})(464.4 \text{ kJ mol}^{-1}) - (1 \text{ mol})(351.5 \text{ kJ mol}^{-1})$$

$$= -226.05 \text{ kJ}$$

The actual value is – 201.2 kJ.

***Exercise* 11.18:** Consider a reaction

$$CH_4 (g) + 2O_2 (g) \rightarrow CO_2 (g) + 2H_2O (g)$$

Calculate $D_{C\text{–}O}$ if $\Delta_r H^0 = -802.24$ kJ, $D_{C\text{–}H} = 414.2$ kJ mol^{-1};

$D_{O_2} = 495.0$ kJ mol^{-1}; $D_{O\text{–}H} = 464.4$ kJ mol^{-1}

Using equation (11.23)

$$-802.24 = (4)(414.2) + (2)(495.0) - (2)(D_{C=O}) - (4)(464.4)$$

$$D_{C=O} = 795.72 \text{ kJ mol}^{-1}$$

The actual value is 732 kJ mol^{-1}.

One can see from the above results that bond energy or $\Delta_r H^0$ of reaction calculated are only approximate values.

11.9. HEAT OF NEUTRALISATION

The heat of reaction which involves the neutralisation of acid by base is known as the heat of neutralisation.

The heat of neutralisation of strong acids and strong bases is constant and is – 55.90 kJ mol^{-1}. But the heat of neutralisation of weak acids and weak bases are less than – 55.90 kJ mol^{-1} because the latter acids or latter bases involve the dissociation of acids into H^+ ions and anion or that of bases into OH^- ions and cations whereas in case of strong acids and strong bases dissociation is already complete and the reaction is only

$$H^+ (aq) + OH^-(aq) = H_2O$$

So

$$\Delta_r H^0 = \Delta_{ioniz} H^0 + \Delta_{neut} H^0 \qquad ...(11.24)$$

***Exercise* 11.19:** Heat of neutralisation of HCN by NaOH is – 12.13 kJ mol^{-1}, calculate $\Delta_{ioniz} H^0$ of HCN.

Using equation (11.24) and $\Delta_{neut} H^0 = -55.90$ kJ mol^{-1}

$$-12.13 = \Delta_{ioniz} H^0 - 55.90$$

$$\Delta_{ioniz} H^0 = +43.77 \text{ kJ mol}^{-1}$$

11.10. HEAT OF SOLUTION

There are two heats of solution, one is 'integral' and the other is the 'differential' heat of solution.

The integral heat of solution is defined as the enthalpy change when 1 mol of solute is dissolved in *n* mol of solvent. The 'differential heat' of solution is defined as the enthalpy change when 1 mol of solute is dissolved in an infinite amount of solution such that its concentration does not change by adding one mol of solute. Mathematically it is defined as $\frac{d\,(m\Delta H)}{dm}$ *i.e.*, heat change is plotted as a function of number of moles of solute and the differential heat of solution can be obtained by finding the slope at any concentration. Thus the differential heat of solution depends on the concentration of the solution.

***Exercise* 11.20:** When different amount of (n_1) water is added to one mol of $HClO_4$ (n_2), the heat evolved is given in the table. Calculate the integral heat at infinite dilution and the differential heat of solution at 0.7401 mol

(mol^{-1}($HClO_4$)$^-$)

$$HClO_4\left(\frac{n_1}{n_2}=2\right)+3H_2O \rightarrow HClO_4\left(\frac{n_1}{n_2}=5\right) \qquad \Delta_{soln}H = -19.58 \text{ kJ}$$

$$HClO_4\left(\frac{n_1}{n_2}=2\right)+13H_2O \rightarrow HClO_4\left(\frac{n_1}{n_2}=15\right) \qquad \Delta_{soln}H = -23.828 \text{ kJ}$$

$$HClO_4\left(\frac{n_1}{n_2}=15\right)+5H_2O \rightarrow HClO_4\left(\frac{n_1}{n_2}=20\right) \qquad \Delta_{soln}H = 0.0544 \text{ kJ}$$

$$HClO_4\left(\frac{n_1}{n_2}=20\right)+30H_2O \rightarrow HClO_4\left(\frac{n_1}{n_2}=50\right) \qquad \Delta_{soln}H = 0.506 \text{ kJ}$$

$$HClO_4\left(\frac{n_1}{n_2}=50\right)+25H_2O \rightarrow HClO_4\left(\frac{n_1}{n_2}=75\right) \qquad \Delta_{soln}H = 0.197 \text{ kJ}$$

$$HClO_4\left(\frac{n_1}{n_2}=75\right)+25H_2O \rightarrow HClO_4\left(\frac{n_1}{n_2}=100\right) \qquad \Delta_{soln}H = 0.159 \text{ kJ}$$

$$HClO_4\left(\frac{n_1}{n_2}=100\right)+100H_2O \rightarrow HClO_4\left(\frac{n_1}{n_2}=200\right) \qquad \Delta_{soln}H = 0.159 \text{ kJ}$$

$$HClO_4\left(\frac{n_1}{n_2}=200\right)+300H_2O \rightarrow HClO_4\left(\frac{n_1}{n_2}=500\right) \qquad \Delta_{soln}H = 0.063 \text{ kJ}$$

$$HClO_4\left(\frac{n_1}{n_2}=500\right)+500H_2O \rightarrow HClO_4\left(\frac{n_1}{n_2}=1000\right) \qquad \Delta_{soln}H = -013 \text{ kJ}$$

$$HClO_4\left(\frac{n_1}{n_2}=1000\right)+1000H_2O \rightarrow HClO_4\left(\frac{n_1}{n_2}=2000\right) \qquad \Delta_{soln}H = -0.034 \text{ kJ}$$

Following table can be made as

n_1	$\frac{n_1+3}{n_2}$	m_2	$\sqrt{m}$	ΔH kJ (mol $HClO_4$)$^{-1}$
1977	2000	0.02775	0.167	–22.737
997	1000	0.05551	0.236	–22.703
497	500	0.1110	0.333	–22.6896
197	200	0.2775	0.527	–22.753
97	100	0.5551	0.745	–22.912
72	75	0.7401	0.86	–23.091
47	50	1.1101	1.054	–23.268
17	20	2.7753		–23.774
13	15	3.7004		–23.828
2	5	11.1012		–19.958
0	2	27.753		0

Plot ΔH (mol^{-1} $HClO_4$) against (molality)$^{1/2}$ in Fig. 11.1. ΔH at infinite dilution is – 23.0 kJ (mol^{-1} $HC1O_4$). The differential heat of dilution at 0.7401 molal solution can be found by drawing a tangent at $\sqrt{07401}$ molal and finding a slope at that point.

$$\Delta_{\text{diff}}H(0.7401 \text{ mol}) = \frac{-21.75-(-25)}{0-(2.25)} = -1.44 \text{ kJ mol}^{-1}.$$

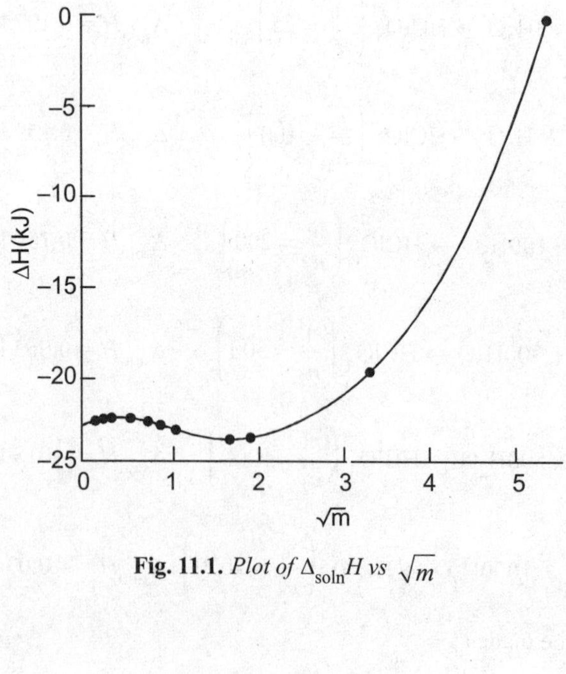

Fig. 11.1. *Plot of* $\Delta_{\text{soln}}H$ *vs* $\sqrt{m}$

PROBLEMS

11.1 2.75 A of constant current and 2.5 V was passed for 164.5 s, 174.2 s and 75.5 s separately. Amount of liquid *A* collected each time is 10 mL and 4.4 mL respectively. If the density of *A* is 1.2835 g ml^{-1} calculate heat of vaporisation.

11.2 10 mL of compound *A* was collected at its boiling point of 80°C. If heat capacity of the apparatus along with liquid *A* is 25 J K^{-1}, calculate the heat required to collect this amount of liquid. Use the data from problem 11.1.

11.3 5 g of acetic acid (*l*) was burnt in excess of O_2 in bomb calorimeter and the products analysed were CO_2 (g) and H_2O (*l*). If the heat capacity of the calorimeter is 1500 J K^{-1}, calculate the rise in temperature of the system: $\Delta_{comb}H = -872.36$ kJ mol^{-1}.

11.4 2 g of benzene (g) was burnt in excess of O_2 in bomb calorimeter.

The reaction is

$$C_6H_6\,(g) + \frac{15}{2}O_2\,(g) \longrightarrow 6\,CO_2\,(g) + 3H_2O\,(l)$$

If heat capacity of the system is 2000 J K^{-1} and rise in temperature is 42.1 deg, calculate the heat of combustion of benzene.

11.5 10 g of *n*-heptane was burnt in excess of O_2 in a constant volume calorimeter. Heat of combustion is – 4811.18 kJ mol^{-1}. If rise in temperature is 25 degree, calculate heat capacity of the calorimeter along with other things.

11.6 Calculate heat of combustion for the following reactions:

$\Delta_f H^0(C_2H_2) = -226\ 73$ kJ mol^{-1}; $\Delta_f H^0(CO_2) = -393.52$ kJ mol^{-1}; $\Delta_f H^0(H_2O, 1) = -285.84$ kJ mol^{-1}; $\Delta_f H^0(C_2O_4H_2) = -487.02$ kJ mol^{-1}; $\Delta_f H^0(CH_3CHO) = -166.76$ kJ mol^{-1}; $\Delta_f H^0(CH_3OH) = -201.17$ kJ mol^{-1}.

(*i*) $C_2H_2(g) + \frac{5}{2}O_2(g) \rightarrow 2CO_2(g) + H_2O(l)$

(*ii*) $CH_3CHO(g) + \frac{5}{2}O_2(g) \rightarrow +2CO_2(g) + 2H_2O(l)$

(*iii*) $(CO_2H)_2\,(s) + \frac{1}{2}O_2(g) \rightarrow 2CO_2(g) + H_2O(l)$

(*iv*) $CH_3OH(l) + \frac{3}{2}O_2(g) \rightarrow CO_2(g) + 2H_2O(l)$

11.7 Consider the reaction *n*-butane (g) → isobutane (g). Calculate $\Delta_r H$ for this reaction. Given: $\Delta_{comb}H$ (*n*-butane) = – 2878.5 kJ mol^{-1}; $\Delta_{comb}H$(isobutane) = – 2871.65 kJ mol^{-1}.

11.8 Calculate $\Delta_f H^0(C_2H_5OH, l)$ by the reaction:

$$2C\,(s) + 3H_2(g) + \frac{1}{2}O_2(g) \rightarrow C_2H_5OH(l).$$

Given: $\Delta_{comb}\bar{H}(C_2H_5OH, l) = -1368.17$ kJ mol^{-1}; $\Delta_{comb}\bar{H}(C, s) = -393$ kJ mol^{-1}; $\Delta_{comb}\bar{H}(H_2) = -285.77$ kJ mol^{-1}.

11.9 Calculate $\Delta_f H^0$ of acetic acid, benzene and *n*-heptane from the data of problems 11.3, 11.4, 11.5, and 11.8.

11.10 Calculate $\Delta_f H^0$ of CS_2 (*l*) from the following data:

$C(s) + O_2(g) \rightarrow CO_2(g); \Delta\bar{H} = -393.3$ kJ mol^{-1}

$S(s) + O_2(g) \rightarrow SO_2(g); \Delta\bar{H} = -293.72$ kJ mol^{-1}

$CS_2(l) + 3O_2(g) \rightarrow CO_2(g) + 2SO_2(g); \Delta\bar{H} = -1108.76$ kJ mol^{-1}

$C(s) + 2S(s) \rightarrow CS_2(l); \Delta\bar{H} = ?$

11.11 Calculate ΔH for the reaction:

$n\text{–}C_3H_8(g) \rightarrow \text{cyclo–}C_3H_6(g) + H_2(g)$

$\Delta\bar{H}_{comb}$ (n–C_3H_8, g) = – 2220.07 kJ mol^{-1}, $\Delta\bar{H}_{comb}$ (cyclo-C_3H_6) = – 2091.16 kJ mol^{-1}; $\Delta\bar{H}_{comb}$ (H_2) = – 285.77 kJ mol^{-1}.

11.12 The formation of ethanol may be considered as

$2C(s) + 2H_2(g) \rightarrow C_2H_4(g); \Delta_r H = 52.28$ kJ mol^{-1}

$C_2H_4(g) + H_2O(g) \rightarrow C_2H_5OH(l); \Delta_r H = -88.09$ kJ mol^{-1}

$H_2(g) + \frac{1}{2}O_2(g) \rightarrow H_2O(g) \; \Delta_r H = -241.83$ kJ mol^{-1}

Calculate $\Delta H_f (C_2H_5OH, l)$

11.13 Calculate $\Delta_r H^0$ for the following reactions using the data of $\Delta_f H^0$ for reactant and products:

(*i*) $C(\text{graphite}) + H_2O(g) \rightarrow H_2(g) + CO(g)$

(*ii*) $Mg(s) + 2HCl(g) \rightarrow MgCl_2(s) + H_2(g)$

(*iii*) $C_2H_6(g) + \frac{7}{2}O_2(g) \rightarrow 2CO_2(g) + 3H_2O(l)$

(*iv*) $NH_3(g) + HCl(g) \rightarrow NH_4Cl(s)$

(*v*) $C_6H_5NH_2(l) + \frac{17}{4}O_2(g) \rightarrow 6CO_2(g) + \frac{7}{2}H_2O(l) + NO_2(g)$

11.14 Calculate $\Delta_r E^0$ for the reactions mentioned in the problem 11.13.

11.15 Calculate $\Delta_r H^0$ for the following atomic reactions:

(*i*) $O(g) + NO_2(g) \rightarrow NO(g) + O_2(g)$

(*ii*) $O(g) + SO_3(g) \rightarrow SO_2(g) + O_2(g)$

(*iii*) $Cl(g) + CH_4(g) \rightarrow CH_3Cl(g) + H(g)$

(*iv*) $H(g) + Br_2(g) \rightarrow HBr(g) + Br(g)$

11.16 The enthalpy of dissociation of Lewis acid-base ($A:B$) complex is 68 kJ mol^{-1}. If $\Delta_f\bar{H}^0$ (A) and $\Delta_f\bar{H}^0$ (B) are – 56.11 kJ mol^{-1} and 220 kJ mol^{-1}, calculate $\Delta_f\bar{H}^0$ (AB).

11.17 A man of 60 kg climbed a tower of height 4 m. If he wants to keep his energy constant, how much glucose he must eat before climbing? Given: $\Delta_{comb}H$ (glucose) = – 2808 kJ mol^{-1} and 30% of this heat is utilized in useful work.

11.18 (*a*) The $\Delta_r H^0$ for the reaction

$C(\text{graphite}) + 2H_2(g) \rightarrow CH_4(g)$ is – 1648.5 kJ mol^{-1}. Calculate the average $D_{C\text{–}H}$.

(*b*) The other data available for the stepwise decomposition of methane is

$$CH_4(g) \rightarrow CH_3(g) + H(g) \qquad D_{CH_3-H} = 430.95 \text{ kJ mol}^{-1}$$
$$CH_3(g) \rightarrow CH_2(g) + H(g) \qquad D_{CH_2-H} = 364.01 \text{ kJ mol}^{-1}$$
$$CH_2(g) \rightarrow CH(g) + H(g) \qquad D_{CH-H} = 523.0 \text{ kJ mol}^{-1}$$
$$CH(g) \rightarrow C(g) + H(g) \qquad D_{C-H} = 338.9 \text{ kJ mol}^{-1}$$

Calculate average D_{C-H} and compare with that obtain in (11.18*a*).

11.19 If $D_{H-H} = 431.79$ kJ mol^{-1}, $D_{C-H} = 412.2$ kJ mol^{-1}, calculate heat of sublimation of C(graphite) and compare with that obtained experimentally (718.39 kJ mol^{-1}). Use data of problem 11.18 if needed.

11.20 $\Delta_r H^0$ for the given reaction is – 2220.07 kJ mol^{-1}:

$$n - C_3H_8(g) + 5O_2(g) \rightarrow 3CO_2(g) + 4H_2O(l)$$

and $D_{C-O} = 494.97$ kJ mol^{-1}; $D_{C-H} = 412.12$ kJ mol^{-1};

$D_{C-O} = 723.83$ kJ mol^{-1}; $D_{O-H} = 464.42$ kJ mol^{-1}

Calculate $D_{C-C.}$

11.21 Consider the reaction ($\Delta_r H^0 = -151.42$ kJ)

$$C_2H_2(g) + H_2O(g) \rightarrow CH_3C{\overset{\displaystyle O}{\underset{\displaystyle H}{\lessgtr}}}$$

Using data of problem 11.20 and $D_{C=C} = 828.43$ kJ mol^{-1}, calculate $D_{C=O}$. Compare with that given in problem 11.20.

11.22 If fluoraffinity of AlF_4^- is defined as

$AlF_3(g) + F^-(g) \rightarrow AlF_4^-(g)$ $Y(AlF_4^-) = ?$

Calculate its value from the following data:

$$Na(s) \xrightarrow[\text{kJ mol}^{-1}]{108.78} Na(g) \xrightarrow[\text{kJ mol}^{-1}]{493.71} Na^+ + e$$

$$\frac{1}{2} F_2(g) \xrightarrow{75.31 \text{ kJ}} F(g) \xrightarrow{e} F^-(g), \qquad EA = -334.72 \text{ kJ mol}^{-1}$$

$$Al(s) + 2F_2(g) \rightarrow AlF_3(g) + F(g) \qquad \Delta H^0 = -1213.36 \text{ kJ}$$
$$AlF_3(g) + F^-(g) \rightarrow AlF_4^-(g) \qquad Y(AlF_4^-) = ?$$
$$Na^+ + AlF_4^- \rightarrow NaAlF_4(s) \qquad U = -606.68 \text{ kJ}$$

$$Na(s) + 2F_2(g) + Al(s) \rightarrow NaAlF_4 \quad \Delta H_f^0 = -2016.69 \text{ kJ}$$

The above reactions can be represented in a cycle as:

$$Na(s) + F_2(g) + Al(g) \xrightarrow[\frac{1}{2}D_{F-F} + S_{Na}]{\Delta H_f^0(AlF_3) +} Na(g) + AlF_3(g) + F(g)$$

$\downarrow \Delta H^0_f(NaAlF_4)$ $\qquad$ $\downarrow I_{Na+}$ $\qquad$ $\downarrow E_F$

$$AlF_3(g) + F\text{–}(g)$$

$\downarrow Y(AlF_4^-)$

$$NaAlF_4(s) \xleftarrow{U(NaAlF_4)} Na^+(g) + AlF\text{–}_4(g)$$

11.23 The fluoraffinity of AlF_4^- can also be represented as

$$\begin{array}{ccc} NaF(g) + AlF_3(g) & \xrightarrow{\Delta H_2} & NaAlF_4 \\ \downarrow \Delta H_1 & \Delta H_3 & \uparrow \\ Na^+(g) + F^-(g) + AlF_3(g) & \xrightarrow[Y(AlF_4^-)]{} & Na^+ + AlF_4^-(g) \end{array}$$

Given: $\Delta H_1 = 635.97$ kJ; $\Delta H_2 = -317.98$ kJ and $\Delta H_3 = -405.8$ kJ. Calculate $Y(AlF_4^-)$.

11.24 The cycle mentioned in problem 11.22 can also be applied to reaction $BF_3(g) + F^-(g) \rightarrow BF_4^-$. Calculate $Y(BF_4^-)$ using data of problem 11.22 and $U(NaBF_4) = -669.44$ kJ mol^{-1}; $\Delta H_f^0(BF_3) = -1133.86$ kJ mol^{-1}; .

$\Delta H_f^0(NaBF_4) = -1778.2$ kJ mol^{-1}.

11.25 Heat of neutralisation of NH_4OH and HCl is -51.46 kJ mol^{-1}. Calculate $\Delta_{ioniz}H^0$ for NH_4OH.

11.26 ΔH^0_{neutr} for CH_3COOH with NaOH is -50.63 kJ mol^{-1}. Calculate $\Delta_{ioniz}H^0$ of acetic acid.

11.27 Calculate heat of neutralisation of NH_4OH and CH_3COOH. Use data of problems (11.25) and (11.26).

Addition Problems

11.28 0.75 g of naphthalene was burnt in a bomb calorimeter in presence of excess of oxygen. The rise in temperature in the calorimeter is by 5.4 degree. If the heat capacity of the calorimeter is 5500 J K^{-1}, calculate the value of ΔE and ΔH per mole.

11.29 Consider the reaction $2O_3\,(g) \rightarrow 3O_2\,(g)$. If $\Delta_f \overline{H}^0 = 142.7$ kJ mol^{-1}, molar heat capacities of O_2 and O_3 are 29.4 J K^{-1} mol^{-1} and 38.2 J K^{-1} mol^{-1} respectively, calculate the value of $\Delta_f \overline{H}^0$ at 227 °C.

11.30 Consider the combustion of ethane as the reaction

$$C_2H_6\,(g) + (7/2)O_2\,(g) \rightarrow 2CO_2\,(g) + 3H_2O\,(g)$$

and calculate the heat of combustion from the given data. $D_{C\text{-}C} = 347$ kJ mol^{-1}, $D_{C\text{-}H} = 414$ kJ mol^{-1}, $D_{O=O} = 498.8$ kJ mol^{-1}, $D_{C=O} = 799$ kJ mol^{-1}, $D_{O\text{-}H} = 460$ kJ mol^{-1}.

11.31 Which of the following standard enthalpy of formation values is not zero at 25 °C? K(*s*), Xe (*g*), CH_4 (*g*), O (*g*), Hg(*l*).

11.32 Consider the reaction as given in Problem (11.30). (*i*) What is value of $\Delta_r E$ if the reaction is multiplied by a factor of 3? (*ii*) What is value of $\Delta_r E$ if the reaction is reversed so that the products become reactants and vice versa? (*iii*) Calculate the value of $\Delta_r E$ if water is in liquid state.

11.33 Consider the combustion of ethanol as mentioned in the reaction

$2CH_3OH\,(l) + 3O_2\,(g) \rightarrow 2CO_2\,(g) + 4H_2O\,(l)$ $\qquad \Delta_r H^0 = -1452.8$ kJ mol^{-1}

Calculate the heat evolved when 55g of methanol is used.

11.34 Consider the following combustion reactions

Fumeric acid $(s) + O_2\,(g) \rightarrow CO_2\,(g) + H_2O\,(l)$ $\qquad \Delta_r H^0 = -1336.0$ kJ mol^{-1}

Maleic acid $(s) + O_2\,(g) \rightarrow CO_2\,(g) + H_2O\,(l)$ $\qquad \Delta_r H^0 = -1359.2$ kJ mol^{-1}

Calculate the enthalpy of the following isomerisation process:

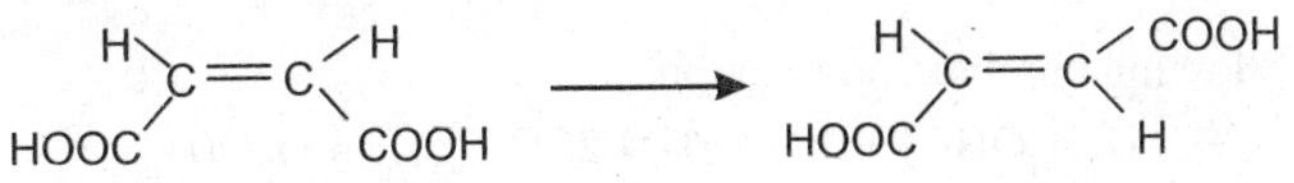

Maleic acid Fumeric acid

11.35 Calculate the enthalpy of formation of molecular nitrogen and hydrogen gas at 127 °C assuming that heat capacities of these gases are independent of temperature. The heat capacities of nitrogen and hydrogen are 19.9 and 20.18 J K^{-1} mol^{-1} respectively.

11.36 Consider the hydrogenation of ethene to form ethane.

$$C_2H_4\ (g) + H_2\ (g) \rightarrow C_2H_6\ (g).$$

Calculate the change in the enthalpy of hydrogenation from 27 °C to 227 °C. The molar heat capacities of hydrogen, ethene and ethane are 20.18, 43.6 and 52.7 J K^{-1} mol^{-1} respectively.

11.37 100 g of ethene has undergone combustion in excess of the oxygen. Calculate the heat of the following reactions and amount of heat evolved. Given data are: $\Delta_f \bar{H}^0(C_2H_4,\ g) = 52.26$ kJ mol^{-1}, $\Delta_f \bar{H}^0(CO_2,\ g) = -393.5$ kJ mol^{-1}, $\Delta_f \bar{H}^0(H_2O,\ g) = -241.8$ kJ mol^{-1},

$$C_2H_4\ (g) + 3O_2\ (g) \rightarrow 2CO_2\ (g) + 2H_2O\ (g)$$

11.38 100 g of ethene has undergone combustion reaction in presence of 288 g of oxygen. Using the data from Problem 11.37, calculate the amount heat evolved.

11.39 Calculate $\Delta_f \bar{H}^0(C_6H_5NH_2,\ l)$ from the following data.

$$2C_6H_5NH_2\ (l) + (33/2)\ O_2\ (g) \rightarrow 12\ CO_2\ (g) + 7\ H_2O\ (g) + 2\ NO\ (g) + 6296.56\ \text{kJ}$$
$$C(s) + 0.5\ CO(g) \rightarrow CO(g) + 110.53\ \text{kJ}$$
$$CO(g) + 0.5\ CO(g) \rightarrow CO_2(g) + 282.98\ \text{kJ}$$
$$H_2(g) + 0.5\ CO(g) \rightarrow H_2O(g) + 24111053\ \text{kJ}$$
$$NO(g) + 0.5\ CO(g) \rightarrow NO_2(g) + 110.53\ \text{kJ}$$

11.40 Esterification reaction of ethanoic acid is given as

$$CH_3COOH\ (l) + C_2H_5OH\ (l) \rightarrow CH_3COOC_2H_5\ (l) + H_2O\ (l)$$

Calculate the heat of the reaction at 298 K as well as 700 K from the data given below:

$\Delta_f \bar{H}^0(CH_3COOH,\ l) = -484.5$ kJ mol^{-1}, $\Delta_f \bar{H}^0(C_2H_5OH,\ l) = -277.69$ kJ mol^{-1},

$\Delta_f \bar{H}^0(CH_3COOC_2H_5,\ l) = -479.0$ kJ mol^{-1}, $\Delta_f \bar{H}^0(H_2O,\ l)$ -285.83 kJ mol^{-1},

$\bar{C}_p(CH_3COOH,\ l) = 124.3$ J K^{-1} mol^{-1}, $\bar{C}_p(C_2H_5OH,\ l) = 111.46$ J K^{-1} mol^{-1},

$\bar{C}_p(CH_3COOC_2H_5,\ l) = 170.1$ J K^{-1} mol^{-1}, $\bar{C}_p(H_2O,\ l) = 75.29$ J K^{-1} mol^{-1}.

11.41 Consider the reaction $C_2H_4\ (g) + H_2O\ (g) \rightarrow CH_3CH_2OH\ (l)$. (*i*) Calculate the heat of the reaction at 298 K, (*ii*) Calculate the heat of the reaction at 500 K and heat evolved if $dT = dp = 0$ and $p = 1$ atm., (*iii*) also heat evolved when $dV = 0$. $\Delta_f \bar{H}^0(C_2H_5OH,\ l) = -277.69$ kJ mol^{-1}, $\Delta_f \bar{H}^0(C_2H_4,\ g) = 52.3$ kJ mol^{-1}, $\Delta_f \bar{H}^0(H_2O,\ l)$ -285.83 kJ mol^{-1}, $\bar{C}_p(C_2H_4,\ g) = 43.56$ J K^{-1} mol^{-1}, $\bar{C}_p(C_2H_5OH,\ l) = 111.46$ J K^{-1} mol^{-1}, $\bar{C}_p(H_2O,\ l) = 75.29$ J K^{-1} mol^{-1}.

11.42. Consider the following combustion reaction.

$$C_2H_5OH(l) + 3\,O_2(g) \rightarrow 2CO_2\,(g) + 3\,H_2O(l)$$

Calculate the standard heat of combustion from the heat of formation of C_2H_5OH (*l*) (–277.69 kJ mol^{-1}), CO_2 (*g*) (–393.509 kJ mol^{-1}) and H_2O (*l*) (–285.835 kJ mol^{-1}) respectively.

11.43 Fluorine is one of the molecule which, can decompose H_2O by the reaction

$$2\,F_2(g) + 2\,H_2O(l) \rightarrow 4\,HF(g) + O_2(g)$$

Data given: $\frac{1}{2}\,H_2(g) + \frac{1}{2}\,F_2(g) \rightarrow HF(g)$ $\quad \Delta_r H_0 = -273.3$ kJ mol^{-1}

$H_2(g) + \frac{1}{2}\,O_2(g) \rightarrow H_2O(l)$ $\quad \Delta_r H_0 = -285.8$ kJ mol^{-1}

11.44 50 g of hexadecane ($C_{16}H_{34}$) undergoes partial combustion in presence of insufficient of oxygen to give the following reactions

$$C_{16}H_{34}(g) + 16.5\,O_2(g) \rightarrow 16CO(g) + 17H_2O(l)$$

and $$C_{16}H_{34}(g) + 24.5\,O_2(g) \rightarrow 16CO_2(g) + 17H_2O(l)$$

At a constant pressure of 1 bar and 298 K, the heat change in the process is found to be –1945.2 kJ. (*i*) Using the data given calculate the percent of hexadecane that reacted to form CO(*g*) in this process, (*ii*) If the same reaction would have been carried out at constant volume, calculate the heat evolved. $\Delta_f \bar{H}^0(H_2O,\,l) = -285.83$ kJ mol^{-1}, $\Delta_f \bar{H}^0(C_{16}H_{34},\,s) = -447.97$ kJ mol^{-1}, =

$\Delta_f \bar{H}^0(CO,\,g) = \Delta_f \bar{H}^0(CO_2,\,g) = -393.5$ kJ mol^{-1}.

11.45 Consider the following reactions:

$2\,P(s) + 3\,Cl_2\,(g) \rightarrow 2\,PCl_3\,(l)$ $\quad \Delta_r H(A) = -640$ kJ (A)

$2\,P(s) + 5\,Cl_2\,(g) \rightarrow 2\,PCl_5\,(s)$ $\quad \Delta_r H(B) = -887$ kJ (B)

Calculate the heat of reaction for the reaction

$$PCl_3(s) + Cl_2\,(g) \rightarrow 2\,PCl_5\,(s)$$

11.46 Molar heat of combustion of isobutene and *n*-butane are –2871 kJ mol^{-1} and –2878 kJ mol^{-1} respectively. Calculate the heat of the reaction for the reaction *n*-Butane → isobutene.

11.47 The standard heat of combustion of *C*(*s*), $H_2(g)$ and $CH_4(g)$ are –393.5 kJ mol^{-1}, –285.8 kJ mol^{-1} and –890.4 kJ mol^{-1} respectively at 298 K. Calculate the standard enthalpy of methane (*g*), *i.e.*, $C(s) + 2\,H_2(g) \rightarrow CH_4(g)$.

11.48 Consider a unimolecular reaction *R* (*g*) → *P*(*g*). $\bar{C}_p\,(R, g) = 7.4 + 2.5 \times 10^{-3}$ T, over the range of 300 to 800 K. Over this range of temperature, the heat of reaction, $\Delta_r \bar{H}^0\,(R,g) = C + 0.25\,T - 2.5 \times 10^{-5}\,T^2$, where C is a constant. Calculate the heat capacity of *P* at constant pressure from the given data.

11.49 Calculate the heat of following reaction at 227 °C from the given data.

$C(s) + H_2O(g) \rightarrow CO(g) + H_2(g)$

$\Delta_f H^o(CO,\,g) = -110.5$ kJ mol^{-1}, $\Delta_f + H^0\,(H_2O,\,g) = -241.8$ kJ mol^{-1},

$$\frac{C_p^0}{R}[H_2O,\,g] = 3.652 + (1.156 \times 10^{-3}\,K^{-1})\,T + (1.42 \times 10^{-7}\,K^{-2})\,T^2$$

$$\frac{C_p^0}{R}[\text{C}, s] = -0.6366 + (7.049 \times 10^{-3}\ \text{K}^{-1})\,T - (5.20 \times 10^{-6}\ \text{K}^{-2})\,T^3 + (1.38 \times 10^{-9}\ \text{K}^{-3})\,T^3$$

$$\frac{C_p^0}{R}[\text{CO}, s] = 3.231 + (8.379 \times 10^{-4}\ \text{K}^{-1})\,T - (9.86 \times 10^{-7}\ \text{K}^{-2})\,T^2$$

$$\frac{C_p^0}{R}[\text{H}_2, g] = 3.496 + (1.006 \times 10^{-4}\ \text{K}^{-1})\,T + (2.42 \times 10^{-7}\ \text{K}^{-2})\,T^2$$

11.50 Similar to unimolecular decomposition reaction of R, consider the dimerisation reaction as: $R \rightarrow R_2$. The heat of this reaction as a function of temperature can be written as: $\Delta_r H = A + B\,T + C\,T^2$, where A, B and C are constants. (*i*) Derive an expression for the heat capacity of this reaction, (*ii*) If C_p of R is independent of temperature and has the value 7.8 J K^{-1} mol^{-1}, express the heat capacity of R_2 as a function of temperature.

11.51 Consider the unimolecular reaction R (*g*) $\rightarrow$ P (*g*) and following data are given for these compounds, *i.e.*,

Quantity	**Compound A**	**Compound B**
C_p (J K^{-1} mol^{-1})	20	35
μ (K bar–1)	1.4	1.9

where μ is the Joule-Thomson coefficient. The reaction is carried at 298 K and 1 bar pressure. When the reaction is carried out at temperature T, the enthalpy change of this reaction at 298 K and at temperature T remains unchanged. Calculate the temperature T.

11.52 Consider the oxidation reaction CO (*g*) + 0.5 O_2 (*g*) $\rightarrow$ CO_2 (*g*) and following data are given for these compounds, *i.e.*,

Quantity	$\Delta_f \bar{H}^0$ **(kJ mol^{-1})**	C_p **(JK^{-1} mol^{-1})**	**μ (K bar^{-1})**
CO (*g*)	–393.5	$6.3423 + 1.8363 \times 10^{-3}\,T$	1.4
O_2 (*g*)		$6.3423 + 1.8363 \times 10^{-3}\,T$	1.9
CO_2 (*g*)	–393.5	$6.369 + 1.01 \times 10^{-2}\,T$	1.10

where μ is the Joule-Thomson coefficient. Assuming that μ and C_p are independent of p, calculate the value ΔH for the given reaction at 298 K with all gases are pressure of 30 bar.

11.53 Xenon reacts with fluorine to form hexafluoroxenon. $\Delta_f \bar{H}^0(\text{XeF}_6, s) = -368.2$ kJ mol^{-1}, $\Delta_f \bar{H}^0(F, g) = 78.99$ kJ mol^{-1}, $\Delta_{\text{sub}} \bar{H}^0(\text{XeF}_6, s) = 62.34$ kJ mol^{-1}. Calculate the bond dissociation of Xe – F bond.

11.54 Show that the work involved in a reversible, adiabatic pressure change of one mole of an ideal gas is given by

$$W = \bar{C}_V T_1 \left[\left(\frac{P_2}{P_1}\right)^{\frac{R}{\bar{C}_P}} - 1\right]$$

where T_1 is the initial temperature and p_1 and p_2 are the initial and final pressures respectively.

CHAPTER 12

SECOND AND THIRD LAWS OF THERMODYNAMICS

12.1. DEFINITIONS OF SECOND LAW OF THERMODYNAMICS

Second law of thermodynamics can be stated in many ways:

(*i*) According to Kelvin, it is impossible to use cyclic process to transfer heat from hot body and convert it to work without transferring a part of heat to the cold body at the same time.

(*ii*) According to Clausius, it is impossible to transfer heat from a cold body to hot body without doing some work.

(*iii*) In terms of entropy, the small change in entropy may be equal to, or greater than DQ/T for any infinitesimal process *i.e.*,

$$dS \geq \frac{DQ}{T}$$

The equality sign is for reversible process and greater than for irreversible process.

12.2. CARNOT CYCLE

Carnot cycle is a hypothetical and ideal heat engine which absorbs heat at higher temperature, does work and returns rest of the heat to other reservoir at lower temperature. It is shown in Fig. 12.1 and consists of the following steps:

(*i*) Isothermal and reversible expansion, *i.e.*,

$$V_1, T_1, p_1 \rightarrow V_2, T_1, p_2 (V_2 > V_1 \text{ and } p_2 < p_1)$$

(*ii*) Adiabatic and reversible expansion, *i.e.*,

$$V_2, T_1, p_2 \rightarrow V_3, T_2, p_3 (V_3 > V_2, T_1 > T_2, p_2 > p_3)$$

(*iii*) Isothermal and reversible compression, *i.e.*,

$$V_3, T_2, p_3 \rightarrow V_4, T_2, p_4 (V_3 > V_4, p_4 > p_3)$$

(*iv*) Adiabatic and reversible compression, *i.e.*,

$$V_4, T_2, p_4 \rightarrow V_1, T_1, p_1 (p_1 > p_4, T_1 > T_2 \text{ and } V_1 > V_4)$$

The values of thermodynamical quantities can be calculated in each step and these are

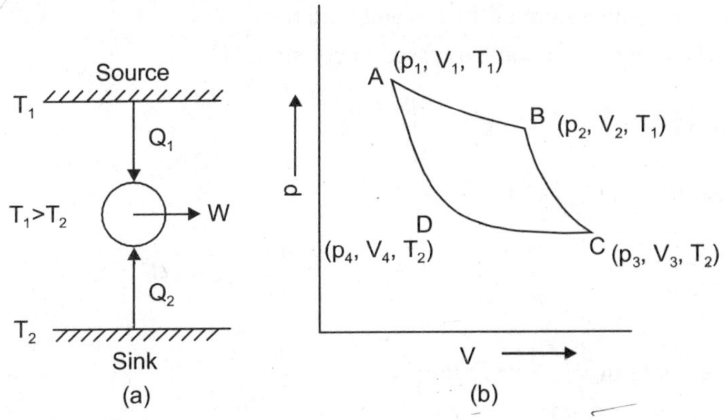

Fig. 12.1 *Carnot cycle*

(*i*) $\Delta E = 0,\ W = Q = nRT_1 \ln \dfrac{V_2}{V_1}$ ...(12.1)

(*ii*) $Q = 0,\ \Delta E = -W = \int_{T_1}^{T_2} C_V\, dT$ ...(12.2)

(*iii*) $\Delta E = 0,\ W = Q = nRT_2 \ln \dfrac{V_4}{V_3}$ ...(12.3)

(*iv*) $Q = 0,\ \Delta E = -W = \int_{T_2}^{T_1} C_V\, dT$ ...(12.4)

Although the Carnot cycle can be applied to any heat engine and containing any material but a fixed amount of ideal gas is considered as working substance in the above engine.

12.3. EFFICIENCY OF HEAT ENGINES

The efficiency of heat engine is defined as the ratio of total amount of work done to the total amount of heat absorbed *i.e.*,

$$\varepsilon = \frac{W}{Q} \quad ...(12.5)$$

In case of Carnot cycle, $Q = Q_1$ and $W = Q_1 + Q_2$

$$\varepsilon = \frac{Q_1 + Q_2}{Q_1} = 1 + \frac{Q_2}{Q_1} \quad ...(12.6)$$

i.e., efficiency can never be greater than unity because the sign of Q_2 *i.e.*, heat given out by the system at cold reservoir is negative. The equation (12.6) can also be written as

$$\varepsilon = 1 - \frac{T_2}{T_1} \quad ...(12.7)$$

where T_1 and T_2 are the temperature of the hot and cold reservoirs.

***Exercise* 12.1:** Derive the relation expressed in equation (12.7).

From equation (12.5) $\varepsilon = \frac{W}{Q_1}$

Now $W = W_{1-2} + W_{2-3} + W_{3-4} + W_{4-1}$

$$= nRT_1 \ln \frac{V_2}{V_1} + \int_{T_1}^{T_2} C_V\, dT + nRT_2 \ln \frac{V_4}{V_3} - \int_{T_2}^{T_1} C_V dT$$

$$= nRT_1 \ln \frac{V_2}{V_1} + nRT_2 \ln \frac{V_4}{V_3} \qquad \text{...}(a)$$

Combining the points *B* and *C* which form two adiabatic ends of Carnot cycle *i.e.*,

$$T_1 V_2^{\gamma-1} = T_2 V_3^{\gamma-1}, \; i.e.,$$

$$\frac{T_1}{T_2} = \left(\frac{V_3}{V_2}\right)^{\gamma-1} \qquad \text{...}(b)$$

Similarly combining the points *A* and *D*, *i.e.*,

$$T_2 V_4^{\gamma-1} = T_1 V_1^{\gamma-1}$$

i.e.,

$$\frac{T_1}{T_2} = \left(\frac{V_4}{V_1}\right)^{\gamma-1} \qquad \text{...}(c)$$

From relations (*b*) and (*c*)

$$\left(\frac{V_3}{V_2}\right)^{\gamma-1} = \left(\frac{V_4}{V_1}\right)^{\gamma-1} \quad \text{or} \quad \frac{V_4}{V_3} = \frac{V_1}{V_2} \qquad \text{...}(d)$$

From (*a*) and (*d*)

$$W = nRT_1 \ln \frac{V_2}{V_1} + nRT_2 \ln \frac{V_1}{V_2}$$

$$W = nR(T_1 - T_2) \ln \frac{V_2}{V_1} \qquad \text{...}(e)$$

Combining the relations (12.5) and (*e*)

$$\varepsilon = \frac{nR(T_1 - T_2) \ln \frac{V_2}{V_1}}{nRT_1 \ln \frac{V_2}{V_1}}$$

$$= \frac{T_1 - T_2}{T_1} = 1 - \frac{T_2}{T_1}$$

***Exercise* 12.2:** An ideal gas is carried through a Carnot cycle. Draw the diagram of the cycle using T and H as variables.

Carnot cycle consists of four steps *i.e.*,

(*a*) Isothermal and reversible expansion *i.e.*,

$$T = \text{Constant}, \ \Delta H = 0$$

(*b*) Reversible adiabatic expansion, *i.e.*,

$$\text{Temperature decreases}, \ \Delta H = \text{Negative}$$

(*c*) Reversible isothermal compression

$$T = \text{Constant}, \ \Delta H = 0$$

(*d*) Reversible adiabatic compression

$$T = \text{positive}, \ \Delta H = \text{positive}$$

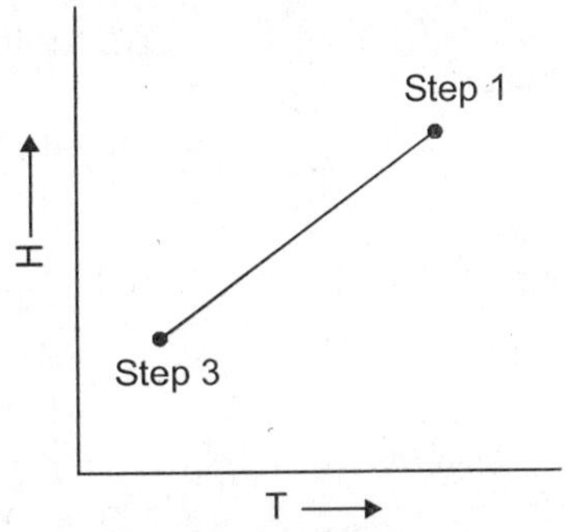

Fig. 12.2

***Exercise* 12.3:** 2 mol of monatomic ideal gas is carried through a Carnot cycle, operating between the temperatures 227°C and 27°C. If $V_1 = 5$ dm^3, $V_2 = 10$ dm^3, calculate the thermodynamic quantities Q, W and ΔE in each step.

Before proceeding further, one should know V_3, p_3, V_4, p_4.

In the step 2, the process is adiabatic, then

$$V_2^{R/C_V} T_1 = V_3^{R/C_V} T_2 = \text{constant} \qquad ...(a)$$

Substituting the values and taking $\bar{C}_V = \frac{3}{2}R$ and $\bar{C}_p = \frac{5}{2}R$ for monatomic gas. $R/C_V = R \times 2/3R = 2/3$, the equation ($a$) reduces,

$$\left(\frac{V_3}{10\ \text{dm}^3}\right)^{2/3} = \frac{(273+227)\ \text{K}}{(273+27)\ \text{K}} = \frac{5}{3}$$

$$V_3 = (10\ \text{dm}^3)(5/3)^{3/2} = 21.516\ \text{dm}^3$$

Step IV is also adiabatic so substituting in equation (a)

$$\left(\frac{V_4}{V_1}\right)^{2/3} = \frac{T_1}{T_2}$$

i.e.,

$$V_4 = (5\ \text{dm}^3)\left(\frac{273+227}{273+27}\right)^{3/2}$$

$$= (5)(5/3)^{3/2} = 10.76\ \text{dm}^3$$

V_4 can also be calculated from the relation $\frac{V_4}{V_3} = \frac{V_1}{V_2}$

so that

$$V_4 = \left(\frac{5}{10}\right)(21.516) = 10.758\ \text{dm}^3$$

Step I

$$\Delta E = 0$$

$$W = Q = nRT \ln \frac{V_{\text{final}}}{V_{\text{initial}}}$$

$$= (2 \text{ mol})(8.314 \text{ J mol}^{-1} \text{ K}^{-1})(500 \text{ K}) \log \frac{10}{5}$$

$$= 5.76 \text{ kJ}$$

Step II

$$Q = 0$$

$$\Delta E = -W = \int_{500}^{300} C_V \, dT = C_V(300 - 500)$$

$$= \frac{3}{2}(8.314 \text{ J mol}^{-1} \text{ K}^{-1})(-200 \text{ K})(2 \text{ mol})$$

$$= -4.99 \text{ kJ}$$

Step III

$$\Delta E = 0$$

$$W = Q = nRT \ln \frac{V_{\text{final}}}{V_{\text{initial}}}$$

$$= (2 \text{ mol})(8.314 \text{ J mol}^{-1} \text{ K}^{-1})(300 \text{ K}) \ln \left(\frac{10.76}{21.516}\right)$$

$$= -3.46 \text{ kJ}$$

Step IV

$$Q = 0$$

$$\Delta E = -W = \int_{300}^{500} C_V \, dT$$

$$= \frac{3}{2}(2 \text{ mol})(8.314 \text{ J mol}^{-1} \text{ K}^{-1})(500 - 300) \text{ K}$$

$$= +4.99 \text{ kJ}$$

$$\text{Efficiency} = \frac{T_1 - T_2}{T_1} = \frac{500 - 300}{500} = 0.4$$

Efficiency can also be calculated as

$$\varepsilon = \frac{Q_1 + Q_2}{Q_1} = \frac{W}{Q}$$

$$= \frac{5.76 - 3.46}{5.76} = 0.399$$

Efficiency calculated in two different ways agrees nicely.

***Exercise* 12.4:** Calculate the efficiency of an engine which is having hot reservoir as the boiling point of water at 1 atm and cold reservoir as the boiling point of liquid nitrogen – 195°C.

Here $T_1 = 100 + 273 = 373\text{ K}$

$T_2 = 273 - 195 = 78\text{ K}$

$$\text{Efficiency} = \frac{(373 - 78)\text{ K}}{373\text{ K}} = 0.79$$

***Exercise* 12.5:** A reversible Carnot cycle does work equivalent to 150 kJ per cycle. If heat supplied by cycle is 225 kJ at 227°C per cycle, calculate, (*a*) the temperature at which heat is rejected, (*b*) the thermal efficiency of the engine.

Using equation (12.5)

$$\varepsilon = \frac{150\text{ kJ}}{225\text{ kJ}} = \frac{2}{3} = 0.667$$

Now
$$\varepsilon = \frac{Q_1 + Q_2}{Q_1} = \frac{T_1 - T_2}{T_1}$$

$$T_1 = 273 + 227 = 500\text{ K},\ \varepsilon = 0.667$$

So,
$$0.667 = \frac{500 - T_2}{500} = 1 - \frac{T_2}{500}$$

or
$$T_2 = 166.5\text{ K}.$$

12.4. HEAT PUMP AND REFRIGERATOR

The difference between heat pump and refrigerator is that in the first case heat is supplied to the body at low temperature by doing work on the system whereas in the second case the heat is removed from a body at low temperature. The coefficient of performance (COP) for each can be defined as

$$\eta = \text{COP (heat pump)} = \frac{Q_1}{-W} = \frac{T_1}{T_1 - T_2} \quad \text{...(12.8)}$$

$$= \frac{Q_1}{-(Q_1 + Q_2)} \quad \text{...(12.9)}$$

$$\eta = \text{COP (refrigerator)} = \frac{Q_2}{-W} = \frac{Q_2}{-(Q_1 + Q_2)} \quad \text{...(12.10)}$$

$$= \frac{T_2}{T_1 - T_2} \quad \text{...(12.11)}$$

***Exercise* 12.6:** A refrigerator working on the reverse Carnot cycle remove 40 kJ min^{-1} of heat from a reservoir at 1°C and rejects heat to reservoir at 41°C. Find (*a*) the coefficient of performance and (*b*) heat rejected to the hot reservoir.

Make use of equation (12.11)

$$\eta = \frac{273 + 1}{(273 + 41) - (273 + 1)} = \frac{274}{40} = 6.85$$

Substitute the value of Q_2 and η in equation (12.10).

$$\eta = \frac{40}{-(Q_1+40)} = 6.85$$

$$Q_1 = -\frac{314}{6.85}\text{ kJ} = -45.84\text{ kJ.}$$

***Exercise* 12.7:** A heat pump is used to draw the water from a well. The temperature of water is 15°C and that of atmosphere 40°C. If the amount of water drawn is 10 kg from the depth of 12 m, calculate the amount of heat supplied to the well.

Coefficient of performance of the heat pump is

$$\eta = \frac{Q_1}{W} = \frac{T_1}{T_1 - T_2}$$

So $$\eta = \frac{(273+40)\text{ K}}{(273+40)-(273+15)\text{ K}} = \frac{313}{25} = 12.52$$

Work done: $$mgh = (10\text{ kg})(9.80\text{ m s}^{-1})(12\text{ m}) = 1176\text{ J} = 1.176\text{ kJ}$$

Heat supplied $$(Q_1) = -\eta \times W = -(12.52)(1.176\text{ kJ}) = -14.72\text{ kJ.}$$

12.5. ENTROPY

Definition: In the Carnot heat engine, efficiency is given as

$$\varepsilon = \frac{Q_1 + Q_2}{Q_1} = \frac{T_1 - T_2}{T_1};\ T_1 > T_2$$

When all the processes are reversible, then

$$1 + \frac{Q_{2(\text{rev})}}{Q_{1(\text{rev})}} = 1 - \frac{T_2}{T_1}$$

or $$\frac{Q_{2(\text{rev})}}{T_1} = -\frac{Q_{2(\text{rev})}}{T_2} \quad ...(12.12)$$

or $$\frac{Q_{1(\text{rev})}}{T_1} + \frac{Q_{2(\text{rev})}}{T_2} = 0$$

$$\sum_i \frac{Q_{i(\text{rev})}}{T_i} = 0 \quad ...(12.13)$$

For any cyclic reversible process ratio of heat absorbed to the temperature of the reservoir is zero and this ratio is called 'Entropy', *i.e.*,

$$dS = \frac{DQ_{rev}}{T} \quad ...(12.14)$$

or

$$\Delta S = S_2 - S_1 = \int_{T_1}^{T_2} \frac{DQ_{rev}}{T} \quad ...(12.15)$$

According to equation (12.13) for a Carnot cycle the net change in the entropy is zero but it can be extended to any other cyclic system. Thus, it is a state function.

Equation (12.14) also requires that heat change in a process must take place reversibly *i.e.*, entropy can only be calculated if the process is reversible. If a process is irreversible between two points, then create a hypothetical reversible process between those points, find the heat change along this process and calculate the entropy change between two points. For reversible process,

$$\Delta S_{system} = -\Delta S_{surr} \quad ...(12.16)$$

$$\Delta S_{universe} = \Delta S_{system} + \Delta S_{surr} \quad ...(12.17)$$

$$= 0$$

For irreversible process $\Delta S_{universe} > 0$ ...(12.18)

The CGS unit of entropy is cal K^{-1} (or gibbs) and denoted as eu. The SI unit is J K^{-1} and denoted as EU (1 eu = 4.184 EU).

12.6. CALCULATION OF ΔS FOR VARIOUS REVERSIBLE PROCESSES

12.6.1. *Entropy Change in Isothermal Transformation*

Under isothermal conditions, *T* is constant and thus can be taken out from integral (12.15). Thus,

$$\Delta S = \frac{1}{T}\int DQ_{rev} = \frac{Q_{rev}}{T} \quad ...(12.19)$$

The phenomena occurring under this heading are mostly the transition processes which take place at constant temperature and constant pressure *e.g.*, sublimation, vaporization, fusion or phase transformation. The above equation can be written as

$$\Delta S = \frac{\Delta H}{T_e} \quad ...(12.20)$$

where T_e is the transition temperature and ΔH is the heat absorbed at T_e.

***Exercise* 12.8:** The entropy of liquid *A* at 0.05 atm and 32.0°C is 0.52 kJ deg^{-1} kg^{-1}. Assume that if vapours of the liquid *A* are ideal, calculate the entropy of the vapours of *A* at 32°C and 0.05 atm. Heat of vapourisation at the above conditions is 2422.12 kJ kg^{-1}.

The above process is a phase transition at constant pressure and at constant temperature,

$$A(l, 0.05 \text{ atm}, 32°C) \rightarrow A(g, 0.05 \text{ atm}, 32°C)$$

$$\Delta S = S_g - S_l = \frac{\Delta H_{vap}}{T}$$

$$= \frac{2422.12 \text{ kJ·kg}^{-1}}{(273+32)\text{K}} = 7.94 \text{ kJ·kg}^{-1}\text{·deg}^{-1}$$

$$S_g = 0.52 \text{ kJ deg}^{-1} \text{ kg}^{-1} + 7.94 \text{ kJ kg}^{-1} \text{ deg}^{-1}$$
$$= 8.46 \text{ kJ kg}^{-1} \text{ deg}^{-1}$$

12.6.2. *Trouton's Rule*

For many liquids or liquids which nearly behave idealy.

$$\Delta_{vap}\bar{S} = 21 \text{ cal mol}^{-1} \text{ deg}^{-1}$$
$$= 87.822 \text{ J mol}^{-1} \text{ deg}^{-1} \quad ...(12.21)$$

From this it follows

$$\frac{\Delta_{vap}\bar{H}}{T_b} = \Delta_{vap}\bar{S} = 87.822 \text{ J mol}^{-1} \text{ deg}^{-1} \quad ...(12.22)$$

where T_b is the boiling point. Thus, either heat of vaporisation or boiling point of a liquid can be calculated approximately. Deviations occur in case of liquids which have strong intramolecular interactions.

***Exercise* 12.9:** The normal boiling points of $CC1_4$, $CHC1_3$ and H_2S are 76.7°C, 61.5°C and –59.6°C respectively. Calculate the molar heat of vapourisation of each liquid if they obey Trouton's rule.

$$\Delta_{vap}\bar{H}(CCl_4) = \Delta_{vap}\bar{S} \times T_b$$
$$= (87.822 \text{ J mol}^{-1} \text{ K}^{-1})(273 + 76.7) \text{ K}$$
$$= 30.71 \text{ kJ mol}^{-1}$$
$$\Delta_{vap}\bar{H}(CHCl_3) = (87.822 \text{ J mol}^{-1} \text{ K}^{-1})(273 + 61.5) \text{ K}$$
$$= 29.376 \text{ kJ mol}^{-1}$$
$$\Delta_{vap}\bar{H}(H_2S) = (87.822 \text{ J mol}^{-1} \text{ K}^{-1})(273 - 59.6) \text{ K}$$
$$= 18.74 \text{ kJ mol}^{-1}$$

The experimental heat of vaporisation of $CC1_4$, $CHCl_3$ and H_2S are 30.18, 29.68 and 18.94 kJ mol^{-1} respectively.

***Exercise* 12.10:** Calculate the vapour pressure of toluene at 77°C if it obeys the Trouton's rule. The boiling point of toluene is 110.6°C.

Since toluene follows Trouton's rule

$$\Delta S = \frac{\Delta_{vap}\bar{H}}{T_b} = 87.822 \text{ J mol}^{-1} \text{ deg}^{-1}$$
$$\Delta_{vap}\bar{H} = (87.822 \text{ J mol}^{-1} \text{ deg}^{-1})(273 + 110.6) \text{ deg}$$
$$= 33688.52 \text{ J mol}^{-1}$$

At boiling point the vapour pressure of toluene is 760 mm. Substituting the proper values in equation, assuming $\Delta_{vap}\bar{H}$ is independent of temperature,

$$\log \frac{p_2}{p_1} = -\frac{\Delta_{vap}\bar{H}}{2.303R}\left[\frac{1}{T_2} - \frac{1}{T_1}\right]$$

$$\log p_2 = \log 760 - \frac{33688.52 \text{ J mol}^{-1}}{(2.303)(8.314 \text{ J K}^{-1} \text{ mol}^{-1})}\left[\frac{1}{350} - \frac{1}{383.6}\right]\text{K}^{-1}$$

$$p_2 = \text{antilog}(2.4405) = 275.7 \text{ mm}$$

12.6.3. *Entropy as a Function of Temperature*

Using equation (12.14), *i.e.*

$$dS = \frac{DQ_{\text{rev}}}{T}$$

at constant volume, $DQ_{\text{rev}} = C_V dT$, and equation (12.14), *i.e.*, entropy at constant volume varies with temperature as

$$dS = C_V dT$$

$$\Delta S = S_2 - S_1 = \int_{T_1}^{T_2} C_V \frac{dT}{T} \quad ...(12.23)$$

Similarly DQ_{rev} at constant pressure, is equal to $C_p dT$. Thus, the change in entropy at constant pressure due to change in temperature will be

$$\Delta S = \int_1^2 dS = \int \frac{DQ_p}{T} = \int_{T_1}^{T_2} \frac{C_p\, dT}{T} \quad ...(12.24)$$

***Exercise* 12.11:** Calculate the increase in the entropy if one mol of krypton is heated from 27°C to 227°C at constant volume and at constant pressure.

As krypton is monatomic gas, the molar heat capacity at constant volume is 3/2R (only the contribution due to translational degrees of freedom) and also independent of temperature, so equation (12.23) becomes

$$\Delta S = n\bar{C}_V \ln \frac{T_2}{T_1}$$

$$= (1 \text{ mol})(3/2 \times 8.314 \text{ J mol}^{-1} \text{ K}^{-1}) \ln \frac{600}{300}$$

$$= 8.64 \text{ J K}^{-1}$$

Similarly ΔS at constant pressure will be

$$S = n\bar{C}p\, ln \frac{T_2}{T_1}$$

$$= (1 \text{ mol}) \left(\frac{5}{2} \times 8.314\, J \text{ mol}^{-1} \text{ K}^{-1}\right) \ln \frac{600}{300}$$

$$= 14.4 \text{ J K}^{-1}.$$

12.6.4. *Entropy Change for Isothermal Expansion*

The general expressions for the entropy change due to expansion at constant temperature are

$$\left(\frac{\partial S}{\partial V}\right)_T = \frac{1}{T}\left[p + \left(\frac{\partial E}{\partial V}\right)_T\right] \quad ...(12.25)$$

$$\Delta S_{\text{sys}} = \int \frac{1}{T}\left[p + \left(\frac{\partial E}{\partial V}\right)_T\right] dV \quad ...(12.26)$$

and $$\Delta S_{\text{sys}} = \int \frac{1}{T}\left[\left(\frac{\partial H}{\partial p}\right)_T - V\right] dp \quad \text{...(12.27)}$$

The above expression can also be expressed as

$$\Delta S_{\text{sys}} = -\int \left(\frac{\partial V}{\partial T}\right)_p dp \quad \text{...(12.28)}$$

For surroundings, it is assumed that the heat change has taken place under isothermal and reversible conditions, hence the entropy change in the surroundings is equal and opposite to the system.

***Exercise* 12.12:** Derive equations (12.25) and (12.27).

Using equation (12.14), *i.e.*,

$$dS = \frac{DQ_{\text{rev}}}{T} \quad \text{...(12.14)}$$

Applying the first law of thermodynamics, *i.e.*, $dE = DQ - pdV$ and substituting the value of DQ in equation (12.14), we get

$$dS = \frac{DQ_{\text{rev}}}{T} = \frac{dE + pdV}{T} \quad \text{...(12.14}a\text{)}$$

Dividing both sides of the equation (12.14 *a*) by dV at constant temperature, we get:

$$\left(\frac{\partial S}{\partial V}\right)_T = \frac{1}{T}\left[\left(\frac{\partial E}{\partial V}\right)_T + p\right] \quad \text{...(12.25)}$$

Similarly variation of entropy with pressure can be derived from equation (12.14) as by substituting E by $H - pV$, we get:

$$dS = \frac{DQ_{\text{rev}}}{T} = \frac{d(H-pV)+pdV}{T} = \frac{dH - pdV - Vdp + pdV}{T} = \frac{dH - Vdp}{T}$$

Dividing both sides of the equation (12.14 *a*) by dp at constant temperature, we get

$$\left(\frac{\partial S}{\partial p}\right)_T = \frac{1}{T}\left[\left(\frac{\partial H}{\partial p}\right)_T - V\right] \quad \text{...(12.27}a\text{)}$$

***Exercise* 12.13:** (*a*) Consider the following Fig. 12.3, where different paths are given. Calculate ΔS for the (*i*) path I, (*ii*) path II, (*iii*) path III, (*iv*) path IV and (*v*) path V. Path I is carried out at constant pressure, path II is carried out reversibly and isothermally, path III is carried out adiabatically and reversibly, paths IV and V are carried out under constant volume. (*b*) Total heat changes and entropy if we follow a cyclic path A → B → C → A and A → B → C → D → A.

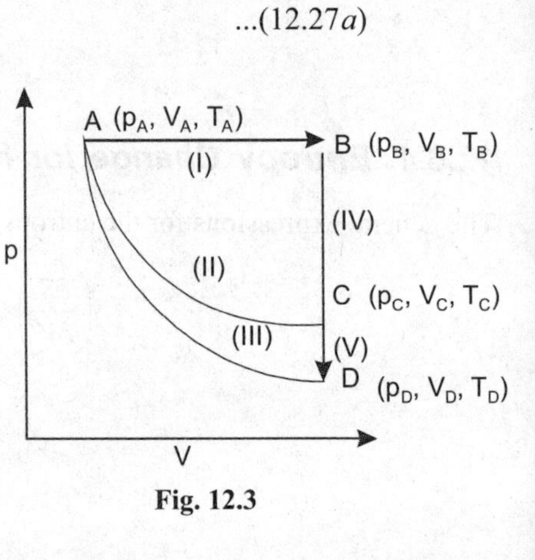

Fig. 12.3

Path (I) is an reversible expansion at constant pressure ($p_{\text{ext}} = p_A = p_B$). So

$$DQ_{\text{rev}}(\text{I}) = dE(\text{I}) - DW_{\text{rev}}(\text{I}) = dE(\text{I}) - p_{\text{ext}}dV(\text{I}) = C_V dT(\text{I}) - p_{\text{ext}}dV(\text{I})$$

So, $$Q_{\text{rev}}(I) = \int_{T_1}^{T_2} C_V(T)dT + p_A(V_B - V_A) \quad ...(1)$$

The entropy change for this process can be evaluated as

$$\Delta S(I) = \frac{Q_{\text{rev}}(I)}{T} = \int_{T_A}^{T_B} \frac{C_V(I)}{T}dT + p\int_{V_A}^{V_B} \frac{dV}{T} \quad ...(2)$$

For ideal gas, $pV = nRT$, replacing T by pV/nR in equation (2), we get

$$\Delta S(I) = \frac{Q_{\text{rev}}(I)}{T} = \int_{T_A}^{T_B} \frac{C_V(I)}{T}dT + p\int_{V_A}^{V_B} \frac{nRdV}{pV} = \int_{T_A}^{T_B} \frac{C_V(I)}{T}dT + nR\int_{V_A}^{V_B} \frac{dV}{V} \quad ...(3)$$

Path (II) is an isothermal reversible expansion, so the expression for ΔS will be:

$$\Delta S(II) = nRT_A \ln \frac{V_C}{V_A} \quad ...(4)$$

Path (III) is an adiabatic reversible expansion, so $DQ_{\text{rev}} = 0$. So the $\Delta S(\text{III}) = 0$. ...(5)

Paths (IV) and (V) are the cooling processes carried out at constant volume. So

$$DW_{\text{rev}}(\text{IV}) + DW_{\text{rev}}(V) = 0$$

$$DQ_{\text{rev}}(IV + V) = dE(IV) + dE(V) = C_V(IV)dT + C_V(V)dT$$

$$Q_{\text{rev}} = \int_{T_B}^{T_C} C_V(IV)dT + \int_{T_C}^{T_D} C_V(V)dT$$

So, $$\Delta S(IV + V) = \frac{Q_{\text{rev}}}{T} = \int_{T_B}^{T_C} \frac{C_V(IV)}{T}dT + \int_{T_C}^{T_D} \frac{C_V(V)}{T}dT \quad ...(6)$$

Since the gas involved is the same in both these processes and entropy is a state function, equation (4) can be written as

$$\Delta S(IV + V) = \int_{T_B}^{T_D} \frac{C_V(IV)}{T}dT \quad ...(7)$$

12.6.5. *For Ideal Gases*

The above equations reduce to

$$\Delta S_{\text{sys}} = \int_{V_1}^{V_2} \frac{p}{T}dV = nR \ln \frac{V_2}{V_1} \qquad \left[\left(\frac{\partial E}{dV}\right)_T = \left(\frac{\partial H}{\partial p}\right)_T = 0\right] \quad ...(12.29)$$

or $$= nR \ln \frac{p_1}{p_2} \quad ...(12.30)$$

and general expression when both expansion and changes in temperature are taking place,

$$\Delta S = n\int_{T_1}^{T_2} \bar{C}_V \frac{dT}{T} + nR \ln \frac{V_2}{V_1} \qquad ...(12.31)$$

$$\Delta S = n\int_{T_1}^{T_2} \bar{C}_p \frac{dT}{T} - nR \ln \frac{p_2}{p_1} \qquad ...(12.32)$$

If $\bar{C}_p$ and $\bar{C}_V$ are assumed to be independent of temperature, the equations (12.31) and (12.32) reduce to

$$\Delta S = n\,\bar{C}_V \ln \frac{T_2}{T_1} + nR \ln \frac{V_2}{V_1} \qquad ...(12.33)$$

$$\Delta S = n\,\bar{C}_p \ln \frac{T_2}{T_1} - nR \ln \frac{p_2}{p_1} \qquad ...(12.34)$$

***Exercise* 12.14:** A 5 dm^3 insulated bulb contains 2 mol of ideal gas at 27°C. This bulb is connected to another insulated bulb of 3 dm^3 capacity by a stop-cock. If the second bulb is evacuated to start with, calculate the entropy change if the stop-cock is opened.

In this case V_1 = 5 dm^3, V_2 = 5 + 3 = 8 dm^3, *i.e.*, gas has expanded from 5 dm^3 to 8 dm^3. Using equation (12.29)

$$\Delta S = (2 \text{ mol})(8.314 \text{ J mol}^{-1} \text{ K}^{-1}) \ln \frac{8}{5}$$

$$= 7.815 \text{ JK}^{-1}.$$

Note: A confusion can arise because the bulbs are insulated so $q = 0 = q_{rev}$ and thus $\Delta S = 0$. But it is not so, as the above process is not reversible. Although $q = 0$, it is only for irreversible process, so $q \neq q_{rev}$ if some reversible path is designed. Here we know the initial and final values of volumes, it does not matter how we have calculated them as it is state function. Yes for *reversible adiabatic* process $\Delta S = 0$ as $q_{rev} = 0$, as mentioned in Section 12.6.6.

***Exercise* 12.15:** 1 mol of CO was in the small cubical container of length $l = a$ at 57°C. The gas was taken out completely from the small container and put into the large container ($l = 2a$) at the same temperature. What is the change in the entropy if it is assumed that process is reversible and gas is behaving as ideal one?

This is a case of isothermal expansion of a CO, so the entropy change is

$$\Delta S = nR \ln \frac{V_2}{V_1}$$

$$V_2 = (2a)^3 = 8a^3$$

$$V_1 = (a)^3 = a^3$$

So $\Delta S = (1 \text{ mol})\,(8.314 \text{ J mol}^{-1} \text{ K}^{-1}) \ln \dfrac{8a^3}{a^3} = 17.29 \text{ J K}^{-1}$.

***Exercise* 12.16:** The molar heat capacity at constant pressure of O_2 is given by the expression

$$\bar{C}_p = 25.723 + 12.979 \times 10^{-3}\, T - 38.618 \times 10^{-7}\, T^2$$

If the temperature of the system containing 2 mol of O_2 is heated from 27°C to 227°C and expanded from 2 dm³ to 8 dm³, calculate the change in the entropy. Assume gas is ideal and the process is carried reversibly.

Substituting the values in equation (12.32)

$$\Delta S = 2 \text{ mol} \int_{T_1}^{T_2} (25.723 + 12.979 \times 10^{-3}\, T - 38.618 \times 10^{-7}\, T^2) \frac{dT}{T}$$

$$+ (2)(8.314 \text{ J mol}^{-1} \text{ K}^{-1}) \ln \frac{8 \text{ dm}^3}{2 \text{ dm}^3}$$

$$\Delta S = (2 \text{ mol})(25.723 \text{ J K}^{-1} \text{ mol}^{-1}) \int_{300}^{500} \frac{dT}{T}$$

$$+ (2 \text{ mol})(12.979 \times 10^{-3} \text{ J K}^{-2} \text{ mol}^{-1}) \int_{300}^{500} dT$$

$$+ (2 \text{ mol})(-38.618 \times 10^{-7} \text{ J K}^{-3} \text{ mol}^{-1}) \int_{300}^{500} T dT$$

$$+ (2 \text{ mol})(8.314 \text{ J mol}^{-1} \text{ K}^{-1}) \ln 4$$

$$= 53.29 \text{ J K}^{-1}.$$

***Exercise* 12.17:** If two heat reservoirs at temperatures T_1 and T_2 $(T_1 > T_2)$ are connected, heat will flow from higher temperature to lower temperature. Let q be the amount of heat transferred from higher to lower temperature. Calculate ΔS of the two reservoirs.

In the above problem, even though the temperature of the reservoirs does not change, the heat transferred q is an irreversible process. We can design the reversible process.

(*i*) Take an ideal gas in a container and place it on the reservoir at T_1. There will be an isothermal expansion and q J are absorbed reversibly.

(*ii*) Remove the container from the heat reservoir and allow the gas to expand adiabatically and reversibly till the temperature is T_2.

(*iii*) Place the container in contact with reservoir at temperature T_2 and carry the isothermal compression in reversible manner such that q amount of heat has transferred to reservoir at temperature T_2. Now entropy change of the gas is

$$\Delta S = \Delta S_1 + \Delta S_2 + \Delta S_3$$

$$= \frac{q}{T_1} + 0 + \frac{-q}{T_2} = \frac{q}{T_1} - \frac{q}{T_2} \qquad ...(12.35a)$$

and entropy change in the reservoirs

$$\Delta S_{res} = -\frac{q}{T_1} + \frac{q}{T_2} \qquad ...(12.35b)$$

***Exercise* 12.18:** Calculate the entropy change in reservoirs when 4000 J of heat is transferred from reservoir at temperature 500 K to 300 K.

$$\Delta S_{res} = -\frac{4000 \text{ J}}{500 \text{ K}} + \frac{4000 \text{ J}}{300 \text{ K}}$$

$$= 13.33 \text{ J K}^{-1} - 8 \text{ J K}^{-1} = 5.33 \text{ J K}^{-1}$$

***Exercise* 12.19:** A piece of alloy weighing 2 kg and at a temperature 500°C is placed in 4 kg of water at 300 K. If the heat capacity of water is 4.184 J K^{-1} g^{-1} and that of alloy 22 J K^{-1} g^{-1}, calculate the entropy change.

The final temperature T can be calculated from the relation,

$$\text{Heat gained} = \text{Heat loss}$$

$$(2 \times 10^3 \text{ g})(22.0 \text{ J K}^{-1}\text{ g}^{-1})(773 - T)\text{ K}$$
$$= (4 \times 10^3 \text{ g})(4.184 \text{ J K}^{-1}\text{ g}^{-1})(T - 300)$$

or $$T = 643 \text{ K}$$

Since heat is not transferred at constant temperature, so using the equation (12.24)

$$\Delta S = \Delta S_1 + \Delta S_2$$
$$= C_p \ln T_2/T_1 + C_p \ln T_2/T_3$$

$$\Delta S = (22 \text{ J K}^{-1}\text{ g}^{-1})(2 \times 10^3 \text{ g}) \ln \frac{643}{773}$$
$$+ (4.184 \text{ J K}^{-1}\text{ g}^{-1})(4 \times 10^3 \text{ g}) \ln \frac{643}{300}$$

$$\Delta S = 4.66 \text{ kJ K}^{-1}.$$

***Exercise* 12.20:** The entropy of liquid water is 0.47 J K^{-1} g^{-1} at 32.5ºC. The vapour pressure is p at this temperature and heat of vaporisation 2422 J g^{-1}. Assuming the vapour to behave as an ideal gas, calculate the entropy of the steam at 200°C and 20 p pressure. The specific heat of steam at pressure p = 1.91 J g^{-1} K^{-1}.

Consider the complete process as sum of

Water	$\xrightarrow{1}$	Water	$\xrightarrow{2}$	Vapour	$\xrightarrow{3}$	Vapour
32.5ºC		vapour 32.5ºC		200ºC		200ºC
		p		p		20 p

$$\Delta S = S(\text{steam at } 200°\text{C}, 20p) - S(\text{liquid}, 32.5, p)$$
$$S(\text{steam}, 200°\text{C}, 20\,p) = \Delta S_1 + \Delta S_2 + \Delta S_3 + 0.47$$

Process (1) is the process at constant temperature and constant pressure, so

$$\Delta S_1 = \frac{2422 \text{ J g}^{-1}}{305.5 \text{ K}} = 7.93 \text{ J g}^{-1}\text{ K}^{-1}$$

ΔS_2 is the change in entropy due to increase of temperature, so

$$\Delta S_2 = (1.91 \text{ J g}^{-1}\text{ K}^{-1}) \ln \frac{473}{305.5} = 0.84 \text{ J g}^{-1}\text{ K}^{-1}$$

ΔS_3 is the change in entropy due to change in pressure

$$\Delta S_3 = -(8.314 \text{ J K}^{-1}\text{ mol}^{-1})\left(\frac{1}{18 \text{ g mol}^{-1}}\right)\ln \frac{20p}{p}$$
$$= -1.38 \text{ J g}^{-1}\text{ K}^{-1}$$

$$S(\text{steam}, 200°\text{C}, 20\,p) = (7.93 + 0.84 - 1.38 + 0.47) = 7.86 \text{ J K}^{-1}\text{ g}^{-1}.$$

12.6.6. *Change in the Entropy for Adiabatic Reversible Process*

Since heat exchange in adiabatic and reversible process is zero *i.e.*, $Q = 0$, so

$$\Delta S_{\text{sys}} = \frac{Q_{\text{rev}}}{T} = \frac{0}{T} = 0 \qquad ...(12.36)$$

Similarly for surroundings (ΔS_{surr}) the entropy change is zero.

12.7. CHANGE IN ENTROPY FOR ISOTHERMAL MIXING

ΔS for mixing of different systems is

$$\Delta S_{\text{sys}} = -R\Sigma n_i \ln x_i \qquad ...(12.37)$$

where n_i is the number of mol, x_i is mole fraction of *i*th component and summation is over the total number of components. For one mol of solution, equation (12.37) becomes,

$$\Delta \bar{S}_{\text{sys}} = -R\Sigma x_i \ln x_i \qquad ...(12.38)$$

Since $x_i < 1$, $\ln x_i$ will be – ve and $\Delta S = +$ ve. This means that entropy always increases on mixing.

***Exercise* 12.21:** Show that entropy change in a binary mixture for isothermal process is maximum when $x_1 = x_2 = \frac{1}{2}$.

Entropy change in binary mixture can be written with the help of equation (12.38).

$$\Delta S_{\text{sys}} = -(R x_1 \ln x_1 + R x_2 \ln x_2)$$

The equation (12.38) can be rewritten as

$$\Delta S_{\text{sys}} = -[R(1-x_2)\ln(1-x_2) + Rx_2 \ln x_2]$$

For maximum condition $\dfrac{d(\Delta S_{\text{mix}})}{dx_2} = 0$

$$\frac{d(\Delta S_{\text{sys}})}{dx_2} = -R\left[(1-x_2)\frac{-1}{1-x_2} - \ln(1-x_2) + \frac{x_2}{x_2} + \ln x_2\right] = 0$$

i.e.,

$$\ln\frac{x_2}{1-x_2} = 0 = \ln 1$$

$$\frac{x_2}{1-x_2} = 1$$

$$x_2 = 1/2$$

Similarly

$$x_1 = 1 - x_2 = 1 - \frac{1}{2} = \frac{1}{2}$$

***Exercise* 12.22:** Calculate the change in entropy if 2 mol of N_2, 3 mol of H_2 and 2 mol of NH_3 are mixed at constant temperature assuming no chemical reaction is occurring.

Mole fraction of each component is

$$x_{N_2} = \frac{2}{2+3+2} = \frac{2}{7}$$

$$x_{H_2} = \frac{3}{2+3+2} = \frac{3}{7}$$

$$x_{NH_3} = \frac{2}{2+3+2} = \frac{2}{7}$$

Substituting the values in equation (12.37)

$$\Delta S = -(8.314 \text{ J mol}^{-1} \text{ K}^{-1}) \left[(2.0 \text{ mol}) \ln \frac{2}{7} + (3.0 \text{ mol}) \ln \frac{3}{7} + 2.0 \text{ mol} \ln \frac{2}{7} \right]$$

$$= 62.80 \text{ J K}^{1}.$$

12.8. STANDARD STATE FOR THE ENTROPY OF IDEAL GAS

$$d\bar{S} = -\frac{R}{p} dp \qquad ...(12.39)$$

is the change in entropy with respect to pressure at constant temperature. It can be integrated to

$$\Delta S = \bar{S} - \bar{S}^0 = -R \ln \frac{p}{1 \text{ atm}} \qquad ...(12.40)$$

$\bar{S}^0$ *i.e.*, molar entropy at 1 atm pressure and at any temperature is the standard state. The equation (12.40) reduces to

$$\bar{S} = \bar{S}^0 - R \ln p \qquad ...(12.40a)$$

where p must be in atmosphere.

***Exercise* 12.23:** (*a*) What is the difference in entropy from standard state at 298 K when the pressure is 600 mm.

The equation to be used is,

$$\Delta\bar{S} = -R \ln p(\text{atm})$$

$$= -(8.314 \text{ J mol}^{-1} \text{ deg}^{-1}) \ln \left(\frac{600 \text{ mm}}{760 \text{ mm}} \right)$$

$$= +1.966 \text{ J mol}^{-1} \text{ deg}^{-1}$$

(*b*) If the standard state entropy of O_2 is 205.03 EU, calculate its entropy under the above conditions.

$$\Delta\bar{S} = \bar{S} - \bar{S}^0 = 1.966 \text{ EU}$$

$$\bar{S} = 205.03 + 1.966 = 206.996 \text{ EU}.$$

12.9. THE THIRD LAW OF THERMODYNAMICS

Similar to second law of thermodynamics, the third law is also defined in number of ways:

(*i*) Nernst heat theorem defines it as

$$\lim_{T \to 0} (-T \Delta S) = 0 \qquad ...(12.41)$$

also $$\lim_{T\to 0}(-\Delta S) = \lim_{T\to 0}\left(\frac{\partial \Delta G}{\partial T}\right)_p = 0 \qquad ...(12.42)$$

(*ii*) Planck stated the third law as "The value of entropy of a pure solid or a pure liquidapproaches zero at 0 K"

$$\lim_{T\to 0} S = 0 \qquad ...(12.43)$$

(*iii*) The most general definition of third law is by Lewis and Randall, who defined as, "If the entropy of each element in some crystalline state be taken as zero at the absolute zero of temperature, then every substance has a finite positive entropy, but at the absolute zero of temperature the entropy may become zero and does so in the case of perfect crystalline substances."

The perfect crystal can be defined as a crystal in which each element or molecule is properly oriented and properly placed. This becomes clear from the example: The solid CO at zero degree absolute will be a perfect crystal if the CO molecules are arranged in the lattice as shown in Fig. 12.4 *a* and the entropy of solid CO will be zero at 0 K but if arranged like Fig. 12.4 *b* and so on, the entropy of this crystal will not be zero.

CO	CO	CO	CO
CO	CO	CO	CO
CO	CO	CO	CO
CO	CO	CO	CO

(a)

CO	CO	OC	CO
OC	CO	CO	OC
OC	CO	CO	CO
CO	OC	CO	OC

(b)

Fig. 12.4 *Arrangements of CO molecules at 0 K.*

The entropy under such circumstances can be calculated theoretically with the help of statistical mechanics. It is given by the expression

$$\bar{S}_0^0 = N_{Av}\, k_b \ln \Omega = R \ln \Omega \qquad ...(12.44)$$

where k_b is the Boltzmann's constant and Ω is the number of complexions in any possible arrangement (see Chapter 9). The value predicted by the equation (12.44) can be greater than zero at 0 K in certain crystal because in some molecules there cannot be complete random orientation due to weak intermolecular forces, size of the molecules, etc.

The deviation in the value of S_0^0, calculated with the help of equation (12.44), can be in case of glassy material. This is because the production of glass depends upon the process employed.

Finally the equation (12.44) predicts the correct value for the crystals in their most stable state at 0 K. S_0^0 will be positive for the metastable crystal even though it is a perfect crystal at 0 K.

12.10. THERMODYNAMIC PROPERTIES AT ABSOLUTE ZERO

With the application of third law, following results can be obtained:

(*i*) $G_0 = H_0 - TS_0 = H_0$ ...(12.45)

(*ii*) $\lim_{T\to 0} \Delta C_p = 0$ ...(12.46)

(*iii*) $\lim_{T\to 0} C_p = 0$ ...(12.47)

(*iv*) $\lim_{T\to 0} C_V = 0$ ...(12.48)

(*v*) $\lim_{T\to 0}\left(\frac{\partial S}{\partial p}\right)_T = \lim_{T\to 0}\left(\frac{\partial S}{\partial V}\right)_T = 0$ ...(12.49)

(*vi*) $\lim_{T\to 0}\left(\frac{\partial p}{\partial T}\right)_V = \lim_{T\to 0}\left(\frac{\partial V}{\partial T}\right)_p = 0$ ...(12.50)

***Exercise* 12.24:** Calculate the entropy of CO crystals at absolute zero.

In the crystal of CO, the random distribution of molecules suggests that either they may be present as shown in Fig. 12.4(*a*) or Fig. 12.4(*b*). That is there are two possibilities. It can be viewed as if the crystal of CO are the mixture of two kinds of molecules. If it is assumed that the mole fraction of each kind is 1/2 the entropy of the crystal of CO can be obtained from equation (12.38).

$$\begin{aligned}\bar{S}_0 &= -R[\frac{1}{2}\ln\frac{1}{2} + \frac{1}{2}\ln\frac{1}{2}]\\ &= (8.314 \text{ J K}^{-1}\text{ mol}^{-1}) \ln 2\\ &= 5.76 \text{ J K}^{-1}\text{ mol}^{-1} = 5.76 \text{ EU}\end{aligned}$$

The actual value is 5.14 EU, so it suggests that mixing of two kinds is not really half-half.

***Exercise* 12.25:** Calculate the value of $\bar{S}_0$ for NO molecule at absolute zero.

Similar to CO, the similarity of nitrogen and oxygen atom suggests that there can be different arrangement of NO like CO molecules in the crystals. But NO is present as a dimer at lower temperature and possible structures can be

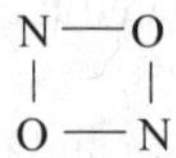

and

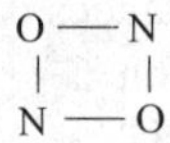

So for each mol of NO, there are only $\frac{1}{2}$ mol of actual molecules and thus the complexions will be half that of CO. Due to this entropy ($\bar{S}_0$) will be half of CO, or

$$\begin{aligned}\bar{S}_0 &= -\frac{R}{2}[\frac{1}{2}\ln\frac{1}{2} + \frac{1}{2}\ln\frac{1}{2}]\\ &= +2.88 \text{ J mol}^{-1}\text{ K}^{-1}\end{aligned}$$

The actual value is 2.76 J mol^{-1} K^{-1}.

12.11. CALCULATION OF ENTROPY AT ANY TEMPERATURE

Since entropy is an additive property so the standard state entropy at any temperature will be due to the contributions of entropies of the phases, transitions of phases which are taking place and the entropy at absolute zero, *i.e.*,

$$S_T^0 = S_0^0 + \sum_i^{\text{phases}} \int_0^T \frac{C_p^0}{T} dT + \sum_j^{\text{transition}} \frac{\Delta H_j}{T_j} \quad ...(12.51a)$$

$$= S_0^0 + \sum_i^{\text{phases}} \int_0^T C_p^0 \, d \ln T + \sum_j^{\text{transition}} \frac{\Delta H_j}{T_j} \quad ...(12.51b)$$

In the equation (12.51*a*), S_0^0 is determined with the help of statistical mechanics, first summation includes the contribution to the entropy due to rise of temperature in all the phases whereas the summation second includes the contributions to the entropy changes due to all the phase transitions which are taking place up to that temperature. The phase transitions can be solid to solid, solid to liquid, liquid to gas etc.

The equation (12.51) can be modified to

$$S_T^0 = S_0^0 + \int_0^{T_D} \frac{C_p^0 \, dT}{T} + \sum_i^{\text{phase}} \int_{T_D}^T \frac{C_p^0 \, dT}{T} + \sum_j^{\text{transition}} \frac{\Delta H_j}{T_j} \quad ...(12.52)$$

where the second term arises due to incapability of determination of heat capacity below temperature T_D and C_p^0 is calculated using Debye's equation, *i.e.*,

$$\int_0^{T_D} \frac{C_p^0 \, dT}{T} = \int_0^{T_D} \frac{CT^3 \, dT}{T} = \frac{1}{3} CT_D^3 = \frac{(C_p)_D}{3} \cdot T_D^3 \quad ...(12.53)$$

where $(C_p)_D$ is the value of C_p at T_D.

In general these integrals are determined graphically but if the analytical expression for C_p^0 is available, it can be used to evaluate the integrals.

***Exercise* 12.26:** Calculate the entropy of methylammonium chloride at 298 K from the following data: $\theta_D = 200.5$ K.

C_p value of β CH_3NH_3Cl can only be obtained experimentally up to 12.04 K.

The above problem can be broken up into number of steps:

(*i*) CH_3NH_3Cl (*s*, 0 K, β form) → CH_3NH_3Cl (*s*, 12.04 K, β form)

$$\Delta \bar{S}_1^0 = \int_0^{12.04} \frac{\bar{C}_p^0 dt}{T} = \int_0^{12.04} \frac{1943.5 \, T^3}{\theta^3} \frac{dT}{T}$$

$$= \frac{1}{3} \left[\frac{(1943.5)(12.04)^3}{(200.5)^3} \right] = 0.14028 \text{ J K}^{-1} \text{ mol}^{-1}$$

(*ii*) CH_3NH_3Cl (*s*, 12.04 K, β form) → CH_3NH_3Cl (*s*, 220.4 K, β form)

$$\Delta \bar{S}_2^0 = \int_{12.04}^{220.4} \frac{\bar{C}_p^0 \, dt}{T} = 93.42 \text{ J K}^{-1} \text{ mol}^{-1}$$

(from numerical integration)

(*iii*) CH_3NH_3Cl (*s*, 220.4 K, β form) → CH_3NH_3Cl (*s*, 220.4 K, γ form)

$$\Delta \bar{S}_3^0 = \frac{\Delta \bar{H}^0}{T} = \frac{1779}{220.4} = 8.07 \text{ J K}^{-1} \text{ mol}^{-1}$$

(*iv*) CH_3NH_3Cl (*s*, 220.4 K, γ form) → CH_3NH_3Cl (*s*, 246.5 K, γ form)

$$\Delta \bar{S}_4^0 = \int_{220.4}^{264.5} \frac{\bar{C}_p^0 \, dT}{T} = 15.44 \text{ J K}^{-1} \text{ mol}^{-1}$$

(from numerical integration)

(*v*) CH_3NH_3Cl (*s*, 264.5 K, γ form) → CH_3NH_3Cl (*s*, 264.5 K, α form)

$$\Delta \bar{S}_5^0 = \frac{\Delta \bar{H}^{0'}}{T'} = \frac{2816}{264.3} = 10.65 \text{ J K}^{-1} \text{ mol}^{-1}$$

(*vi*) CH_3NH_3Cl (*s*, 264.5 K, α form) → CH_3NH_3Cl (*s*, 298.15 K, α form)

$$\Delta \bar{S}_6^0 = \int_{2645.5}^{298.15} \frac{\bar{C}_p^0 \, dT}{T} = 10.69 \text{ J K}^{-1} \text{ mol}^{-1}$$

$$\Delta \bar{S}_{298.15}^0 = \Delta \bar{S}_1^0 + \Delta \bar{S}_2^0 + \Delta \bar{S}_3^0 + \Delta \bar{S}_4^0 + \Delta \bar{S}_5^0 + \Delta \bar{S}_6^0$$

$$= 138.41 \text{ J K}^{-1} \text{ mol}^{-1}.$$

12.12. ΔS_T^0 FOR A CHEMICAL REACTION AND ITS DEPENDENCE ON TEMPERATURE

Let a chemical reaction be

$$aA + bB + ... \rightarrow l\,L + mM + ... \qquad ...(12.54)$$

Since entropy is an additive function, the ΔS_T^0 for the above chemical reaction will be

$$\Delta S_T^0 = l\, S_{T,L}^0 + m S_{T,M}^0 + ... - a S_{T,A}^0 - b S_{T,B}^0 \qquad ...(12.55)$$

$$= \sum_i^{\text{products}} n_i \, S_{T,i}^0 - \sum_j^{\text{reactant}} n_j \, S_{T,j}^0 \qquad ...(12.56)$$

where n_i and n_j are the stoichiometric coefficients of the products and reactants in a balanced chemical equation.

If ΔS_{298}^0 of any chemical reaction is known, ΔS_T^0 at any temperature can be evaluated as:

$$\Delta S_T^0(\text{reaction}) = \Delta S_{298}^0 (\text{reaction}) + \int_{298}^{T} \frac{\Delta C_p^0 \,(\text{reaction})}{T} dT \qquad ...(12.57)$$

***Exercise* 12.27:** Calculate the entropy (ΔS_{298}^0) of the following reaction:

$$2\, SO_2 + O_2 \rightarrow 2\, SO_3$$

if $\bar{S}_{298}^0 = 248.53$ EU for SO_2, $\bar{S}_{298}^0 (O_2) = 205.03$ EU and $\bar{S}_{298}^0 (SO_3) = 256.23$ EU

$$\Delta \bar{S}_{298}^0 = \sum_i^{\text{products}} n_i \, \bar{S}_{i,298}^0 - \sum_j^{\text{reactants}} n_j, \bar{S}_{298,j}^0$$

$$= 2 \times \bar{S}_{298}^0 (SO_3) - 2 \times \bar{S}_{298}^0 (SO_2) - \bar{S}_{298}^0 (O_2)$$

$$= (2 \text{ mol}) (256.2 \text{ J deg}^{-1} \text{ mol}^{-1}) - (2 \text{ mol}) (248.53 \text{ J deg}^{-1} \text{ mol}^{-1})$$

$$- (1 \text{ mol}) (205.03 \text{ J deg}^{-1} \text{ mol}^{-1})$$

$$= - 189.63 \text{ J deg}^{-1}.$$

***Exercise* 12.28:** Calculate ΔS^0 of the reaction mentioned in exercise 12.27 at 504 K. Given data is

$$\bar{C}_p^0(O_2) = 25.72 + 12.98 \times 10^{-3}\, T - 38.62 \times 10^{-7}\, T^2$$

$$\bar{C}_p^0\,(SO_2) = 49.77 + 4.56 \times 10^{-3}\, T$$

$$\bar{C}_p^0\,(SO_3) = 24.43 + 98.48 \times 10^{-3}\, T - 405.30 \times 10^{-7}\, T^2$$

ΔC_p^0 for reaction can be calculated with equation $\Delta C_p^0 = \Sigma_i\, n_i\, C_p^0$ (products) – $\Sigma_j\, n_j\, C_p^0$ (reactants)

$$\begin{aligned}\Delta C_p^0 &= (2\text{ mol})\,(24.43 + 98.48 \times 10^{-3}\, T - 405.30 \times 10^{-7}\, T^2)\\ &\quad - (2\text{ mol})\,(49.77 + 4.56 \times 10^{-3}\, T)\\ &\quad - (1\text{ mol})\,(25.72 + 12.98 \times 10^{-3}\, T - 38.62 \times 10^{-7}\, T^2)\\ &= -76.4 + 174.86 \times 10^{-3}\, T - 771.99 \times 10^{-7}\, T^2\end{aligned}$$

ΔS^0 (500) can be calculated using data of exercise 12.27 and equation (12.57):

$$\begin{aligned}\Delta S^0_{500} &= \Delta S^0_{298} + \int_{298}^{500} (-76.4 + 174.86 \times 10^{-3}\, T - 771.99 \times 10^{-7}\, T^2)\frac{dT}{T}\\ &= -189.62 + (-76.4)\left[\ln T\right]_{298}^{500} + (174.86 \times 10^{-3})\,(500 - 298)\\ &\quad + (-771.99 \times 10^{-7})\left(\frac{1}{2}\right)(500^2 - 298^2)\\ &= 200.06 \text{ EU.}\end{aligned}$$

PROBLEMS

12.1 A reversible engine, operating in a rectangular cycle, whose sides are parallel to the p and V axis. The material used in the engine is monatomic ideal gas. If $p_1 = 3$, p_3 and $V_2 = 5V_1$, where p_1 and p_3 are the higher and lower pressure and V_1 and V_2 the lower and the higher volumes. Calculate the efficiency of the engine.

12.2 Calculate the efficiency of a heat engine if it is working between 100°C and 20°C. How one can increase the efficiency of heat engine?

12.3 Show that it is not possible for two adiabats to cross on a two dimensional p–V diagram using second law of thermodynamics. Is it possible for the isotherm to cross in such a diagram?

12.4 1 mol of an ideal gas is allowed to expand reversibly when in contact with heat bath at temperature 400 K. The initial and final volumes are 4 dm^3 and 12 dm^3. This work done is used to derive a refrigerator working between a reservoir at 250 K and the one at 320 K. What is the maximum amount of heat withdrawn from the 250 K reservoir?

12.5 An ideal gas is carried through a Carnot cycle. Draw the diagram of the cycle using the following coordinates:

(*a*) *p*, *T*, (*b*) *T*, *H*, (*c*) *T*, *S*, (*d*) *E*, *S* and (*e*) *S*, *V*.

12.6 Two mol of diatomic ideal gas is carried through a Carnot cycle, operating between the temperature 400 K and 300 K. If V_1 = 5 dm^3, V_2 = 25 dm^3, calculate the thermodynamic quantities Q, W and ΔE in each step. Calculate, also, the efficiency of the cycle.

12.7 A gas which is carried through a Joule cycle, obeys the following equation of state

$$p(\bar{V} - b) = RT$$

where b is a constant with + ve value. If the Joule cycle consists, (*a*) increase of volume at constant pressure (*b*) adiabatic expansion, (*c*) decrease in volume under constant pressure, and (*d*) adiabatic compression steps. Draw the diagrams for the Joule cycle, using the following coordinates, (*i*) p, V; (*ii*) T, S; (*iii*) E, V.

12.8 Two experiments were performed, containing similar and same amount of a gas. The enclosed gas, initially at the conditions shown, is supplied an amount of energy E, in one case by slowly pushing in piston B and in other by turning the paddle wheel. Determine for each experiment:

(*a*) the work output via piston A;

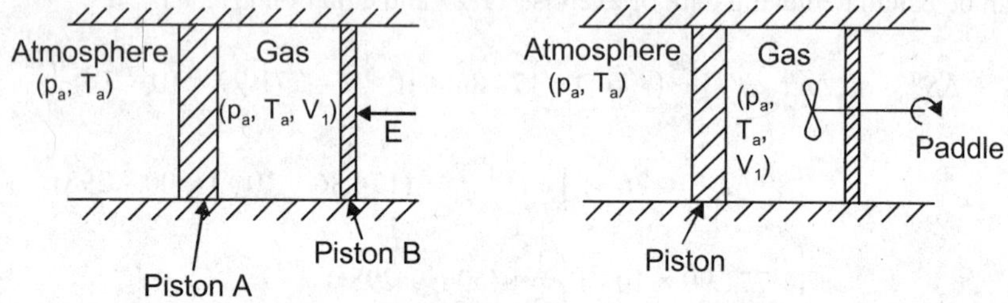

(*b*) the change in the entropy of the enclosed gas. Assume that all the pistons are moveable, massless and frictionless. All the boundaries and pistons are adiabatic.

12.9 The door of the refrigerator was left open. The temperature of the room is 40°C and that of refrigerator is 0°C. Assume that heat capacity of room is 500 kJ deg^{-1}, $\Delta_{fus}\bar{H}$ of water 334.72 J g^{-1} and operation of cycle is ideal. If machine takes 1.5 hr to freeze 1 kg of ice with the above operating temperatures, calculate the change in the temperature of the room.

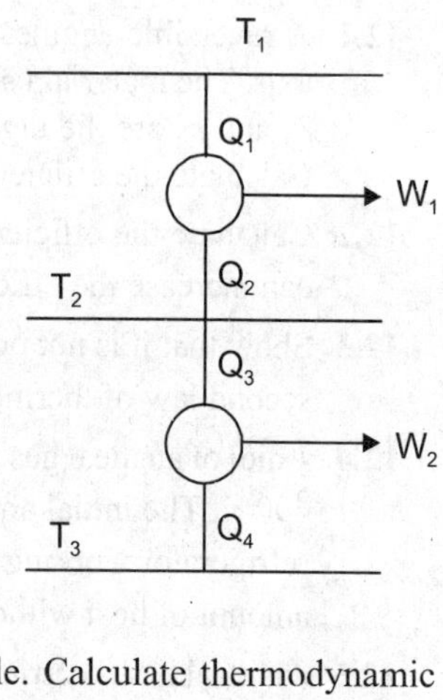

12.10 Two heat engines are working between three temperatures as shown in diagram: (*a*) find the relations between the temperatures, T_1, T_2 and T_3 if (*a*) two engines have the same efficiencies, (*b*) two engines do the same amount of work, (*c*) If T_1 = 1500 K, and T_3 = 300 K calculate T_2 in each case.

12.11 How the efficiency of heat engine can be increased by (*a*) decreasing the sink temperature or (*b*) increasing the temperature of the source.

12.12 1 mol of monatomic ideal gas is carried through a Carnot cycle. Calculate thermodynamic quantities, Q, W and ΔE, in each step. Also calculate the efficiency. Given: V_1 = 8 dm^3: T_1 = 500 K, V_2 = 15 dm^3, V_3 = 25 dm^3.

12.13 A Carnot cycle is operating between the temperatures 100 °C and 27°C. If the heat absorbed from hot bath is 10^{3}J, calculate the heat rejected at the cold bath.

12.14 A reversible heat engine absorbs 4×10^3 kJ of heat at 200°C and rejects the remaining heat to sink at 60°C after doing 1184 kJ of work. Calculate the change in the entropy of (*i*) heat source, (*ii*) heat sink, (*iii*) the universe as a whole.

12.15 Calculate the increase in the entropy of 2 kg of ice when it changes to liquid at 0°C by supplying the heat reversibly to a mixture of ice and liquid at 0 °C. The specific heat of fusion of ice at 0°C is 336.86 J g^{-1}.

12.16 A heat engine absorbs 500 kJ of heat from a bath maintained at 100°C and rejects (*i*) 450 kJ, (*ii*) 433 kJ and (*iii*) 400 kJ heat to sink kept at 40°C. Which of the respective cases represent (*a*) reversible cycle, (*b*) irreversible cycle and (*c*) impossible cycle.

12.17 Show that (*i*) $\left(\frac{\partial S}{\partial V}\right)_T = \frac{1}{T}\left[p + \left(\frac{\partial E}{\partial V}\right)_T\right]$;

(*ii*) $\left(\frac{\partial S}{\partial V}\right)_T = -\left(\frac{\partial p}{\partial T}\right)_V$; and (*iii*) $\left(\frac{\partial S}{\partial V}\right)_T = \frac{\alpha}{\beta}$,

where α is the coefficient of expansion and β is compressibility factor.

12.18 Prove the following relations:

(*a*) $\left(\frac{\partial S}{\partial p}\right)_T = -\alpha V$

(*b*) $E = -T^2\left[\frac{\partial(A/T)}{\partial T}\right]_V = \left[\frac{\partial(A/T)}{\partial(1/T)}\right]_V$

(*c*) $H = -T^2\left[\frac{\partial(G/T)}{\partial T}\right]_p$

(*d*) $\Delta V = -nF\left(\frac{\partial \varepsilon}{\partial p}\right)_T$, where n is the number of mol of electrons, V is the change in the volume of chemical reaction, F is the Faraday constant and $\left(\frac{\partial \varepsilon}{\partial p}\right)_T$ is the variation of voltage with pressure.

(*e*) $dS = \frac{C_p}{T}dT - \left(\frac{\partial V}{\partial T}\right)_p dp$

$= \frac{C_p}{T}dT - V\alpha\, dp$

(*f*) $dS = \frac{\gamma}{\gamma-1}\left(\frac{\partial p}{\partial T}\right)_V dV + \frac{1}{\gamma+1}\left(\frac{\partial V}{\partial T}\right)_p dp$

where $\gamma = \frac{C_p}{C_V}$

12.19 Calculate the change in entropy at 25°C when the copper metal is compressed by applying pressure of 2000 atm. Assume that α and β are constant in this range. Use relation (*i*) of problem 12.18. The density of copper is 8.96 kg dm^{-3} and $\alpha = 0.48$ deg^{-1}.

12.20 Two metal blocks, each of 5 kg, are at 1000 K and at 500 K. They are brought together and thermal equilibrium is reached. Now if a heat engine is formed by choosing hot reservoir as two metal piece together and cold reservoir at 27°C: (*a*) calculate the work obtained by this engine till the temperature of the blocks came down to 350 K; (*b*) calculate the work obtained if the engines were formed with the metal blocks separately but with lower temperature of blocks as 350 K. Specific heat of metal is 25 J g^{-1} deg^{-1}.

12.21 (*a*) 10 mol of water was heated very slowly establishing reversible conditions, from 20°C to 90°C. Calculate ΔS for water, reservoir and the total.

(*b*) If the same amount of water is heated by placing the water in the reservoir at 90°C, calculate ΔS for the water, reservoir and the total.

12.22 A resistor of 25 ohm resistance is kept at a constant temperature of 20°C. Calculate the entropy change in the resistor if an electric current of 15 amp is passed for 5 sec. Find the entropy change in the surroundings.

12.23 A resistor of 25 ohm resistance is thermally insulated. If a current of 15 amp is passed for 5 sec, calculate the entropy change in the surroundings and resistor. Mass of the resistor is 50 g and specific heat is 1.04 J g^{-1} des^{-1}.

12.24 The temperature of 200 g of water at 40°C was increased to 100°C when desired amount of steam at 100°C and 1 atm pressure was condensed into water. If heat of vaporisation of water is 2.26 kJ g^{-1} and heat capacity of water is 4.184 J g^{-1} deg^{-1}, calculate the total entropy change.

12.25 Thermodynamic quantities are generally expressed at constant pressure. Show that entropy of fusion at constant pressure and at constant volume are related by the expression:

$$\Delta S_{fus}(V) = \Delta S_{fus}(p) + \frac{\alpha}{\beta}(V_s - V_l)$$

where α and β are coefficient of expansion and compression, V_s and V_l are the volume of solid and liquid phase.

12.26 Two mol of ideal gas was heated from 25°C to 127°C. If $C_V = 25.2 + 0.02T + 0.00015T^2$, calculate the value of S at 127°C if S at 25°C = 212.6 J deg^{-1} mol^{-1}.

12.27 Calculate the value of ΔS when 2 mol of monatomic ideal gas at 10 atm pressure and 200°C is expanded to 10 times its original volume by (*a*) reversible isothermal and (*b*) reversible adiabatic.

12.28 The standard state entropy of SO_2 at 25°C is 248.53 J deg^{-1} mol^{-1}. Calculate the molar entropy of SO_2 at 5 atm at 100°C if the $\bar{C}_p$ (J deg^{-1} mol^{-1}) varies as

$$\bar{C}_p = 25.72 + 57 \times 10^{-3}T - 380.87 \times 10^{-7}\, T^2$$

12.29 If 5 mol of ideal monatomic gas at temperature 2000°C and at 10 atm pressure expanded adiabatically and irreversibly against pressure of 5 atm, calculate the value of ΔS for gas and universe.

12.30 (*a*) 1 mol of monatomic ideal gas at 100°C and 2 atm pressure is present in 15.3 dm^3 of cylinder. If the gas is compressed adiabatically and irreversibly at a pressure of 4 atm, calculate ΔV, ΔS and ΔT for the process.

(*b*) What is the entropy change of the surroundings?

(*c*) If the above process is carried out isothermally by placing the cylinder in a bath maintained at 100°C, calculate ΔS for the system and surroundings.

12.31 40 g of neon gas is heated from 20°C to 500°C at a constant volume of 20 dm^3. Calculate ΔS if it is assumed that the gas is ideal and heat capacity remains constant in this temperature range.

12.32 Repeat the above problem (12.31) if the process is carried out at constant pressure of 700 mm of Hg.

12.33 Repeat the problems (12.28) if the gas under consideration is N_2.

(*a*) $C_p = 29.04$ J K^{-1} mol^{-1};

(*b*) $C_p = 27.30 + 5.23 \times 10^{-3}\, T - 0.042 \times 10^{-7}\, T^2$.

12.34 200 g of Fe at – 10°C is brought into contact with 400 g of Fe at 420°C. Calculate the entropy change of the process when the equilibrium is reached and also the system is thermally insulated from surroundings. The molar heat capacity of Fe is 24.98 J deg^{-1} mol^{-1}.

12.35 If Trouton's law is followed by (*i*) benzene, (*ii*) *n*-octane, (*iii*) chloroform, calculate the heat of vaporisation of each molecule. Boiling points of benzene, *n*-octane and chloroform are 80.10°C, 125.66°C and 61.2°C respectively.

12.36 The heat of vapourisation of water, ethanol, acetic acid and methane are 40.66 kJ mol^{-1}, 38.58 kJ mol^{-1}, 24.35 kJ mol^{-1} and 8.18 kJ mol^{-1}. If the boiling points of water, ethanol, acetic acid and methane are 100°C, 78.5°C, 118.3°C and –161.49°C respectively. Calculate the entropy of vapourisation. Is the Trouton's rule valid? If not, explain.

12.37 2 mol of ideal gas, initially at 67°C and 4 atm pressure, is expanded irreversibly to 27°C and 1 atm pressure. Calculate the ΔS of this process considering that this process is not carried at constant volume, constant temperature or at constant pressure. $C_v = \frac{3}{2} R$.

12.38 A metal piece of mass 70 kg at 600°C is quenched in 300 kg of oil at 40°C. If the specific heat of oil is 5.02 J deg^{-1} g^{-1} and that of metal 22.59 J g^{-1} K^{-1} and heat loss to surroundings is negligible, calculate the change in the entropy of (*a*) the metal piece, (*b*) the oil and the total.

12.39 A vessel of capacity 2 dm^3 contains O_2 at a pressure of 4 atm and at temperature 27°C. It is stirred with a paddle wheel till the pressure increased to 10 atm. The whole system is adiabatic and paddle wheel was driven by electric motor. If the heat capacity of O_2 is 29.50 J mol^{-1} deg^{-1} and considering that it behaves as an ideal gas calculate the work done and change in the entropy.

12.40 A liquid of mass m_1 and at temperature T_1 is mixed with an equal amount of the same liquid at temperature T_2. Show that entropy change of the universe is

$$2 \text{ mc} \ln \frac{(T_1 + T_2)/4}{\sqrt{T_1 T_2}} \quad \text{or} \quad 2 \text{ mc} \ln \frac{(\sqrt{T_1} + \sqrt{T_2}}{\sqrt{T_1 T_2}}$$

where c is the heat capacity (J deg^{-1} g^{-1}).

12.41 Prove that coefficient of compressibility at constant entropy (at adiabatic process) and coefficient of compressibility at constant temperature are related as

$$\left(\frac{\partial V}{\partial p}\right)_T = \gamma \left(\frac{\partial V}{\partial p}\right)_S$$

where $\gamma = \dfrac{C_p}{C_V}$.

12.42 Calculate the vapour pressure of the molecules mentioned in the problem (12.36) at 400 K from the data given in problem (12.36) assuming that vapours are acting as ideal gas.

12.43 Show that the entropy of solids or liquids can be calculated from the following approximate relation:

$$S = C_p \ln T - V_0 \beta p + V_0 \beta p_0 - C_p \ln T_0 + S_0$$

where β is the coefficient of compressibility, V_0 is the volume at T_0, S_0 is a constant and assume that C_p is independent of temperature.

12.44 Show that

(a) $$\left(\frac{\partial S}{\partial p}\right)_T = \frac{M}{\rho^2}\left(\frac{\partial \rho}{\partial p}\right)_p$$

where ρ is the density of the material.

(b) If the density of any compound between 100ºC to 200°C can be expressed as

$$\rho = 4.186 - 1.624 \times 10^{-3}t - 4.56 \times 10^{-6}t^2$$

where t is in °C, calculate the change in the entropy of 1 mol of the compound per unit pressure at 150ºC.

12.45 Calculate the entropy of metal Ag at 500ºC and 200 atm if the standard state entropy at 298°C is 42.70 J mol^{-1} deg^{-1}; $\bar{C}_p$ = 25.48 J mol^{-1} deg^{-1}; β = 9.9 × 10^{-13} cm^2 dyne^{-1}, d = 10.5 kg dm^{-3}. Assume that $\bar{C}_p$, β and V_0 remain constant under these conditions. (Use equation mentioned in problem 12.43.)

12.46 Calculate the entropy change for the irreversible process at 1 atm pressure

$$A(1, -10.5°C) \rightarrow A(s, -10.5°C)$$

$\bar{C}_p(1)$ = 10.0 J deg^{-1} mol^{-1}, $\bar{C}_p(s)$ = 18.0 J deg^{-1} mol^{-1}, $\Delta\bar{H}_{fus}$ = 40 kJ mol^{-1}. Calculate the entropy change of the surroundings if the above process is taking place in a large thermal bath. The freezing point of the liquid is 5°C and $\Delta\bar{H}_{fus}(-10.5°C)$ = 39.53 kJ.

12.47 Find the relation between the dependence of entropy on the volume for a van der Waals gas at constant temperature.

12.48 The volume occupied by 2 mol of CH_4 at 27°C is 5 dm^3. Calculate the entropy change if the gas is expanded to 9.6 dm^3 isothermally and reversibly. Given: b = 0.0428 dm^3 mol^{-1}.

12.49 Prove the following relation for a van der Waals gas if it is following isoentropic process

$$T(V-b)^{R/C_V} = \text{Constant}$$

12.50 In atomic recombination reactions

$$A + A \rightarrow A_2$$

the diatomic molecules are formed at room temperature and at atmospheric pressure where ΔS = – ve for the process. Is the second law of thermodynamics violated?

Entropy of Mixing

12.51 4 dm^3 of ethane at pressure 2 atm and 27°C and 6 dm^3 of O_2 at pressure 4 atm and 27°C are forced into 4 dm^3 evacuated flask. If the temperature is maintained at 27°C, calculate ΔS for this process assuming that the gases are ideal.

12.52 A 5 dm^3 bulb contains N_2 at 27°C and 2 atm pressure. Another 2 dm^3 bulb contains H_2 at the same temperature but at 5 atm pressure. Two bulbs are connected by a valve and allowed to mix isothermally. Calculate the entropy change assuming that gases behave ideally and equilibrium is reached. $\bar{C}_p(O_2)$ = 29.50 J mol^{-1} deg^{-1}; $\bar{C}_V(O_2)$ = 21.13 J mol^{-1} deg^{-1}; $\bar{C}_p(N_2)$ = 29.04 J mol^{-1} deg^{-1}; $\bar{C}_V(N_2)$ = 20.71 J mol^{-1} deg^{-1}.

12.53 One litre each of two ideal gases at temperature T_1 and pressure p_1 are mixed together. Show that entropy change per mol of mixing is

$$\Delta S = R \ln 2$$

12.54 One litre of gas 1 and 4 litres of gas 2 at constant temperature T and pressure p are forced into 4 litre vessel at the temperature T. Calculate the entropy change.

Third Law

12.55 Calculate the entropy of SO_2, (g) at 400 K from the following data. The contribution to the entropy by

(a) Debye function from 0 to 15 K = 1.255 EU,

(b) From 15 K to 197.64 K, solid = 84.182 EU.

(c) From 197.64 to 263.08 K, liquid = 24.937 EU.

(d) Heat of fusion and heat of vapourisation are 7.402 kJ mol^{-1} and 24.937 kJ mol^{-1} respectively.

(e) The melting point is 197.64 K and boiling point 263.08 K.

(f) The variation of heat capacity of SO_2, (g) is

$\bar{C}_p = 49.769 + 4.556 \times 10^{-3}\, T$

12.56 Show that the entropy of any substance at low temperature where Debye's relation of heat capacity is valid is one-third of the heat capacity.

12.57 Consider the reaction

$$A\,(s) + 7\,H_2O\,(g) \rightarrow A, 7\,H_2O\,(s)$$

The absolute entropy of A, 7 H_2O (s), computed with the help of third law of thermodynamics at 298 K is 526.6 J deg^{-1} mol^{-1}. The other thermodynamic data available for the above reaction is

$$\Delta G^0_{298} = -\,52.52 \text{ kJ mol}^{-1}$$
$$\Delta H^0_{298} = -\,252.64 \text{ kJ mol}^{-1}$$

Further, the entropies of A and water at 298 K are 80 J deg^{-1} mol^{-1} and 188.72 J deg^{-1} mol^{-1}. Is A 7H_2O is a perfect crystal at 0 K?

12.58 The dissociation pressure of Ag_2O (s) at 173°C is 0.12 mm, *i.e.*, $Ag_2O\,(s) \rightleftharpoons 2\,Ag\,(s) + \frac{1}{2}\,O_2$ (g). Also the third law entropies of Ag_2O, Ag and O_2 at 298 K are 121.75 EU, 42.68 EU and 205.0 EU respectively. If ΔH^0_{298} of the above reaction is 30.54 kJ, find out whether third law of thermodynamics is valid or not.

Entropy Changes in Reactions and Variation of Entropy with Temperature

12.59 In the following reactions, calculate the molar entropy of each component at 127°C and the entropy of each reaction at 127°C.

(a) $CH_4\,(g) + 2\,O_2\,(g) \rightarrow CO_2\,(g) + 2\,H_2O\,(g)$

(b) $N_2\,(g) + 3\,H_2\,(g) \rightarrow 2\,NH_3\,(g)$

(c) $H_2\,(g) + Br_2\,(g) \rightarrow 2\,HBr\,(g)$

Given:
$\bar{C}_p\,(H_2) = 29.066 - 0.833 \times 10^{-3}\, T + 20.117 \times 10^{-7}\, T^2$
$\bar{C}_p\,(Br_2) = 35.241 - 4.075 \times 10^{-3}\, T - 14.853 \times 10^{-7}\, T^2$
$\bar{C}_p\,(HBr) = 27.521 + 3.395 \times 10^{-3}\, T + 6.615 \times 10^{-7}\, T^2$
$\bar{C}_p\,(N_2) = 26.983 + 5.91 \times 10^{-3}\, T - 3.377 \times 10^{-7}\, T^2$
$\bar{C}_p\,(NH_3) = 25.895 + 32.999 \times 10^{-3}\, T - 30.46 \times 10^{-7}\, T^2$
$\bar{C}_p\,(O_2) = 25.503 + 13.612 \times 10^{-3}\, T - 42.556 \times 10^{-7}\, T^2$

$\bar{C}_p\,(CH_4) = 14.318 + 74.664 \times 10^{-3}\,T - 174.264 \times 13^{-7}\,T^2$
$\bar{C}_p\,(CO_2) = 26.76 + 42.649 \times 10^{-3}\,T - 147.82 \times 10^{-7}\,T^2$
$\bar{C}_p\,(H_2O) = 30.072 + 9.93 \times 10^{-3}\,T + 8.721 \times 10^{-7}\,T^2$
$\bar{S}^0\,(H_2, g) = 130.58$ EU, $S^0\,(Br_2, g) = 245.35$ EU,
$\bar{S}^0\,(HBr, g) = 198.49$ EU, $\bar{S}^0\,(N_2, g) = 191.49$ EU, $\bar{S}^0\,(NH_3, g) = 192.51$ EU,
$\bar{S}^0\,(O_2, g) = 205.03$ EU, $\bar{S}^0(CO_2, g) = 213.64$ EU,
$\bar{S}^0\,(H_2O, g) = 188.72$ EU, $\bar{S}^0\,(CH_4, g) = 186.19$ EU.

12.60 Calculate the entropy changes of the following reactions at 25°C:

(*a*) $Ca^{2+}\,(aq) + CO_3^{2-}\,(aq) = CaCO_3\,(s)$

(*b*) $H^+\,(aq) + S^{2-}\,(aq) = HS^-(aq)$.

$\bar{S}°\,(CaCO_3, s) = 92.89$ EU, $\bar{S}^0\,(Ca^{2+}\,aq) = -55.23$ EU, $\bar{S}^0(CO_3^-\,aq) = -53.14$ EU, $\bar{S}^0\,(H^+\,aq) = 65.30$ EU, $\bar{S}^0\,(S^{2-}\,aq) = 22.18$ EU, $\bar{S}^0\,(HS^-\,aq) = 61.09$ EU.

12.61 The formation of ozone can be written as

$O_2\,(g) + O\,(g) + O_2\,(g) \rightarrow O_s\,(g) + O_2\,(g)$

Calculate the entropy change for this reaction at 25°C. Given:

$\bar{S}^0\,(O_2) = 205.03$ EU, $\bar{S}^0\,(O_3) = 237.65$ EU, $\bar{S}^0\,(O) = 160.95$ EU.

Additional Problems

12.62 In the Carnot cycle 15 kJ of heat is transferred to the system during the isothermal expansion stroke and produced 4.5 kJ of useful work. If the compression stroke of the engine works at 350 K, calculate the temperature during the isothermal expansion stroke. How much energy if lost to the engine during the compression stroke?

12.63 We have learnt in Section 12.3 that the efficiency of the Carnot heat engine depends on the two temperatures, *i.e.*, heat absorbed at high temperature, T_h and heat lost at lower temperature T_l. (*i*) Derive an expression for the total differential efficiency (ε) which shows the rate at which engine efficiency changes as we vary T_h and T_l, (*ii*) Predict whether ε increases by raising T_h or by lowering T_l? (*iii*) Take the source temperature as 700 K and sink temperature as 300 K. Apply the above expression to suggest how the efficiency of the engine can be increased either by increasing the source temperature of decreasing the sink temperature?

12.64 Liquid helium is produced by pre-cooling it from room temperature to certain lower temperature. Suppose we want to carry out this process. Consider 200 moles of helium in a closed container in equilibrium with the surroundings at 300 K. We would like to cool it to 150 K by connecting the container to Carnot refrigerator. The molar heat capacity of liquid helium is 1.5 R and the total heat capacity at constant volume for the helium container is 840 J K^{-1}. If we assume that these heat capacities do not depend on temperature and high temperature site is maintained at constant temperature of 300 K, calculate the amount of work that must be done to cool the helium to 150 K.

12.65 Two moles of an ideal gas at 300 K and 6 atm pressure is expanded isothermally against a constant pressure of 2.5 atm. till the volume is 30 L. Calculate the values of Q, W, ΔE and ΔS. (*ii*) Suppose the same above process is carried out isothermally and reversibly to the final volume of 30 L, calculate the values of Q, W, ΔE and ΔS. Compare the values of each quantity obtained in two manners.

12.66 On a hot summer day, a person tries to cool himself by opening the door of a refrigerator. Thermodynamically speaking is this a wise action?

12.67 Supercooled liquid water has been cooled below its normal freezing point. This state is thermodynamically unstable and tends to freeze into ice spontaneously. Now take 2 moles of supercooled liquid water turning into ice at –10 °C and 1 atm. pressure. Calculate the values of $\Delta_{sys} S$, $\Delta_{sur} S$ and $\Delta_T S$ for this process. The molar heat capacities of water and ice for this temperature are 75.3 J K^{-1} mol^{-1} and 37.7 J K^{-1} mol^{-1} respectively and molar heat fusion is 6.01 kJ mol^{-1}.

12.68 Suppose an gas follows the equation of state as $p(V - nb) = nRT$, where b is a constant reflecting the size of the molecules. Considering that this gas is allowed to follow the paths (II) and (III) + (V) and E is a function of temperature only (Example 12.13), calculate the values of Q_{rev} and ΔS along these paths. **[Ans.:** ΔS(II) = ΔS(III + V)]

12.69 A gas following the equation of state as $p(V - nb) = nRT$, where b is a constant reflecting the size of the molecules and energy is a function of temperature only. Derive an expression for the change in entropy for the process undergoing expansion against vacuum for a one mole of gas at T and molar volume V_1 to molar V_2.

12.70 In one reversible path the volume is expressed by the relation $V = AT\,e^{0.001T}$, where A is a constant having units L K^{-1}. Calculate the value of ΔS for two moles of monoatomic ideal gas heated from 27 °C to 250 °C.

12.71 For 2 mol of monoatomic ideal gas initial state at 27 °C is defined as $V_1 = 300\,A\,e^{0.3}$ L and $p = \frac{R}{A}e^{-0.3}$ atm. where A is a constant. The ideal gas is first heated to 500 K at constant volume, after which it is expanded isothermally against a constant pressure $p_{ext} = p = \frac{R}{A}e^{-0.4}$ until the internal pressure and the external pressure becomes equal. Calculate the values of Q, W, ΔE and ΔS.

12.72 A quantity of 5.0 moles of an ideal gas is reversibly heated at constant volume from 27 °C to 227 °C. Calculate the entropy change. What would be the value of entropy change if the process is carried out irreversibly?

12.73 Two moles of an ideal gas is first heated at constant pressure from T to $4T$ and second cooled back to T at constant volume. (*i*) Derive expression for ΔS for the overall process. (*ii*) Show that the overall process is equivalent to isothermal expansion of the gas at T from V to $3V$, where V is original volume. (*iii*) Show that the value of ΔS for the process in (*i*) is the same as that in (*ii*).

12.74 Two moles of an ideal gas at 27 °C expands isothermally from 1.5 L to 3.5 L. (*i*) reversibly and (*ii*) against constant pressure of 15 atm. Calculate the values of $\Delta_{sys} S$, $\Delta_{sys} S$ and the entire system.

12.75 50 g of liquid water at 27 °C (called it A) is mixed with 250 g of water (called B) at 250 °C. (*i*) Calculate the final temperature of the system assuming that is mixing is carried out adiabatically. (*ii*) Calculate the entropy change of systems A, B and the entire system.

12.76 Consider a two compartment system in which A and B are large and form a one component system. Each system is at equilibrium but they themselves are not equilibrium with each other. If the two systems are separated from each other by a rigid, heat conducting wall and form the isolated system, show that ΔS of the system will be positive. Given $T_A > T_B$,

12.77 Molar heat capacity of a gas at constant pressure is given as:

$$\bar{C}p = 25.72 + (12.98 \times 10^{-3}\ \text{K}^{-1})T - (38.62 \times 10^{-7}\ \text{K}^{-2})T^2$$

Calculate the value of ΔS when one mole of the gas is heated from 27 °C to 827 °C.

12.78 Consider the following process taking place. One mole of an ideal gas undergoes reversible cooling at constant volume V_1 from p_1, V_1, T_1 to p_2, V_1, T_2. It is followed by a reversible expansion at constant pressure p_2 from p_2, V_1, T_2 to p_2, V_2, T_1. Calculate the value of ΔS for the both processes and total also.

12.79 Starting with the equation $dE = TdS - pdV$ and taking E as a function of $E(V, T)$ and S as a function of $S(T, V)$ show that $\left(\frac{\partial S}{\partial T}\right)_V = \frac{C_V}{T}$ and $\left(\frac{\partial S}{\partial V}\right)_T = \frac{1}{T}\left[p + \left(\frac{\partial E}{\partial V}\right)_T\right]$

12.80 According to Debye theory, the low temperature molar heat capacity of non-metallic solids is given by

$$\bar{C}p(T) = \frac{12\pi^4}{5} R\left(\frac{T}{\Theta_D}\right)^3$$

where Θ_D is the Debye temperature of the solid and has the units of temperature. Calculate the expression for the entropy. The above equation is generally obeyed in the temperature of 0 to 10-20 K.

12.81 The molar heat capacity of C_2H_2 (g) over the range of 300 K < T < 1000 K can be written as

$$\frac{\bar{C}_V}{R} = 16.4105 - \frac{6085.929 \text{ K}}{T} + \frac{822826 \text{ K}^2}{T^2}$$

Calculate ΔS if two moles of ethene is heated over the range of 300 K to 800 K.

12.82 Using the data of Problem 12.81, calculate ΔS if two moles of ethene is heated over the range of 300 K to 800 K at constant pressure. We can assume that ethene gas to behave ideally.

12.83 We know that $\bar{C}_p - \bar{C}_V = R$ for ideal gas. The difference in the results of Problems 12.81 and 12.82 by using this expression is

$$\Delta\bar{S}_p - \Delta\bar{S}_V = R\ln\frac{T_2}{T_1}$$

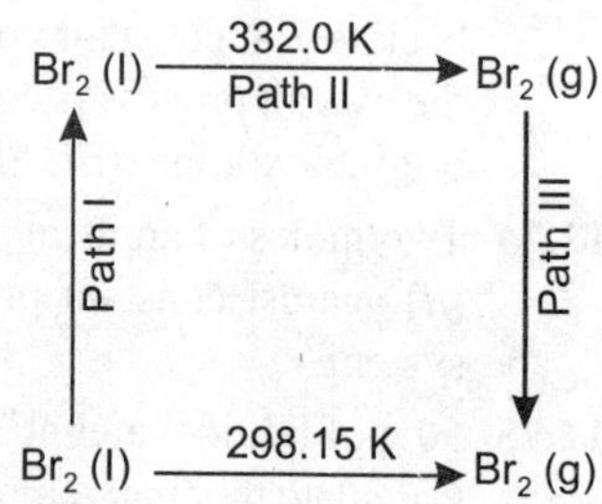

12.84 Calculate ΔS for path followed by bromine molecule as depicted at 298.15 K and same thing by following the paths I, II and III.

Given $\bar{C}_p(l) = 75.69 \text{ J K}^{-1} \text{ mol}^{-1}$, $\bar{C}_p(l) = 75.69 \text{ J K}^{-1} \text{ mol}^{-1}$,

$\Delta_{vap}H(332 \text{ K}) = 29.54 \text{ kJ mol}^{-1}$, Br_2 is a liquid at 298 K

and boiling point is 332 K, $S^0_{298} = 152.2 \text{ J K}^{-1} \text{ mol}^{-1}$.

12.85 Use the following data to calculate the standard molar entropy of $N_2(g)$ at 298 K.

$$C_p^0[N_2(s_1)] = -0.03165 + (0.05460 \text{ K}^{-1})T + (3.520 \times 10^{-3} \text{ K}^{-2})T^2 - (2.064 \times 10^{-5} \text{ K}^{-3})T^3$$

$$(10 \text{ K} \le T \le 35.61 \text{ K})$$

$$C_p^0[N_2(s_2)] = -0.1696 + (0.2379 \text{ K}^{-1})T - (4.214 \times 10^{-3} \text{ K}^{-2})T^2 + (3.036 \times 10^{-5} \text{ K}^{-3})T^3$$

$$(35.61\ \text{K} \le T \le 63.15\ \text{K})$$

$$C_p^0[\text{N}_2(l)] = -18.44 + (1.053\ \text{K}^{-1})T - (0.0148 \times 10^{-3}\ \text{K}^{-2})T^2 + (7.064 \times 10^{-5}\ \text{K}^{-3})T^3$$

$$(63.15\ \text{K} \le T \le 77.36\ \text{K})$$

$C_p^0[N_2(g)]/R = 3.500$ from $77.36\ \text{K} \le T \le 1000\ \text{K})$, $Cp(\text{T} = 10.0\ \text{K}) = 6.151\ \text{J K}^{-1}\ \text{mol}^{-1}$, T_{trs} = 35.61 K, $\Delta_{\text{trs}}\bar{H} = 0.2289\ \text{kJ mol}^{-1}$, T_{fus} = 63.15 K, $\Delta_{\text{fus}}\bar{H} = 0.71\ \text{kJ mol}^{-1}$, T_{vap} = 77.36 K, $\Delta_{\text{vap}}\bar{H} = 5.57\ \text{kJ mol}^{-1}$.

12.86 Using the data of the Problem 12.11 and $C_p^0[N_2(g)]/R = 3.307 + (6.29 \times 10^{-4}\ \text{K}^{-1})T$ for $T \ge 77.36\ \text{K}$. Calculate the standard molar entropy of N_2 at 1000 K.

12.87 Use the following data calculate the standard molar entropy of $Cl_2(g)$ at 298 K.

$$C_p^0[\text{Cl}_2(s)] = -1.545 + (0.1502\ \text{K}^{-1})T - (1.179 \times 10^{-3}\ \text{K}^{-2})T^2 + (3.441 \times 10^{-6}\ \text{K}^{-3})T^3$$

$$(15\ \text{K} \le T \le 172.12\ \text{K})$$

$$C_p^0[\text{Cl}_2(l)] = 7.689 + (5.582 \times 10^{-3}\ \text{K}^{-1})T - (1.954 \times 10^{-5}\ \text{K}^{-2})T^2\ (172.12\ \text{K} \le T \le 239.0\ \text{K})$$

$$C_p^0[\text{Cl}_2(g)]/R = 3.812 + (1.220 \times 10^{-3}\ \text{K}^{-1})T - (4.856 \times 10^{-7}\ \text{K}^{-2})T^2 (239.0\ \text{K} \le T \le 1000\ \text{K})$$

T_{fus} = 172.12 K, $\Delta_{\text{fus}}\bar{H} = 6.406\ \text{kJ mol}^{-1}$, T_{vap} = 239.0 K, $\Delta_{\text{vap}}\bar{H} = 20.40\ \text{kJ mol}^{-1}$.

12.88 Arrange the following molecules in order of increasing standard molar entropy: CH_2Cl_2 (*g*), $CHCl_3$ (*g*), CH_3Cl (*g*).

12.89 Without referring to the Tables for the data, predict whether the entropy change is positive, nearly zero or negative for each of the following reactions:

(*i*) $N_2(g) + O_2(g) \rightarrow 2NO(g)$

(*ii*) $Mg(s) + O_2(g) \rightarrow 2MgO(s)$

(*iii*) $H_2(g) + CO_2(g) \rightarrow H_2O(g) + CO(g)$

(*iv*) $CO_2(g) + 2NH_3(g) \rightarrow (NH_2)_2CO(s) + H_2O(l)$

CHAPTER 13

FREE ENERGY AND CHEMICAL EQUILIBRIUM

13.1. FREE ENERGY FUNCTIONS (DEFINITIONS)

13.1.1. *Helmholtz Free Energy*

It is denoted by A and is also known as 'Total Work Function'. It is defined as

$$A = E - TS \quad ...(13.1)$$

and for small change

$$dA = dE - TdS - SdT \quad ...(13.2)$$

Now $dS = \frac{DQ_{rev}}{T} = \frac{dE + pdV}{T}$ (using first law of thermodynamics)

or

$$dE = TdS - pdV \quad ...(13.3)$$

$$dA = -pdV - SdT \quad ...(13.4)$$

It can also be shown that at constant temperature and constant volume

$$dA = -DW \quad ...(13.5)$$

i.e., change in the Helmholtz free energy defines the total work that can be obtained. The total work includes all kinds of works.

13.1.2. *Gibb's Free Energy*

It is denoted as G and is defined as

$$G = E + pV - TS \quad ...(13.6)$$

$$= H - TS = A + pV \quad ...(13.7)$$

and for small change

$$dG = dE + pdV + Vdp - TdS - SdT \quad ...(13.8)$$

Combing this equations (13.8) with (13.3)

$$dG = Vdp - SdT \quad ...(13.9)$$

Equation (13.9) can be derived as

$$H = E + pV$$
$$dH = dE + pdV + Vdp$$

Combing this equation with (13.3), we get

$$dH = TdS - pdV + pdV + Vdp = TdS + Vdp \quad ...(13.7a)$$

Now $$dG = dH - TdS - SdT$$

or $$= TdS + Vdp - TdS - SdT = Vdp - SdT \quad ...(13.9a)$$

It can also be shown that at constant temperature and pressure *i.e.* from second law of thermodynamics

$$dS \geq \frac{DQ_{rev}}{T} = \frac{dE + DW}{T} = \frac{dE + pdV + DU}{T}$$

where W is the total work, pdV the work of expansion and U the useful work. After rearranging

$$-dE - pdV + TdS \geq DU$$

At constant temperature and pressure,

$$-d[E + pV - TS] = -dG \geq DU \quad ...(13.10)$$

where U is the work apart from the work of expansion that can be obtained by the decrease in free energy. Or it is also known as 'useful work function' *i.e.* useful work that can be obtained at the expense of decrease in the free energy.

***Exercise* 13.1:** Calculate the value of ΔA for the process

$$2\text{He}(g, 27°\text{C}, 1\text{ atm}) \rightarrow 2\text{He}(g, 27°\text{C}, 5\text{ atm})$$

Consider the process to be reversible.

Since, the above process is carried at constant temperature and in a reversible manner.

$$\Delta A = -W_{max} = -\text{(Work of compression)}$$
$$= -nRT \ln \frac{p_1}{p_2}$$
$$= -(2\text{ mol})(8.314\text{ J K}^{-1}\text{ mol}^{-1})(300\text{ K})\left(\ln \frac{1}{5}\right)$$
$$= 8.029\text{ kJ}$$

***Exercise* 13.2:** Calculate the value of ΔA for the following process

$$C_8H_{18}(g) + \frac{25}{2}O_2(g) \rightarrow 8CO_2(g) + 9H_2O(l)$$

given $\Delta H^0 = -5.93 \times 10^3$ kJ. The standard entropies of n C_8H_{18}, O_2, CO_2 and H_2O are 463.67 EU, 205.03 EU, 213.64 EU and 69.94 EU respectively and the reaction is carried out in a bomb calorimeter. The reactants were at 1 atm and 25°C to start with and the reaction is complete.

ΔA can be calculated from the expression

$$\Delta A = \Delta E - T\Delta S$$
$$\Delta E = \Delta H - \Delta(pV) = \Delta H - \Delta nRT$$
$$= -5.93 \times 10^3\text{ kJ} - \left(8 - \frac{25}{2} - 1\right)(8.314\text{ J K}^{-1}\text{ mol}^{-1})(298\text{ K})$$
$$= -5915.67\text{ kJ}$$

ΔS cannot be calculated as simply the difference in entropies of products and reactants because there is a change in the number of mol in forming products, so the pressure of the products will not be the same as that of reactants. Now the final pressure will be

$$p_{\text{fin}} = p_{\text{ini}} \frac{n_f}{n_i} = (1 \text{ atm})\left(\frac{8 \text{ mol}}{13.5 \text{ mol}}\right)$$
$$= 0.593 \text{ atm}$$

The entropies of liquids remain nearly constants unless the pressure is very high. But the entropy of CO_2 at 0.593 atm

$$\overline{S} = \overline{S}^0 + R \ln \frac{p_1}{p_2}$$
$$= 213.64 + (8.314 \text{ J K}^{-1} \text{ mol}^{-1}) \ln \left(\frac{1 \text{ atm}}{0.593 \text{ atm}}\right)$$
$$= 217.99 \text{ EU}$$

$$\Delta S \text{ (of the reaction)} = (8 \text{ mol})(217.99 \text{ EU}) + (9 \text{ mol})(69.94 \text{ EU}) - (463.67 \text{ EU}) - (12.5 \text{ mol})(205.03 \text{ EU})$$
$$= -653.165 \text{ EU}$$
$$\Delta A = \Delta E - T\Delta S$$
$$= -5916.67 - (298 \text{ K})(-653.165 \text{ J deg}^{-1})(10^{-3} \text{ kJ J}^{-1})$$
$$= -5722.03 \text{ kJ}$$

***Exercise* 13.3:** Calculate $\Delta G - \Delta A$ for the combustion reaction of *n*-propane at 300 K.

$$C_3H_8(g) + 5O_2(g) \rightarrow 3CO_2(g) + 4H_2O(l)$$
$$\Delta G - \Delta A = \Delta(pV) \text{ and for ideal gases}$$
$$\Delta G - \Delta A = \Delta nRT$$
$$= (3\text{–}5\text{–}1 \text{ mol})(8.314 \text{ J mol}^{-1} \text{ K}^{-1})(300 \text{ K})$$
$$= -7482.6 \text{ J}$$

***Exercise* 13.4:** In the cell

$$\text{Pt} \mid H_2(1 \text{ atm}) \mid H^+(1 \text{ mol}) \| \; Zn^{++}(1 \text{ mol}) \mid \text{Zn}$$

the cell reaction is

$$Zn^{2+}(aq) + H_2(g) \rightarrow \text{Zn}(s) + 2H^+(aq)$$

and e.m.f. is – 0.763 V at 25°C. Calculate the free energy change of this reaction.

The change in the free energy is related to electrical work as

$$\Delta G^0 = -nFE_0 \qquad \text{...(13.11)}$$

where n is the number of mol of electrons. F the Faraday constant (9.65×10^4 C mol^{-1}) = 9.65×10^4 J mol^{-1} V^{-1} and E_0 is the standard cell potential. Substituting the values in equation (13.11).

$$\Delta G^0 = -(2 \text{ mol})(9.65 \times 10^4 \text{ J mol}^{-1} \text{ V}^{-1})(-0.763 \text{ V})$$
$$= 14.73 \times 10^4 \text{ J}$$

Since ΔG^0 is positive the reaction is not spontaneous.

***Exercise* 13.5:** Calculate the value of ΔG for the process mentioned in exercise 13.1.

Using equation (13.7)

$$\Delta G = \Delta H - T\,\Delta S$$

Since temperature is constant, $\Delta H = 0$ and

$$\Delta S = nR\ln\frac{p_1}{p_2}$$

$$= (8.314\ \text{J mol}^{-1}\ \text{K}^{-1})\ (2\ \text{mol})\ \ln\left(\frac{1}{5}\right)$$

$$= -26.76\ \text{J K}^{-1}$$

$$\Delta G = -(300\ \text{K})\ (-26.76\ \text{J K}^{-1})$$

$$= +8.029\ \text{kJ}$$

Note: The above problem can also be solved with the help of equation (13.7) *e.g.*

$$\Delta G = \Delta A - \Delta(pV)$$

Since temperature is constant and $\Delta n = 0$

$$\Delta G = \Delta A = -W_{max}$$

$$= -nRT\ln\frac{V_2}{V_1} = -nRT\ln\frac{p_1}{p_2}$$

$$= +8.029\ \text{kJ}$$

***Exercise* 13.6:** Calculate the value of ΔG for the process mentioned in exercise (13.2), using the data mentioned there and also using the equation (13.7).

In exercise 13.2, ΔG can be calculated as

$$\Delta G = \Delta H - T\Delta S$$

Substituting the respective values

$$\Delta G = -5930\ \text{kJ} - (298\ \text{K})\ (-653.165\ \text{EU})$$

$$= -5930\ \text{kJ} + 194.64\ \text{kJ}$$

$$= -5735.36\ \text{kJ}$$

For ideal gases

$$\Delta G = \Delta A + \Delta nRT$$

$$= -5722.03 + (-5.5\ \text{mol})\ (8.314\ \text{J mol}^{-1}\text{K}^{-1})\ (298\ \text{K})$$

$$= -5722.03 - 13.63$$

$$= -5735.66\ \text{kJ}$$

In both the cases the answer is same.

***Exercise* 13.7:** For the reaction

$$4\text{C (graphite)} + 5\text{H}_2\ (g) \rightarrow n - \text{C}_4\text{H}_{10}\ (g),\ \Delta \bar{H}^0 = -124.73\ \text{kJ}$$

and

$$\Delta S^0 = -365.8\ \text{EU}$$

For reaction

$$4\text{C (graphite)} + 5\text{H}_2\ (g) \rightarrow \text{iso C}_4\text{H}_{10}\ (g)$$

$$\Delta H^0 = -131.6\ \text{kJ and}\ \Delta S^0 = -381.079\ \text{EU}$$

Calculate ΔG^0 for the reaction

$$n\text{C}_4\ \text{H}_{10}\ (g) \rightarrow \text{iso C}_4\text{H}_{10}\ (g)$$

The free energy of formation of *n*-butane is

$$\Delta G^0_n = \Delta H^0 - T\Delta S^0$$

$$= -124.73 - (298\text{ K})(-365.8\text{ EU}) \times 10^{-3} = -15.72\text{ kJ}$$

ΔG^0 (iso-butane) = – 131.6 – (298 K) (– 381.079 EU) × 10^{-3} = –18.04 kJ

The final reaction can be written as

$$\Delta G^0 = \Delta G_i^0 - \Delta G_n^0$$
$$= -18.04 - (-15.72) = -2.32\text{ kJ}$$

The transformation of isobutane from *n*-butane is spontaneous.

13.2. COUPLED REACTIONS

It can be seen quite often that certain reactions are not spontaneous *i.e.*, $\Delta G > 0$. But if the same reactions are coupled to other reactions, which are spontaneous, the original reactions become feasible. This can only be possible if the total free energy change *i.e.*, sum of the free energies of the second reaction and the original reaction is negative. So in the presence of reactions where ΔG is very negative, the non-spontaneous reactions becomes spontaneous.

***Exercise* 13.8:** Show that the reaction

$$\text{Glucose} + \text{Fructose} \rightarrow \text{Sucrose} + H_2O \quad \Delta G_1^0 = 22.99\text{ kJ mol}^{-1} \quad ...(a)$$

is spontaneous if coupled with the reaction

$$\text{ADP} + HPO_4^{2-} + H^+ \rightarrow \text{ATP} + H_2O \quad \Delta G_2^0 = +29.26\text{ kJ mol}^{-1} \quad ...(b)$$

where ADP and ATP are adinosine diphosphate and adinosine triphosphate respectively.

The mechanism for the formation of sucrose from glucose and fructose in presence of ATP is

$$\text{Glucose} + \text{ATP} \rightarrow \text{ADP} + \text{Gluco 1-phosphate} \quad ...(c)$$

$$\text{Gluco 1-phosphate} + \text{Fructose} \rightarrow \text{Sucrose} + HPO_4^{2-} + H^+ \quad ...(d)$$

Adding (*c*) and (*d*) gives

$$\text{Glucose} + \text{ATP} + \text{Fructose} \rightarrow \text{Sucrose} + HPO_4^{2-} + H^+ + \text{ADP} \quad ...(e)$$

Similarly (*a*) – (*b*) gives

$$\text{Glucose} + \text{Fructose} + \text{ATP} \rightarrow \text{Sucrose} + HPO_4^{2-} + H^+ + \text{ADP} \quad ...(f)$$

Now (*e*) and (*f*) are same, so

$$\Delta G^0 = \Delta G_1^0 - \Delta G_2^0 = (22.99 - 29.26)\text{ kJ mol}^{-1}$$
$$= -6.27\text{ kJ mol}^{-1}$$

Coupling a reaction (*d*) and reverse of (*b*) makes the complete reaction feasible.

13.3. CRITERIA FOR EQUILIBRIUM AND SPONTANEITY

After defining thermodynamic quantities, one can list the conditions under which whether a reaction can procede further or not or the condition for equilibrium. These conditions have been mentioned in Table 13.1.

TABLE 13.1. Criteria for Equilibrium

Constraints	Condition for spontaneity		Equilibrium condition	
None	$-(dE + pdV - TdS) - DU = +ve$		$-(dE + dpV - TdS) - DU = 0$	
	Small change	Finite change	Small change	Finite change
Isolated system	$dS = +ve$	$\Delta S = +ve$	$dS = 0$	$\Delta S = 0$
T constant	$dA + DW = -$	$\Delta A + W = -$	$dA + DW = 0$	$\Delta A + W = 0$
T, p constant	$dG + DU = -$	$\Delta G + U = -$	$dG + DU = 0$	$\Delta G + U = 0$
$U = 0$, T, p constant	$dG = -$	$\Delta G = -$	$dG = 0$	$\Delta G = 0$
$U = 0$, T, V constant	$dA = -$	$\Delta A = -$	$dA = 0$	$\Delta A = 0$

13.4. CALCULATION OF FREE ENERGY

Similar to E and H, it is not possible to calculate the absolute value of Gibb's free energy with the help of thermodynamics, so change in the free energy is calculated

$$\Delta G = \Delta H - \Delta(TS) \qquad ...(13.12)$$

At constant temperature and pressure

$$\Delta G = \Delta H - T\Delta S \qquad ...(13.13a)$$

or

$$dG = dH - TdS \qquad ...(13.13b)$$

The value of free energy can be calculated by three different experimental methods.

(*i*) Calorimetric calculation of ΔH and ΔS for the reactants and products and using equation (13.13*a*).

(*ii*) Knowing the equilibrium constant, one can calculate ΔG.

(*iii*) If one can form e.m.f. cell, the ΔG can be calculated from the e.m.f. of the cell.

The last two methods will be discussed in Section 13.16 and Chapter 16.

***Exercise* 13.9:** Calculate ΔG^0 of the reaction at 25°C

$$H_2S(g) + 3/2\ O_2(g) \rightarrow H_2O(l) + SO_2(g)$$

from the following data.

$$\Delta\bar{H}_f^0(H_2S) = -20.146 \text{ kJ}, \quad \Delta\bar{H}_f^0(H_2O, l) = -285.838 \text{ kJ},$$

$$\Delta\bar{H}_f^0(SO_2, g) = -296.897 \text{ kJ}, \quad \bar{S}^0(O_2) = 205.029 \text{ EU},$$

$$\bar{S}^0(H_2S, g) = 248.53 \text{ EU}, \quad \bar{S}^0(H_2O) = 69.915 \text{ EU},$$

$$\bar{S}^0(SO_2, g) = 248.53 \text{ EU. All these data are at 25° C.}$$

The free energy of the above reaction will be given as

$$\Delta G^0 = \Delta H^0 \text{ (reaction)} - T\Delta S^0 \text{ (reaction)}$$

Now

$$\Delta H^0 = \sum_i^{\text{Products}} n_i \Delta H_f^0 - \sum_j^{\text{Reactants}} n_j \Delta H_f^0$$

$$\Delta H^0 = (-285.838 \text{ kJ mol}^{-1})(1 \text{ mol}) + (-296.897 \text{ kJ mol}^{-1})(1 \text{ mol}) - (-20.146 \text{ kJ mol}^{-1})(1 \text{ mol})$$

$$= -562.589 \text{ kJ}$$

Similarly ΔS^0 (reaction)

$$\Delta S^0 = (1 \text{ mol})(69.915 \text{ EU mol}^{-1}) + (1 \text{ mol})(248.53 \text{ EU mol}^{-1}) - \left(\frac{3}{2}\text{mol}\right)(205.029 \text{ EU mol}^{-1}) - (1 \text{ mol})(205.644 \text{ EU mol}^{-1})$$

$$= -194.743 \text{ EU}$$

Substituting the values in equation (13.13 *a*)

$$\Delta G^0 = (-562.589 \text{ kJ}) - (298 \text{ K})(-194.743 \text{ J K}^{-1})$$

$$= -504.56 \text{ kJ}$$

***Exercise* 13.10:** 5 mol of ideal gas is compressed from 1 atm to 6 atm at 27°C. Calculate the ΔA.

Using equation (13.4) at constant temperature

$$\left(\frac{\partial A}{\partial V}\right)_T = -p$$

At constant temperature

$$\Delta A = -\int p dV = -nRT \int \frac{dV}{V} = -nRT \ln \frac{V_2}{V_1}$$

$$= -nRT \ln \frac{p_1}{p_2}$$

$$\Delta A = -(5 \text{ mol})(8.314 \text{ J K}^{-1} \text{ mol}^{-1})(300 \text{ K}) \ln \frac{1}{6}$$

$$= 22.345 \text{ kJ}$$

***Exercise* 13.11:** 5 mol of O_2 at 1 atm pressure and 25°C is heated to 127°C at constant volume. Calculate the value of ΔA for the above process if $\overline{S}^0 (O_2) = 205.029$ EU and $C_V = 21.13$ J deg^{-1} mol^{-1}.

Assume that C_V for O_2 remains constant in this temperature range.

Equation (13.4) at constant volume can be written as

$$dA = -SdT$$

$$\Delta A = -\int_{298}^{400} SdT \qquad ...(13.14)$$

S is also a function of temperature and at constant volume

$$dS = \int C_V \frac{dT}{T}$$

So

$$S = C_V \ln T + \text{Constant} \qquad ...(13.15)$$

At T = 298 K, $S = S_{298}$, so

$$\text{Constant} = S_{298} - C_V \ln 298$$

Thus

$$S = S_{298} - C_V \ln 298 + C_V \ln T \qquad ...(13.16)$$

Substitute the value of S in equation (13.14)

$$\Delta A = -\int_{298}^{400} S_{298}\, dT + \int_{298}^{400} C_V \ln 298\, dT - \int_{298}^{400} C_V \ln T\, dT$$

$$\Delta A = -S_{298}\,[T]_{298}^{400} + C_V \ln 298\,[T]_{298}^{400} - C_V\,[T \ln T - T]_{298}^{400}$$

$$= -(205.029 \text{ J K}^{-1} \text{ mol}^{-1})(400 - 298) \text{ K} + (21.13 \text{ J K}^{-1} \text{ mol}^{-1})$$
$$(\ln 298)(400 - 298) \text{ K} - (21.13 \text{ J K}^{-1} \text{ mol}^{-1})(400 \ln 400 - 298 \ln 298 - 400 + 298)$$
$$= -209130 + 12278.72 - 12611.482 = -21.25 \text{ kJ}.$$

***Exercise* 13.12:** Consider the reaction in exercise 13.9. Considering work function as the criteria for the feasibility of the reaction (*a*) indicate whether the reaction is spontaneous or not and (*b*) calculate the temperature at which reaction will reverse.

(*a*) From the data of exercise (13.9)

$$\Delta H^0 = -562.589 \text{ kJ, so}$$
$$\Delta E^0 = \Delta H^0 - \Delta nRT$$
$$= -562.589 - (1 - 1 - 3/2 \text{ mol})(8.314 \text{ J K}^{-1} \text{ mol}^{-1})(298 \text{ K})$$
$$= -558.87 \text{ kJ}$$
$$\Delta S^0 = -194.743 \text{ EU}$$

Using equation (13.1)

$$\Delta A^0 = \Delta E^0 - T\Delta S^0$$
$$= -558.87 - (-194.743)(298)(10^{-3})$$
$$= -500.84 \text{ kJ}$$

The reaction is spontaneous

(*b*) The reaction will go in reverse direction at a temperature when ΔA^0 just becomes +ve. The minimum temperature will be thus (assuming ΔE^0 and ΔS^0 remain same)

$$T_{\min} = \frac{\Delta E^0}{\Delta S^0} = \frac{-558870 \text{ J}}{-194.743 \text{ J K}^{-1}} = 2870 \text{ K}$$

13.5. PROPERTIES OF HELMHOLTZ FREE ENERGY

The properties of A can be derived from the equation (13.4) *i.e.*

$$dA = -pdV - SdT \qquad ...(13.4)$$

i.e.

$$\left(\frac{\partial A}{\partial V}\right)_T = -p \qquad ...(13.17)$$

and

$$\left(\frac{\partial A}{\partial T}\right)_V = -S \qquad ...(13.18)$$

Equation (13.18) shows that increase of temperature at constant volume decreases the work function because S is always a positive quantity. Greater the value of entropy, greater will be decrease in A.

Similarly increase of volume will decrease the work function because p is also a positive quantity.

The equation (13.18) can also be written as

$$\left(\frac{\partial A}{\partial T}\right)_V = \frac{A-E}{T} \qquad ...(13.19)$$

and

$$\left(\frac{\partial (A/T)}{\partial T}\right)_V = -\frac{E}{T^2} \qquad ...(13.20)$$

The above equations can also be applied to the chemical reactions and resulting equations are

$$\left(\frac{\partial (\Delta A)}{\partial T}\right)_V = -\Delta S \qquad ...(13.21a)$$

$$\left(\frac{\partial (\Delta A)}{\partial T}\right)_V = \frac{\Delta A - \Delta E}{T} \qquad ...(13.21b)$$

$$\left(\frac{\partial (\Delta A/T)}{\partial T}\right)_V = -\frac{\Delta E}{T^2} \qquad ...(13.21c)$$

The temperature at which reactions become spontaneous from non-spontaneous can be calculated using above equations.

13.6. PROPERTIES OF GIBB'S FREE ENERGY

The properties of Gibb's free energy can be derived from the equation (13.9)

$$dG = Vdp - SdT \qquad ...(13.9)$$

i.e.,

$$\left(\frac{\partial G}{\partial p}\right)_T = V \qquad ...(13.22a)$$

$$\left(\frac{\partial G}{\partial T}\right)_p = -S \qquad ...(13.22b)$$

Similar to work function, increase of temperature at constant pressure will decrease the free energy. The decrease is greater for the gases and less for the solids.

Increase of pressure at constant temperature increases the free energy. The larger the volume of the system, greater will be increase in G.

Similar to the enthalpy and internal energy, the standard state for the free energy functions is the most stable state at 1 atmosphere pressure at any temperature. Generally all the tables list the values of thermodynamic quantities at 298 K. The values at other temperatures can also be calculated.

13.6.1. *Variation of G with Pressure*

Integrating equation (13.22*a*)

$$\int_1^p dG = \int_1^p V dp$$

or

$$\Delta G = G - G^0 = \int_1^p V dp$$

$$G = G^0 + \int_1^p V dp \qquad ...(13.23)$$

where G^0 is the standard free energy at a given temperature.

For solids and liquids, the change of volume is very small with the increase of pressure unless the pressure is very large, the equation (13.23) becomes

$$G = G^0 + V(p - 1) \qquad ...(13.24)$$

Unless the pressure is very large, the second term is negligible *i.e.*,

$$G = G^0 \qquad ...(13.25)$$

For ideal gases

$$G = G^0 + nRT \ln\left(\frac{p \text{ atm}}{1 \text{ atm}}\right) \qquad ...(13.26)$$

Since $\mu = \frac{G}{n}$ *i.e.*, free energy per mol

$$\mu = \mu^0 + RT \ln p \qquad ...(13.27)$$

The p is expressed in atmosphere, μ is also called chemical potential.

***Exercise* 13.13:** Calculate the value of $\mu - \mu^0$ for one mol of ideal gas at 25°C and 400 mm of pressure.

$\mu - \mu^0$ is given by equation (13.27) as

$$\mu - \mu^0 = RT \ln p \text{ atm/1 atm}$$

$$\mu - \mu^0 = (8.314 \text{ J mol}^{-1} \text{ K}^{-1}) (1 \text{ mol}) (298 \text{ K}) \ln \frac{400}{760}$$

$$= -1590 \text{ J.}$$

***Exercise* 13.14:** (*a*) Calculate free energy of an isothermal reversible expansion process if a pressure of one mol of ideal gas drops from 100 atm to 20 atm at a temperature 25°C. (*b*) Calculate the ΔG if the expansion is carried against a constant pressure of 2 atm.

(*a*) Substituting the proper values in

$$\Delta G = nRT \ln \frac{p_2}{p_1}$$

$$\Delta G = (1 \text{ mol})\ (8.314 \text{ J mol}^{-1} \text{ K}^{-1})\ (298 \text{ K}) \ln \frac{20 \text{ atm}}{100 \text{ atm}}$$

$$= -3.988 \text{ kJ}$$

(*b*) The free energy (*G*) is a state function and does not depend on path. The value of ΔG will be same as calculated in part (*a*). It is because the initial and final states are defined.

***Exercise* 13.15:** What will be the free energy change of the following process

$$(1 \text{ mol})\ H_2O\ (l, 100°C) \rightarrow (1 \text{ mol})\ H_2O\ (g, 100°C)$$

at 100°C?

The above process is a phase change process at constant temperature, so entropy change will be

$$\Delta S = \frac{\Delta H_{vap}}{T_{boil}}$$

Substituting in equation (13.7)

$$\Delta G = \Delta H_{vap} - T_b \left(\frac{\Delta H_{vap}}{T_b} \right)$$

$$= 0$$

i.e., $\Delta G = 0$ for any phase transition. This is because phase change is an equilibrium process at the phase transition temperature and pressure.

13.6.2. *Variation of G with Temperature*

The change of free energy with temperature at constant pressure is given by

$$\left(\frac{\partial G}{\partial T} \right)_p = -S \qquad ...(13.22b)$$

The above equation can also be expressed as

$$\left(\frac{\partial G}{\partial T} \right)_p = \frac{G - H}{T} \qquad ...(13.28)$$

$$\left(\frac{\partial (G/T)}{\partial T} \right)_p = -\frac{H}{T^2} \qquad ...(13.29)$$

$$\frac{\partial (G/T)}{\partial (1/T)_p} = H \qquad ...(13.30)$$

Equation (13.29) is used more frequently and is known as Gibbs-Helmholtz equation. It can be expressed as

$$\frac{G(T_2)}{T_2} = \frac{G(T_1)}{T_1} - \int_{T_1}^{T_2} \frac{H}{T^2} dT \qquad ...(13.31)$$

or change in the free energy of reaction with temperature can be written as

$$\frac{\Delta G(T_2)}{T_2} - \frac{\Delta G(T_1)}{T_1} = -\int_{T_1}^{T_2} \frac{\Delta H}{T^2} dT \qquad \text{...(13.32)}$$

If ΔH is constant within the temperature range T_1 and T_2 the above integral can be integrated. Otherwise temperature dependence of ΔH should be substituted and then either analytical or graphical evaluation of integral can be made.

Generally the standard state values of thermodynamic functions are given at 25°C (298 K), so to calculate the values at any other temperature following procedure is employed,*e.g.* ΔC_p, can be expressed as

$$\Delta C_p = \Delta a + \Delta bT + \Delta cT^2 + \Delta dT^3 \qquad \text{...(13.33}a\text{)}$$

or

$$\Delta C_p = \Delta a' + \Delta b'T + \Delta c'T^{-2} \qquad \text{...(13.33}b\text{)}$$

Now

$$d\Delta H = \int_{298}^{T} \Delta C_p dT \qquad \text{...(13.34)}$$

or

$$\Delta H(T) = \Delta H^0_{298} + \Delta a(T - 298) + \frac{\Delta b}{2}(T^2 - 298^2) + \frac{\Delta c}{3}(T^3 - 298^3) + \frac{\Delta d}{4}(T^4 - 298^4) \qquad \text{...(13.35)}$$

or

$$\Delta H(T) = \Delta H^0_{298} + \Delta a'(T - 298) + \frac{\Delta b'}{2}(T^2 - 298^2) - \Delta c'\left(\frac{1}{T} - \frac{1}{298}\right) \qquad \text{...(13.36)}$$

It can also be written as

$$\Delta H_T = J + (\Delta a)T + \left(\frac{\Delta b}{2}\right)T^2 + \tfrac{1}{3}(\Delta c)\, T^3 + \frac{1}{4}(\Delta d)\, T^4 \qquad \text{...(13.37)}$$

or

$$\Delta H_T = J' + (\Delta a')\, T + \frac{1}{2}(\Delta b')T^2 - \frac{1}{T}(\Delta c') \qquad \text{...(13.38)}$$

$$\text{where } J = \Delta H^0_{298} - (\Delta a)(298) - \frac{1}{2}(\Delta b)(298^2) - \frac{1}{3}(\Delta c)(298^3) + \frac{1}{4}(\Delta d)(298^4) \qquad \text{...(13.39)}$$

$$J' = \Delta H^0_{298} - (\Delta a')(298) - \frac{1}{2}(\Delta b')(298^2) + (\Delta c')\frac{1}{298} \qquad \text{...(13.40)}$$

Now substituting this expression in (13.32) and integrating

$$\Delta G_T = J + KT - \Delta a\, T \ln T - \frac{\Delta b}{2} T^2 - \frac{\Delta c}{6} T^3 - \frac{\Delta d}{12} T^4 \qquad \text{...(13.41)}$$

$$\Delta G_T = J' + K'T - \Delta a'\, T \ln T - \frac{\Delta b'}{2} T^2 - \frac{\Delta c'}{2} T^{-1} \qquad \text{...(13.42)}$$

where

$$K = \frac{\Delta G_{298} - J'}{298} + \Delta a \ln 298 + \frac{\Delta b}{2}(298) + \frac{\Delta c}{6}(298^2) + \frac{\Delta d}{12}(298^3) \quad ...(13.43)$$

$$K' = \frac{\Delta G_{299} - J}{298} + \Delta a' \ln 298 + \frac{\Delta b'}{2}(298) + \frac{\Delta c'}{2}(298^{-2}) \quad ...(13.44)$$

Generally the free energy functions are tabulated as $(G_T^0 - E_0^0)/T$ or $(G_T^0 - H_0^0)/T$ because at absolute zero $H_0^0 = E_0^0$ and thus ΔG_T^0 can be determined as

$$\frac{\Delta(G_T^0 - H_0^0)}{T} = \sum_i^{\text{products}} n_i \left(\frac{G_T^0 - H_0^0}{T}\right)_i - \sum_j^{\text{reactants}} n_j \left(\frac{G_T^0 - H_0^0}{T}\right)$$

and temperature dependence of ΔG^0 (reaction) by

$$\Delta G^0 = T\left(\frac{\Delta(G_T^0 - H_0^0)}{T}\right) + \Delta H_0^0(\text{reaction})$$

Free energy functions have been discussed in Chapter 9.

***Exercise* 13.16:** Calculate ΔG_{500}^0 for the reaction

$$H_2S(g) + \frac{3}{2}O_2(g) \rightarrow H_2O\,(g) + SO_2\,(g)$$

if

$$C_p\,(H_2S) = 26.715 + 23.866 \times 10^{-3}\,T - 50.626 \times 10^{-7}\,T^2$$
$$C_p\,(O_2) = 25.503 + 13.612 \times 10^{-3}\,T - 42.556 \times 10^{-7}\,T^2$$
$$C_p\,(H_2O) = 30.204 + 9.933 \times 10^{-3}\,T + 11.171 \times 10^{-7}\,T^2$$
$$C_p\,(SO_2) = 25.719 + 57.923 \times 10^{-3}\,T - 380.87 \times 10^{-7}\,T^2$$

All the values are expressed as J mol^{-1} K^{-1} and

$$\Delta H_{298}^0 = -518.619\text{ kJ},\ \Delta G_{298}^0 = -495.946\text{ kJ}$$

From the above data, heat capacity of the reaction can be calculated from (13.33)

$$\Delta C_p^0 = (30.204 + 9.933 \times 10^{-3}\,T + 11.171 \times 10^{-7}\,T^2) + (25.719 + 57.923 \times 10^{-3}\,T - 380.87 \times 10^{-7}\,T^2)$$
$$- \frac{3}{2}(25.503 + 13.612 \times 10^{-3}\,T - 42.556 \times 10^{-7}\,T^2) - (26.715 + 23.866 \times 10^{-3}\,T - 50.626 \times 10^{-7}\,T^2)$$
$$= -9.0465 + 23.572 \times 10^{-3}\,T - 255.239 \times 10^{-7}\,T^2$$

i.e.

$$\Delta a = -9.0465\text{ J K}^{-1}\text{ mol}^{-1},\ \Delta b = 23.572 \times 10^{-3}\text{ J K}^{-2}\text{ mol}^{-1}$$
$$\Delta c = -255.239 \times 10^{-7}\text{ J K}^{-3}\text{ mol}^{-1}$$

Substitute the respective values in (13.39) to calculate J

$$J = -518.619 - (-9.0465 \times 10^{-3}\text{ kJ K}^{-1}\text{ mol}^{-1})(298\text{ K}) - \frac{1}{2}(23.572 \times 10^{-6}\text{ kJ K}^{-2}\text{ mol}^{-1})(298\text{ K})^2 -$$
$$\frac{1}{3}(-255.239 \times 10^{-10}\text{ kJ mol}^{-1}\text{ K}^{-3})(298\text{ K})^3$$
$$= -520.043\text{ kJ}$$

K can be calculated using relation (13.43)

$$K = \frac{-495.946 - (-518.619)}{298} + (-9.0465)\ \ln 298 + \frac{1}{2}(23.572 \times 10^{-3})$$

$$(298) + \frac{1}{6}(-255.239 \times 10^{-7})\ (298)^2$$

$= 27.679\ \text{J K}^{-1} = 27.679 \times 10^{-3}\ \text{kJ K}^{-1}$

Substitute the values of $J, K, \Delta a, \Delta b, \Delta c$ and $T = 500$ K in (13.41)

$$\Delta G_{500} = -518.69\ \text{kJ} + (27.679 \times 10^{-3}\ \text{kJ K}^{-1})\ (500\ \text{K}) - (-9.0465 \times 10^{-3}\ \text{kJ K}^{-1})\ (500\ \text{K})\ \ln 500$$

$$-\frac{1}{2}(23.572 \times 10^{-6}\ \text{kJ K}^{-2})\ (500\ \text{K})^2 - \frac{1}{6}(255.239 \times 10^{-10}\ \text{kJ K}^{-3})\ (500\ \text{K})^3$$

$$= -479.16\ \text{kJ}$$

***Exercise* 13.17:** Consider the following reaction.

$$CO(g) + H_2O(g) \rightarrow CO_2(g) + H_2(g)$$

From the following data, indicate whether the reaction is feasible or not at 298 K. Calculate the temperature at which the reaction will go in the reverse direction.

$\Delta H_f^0(CO, g) = -110.52\ \text{kJ mol}^{-1}$, $\Delta H_f^0(H_2O, g) = -241.83\ \text{kJ mol}^{-1}$,

$\Delta H_f^0(CO_2, g) = -393.54\ \text{kJ mol}^{-1}$, $\overline{S^0}(CO) = 197.91$ EU,

$S^0(H_2O) = 188.72$ EU, $\bar{S}^0(CO_2) = 213.64$ EU, $\overline{S^0}(H_2) = 130.59$ EU

The reaction will be feasable if ΔG will be negative. Using equation (13.7 *b*)

$$\Delta G^0 = \Delta H^0 - 298\Delta S^0$$

$$\Delta H^0 = -393.54 - (-110.52) - (-241.83) = -41.19\ \text{kJ}$$

$$\Delta S^0 = 213.64 + 130.59 - 197.91 - 188.72 = -42.4\ \text{J mol}^{-1}\ \text{deg}^{-1}$$

$$\Delta G^0 = -41.19\ \text{kJ} - (298\ \text{K})(-42.4 \times 10^{-2}\ \text{kJ deg}^{-1})$$

$$= -41.19 + 12.64 = -28.55\ \text{kJ}$$

The free energy change of the reaction is – 28.55 kJ, so the reaction will be feasable.

The reaction will go in the opposite direction as soon as the free energy of the reaction will be positive or starts changing the sign. So the minimum temperature at which ΔG^0 becomes zero can be calculated as (assume that ΔH^0 and ΔS^0 do not change appreciably in this temperature range)

$$0 = \Delta H^0 - T_{\min}\ \Delta S^0$$

$$T_{\min} = \frac{\Delta H^0}{\Delta S^0} = \frac{-41.19 \times 10^3\ \text{J}}{-42.4\ \text{J K}^{-1}} = 971.5\ \text{K}$$

i.e. at temperature greater than 971.5 K, the reaction will go in opposite direction.

***Exercise* 13.18:** One mol of CO_2 at 1 atm pressure and 25°C was heated to 127°C at constant pressure. Calculate ΔG for the above process if $\overline{S^0}(CO_2) = 213.64$ EU, $C_p(CO_2) = 37.49\ \text{J mol}^{-1}\ \text{deg}^{-1}$. Assume C_p remains constant over the above range.

Using equation (13.22) at constant pressure

$$dG = -S\,dT$$

So
$$\Delta G = -\int_{298}^{400} S\,dT$$

Further entropy is also a function of temperature and at constant pressure

$$dS = \frac{C_p dT}{T}$$

$$S = C_p \ln T + \text{constant}$$

at 298 K, $S = 213.64$ J mol^{-1} deg^{-1}, So

Constant = (213.64 – 37.49 ln 298) EU = 0.056 EU

$$S = C_p \ln T + 0.056$$

Now

$$\Delta G = -\int_{298}^{400} (0.056)\,dT - \int_{298}^{400} C_p \ln T\,dT$$

$$= -(0.056)(400 - 298) - \left[C_p T \ln T - C_p T\right]_{298}^{400}$$

$$= -(0.056)\,(102\text{ K}) - (37.49\text{ J K}^{-1})\,(400 \ln 400 - 298 \ln 298 - 400 + 298)\,(\text{K}) = -22.38\text{ kJ}$$

13.7. SOME MATHEMATICAL RESULTS

13.7.1. *Cross-Derivatives*

Consider a function of two variables *i.e.*, $f(x, y)$. If the first derivative with respect to each variable $\left(\frac{\partial f(x,y)}{\partial x}, \frac{\partial f(x,y)}{\partial y}\right)$ and the function itself $f(x, y)$ are exact differentials, then cross-derivatives of the first derivatives $\left(i.e., \frac{\partial^2 f(x,y)}{\partial y\,\partial x}, \frac{\partial^2 f(x,y)}{\partial x\,\partial y}\right)$ are equal.

$$df = \left(\frac{\partial f}{\partial x}\right)_y dx + \left(\frac{\partial f}{\partial y}\right)_x dy \qquad ...(13.45)$$

Since $\left(\frac{\partial f}{\partial x}\right)_y$ and $\left(\frac{\partial f}{\partial y}\right)_x$ are also functions of x, y, and also exact differentials then

$$\frac{\partial^2 f}{\partial y \partial x} = \frac{\partial^2 f}{\partial x \partial y} \qquad ...(13.46)$$

Sometimes equation (13.45) can be written as

$$df = M(x, y)dx + N(x, y)dy \qquad ...(13.47)$$

and then with the help of cross-derivatives

$$\left(\frac{\partial M}{\partial y}\right)_x = \left(\frac{\partial N}{\partial x}\right)_y \qquad ...(13.48)$$

where $$M = \left(\frac{\partial f}{\partial x}\right)_y \text{ and } N = \left(\frac{\partial f}{\partial y}\right)_x$$

***Exercise* 13.19:** Apply the cross-derivative rule to the following fundamental equation

$$dH = TdS + Vdp \qquad ...(13.49)$$

In this case H is a function of four variables T, S, V and p and further comparing the equation (13.49) with equation (13.47)

$$T = M = \left(\frac{\partial f}{\partial x}\right)_{y'}, \ dS = dx$$

$$V = N = \left(\frac{\partial f}{\partial y}\right)_{x'}, \ dp = dy$$

So with the application of cross-derivatives

$$\left(\frac{\partial T}{\partial p}\right)_S = \left(\frac{\partial V}{\partial S}\right)_p$$

***Exercise* 13.20:** The differential of the pressure of a certain gas is given by one of the following relations:

$$dp = \frac{2RT}{(V-b)^2}dV + \frac{R}{V-b}dT \qquad ...(13.50a)$$

$$dp = -\frac{RT}{(V-b)^2}dV + \frac{R}{V-b}dT \qquad ...(13.50b)$$

Identify the exact equation.

Since the function as well as the first derivatives are exact differentials, the relation (13.48) must be satisfied.

In equation (13.50a)

$$M = \frac{2RT}{(V-b)^2}, \text{ So } \left(\frac{\partial M}{\partial T}\right)_V = \frac{2R}{(V-b)^2}$$

$$N = \frac{R}{V-b}, \text{ So } \left(\frac{\partial N}{\partial V}\right)_T = -\frac{R}{(V-b)^2}$$

If the equation (13.50a) is a correct one then

$$\left(\frac{\partial M}{\partial T}\right)_V = \left(\frac{\partial N}{\partial V}\right)_T$$

But it is not, so first equation is not the correct one. In second equation

$$M = -\frac{RT}{(V-b)^2}, \text{ So } \left(\frac{\partial M}{\partial T}\right)_V = -\frac{R}{(V-b)^2}$$

$$N = \frac{R}{V-b}, \text{ So } \left(\frac{\partial N}{\partial V}\right)_T = -\frac{R}{(V-b)^2}$$

Here $\left(\frac{\partial M}{\partial T}\right)_V = \left(\frac{\partial N}{\partial V}\right)_T$

So second equation is a correct one.

13.7.2. *Cyclic Rule*

If any three variables (x, y, z) are connected by a functional relationship, then the three partial derivatives are connected by

$$\left(\frac{\partial x}{\partial y}\right)_z \left(\frac{\partial y}{\partial z}\right)_x \left(\frac{\partial z}{\partial x}\right)_y = -1 \qquad ...(13.51)$$

and is known as cyclic rule. In the above relation variables x, y, z in the numerator are related to $y, z, x,$ in the denominator and to the subscripts z, x, y in a cyclic manner.

The above relation can be written as

(*i*) Write the three variables in any manner.

(*ii*) Write the same three variable below the already written one in such a manner that no vertical column should have the same variables. For example, in the above case, two combinations are possible *i.e.*

$$\begin{matrix} x, y, z & \quad x, y, z \\ y, z, x & \quad z, x, y \end{matrix}$$

Cyclic relations are

$$\left(\frac{\partial x}{\partial y}\right)_z \left(\frac{\partial y}{\partial z}\right)_x \left(\frac{\partial z}{\partial x}\right)_y = -1$$

and

$$\left(\frac{\partial x}{\partial z}\right)_y \left(\frac{\partial y}{\partial x}\right)_z \left(\frac{\partial z}{\partial y}\right)_x = -1$$

Both these relation are the same because $\frac{1}{-1} = -1.$

***Exercise* 13.21:** The Joule Thomson coefficient is given by the relation

$$\mu_{JT} = -\frac{1}{C_p}\left(\frac{\partial H}{\partial p}\right)_T$$

Derive this expression with the help of cyclic rule.

In the above equation, there are three variables, $H, p, T,$ so applying cyclic rule *i.e.*

$$H, p, T$$
$$p, T, H$$

We get

$$\left(\frac{\partial H}{\partial p}\right)_T \left(\frac{\partial p}{\partial T}\right)_H \left(\frac{\partial T}{\partial H}\right)_p = -1 \qquad ...(13.52)$$

Now $$\mu_{JT} = \left(\frac{\partial T}{\partial p}\right)_H \text{ and } C_p = \left(\frac{\partial H}{\partial T}\right)_p$$

Substituting these two in equation (13.52)

$$\left(\frac{\partial H}{\partial p}\right)_T \left(\frac{1}{\mu_{JT}}\right) \frac{1}{C_p} = -1$$

or $$\mu_{JT} = -\frac{1}{C_p}\left(\frac{\partial H}{\partial p}\right)_T$$

after rearrangement.

***Exercise* 13.22:** Derive the following relation

$$C_p - C_V = \frac{TV\alpha^2}{\beta}$$

where α and β are the coefficient of expansion and compressibility.

Taking three variables as, p, V, T and applying cyclic rule, one gets

$$\left(\frac{\partial V}{\partial p}\right)_T \left(\frac{\partial p}{\partial T}\right)_V \left(\frac{\partial T}{\partial V}\right)_p = -1 \qquad ...(13.53)$$

Now

$$\left(\frac{\partial V}{\partial p}\right)_T = -\beta V$$

$$\left(\frac{\partial V}{\partial T}\right)_p = \alpha V$$

Substituting these results in equation (13.53)

$$(-\beta V)\left(\frac{\partial p}{\partial T}\right)_V \left(\frac{1}{V\alpha}\right) = -1$$

or

$$\left(\frac{\partial p}{\partial T}\right)_V = \frac{\alpha}{\beta}$$

Using the following relation

$$C_p - C_V = \left[p + \left(\frac{\partial E}{\partial V}\right)_T\right]\left(\frac{\partial V}{\partial T}\right)_p \quad ...(13.54)$$

The value of $\left(\frac{\partial E}{\partial V}\right)_T$ can be calculated from the relation

$$dE = T\,dS - p\,dV$$

by rearranging as

$$\left(\frac{\partial E}{\partial V}\right)_T = T\left(\frac{\partial S}{\partial V}\right)_T - p$$

But from Maxwell's relations

$$\left(\frac{\partial S}{\partial V}\right)_T = \left(\frac{\partial p}{\partial T}\right)_V$$

So $$\left(\frac{\partial E}{\partial V}\right)_T = T\left(\frac{\partial P}{\partial T}\right)_V - p = T\frac{\alpha}{\beta} - p \quad ...(13.55)$$

Combining equations (13.53), (13.54) and (13.55)

$$C_p - C_V = \left(p + T\frac{\alpha}{\beta} - p\right)\alpha V$$

$$= \frac{\alpha^2 VT}{\beta}.$$

13.8. THE FUNDAMENTAL EQUATIONS OF THERMODYNAMICS AND MAXWELL'S RELATIONS

If the work against expansion is the only work then these four fundamental equations can be written as

$$dE = T\,dS - p\,dV \quad ...(13.3)$$
$$dH = T\,dS + V\,dp \quad ...(13.7a)$$
$$dA = -S\,dT - p\,dV \quad ...(13.4)$$
$$dG = -S\,dT + V\,dp \quad ...(13.9)$$

There are many Maxwell's relations which can be derived from the four fundamental equations. Actually the above relations involves four variables S, p, V and T. These variables can be arranged in number of ways to get the Maxwell's relations and the general expression is

$$\left(\frac{\partial w}{\partial x}\right)_z = \pm\left(\frac{\partial Y}{\partial z}\right)_x \quad ...(13.56)$$

where w, x, y and z can be S, V, p and T respectively.

The following two rules are used:

(*i*) The cross products *xy* and *wz* must be work terms *i.e.*, pV and ST.

(*ii*) The negative sign is chosen if *w* and *x* or *z* and *y* are *T* and *V*.

The commonly used Maxwell's relations are

$$\left(\frac{\partial T}{\partial V}\right)_S = -\left(\frac{\partial p}{\partial S}\right)_V \qquad ...(13.57a)$$

$$\left(\frac{\partial T}{\partial p}\right)_S = \left(\frac{\partial V}{\partial S}\right)_p \qquad ...(13.57b)$$

$$\left(\frac{\partial S}{\partial p}\right)_T = -\left(\frac{\partial V}{\partial T}\right)_p \qquad ...(13.57c)$$

$$\left(\frac{\partial S}{\partial V}\right)_T = \left(\frac{\partial p}{\partial T}\right)_V \qquad ...(13.57d)$$

***Exercise* 13.23:** From equation (13.9) derive the Maxwell's relation $\left(\frac{\partial S}{\partial p}\right)_T = -\left(\frac{\partial V}{\partial T}\right)_p$ and compare this with that derived from equation (13.56).

At constant pressure, equation (13.9) gives

$$\left(\frac{\partial G}{\partial T}\right)_p = -S \qquad ...(13.58a)$$

and at constant temperature

$$\left(\frac{\partial G}{\partial p}\right)_T = V \qquad ...(13.58b)$$

Differentiate (13.58*a*) with respect to *p* at constant temperature and (13.58*b*) with respect to temperature at constant pressure *i.e.*

$$\frac{\partial}{\partial p}\left[\left(\frac{\partial G}{\partial T}\right)_p\right]_T = -\left(\frac{\partial S}{\partial p}\right)_T$$

and

$$\frac{\partial}{\partial T}\left[\left(\frac{\partial G}{\partial p}\right)_T\right]_p = \left(\frac{\partial V}{\partial T}\right)_p$$

According to cross-derivative rules

$$\left[\frac{\partial}{\partial p}\left(\frac{\partial G}{\partial T}\right)_p\right]_T = \frac{\partial}{\partial T}\left[\left(\frac{\partial G}{\partial p}\right)_T\right]_p$$

So

$$\left(\frac{\partial S}{\partial p}\right)_T = -\left(\frac{\partial V}{\partial T}\right)_p$$

If $w = S, x = p$, in equation (13.56) the first rule says that $y = V, z = T$, giving

$$\left(\frac{\partial S}{\partial p}\right)_T = \pm\left(\frac{\partial V}{\partial T}\right)_p$$

By the second rule, the nagative sign will be chosen and the final relation

$$\left(\frac{\partial S}{\partial p}\right)_T = -\left(\frac{\partial V}{\partial T}\right)_p$$

agrees with the relation derived above.

***Exercise* 13.24:** Using Maxwell's relations show that

(*i*) $$\left(\frac{\partial C_V}{\partial V}\right)_T = T\left(\frac{\partial^2 p}{\partial T^2}\right)_V = T\left[\frac{\partial}{\partial T}\left(\frac{\alpha}{\beta}\right)_V\right]$$

(*ii*) $$\left(\frac{\partial C_p}{\partial p}\right)_T = -T\left(\frac{\partial^2 V}{\partial T^2}\right)_p = -T\left[\frac{\partial}{\partial T}(V\alpha)\right]$$

(*i*) $T\,dS = C_V\,dT + P\,dV$

At constant volume,

So $$C_V = T\left(\frac{\partial S}{\partial T}\right)_V$$

Differentiate w.r.t. volume at constant temp.

$$\left(\frac{\partial C_V}{\partial V}\right)_T = \left[\frac{\partial}{\partial V}\left(T\frac{\partial S}{\partial T}\right)_V\right]_T = T\left[\frac{\partial}{\partial V}\left(\frac{\partial S}{\partial T}\right)_V\right]_T$$

$$= T\left[\frac{\partial}{\partial T}\left(\frac{\partial S}{\partial V}\right)_T\right]_V$$

Now from Maxwell's relation $$\left(\frac{\partial S}{\partial V}\right)_T = \left(\frac{\partial p}{\partial T}\right)_V$$

$$\left(\frac{\partial C_V}{\partial V}\right)_T = T\left[\frac{\partial}{\partial l}\left(\frac{\partial p}{\partial T}\right)_V\right]$$

$$= T\left(\frac{\partial^2 p}{\partial T^2}\right)_V$$

$$= T\left[\frac{\partial}{\partial T}\left(\frac{\alpha}{\beta}\right)\right]_V \quad \because \left(\frac{\partial p}{\partial T}\right)_V = \frac{\alpha}{\beta}$$

(*ii*) Similarly

$$C_p = T\left(\frac{\partial S}{\partial T}\right)_p$$

Differentiate with respect to p at constant temp.

$$\left(\frac{\partial C_p}{\partial p}\right)_T = \frac{\partial}{\partial p}\left[T\left(\frac{\partial S}{\partial T}\right)_p\right]_T$$

$$= T\left[\frac{\partial}{\partial p}\left(\frac{\partial S}{\partial T}\right)_p\right]_T = T\left[\frac{\partial}{\partial T}\left(\frac{\partial S}{\partial p}\right)_T\right]_p$$

But
$$\left(\frac{\partial S}{\partial p}\right)_T = -\left(\frac{\partial V}{\partial T}\right)_p$$

$$\left(\frac{\partial C_p}{\partial p}\right)_T = T\left[\frac{\partial}{\partial T}\left(-\frac{\partial V}{\partial T}\right)_p\right]_p = -T\left(\frac{\partial^2 V}{\partial T^2}\right)_p$$

$$= -T\left[\frac{\partial}{\partial T}(Va)\right]_p \quad \because \frac{\partial V}{\partial T} = V\alpha$$

13.9. SYSTEMS OF VARIABLE COMPOSITION

The free energy of a system of fixed composition is given by equation (13.9)

$$dG = -S\,dT + V\,dP \qquad \text{...(13.9)}$$

but for system where the number of mol of components vary the free energy is given as

$$dG = -S\,dT + V\,dp + \left(\frac{\partial G}{\partial n_1}\right)_{n_i \neq n_1, p, T} dn_1 + \left(\frac{\partial G}{\partial n_2}\right)_{n_i \neq n_2, p, T} dn_2 + \ldots \qquad \text{...(13.59)}$$

The term $\left(\frac{\partial G}{\partial n_i}\right)_{n_j \neq n_i, p, T} = \mu_i$ is known as chemical potential or partial molal free energy. So

$$dG = -S\,dT + V\,dp + \mu_1 dn_1 + \mu_2 dn_2 + \ldots \qquad \text{...(13.60)}$$

where $dn_1, dn_2, \ldots$ are the changes in the number of mol of the components.

13.9.1. *Chemical Potential*

The chemical potential of any component is defined as the change in the free energy of the system if one mol of the component is added to system of infinite amount such that no change in the composition takes place in the system. Mathematically it is defined as

$$\mu_i = \left(\frac{\partial G}{\partial n_i}\right)_{n_j \neq n_i, p, T} \qquad \text{...(13.61)}$$

13.9.2. *Chemical Potential of Pure Ideal Gas*

As derived earlier (equation 13.27) the chemical potential of pure ideal gas is

$$\mu = \mu^0(T) + RT \ln p \qquad ...(13.27)$$

where $\mu^0(T)$ is the standard chemical potential (*i.e.* potential at one atmosphere of gas). The chemical potential of ideal gas in the mixture of gases is

$$\mu_i = \mu_{i(\text{pure})} + RT \ln x_i \qquad ...(13.62)$$

where $\mu_{i(\text{pure})}$ is the chemical potential of pure gas at the same temperature and pressure as that of mixture and x_i is the mole fraction. It is clear from the equation (13.62) that chemical potential of any gas in mixture is less than that of pure gas at the same temperature and pressure because x_i is less than unity and $\ln x_i$ will be negative.

***Exercise* 13.25:** (*a*) Calculate the chemical potential, over and above the standard chemical potential, of O_2 present in a mixture where the partial pressure is (*i*) 2 atm (*ii*) 10 atm. (*b*) Calculate the free energy change when 2 mol of O_2 is transferred from partial pressure 10 atm to partial pressure 2 atm at 27°C.

(*a*) Using equation (13.26)

$$\mu = \mu^0 + RT \ln p_i$$

(*i*)
$$\mu_2 - \mu^0 = (8.314 \text{ J K}^{-1} \text{ mol}^{-1})(300 \text{ K}) \ln 2$$
$$= 1.729 \text{ kJ}$$

(*ii*)
$$\mu_{10} - \mu^0 = (8.314 \text{ J K}^{-1} \text{ mol}^{-1})(300 \text{ K}) \ln 10$$
$$= 5.741 \text{ kJ}$$

(*b*) Change in free energy when 1 mol of O_2 is transferred from 10 atm to 2 atm is

$$\mu_2 - \mu_{10} = (\mu_2 - \mu^0) - (\mu_{10} - \mu^0)$$
$$= 1.729 \text{ kJ} - 5.743 \text{ kJ} = -4.014 \text{ kJ mol}^{-1}$$
$$\Delta G = (2 \text{ mol})(\mu_2 - \mu_{10})$$
$$= (2 \text{ mol})(-4.014 \text{ kJ mol}^{-1})$$
$$= -8.028 \text{ kJ.}$$

13.10. THE FREE ENERGY OF MIXTURE

If the mixture contains $n_1, n_2, n_3,...$ mol, μ_1, μ_2, μ_3, are the chemical potentials of the component 1, 2, 3, ... the free energy of the mixture at constant temperature and pressure is

$$G = n_1 \mu_1 + n_2\mu_2 + ... \qquad ...(13.63)$$
$$= \sum_i n_i \mu_i$$

Free Energy of Ideal Mixing

With the help of equations (13.62) and (13.63) the free energy of mixing can be derived ($\Delta_{\text{mix}} G$) as

$$\Delta_{\text{mix}} G = nRT \sum x_i \ln x_i \qquad ...(13.64)$$

where n = total number of mol of a system and x_i is the mole fraction of the *i*th component. The summation is over the total number of components.

Since the equation (13.64) gives negative value of free energy of mixing of ideal gases, it is clear that mixing is a spontaneous process.

Combining equations (13.64) and (13.13 *b*), the heat of mixing of ideal gases can be calculated from

$$\Delta_{mix}G = \Delta_{mix}H - T\Delta_{mix}S \quad ...(13.13b)$$

i.e.

$$\Delta_{mix}H = nRT\sum_i x_i \ln x_i + (T)(-NR\Sigma x_i \ln x_i)$$
$$= 0$$

Similarly it can also be shown that volume change in mixing of ideal gases is also zero *i.e.*,

$$\Delta_{mix}V = \left(\frac{\Delta_{mix}G}{\partial p}\right)_{T,x_i} = 0$$

Since $\Delta_{mix}G$ is independent of pressure.

Note: If $\Delta_{mix}H = \Delta_{mix}V = 0$, does not necessarily means that mixture formed is an ideal one unless other criterian are observed.

***Exercise* 13.26:** Show that in a binary system, the decrease in free energy of mixing is maximum if $x_1 = x_2 = 1/2$.

The free energy of mixing of two components is

$$\Delta_{mix}G = NRT x_1 \ln x_1 + NRT x_2 \ln x_2$$

Differentiating with respect to x_1 after rearranging the above equation as

$$\Delta_{mix}G = NRT x_1 \ln x_1 + NRT(1-x_1)\ln(1-x_1)$$

because $x_1 + x_2 = 1$

$$\left(\frac{\Delta_{mix}G}{\partial x_1}\right)_{T,p} = NRT\left[\ln x_1 + x_1\frac{1}{x_1} - (1-x_1)\frac{1}{1-x_1} - \ln(1-x_1)\right] = 0$$

$$\ln\frac{x_1}{1-x_1} = 0 = \ln 1$$

$$\frac{x_1}{1-x_1} = 1 \text{ or } x_1 = 1/2$$

Similarly $x_2 = 1 - 1/2 = 1/2$

***Exercise* 13.27:** The heat of mixing of two components is not zero if it is not ideal solution. What should be the maximum positive value of $\Delta_{mix}H$ at 25°C so that the mixing is possible?

In case of ideal solutions,

$$\Delta_{mix}H = 0, \text{ so the equation (13.13}b\text{)}$$

becomes

$$\Delta_{mix}G = -T\Delta_{mix}S$$

This equation shows that mixing is only possible due to increase in the entropy of mixing. It was shown in exercise (12.19) for binary system, the entropy of mixing is maximum if $x_1 = x_2 = 1/2$. So entropy of mixing per mol of mixture is

$$\Delta_{mix}S = -(1\text{ mol})(8.314\text{ J mol}^{-1}\text{ K}^{-1})\left(\frac{1}{2}\ln\frac{1}{2} + \frac{1}{2}\ln\frac{1}{2}\right)$$

$$= 5.76\text{ J mol}^{-1}\text{ K}^{-1} = 5.76\text{ EU mol}^{-1}$$

Substituting the value of ΔS in equation above

$$\Delta_{mix}G = -(298\text{ K})(5.76\text{ J mol}^{-1}\text{ K}^{-1})$$
$$= -1717.32\text{ J mol}^{-1}$$

For the mixing to be spontaneous,

$$\Delta_{mix}G = -\text{ve. Thus from equation (13.13}\,b)$$

$$\Delta_{mix}G = \Delta_{mix}H - T\Delta_{mix}S$$

the value of $\Delta_{mix}G$ at 298 K can vary from 0 to – 1717.32 J mol^{-1} for spontaneous mixing. If $\Delta_{mix}H$ is more positive than 1717.32 J mol^{-1}, $\Delta_{mix}G$ will be positive and mixing becomes non-spontaneous. Thus maximum positive value of

$$\Delta_{mix}H = 1717.32 \text{ J mol}^{-1}$$

***Exercise* 13.28:** 5 mol of benzene and 3 mol of toluene are mixed together at 1 atm and 25°C. (*i*) Calculate $\Delta G, \Delta A, \Delta H, \Delta E, \Delta V$ of mixing if the two form ideal solution (*ii*) if the standard entropies of benzene and toluene are 269.20 EU and 319.74 EU respectively, calculate the entropy of the mixture.

(*i*) In the mixture

$$x\text{ (benzene)} = \frac{5}{3+5} = 0.625$$

$$x\text{ (toluene)} = \frac{3}{8} = 0.375$$

Substituting the values in equation (13.64)

$$\Delta_{mix}G = (8.314 \text{ J K}^{-1} \text{ mol}^{-1})(298 \text{ K})(0.625 \ln 0.625 + 0.375 \ln 0.375)$$
$$= -1.639 \text{ kJ mol}^{-1}$$

$$\Delta_{mix}S = -(8.314 \text{ J K}^{-1} \text{ mol}^{-1})(0.625 \ln 0.625 + 0.375 \ln 0.375)$$
$$= 5.50 \text{ J mol}^{-1} \text{ K}^{-1}$$

$$\Delta_{mix}S = (8 \text{ mol})\ \Delta_{mix}S = (8 \text{ mol})(5.50 \text{ J mol}^{-1} \text{ K}^{-1})$$
$$= 44 \text{ J K}^{-1}$$

$$\Delta_{mix}G = \Delta_{mix}A = -1.639 \text{ kJ mol}^{-1}$$

$$\Delta_{mix}H = \Delta_{mix}E = \Delta_{mix}V = 0$$

(*ii*) Total entropy of the components before mixing

$$= (5 \text{ mol})(269.20 \text{ J deg}^{-1} \text{ mol}^{-1}) + (3 \text{ mol})(319.74 \text{ J mol}^{-1} \text{ deg}^{-1})$$
$$= 1346 \text{ J deg}^{-1} + 959.22 \text{ J deg}^{-1} = 2305.22 \text{ J deg}^{-1}$$

Total entropy of the solution after mixing

$$= 2305.22 + 44 = 2349.22 \text{ J deg}^{-1}$$

***Exercise* 13.29:** The partial molal volumes of ethanol and H_2O in a solution containing 5.0 mol of H_2O and 1.05 mol of ethanol are 17.839 ml and 55.10 mL respectively. Calculate the excess volume if density of ethanol is 0.7893 kg dm^{-3}.

If the solution is ideal, the total volume calculated by any procedure will be same. Let the density of water be 1 kg dm^{-3} so

$$V^0(H_2O) = \frac{\text{Molecular wt.}}{\text{density}} = \frac{(18 \text{ g mol}^{-1})(10^{-3} \text{ kg g}^{-1})}{1 \text{ kg dm}^{-3}}$$
$$= 18.0 \times 10^{-3} \text{ dm}^3 \text{ mol}^{-1}$$

$$V^0(C_2H_5OH) = \frac{(46 \text{ g mol}^{-1})(10^{-3} \text{ kg g}^{-1})}{0.7893 \text{ kg dm}^{-3}} = 58.28 \times 10^{-3} \text{ dm}^3 \text{ mol}^{-1}$$

$$V_{\text{total}} = (18.0 \times 10^{-3}\ \text{dm}^3\ \text{mol}^{-1})\ (5\ \text{mol}) + (1.05\ \text{mol})\ (58.28 \times 10^{-3}\ \text{dm}^3\ \text{mol}^{-1})$$
$$= 151.19 \times 10^{-3}\ \text{dm}^3$$

Now by second procedure

$$V_{\text{total}} = \overline{V}_{\text{(ethanol)}}\ n_{\text{etha}} + \overline{V}_{H_2O}\ n_{H_2O}$$
$$= (55.10 \times 10^{-3}\ \text{dm}^3\ \text{mol}^{-1})(1.05\ \text{mol}) + (17.839 \times 10^{-3}\ \text{dm}^3\ \text{mol}^{-1})\ (5\ \text{mol})$$
$$= 147.05 \times 10^{-3}\ \text{dm}^3$$

The value of total volume obtained by two methods are not same. The excess volume

$$V_{\text{excess}} = (147.05 - 151.19)\ 10^{-3}\ \text{dm}^3$$
$$= -4.14 \times 10^{-3}\ \text{dm}^3$$

and

$$\overline{V}_{\text{excess}} = \frac{-4.14 \times 10^{-3}}{5.0 + 1.05} = -0.68 \times 10^{-3}\ \text{dm}^3\ \text{mol}^{-1}$$

***Exercise* 13.30:** The equations (13.26) can also be applied to calculate ΔG even if the solutions are non-ideal provided (*i*) their vapour are ideal (*ii*) p_i and p'_i are the partial pressures of the vapours that are in equilibrium with the liquid solution *i.e.*, the problem becomes as:

Find the free energy change for the following transfer process

$$A\ (\text{soln}, x_A = x_A) \rightarrow A\ (\text{soln}, x_A = x_{A'})$$

where A is volatile and none of the solution is ideal.

Let P_A and $P_{A'}$ be the partial vapour pressure of the vapours in equilibrium with the respective solution, consider the following steps

$$A\ (\text{soln}, x_A = x_A) \xrightarrow{1} A\ (g, P_A = P_A) \xrightarrow{2} A\ (g, P_A = P_{A'}) \xrightarrow{3} A\ (\text{soln}, x_A = x_{A'})$$

So,
$$\Delta G = \Delta G_1 + \Delta G_2 + \Delta G_3$$

Since the steps (1) and (3) are the equilibrium steps $\Delta G_1 = \Delta G_3 = 0$. The change in the free energy for the above process is

$$\Delta G = \Delta G_2 = nRT \ln (p'_{,}(A)/p_{(A)})$$

Note: If the data on equilibrium partial pressure is available, it helps in calculating the ΔG even if the solutions are non-ideal.

***Exercise* 13.31:** Calculate the free energy change for the following process at 30°C.

$$\text{Ether (soln } x_A = 0.2) \rightarrow \text{Ether (soln } x_A = 0.8)$$

The partial pressure of ether at $x_A = 0.2$ and $x_A = 0.8$ are 202 mm and 535 mm respectively. Using the results derived in exercise 13.30 *i.e.*

$$\Delta G = nRT \ln \frac{p'(x_A)}{p(x_A)}$$

$$p(x_A) = p'(x_A = 0.8) = 535\ \text{mm}$$
$$p(x_A) = p(x_A = 0.2) = 202\ \text{mm}$$

$$\Delta G = (1\ \text{mol})\ (8.314\ \text{J deg}^{-1}\ \text{mol}^{-1})\ (303\ \text{K}) \ln \left(\frac{535}{202}\right)$$
$$= 2429.35\ \text{J or } 2.43\ \text{kJ}$$

13.11. CHEMICAL EQUILIBRIUM IN A MIXTURE

In a closed system, the free energy of general reaction of type

$$n_a A + n_b B + \cdots \rightarrow n_1 L + n_m M + \ldots \qquad \ldots(13.65)$$

at constant temperature and pressure is

$$\Delta G = \sum_{i}^{\text{Product}} n_i \mu_i - \sum_{j}^{\text{Reactant}} n_j \mu_j$$

At equilibrium, $\Delta G = 0$ so equation (13.65) reduces to

$$\sum_{i}^{\text{Product}} n_i \mu_i - \sum_{j}^{\text{Reactant}} n_j \mu_j = 0 \qquad \ldots(13.66)$$

13.11.1. *Chemical Equilibrium in a Mixture of Ideal Gases*

If the reaction (13.65) involves ideal gases the free energy of the reaction at constant temperature and pressure is

$$\Delta G = \Delta G^0 + RT \ln \frac{p_L^{n_L} \cdot p_M^{n_M}}{p_A^{n_A} \cdot p_B^{n_B}} \ldots \qquad \ldots(13.67a)$$

where G^0 is standard state free energy and p_i are the partial pressures of the gases. At equilibrium, $\Delta G = 0$ and the partial pressures of the gases will be equilibrium partial pressures *i.e.*,

$$\Delta G^0 = -RT \ln \frac{p_{L(\text{eq})}^{n_L} \cdot p_{M(\text{eq})}^{n_M}}{p_{A(\text{eq})}^{n_A} \cdot p_{B(\text{eq})}^{n_B}} \ldots \qquad \ldots(13.67b)$$

$$= -RT \ln K_p \qquad \ldots(13.67c)$$

where K_p is the equilibrium constants or ratio of the products of partial pressure of the products and reactants at equilibrium.

Though the value of K_p is just a number *i.e.*, unitless but one thing must be kept in mind, that is, unit of partial pressures of the components. If one is talking of standard free energy at 1 atm pressure, the unit of partial pressure should be in atm. If the partial pressure is expressed in mm, the standard free energy will be the free energy at 1 mm pressure. One can also write the values of equilibrium constant as K_p (atm) or K_p (mm).

The ΔG^0 is independent of total pressure so is the equilibrium constant of total pressure. Again, it does not mean that partial pressures of the components, degree of dissociation, course of reaction etc. are independent of total pressure. It is only the equilibrium constant K_p as well as ΔG^0 do depend upon temperature and will be discussed later.

Following are some important aspects of chemical equilibrium:

(*i*) The value of K_p is always associated with a particular reaction even at given temperature *e.g.* K_p of reaction

$$aA(g) + bB(g) \rightleftharpoons c\,C(g) + dD(g)$$

is different than the K_p of the reverse reaction

(*ii*) The value of K_p depends upon the values of the stoichiometric coefficients *e.g.* K_p of reaction

$$N_2 + 3H_2 \rightleftharpoons 2NH_3$$

$$K_p = \frac{[NH_3]^2}{[N_2][H_2]^3}$$

is different than the K_p of reaction

$$\frac{1}{2} N_2 + \frac{3}{2} H_2 \rightleftharpoons NH_3$$

$$K'_p = \frac{[NH_3]}{[N_2]^{\frac{1}{2}}[H_2]^{\frac{3}{2}}}$$

$$K_p = K'^2_p$$

***Exercise* 13.32:** The ΔG_{298} for the reaction

$$2H_2(g, 1 \text{ atm}) + O_2(g, 1 \text{ atm}) \rightarrow 2H_2O(g, 1 \text{ atm})$$

is – 228.60 kJ. Calculate the value of ΔG_{298} for the reaction

$$2H_2(g, 0.1 \text{ atm}) + O_2(g, 0.2 \text{ atm}) \rightarrow 2H_2O(g, 0.3 \text{ atm})$$

The ΔG_{298} for the first reaction is the standard free energy of reaction as each species is at 1 atm pressure *i.e.* ΔG^0_{298}. So

$$\Delta G_{298} = \Delta G^0_{298} + RT \ln \frac{p^2_{H_2O}}{p^2_{H_2} p_{O_2}}$$

$$= -228.60 \text{ kJ} + (8.314 \text{ J deg}^{-1})(298 \text{ K}) \ln \frac{(0.3)^2}{(0.1)^2(0.2)}$$

$$= -219.17 \text{ kJ}$$

***Exercise* 13.33:** Consider the exercise 13.16 Calculate the value of equilibrium constant of the reaction

$$H_2S(g) + \frac{3}{2} O_2(g) \rightleftharpoons H_2O(g) + SO_2(g)$$

at 298 K and 500 K. Use the data mentioned in exercise 13.16.

This can be done in number of ways:

(*i*) ΔG^0_{298} can be calculated either from the standard free energy of formation of each compound. This is given to be – 495.946 kJ.

Use equation (13.65) $$\log K_p = -\frac{\Delta G^0}{2.303RT} = -\frac{-495946 \text{ J}}{(2.303)(8.314 \text{ J K}^{-1})(298 \text{ K})}$$

$$= 86.9189$$

$$K_p = 10^{86.92} = 8.295 \times 10^{86}$$

(*ii*) ΔG^0_{298} of the reaction can also be calculated from heat of formation and entropies of the individual components.

$$\Delta G^0 = \underset{\text{Prod.}}{\Sigma \Delta H_f^0} - \underset{\text{Reac.}}{\Sigma \Delta H_f^0} - T(\underset{\text{Prod.}}{\Sigma S^0} - \underset{\text{Reac.}}{\Sigma S^0})$$

$$= [(-241.82 \text{ kJ mol}^{-1})(1 \text{ mol}) + (-296.90 \text{ kJ mol}^{-1})(1 \text{ mol}) - (-20.815 \text{ kJ mol}^{-1})(1 \text{ mol})] - (298 \text{ K})[(188.72 \text{ J mol}^{-1} \text{ K}^{-1})(1 \text{ mol}) + (248.53 \text{ J K}^{-1} \text{ mol}^{-1})(1 \text{ mol}) - (205.64 \text{ J mol}^{-1} \text{ K}^{-1})(1 \text{ mol}) - (205.03 \text{ J mol}^{-1} \text{ K}^{-1})(\frac{3}{2} \text{ mol})]$$

$$= -495275 \text{ J}$$

i.e. quite close to that obtained in part 1

$$\log K_p = -\frac{\Delta G^0}{2.303\, RT} = -\frac{-495275 \text{ J}}{(2.303)(8.314 \text{ J K}^{-1})(298 \text{ K})}$$

$$= 86.80$$

$$K_p = 1086.10^{0.80} = 6.30 \times 10^{86}$$

(*iii*) Equilibrium constant at 500 K can be calculated by using the value of ΔG^0 obtained in exercise 13.16 and the procedure is already described there.

$$\log K_p = -\frac{\Delta G^0_{500}}{2.303\, RT} = -\frac{-479155 \text{ J}}{(2.303)(8.314 \text{ J K}^{-1})(298 \text{ K})}$$

$$= 50.0498$$

$$K_p = 10^{.0498} \times 10^{50} = 1.12 \times 10^{50}$$

(*iv*) The other procedure is to use the equation (13.71), using the data of the exercise 13.16 *i.e.*

$$A' = \Delta a = -9.0465 \text{ J K}^{-1} \text{ mol}^{-1}$$

$$B' = \Delta b = 23.572 \times 10^{-3} \text{ J K}^{-2} \text{ mol}^{-1}$$

$$C' = \Delta c = -255.239 \times 10^{-7} \text{ J K}^{-3} \text{ mol}^{-1}$$

$$\Delta H^0_{298} = -518.629 \text{ kJ}$$

$\log K_p(298 \text{ K}) = 6.30 \times 10^{86}$ *i.e.*

$\ln K_p (500 \text{ K}) - 2.303 \log (6.30 \times 10^{86})$

$$= -\frac{518619 \text{ J}}{8.314 \text{ J K}^{-1}}\left(\frac{1}{500} - \frac{1}{298}\right)\text{K}^{-1} + \frac{(-9.0465 \text{ J K}^{-1} \text{ mol}^{-1})}{(8.314 \text{ J K}^{-1} \text{ mol}^{-1})} \ln \frac{500 \text{ K}}{298 \text{ K}}$$

$$+ \frac{23.572 \times 10^{-3} \text{ J K}^{-2} \text{ mol}^{-1}}{8.314 \text{ J K}^{-1} \text{ mol}^{-1}}(500\text{–}298) \text{ K}$$

$$+ \frac{(-255.239 \times 10^{-7} \text{ J K}^{-3})(500^2 - 298^2)\text{K}^2}{2 \times 8.314 \text{ J K}^{-1}}$$

$$\ln K_p = 115.093$$

$$K_p = 9.45 \times 10^{49}$$

***Exercise* 13.34:** The ozone is obtained from oxygen by the reaction

$$\frac{3}{2}O_2 \rightleftharpoons O_3$$

If the ΔG_f^0 of ozone is 163.43 kJ mol^{-1}, calculate K_p at 25°C.

The standard free energy of the reaction (13.66) will be

$$\Delta G^0_{\text{(reaction)}} = \sum_i^{\text{Products}} n_i \Delta G^0_{f(i)} - \sum_j^{\text{Reactants}} n_j \Delta G^0_{f(j)}$$

$$= 163.43 - 3/2 \times 0$$

$$= 163.43 \text{ kJ}$$

Substituting the values in equation (13.67 *c*)

$$\ln K_p = -\frac{\Delta G^0}{RT} = \frac{-163.43\times10^3 \text{ J mol}^{-1}}{(8.314 \text{ J mol}^{-1}\text{ K}^{-1})(298 \text{ K})}$$

or

$$\log K_p = \frac{-163.43\times10^3}{(8.314)(298)(2.303)} = -28.67$$

$$K_p = 10^{-28.69} = 0.20 \times 10^{-28}$$

13.12. DIFFERENT FORMS OF EQUILIBRIUM CONSTANTS

The concentration of gases can be expressed in different ways, the following relations exist beween the equilibrium constants:

$$K_p = K_x P^{\Delta\nu} \quad ...(13.68a)$$

$$K_p = K_e (RT)^{\Delta\nu} \quad ...(13.68b)$$

where K_x and K_c are the equilibrium constant when the concentration terms are expressed as mole fraction and mol dm^{-3}. P is the total pressure of the system excluding inert gas and $\Delta\nu$ is the change in the number of mol of the reaction.

Since K_p is independent of pressure, K_x will depend on pressure if $\Delta\nu \neq 0$ and K_c is only a function of temperature.

If $\Delta\nu = 0$, $K_p = K_c$ and the equation can be written as

$$\Delta G^0 = -RT \ln K_c \quad ...(13.67c)$$

where ΔG^0 is the free energy change of the reaction but each gas is not at a partial pressure of 1 atm but at a concentration of 1 mol dm^{-3}.

Note: Looking at the equation (13.68*b*), the unit of gas constant should be in dm^3 atm mol^{-1} deg^{-1} because the unit of pressure is in atm and that of concentration is mol dm^{-3}, whereas in equations (13.67) and (13.68), the unit of R can be in Joules mol^{-1} deg^{-1} if ΔG^0 is expressed in joules.

***Exercise* 13.35:** In the exercise 13.34 if the reaction is carried at 1 atm pressure and 25°C, calculate K_x and K_c.

The value of K_p obtained in exercise 13.34 is

$$K_p = 10^{-28.69} = 2.04 \times 10^{-29}$$

Now substituting the respective values in relation (13.68 *a*)

$$K_x = K_p(P)^{-\Delta\nu} = (2.04 \times 10^{-29})(1 \text{ atm})^{-1/2} = 2.04 \times 10^{-29} \text{ atm}^{-1/2}$$

Similarly

$$K_c = K_p(RT)^{-\Delta\nu} = \frac{2.04\times10^{-29}}{\left[(0.0821\text{ atm dm}^3\text{ mol}^{-1}\text{ K}^{-1})(298\text{ K})\right]^{1/2}}$$

$$= 4.13 \times 10^{-30}$$

13.13. TEMPERATURE DEPENDENCE OF EQUILIBRIUM CONSTANT

Differentiating equation (13.67 *c*) with respect to temperature

$$\frac{d\ln K_p}{dT} = -\frac{1}{R}\frac{d}{dT}\left(\frac{\Delta G^0}{T}\right) \qquad \text{...(13.69}a\text{)}$$

$$= \frac{\Delta H^0}{RT^2}$$

$$\ln K_p(T_2) - \ln K_p(T_1) = \int_{T_1}^{T_2}\frac{\Delta H^0}{RT^2}\,dT \qquad \text{...(13.69}b\text{)}$$

If ΔH^0 is independent of temperature, specially for small range of temperatures, the integral can be integrated to

$$\ln K_p(T_2) - \ln K_p(T_1) = \frac{\Delta H^0}{R}\left[\frac{1}{T_1}-\frac{1}{T_2}\right] \qquad \text{...(13.69}c\text{)}$$

If ΔH^0 is not independent of temperature, then either the integral is evaluated graphically or using analytical form of ΔH^0. ΔH^0 can be expressed as a power series as

$$\Delta H^0 = \Delta H_0^0 + A'T + B'T^2 + C'T^3 + \ldots \qquad \text{...(13.70)}$$

So the above integral (13.68 *b*) becomes

$\ln K_p(T) - \ln Kp(T_0)$

$$= -\frac{\Delta H_0^0}{R}\left[\frac{1}{T}-\frac{1}{T_0}\right] + \frac{A'}{R}\ln\frac{T}{T^0} + \frac{B'}{R}(T-T_0) + \frac{C'}{2R}(T^2 - T_0^2) + \ldots. \qquad \text{...(13.71)}$$

Similarly

$$\frac{d\ln K_c}{dT} = \frac{\Delta E^0}{RT^2} \qquad \text{...(13.72)}$$

and this equation can be used in the same manner as equation (13.69 *a*).

***Exercise* 13.36:** The equilibrium constant for the reaction

$$2\text{ HI} \rightleftharpoons H_2 + I_2$$

at temperatures 393.8°C and 490.8°C are 1.643×10^{-2} and 2.184×10^{-2} respectively. If ΔH^0 remains constant during this temperature interval, calculate its value.

Substituting the values of respective terms in equation (13.69) one gets

$$\ln\frac{2.184\times10^{-2}}{1.643\times10^{-2}} = \frac{-\Delta H^0}{8.314\text{ J mol}^{-1}\text{ K}^{-1}}\left(\frac{1}{763.8}-\frac{1}{666.8}\right)K^{-1}$$

i.e.,
$$\Delta H^0 = \left(\frac{8.314 \times 763.8 \times 666.8}{97.0}\right)\left(\ln \frac{2.184}{1.643}\right)$$
$$= 12.42 \text{ kJ}$$

13.14. THE LE CHATELIER PRINCIPLE

If the external constraints (or conditions) under which the chemical equilibrium is established, are changed, the concentration of the reactants and products will change in such a manner as to oppose the change. Mathematically the following equations can be established *i.e.*,

$$\left(\frac{\partial x_e}{\partial T}\right)_p = \frac{\Delta H_0}{TG''_e} \qquad \text{...(13.73}a\text{)}$$

and

$$\left(\frac{\partial x_e}{\partial p}\right)_T = -\frac{\Delta v}{p}\frac{RT}{G''_e} \qquad \text{...(13.73}b\text{)}$$

where x_e is the degree of advancement or the change of concentration from equilibrium position, G''_e is the $\left(\frac{\partial^2 \Delta G}{\partial x^2}\right)_{T,p}$ and is positive quantity, Δv is the change in the number of mol in a reaction.

From equation (13.73*a*) it is clear that if ΔH^0 is positive (endothermic reactions) increase of temperature will shift the equilibrium towards right and vice versa.

Similarly from equation (13.73 *b*) increase of pressure will shift the equilibrium to right if Δv is – ve and vice versa.

***Exercise* 13.37:** Consider the reactions

(*i*) $$H_2S\,(g) + \frac{3}{2}O_2\,(g) \rightleftharpoons SO_2\,(g) + H_2O\,(g)$$
$$\Delta H^0 = -518.62 \text{ kJ}$$

(*ii*) $$2HI \rightleftharpoons H_2 + I_2$$
$$\Delta H^0 = 12.42 \text{ kJ}$$

What is the effect of increase of temperature and pressure on the course of above reactions.

(*i*) Consider the equation (13.73 *a*), T is + ve, G''_e is + ve, ΔH^0 = – ve

so

$$\left(\frac{\partial x_e}{\partial T}\right)_p = -\text{ve}$$

i.e., increase of temperature (ΔT = + ve) will retard the advancement of reaction. In other words, increase of temperature shift the equilibrium in backward direction. Consider the equation (13.73 *b*), p is + ve, G''_e = + ve

$$\Delta v = 1 + 1 - 1 - 3/2 = -1/2 \text{ so}$$

$$\left(\frac{\partial x_e}{\partial T}\right)_T = +\text{ve}$$

i.e., increase of pressure will accelerate the degree of advancement. In other words, the increase of pressure will shift the reaction towards right.

(*ii*) In this reaction, $\Delta H^0 = +$ ve, $\Delta v = 0$ so increase of temperature will shift the equilibrium to right whereas pressure has no effect on the course of reaction.

13.15. EFFECT OF ADDITION OF INERT GAS

The effect of adding inert gas at constant temperature on the course of reaction can be studied from two points of view, (*i*) Addition of inert gas, if it does not change the volume (*i.e.*, at constant volume), will have no effect on the position of equilibrium because the partial pressure of each gas will remain the sam e *i.e.*, $p_i = \left(\frac{n_i}{V}\right) RT$ and each term on right hand side does not change, (*ii*) Addition at constant total pressure can shift the position of equilibrium though K_p and P remain constant. The following observations can be made with the help of equation (13.73 *b*):

(*a*) If $\Delta v > 0$, an increase in K_x results with the corresponding shift in the equilibrium from left to right.

(*b*) If $\Delta v < 0$ the effect will be opposite.

13.16. CALCULATIONS OF K AND CONCENTRATION OF SPECIES AT EQUILIBRIUM FOR IDEAL GASES

***Exercise* 13.38:** Calculate the partial pressure of NO_2 (*g*) at 1 atmosphere of total pressure and 450°C in the following equilibrium reaction

$$N_2O_4\,(g) \rightleftharpoons 2\,NO_2(g)$$

Given ΔG_f^0 (N_2O_4) = 98.286 kJ and ΔG_f (NO_2) = 51.84 kJ and initially 2 mol of N_2O_4 was present. Assume ΔG^0 does not change.

Let a_e be the degree of dissociation of N_2O_4 at this temperature and pressure then at equilibrium

$$n\,(N_2O_4) = 2(1-a_e) \text{ and } n\,(NO_2) = 4a_e$$

$$n\,(N_2O_4) + n\,(NO_2) = 2 - 2a_e + 4a_e = 2\,(1 + a_e)$$

$$x_{N_2O_4} = \frac{2(1-a_e)}{2(1+a_e)} = \frac{1-a_e}{1+a_e}$$

$$x_{NO_2} = \frac{2.2a_e}{2(1+a_e)} = \frac{2a_e}{1+a_e}$$

$$p_{N_2O_4} = \frac{1-a_e}{1+a_e} P \text{ atm and } p_{NO_2} = \frac{2a_e}{1+a_e} P \text{ atm}$$

So equilibrium constant will be

$$K_p = \frac{p^2_{NO_2}}{p_{N_2O_4}} = \frac{(2a_e/1+a_e)^2 \;\; P^2 \text{ atm}^2}{(1-a_e/1+a_e) \;\; P \text{ atm}}$$

$$= \frac{4a_e^2 P}{1-a_e^2} \text{ atm} = \frac{4a_e^2}{1-a_e^2} \text{ as } P = 1 \text{ atm}$$

K_p can be calculated from the relation

$$\Delta G^0 = -RT \ln K_p$$

$$\Delta G^0_{\text{reaction}} = \sum_i^{\text{Product}} n_i \Delta G_f^0 - \sum_j^{\text{Reactant}} n_j \Delta G_f^0$$

$$= 2 \times 51.84 - 98.286 = 5.395 \text{ kJ}$$

Substituting this value in equation (13.67 *c*)

$$\log K_p = \frac{-(5.395 \times 10^3 \text{ J})}{(2.303)(8.314 \text{ J mol}^{-1} \text{ K}^{-1})(723 \text{ K})}$$

$$= -0.3897$$

$$K_p = 0.4077 = 0.408$$

Note: In above types of reactions and reactions $A \rightleftharpoons B + C$, the original concentration of A cancels out. The above relations are only valid when original amount of B or B and C are zero.

Substituting the value of K_p in equation

$$0.408 = \frac{4a_e^2}{1 - a_e^2}$$

$$0.408 - 0.408\, a_e^2 = 4a_e^2$$

$$a_e = \left(\frac{0.408}{4.408}\right)^{1/2} = 0.304$$

Percentage dissociation = 30.4

Amount of NO_2 at equilibrium $= 4a_e = 4 \times 0.304 = 1.216$ mol

Amount of $N_2O_4 = 2(1 - 0.304)$

$= 2 \times .696 = 1.392$ mol

13.17. HETEROGENEOUS SYSTEM

Equilibrium between Ideal Gas and Pure Condensed Phases

If the substances taking part in the chemical equilibrium is of more than one phase, it is called heterogeneous equilibrium. Consider a reaction

$$CaCO_3(s) \rightleftharpoons CaO(s) + CO_2(g)$$

It is said earlier that chemical potentials of condensed phases are insensitive to the pressure change, unless the pressure is very large. So at equilibrium

$$\Delta G = \mu(CaO) + \mu(CO_2) - \mu(CaCO_3) = 0$$

or $$\mu^0(CaO) + \mu^0(CO_2) + RT \ln p_e(CO_2) - \mu^0(CaCO_3) = 0$$

or $$\Delta G^0 = -RT \ln p_e(CO_2) = -RT \ln K_p$$

and $$K_p = p_e(CO_2)$$

***Exercise* 13.39:** Calculate the pressure of CO_2 at 427°C in the reaction under equilibrium

$$CaCO_3(s) \rightleftharpoons CaO(s) + CO_2(g)$$

condition if $\Delta G_f^0(CaCO_3) = -1129$ kJ, $\Delta G_f^0(CaO) = -604.2$ kJ,

$$\Delta G_f^0(CO_2) = -394.6 \text{ kJ}$$

The free energy change of the reaction is

$$\Delta G^0_{(\text{reaction})} = (-604.2) + (-394.6) - (-1129) = 130.2 \text{ kJ}$$

Substituting the values in equation (13.67 *c*)

$$\log K_p = \log p_e(CO_2) = -\frac{130.2 \times 10^3 \text{ J}}{(2.303)(8.314 \text{ J mol}^{-1} \text{ K}^{-1})(700 \text{ K})}$$

$$= -9.712$$

$$P_e(CO_2) = 1.94 \times 10^{-10} \text{ atm}$$

***Exercise* 13.40:** The equilibrium constant for the reaction

$$C(s) + CO_2(g) \rightleftharpoons 2CO(g)$$

as a function of temperature can be expressed as

$$RT \ln K_p = -40900 + 4.9T \ln T - 4.95 \times 10^{-2}T^2 + 5.1 \times 10^{-7}T^3 + 12.66T$$

in the range of 1100 to 1500 K. If the $p_e(CO_2)$ at 1200 K is 1.2 atm, calculate $p_e(CO)$.

For the given reaction

$$K_p = \frac{p_{CO}^2}{p_{CO_2}} \quad \text{or } p_{CO} = (K_p p_{CO_2})^{1/2}$$

K_p at 1200 K can be calculated from the given relation

$$\ln K_p = \frac{-1}{(1200 \text{ K})(8.314 \text{ J K}^{-1})}[-40900 + (4.9)(1200) \ln 1200 - (4.95 \times 10^{-2})(1200)^2 + (5.1 \times 10^{-7})(1200)^3 + (12.66)(1200)]$$

$$\ln K_p = -5.454$$

$$K_p = 4.28 \times 10^{-3}$$

Using the above derived equation

$$p_{CO} = (4.28 \times 10^{-3} \times 1.2)^{1/2}$$

$$= 0.072 \text{ atm.}$$

13.18. REAL GASES

13.18.1. *Fugacity*

The chemical potential of ideal gas is a function of the pressure of gas, whereas that for *real gases,* it is given by the relation

$$\mu = \mu^0 + RT \ln f \quad \text{...(13.74)}$$

where *f* is known as fugacity and defined mathematically as

$$\lim_{p \to 0} \frac{f}{p} = 1 \quad \text{...(13.75)}$$

i.e. as the pressure approaches zero, fugacity approaches pressure. In other words for ideal gases, the pressure and fugacity are equal and physically fugacity is the measure of pressure of real gas.

μ^0 is the standard chemical potential *i.e.* chemical potential when the fugacity is *unity.*

13.18.2. *Pressure Dependence of Fugacity*

The change of fugacity with pressure is given by

$$\ln f_2 - \ln f_1 = \int_{p_1}^{p_2} \frac{\bar{V}}{RT} dp \qquad ...(13.76)$$

Knowing the value of fugacity at one pressure, the value at other can be obtained either by evaluating the integral graphically or analytically.

***Exercise* 13.41:** Derive equation 13.76 expressing fugacity as a function of pressure.

Using equation (13.22*a*) in terms of molar quantities, we get

$$\left(\frac{\partial \mu}{\partial p}\right)_T = \bar{V} \qquad ...(1)$$

or at constant temperature, we can write equation (1) as

$$d\mu = \bar{V} dp \qquad ...(2)$$

Similarly differential form of equation (13.74) at constant temperature can be written as

$$d\mu = RTd \ln f \qquad ...(3)$$

Comparing equations (2) and (3) at constant temperature, we get

$$d \ln f = \frac{\bar{V}}{RT} dp \qquad ...(4)$$

Equation (4) can be written in integrated form by using the reference as $f = 1$ at $p = 1$ atm to any arbitrary pressure as

$$\ln f_2 - \ln f_1 = \int_{1}^{p} \frac{\bar{V}}{RT} dp \qquad ...(5)$$

For condensed phase, *i.e.*, solids or liquids and assuming that these forms are incompressible, we can write equation (5) as

$$\ln f - \ln 1 = \ln f = \frac{\bar{V}}{RT}(p-1) \qquad ...(13.76a)$$

13.18.3. *Temperature Dependence of Fugacity*

The change of fugacity with temperature is given by

$$\left(\frac{\partial \ln f}{\partial T}\right)_p = \frac{H^* - H}{RT^2} \qquad ...(13.77)$$

where H^* is the partial molar enthalpy of the substance in state A^* *i.e.* at zero pressure. Thus the difference $(H^* - H)$ is the change in the molar enthalpy when the substance is brought from state A to its state of zero pressure. Sometimes this is also called the "ideal molar heat of vapourisation" for the given state. If the given state is also a gas, it is called *Joule Thomson Heat.*

***Exercise* 13.42:** Using equation (13.74) show that

$$x_1\left(\frac{\partial \ln f_1}{\partial x_1}\right)_{l,T} = x_2\left(\frac{\partial \ln f_2}{\partial x_2}\right)_{p,T}$$

From Gibbs-Duhem relation *i.e.*,

$$n_1 d\mu_1 = -n_2 d\mu_2$$

or

$$x_1 d\mu_1 = -x_2 d\mu_2$$

At constant temperature and pressure, equation (13.74) becomes

$$d\mu = RTd\ln f$$

So

$$x_1\, RTd\ln f_1 = -x_2\, RTd\ln f_2$$

Divide both sides by $RTdx_1$ at constant T and p

$$x_1\left(\frac{\partial \ln f_1}{\partial x_1}\right)_{l,T} = -x_2\left(\frac{\partial \ln f_2}{\partial x_1}\right)_{p,T}$$

Now $x_1 = 1 - x_2$ *i.e.*, $dx_1 = -dx_2$

The above relation becomes

$$x_1\left(\frac{\partial \ln f_1}{\partial x_1}\right)_{l,T} = x_2\left(\frac{\partial \ln f_2}{\partial x_2}\right)_{p,T}$$

13.19. CALCULATION OF FUGACITY OF REAL GASES

Fugacity of real gases can be evaluated either graphically or analytically.

13.19.1. *By Graphical Methods*

(*i*) *Using* α *function:* The fugacity of any real gas at a pressure p is given as

$$\ln f = \ln p + \frac{1}{RT}\int_0^p \alpha\, dp \qquad ...(13.78a)$$

where $\alpha = \left(-\frac{RT}{p} + \overline{V}\right)$, p is the pressure of the real gas, $\overline{V}$ is the volume of 1 mol of real gas and α is the difference in the molar volume of ideal and real gas.

The integral can be evaluated graphically *i.e.*, area under the curve, which is a plot of α vs p.

(*ii*) *Based on Compressibility Factor*: The equation (13.78*a*) can be written as

$$\ln f = \ln p - \int_0^p \frac{1-Z}{p}\,dp \qquad ...(13.78b)$$

The integral can be evaluated graphically by plotting $(1 - Z)/p$ vs p and measuring the area under the curve. For gases below their Boyle temperatures, $(1 - Z)$ is positive at moderate temperature and fugacity will be thus less than the pressure. For gases above the Boyle temperatures, the fugacity is greater than the pressure.

13.19.2 *Analytical Method*

The behaviour of real gases can be expressed by different equations of states. Using any particular equation of state, the above integral can be evaluated and thus the fugacity can be calculated.

***Exercise* 13.43:** Using the following data calculate the value of f and γ for H_2 at 100°C and 300 atm pressure, using graphical method.

The value of V_{id} at any pressure (say 10 atm) can be calculated from ideal gas equation

$$\overline{V}_{id} = \frac{RT}{p} = \frac{(0.0821\text{ atm dm}^3\text{ deg}^{-1}\text{ mol}^{-1})(373\text{K})}{10\text{ atm}}$$

$$= 3.0623\text{ dm}^3\text{ mol}^{-1}$$

and $$\alpha = \overline{V} - \frac{RT}{p} = 3.0776 - 3.0623$$

$$= 0.0153\text{ dm}^3\text{ mol}^{-1}$$

The values of α at different pressures are calculated and listed in the table on next page. The values of α are plotted vs pressure in Fig. 13.1. The area under the curve = 4.92 atm dm^3 mol^{-1}, using equation (13.78a)

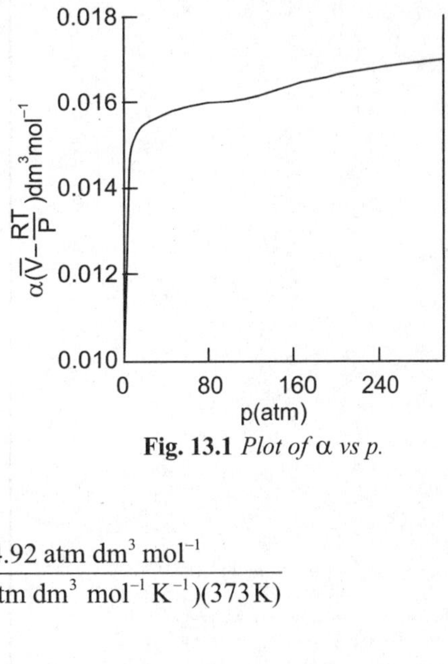

Fig. 13.1 *Plot of* α *vs p.*

$$\ln f = \ln 300 + \frac{4.92\text{ atm dm}^3\text{ mol}^{-1}}{(0.821\text{ atm dm}^3\text{ mol}^{-1}\text{ K}^{-1})(373\text{K})}$$

So $$\ln f = 5.8645$$

$$f = 352.3$$

and $$\gamma = \frac{352.3}{300} = 1.174$$

***Exercise* 13.44:** The value of f or γ for H_2 can also be calculated by using the equation (13.78 b) *i.e.* from the plot of $Z - 1$ against ln p_r. Prepare such a plot for H_2 at 100°C and by selecting various pressures up to 300 atm, calculating p_r and finding Z from the Fig. 8.3. Calculate f and v for H_2 at 100°C and 300 atm. $T_c(H_2) = 33.3$ K and $p_c(H_2) = 12.8$ atm.

Prepare the table as follows, taking pressure = 300 atm *i.e.*

$$p_r = \frac{300\text{ atm}}{12.8\text{ atm}} = 23.44$$

p (atm)	1	10	20	50	60	80	100	120	140	180	200	240	280	300
$\bar{V}$ ($dm^3\ mol^{-1}$)	30.6347	3.0776	1.5467	0.6283	0.5263	0.3988	0.3223	0.2712	0.2350	0.1866	0.1699	0.1443	0.1263	0.1191
$\bar{V}_{id}$ ($dm^3\ mol^{-1}$)	30.6233	3.0623	1.5312	0.6125	0.5104	0.3828	0.3062	0.2552	0.2187	0.1701	0.1531	0.1276	0.1094	0.1021
α ($dm^3\ mol^{-1}$)	0.0114	0.0153	0.0155	0.0158	0.0159	0.0160	0.0161	0.0160	0.0163	0.0165	0.0168	0.0167	0.0169	0.0170

and $$T_r = \frac{373\ \text{K}}{33.3} = 11.20$$

From Fig. 8.3, the value of Z at $p_r = 23.44$ and $T_c = 11.20$ is 0.0125. So $Z - 1 = -0.9875$. Similarly other data in the table is prepared.

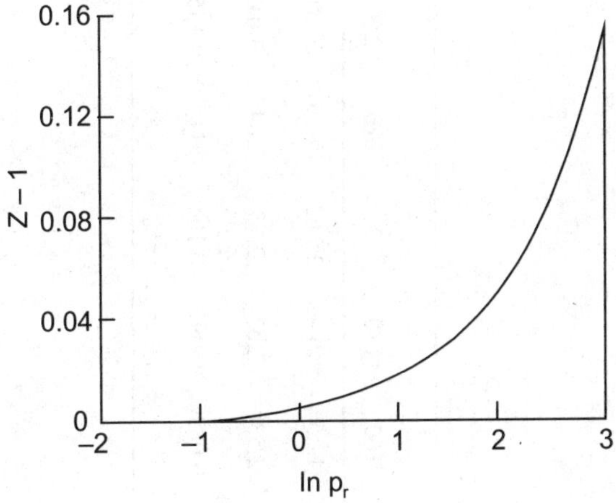

Fig. 13.2 *Plot of Z–1 vs ln p_r.*

The $(Z-1)$ is plotted vs ln p_r in Fig. 13.2. The value

$$\int_0^{300} (Z-1)\,d\ln p_r = 0.1534$$

Using equation (13.78*b*)

$$\ln f = \ln 300 + 0.1534 = 5.8572$$
$$f = e^{5.8572} = 349.74$$

and $$\gamma = \frac{349.74}{300} = 1.166$$

The values of f and γ agree quite nicely with those obtained in exercise 13.43.

***Exercise* 13.45:** Derive the relation for calculating the fugacity of gas obeying van der Waals equation of state.

The equation (13.78*a*) can be rewritten as

$$RT(\ln f - \ln p) = -\int_0^p \left(\frac{RT}{p} - \overline{V}\right) dp$$

$$= \int_0^p \overline{V}\,dp - RT\int_0^p \frac{dp}{p}$$

The van der Waals equation can be written as

$$p = \frac{RT}{\overline{V} - b} - \frac{a}{\overline{V}^2}$$

p(atm)	1	10	20	50	60	80	100	120	140	160	180	200	220	260	280	300
p_r	0.078	0.78	1.56	3.91	4.69	6.25	7.81	9.38	10.94	12.5	14.4	15.63	17.19	20.31	21.88	23.44
$\ln p_r$	–2.55	–0.25	0.45	1.36	1.55	1.83	2.06	2.24	2.39	2.53	2.67	2.75	2.84	3.01	3.09	3.15
$Z-1$	.005	.005	0.01	0.026	0.031	0.042	0.052	0.063	0.073	0.083	0.096	0.104	0.115	0.135	0.146	0.156

So at constant temperature

$$dp = -\frac{RTd\overline{V}}{(\overline{V}-b)^2} + \frac{2ad\overline{V}}{\overline{V}^3}$$

Substituting the value of dp in the first integral

$$RT(\ln f - \ln p) = \int_{\infty}^{\overline{V}}\left[-\frac{RT\overline{V}\,d\overline{V}}{(\overline{V}-b)^2} + \frac{2a\overline{V}d\overline{V}}{V^3}\right] - RT\left[\ln p\right]_0^p$$

Add and subtract $\frac{RTb}{(\overline{V}-b)^2}$ in the first integral and eassange

$$= -\int_{\infty}^{\overline{V}}\frac{RT(\overline{V}-b)d\overline{V}}{(\overline{V}-b)^2} - \int_{\infty}^{\overline{V}}\frac{RTb}{(\overline{V}-b)^2}d\overline{V} + \int_{\infty}^{\overline{V}}\frac{2ad\overline{V}}{\overline{V}^2} - RT\left[\ln p\right]_0^p$$

$$= -RT\left[\ln \overline{V}-b\right]_{\infty}^{\overline{V}} + \left[\frac{RTb}{\overline{V}-b}\right]_{\infty}^{\overline{V}} - \left[\frac{2a}{\overline{V}}\right]_{\infty}^{\overline{V}} - RT\left[\ln p\right]_0^p$$

$$= -\left[RT\ln[p(V-B)]\right]_{\substack{p=0\\ \overline{V}=\infty}}^{p,\overline{V}} + \left[\frac{RTb}{\overline{V}-b}\right]_{\infty}^{\overline{V}} - \left[\frac{2a}{\overline{V}}\right]_{\infty}^{\overline{V}}$$

Now

$$\lim_{\substack{p\to 0\\ \overline{V}\to\infty}} p(\overline{V}-b) = p\overline{V} = RT$$

$$RT\ln f - RT\ln p = -RT\ln\left[p(\overline{V}-b)\right] + RT\ln RT + \frac{RTb}{\overline{V}-b} - \frac{2a}{\overline{V}}$$

Thus after reassanging we get

$$RT\ln f = RT\ln\left(\frac{RT}{\overline{V}-b}\right) + \frac{RTb}{\overline{V}-b} - \frac{2a}{\overline{V}}$$

or

$$\ln f = \ln\frac{RT}{\overline{V}-b} + \frac{b}{\overline{V}-b} - \frac{2a}{RT\overline{V}} \qquad \text{...(13.79)}$$

13.19.3. *Approximate Relations*

The following approximate relations can also be used to calculate the fugacities of gases obeying van der Waals equation and at low pressure.

At low pressures where $\frac{bp}{RT} \ll \frac{p\overline{V}}{RT} \cong 1$ for van der Waals gas, $a \cong \frac{a}{RT} - b$. So the expression (13.79) can be written as

$$\ln f = \ln p - \frac{ap}{(RT)^2} + \frac{bp}{RT} + \frac{abp^2}{2(RT)^3} \qquad \text{...(13.80)}$$

It is generally found that α is roughly constant at relatively low pressure *e.g.* H_2 is good case. Then equation (13.78) becomes

$$\ln f = \ln p - \frac{a}{RT} p \qquad ...(13.81)$$

This can further be approximated as

$$\frac{f}{p} = e^{-(\alpha p/RT)}$$

$$= 1 - \frac{\alpha p}{RT} \text{ (neglecting higher terms)}$$

$$= 1 - \frac{p}{RT}(-\overline{V} + \overline{V}_{\text{id}})$$

$$= 1 + \frac{p\overline{V}}{RT} - 1 = \frac{p\overline{V}}{RT} = Z$$

$$\frac{f}{p} = Z \qquad ...(13.82)$$

or

$$f = pZ$$

The equation (13.82) can also be written as

$$\frac{f}{p} = \frac{p}{p_{\text{id}}} \qquad ...(13.83)$$

If a gas obeys the Berthelot equation of state *e.g.*

$$p\overline{V} = RT + \left(\frac{aRT_c p}{128 p_c}\right)\left(1 - \frac{6T_c^2}{T^2}\right) \qquad ...(13.84)$$

The approximate value of fugacity can be calculated from the equation

$$\ln f = \ln p + \left(\frac{2T_c p}{128 p_c T}\right)\left(1 - \frac{6T}{T^2}\right) \qquad ...(13.85)$$

***Exercise* 13.46:** Calculate the fugacity of H_2 at 100°C and at 300 atm using the equation (13.79) $a = 0.244$ dm^6 atm mol^{-2}, $b = 0.0266$ dm^3 mol^{-1} and $\overline{V} = 0.119$ dm^3 mol^{-1}.

Substituting the values of various terms in equation (13.79)

$$\ln f = \ln\left[\frac{(0.0821 \text{ dm}^3 \text{ atm deg}^{-1} \text{ mol}^{-1})(373\text{K})}{(0.119 - 0.0266 \text{ dm}^3 \text{ mol}^{-1})}\right]$$

$$+ \frac{0.0266 \text{ dm}^3 \text{ mol}^{-1}}{(0.119 - 0.0266) \text{ dm}^3 \text{ mol}^{-1}}$$

$$- \frac{(2)(0.2244 \text{ dm}^6 \text{ atm mol}^{-2})}{(0.0821 \text{ dm}^3 \text{ atm K}^{-1} \text{ mol}^{-1})(373 \text{ K})(0.119 \text{ dm}^3 \text{ mol}^{-1})}$$

$$= 5.9574$$

$$f = 386.6$$

***Exercise* 13.47:** Calculate the fugacity of H_2 at 100°C at 300 atm using equation (13.80). $a = 0.244$ dm^6 atm mol^{-2} and $b = 0.0266$ dm^3 mol^{-1}.

Substituting the values of various terms in equation (13.80)

$$\ln \gamma = -\frac{(0.244 \text{ dm}^6 \text{ atm mol}^{-2})(300 \text{ atm})}{[(0.082 \text{ atm mol}^{-1} \text{K}^{-1} \text{dm}^3)(375 \text{ K})]^2}$$

$$+ \frac{(0.0266 \text{ dm}^3 \text{ mol}^{-1})(300 \text{ atm})}{(0.082 \text{ atm dm}^3 \text{ mol}^{-1} \text{K}^{-1})(373 \text{ K})}$$

$$+ \frac{(0.244 \text{dm}^6 \text{ atm mol}^{-2})(0.0266 \text{dm}^3 \text{ mol}^{-1})(300 \text{atm})^2}{(2)[(0.082 \text{atm mol}^{-1} \text{K}^{-1} \text{dm}^3)(373 \text{K})]^3}$$

$$= -0.0781 + 0.2606 + 0.0101$$

$$= 0.2028$$

$$\gamma = e^{0.2028} = 1.225$$

So $$f = p\gamma = 1.255 \times 300 = 367.5$$

***Exercise* 13.48:** Calculate the fugacity of H_2 at 100°C and 300 atm, using equation (13.83). The density of H_2 at the above conditions is 16.79 g dm^{-3}. The molecular weight of H_2 is 2.0 g mol^{-1}.

Now for ideal gas

$$d = \frac{pM}{RT}$$

So $$p_{id} = \frac{d.RT}{M}$$

$$= \frac{(16.79 \text{ g dm}^{-3})(0.0821 \text{ dm}^3 \text{ atm mol}^{-1} \text{ K}^{-1})(375 \text{K})}{(2 \text{ g mol}^{-1})}$$

$$= 257.08 \text{ atm}$$

Substituting the value of p and p_{id} in equation (13.83)

$$f = \frac{p^2}{p_{id}} = \frac{(300 \text{ atm})^2}{(257.08 \text{ atm})}$$

$$= 350.08 \text{ atm}$$

***Exercise* 13.49:** Calculate the fugacity of H_2 at 100°C and at 300 atm, using equation (13.85). Given $p_c = 12.8$ atm, $T_c = 33.3$ K.

The equation (13.85) is

$$\ln \gamma = \left(\frac{9T_c p}{128 p_c T}\right)\left(1 - \frac{6T_c^2}{T^2}\right)$$

$$= \frac{(9)(33.3\,\text{K})(300\,\text{atm})}{(128)(12.8\,\text{atm})(373\,\text{K})}\left(1 - \frac{(6)(33.3\,\text{K})^2}{(373\,\text{K})^2}\right)$$

$$\ln \gamma = 0.14012$$

$$\gamma = e^{+0.14012}$$

$$= 1.15$$

$$f = p \times \gamma = 300 \times 1.15 = 345 \text{ atm}$$

13.20. MIXTURE OF REAL GASES AND CERTAIN RELATIONS

(*i*) According to Lewis and Randall, the following approximate relations exist

$$f_i = x_i f_i^0 \qquad ...(13.86)$$

where f_i^0 is the fugacity of the pure component and f_i that of *i*th component in mixture.

Fugacity coefficient (γ_i) is defined as

$$\gamma_i = \frac{f_i}{p_i} \qquad ...(13.87)$$

(*ii*) The approximation that fugacity coefficient of *i*th component in mixture is equal to the fugacity coefficient of the pure component at the same temperature and pressure *i.e.*,

$$\gamma_i = \gamma_i^0 \qquad ...(13.88)$$

13.21. EQUILIBRIUM CONSTANTS FOR REACTIONS INVOLVING REAL GASES

Considering the general reaction (13.63) and the relations involving chemical potential and fugacity, equilibrium constant can be written as

$$K = K_f = \frac{f_L^{n_L}\, f_M^{n_M} \ldots}{f_A^{n_A}\, f_B^{n_B} \ldots} \qquad ...(13.89)$$

where f_i's are the respective fugacities. The change in standard free energy of the reaction and equilibrium constant are related by

$$\Delta G^0 = -RT \ln K_f \qquad ...(13.90)$$

Combining the equation (13.87), $K = K_f$ will be

$$K = K_f = \frac{(p_L\,\gamma_L)^{n_L}(p_M\gamma_M)^{n_M}\ldots}{(p_A\gamma_A)^{n_A}(p_B\gamma_B)^{n_B}\ldots}$$

$$= \frac{(p_L^{n_L})(p_M^{n_M})\ldots(\gamma_L)^{n_L}(\gamma_M)^{n_M}\ldots}{(p_A)^{n_A}(p_B)^{n_B}\ldots(\gamma_A)^{n_A}(\gamma_B)^{n_B}\ldots} \qquad ...(13.91)$$

$$= K_p K_\gamma$$

where K_p is the partial pressure equilibrium constant, K_γ is the corresponding ratio of the fugacity coefficients for the respective pure gases at the specified total pressure.

For ideal gases $\gamma_i = 1$. So $K = K_p$, same as equation (13.69).

***Exercise* 13.50:** The equilibrium constant for the reaction

$$N_2(g) + 3H_2(g) \rightleftharpoons 2\,NH_3\,(g)$$

at 450°C and 600 atm is 4.516×10^{-5}. (*a*) If the fugacity coefficients of pure gases at 450°C and 600 atm are $\gamma_{H_2} = 1.2874$, $\gamma_{N_2} = 1.3238$ and $\gamma_{NH_3} = 0.8548$ in atm, calculate the amount of NH_3 formed under these conditions, (*b*) Repeat the calculations assuming that gases are ideal.

(*a*) The equilibrium constant in thermodynamic sense is

$$K = \frac{f_{NH_3}^2}{f_{N_2}\, f_{H_2}^3}$$

$$= \frac{p_{NH_3}^2}{p_{N_2}\, p_{H_2}^3} \frac{\gamma_{NH_3}^2}{\gamma_{N_2}\, \gamma_{H_2}^3} = K_p\, K_\gamma$$

Let a mol of N_2 react with $3a$ mol of H_2 to give $2a$ mol of NH_3. To start with if N_2 and H_2 are in the ratio of 1:3, the amount of each species present at equilibrium is $1 - a$, $3(1 - a)$, $2a$. The partial pressure of N_2, H_2 and NH_3 will be $\frac{(1-a)\,600}{2(2-a)}$ atm, $\frac{3(1-a)}{2(2-a)} \times 600$ atm and $\frac{2a}{2(2-a)}$ 600 atm. Substituting the respective values

$$4.516 \times 10^{-5} = \frac{4a^2\,(600)^2}{4(2-a)^2} \cdot \frac{2(2-a)}{(1-a)\,600} \times \frac{8\,(2-a)^3}{27\,(1-a)^3(600)^3} \frac{(0.8548)^2}{(1.2874)^3(1.3238)}$$

Rearranging and solving the quadratic equation in a, the values of a are 0.43 and 1.56. The second value is not acceptable as the degree of conversion to NH_3 cannot be more than 1. So the degree of conversion of N_2 and H_2 to NH_3 at the above conditions is 43%.

(*b*) If one considers, the gases are ideal, under these conditions

$$K_{\text{therm}} = K_p$$

$$4.516 \times 10^{-5} = \frac{a^2(2-a)^2}{(1-a)^4} \frac{16}{(27)(600)^2}$$

Arranging the above equation and solving for a, gives $a = 0.55$.

***Exercise* 13.51:** Consider the Haber process for formation of NH_3 as $1/2\,N_2\,(g) + 3/2\,H_2\,(g) \rightleftharpoons NH_3\,(g)$. If the fractions of N_2, H_2, NH_3 and Ar present at equilibrium and at 300 atm and 450°C are 0.1407, 0.50, 0.355 and 0.0041 respectively, calculate the equilibrium constant (*b*) If γ_{N_2}, γ_{H_2} and γ_{NH_3} at this pressure are 1.153, 1.1163 and 0.9354 respectively, calculate the equilibrium constant. Compare the two values.

(*a*) Here using the relation (13.68 *a*), $\Delta v = -2$

$$K_p = K_x\,(P)^{\Delta v} = \frac{(0.355)}{(0.1407)^{1/2}\,(0.5)^{3/2}} \times \frac{1}{(300)^2} = 8.92 \times 10^{-3}$$

(*b*) Using the equation (13.91)

$$K_{\text{ther}} = K_p \times K_\gamma = 8.92 \times 10^{-3} \frac{0.9354}{(1.153)^{1/2}\,(1.1163)^{3/2}} = 6.59 \times 10^{-3}$$

Note: One can clearly see that at high pressure, K_p is not good thermodynamic equilibrium constant for real gases. One should use thermodynamic equilibrium constant.

PROBLEMS

Helmholtz and Gibb's Free Energy Calculation

13.1 Calculate the value of $\Delta G - \Delta A$ at 100°C for the following reactions

(*i*) $2C_2H_6(g) + 7\,O_2(g) \rightarrow 4\,CO_2(g) + 6H_2O(l)$

(*ii*) $CO_2\,(g) + H_2\,(g) \rightarrow CO\,(g) + H_2O\,(g)$

(*iii*) $H_2O_2(g) \rightarrow H_2O\,(g) + \frac{1}{2}O_2\,(g)$

(*iv*) $C_6H_6\,(g) + \frac{15}{2}O_2\,(g) \rightarrow 6\,CO_2\,(g) + 3\,H_2O\,(g)$

(*v*) $CH_3COOH\,(l) + 2\,O_2\,(g) \rightarrow 2\,CO_2\,(g) + 2\,H_2O\,(g)$

13.2 Calculate the change in Helmholtz free energy when two mol of ideal gas at 50°C and at a pressure of 1 atm is compressed isothermally and reversibly to 5 atm.

13.3 Calculate the value of ΔA for the reaction mentioned in problem 13.1 (*v*) from the following data. This reaction was carried out in bomb calorimeter. Initially the reactants were at one atmosphere and the reaction is complete. $\Delta H^0_{298} = -\,872.36$ kJ. The standard entropies of CH_3COOH, O_2, CO_2 and H_2O at 25°C are 159.83 J mol^{-1} deg^{-1}, 205.02 J mol^{-1} deg^{-1}, 43.64 J deg^{-1} mol^{-1} and 69.96 J deg^{-1} mol^{-1} respectively.

13.4 Calculate the value of ΔA, ΔE and ΔS, when one mol of an ideal gas is heated from 25°C to 125°C at constant volume. The $C_v = 20.90$ J deg^{-1} mol^{-1}, $S^0_{298} = 126.05$ J K^{-1} mol^{-1}.

13.5 Calculate the value of ΔG for the reaction mentioned in problem 13.3.

13.6 Calculate the value of ΔG for the following process

$$2\,A\text{ (gas, 298 K, 1 atm)} \rightarrow 2\,A\text{ (g, 227°C, 1 atm)}$$

Given, gas is ideal, $C_p = 29.29$ J deg^{-1} mol^{-1} and $S^0_{298} = 107$ J deg^{-1} mol^{-1}.

13.7 Calculate the value of ΔA for the reaction

$$2\,C_2H_6\,(g) + 7\,O_2\,(g) \rightarrow 4\,CO_2\,(g) + 6\,H_2O\,(l)$$

from the data of problem 13.3, $S^0_{298}\,(C_2H_6) = 229.49$ J deg^{-1} mol^{-1}.

13.8 Calculate the value of ΔG^0 of the reactions mentioned in the problem 13.1 (*i*) and 13.1 (*ii*) from the following data:

$\Delta H^0_f\,(H_2O, l) = -\,285.86$ kJ mol^{-1},

$\Delta H^0_f\,(CO, g) = -\,110.50$ kJ mol^{-1},

$\Delta H^0_f\,(C_2H_6 g) = -\,84.68$ kJ mol^{-1},

$\Delta H^0_f\,(CO_2, g) = -\,393.51$ kJ mol^{-1},

ΔH_f^0 (H_2O, g) = – 241.79 kJ mol^{-1},

ΔH_f^0 (H_2O_2, g) = – 136.11 kJ mol^{-1},

$\overline{S}^0$ (H_2O, g) = 188.7 EU, $\overline{S}^0$ (O_2, g) = 205.02 EU,

$\overline{S}^0$ (CO, g) = 197.9 EU, $\overline{S}^0$ (H_2, g) = 130.58 EU,

$\overline{S}^0$ (CO_2, g) = 213.80 EU, $\overline{S}^0$ (H_2O, e) = 69.94 EU,

$\overline{S}^0$ (C_2H_6, g) = 229.28 EU.

13.9 Calculate ΔG^0 of the relations mentioned in problem 13.1 with the following data and see whether the two values agree or not.

ΔG_f^0 (C_2H_6, g) = – 32.89 kJ mol^{-1},

ΔG_f^0 (CO_2, g) = – 394.38 kJ mol^{-1},

ΔG_f^0 (H_2O, l) = 237.19 kJ mol^{-1},

ΔG_f^0 (CO, g) = – 137.28 kJ mol^{-1},

ΔG_f^0 (H_2O, g) = – 228.53 kJ mol^{-1},

ΔG_f^0 (H_2O_2, g) = – 103.34 kJ mol^{-1}.

13.10 (*a*) Consider the formation of NH_3 *i.e.*,

$$H_2 (g) + 3 H_2 (g) \rightarrow 2 NH_3 (g)$$

Calculate ΔH^0, ΔS^0, ΔA^0 and ΔG^0 at 25° from the following data:

ΔH_f^0 (NH_3) = – 46.19 kJ mol^{-1}, $\overline{S}^0$ (N_2) = 191.50 EU, $\overline{S}^0$ (H_2) = 130.58 EU, $\overline{S}^0$ (NH_3) = 192.51 EU, ΔG_f^0 (NH_3) = – 16.64 kJ mol^{-1}. Also indicate which of the factors *i.e.*, ΔH, or ΔS are having more weightage for the spontaneity,

(*b*) Consider the reverse reaction. Calculate the temperature at which this reaction becomes spontaneous.

13.11 Calculate the following thermodynamic quantities (w, q, ΔE, ΔH, ΔA, ΔG, ΔS) when 2 mol of A are vapourised at its boiling point at 73.5°C and 1 atm pressure. ΔH_{vap} = 38.57 kJ mol^{-1}, d (A, l) = 0.7893 kg dm^{-3}, d (A vapour) = 1.24 × 10^{-4} kg dm^{-3}. The molecular weight is 36 g mol^{-1}.

13.12 2 mol of ethane initially at 27°C and 1 atm pressure are heated at constant pressure in such a manner that its volume increases 4 times. If we assume that ethane behaves as an ideal gas and $C_p = 5.753 + 175.11 \times 10^{-3}\, T - 578.52 \times 10^{-7}\, T^2$ (J mol^{-1} deg^{-1}), calculate w, q, ΔE, ΔH, ΔS, ΔA and ΔG.

13.13 Calculate the value of ΔG_{298}^0 for the reaction

$$3\, C_2H_2 (g) \rightarrow C_6H_6 (g)$$

from the data of following two reactions

(*i*) $2\,C\,(graphite) + H_2\,(g) \rightarrow C_2H_2\,(g)$

$$\Delta H^0_{298} = 226.75 \text{ kJ}, \ \Delta S^0_{298} = 58.84 \text{ EU}$$

(*ii*) $6\,C\,(graphite) + 3\,H_2\,(g) \rightarrow C_6H_6\,(g)$

$$\Delta H^0 = 82.93 \text{ kJ}, \ \Delta S^0_{298} = -\,152.51 \text{ EU}$$

13.14 6 g of liquid *A* enclosed in glass bulb was placed in an evacuated container maintained at 40°C. Now the glass bulb is broken and the liquid evaporated. When the equilibrium is reached at 40°C, it was found that 1 g of liquid has vaporised and the pressure is 100 mm. If the heat of vaporisation is 600 J g^{-1}, calculate *w, q,* ΔH, ΔG and ΔS for this process.

Coupled Reactions

13.15 The equilibrium vapour pressure of $CdBr_2 4H_2O\,(s)$ and $CdBr_2(s)$ at 25°C is 30 mm whereas that of $Mg(OH)_2$ and MgO at 25° is 4.5×10^{-4} mm. Calculate the free energy change for the process.

$$CdBr_2 \cdot 4H_2O + 4\,MgO \rightarrow CdBr_2(s) + 4\,Mg(OH)_2$$

13.16 The equilibrium vapour pressure of $CuSO_4 \cdot H_2O$ and $CuSO_4$ at 25°C is 2.94 mm and that of $Ca(OH)_2$ and CaO at 25°C is 4.7×10^{-9} mm. Calculate the free energy change for the process.

$$CuSO_4 \cdot H_2O\,(s) + GaO\,(s) \rightarrow CuSO_4\,(s) + Ca\,(OH)_2\,(s)$$

13.17 The standard free energy change for the reaction

$$TiO_2\,(s) + 2\,Cl_2\,(g) \rightarrow TiC1_4\,(l) + O_2\,(g) \ \Delta G^0 = 161.92 \text{ kJ}$$

and that for the reaction

$$C\,(s) + O_2\,(g) \rightarrow CO_2\,(g) \ \Delta G^0 = -\,394.55 \text{ kJ}$$

If the two reactions are coupled as

$$C\,(s) + TiO_2\,(s) + 2\,Cl_2\,(g) \rightarrow TiC1_4\,(l) + CO_2\,(g)$$

(*i*) will the reaction be feasible?

(*ii*) calculate the equilibrium constant at 298 K.

(*iii*) if the partial pressure of Cl_2 at equilibrium is 0.5 atm, calculate that of CO_2.

Criteria for Spontaneity

13.18 What are the criteria for equilibrium and spontaneity under the following conditions: (*a*) at constant *E* and *V* (*b*) at constant *p* and *S* (*c*) at constant *T* and *V*.

13.19 Heat of vapourisation of SO_2 is 24.916 kJ mol^{-1} and entropy of vapourisation is 94.684 EU. Calculate the temperature at which both the phases can co-exist.

13.20 It has been found that Al can be used for the reduction of some of the metallic oxides. Just from the free energies of the reduction reactions of the oxides BaO, Cu_2O and Fe_2O_3, show which of the reactions are spontaneous. $\Delta G^0_f\,(Al_2O_3) = -\,1576.53$ kJ mol^{-1},

$$\Delta G^0_f\,(BaO) = -\,528.44 \text{ kJ mol}^{-1},$$

$$\Delta G^0_f\,(Fe_3O_4) = -\,1014.20 \text{ kJ mol}^{-1},$$

$$\Delta G^0_f\,(Fe_2O_3) = -\,740.99 \text{ kJ mol}^{-1},$$

$$\Delta G_f^0 (Cu_2O) = -146.36 \text{ kJ mol}^{-1}.$$

13.21 If anyone of the above reactions is not feasible, calculate the temperature at which each of the above reaction becomes spontaneous. Given $\overline{S}^0$ (Al) = 28.32 EU, $\overline{S}^0$ (BaO, S) = 70.29 EU,

$$\overline{S}^0 (Ba) = 66.95 \text{ EU}, \ \overline{S}^0 (Cu) = 28.32 \text{ EU},$$

$$\overline{S}^0 (Fe_2O_3) = 89.96 \text{ EU}, \ \overline{S}^0 (Al_2O_3) = 50.99 \text{ EU},$$

$$\overline{S}^0 (Cu_2O) = 100.83 \text{ EU}, \ \Delta H_f^0 (Fe_2O_3) = -822.16 \text{ kJ mol}^{-1},$$

$$\Delta H_f^0 (Al_2O_3) = -1669.79 \text{ kJ mol}^{-1},$$

$$\Delta H_f^0 (Fe_3O_4) = -1117.13 \text{ kJ mol}^{-1},$$

$$\Delta H_f^0 (Cu_2O) = -166.69 \text{ kJ mol}^{-1}.$$

13.22 Consider the following processes

(*i*) A(*l*, 10°C) → A(*s*, 10°C)

(*ii*) A(*l*, 15°C) → A(*s*, 15°C)

(*iii*) A(*s*, 20°C) → A(*l*, 20°C)

and 15°C is the fusion temperature: (*a*) Which of the above process is spontaneous and why? (*b*) Which of the phase will be more stable?

Variation of *A* with Temperature and Pressure

13.23 The entropy of C_2H_6 at 25°C and 1 atm is 229.49 EU. If the heat capacity at constant volume is 44.56 J mol^{-1} deg^{-1}, calculate ΔA when C_2H_6 is heated to 127°C at constant volume. Assume that C_V remain constant in this range.

13.24 Consider the reaction

$$MO(s) = CO_2(g) \rightarrow MCO_3(s)$$

with $\Delta A_{298} = -20.5$ kJ and $\Delta E_{298} = -80.4$ kJ. If the stoichiometric amount of MO and CO_2 are sealed in the constant volume vessel at 25°C the value of ΔA_{298} shows that the above process is spontaneous. Find out whether the process is spontaneous even at 600°C.

13.25 Calculate the values of ΔA for the following processes,

$$d_{CCl_4} = 1.594 \text{ kg dm}^{-3}$$

(*i*) CCl_4(*l*,76.75°C, 1 atm) → CCl_4 (*g*, 76.75°C, 1 atm)

(*ii*) CCl_4(*l*, 76.75°C, 1 atm) → CCl_4(*g*, 76.75°C, 100 mm)

where 76.75°C is the boiling point of CCl_4 and vapours are ideal.

Variation of *G* with Pressure and Temperature

13.26 If 20 g of liquid ethanol is converted into vapours at 34.9°C and pressure of 2 mm, calculate the free energy change. $\Delta H_{vap} = 41.46$ kJ mol^{-1} and boiling point is 78°C. The equilibrium vapour pressure of ethanol at 34.9°C is 100 mm. Will this transformation of liquid ethanol to vapour ethanol be a spontaneous?

13.27 The vapour pressures of solid and liquid carbon dioxide at – 57°C are 3294.6 and 3495.0 mm respectively. Calculate the change in the free energy of this process.

13.28 Consider the process

$$H_2O(l, -15°C) \rightarrow H_2O(s, -15°C)$$

(*a*) If the vapour pressures of liquid water and ice at – 15°C are 1.241 and 1.436 mm of Hg, calculate the free energy change for this process.

(*b*) Suppose that the process is carried out in the thermostat maintained at –15°C and both form isolated system. Show that this process is spontaneous $\Delta H_{fus}(H_2O) = 333.46 \text{ J g}^{-1}$ at 0°C and heat capacities of ice and water at –15°C are 2.00 J deg $^{-1}$ mol^{-1} and 4.184 J deg^{-1} mol^{-1} respectively.

13.29 Compute the free energy change for the following process if the equilibrium vapour pressure of water at 50°C is 92.51 mm.

$$H_2O(l, 50°C\ 1\ atm) = H_2O(g, 50°C, 1\ atm)$$

13.30 The equilibrium vapour pressure of water at 25ºC over the mixture of $CuSO_4$ and $CuSO_4 \cdot H_2O$ is 2.94 mm of Hg. Calculate the free energy of the following processes:

(*a*) $CuSO_4 \cdot H_2O(s) \rightarrow CuSO_4(s) + H_2O(g)$ $\quad p = 2.94$ mm

(*b*) $CuSO_4 \cdot H_2O(s) \rightarrow CuSO_4(s) + H_2O(g)$ $\quad p = 1$ atm.

13.31 The equilibrium vapour pressure of $CuSO_4 \cdot 5H_2O$ and $CuSO_4 \cdot 3H_2O$ at 25°C is 7.8 mm. Calculate the free energy changes of the processes.

(*a*) $CuSO_4 \cdot 3H_2O(s, 100\text{ mm}, 25°C) + H_2O(g, 100\text{ mm}, 25°C)$
$\rightarrow CuSO_4 \cdot 5H_2O(s, 100\text{ mm}, 25°C)$

13.32 Using the data given in problems (13.1) and (13.8) and the following data, calculate the ΔG for reactions 13.1 (*i*), 13.1 (*ii*) at 1000 K.

$$C_p(O_2) = 25.72 + 12.979 \times 10^{-3}T - 38.62 \times 10^{-7}T^2$$
$$C_p(H_2) = 29.066 - 0.836 \times 10^{-3}T + 20.117 \times 10^{-7}T^2$$
$$C_p(CO_2) = 26.00 + 43.50 \times 10^{-3}T - 148.32 \times 10^{-7}T^2$$
$$C_p(CO) = 26.86 + 6.966 \times 10^{-3}T - 8.20 \times 10^{-7}T^2$$
$$C_p(H_2O) = 30.36 + 9.615 \times 10^{-3}T + 11.84 \times 10^{-7}T^2$$
$$C_p(C_2H_6) = 9.40 + 159.83 \times 10^{-3}T - 462.25 \times 10^{-7}T^2$$

The values of C_p are J mol^{-1} deg^{-1}.

13.33 If it is assumed that ΔH is independent of temperature in the range mentioned in the problem 13.32, calculate the value of ΔG at 1000 K for each reaction. Compare the results with those obtained in the problem 13.32 and calculate the error in each case.

13.34 The heat capacity (C_p) of O_2 at 25°C is 29.14 J deg^{-1} mol^{-1} and standard entropy at 25°C is 205.1 J deg^{-1} mol^{-1}. If it is assumed that heat capacity remains constant in the range 298 K to 500 K find ΔG when 1 mol of O_2 is heated from 298 K to 500 K at constant pressure of 1 atm.

13.35 Repeat the problem 13.34 if C_p of O_2 is given by

$$C_p = 29.96 + 4.18 \times 10^{-3}T - 1.67 \times 10^{+5}T^{-2}$$

13.36 The standard free energies of formation at 25°C of Cis-2-butene and *n* butane are 65.86 kJ mol^{-1} and –15.69 kJ mol^{-1}. Consider the reaction

$$H_3C - CH = CH - CH_3(g) + H_2(g) \rightarrow nC_4H_{10}(g)$$

(*a*) Calculate the free energy change for this reaction,

(*b*) Knowing the standard free energies of the species, calculate the free energies of cis 2-butene at pressure of 20 mm, H_2 at 100 mm and butane at 600 mm, (*c*) Calculate the free energy change of the above reaction if the gases are present at the above partial pressures,

(*d*) Compare this value with that obtained in part (*a*) and discuss the criteria of spontaneity in each case.

Maxwell's Relations

13.37 Show that the equation

$$TdS = dE + p\,dV$$

can be written also in the form

$$TdS = C_V dT + T\frac{\alpha}{\beta}dV$$

$$TdS = \frac{\beta C_V}{\alpha}dp + \frac{C_p}{\overline{V}\alpha}dV$$

$$TdS = C_p\,dT - VT\,a\,dp$$

13.38 Prove the following relations

(*i*) $\left(\frac{\partial E}{\partial T}\right)_V = C_p - \frac{VT}{\beta}\alpha$

(*ii*) $\left(\frac{\partial E}{\partial T}\right)_p = C_p - p\overline{V}\alpha$

(*iii*) $\left(\frac{\partial E}{\partial V}\right)_T = \frac{Ta}{\beta} - p$

(*iv*) $\left(\frac{\partial E}{\partial p}\right)_T = p\overline{V}\beta - \overline{V}Ta$

13.39 The term *TdS* can also be written as

$$TdS = C_V dT + T\left(\frac{\partial p}{\partial T}\right)_V dV$$

Using the above expression calculate the heat absorbed when 2 mol of gas obeying van der Waals equation of state is expanded reversibly and isothermally from 2 dm ^{3} to 10 dm^3.

$$a = 4.39 \text{ dm}^6 \text{ atm mol}^{-2}, b = 0.52 \text{ dm}^3 \text{ mol}^{-1}$$

13.40 Using the expression mention in problem 13.37, calculate the heat transferred when 0.200 dm ^{3} of liquid *A* at 5°C is compressed isothermally and reversibly by increasing the pressure from 1 atm to 200 atm. The value of $\alpha = 2.50 \times 10^{-4}$ deg^{-1} and is assumed to be constant in this pressure range. Change in volume is also negligible.

13.41 The values of α and β for $CC1_4$ are 12.4×10^{-4} deg^{-1} and 107×10^{-6} atm^{-1} respectively, $d = 1.594$ kg dm^{-3}. Using the relations from problems 13.39 and 13.40 or otherwise calculate *Q, W* and Δ*E* when 200 g of $CC1_4$ is compressed isothermally and reversibly from 0 to 1000 atm at 27°C.

13.42 Prove the following relations

(i) $\left(\dfrac{\partial H}{\partial p}\right)_T = \overline{V} - T\left(\dfrac{\partial V}{\partial T}\right)_p$

(ii) $\left(\dfrac{\partial H}{\partial V}\right)_T = T\left(\dfrac{\partial p}{\partial T}\right)_V + \overline{V}\left(\dfrac{\partial p}{\partial V}\right)_T$

(iii) $\left(\dfrac{\partial H}{\partial S}\right)_T = T - \overline{V}\left(\dfrac{\partial T}{\partial V}\right)_p$

(iv) $\left(\dfrac{\partial H}{\partial S}\right)_V = T - \overline{V}\left(\dfrac{\partial V}{\partial T}\right)_S$

(v) $\overline{H} = \overline{G} - T\left(\dfrac{\partial G}{\partial T}\right)_p$

(vi) $C_p = -T\left(\dfrac{\partial^2 G}{\partial T^2}\right)_p$

(vii) $dH = C_p dT + (\overline{V} - \overline{V}\, T\,\alpha)\, dp$

(viii) $\left(\dfrac{\partial H}{\partial T}\right)_V = C_V + V\left(\dfrac{\partial p}{\partial T}\right)_V$

(ix) $\dfrac{\partial^2 H}{\partial T \partial p} = T\dfrac{\partial^2 S}{\partial T \partial p}$

13.43 With the help of Maxwell's equations, show that

$\left(\dfrac{\partial E}{\partial V}\right)_T = \left(\dfrac{\partial E}{\partial p}\right)_T = 0$ for ideal gas ($p\overline{V} = RT$).

(**Hint:** Start from $dE = TdS - pdV$)

13.44 Prove that C_p is never less than C_V for every material.

$$\left[\textbf{Hint: } \text{Prove the relation } C_p - C_V = -T\left(\frac{\partial V}{\partial T}\right)_p^2 \left(\frac{\partial p}{\partial T}\right)_V = \frac{\overline{V}\, T\alpha^2}{\beta}\right]$$

13.45 Show that

(i) $\left(\dfrac{\partial C_V}{\partial V}\right)_T = T\left[\dfrac{\partial}{\partial T}\left(\dfrac{\alpha}{\beta}\right)\right]_V$

(ii) $\left(\frac{\partial C_p}{\partial p}\right)_T = -T\left[\frac{\partial}{\partial T}(\bar{V}\alpha)\right]_p$

(iii) C_p or C_V is only a function of temperature for ideal gas,

i.e., $\left(\frac{\partial C_V}{\partial V}\right)_T = \left(\frac{\partial C_p}{\partial p}\right)_T = 0$

$\left[\textbf{Hint: Get } \left(\frac{\partial C_V}{\partial V}\right)_T = T\left(\frac{\partial^2 p}{\partial T^2}\right)_V\right]$

(iv) $\left(\frac{\partial C_V}{\partial V}\right)_T = 0$ for van der Waals gas.

13.46 (*a*) Show that the Joule-Thomson coefficient can also be written as

$$\mu_{JT} = \frac{T^2}{C_p}\left[\frac{\partial}{\partial T}\left(\frac{\partial \bar{V}}{\partial T}\right)\right]_p$$

(*b*) Find the expression for μ for an ideal gas as well as for the gas obeying $p(\bar{V} - b) = RT$.

13.47 If the isothermal compressibility and adiabatic compressibility are defined as

$$\beta_T = -\frac{1}{V}\left(\frac{\partial V}{\partial p}\right)_T, \beta_S = -\frac{1}{V}\left(\frac{\partial V}{\partial p}\right)_S$$

(*a*) Show that adiabatic compressibility is always less than or equal to isothermal compressibility.

(*b*) Show $\frac{\beta_T}{\beta_S} = \gamma$ where γ is the ratio of heat capacities at constant pressure and at constant volume.

Free Energy of Mixing

13.48 Calculate free energy change *i.e.* ΔG^0 for each of the process

(*i*) Acetone (*l*) → Acetone ($l, x_{\text{Acetone}} = 0.7$)

(*ii*) Acetone ($l, x_{\text{Acetone}} = 0.2$) + $CHCl_3$ ($l, x_{\text{CHC13}} = 0.8$)
→ Soln

(*iii*) when 2 mol of Acetone, 5 mol of $CHC1_3$ and 2 mol of ethyl alcohol are mixed

(*iv*) find $\Delta_{\text{mix}}S^0$ for each of the above process.

13.49 If the standard free energy of formation of carbontetrachloride and ethanol are –68.62 kJ mol^{-1} and –174.77 kJ mol^{-1}, calculate the free energy of the mixture when mole fractions of carbontetrachloride and ethanol in the solution are 0.44 and 0.56 respectively.

13.50 Consider the process

$$\text{Cispentene} \rightleftharpoons \text{Cyclopentene}$$

ΔH_f^0 (cispentene) = – 77.40 kJ mol^{-1}, ΔH_f^0 (cyclo-pentene) = 32.98 kJ mol^{-1}, $\bar{S}_f^0$ (cispentene) = 292.88 EU, $\bar{S}^0$ (cyclopentene) = 289.53 EU. (*i*) Calculate the free energy of formation of each species and the process, (*ii*) Suppose that gases do not transform into each other. Using the data from part (*i*), calculate the free energy of each species in a mixture containing $x_{\text{cia-pentene}} = 0.728$ and $x_{\text{cypentene}} = 0.272$ and each gas at a pressure of 1 atm. (*iii*) Calculate the free energy of the mixture formed in part (*ii*) but still unreactive and total pressure is 1 atm. (*iv*) If the equilibrium is reached quickly, calculate K_p. Will it depend on total pressure? (*v*) Calculate the composition of the mixture at equilibrium, (*vi*) Calculate the chemical potential of each species at equilibrium.

Equilibrium Constant

13.51 Calculate the equilibrium constants for the reactions mentioned in the problem 13. 1(*i*), using the data either obtained in problem 13.8 or 13.9.

13.52 Calculate the equilibrium constants expressed as K_x and K_c for the reactions mentioned in problem 13.1 using the data from problem 13.51.

13.53 Consider the reaction

$$H_2(g) + Br_2(g) \rightarrow 2HBr(g)$$

Calculate (*a*) ΔG^0 of the reaction. Given ΔH_f^0 (HBr) = – 36.23 kJ mol^{-1}, $\overline{S^0}(H_2)$ = 130.58 EU, $\bar{S}^0$ (Br_2, g) = 245.18 EU, $\bar{S}^0$(HBr) = 198.32 EU. (*b*) K_p, K_c and K_x at 25°C. (*c*) If 2 mol of H_2 and Br_2 each, were mixed together, calculate the partial pressure of each species at equilibrium. (*d*) Assume that ΔH^0 of the above reaction remains constant over the range of temperature at which the reaction is studied. Calculate the value of K_p at 1000 K.

13.54 Consider the reaction

$$2H_2S(g) + 3O_2(g) \quad = \quad 2H_2O(g) + 2SO_2(g)$$

(*a*) Calculate, the value of ΔH^0.

(*b*) Calculate ΔG^0 at 298 K.

(*c*) What will happen to K_p if the temperature is increased or decreased.

(*d*) If the total pressure is increased or decreased what will be the effect on K_p, K_c or K_x.

(*e*) Calculate the value of K_p at 298 K.

13.55 Consider the reaction

$$H_2O_2(g) \quad = \quad H_2O(g) + 1/2\, O_2(g)$$

(*a*) Calculate ΔG^0 of this reaction, given;

$$\Delta G_f^0\,(H_2O, g) \quad = \quad -\,227.58 \text{ kJ mol}^{-1},$$

$$\Delta G_f^0\,(H_2O_2, g) \quad = \quad -\,103.34 \text{ kJ mol}^{-1}.$$

(*b*) Calculate K_p, K_x and K_c at 25°C.

(*c*) Calculate the value of K_p at 227°C assuming ΔH^0 remains constant.

(*d*) If the pressure of $H_2O_2(g)$ is 1 atm in beginning where as $p_{O2} = 0$ and $p_{H2O} = 0$, calculate the partial pressures of $H_2O_2(g)$ and O_2 at the equilibrium temperatures 25°C and 227°C.

13.56 The equilibrium constants for the following reactions at 1395 K

$$2H_2O \rightleftharpoons 2H_2 + O_2 \qquad K_1 = 2.1 \times 10^{-13}$$

$$2CO_2 \rightleftharpoons 2CO + O_2 \qquad K_2 = 1.4 \times 10^{-12}$$

Calculate the value of K for the reaction

$$H_2 + CO_2 \rightleftharpoons CO + H_2O$$

13.57 Calculate the equilibrium constant for the reaction

$$O_2 + 4HCl \rightleftharpoons 2H_2O + 2Cl_2 \text{ at 1565 K from the data}$$

$$2H_2O \rightleftharpoons 2H_2 + O_2 \quad K_1 = 1.6 \times 10^{-11}$$

$$2HCl \rightleftharpoons H_2 + Cl_2 \quad K_2 = 2.22 \times 10^{17}$$

13.58 Consider the equilibrium reaction

$$2A + B \rightleftharpoons A_2B$$

where A, B and A_2B are stable gaseous molecules. The temperature at the equilibrium is 800°C and the total pressure is 6 atm. The analysis of the equilibrium mixture was carried out by drawing 1 dm^3 of sample at STP. It is found that mixture contains 0.6 dm^3 (at STP) of compounds A and A_2B and rest of B. Calculate the equilibrium constant.

13.59 Consider the equilibrium reaction

$$A(g) + B(g) \rightleftharpoons C(g) + D(g)$$

at 500°C. To start with, 2 g of A and 5 g of B were introduced in the evacuated flask maintained at above temperature. At equilibrium the total pressure is 0.8 atm. The analysis shows that mole fraction of compound D in the equilibrium mixture is 0.2. If the molecular weights of A, B, C and D are 32 g, 40 g, 56 g and 64 g per mol respectively, calculate the equilibrium constant.

13.60 It is found that equilibrium constant increases by a factor of four when the temperature is increased from 25°C to 40°C. Calculate the value of ΔH^0.

13.61 Following values of equilibrium constants for the inhibitor binding to the enzyme carbonic anhydrase are obtained. Calculate ΔG^0, ΔH^0 and ΔS^0 at 25°C for this process.

T°C	15	20	25	30	35
$K(\times 10^{-7})$(M)	7.26	5.11	4.19	3.60	2.32

13.62 The variation of equilibrium constant with temperature for the reaction

$$H_2(g) + I_2(g) \rightleftharpoons 2HI(g)$$

is

$T(K)$	457.6	490.8	393.8
K	49.54	45.62	60.80

Calculate the value of ΔH if it is assumed to be constant in this range.

13.63 In the manufacture of SO_3 from SO_2 by the reaction

$$2SO_2(g) + O_2(g) \rightleftharpoons 2SO_3(g)$$

it has been found that variation of equilibrium constant with temperature in the range 800 to 1170°C can be expressed as

$$\log K_p = -9.346 + \frac{9888}{T}$$

Calculate ΔH of this reaction.

13.64 The variation of equilibrium constant with temperature for the formation of ammonia at total pressure of 10 atm are

$$N_2 + 3H_2 \rightleftharpoons 2NH_3$$

$T(K)$	623	673	723	773
$10^4 \times$ K(atm)	7.08	1.66	0.434	0.145

Calculate the heat of formation of this from the data of C_p

$C_p(H_2) = 29.07 - 0.84 \times 10^{-3}\,T + 20.12 \times 10^{-7}\,T^2$
$C_p(N_2) = 26.98 + 5.91 \times 10^{-3}\,T - 3.38 \times 10^{-7}\,T^2$
$C_p(NH_3) = 25.90 + 33.00 \times 10^{-3}\,T - 30.46 \times 10^{-7}\,T^2$

13.65 In the water gas reaction

$$CO_2(g) + H_2(g) \rightleftharpoons CO(g) + H_2O(g)$$

the equilibrium constants calculated at temperature 959 K and 1259 K are 0.534 (atm) and 1.571 (atm) respectively. Calculate the value of ΔH^0 by using the data of C_p from problems 13.64 and 13.66. Also calculate the constant of integration.

13.66 In the following reaction

$$C(s) + CO_2(g) \rightleftharpoons 2CO(g)$$

Calculate the value of K_p at 1100°C when, $\Delta H^0 = 186.46$ kJ

C_p(graphite) $= 1.1 + 4.8 \times 10^{-3}\,T + 1.2 \times 10^{-6}\,T^2$
$C_p(CO_2) = 26.00 + 43.50 \times 10^{-3}\,T - 148.32 \times 10^{-7}\,T^2$
$C_p(CO) = 26.86 + 6.966 \times 10^{-3}\,T - 8.20 \times 10^{-7}\,T^2$
$\Delta G^0(298) = 120.003$ kJ

13.67 What will be the effect of increasing pressure over the following reaction?

(*i*) $H_2(g) + Br_2(g) \rightleftharpoons 2HBr(g)$ $\quad \Delta H^0 = -106.36$ kJ

(*ii*) $H_2O_2(g) \rightleftharpoons H_2O(g) + \frac{1}{2}O_2(g)$ $\quad \Delta H^0 = -105.68$ kJ

(*iii*) $2H_2S\,(g) + 3\,O_2 \rightleftharpoons 2H_2O\,(g) + 2\,SO_2\,(g)$ $\quad \Delta H^0 = -237.07$ kJ

(*iv*) $Br_2\,(g) \rightleftharpoons 2\,Br\,(g)$ $\quad \Delta H^0 = 223.5$ kJ

(*v*) $CO_2\,(g) + H_2\,(g) \rightleftharpoons CO\,(g) + H_2O\,(g)$ $\quad \Delta H^0 = 41.22$ kJ

(*vi*) $2\,SO_2(g) + O_2(g) \rightleftharpoons 2\,SO_3\,(g)$ $\quad \Delta H^0 = -140$ kJ

(*vii*) $C\,(s) + CO_2\,(g) \rightleftharpoons 2\,CO\,(g)$ $\quad \Delta H^0 = -172.47$ kJ

13.68 What will be the effect of increasing the temperature over the reactions mentioned in the problem 13.67? The heat of the reaction is mentioned along with the reaction.

13.69 Consider the dissociation of A_2 gas into atoms at 298 K.

$$A_2\,(g) \rightarrow 2\,A\,(g)$$

If the heat of dissociation of A_2 (*g*) is 82.38 kJ, calculate the degree of dissociation (*a*) when the pressure is 40 mm and (*b*) when the pressure is 200 mm.

13.70 Calculate the amount of HI formed at equilibrium when 4.58×10^{-3} mol of H_2 and 0.9733 mol of I_2 were mixed. The equilibrium constant at 457.6°C is 48.7.

13.71 A one litre reaction vessel containing 0.015 mol of N_2 and 0.02 mol of PCl_5 are heated to 227°C. The total pressure at equilibrium is 1.843 atm. Assuming all gases are ideal, calculate K_p for the following reaction

$$PCl_5\,(g) \rightleftharpoons PCl_3\,(g) + Cl_2\,(g)$$

13.72 Calculate percentage of ammonia at the equilibrium at 400°C and total pressure of 10 atm, using the data from the problem 13.64.

13.73 The ΔG^0 for the reaction $SO_2(g) + \frac{1}{2}O_2\,(g) \rightarrow SO_3\,(g)$ is $\Delta G^0 = -22600 + 21.41\,T$ (cal).

Calculate the temperature at which the conversion of SO_2 to SO_3 is 80% if the initial mixture contains 15% SO_2 and 20% O_2 and $P = 1$ atm.

13.74 Ammonium chloride decomposes as

$$NH_4Cl\,(s) \rightleftharpoons NH_3\,(g) + HCl\,(g)$$

(*a*) If the total pressure observed at equilibrium and at 25°C is 0.618 atm, calculate the equilibrium constant. Given $P_{NH_3} = P_{HCl} = 0$ at zero time.

(*b*) If, in the above experiment, $P_{NH_3} = 0.108$, is the initial pressure of NH_3 and the equilibrium constant is 9.40×10^{-2}, calculate the partial pressures of NH_3 and HCl if the total pressure in the system is 0.622 atm.

13.75 The total pressure observed at equilibrium in the dissociation reaction of ammonium carbamate is 1.35 atm at certain temperature.

$$NH_4CO_2NH_2\,(s) \rightleftharpoons 2\,NH_3\,(g) + CO_2\,(g)$$

Calculate the value of equilibrium constant.

13.76 In the following reaction

$$CuSO_4 \cdot 3H_2O\,(s) + 2\,H_2O(g,\,23.7\text{ mm}) \rightarrow CuSO_4 \cdot 5H_2O\,(s)$$

the equilibrium vapour pressure of water over $CuSO_4 \cdot 3H_2O$ (*s*) and $CuSO_4 \cdot 5H_2O$ (*s*) at 25°C is 7.8 mm.

(*a*) Calculate the free energy for the above reaction.

(*b*) Calculate the equilibrium constant.

Fugacity

13.77 The virial equation for N_2 at 0°C is

$$p\bar{V} = 22.414 - 1.027 \times 10^{-2}p + 6.3844 \times 10^{-5}p^2$$

Calculate the value of fugacity at 0°C and 100 atm pressure.

13.78 The virial equation for N_2 at 100°C is

$$p\bar{V} = 30.619 + 0.656 \times 10^{-2}p + 3.3835 \times 10^{-5}p^2$$

Calculate the value of fugacity at 100°C and 100 atm.

13.79 Calculate the value of *a* ($a = \bar{V} - \bar{V}_{id}$) for N_2 at 1 atm, 5 atm, 10 atm, 50 atm, 100 atm and finally the value of fugacity at 100 atm and temperature equal to 0°C and 100°C. Compare the values obtained in problems 13.77 and 13.78.

13.80 The van der Waals constants for N_2 are, $a = 1.39$ dm^{-6} atm mol^{-2} and $b = 0.0442$ dm^3 mol^{-1}. Calculate the values of fugacities of N_2 at (*a*) 0°C and 100 atm (*b*) at 100°C and 100 atm. Compare the values obtained in earlier problems 13.77, 13.78 and 13.79.

13.81 The van der Waals constants for CO_2, H_2, CO and H_2 are

	a (*atm dm*$^{-6}$ *mol*$^{-2}$)	*b* (*dm*^{3} *mol*$^{-1}$)
CO_2	3.59	0.0427
CO	1.49	0.0399
H_2O	5.46	0.0305
H_2	0.244	0.0266

Calculate the values of fugacities of each species at 400°C and 100 atm pressure using problems 13.79 and 13.80.

13.82 The molecular wt. of N_2 is 28.0 g and densities at 0°C and 100 atm and 100°C and 100 atm are 1.273×10^{-1} kg dm^{-3} and 0.093 kg dm^{-3} respectively. Calculate the fugacities under the above conditions using problem 13.83. Find the percentage error if we assume that values of fugacities of N_2 obtained in problems 13.77 and 13.78 are correct.

13.83 Calculate the fugacity of liquid chlorine at 0°C if the vapour pressure at 0°C is 3.66 atm and the density of Cl_2 vapours at 0°C and 3.66 atm is 11.81 kg m^{-3}.

13.84 The reduced temperature and reduced pressure of N_2 are $T_c = 126.1$ K and $p_c = 33.5$ atm. Calculate the value of fugacity and fugacity coefficient of N_2 at (*i*) 100°C and 100 atm, and (*ii*) 0°C and 100 atm from the Fig. 13.3.

13.85 The critical constants of H_2, CO, CO_2 and H_2O are:

	T_c (°K)	p_c (atm)
H_2	33.3	12.8
CO	134.6	34.4
CO_2	304.2	72.8
H_2O	647.3	217.7

Calculate the fugacities of the above gases at 400°C and 100 atm from the chart.

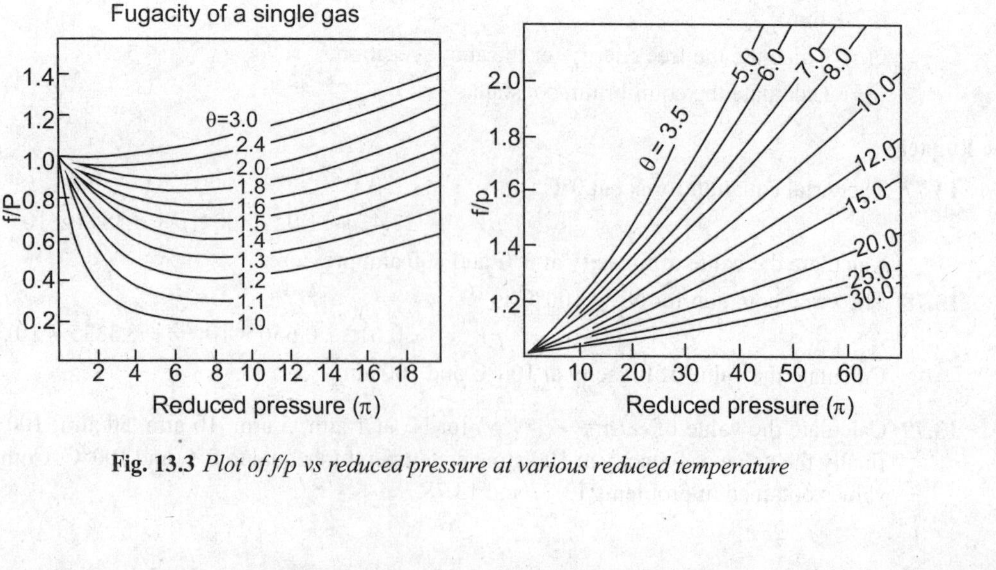

Fig. 13.3 *Plot of f/p vs reduced pressure at various reduced temperature*

13.86 The following is the data of volume at different pressures. Calculate the fugacities of CO_2 at 100 atm and 100°C by using problem 13.78 *i.e.*, by graphical method.

p(atm)	1	20	10	60	80	100	40
$\overline{V}$ (dm^3 mol^{-1})	26.2	1.20	2.52	0.32	0.18	0.098	0.54

13.87 Using the Fig. 8.3 calculate the values of Z at number of pressures for H_2. Critical temperature and pressure are 33.3 K and 12.8 atm. Calculate the value of fugacity at 100°C and 100 atm.

13.88 The fugacity of gas O_2 at 0°C and 50 atm pressure is 48.0 atm. Calculate the amount of correction to be made to ΔG^0 for O_2 at the above conditions. Whether the correction is to be added or subtracted?

13.89 The fugacities of O_2 at 0°C and at pressures 100 atm and 600 atm are 92.5 and 540 respectively. Calculate the free energy change for the process.

$$O_2\ (g,\ 600\ \text{atm}) \rightarrow O_2\ (g,\ 100\ \text{atm})$$

Calculate, also, the free energy change for the above process if O_2 is considered ideal gas and the percentage error.

13.90 Calculate the equilibrium constant for the reaction

$$CO_2\ (g) + H_2\ (g) \rightleftharpoons CO\ (g) + H_2O\ (g)$$

at 400°C and 100 atm, using the data from the problem 13.81.

Addition Problems

13.91 Dissolution of some ionic salts in water leads to cooling and it is termed as spontaneous process. Explain whether enthalpy of reaction or entropy of reaction contribute to this process?

13.92 Express variation of enthalpy of reaction with pressure at constant temperature by taking variables as p, V and T only.

13.93 Express Joule-Thomson coefficient in terms of $\overline{C}_p$ and the variables p, V and T only.

13.94 Derive the general expression for the variation of Gibbs free energy with temperature in terms of heat capacity at constant pressure and temperature.

13.95 Two moles of an ideal gas is expanded from 2.5 L to 5.4 L at 27 °C. (*i*) Calculate the values of Q, W, ΔE, ΔH, ΔA and ΔG, if the process is carried out isothermally and reversibly, (*ii*) isothermally and irreversibly against the pressure of 1.2 atm.

13.96 Water gas is obtained by the following reaction:

$$H_2O(g) + C(s) \rightarrow CO(g) + H_2(g)$$

From the given data suggest whether the reaction is feasible at 298k? If not calculate the temperature at which the given reaction will be feasible? $\Delta_f H^0$ [H_2O, g] = –241.82 kJ K^{-1} mol^{-1}, $\Delta_f H^0$[CO, g], = –110.53 kJ K^{-1} mol^{-1}, S^0 (C, s) = 5.470 J K^{-1} mol^{-1}, S^0 (CO, g) = 197.67 J K^{-1} mol^{-1}, S^0 (H_2, g) = 130.684 J K^{-1} mol^{-1}, S^0 (H_2O, g) = 109.6 J K^{-1} mol^{-1}.

13.97 5 moles ideal gas is expanded isothermally from 45 L to 80 L at 298 K. Calculate the values of ΔG and ΔH for this process.

13.98 Certain bacteria in the soil obtain the necessary energy for the growth by oxidising nitrite to nitrate as

$$2NO_2^-(aq) + O_2(g) \rightarrow 2NO_3^-(aq)$$

$\Delta_f G^0[NO_2^-\ (aq)] = -34.6$ kJ mol^{-1}, $\Delta_f G^0[NO_3^-\ (aq)] = -110.5$ kJ mol^{-1} respectively.

Calculate the amount of Gibb's energy released when one mole of NO_2^- is oxidised to one mole of NO_3^-.

13.99 Calculate the value of $\Delta_r G^0$ for the fermentation reaction of α-*D*-glucose:

$$\alpha\text{-D-glucose } (aq) \rightarrow 2C_2H_5OH\ (l) + 2CO_2(g)$$

Given $\Delta_f G^0$ [α-D-glucose (*aq*)] = – 914.5 kJ mol^{-1}, $\Delta_f G^0$ [C_2H_5OH (*l*)] = –174.78 kJ mol^{-1}, $\Delta_f G^0$ [$CO_2(g)$] = –394.36 kJ mol^{-1}.

13.100 Urea is obtained from the reaction: $CO_2(g) + 2NH_3(g) \rightarrow (NH_2)_2CO(s) + H_2O(l)$. Calculate the value of $\Delta_r G^0$ for the reaction at 298 K. If it is assumed that the gases behave as ideal gases, calculate the value of $\Delta_r G^0$ for the reaction at 10. 0 atm. Given $\Delta_f G^0$ [CO_2 (*g*)] = –394.36 kJ mol^{-1}, $\Delta_f G^0$ [H_2O (*l*)] = –228.57 kJ mol^{-1}, $\Delta_f G^0$ [$NH_3(g)$] = –16.45 kJ mol^{-1}, $\Delta_f G^0$ [urea (*s*)] = –197.15 kJ mol^{-1}.

13.101 Crystallisation of the compounds from their supersaturated solution is a spontaneous process. Guess about the signs of ΔS and ΔH.

13.102 Give the conditions under which each of the following equations may be applied: (*i*) $dA \geq 0$ (equilibrium and nonspontaneous process), (*ii*) $dG \leq 0$ (equilibrium and spontaneous process), (*iii*) $\ln\frac{p_2}{p_1} = -\frac{\Delta\bar{H}}{R}\left(\frac{T_1 - T_2}{T_1T_2}\right)$ and (iv) $\Delta A = nRT\ln\frac{V_2}{V_1}$

13.103 A person heated water in a closed bottle in a microwave oven for tea. After removing the bottle from the oven, he added a tea bag or coffee to the hot water. To his surprise the water started to boil violently. Explain what happened?

13.104 It is well known that protein molecules are polypeptide chains made up of amino acids. In their physiological functioning or native state, these chains fold in a unique manner such that the nonpolar groups of the amino acids are usually buried in the interior regions of the proteins, where there is little or no contact with water. When protein denatures, the nonpolar chains unfold so that these nonpolar groups are exposed to water. A useful estimate of the changes in thermodynamic quantities as a result of denaturation is to transfer of a hydrocarbon such as methane (or a nonpolar substance) from an inert solvent (such as benzene or CCl_4) the aqueous environment:

(*i*) CH_4 (inert solvent) $\rightarrow$ CH_4 (*g*)

(*ii*) CH_4 (*g*) $\rightarrow$ CH_4 (*aq*)

If the values of ΔH^0 and ΔG^0 are approximately 2.0 kJ mol^{-1} and –14.5 kJ mol^{-1} respectively for (*i*) and –13.5 kJ mol^{-1} and 26.5 kJ mol^{-1} respectively for (*ii*), calculate the values of ΔH^0 and ΔG^0 for the transfer of 1 mole of CH_4 according to equation:

$$CH_4 \text{ (inert solvent)} \rightarrow CH_4\ (aq)$$

Explain your results when we assume that the temperature is kept at 298 K.

13.105 If the standard free energy of the reaction is negative, does it mean that when the reactants are mixed together the reaction will lead to products spontaneously? Explain briefly.

13.106 Instead of ideal gas in the Problem 13.104, it is a real gas which obeys the equation of state as: $p(V - nb) = nRT$, where b is a function of temperature of only. Calculate the values of ΔG and ΔH for this process.

13.107 Show that: (*i*) $\sum_{i=1}^{i=k} \bar{A}_i dn_i = \sum_{i=1}^{i=k} \bar{G} dn_i$, (*ii*) $\sum_{i=1}^{i=k} \bar{H}_i dn_i = \sum_{i=1}^{i=k} \bar{G} dn_i$.

13.108 The molar enthalpy of vapourisation of benzene at it normal boiling point (80.09 °C) is 30.72 kJ mol^{-1}. If we assume that $\Delta_{vap}\bar{H}$ and $\Delta_{vap}\bar{S}$ are independent of temperature, calculate the value of $\Delta_{vap}\bar{G}$ at 60 °C, 80.09 °C and 100 °C.

13.109 Repeat the Problem 13.108 by considering that molar heat capacities of liquid and gaseous benzene as 136.3 J K^{-1} mol^{-1} and 82.4 J K^{-1} mol^{-1} respectively.

13.110 Consider a gas obeying the van der Waals equation of state. Two moles of this gas is expanded isothermally from volume V_1 to V_2 at temperature T. Derive an expression giving ΔS, ΔA, and ΔG, for the given process in terms of V_1, V_2, T and van der Waals parameters.

13.111 The equilibrium constant (K_p) for the reaction 2 NO_2 (*g*) $\rightleftarrows$ N_2O_4 (*g*) at 298 K is 6.739. (*i*) Calculate the value of K_c at 298 K. (*ii*) If the total pressure at equilibrium condition and at temperature 298 K is p_T and K_p is the equilibrium constant the general expression for the partial pressure of N_2O_4 (*g*) can be written as

$$K_p\, p^2_{N_2O_4(g)} - [2Kp\, p_T + 1]p_{N_2O_4(g)} + K_p\, p_T^2 = 0$$

(*iii*) If the total pressure at 298 K at the equilibrium is 1.00 atm., calculate the partial pressures of NO_2 (*g*) and N_2O_4 (*g*) and the number of moles for each component, (*iv*) if the initial pressure of NO_2(*g*) at 298 K is 2 atm., calculate the partial pressures of and NO_2 (*g*) and N_2O_4 (*g*) at equilibrium condition.

13.112 Consider the reaction: $2SO_2\,(g) + O_2(g) \rightleftarrows 2SO_3(g)$ and equilibrium constant at 960 K is 10. Calculate $\Delta_r G$. In case of the same reaction and under the given conditions, find out whether is reaction is spontaneous or not?

$$2SO_2(1.0 \times 10^{-3}\text{bar}, g) + O_2(1.0\text{ bar}, g) \rightleftarrows 2SO_3(5.0 \times 10^{-5}\text{ bar}, g)$$

13.113 Prove equations (13.69*a*), (13.71) and (13.72).

13.114 Prove equations (13.68*a*) and (13.68*b*). Under what conditions $K_p = K_x = K_c$.

13.115 Mathematically fugacity is defined as: $\lim_{p\to 0} \frac{f}{p} = 1$ and let us assume that there exists a nonideal gas for which $f = \frac{p}{(1+bp)}$, where *b* is equal to a constant for this choice of reference point. (*i*) If we define the fugacity $\lim_{p\to 0} \frac{f}{p} = 2$, as our reference point, determine the form fugacity will have for this gas (*ii*) Show that ΔG for the isothermal expansion of the gas from pressure p_1 to p_2 is the same for either choice of reference.

13.116 Consider two gases *A* and *B* which are in equilibrium as: $A(g) \rightleftarrows B(g)$. A container containing total of *n* moles of *A*(*g*) and *B*(*g*) are heated at constant pressure in the presence of a catalyst so that the system is always in equilibrium. If heating the above system by an amount dT resulted in the change in number of moles of *B*(*g*) at equilibrium by amount dn_B, show that the total heat associated with this change due to enthalpy of reaction is given by:

$$\Delta H dn_B = \frac{\Delta H^2 nK}{RT^2(K+1)^2} dT$$

where *K* thermodynamic is the equilibrium constant and $\Delta \bar{H}$ is molar enthalpy of the reaction.

13.117 Calculate the activity of $C(s)$ in the form of coke at 100 atm. and 1000 K.

13.118 Consider the conversion of graphite carbon to diamond carbon as

$$C(\text{graphite}) \rightleftarrows C(\text{diamond})$$

The standard molar Gibbs energy for the conversion of graphite into diamond is 2.9 kJ mol $^{-1}$ at 298 K. The density of graphite is 2.27 g cm^{-3} and that of diamond is 3.52 g cm^{-3} at 298 K. At what pressure will these two forms of carbon be at equilibrium at 298 K. Note that the activity of solid substance is given as

$$\ln a = \frac{\bar{V}}{RT}(p-1)$$

where $\bar{V}$ is the molar volume of the condensed phase and is essentially constant over a moderate range of pressure.

13.119 The decomposition of silver oxide gave the following data at different temperatures:

$$Ag_2O(s) \rightleftarrows 2Ag(s) + \frac{1}{2} O_2(g)$$

t (°C)	172	177	182	187
p (torr)	402.8	490.6	582.7	688.0

Calculate the values of K_p at different temperature. From the plot of ln K_p versus $1/T$ formulate an analytical expression. Using the expression obtained between ln K_p versus $1/T$, calculate the value of $\Delta_r\bar{H}$. Using heat capacity data given below, calculate the values of $\Delta_r\bar{H}^0$, $\Delta_r\bar{S}^0$ and $\Delta_r\bar{G}^0$ at 298 K.

13.120 Calculate the activity of $H_2O(l)$ as a function of pressure from 1 atm. to 100 atm. at 20 °C. Given density of water at 20 °C to be 0.9982 g mL^{-1} and assume that it is independent of pressure.

13.121 Repeat the Problem 13.120 for the following reaction

$$CaCO_3(\text{calcite}) \rightleftarrows CaCO_3(\text{aragonite})$$

where calcite and aragonite are two crystalline form of calcium carbonate. The $\Delta_r G^0$ at 298 K is 1.04 kJ mol^{-1}. The density of calcite and aragonite are 2.710 g cm^{-3} and 2.930 g cm^{-3} respectively. Calculate the pressure at which the two forms are at equilibrium.

13.122 The equilibrium constant as a function of temperature for the reaction

$$2C_3H_6(g) \rightleftarrows C_2H_4(g) + C_4H_8(g)$$

can be given by the equation

$$\ln Kp(T) = -2.395 - \frac{2505}{T} + \frac{3.477 \times 10^6 K^2}{T^2}$$

between temperature in range of 300 to 600 K. Calculate the values of $\Delta_r\bar{H}^0$, $\Delta_r\bar{S}^0$ and $\Delta_r\bar{G}^0$ at 298 K.

CHAPTER 14

PHASE EQUILIBRIA AND PHASE DIAGRAMS

14.1. EQUILIBRIUM CONDITION

Equilibrium condition for any system is that chemical potential of each constituent throughout the system must be same. If there are several phases of each constituent, then the chemical potential of each constituent in every phase must have the same value. For example, if at any temperature and pressure liquid water is in equilibrium with water vapours and solid ice then

$$\mu_{(ice)} = \mu_{H_2O(l)} = \mu_{H_2O(vapour)} \qquad ...(14.1)$$

***Exercise* 14.1:** In Fig. 14.1, the chemical potential of each phase (solid, liquid and gas) at constant pressure is plotted vs. temperature. Predict the phases which are present or more stable in the three boundaries.

It is clear from Fig. 14.1, that at temperature less than melting point (T_m), the chemical potential of solid phase is the least. Criterion of spontaneity says that system is stable if the free energy decreases. Thus below T_m, solid phase is stable, between temperature T_m and T_b liquid phase and above T_b vapour phase will be stable.

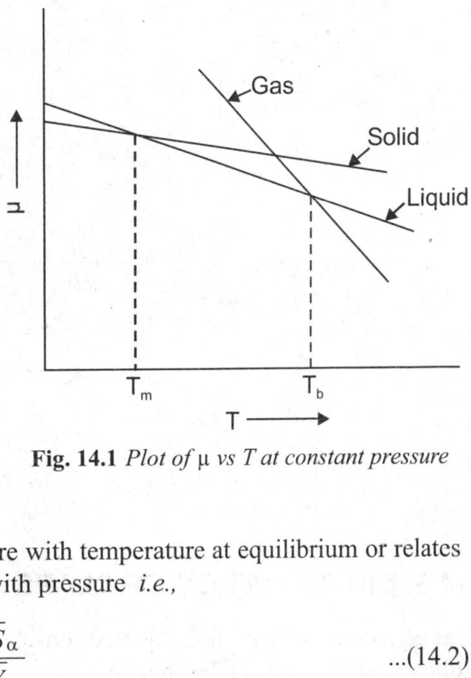

Fig. 14.1 *Plot of* μ *vs T at constant pressure*

14.2. THE CLAPEYRON EQUATION

The Clapeyron equation describes the variation of pressure with temperature at equilibrium or relates the quantitative dependence of equilibrium temperature with pressure *i.e.,*

$$\frac{dp}{dT} = \frac{\Delta S}{\Delta \overline{V}} = \frac{\overline{S}_\beta - \overline{S}_\alpha}{\overline{V}_\beta - \overline{V}_\alpha} \qquad ...(14.2)$$

where ΔS is the change in entropy and $\Delta \overline{V}$ is the change in volume accompanying in transfer of one mol of substance from phase α to phase β. Thus, with the help of above equation (14.2) a graph can be constructed between the equilibrium pressure and temperature for any phase transformation.

***Exercise* 14.2:** Calculate the change in pressure per degree in the solid-liquid equilibrium of water at freezing point. d(ice) = 0.9168 kg dm^{-3}, d(water) = 0.9998 kg dm^{-3}, ΔS_{fusion} = 22 EU

$$\bar{V}(\text{ice}) = \frac{(18\text{ g mol}^{-1})(10^{-3}\text{ kg g}^{-1})}{0.9168\text{ kg dm}^{-3}} = 19.634 \times 10^{-3}\text{ dm}^3\text{ mol}^{-1}$$

$$\bar{V}(\text{water}) = \frac{(18\text{ g mol}^{-1})(10^{-3}\text{ kg g}^{-1})}{0.9998\text{ kg dm}^{-3}} = 18.004 \times 10^{-3}\text{ dm}^3\text{ mol}^{-1}$$

$$\begin{aligned}\Delta p &= \frac{\Delta S_{fus}}{\bar{V}_l - \bar{V}_s}\Delta T = \frac{(22\text{ J K}^{-1}\text{mol}^{-1})(-1\text{ K})}{(18.004 - 19.634)\times 10^{-3}\text{dm}^3\text{mol}^{-1}} \\ &= 13.494 \times 10^3\text{ J dm}^{-3} = 13.494 \times 10^6\text{ N m}^{-2} \\ &= 133\text{ atm}\end{aligned}$$

***Problem* 14.3:** The densities of liquid water and water vapour at its normal boiling point are 0.958 and 5.98 × 10^{-4} kg dm^{-3} and the change in the entropy of vaporisation is 108.99 EU. Calculate the change of pressure for the change in temperature by one degree.

Using equation (14.2), one has to calculate $\bar{V}_l$, $\bar{V}_v$

$$\bar{V}_l = \frac{\text{M.W.}}{d} = \frac{18 \times 10^{-3}\text{kg mol}^{-1}}{0.958\text{ kg dm}^{-3}} = 18.79 \times 10^{-3}\text{ dm}^3\text{ mol}^{-1}$$

$$\bar{V}_v = \frac{18 \times 10^{-3}\text{kg mol}^{-1}}{5.98 \times 10^{-4}\text{ kg dm}^{-3}} = 30.100\text{ dm}^3\text{ mol}^{-1}$$

$$\begin{aligned}\frac{dp}{dT} &= \frac{\Delta p}{\Delta T} = \frac{\bar{S}_v - \bar{S}_l}{\bar{V}_v - \bar{V}_l} = \frac{\Delta S_{vap}}{\bar{V}_v - \bar{V}_l} = \frac{108.99\text{ J K}^{-1}\text{mol}^{-1}}{(30.100 - .0188)\text{ dm}^3\text{mol}^{-1}} \\ &= 3.6233\text{ J dm}^{-3}\text{ K}^{-1} \\ \Delta p &= (3.6233\text{ J dm}^{-3}\text{ K}^{-1})(1\text{ K}) \\ &= 3.6233\text{ J dm}^{-3} \\ &= (3.6233 \times 10^3)(\text{kg m}^2\text{ s}^{-2}\text{ m}^{-3}) \\ &= (3.6233 \times 10^3)\text{ kg m s}^{-2}\text{ m}^{-2} \\ &= 3.6233 \times 10^3\text{ N m}^{-2}\end{aligned}$$

14.3. INTEGRATION OF CLAPEYRON EQUATION

The equation (14.2) can be integrated considering the various phase transformation equilibrium and the following equations can be obtained.

14.3.1. *Solid-liquid Equilibrium*

In case of solid-liquid equilibrium, the equation (14.2) can be written as

$$\frac{dp}{dT} = \frac{\Delta S_{\text{fus}}}{\Delta V_{\text{fus}}} = \frac{\Delta H_{\text{fus}}}{T_m \Delta V_{\text{fus}}} \quad ...(14.3)$$

and can be integrated to

$$p_2 - p_1 = \frac{\Delta H_{\text{fus}}}{\Delta V_{\text{fus}}} \ln \frac{T'_m}{T_m} \quad ...(14.4)$$

where T'_m is the melting point at pressure p_2 and T_m at p_1 and also assuming that ΔH_{fus} and ΔV_{fus} are independent of temperature in this range. The approximate equation can also be used as

$$p_2 - p_1 = \frac{\Delta H_{\text{fus}}}{\Delta V_{\text{fus}}} \frac{\Delta T}{T_m} \quad ...(14.5)$$

where ΔT is the change in the melting point and is quite small relative to the magnitude of the melting point.

14.3.2. *Clausius-Clapeyron Equation*

This equation relates the variation of pressure at the condensed phase—vapour equilibria with temperature. The relation can be derived from Clapeyron equation, with the assumptions that *molar volume of vapours is much larger than that of solids*, *vapours behave ideally* and $\Delta\bar{H}$ is independent of temperature. The equation is

$$\frac{d \ln p}{dT} = \frac{\Delta \bar{H}}{RT^2} \quad ...(14.6)$$

where $\Delta\bar{H}$ is molar heat of vaporisation of liquid or molar heat of sublimation. The above equation can be integrated to

$$\ln p_2 = \ln p_1 + \frac{\Delta \bar{H}}{R}\left[\frac{1}{T_1} - \frac{1}{T_2}\right] \quad ...(14.7)$$

If $T_1 = T_0$ the normal boiling point *i.e.*, $p_1 = 1$ atm or $T_1 = T_0$, the sublimation temperature at $p_1 = 1$ atm, the above equation becomes

$$\ln p = \frac{\Delta H}{RT_0} - \frac{\Delta H}{RT} \quad ...(14.8)$$

From the plot of $\ln p$ vs $(1/T)$, the molar heat of vaporisation or molar heat of sublimation can be obtained.

The equation (14.7) has been obtained with the following assumptions. (1) $\bar{V}_g \gg \bar{V}_l$ or $\bar{V}_s$ (2) vapours behave ideally and (3) $\Delta\bar{H}$ is independent of temperature.

The assumption (1) is acceptable but (2) and (3) are not true *i.e.,* vapours never behave ideally and ΔH decreases as temperature increases and is zero at critical temperature.

In general a straight line is obtained for the plot of log p vs $1/T$. This has been explained by O.L.I. Brown (*J. Chem. Educ.,* **28,** 428, 1951) as follows:

$$\Delta \bar{V} = \bar{V}_g - \bar{V}_l = \bar{V}_g\left(1 - \frac{\bar{V}_l}{\bar{V}_g}\right) = \frac{ZRT}{p}\left(1 - \frac{\bar{V}_l}{\bar{V}_g}\right) \quad ...(14.9)$$

where Z is the compressibility factor of vapours under given conditions, *i.e.*, a function of T and P. Now

$$\frac{dp}{dT} = \frac{\Delta H}{T} \times \frac{p}{ZRT\left(1-\bar{V}_l/\bar{V}_g\right)} \quad ...(14.10a)$$

$$\frac{1}{p}dp = \frac{\Delta \bar{H}}{ZR\left(1-\bar{V}_l/\bar{V}_g\right)} \cdot \frac{dT}{T^2} \quad ...(14.10b)$$

or

$$\frac{d \ln p}{d(1/T)} = -\frac{\Delta \bar{H}}{ZR\left(1-\bar{V}_l/\bar{V}_g\right)} \quad ...(14.10c)$$

Z varies from unity at $p = 0$ to about 0.2 at p = critical pressure. Similarly $1 - \bar{V}_l/\bar{V}_g$ varies from 1 to 0 and $\Delta \bar{H}$ drops from sizable value to zero. Thus with increase in T both the numerator and denominator decreased and thus cancel the effect and straight line is observed.

***Exercise* 14.4:** The following data is obtained for (vapour pressure–temperature) H_2O. Calculate the molar heat of vaporisation.

t^0(c)	25.2	30.6	40.4	46.4	52.0
p(mm)	24.04	32.93	56.51	77.21	102.09

The above data can be written as

$\frac{1}{T}\times 10^3$	3.353	3.294	3.191	3.131	3.077
$\log p$	1.381	1.518	1.752	1.888	2.009

$\log p$ vs $\frac{1}{T}$ is plotted in Fig. 14.2.

$$\text{slope} = \frac{1.91-1.36}{(3.12-3.36)10^{-3}} = -\frac{55}{24}\times 10^3 \text{K}$$

$$\Delta \bar{H} = -2.303\ R\ \text{slope}$$

$$= -(2.303)(8.314 \text{ J K}^{-1}\text{ mol}^{-1})\left(-\frac{55}{24}\times 10^3\text{K}\right)$$

$$= 43.88 \times 10^3 \text{ J mol}^{-1}$$

$$= 43.88 \text{ kJ mol}^{-1}$$

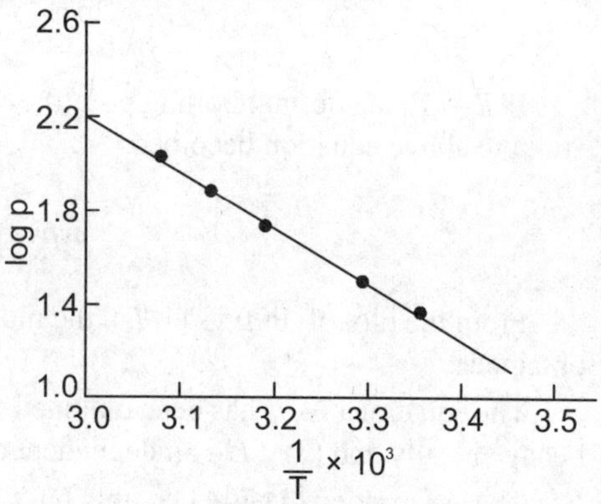

Fig. 14.2 *Plot of log p vs 1/T*

Note: Equation (14.8) says that p should be in atm. But when the method of calculation is graphical, the constant terms are collected together and summed as intercept *e.g.* here equation (14.8) becomes

$$\ln p = \left(\ln 760 + \frac{\Delta \bar{H}^0}{RT^0}\right) - \frac{\Delta \bar{H}^0}{RT}$$

***Exercise* 14.5:** Calculate the heat of sublimation of solid CO_2 from the following data:

p(mm)	10	40	100	400	760
t(°C)	– 119.5	– 108.6	– 102.2	– 85.7	– 78.2

Using the relation (14.7)

$$\log \frac{p_2}{p_1} = \frac{\Delta \overline{H}}{2.303R}\left(\frac{1}{T_1} - \frac{1}{T_2}\right)$$

or

$$\Delta \overline{H}_{\text{sub}} = \frac{2.303RT_1T_2}{T_2 - T_1}\log \frac{p_2}{p_1}$$

$$= \frac{(2.303)\left(8.316\text{J K}^{-1}\text{mol}^{-1}\right)(153.5\text{ K})(174.4\text{ K})}{(174.4 - 153.5)\text{K}}\log \frac{40\text{mm}}{10\text{mm}}$$

$$= 13018.44 \text{ J mol}^{-1} = 13.018 \text{ kJ mol}^{-1}$$

or it can also be solved by plotting ln p vs $\frac{1}{T}$ and from slope $\Delta \overline{H}_{\text{sub}}$ can be calculated.

***Exercise* 14.6:** Calculate vapour pressure of toluene at 100°C if it is assumed that toluene obeys Truton's rule and the normal boiling point is 110.62°C.

Use equation (12.22) to calculate molar heat of vaporisation and equation (14.7) to calculate the vapour pressure at 100°C

$$\Delta \overline{H}_{\text{vap}} = (87.86 \text{ J K}^{-1} \text{ mol}^{-1})(383.62 \text{ K})$$

$$= 33704.85 \text{ J mol}^{-1}$$

$$\log \frac{p_2}{1\text{ atm}} = \frac{33704.85 \text{ J mol}^{-1}}{(2.303)\left(8.316 \text{ J K}^{-1} \text{ mol}^{-1}\right)}\left(\frac{1}{383.62} - \frac{1}{373}\right)\text{K}^{-1}$$

$$= \frac{(33704.85)(-10.62)}{(2.303)(8.316)(383.62)(373)}$$

$$= -0.1306$$

$$p_2 = e^{-0.1306} = 0.74 \text{ atm}$$

***Exercise* 14.7:** The vapour pressure of a liquid A can be expressed as

$$\log p = 6.2512 - 1526.2T^{-1} - 41264.4T^{-2}$$

where p is mm. The densities of liquid and vapour phase are 0.752 kg dm^{-3} and 2.14 × 10^{-4} kg dm^{-3} respectively. If the molecular weight of A is 40 g mol^{-1}, calculate heat of vaporisation at 478.5 K, the normal boiling point,

One needs $\frac{dp}{dT}$, $\overline{V}_l$, $\overline{V}_v$ and p at 478.5 K before one uses the similar equation (14.3) for liquid vapour equilibrium *i.e.*,

$$\frac{dp}{dT} = \frac{\Delta \bar{H}_{vap}}{T_b \Delta \bar{V}} \qquad ...(14.3a)$$

where $\Delta \bar{H}_{vap}$ is heat of vapourisation, $\Delta \bar{V}$ is change in the molar volumes of vapours and liquid and T_b is the boiling temperatue.

Since pressure as well as the standard state are expressed in atm, it will be nice to convert the above relation in atm.

$$\log p(\text{atm}) = \log p(\text{mm}) - \log 760\ (\text{mm atm}^{-1})$$

$$\frac{d \log p(\text{atm})}{dT} = \frac{d \log p(\text{mm})}{dT}$$

i.e., it is immaterial whether the pressure is expressed in atm or in mm so long as change of pressure with temperature is considered.

Now

$$\frac{d \log p}{dT} = \frac{1}{p}\frac{dp}{dT} = \frac{1526.2}{T^2} + \frac{82528.8}{T^3}$$

At boiling point $p = 1$ atm, $T_b = 478.5$ K

$$\frac{dp}{dT} = (1\ \text{atm})\left(\frac{1526.2}{(478.5)^2} + \frac{82528.8}{(478.5)^3}\right) = 7.419 \times 10^{-3}\ \text{atm K}^{-1}$$

$$\bar{V}_l = \frac{(40\ \text{g mol}^{-1})(10^{-3}\ \text{kg g}^{-1})}{0.752\ \text{kg dm}^{-3}} = 53.19 \times 10^{-3}\ \text{dm}^3\ \text{mol}^{-1}$$

$$\bar{V}_V = \frac{(40\ \text{g mol}^{-1})(10^{-3}\ \text{kg g}^{-1})}{2.14 \times 10^{-4}\ \text{kg dm}^{-3}} = 186.92\ \text{dm}^3\ \text{mol}^{-1}$$

$$\Delta \bar{V} = \bar{V}_V - \bar{V}_1 = 186.92 - 0.053 = 186.867\ \text{dm}^3\ \text{mol}^{-1}$$

Now using the equation (14.3*a*)

$$\Delta \bar{H}_{vap} = T \Delta \bar{V} \frac{dp}{dT}$$

$$= (478.5\ \text{K})\ (186.867\ \text{dm}^3\ \text{mol}^{-1})\ (7.419 \times 10^{-3}\ \text{atm K}^{-1})$$

$$= 663.38\ \text{dm}^3\ \text{atm mol}^{-1}$$

$$= (663.38 \times 10^{-3}\ \text{m}^3\ \text{atm mol}^{-1})\ (101325\ \text{J atm}^{-1}\ \text{m}^{-3})$$

$$= 67.22\ \text{kJ mol}^{-1}$$

14.4. EFFECT OF PRESSURE ON THE VAPOUR PRESSURE

The equilibrium pressure in case of liquid-vapour equilibrium changes with external pressure and the relation is

$$\frac{dp}{dP} = \frac{\bar{V}_l}{\bar{V}_g} \qquad ...(14.11)$$

where p is the vapour pressure of liquid, P is total pressure ($p + p_{\text{foreign gas}}$), $\bar{V}_l$ and $\bar{V}_g$ are the molar volumes of the liquid and vapours. If vapours are ideal and $\overline{V}_l$ is constant during the range of pressure used, the above equation can be integrated as

$$\ln \frac{p_2}{p_1} = \frac{\overline{V}_l}{RT}(p_2 - p_1) \qquad \text{...(14.12)}$$

***Exercise* 14.8:** The density of liquid ethanol at 35°C is 0.7767 kg dm^{-3} and its vapour pressure at the same temperature is 100 mm. Calculate the vapour pressure of ethanol when N_2 is added till the pressure is 100 atm.

Equation (14.12) can be used to calculate the vapour pressure of ethanol at 100 atm total pressure

$$p_1 = p_1 = \frac{100 \text{ mm}}{760 \text{ mm atm}^{-1}} = 0.13 \text{ atm}$$

$$\overline{V}_l = \frac{(46 \text{ g mol}^{-1})(10^{-3} \text{ kg g}^{-1})}{0.7767 \text{ kg dm}^{-3}} = 59.225 \times 10^{-3} \text{ dm}^3 \text{ mol}^{-1}$$

$$\ln \frac{p_2}{p_1} = \frac{59.225 \times 10^{-3} \text{ dm}^3 \text{mol}^{-1}}{(308 \text{ K})(0.0821 \text{ atm dm}^3 \text{ K}^{-1} \text{ mol}^{-1})} (100 - 0) \text{ atm}$$

Since $p_1 = p_1$ is very small $= 0.13$ atm

$$\ln \frac{p_2}{p_1} = 0.234$$

$$\frac{p_2}{p_1} = e^{0.234} = 1.2636$$

$$p_2 = 100 \times 1.2636 = 126.36 \text{ mm}$$

14.5. PHASE RULE

14.5.1. *Phase*

A phase is defined as that portion of the system which is uniform or homogeneous among its submacroscopic state but is definitely separated from other such portions of the system by well defined boundaries. Mixture of the solids or two immiscible liquids can form separatc phases whereas mixture of gases is a one phase as it is homogeneous system. The common symbol for number of phases is P.

***Exercise* 14.9:** How many phases are present in the following equilibria.

$$CaCO_3 (s) \rightleftharpoons CaO (s) + CO_2 (g)$$

In the above equation, two solids are of different structure and are separated by definite boundaries. So in all there are *three* phases, *i.e.,* two solids and one gas.

***Exercise* 14.10:** An iceberg is floating in the lake. If one considers the lake, iceberg and atmosphere as a one system, what are the number of phases?

Iceberg is a solid form of water, lake is liquid water and atmosphere contains water vapours and other gases. So there are three phases (solid, liquid and vapour).

14.5.2. *Component*

The number of components in a system is defined as the minimum number of "chemically independent" species which are needed to describe the composition of each phase in a system.

***Exercise* 14.11:** In the exercise 14.9, find the number of components.

In the exercise 14.9, there are three chemically distinct species, $CaCO_3$, CaO and CO_2. But the system is under equilibrium and thus the number of components are two. It is because if the composition of two species are fixed or defined, that of third one is automatically defined.

***Exercise* 14.12:** Calculate the number of components in exercise 14.10 if atmosphere only contains water vapours.

Since iceberg, liquid water and water vapours, all consist of one chemical species, there is only one component *i.e.*, H_2O.

14.5.3. *Degrees of Freedom*

The degrees of freedom are defined as the minimum number of intensive variables which must be specified such that the remaining intensive variables can be fixed. The minimum number of intensive variables can be temperature, pressure, concentration. The symbol for degree of freedom is 'F' and it is invariant if $F = 0$, univariant if $F = 1$, bivariant if $F = 2$ and so on.

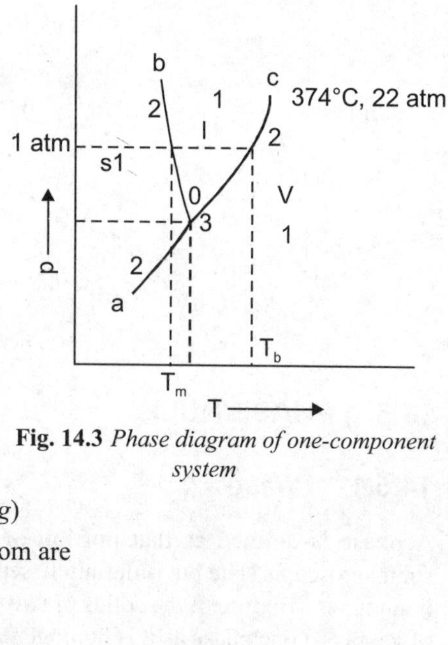

Fig. 14.3 *Phase diagram of one-component system*

14.5.4. *Gibbs Phase Rule*

Gibbs phase rule gives a relation between the degrees of freedom in a system of 'C' components and P phases. The relation is

$$F = C - P + 2 \qquad ...(14.13)$$

***Exercise* 14.13:** Calculate the degrees of freedom in system described by the exercises (14.9) and (14.11).

In the system

$$CaCO_3\,(s) \rightleftharpoons CaO\,(s) + CO_2\,(g)$$

there are *3* phases and *2* components, so the degrees of freedom are

$$F = 2 - 3 + 2 = 1$$

i.e., the system is univariant.

***Exercise* 14.14:** In Fig. 14.3, calculate the number of degrees of freedom at the points labelled 1, 2 and 3.

The phase diagrams is of one component, system *i.e.*, $C = 1$,

At point 1, $P = 1$, So $F = 1 - 1 + 2 = 2$

At point 2, $P = 2$, So $F = 1 - 2 + 2 = 1$

At point 3, $P = 3$, So $F = 1 - 3 + 2 = 0$

i.e., point 1 is bivariant, point 2 is univariant and point 3 is invariant called Triple point.

14.6. PHASE DIAGRAMS FOR ONE-COMPONENT SYSTEM

For one-component system, the Gibbs phase rule equation (14.13) becomes

$$F = 1 - P + 2 = 3 - P$$

Since there will always be at least one phase in the system, so the minimum degrees of freedom will be two. These can be considered as temperature and pressure and phase diagrams can be constructed with these two variables.

Equations depicting the equilibrium vapour pressure of the solid $\rightarrow$ vapour and liquid $\rightarrow$ vapour are expressed as

$$p_s^{\text{eq}} = p_0 e^{\left\{-\frac{\Delta_{\text{sub}}\bar{H}}{R}\left[\frac{1}{T^{\text{eq}}} - \frac{1}{T^o}\right]\right\}} \qquad \text{...(14.1)}$$

$$p_2^{\text{eq}} = p_1^{\text{eq}} e^{\left\{-\frac{\Delta_{\text{vap}}\bar{H}}{R}\left[\frac{1}{T_2^{\text{eq}}} - \frac{1}{T_1^{\text{eq}}}\right]\right\}} \qquad \text{...(14.2)}$$

where $\Delta_{\text{vap}}\bar{H}$ and $\Delta_{\text{sub}}\bar{H}$ heat of vapourisation and sublimation of liquid respectively,, p_2^{eq} and p_1^{eq} are vapour pressure of the liquid at temperature T_2 and T_1, p^0 is the equilibrium vapour pressure over the solid at temperature T^0. At all points, (p_s^{eq}, T^{eq}) the solid and vapour are both present in equilibrium. Under all the approximations, equation (14.1) gives a curve depicting the solid-vapour equilibrium. Similarly equation (2) gives a curve depicting the vapour pressure at the liquid-vapour equilibrium. Intersection of two curves gives a point at which all the three phases (solid, liquid and vapour) are in equilibrium. This is a unique point and is called *triple point*.

***Exercise* 14.15:** In liquid-vapour phase equilibrium, the liquid-vapour curves end at the critical point. Explain some of its characteristics.

In the liquid-vapour phase equilibrium, a temperature is reached, above which a liquid can be transformed to vapour phase without ever passing through a two phase state, *i.e.*, in other words, the gas passes gradually and continuously transform into the liquid state without a two phase region appearing and without any apparent condensation. Smooth passage of liquid to vapour state without appearance of two phase region is because both liquid and vapour phases are fluids and the difference between them is purely one of degree rather than actual structure. The fluctuations in the coexistence of one phase or the other are sensitive that it requires very sensitive techniques to study them. At the critical temperature the fluctuations in the scattering of light are very strong and the system appears milky, *i.e.*, somewhat like a finely dispersed fog. This effect is known as *critical opalescence*. $\Delta_{\text{vap}}\bar{H}$ is zero near the critical temperature and densities of the liquid phase and vapour phase are the same. These densities at the critical temperature are called *orthobaric densities*. The liquid and vapour phase simply submerge with each other.

***Exercise* 14.16:** Explain briefly whether you would expect a critical temperature in solid-liquid phase diagram curve.

As learnt in Exercise 14.15, it is essential to have similar structure of the two phases if critical temperature is to be observed. In case of solid-liquid or in general if the structures of the phases (even solid-solid) are different and remain so, it is not possible to pass from one phase to another phase in a gradual continuous manner. In such cases, the critical temperature will not be observed. For such phases, the coexistence curve separating the two phases must continue indefinitely or intersect the coexistence curves of other phases and in reality such curves are present in many cases.

***Exercise* 14.17:** In Fig. 14.3 the phase diagram for water is given. Discuss this.

The areas *S*, *L* and *V* are the spaces where only one phase is present *i.e.,* solid, liquid and vapours.

The lines *co*, *bo* and *ao* give a variation of equilibrium vapour pressure with temperature in the system $l - v$, $s - l$ and $s - v$ equilibria.

The slope of line *bo* is negative because the $\Delta V = \bar{V}_l - \bar{V}_s$ is negative for water system, so Clapeyron equation predicts decrease of vapour pressure with increase of temperature for liquid water and ice equilibrium. Slope of curve *ao* is steeper than *co* because $\Delta \bar{H}_{sub} = \Delta \bar{H}_{fus} + \Delta \bar{H}_{vap}$ near *b*.

Point *o* is a triple point, *i.e.,* ice, water and water vapours are present together so phase rule gives invariant system. $p = 4.58$ mm and $t = 0.0098°C$ at this point. Point C and dotted line are critical point and critical isotherm and the values of pressure and temperature are 22 atm and 374°C. Above this temperature and pressure liquid and vapour phase cannot be distinguished by specified boundaries.

***Exercise* 14.18:** Construct the phase diagram for a compound *A* (*i.e.* one component system) from the following data: (*i*) Compound *A* exists in two solid forms, the densities of both the solid forms are greater than that of liquid phase (*ii*) The melting point of *A* is 20.2°C under its own vapour pressure of 5 mm of Hg (*iii*) Phases A_1, A_2 and liquid are at equilibrium at 1000 atm and 70.2°C (*iv*) The transition temperature of solid phases increases with the increase of pressure (*v*) the boiling point of liquid is 140°C and (*vi*) if the solid form A_1 is heated slowly under its vapour pressure it is converted to S_2 at 15°C.

The phase diagram is constructed with the help of above data and shown in Fig. 14.4.

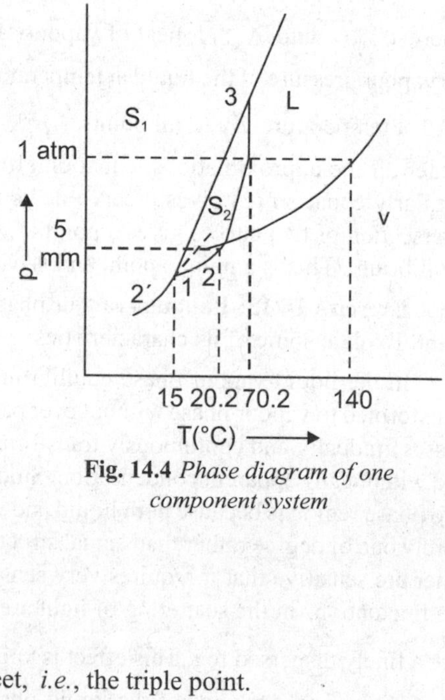

Fig. 14.4 *Phase diagram of one component system*

***Exercise* 14.19:** The vapour pressure of naphthalene in mm can be expressed as

$$\log p = -\frac{0.05223}{T}a + b$$

between 0 – 80°C, $a = 71401$, $b = 11.450$

between 120 – 200°C, $a = 47362$, $b = 7.927$

Extrapolate the two curves to find out where these meet, *i.e.*, the triple point.

The vapour pressure at different temperatures are calculated with above relation and are given below in tabular form

Temp. (K)	273	285	300	325	340	350	353	375	400	450	475	500	550	575
p (mm)	6.16×10^{-3}	0.0232	0.1045	0.9447	3.031	6.236	7.682	21.4	55.3	269.06	523.81	954.03	2687.4	4072
log *p* (mm)	–2.2103	–1.6352	–0.9809	–0.0247	0.4815	0.7949	0.8825	1.33	1.74	2.43	2.72	2.98	3.43	3.61

log p vs T is plotted in Fig. 14.5. The two curves when extrapolated, meet at point $T = 358$ K, log p = 0.96 or $p = 9.12$ mm.

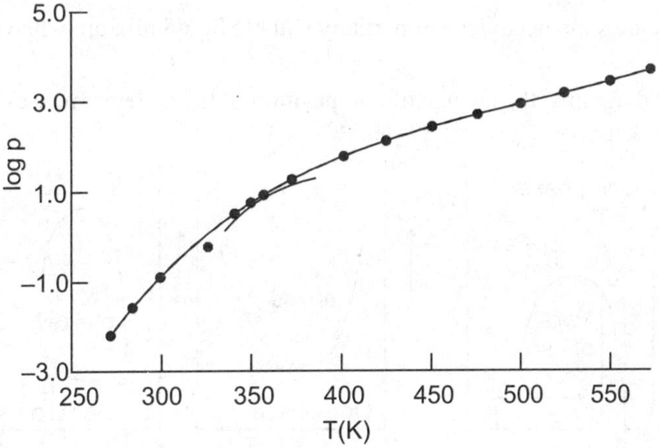

Fig. 14.5 *Plot of log p vs T*

14.7. ▌TWO COMPONENT SYSTEM

In the two-component system the phase rule equation (14.13) reduces to

$$F = 4 - P$$

Now since there will be at least one phase always present, we need three variables to describe the system completely and this is difficult to grasp or visualise. Thus always one of the variable, pressure or temperature is held constant and phase diagrams are expressed in terms of temperature and composition or pressure and composition.

14.7.1. *Lever's Rule*

In the two-component phase diagrams there are regions where two phases, *e.g.* liquid-vapour (see Fig. 14.6) are present, the composition of the entire system *i.e.,* of liquid phase and vapour phase, can be calculated with the help of Lever's rule. The procedure is as follows. Take any state point, for example, '*a*' in two phase region. Draw a horizontal line, called "tic line" and let it meet liquid-liquid vapour equilibrium curve at *l* and vapour-liquid vapour curve at v. Then ratio of the amount of liquid phase to vapour phase is given as

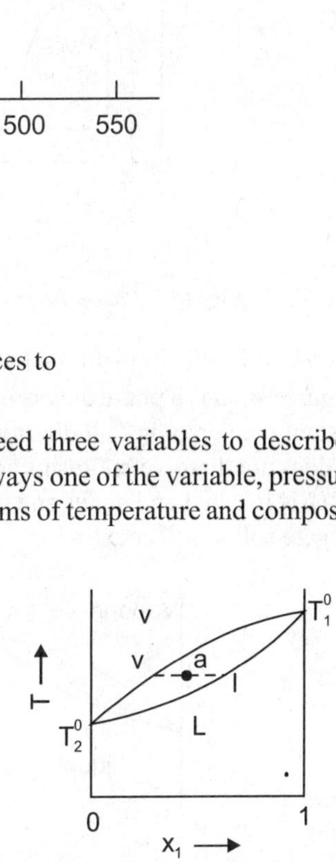

Fig. 14.6 *Phase diagram for liquid-vapour equilibria*

$$\frac{n_{\text{liq}}}{n_{\text{vap}}} = \frac{av}{al} \quad \text{...(14.14)}$$

14.7.2. *Liquid-Liquid Phase Diagrams*

Liquids can form various kind of mixtures with other liquids. So different kinds of phase diagrams can be obtained and they are shown below.

(*a*) Figure 14.7 shows the phase diagrams for the partially miscible liquids.

(*b*) Figure 14.8 shows the phase diagrams of the miscible liquids which form either ideal or real solutions.

(*c*) Figure 14.9 shows the behaviour of partially soluble liquid mixture when distillation is carried out.

In all these phase diagrams, the change of composition of the different phases with temperature is plotted and pressure is kept constant.

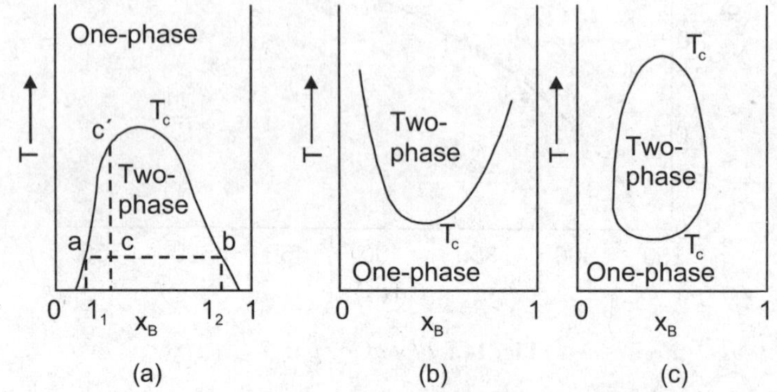

Fig. 14.7 *Phase diagrams of two partially miscible liquids (pressure is constant)*

***Exercise* 14.20:** Explain the phase diagram drawn in Fig. 14.7 *a.*

Figure 14.7*a* is a phase diagram for two partially miscible liquids. If a solute is added to the solvent at constant temperature T_1 it dissolves in the beginning, forming only one phase. After the point *a*, the solute does not dissolve but forms another layer and thus two phases are formed, till the composition of point '*b*' is reached and again one phase is obtained. In the region between *a* and *b* two phases called "conjugate solutions" coexist.

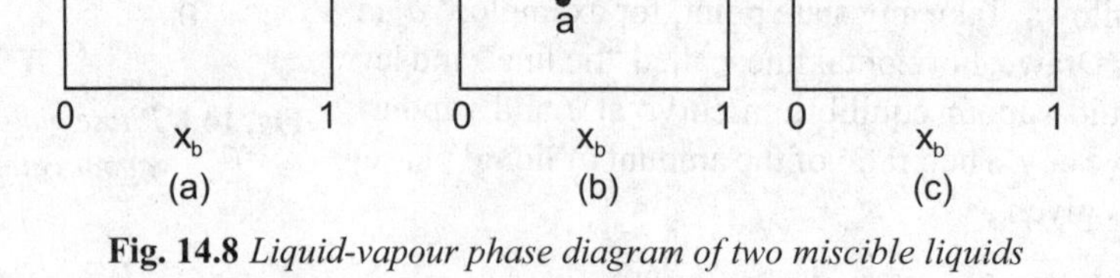

Fig. 14.8 *Liquid-vapour phase diagram of two miscible liquids*

If the temperature is raised, the solubility also changes. It increases with increase of temperature and above the temperature T_c, the liquids are completely miscible and one phase is obtained. Temperature T_c is called the "critical solution temperature" and is also called the "upper consolute temperature".

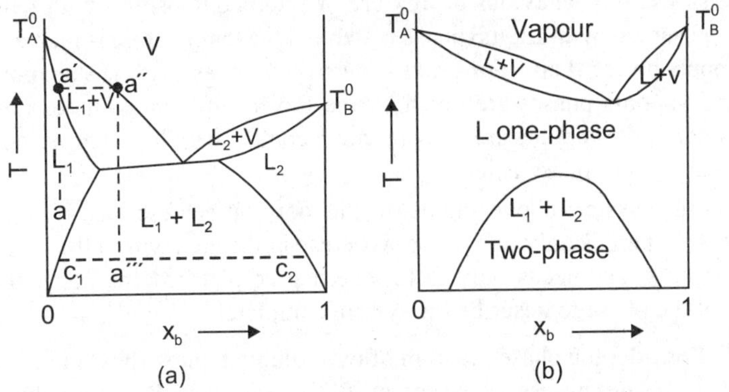

Fig. 14.9 *Liquid-vapour phase diagram of partially soluble liquids*

Any point within the loop is a state point of the system, consisting of two phases and relative mass of two layers is given by Lever's rule. For example, at state point c, there will be two layers L_1 of composition l_1 and L_2 of composition l_2. The relative composition of the two layers is given by Lever's rule and is

$$\frac{\text{mol of } l_1}{\text{mol of } l_2} = \frac{bc}{ac}$$

As the temperature increases, the ratio bc/ac increases. In other words the layer L_1 gets richer in component 2 and L_2 gets richer in component 1. This is the behaviour shown by dashed line, till the point c is reached where the solution is homogeneous once again.

According to phase rule, the equation (14.13) becomes $F = 3 - P$ under the present system. At any point outside the loop, there is only one phase, one needs two degrees of freedom (*i.e.,* temperature and composition) to describe the system completely. But inside the loop, there are two phases, the degree of freedom is only one *i.e.,* if the temperature is fixed, then by drawing a tie line, one can describe the composition of the system completely with the help of Lever's rule. Similarly if one knows the composition of one of the conjugate solution, the composition of the other conjugate solution and the temperature can be obtained with the help of tie line.

***Exercise* 14.21:** Explain the phase diagram shown in Fig. 14.8 *b*.

The phase diagram depicted in Fig. 14.8 *b* is that of two liquids which are completely soluble and form real solution (non-ideal solution).

The area above the loop is one-phase area, having vapours of components C_1 and C_2. The area below the loop is also a one-phase area, having liquid solution of two components. The area between the loop depicts two phases, liquid and vapour at equilibrium with each other.

The phase diagram shows the variation of boiling point of the liquid mixture with the change of its composition. T_1^0 and T_2^0 are the normal boiling points of pure components 1 and 2. T_{max} is the maxima in the boiling point of the solution and this point is called 'Azeotrope'. The composition at this point is known as azeotropic composition. The composition of liquid and vapour phase is same and the mixture of this composition behaves like a pure compound. The two liquids cannot be separated by distillation and boiling point of this mixture at one atmosphere is a sharp boiling point.

The dotted line shows the behaviour of mixture of fixed composition with temperature. Let the state of system be represented by *a*. Liquid phase is stable if the temperature is increased till one reaches *a*'. At this point, vapours just start appearing and system is two phase. At the state point *a*", the relative amount of liquid and vapours phases are given by the Lever's rule and vapour phase gets richer in component *B*. At point *a*", the vapours are just present in contact with liquid phase and if the mixture is heated further, the only vapour phase is left.

The description of the system above and below the loop can be described if both temperature and composition are fixed as there is only one phase. Whereas in the area within the loop and on the lines, two phases are present, so one needs only to fix one degree of freedom, *i.e.,* either temperature or composition of one of the phase to describe the system completely.

***Exercise* 14.22:** Consider the phase diagram of two soluble liquids shown in Fig. 14.10. Calculate the number of theoretical plates necessary to separate the pure form of *B* from a solution of composition x_1.

If liquid of composition x_1 is heated up to a temperature T_1, the vapours of composition x_2 is obtained and are richer in *B* relatively. Cool these vapours to temperature T_2 and the liquid phase of composition x_2 is obtained. At temperature T_2, vapours of composition x_3 are obtained. Cooling these vapours to temperature T_3, liquid solution richer in *B* is obtained. In this way obtaining vapours at temperature T_3 and subsequent cooling will give pure liquid *B* if it meets the point T_B^0.

Now 'Theoretical plate' is defined as a simple distillation step in which an equilibrium between the solution and vapour is established and the vapour is condensed to liquid of different composition. The four theoretical plates, as shown in Fig. 14.10, will be enough to get the pure liquid *B*.

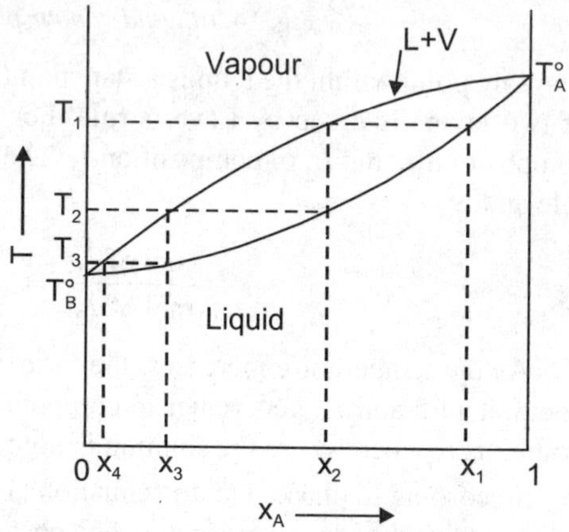

Fig. 14.10 *Principle of distillation*

***Exercise* 14.23:** Consider the following liquid-liquid phase diagram (14.7 *a*). In this 25 g of *A* and 60 g of *B* are mixed at 25°C and at this temperature the system forms two phases. If the wt. of *B* in phase I is 15% and that of *B* in phase II, 80%, calculate the masses of each phase in equilibrium.

Applying Lever's rule to Fig. 14.7 *a*

$$\frac{m_1}{m_2} = \frac{cb}{ac}$$

$$= \frac{(\text{wt. \% } B)_2 - (\text{wt. \% } B)_{AB}}{(\text{wt. \% } B)_{AB} - (\text{wt. \% } B)_1}$$

Now $$(\text{wt. \% } B)_{AB} = \frac{60 \text{ g}}{25 \text{ g} + 60 \text{ g}} = 0.706 = 70.6\%$$

Also $$m_1 + m_2 = 25 + 60 = 85 \text{ g}$$

$$\frac{m_1}{m_2} = \frac{80-70.6}{70.6-15} = \frac{9.4}{55.6} = 0.169$$

i.e.,

$$m_1 = 0.169\, m_2$$

$$0.169\, m_2 + m_2 = 95 \text{ g}$$

$$m_2 = \frac{85 \text{ g}}{1.169} = 72.79$$

$$m_1 = 85 - 72.79 = 12.21 \text{ g}$$

Note: (wt. % B)$_{AB}$ means weight % of B in total amount of A and B (wt. % B)$_I$ means weight % of B in layer I and so on.

***Exercise* 14.24:** The following data is given about the liquid-vapour phase diagram *i.e.,* the boiling point of various composition of liquids A and B.

T_b(°C)	53.4	54.0	55.5	58.6	60	62.8	63	62.2
$10^2 \times x_A(l)$	0.0	3.5	12.0	30.65	45.1	61.1	70.0	82.3
$10^2 \times x_A(\text{v})$	0.0	2.5	7.9	25.42	34.8	59.5	77.5	84.1
T_A(°C)	61.2	60.0	59.2					
$10^2 \times x_A(l)$	86.7	95.1	100					
$10^2 \times x_A(\text{v})$	91.5	97	100					

(*a*) Draw the liquid-vapour phase diagram as boiling point vs mole fraction. (*b*) Find the azeotropic temperature and azeotropic composition. (*c*) Take a mixture containing liquid as 0.35 mole fraction and heat up to 59.5°C in a closed system at 1 atm. What will be the composition of the liquid phase and that of vapour phase. (*d*) If 60 g of A (mol. wt. 48 g mol^{-1}) and 120 g of B (mol. wt. 80) are heated up to 62°C in a closed vessel at *a* atm, calculate the amount of liquid phase and vapour phase.

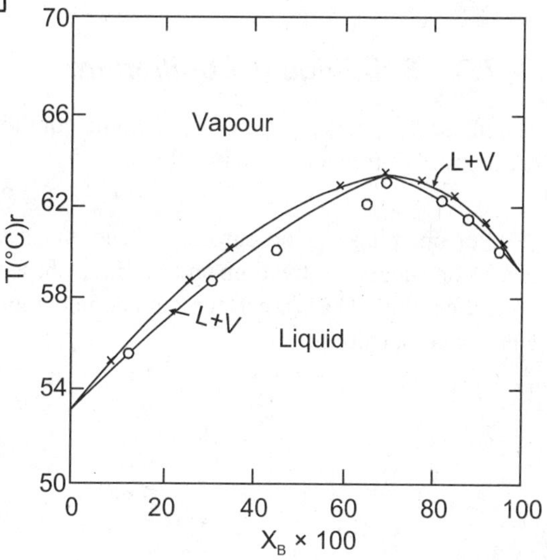

Fig. 14.11 *Plot of boiling points vs composition of binary mixture*

(*a*) Phase diagram is drawn in Fig. 14.11.

(*b*) Azeotropic temperature is 63.3°C and azeotropic composition is $x_B = 0.69$, $x_A = 0.31$.

(*c*) When this mixture is heated in a closed vessel the equilibrium is established at 59.5°C. The composition of liquid phase is $x_A(\text{v}) = 0.695$, $x_B(\text{v}) = 0.305$, $x_A(1) = 0.653$, $x_B(1) = 0.347$.

(*d*) $n_A = \dfrac{60 \text{ g}}{48 \text{ g mol}^{-1}} = 1.25 \text{ mol}, \quad n_B = \dfrac{120 \text{ g}}{80 \text{ g mol}^{-1}} = 1.5 \text{ mol}$

$$x_A \text{ (initial)} = \frac{1.25}{1.25+1.5} = 0.455$$

$$x_B \text{ (initial)} = \frac{1.5}{1.25+1.5} = 0.545$$

When this mixture is heated up to 62°C, the point is depicted by 'O'. So by the Lever's rule

$$\frac{n(1)}{n(\text{v})} = \frac{Oa}{bO} = \frac{0.545-0.505}{0.575-0.545} = 1.33$$

$$n_1 = 1.33\, n(\text{v})$$

Now total number of mol of both components including liquid and vapours are

$$n_{(1)} + n_{(\text{v})} = 2.75 \text{ mol}$$

Using the above equations

$$n_{(\text{v})} = \frac{2.75}{2.33} = 1.18 \text{ mol}$$

$$n_{(1)} = 2.75 - 1.18 = 1.57 \text{ mol}$$

14.7.3. *Solid-liquid Equilibrium*

The phase diagrams in case of solid-liquid equilibrium are expressed generally as constant pressure so the degrees of freedom are given by

$$F = 3 - P$$

For one phase, two degrees of freedoms are required to express the system completely, for two phases one degree of freedom and for three phases, the degrees of freedom are zero. This point is called 'eutectic point'. The temperature and composition at this point are known as eutectic temperature and eutectic composition.

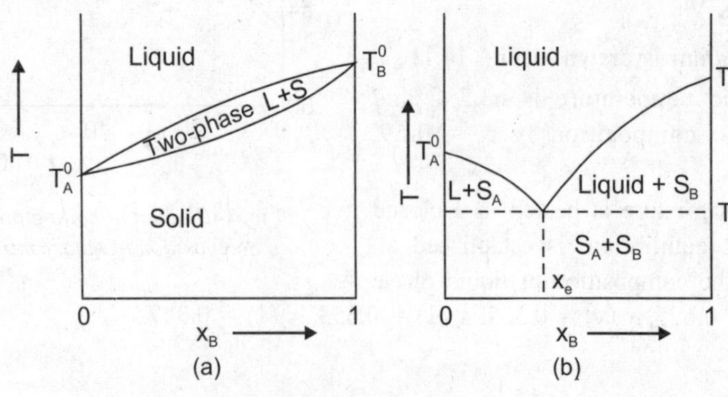

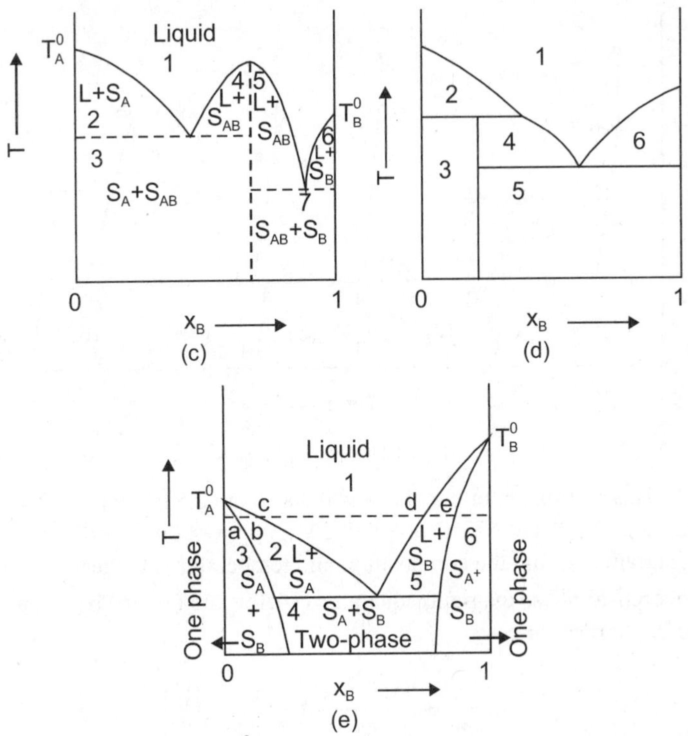

Fig. 14.12 *Solid-liquid phase diagrams at constant pressure*

There are many kinds of phase diagrams. Following are some common ones.

(*i*) Figure 14.12*a* is the phase diagram when two components are completely soluble in solid and liquid phase.

(*ii*) Figure 14.12*b* represents two components which are completely immiscible in solid phase but soluble in liquid phase.

(*iii*) Figure 14.12*c* represents a stable compound formation between two components and solid phases insoluble with each other. This is also known as congruent melting point compound formation.

(*iv*) Figure 14.12*d* represents when two components form unstable compound, called incongruent melting point compound formation.

(*v*) Figure 14.12*e* represents diagram for two components which show partial miscibility with each other in solid phase.

Phase diagrams for solid-liquid equilibrium are determined by the cooling curves method. These curves are constructed by taking a solid of certain composition, melting them completely into liquid phase and then noting the temperature at various times till the system completely solidifies. Thus cooling curve is a plot between temperature and time. The cooling curve for pure component and at the eutectic composition will consist of a 'plateau' or 'arrest'. At other compositions the cooling curve will consist of (1) a break where the solid begins to solidify and the liquid changes composition and (2) an 'arrest' at which the remaining liquid phase solidifies at the eutectic composition. The cooling curves are shown in Fig. 14.13.

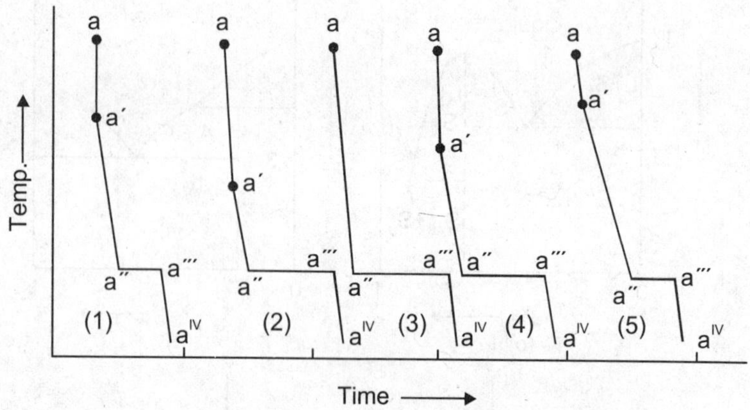

Fig. 14.13 *Cooling curves*

***Exercise* 14.25:** Heats of fusion of metal A and metal B are 2.5 kJ mol^{-1} and 4.0 kJ mol^{-1} and their fusion temperatures are 527°C and 672°C respectively. Suppose both form ideal solution, construct a theoretical phase diagram and find the composition and temperature at eutectic point.

To draw the theoretical phase diagram, equations (14.15) and (14.16) can be used to calculate the solubilities at different temperature.

$$\ln x_A = \frac{\Delta \overline{H}_{\text{fus}}(A)}{R}\left(\frac{1}{T_m(A)} - \frac{1}{T}\right) \qquad \text{...(14.15)}$$

and

$$\ln x_B = \frac{\Delta \overline{H}_{\text{fus}}(B)}{R}\left(\frac{1}{T_m(B)} - \frac{1}{T}\right) \qquad \text{...(14.16)}$$

The above equations can be rewritten in the following form after making substitutions.

$$\log x_A = 0.163 - \frac{130.344}{T}$$

and

$$\log x_B = 0.221 - \frac{208.86}{T}$$

Taking different values of T, smaller than $T_m\,(i)$, $x\,(i)$ can be calculated.

As model calculations, say $T = 400$ K

$$\log x_A = 0.163 - \frac{130.344}{400}$$

$$= -0.1629 = \bar{1}.8371$$

$$x_A = 0.6871$$

Similarly

$$\log x_B = 0.221 - \frac{208.86}{400}$$

$$= -0.3011 = \bar{1}.6989$$

$$x_B = 0.5$$

Following table can be obtained.

Temperature (T K)	300	400	500	600	700	800	900
x_A	0.535	0.6871	0.798	0.882	0.948		
x_B	0.335	0.5	0.635	0.746	0.837	0.912	0.975

Temperature vs x_A and Temperature vs x_B are plotted in Fig. 14.14 x_A(eut) = 0.588, T_e = 331 K.

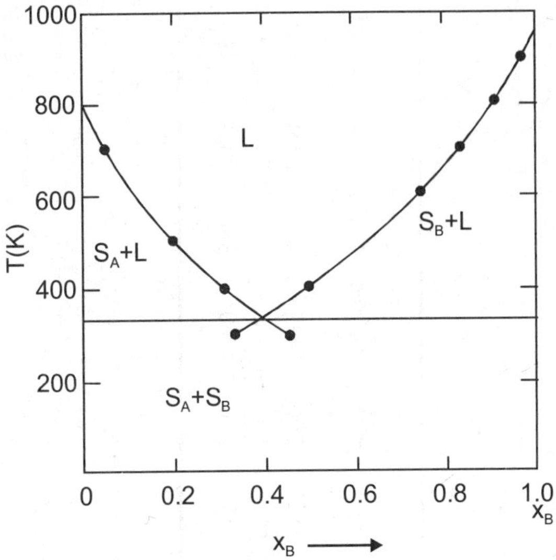

Fig. 14.14 *Plot of melting points of binary mixture at various compositions*

***Exercise* 14.26:** The results of the thermal analysis of Au and Sb are given below: (*a*) Construct the phase diagram (*b*) Indicate if there is any compound formation. If yes, find the molecular formula.

The data of next page is plotted in Fig. 14.15. It looks a compound formation but unstable one because there is no well defined maximum in the plot. The temperature where the curve for compound formation and the separation of Sb from liquid melt is meeting, is at 460°C.

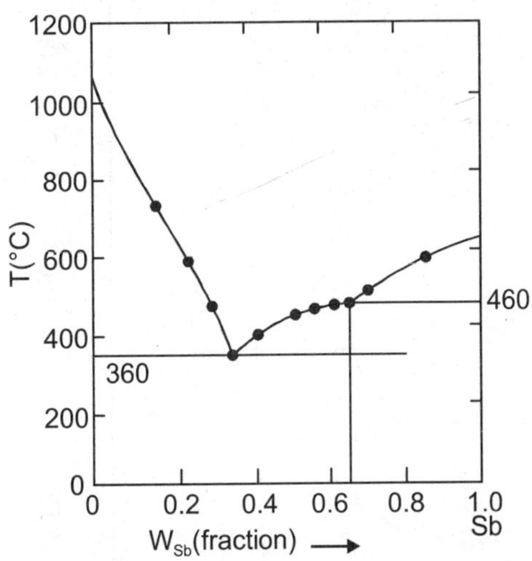

Fig. 14.15 *Solid-liquid phase diagram of Au–Sb system*

Atomic % Sb	0	15.12	22.06	28.61	34.83	40.93	51.66	56.74	61.58	66.21	70.64	86.51	100
1st separation of solid (°C)	1064	728	581	472	357	396	443	455	458	460	495	580	631
Invariant equilibrium (°C)	1064	360	361	362	357	360	360	360	359	460	460	460	631

$$\text{Wt. \% Sb} = 66\%$$
$$\text{Wt. \% Au} = 34\%$$

If one takes 100 g of total material

$$\text{Amount of Sb} = 66 \text{ g}$$

$$\text{Number of mol (Sb)} = \frac{66 \text{ g}}{122 \text{ g mol}^{-1}} = 0.541$$

$$\text{Amount of Au} = 34 \text{ g}$$

$$\text{Number of mol (Au)} = \frac{34 \text{ g}}{197 \text{ g mol}^{-1}} = 0.173$$

The ratio of the number of mol of Au : Sb :: 1 : 3. So the molecular formula could be $AuSb_3$.

***Exercise* 14.27:** (*a*) Label the areas mentioned in the following diagram which depict the solubility of a salt MX and its hydrates.

(*b*) Starting from point *a* mention the changes that would take place if the water vapours are removed at constant temperature.

(*a*) (1) Salt solution (2) ice and solution
(3) ice + $MX \cdot 7H_2O$ (4) solution + $MX \cdot 7H_2O$
(5) $MX \cdot 7H_2O$ + solution (6) $MX \cdot 7H_2O + MX \cdot 5H_2O$
(7) $MX \cdot 5H_2O$ + solution (8) $MX \cdot 2H_2O$ + solution
(9) $MX \cdot 2H_2O + MX \cdot 5H_2O$ (10) $MX \cdot 2H_2O$ + MX

(*b*) (*i*) Removal of water vapours from *a* will concentrate the solution in $MX \cdot 7H_2O$ till one reaches the point *b* on horizontal line.

(*ii*) At *b* $MX \cdot 7H_2O$ solid phase will separate out till the system is pure $MX \cdot 7H_2O$.

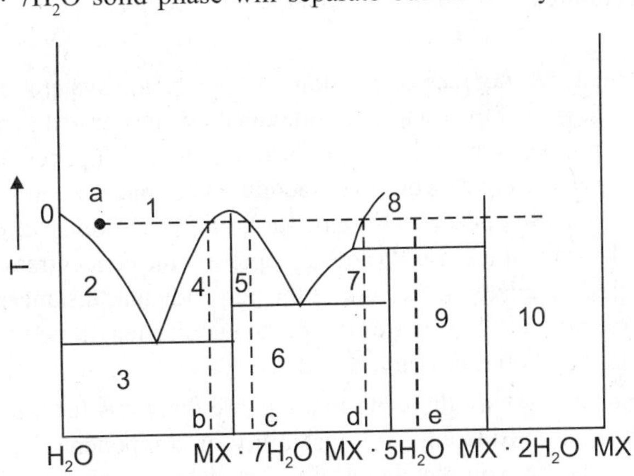

Fig. 14.16 *Phase diagram of $MX{\cdot}H_2O$ system*

(*iii*) Further removal of water will lead to complete solution phase after the point *c*.

(*iv*) At point *d* crystals of $MX \cdot 2H_2O$ will separate till the system is pure, because $MX \cdot 5H_2O$ is unstable above the line *xy*.

(*v*) At point *d* again $MX \cdot 5H_2O$ will completely dissolve to form solution.

(*vi*) At point *e* $MX \cdot 2H_2O$ will separate out till it is pure. Further removal of water will lead to the formation of MX.

***Exercise* 14.28:** Explain the phase diagram shown in Fig. 14.12*e*.

Phase diagram depicted in Fig. 14.12*e* gives the partial solubility of component *A* in component *B* in solid phase. Area 1 is a liquid phase. Area 2 saturated solid solution of *B* in *A* and liquid melt. Area 3, a single solid phase consisting of unsaturated solid solution of *B* in *A*, Area 4, saturated solid solution of *A* in *B* and that of *B* in *A*, Area 5, liquid and saturated solid solution of *A* in *B* and Area 6, single phase, unsaturated solution of *A* in *B*.

The degrees of freedom in regions 1, 3 and 6 are two because there is only one phase. In the region 2, 4 and 5 the degrees of freedom is only one because it is two phase region, *i.e.*, solid, solid-solid and liquid. The degrees of freedom on the lines are one and the point '0' is the eutectic point, no degree of freedom.

If one starts with point *a i.e.*, some known amount of component *B* in *A* in solid phase and start adding component *B* at constant temperature, the solid component *A* will be saturated in *B* and saturated solid solution is obtained till one reaches *b*. Further addition of *B* will produce two phase region, saturated solid solution *B* in *A* and liquid melt. After point *c* the complete system is in liquid phase till one reaches the point *d*. At point *d* the solid solution of *B* saturated with *A* crystallise out. Further addition will bring out more solid solution of *A* in *B* till the point *e* and after this the whole system freezes to unsaturated solid solution of *A* in *B*.

14.8. THREE COMPONENT SYSTEM

According to phase rule, the degrees of freedom are given by

$$F = C - P + 2$$
$$= 5 - P$$

and if pressure and temperature are fixed, the above equation becomes

$$F = 3 - P$$

For one phase, we need two degrees of freedom to specify the system completely and for two phases in equilibrium, one degree of freedom. Thus one can draw the phase diagram in a plane. The best way to represent three component system is to have a triangular graph paper. The concentrations can either be represented in terms of weight % or mole fractions. The apices are joined to the middle points of the opposite sides, *i.e.*, *Aa*, *Bb* and *Cc*. The zero starts from the points *a*, *b*, *c* and points *A*, *B*, *C* represent 100% or unity. Thus the lines *Aa*, *Bb* and *Cc* represent the concentrations of components *A*, *B* and *C*. Further, the triangle is equilateral, the sum of the perpendicular distances from any point in the triangle to the sides is constant and is equal to the perpendicular length between the corner and the centre of the opposite side, *i.e.*, 100% or unity.

Similar to two component system, there are many phase diagrams for three component systems. Here only simple one will be considered, *i.e.*, miscibility of components, *A*, *B*, *C* with each other. Specially let *A* and *B*, *B* and *C* are completely miscible but *A* and *C* only partially.

***Exercise* 14.29:** Locate the following points on the triangular graph paper

(*a*) $x_A = 0.2$, $x_B = 0.3$, $x_c = 0.5$

(*b*) $x_A = 0.15$, $x_B = 0.45$, $x_c = 0.4$

(*a*) In the triangular graph paper, the lines parallel to side *BC* represent the composition of *A*, parallel to *CA* the composition of *B* and parallel to *AB*, that of *C*. Starting from point *a*, move along *aA* uptill 0.2. Draw a horizontal line parallel to *BC*. Starting from point *b*, move along *bB* uptill 0.3 and draw a line parallel to *CA*. Starting from *c* move along *Cc* till 0.5 and draw a line parallel to *AB*. The intersection of these three parallel lines is point $x_A = 0.2$, $x_B = 0.3$, $x_c = 0.5$ *i.e.*, point 0.

(*b*) Proceed in the manner as mentioned above and the point 0 represents $x_A = 0.15$, $x_B = 0.45$ and $x_c = 0.4$.

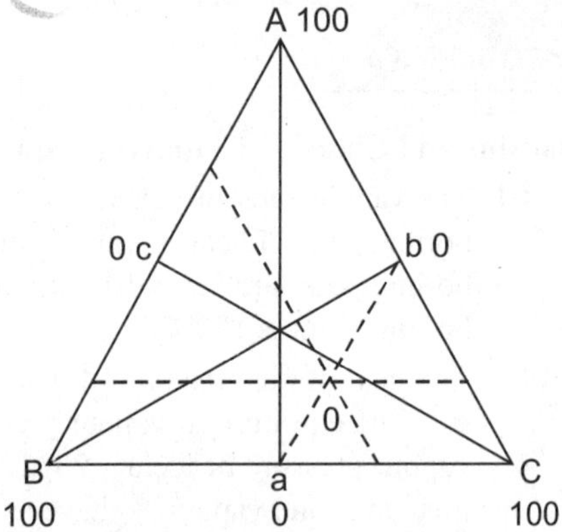

Fig. 14.17 *Location of points in three component system*

***Exercise* 14.30:** In the miscibility experiments of three components, *i.e.*, acetic acid, benzene and water, the M.Sc. students took known amount of water and benzene and kept on adding acetic acid till the solution is one phase. The following data is obtained at 35°C and atmospheric pressure.

Construct the phase diagram.

Benzene			Water			Acetic Acid		
$10^{+3} \times v_1$ (dm^3)	$10^3 \times m_1$ (kg)	wt. %	$v_2 \times 10^{+3}$ (dm^3)	$m_2 \times 10^3$ (kg)	wt. %	$10^3 \times v_3$ (dm^3)	$10^3 \times m_3$ (kg)	wt. %
5.0	4.35	20.8	3.0	3.0	14.4	12.9	13.55	64.9
5.0	4.35	6.28	21.0	21.0	30.3	41.8	44.0	63.5
5.0	4.35	35.8	1.0	1.0	8.24	6.5	6.82	56.0
20.0	17.40	56.7	1.0	1.0	3.26	11.7	12.3	40.2
35.0	30.45	63.2	1.0	1.0	2.07	16.0	16.8	34.8
3.0	2.61	4.45	15.0	15.0	25.6	39.0	41.0	69.9
2.0	1.74	3.18	25.0	25.0	45.6	26.85	28.2	14.8
2.0	1.74	2.07	75.0	75.0	89.3	7.05	7.42	8.81
2.0	1.74	1.82	100.0	100.0	95.8	2.4	2.52	2.31

The density of benzene, acetic acid and water at this temperature can be taken as 0.87 kg dm^{-3}, 1.05 kg dm^{-3} and 1.0 kg dm^{-3} respectively.

Convert the volume into mass for each component and calculate the percentage weight composition for each component. These have been mentioned in columns 3, 6 and 9 of the Table. Plot each point and procedure is mentioned in exercise 14.27. The desired phase diagram is depicted in Fig. 14.18.

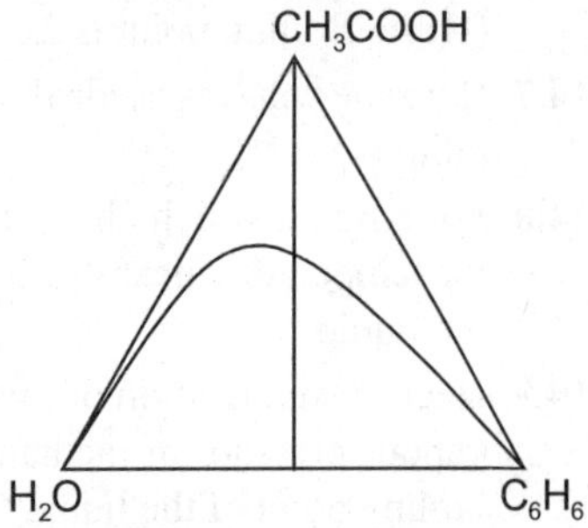

Fig. 14.18 *Phase diagram of H_2O, C_6H_6 and CH_3COOH system*

PROBLEMS

Clausius and Clausius-Clapeyron Equation

14.1 The vapour pressure exerted over a liquid A is 5×10^6 dynes cm^{-2}. Calculate the change in boiling point if heat of vaporisation is 100 kJ mol^{-1} and densities of vapours and liquid at its boiling point are 3.5×10^{-4} kg dm^{-3} and 1.25 kg dm^{-3}. Molecular weight is 100 g mol^{-1} and boiling point is 150°C.

14.2 The heat of vaporisation of water at 0.0°C is 2.20 kJ g^{-1} and heat of fusion at 0°C 334.72 J g^{-1}. The equilibrium vapour pressure of water at 0°C is 4.95 mm. Calculate the sublimation vapour pressure of ice at –23°C, assuming that heat of sublimation of water is constant over this range and vapours behave as ideal gas.

14.3 The following data is obtained when M.Sc. students performed the variation of vapour pressure of acetophenone with temperature experiment.

t(°C)	78.0	109.4	133.6	178.0	202.4
p(mm)	10	40	100	400	760

Calculate the heat of vaporisation and the boiling point of acetophenone. The actual boiling point is 202.4°C. Explain if there is any discrepancy.

14.4 The densities of ice and water at its freezing point are 0.9168 kg dm^{-3} and 0.9998 kg dm^{-3} respectively. If the heat of fusion of water is 334.72 J g^{-1}, calculate the melting point of water at (*a*) 50 atm, (*b*) 500 atm.

14.5 The vapour pressure (mm) of liquid A can be expressed as $\log p = 7.6546 - \dfrac{1686.77}{T}$. The density of liquid phase at 80.2°C is 0.878 kg dm^{-8} and molecular weight is 80 g mol^{-1}. If heat of vaporisation is 30 kJ mol^{-1}, calculate the density of vapour phase of A at 80.2°C (normal boiling point).

14.6 A liquid of molecular weight 80 g mol^{-1} melts at 10°C and has a critical temperature of 350°C. The density of solid and liquid phases are 0.95 and 0.87 kg dm^{-3} respectively. If the melting point of liquid at the triple point is 9.95°C (where the system is subjected only to its own low vapour pressure), calculate, $\Delta\bar{H}_v$, $\Delta\bar{H}_f$, $\Delta\bar{H}_{sub}$ and the vapour pressure at the triple point. Assume the liquid is ideal.

(**Hint:** Boiling point is 2/3 of the critical temperature expressed in K.)

14.7 Benzene behaves as ideal liquid. What will be its boiling point at 5 mm when its normal boiling point is 80.1°C?

14.8 For a liquid which obeys Trouton's law, the vapour pressure increases by 8 mm per degree at the temperature near its boiling point. Find out the heat of vaporisation and the boiling point of the liquid.

14.9 Heat of vaporisation of methanol is 35.27 kJ mol^{-1} and that of a liquid A is 25 kJ mol^{-1}. If the vapour pressure of methanol at 50°C is the same as that of liquid A (*a*) calculate the normal boiling point of the liquid (*b*) which of the two liquid obey Trouton's law nicely. Boiling point of methanol is 65°C.

14.10 The boiling points of carbontetrachloride and ethanol are 76.7 and 78.5°C. The heat of vaporisation of CCl_4 and ethanol are 30 kJ mol^{-1} and 38.58 kJ mol^{-1} respectively. Calculate the temperature at which both the liquids have the same vapour pressure.

14.11 The vapour pressure of a liquid increases 6 % near the temperature point of 40°C. Calculate $\Delta\bar{H}_{vap}$ and the normal boiling of the liquid. Mention the assumptions made and the approximation used.

14.12 Following are the data for a compound: melting point at 1 atm = 10.5°C, ΔH_{fus} = 410 J g^{-1}, d(solid) = 1.462 kg dm^{-3}, d(liquid) = 1.245 kg dm^{-3}, molecular wt. = 54 g mol^{-1}, normal boiling point = 140°C. Calculate (a) Heat of sublimation from the above data or other (b) melting point at its own vapour pressure.

14.13 Assume that $\Delta\bar{H}_{vap}$ is not independent of temperature and can be expressed as

$$\Delta\bar{H}_{vap} = \Delta\bar{H}_{vap,0} + \Delta C_p T$$

where $\Delta H_{0,\ vap}$ is the heat of vaporisation at 0°K and $\Delta C_p = C_p$ (vap) – C_p(liq), is generally negative, show that the Clausius-Clapeyron equation reduces to

$$\ln p = -\frac{\Delta H_{0+vap}}{RT} + \frac{\Delta C_p}{RT} \ln T + \text{constant for ideal vapour.}$$

14.14 The vapour pressure of ethylchloride can be expressed as $\log\ p = -\dfrac{0.05223\ a}{T} + b$

where a = 26319 K, b = 7.691 and p is measured in mm. Calculate the heat of vaporisation at 40°C and at its boiling point.

14.15 The vapour pressure of CCl_4 (liq.) and that of solid CCl_4 can be expressed as

$$\log p_s = -\frac{0.05233\ a}{T} + b,\ a = 34608,\ b = 8.05$$

$$\log p_l = -\frac{0.05233\ a}{T} + b,\ a = 33914,\ b = 8.004$$

where p_i is expressed in mm. Calculate the pressure and temperature at the triple point.

14.16 Calculate the sublimation vapour pressure of SO_2 at –100°C assuming that the heat of sublimation is constant. The ΔH_{vap}= 24.95 kJ mol^{-1} and $\Delta\bar{H}_{fus}$= 7.40 kJ mol^{-1}. The vapour pressure of SO_2 at – 76.8°C is 10 mm.

14.17 What conclusion you would draw about the vapour pressure at $T = \infty$ of the liquids those obey Trouton's rule?

14.18 The vapour pressure of liquid A is 60 mm at temperature 35°C and that of solid is 1 mm less than that of liquid at the same temperature. The difference between the vapour pressures of solid and liquid (*i.e.*, $p_l - p_s$) is 2 mm when the temperature is 33°C. If $\Delta\bar{H}_{vap}$ is 40 kJ, calculate: (a) $\Delta\bar{H}_{fus}$, (b) $\Delta\bar{H}_{sub}$ and (c) melting point of A.

14.19 The heat of transition for the compound A to B in the solid phase is 9.0 kJ mol^{-1}. The equilibrium between the two form is established at (a) 2 atm pressure and 140°C and (b) 60 atm pressure and 144°C. Find out which form will be having higher density.

14.20 If the changes in pressure and temperature are very small, by using Trouton's rule, it can be shown that

$$T = 1.25 \times 10^{-4} T_{obs}(p_{(mm)} - 760)$$
$$= 9.46 \times 10^{-2} T_{obs}(p_{(atm)} - 1)$$

14.21 The vapour pressure of water at 0.0°C is 4.58 mm and its density is 1 kg dm^{-3}. Calculate the pressure applied such that its vapour pressure at this temperature is increased to 6.6 mm.

14.22 The density of liquid mercury at its melting point 126.2°C is 13.30 kg dm^{-3}. If the vapour pressure at 126.2°C temperature is 1 mm, calculate the vapour pressure of mercury at (*a*) 100 atm and (*b*) 10 atm.

14.23 The vapour pressure (mm) of methanol can be nicely expressed as

$$\log p = -\frac{A}{T} + B$$

in the temperature range of – 10 to 80°C. Calculate the molar volume of methanol vapour at 20°C.

$A = 38324$, $B = 8.8017$ and density of liquid methanol is 0.7914 kg dm^{-3} at 20°C.

Phase Rule

14.24 In the following systems, calculate the number of phases.

(*a*) $N_2(g) + 3H_2(g) \rightleftharpoons 2NH_3(g)$

(*b*) I_2 dissolved in CCl_4

(*c*) $NH_4CO_2NH_2(s) \rightleftharpoons 2NH_3(g) + CO_2(g)$

(*d*) $C(s) + H_2O(g) \rightleftharpoons CO(g) + H_2(g)$

(*e*) $2Pb_3O_4(s) \rightleftharpoons 6PbO(s) + O_2(g)$

(*f*) $CuSO_4.5H_2O(s) \rightleftharpoons CuSO_4.3H_2O(s) + 2H_2O(g)$

(*g*) $Fe(s) + H_2O(g) \rightleftharpoons FeO(s) + H_2(g)$

14.25 Calculate the number of components in each case of the problem 14.24.

14.26 Calculate the degrees of freedom for the systems mentioned in problem 14.24.

14.27 Is it possible to have quadrupole points for one component system?

One-Component System

14.28 Draw the phase diagram (*p* vs *T*) for the solid-liquid equilibrium for univalent system when $\bar{V}_l - \bar{V}_s = -$ve and $\bar{V}_l - \bar{V}_s = 0$. Give some example if possible.

14.29 In the one-component phase diagram (Fig. P14.29) list the degrees of freedom, and the phases present in the areas, lines and at the points.

14.30 In the one-component phase diagrams, generally, the slope of the line representing the equilibrium of *s* – v is more than that for the line representing the *l* – v equilibrium near the tripple point. Explain this.

14.31 In the diagram mentioned in the problem 14.29 describe the behaviour if you start at a temperature T_1 and start increasing the pressure.

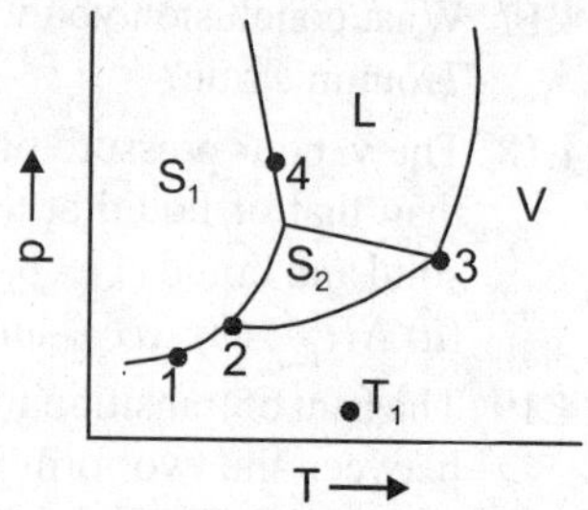

Fig. P14.29 *Phase diagram of one component system*

14.32 Heat of fusion and heat of vaporisation of water are 6.01 kJ mol^{-1} and 40.66 kJ mol^{-1}, vapour pressure of ice at – 5°C and that of liquid water at 10°C are 3.013 mm and 9.209 mm. Calculate the values of vapour pressures at various temperatures (–40°C, –25°C, –15°C, 0°C, 15°C, 25°C, 40°C) of ice and liquid water. Draw a plot p vs T and find the temperature and pressure where the two curves meet. Is this matches with tripple point?

14.33 The vapour pressure (mm) of benzene can be represented by the equation $\log p = -\frac{0.05223a}{T} + b$ where a and b have different values in different temperature ranges.

–58°C to –30°C solid,	$a = 42904$,	$b = 9.556$
–30°C to –5°C solid,	$a = 44222$,	$b = 9.846$
0°C to 42°C liquid,	$a = 34172$,	$b = 7.9622$
42°C to 100°C liquid,	$a = 32295$,	$b = 7.6546$

Compute the vapour pressure for different temperatures, plot p vs T and calculate the pressure and temperature of triple point.

14.34 Draw the phase diagram for one-component system from the following data: (i) compound A exists in two solid forms A_1, A_2, density of A_1 as solid is greater than that of liquid while solid A_2 is less dense than that of liquid (ii) the solid A_1, solid A_2 and vapours are at equilibrium at 2.5 mm at 5.5°C (iii) solid A_1, liquid and solid A_2 are at equilibrium at 15°C and 5 atm (iv) solid A_2, L and vapours at equilibrium at 30°C and 100 mm (v) the transition temperature of solids A_1 and A_2 increases with increase of pressure.

14.35 Draw the phase diagram for one-component system A from the following data: (i) compound A exists in two solid forms A_1 and A_2, densities of solids A_1 and A_2 are more than that of liquid phase (ii) the transition temperature of A_1 and A_2 decreases with the increase of pressure (iii) the phase A_1 is stable at low pressure (iv) the melting point of solid at its own vapour pressure (10 mm) is 8.2°C (v) the boiling point is 90°C at 1 atm and (vi) A_1, A_2 and liquid coexist at 100 atm and 40°C. Explain the areas and the lines.

Two Component System: Solid-liquid Diagram

14.36 Construct the solid-liquid phase diagram for cinnamic acid (A) and benzoic acid (B) system from the following data:

wt. of A (g)	2	4	6	5	5	5	7.04	0
wt. of B (g)	5	5	5	2	4	6	4.82	5
Freezing point (°C)	97	95	85.5	99	78	75.5	100	115

Calculate the composition at the eutectic point and also find the temperature.

14.37 Construct the phase diagram for Zn and Mg system as melting point plotted against mole fraction from the following data T_f(Mg) = 651°C, T_f(Zn) = 419.5°C, $T_f(Mg_aZn_b)$ = 590°C. The composition of the compound is 15% by weight of Mg and 85% by weight of Zn. The lowest freezing point of Zn observed is 368°C for a composition containing 3.2% of Mg by weight and that for Mg (347°C) for a composition containing 49% by weight of Mg. Find the molecular formula of the compound. Also label the phases, components and degrees of freedom at each line, point and area.

14.38 Indicate the behaviour of a cooling curve for the system containing 60% by weight of Mg and starting at a temperature of 700°C in problem 14.37.

14.39 In a two-component solid-liquid phase equilibrium study, draw phase diagram with the following data. $T_f(\text{Mg}) = 650°\text{C}$, $T_f(\text{Ge}) = 1115°\text{C}$ and $T_f(\text{Mg}_x\text{Ge}_y) = 936°\text{C}$. The lowest melting point observed for Mg is at 635°C with having Ge percentage is 3.4% by weight and the lowest melting point of Ge observed is at 680°C with Ge 82.3% by weight. The composition of the mixture at the highest temperature is 61 % weight of Ge.

14.40 The Ag and Ge nearly form ideal system and are completely miscible in both the phases. If the melting points of Ag and Ge are 1700 K and 1210 K respectively, draw the phase diagram.

14.41 Explain the line, points and regions labelled in Fig. 14.10.

14.42 Construct the solid-liquid phase diagram for Bi and Cd system from the following data, $T_f(\text{Bi}) = 271°\text{C}$, $T_f(\text{Cd}) = 321°\text{C}$. The eutectic composition is 60% by weight of Bi and eutectic temperature is 144°C. Also draw the phase diagram as the atomic weight fraction.

14.43 The heat of fusion of benzene and naphthalene are 9.832 kJ mol^{-1} and 19.08 kJ mol^{-1}. The fusion temperatures of benzene and naphthalene are 5.6°C and 80°C respectively. Using the equations (14.15) and (14.16), calculate the freezing point of mixture for different compositions of benzene in naphthalene and naphthalene in benzene. Plot T_f vs x (benzene) and find the eutectic composition and temperature, compare them with the experimental ones. $T_e = -3.6°\text{C}$ and x_e (benzene) = 0.868.

14.44 The heat of fusion of *p*-xylene is 16.92 kJ mol^{-1}. $T_f(\text{para}) = 13.5°\text{C}$, $T_f(\text{meta}) = -48.6°\text{C}$. If the eutectic temperature is $-53.1°\text{C}$ and assuming that two compounds form ideal solution overall composition, (*a*) calculate the eutectic composition and heat of fusion of meta xylene, (*b*) Calculate the heat required to melt 1 mol of eutectic composition.

14.45 Calculate the heats of fusion of cinnamic acid and benzoic acid from the data given in the problem 14.36.

14.46 The heat of fusion of naphthalene and *p*-dichlorobenzene can be expressed as

$$\Delta\bar{H}_{\text{fus}}(\text{C}_{10}\text{H}_8) = 17844.8 - 110.\,1T + 0.64T^2$$

$$\Delta H_{\text{fus}}(\text{C}_6\text{H}_4\text{Cl}_2) = -42886 + 393.6T - 0.63T^2$$

Construct the solid-liquid phase diagram if melting points of naphthalene and*p*-dichlorobenzene are 80.55°C and 53.1°C. Find eutectic temperature and composition.

Partially Miscible Liquids

14.47 (*a*) Construct the liquid-liquid phase diagram for two partially immiscible liquids A and B with the following data:

T(°C)	80	70	52	38	26
wt. % $(B)_1$	15	17	21	28	39
wt. % $(B)_2$	92	86	76	68	58

(*b*) Calculate the temperature and composition of the lower consolute.

(*c*) If the molecular weights of *A* and *B* are 100 g mol^{-1} and 150 g mol^{-1}, draw the phase diagram as *x* vs temperature.

(*d*) Take 10 g of *A* and 20 g of *B* and mix them at 45°C. Calculate the amount of *B* present in both the layers.

14.48 (*a*) Construct the liquid-liquid phase diagram for two partially miscible liquids *A* and *B* with the following data:

T(°C)	9	22	33	40	45	51	58
wt.%$(A)_1$	9	14	20	26	31	39	55
wt.%$(A)_2$	100	100	100	100	96	91	82

(*b*) Calculate the temperature and composition of the upper consolute.

(*c*) Label the two-phase and one-phase region.

(*d*) Take 40 g of *A* and 60 g of *B*, mix them together and calculate the relative proportions of *A* in phase I and phase II at 50°C.

14.49 In the phase diagram (14.9*a*) of liquid-vapour equilibrium drawn as *T* vs x_A at constant pressure, identify the phases and calculate degrees of freedom of the regions marked.

14.50 10 g of A and 50 g of B are mixed together at 25°C. These two are partially soluble and form two-phase system. If the weight % of *A* in phase I is 5% and in phase II is 60 %, calculate the masses of the two phase in equilibrium.

14.51 Take point *a* as mentioned in Fig. 14.9*a*. Explain the behaviour of the system if you move on the dotted line.

14.52 Explain the Fig. 14.9*b*. Under what conditions the Fig. 14.9*b* will become Fig. 14.9*a*.

Liquid Vapour Equilibrium

14.53 Following data is obtained for the methanol and chloroform system.

wt. %		B.pt. °C	wt. %		B.pt. °C
$CHCl_3(l)$	$CHCl_3(v)$		$CHCl_3(l)$	$CHCl_3(v)$	
100	0	54	62.98	79.61	54.5
91.73	90.80	53	58.15	79.97	55.5
83.21	86.21	52.5	51.40	70.53	56.5
76.20	82.51	54	42.84	53.01	57.5
67.51	81.44	54	27.09	41.02	60
			0	0	68

(*a*) Construct the liquid-vapour phase diagram as boiling point vs mole fraction of $CHCl_3$.

(*b*) Find if there is any minimum or maximum in boiling point.

(*c*) If there is, find the composition at that point.

(*d*) Is the system exhibit a positive or negative deviation?

14.54 (*a*) Construct the liquid-vapour phase diagram (boiling point vs mole fraction of $CHCl_3$) for the $CHCl_3$– $(CH_3)_2CO$ system from the data given on page 736 (Table given next page).

(*b*) At 58°C what is the composition of liquid phase and vapour phase.

(*c*) Find the composition and temperature at the azeotropic point.

(*d*) Take 50 g of $(CH_3)_2CO$ and 50 g of $CHCl_3$, mix together and heat the mixture to 61°C. Which component will be richer in vapour phase.

(*e*) Cool the vapours obtained in part (*d*) and find its temperature.

(*f*) Carry out the steps (*d*) and (*e*) for a solution containing 100 g of $CHCl_3$ and 20 g of $(CH_3)_2CO$.

(*g*) Calculate the relative amount of liquid and vapour phase at $T = 57°C$ and $X(CHCl_3) = 12\%$.

B. pt. (°C)	55.1	56.2	57.1	58.3	60.1	60.9	61.6	62.3	63.3	63.4	62.9	61.1	60.4
x_{CHCl_3}(l)	0	7	12.17	21.2	31.0	35.75	42.0	46.6	58.98	70.81	77.96	95.05	100
y_{CHCl_3}(v)	0	4	7	12.06	21.05	25.95	31.0	36.95	52.0	76.86	85.28	97.33	100

14.55 The following data is obtained for the liquid-vapour equilibrium in case of benzene and CS_2 system at 25°C.

(*a*) Construct the liquid-vapour phase diagram as P_T vs x_{CS_2} as well as P_T vs y_{CS_2}.

(*b*) Calculate the azeotropic composition and the pressure.

(*c*) Plot p_{CS_2} *vs* x_{CS_2} and p_{ben} vs x_{ben}. Calculate the Henry's constants for each solute in the solvent.

(*d*) Calculate the relative amount of liquid phase and vapour phase at $P_T = 200$ mm and $x_{CS_2} = 0.8$.

$x_{CS_2}(10^{-2})$	100	97.89	95.32	88.47	81.76	69.98	50.14	36.98	23.37	12.86	5.51	0.00
p_{CS_2}(mm)	361.1	352.7	344.1	321.4	299.1	263.6	197.4	156.6	102.5	60.2	27.9	0
p_{ben}(mm)	0	3.5	7.1	16.0	24.0	35.1	53.7	63.5	75.3	83.8	89.5	94.9

14.56 Benzene and toluene form ideal solution, (*a*) Calculate the total vapour pressure for different compositions of benzene and toluene (*b*) Calculate also the composition of the vapour phase at each total pressure and calculated above (*c*) Plot P_T vs x_{ben} and P_T vs y_{ben}. p^0(ben) and p^0_{tol} at 40°C are 186 mm and 64 mm respectively (*d*) The following is the experimental data at 40°C. Compare the theoretical total vapour pressure vs experimental

$x_{benzene}$	0	0.1	0.2	0.3	0.4	0.5	0.6	0.7	0.8	0.9	1.0
P_T(mm)	64	78	94	109	122	136	148	158	168	178	186

14.57 The boiling point diagram for two miscible liquids is shown in Fig. 14.8. (*a*) Take a solution containing $x_A = 0.4$ and boil in the open. Find the composition of the vapours when they were formed in the beginning and that of vapours when they were formed in the end.

(*b*) Heat the same solution in a closed vessel at 1 atm. Calculate the temperature and the composition of the liquid when 3/4 of the solution is vaporised.

14.58 The liquid-vapour phase diagram for two liquids *A* and *B* is given in Fig 14.8. The normal boiling points of liquids are 80°C and 35°C and they form ideal solutions. A solution containing 0.3 mol of *A* and 0.7 mol of *B* was heated upto 60°C in a closed system and at 1 atm. Thus n_v mol of vapours of composition y_A and n_l mol of liquid remaining with composition x_A were obtained. The vapour pressure of the condensate in equilibrium with liquid phase is 800 mm at 60°C with the composition $y_A^1 = 0.70$. Calculate (*a*) y_A, x_A; (*b*) n_v and n_l and (*c*) using the above data construct the liquid-vapour phase diagram in terms of total pressure and composition.

14.59 The liquid-vapour phase diagram for liquids A and B is drawn as temperature vs composition. A solution containing 0.4 mol of A and 0.6 mol of B is heated in open till its boiling point is increased by 10°C. From the phase diagram calculate: (a) initial and final boiling points (b) composition of the liquid when boiling stopped (c) initial and final and thus average composition of vapour phase (d) the mol of liquid remaining in liquid when boiling stopped and (e) draw the vapour pressure composition diagram for this system.

14.60 Redraw the phase diagrams 14.8 and 14.9 in terms of vapour pressure composition diagrams.

Three Component System

14.61 Plot the following points on a triangular coordinate system.

(a) $x_A = 0.75,\quad x_B = 0,\quad x_C = 0.25$

(b) $x_A = 0.65,\quad x_B = 0.10,\quad x_C = 0.25$

(c) $x_A = 0.45,\quad x_B = 0.20,\quad x_C = 0.35$

(d) $x_A = 0.10,\quad x_B = 0.20,\quad x_C = 0.70$

(e) $x_A = 0.50,\quad x_B = 0.10,\quad x_C = 0.40$

(f) $x_A = 0.00,\quad x_B = 0.20,\quad x_C = 0.80$

14.62 Construct the phase diagram for ethyl alcohol, water and benzene from the data. The data below is when a single phase obtained.

wt%			wt%		
Benzene	**Water**	**EtOH**	**Benzene**	**Water**	**EtOH**
60.8	7.1	32.1	45.1	12.6	42.3
65.9	5.6	28.5	36.5	16.7	46.8
75.7	3.4	21.1	20.9	26.6	52.5
82.5	2.0	15.5	10.9	37.2	51.9
87.6	1.2	11.2	5.9	46.2	47.9
91.2	0.7	8.1	2.8	55.6	41.6

14.63 The data given below is for three-component system when they form uniform system. Construct the phase diagram.

wt%			wt%		
$CHC1_3$	H_2O	$(CH_3)_2CO$	$CHCI_3$	H_2O	$(CH_3)_2CO$
1.23	82.97	15.8	70.0	1.3	28.7
1.29	73.11	25.6	55.7	2.2	42.1
1.71	62.29	36.0	42.9	4.4	52.7
3.20	54.1	42.7	35.8	7.7	56.5
5.10	45.6	49.3	28.4	10.3	61.3
9.8	34.5	55.7	20.4	18.6	61.0

Addition Problems

14.64 Derive Clapeyron equation (14.2)

14.65 Derive equation (14.7) assuming that molar heat of vapourisation is independent of temperature as well as assuming that molar heat of vapourisation is dependent of temperature.

14.66 Using the data given in Exercise 14.1, calculate the differential change of temperature with pressure and calculate the melting point of water at 500 atm.

14.67 Calculate the value of $\frac{dT}{dp}$ for water at its normal boiling point of 373 K. Molar enthalpy of vapourisation of water at 373 K is 40.65 kJ mol^{-1} and the densities of the liquid and vapour are 0.9584 g mL^{-1} and 0.6010 g L^{-1} respectively. Estimate the boiling point of water at 5 atm.

14.68 Vapour pressure of benzene can be expressed as

$$\ln p(torr) = 16.725 - \frac{3229.86\ \text{K}}{T} - \frac{118345\ \text{K}^2}{T^2}$$

in the range of 298.2 K to its normal boiling point 353.24 K. Calculate molar volume of the vapour at its equilibrium pressure at 353.24 K. Compare this value of molar volume assuming that the vapours of benzene are behaving in ideal manner. Given: $\Delta_{\text{vap}}\bar{H}$ at 353.24 K is 30.8 kJ mol^{-1} and molar volume of liquid benzene at 353.24 K is 96.0 cm^{3} mol^{-1}.

14.69 If the vapour pressure of benzene at 373.2 K is 1360 torr, calculate the value of $\Delta_{\text{vap}}\bar{H}$. Compare this value with that of experimental value 30.8 kJ mol^{-1}.

14.70 Vapour pressure of ammonia in torr can be expressed as:

$\ln p(\text{torr}) = 34.4834 - 1.81630 \ln T(\text{K}) - \frac{4124.4\ \text{K}}{T}$ in the temperature range of 146 K to 195 K.

Calculate $\Delta_{\text{sub}}\bar{H}$ in the above temperature range.

14.71 Following data exit for oxygen molecule. At triple point, T_{trip} = 54.3 K and p_{trip} = 1.14 torr; critical point, T_{critic} = 154.6 K and p_{critic} = 37828 torr, normal boiling point –182.9 °C and normal melting point –218.4 °C. Construct the phase diagram qualitatively. On comparing with phase diagram for water, does it melt under applied pressure?

14.72 Repeat the Problem 14.71 for iodine from the following data: T_{trip} = 113 °C and p_{trip} = 91.2 torr; critical point, T_{critc} = 512 °C and p_{critic} = 116 atm., normal boiling point 184 °C and normal melting point 114 °C, density of liquid iodine > density of solid iodine.

14.73 Consider equilibrium between diamond and graphite, *i.e.*, Graphite(s) → Diamond(s). If standard heat of formation of graphite and diamond are 0 kJ mol^{-1} and 1.895 kJ mol^{-1} respectively, (*i*) calculate the heat of reaction for the above process, (*ii*) Density of graphite and diamond are 2.25 g cm^{-3} and 3.51 g cm^{-3} respectively. Calculate the ΔV of this phase transformation. (*iii*) Graphite and diamond are found to be at equilibrium at 3000 K and 235 bar of pressure. If we assume that heat of reaction of the above reaction is independent of temperature, calculate the rate at which the equilibrium pressure changes temperature at that temperature and pressure.

14.74 The normal boiling point of water is 373 K. Calculate the vapour pressure of water at 60 ° C. Given $\Delta_f H^0(l) = -285.83$ kJ mol^{-1}, $\Delta_f H^0(g) = -241.82$ kJ mol^{-1} and $\Delta C_p = 0$.

14.75 Least square analysis of the plot between the vapour pressure and temperature gave the following expression: $\ln(p^{eq}) = 21.120 - \dfrac{5097.4 \text{ K}}{T}$. Calculate $\Delta_{vap}H$ and normal boiling.

14.76 We know when a wire tied some weights on both ends are hung over the ice cube, the wire pass through the ice cube without cutting it. Now suppose the weight on each end is 20 kg and other given data are $\Delta_{fus}H^0(H_2O, s) = -6.010$ kJ mol^{-1}, densities of water and ice are 0.9998 g cm^{-3} and 0.9168 g cm^{-3} respectively. Using given data, calculate value of T^{eq}.

14.77 In a five litre container at 300 K holds 250 mL of liquid water. The equilibrium vapour pressure at this temperature is found to be 25 mm of Hg. What pressure of Xe may be added in the container so that the vapour pressure of water is doubled?

14.78 Consider the equilibrium, $Al(s) \rightleftarrows Al(l)$. Given S^0 $(Al, s) = 28.33$ J K^{-1} mol^{-1} and $S^0(Al, l) =$ 39.55 J K^{-1} mol^{-1}. These values are at 298 K and 1 atm. pressure and assume that these values are independent of temperature. Calculate the temperature at which the molar free energies of the two phases are equal.

14.79 Consider the equilibrium, $Al(s) \rightleftarrows Al(l)$. Given densities of solid Al and liquid Al are 2.70 g cm^{-3} and 2.357 g cm^{-3} respectively. S^0 $(Al, s) = 28.33$ J K^{-1} mol^{-1} and $(Al, s) = 28.33$ J K^{-1} mol^{-1}. Assuming that the partial molal volumes of the liquid and solid forms are independent of pressure. calculate the chemical potentials of $Al(s)$ and $Al(l)$ as a function of pressure at 298 K.

14.80 The normal boiling point of bromine is 58.78 °C. Calculate its vapour pressure at 35 °C. Given $\Delta_f H_0$ $(Br_2, g) = 30.907$ kJ mol^{-1}. Assume that $\Delta C_p = 0$ for the vaporisation process.

14.81 The equilibrium vapour pressure of a solid at 300 K is 20 mm of Hg. If the heat of sublimation is 32.5 kJ mol^{-1}, calculate the vapour pressure of the solid at 127 °C.

14.82 The vapour pressures of a liquid at 285 K and 330 K are 25.2 and 292.4 mm of Hg. (*i*) Calculate the heat of vapourisation and normal boiling point by assuming that $\Delta C_p = 0$, *i.e.*, heat of vapourisation is independent in this range of temperature. (*ii*) If the standard heat of formation the liquid is 45.65 kJ mol^{-1}, calculate the heat of formation of the vapour from of liquid.

14.83 In literature there are two kinds of boiling points. One is the *normal boiling point* and the second is the *standard boiling point*. The former is at 760 torr and the second is at 1 bar = 750.6 torr. Using the data of the Problem 14.10 (heat of vapourisation and $\Delta C_p = 0$), calculate the standard boiling point and the percent difference in the two.

14.84 Vapour pressure of CCl_4 (l) at 330.95 K and 277.45 K are 400 and 40 torr respectively, (*i*) calculate the normal boiling point and molar heat of vapourisation ($\Delta_{vap}\bar{H}$) of CCl_4, (*ii*) if we assume that heat capacities of CCl_4 for solid, liquid and vapour phases respectively are identical then $\Delta_{sub}\bar{H} = \Delta_{fus}\bar{H} + \Delta_{vap}\bar{H}$. Calculate the melting point of CCl_4 from the given data: $\Delta_{fus}\bar{H} = 2.677$ kJ mol^{-1} and vapour pressure of $CCl_4(s)$ at 223.15 K is 1 torr.

14.85 Standard heats of formation are given for the liquid and gaseous states and it is assumed that $\Delta C_p = 0$. $\Delta_f\bar{H}(H_2O, l) = -285.83$ kJ mol^{-1}, $\Delta_f\bar{H}(H_2O, g) = -241.82$ kJ mol^{-1}. (*i*) Calculate $\Delta_{vap}\bar{H}$ at 298 K for the process $H_2O\,(l) \rightarrow H_2O(g)$. (*ii*) Calculate the normal boiling point of water if the vapour pressure of water at 298 K is 23.756 mm Hg. (*iii*) Taking the data from the parts (*i*) and (*ii*), calculate the vapour pressure of water at 373 K and calculate the percent difference, if any.

14.86 In all the problems given above we have assumed that the heat capacities of all the phases are identical. In this problem, let us take $C_p(l) = 65$ J K^{-1} mol^{-1}, and $C_p(g) = 30$ J K^{-1} mol^{-1}, $\Delta_{vap}\bar{H}$ = 8.5 kJ mol^{-1} at 300 K. If the equilibrium vapour pressure of liquid A is 50 mm Hg at 300 K, calculate the vapour pressure of liquid A at 400 K, considering heat of vapourisation is not independent of temperature.

14.87 Empirical relation is given below for the vapour pressure of methanol as

$$\ln p(bar) = -\frac{10.752849T_c}{T} + 16.758207 - \frac{3.603425T}{T_c} + \frac{4.37323T^2}{T_c^2} - \frac{2.381377T^3}{T_c^3}$$

$$+4.572199(1-x)^{1.70}$$

Given $T_c = 512.60$ K, (*i*) Calculate the normal boiling point of methanol, (*ii*) also calculate the standard boiling point of methanol.

CHAPTER 15

ELECTROCHEMISTRY

15.1. DEFINITIONS

15.1.1. *Current*

The current, generally denoted by I, is defined as the rate at which a charge dQ is conducted through the medium, *i.e.* current is depicted as

$$I = \frac{dQ}{dt} \qquad ...(15.1)$$

The SI unit for current is ampere and is abbreviated as *amp. The ampere is defined to be the current flowing through two parallel wires of infinite length separated by 1 meter that produces a force per unit length between the wires of exactly* 2×10^{-7} N m^{-1}.

15.1.2. *Charge*

The unit of charge is called *coulomb* which is defined to be the total charge passing a point when a constant current of 1 ampere flows for 1 second. The total current Q is defined as

$$Q = \int_0^{\Delta 1} I dt \qquad ...(15.2)$$

If the current I is constant, intergral (15.2) can be written as

$$Q = I\Delta t = It \qquad ...(15.3)$$

and if current is measured from zero value, $\Delta t = t_2 - t_1 = t_2 = t$.

15.1.3. *Current Density*

The current density L is defined as current per unit cross-sectional area, *i.e.*

$$L = \frac{I}{A} \qquad ...(15.4)$$

where A is the cross-sectional area of the medium through which the current passes. SI unit of current density is thus amp m^{-2}.

Unless we are dealing with super conductors, the flow of the current is resisted and thus for the flow of electric current there must be an electric field E present to derive the charge carrier through the medium. SI unit of electric field is N C^{-1}, *i.e.* newton per coulomb. Current density is proportional to the electric field and is related by:

$$L = \kappa E \qquad ...(15.5)$$

where κ is called the specific conductivity or specific conductance of the medium and inverse of specific conductivity is called resistivity (ρ). SI unit of specific conductivity is C N^{-1} m^{-2} and thus the unit of ρ is N m^2 C^{-1}.

15.1.4. *Resistance*

Resistance is nothing but the obstructions present in the system to the flow of current. The relation between the current, e.m.f. and resistance is

$$R = \frac{E}{I} \qquad ...(15.6)$$

where E is in volts, I is in amperes and R in ohms. This is known as Ohm's law. Electric field is measured in nowtons of force per coulomb (N C^{-1}). Superconductors are exceptions and offers zero resistance, *i.e.* once initiated, current can continue to flow indefinitely in a superconductor without the presence of driving force, so long as no work is done by the system

The resistance of any system is measured with the help of Wheatstone bridge as mentioned in Fig. 15.1. The resistance of the unknown system is obtained by the relation

$$X = R_3 \frac{R_1}{R_2} \qquad ...(15.7)$$

when the current flowing through galvanometer G is zero and the values of R_1, R_2 and R_3 are known.

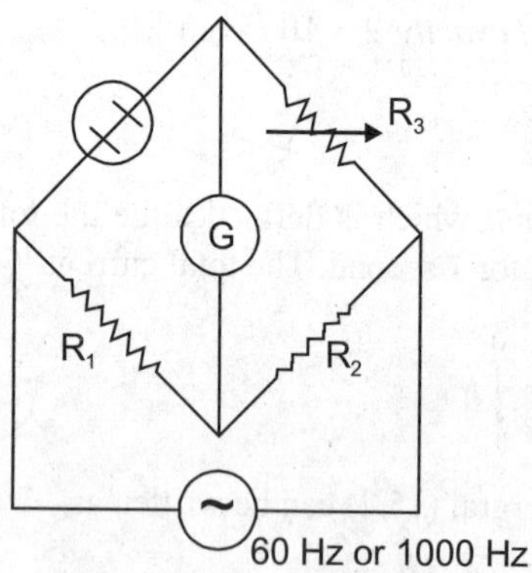

Fig. 15.1 *Wheatstone bridge*

15.1.5. *Conductance*

The current carried by the ions, the electrons etc. is termed as conductance and is reciprocal of resistance *i.e.*

$$L = 1/R \qquad ...(15.8)$$

The unit for conductance is mho. The SI units of conductance is siements (S), named after Ernst Werner von Siemens. S is inverse of ohm, *i.e.* mho (Ω^{-1}).

15.1.6. *Specific Conductance*

Specific conductance is defined as

$$L = \kappa \frac{A}{l} \text{ or } \kappa = L.\frac{l}{A} = LK \qquad ...(15.9)$$

where A is the area of the electrodes, l is the distance between the electrodes and κ is the specific conductance. It is defined as the conductance of a 1 m cube of the solution. The ratio of l/A (K) is known as the cell constant and it is not easy to find from the geometrical dimensions of the cell. Its value is calculated by measuring the conductance of KCl(aq) solution of known specific conductance.

A.C. current is used to calculate the conductance of an electrolyte. For weak electrolytes, the frequency of a current is generally 50 cycles s^{-1} and for strong electrolytes, it is 1000 cycles s^{-1}. This is to avoid deposition of materials on the electrodes. The electrodes used in the conductance cell are made of platinised platinum.

15.1.7. *Molar Conductance*

Molar conductance (Λ) is defined as the conductance of solution containing one mol of electrolyte and placed between two parallel electrodes 1 metre apart, Λ is defined as

$$\Lambda = \frac{\kappa}{C} \qquad ...(15.10)$$

where C is the concentration of the electrolytes expressed as mol per metre cube (Molarity), κ is the specific conductance expressed as mho m^{-1}. Its unit is mho mol^{-1} m^2. The other commonly used unit of conductance is siemens(S), so unit for conductivity will be S mol^{-1} m^2. Further, Λ will be independent of concentration of the electrolyte and therefore it is a true measure of the conductivity of a particular combination of charge carriers. This is only possible if there is no interaction between the charge carriers. As the concentration increases, the molar conductance decreases, because increase of concentration leads to the ion-pair formation and thus a decrease in the concentration of charge carriers and molar conductance. The Debye-Hückel theory indicates that these interactions exhibit a dependence on $\sqrt{c}$ at low concentrations, *i.e.*

$$\mathring{\Lambda} = \mathring{\Lambda} + a\sqrt{c} \qquad ...(15.11)$$

where $\mathring{\Lambda}$ is the molar conductance at infinite dilution and a is the slope of the plot between Λ and $\sqrt{c}$.

Equation (15.7) can be written:

$$X = \frac{R_1 R_3}{R_2} \qquad ...(15.7a)$$

Taking inverse of equation (15.9) , *i.e.*

$$X = \frac{1}{L} = \frac{l}{\kappa A} = \frac{R_1 R_3}{R_2} \qquad ...(15.12)$$

Substituting the value of κ from (15.10) in equation (15.12), we get

$$X = \frac{l}{C\Lambda A} = \frac{R_1 R_3}{R_2}$$

or
$$\frac{l}{A} = \frac{C\Lambda R_1 R_2}{R_2} \quad ...(15.13)$$

In Table 15.1, we have summarised the electrical quantities, their symbols, units and unit conversions.

TABLE 15.1. Electrical quantities, their symbols, units and conversion factors

Electrical quantities	Symbol	Unit names, Abbreviations and unit conversions
Current	I	ampere (amp); 1 amp = 1 C s^{-1}
Charge	Q	coulomb (C); 1 C = 1 amp s
Electric field	E	1 N C^{-1} = 1 kg m s^{-2} C^{-1} = 1 V m^{-1}
Current density	J	amp m^{-2}
Specific conductance	κ	1 amp C m^{-2} N^{-1} = 1 C^2 s kg^{-1} m^{-3}
Resistivity	ρ	1 amp^{-1} $C^{-1}m^2$ N = 1 C^{-2} s^{-1} kg m^3
Electric potential	Φ	1 volt(V) 1 J C^{-1} = 1 N m C^{-1}
Resistance	R	1 ohm (Ω) = 1 C^{-2} s^{-1} kg m^2 = 1 V amp^{-1}
Conductivity	S	1 siemens (S) = 1 $Ω^{-1}$ = C^2 s kg^{-1} m^{-2} = 1 V^{-1} amp

***Exercise* 15.1:** Derive the base units of (*i*) resistivity (ρ), (*ii*) specific conductivity and (*iii*) resistance.

(*i*) We know that ρ = N m^2 C^{-1} and N = kg m s^{-2}.

Substituting these dimensions in ρ we get:

$$\rho = (\text{kg m s}^{-2})(\text{m}^2)(\text{C}^{-1}) = \text{kg m}^3 \text{ s}^{-2} \text{ C}^{-1}.$$

(*ii*) Similarly the base unit of specific conductivity, *i.e.* 1/ρ is C s^2 kg^{-1} m^{-3}.

(*iii*) The relation between resistance and resistivity is given as:

$$R = \frac{\rho L}{A} = \frac{(\text{kg m}^3\text{C}^{-2}\text{s}^{-1})(\text{m})}{\text{m}^2} = \text{kg m}^2\text{C}^{-2}\text{s}^{-1}$$

In terms of electrical SI units, one ohm = V amp^{-1}.

κ depends on the concentration, (*i.e.* increasing the concentration increases the conductance) whereas molar conductance is independent of the concentration of electrolyte as it is obtained by dividing the conductance by concentration. In reality it does not happen so because it is well known that as the concentration increases, the complete dissociation of the electrolyte is not possible and leads to ion-pair formation. In other words, it decreases the molar conductance when concentration increases.

***Exercise* 15.2:** Derive the equation $R = \dfrac{pl}{A}$.

Potential is defined as the function whose negative derivative with respect to coordinate x gives rise to force in the direction of x, *i.e.*

$$E_x = -\frac{\partial \Phi}{\partial x} \quad ...(1)$$

Its value between the two limits can be obtained as

$$\int_{\Phi_1}^{\Phi_2} d\Phi = -\int_{x_1}^{x_2} E_x dx \qquad ...(2)$$

If we assume that the value of the electric field over a small distance is constant, equation (2) can be integrated to give

$$\Phi_2 - \Phi_1 = \Delta\Phi = -E_x(x_2 - x_1) = -E_x \Delta x$$

or

$$E_x = -\frac{\Delta\Phi}{\Delta x} \qquad ...(3)$$

Equation (3) gives the units of electric potential as N m C^{-1}. Since N m is the energy units, *i.e.* J, so the electric potential units can be J C^{-1}. Substituting the value of electric field in equation (15.5), we get

$$L = \kappa E = -\kappa \frac{\Delta\Phi}{\Delta x} = \frac{l}{A} \qquad ...(4)$$

The magnitude of the electric potential can be written as

$$|\Delta\Phi| = \frac{I\Delta x}{\kappa A} = \left|\frac{\rho\Delta x}{A}\right| I \qquad ...(5)$$

The quantity in the brackets in equation (5) is called the resistance R of the conductor, *i.e.* by definition

$$R = \frac{\rho\Delta x}{A} \qquad ...(6)$$

Thus equation (5) can be written as

$$|\Delta\Phi| = RI \qquad ...(15.14)$$

The equation (15.14) indicates that the electric field is proportional to distance and resistance and will be a function of position along the conductor. If our conductor is of homogeneous composition whose cross-sectional area A is constant then the current density and the electric field will be constant. For such a system, we may take Δx to be the total length of the conductor, l and we get

$$R = \frac{\rho l}{A} \qquad ...(15.15)$$

It may be mentioned here that in many materials, the resistivity is independent of the electric field. This is true for those materials and for electrolytic solutions if the current density is not too high and steady state conditions are maintained.

***Exercise* 15.3:** The conductance of solutions of different concentrations of strong electrolyte is given below. Calculate the molar conductance at infinite dilution.

C (10^{-4} mol L^{-1})	1.0	5.0	10.0	50.0	100.0
κ (S cm^{-1})	0.0425	0.2113	0.4212	2.0785	4.119

Using equation 17.11 and the data given above, following Table can be constructed, *i.e.*:

$C\,(10^{-4}\text{ mol L}^{-1})$	1.0	5.0	10.0	50.0	100.0
$\kappa\,(S\text{ cm}^{-1})$	0.0425	0.2113	0.4212	2.0785	4.119
$\sqrt{C}\ (\sqrt{\text{mol}}\sqrt{L})$	0.01	0.02236	0.03162	0.07071	0.1000
$\Lambda\,(S\text{ cm}^2\text{ mol}^{-1})$	424.5	422.6	421.2	415.7	411.9

When Λ is plotted against $\sqrt{C}$ Λ. The above data can be fitted into an equation

$$\Lambda = 425.74 - 139.62\sqrt{C}$$

So $\Lambda_0 = 425.74\text{ S cm}^2\text{ mol}^{-1}$.

***Exercise* 15.4:** The conductance of 0.0075 mol dm^{-3} solution of KCl is 1.49×10^3 micro mho. If the cell constant is 105 m^{-1}, calculate the specific conductance as well as the molar conductance.

The conductance and specific conductance are related by the expression.

$$\begin{aligned}\kappa &= (\text{cell const})(L)\\ &= (1.49 \times 10^3 \mu\Omega^{-1})(105\text{ m}^{-1})\\ &= 1.565 \times 10^5 \mu\text{ mho m}^{-1} = 0.1565\text{ mho m}^{-1}\end{aligned}$$

Substituting the values in (15.10)

$$\Lambda = \frac{1.565 \times 10^5 \times 10^{-6}}{(0.0075)(1000)} = 0.021\text{ mho m}^2\text{ mol}^{-1}$$

***Exercise* 15.5:** The dimensions of the electrodes used in the conductance cell are 0.95 cm and 1.015 cm. If the two electrodes are held at a distance of 0.45 cm, calculate the cell constant.

Cell constant is defined as the ratio of the distance between the electrodes to the area cross-section of the electrode, *i.e.*

$$K = \frac{0.45 \times 10^{-2}\text{ m}}{(0.95 \times 10^{-2}\text{ m})(1.015 \times 10^{-2}\text{ m})} = 46.8\text{ m}^{-1}$$

***Exercise* 15.6:** The resistance of 0.1 M KCl solution in a conductance cell is 325 ohms and the specific conductance of the same solution is 1.29 mho m^{-1}. If the resistance of 0.05 M NaCl solution in the same cell is 752.4 ohms, calculate the molar conductance of the same solution.

Substituting the values of R and κ for KCl solution the cell constant is given as

$$\begin{aligned}K &= \kappa R = (1.29\text{ ohm}^{-1}\text{ m}^{-1})(325\text{ ohm})\\ &= 419\text{ m}^{-1}\end{aligned}$$

The specific conductance of NaCl solution can be calculated as

$$\kappa = \frac{K}{R} = \frac{419\text{ m}^{-1}}{752.4\text{ ohm}} = 0.557\text{ mho m}^{-1}$$

Substituting the values of κ and C in equation (15.10) Λ can be calculated as

$$\begin{aligned}\Lambda &= \frac{0.557\text{ mho m}^{-1}}{(0.05\text{ mol dm}^{-3})(1000\text{ dm}^3\text{ m}^{-3})}\\ &= 0.011\text{ mho m}^2\text{ mol}^{-1}\end{aligned}$$

15.2. KOHLRAUSCH'S LAW

Plot of molar conductance vs $\sqrt{c}$ is linear for strong electrolytes and highly curved for weak electrolytes. But the plot in latter case reaches the limiting value. This led Kohlrausch's to define law of independent migration of ions *i.e.* the molar conductance of any electrolyte at infinite dilution is the sum of the molar conductances of the ions at infinite dilution *i.e.*

$$\mathring{\Lambda} = \lambda_+^0 + \lambda_-^0 \qquad ...(15.16)$$

The main application of Kohlrausch's law is to find out the limiting value of the molar conductance of weak electrolytes *i.e.* It can be shown that

$$\mathring{\Lambda}\,(AD) = \mathring{\Lambda}\,(AB) + \mathring{\Lambda}\,(CD) - \mathring{\Lambda}\,(CB) \qquad ...(15.17)$$

where $\mathring{\Lambda}$'s are the molar conductances of species AD, AB, CD and CB respectively and these are strong electrolytes generally (except AD).

Now using Kohlrausch's law

$$\mathring{\Lambda}\,(AD) = \lambda_{A+}^0 + \lambda_{B-}^0 + \lambda_{C+}^0 + \lambda_{D-}^0 - \lambda_{C+}^0 - \lambda_{B-}^0$$

$$= \lambda_{A+}^0 + \lambda_{D-}^0$$

***Exercise* 15.7:** Calculate the molar conductance of NH_4OH at infinite dilution from the following data.

$\mathring{\Lambda}$ (NaOH) = 247.8×10^{-4}, $\mathring{\Lambda}$ (NaCl) = 126.45×10^{-4} and

$\mathring{\Lambda}$ (NH_4Cl) = 149.7×10^{-4}. These values are at 25°C and unit is mho mol^{-1} m^2.

The limiting molar conductance of NH_4OH at 298 K is

$$\mathring{\Lambda} = \lambda^0(NH_4^+) + \lambda^0(OH^-)$$

The desired equation (15.17) can be obtained

$$\mathring{\Lambda}\,(NH_4OH) = \lambda^0(NH_4^+) + \lambda^0(Cl^-) + \lambda^0(Na^+) + \lambda^0(OH^-) - \lambda^0(Na^+) - \lambda^0(Cl^-)$$

$$= \mathring{\Lambda}\,(NH_4Cl) + \mathring{\Lambda}\,(NaOH) - \mathring{\Lambda}\,(NaCl)$$

$$= (149.7 + 247.8 - 126.45) \times 10^{-4}$$

$$= 271.05 \times 10^{-4} \text{ mho } m^2 \text{ mol}^{-1}$$

15.3. THE ARRHENIUS LAW OF DISSOCIATION

The degree of dissociation 'α' can be given by the expression

$$\alpha = \frac{\Lambda}{\mathring{\Lambda}} \qquad ...(15.18)$$

where Λ is the molar conductance at some concentration C and $\mathring{\Lambda}$ is the molar conductance at infinite dilution. The above relation only holds good for weak electrolytes, where the ionic interaction effects are minimum.

If one knows the degree of dissociation, one can easily calculate dissociation constant of a weak electrolyte *e.g.* consider 1:1 electrolyte

$$AB \rightleftharpoons A^+ + B^-$$

If α is the degree of dissociation and m is the molarity of the electrolyte AB, then concentrations of the species AB, A^+ and B^- will be, $m(1 - \alpha)$, $m\alpha$ and $m\alpha$ respectively. The dissociation constant will be

$$K = \frac{[A^+][B^-]}{[AB]}$$

$$= \frac{(m\alpha)(m\alpha)}{m(1-\alpha)} = \frac{m\alpha^2}{1-\alpha} \qquad ...(15.19)$$

Similar relations can be derived for other kinds of electrolytes.

***Exercise* 15.8:** The 0.0185 mol dm^{-3} solution of acetic acid has the conductance 2.34 × 10^2 μmho. If the cell constant is 105 m^{-1} and $\mathring{\Lambda}$ (CH_3COOH) = 391 × 10^{-4} mho m^2 mol^{-1}, calculate the equilibrium constant of acetic acid.

To calculate the dissociation constant, one needs (*a*) specific conductance, (*b*) molar conductance and (*c*) degree of dissociation besides the given data.

Substituting the desired values in equation (15.9)

$$\kappa = (2.34 \times 10^2 \text{ μmho})(105 \text{ m}^{-1}) = 2.45 \times 10^{-2} \text{ mho m}^{-1}$$

Substituting the values of κ and *C* in the equation (15.10), we get

$$\Lambda = \frac{(2.457 \times 10^{-2} \text{ mho m}^{-1})}{(1.85 \times 10^{-2} \text{mho dm}^{-3})(10^3 \text{dm}^3 \text{ m}^{-3})}$$

$$= 1.33 \times 10^{-3} \text{ mho m}^2 \text{ mol}^{-1}$$

The degree of dissociation α, can be

$$\alpha = \frac{\Lambda}{\mathring{\Lambda}} = \frac{1.33 \times 10^{-3} \text{mho m}^2 \text{ mol}^{-1}}{390 \times 10^{-4} \text{mho m}^2 \text{ mol}^{-1}}$$

$$= 0.034$$

The acetic acid dissociates as

$$CH_3COOH \rightleftharpoons CH_3COO^- + H^+$$

The equilibrium constant will be given by the similar expression as (15.19)

$$K = \frac{m\alpha^2}{1-\alpha} = \frac{(1.85 \times 10^{-2} \text{ M})(0.034)^2}{1-0.034}$$

$$= 2.221 \times 10^{-5} \text{ M}$$

***Exercise* 15.9:** The dissociation constant of *n* butyric acid at 25°C is 1.515 × 10^{-5} and the molar conductance at infinite dilution 382.42 × 10^{-4} mho m^2 mol^{-1}. Calculate the specific conductance of the acid at 0.01 M solution.

The normal butyric acid dissociates as

$$C_3H_7COOH \rightleftharpoons C_3H_7COO^- + H^+$$

and
$$K = \frac{m\alpha^2}{1-\alpha}$$

$$1.515 \times 10^{-5} = \frac{0.01 \times \alpha^2}{1-\alpha} \approx 0.01 \times \alpha^2$$

Because α is small and 1 – α ≈ 1

So
$$\alpha^2 = \frac{1.515 \times 10^{-5}}{1 \times 10^{-2}} = 1.515 \times 10^{-3}$$

$$\alpha = \sqrt{1.515 \times 10^{-3}} = 3.89 \times 10^{-2}$$

From Arrhenius law $\alpha = \dfrac{\Lambda}{\overset{\circ}{\Lambda}}$

or $\Lambda = (3.89 \times 10^{-2})(382.42 \times 10^{-4} \text{ mho m}^2 \text{ mol}^{-1})$

$$= 1.488 \times 10^{-3} \text{ mho m}^2 \text{ mol}^{-1}$$

Substituting this value and the concentration in equation (15.10)

$$\kappa = (1.488 \times 10^{-3} \text{ mho m}^2 \text{ mol}^{-1})(10 \text{ mol m}^{-3})$$

$$= 1.488 \times 10^{-2} \text{ mho m}^2 \text{ mol}^{-1}$$

15.4. ELECTROLYSIS AND ELECTRODE REACTIONS

Electrolysis is a process in which chemical reactions occur at the electrodes, dipping in the electrolytes, when voltage is applied across them. The electrode that is charged positively is called anode and the electrode that is charged negatively is called cathode. The electrodes, like platinum, which only transfer electrons to and from the solutions are termed as "inert electrodes". The "reactive electrodes" are those which enter chemically into the electrode reactions. During electrolysis, the reduction takes place at the cathode and oxidation takes place at the anode.

There can be many types of electrode reactions but general features are summarised below:

(*i*) Current carrying ions are discharged at the electrodes.

(*ii*) A negative ion which is difficult to get discharged at the anode, leads to the decomposition of H_2O and formation of O_2, H^+ and electron.

(*iii*) A positive ion which is difficult to get discharged at cathode, leads to the decomposition of water, with the formation of H_2, OH^- and absorption of electron.

The amount of material accumulated at or dissolved of the electrodes is given by Faraday law *i.e.* "when 96543 coulombs (one Faraday) of charge is passed through an electrolyte, one equivalent of product is deposited at the electrodes or one equivalent of chemical reaction takes place at the electrodes". For approximate calculations one Faraday is taken as 96500 coulombs.

***Exercise* 15.10:** If 50×10^{-3} amp of current is passed through copper coulometer for 60 min., calculate the amount of copper deposited.

Charge $= i \times t = (50 \times 10^{-3} \text{ A})(60 \times 60 \text{ s})$

$= 180$ C

The chemical reaction is

$$Cu^{2+} + 2e^- \rightarrow Cu(s)$$

Since two electrons are involved for the deposition of 1 mole of Cu, so 2 F will deposit 1 mol of copper *i.e.*

$$\text{Amount of copper deposited} = \frac{(63.5 \text{ g mol}^{-1}(180 \text{ C})}{2 \times 96500 \text{ C mol}^{-1}}$$

$$= 0.118 \text{ g}$$

***Exercise* 15.11:** Cell *A* contains $CuSO_4$ solution in water and copper electrodes. Cell *B* consists of Ag_2SO_4 solution in water and platinum electrodes. These are connected in series and current is passed till 4 g of O_2 is collected at cell *B*. (*i*) Give the electrode reactions, (*ii*) Calculate the amount of material deposited at the two cathodes.

Cell A: The reactions at anode and cathode are

$$Cu(s) \rightarrow 2e^- + Cu^{2+}$$

$$Cu^{2+} + 2e^- \rightarrow Cu(s)$$

Cell B:

$$2H_2O(l) \rightarrow O_2(g) + 4H^+ + 4e^-$$
$$Ag^+ + e^- \rightarrow Ag(s)$$

From above reactions it is clear that 4 F give 32 g of O_2. So 4 g of O_2 will need only

$$\frac{(4)(96500\,C)\,(4\,g)}{32\,g} = 0.5\ F$$

$$\text{Amount of Ag deposited in cell } B = (0.5\ F)(108\ g\ F^{-1}) = 54\ g$$

$$\text{Amount of Cu deposited in cell } A = \frac{(63.5\ g)(0.5\ F)}{2\ F} = 15.875\ g$$

15.5. IONIC MOBILITIES

Ionic mobility can be calculated by actual measuring the distance travelled by any ion in given time. This process is called as a moving boundary method. The relation used to calculate ionic mobility is:

$$U_{\neq} = \frac{x}{t\left(\frac{d\varepsilon}{dx}\right)} \qquad ...(15.20)$$

where x is the distance measured in m, t is the time measured in seconds and $\frac{d\varepsilon}{dx}$ is the field strength measured in volts m^{-1}. So the unit of ionic mobility is $m^2\ V^{-1}\ s^{-1}$. The field strength can be calculated as

$$E = \frac{d\varepsilon}{dx} = \frac{I}{A\kappa} \qquad ...(15.21)$$

where I is the current in amp, A is the cross-sectional area (m^2) and κ is the specific conductance expressed as mho m^{-1}.

Ionic mobility of any ion can also be calculated if one knows the conductance of that ion at a given concentration *i.e.*

$$U_{\neq} = \frac{\lambda_{\neq}}{ZF} \qquad ...(15.22)$$

where F is Faraday and $\lambda_{\neq}$ is the molar conductance at given concentration and Z is the valence of cation.

Knowing ionic mobilities of the ions, one can relate them to the transport number *i.e.*

$$t_+ = \frac{U_+}{U_+ + U_-} \qquad ...(15.23)$$

and

$$t_- = \frac{U_-}{U_+ + U_-} \qquad ...(15.24)$$

such that

$$t_+ + t_- = 1 \qquad ...(15.25)$$

***Exercise* 15.12:** Calculate the ionic mobilities of K^+ and OH^- ions at infinite dilution. $\lambda^0_{K+} = 73.50 \times 10^{-4}$ mho m^2 mol^{-1} and $\lambda^0_{OH^-} = 197.6 \times 10^{-4}$ mho m^2 mol^{-1}.

Ionic mobilities can be calculated by using the equation (15.22).

$$U_+ = \frac{\lambda^0_{K+}}{Z_{K+}F} = \frac{73.50\times10^{-4}\text{ mho m}^2\text{ mol}^{-1}}{(96500\text{ C eq}^{-1})(1\text{ eq mol}^{-1})}$$

$$= 7.62 \times 10^{-8}\text{ m}^2\text{ V}^{-1}\text{ s}^{-1}$$

$$U\_(OH^-) = \frac{197.6\times10^{-4}\text{ mho m}^2\text{ mol}^{-1}}{(96500\text{ C eq}^{-1})(1\text{ eq mol}^{-1})}$$

$$= 2.05 \times 10^{-7}\text{m}^2\text{ V}^{-1}\text{ s}^{-1}$$

***Exercise* 15.13:** In the moving boundary experiment, the boundary moved only 38.7 mm in 20 min when a steady current of 1.82×10^{-3} A was passed. The strength of the solution is 0.02 mol dm $^{-3}$, $\kappa = 0.278$ mho m^{-1} and tube has a bore 4.146 mm. Calculate the field strength and the mobility of the K^+.

The field strength can be calculated using equation (15.21) whereas the ionic mobility by using equation (15.20).

$$\frac{d\varepsilon}{dx} = \frac{I}{A\kappa} = \frac{1.82\times10^{-3}\text{ A}}{\pi\left(\frac{4.146}{2}\times10^{-3}\text{ m}\right)^2(0.2788\text{ mho m}^{-1})}$$

$$= 483.3\text{ V m}^{-1}$$

$$U_{K+} = \frac{x}{t\left(\frac{d\varepsilon}{dx}\right)} = \frac{38.7\times10^{-3}\text{m}}{(1200\text{ s})(483.3\text{V m}^{-1})}$$

$$= 6.673 \times 10^{-8}\text{ m}^2\text{ V}^{-1}\text{ s}^{-1}$$

15.6. ▌TRANSPORT NUMBER

Transport number of any ion is defined as the fraction of the total current carried by that particular ion. It is also called "transference number".

The transport number is calculated either by (*a*) Hittorf method or by (*b*) moving boundary method. The apparatus used in Hittorf method or in moving boundary method is shown in Figs. 15.2 and 15.3.

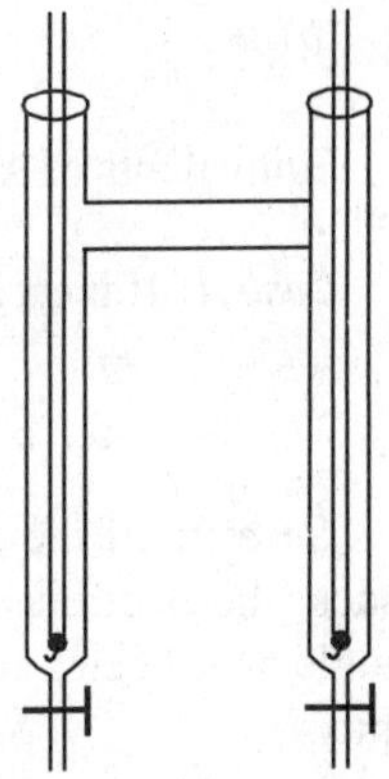

Fig. 15.2 *Hittorf cell*

Hittorf Method

The basis of Hittorf method is that change in the amount of electrolyte in either end compartment depends both on the electrolysis reaction and on the number of ions that have migrated in or out in the process of carrying current *i.e.*, speeds of the ions. Thus it needs determination of the concentration of the solution in each tube as well as the amount of product deposited at the cathode of coulometer (if used for measuring the current passed). The t_i can be calculated from

$$t_+ = \frac{\text{No. of } g \text{ equivalent of electrolyte lost from anode}}{\text{No. of } g \text{ equiv. of metal deposited in coulometer}} \quad ...(15.26a)$$

$$t_- = \frac{\text{No. of } g \text{ equiv. of electrolyte lost from cathode comp.}}{\text{No. of } g \text{ equiv. of metal deposited in coulometer}} \quad ...(15.26b)$$

Such that $t_+ + t_- = 1$ ...(15.27)

The experimental data can be used as follows to calculate the transport number. Let AB be the electrolyte and E be the equivalent weight.

(*i*) If silver coulometer is used to measure the amount of current passed, then equivalent of silver

$$\text{deposited in silver coulometer} = \frac{W}{108}$$

$$= z$$

(If copper coulometer is used and W is amount of copper deposited, the number of equivalent copper deposited $= \frac{W}{31.8} = z$)

(*ii*) Let a g of anode solution contain b g of AB (before electrolysis), wt. of water = $(a - b)$ g,

i.e. $(a - b)$ g of water contains b g of AB, equivalents of AB, before electrolysis $= \frac{b}{E} = y$

(*iii*) Let c g of anode solution contain d g of AB after electrolysis.

$$\text{wt. of water} = (c - d)\ g$$

i.e. $(c - d)$ g of water contains d g of AB after electrolysis. So $(a - b)$ g of water will have $= \frac{d}{c-d}$ $(a - b)$ g

$$\text{Equivalents of AB after electrolysis} = \frac{(d)(a-b)}{(c-d)(E)} = x$$

***Case A*:** If inert electrodes are used in H cell. Decrease in concentration around anode = $(y - x)$

$$\therefore \quad t_{A+} = \frac{y-x}{z}$$

***Case B*:** If the attackable electrodes are used the concentration of A^+ should increase before B^- attacks the electrode so $x > y$. Increase in concentration around anode = $(x - y)$ equi. But actual increase should have been z equivalent if no A^+ ions move out. So the amount of A^+ ions moved out = $z - (x - y)$ equiv.

$$t_{A+} = \frac{z-(x-y)}{z}$$

***Exercise* 15.14:** Consider an electrolyte MA, giving M^+ and A^- ions. If t_+ and t_- are the transport numbers of M^+ and A^-, derive relations (15.26*a*) and (15.26*b*).

Let us assume that 1 Faraday of electricity is passed. The changes observed in different compartments are:

Compartment I cathode	Compartment II	Compartment III anode
1 g equiv. of M^+ is discharged t_+ g equiv. of M^+ migrate in t_- g equiv. of A^- migrate out	t_+ g equiv. of M^+ migrate to 1 t_- g equiv. of A^- migrate from I t_+ g equiv. of M^+ migrate from III t_- g equiv. of A^- migrate to III	1 g equiv. of A^- is discharged t_- equiv. of A^- migrate in t_+ g equiv, of M^+ migrate out.
Net result Loss of $1 - t_+ = t_-$ g equivalent of M^+ Loss of t_- g equiv. of A^- *i.e.* net loss of t_- g equiv. of MA	No change	Loss of $1 - t_- = t_+$ g equivalent of M^+ Loss of t_+ g equiv. of M^+ net loss of t_+ g equiv. of MA

From above results

$$\frac{\text{No. of } g \text{ equvi. of electrolyte lost from cathode}}{\text{No. of } g \text{ equiv. of electrolyte lost from anode}} = \frac{t_-}{t_+}$$

or it can be written as

$$\frac{\text{No. of } g \text{ equiv. of electrolyte lost from cathode}}{\text{No. of } g \text{ equiv. deposited at each electrode}} = \frac{t_-}{1}$$

If coulometer is used to measure current then the denominator is replaced by the number of equiv. of metal deposited in coulometer.

***Exercise* 15.15:** In the Hittorf cell using the silver electrodes and $AgNO_3$ as an electrolyte, certain amount of current was passed. The anode compartment contains 0.2157 g of $AgNO_3$ in 24.96 g of water before electrolysis and 27.5 mL (27.8538 g) solution containing $AgNO_3$ required 49.8 mL of 0.0331 N KCNS solution after the electrolysis. Coulometer attached in series in H-cell deposited 0.014 g of copper by the same amount of current. Calculate transport number of Ag^+.

$$\text{No. of equivalent of Cu deposited} = \frac{0.014 \text{ g}}{31.8 \text{ g equiv.}^{-1}}$$

$$= 4.40 \times 10^{-4} \text{ equivalent}$$

i.e. theoretical increase of Ag^+ in anode

Before analysis: 24.96 g containing 0.2157 g of $AgNO_3$.

$$\text{After analysis:} \quad N(AgNO_3) = \frac{(49.8 \text{ mL})(0.0331 \text{ N})}{(27.5 \text{ mL})} = 0.0599 \text{ N}$$

$$\text{Amount of } AgNO_3 \text{ in 27.5 mL} = \frac{(0.0599 \text{ N})(27.5 \text{ mL})(170 \text{ g N}^{-1})}{(1000 \text{ mL})}$$

$$= 0.28 \text{ g}$$

$$\text{Amount of water in 27.8538 g of electrolyte} = 27.8538 - 0.28$$

$$= 27.5738 \text{ g}$$

Now 27.5738 g of water before electrolysis will have

$$AgNO_3 = \frac{0.2157}{24.96} \times 27.5738 = 0.238 \text{ g}$$

$$\text{Difference in amount} = 0.28 \text{ g} - 0.238 \text{ g} = 0.042 \text{ g}$$

$$= \frac{0.042 \text{ g}}{170 \text{ g equiv.}^{-1}} = 2.47 \times 10^{-4} \text{ equiv.}$$

Now theoretical increase in anode compartment of Ag^+ should be 4.40×10^{-4} equiv. but actually it is only 2.47×10^{-4} equiv. So $(4.40 - 2.47) \times 10^{-4} = 1.93 \times 10^{-4}$ equiv. has migrated.

$$t_{Ag+} = \frac{1.93 \times 10^{-4} \text{ equiv.}}{4.40 \times 10^{-4} \text{ equiv.}} = 0.44$$

***Exercise* 15.16:** 0.2% solution of NaOH was electrolysed using Pt electrodes in the Hittorf cell. The current of 120 mA was passed for 30 min. After the electrolysis, the 25.64 g cathode solution contains 0.0560 g of NaOH. Calculate the t_{Na+} and t_{OH^-}.

Since inert electrode are used case A is applicable here.

$$\text{No. of equivalents current passed} = \frac{(i \text{ A})(t \text{ s})}{(96500 \text{ C equiv.}^{-1})}$$

$$= \frac{(120 \times 10^{-3} \text{ A})(30 \times 60 \text{ s})}{(96500 \text{ C equiv.}^{-1})}$$

$$= 2.238 \times 10^{-3} \text{ equiv.}$$

Before electrolysis 100 g of cathode soln. contains = 0.2 g NaOH

After electrolysis:

$$25.64 \text{ g cathode solution} = 0.056 \text{ g}$$

$$100 \text{ g cathode solution} = \frac{0.056 \text{ g}}{25.6 \text{ g}} \times 100 \text{ g} = 0.2188 \text{ g}$$

$$\text{Increase in the concentration in cathode} = 0.2188 - 0.2 = 0.0188 \text{ g}$$

$$= \frac{0.0188 \text{ g}}{40 \text{ g equiv.}^{-1}}$$

$$= 4.7 \times 10^{-4} \text{ equiv.}$$

$$t_{Na+} = \frac{4.7 \times 10^{-4} \text{ equiv.}}{2.238 \times 10^{-3} \text{ equiv.}} = 0.21$$

$$t_{OH-} = 1 - 0.21 = 0.79$$

Moving Boundary Method

In the moving boundary method, the transport number is calculated by

$$t_i = \frac{F\,1000\,c_i}{I}\frac{dV}{dt} \qquad ...(15.28)$$

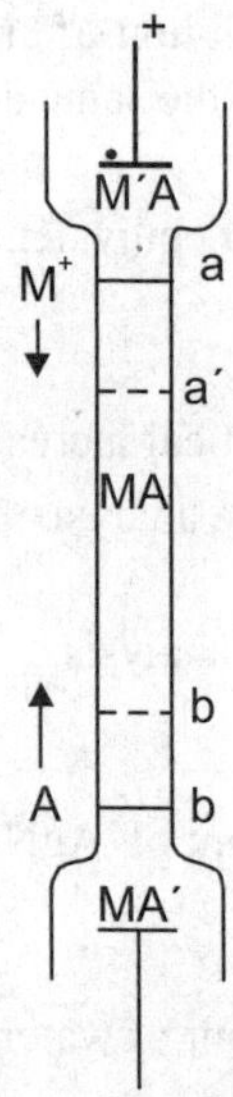

Fig. 15.3 *Moving boundary method*

where c_i is the concentration of ion i in equivalents dm^{-3}, I is the current in amperes, V is the volume in m^3 through which the boundary passes and t is the time in sec.

***Exercise* 15.17:** Calculate transference number of K^+ and Cl^- ions from the data of exercise 15.13.

Making use of the equation (15.28)

where A is the area cross, c is the concentration, l is the length of the boundary moved in time t under the electric current, F is the Faraday constant.

$$t_{K+} = \frac{ACFl \times 1000}{I \times t}$$

$$= \frac{\pi(2.073 \times 10^{-3}\ m)^2(20\ equiv.\ m^{-3})(96500\ A\ s\ equiv^{-1})(38.7 \times 10^{-3}\ m)}{(1.82 \times 10^{-3}\ A)(1200\ s)}$$

$$= 0.462$$

$$t_{Cl-} = 1 - t_{K+} = 1 - 0.462 = 0.538$$

***Exercise* 15.18:** In the moving boundary method a current of 25 mA was passed for 10 min. If the radius of the capillary is 3 mm, calculate the distance moved by H^+ in a solution of concentration 50 equiv. m^{-3} and $t_{H+} = 0.8292$.

Equation (15.28) can be rearranged as

$$l = \frac{t_+ It}{F \times C \times A}$$

$$= \frac{(0.8292)(25 \times 10^{-3}\ A)(10 \times 60\ s)}{(96500\ C\ equiv.^{-1})(50\ equiv.\ m^{-3})(\pi)(3 \times 10^{-3}\ m)^2}$$

$$= 9.1\ cm$$

15.7. IONIC CONDUCTANCES

The molar conductance of any ion at infinite dilution is given by

$$v^+\lambda^{0+} = (t_+)^0 \mathring{\Lambda} \qquad ...(15.29)$$

$$v^-\lambda^{0-} = (t_-)^0 \mathring{\Lambda} \qquad ...(15.30)$$

where $(t_i)^0$ is the transport number of ion extrapolated to infinite dilution and v^+ or v^- are the number of ions associated in the molecule.

This can be extended to calculate the equivalent (or molar) conductance of ions at any other concentrations by the relations

$$v^+\lambda^+ = t_+\Lambda \qquad ...(15.31)$$

$$v^-\lambda^- = t_-\Lambda \qquad ...(15.32)$$

where each term has been defined earlier. But one thing must be kept in mind that Kohlrausch's law of independent migration of ions are not valid at such concentrations of electrolytes. So the ionic conductances calculated in such a way depend upon the nature of the electrolyte rather than a property of ion *i.e.* the conductance under such circumstances is the property of the electrolyte rather than that of individual ions.

***Exercise* 15.19:** The molar conductance of HCl at 0.05 M is $399.09 \times 10^{-4}\ m^2$ mho mol^{-1} and the transport number of H^+ ion at this concentration is 0.829. Calculate the ionic molar conductances of H^+ and Cl^-.

Using equations (15.31) and (15.32)

$$\lambda_{H^+} = (t_{H+})(\Lambda_{HCl})$$
$$= (0.829)(399.09 \times 10^{-4}\ m^2\ mho\ mol^{-1})$$
$$= 330.85 \times 10^{-4}\ m^2\ mho\ mol^{-1}$$

Similarly

$$\lambda_{Cl^-} = (1 - 0.829)(399.09 \times 10^{-4}\ m^2\ mho\ mol^{-1})$$
$$= 68.24 \times 10^{-4}\ m^2\ mho\ mol^{-1}$$

***Exercise* 15.20:** Calculate the molar conductance of $CaCl_2$ at infinite dilution from the following data.

$$U^0_{Ca^{2+}} = 6.16 \times 10^{-2}\ m^2\ V^{-1}\ s^{-1}$$
$$U^0_{Cl^-} = 7.91 \times 10^{-8}\ m^2\ V^{-1}\ s^{-1}$$

The molar conductance at infinite dilution of an electrolyte can be obtained by finding, first, the molar conductance of individual ions and then adding up. Using equations (15.22) and (15.16)

$$\lambda^0_{Ca^{2+}} = (6.16 \times 10^{-8}\ m^2\ V^{-1}\ s^{-1})(96500\ C\ equiv^{-1})(2\ equiv\ mol^{-1})$$
$$= 118.88 \times 10^{-4}\ m^2\ mho\ mol^{-1}$$
$$\lambda^0_{Cl^-} = (7.91 \times 10^{-8}\ m^2\ V^{-1}\ s^{-1})(96500\ A\ s\ equiv^{-1})(1\ equiv\ mol^{-1})$$
$$= 77.33 \times 10^{-4}\ m^2\ mho\ mol^{-1}$$
$$\mathring{\Lambda}_{(CaCl_2)} = \lambda^0_{Ca^{2+}} + 2\lambda^0_{Cl^-} = 118.88 \times 10^{-4} + 2 \times 77.33 \times 10^{-4}$$
$$= 273.54 \times 10^{-4}\ m^2\ mho\ mol^{-1}$$

15.8. COLLIGATIVE PROPERTIES OF ELECTROLYTES

As mentioned in Section (17.5.2) the colligative properties depend on the number of particles present in the system and not on their nature. So these properties mentioned in Section (17.5.2) get modified for electrolytes *i.e.*

$$\Delta T_{F.P.} = iK_{FP}m \qquad ...(15.33)$$
$$\Delta T_{B.P.} = iK_{BP}m \qquad ...(15.34)$$

where i is the van't Hoff factor and is defined as the average number of ions produced per mole of electrolyte. It is given by the relation

$$i = 1 - \alpha + \alpha\nu \qquad ...(15.35)$$

where α is the degree of dissociation and ν is the total number of ions if there is complete dissociation. The equation (15.35) can be rearranged to give

$$\alpha = \frac{i-1}{\nu-1} \qquad ...(15.36)$$

***Exercise* 15.21:** The elevation in boiling points observed for water containing 2.84 g, 4.84 g and 5.89 g of KCl in 125 g of water are 0.27°, 0.44° and 0.59° respectively. Calculate the values of 'i' in each case and explain the results. The molal elevation in boiling point constant is 0.513 degree mol $^{-1}$.

Calculate the molality of KCl for each solution and then using the equation (15.34) the i factor can be calculated.

$$\text{Molality of solution I} = \frac{2.84}{125} \times \frac{1000}{74.5} = 0.305\ M$$

and molalities of solutions II and III are 0.52 M and 0.632 M respectively.

From equation (15.34)

$$i = \frac{\Delta T}{K_b m} = \frac{0.27^\circ}{(0.513 \text{ deg mol}^{-1})(0.305 \text{ mol})}$$
$$= 1.726$$

$$i(\text{solution II}) = \frac{(0.44 \text{ deg})}{(0.513 \text{ deg mol}^{-1})(0.52 \text{ mol})}$$
$$= 1.649$$

and i(solution III) is 1.60.

The value of i decreases as the concentration of KCl increases. This shows that at higher concentration the molecules are held together and can be considered as ion pairs *i.e.* one species.

***Exercise* 15.22:** With the help of data of exercise 15.21, calculate the apparent molecular weight of KCl.

Using the equation (15.34), the molecular weight of solute is given as

$$M_2 = \frac{K_b}{\Delta T} \times \frac{w_2 \times 1000}{w_1}$$

For solution I

$$M_2 = \frac{(0.513 \text{ deg mol}^{-1})(2.84 \text{ g})(1000 \text{ g})}{(0.27 \text{ deg})(125 \text{ g})}$$
$$= 43.17 \text{ g mol}^{-1}$$

Similarly for solution II and solution III, the molecular weights of KCl are 45.14 and 46.49 g mol^{-1} respectively.

Note: The value of i can also be calculated from the relation (15.37)

$$i = \frac{M_{ac}}{M_{app}} \quad ...(15.37)$$

where M_{ac} and M_{app} are actual and apparent molecular weights.

15.9. APPLICATIONS OF CONDUCTANCE MEASUREMENTS

Following are some of the applications of conductance measurements.

15.9.1. *Degree of Ionisation of Weak Electrolytes*

Using Arrenius equation (15.18) one can calculate the degree of dissociation and thus the equilibrium constant. This equation is only valid for weak electrolytes but for strong electrolytes erroneous results are obtained.

15.9.2. *Solubility Product of Sparingly Soluble Salts*

The basis involved is that molar conductance calculated by Kohlrausch's law (at infiuite dilution) should not be different from the actual value because the concentration of these ions are very small. So equations (15.15) and (15.16) can be used to calculate K_{sp}.

***Exercise* 15.23:** The vant Hoff factor i for $Pb(NO_3)_2$ at 0.1 M concentration is 2.13. Calculate the degree of dissociation and the dissociation constant for the reaction

$$Pb(NO_3)_2 \rightleftharpoons Pb^{2+} + 2NO_3^-$$

Substitute the values of i and v in equation (15.36) so that

$$\alpha = \frac{2.13-1}{3-1} = \frac{1.13}{2} = 0.565$$

The equilibrium constant for the above reaction can be written as

$$K = \frac{[Pb^{2+}][NO_3^-]^2}{[Pb(NO_3)_2]}$$

If α is the degree of dissociation and m is the molarity, the concentrations of Pb^{2+}, NO_3^- and $Pb(NO_3)_2$ are $m\alpha$, $2m\alpha$ and $m(1-\alpha)$.

So

$$K = \frac{(m\alpha)(2m\alpha)^2}{m(1-\alpha)} = \frac{4m^2\alpha^3}{1-\alpha}$$

$$= \frac{(4)(0.1)^2(0.565)^3}{1-0.565}$$

$$= 1.66 \times 10^{-2}$$

***Exercise* 15.24:** The solubility product of $BaSO_4$ is 1.1×10^{-10} mol^2 dm^{-6}. If the ionic conductance at infinite dilution at 25°C of $Ba^{2+} = 127.28 \times 10^{-4}$ m^2 mho mol^{-1} and $SO_4^{2-} = 160.0 \times 10^{-4}$ m^2 mho mol^{-1}, calculate the specific conductance of $BaSO_4$.

$$BaSO_4 \rightleftharpoons Ba^{2+} + SO_4^{2-}$$

$$1.1 \times 10^{-10} = K_{sp} = [Ba^{2+}][SO_4^{2-}] = m^2$$

or

$$m = (1.1 \times 10^{-10})^{1/2} = [1.049 \times 10^{-5} \text{ mol dm}^{-3}]$$

$$= 1.049 \times 10^{-2} \text{ mol m}^{-3}$$

$$\mathring{\Lambda}(BaSO_4) = (127.28 + 160) \times 10^{-4} \text{ m}^2 \text{ mho mol}^{-1} = 287.28 \times 10^{-4}$$

$$\kappa = \mathring{\Lambda} C = (287.28 \times 10^{-4} \text{ m}^2 \text{ mho mol}^{-1})(1.049 \times 10^{-2} \text{ mol m}^{-3})$$

$$= 3.01 \times 10^{-4} \text{ mho m}^{-1}$$

PROBLEMS

15.1 The resistance of 0.01974 molar KCl solution is 1/2.505 ohms. If the specific conductance of the above solution of KCl is 0.2768 mho m^{-1}, calculate the cell constant of the cell.

15.2 The dimensions of the cell electrodes were found to be 1.052 cm and 0.995 cm. If the distance between the two electrodes in the conductance cell is 1.15 cm, calculate the cell constant.

15.3 Calculate the distance between the electrodes if the conductance of the solution containing 0.0745 g of KCl in 100 mL of conductivity water is 0.0011 mho. Given that the electrodes are 1 cm × 1 cm. The specific conductance of the above solution at 18°C is 0.1227 mho m^{-1}.

15.4 The resistance of the solution containing 0.368 g of KCl in 0.25 dm^3 of water is $\frac{1}{2.505}$ ohm and the specific conductance is 0.28 mho m^{-1}. In the same conductivity cell the resistance of aqueous solution of sodium acetate containing 0.164 g in 0.100 dm^3 is $1/0.92 \times 10^{-3}$ ohm, calculate the specific conductance as well as the molar conductance of sodium acetate solution.

15.5 The specific conductance of a solution containing 0.745 g of KCl in 0.1 dm^3 water at 25° is 1.288 mho m^{-1}. If the dimensions of the electrodes are 1.12×1.35 cm^2 and the distance between the electrodes is 1.45 cm, calculate the resistance, conductance and the molar conductance of the solution.

15.6 Calculate molar conductance of NaCl solution at infinite dilution graphically from the following data if the cell constant is 110.5 m^{-1}.

concn $\times 10^{-3}$ (mol m^{-3})	0.02	0.0133	0.01	0.008	0.0066	0.0057
$10^3 \times$ conductance (mho)	2.00	1.50	1.08	0.87	0.73	0.667

15.7 What will be resistance of 1000 ppm solution of KCl in water at 25°C when the electrodes used are 1 sq cm each and are kept 1 cm apart. The molar conductance of KCl at infinite dilution is 149.9×10^{-4} mho m^2 mol^{-1}.

15.8 Assume that water is completely ionised. Calculate the limiting molar conductance at 25°C. Given $\mathring{\Lambda}_{HCl} = 426.16 \times 10^{-4}$ m^2 mol^{-1} mho, $\mathring{\Lambda}_{NaOH} = 247.8 \times 10^{-4}$ m^2 mol^{-1} mho and $\mathring{\Lambda}_{NaCl} = 126.45 \times 10^{-4}$ mho m^2 mol^{-1}.

15.9 The dissociation constant of acetic acid is 1.75×10^{-5} and $\mathring{\Lambda}_{CH_3COOH} = 390.6 \times 10^{-4}$ mho m^2 mol^{-1}. Calculate the specific conductance of 0.01 molar acetic acid solution.

15.10 The molar conductances (mho m^2 mol^{-1}) of various electrolytes at infinite dilution are given as follows.

$\mathring{\Lambda}_{BaCl_2} = 279.96$ $\qquad$ $\mathring{\Lambda}_{Ba(NO_3)_2} = 270.08$

$\mathring{\Lambda}_{BaBr_2} = 284.08$ $\qquad$ $\mathring{\Lambda}_{BaSO_4} = 286.88$

$\mathring{\Lambda}_{MgCl_2} = 258.80$ $\qquad$ $\mathring{\Lambda}_{Mg(NO_3)_2} = 248.92$

$\mathring{\Lambda}_{MgBr_2} = 262.92$ $\qquad$ $\mathring{\Lambda}_{MgSO_4} = 265.72$

If the Kohlrausch law of independent migration of ions are valid, how will you prove it from the above data.

15.11 Calculate the molar conductance of $Ba(OH)_2$ at infinite dilution and at 25°C using Kohlrausch law of independent migration. Given $\mathring{\Lambda}_{NaOH} = 248.61 \times 10^{-4}$ mho m^2 mol^{-1}, $\mathring{\Lambda}_{BaCl_2} = 279.96 \times 10^{-4}$ mho m^2 mol^{-1} and $\mathring{\Lambda}_{NaCl} = 126.45 \times 10^{-4}$ mho m^2 mol^{-1}.

15.12 If the specific conductance of a solution containing 0.745 g of KCl in 100 mL of water at 25°C is 1.288 mho m^2, calculate the molar conductance at 25°C as well as at 50°C. The temperature coefficient is 1.9 % of value of Λ at 25°C ($\mathring{\Lambda}_{t0} = \mathring{\Lambda}_{250} [1 + x (t - 25)]$ where x the coefficient of increase of $\mathring{\Lambda}$).

15.13 The molar conductance of Na_2SO_4 at infinite dilution and at temperatures 25°C and 18°C are

259.8×10^{-4} mho m^2 mol^{-1} and 223.8×10^{-4} mho m^2 mol^{-1}, calculate the temperature coefficient for Na_2SO_4.

15.14 Calculate the molar conductance of HIO_4 at infinite dilution from the following data observed at 25°C. $\overset{\circ}{\Lambda}_{KIO_4} = 127.92 \times 10^{-4}$ mho m^2 mol^{-1}, $\overset{\circ}{\Lambda}_{HCl} = 426.16 \times 10^{-4}$ mho m^2 mol^{-1} and $\overset{\circ}{\Lambda}_{KCl} = 149.86 \times 10^{-4}$ mho m^2 mol^{-1}.

15.15 The transference numbers of H^+ ions and Cl^- ions are measured by moving boundary method. The diameter of the capillary was 5.2 mm, the concentration of acid is 0.05 M, the current passed is 2.3 mA and the distance travelled by the boundary of H^+ ions is 0.0186×10^{-3} m s^{-1}. Calculate the transport numbers of H^+ and Cl^- ions.

15.16 If the diameter of the capillary tube used in moving boundary method is 4.8 mm, the steady current passed is 6.4 mA, the concentration of $LaCl_3$ is 0.05 N and the distance covered in two minutes is 3.93×10^{-3} m, calculate the transport number of La^{3+} and Cl^- ions.

15.17 If the ionic mobilities of La^{3+} and Cl^- at infinite dilution are 7.21×10^{-8} m^2 V^{-1} s^{-1} and 7.91×10^{-8} m^2 V^{-1} s^{-1} respectively, calculate the transference numbers of La^{3+} and Cl^- ions at infinite dilution and see that these numbers are different than those obtained in preceding problem. This indicates that the transport numbers depend upon the concentration of the solution.

15.18 Following data is observed in the moving boundary method when a steady current of 12×10^{-3} A was passed through 0.0425 M H_2SO_4 solution. Calculate the transference number of H^+ and SO_4^{2-} ions.

t(s)	85	172	256	343	426
Volume $\times 10^3$ moved (dm^3)	0.1	0.2	0.3	0.4	0.5

15.19 In the Hittorf's method for finding the transport number of ions, it was found that normality of $AgNO_3$ before and after passing current is 4.88×10^{-2} N and 5.60×10^{-2} N when 27 mL of the solution was withdrawn. The same amount of current deposited 0.01307 g of copper in copper coulometer calculate the transport number of Ag^+ and NO_3^- ion.

15.20 In the Hittorf method for finding the transport number of ions, before electrolysis the anode compartment solution was 0.5% $CuSO_4$ solution. 25 mA of current was passed for 100 min. If the $t_{Cu^{2+}}$ is 0.31, calculate the change in the concentration of anodic compartment when the (*a*) copper electrodes were used (*b*) when platinum electrodes were used.

15.21 Calculate the ionic conductance of Na^+ and SO_4^{2-} ions at infinite dilution from the following data.

$\overset{\circ}{\Lambda}_{NaCl} = 126.45 \times 10^{-4}$, $\overset{\circ}{\Lambda}_{HCl} = 426.16 \times 10^{-4}$ and $\overset{\circ}{\Lambda}_{H_2SO_4} = 859.24 \times 10^{-4}$ mho m^2 mol^{-1}. $t_{Na^+} = 0.386$.

15.22 The molar conductance of HCl at infinite dilution is 426.10×10^{-4} mho m^2 mol^{-1} and the transport number at infinite dilution of H^+ is 0.821. Calculate the ionic conductances of the individual ions.

15.23 The transport number of La^{3+} in $LaCl_3$ is given as a function of concentration

Normality	0.01	0.02	0.05	0.1	0.2
$t_{La^{3+}}$	0.462	0.458	0.448	0.438	0.423

Calculate the transport number of La^{3+} at infinite dilution as well as the ionic conductance of individual ions. Compare the value of $\lambda^0_{Cl^-}$, calculated in problems 15.22 and 15.23. $\mathring{\Lambda}_{LaCl_3}$ $= 434.49 \times 10^{-4}$ mho m^2 mol^{-1}.

15.24 From the following data, show that conductance of ion is a property of the ion only at infinite dilution:

conc.	HCl		KCl		NaCl	
($10^{-3} \times$ mol m^{-3})	$\lambda_{HCl \times 10^4}$	t_{Cl^-}	$\lambda_{KCl \times 10^4}$	t_{Cl^-}	$\lambda_{NaCl \times 10^4}$	t_{Cl^-}
0.1	391.32	0.169	128.96	0.51	106.74	0.615

The unit of molar conductance is mho m^2 mol^{-1}.

15.25 Using the data of the problem 15.23, show that the ionic mobilities of ions are the property of the ions only at infinite dilution.

15.26 Using the data of problem 15.22, calculate the ionic mobilities of both the ions.

15.27 In the electrolysis cell, the two electrodes are kept at a distance of 10 cm and voltage applied between the two electrodes is 50 V. If the solution is considered dilute enough to be at infinite dilution, calculate the time taken by La^{3+} and Cl^- ions to reach the respective electrodes if they start from the opposite electrodes $U_{Cl^-} = 7.91 \times 10^{-8}$ m^2 V^{-1} s^{-1}, $U_{La^{3+}} = 7.21 \times 10^{-8}$ m^2 V^{-1} s^{-1}.

15.28 Calculate the ionic mobilities of the following ions. Given:

$$\lambda^0_{Ag^+} = 61.92 \times 10^{-4} \text{ mho m}^2 \text{ mol}^{-1},$$
$$\lambda^0_{Ca^{2+}} = 119.0 \times 10^{-4} \text{ mho m}^2 \text{ mol}^{-1},$$
$$\lambda^0_{Li^+} = 38.69 \times 10^{-4} \text{ mho m}^2 \text{ mol}^{-1},$$
and
$$\lambda^0_{NH_4^+} = 73.4 \times 10^{-4} \text{ mho m}^2 \text{ mol}^{-1},$$

15.29 In a certain experiment, a constant current of 2.05×10^{-2} A was passed and the ions travelled only a distance of 40 nm s^{-1}. If the strength of the solution is 0.01 mol dm^{-3}, $\kappa = 4.84 \times 10^{-1}$ mho m^{-1} and the tube has bore of 5.0 mm, calculate the field strength and the mobility of the respective ion.

15.30 In a moving boundary experiment, the constant current of 4.5×10^{-3} A was applied. If the Λ of the 0.01 M solution is 247.88×10^{-4} Λ^{-1} mol^{-1} m^2 and area cross-section 0.2 mm^2, calculate the time taken by the ions to travel 10 cm.

$$U^{A-1} = 5.4 \times 10^{-8} \text{ m}^2 \text{ V}^{-1} \text{ s}^{-1}$$

Colligative Properties of Electrolytes

15.31 0.8255 g of $K_3[Fe(CN)_6]$ was dissolved in 250 g of water. Freezing point of the solution was observed to be –0.063°C. Calculate the apparent molecular weight of the compound if the freezing point constant of water is 1.86 degree mol^{-1}.

15.32 0.331 g of lead nitrate was dissolved in 1000 g of water and the freezing point of water observed was –0.0054°C. Similarly when 0.83 g of lead nitrate dissolved in 250 g of water, the freezing point observed was –0.049°C. Calculate the freezing point depression constant in each case and compare the values. Explain if these values are different.

15.33 Calculate the vant Hoff factors from the data of problems 15. 31 and 15.32. Explain the difference assuming that there is no experimental error.

Equilibrium Constant

15.34 The values of vant Hoff's factor obtained for different concentration of $MgSO_4$ from the depression in the freezing point experiment are

m	0.001	0.01	0.1	1.0
i	1.82	1.53	1.21	1.09

Calculate the degree dissociation assuming Arrhenius theory and also the dissociation constant of $MgSO_4$.

15.35 The following data is observed for different concentration of $BaCl_2$ at 25°C:

(mol dm^{-3}) $\Lambda \times 10^4$	0.000	0.001	0.002	0.02	0.2	2.0
(mho m^2 mol^{-1})	279.96	271.92	268.68	247.88	210.38	161.5

Calculate the degree of dissociation and dissociation constant assuming that Arrhenius law of dissociation is valid and cell constant is 1.15 m^{-1}.

15.36 The molar conductance measured for different solutions of H_2SO_4 are

(mol dm^{-3})	0.0	0.001	0.002	0.02	0.2
$\Lambda \times 10^{+4}$ (mho m^2 mol^{-1})	859.2	826.2	799.0	672.8	501.6

Calculate the degree of dissociation and dissociation constant up to 1 molar solution.

15.37 The decrease in the freezing point of water observed when different amount of sulphuric acid is added to water are

M	0.01	0.1	1.0
ΔT_f	0.04584	0.394	4.04

If the freezing point depression constant for water is 1.86 degree mol^{-1}, calculate the van't Hoff factor as well as the dissociation constant of H_2SO_4. Compare it with that calculated in Problem 15.36. Also explain why the van't Hoff's factor decreases with increase in concentration.

15.38 If the specific conductance of pure distilled water at 25°C is 0.58×10^{-5} mho m^{-1} and the molar conductance of water at infinite dilution and at 25°C is 548.3×10^{-4} mho m^2 mol^{-1}, calculate the dissociation constant of water.

15.39 The specific conductance of pure water at 25°C is 0.58×10^{-5} mho m^{-1} and of saturated solution of AgCl prepared in same water at 25°C is 2.83×10^{-5} mho m^{-1}. If λ^0_{Ag+} and λ^0_{Cl-} are 61.92×10^{-4} and 75.23×10^{-4} mho m^2 mol^{-1} at 25°C, calculate the solubility product of AgCl.

15.40 The solubility product of AgBr at 25°C is 4.8×10^{-13} and λ^0_{Ag+} and λ^0_{Br-} are 61.92×10^{-4} and 78.4×10^{-4} mho m^2 mol^{-1} respectively. Calculate the specific conductance of the saturated solution of AgBr.

15.41 The solubility product of Mg $(OH)_2$ is $10^{-10.6}$ mol^3 dm^{-9} and $\lambda^0_{Mg^{2+}}$ and $\lambda^0_{OH^-}$ are 106.12×10^{-4} and 198.5×10^{-4} Ω^{-1} m^2 mol^{-1} at 25°C. Calculate the specific conductance of the saturated solution of Mg $(OH)_2$.

15.42 The equivalent conductance of 0.1 equi L^{-1} of H_3PO_4 at 18°C is 96.5 mho cm^2 equi^{-1}. If $\mathring{\Lambda}_{HCl}$ = 378.3, $\mathring{\Lambda}_{NaCl}$ = 109.0, $\mathring{\Lambda}_{NaH_2PO_4}$ = 70.0 mho cm^2 equi^{-1} respectively, calculate the degree of dissociation and dissociation constant for the reaction

$$H_3PO_4 \rightleftharpoons H^+ + H_2PO_4^-$$

Additional Problems

15.43 Show that the base unit of ohm is kg m^2 C^{-2} s^{-1}.

15.44 The specific conductance of a 0.01 molar aqueous KCl solution at 298 K is 0.00141 S cm^{-1}. This solution was placed in a conductance cell and its resistance is measured using a bridge it is found that a current null point in the bridge is produced whenever R_1 = 100 Ω, R_2 = 52.796 Ω and R_3 = 150 Ω. Calculate the cell constant.

15.45 The ionic mobilities of K^+ and Cl^- ions in a KCl solution of concentration 0.20 mol L^{-1} are 62.9×10^{-5} and 65.6×10^{-5} cm^2 V^{-1} s^{-1} respectively at 298 K. Calculate the transport numbers and ionic conductances of $K^+(aq)$ and $Cl^-(aq)$. What is the molar conductivity of KCl at this concentration and temperature? Assume that there is complete ionisation and no ion pairing.

[**Ans.** t_{K^+} = 0.489, t_{Cl^-} = 0.511, Λ_{KCl} = 124.0 S cm^2 mol^{-1}]

15.46 The solution of KCl of concentration 0.40 mol L^{-1} at 298 K is placed in the conductivity cell. If A = 30 cm^2, l = 10 cm and the potential applied across the cell is 60 volts, calculate the expected current density and the current in the cell. Again assume that KCl is completely dissociated.

15.47 The specific conductance of a 0.015 M aqueous solution of KCl solution at 298 K is 0.002 S cm^{-1}. When the solution is placed in a conductance cell and its resistance is measured using a wheatstone bridge. It is found that null point in bridge is produced whenever R_1 = 140 Ω, R_2 = 60.5 Ω and R_3 = 170 Ω. Calculate the cell constant.

15.48 Using the cell constant value obtained in problem 15.47 and the same wheatstone bridge, calculate the molar conductivity of 0.0016 M solution of HCl.

15.49 Conductance of different solutions of $AgNO_3$ at 298 K is given below. Calculate the value $\mathring{\Lambda}$.

C (10^{-4} mol L^{-1})	5.0	10.0	50.0	100.0	200.0
κ (S cm^{-1})	0.0656	013	0.6235	1.247	2.427

15.50 A moving boundary apparatus with a cross-sectional area of 9 cm^2 contains a 1.00×10^{-2} M $BaCl_2$ solution above a $CaCl_2$ solution. When the $BaCl_2$ solution is placed in a conductance cell whose cell constant is 0.35 cm^{-1}, the resistance of the solution is found to be 151 41. 24 Ω. An electric potential is applied across the moving boundary apparatus such that a current of 5 A is flowing through the cell. After 200 s the $BaCl_2$-$CaCl_2$ boundary has moved 26.17 cm. Calculate electric mobility of Ba^{2+} from these data. [**Ans.** 0.0005836 cm^2 V^{-1} s^{-1}]

15.51 The ionisation constant of HF in aqueous solution at 298 K is found to be 3.5×10^{-4}. Use the Debye-Huckel theory and the Debye-Huckel-Onsager equation to estimate the specific conductivity of a 0.02 molar HF solution at 298 K, the molar ionic conductance of H_3O^+ and F^-.

15.52 The resistance of 0.01 mol L^{-1} aqueous solution of formic acid (HCOOH) at 298 is found to be 884.11 Ω. If the cell constant of the cell is 0.460 cm^{-1}, (*i*) calculate specific conductance of the solution, (*ii*) the limiting molar ionic conductance of H_3O^+ and formate ion are 349.65 S cm^2 mol^{-1} and 54.6 S cm^2 mol^{-1} respectively. Use the Debye –Huckel-Onsager equation to calculate the concentration of H_3O^+ and $HCCO^-$ ions and (*iii*) calculate the ionisation constant of formic acid.

15.53 In the Hittorf method following data is obtained regarding the ionic mobilities of K^+ and Cl^- at different concentrations of KCl at 298 K. Calculate transport number of K^+ at each concentration.

C (mol L^{-1})	0.00	0.01	0.10	0.20	0.3
$U(K^+)$. C m^2 V^{-1} s^{-1}	0.000762	0.000718	0.000654	0.000629	0.000566
$U(K^+)$. C m^2 V^{-1} s^{-1}	0.000791	0.000746	0.000682	0.000656	0.000593

CHAPTER 16

ELECTROCHEMICAL CELLS

16.1. INTRODUCTION

16.1.1. *Galvanic Cell*

It is a cell in which free energy of a chemical reaction is converted into electrical energy. It is also called electrochemical cell. The relation between the free energy of the chemical reaction and the voltage of the cell is

$$\Delta G = -nFE \qquad ...(16.1)$$

where F is the faraday, constant, E is the e.m.f. of cell in volts and n is the number of mol of electrons involved in the balanced reaction.

16.1.2. *Electrolysis Cell*

It is the cell in which electrical energy is used to carry out a chemical reaction. It is reverse of the galvanic cell. The e.m.f. required to carry out this process will be little higher than the e.m.f. produced in chemical reaction and is supplied from surroundings.

The spontaneity of chemical reaction requires ΔG to be negative. If the e.m.f. of the cell is +ve, it is a galvanic cell. There will be equilibrium if ΔG and E are zero. The reaction with more positive value of E will occur first in preference to other chemical reaction with low positive e.m.f.

***Exercise* 16.1:** The overall reaction of the discharge of storage battery is

$$Pb\,(s) + PbO_2\,(s) + 2H_2SO_4\,(l) \rightarrow 2PbSO_4\,(s) + 2H_2O\,(l)$$

If the standard free energies of formation of $PbO_2(s)$, H_2SO_4, $PbSO_4$ and water are –217.38 kJ mol^{-1}, –909.27 kJ mol^{-1}, – 813.20 kJ mol^{-1}, – 237.18 kJ mol^{-1} respectively (*a*) calculate ΔG^0 for the reaction, (*b*) write the cell formed and the half cell reactions and (*c*) calculate the e.m.f. of the cell and predict whether the reaction is spontaneous or not.

(*a*) The free energy of the reaction can be calculated by the use of equation taking ΔG_f^0 of pure elements as zero.

$$\Delta G^0 = \sum_i^{\text{Products}} n_i G_i^0 - \sum_j^{\text{Reactants}} n_j G_j^0$$

$$\Delta G^0 = (2 \text{ mol})(-813.20 \text{ kJ mol}^{-1}) + (2 \text{ mol})(-237.18 \text{ kJ mol}^{-1}) - (1 \text{ mol})(-217.38 \text{ kJ mol}^{-1}) - (2 \text{ mol})(-909.27 \text{ kJ mol}^{-1})$$

$$= -64.84 \text{ kJ}$$

(*b*) Since the free energy of the reaction is negative, the reaction is spontaneous.

The cell formed from the given reaction is

$$Pb \mid PbSO_4 \mid H_2SO_4 \mid PbO_2$$

the half cell reactions are:

At right hand electrode

$$\frac{1}{2}PbO_2\,(s) + \frac{1}{2}\,SO_4^{2-}\,(aq) + 2H^+\,(aq) + e- \rightarrow \frac{1}{2}PbSO_4\,(s) + H_2O\,(l)$$

At left hand electrode

$$\frac{1}{2}\,PbSO_4\,(s) + e^- \rightarrow \frac{1}{2}\,Pb\,(s) + \frac{1}{2}\,SO_4^{2-}\,(aq)$$

The balanced reaction is, *i.e.* R – L

$$\frac{1}{2}Pb\,(s) + \frac{1}{2}\,PbO_2\,(s) + H_2SO_4\,(aq) \rightarrow PbSO_4\,(s) + H_2O\,(l)$$

or $$Pb\,(s) + PbO_2\,(s) + 2H_2SO_4\,(aq) \rightarrow 2PbSO_4\,(s) + 2H_2O\,(l)$$

(*c*) The standard e.m.f. of the cell can be calculated from the equation (16.1).

$$\Delta G^0 = -nFE^0$$

$$E^0 = \frac{-64.84 \text{ kJ}}{-(2 \text{ mol})(96.5 \text{ kJ mol}^{-1}\text{ V}^{-1})} = 0.336 \text{ V}$$

***Exercise* 16.2:** In the copper coulometer 20.0 A of current was passed for 1 hour. Calculate the amount of copper deposited.

The chemical reaction is

$$Cu^{2+}\,(aq) + 2e^- \rightarrow Cu\,(s)$$

i.e. 2 mol of electrons, are required to deposit one mol of copper

$$q = (20.0 \text{ A})(1 \text{ hr})(3600 \text{ s hr}^{-1})$$

This amount of electricity corresponds to

$$= \frac{72\times10^3 \text{ C}}{96.5\times10^3 \text{C mol}^{-1}} = 0.746 \text{ mol}$$

i.e. 0.746 N_{av} electronic charges.

The above reaction consumes 2 mol electrons to give 1 mol of Cu

So amount of copper deposited $$= \frac{(63.5\times10^{-3} \text{ kg mol}^{-1})(0.746 \text{ mol})}{2}$$

$$= 23.68 \times 10^{-3} \text{ kg}$$

***Exercise* 16.3:** The standard reduction potentials of the electrodes $Fe^{3+} \mid Fe$ and $Fe^{2+} \mid Fe$ are – 0.035, – 0.440 V respectively. Is it easy to oxidise Fe to Fe^{2+} or Fe to Fe^{3+}?

Since the standard reduction potentials are given the standard oxidation potentials will be reverse of the given values *i.e.* 0.036 V and 0.44 V. The standard oxidation potential of Fe | Fe^{2+} is more positive than that of Fe | Fe^{3+} electrode. So it is easy to oxidise Fe to Fe^{2+}.

16.2. ELECTRODES

There are two types of electrodes:

(a) *Anode*: In galvanic cell, the anode, where the oxidation is taking place is negatively charged because due to spontaneous chemical reaction, the electrons will be released to this electrode.

In electrolysis cell, an external source of voltage is supplied and due to this anode is positively charged with respect to cathode. Thus the negatively charged ions come to anodes to get oxidised.

(b) *Cathode*: Cathodes are the electrodes where reduction of chemical species takes place.

In galvanic cell, it is positively charged with respect to anode. Here the positively charged species come to get reduced by the electrons coming from anode.

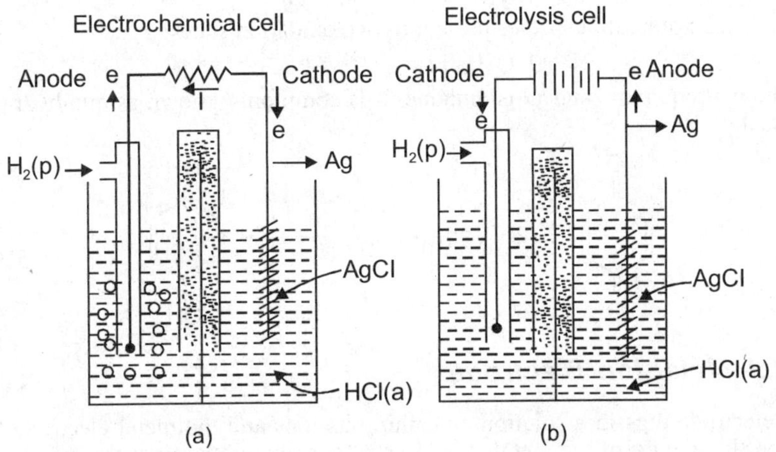

Fig. 16.1 *(a) Electrochemical cell (b) Electrolysis cell*

In electrolytic cell, the cathode is negatively charged electrode. The positively charged ions (cations) come to this electrode to get reduced.

Thus in the galvanic cell, the electrons move from anode to cathode in external circuit. Whereas in electrolytic cell, the electrons are supplied by the external battery to enter through cathode and come out through anode. It is shown in Fig. 16.1.

16.3. NATURE OF ELECTRODES

In any cell, the electrodes are generally either dipped directly in the solution or are connected through the salt bridge which allows the passage of electrons through it. Salt bridge is generally used when the electrodes have to be dipped in different solutions and they are miscible. The various types of electrodes used and the chemical reactions which take place are:

16.3.1. *Gas Electrodes*

Here the inert electrode is present in contact with the gas at 1 atm pressure (unless specified) and its ions in the solution *e.g.* Hydrogen gas electrode, etc. It is denoted as $H^+ | H_2$ (1 atm) | Pt and the chemical reaction is

$$H^+ (aq) + e^- \rightarrow \frac{1}{2} H_2(g)$$

The nature of inert electrode should be such that, when the external voltage is changed even by small amount from the equilibrium value, the reaction will proceed one way or the other. Most suitable electrode is platinum with fine coat of platinum powder or platinum black.

16.3.2. *Oxidation–Reduction Electrode*

Here the inert electrode (generally platinum) is dipping in the solution which contains ions of species having two oxidation states, *e.g.* platinum electrode dipping in solution containing stannous and stannic ions. It is denoted as Sn^{4+}, Sn^{2+} | Pt and chemical reaction is

$$Sn^{4+} (aq) + 2e^- \rightarrow Sn^{2+} (aq)$$

The solution can be of neutral species present in two oxidation states *e.g.*,

$$H^+, Q, QH_2 | Pt$$

where QH_2 is the hydroquinone and *Q* is quinone. It is commonly known as quinhydrone electrode. The chemical reaction is

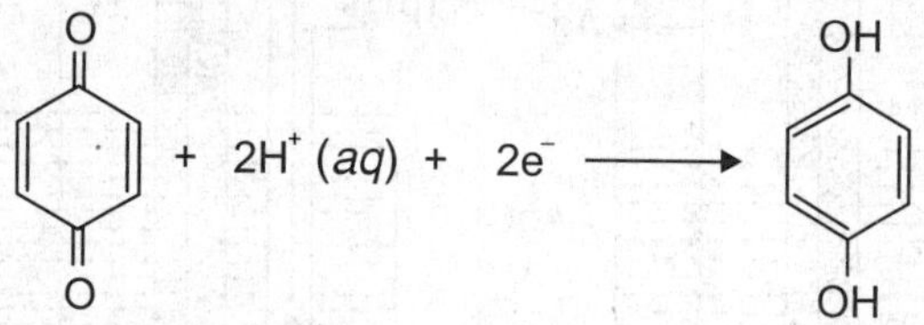

16.3.3. *Metal–Metal Ion Electrode*

Here the metal electrode dips in a solution containing its ions and the metal electrode takes part in chemical reaction. Reactivity of the metal should be intermediate otherwise the reactive metals will react with water and cannot act as electrodes. The example is the copper electrode dipping in copper sulphate solution. It is denoted as M^{n+} | M in general and in particular Cu^{2+} (C) | Cu and C is the concentration. The chemical reaction is

$$Cu^{2+} (aq) + 2e^- \rightarrow Cu(s)$$

16.3.4. *Metal–Insoluble Salt Electrode*

Here the metal electrode is in contact with its insoluble salt which is further in contact with its ions (generally the anion). The most common is the calomal electrode. This consists of mercury, mercurous chloride (Hg_2Cl_2) and saturated solution of KCI. But the concentration of Cl^- ions can be different. The chemical reaction taking place is

$$\frac{1}{2} Hg_2Cl_2(s) + e^- \rightarrow Hg(s) + Cl^- (aq)$$

and depicted as $Cl^- | Hg_2Cl_2 | Hg | Pt$.

Similarly lead amalgam lead sulphate electrode is depicted as $SO_4^{2-} | PbSO_4 | Pb(Hg)$ and the chemical reaction is

$$PbSO_4(s) + 2e^- \rightarrow Pb(s) + SO_4^{2-} (aq)$$

***Exercise* 16.4:** Write cell electrode for the following reactions:

(*i*) $2MnO_4^-(aq) + 6H^+(aq) + 5H_2C_2O_4(aq) \rightleftharpoons 2Mn^{2+}(aq) + 8H_2O\ (l) + 10CO_2(g)$

(*ii*) $Cu(OH)_2(s) \rightarrow Cu^{2+}(aq) + 2OH^-(aq)$

(*iii*) $AgCl(s) + e^- \rightarrow Ag(s) + Cl^-(aq)$

(*i*) Half-cell reactions are

$$MnO_4^-(aq) + 8H^+(aq) + 5e^- \rightarrow Mn^{2+}(aq) + 4H_2O\ (l)$$
$$2CO_2(g) + 2H^+(aq) + 2e^- \rightarrow H_2CO_4$$

and electrodes

$$MnO_4^-, H^+, Mn^{2+} \mid Pt \text{ and } Pt \mid CO_2 \mid H^+$$

(*ii*) Half-cell reactions are

$$Cu(OH)_2(s) + 2e^- \rightarrow Cu(s) + 2OH^-(aq)$$
$$Cu^{2+}(aq) + 2e^- \rightarrow Cu(s)$$

and electrodes are $Cu^{2+} \mid Cu$ and $Pt \mid Cu(OH)_2 \mid OH^-$

(*iii*) Electrode is $Cl^- \mid AgCl \mid Ag$.

***Exercise* 16.5:** Write the cell reactions for the following cells

(*i*) $Zn \mid ZnO_2^-, OH^- \mid HgO \mid Hg$

(*ii*) $Pb \mid PbSO_4 \mid H_2SO_4 \mid PbSO_4 \mid PbO_2 \mid Pt$

(*iii*) $Pt \mid H_2 \mid HC1 \mid Hg_2Cl_2 \mid Hg \mid Pt$

(*i*) The electrode reactions are

$$ZnO_2^{2-}(aq) + 2H_2O\ (l) + 2e^- \rightarrow Zn(s) + 4OH^-(aq) \quad ...(16.2a)$$
$$HgO(s) + H_2O\ (l) + 2e^- \rightarrow Hg(s) + 2OH^-(aq) \quad ...(16.2b)$$

Cell reaction is (16.2*a*) – (16.2*b*) *i.e.*

$$Zn(s) + HgO(s) + 2OH^-(aq) \rightarrow ZnO_2^{2-}(aq) + Hg(s) + H_2O\ (l)$$

(*ii*) The electrode reactions are

$$PbSO_4(s) + 2e^- \rightarrow Pb(s) + SO_4^{2-}(aq) \quad ...(16.2c)$$
$$PbO_2(s) + SO_4^{2-}(aq) + 4H^+(aq) + 2e^- \rightarrow PbSO_4(s) + 2H_2O\ (l) \quad ...(16.2d)$$

Cell reaction is (16.2*d*) – (16.2*c*)

$$Pb(s) + PbO_2(s) + 4H^+(aq) + 2SO_4^{2-}(aq) \rightarrow 2PbSO_4(s) + 2H_2O\ (l)$$

(*iii*) The electrode reactions are

$$2H^+(aq) + 2e^- \rightarrow H_2(g)$$
$$Hg_2Cl_2(s) + 2e^- \rightarrow 2Hg(l) + 2Cl^-(aq)$$

The cell reaction is

$$Hg_2Cl_2(s) + H_2 \rightarrow 2Hg(l) + 2H^+(aq) + 2C1^-(aq)$$

16.4. ELECTRODE POTENTIAL

It is very difficult to find the voltage of an individual electrode. So it is combined with the standard hydrogen electrode (*i.e.* platinum electrode in the atmosphere of H_2 at 1 atm pressure and in contact with H^+ ion at unit activity) and the e.m.f. of the cell is measured. It is assumed that the standard e.m.f. of standard hydrogen electrode is zero. So the e.m.f. of such a cell is the e.m.f. of the electrode, if the following conventions are used regarding the nature of electrode potentials.

(1) The standard electrode potential is used to designate the potential which can be obtained if the constituents are present at unit activity. Now, standard electrode potential is positive if its value is more positive than that of standard hydrogen electrode and negative value if its electrode potential is more negative than that of standard hydrogen electrode potential.

(2) The electrode is considered positive if reduction reaction is taking place and it is negative if oxidation reaction is taking place.

(3) The IUPAC (International Union of Pure and Applied Chemistry) convention always mention standard electrode potential as the reduction potential.

The standard reduction potentials of many single electrode potentials have been compiled in Table 16.1.

TABLE 16.1. Standard reduction potentials E^0, for half-cells at 298 K (pH = 0) Electrochemical

Electrode	Electrode reaction	E^0 (V)
$Pt \mid F_2 \mid F^-$	$F_2(g) + 2e^- \rightarrow 2F^-(aq)$	+2.87
$Pt \mid Co^{3+}, Co^{2+}$	$Co^{3+}(aq) + e^- \rightarrow Co^{2+}(aq)$	+1.92
$Pt \mid Ce^{4+}, Ce^{3+}$	$Ce^{4+}(aq) + e^- \rightarrow Ce^{3+}(aq)$	+1.72
$Pt \mid MnO_4^-, Mn^{2+}$	$MnO_4^-(aq) + 8H^+(aq) + 5e^- \rightarrow Mn^{2+}(aq) + 4H_2O(l)$	+1.507
$Pt \mid Mn^{3+}, Mn^{2+}$	$Mn^{3+}(aq) + e^- \rightarrow Mn^{2+}(aq)$	+1.54
$Au \mid Au^{3+}$	$Au^{3+}(aq) + 3e^- \rightarrow Au(s)$	+1.498
$Pt \mid Cl_2 \mid Cl^-$	$Cl_2(g) + 2e^- \rightarrow 2Cl^-(aq)$	+1.36
$Pt \mid Cr_2O_7^{2-}, Cr^{2+}$	$Cr_2O_7^{2-}(aq) + 14H^+(aq) + 6e^- \rightarrow 2Cr^{2+}(aq) + 7H_2O(l)$	+1.23
$Pt \mid Tl^{3+}, Tl^{2+}$	$Tl^{3+}(aq) + 2e^- \rightarrow Tl^+(aq)$	+1.252
$Pt \mid O_2, H_2O$	$O_2(g) + 4H^+(aq) + 4e^- \rightarrow 2H_2O(l)$	+1.229
$Pt \mid Br_2 \mid Br^-$	$Br_2(g) + 2e^- \rightarrow 2Br^-(aq)$	+1.087
$Pt \mid Hg^{2+}, Hg_2^{2+}$	$2Hg^{2+}(aq) + 2e^- \rightarrow Hg_2^{2+}(aq)$	+0.92
$Hg \mid Hg^{2+}$	$Hg^{2+}(aq) + 2e^- \rightarrow Hg(l)$	+0.851
$Ag \mid Ag^+$	$Ag^+(aq) + e^- \rightarrow Ag(s)$	+0.800

(Contd...)

$Pt \mid Fe^{3+}, Fe^{2+}$	$Fe^{3+}(aq) + e^- \rightarrow Fe^{2+}(aq)$	+0.771
$Pt \mid I_2 \mid I^-$	$I_2(g) + 2e^- \rightarrow 2I^-(aq)$	+0.536
$Pt \mid O_2, OH^-$	$O_2(g) + 2H_2O(l) + 4e^- \rightarrow 4OH^-(aq)$	+0.401
$Pt \mid Fe(CN)_6^{3-}, Fe(CN)_6^{4-}$	$Fe(CN)_6^{3-}(aq) + e^- \rightarrow Fe(CN)_6^{4-}(aq)$	+0.36
$Cu \mid Cu^{2+}$	$Cu^{2+}(aq) + 2e^- \rightarrow Cu(s)$	+0.342
$Pt \mid Hg \mid Hg_2Cl_2 \mid Cl^-$	$Hg_2Cl_2(s) + 2e^- \rightarrow 2Hg(s) + 2Cl^{-1}(aq)$	+0.268
$Ag \mid AgCl \mid Cl^-$	$Ag\,Cl(s) + e^- \rightarrow Ag(s) + Cl^{-1}(aq)$	+0.222
$Pt \mid Sn^{4+}, Sn^{2+}$	$Sn^{4+}(aq) + 2e^- \rightarrow Sn^{2+}(aq)$	+0.151
$Pt \mid Cu^{2+}, Cu^+$	$Cu^{2+}(aq) + e^- \rightarrow Cu^+(aq)$	+0.153
$Ag \mid AgBr \mid Br^-$	$AgBr(s) + e^- \rightarrow Ag(s) + Br^{-1}(aq)$	+0.0713
$Pt \mid H_2 \mid H^+$	$2H^+(aq) + 2e^- \rightarrow H_2(g)$	0.0
$Pb \mid Pb^{2+}$	$Pb^{2+}(aq) + 2e^- \rightarrow Pb(s)$	–0.126
$Sn \mid Sn^{2+}$	$Sn^{2+}(aq) + 2e^- \rightarrow Sn(s)$	–0.138
$Co \mid Co^{2+}$	$Co^{2+}(aq) + 2e^- \rightarrow Co(s)$	–0.277
$Tl \mid Tl^+$	$Tl^+(aq) + e^- \rightarrow Tl(s)$	–0.336
$Pb \mid PbSO_4 \mid SO_4^{2-}$	$PbSO_4(s) + 2e^- \rightarrow Pb(s) + SO_4^{2-}(aq)$	–0.359
$Cd \mid Cd^{2+}$	$Cd^{2+}(aq) + 2e^- \rightarrow Cd(s)$	–0.403
$Pt \mid Cr^{3+}, Cr^{2+}$	$Cr^{3+}(aq) + e^- \rightarrow Cr^{2+}(aq)$	–0.41
$Fe \mid Fe^{2+}$	$Fe^{2+}(aq) + 2e^- \rightarrow Fe(s)$	–0.447
$Zn \mid Zn^{2+}$	$Zn^{2+}(aq) + 2e^- \rightarrow Zn(s)$	–0.762
$Pt \mid H_2O \mid H_2, OH^-$	$2H_2O(l) + 2e^- \rightarrow H_2(g) + 2OH^-(aq)$	–0.828
$Mn \mid Mn^{2+}$	$Mn^{2+}(aq) + 2e^- \rightarrow Mn(s)$	–1.180
$Al \mid Al^{3+}$	$Al^{3+}(aq) + 3e^- \rightarrow Al(s)$	–1.662
$Mg \mid Mg^{2+}$	$Mg^{2+}(aq) + 2e^- \rightarrow Mg(s)$	–2.372
$Na \mid Na^+$	$Na^+(aq) + e^- \rightarrow Na(s)$	–2.714

(Contd...)

$Ca \mid Ca^{2+}$	$Ca^{2+}(aq) + 2e^- \rightarrow Ca(s)$	–2.868
$Sr \mid Sr^{2+}$	$Sr^{2+}(aq) + 2e^- \rightarrow Sr(s)$	–2.899
$Ba \mid Ba^{2+}$	$Ba^{2+}(aq) + 2e^- \rightarrow Ba(s)$	–2.905
$K \mid K^+$	$K^+(aq) + e^- \rightarrow K(s)$	–2.931
$Li \mid Li^+$	$Li^+(aq) + e^- \rightarrow Li(s)$	–3.05

***Exercise* 16.6:** Calculate the standard reduction potential of the sodium $Na^+(aq) + e^- \rightarrow Na(s)$. Given $\Delta_f \bar{H}^0[Na^+(aq)] = -240.12$ kJ mol^{-1}, $\Delta_f \bar{H}^0[H^+(aq)] = 0$ kJ mol^{-1}, $S^0[Na(s)] = 51.3$ J K^{-1} mol^{-1}, $S^0[Na^+(aq)] = 59.0$ J K^{-1} mol^{-1}.

As we have learnt that we cannot have a single electrode, we have to combine it with either standard hydrogen electrode or with an electrode whose standard electrode potential is known. To make the problem simpler, connect the above electrode with standard hydrogen electrode and reaction will be

$Na^+(aq) + 1/2\ H_2(g) \rightarrow Na(s) + H^+(aq)$

In this cell Na^+ ions are reduced at the sodium electrode and hydrogen molecules are oxidised at the hydrogen electrode. Heat of the reaction can be calculated from

$$\Delta_r H^0 = \Delta_f \bar{H}^0[Na(s)] + \Delta_f \bar{H}^0[H^+(aq)] - \Delta_f \bar{H}^0[Na^+(aq)] - 1/2[\Delta_f \bar{H}^0[H_2(g)]]$$

$$= 0 + 0 - \Delta_f \bar{H}^0[-240.12 \text{ kJ mol}^{-1}] - 1/2[0] = 240.12 \text{ kJ mol}^{-1}$$

$$\Delta r S^0 = \bar{S}^0[Na(s)] + \bar{S}^0[H^+(aq)] - \bar{S}^0[Na^+(aq)] - 1/2[\bar{S}^0[H_2(g)]]$$

$$\Delta r S^0 = [51.21] + 0 - [59.0] - 1/2[130.6] = -73.1 \text{ J K}^{-1} \text{ mol}^{-1}$$

$$\Delta_r G^0 = \Delta_r H^0 - T\Delta_r S^0 = 240.12 \text{ kJ mol}^{-1} - \frac{(298\text{K})(-73.1 \text{ J K}^{-1} \text{ mol}^{-1})}{1000 \text{ J kJ}^{-1}}$$

$$= 261.90 \text{ kJ mol}^{-1}$$

Using equation (16.6) and substituting the variables, we get

$$E_0 = -\frac{(261.90 \text{ kJ mol}^{-1})(1000 \text{ J kJ}^{-1})}{96485 \text{ C mol}^{-1}} = 2.7144 \text{ V}$$

16.5. CONVENTION FOR ELECTROCHEMICAL CELLS AND ELECTRODE REACTIONS

The following conventions are used to calculate the e.m.f. of the electrochemical cells and to write the electrode as well as overall cell reactions.

(*i*) The reaction taking place at the left electrode is written as an oxidation reaction and the reduction reaction occurs at right hand electrode. The overall cell reaction is sum of these two electrode reactions.

(*ii*) The standard e.m.f. of the cell is given as the standard electrode potential of the right electrode minus that of the left electrode *i.e.*

$$E^0 = \varepsilon^0_{right} - \varepsilon^0_{left} \qquad ...(16.3)$$

(*iii*) If the E^0 calculated by step (*ii*) is positive, then the cell reaction taking place in step (*i*) is a spontaneous reaction where the activities of the reactants and products are unity.

Important thing to be kept in mind is that if in a cell reaction, more than one electron is involved, the standard electrode potential does not change but the standard free energy of the chemical reaction will be multiple of FE_0 *i.e.*, the actual value of n should be used.

***Exercise* 16.7:** (*i*) Calculate the e.m.f. of the cell mentioned in exercise 16.5 *i*.

$$\varepsilon^0\,(OH^-, ZnO_2^- |Zn) = -1.216 \text{ V}, \ \varepsilon^0\,(OH^- | HgO | Hg) = 0.098 \text{ V}$$

(*ii*) Calculate the free energy change and (*iii*) indicate whether the reaction is spontaneous?

(*i*) Cell reaction is written in exercise 16.5 *i*. The e.m.f. of the cell is

$$E^0 = \varepsilon^0_{right} - \varepsilon^0_{left} = 0.098 - (-1.216) = 1.314 \text{ V}$$

(*ii*)
$$\Delta G^0 = -nFE^0 = -(2)\,(96.5 \text{ kJ mol}^{-1} \text{ V}^{-1})\,(1.314 \text{ V})$$
$$= -253.6 \text{ kJ}$$

(*iii*) Reaction is spontaneous.

***Exercise* 16.8:** Calculate the standard e.m.f. of the reaction

Given
$$Fe^{3+}(aq) + 3e^- \rightarrow Fe\,(s) \quad \varepsilon^0_3 = ?$$
$$Fe^{3+}(aq) + e^- \rightarrow Fe^{2+}(aq) \quad \varepsilon^0_1 = 0.771 \text{ V}$$
$$Fe^{2+}(aq) + 2e^- \rightarrow Fe\,(s) \quad \varepsilon^0_2 = -0.44 \text{ V}$$

The above problem can be tackled by two ways

(*i*) Calculation of free energy

$$\Delta G^0_1 = -nF\varepsilon^0 = -(1)\,(F)\,(0.771 \text{ V}) = -0.771\, F\text{V}$$

$$\Delta G^0_2 = -(2)\,(F)\,(-0.44) = 0.88\, F\text{V}$$

$$\Delta G^0_3 = -(3)\,(F)\,(\varepsilon^0) = -3\,F\varepsilon^0 \text{ V}$$

The free energy change for the unknown process can be obtained as

$$\Delta G^0_3 = \Delta G^0_1 + \Delta G^0_2$$

$$-3F\varepsilon^0 = -0.771\,F + 0.88\,F = 0.109\,F$$

or
$$\varepsilon^0 = -\frac{0.109}{3} = -0.036 \textbf{ V}$$

(*ii*) Making use of Nernst equation, ε_1, ε_2 and ε_3 can be written as

$$\varepsilon_1 = \varepsilon^0_1 - \frac{0.0591}{1} \log \frac{a_{Fe^{2+}}}{a_{Fe^{3+}}}$$

$$\varepsilon_2 = \varepsilon^0_2 - \frac{0.0591}{2} \log \frac{1}{a_{Fe^{2+}}}$$

$$\varepsilon_3 = \varepsilon^0_3 - \frac{0.0591}{3} \log \frac{1}{a_{Fe^{3+}}}$$

Now ε_3^0 can be calculated from ε_1^0 and ε_2^0 by making the coefficient of log terms equal *i.e.*

$$\varepsilon_3 = \frac{1}{3}\varepsilon_1 + \frac{2}{3}\varepsilon_2 = \frac{\varepsilon_1^0}{3} + \frac{2}{3}\varepsilon_2^0 - \frac{0.0591}{3}\log \frac{a_{Fe^{2+}}}{a_{Fe^{3+}}} \cdot \frac{1}{a_{Fe}^{2+}}$$

Now if activities are unity

$$\varepsilon_3^0 = \frac{\varepsilon_1^0 + 2\varepsilon_2^0}{3} = \frac{0.771 - 0.88}{3} = -0.036 \text{ V}$$

***Exercise* 16.9:** Calculate ε^0 for the reaction

Given $Cr^{2+}(aq) + 2e^- \rightarrow Cr(s)$ $\varepsilon_1^0 = ?$

$Cr^{3-}(aq) + 3e^- \rightarrow Cr(s)$ $\varepsilon_2^0 = 0.5$ V

$Cr^{3+}(aq) + e^- \rightarrow Cr^{2+}(aq)$ $\varepsilon_3^0 = 0.41$V

From the above reactions the ΔG_1^0 can be obtained as

$$\Delta G_1^0 = -2F\,\varepsilon_1^0 = \Delta G_2^0 - \Delta G_3^0$$
$$= -(3)(F)(0.5) - [-(1)(F)(-0.41)]$$

or

$$\varepsilon_1^0 = \frac{3 \times 0.5 + 0.41}{2} = \frac{1.5 + 0.14}{2} = 0.955 \text{ V}$$

It can also be written as

$$\varepsilon_1^0 = \frac{3\,\varepsilon_2^0 - \varepsilon_3^0}{2} = \frac{(3)(0.5) - (1)(-0.41)}{2}$$
$$= 0.955 \text{ V}$$

16.6. DEPENDENCE OF E.M.F. OF A CELL ON ACTIVITIES OF REACTION SPECIES, i.e. NERNST EQUATION

The free energy change for the cell reaction

$$aA + bB \rightarrow cC + dD$$

is given as

$$\Delta G = \Delta G^0 + RT \ln \frac{a_C^c\, a_D^d}{a_A^a\, a_B^b} \qquad ...(16.4)$$

After substitutions $\Delta G = -nFE$ and $\Delta G^0 = -nFE^0$ in equation (16.4) and rearranging

$$E = E^0 - \frac{RT}{nF} \ln \frac{a_C^c\, a_D^d}{a_A^a\, a_B^b} \qquad ...(16.5)$$

The equation (16.5) can also be written as

$$E = E^0 - \frac{0.0591}{n} \log \frac{a_C^c\, a_D^d}{a_A^a\, a_B^b} \qquad ...(16.5a)$$

at 298 K.

The equations (16.5) and (16.5*a*) are often called Nernst equations. If the solution is behaving as ideal, the activities can be replaced by concentrations *i.e.*

$$E = E^0 - \frac{0.0591}{n} \log \frac{C_C^c \, C_D^d}{C_A^a \, C_B^b} \quad ...(16.6)$$

***Exercise* 16.10:** Consider the following cell

$$Ag \mid AgCl \mid HCl\ (0.02M) \mid H_2\ (0.56\ atm) \mid Pt$$

Given $\varepsilon^0(Cl^- \ AgCl\ Ag) = 0.2225$ V.

(*a*) Write the electrode reactions and overall reaction, (*b*) Calculate the e.m.f. of the cell at 25°C neglecting the activity coefficients, (*c*) Calculate the e.m.f. of the cell using activities, obtaining activity coefficients by using Debye-Hückel limiting law.

(*a*) At left electrode

$$Ag(s) + Cl^-(aq) \rightarrow AgCl(s) + e^- \qquad \varepsilon_1^0 = 0.2225 \text{ V}$$

At right electrode

$$H^+(aq) + e^- \rightarrow \frac{1}{2} H_2(g) \qquad \varepsilon_2^0 = 0$$

Overall reaction

$$Ag(s) + H^+(aq) + Cl^-(aq) \rightarrow AgCl(s) + \frac{1}{2} H_2(g)$$

$$E^0 = \varepsilon_2^0 - \varepsilon_1^0 = -0.2225 \text{ V}$$

i.e. standard e.m.f. of the cell at 25°C is –0.2225 V.

Since the e.m.f. is negative, the overall reaction is not spontaneous. So the left hand electrode is cathode and the right hand electrode is anode.

(*b*) Using Nernst equation (16.6) the e.m.f. at 25°C is

$$E = -0.2225 - \frac{0.0591}{1} \log \frac{p_{H_2}^{1/2}}{[H^+][Cl^-]}$$

$$= -0.2225 - (0.0591) \log \frac{(0.56)^{1/2}}{(0.02)(0.02)}$$

$$= -0.4159 \text{ V}$$

(*c*) Writing the Nernst equation (16.5*b*) as

$$E = E^0 - \frac{0.0591}{1} \log \frac{p_{H_2}^{1/2}}{[H^+][Cl^-]} \frac{\gamma_{H_2}^{1/2}}{\gamma_{H^+} \gamma_{Cl^-}}$$

Considering hydrogen as ideal gas

$$\gamma_{H_2} = 1 \text{ and } \gamma_{H^+} \gamma_{Cl^-} = \gamma_{\pm}^2 (HCl)$$

The mean activity coefficient of HCl by Debye-Hückel limiting law is given as

$$\log \gamma_{\pm} = 0.591\, z_+ z_- \mu^{1/2}$$

$$= (0.591)(1)(-1)[\frac{1}{2}(0.02)(1)^2 + \frac{1}{2}(0.02)(-1)^2]^{1/2}$$

$$= (0.591)(-1)(0.02) = -0.0118$$
$$= \bar{1}.9882$$
$$\gamma_{\pm} = 0.973$$

With the help of results derived in part (*b*) and in (*c*) the equation (16.5*b*) can be written as

$$E = E^0 - \frac{0.0591}{1} \log \frac{p_{H_2}^{1/2}}{[H^+][Cl^-]} - \frac{0.0591}{1} \log \frac{1}{\gamma^2_{\pm(HCl)}}$$
$$= -0.4159 + (2)(0.0591) \log 0.973$$
$$= -0.4159 - 0.0014$$
$$= -0.4173 \text{ V}$$

Note: The e.m.f. of the above cell can also be calculated by calculating the standard potential of the individual electrode at the given concentration and then taking the difference *i.e.*

$$\varepsilon_1 = \varepsilon_1^0 - \frac{(0.0591)}{1} \log (0.02)$$
$$= 0.2225 + 0.1004 = 0.3229 \text{ V}$$

for $Cl^- | AgCl | Ag$ electrode. For electrode $H^+ | H_2 | Pt$

$$\varepsilon_2 = \varepsilon_2^0 - (0.0591) \log \frac{0.56^{1/2}}{0.02}$$
$$= 0 - 0.093 \text{ V} = -0.093 \text{ V}$$
$$E = \varepsilon_2 - \varepsilon_1 = -0.093 - 0.3229$$
$$= -0.4159 \text{ V}$$

i.e. The voltage is same as calculated above.

***Exercise* 16.11:** Consider the cell $Ag | AgBr(s) | Br^- (a = 0.32), Cu^{2+} (a = 0.42), Cu^+ (a = 0.36) |$ Pt. The e.m.f. of this cell at 25°C is 0.0565 V. (*a*) Write the cell reactions (*b*) Calculate the standard e.m.f. of the cell,

(*a*) At left electrode

$$Ag(s) + Br^-(aq) \rightarrow AgBr(s) + e^-$$

At right electrode

$$Cu^2(aq) + e^- \rightarrow Cu^+(aq)$$

Overall reaction

$$Ag(s) + Cu^{2+}(aq) + Br^-(aq) \rightarrow AgBr(s) + Cu^+(aq)$$

The Nernst equation (16.5*a*) can be written as

$$E = E^0 - \frac{(0.0591)}{(1)} \log \frac{a_{Cu^+}}{a_{Cu^{2+}} a_{Br^-}}$$

i.e.
$$0.0565 = E^0 - (0.0591) \log \frac{0.36}{0.42 \times 0.32}$$
$$= E^0 - 0.0253$$
$$E^0 = 0.0565 + 0.0253 = 0.0818 \text{ V}$$

***Exercise* 16.12:** The standard electrode potentials of the electrodes $Cu^{2+} | Cu$ and $Ag^+ | Ag$ are 0.337 V and 0.7991 V. What would be the concentration of Ag^+ in a solution containing 0.06 M of Cu^{2+}

ion such that both the metals can be deposited together? Assume that activity coefficients are unity and both silver and copper do not dissolve among themselves.

The individual reactions are

$$Cu^{2+}(aq) + 2e^- \rightarrow Cu(s)$$

and

$$Ag^+(aq) + e^- \rightarrow Ag(s)$$

The electrode potentials given by Nernst equation (16.5 *a*)

$$\varepsilon(Cu^{2+} \mid Cu) = \varepsilon^0 - \frac{0.0591}{2} \log \frac{1}{[Cu^{2+}]}$$

$$= 0.337 - \frac{(0.0591)}{2} \log \frac{1}{0.06}$$

$$= 0.337 - 0.036 = 0.301$$

$$\varepsilon(Ag^+|Ag) = 0.7991 - \frac{0.0591}{1} \log \frac{1}{[Ag^+]}$$

Two metals will be deposited together when the electrode potentials are equal *i.e.*

$$0.301 = 0.7991 - 0.0591 \log \frac{1}{[Ag^+]}$$

i.e.

$$\log \frac{1}{[Ag^+]} = \frac{0.7991 - 0.301}{0.0591} = 8.428$$

$$\frac{1}{[Ag^+]} = 10^{8.428}$$

or

$$[Ag^+] = 10^{-8.428} = 0.37 \times 10^{-8} \text{ mol dm}^{-3}$$

16.7. CONCENTRATION CELLS

In concentration cell, the two electrodes are of the same material and they are dipping in the solutions of their respective ions at different concentrations. The two solutions are separated either by liquid junction or by salt bridge. The electrode dipping in a solution of higher concentration of positive ions is positive and the electrode dipping in lower concentration of positive ions is negative. The overall cell reaction is nothing but a transference of material from higher activity to lower activity. The e.m.f. of the cell is given by

$$E = -\frac{RT}{nF} \ln \frac{a_1}{a_2} \qquad ...(16.7)$$

where a_1 and a_2 are the activities of the ion in two solutions. The term E^0 cancels out because both the electrodes are same and will have the same standard electrode potential at unit activity. Thus E is positive if $a_2 > a_1$ and negative if $a_1 > a_2$. The above equation holds good only if the liquid junction potential is eliminated by using KCl salt bridge.

In case if the liquid junction is present, the e.m.f. of the electrochemical cell is given by

$$E = -\frac{RT}{nF} t_i \ln \frac{a_1}{a_2} \qquad ...(16.8)$$

where t_i is the transport number of the concerned ion. This is also one of the nice methods to calculate the transport number provided a good liquid junction can be set up.

***Exercise* 16.13:** In the following cell

$$Zn - Hg\ (c_1)\ |\ Zn^{2+}\ |\ Zn - Hg\ (c_2)$$

if the concentration c_1 and c_2 are 2 g and 0.5 g per 100 g of mercury, calculate the e.m.f. of the cell at 25°C.

The cell reaction is

$$Zn\ (c_1) \rightarrow Zn^{2+}(aq) + 2e^-$$

$$\frac{Zn^{2+}(aq) + 2e^- \rightarrow Zn\ (c_2)}{Zn\ (c_1) \rightarrow Zn\ (c_2)}$$

Using equation (16.7)

$$E = -\frac{0.05915}{2} \log \frac{0.5}{2.0}$$

$$= +0.0178 \text{ V}$$

In general the amalgams are very close to ideal solutions so the concentration can be used.

***Exercise* 16.14:** Assuming that hydrogen behaves as an ideal gas, calculate the e.m.f. of the cell at 25°C

$$Pt\ |\ H_2(p_1)\ |\ HCl\ |\ H_2(p_2)\ |\ Pt$$

if $p_1 = 640$ mm, $p_2 = 425$ mm

The overall reaction is

$$H_2\ (p_1) \rightarrow H_2\ (p_2)$$

involving two electrons. So using equation (16.7) with activity being equal to pressure.

$$E = -\frac{0.05915}{2} \log \frac{425 \text{ mm}}{640 \text{ mm}}$$

$$= 5.258 \times 10^{-3} \text{ V}$$

***Exercise* 16.15:** Consider the cell

$$Pt\ |\ H_2\ |\ HCl\ (0.2\ m)\ |\ AgCl\ |\ Ag - Ag\ |\ AgCl\ |\ HCl\ (0.015\ m)\ |\ H_2\ |\ Pt$$

(*a*) Write the overall cell reaction as well as the individual cell reactions.

(*b*) Calculate the e.m.f. of the cell assuming no transference or without liquid junction and $\gamma_i = 1$

(*c*) Repeat the calculation (*b*) with $\gamma_i \neq 1$ and calculate γ_i with Debye-Hückel limiting law.

(*a*) The above cell is a combination of two cells. Individual cell reactions are:

At left cell

$$\frac{1}{2}H_2\ (g) \rightarrow H^+(aq) + e^-$$

$$\frac{AgCl\ (s) + e^- \rightarrow Ag(s) + Cl^-\ (aq)}{\frac{1}{2}H_2(g) + AgCl\ (s) \rightarrow Ag\ (s) + H^+\ (aq,\ c_1) + Cl^-\ (aq,\ c_1)}$$

At right cell

$$Cl^-(aq) + Ag(s) \rightarrow AgCl\ (s) + e^-$$

$$\frac{H^+(aq) + e^- \rightarrow \frac{1}{2}H_2(g)}{Ag(s) + H^+ \ (aq, c_2) + Cl^- \ aq, (c_2) \rightarrow AgCl\,(s) + \frac{1}{2}H_2(g)}$$

Overall reaction of the two cells

$$H^+ \ (aq, c_2) + Cl^- \ (aq, c_2) \rightarrow H^+ \ (aq, c_1) + Cl^- \ (aq, c_1)$$

(*b*) Assuming the solution is ideal the equation (16.7) becomes

$$E = -\frac{0.05915}{1}\log\frac{c_1\ (HCl)}{c_2\ (HCl)}$$

$$= -0.05915 \log\frac{0.015\ m}{0.2\ m}$$

$$= +0.0665\ V$$

(*c*) By Debye-Hückel limiting law

$$\log \gamma_{\pm} = 0.5091\ z_+z_- \sqrt{\mu}$$

The ionic strength in the two cells can be calculated by assuming that concentration of ions due to the dissociation AgCl will be small. So

$$\mu_1 = \frac{1}{2}[(0.2)\,(1)^2 + (0.2)\,(-1)^2] = 0.2\ m$$

and $$\mu_2 = 0.015\ m$$

So

$$\log \gamma_{\pm}(\mu_1) = (0.5091)\,(1)\,(-1)\,(0.2)^{1/2}$$

$$= -0.2277$$

$$\gamma_{\pm}(\mu_1) = 0.592$$

$$\log \gamma_{\pm}(\mu_2) = (0.5091)\,(1)\,(-1)\,(0.015)^{1/2}$$

$$= -0.0624$$

$$\gamma_{\pm}(\mu_2) = 0.866$$

Now $$a_{\pm}(1) = c_{\pm} \times \gamma_{\pm}$$

Since here $c_{\pm}$ is same as c because the electrolyte is 1:1. So

$$a_{\pm}(1) = (0.2)\,(0.592) = 0.1184$$

$$a_{\pm}(2) = (0.015)\,(0.866) = 0.013$$

and the e.m.f. of the cell is

$$E = \frac{0.05915}{1}\log\frac{0.1184}{0.013} = 0.0568\ \textbf{V}$$

***Exercise* 16.16:** Consider the cell

$$Pt\,|\,H_2(1\ atm)\,|\,HC1(0.25\ m)\,|\,HC1(0.15\ m)\,|\,H_2(0.72\ atm)\,|\,Pt$$

Write the electrode reactions, transference changes and thus overall changes. Calculate the e.m.f. if the transference number of H^+ is 0.8.

At left electrode $\frac{1}{2}H_2(g, p_1) \rightarrow H^+(aq, c_1) + e^-$

At right electrode $H^+(aq, c_2) + e^- \rightarrow \frac{1}{2}H_2(aq, p_2)$

Overall reaction $\frac{1}{2}H_2(aq, 1\text{ atm}) + H^+(aq, 0.15\text{ m})$

$$\rightarrow \frac{1}{2}H_2(g, 0.72\text{ atm}) + H^+(aq, 0.25\text{ m}) \quad ...(16.9a)$$

The e.m.f. of the cell without transference can be calculated by using Nernst equation.

$$E = 0 - 0.059 \log \frac{(0.72)^{1/2}(0.25)}{(1)^{1/2}(0.15)}$$

$$= -8.895 \times 10^{-3}\text{ V}$$

When transference is considered, the junction reaction are

$$t_+ H^+(aq, 0.25\text{ m}) \rightarrow t_+ H^+(aq, 0.15\text{ m})$$

$$(1 - t_-) H^+(aq, 0.25\text{ m}) \rightarrow (1 - t_-)H^+(aq, 0.15\text{ m}) \quad ...(16.9b)$$

$$t_- Cl^-(aq, 0.15\text{ m}) \rightarrow t_- Cl^-(aq, 0.25\text{ m}) \quad ...(16.9c)$$

or $$(1 - t_+) Cl^-(aq, 0.15\text{ m}) \rightarrow (1 - t_+) Cl^-(aq, 0.25\text{ m}) \quad ...(16.9d)$$

Overall reaction is *i.e.* adding the reactions (16.9*a*), (16.9*b*) and (16.9*c*)

$$\frac{1}{2}H_2(g, 1\text{ atm}) + H^+(aq, 0.15\text{ m}) + H^+(aq, 0.25\text{ m}) - t_- H^+(aq, 0.25\text{ m}) + t_- Cl^-(aq, 0.15\text{ m})$$

$$\rightarrow \frac{1}{2}H_2(g, 0.72\text{ atm}) + H^+(aq, 0.25\text{ m}) + H^+(aq, 0.15\text{ m}) - t_- H^+(aq, 0.15) + t_- Cl^-(aq, 0.25)$$

i.e. $$\frac{1}{2}H_2(g, \text{atm}) + t_- H^+(aq, 0.15) + t_- Cl^-(aq, 0.15)$$

$$\rightarrow \frac{1}{2}H_2(g, 0.72\text{ atm}) + t_- H^+(aq, 0.25) + t_- Cl^-(aq, 0.25)$$

So using Nernst equation

$$E = 0 - \frac{0.0591}{1} \log \left[\frac{(0.72)^{1/2}(0.25)^{t-}(0.25)^{t-}}{1^{1/2}(0.15)^{t-}(0.15)^{t-}}\right]$$

$$= -(0.0591) \log \left[\frac{(0.72)^{1/2}(0.25)^{0.2}(0.25)^{0.2}}{(0.15)^{0.2}(0.15)^{0.2}}\right]$$

$$= -(0.0591) \log (0.72)^{1/2} \left[\frac{0.25 \times 0.25}{0.15 \times 0.15}\right]^{0.2}$$

$$= -1.03 \times 10^{-3}\text{ V}$$

16.8. THERMODYNAMICS OF THE CELLS

The e.m.f. of the cell is related to free energy by equation (16.1) *i.e.*

$$\Delta G = -nFE \quad ...(16.1)$$

Now

$$\left(\frac{\partial \Delta G}{\partial T}\right)_p = -\Delta S \qquad ...(16.10)$$

So

$$\left(\frac{\partial \Delta G}{\partial T}\right)_p = -nF\left(\frac{\partial E}{\partial T}\right)_p = -\Delta S$$

or

$$\Delta S = nF\left(\frac{\partial E}{\partial T}\right)_p \qquad ...(16.11)$$

The enthalpy of the cell reaction will be

$$\Delta H = \Delta G + T\Delta S$$

$$= -nFE + TnF\left(\frac{\partial E}{\partial T}\right)_p \qquad ....(16.12)$$

The thermodynamic quantities of the cell reaction can be calculated by equations (16.1), (16.11) and (16.12) provided one knows the e.m.f. of cell and its dependence on temperature.

The heat effects in the system can be calculated as follows. If the process is irreversible (*i.e.* by mixing the reactants together), heat flow to the system can be given by the reaction $\Delta H = Q_p$. If the process is reversible the heat flow to system is given by

$$Q_p = T\Delta S$$

***Exercise* 16.17:** Consider the cell

$$Cd \mid CdCl_2{\cdot}25H_2O \mid AgCl\,(s) \mid Ag$$

The e.m.f. of the cell at 15°C is 0.67531 V and temperature coefficient of e.m.f. is –0.00065 V deg^{-1}. Calculate the value of ΔH at 15°C and heat flow if the process is carried reversibly.

The cell reaction is

At anode $Cd(s) + 2Cl^-(aq) \rightarrow CdCl_2(s) + 2e^-$

At cathode $2AgCl(s) + 2e^- \rightarrow 2Ag(s) + 2Cl^-(aq)$

$$Cd(s) + 2AgCl(s) \rightarrow CdCl_2(s) + 2Ag(s)$$

The number of electrons involved are two.

So

$$\Delta G = -nFE$$
$$= -(2)(96500\text{ C})(-0.67531\text{ V})$$
$$= -130.335\text{ kJ}$$
$$\Delta S = (2)(96500\text{ C})(-0.00065\text{ V deg}^{-1})$$
$$= -125.45\text{ J deg}^{-1}$$
$$\Delta H = \Delta G + T\Delta S$$
$$= (-130.335\text{ kJ}) + (288\text{ K})(-125.45\text{ J K}^{-1})(10^{-3}\text{ kJ J}^{-1})$$
$$= -166.465\text{ kJ}$$
$$Q_p = (288\text{ K})(-125.45\text{ J K}^{-1}\text{ mol}^{-1})$$
$$= -36.13\text{ kJ mol}^{-1}$$

***Exercise* 16.18:** Consider the cell

$$Pb \mid PbSO_4 \mid Na_2SO_4\,.10H_2O(aq) \mid Hg_2SO_4(s) \mid Hg$$

The temperature coefficient of the e.m.f. of above cell is 0.000174 V deg^{-1} and heat of reaction is –176.146 kJ. Calculate e.m.f. of the cell. The cell reaction is

At anode $Pb(s) + SO_4^{2-}(aq) \rightarrow PbSO_4(s) + 2e^-$

$$\text{At cathode} \quad \frac{Hg_2SO_4(s) + 2e^- \rightarrow 2Hg(l) + SO_4^{2-}(aq)}{Pb(s) + Hg_2SO_4(s) \rightarrow PbSO_4(s) + 2Hg(l)}$$

i.e. the number of electrons involved are two, substituting the values of known terms in equation (16.12) *i.e.*

$$-176146 \text{ J} = -(2)(96500 \text{ J V}^{-1})E^0 + (298 \text{ K})(96500 \text{ J V}^{-1})(2)(0.000174 \text{ V K}^{-1})$$

$$E^0 = 0.9645 \text{ V}$$

16.9. APPLICATIONS

16.9.1. *Calculation of Equilibrium Constants*

Equation (16.5) gives the e.m.f. of the cell for the overall cell reaction at any concentration of the species

$$aA + bB \rightarrow cC + dD$$

i.e.
$$E = E^0 - \frac{RT}{nF} \ln \frac{a_C^c \, a_D^d}{a_A^a \, a_B^b} \quad \text{...(16.5)}$$

If the reaction (16.3) represents equilibrium then under these condition $\Delta G = 0$ and in turn $E = 0$. Whereas the quotient

$$\frac{(a_C^c)_e \, (a_D^d)_e}{(a_A^a)_e \, (a_B^b)_e} = K \text{ the equilibrium constant}$$

i.e.

$$\ln K = \frac{nE}{RT} E^0 \quad \text{...(16.13)}$$

or

$$\log K = \frac{n}{0.059} E^0 \text{ at 298 K} \quad \text{...(16.13}a\text{)}$$

Thus knowing E^0, K can be calculated.

***Exercise* 16.19:** The standard electrode potentials of the electrodes $Ag^+ | Ag$ and $Fe^{3+}, Fe^{2+} | Pt$ are 0.7991 V and 0.771 V respectively. Calculate the equilibrium constant for the reaction

$$Ag(s) + Fe^{3+}(aq) \rightleftharpoons Ag^+(aq) + Fe^{2+}(aq)$$

The cell formed for the above reaction is $Ag | Ag^+; Fe^{3+}, Fe^{2+} | Pt$.

At anode $\quad Ag(s) \rightarrow Ag^+(aq) + e^-$

At cathode $\quad Fe^{3+}(aq) + e^- \rightarrow Fe^{2+}(aq)$

Overall reaction

$$Ag(s) + Fe^{3+}(aq) \rightleftharpoons Ag^+(aq) + Fe^{2+}(aq)$$

The e.m.f. of the cell can be written as

$$E^0 = 0.771 - 0.7991 = -0.0281 \text{ V}$$

i.e.

$$E^0 = \frac{RT}{nF} \ln \frac{a_{Ag^+} a_{Fe^{2+}}}{a_{Fe^{3+}}} = \frac{RT}{nF} \ln K$$

since *'a'* for pure solid is unity.

At 25°C the above equation becomes

$$E^0 = \frac{0.0591}{n} \log K$$

i.e.

$$\log K = -\frac{(0.0281 \text{ V})(1)}{(0.0591 \text{ V})} = -0.4751$$

$$K = 0.335$$

***Exercise* 16.20:** If zinc metal piece is placed in a solution containing 0.1 M Cu^{2+} ion, calculate the concentration of Zn^{2+} and Cu^{2+} ions at equilibrium. Assume the activity coefficients are unity and no Zn^{2+} ions are present in the beginning. The standard electrode potentials at 25°C of $Zn^{2+} | Zn$ and $Cu^{2+} |$ Cu are –0.7628 V and 0.337 V respectively.

At equilibrium, the e.m.f. of two electrodes are equal

$$E_{Zn^{2+}/Zn} = E^0_{Zn^{2+}/Zn} + \frac{RT}{2F} \ln Zn^{2+}$$

$$E_{Cu^{2+}/Cu} = E^0_{Cu^{2+}/Cu,} + \frac{RT}{2F} \ln Cu^{2+}$$

Since at equilibrium, $E_{Zn^{2+}/Zu} = E_{Cu^{2+}/Cu,}$ so

$$E^0_{Zn^{2+}/Zn} + \frac{RT}{2F} \ln Zn^{2+} = E^0_{Cu^{2+}/Cu} + \frac{RT}{2F} \ln Cu^{2+}$$

or

$$\frac{RT}{nF} \ln \frac{Zn^{2+}}{Cu^{2+}} = -E^0_{Zn^{2+}/Zn} + E^0_{Cu^{2+}/Cu}$$

At 25°C

$$\frac{0.0591}{2} \log \frac{Zn^{2+}}{Cu^{2+}} = -(-0.7628) + .337$$

$$= 1.0998$$

or

$$\log \frac{Zn^{2+}}{Cu^{2+}} = \frac{(1.0098)(2)}{0.0591} = 37.22$$

$$\frac{Zn^{2+}}{Cu^{2+}} = 10^{37.22}$$

i.e. concentration of Cu^{2+} will be very small in comparison to Zn^{2+} ions *i.e.* complete displacement of Cu^{2+} by Zn metal.

***Exercise* 16.21:** Calculate the equilibrium constant for the reaction

$$Zn^{2+}(aq) + 4NH_3\,(aq) \rightarrow [Zn(NH_3)_4]^{2+}(aq)$$

$\varepsilon^0(Zn^{2+}/Zn) = -\,0.763$ V and $\varepsilon^0[Zn(NH_3)_4]^{2+}/Zn + NH_3) = -\,1.03$ V

The electrode reactions for the given electrodes, can be written as

$$Zn^{2+}\,(aq) + 2e^- \rightarrow Zn(s) \qquad \varepsilon_1^0 = -\,0.763 \text{ V}$$

$$[Zn(NH_3)_4]^{2+}(aq) + 2e^- \rightarrow Zn(s) + 4NH_3\,(aq) \qquad \varepsilon_2^0 = -\,1.03 \text{ V}$$

Reactions (1) – (2)

$$Zn^{2+}(aq) + 4NH_3(aq) \rightarrow [Zn(NH_3)_4]^{2-}(aq)$$

The standard e.m.f. of this reaction $= \varepsilon_1^0 - \varepsilon_2^0$

$$= -\,0.763 \text{ V} - (-\,1.03) = 0.267 \text{ V}$$

According to Nernst equation

$$E = E^0 - \frac{RT}{2F} \ln \frac{a_{[Zn(NH_3)_4]^{2+}}}{a_{Zn^{2+}} \cdot a^4_{NH_2}}$$

If the process is equilibrium, $E = 0$ at 25°C

$$\frac{0.0591}{2} \log K = 0.267$$

$$\log K = \frac{(0.267)(2)}{(0.059)} = 9.036$$

$$K = 10^{\cdot 036}\ 10^9 = 1.09 \times 10^9$$

This equilibrium constant is also called the "Stability Constant" of the complex. If the reaction is written in a reverse manner, it will be "instability constant".

16.9.2. *Solubility Product*

The solubility product of sparingly soluble salt MX can be discussed in terms of the equilibrium of the kind

$$MX\,(s) \rightleftharpoons M^+\,(aq) + X^-\,(aq) \qquad ...(16.14)$$

Since, the activity of the pure solid is always unity, the equilibrium constant of the solubility product can be written

$$K_{sp} = a_{M^+}\, a_{X^-} \qquad ...(16.15a)$$

When the solubility of the salt is very small, the solution is ideal and thus $a_i = C_i$ of ions. The equation (16.15*a*) can be written as

$$K_{sp} = C_{M^+} \cdot C_{X^-} \qquad ...(16.15b)$$

Using Nernst relation, the e.m.f. of the cell will be

$$E = E^0 - \frac{RT}{nF} \ln \frac{a_{M^+}\, a_{X^-}}{a_{M_X}} \qquad ...(16.16)$$

Under equilibrium conditions the e.m.f. of the cell will be zero *i.e.,* $E = 0$ and also the activity of pure solid is unity. Further $K_{sp} = a_{M^+} \cdot a_{X^-}$, the above equation at 25°C can be written as

$$\log K_{sp} = \frac{E^0 . n}{0.0591} \quad ...(16.17)$$

The solubility product thus can be calculated from the standard e.m.f. of the cell, formed in such a way that the final reaction is of the type (16.14).

***Exercise* 16.22:** The solubility product of $Fe(OH)_3$ at 25°C is $10^{-36.4}$ mol^4 dm^{-12} and ε^0 (Fe^{3+} | Fe) = – 0.036 V. Calculate the standard e.m.f. of the reaction

$$Fe(OH)_3\,(s) \rightarrow Fe^{3+}(aq) + 3\,OH^-(aq)$$

The cell can be formed as

$$Fe \mid Fe^{3+};\ OH^- \mid Fe(OH)_3\,(s) \mid Fe$$

The electrode reactions can be written as

$$Fe(s) \rightarrow Fe^{3+}(aq) + 3e^-$$

$$Fe\,(OH)_3\,(s) + 3e^- \rightarrow Fe\,(s) + 3\,OH^-(aq)$$

Overall reaction is *i.e.*, by adding

$$Fe(OH)_3\,(s) \rightarrow Fe^{3+}(aq) + 3\,OH^-(aq)$$

i.e.

$$E^0 = \varepsilon^0_{OH-/Fe(OH)_3\,/Fe} - \varepsilon^0_{Fe^{2}+/Fe} \quad ...(16.18)$$

$\varepsilon^0_{OH-/Fe(OH)_3\,/Fe}$ can be calculated if we know E^0 and E^0 can be calculated as follows

$$E^0 = \frac{0.0591}{3} \log 10^{-36.4}$$

$$= \frac{0.0591 \times (-36.4)}{3} = -\,0.75\ \text{V}$$

Substitute the values in relation (16.18) *i.e.*

$$-\,0.75 = \varepsilon^0_{OH-/Fe(OH)_3\,/Fe} - (-\,0.036\ \text{V})$$

or

$$\varepsilon^0_{OH-/Fe(OH)_3\,/Fe} = -\,0.75 - 0.036$$

$$= -\,0.786\ \text{V}$$

***Exercise* 16.23:** Calculate the K_{sp} of AgI by forming a proper cell. Given $\varepsilon^0_{I-/\,Ag\,I(s)/\,Ag} = -\,0.151$ V and $\varepsilon^0_{Ag+/\,Ag} = 0.7991$ V.

The cell can be written as

$$Ag \mid Ag^+;\ I^- \mid AgI \mid Ag$$

At left electrode

$$Ag(s) \rightarrow Ag^+(aq) + e^- \quad \varepsilon^0 = 0.7991\ \text{V}$$

At right electrode

$$AgI(s) + e^- \rightarrow Ag(s) + I^-(aq) \quad \varepsilon^0 = -0.151 \text{ V}$$

$$AgI(s) \rightarrow Ag^+(aq) + I^-(aq)$$

The standard e.m.f. of the cell is

$$E^0 = \varepsilon_R^0 - \varepsilon_L^0 = -0.151 - 0.7991$$

$$= -0.9501 \text{ V}$$

Using relation is (16.17) at 25°C

$$\log K_{sp} = -\frac{(0.9501)(1)}{0.059} = -16.1$$

$$K_{sp} = 10^{-16.1} = 7.94 \times 10^{-17}$$

PROBLEMS

16.1 Write the half-cell reactions for the following cells:

(*a*) $Pt \mid Ti^{3+}, Ti^{4+}; Fe^{3+}, Fe^{2+} \mid Pt$

(*b*) $Pb \mid PbSO_4 \mid SO_4^{2-}; Zn^{2+} \mid Zn$

(*c*) $Pt \mid Hg \mid Hg_2Cl_2 \mid Cl^-; H^+ \mid H_2 \mid Pt$

(*d*) $Pt \mid QH_2, Q; H^+; Cl^- \mid AgCl \mid Pt$

(*e*) Pt | fumarate^{2-}, 2H$^+$, succinate

(*f*) Pt | cytochrome C(Fe^{3+}) cytochrome C(Fe^{2+}).

16.2 Write the electrochemical cells for the following reactions.

(*a*) $H_2(g) + Cu^{2+}(g) \rightarrow Cu(s) + 2H^+(g)$

(*b*) $Cl_2(g) + 2Fe^{2+}(g) \rightarrow 2Fe^{3+}(g) + 2Cl^-(g)$

(*c*) $Zn(s) + 2AgCl(s) \rightarrow 2Ag(s) + ZnCl_2(s)$

(*d*) $[Fe(CN)_6]^{3-}(aq) + \frac{1}{2}H_2(g) \rightarrow [Fe(CN)_6]^{4-}(aq) + H^+(aq)$

(*e*) $\text{Lactate}^-(aq) + \text{NAD}(aq) \rightleftharpoons \text{pyruvate}^{2-}(aq) + \text{NADH}(aq) + H^+(aq)$

16.3 Calculate the standard e.m.f. and the standard free energy change for each cell mentioned in problems (16.1*a–d*) and (16.2). Calculate also the equilibrium constants.

16.4 The half-cell reactions and their e.m.f. are mentioned as below

$$Ag(s) \rightarrow Ag^+(aq) + e^- \qquad \varepsilon^0 = -0.799 \text{ V}$$

$$2I^-(aq) \rightarrow I_2(g) + 2e^- \qquad \varepsilon^0 = -0.5335 \text{ V}$$

Check whether free iodine will react with Ag or not.

16.5 The standard electrode potential of $Fe^{3+}, Fe^{2+} \mid Pt$ is 0.771V. This electrode is connected with another electrode but the e.m.f. of the cell is still 0.771 V. (*a*) Name the second unknown electrode (*b*) Write the complete cell (*c*) Overall chemical reaction.

16.6 (*a*) Write the cell for the following reaction

$$2MO_4^-(aq) + 6H^+(aq) + 5H_2C_2O_4(aq) \rightarrow 2Mn^{2+}(aq) + 8H_2O + 10CO_2(g)$$

(*b*) Name the anode and cathode.

(*c*) Calculate the standard e.m.f. of the cell.

16.7 Explain the following behaviour. In the Daniel cell if Na_2S is added to Cu—Cu^{2+} side of the cell, the cell e.m.f. drops as predicted qualitatively by Le-chatelier's principle. However if lot of Na_2S is added, the reversal of the voltage takes place. Given, $K_{sp}(CuS) = 1.0 \times 10^{-28}$

$$Zn^{2+}(aq) + 2e^- \rightarrow Zn(s) \qquad \varepsilon^0 = -0.763 \text{ V}$$
$$Cu^{2+}(aq) + 2e^- \rightarrow Cu(s) \qquad \varepsilon^0 = 0.337 \text{ V}$$
$$Cu(s) + S^{2-}(aq) \rightarrow CuS(s) + 2e^- \qquad \varepsilon^0 = 0.78 \text{ to } 1.11 \text{ V}$$
$$\text{or } \varepsilon^0 = 0.8$$

e.m.f. of the Reactions

16.8 The reduction potentials of $Cu^{2+}|Cu$ and $Cu^+|Cu$ at 25°C are 0.337 and 0.521 volts respectively.

(*a*) Calculate the standard e.m.f. for the reaction

$$Cu^{2+}(aq) + e^- \rightarrow Cu^+(aq)$$

(*b*) Calculate ΔG^0 for the above reaction

(*c*) Is it easy to oxidise Cu to Cu^{2+} or Cu to Cu^+ ?

16.9 The standard reduction potential for the reactions

$$Sn^{2+}(aq) + 2e^- \rightarrow Sn(s)$$
$$Sn^{4+}(aq) + 2e^- \rightarrow Sn^{2+}(aq)$$

are –0.136 V and 0.15 V respectively

(*a*) Calculate the standard e.m.f. of the reaction.

$$Sn^{4+}(aq) + 2e^- \rightarrow Sn(s)$$

(*b*) Calculate ΔG^0 for the reaction

(*c*) Is it easy to reduce Sn^{2+} to Sn or Sn^{4+} to Sn?

16.10 In the following sequence of reactions, the standard reduction potentials in volts in acidic medium (ε_A^0) for individual reactions are mentioned on the arrows

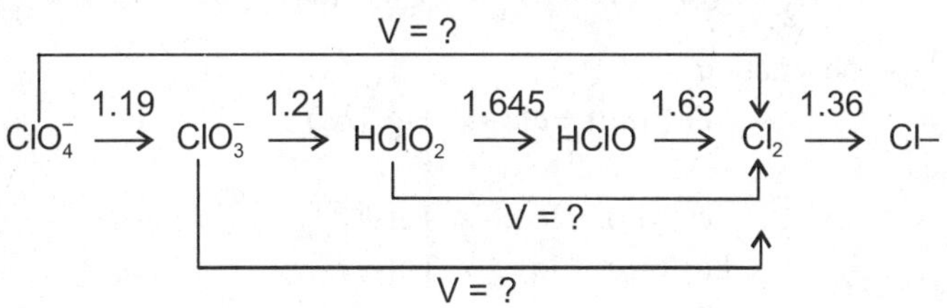

From the above diagram

(*a*) Calculate ε_A^0 for the reaction $ClO_3^-(aq) \rightarrow Cl_2(g)$

(*b*) Calculate ε_A^0 for the change $ClO_4^-(aq) \rightarrow HClO$

(*c*) Calculate ΔG^0 for the changes mentioned in parts (*a*) and (*b*)

(*d*) Is it easy to reduce $ClO_3^-(aq)$ to $Cl_2(g)$ or $ClO_4^-(aq)$ to $HClO(aq)$?

(*e*) Calculate ε_A^0 and ΔG^0 for the reaction

$$2HClO_2(aq) \rightarrow HClO(aq) + ClO_3^-(aq) + H^+(aq)$$

(*f*) Find the other missing values mentioned in the above scheme.

16.11 From the following sequence of reactions in acidic media (ε_A^0 are at 25°C and expressed in volts).

$$MnO_4^- \xrightarrow{0.564} MnO_4^{-2} \xrightarrow{2.26} MnO_2 \xrightarrow{0.95} Mn^{3+} \xrightarrow{1.51} Mn^{2+} \xrightarrow{-1.18} Mn$$

(*a*) Calculate ε_A^0 for $MnO_4^- \rightarrow Mn^{2+}$

(*b*) Which will be better oxidising agent in acidic media, $MnO_4^-(aq)$ or $MnO_4^{2-}(aq)$ when the final form is Mn^{2+}.

(*c*) Calculate the standard e.m.f. for the following reaction

$$3MnO_4^{2-}(aq) + 4H^+(aq) \rightarrow MnO_2(s) + 2MnO_4^-(aq) + 2H_2O$$

(*d*) Calculate also ΔG^0.

16.12 The standard free energy change for the reaction

$$H_2(g, 1\text{ atm}) + \frac{1}{2}O_2(g, 1\text{ atm}) \rightarrow H_2O(l)$$

is –237.19 kj mol^{-1} at 25°C (*a*) Write the cell and the electrode reactions, (*b*) Calculate the standard oxidation potential of Pt | $\frac{1}{2}O_2$(l atm) | OH^-. (*c*) Calculate the standard reduction potential of Pt | O_2(1 atm) | H^+ $a_{H^+} = 1$).

16.13 The standard e.m.f. for the half-cell reactions

$$Cl^-(aq) + 3H_2O(l) \rightarrow ClO_3^-(aq) + 6H^+ + 6e^- \quad \varepsilon^0 = -1.45\text{ V}$$

$$\tfrac{1}{3}Cl_2(g) + e^- \rightarrow Cl^-(aq) \quad \varepsilon^0 = -1.36\text{ V}$$

Calculate the standard e.m.f. for the cell reaction

$$\frac{1}{2}Cl_2(g) + 3H_2O(l) \rightarrow ClO_3^-(aq) + 6H^+(aq) + 5e^-$$

16.14 Calculate ε^0 for the reaction

$$Fe^{3+}(aq) + e^- \rightarrow Fe^{2+}(aq)$$

From the data

$$Fe^{2+}(aq) + 2e^- \rightarrow Fe(s) \quad \varepsilon^0 = -0.44\text{ V}$$
$$Fe^{3+}(aq) + 3e^- \rightarrow Fe(s) \quad \varepsilon^0 = -0.036\text{ V}$$

Nernst Equation

16.15 Calculate the electrode potential of hydrogen electrode when the solution containing H^+ is at $a_{H^+} = 0.5$ and also in contact with hydrogen gas having fugacity 0.99.

16.16 Calculate the e.m.f. of each cell and also write the cell reactions

(*a*) Pt | Pb(*s*), Pb^{2+} (0.1) || Sn^{4+} (0.02), Sn^{2+} (*a* = 0.1) | Pt

(*b*) Bi | $Bi_2O_3(s)$ | $OH^-(a = 1.2)$ || $Fe(OH)_3$ | Fe

(*c*) Pt | $\frac{1}{2}H_2(p_{H2} = 0.3\text{ atm})$ | H^+ ($_{[H+]}$ = 1M); Cl^- | AgCl | Ag

(*d*) Pt | NAD^+, H^+, at pH = 7, $_{[NAD^+]}$ = 0.1M, $_{[NADH]}$ = 0.2M

16.17 If the e.m.f. of the cell

$$Pt \mid \frac{1}{2} H_2(1 \text{ atm}) \mid HCl(xM) \mid \frac{1}{2} Hg_2Cl_2 \mid Hg$$

is 0.70 V at 25°C, calculate the pH of the solution. Given, ε^0 for calomal electrode is 0.2 V.

16.18 Consider the half cell Fe^{3+}, Fe^{2+} | Pt. Write the Nernst equation and calculate the ratio of oxidised to the reduced form at half cell potential 0.1, 0.2, 0.3 V and 0.6 V given $\varepsilon^0 = 0.7591$ V.

16.19 The standard hydrogen cell is described as when its $a_{H^+} = 1$, $p_{H_2} = 1$ atm and $\varepsilon^0 = 0$. If the standard state is considered when $a_{OH^-} = 1$ or $a_{H^+} = 7$, calculate the standard hydrogen electrode potential under these conditions.

16.20 For half cell reaction $MnO_4^- \rightarrow Mn^{2+}$ (in acidic media) $\varepsilon^0 = 1.51$ V. Assuming that $[MnO_4^-] = [Mn^{2+}]$, calculate the redox potential at pH = 0, 2, 4 and 6. Also explain why the ability of permagnate ion to oxidise halide ions to free halogens depends upon the acidity of the solution. The ε^0 for $Cl^- \mid Cl_2 \mid Pt$, $Br^- \mid Br_2 \mid Pt$ and $I^- \mid I_2 \mid Pt$ are 1.36 V, 1.07 V and 0.54 V respectively.

16.21 The potential for the reaction

$$O_2(g) + 4H^+(aq) + 4e^- \rightarrow 2H_2O(l)$$

is 1.23 in 0.1 N acid solution. Calculate the potential of this couple in aqueous solution having pH = 10,14.

16.22 ε^0_{rec} for the reaction $VO^{2+}(aq) + 2H^+(aq) + e^- \rightarrow V^{3+}(aq) + H_2O(l)$ is –0.36 V. Is the solution of V^{3+} subject to air oxidation at pH=7? Give reasons for your answer.

16.23 For the reaction $NO_3^-(aq) + 4H^+(aq) + 3e^- \rightarrow NO(g) + 2H_2O(l)$ ε^0 is 0.96 V. (*a*) What will be the value of ε^0 for the half reaction at pH = 7, (*b*) Will the NO_3^- ion at pH = 0 be able to oxidise chlorine to ClO_4 if ε^0 for the reaction

$$ClO_4^-(aq) + 8H^+(aq) + 7e^- \rightarrow \frac{1}{2} Cl_2(g) + 4H_2O(l)$$

is 1.38 V.

16.24 The standard state in biochemical system is defined in the same manner as for normal systems except the pH = 7. If $\varepsilon^{0'}$ (pH = 7) for the half cell NAD^+, H^+ NADH is 0.32 V, calculate the value of ε^0 at pH = 0 and 4. The standard state for NAD^+ and NADH is 1 M.

16.25 The standard electrode potentials of Cu^{2+} | Cu and Zn^{2+} | Zn are 0.337 V and –0.7628 V respectively. A solution contains 0.05 m of Cu^{++} and *x* m of Zn^{2+} ions. What would be the value of *x* such that both Cu^{2+} and Zn^{2+} are deposited simultaneously. Assume that zinc and copper do not dissolve with each other and $\gamma_i = 1$.

Concentration Cell

16.26 Consider the cell

$$Pb \mid PbSO_4(s) \mid H_2SO_4\,(m = 1.0) \mid H_2SO_4(m = 0.1)/PbSO_4/Pb$$

(*a*) Write the anode, cathode and overall reaction when the transference or no transference of ions are taking place.

(*b*) Calculate the e.m.f. of the cell without transference of ions.

(*c*) Calculate the e.m.f. of the cell when transference is taking place, $t_{H^+} = 0.831$.

16.27 Consider the cell

$$Hg(l) \mid Hg_2Cl_2(s) \mid KCl(0.06\ m) \mid KCl(0.025\ m) \mid Hg_2Cl_2(s) \mid Hg(l)$$

(*a*) Write the anode, cathode and overall reaction of the cell without transference and thus calculate the e.m.f. of the cell.

(*b*) Consider the transference of both the ions and write overall reaction. Calculate the e.m.f. of the cell if $t_{K^+} = 0.49$.

16.28 Consider the cell

$$Pt \mid Zn\text{–}Hg(m_1) \mid SO_4^{2-} \mid Zn\text{–}Hg(0.2\ m) \mid Pt$$

Write the overall reaction and calculate the value of m_1 if the e.m.f. of the cell at 25°C is 0.045 V.

16.29 Given the cell

$$Pt \mid Cl_2(f = 0.5\ atm) \mid H^+ \mid Cl_2(f = 0.25\ atm) \mid Pt$$

Name the anode and the cathode and thus calculate the e.m.f. of the cell.

Thermodynamic of Cells

16.30 The e.m.f. measured of the cell Cd | $CdCl_2.2.5H_2O(aq)$ AgCl(*s*) | Ag at 25°C and 0°C are 0.6753 V and 0.6915 V respectively.

Calculate $\left(\frac{\partial E}{\partial T}\right)$, ΔG^0, ΔS^0, ΔH^0 for the reaction at 25°C.

$$Cd(s) + 2AgCl(s) + aq \rightarrow CdCl_2 \cdot 2 \cdot 5H_2O(sat) + 2Ag(s)$$

16.31 The e.m.f. of the standard western cell at any temperature (°C), can be calculated from the relation

$$E = 1.0183 - 4.06 \times 10^{-5}(t - 20) - 9.5 \times 10^{-7}(t - 20)^2 + 1 \times 10^{-9}(t - 20)^3$$

where E^0 at 20°C is 1.0183 V. Calculate dE/dt. If the cell is Hg | Hg_2SO_4 | SO_4^{2-}, $CdSO_4{\cdot}8H_2O(sat)$ | Cd, calculate ΔH^0.

16.32 The voltages of certain cell at 25°C and 20°C are 0.35252 V and 0.35333 V respectively. If the number of electrons involved in the overall reaction are two, calculate ΔG^0, ΔS^0 and ΔH^0 at 25°C.

16.33 The standard e.m.f. of the cell Zn | Zn^{2+} || Fe^{3+}, Fe^{2+} | Pt at 25°C is 1.53 V and at 50°C is 1.55 V. Calculate ΔH^0_{298} for the overall reaction.

16.34 The e.m.f. of the cell Pb | $PbSO_4$ | $Na_2SO_4 10H_2O$(sat) | $HgSO_4$ | Pt is 0.9647 V at 25°C. The temperature coefficient is 1.74×10^{-4} V deg^{-1}. Calculate ΔH^0 for the reaction

$$Pb(s) + Hg_2SO_4(s) \rightarrow PbSO_4(s) + Hg(l)$$

Equilibrium Constant

16.35 (*a*) Calculate the equilibrium constant for the reaction

$$Cu(s) + Cu^{2+}(aq) \rightarrow 2Cu^+(aq)$$

Using the data given in problems 16.8.

(*b*) If excess of Cu metal is added to 0.2 M Cu^{2+} ion solution, calculate the concentration of Cu^+ ion.

16.36 Calculate the equilibrium constant at 25°C for the reaction

$$3MnO_4^{2-}(aq) + 4H^+(aq) \rightarrow MnO_2(s) + 2MnO_4^-(aq) + 2H_2O(l)$$

using the data of the problem 16. 11 *d*.

16.37 Calculate the equilibrium constant for the reaction

$$Sn^{2+}(aq) + Cl_2(g) \rightarrow Sn^{4+}(aq) + 2Cl^-(aq)$$

Given

$$Sn^{4+}(aq) + 2e^- \rightarrow Sn^{2+}(aq) \qquad \varepsilon^0 = 0.15 \text{ V}$$

$$Cl^-(aq) + e^- \rightarrow \frac{1}{2}Cl_2(g) \qquad \varepsilon^0 = 1.3595 \text{ V}$$

16.38 (*a*) Calculate the equilibrium constant for the reaction of metallic cobalt and Ni^{2+} ion. ε^0 for $Ni^{2+}/Ni = -0.25$ V and for $Co^{2+}/Co = -0.277$ V.

(*b*) The above values of ε^0 are in 1 M H^+ ions. If the pH of the solution is raised to 4, what will be the effect on the values of *K*. Assume that at pH = 4, the only species present in the solution are Ni^{2+} and Co^{2+}.

16.39 Calculate the equilibrium constant for the following reactions

(*a*) $Ni^{2+}(aq) + 2Fe^{2+}(aq) \rightarrow Ni(s) + 2Fe^{3+}(aq)$

(*b*) $Zn^{2+}(aq) + Pb(s) + SO_4^{2-}(aq) \rightarrow Zn(s) + PbSO_4(s)$

16.40 Calculate the equilibrium constant for the reaction

$$2Fe^{2+}(aq) + Au^{3+}(aq) \rightarrow 2Fe^{3+}(aq) + Au^+(aq)$$

The half-cell potentials (ε^0) for electrodes Au^+/Au, Au^{3+}/Au and Fe^{3+}, Fe^{2+}/Pt are + 1.68 V, + 1.50 V and +0.771 V respectively.

16.41 In the problem 16.40, if excess of metal gold is added to 0.4 M solution containing only Fe^{3+} (*i.e.* no Fe^{2+} or Au^{3+} ions) calculate the concentration of Fe^{2+} and Au^{3+} ions as well as the equilibrium constant for the reaction

$$3Fe^{3+}(aq) + Au(s) \rightarrow 3Fe^{2+}(aq) + Au^{3+}(aq)$$

16.42 Using the data of the problem (16.10 *c*), calculate the equilibrium constant of the reaction at 25°C.

$$2HClO_2^-(aq) \rightarrow HClO^-(aq) + ClO_3^-(aq) + H^+(aq)$$

16.43 Calculate the solubility product of the reaction

$$Fe(OH)_3(aq) \rightarrow Fe^{3+}(aq) + 3(OH^-)(aq)$$

Given

$$Fe(OH)_3(s) + 3e^- \rightarrow Fe(s) + 3OH^-(aq) \qquad \varepsilon^0 = -0.77 \text{ V}$$

$$Fe^{3+}(aq) + 3e^- \rightarrow Fe(s) \qquad \varepsilon^0 = -0.036 \text{ V}$$

16.44 The K_{sp} of $PbSO_4$ is 1.06×10^{-8} and the given data is

$$PbSO_4(s) + 2e^- \rightarrow Pb(s) + SO_4^{2-}(aq) \qquad \varepsilon^0 = -0.356 \text{ V}$$

$$Pb^{2+}(aq) + 2e^- \rightarrow Pb(s) \qquad \varepsilon^0 = -0.126 \text{ V}$$

(*a*) Calculate the e.m.f. of the cell at 25°C if molarity of sulphuric acid is 0.02 and neglecting activity coefficients in the cell

$$Pb \mid PbSO_4(s) \mid 0.02 \text{ M } H_2SO_4 \mid H_2(1 \text{ atm}) \mid Pt$$

(b) Calculate values of activity coefficient using Debye-Hückel limiting law and also the e.m.f. using activities.

16.45 The solubility product of CuS is 1.0×10^{-28}. (*a*) Calculate the standard potential of the Cu^{2+}/Cu electrode in a cell formed by the saturated CuS solution (*b*) Calculate the standard potential of S^{2-} | CuS | Cu electrode.

16.46 The following cell is given

$$Cd \mid Cd(OH)_2 \mid NaOH(0.04\ m) \mid H_2(1\ atm) \mid Pt$$

The e.m.f. and the standard e.m.f. of the cell at 50°C are 0.05 V and 0.45 V respectively. (*a*) Write anode and cathode as well as the overall reaction (*b*) Calculate the solubility product of $Cd(OH)_2$ and 25°C.

16.47 Calculate the stability constant for the complex formed in the reaction

$$Cu^{2+}(aq) + 4NH_3(aq) \rightarrow Cu(NH_3)_4^{2+}(aq)$$

Given

$$Cu^{2+}(aq) + 2e^- \rightarrow Cu(s) \qquad \varepsilon^0 = 0.337\ V$$

$$Cu(NH_3)_4^{2+}(aq) + 2e^- \rightarrow Cu(s) + 4NH_3(g) \qquad \varepsilon^0 = -0.12\ V$$

16. 48 The stability constant for the complex $[Ag(NH_3)_2]^{1+}$ is 1.61×10^7.
If Ag^+/Ag is 0.7991 V and E^0 for

$$Ag(NH_3)_2^+(aq) + e^- \rightarrow Ag(s) + 2NH_3(g) \text{ is } 0.373\ V.$$

calculate the amount of $[Ag(NH)_2]^+$ ions when excess of metal silver is added to liquid ammonia.

Additional Problems

16.49 Calculate the half cell reduction potential for the electrode reaction: $Al^{3+}(aq) + 3e^- \rightarrow Al(s)$.

Given $\Delta_f \bar{H}^0[Al^{3+}(aq)] = -531.0\ kJ\ mol^{-1}$, $\Delta_f \bar{H}^0[Al(s)] = 0\ kJ\ mol^{-1}$. $S^0[Al(s)] = 28.32\ J\ K^{-1}\ mol^{-1}$, $S^0[Al^{3+}(aq)] = -321.7\ J\ K^{-1}\ mol^{-1}$.

16.50 Repeat the problem 16.49 for the reaction: $Fe^{3+} + e^- \rightarrow Fe^{2+}$. Given: $\Delta_f \bar{H}^0[Fe^{3+}(aq)] = -48.5.0\ kJ\ mol^{-1}$, $\Delta_f \bar{H}^0[Fe^{2+}(aq)] = -89.1\ kJ\ mol^{-1}$. $S^0[Fe^{2+}(aq)] = -137.7\ J\ K^{-1}\ mol^{-1}$, $S^0[Fe^{3+}(aq)] = -315.9\ J\ K^{-1}\ mol^{-1}$.

16.51 Consider a Daniel cell operating under the standard state conditions. Suppose the cell's reaction is multiplied by 3. What will be effect of this have on each of the following quantities in the Nernst equation?

(*i*) Q, (*ii*) $\ln Q$, (*iii*) E and (*iv*) E^0

16.52 Calculate the standard electrode potential of the reaction; $Li^+(aq) + e^- \rightarrow Li(s)$. $\Delta_f \bar{H}^0[Li^+(aq)] = -278.5\ mol^{-1}$, $\Delta_f \bar{H}^0[Li(s)] = 0\ kJ\ mol^{-1}$. $S^0[Li(s)] = 28.03\ J\ K^{-1}\ mol^{-1}$, $S^0[Li^+(aq)] = 14.23\ J\ K^{-1}\ mol^{-1}$.

16.53 Using the half cell electrode data from the Table 16.1, calculate the equilibrium constant for the reaction: $Sn^{2+}(aq) + 2Ag(s) \rightleftharpoons Sn(s) + 2Ag^+(aq)$.

16.54 Calculate the standard e.m.f of the reaction: $Cu^{2+}(aq) + Fe(s) \rightarrow Cu(s) + Fe^{3+}$. Which electrode will be anode and which one will be cathode? Calculate the equilibrium constant.

16.55 The half cell reaction for an electrode is: $Cu^{2+}(aq) + 2e^- \rightarrow Cu(s)$. Calculate the number of grams of copper that can be obtained when 2 faradays of current is passed.

16.56 The equilibrium constant for the reaction: $Sr^{2+}(aq) + Mg(s) \rightleftharpoons Sr(s) + Mg^{2+}(aq)$ is 1.52×10^{-18} at 298 K. Calculate the value of E^0 for the cell made up of $Mg | Mg^{2+}$ and $Sr^{2+} | Sr$ half cell.

16.57 Calculate the value of E^0, $\Delta_r G^0$ and K for the following reactions:

(*i*) $5Fe^{2+}(aq) + MnO_4^-(aq) + 8H^+(aq) \rightleftharpoons Mn^{2+}(aq) + 4H_2O(l) + 5Fe^{3+}(aq)$

(*ii*) $Zn^{2+}(aq) + Sn^{2+}(aq) \rightleftharpoons Zn(s) + Sn^{4+}(aq)$

Given data are: $Pt | Fe^{3+}, Fe^{2+} = 0.771$ V, $Pt | O_2, H_2O = 1.229$ V, $Pt | MnO_4^-, Mn^{2+} = 1.507$ V, $Zn | Zn^{2+} = -0.762$ V, $Pt | Sn^{4+}, Sn^{2+} = 0.151$ V.

16.58 Consider the following cell:

$$Ag(s) | AgCl(s) | NaCl(aq) | Hg_2Cl_2(s) | Hg(l).$$

(*i*) Calculate the values of $\Delta_r G^0$, $\Delta_r H^0$, $\Delta_r S^0$, for the reaction at 298 K, (*ii*) write the half cell reaction, (*iii*) the standard emfs of the cell at several temperatures are given below:

T (K)	291	298	303	311
E^0 (10^{-3} V)	0.043	0.0454	0.0471	0.0501

16.59 Calculate the ratio of $[Pb^{2+}]$ to $[Sn^{2+}]$ value for the reaction at equilibrium. Use standard reduction potential of $Sn^{2+} | Sn$ and $Pb^{2+} | Pb$ from Table 16.1.

16.60 Consider the following cell $Al(s) | Al^{3+}(aq, 0.35M) || Al^{3+}(aq, 0.65) | Al(s)$. Calculate the e.m.f. of the cell.

16.61 Consider the following reactions:

$$Ag^+(aq) + e^- \rightarrow Ag(s)$$
$$AgCl(s) + e^- \rightarrow Ag(s) + Cl^-(aq)$$

Calculate the solubility product of AgCl, suing the data from Table 16.1.

16.62 Consider a very important reduction reaction

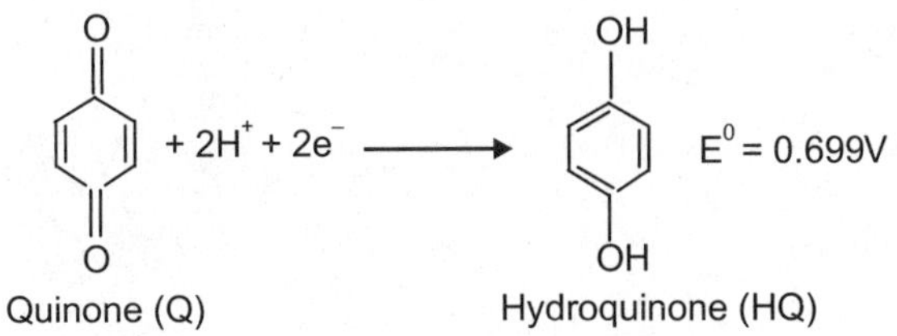

A quinhydrone electrode can be made by immersing a piece of platinum wire in a quinhydrone solution. (*i*) Derive an expression for the electrode potential of this couple in terms E^0 and the hydrogen ion concentration, (*ii*) When the quinine-hydroquinone couple is joined to a saturated calomel electrode, the emf of the cell is found to be 0.24 V. In this cell the saturated calomel electrode acts as the anode. Calculate the pH of the quinihydrone solution at 298 K.

16.63 The $\Delta_r S^0$ of the Daniell cell is 22.7 J K^{-1} mol^{-1}. Calculate the temperature coefficient $(\frac{\partial E^0}{\partial T})_p$ of the cell and the emf of the cell at 55 ^{0}C.

16.64 The standard electrode potentials of the following half cells are:

$$2H^+(aq) + 2e^- \rightarrow H_2(g)$$

$$2H_2O(l) + 2e^- \rightarrow H_2(g) + 2OH^-(aq)$$

0.0V and – 0.828 V. Calculate the value of K_w, *i.e.* $[H^-][OH^-]$ at 298.

16.65 Consider the following disproportionation reaction and calculate the values of $\Delta_r G^0$ and K:

$$Hg_2^{2+}(aq) \rightarrow Hg^{2+}(aq) + Hg(l)$$

Given: $2Hg^{2+}(aq) + 2e^- \rightarrow Hg_2^{2+}(aq) \qquad E_0 = 0.920\ V$

$Hg_2^{2+}(aq) + 2e^- \rightarrow Hg(l) \qquad E_0 = 0.797\ V$

16.66 Consider the following disproportionation reaction and calculate the values of $\Delta_r G^0$ and K:

$$3Au^+(aq) \rightarrow Au^{3+} + 2Au(s)$$

Given: $Au^+(aq) + e^- \rightarrow Au(s) \qquad E_0 = 1.69\ V$

$Au^{3+}(aq) + 3e^- \rightarrow Au(s) \qquad E_0 = 1.498\ V$

CHAPTER 17

IDEAL DILUTE SOLUTIONS

17.1. INTRODUCTION : CONCENTRATION UNITS

In any quantitative study it is necessary to know the amounts of solute present in a given amount of solvent or concentration of the solution. Although chemists use different concentration units we will discuss only four in the coming sections.

17.1.1. *Percent by Weight*

The percent by weight (also called percent by mass) of a solute in a solution is defined as;

$$\text{Percent by weight} = \frac{\text{Weight of solute}}{\text{Weight of solute + Weight of solvent}} \times 100$$

$$\text{Percent by weight} = \frac{\text{Weight of solute}}{\text{Weight of solution}} \times 100 \qquad ...(17.1)$$

17.1.2. *Mole Fraction*

Mole fraction of any component *i* in a solution is defined as:

$$\chi_i = \frac{\text{Number of moles of component } i}{\text{Number of moles of all the components}}$$

$$\chi_i = \frac{n_i}{\sum_i n_i} \qquad ...(17.2)$$

where n_i is the number of moles of component *i* and summation is the number of moles over all the components, χ_i is the mole fraction of component *i*.

17.1.3. *Molarity (M)*

Molarity is defined as the number of moles of solute dissolved in 1 liter of solution, *i.e.*

$$\text{Molarity (M)} = \frac{\text{Number of moles of solute}}{L \text{ solution}} \qquad ...(17.3)$$

The units of molarity, thus, are moles per liter (mol L^{-1}). By convention we use square brackets [ ] to represent concentration.

17.1.4. *Molality* (m)

Molality is defined as the number of moles of solute dissolved in 1 kg (1000 g) of solvent, *i.e.*

$$\text{Molality (m)} = \frac{\text{Number of moles of solute}}{\text{Weight of solvent in kg}} \qquad ...(17.4)$$

The units of molality, thus, are moles per kg of solvent (mol kg^{-1}).

Advantage of using weight percent is that we do not know the molar mass of the solute. This unit is mostly used by biochemists, who frequently work with macromolecules either unknown molar mass or of unknown purity. For example, the common unit for protein or DNA solution is mg m L^{-1} or mg per milliliter. Further, this unit is independent of temperature. Mole fraction is useful for calculating the partial pressures of gases and in the study of partial pressures of solutions. Molarity is the most commonly used units and it is easy to measure the volume of the solvents than weighing the solvents. The main drawback is that this unit is dependent on temperature and it is difficult to know the exact volume of solvent present in the solution. Molality is independent of temperature is used in the studies which involve the changes in temperature, *e.g.* colligative properties of solutions.

***Exercise* 17.1:** Calculate the amount of water that must be added to prepare to 40 g of glucose to prepare the 7.5% aqueous glucose solution by weight.

Using equation (17.1), and substituting the values of different parameters, we get

$$7.5 = \frac{40 \text{ g}}{40 \text{ g} + \text{Weight of solvent}} \times 100$$

or $\quad 7.5 \times 40 + 7.5 \times \text{Weight of solvent} = 4000$

or $$\text{weight of solvent} = \frac{(4000 - 300) \text{ g}}{7.5} = 493.3 \text{ g.}$$

***Exercise* 17.2:** The commercially available H_2SO_4 is 98% by weight. If the density of H_2SO_4 is 1.83 g cm^{-3}, calculate the molarity and molality of the concentrated H_2SO_4.

Molar mass of $H_2SO_4 = 2 \times 1 + 1 \times 32 + 4 \times 16 = 98$ g mol^{-1}.

98% H_2SO_4 means that it contains 98 g of H_2SO_4 and 2 g of H_2O.

$$\text{No. of moles } H_2SO_4 = \frac{\text{Mass of } H_2SO_4}{\text{Molar of mass of } H_2SO_4} = \frac{98 \text{ g}}{98 \text{ g mol}^{-1}} = 1 \text{ mol}^{-1}$$

$$m\,(H_2SO_4) = \frac{\text{No. of moles of } H_2SO_4}{\text{Mass of } H_2O} \times 1000 \text{ g}$$

$$= \frac{1 \text{ mol of } H_2SO_4}{2 \text{ g of } H_2O} \times 1000 \text{ g} = 500 \text{ mol kg}^{-1}$$

$$\text{Volume of solution} = \frac{\text{Mass of solution}}{\text{Density of solution}} = \frac{100 \text{ g of solution}}{1.83 \text{ g cm}^{-3}} = 54.64 \text{ mL}$$

$$\text{M }(H_2SO_4) = \frac{\text{No. of moles of } H_2SO_4}{\text{Volume of solution}} \times 1000 \text{ mL}$$

$$= \frac{1 \text{ mol of } H_2SO_4}{54.46 \text{ mL}} \times 1000 \text{ mL} = 18.3 \text{ M.}$$

***Exercise* 17.3:** Calculate the molality of the 2.75 M solution of methanol if the density of the solution is 0.851 g cm^{-3}.

2.75 M solution means that it contains 2.75 moles of methanol 1 L solution, *i.e.* it contains = (2.75 mol) (32 g mol^{-1}) = 88 g

$$\text{Mass of 1 L solution} = \frac{1000 \text{ mL}}{0.851 \text{ gm L}^{-1}} = 1175.88 \text{ g}$$

Mass of solvent in 1 L solution = 1175.88 – 88.00 = 1087.88 g

$$m\ (CH_3OH) = \frac{\text{No. of moles of } CH_3H}{\text{mass of } H_2O} \times 1000 \text{ g}$$

$$= \frac{2.75 \text{ mol of } CH_3OH}{1087.88 \text{ g of } CH_3OH} \times 1000 \text{ g} = 2.53 \text{ m.}$$

17.2. IDEAL SOLUTIONS

17.2.1. *Definition*

A solution is said to be an ideal if it obeys Raoult's law throughout the composition range of the system.

Raoult's law in its more generalised form is defined as the fugacity of any component in solution being equal to the product of its fugacity in its pure state at the same temperature and pressure and its mole fraction in the solution *i.e.,*

$$f_i = x_i f_i^* \qquad ...(17.5)$$

whereas the relation between the partial pressure and its composition in solution is the approximation in case of solutions having components of small partial pressure

$$p_i = x_i p_i^0 \qquad ...(17.6)$$

The chemical potential of any component in solution is defined as

$$\mu_i = \mu_i^0 + RT \ln x_i \qquad ...(17.7)$$

where μ_i^0 is the chemical potential of the pure component i at the same pressure and temperature of the solution.

17.2.2. *Consequences of the Definition*

In case of ideal solutions, the volume of mixing and heat of mixing are zero *i.e.,*

$$\Delta V_{mix} = 0 \qquad ...(17.8)$$

$$\Delta H_{mix} = 0 \quad ...(17.9)$$

Similarly, the free energy and entropy of mixing is given as

$$\Delta S_{mix} = -nR\,\Sigma\, x_i \ln x_i \quad ...(17.10)$$

$$\Delta G_{mix} = nRT\,\Sigma\, x_i \ln x_i \quad ...(17.11)$$

Note: If a solution follows Raoult's law consequences are that heat and volume of mixing are zero but the reverse is not true *i.e.,* if heat of mixing or volume of mixing is zero, the solution may or may not follow Raoult's law.

17.3. VAPOUR PRESSURE OF MIXTURE OF VOLATILE COMPONENTS

Raoult's law relates the vapour pressure of a component to its composition (mole fraction) in the solution *i.e.,*

$$p_i = x_i p_i^0 \quad ...(17.6)$$

Total vapour pressure of the volatile components of solutions is thus

$$P = p_1 + p_2 + p_3 \ldots \quad ...(17.12)$$

In case of *binary mixture,* the total vapour pressure can be expressed in terms of a mole fraction of one of the components in the solution as

$$P = p_2^0 + (p_1^0 - p_2^0)x_1 \quad ...(17.13)$$

Similarly, the total vapour pressure can also be expressed in terms of the composition of the vapours of the components *i.e.,*

$$P = \frac{p_1^0 - p_2^0}{p_1^0 + (p_2^0 - p_1^0)y_1} \quad ...(17.14)$$

or

$$y_1 = \frac{p_1^0 x_1}{p_2^0(1 - x_1) + p_1^0 x_1} \quad ...(17.15)$$

Thus the variation of total vapour pressure with the change in composition of the liquid phase and vapour phase can be plotted and shown in the Fig. 17.1. where as vapour pressure of individual components and total vapour pressure is plotted in Fig. 17.2.

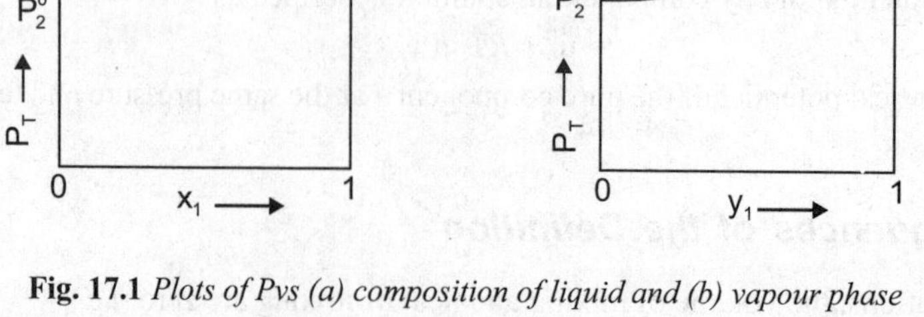

Fig. 17.1 *Plots of P vs (a) composition of liquid and (b) vapour phase*

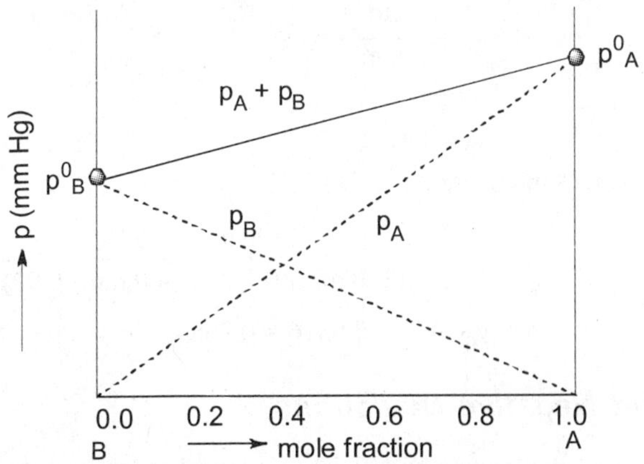

Fig. 17.2 *Total vapour pressure of liquids A and B as a function of the mole fraction of liquid A. Dotted lines represents the partial pressures of the components A and B*

***Exercise* 17.4:** 40 g of liquid A and 70 g of liquid B are mixed together. If two form ideal solution, calculate the composition of the vapours. p_A^0 = 200 mm, p_B^0 = 700 mm, M_A = 56 g mol^{-1}, M_B = 90 g mol^{-1}.

Make use of the equations (17.13) and (17.15)

$$n_A = \frac{40 \text{ g}}{56 \text{ g mol}^{-1}} = 0.714 \text{ mol}$$

$$n_B = \frac{70 \text{ g}}{90 \text{ g mol}^{-1}} = 0.778 \text{ mol}$$

$$x_A = \frac{0.714 \text{ mol}}{1.492 \text{ mol}} = 0.479$$

$$x_B = \frac{0.778 \text{ mol}}{1.492 \text{ mol}} = 0.521$$

$$y_A = \frac{(250 \text{ mm})(0.479)}{(700 \text{ mm})(0.52) + (200 \text{ mm})(0.479)} = 0.208$$

$$y_B = 1 - 0.208 = 0.792$$

Note: This can also be solved by first finding the vapour pressure of each component and then mole fraction of vapours.

***Exercise* 17.5:** If the vapour pressure of pure liquids A and B are 300 mm and 800 mm at 75°C, calculate the composition of the mixture such that it boils at 75°C. Also find the composition of vapour phase.

Let x_A be the mole fraction of liquid A in the solution which boils at 75°C. Since the total vapour pressure should be 760 mm at the boiling point so,

$$760 = (x_A)(300 \text{ mm}) + (1 - x_A)(800 \text{ mm})$$

$$760 = 300\,x_A + 800 - 800\,x_A$$

$$500\,x_A = 40$$

$$x_A = \frac{40}{500} = 0.08$$

and

$$x_B = 0.92$$

Substitute respective values in equation (17.15)

$$y_A = \frac{(300 \text{ mm})(0.08)}{(300 \text{ mm})(0.08) + (800 \text{ mm})(0.92)} = 0.0316$$

$$y_B = 1 - 0.0316 = 0.9684.$$

17.3.1. *Positive and Negative Deviations*

Most of the solutions do not obey equations (17.5), (17.8) and (17.9) and thus are called non-ideal. These solutions can have positive deviations or negative deviations. In case of *positive deviations*, the vapour pressure of the solution is greater than the sum of the vapour pressures of the components forming ideal solutions. This is because in this case the intermolecular forces between unlike molecules are weaker than those between the like molecules. Thus there is a greater tendency for these molecules to leave the solution than in case of ideal solution. In case of negative deviation, it is reverse, *i.e.* the vapour pressure of the solution is less than that of the ideal solution obtained from the same components. The positive deviations are depicted in Fig. 17.2.

TABLE 17.1. Henry's constants for some of the gases in water

Gas	K (torr)	K' (atm. mol^{-1} kg H_2O)
H_2	5.54×10^7	1311
He	1.12×10^7	2649
Ar	2.80×10^7	662
N_2	6.80×10^7	1610
O_2	3.27×10^7	773
CO_2	1.24×10^6	29.3
H_2S	4.27×10^5	10.1

Deviations from Henry's law can occur: (*i*) The Henry's law holds good only for dilute solutions, (*ii*) If the dissolved gas interacts chemically with the solvent and thus the solubility of gases can be enhanced. The examples can be CO_2, NH_3, H_2S, HCl etc, (*iii*) Dissolution of gases and interacting, for example, O_2 is only sparingly soluble in water but its solubility increases dramatically if the solutions contain haemoglobin or myoglobin. Henry's constants in different units for few gases are given in table 17.1.

Decrease in vapour pressure of solution is not due to modification of intermolecular forces but it is because of entropy effect. When a solvent evaporates the entropy of the universe increases because the entropy of the substance in gaseous state is greater than that in the liquid state at the same temperature. The entropy of the solution itself is accompanied with the increase in the entropy because the solution has greater entropy than the pure solvent. Therefore, the solution has a smaller driving force for evaporation. In other words, evaporation from a solution will result in a smaller increase in entropy than the case for a pure solvent. Consequently, the solvent has a less tendency to leave the solution and the solution will have a lower vapour pressure than that of pure solvent.

17.4. GIBB'S-DUHEM EQUATION

If the temperature and pressure are kept constant the chemical potentials of the components of the solution do not vary independently with the change in composition of the solution. They are related by the equation

$$\Sigma n_i d\mu_i = 0 \qquad ...(17.16)$$

***Exercise* 17.6:** Prove the Gibb's-Duhem equation.

At constant temperature and pressure the free energy of the mixture is given by the equation (17.17)

$$G = \Sigma n_i \mu_i \qquad ...(17.17)$$

Differentiating

$$dG = \Sigma (n_i d\mu_i + \mu_i dn_i) \qquad ...(17.17a)$$

The fundamental equation under these conditions is

$$dG = \sum_i \mu_i dn_i \qquad ...(17.17b)$$

Subtracting (17.17*b*) from (17.17*a*)

$$\sum_i n_i d\mu_i = 0$$

***Exercise* 17.7:** If a solute follows Henry's law, then the solvent must follow Raoult's law.

Henry's law states

$$f_2 = K_2 x_2 \qquad ...(17.18)$$

$$\ln f_2 = \ln K_2 + \ln x_2$$

and

$$\frac{d \ln f_2}{dx_2} = \frac{1}{x_2} \qquad ...(17.19)$$

Gibb's-Duhem equation in case of binary mixture is

$$n_1 d\mu_1 + n_2 d\mu_2 = 0 \qquad ...(17.16)$$

The chemical potential of any component is

$$\mu_i = \mu_i^0 + RT \ln f_i$$

At constant temperature

$$d\mu_i = RT\, d \ln f_i$$

Substituting in equation (17.16)

$$n_1 RT\, d \ln f_1 + n_2\, RT\, d \ln f_2 = 0$$

Dividing it by $(n_1 + n_2)$

$$x_1 d \ln f_1 + x_2 d \ln f_2 = 0$$

or

$$x_1 \frac{d \ln f_1}{dx_1} + x_2 \frac{d \ln f_2}{dx_1} = 0 \qquad ...(17.20)$$

Now for binary mixture

$$x_1 + x_2 = 1$$

or

$$dx_1 = -dx_2 \qquad ...(17.21)$$

Combining equations (17.21) and (17.20)

$$x_1 \frac{d \ln f_1}{dx_1} = x_2 \frac{d \ln f_2}{dx_2}$$

Substituting the value of $\frac{d \ln f_2}{dx_2}$ from (17.19)

$$x_1 \frac{d \ln f_1}{dx_1} = x_2 \cdot \frac{1}{x_2} = 1$$

or

$$d \ln f_1 = \frac{dx_1}{x_1}$$

or

$$\ln f_1 = \ln C + \ln x_1$$

Removing logarithms

$$f_1 = C x_1$$

Now when $x_1 = 1, f_1 = f_1^0$ and $C = f_1^0$

Thus

$$f_1 = f_1^0 x_1$$

and this is the definition of Raoult's law.

17.5. IDEAL DILUTE SOLUTIONS

17.5.1. *Henry's Law*

At a fixed temperature, the general form of Henry's law is

$$f_2 = K_2 x_2 \qquad ...(17.18)$$

where f_2 is the fugacity of solute, x_2 is its mole fraction in solution and K_2 is the Henry's constant. The above equation can also be written for dilute solutions as

$$f_2 = K_2' \frac{n_1}{n_2}$$

$$= K_2'' m_2 \qquad ...(17.22)$$

and m_2 is the molarity of the solute.

The original form of Henry's law states that at constant temperature the solubility of gas in a given quantity of solvent is directly proportional to its partial pressure (p_2) above the solution *i.e.*,

$$p_2 = K_2 x_2 \qquad ...(17.23)$$

The chemical potential of the solute, then, can be expressed as

$$\mu_2 = \mu_2^0 + RT \ln x_2 \qquad ...(17.24)$$

where μ_2^0 is chemical potential of the solute when its mole fraction is unity if the Henry's law is obeyed over the whole range. If not, then it is a hypothetical state obtained by the extension of Henry's law, shown in Fig. 17.3.

For dilute solution the equation (17.24) can be written as

$$\mu_2 = \mu_2^0 + RT \ln m_2 \qquad ...(17.25)$$

where m_2 is molality and μ_2^0 is the chemical potential at $m_2 = 1$ if the equation (17.25) is valid at unit molality. Generally this state is also obtained as a hypothetical state as an extension of Henry's law up to 1 molal solution.

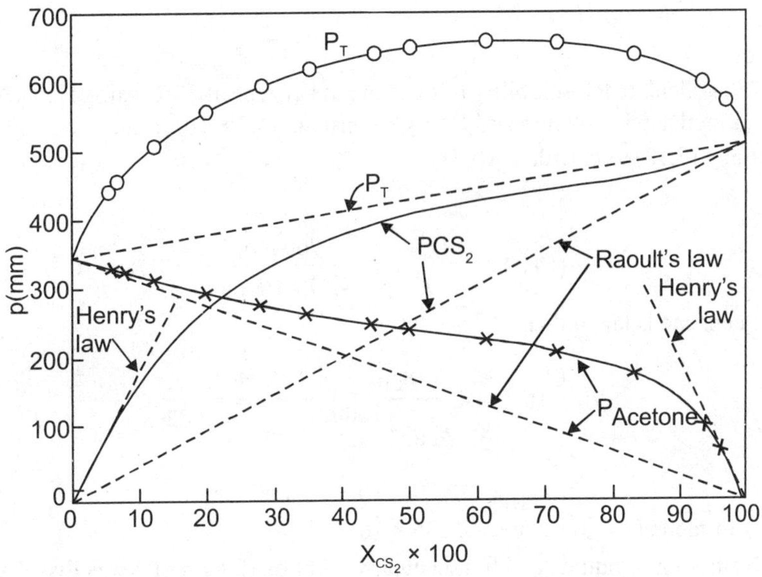

Fig. 17.3 *Plot of partial pressure of CS_2, acetone and P_T vs mole fraction of CS_2*

***Exercise* 17.8:** If we assume that air contains 80% N_2 and 20% O_2, calculate the proportion of N_2 and O_2 in water. Henry's constant of N_2 and O_2 in H_2O are 6.51×10^7 mm and 3.30×10^7 mm respectively.

If the pressure of air over water is 760 mm, the partial pressures of N_2 and O_2 will be

$$p_{N_2} = \frac{760 \times 80}{100} = 608 \text{ mm}$$

$$p_{O_2} = \frac{760 \times 20}{100} = 152 \text{ mm}$$

Applying Henry's law

$$p_{N_2} = K_{N_2} x_{N_2}$$

$$x_{N_2} = \frac{p_{N_2}}{K_{N_2}} = \frac{608 \text{ mm}}{6.51 \times 10^7 \text{ mm}} = 9.34 \times 10^{-6}$$

$$x_{O_2} = \frac{p_{O_2}}{K_{O_2}} = \frac{152 \text{ mm}}{3.30 \times 10^7 \text{ mm}} = 4.61 \times 10^{-6}$$

i.e., Mole fractions of N_2 and O_2 in water are 9.34×10^{-6} and 4.61×10^{-6} respectively. Relative proportion of N_2 and O_2 in water are:

$$\frac{9.34 \times 10^{-6}}{(9.34 + 4.61) \times 10^{-6}} = 0.67$$

and

$$\frac{4.61 \times 10^{-6}}{(9.34 + 4.61) \times 10^{-6}} = 0.33$$

i.e., in the ratio of 2 : 1.

***Exercise* 17.9:** Calculate the solubility of methane in benzene at 25°C using Henry's law when the partial pressure of methane is 760 mm and Henry's constant 4.27×10^{5} mm.

Using equation (17.18) to calculate $x(CH_4)$

i.e.,

$$x(CH_4) = \frac{p_{CH_4}}{K} = \frac{760 \text{ mm}}{4.27 \times 10^5 \text{ mm}} = 1.78 \times 10^{-3}$$

If 1000 g of benzene is taken then

$$1.78 \times 10^{-3} = \frac{n_{CH_4}}{n_{CH_4} + \frac{1000}{78}} = \frac{n_{CH_4}}{n_{CH_4} + 12.82}$$

So $$n_{CH_4} = 2.29 \times 10^{-2} = \text{molality}$$

The molality of methane in benzene = 2.286×10^{-2}

***Exercise* 17.10:** 80 g of liquid A with molecular weight of 120 g mol^{-1} was dissolved in 1000 g of liquid B (molecular weight 160 g mol^{-1}). The vapour pressure of liquid B of this solution is 400 torr. Calculate Henry's constant, vapour pressure of liquid A (pure and in the solution) if the total vapour pressure is 475 torr.

The above problem can be solved as

$$n_A = \frac{80}{120 \text{ g mol}^{-1}} = 0.667 \text{ mol}$$

$$n_B = \frac{1000 \text{ g}}{160 \text{ g mol}^{-1}} = 6.250 \text{ mol}$$

$$x_A = \frac{0.667 \text{ mol}}{6.917 \text{ mol}} = 0.096$$

$$x_B = \frac{6.25 \text{ mol}}{6.917 \text{ mol}} = 0.904$$

Now

$$p_B = x_B p_B^0$$

i.e.,

$$p_B^0 = \frac{p_B}{x_B} = \frac{400 \text{ mm}}{0.904} = 442.48 \text{ mm}$$

Now

$$P_T = x_A p_A^0 + x_B p_B^0$$

$$475 = 0.096 p_A^0 + 400$$

$$p_A^0 = \frac{(475 - 400) \text{ torr}}{0.096} = 781 \text{ mm}$$

The partial pressure of liquid A above the solution is 75 mm. So using equation (17.18)

$$K = \frac{p}{x_A} = \frac{(475 - 400) \text{ mm}}{0.096} = 781 \text{ mm}$$

***Exercise* 17.11:** Vapour pressure of CS_2 and acetone is given as a function x_1 (CS_2) at 35.17°C

x_1	p_1 (mm)	p_2 (mm)	P_T	x_1	p_1 (mm)	p_2 (mm)	P_T
0	0	343.8	343.8	49.74	404.1	242.1	646.2
6.24	110.7	331.0	441.7	61.24	426.9	227.0	653.9
7.11	123.1	328.7	451.8	67.13	438.0	217.0	655.0
12.12	191.7	313.5	505.2	71.97	447.5	207.1	654.6
19.91	271.9	290.6	562.5	82.80	464.9	180.2	645.1
27.61	323.3	275.2	598.5	93.50	491.9	109.2	601.1
35.02	358.3	263.9	622.2	96.20	500.8	73.4	574.2
44.74	390.4	250.2	640.6	100.00	512.3	0	512.3

(*i*) Plot x_1 $(CS)_2$ vs p_1 (CS_2), p_2 (acetone) and P_T. (*ii*) Indicate whether there is +ve or –ve deviation from Raoult's law. (*iii*) Calculate Henry's constant.

(*i*) The above data has been plotted in Fig. 17.3.

(*ii*) There is a positive deviation from Raoult's law because vapour pressure of the system is greater than that predicted by Raoult's law.

(*iii*) Using equation (17.23)

$$K\,(CS)_2 = \frac{280 \text{ mm}}{0.15} = 1.87 \times 10^3 \text{ mm}$$

$$K\,(\text{Acetone}) = \frac{100 \text{ mm}}{0.15} = 2.0 \times 10^3 \text{ mm}$$

17.5.2. Nernst's Distribution Law

If a solute is dissolved in two immiscible solvents, the ratio of the mole fractions of solute in two solvents at equilibrium in two solvents is a constant K, *i.e.*,

$$\frac{x_2'}{x_2} = \frac{K_2}{K_2'} = K \quad ...(17.26)$$

where K_2, K_2' are Henry's constants in two solvents, x_2 and x_2' the respective mole fractions of solute in two solvents. The above relation can also be expressed in terms of concentration unit *e.g.* in molality unit.

$$\frac{m_2'}{m_2} = \frac{K_2}{K_2'} = K \quad ...(17.27)$$

K and K' are two different constants.

***Exercise* 17.12:** To 0.100 dm^3 of water containing 0.4 mol dm^{-3} SO_2, 0.100 dm^3 of $CHCl_3$ was added. After the equilibrium is reached, calculate the amount of SO_2 present in two solvents which were completely immiscible and the Nernst constant

$$\frac{C_{H_2O}}{C_{CHCl_3}} = 0.98$$

The amount of SO_2 present in 100 mL of H_2O

$$= (0.4 \text{ mol dm}^{-3})(0.1 \text{ dm}^3)(64 \text{ g mol}^{-1})$$

$$= 2.56 \text{ g}$$

Let x g be present in H_2O and $(2.56 - x)$ g be present in $CHCl_3$ after equilibrium is reacted. Concentration of SO_2 in

$$H_2O = \left(\frac{x \text{ g}}{64 \text{ g mol}^{-1}}\right)\left(\frac{1000 \text{ mL dm}^{-3}}{100 \text{ mL}}\right)$$

$$= \frac{x}{6.4} \text{ mol dm}^{-3}$$

Concentration of SO_2 in $CHCl_3$ $= \left(\frac{2.56 - x}{64 \text{ g mol}^{-1}}\right)\left(\frac{1000 \text{ mL dm}^{-3}}{100 \text{ mL}}\right)$

$$= \frac{2.56 - x}{6.4} \text{ mol dm}^{-3}$$

According to Nernst's law

$$\frac{C_{H_2O}}{C_{HCl_3}} = 0.98$$

i.e., $$\frac{x}{6.4} \times \frac{6.4}{2.56 - x} = 0.98$$

$$x = \frac{2.509}{1.98} = 1.267 \text{ g}$$

i.e., amount of SO_2 in 100 dm^{-3} of water is 1.267 g and in $CHCl_3$ 1.293 g.

***Exercise* 17.13:** If C_1 and C_2 are the concentrations (mol dm^{-3}) of the solute x in two immiscible solvents and solute exists as a polymer in solvent II, show that

$$\frac{C_1}{n\sqrt{C_2}} = \text{Constant}$$

The Nernst distribution law is applicable only to the species which have the same form in both the solvents. Now let the equilibrium in solvent II be

$$nx \rightleftharpoons x_n \qquad ...(17.28)$$

If α_a is the degree of association, the concentration of associated form will be $\alpha_a C_2/n$ and that of monomer is $(1-\alpha_a) C_2$. The association constant is defined as

$$K_a = \frac{\alpha_a C_2}{n(1-\alpha_a)^n C_2^n} \qquad ...(17.29)$$

or

$$(1-\alpha_a) C_2 = \sqrt[n]{\alpha_a C_2 / K_a n} \qquad ...(17.30)$$

So the Nernst's distribution law becomes

$$\frac{C_1}{(1-\alpha_a)C_2} = \frac{C_1}{\sqrt[n]{\alpha_a C_2 / K_a n}} = K \qquad ...(17.31)$$

If the compound x is completely present as polymer in solvent II, α_a will be unity and the equation (17.31) becomes

$$\frac{C_1}{\sqrt[n]{C_2}} = \text{Constant} \qquad ...(17.32)$$

because n, K_a and K are all constants.

17.6 ▌COLLIGATIVE PROPERTIES

Colligative properties are those properties which only depend upon the number of molecules (or species) of solute relative to total number of molecules present in the solution but not on the nature of the solute particles *i.e.* on the size or molar mass of the molecules. Two broad assumptions made are: (*i*) The solutions are dilute so that the solvent obeys Raoult's law and (*ii*) the solutions contain nonelectrolytes. Figure 17.4 represents the plot of chemical potential as a function of temperature of pure solid, liquid and gaseous phases as well as that of liquid solution. Intersection of respective curves define the melting and boiling points. These are:

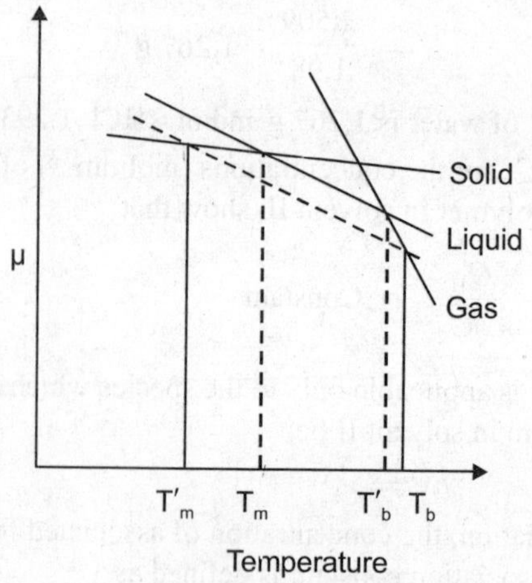

Fig. 17.4 *Plot of chemical potential versus temperature to illustrate colligative properties. The dashed line denotes the solution phase.* T'_m *and* T_m *stands for the melting points of solution and pure solvents, whereas* T'_b *and* T_b *boiling points of solution and pure solvent.*

17.6.1. *Relative Lowering of Vapour Pressure*

The vapour pressure of solvent is lowered when any solute is dissolved in it and expression is

$$p_1 = p_1^0 x_1$$

$$= p_1^0(1 - x_2) \qquad (x_1 + x_2 = 1)$$

or

$$\frac{p_1^0 - p_1}{p_1^0} = x_2 \qquad \text{...(17.33)}$$

where x_1 and x_2 are the mole fractions of solvent and solute.

***Exercise* 17.14:** The vapour pressure of ethanol at 20°C is 44.5 mm. When 15 g of nonvolatile compound A is dissolved in 500 g of ethanol, the vapour pressure is 43.52 mm. Calculate the molecular weight of compound A.

Since the solution is dilute so the Raoult's law will be applicable and equation (17.33) can be used.

$$x_2 = \frac{(44.50 - 43.52)\text{ mm}}{44.50\text{ mm}} = 0.022$$

Now

$$x_2 = \frac{m_2 / M_2}{(m_1 / M_1) + (m_2 / M_2)}$$

$$\frac{m_1}{M_1} = \frac{500\text{ g}}{46\text{ g mol}^{-1}} = 10.87\text{ mol}^{-1}$$

So

$$0.022 = \frac{15 / M_2}{10.87 + 15 / M_2}$$

or

$$(0.022)(10.87) + (0.022)\left(\frac{15}{M_2}\right) = \frac{15}{M_2}$$

$$M_2 = \frac{15 \times 0.978}{0.239} = 61.34 \text{ g mol}^{-1}.$$

***Exercise* 17.15:** 15 g of glucose is added to 250 g of water at 20°C. If the vapour pressure of water at 20°C is 17.535 mm, calculate the relative lowering of vapour pressure of water at above temperature.

Since the solution is dilute, the Raoult's law is applicable

$$n_2 = \frac{15 \text{ g}}{180 \text{ g mol}^{-1}} = 0.083 \text{ mol}$$

$$n_1 = \frac{250 \text{ g}}{18 \text{ g mol}^{-1}} = 13.889 \text{ mol}$$

$$x_2 = \frac{0.083}{13.889 + 0.083} = 5.94 \times 10^{-3}$$

Using equation (17.33) $\quad p_1^0 - p_1 = (5.94 \times 10^{-3})(17.535 \text{ mm}) = 0.104 \text{ mm}.$

17.6.2. *Depression in Freezing Point*

When a nonvolatile solute is dissolved in a solvent, the freezing point of solvent decreases. The decrease ΔT_f is given as (17.34) if the ΔT_f is not very large and the solution is ideal.

$$\Delta T_f = \frac{RT_f^2 \ln x_2}{\Delta \bar{H}_{\text{fusion}}} \qquad \text{...(17.34)}$$

$\Delta \bar{H}_{\text{fusion}}$ is the molar heat of fusion of solvent, x_2 is the mole fraction of solute and T_f is the true freezing point. For very dilute solution and behaving ideally, the above expression reduce to

$$\Delta T_f = K_f m \qquad \text{...(17.35)}$$

where K_f is molal freezing point depression constant and expressed as

$$K_f = \frac{MRT^2}{\Delta \bar{H}_{\text{fus}} 1000} \qquad \text{...(17.36)}$$

where M is the molecular weight of solvent and m is the molality of solute.

With the help of depression in freezing point, the quantities like molecular weight of solute, activity and activity coefficients, dissociation constants of weak electrolytes and van't Hoff factor can be determined.

***Exercise* 17.16:** If heat of fusion of water is 333.86 J g^{-1} at 0°C (at its freezing point), calculate the molal freezing point constant.

Make use of the equation (17.36)

$$T = 0 + 273 = 273 \text{ K}$$

$$\Delta \bar{H}_{\text{fus}} = 333.86 \times 18 \text{ J mol}^{-1}$$

So

$$K_f = \frac{(18 \text{ g mol}^{-1})(8.314 \text{ J K}^{-1}\text{mol}^{-1})(273 \text{ K})^2}{(333.86 \times 18 \text{ J mol}^{-1})(1000 \text{g kg}^{-1})}$$

$$= 1.856 \text{ K kg mol}^{-1}$$

***Exercise* 17.17:** 4 g of nonvolatile solute is dissolved in 250 g of water. The depression in the freezing point of water observed is 0.087. Calculate the molecular weight of the solute. K_f= 1.856 deg kg mol^{-1}.

Assuming that solution is in equilibrium only with ice, the equation (17.35) is valid

$$\Delta T_f = K_f m_2$$

So $$m_2 = \frac{0.087\,\text{deg}}{1.856\,\text{deg kg mol}^{-1}} = 0.047\ \text{mol kg}^{-1}$$

Now $$m_2 = \frac{w_2 / M_2}{w_1} \times (1000\ \text{g kg}^{-1})$$

So $$M_2 = \frac{w_2 \times 1000}{w_1 \times m_2} = \frac{(4.05\ \text{g})(1000\,g\ \text{kg}^{-1})}{(250\ \text{g})(0.047\ \text{mol kg}^{-1})}$$

$$= 344.69\ \text{g mol}^{-1}$$

where w_2 is the amount of solute, w_1 amount of solvent.

Molal freezing point-depression constants and molal *f* boiling point-elevation constants are compiled in Table 17.2.

TABLE 17.2. Molal freezing point-depression constants and molal boiling point elevation constants of some common solvents

Solvent	K_f (K mol^{-1} kg)	K_b (K mol^{-1} kg)
H_2O	1.86	0.512
C_2H_5OH	–	1.22
C_6H_6	5.12	2.53
$CHCl_3$	–	3.63
CH_3COOH	3.90	2.93
CCl_4	–	5.03

17.6.3. *Elevation of Boiling Point*

The normal boiling point of a solution is the temperature at which its vapour pressure is equal to the external pressure (equal to 1 atm.). Lowering of vapour pressure of solution with addition of nonvolatile solute will lead to increase of boiling point. Similar to depression in freezing point, if nonvolatile solute is added to a solvent, its boiling point is increased and the increase is given by

$$\Delta T_b = -\frac{RT_b^2 \ln x}{\Delta \bar{H}_{\text{vap}}} \qquad \text{...(17.37)}$$

where T_b is the normal boiling point and increase is not very large. The above relation for dilute solution is

$$\Delta T_b = K_b m_2 \qquad \text{...(17.38)}$$

where K_b is the molal boiling point elevation constant and is given as

$$K_b = \frac{M\ RT_b^2}{1000\ \Delta \bar{H}_{\text{vap}}} \qquad \text{...(17.39)}$$

The assumptions involved in deriving equations (17.35) or (17.38) are:

(*i*) Only the solvent is transformed into solid form or changed into vapours.

(*ii*) The solution is dilute so Raoult's law is applicable and other approximations of dilute solutions are valid *i.e.*,

$$x_2 = \frac{n_2}{n_1 + n_2} = \frac{n_2}{n_1} = m_2 \text{ if } n_1 = \frac{1000}{M_1}$$

***Exercise* 17.18:** 5 g of nonvolatile compound was dissolved in 250 g of water. If the molecular weight of the compound is 140 g mol^{-1}, calculate the elevation in the boiling point of water using the equations (17.37) and (17.38) $\Delta \bar{H}_{\text{vap}} = 40.66$ kJ mol^{-1}

$$n_1 = \frac{250 \text{ g}}{18 \text{ g mol}^{-1}} = 13.889 \text{ mol}$$

$$n_2 = \frac{5 \text{ g}}{140 \text{ g mol}^{-1}} = 0.036 \text{ mol}$$

$$x_1 = \frac{(13.889) \text{ mol}}{(13.889 + 0.036) \text{ mol}} = 0.9974$$

Using equation (17.37)

$$\Delta T_b = -\frac{(8.314 \text{ J K}^{-1} \text{ mol}^{-1})(373 \text{ K})^2}{40660 \text{ J mol}^{-1}} \ln 0.9974$$

$$= 0.0741 \text{ degree}$$

$$m_2 = \frac{5 \text{ g}}{140 \text{ g mol}^{-1}} \times \frac{1000 \text{ g kg}^{-1}}{250 \text{ g}} = 0.143 \text{ mol kg}^{-1}$$

Using equation (17.38)

$$\Delta T_b = (0.514 \text{ K kg mol}^{-1})(0.143 \text{ mol kg}^{-1}) = 0.0734 \text{ K}$$

The difference is small *i.e.*, 0.0007 degree but at the same time solution is also very dilute. The difference becomes large if concentration increases (see Problem 17.54).

***Exercise* 17.19:** 1.51 g of NaCl was dissolved in 500 g of water and the elevation in boiling point observed is 0.0514 deg. Calculate the van't Hoff factor. $K_b = 0.514$ deg kg mol^{-1}.

If the solution is dilute and the compound dissociates, the equation (17.38) modified to

$$\Delta T_b = vK_b m = m' K_b \qquad \text{...(17.40)}$$

where v is number of species to which an undissociated molecule dissociates, m' is apparent molality.

Using above equation

$$m' = \frac{0.0514 \text{ deg}}{0.514 \text{ deg kg mol}^{-1}} = 0.1 \text{ mol kg}^{-1}$$

$$M_{\text{app}} \text{ (solute)} = \frac{w_2}{m} \times \frac{1000}{w_1}$$

$$= \frac{(1.51 \text{ g})(1000 \text{ g kg}^{-1})}{(0.1 \text{ mol kg}^{-1})(500 \text{ g})} = 30.2 \text{ g mol}^{-1}$$

van't Hoff factor $i = \dfrac{M_{\text{theo}}}{M_{\text{app}}}$ ...(17.41)

$$= \frac{58.5 \text{ g mol}^{-1}}{30.2 \text{ g mol}^{-1}} = 1.937$$

Degree of dissociation $\alpha = \dfrac{i-1}{\gamma-1}$ ...(17.42)

$$= \frac{1.937-1}{2-1} = 0.937.$$

17.6.4. *Dependence of Solubility on Pressure*

As in case of depression in freezing point, there is an equilibrium between the solid solvent and the solvent in solution, on this case there is an equilibrium between the solid solute and solute in solution at the saturated point. If we assume that solution is behaving as an ideal one, then solubility is also an colligative property and change of solubility with pressure is given by

$$\left[\frac{\partial \ln x_2(\text{satd})}{\partial p}\right]_T = -\frac{\Delta \bar{V}_{\text{fusion}}}{RT} \quad ...(17.43)$$

where $\Delta \bar{V}$ is the molar volume change for the transition from the pure solid solute to the supercooled liquid state, x_2 is the mole fraction of solute at saturated point.

17.6.5. *Dependence of Solubility on Temperature*

Under the similar situation as above, the change in solubility with temperature at constant pressure and at saturation point is given by

$$\left[\frac{\partial \ln x_2(\text{satd})}{\partial T}\right]_P = \frac{\Delta \bar{H}_{\text{fusion}}}{RT^2} \quad ...(17.44a)$$

where $\Delta \bar{H}_{\text{fusion}}$ is the molar heat of fusion of solute. The equation (17.44 *a*) can be written as

$$\ln x_2 = -\frac{\Delta \bar{H}_{\text{fus}}}{R}\left[\frac{1}{T}-\frac{1}{T_0}\right] \quad ...(17.44b)$$

where T_0 is the melting point of pure solute. It can also be written as

$$\ln x_2 = \frac{\Delta \bar{H}_{\text{fus}}}{T_0 R}\left[1-\frac{T_0}{T}\right] = \frac{\Delta \bar{S}_{\text{fus}}}{R}\left(1-\frac{T_0}{T}\right) \quad ...(17.44c)$$

***Exercise* 17.20:** Calculate the ideal solubility of naphthalene at 20°C. The actual solubility of naphthalene expressed as mole fraction at 20°C in benzene and methyl alcohol is 0.241 and 0.018 respectively. Explain the differences if any. $\Delta \bar{H}_{\text{fus}}$ = 19.08 kJ mol^{-1}. Melting point of naphthalene is 80°C.

Using equation (17.44*a*)

$$\ln x_2 = -\frac{(19080 \text{ J mol}^{-1})}{(8.314 \text{ J mol}^{-1}\text{K}^{-1})}\left(\frac{1}{293} - \frac{1}{353}\right) \text{K}^{-1} = -1.331$$

$$\ln x_2 = 0.264$$

The solubility (x_2) of naphthalene in benzene is 0.241 and in methyl alcohol is 0.018. The differences are due to the assumptions involved in deriving equation (17.44 *b*). (*a*) $\Delta\overline{H}_{fus}$ remains constant over the wide temperature range, which is a function of temperature, (*b*) the law is never good when the solvents are hydrogen bonding but solute are not and (*c*) the law is never accurate for solutions of ionic materials in water. This is because the solutions of such solutes are not ideal at such concentrations and are far below their melting points.

17.7. OSMOTIC PRESSURE

Osmotic pressure (represented as Π) is that external pressure which must be applied in order to stop the flow of pure solvent into the solution through a semipermeable membrane. The general formula to determine osmotic pressure is

$$\Pi = -\frac{RT}{\overline{V}_1}\ln\frac{p_{soln}}{p_1^0} \quad ...(17.45)$$

where $\overline{V}_1$ is the molar volume of solvent, p_{soln} is the vapour pressure of the solvent in solution and p_1^0 is that of pure solvent. If the solution is ideal

$$\Pi = -\frac{RT}{\overline{V}_1}\ln x_1 \quad ...(17.46)$$

where x_1 is the mole fraction of solvent.

If the solution is dilute and ideal the equation (17.46) reduces to

$$\Pi = \frac{RT\, x_2}{\overline{V}_1} \quad ...(17.47)$$

$$= C\,RT \quad ...(17.48)$$

where C is the molarity of the solute in solution and R is expressed as atm dm^3 K^{-1} mol^{-1}. Similar to other colligative properties, the osmotic pressure experiments are very useful in the determination of molecular weight of high molecular weight polymers and biological molecules.

***Exercise* 17.21:** Using the data given in exercise 17.14, calculate the osmotic pressure of the aqueous solution at 20°C. The density of water at this temperature is 0.998 kg dm^{-3}.

The equation (17.45) can be used to calculate the osmotic pressure.

$$\overline{V}_{H_2O} = \frac{(18.0 \text{ g mol}^{-1})(10^{-3} \text{ kg g}^{-1})}{0.998 \text{ kg dm}^{-3}} = 18.04 \times 10^{-3} \text{ dm}^3 \text{ mol}^{-1}$$

$$\Pi = \frac{RT}{\overline{V}_{H_2O}}\ln\frac{p^0_{H_3}}{p_{H_2O}}$$

$$= \frac{(0.0821\ \text{dm}^3\ \text{atm K}^{-1}\text{mol}^{-1})\,(293\ \text{K})}{(18.04\times10^{-3}\ \text{dm}^3\text{mol}^{-1})}\ln\frac{17.535\ \text{mm}}{16.075\ \text{mm}}$$

$$= 115\ \text{atm.}$$

***Exercise* 17.22:** 5.0 g of polymer of molecular weight 50.0 kg mol^{-1} is dissolved in 1 dm^3 of water. If the density of this solution is 0.96 kg dm^{-3}, calculate the height of water that will represent this pressure.

The osmotic pressure exerted by the above solution can be calculated by the equation (17.48)

$$\Pi = C\,RT$$

$$= \left(\frac{5.0\times10^{-3}\text{kg}}{50.0\ \text{kg mol}^{-1}}\right)\left(\frac{0.0821\ \text{atm dm}^3\ \text{K}^{-1}\ \text{mol}^{-1}\ (300\ \text{K}}{1\ \text{dm}^3}\right)$$

$$= 2.463\times10^{-3}\ \text{atm} = 1.872\ \text{mm}$$

Using the relation $\frac{h_{soln}}{h_{Hg}} = \frac{d_{Hg}}{d_{soln}}$

$$h_{sol} = \frac{13.6\ \text{kg dm}^{-3}}{0.96\ \text{kg dm}^{-3}}\,(1.872\ \text{mm}) = 26.52\ \text{mm}$$

Some definitions commonly used in the study of osmotic pressures are:

***(i)* Isotonic solutions:** If two solutions are of equal concentrations and hence have the same osmotic pressure, they are called isotonic solutions.

***(ii)* Hypotonic solutions:** For two solutions of unequal osmotic pressures, the more concentrated solution is called hypertonic and the less concentrated solution is said to be hypotonic.

***(iii)* Hemolysis:** Hemolysis is a technique used by biochemists to study the contents of red blood cells. The red blood cells are placed in hypotonic solution which causes the water to move into the cell. The cells swell and eventually burst. Thus haemoglobin and other protein molecules are released in to the solution.

***(iv)* Crenation:** When a solution is placed in a hypertonic solution, the intracellular water tends to move out of the cell by osmosis to the more concentrated, surrounding solution. This process is known as crenation, which causes the cell to shrink and eventually cease functioning.

17.8. EQUILIBRIUM IN SOLUTION

Generally the reactions taking place in solutions are of the type:

(*a*) *Ionisation:* The ionisation could be of acids, bases and the salts. Generally the strong acids like HCl, HNO_3, $HClO_4$ and H_2SO_4, the strong bases like NaOH, KOH, LiOH and $Ba(OH)_2$ and salts of strong acids and strong bases are completely ionised in the solution. This is further true since we generally deal with dilute solutions. For other acids, bases and salts the following equilibrium exists.

$$AB \rightleftharpoons A^+ + B^-$$

where A and B can be any kind of species.

(*b*) The dissolution of a precipitate *i.e.*,

$$AgBr(s) \rightleftharpoons Ag^+ + Br^-$$

and the effect of common ion on the solubility.

(*c*) An oxidation-reduction reaction

$$2Fe^{3+} + 2I^- \rightleftharpoons 2Fe^{2+} + I_2$$

(*d*) Or, in general, a reaction which can be represented as

$$aA + bB \rightleftharpoons cC + dD$$

Problems concerning each type of reactions will be briefly discussed with few solved problems. The oxidation-reduction reactions and stability of the complex formation has already been discussed in Chapter 16. Here *it will be assumed that the solutions are ideal such that* $\gamma_i = 1$, *i.e., activity coefficients are unity.*

17.8.1. *Calculations of Ionic Concentration in Complicated Ionic Equilibria*

The problems generally considered in ionic equilibrium is to help students understand the acid-base titrations (18.11) and precipitations encountered in analytical laboratory. This can be easily extended to complicated ionic equilibria and steps involved are listed as follows:

1. Set up expressions for all the independent equilibria involved in the system concerned.
2. List all the concentrations of interest. These must include all concentrations in the equilibrium expressions and may also contain additional species (such as Na^+, NO_3^- etc.) which are present but do not react.
3. If the ions of the solvent are involved in the equilibria (acid-base systems) set up the conditions for electroneutrality.
4. Use mass balance relations to obtain enough independent relations to define the system completely. The total unknown number of concentrations terms should be equal to the total number of relations.
5. Make as many approximations in the mass balance and electroneutrality relations as is consistent with qualitative chemical knowledge of system. In particular neglect $[OH^-]$ compared to $[H^+]$ in acid solutions, neglect $[H^+]$ compared to $[OH^-]$ in basic solutions; assume precipitation or complex formation reactions go to completion etc.
6. Solve the approximate equations.
7. Check the results obtained by the approximate method to see if they are consistent approximations within reasonable limit of error; say 5%. Discard those approximations which are not valid and if necessary re-calculate the results.

***Exercise* 17.23:** Calculate the concentration of each ion in the mixture; 0.2 mol of $Ba(NO_3)_2$, 0.4 mol of $NaNO_3$ and 0.8 mol of NaCl dissolved in 0.4 dm^3.

Since all the electrolytes are ionised, the concentration of each ion will be

$$[Ba^{2+}] = 0.2 \text{ mol}/0.4 \text{ dm}^3 = 0.5 \text{ mol dm}^{-3}$$

$$[NO_3^-] = \frac{1}{0.4 \text{ dm}^3} (NO_3^- \text{ from } Ba(NO_3)_2 + NO_3^- \text{ from } NaNO_3)$$

$$= \frac{1}{0.4 \text{ dm}^3} (2 \times 0.2 \text{ mol} + 0.4 \text{ mol}) = 2 \text{ mol dm}^{-3}$$

$$[\mathrm{Na^+}] = \frac{1}{0.4\ \mathrm{dm^3}}(\mathrm{Na^+\ from\ NaNO_3 + Na^+\ from\ NaCl})$$

$$= \frac{1}{0.4\ \mathrm{dm^3}}(0.4\ \mathrm{mol} + 0.8\ \mathrm{mol}) = 3\ \mathrm{mol\ dm^{-3}}$$

$$[\mathrm{Cl^-}] = \frac{0.8\ \mathrm{mol}}{0.4\ \mathrm{dm^3}} = 2\ \mathrm{mol\ dm^{-3}}$$

***Exercise* 17.24: Is the above solution electrically neutral?**

$$\text{Charge of } [\mathrm{Ba^{2+} + Na^+}] = \text{Charge of } [\mathrm{NO_3^- + Cl^-}]$$

$$(2 \times 0.5 + 3 \times 1)\ \text{mol} = (2 + 2)\ \text{mol}$$

i.e., $$4 \text{ mol of +ve charge} = 4 \text{ mol of } -\text{ve charge.}$$

***Exercise* 17.25: Calculate the concentration of H^+, CN^- and HCN when 2 mol of HCN is dissolved in 1 dm^3 of water and thus find pH of the solution. The dissociation constant of HCN is 7.2×10^{-10} at 25°C.**

The equilibrium reaction is

$$\mathrm{HCN} \rightleftharpoons \mathrm{H^+ + CN^-}$$

If α is the degree of dissociation, then at equilibrium $[H^+] = 2\alpha$ mol, $[CN^-] = 2\alpha$ mol and [HCN] = $2(1 - \alpha)$ mol. So

$$\frac{[\mathrm{H^+}][\mathrm{CN^-}]}{[\mathrm{HCN}]} = K$$

or

$$\frac{(2\alpha)(2\alpha)}{2(1-\alpha)} = 7.2 \times 10^{-10}$$

or

$$2\alpha^2 + 7.2 \times 10^{-10}\alpha - 7.2 \times 10^{-10} = 0$$

$$\alpha = \frac{-7.2\times10^{-10} \pm [(7.2\times10^{-10})^2 + 4(2)(7.2\times10^{-10})]^{1/2}}{4} = 1.89 \times 10^{-5}$$

So

$$[\mathrm{H^+}] = (1.89 \times 10^{-5})(2\ \mathrm{mol}) = 3.78 \times 10^{-5}\ \mathrm{mol\ dm^{-3}} = [\mathrm{CN^-}]$$

$$\mathrm{HCN} = 2(1-1.89 \times 10^{-5}) = 1.99996\ \mathrm{mol\ dm^{-3}}$$

pH is defined as the negative of logarithm of hydrogen ion concentration *i.e.,* pH = – log $[H^+]$. Actually it is defined in terms of measurements on standard solutions and more closely defined by the relation (for detailed description see Section 18.11.3)

$$\mathrm{pH} = -\log a_{\mathrm{H^+}}$$

where a_{H^+} is activity of hydrogen ion.

The pH of above solution is

$$\mathrm{pH} = -\log(3.78\times10^{-5}) = 4.42.$$

Note: The above problem can be solved by assuming $2(1-\alpha) = 2$ as the value of K is very small. So

$$\frac{[H^+][CN^-]}{[HCN]} = K$$

$$\frac{(2\alpha)(2\alpha)}{2(1-\alpha)} = 2\alpha^2 = 7.2 \times 10^{-10}$$

or $$\alpha = 6.0 \times 10^{-5}$$

i.e., very close to 3.78×10^{-5}. One can also use the second method.

***Exercise* 17.26:** (*a*) 5 mL of 0.05 M NH_4Cl is mixed with 2 mL of 0.02 M NH_3. Calculate $[H^+]$. (*b*) If the solution is diluted to 70 dm^3 what is the $[H^+]$?

(*a*) The chemical equations involved are

$$H_2O \rightleftharpoons H^+ + OH^- \qquad K_w = 10^{-14} \qquad ...(17.49a)$$

$$NH_3 + H_2O \rightleftharpoons NH_4^+ + OH^- \qquad K_b = \frac{[NH_4^+][OH^-]}{[NH_3]} = 1.8 \times 10^{-5} \qquad ...(17.49b)$$

After mixing, the concentration of NH_4Cl is $0.05 \times \frac{5}{7} = 0.0357$ M and that of

$$NH_3 = \frac{0.02 \times 2}{7} = 0.0057 \text{ M.}$$

The electroneutrality relation shows

$$[H^+] + [NH_4^+] = [OH^-] + [Cl^-]$$
$$= [OH^-] + 0.0357 \text{ M} \qquad ...(17.49c)$$

Assuming NH_4Cl is completely dissociated.

A mass balance on NH_3 gives

$$[NH_3] + [NH_4^+] = 0.0057 + 0.0357 = 0.0414 \qquad ...(17.49d)$$

We have four equations (17.49*a*), (17.49*b*), (17.49*c*) and (17.45*d*). Let us make approximation: (*i*) Solution will be basic so $[H^+]$ will be small, (*ii*) Since NH_4Cl will be completely dissociated so the concentration of OH^- will also be small. This is not a good assumption but can be justified later. So electroneutrality gives $[NH_4^+] = 0.0357$ M and mass balance gives $[NH_3] = 0.0057$ M. Substituting in equation (17.49*b*)

$$\frac{(0.0357)[OH^-]}{0.0057} = 1.8 \times 10^{-5} \text{ or } [OH^-] = 2.87 \times 10^{-6} \text{ M}$$

Using equation (17.49*a*) $$[H^+] = \frac{10^{-14}}{2.87 \times 10^6} = 3.46 \times 10^{-9} \text{ M}$$

If the approximation (*ii*) is not used, the $[OH^-]$ can be calculated using equations (17.49*a*), (17.49*b*) and (17.49*c*). The quadratic equation will give $[OH^-] = 3.4 \times 10^{-6}$ M. The error is not much and one is justified in using approximation (*ii*).

(*b*) In the part (*b*) the solution is diluted to 10^4 times and one cannot neglect the concentration of $[H^+]$ and $[OH^-]$. These will be comparable.

Now electroneutrality gives.

$$[H^+] + [NH_4^+] = [OH^-] + 3.57 \times 10^{-6}\ M \quad ...(17.49c')$$

and mass balance gives

$$[NH_3] + [NH_4^+] = 4.14 \times 10^{-6}\ M \quad ...(17.49d')$$

Using equations (17.49*b*) and (17.49*d'*), *i.e.*

form (17.49*b*), $$[NH_3] = \frac{[NH_4^+][OH^-]}{1.8\times10^{-5}}$$

further $$[NH_4^+] + \frac{[NH_4^+][OH^-]}{1.8\times10^{-5}} = 4.14 \times 10^{-6}$$

or $$1.8 \times 10^{-5}[NH_4^+] + [NH_4^+][OH^-] = (4.14 \times 10^{-6})(1.8 \times 10^{-5}) = 7.452 \times 10^{-11}$$

or $$(1.8 \times 10^{-5} + [OH^-])[NH_4^+] = 7.452 \times 10^{-11}$$

$$[NH_4^+] = \frac{7.452\times10^{-11}}{[OH^-]+1.8\times10^{-5}}$$

From (17.49*a*) $$[H^+] = \frac{10^{-14}}{[OH^-]}$$

Substituting in relation (17.49*c'*)

$$\frac{10^{-14}}{[OH^-]} + \frac{7.452\times10^{-11}}{[OH^-]+1.8\times10^{-5}} = [OH^-] + 3.57 \times 10^{-6}$$

Let $[OH^-] = 10^{-7}\,x$ so the above equation becomes

$$\frac{10^{-7}}{x} + \frac{7.452\times10^{-4}}{x+180} = x\cdot10^{-7} + 3.57 \times 10^{-6}$$

$$x\,10^{-7} + 180 \times 10^{-7} + 7.452 \times 10^{-4}\,x = (x^2 + 180\,x)(10^{-7}\,x + 3.57 \times 10^{-6})$$

Multiply by 10^{-7} and rearrange

$$x^3 + 315.7\,x^2 - 1027\,x - 180 = 0$$

Since the concentration of OH^- is small one can neglect the x^3 term.

The second order equation

$$315.7x^2 - 1027\,x - 180 = 0$$

give $x = 3.42$ and one negative value which can be neglected. So

$$[OH^-] = 3.42 \times 10^{-7}\ M \text{ and } [H^+] = 2.9 \times 10^{-8}\ M$$

$$pH = 7.54$$

If one does not neglect x^3, the improved value of x obtained by Newton's method will be 3.32, not much different from 3.42.

From the above results one can see that the concentrations of H^+ and OH^- cannot be neglected as these are comparable.

***Exercise* 17.27:** Calculate $[Pb^{2+}]$ (*a*) for a saturated solution of $PbCl_2$ in water and (*b*) for saturated solution of $PbCl_2$ in 0.08 M NaCl. $K_{sp} = 1.8 \times 10^{-4}$ mol^3 dm^{-9}

(*a*) The reaction is

$$PbCl_2 \rightleftharpoons Pb^{2+} + 2\,Cl^-$$

Neglecting the concentration of ions of water since they do not affect the equilibrium, the electroneutrality of the solution gives

$$2\,[Pb^{2+}] = [Cl^-]$$

or according to mass balance, for each mol of precipitate dissolved it gives 1 mol of Pb^{2+} ions and 2 mol of Cl^- ions *i.e.*,

$$[Pb^{2+}]\,[Cl^-]^2 = K_{sp} = 1.8 \times 10^{-4}\ mol^3\ dm^{-9}$$

or

$$[Pb^{2+}]\,[2Pb^{2+}]^2 = 4\,[Pb^{2+}]^3 = 1.8 \times 10^{-4}$$

$$[Pb^{2+}] = 0.036\ mol\ dm^{-3}.$$

(*b*) When NaCl is added, the $[Cl^-]$ is increased and that of $[Pb^{2+}]$ decreased. According to electroneutrality

$$2\,[Pb^{2+}] + [Na^+] = [Cl^-]$$

or mass balance can also give the above results. The concentration of

$[Na^+] = [NaCl] = 0.08$ M as NaCl is completely dissociated. So

$$[Pb^{2+}]\,([2Pb^{2+}] + [0.08])^2 = 1.8 \times 10^{-4}$$

$$f(x) = 4[Pb^{2+}]^3 + 0.32\,[Pb^{2+}]^2 + 6.4 \times 10^{-3}\,[Pb^{2+}] - 1.8 \times 10^{-4} = 0$$

$$f'(x) = 12\,[Pb^{2+}]^2 + 0.64\,[Pb^{2+}] + 6.4 \times 10^{-3} = 0$$

The cubic equation in $[Pb^{2+}]$ can be solved by Newton's method of approximate root and the approximate root $[Pb^{2+}]$ can be taken M 0.036 *i.e.*,

$$x_1 = x_0 - \frac{f(x_0)}{f'(x_0)}$$

Substituting the values of $x_0 = 0.036$ in above two equations.

$$x_1 = 0.036 - \frac{4\,(0.036)^3 + 0.32\,(0.036)^2 + 6.4 \times 10^{-3}(0.036) - 1.8 \times 10^{-4}}{12\,(0.036)^2 + 0.64\,(0.036) + 6.4 \times 10^{-3}}$$

$$= 0.0204$$

$$x_2 = 0.0204 - \frac{4\,(0.0204)^3 + 0.32(0.0204)^2 + 6.4 \times 10^{-3}(0.0204) - 1.8 \times 10^{-4}}{12\,(0.0204)^2 + 0.64\,(0.0204) + 6.4 \times 10^{-3}}$$

$$= 0.0156$$

Similarly $x_3 = 0.015$ which is quite close to x_2 so the concentration of Pb^{2+} in presence of 0.08 M NaCl is 0.015 M.

17.8.2. *Calculation of Molar Solubilities from Equilibrium Constants (Solubility Product)*

In general, the calculation of molar solubilities of sparingly soluble salts in water involves the following equilibria:

$$M_aX_b(s) \rightleftharpoons aM^{b+}(aq) + bX^{a-}(aq) \qquad ...(17.50a)$$

and

$$K_{sp} = [M^{b+}]^a [X^{a-}]^b \qquad ...(17.50b)$$

The assumption involved here is that only M^{b+} and X^{a-} ions exist in solution and if the salt is dissolved in water, there is no other source of the ions. If s is the solubility in mol dm^{-3} (*g* formula weight dm^{-3})

then

$$M^{b+} = as, X^{a-} = bs \text{ and } K_{sp} = (as)^a (bs)^b. \qquad ...(17.51)$$

Rarely any other reactions of the ions are discussed further but there are three classes of reactions which can take place with these ions: (*i*) Hydrolysis of anion (only salts of strong acids will be free from this difficulty and most of them are very soluble). (*ii*) Hydrolysis of cation which is the same as the formation of hydroxide complexes, (*iii*) Formation of complexes, particularly neutral molecules of salt or ion pairs by often complicated ions. It is difficult to find the salts which are free from these kinds of reactions but there are few examples e.g. sparingly soluble salts of strong acids *i.e.,* sulphates of Sr, Ba, Pb, Hg (*I*) and Ag; the fluorides of IIa; chlorides of Cu (I), Hg (I), Ag and Tl (I); the bromides and iodides of Ag and Tl (I); the iodates of Ca, Ba, La (III), Ce (IV), Cu (II), Ag (I), Hg (I), Pb, Tl (I); $AgBrO_3$ and Mg $(OH)_2$. These salts are free from the above reactions.

***Exercise* 17.28:** Calculate the concentration of all the species in saturated AgCl. The following data can be used:

$$AgCl(s) \rightleftharpoons Ag^+(aq) + Cl^-(aq) \quad K_1 = 1.8 \times 10^{-10} = [Ag^+][Cl^-] \qquad ....(17.52a)$$

$$Ag^+(aq) + OH^-(aq) \rightleftharpoons AgOH \quad K_2 = 2.0 \times 10^2 = \frac{[AgOH]}{[Ag^+][OH^-]} \qquad ...(17.52b)$$

$$Ag^+(aq) + Cl^-(aq) \rightleftharpoons AgCl \quad K_3 = 1.1 \times 10^3 = \frac{[AgCl]}{[Ag^+][Cl^-]} \qquad ...(17.52c)$$

$$AgCl + Cl^-(aq) \rightleftharpoons AgCl_2^-(aq) \quad K_4 = 1.0 \times 10^2 = \frac{[AgCl_2^-]}{[AgCl][Cl^-]} \qquad ...(17.52d)$$

$$AgCl + Ag^+(aq) \rightleftharpoons Ag_2Cl^+(aq) \quad K_5 = 1.5 \times 10^2 = \frac{[Ag_2Cl^+]}{[AgCl][Ag^+]} \qquad ...(17.52e)$$

$$AgCl_2^-(aq) + Cl^-(aq) \rightleftharpoons AgCl_3^{2-}(aq) \quad K_6 = 1.0 = \frac{[AgCl_3^{2-}]}{[AgCl_2^-][Cl^-]} \qquad ...(17.52f)$$

$$Ag_2Cl^+(aq) + Ag^+(aq) \rightleftharpoons Ag_3Cl^{2+}(aq) \quad K_7 = 1.8 = \frac{[Ag_3Cl^{2+}]}{[Ag_2Cl^+][Ag^+]} \qquad ...(17.52g)$$

$$H_2O(l) \rightleftharpoons H^+(aq) + OH^-(aq) \quad K_w = 10^{-14} = [H^+][OH^-]$$

As said earlier, here the extent of other reactions of Ag^+ and Cl^- ions are very small and can be shown below by calculating their concentrations.

Mass balance on Ag, on chloride and on OH^- give

$$[Ag^+] + [AgOH] + [AgCl] + [AgCl_2^-] + [AgCl_3^{2-}] + 2[Ag_2Cl^+] + 3[Ag_3Cl^{2+}] = s \quad ...(17.52h)$$

$$[Cl^-] + [AgCl] + 2[AgCl_2^-] + 3[AgCl_3^{2-}] + [Ag_2Cl^+] + [Ag_3Cl^{2+}] = s \quad ...(17.52i)$$

and $$[H^+] = [AgOH] + [OH^-] \quad ...(17.52j)$$

where s is the solubility of AgCl in water. If Ag^+ and Cl^- are predominent, from reaction (17.52 a)

$$[Ag^+][Cl^-] = s^2 = 1.8 \times 10^{-10}\ M^2$$

$$s = 1.35 \times 10^{-5}\ M \quad ...(17.52k)$$

Combining equations (17.52c) and (17.52k)

$$[AgCl] = K_3[Ag^+][Cl^-]$$
$$= (1.1 \times 10^3)(1.35 \times 10^{-5})(1.35 \times 10^{-5}) = 2.0 \times 10^{-7}\ M$$

From equation (17.52b)

$$\frac{[AgOH]}{[OH^-]} = K_2 \times [Ag^+] = (2.0 \times 10^2)(1.35 \times 10^{-5}) = 2.7 \times 10^{-3}$$

i.e., in equation (17.52j) the amount of AgOH is negligible.

From K_w: $$[OH^-] = [H^+] = 10^{-7}\ M$$

So $$[AgOH] = (2.7 \times 10^{-3})(10^{-7}\ M) = 2.7 \times 10^{-10}\ M$$

From relation (17.52d)

$$[AgCl_2^-] = K_4[AgCl][Cl^-]$$
$$= (1.0 \times 10^2)(2.0 \times 10^{-7})(1.35 \times 10^{-5}) = 2.7 \times 10^{-10}\ M$$

From relation (17.52f)

$$[Ag_2Cl^+] = K_5[AgCl][Ag^+]$$
$$= (1.5 \times 10^2)(2.0 \times 10^{-7})(1.35 \times 10^{-5}) = 4.05 \times 10^{-19}\ M$$

Similarly from relations (17.52g) and (17.52h)

$$[AgCl_3^{2-}] = 3.6 \times 10^{-15}\ M, [Ag_3Cl^{2+}] = 9.7 \times 10^{-15}\ M$$

***Exercise* 17.29:** Calculate the concentration of all the species in the saturated solution of HgS. Use the following data:

$$HgS(s) \rightleftharpoons Hg^{2+}(aq) + S^{2-}(aq) \quad K_1 = 1.0 \times 10^{-51} = [Hg^{2+}][S^{2-}] \quad ...(17.53a)$$

$$Hg^{2+}(aq) + H_2O\ (l) \rightleftharpoons HgOH^+(aq) + H^+(aq) \quad K_2 = 3.2 \times 10^{-3} = \frac{[HgOH^+][H^+]}{[Hg^{2+}]} \quad ...(17.53b)$$

$$HgOH^+(aq) + H_2O\ (l) \rightleftharpoons Hg(OH)_2(s) + H^+(aq) \quad K_3 = 1.4 \times 10^{-3} = \frac{[Hg(OH)_2][H^+]}{[HgOH^+]} \quad ...(17.53c)$$

$$S^{2-}(aq) + H_2O\ (l) \rightleftharpoons HS^-(aq) + OH^-(aq) \quad K_4 = 8.0 \times 10^{-2} = \frac{[HS^-][OH^-]}{[S^{2-}]} \quad ...(17.53d)$$

$$HS^-(aq) + H_2O\ (l) \rightleftharpoons H_2S\ (g) + OH^-(aq) \quad K_5 = 1.0 \times 10^{-7} = \frac{[H_2S][OH^-]}{[HS^-]} \quad ...(17.53e)$$

$$HgS(l) + H_2S\ (g) \rightleftharpoons Hg(SH)_2\ (s) \quad K_6 = 6.3 \times 10^{-7} = \frac{[Hg(SH)_2]}{[HgS][H_2S]_2} \quad ...(17.53f)$$

$$Hg(SH)_2\,(s) \rightleftharpoons HgS_2^{2-}\,(aq) + 2H^+\,(aq) \quad K_7 = 1.0 \times 10^{-7} = \frac{[HgS_2^{2-}][H^+]^2}{[Hg\,(SH)_2]} \quad ...(17.53g)$$

$$H_2O\,(l) \rightleftharpoons H^+\,(aq) + OH^-\,(aq) \quad K_w = 10^{-14} = [H^+][OH^-]$$

The mass balance on Hg and sulphur gives

$$[Hg^{2+}] + [HgOH^+] + [Hg(OH)_2] + [Hg(SH)_2] + [HgS_2^{2-}] = s \quad ...(17.53h)$$

$$[S^{2-}] + [HS^-] + [H_2S] + 2[Hg(SH)_2] + 2[HgS_2^{2-}] = s \quad ...(17.53i)$$

and electroneutrality gives

$$[H^+] + [HgOH^+] + 2[Hg^{2+}] = [OH^-] + 2[S^{2-}] + [HS^-] + 2[HgS_2^{2-}] \quad ...(17.53j)$$

Make assumptions (it will be shown later that these assumptions are right) that main species present are H^+, OH^- and then HgS_2^{2-} and $Hg(OH)_2$. With this, equations (17.53*h*), (17.53*i*) and (17.53*j*) become

$$[Hg(OH)_2] + [HgS_2^{2-}] = s \quad ...(17.53k)$$

$$2[HgS_2^{2-}] = s \quad ...(17.53l)$$

and

$$[H^+] = [OH^-] = 10^{-7} \quad ...(17.53m)$$

Combining equations (17.53*i*) and (17.53*k*)

$$[Hg(OH)_2] = [HgS_2^{2-}] = s/2 \quad ...(17.53n)$$

From (17.53*c*),

$$\frac{[Hg(OH)_2]}{[HgOH^+]} = \frac{K_3}{[H^+]} = \frac{1.4 \times 10^{-3}}{10^{-7}} = 1.4 \times 10^4 \quad ...(17.53o)$$

From (17.53*b*),
$$\frac{[HgOH^+]}{[Hg^{2+}]} = \frac{K_2}{[H^+]} = \frac{3.2 \times 10^{-3}}{10^{-7}} = 3.2 \times 10^4 \quad ...(17.53p)$$

From (17.53*g*),
$$\frac{[HgS_2^{2-}]}{[Hg(SH)_2]} = \frac{K_7}{[H^+]^2} = \frac{10^{-7}}{(10^{-7})^2} = 10^7 \quad ...(17.53q)$$

From (17.53*f*),
$$\frac{[Hg(SH)_2]}{[H_2S]} = \frac{K_6}{[HgS]} = K_6 = 6.3 \times 10^{-7} \quad (17.53r)$$

Multiply (17.53*o*) by (17.53*p*) and use (17.53*n*)

$$[Hg^{2+}] = \frac{[Hg(OH)_2]}{(1.4 \times 10^4)(3.2 \times 10^4)} = \frac{s}{9 \times 10^8}\text{ M} \quad (17.53s)$$

Multiply (17.53*q*) and (17.53*r*) gives

$$[HgS_2^{2-}] = 6.3[H_2S] \quad ...(17.53t)$$

From (17.53*d*),
$$\frac{[HS^-]}{[S^{2-}]} = 8.0 \times 10^5 \quad ...(17.53u)$$

From (17.53*e*)
$$\frac{[H_2S]}{[HS^-]} = 1.0 \quad ...(17.53v)$$

Combining (17.53*t*), (17.53*u*) and (17.53*v*)

$$[S^{2-}] = \frac{s}{1.0 \times 10^7} \quad \text{...(17.53}w\text{)}$$

Finally using (17.53*s*) and (17.53*w*) in (17.53*a*)

$$[Hg^{2+}][S^{2-}] = \frac{s^2}{(9 \times 10^8)(1 \times 10^7)} = 1.0 \times 10^{-51}$$

or

$$s = 3.0 \times 10^{-18}$$

Now if we completely neglect the other reactions mentioned above, the solubility of HgS will be

$$[Hg^{2+}][S^{2-}] = s^2 = 1.0 \times 10^{-51} \quad \text{or} \quad s = 3.2 \times 10^{-26}$$

Two values clearly show that some other reactions are taking place and that enhance the solubility of HgS by a factor of 10^8. Concentrations of the other species can be calculated by using the respective relations and are

$$[H^+] = [OH^-] = 10^{-7}\text{ M}; \quad [Hg(OH)_2] = [HgS_2^{2-}] = 1.5 \times 10^{-18}\text{ M}$$

$$[H_2S] = [HS^-] = 2.4 \times 10^{-19}\text{ M}, [HgOH^+] = 1.1 \times 10^{-22}\text{ M}$$

$$[Hg(SH)_2] = 1.5 \times 10^{-25}\text{ M}, \quad [S^{2-}] = 3.0 \times 10^{-25}\text{ M}$$

$$[Hg^{2+}] = 3.3 \times 10^{-27}$$

Note: One can clearly see that the concentration of actual species (*i.e.*, S^{2-} and Hg^{2+}) are quite small and hydrolysis Hg^{2+} and S^{2-} ions are taking place. Further it is better to think that HgS dissolves in H_2O as

$$2HgS + 2H_2O \rightleftharpoons HgS_2^{2-} + Hg(OH)_2 + 2H^+$$

and Hg^{2+} and S^{2-} are formed by the dissociation of these species. Lastly the K_{sp} of HgS is still defined by the reaction (17.53*a*) as it is the experimentally determined quantity but only solubility of HgS in H_2O changes.

PROBLEMS

17.1 60 g of benzene and 80 g of naphthalene were mixed together. Calculate (*a*) heat of mixing (*b*) ΔG_{mix} at 298 K and (*c*) ΔS_{mix}.

17.2 Three cylinders of N_2, H_2 and O_2 contain gases at $V_{N2} = 10\text{ dm}^3$, $p_{N_2} = 20$ atm; $V_{H2} = 3\text{ dm}^3$, $p_{H_2} = 5$ atm, $V_{O2} = 15\text{ dm}^3$, $p_{O_2} = 2$ atm at 27°C. These gases are mixed while keeping the temperature constant but the final total pressure is 10 atm. Calculate ΔE, ΔH, ΔS and ΔG of mixing when the equilibrium is reached.

17.3 10^4 dm^3 of air at 27°C and 1 atm was separated into O_2, N_2 and CO_2. These gases were stored in bottle at the above temperature but at 20 atm. If the gases were considered ideal and the proportions of O_2, N_2 and CO_2 is = 0.20:0.77:0.03, calculate ΔS and ΔG for this process.

17.4 Show that in a ternary mixture ΔS_{mix} and ΔG_{mix} is maximum when $x_1 = x_2 = x_3 = \dfrac{1}{3}$.

Vapour Pressure of Mixture of Volatile Components

17.5 60 g of benzene and 80 g of naphthalene were mixed together. If the vapour pressure of benzene at 27°C is 50.71 mm and that of naphthalene is 32.06 mm, calculate the mole fraction of benzene in vapour phase. The two form ideal solution.

17.6 80 g of ethanol was mixed with 50 g of compound A and both form ideal solution. If $p^0_{C_2H_5OH}$ and p^0_A at 25°C are 44.5 mm and 88.7 mm, $p_{C_2H_5OH}$ and p_A are 23.45 mm and 41.96 mm respectively, calculate the molecular weight of compound A.

17.7 Two compounds A and B form ideal solutions at all compositions. If the vapour pressure of A at 80°C is 860 mm and that of B at 80°C is 350 mm, calculate the composition of the solution such that it boils at 80°C. Also calculate the composition of the vapour phase.

17.8 The vapour pressure of pure acetone at 55.1°C is 632.8 mm and that of $CHCl_3$ at 55.1°C is 741.8 mm. Assuming that they form ideal solution over the entire range of composition, plot P_T, p_{cs_2}, $p_{Acetone}$ as a function of x_{cs_2}. The actual data observed for different composition of mixture is

$100 \times x_{Acetone}$	0	11.8	23.4	36	50.8	58.2	64.5	72.1
$p_{Acetone}$(mm)	0	54.9	110.1	202.4	322.7	405.9	454.1	521.1
p_{CHCl_3}	632.8	548.1	469.4	359.7	257.7	193.6	161.2	120.7

Plot them on the same graph paper. Indicate whether it has positive deviation or negative deviation from the ideal solution. Find P_T when the $x_{CHCl_3} = 0.5$ in both cases. $p^0_{Acetone}$ at 55.1°C is 741 mm.

17.9 Suppose that acetone and $CHCl_3$ form ideal solution over the entire range of compositions.

$p^0_{CHCl_3} = 143$, and $p^0_{(CH_3)_2CO} = 397$ mm at 17°C. Plot P_T, p_{CHCl_3} and $p_{Acetone}$ vs mole fraction of acetone. The following is the experimental data. Plot this also on the same graph paper. Calculate p_{CHCl_3}, when its $x_{CHCl_3} = 0.75$. Calculate y_{CHCl_3} and indicate whether there is positive deviation or negative one.

$100 \times x_{CHCl_3}$	0	10	20	30	40	50	60	70	80	90	100
p_{CHCl_3} (mm)	0	6.0	13	22	34	50	71	91	111	129	143
$p_{(CH_3)_2CO}$ (mm)	397	354	304	251	196	143	95	60	32	14	0

17.10 From the plots of problems (17.8) and (17.9): (*a*) calculate Henry's constants when acetone and $CHCl_3$ are solutes and (*b*) indicate the regions where Raoult's law or Henry's laws are valid.

17.11 If x_1 and y_1 are the mole fractions of component 1 in solution and vapour phase, calculate the value of x_1 for which $y_1 - x_1$ has the minimum value. Calculate the vapour pressure at this composition.

17.12 It is generally found that in a binary mixture if one component shows the negative behaviour from the Raoult's law, the other component will also show the same behaviour. Explain this behaviour on molecular basis.

17.13 Liquids A and B form ideal solution. The solution of liquids A and B of composition $x_A = 0.3$ is placed in a closed system and the vapour pressures of pure liquid A and B at temperature T are 600 mm and 1000 mm. The pressure above the system is decreased till the vapours are just formed. Calculate the total pressure and the composition of the vapours. If the liquid boils at this temperature, calculate the composition of the solution.

17.14 The vapour pressure of a solution containing 2 mol of liquid A and 3 mol of liquid B at 60°C is 500 mm Hg. When 2 mol of A is further added to the above solution, the vapour pressure decreases to 400 mm Hg at the same temperature. (*a*) Calculate the vapour pressures of the pure liquids if the solution formed by liquids A and B is ideal, (*b*) The vapours of the second solution is condensed. Calculate the composition of the vapours in equilibrium with the condensate at 60°C.

17.15 The two liquids A and B have the same molecular weights and forms ideal solution. The solution of composition x_A has the vapour pressure 700 mm Hg at 80°C. The above solution is distilled without reflux till 3/4 of the solution is collected as condensate. The composition of the condensate is $x'_A = 0.75$ and that of residue $x_A = 0.3$. If the vapour pressure of the residue at 80°C is 600 mm, calculate x_A, p^0_A and p^0_B.

17.16 A solution containing compound *x* in water and a solution containing urea in water were put in a closed system. By doing this some water vapours were removed from one solution and got condensed in other. It is found, when both the solutions were at equilibrium vapour pressure, that one solution contains 2% of compound *x* and the other 5% urea by weight. Calculate the molecular weight of *x*.

Henry's Law

17.17 The Henry's constant of CH_4 in water is 31.4×10^6 mm. Calculate the solubility of CH_4 in H_2O at 25°C when the partial pressure of CH_4 over H_2O is 760 mm.

17.18 The water is in equilibrium with air at a pressure of 10 atm and at 25°C. If we suppose that air contains 20% of O_2 and 80% N_2 by volume at this temperature and pressure, calculate the composition of these gases dissolved in water.

$$K_{O_2} = 3.30 \times 10^7 \text{ mm}, \qquad K_{N_2} = 6.51 \times 10^7 \text{ mm}$$

17.19 It has been found that 1000 g of water contains 5.51×10^{-2} g of gas *x* dissolved in it at 25°C and partial pressure of 760 mm. Calculate the molecular weight of the gas if Henry's constant is 23.0×10^6 mm. If the gas is saturated hydrocarbon, calculate its molecular formula.

17.20 Calculate the amount of NaOH required to neutralise the amount of CO_2 dissolved in 1000 g of aerated water. The partial pressure of CO_2 at 25°C is 800 mm and Henry's constant 1.25×10^6 mm.

17.21 5% (by weight) solution of volatile compound in water as solvent has a vapour pressure of 60 mm Hg at 25°C. The vapour pressure of pure water at 25°C is 23.76 mm Hg.

(*a*) Calculate the partial pressures of two components by applying Raoult's law.

(*b*) Calculate the total pressure if the concentration of the solute is increased to 8% by weight. Take molecular weight of the compound as 50 g mol^{-1}.

17.22 Consider the equilibrium reaction

$$A \rightarrow B$$

The molecular weight of A and B are same *i.e.*, 156 g mol^{-1}. The standard free energy of formation at 300 K of A and B are – 480.6 kJ mol^{-1} and – 450.34 kJ mol^{-1}. The saturation solubility of A and B in a solvent is 50 g kg^{-1} and 90 g kg^{-1} of the solvent respectively. If the Henry's law is observed up to this limit of solubility and standard state is that of 1 molal solution, calculate the equilibrium constant of the above reaction in this solvent.

Nernst Distribution Law

17.23 50 mL of H_2O and 25 mL of CCl_4 were taken in a cylinder. 4 g of I_2 was added. Calculate the amount of I_2 present in each layer by using Nernst law. Nernst coefficient $C_{CCl_4}/C_{H_2O} = 86$.

17.24 In a cylinder, which contains 50 g of water and 70 g of benzene, some $HgCl_2$ was added. Calculate the Nernst constant $C_{H_2O}/C_{C_6H_6}$ if after equilibrium it is found that water contains 0.204 g and benzene contains 0.0217 g of $HgCl_2$. Density of benzene is 1.50 kg dm^{-3}.

17.25 Show that if a compound dissociates in one solvent and associates in another, the value of K, the Nernst distribution coefficient is given by

$$K = \frac{\alpha_d^2 C_A^2 / K_d}{\sqrt[n]{\alpha_a C_{B/nKa}}}$$

where α_d and α_a are degree of dissociation and association, K_a and K_d are association and dissociation constants.

17.26 Benzoic acid forms dimer in benzene and dissociates in water. If the concentration of benzoic acid in water $H_2O(C_1)$ and in benzene (C_2) are:

$C_1 = 1.758 \times 10^{-4}$ mol dm^{-3}, 11.53×10^{-3} mol dm^{-3}

$C_2 = 5.937 \times 10^{-3}$ mol dm^{-3}, 224.8×10^{-3} mol dm^{-3}

calculate the Nernst distribution coefficient. The $K_d = 6.0 \times 10^{-5}$ in mol dm^{-3} at 10°C.

17.27 The equilibrium of the type in solvent 1 can be

$$A + B \rightleftharpoons AB$$

studied with the help of Nernst distribution law provided a second solvent can be found which is immiscible with the solvent I, dissolves only one of the species without dissociation or association. Show that equilibrium constant K_a for the above reaction is

$$K_a = \frac{C_{AB}}{C_A C_B} = \frac{C_A^0 - C_{II}K}{(C_{II}K)(C_B^0 - C_A^0 + C_{II}K)}$$

where C_A, C_B, C_{AB} are the concentration of free A, free B and AB, C_B^0, C_A^0 are the total concentration of B, A in solvent II and C_I and C_{II} the total concentration in solvent I and II. K is the distribution coefficient.

17.28 Using the results of the above problem (17.27) calculate K_a for $I_2 + I^- \rightleftharpoons I_3^-$.

Given $C_A^0 = 0.02832$ mol dm^{-3}, $C_B^0 = 0.125$ mol dm^{-3}

$$C_{II} = 0.1896 \text{ mol dm}^{-3}, \quad K = \frac{C_{H_2O}}{C_{cs_2}} = \frac{1}{625}.$$

17.29 The extraction process where a compound is extracted by one solvent from the other solvent (two solvents are immiscible with each other) the amount extracted by the second solvent is given as

$$w_n = w\left(\frac{KV_1}{KV_1 + V_2}\right)^n$$

where w is amount dissolved in V_1 mL of given solvent, w_n the amount extracted by V_2 mL of solvent used for extraction and n is number of extractions.

17.30 The distribution coefficient of lactic acid between water and chloroform $\frac{C_{H_2O}}{C_{CHCl_3}}$ is 1/0.203 at 25°C. 200 mL of chloroform contains 1 molar solution of lactic acid. Calculate the amount of lactic acid withdrawn when 100 mL of water is shaken with solution (*a*) once, (*b*) twice by taking 50 mL each time and (*c*) four times by taking 25 mL each time.

Relative Lowering of Vapour Pressure

17.31 The vapour pressure of water at 25°C is 23.76 mm.

(*a*) Calculate the vapour pressure of water at the same temperature when 20 g of glucose is dissolved in 100 g of water.

(*b*) Calculate the vapour pressure of 20% glucose solution in water.

(*c*) Perform the same calculations as in part (*a*) when 38 g of sucrose is dissolved in 100 g of water. Is the results in this case are different? If not explain this.

17.32 15 g of benzoic acid (C_6H_5COOH) is dissolved in 100 g of water and 100 g of benzene separately at 25°C. The vapour pressures of pure water and pure benzene are 23.756 mm and 72.5 mm respectively and those of solutions 23.45 mm and 70.2 mm. If it is assumed that benzoic acid is non-volatile, can you guess about the bonding between the different molecules of benzoic acid in different solvents.

17.33 The vapour pressure of water at 20°C is 17.54 mm and that of 10% urea is 16.93 mm. Calculate its molecular weight.

17.34 126.28 g of compound A was dissolved in 1000 g of water. Calculate the relative lowering in the vapour pressure of water at 20°C when the molecular weight of A is 250 g and $p^0_{H_2O}$ is 17.54 mm at 20°C. Also calculate the decrease in the vapour pressure if the compound A dissociates into B and C.

Depression in Freezing Point

17.35 Calculate the molal freezing point constant for benzene if the freezing point of benzene is 5.45°C and heat of fusion 9.837 kJ mol^{-1}.

17.36 It has been noticed that the minimum temperature recorded in hill station is –10°C. Since water is used in car radiator and it freezes at 0°C, calculate the amount of glycerine added in 40 dm^3 of water so that water does not freeze. ΔH_{fus} = 6.01 kJ mol^{-1} and the molecular weight of glycerine is 82 g mol^{-1}. Use equations (17.34) and (17.35).

17.37 1 g of urea when dissolved in 100 g of certain solvent decreases its freezing point by 0.2°C. 1.6 g of unknown compound when dissolved in 80 g of the same solvent depresses the freezing point by 0.36°C. Calculate the molecular weight of unknown compound.

17.38 2 g of KCl is dissolved in 100 g of water. If the dissociation of KCl in such solution is only 90% (means existence of separate ions) calculate the freezing point of the solution.

17.39 A compound with molecular weight 80 g mol^{-1} forms polymer when dissolved in benzene. If the freezing point of benzene is 5.1°C when 4 g of compound A is dissolved in 100 g of benzene, calculate the apparent molecular weight and degree of association. The freezing point of benzene is 5.45°C and molal freezing point constant is 5.1 K kg mol^{-1}.

17.40 The Henry's constant for methane in benzene at 25°C is 31.4 × 10^6 torr. If the partial pressure of CH_4 over 1 kg of benzene at 25°C is 2 atm, calculate the depression in the freezing point of benzene by the dissolved methane. Assume that Henry's constant has the same value at 25°C and at 5°C.

17.41 A bunch of students performed the depression in the freezing point experiment using acetic acid as, solute and water as solvent. The concentration of acetic acid in water was determined by titrating the solution with 0.0705 M NaOH. From the following data calculate the molecular weight of acetic acid for each concentration.

V_{NaOH} for 25 mL of solution	7.2	13.4	25.1
ΔT	0.06	0.1	0.18

Is the molecular weight obtained consistent? If not explain.

17.42 Calculate the vant Hoff factor in each case from the problem 17.41 and the degree of dissociation.

17.43 The depression in the freezing point of benzene observed when different amount of CCl_4 was dissolved in 50 g of benzene are as follows:

m	0.1184	0.1738	0.3499	0.563	0.816	1.166
ΔT_f	0.603	0.883	1.761	2.801	4.005	5.6

Calculate the molar freezing point constant in each case. Explain if the results obtained are not same. Plot K_f vs m and get the value of K_f at $m = 0$. Compare this value with that obtained theoretically.

17.44 The decrease in the freezing point of benzene observed by different amount of acetic acid in 1000 g of benzene is as follows.

w_{Acetic} acid (g)	0.201	0.399	0.895	2.894	5.802	14.25
ΔT	0.0156	0.0277	0.0539	0.1472	0.253	0.608

Calculate the degree of association and thus the equilibrium constant for the association.

17.45 0.4 g of glucose was dissolved in 100 g of water and the depression in the freezing point is – 0.041. Calculate the heat of fusion of water.

17.46 It is found that compound A is contaminated by B but both are nonvolatile. By cooling curve method it is found that the solid phase starts appearing at a temperature of 125°C and this is 30°C less than the melting point of A. When 2 g of impure compound is dissolved in 100 g of water, the freezing point observed is – 0.27°C. If the molal freezing point constant for water is 1.86 K molal^{-1}, calculate (*a*) the composition of the impure compound and (*b*) ΔH_{fus} of compound A. The molecular weights of A and B are 150 g mol^{-1} and 120 g mol^{-1}.

Elevation in Boiling Points

17.47 When 2.1035 g of nonvolatile compound x is dissolved in 70 g of H_2O, the boiling point is increased by 0.23°C. Calculate the molecular weight of the compound $K_b = 0.5$ deg kg mol^{-1}.

17.48 An ionic compound x of molecular weight 140 g mol^{-1} is dissolved in 70 g of water. The boiling point of the solution observed for different concentrations are as follows:

w_2 (g)	1.947	3.9958	5.5006
ΔT	0.13	0.25	0.33

Calculate the molecular weight in each case and see whether it is consistent with the actual. If not explain.

17.49 Calculate the vant Hoff factor in each case from the data of problem 17.48.

17.50 0.65 g of naphthalene was dissolved in 100 g of methyl acetate. The elevation in the boiling point of methyl acetate solution observed was 0.103°C. If the actual boiling point of methyl acetate is 57°C, calculate its molar heat of vapourisation. Also calculate the relative lowering of vapour pressure of methyl acetate at its boiling point.

17.51 The heat of vapourisation of water is 2.258 kJ g^{-1} at its boiling point (100°C) and at 1 atm. Calculate the boiling point of water in solution when it has the mole fraction 0.8, 0.6, 0.4 and 0.2. Plot T against mole fraction and comment on the plot.

17.52 40 g of nonvolatile compound A when dissolved in 100 g of solvent, its boiling point increased from 125°C to 128°C, whereas its freezing point decreased from –20°C to –24.5°C. If the molecular weight of the solvent is 200 g mol^{-1} and its ΔH_{sub} is 500 kJ mol^{-1}, calculate the molecular weight of the solute as well as ΔH_{vap} and ΔH_{fus} of the solvent. Assume that the solution so formed is ideal and dilute.

17.53 2% (by weight) solution of unknown nonvolatile compound in CCl_4 (as solvent) boils at 79°C whereas pure CCl_4 boils at 76.7°C. The vapour pressure of the above solution at 76.7°C is 740 mm. Calculate the molecular weight of the compound as well as ΔH_{vap} of CCl_4.

17.54 30 g of urea ($CO(NH_2)_2$) was dissolved in 100 g of water. Calculate its boiling point using equations (17.37) and (17.38). Explain the difference in results if any.

Solubility

17.55 Calculate the solubility of naphthalene in chlorobenzene at 40°C assuming that it forms ideal solution. Melting point of naphthalene is 80°C and heat of fusion 149.06 J g^{-1}. The actual value of solubility is 0.256(x_2).

17.56 The saturated solution of naphthalene in nitrobenzene at 20°C contains 334 g of naphthalene in 1000 g of nitrobenzene. If the ideal law of solubility is valid, calculate the entropy of fusion of naphthalene. Molecular weights of naphthalene 128 g mol^{1} and that of nitrobenzene 123 g mol^{-1}.

17.57 The molar heat of fusion of compound A is 27120.6 J mol^{-1}. Calculate the ideal solubility at 57°C if the melting point is 86°C.

17.58 The entropy of fusion of benzene at its melting point (5.533°C) is 35.30 J deg^{-1} mol^{-1}. Calculate its ideal solubility in toluene at 25°C.

17.59 A compound x is dissolved in solvent 'B' till it forms a saturated solution. The boiling point of this saturated solution is 95°C whereas that of pure solvent 92°C. The melting points of x and B are 125°C and –80°C respectively. If the heat of vapourisation of solvent is 80 kJ mol^{-1}, molecular weight is 58 g mol^{-1} and assuming the behaviour of ideality, calculate (*a*) composition of the saturated solution and (*b*) heat of fusion of x.

Osmotic Pressure

17.60 20 g of glucose was dissolved in 100 g of water. Calculate the osmotic pressure at 25°C.

17.61 A compound x was dissolved in water at 27°C. It is found that vapour pressure lowering at 27°C is 0.72 mm. If the vapour pressure of water at 27°C is 26.74 mm, calculate the osmotic pressure of the solution.

17.62 0.2 g of polymer x was dissolved in 100 mL of water and osmotic pressure observed was 400 mm at 25°C. Calculate the molecular weight of polymer x.

17.63 The osmotic pressure observed when 0.2 g of compound xy_2 dissolved in 500 mL of water at 27°C is 0.4 atm. If it is assumed that the compound xy_2 is completely dissociated into x and y, calculate the apparent molecular weight.

17.64 The osmotic pressure observed is 3 atm at 25°C when 7.36 g of urea is dissolved in 1 dm ^{3} of water, calculate the ΔG for the process.

$$H_2O \text{ (pure, 25°C)} = H_2O \text{ (soln. 25°C)}$$

17.65 300 g of sucrose was dissolved in 1 dm^{+3} of water. Calculate the osmotic pressure of this solution at 0°C using equations (17.46) and (17.48). The experimental value observed is 26.8 atm. Explain the results.

Chemical Equilibrium in Ideal Solutions

17.66 Calculate the concentration of each ion in the following solutions diluted to 2 dm ^{3}:

(*a*) 0.2 mol $Ba(OH)_2$, 0.6 mol KOH and 0.5 mol LiOH

(*b*) 0.45 mol H_2SO_4, 0.15 mol K_2SO_4 and 0.45 mol KNO_3

(*c*) 0.25 g $NaNO_3$, 0.54 g of K_2SO_4 and 2.54 g $Pb(NO_3)_2$

(*d*) 5 g $Fe(C1O_4)_3$, 10 g $Ca(NO_3)_2$ and 0.5 mol HNO_3.

17.67 Show that in each of the solution mentioned in 17.66 the electroneutrality exists.

17.68 Calculate the concentration of H^+, F^- and HF when 0.6 mol of HF is dissolved in 1 dm^3 of H_2O. Also calculate the pH. Dissociation constant of HF is 6.7×10^{-4}.

17.69 Calculate the concentration of H^+, F^- and HF when 0.2 mol of KF is added to the solution mentioned in problem 17.68.

17.70 If 0.25 mol of KCN is added to 2 mol of HCN dissolved in 1 dm ^{3}, calculate the concentration of each ion. Given

$$K_a = \frac{[H^+][CN^-]}{[HCN]} = 7.2 \times 10^{-10}, \; CN^- + H_2O \rightleftharpoons HCN + OH^-; \quad K = 1.4 \times 10^{-5}$$

17.71 16.7 g of o-phthalic acid is dissolved in 1 dm^3 of water. Calculate the pH of the solution. Given

$$C_6H_4(COOH)_2 \rightleftharpoons C_6H_4COOH(COO)^- + H^+ \quad K_1 = 1.1 \times 10^{-3}$$

$$C_6H_4COOH(COO)^- \rightleftharpoons C_6H_4(COO)^{2-} + H^+ \quad K_2 = 3.9 \times 10^{-6}$$

17.72 Calculate the pH of the solution mentioned in problem 17.71 when 21.1 g of disodium phthalate is added.

17.73 A buffer is prepared by mixing 0.05 mol of benzoic acid (pk_a = 4.20) and 0.02 mol of sodium benzoate per dm^3.

(*a*) Calculate the pH of this solution

(*b*) Calculate the pH of the above solution when 10 mL of 0.2 molar HCl is added to it.

17.74 Calculate the concentration of $[Ba^{2+}]$ (*a*) for saturated solution of $BaSO_4$ in H_2O

(*b*) for saturated solution of $BaSO_4$ in 0.16 M H_2SO_4. $K_{sp}(BaSO_4) = 1.0 \times 10^{-10}$.

17.75 Calculate the concentration of each species (*a*) for saturated solution of $BaCO_3$($K_s = 5 \times 10^{-9}$) assuming no hydrolysis of any ion.

(*b*) for saturated solution of $BaCO_3$ in water assuming the following reactions:

$$CO_3^{2-} + H_2O \rightleftharpoons HCO_3^- + OH^- \quad K_1 = 2.13 \times 10^{-4}$$

$$HCO_3^- + H_2O \rightleftharpoons H_2CO_3 + OH^- \quad K_2 = 2.22 \times 10^{-8}$$

17.76 Consider the problem 17.73(*a*) with following more data:

$$Ba^{2+} + OH^- \rightleftharpoons BaOH^+ \qquad K_1 = 4.8$$

$$SO_4^{2+} + H^+ \rightleftharpoons HSO_4^- \qquad K_2 = 9.8$$

$$H_2O \rightleftharpoons H^+ + OH^- \qquad K_w = 10^{-14}$$

Calculate the concentration of each species and show that assumption made in problem 17.73 *a* is valid one.

17.77 Saturated solution of CaF_2 in water can undergo following reactions:

$$CaF_2(s) \rightleftharpoons Ca^{2+} + 2F^- \qquad K_1 = 4.0 \times 10^{-11}$$

$$Ca^{2+} + OH^- \rightleftharpoons CaOH^+ \qquad K_2 = 20$$

$$F^- + H^+ \rightleftharpoons HF \qquad K_3 = 1.5 \times 10^3$$

$$Ca^{2+} + F^- \rightleftharpoons CaF^+ \qquad K_4 = 10$$

Calculate the concentration of each species.

17.78 0.200 dm^3 of 0.6 M NaCl is added to 0.1 dm^3 solution containing 2.4×10^{-3} mol of $AgNO_3$ and 3×10^{-3} mol of $Na_2S_2O_3$. Given

$$K_{sp}(AgCl) = 1.2 \times 10^{-10}$$

and $$K = \frac{[Ag^+][S_2O_3^{2-}]^2}{[Ag(S_2O_3)_2^{3-}]} = 10^{-13} \text{ for the reaction}$$

$$[Ag(S_2O_3)_2]^{3-} \rightleftharpoons Ag^+ + 2S_2O_3^{2-}$$

Calculate the concentration of each species.

17.79 To a solution containing 0.01 M Cd^{2+} and 0.05 M Pb^{2+}, sodium oxalate is added dropwise. (*a*) Calculate $[C_2O_4^{2-}]$ such that lead oxalate precipitate. (*b*) Calculate $[C_2O_4^{2-}]$ such that Cd^{2+} precipitate. (*c*) Calculate the $[Pb^{2+}]$ in solution when Cd^{2+} will start precipitating K_{sp} (lead oxalate) $= 8.3 \times 10^{-12}$, K_{sp} (cadmium oxalate) $= 1.8 \times 10^{-8}$.

17.80 Calculate the pH needed to dissolve 0.002 mol of AgCN completely in 1 dm^3 of solution K_{sp} (AgCN) $= 1.6 \times 10^{-14}$,

$$K = \frac{[H^+][CN^-]}{[HCN]} = 7.2 \times 10^{-10}$$

17.81 Calculate the concentration of each species present in the saturated solution of MnS. Given

$$MnS(s) \rightleftharpoons Mn^{2+} + S^{2-} \qquad K_1 = 2.5 \times 10^{-10}$$

$$Mn^{2+} + OH^- \rightleftharpoons MnOH^+ \qquad K_2 = 2.5 \times 10^3$$

$$Mn(OH)_2(s) \rightleftharpoons Mn^{2+} + 2OH^- \qquad K_3 = 1.6 \times 10^{-13}$$

$$Mn(OH)_2 + OH^- \rightleftharpoons Mn(OH)_3^- \qquad K_4 = 1.0 \times 10^{-5}$$

$$S^{2-} + H_2O \rightleftharpoons HS^- + OH^- \qquad K_5 = 8.0 \times 10^{-2}$$

$$HS^- + H_2O \rightleftharpoons H_2S + OH^- \qquad K_6 = 1.0 \times 10^{-7}$$

Additional Problems

17.82 In an experiment one needs 10 % NaCl solution. How grams of water is required for 30 grams of NaCl?

17.83 Calculate the molarity of 3.4 mol kg^{-1} sucrose solution if the density of the solution is 1.5 g cm^{-3}.

17.84 (*i*) Calculate the molarity and (*ii*) molality of concentrated H_2SO_4 if the density of the 38% (weight by weight) solution is 1.1885 g mol^{-3}.

17.85 If molality of a solution is 2.5 mol kg^{-1}, calculate its molarity. The density of the solution is 0.95 g cm^{-3}.

17.86 If the molarity of the solution is 1.25 mol L^{-1}, calculate its molality of the solution. Density of the solution is 0.9 g cm^{-3}.

17.87 The strength of the alcoholic beverages is usually described in terms of proof which is defined as twice the percentage by volume of ethanol. Calculate the number of grams of alcohol in 2.5 L of 90% proof whisky. Calculate the molality of whisky. Density of ethanol is 0.80 g cm^{-3}.

17.88 In case of non-ideal solution, explain briefly the following observations, *i.e.* $\Delta_{mix}H > 0, \Delta_{mix}H < 0$, $\Delta_{mix}V > 0$ and $\Delta_{mix}V < 0$.

17.89 Explain briefly a nice observation. A miner took a beverage with him in a mine which 1000 ft deep. While taking lunch he opened the beverage and found that the bottle seems to be flat, *i.e.* not much effervescence was observed on opening the bottle. Shortly after lunch the miner came up and started feeling to belch.

17.90 A beverage bottle at room temperature was allowed to freeze. Now when the bottle was opened while frozen or at very low temperature, the effervescence occurs more vigorously as compared to that when the bottle is opened at room temperature.

17.91 The freezing point of water in presence of 0.02 m acetic acid is found to be 0.025 K. Calculate the degree of dissociation of acetic acid.

17.92 Assuming that $K_3[Fe(CN)_6]$ is strong electrolyte, calculate the freezing point of water for 0.015 m solution.

17.93 The osmotic pressure of blood plasma is approximately 7.5 atm at 37°C. Estimate the total concentration of dissolved species and the freezing point of blood plasma.

17.94 Calculate the osmotic pressure of a solution containing 200 g of haemoglobin in one liter of water at 298 K. Molar mass of haemoglobin is 65,000 g mol^{-1}.

17.95 There are two compartments. Compartment one which we can refer as the solvent compartment contains 0.0002 M concentration of Na_3OP protein. Calculate the osmotic pressure of the solution at 298 K if the second compartment which we can refer as solution contains 0, 0.10 M, 0.02 M and 4 M NaCl solution.

17.96 The solubility of nitrogen in blood at 37°C and a partial pressure of 0.80 atm. is 5.6 × 10^{-4} mol L^{-1}. A deep sea diver breaths compressed air with the partial pressure of N_2 equal to 4.0 atm. Assume that the total volume of blood in the body is 5.0 L. Calculate the amount N_2 gas released (in liters) when the diver returns to the surface of water, where the partial pressure of N_2 is 0.8 atm.

17.97 Based on the results of problem 17.96, when the diver swims from under water to top of the water, he feels dizzy under mild conditions and can lead to death under strong conditions. In order to dilute the oxygen, helium is mixed. Explain briefly.

17.98 Two liquids *A* and *B* of normal boiling points T_A^0 and T_B^0 respectively form ideal solution when different amounts were added. Predict the range of boiling points of solutions formed by mixing different amounts of liquids *A* and *B*.

17.99 The vapour pressures of ethanol and *n*-propanol at 36.4°C are 108 mm Hg and 40.0 mm Hg respectively. (*i*) Calculate the mole fraction of ethanol and *n*-propanol in the mixture of ethanol and *n*-propanol to be 72 mm Hg at 36.4°C, (*ii*) calculate the vapour pressure of the mixture in which the mole fraction of ethanol is 0.4 and (*iii*) the composition of the vapours over the solution.

17.100 Two liquids *A* and *B* form ideal mixture at 75°C. Mixture of 2.4 moles of *A* and 4.0 moles of *B* possess vapour pressure of 450 mm Hg at 75°C. Suppose we add 1 mole of *A* to this mixture the vapour pressure decreases to 400 mm Hg. Calculate the vapour pressures of the pure liquids *A* and *B*. (*ii*) In both cases, calculate the mole fractions of the vapour phase.

17.101 Consider two liquids *A* and *B* which form an ideal mixture over the entire concentration range. The vapour pressures of pure liquids *A* and *B* at 300 K are 250 and 150 mm Hg. (*i*) Calculate the partial pressure of each liquid over the range of concentration in the form of mole fraction and construct the plot between partial pressures of *A* and *B* as well as total pressure as a function of mole fraction, (*ii*) Calculate the total pressure over the binary solution if the mole fraction of liquid *B* is 0.35, (*iii*) Is it possible to have a concentration when $p_A = p_B$ and calculate the mole fraction of each liquid at this condition? (*Hint:* The point where $p_A = p_B$ corresponds to the crossing point of the two lines representing Raoult's law on a plot of ideal vapour pressure *versus* mole fraction). (*iv*) Show that the total equilibrium vapour pressure over the solution at the point where $p_A = p_B$ is given by

$$p_T = \frac{2p_A^0 p_B^0}{p_A^0 + p_B^0}$$

17.102 Continuing with the problem 17.101, let the mole fraction of the liquid *B*, *i.e.* $X_B = 0.357$ when $p_A = p_B$ and p_T is 357.5 mm Hg. Calculate the value of p_A^0 and p_B^0.

17.103 Vapour pressure of water at 45°C is 71.9 mm Hg. Suppose that 12.5 g of $Ca(NO_3)_2$ was dissolved in 400 g of water. (*i*) Calculate the vapour pressure of water at 45°C assuming that there is no dissociation of $Ca(NO_3)_2$, (*ii*) If the vapour pressure of water under these is 55.4 mm Hg, calculate the percentage dissociation, (*iii*) Suppose $Ca(NO_3)_2$ is completely dissociated. Calculate the vapour pressure of water under this condition.

17.104 Using the data from problem 17.103, (*i*) calculate the freezing point of water assuming there is no dissociation of $Ca(NO_3)_2$, (*ii*) if the freezing point of water under these conditions is –0.55° C, calculate the percentage dissociation, (*iii*) suppose $Ca(NO_3)_2$ is completely dissociated. Calculate the freezing point of water under this condition. K_f = 1.86 K mol^{-1} kg.

17.105 Using the data from problem 17.103, (*i*) calculate the boiling point of water assuming there is no dissociation of $Ca(NO_3)_2$, (*ii*) if the boiling point of water under these conditions is 100.35°C, calculate the percentage dissociation of $Ca(NO_3)_2$, (*iii*) suppose $Ca(NO_3)_2$ is completely dissociated. Calculate the boiling point of water under this condition. K_b = 0.512 K mol^{-1} kg.

17.106 0.2 g of lysozyme protein is dissolved in 150 g of water at 298 K. If the molecular weight of lysozyme is 13,930 g mol^{-1}, calculate (*i*) decrease in the relative vapour pressure of water, (*ii*) depression in freezing point, (*iii*) elevation in the boiling points and, (*iv*) the osmotic pressure. The vapour pressure of pure water is 23.76 mm Hg at 298 K.

17.107 The vapour pressure of sea water at 298 K is 21.98 mm Hg. If we assume that sea water contains only NaCl (a strong electrolyte), calculate the concentration of NaCl. Take vapour pressure of pure water at 298 K is 23.76 mm Hg.

17.108 The water rises from the roots to top of the trees by the action of osmosis phenomenon. If the height of tree is 200 ft calculate the osmotic pressure required to raise the water to the top from the root.

17.109 22 g of haemoglobin in 1 L of the solution raises the height of the solution as compared to pure solvent by 7.88 mm. Calculate the molar mass of haemoglobin.

17.110 The solubility of barium sulphate at 20 and 100°C in water are 2.3×10^{-4} and 4.1×10^{-4} g per 100 mL of water. Calculate $\Delta_{fus}\bar{H}$ and $\Delta_{fus}\bar{S}$.

17.111 The solubility of oxygen at 20 and 100°C in water are 4.5×10^{-3} and 3.3×10^{-3} g per 100 mL of water. Calculate $\Delta_{fus}\bar{H}$ and $\Delta_{fus}\bar{S}$.

17.112 The solubility of CO_2 (*g*) in [g of CO_2 per 100 g of H_2O] in water at different temperatures when the partial pressure of $CO_2(g)$ over the solution is 760 mm Hg.

T (K)	273	283	288	293	298	303	313	323	333
Solubility	0.3346	0.2318	0.1970	0.1688	0.1449	0.1257	0.0973	0.0761	0.0576

(*i*) Plot In $\chi^{sat}_{CO_2}$ versus T^{-1}. Does the degree of linearity indicate that the solution is ideal?

(*ii*) Assuming the solution to be ideal, what is the predicted molar enthalpy of vaopurisation of CO_2.

17.113 Repeat the problem 17.112 for Li_2SO_4 from the data given below:

T (K)	273	293	313	333	353	373
Mass % $LiSO_4$	26.3	25.6	25.3	24.8	24.0	23.6

17.114 Calculate the solubility of $Ni_3(PO_4)_2$ at 298 K. K_{sp} for this molecule is 4.74×10^{-32}.

17.115 Repeat the problem 17.114 for $Ni_3(PO_4)_2$ in the absence of 0.1 M $BaCl_2$ and in presence of 0.01 M $BaCl_2$. Assume that $Ba_3(PO_4)_2$ does not precipitate and $BaCl_2$ ionises completely. K_{sp} $[Ni_3(PO_4)_2] = 4.74 \times 10^{-32}$.

CHAPTER 18

NON-IDEAL SOLUTIONS

18.1. INTRODUCTION

It has been assumed in previous chapters that the systems behave in an ideal manner and all the thermodynamical relations have been derived with this model. Due to departure from the ideal behaviour, the derived thermodynamic relations have to be modified and thus true picture of the systems can be described, although it is difficult. Before we discuss the main topics, let us learn about the thermodynamics of dissolution of ionic compounds in water. This chapter is mainly divided into two sections:

(*i*) Partial molal quantities *e.g.* partial molal volume, enthalpy etc.

(*ii*) Activity and activity coefficients, application of Debye-Hückel limiting law.

18.2. DISSOLUTION OF IONS IN WATER

We have learnt in Chapter 4 that the ions are held together by coulombic forces. For example, in solid NaCl, Na^+ cation and Cl^- anion are held together by electrostatic forces in a crystal lattice. Two kinds of forces are responsible for the solution of electrolytes in solvents. We will take the solvent as water as it is very commonly used one.

18.2.1. *Hydration*

Each ion is surrounded by a number of water molecules, called hydration number of an ion. Hydration number is directly proportional to the charge density of an ion and inversely proportional to the size of the ion. It may be pointed out that the water in hydration shell has different properties than in bulk water. For example, water molecules in the hydration sphere of an ion do not exhibit individual translation motion. They move with the ion as whole. Because of this reason size of the solvated ion is greater than that of the ion in the gas phase or their crystal or ionic radii. For example, the radii of the hydrated Li^+, Na^+ and K^+ ions are estimated to be 0.366 nm, 0.280 nm and 0.187 nm respectively. Water molecules in the hydration sphere and water molecules in the bulk water are in dynamical equilibrium, *i.e.* water molecules from the hydration sphere moves into the bulk water and water molecules from bulk water moves into the hydration sphere. Because of the different interactions with ions, the mean life time of water molecule in the hydration sphere can have different values, *e.g.* for Br^- ion it is 10^{-11} s, Na^+, 10^{-9} s, Cu^{2+}, 10^{-7} s; Fe^{3+}, 10^{-5} s, Al^{3+}, 7 s and Cr^{3+}, 1.5×10^5 s or 42 hr.

As learnt in Chapter 4, ion-dipole interactions can affect many of the bulk properties of the water molecules. Depending on the magnitude of these interactions these ions can be clubbed into either structure making or structure breaking. Small ions or multi-charged ions can exert strong electric fields, which can polarise the water molecules, producing additional order beyond the first hydration layer. This interaction can increase the viscosity of the solution. The examples can be: Li^+, Na^+, Mg^{2+}, Al^{3+}, Er^{3+}, OH^- and F^-.

On the other hand, large monovalent ions such as K^+, Rb^+, Cs^+, NH_4^+, Cl^-, NO_3^-, ClO_4^-, possess diffuse surface charges and hence weak electric field. These weak ions fail to polarise the water molecules beyond the first layer of hydration. Consequently the viscosities of solutions containing these ions are usually lower than that of pure water. These ions are called *structure breaking* solvents.

18.2.2. *Lattice Energy*

Ionic interactions between the ions due to coulombic interactions in vacuum is given by

$$F = \frac{q_{M^{n+}} q_{M^{m-}}}{4\pi\varepsilon_0 r^2} \qquad \text{...(18.1)}$$

where $q_{M^{n+}}$ and $q_{M^{m-}}$ are electronic charge on the cations and anions and r is the distance between the two ions. $4\pi\varepsilon_0$ is the permittivity constant to convert *cgs* units to SI unit and thus the force between the two species is given in newton. If the solvent molecules are present in between the two ions, electrostatic interaction decreases and it is given by equation (18.2) as

$$F = \frac{q_{M^{n+}} q_{M^{m-}}}{4\pi\varepsilon_0 \varepsilon r^2} \qquad \text{...(18.2)}$$

where ε is the dielectric constant of the medium. Dielectric constant depends on temperature. For example, the dielectric constant of water at 298 K is 78.54 and at 343 K is nearly 64. Dielectric constants of some commonly used solvents are compiled in Table 18.1.

TABLE 18.1. Dielectric constants of some commonly used solvents at 298 K

Liquid	Dielectric constant (ε)	Liquid	Dielectric constant (ε)
H_2SO_4	101	CH_3OH	32.6
H_2O	78.54	C_2H_5OH	24.3
$(CH_3)_2SO$	49	CH_3COCH_3	20.7
$C_3H_8O_2$	42.5	CH_3COOH	6.2
CH_3NO_2	38.6	C_6H_6	4.6
$HOCH_2CH_2OH$	37.7	$C_2H_5OC_2H_5$	4.3
CH_3CN	36.2	CS_2	2.6

***Exercise* 18.1:** Calculate the force between Na^+ and Cl^- ions at $r = 1$ nm distance. (*i*) in vacuum and (*ii*) in water at 298 K. ε (H_2O) = 78.54, $q_{Na^+} = 1.6 \times 10^{-19}$ C and $q_{Cl^-} = -1.6 \times 10^{-19}$ C.

Using equation (18.1) and substituting the values, we get

$$F = \frac{q_{M^{n+}} q_{M^{m-}}}{4\pi\varepsilon_0 r^2}$$

$$= \frac{(1.6 \times 10^{-19}\,\text{C})(-1.6 \times 10^{-19}\,\text{C})}{(4)(3.14)(8.854 \times 10^{-12}\ \text{C}^2\ \text{N}^{-1}\ \text{m}^{-2})(1 \times 10^{-9}\,\text{m})^2}$$

$$= -2.31 \times 10^{-10}\ \text{N}$$

(*ii*) Using equation (18.2) and substituting the values, we get

$$F = \frac{q_{M^{n+}} q_{M^{m-}}}{4\pi\varepsilon_0 \varepsilon r^2}$$

$$= \frac{(1.6 \times 10^{-19}\,\text{C})(-1.6 \times 10^{-19}\,\text{C})}{(4)(3.14)(8.854 \times 10^{-12}\ \text{C}^2\ \text{N}^{-1}\ \text{m}^{-2})(78.54)(1 \times 10^{-9}\,\text{m})^2}$$

$$= -2.94 \times 10^{-12}\ \text{N}.$$

18.2.3. *Thermodynamics of Ions in Solution*

Thermodynamics of dissolution of electrolytes can be represented as: (again taking example of NaCl)

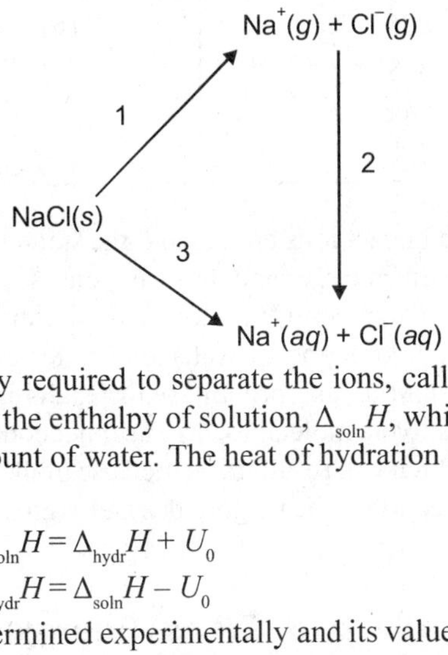

The step 1 involves the energy required to separate the ions, called as *lattice energy* (U_0). The enthalpy change for the process 3 is the enthalpy of solution, $\Delta_{soln}H$, which is heat absorbed or released when NaCl dissolved in a large amount of water. The heat of hydration of ions $\Delta_{hydr}H$ can be obtained with the help of Hess's law, *i.e.*

$$\Delta_{soln}H = \Delta_{hydr}H + U_0$$

$$\Delta_{hydr}H = \Delta_{soln}H - U_0 \qquad \text{...(18.3)}$$

The quantity $\Delta_{soln}H$ can be determined experimentally and its value is $\Delta_{soln}H = 3.8$ kJ mol^{-1} and the value of U_0 can be determined if the structure of the lattice is known. In case of NaCl, the value of U_0 is 787 kJ mol^{-1}. So using equation (18.3) we can calculate the value of $\Delta_{hydr}H$ as

$$\Delta_{hydr}H = 3.8 - 787 = -783.2\ \text{kJ mol}^{-1}$$

From the above results it is clear that the hydration energy is very large. It may be mentioned here that the hydration energy is obtained from the contribution of hydration from each ion. It is not possible to calculate experimentally the hydration energy of individual ion. The hydration energy value of the process

$$H^+(g) \rightarrow H^+(aq) \quad ...(18.4)$$

has been studied theoretically and found to be –1089 kJ mol^{-1}. Using this value of $\Delta_{hydr}H$ for H^+ ion, the values of other ions can be obtained from the hydration energy of the electrolytes. Values of $\Delta_{hydr}H^0$, $\Delta_{hydr}S^0$ and ionic radii for some occurring ions are compiled in Table 18.2.

TABLE 18.2. Values of standard $\Delta_{hydr}H^0$, $\Delta_{hydr}S^0$ and ionic radii for few common ions at 1 bar

Ion	$-\Delta_{hydr}H^0$ (kJ mol^{-1})	$-\Delta_{hydr}S^0$ (Jk^{-1} mol^{-1})	Ionic Radius (nm)
H^+	1089	132	–
Li^+	520	119	0.060
Na^+	405	89	0.095
K^+	314	51	0.133
Ag^+	468	94	0.126
Mg^{2+}	1926	268	0.065
Ca^{2+}	1579	209	0.099
Ba^{2+}	1309	159	0.135
Mn^{2+}	1832	243	0.080
Fe^{2+}	1950	272	0.076
Cu^{2+}	2092	259	0.072
Fe^{3+}	4355	460	0.064
F^-	506	151	0.136
Cl^-	378	96	0.181
Br^-	348	80	0.195
I^-	308	60	0.216

Hydration energy of the compact ions are much larger than the ions of bigger ions as the charge density of the compact ions are much larger and thus can polarise the water molecules beyond the first hydration shell of water. Two factors contribute to standard entropy of hydration. (*i*) The hydration process can result in considerable ordering of water molecules around the ion and thus will lead to reduction in entropy of the system as this ordering will results in the considerable reduction of the various kinds of motions of the solvent molecules. (*ii*) The other contribution results from the dissociation of the ionic molecules to ions which will result in the increase of the entropy. Thus the two contributions are occurring in the opposite directions. The magnitudes and sign of $\Delta_{hydr}S^0$ depend on the magnitudes of these two factors.

18.2.4. *Enthalpy, Entropy and Gibbs Energy of Formation of Ions in Solution*

We have learnt above how to calculate the hydration energy of the individual ions. The enthalpy of formation of ions cannot be measured individually. To solve this problem it is assumed and assigned arbitrary value of enthalpy of formation to hydrogen ion, *i.e.* $\Delta_f H^0[H^+(aq)] = 0$ and thus the values of $\Delta_f H^0$ for other ions can be obtained relative to this value. For example, see the Exercise 18.2.

Exercise **18.2:** Consider the reaction:

$$\frac{1}{2}H_2(g)+\frac{1}{2}Cl_2(g) \rightarrow H^+(aq)+Cl^-(aq) \quad \Delta_r H^0 = -167.2 \text{ kJ mol}^{-1}$$

Calculate the $\Delta_f \bar{H}^0[Cl^-(aq)]$

The enthalpy of reaction ($\Delta_r H^0$), which can be determined experimentally, can be written as:

$$\Delta_r H^0 = \Delta_r \bar{H}^0[H^+(aq)] + \Delta_r \bar{H}^0[Cl^-(aq)] - \frac{1}{2}(0) - \frac{1}{2}(0)$$

so that

$$\Delta_r H^0 = \Delta_r \bar{H}^0[Cl^-(aq)] = \Delta_f \bar{H}^0[Cl^-(aq)] = -167.2 \text{ kJ mol}^{-1}$$

Once $\Delta_f \bar{H}^0[Cl^-(aq)]$ is obtained, $\Delta_f \bar{H}^0$ of other ions can be obtained. Values of standard $\Delta_f \bar{G}^0$ and $\bar{S}^0$ can also be determined on the same lines as we have obtained $\Delta_f \bar{H}^0$ for the ions, *i.e.* assigning $\Delta_f \bar{G}^0$ and $\bar{S}^0$ the values to zero. Because $\bar{S}^0$ values of ions in aqueous solution are relative to that of the H^+ ion, $\bar{S}^0$ may be positive or negative. The magnitude and sign of the entropies reinfluenced by the extent to which they can order the water molecules around themselves in solution as compared to $H^+(aq)$. The values of $\Delta_f \bar{H}^0$, $\Delta_f \bar{G}^0$ and $\bar{S}^0$ some ions are compiled in Table 18.3. Small, highly charged ions have negative entropy values, whereas large, singly charged ions have positive entropy.

TABLE 18.3. Values of standard $\Delta_r H^0$, $\Delta_{hydr} S^0$ and ionic radii for few common ions at 1 bar

Ion	$\Delta_f \bar{H}^0$ (kJ mol^{-1})	$\Delta_f \bar{G}^0$ (kJ mol^{-1})	$\bar{S}^0$ (J K^{-1} mol^{-1})
H^+	0	0	0
Li^+	–278.5	–293.8	14.23
Na^+	–239.7	–261.9	60.25
K^+	–251.2	–282.3	102.5
Mg^{2+}	–462.0	–456.0	–138.1
Ca^{2+}	–543.0	–553.0	–55.23
Fe^{2+}	–87.9	–84.9	–137.7
Zn^{2+}	–152.4	–147.2	–112.1
Fe^{3+}	–47.7	–4.7	–293.3
OH^-	–229.9	–157.3	–10.54
F^-	–329.1	–276.5	–13.8
Cl^-	–167.2	–131.2	56.5
Br^-	–120.9	–102.8	80.71
I^-	–55.9	–51.7	109.37
CO_3^{2-}	–676.3	–528.1	–53.14
NO_3^-	–206.6	–110.5	146.4
PO_4^{3-}	–1284.1	–1025.6	–217.6

***Exercise* 18.3:** The standard $\Delta_r \bar{H}^0$ for the reaction

$$Na(s) + 1/2\ Cl_2(g) \rightarrow Na^+(aq) + Cl^-(aq)$$

is -406.9 kJ mol^{-1}. Calculate the $\Delta_f \bar{H}^0$ for $Na^+(aq)$ if $\Delta_f \bar{H}^0$ for $Cl^{-1}(aq) = -167.2$ kJ mol^{-1}.

Using the desired equation for the heat or reaction and substituting the desired value we get:

$$\Delta_r \bar{H}^0 = \Delta_f \bar{H}^0 (Na^+) + \Delta_f \bar{H}^0 (Cl^-) - \Delta_f \bar{H}^0 [Na(s)] - 1/2\ \Delta_f \bar{H}^0\ [Cl_2(g)]$$

$$-406.9 \text{ kJ mol}^{-1} = \Delta_f \bar{H}^0\ (Na^+) \text{ kJ mol}^{-1} - 167.2 \text{ kJ mol}^{-1} - 0 - 0$$

$$\Delta_f \bar{H}^0\ (Na^+) = -406.9 \text{ kJ mol}^{-1} + 167.2 \text{ kJ mol}^{-1} = -239.7 \text{ kJ mol}^{-1}$$

18.3. PARTIAL MOLAL PROPERTIES

Mathematically it is defined as

$$\left(\frac{\partial J}{\partial n_i}\right)_{T,p,n_j} = \bar{J}_i \quad ...(18.5)$$

where $\bar{J}_i$ is partial molal properties of *i*th component. Physically it means that increase in thermodynamic quantity J observed when one mol of component i is added to such a large system that its composition remains constant.

At constant pressure and temperature equation (18.5) can be written as

$$dJ = \sum_i \bar{J}_i dn_i \quad ...(18.6)$$

and can be integrated to

$$J = \sum_i n_i \bar{J}_i \quad ...(18.7)$$

The physical meaning of this integration is that to a solution of fixed composition, a solution of components n_1, n_2, ..., n_i (whose composition also resembles with the old solution) are added further so that the relative composition of each species remains constant. Due to this partial molal quantity remains same and integration is carried out over the number of mol.

There are three main partial molal thermodynamic properties *i.e.* (*i*) partial molal volumes of components in solution, (*ii*) partial molal enthalpies (also called as differential heat of solution) and (*iii*) partial molal free energy (called chemical potential). These properties can be determined with the help of (*i*) graphical method, (*ii*) using analytical relations representing J and n_i and (*iii*) using a function called apparent molal quantity defined as

$$\phi_{J_i} = \frac{J - n_i \bar{J}_i^0}{n_i} \quad ...(18.8)$$

where $\tilde{J}_i^0$ is the molal value for pure component and using methods of intercept. One thing must be kept in mind that partial molal property of a component in a solution and molal property for pure component is same if the solution is ideal.

18.4. CALCULATION OF PARTIAL MOLAL QUANTITIES

18.4.1. *Graphical Method*

In this case J is plotted as a function of composition of solution, keeping the amount of all the other components fixed except one.

(*i*) If this plot is linear, the slope of the line will be the partial molal quantity of that component. This also suggests that the partial molal properties of the components are also independent of concentration.

(*ii*) If the plot of J in the above case is not linear then

$$\bar{J}_i = \left(\frac{\Delta J}{\Delta n_i}\right)_{T,p,n_j}$$

$$\Delta n \to 0$$

i.e. $\left(\frac{\partial \Delta J}{\partial n_j}\right)_{T,p,n_j}$ vs n_i is again plotted and intercept at $n_i = 0$ is $\tilde{J}_i$.

(*a*) *Partial Molal Volume:* The above relations hold good for partial molal volume.

(*b*) *Partial Molal Enthalpy:* Since absolute value of enthalpy cannot be determined with the help of thermodynamics, one defines relative partial molal enthalpy as

$$L_i = \bar{H}_i - \bar{H}_i^0 \qquad ...(18.9)$$

where $\bar{H}_i^0$ is the partial molal enthalpy of component i in the standard state *i.e.* the infinitely dilute solution. Thus for binary mixture above relations get modified as

$$\left(\frac{\partial \Delta H}{\partial n_1}\right)_{n_2} = \bar{L}_1 \qquad ...(18.10)$$

for solvent and

$$\left(\frac{\partial \Delta H}{\partial n_2}\right)_{n_1} = \bar{L}_2 - L_2 \qquad ...(18.11)$$

where L_2 is the relative molal enthalpy of solute.

Since $\lim_{n_2 \to 0} \bar{L}_2 = 0$, so

$$\lim_{n_2 \to 0}\left(\frac{\partial \Delta H}{\partial n_2}\right)_{n_1} = -L_2 \qquad ...(18.12)$$

Thus the process involves the calculation of first L_2 and then $\bar{L}_2$.

***Exercise* 18.4:** Different amount of glycine was dissolved in 1 kg of water and density of each solution as a function of molality obtained is as follows:

m_2	0	0.1009	0.20348	0.619	1.9629	1.0496
d (kg dm^{-3})	0.997074	1.00026	1.003489	1.01598	1.05167	1.02821
V (dm^3)	1.0029	1.0073	1.0117	1.0299	1.0909	1.0491

Calculate the partial molal volume of glycine.

From the given data, the total volume of the solution can be obtained by the relation

$$V = \frac{1\text{ kg of water} + \text{Amount of solute}}{\text{Density of solution}}$$

i.e.
$$V = \frac{1\text{ kg} + (0.1009\text{ mol})(75\text{ g mol}^{-1})(10^{-3}\text{ kg g}^{-1})}{1.00026\text{ kg dm}^{-3}}$$

$$= 1.0073\text{ dm}^3$$

Similarly other values of V are calculated and listed in the above table. The values of V are plotted against m_2 in Fig. 18.1. The plot is linear and is the case I *i.e.* slope is the value of partial molal volume of glycine.

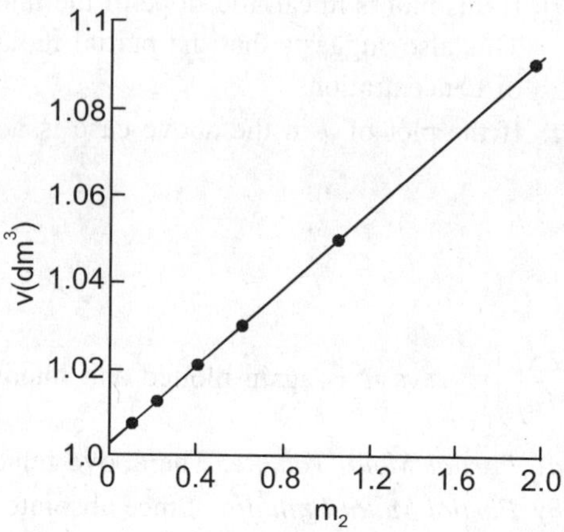

Fig. 18.1

$$\text{Slope} = \frac{(1.074 - 1.025)\text{ dm}^3}{(1.6 - 0.5)\text{ mol}} = 0.445\text{ dm}^3\text{ mol}^{-1}$$

The partial molal volume of glycine is 0.0445 dm^3 mol^{-1} and is independent of concentration in this range.

***Exercise* 18.5:** Densities of formic acid and water mixtures at different proportions are given below. Calculate the partial molal volume of formic acid at 46 molal solution.

wt % formic acid	$\frac{[HCO_2H]}{[H_2O]} = \frac{n_2}{n_1}$	$\frac{[H_2O]}{[HCO_2H]} = \frac{n_1}{n_2}$	d (kg dm^{-3})	$V \times 10^3$ dm^3 (mol H_2O)$^{-1}$	$\frac{\Delta V}{\Delta n_2}$
1	2	3	4	5	6
10	0.0434	23.041	1.0246	19.516	35.596
20	0.0979	10.216	1.0488	21.456	35.974
30	0.1677	5.965	1.0727	23.967	36.474
40	0.2610	3.831	1.0995	27.370	36.477
50	0.3913	2.556	1.1207	32.123	36.932
55	0.478	2.090	1.1320	35.325	37.309
60	0.5869	1.704	1.1424	39.388	36.959
65	0.727	1.376	1.1543	44.566	36.677
70	0.9130	1.095	1.1655	51.388	37.521
75	1.174	0.852	1.1769	61.18	37.588
80	1.565	0.639	1.1860	75.877	37.58

85	2.218	0.450	1.1953	100.417	37.60
90	3.518	0.284	1.2044	149.30	37.59
95	7.428	0.135	1.2140	296.28	
100			1.2212		

The partial molal volume of formic acid can be calculated from the above data if one knows how the volume increases by adding different amount of formic acid to a fixed quantity of water (let it be one mol).

(*i*) The column 2 gives the number of mol of formic acid per mol of water. This had been calculated as: Consider 10% HCO_2H solution.

$$n_2 = \frac{10\text{ g}}{46\text{ g mol}^{-1}} = 0.2172\text{ mol},\ n_1 = \frac{90\text{ g}}{18\text{ g mol}} = 5\text{ mol}$$

$$\frac{n_2}{n_1} = \frac{0.2172\text{ mol}}{5\text{ mol}} = 0.0434\text{ mol of HCO}_2\text{H per mol of H}_2\text{O.}$$

Similarly the other data has been calculated in column 2.

(*ii*) Column 3 gives the number of mol of water per mol of formic acid *i.e.*

$$\frac{1}{n_2/n_1} = \frac{n_1}{n_2} = \frac{1}{0.0434} = 23.041$$

mol of water per mol of formic acid.

(*iii*) Column 5 list the volume of n_2 mol of formic acid and one mol of water. This has been calculated as follows:

Consider 10% HCO_2H solution. The amount (W) containing n_2 mol of HCO_2H and 1 mol of water will be

$$\begin{aligned} W &= n_2 \times M_{HCO_2H} + 1 \times M_{H_2O} \\ &= (0.0434\text{ mol})(46\text{ g mol}^{-1}) + (1\text{ mol})(18.0\text{ g mol}^{-1}) = 19.996\text{ g} \\ &= 19.996 \times 10^{-3}\text{ kg} \end{aligned}$$

$$\text{So volume } V = \frac{W}{d} = \frac{19.966 \times 10^{-3}\text{ kg}}{1.0246\text{ kg dm}^{-3}} = 19.516 \times 10^{-3}\text{ dm}^3$$

(*iv*) Column 6 list the $\Delta V/\Delta n_2$ at $n_1 = 1$ mol. This has been calculated by taking the two successive observations: *e.g.* 10% and 20% HCO_2H solution.

$$\frac{\Delta V}{\Delta n_2} = \frac{V(20\%) - V(10\%)}{n_2/n_1(20\%) - n_2/n_1(10\%)} = \frac{(21.456 - 19.516) \times 10^{-3}}{0.0979 - 0.0434}$$

$$= 35.596 \times 10^{-3}\text{ dm}^3\text{ mol}^{-1}\ (\text{HCO}_2\text{H})$$

Similarly the other data is calculated and listed in the column 6.

(*v*) $\dfrac{\Delta V}{\Delta n_2} = \dfrac{\Delta V}{\Delta(n_2/n_1)}$ is plotted vs n_2 in Fig. 18.2. The plot is not a straight line *i.e.* case II of

18.4.1. The V_2 at 6 molal is calculated by drawing the tangent at the point $\left(\frac{n_2}{n_1} = \frac{6 \text{ mol}}{55.556 \text{ mol}}\right) = 0.108$ mol (mol $H_2O)^{-1}$ and slope is 45.15×10^{-3} dm^3 mol^{-1} *i.e.* the partial molal volume of HCO_2H.

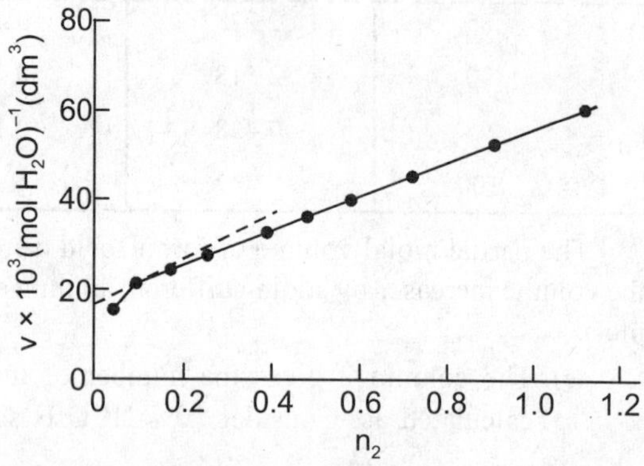

Fig. 18.2

18.4.2. *Analytical Methods*

If the value of extensive property can be expressed as an algebraic function of the composition, the partial molal property can be calculated analytically.

***Exercise* 18.6:** Consider a hypothetical solute dissolved in 1 kg of water. The volume V (mL) at 25°C and 1 atm can be expressed as

$$V = 1000.38 + 20.563\, m_2 + 2.024\, m_2^2 - 0.24\, m_2^3$$

(*i*) Derive the relation for the partial molal volume of solute A and calculate its value at 1 molal solution.

(*ii*) Calculate the partial molal volume of water also. Since the amount of solvent is fixed, using equation (18.5)

$$\bar{V}_2 = \left(\frac{\partial V}{\partial n_2}\right)_{n_1} = \left(\frac{\partial V}{\partial m_2}\right)_{n_1} = 20.563 + 4.048\, m_2 - 0.72\, m_2^2$$

Substituting $m_2 = 1$, $\bar{V}_2 = 25.331$ dm^3.

(*iii*) $n_1 = \dfrac{1000 \text{ g}}{18.0 \text{ g mol}^{-1}} = 55.55$ mol

Using equation $d\tilde{J}_1 = -\dfrac{n_2}{n_1} d\tilde{J}_2$

and $\qquad V_1 = 18.07 \text{ mL mol}^{-1} = 18.07 \times 10^{-3} \text{ dm}^3 \text{ mol}^{-1}$

i.e.,

$$\tilde{J}_1 - \tilde{J}_1^0 = \bar{V} - \bar{V}_1^0 = -\int \frac{n_2}{n_1} d\bar{V} \qquad \text{...(18.13)}$$

n_1 is constant and $\bar{V}_1^0$ is the volume of pure solvent. The molal volume of the solvent can be calculated by integrating the integral between limits $n_2 = 0$ and $n_2 = n_2$

$$\bar{V}_1 - \bar{V}_1^0 = -\frac{1}{n_1}\int_0^{n_2} n_2\, dV_2 \qquad \text{...(18.14)}$$

Now $d\bar{V}_2 = 4.048 - 1.44\, m_2^2$

$$\bar{V}_1 - 18.07 = -\frac{1}{55.55}\int_0^{m_2} (4.048\, m_2 - 1.44\, m_2^2)\, dm_2$$

$$\bar{V}_1 = 18.07 - \frac{1}{55.55}(2.024\, m_2^2 - 0.48\, m_2^3)$$

for $m_2 = 1$

$$\bar{V}_1 = 18.07 - 0.045 = 18.12 \times 10^{-3}\ \text{dm}^3$$

***Exercise* 18.7:** Consider a hypothetical solute A and when n_2 mol of A is dissolved in 1000 g of water, the ΔH can be expressed as

$$\Delta H = 20.5\, m_2 + 8.4\, m_2^2$$

Calculate $\bar{L}_2$ and $\bar{L}_1$ when $m_2 = 2$ M.

Using equation (18.12) and here $n_2 = m_2$

$$\bar{L}_{2_{m_2 \to 0}} = \left(\frac{\partial \Delta H}{\partial m_2}\right)_{n_1} = 20.5 + 16.8\, m_2 = 20.5$$

Using equation (18.11)

$$\bar{L}_2 = 20.5 + 16.8\, m_2 - 20.5 = 16.8\, m_2$$

$$\bar{L}_2 = (16.8)(2) = 33.6\ \text{J mol}^{-1}$$

Using equation (18.14) and substituting proper values of m_2

$$\bar{L}_1 = -\frac{1}{55.55}\int_0^{m_2}(m_2)(16.8\, dm_2) = -\frac{8.4\, m_2^2}{55.55}$$

for $\quad m_2 = 2$ M

$$\bar{L}_1 = -\frac{(8.4)(2)^2}{55.55} = -0.605\ \text{J mol}^{-1}$$

18.4.3. *Partial Molal Quantities from Apparent Molal Quantities*

Rewriting the equation (18.8) for binary system

$$\tilde{J} = n_1 \tilde{J}_1^0 + n_2 \phi J_2 \qquad \text{...(18.15)}$$

$$\tilde{J}_2 = \left(\frac{\partial J}{\partial n_2}\right)_{n_1} = \phi J_2 + n_2\left(\frac{\partial\, \phi J_2}{\partial n_2}\right)_{n_1} \qquad \text{...(18.16)}$$

and

$$\tilde{J}_1 = \left(\frac{\partial J}{\partial n_1}\right)_{n_2} = \tilde{J}_1^0 + n_2\left(\frac{\partial\, \phi J_2}{\partial n_1}\right)_{n_2} \qquad \text{...(18.17)}$$

For electrolytes the equations (18.16) and (18.17) are modified as

$$\tilde{J}_2 = \phi\, \tilde{J}_2 + \frac{\sqrt{n_2}}{2}\left(\frac{\partial \phi J_2}{\partial \sqrt{n_2}}\right)_{n_1} \qquad \text{...(18.18)}$$

$$\tilde{J}_1 = \tilde{J}_1^0 - \frac{n_2^{3/2}}{2}\left(\frac{\partial \phi J_2}{\sqrt{n_1}}\right)_{n_2} \qquad \text{...(18.19)}$$

***Exercise* 18.8:** If d and d_1 are the densities of pure water and water in which w_2 kg of solute is dissolved in 1000 g of water respectively, calculate ϕV_2.

If $J = V$, the equation (18.8) becomes

$$V = n_1 \bar{V}_1^0 + n_2 \phi V_2$$

Now

$$\bar{V}_1^0 = \frac{1\text{ kg}}{d\text{ kg dm}^{-3}} = \frac{1}{d}\text{ dm}^3$$

$$V = \frac{1 + w_2}{d_1}\text{ dm}^3 \text{ and } n_2 = \frac{w_2}{M_2} \times 10^3 = m_2$$

Equation (18.8) becomes

$$\phi V_2 = \frac{1}{n_2}\left[\frac{1+w}{d_1} - \frac{1}{d}\right]$$

It can be rearranged as

$$\phi V_2 = \frac{1}{d_2}\left[\frac{M_2}{10^3} + \frac{1}{dm_2}(d - d_1)\right]\text{dm}^3 \quad \text{...(18.20)}$$

***Exercise* 18.9:** Derive general expression for apparent molal volume utilising the data of exercise 18.6 and calculate $\bar{V}_2$ for 1 molal solution

$$V_1 = 18.07 \times 10^{-3}\text{ dm}^3\text{ mol}^{-1}, \quad n_1 = \frac{1\text{ kg}}{(18\text{ g mol}^{-1})(10^{-3}\text{ kg g}^{-1})}$$

$$= 55.555\text{ mol}$$

and $n_2 = m_2$, use equation (18.8)

$$\phi V_2 = \frac{1001.38 + 20.563\, m_2 + 2.024\, m_2^2 - 0.24\, m_2^3 - (55.556)(18.07)}{m_2}$$

$$= 20.563 + 2.024\, m_2 - 0.24\, m_2^2$$

$$\left(\frac{\partial \phi V_2}{\partial m_2}\right)_{n_1} = 2.024 - 0.48\, m_2$$

Using equation (18.16)

$$\bar{V}_2 = 20.563 + 2.024\, m_2 - 0.24\, m_2^2 + m_2(2.024 - 0.48\, m_2)$$

$$= 20.563 + 4.048\, m_2 - 0.72\, m_2^2$$

If $m_2 = 1$

$$\bar{V}_2 = 25.33\text{ ml mol}^{-1} = 25.33 \times 10^{-3}\text{ dm}^3\text{ mol}^{-1}$$

***Exercise* 18.10:** Following data is given for the glycine in water at 25°C. (*a*) Calculate the apparent molal volume of glycine, (*b*) partial molal volume of glycine and (*c*) the partial volume of water at 1 molal solution.

m (mol kg^{-1})	0	0.10129	0.20348	0.619	1.0496	1.9629
d (kg dm^{-3})	0.99704	1.000287	1.003489	1.01598	1.02821	1.05167
$\phi V_2 \times 10^{-3}$ (dm^3 mol^{-1})		43.24	43.30	43.73	44.07	44.85

Using equation (18.20), the apparent molal volumes at different concentrations can be calculated. As a sample one value is calculated as:

$$\phi V_2 = \frac{1}{(1.000287 \text{ kg dm}^{-3})}\left[(75 \text{ g mol}^{-1})(10^{-3} \text{ kg g}^{-1}) + \frac{(1 \text{ kg})(0.997074 - 1.00287)(\text{kg dm}^{-3})}{(0.10129 \text{ mol})(0.997074 \text{ kg dm}^{-3})}\right]$$

$$= \frac{1}{1.000287}[0.075 - 0.0318] \text{ dm}^3 \text{ mol}^{-1} = 0.0432 \text{ dm}^3 \text{ mol}^{-1}$$

The other values of ϕV_2 are mentioned in the above table. ϕV_2 vs m is plotted in Fig. 18.3. The plot is a straight line and slope

$$= \frac{(10^{-3})(44.8 - 44.22) \text{ dm}^3 \text{ mol}^{-1}}{(1.86 - 1.20) \text{ mol}} = (0.8788 \times 10^{-3}) \text{ dm}^3 \text{ mol}^{-2}$$

From plot ϕV_2 at 1 mol = 44.042×10^{-3} dm^3 mol^{-2} and using equation (18.16)

$$\bar{V}_2 = 44.042 \times 10^{-3} \text{ dm}^3 \text{ mol}^{-1} + (1 \text{ mol})(0.8788 \times 10^{-3} \text{ dm}^3 \text{ mol}^{-2})$$

$$= 44.92 \times 10^{-3} \text{ dm}^3 \text{ mol}^{-1}$$

Using equation (18.8)

$$\bar{V}_1 = \frac{V - n_2\bar{V}_2}{n_1}, \; n_1 = \frac{1000 \text{ g}}{18 \text{ g mol}^{-1}} = 55.556 \text{ mol}$$

Now

$$V = \frac{1 \text{ kg} + (0.1009 \text{ mol})(75 \text{ g mol}^{-1})(10^{-3} \text{ kg g}^{-1})}{1.00026 \text{ kg dm}^{-3}}$$

$$= 1.00731 \text{ dm}^3$$

$$\bar{V}_1 = \frac{1.00731 \text{ dm}^3 - (0.1009 \text{ mol})(44.92 \times 10^{-3} \text{ dm}^3 \text{ mol}^{-1})}{55.556 \text{ mol}}$$

$$= 0.01805 \text{ dm}^3 \text{ mol}^{-1}$$

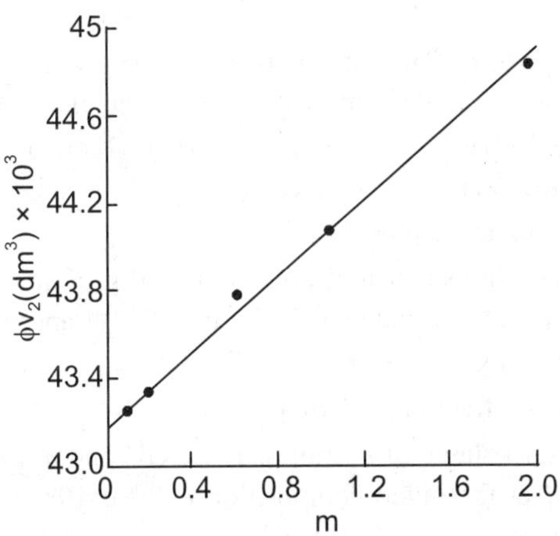

Fig. 18.3

Apparent Molal Enthalpies: The apparent molar enthalpy and relative molar enthalpy can be defined as

$$\phi H_2 = \frac{H - n_1\bar{H}_1^0}{n_2} \qquad ...(18.21)$$

$$\phi L_2 = \frac{L - \bar{L}^0}{n_2} \qquad ...(18.22)$$

where H is the heat content of the solution, $\bar{H}_1^0$ heat content of pure solvent, L the relative heat content of the solution and $\bar{L}^0$ heat content of the solution at zero molality. The relative partial molal heat content of the solute and solvent are given by the expressions:

$$\bar{L}_2 = \phi L_2 + n_2\left(\frac{\partial \phi L_2}{\partial n_2}\right)_{n_1} \qquad ...(18.23)$$

$$\bar{L}_1 = -\frac{n_2^2}{n_1}\left(\frac{\partial \phi L_2}{\phi n_2}\right)_{n_1} \qquad ...(18.24)$$

18.4.4. *Methods of Intercepts*

A new quantity *i.e.* "mean value" of the extensive property per mol of mixture is defined as

$$\tilde{J}_1 = \frac{J}{n_1 + n_2} \qquad ...(18.25)$$

or

$$\tilde{J}_2 = \left(\frac{\partial J}{\partial n_2}\right)_{n_1} = \tilde{J}' + (n_1 + n_2)\left(\frac{\partial \tilde{J}'}{\partial n_2}\right)_{n_1} \qquad ...(18.26)$$

The above equation can be rearranged as

$$\tilde{J}_2 = \tilde{J}' - x_1\frac{d\tilde{J}'}{dx_1} \qquad ...(18.27)$$

Value of J' is plotted against x_1. The intercepts at $x_1 = 0$ gives $\tilde{J}_2$ and at $x_1 = 1$ gives $\tilde{J}_1$. The advantage of this method is that it simultaneously gives the value of $\tilde{J}_1$ and $\tilde{J}_2$.

***Exercise* 18.11:** Calculate the partial molal volume of HCO_2H and H_2O at 6 molal HCO_2H solution by the intercept method. Use the data of exercise 18.5.

The following procedure can be adopted:

(*i*) Column 2 list the total number of mol present in 100 g of solution of different composition *e.g.* 10% solution of HCO_2H contains 0.01 kg of HCO_2H and 0.09 kg of H_2O *i.e.* 0.217 mol $HCO_2H(n_2)$ and 5 mol $H_2O(n_1)$, $n_1 + n_2 = 5.217$ mol.

(*ii*) Column 3 lists the mole fraction (x_1) of H_2O.

(*iii*) Column 4 lists the total volume of n_2 mol of H_2COOH and n_1 mol of H_2O, 0.1 kg of solution divided by its density at the given composition *e.g.* for 10% solution. $d = 1.0246$ kg dm^{-3}, so

$$V = \frac{0.10 \text{ kg}}{1.0246 \text{ kg dm}^{-3}} = 0.097599 \text{ dm}^3 = 97.599 \times 10^{-3} \text{ dm}^3$$

wt % HCO_2H	$n_1 + n_2$ mol	x_1	$V \times 10^3$ (dm^3)	$\bar{V} = \frac{V}{n_1+n_2} \times 10^3$ dm^3 (mol soln)
1	2	3	4	5
1	5.556	1	100.18	18.031
10	5.217	0.96	97.599	18.708
20	4.879	0.91	95.347	19.542
30	4.541	0.86	93.205	20.525
40	4.203	0.79	91.216	21.706
50	3.865	0.69	89.230	23.087
55	3.696	0.676	88.339	23.901
60	3.526	0.63	87.535	24.826
65	3.357	0.58	86.633	25.807
70	3.189	0.52	85.800	26.905
75	3.019	0.46	84.968	28.145
80	2.850	0.39	84.317	29.585
85	2.681	0.31	83.661	31.205
90	2.512	0.22	83.029	33.052
95	2.343	0.12	82.372	35.157
100	2.174	0.0	81.887	37.666

(*iv*) Column 5 lists the volume per mol of the solution *i.e.*

$$\bar{V} = \frac{97.599 \times 10^{-3} \text{ dm}^3}{5.217 \text{ mol}} = 18.708 \times 10^{-3} \text{ dm}^3 \text{ (mol soln)}^{-1}$$

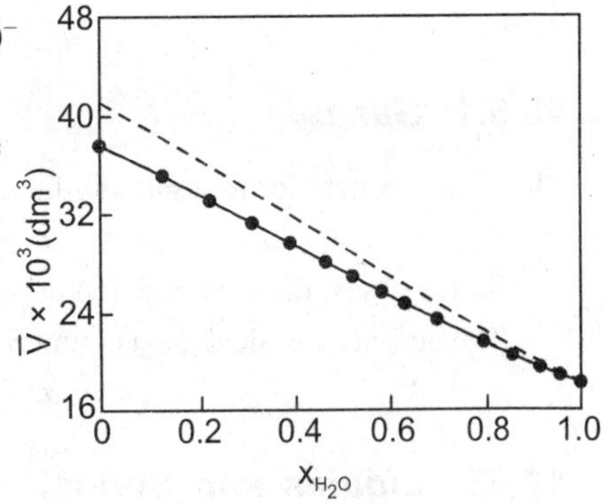

Fig. 18.4

Similarly the other data is obtained and listed in the above table.

(*v*) $\bar{V}$ is plotted against x_1 in Fig. 18.4. The V_2 at 6 molal HCO_2H solution is obtained by drawing a tangent at

$$x_1 = \frac{55.556 \text{ mol}}{(55.556 + 6) \text{ mol}} = 0.9$$

Intercepts on the ordinates gives $V_2 = 41.6 \times 10^{-3}$ dm^3 and $V_1 = 17.4 \times 10^{-3}$ dm^3.

18.5. ▌ACTIVITY AND ACTIVITY COEFFICIENT

Introduction

The term fugacity (13.78*a*) has been defined to relate the deviation of the real gases from ideal behaviour. The same term can be used for the liquids and solids to relate their behaviour in comparison with ideal one. But in some cases, specially the solids, the vapour pressure is so small that it cannot be measured. A new term 'activity' is used to tell the departure of the system from the ideal behaviour.

Definitions

Activity: Activity (also called relative fugacity) is defined as the ratio of the fugacity f_i in any given state to the fugacity f_i^0 in some standard state at the same temperature *i.e.*

$$a_i = \frac{f_i}{f_i^0} \quad ...(18.28)$$

From the definition of activity it is clear that it directly gives the difference in chemical potential of the chosen and the standard state *i.e.*

$$\begin{aligned} \mu - \mu_0 &= RT \ln f_i + \beta_i(T) - RT \ln f_i^0 - \beta_i(T) \\ &= RT \ln (f_i/f_i^0) \\ &= RT \ln a_i \end{aligned} \quad ...(18.29)$$

Activity Coefficient: Activity coefficient is defined as

$$\gamma_i = \frac{a_i}{x_i} \quad ...(18.30a)$$

or

$$\gamma_i = \frac{a_i}{m_i} \quad ...(18.30b)$$

where x_i and m_i are the mole fraction and molality of component *i.* The first definition holds good where Raoult's law is applicable and second definition when Henry's law is applicable. Thus activity coefficient measures the deviation of a solution from ideal behaviour. When the solution is ideal, $\gamma = 1$

$$a_i = x_i \quad ...(18.31a)$$

or

$$a_i = m_i \quad ...(18.31b)$$

18.6. CHOICE OF STANDARD STATE

It is necessary to define a standard state and under different conditions different standard states are used.

18.6.1. *Gases*

The standard state for gases at any fixed temperature is that when the gas has fugacity of 1 atm. *i.e.*

$$a_i = f_i \quad ...(18.32)$$

So for gases, fugacity and activity are the same thing.

The activity for ideal gas is numerically equal to pressure

$$a_i = f_i = p_i \quad ...(18.33)$$

18.6.2. *Liquids and Solids*

(*a*) ***Pure substances:*** The state of pure solid or of pure liquid at any temperature and at a pressure of 1 atm is the standard state. Thus the activities of the pure solids and liquids are unity at a pressure of 1 atm.

(*b*) ***Solvent in solution:*** Solvent is generally the component in a solution which has higher mole fraction or larger concentration. For solvent the standard state is the pure substance at the same temperature and pressure as the solution, *i.e.,*

$$f^{\bullet}_{i(x_1 = 1)} = f_1 \text{ (standard state)} = f_1^0 \quad ...(18.34)$$

So from this the activity of pure components will always be unity

$$a_1 = \frac{f_i}{f_1^0} = \frac{f_i^\bullet}{f_1^0} = \frac{f_1^0}{f_1^0} = 1$$

Under these circumstances the activity of any component in a solution will always be less than that of pure component. The value of activity coefficient will be greater than unity for those component which has positive deviation from Raoult's law and less than unity which has negative deviation from Raoult's law.

For solvent in ideal solutions, the value of $\gamma_i = 1$ *i.e.*

$$a_1 = x_1 \qquad \text{...(18.35)}$$

(c) ***Solute in solution:*** Standard state for solute can be either based on mole fraction or molality. (*i*) The standard state for solute is a hypothetical state in which the solute would exist if it followed Henry's law up to mole fraction x_2 equal to unity. So one should keep in mind that actual fugacity of pure solute is not equal to the fugacity of the standard state chosen

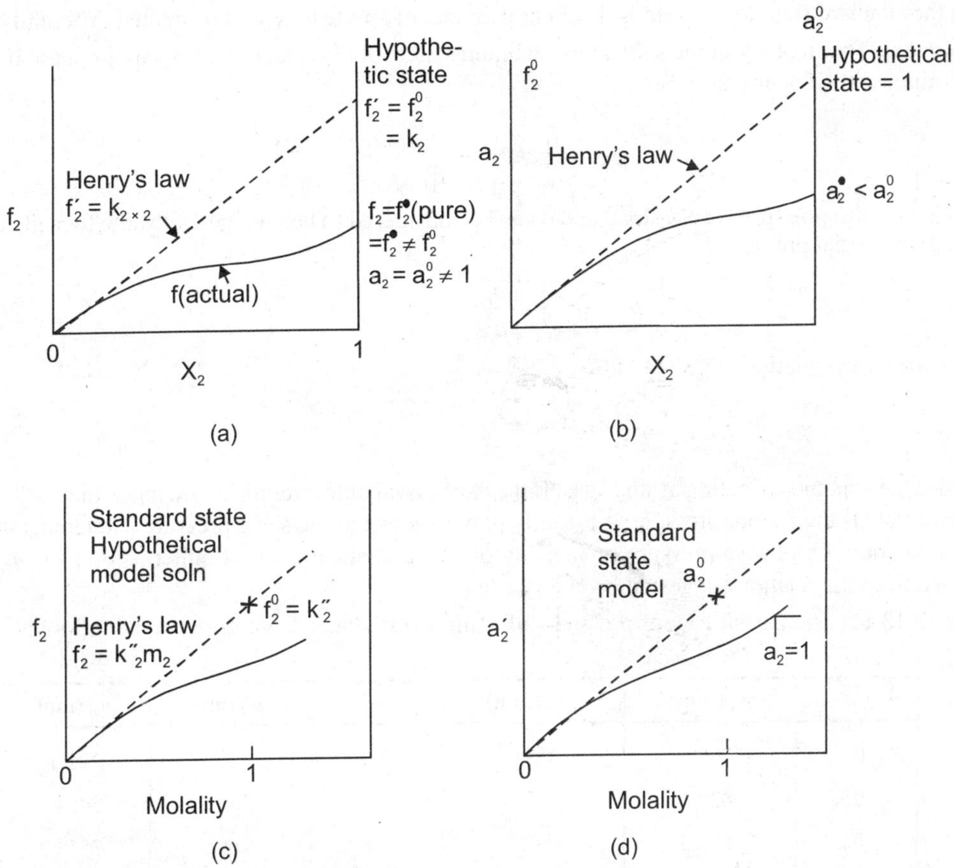

Fig. 18.5 *(a) Establishment of standard state for solute, (b) activity and activity coefficient when standard state is set by extrapolation of Henry's law, (c) establishment of standard state for solute on molality basis, (d) activity and activity coefficient when standard state is established on molality basis.*

(Fig. 18.5*a* and *b*). (*ii*) In some cases where the data of fugacity of the solute is available over the entire range of concentration upto $x_2 = 1$, the pure solute can be selected as the standard state, as in case of solvent, (*iii*) The reference state is a hypothetical 1 molal solution *i.e.* a state that has a fugacity which 1 molal solution would have if the solution obeyed the limiting Henry's law, as shown in the Fig. 18.5 (*c* and *d*). (One thing must be kept in mind that value of activity and activity coefficients may be different depending upon the standard state but the change in the free energy and in other thermodynamic quantities remain the same).

18.7. METHODS OF FINDING ACTIVITY AND ACTIVITY COEFFICIENTS

There are many methods to calculate the activities but only few important ones will be discussed.

18.7.1. *From Vapour Pressure Measurements*

This method is only useful if the substances are volatile and their vapour pressure can be measured. Since the standard state for solvent is different than that of solute they will be treated differently.

Solvent: The fugacity of the solvent in the liquid state will be equal to that in vapour phase if it is in equilibrium with vapour phase. So,

$$a_1 = \frac{f_1}{f_1^0} = \frac{f_1}{f_1^*} = \frac{f_{1(\text{vapour})}}{f^*_{1\,(\text{vapour})}} \qquad ...(18.36)$$

If the vapour pressure is very small and it can be considered as behaving ideally, the fugacity can be replaced by partial pressure

$$a_1 = \frac{p_1}{p_1^0} = \frac{p_1}{p_1^*} \qquad ...(18.37)$$

Solute: The equation (18.28) can be written as

$$a_2 = \frac{f_2}{f_2^0} = \frac{p_2}{p_2^0} \qquad ...(18.38)$$

provided the vapours are acting as ideal gas. If the data is available over the entire range then $p_2^0 = p_2^*$ *i.e.* standard state is the vapour pressure or fugacity of pure solute at the same pressure and temperature as that of solution. Otherwise pressure or fugacity of a hypothetical state of either $x_2 = 1$ or $m_2 = 1$ is obtained from the relation suggested by Henry's law.

***Exercise* 18.12:** The partial vapour pressures of compounds *A* and *B* are given in the following table:

x_B	p_A (mm)	p_B (mm)	x_B	p_A (mm)	p_B (mm)
0	700	0	0.6	420	325.5
0.1	630	72.5	0.7	352.5	345
0.2	572.5	140	0.8	260	355
0.3	540	190	0.9	145	370
0.4	505	240	1.0		400
0.5	470	285			

(*a*) Plot the partial pressures of *A* and *B* as the composition and indicate whether the liquids have negative or positive deviation from Raoult's law. (*b*) Calculate the Henry's constant for both *A* and *B*. (*c*) Calculate the activity coefficients of *B* at mole fraction $x_B = 0.8$, considering the standard state as (*i*) pure component and (*ii*) hypothetical state of Henry's law at unit mole fraction of *B*, assuming the vapours are ideal, (*d*) Calculate the activity coefficient of *A* at $x_A = 0.8$ and considering it as solvent.

(*a*) The partial pressures of *A* and *B* are plotted in Fig. 18.6. Both the component have positive deviations from Raoult's law.

(*b*) Henry's law states

$$p_i = K_i x_i$$

Applying to component *B*, $K_B = 720$ mm when $x_B = 1$, $p_B = 720$ mm.

Applying to component *A*

$$K_A = \frac{p_A(2) - p_A(1)}{x_A(2) - x_A(1)} = \frac{(760.5 - 460)\text{ mm}}{0.5 - 0.3}$$

$$= \frac{300.5}{0.2} = 1502.5 \text{ mm}$$

(*c*) (*i*)

$$a_B = \frac{p_B}{p_B^{\circ}} = \frac{p_B}{p_B^{\bullet}}$$

$$p_B^* = 400 \text{ mm}, p_B = 355 \text{ mm at } x_B = 0.8$$

$$a_B = \frac{355 \text{ mm}}{400 \text{ mm}} = 0.8875$$

$$\gamma_B = \frac{a_B}{x_B} = \frac{0.8875}{0.8} = 1.109$$

(*ii*) The partial pressure of hypothetical Henry's law state of $x_B = 1$ is 720 mm

$$a_B' = \frac{355 \text{ mm}}{720 \text{ mm}} = 0.493$$

$$\gamma_B' = \frac{a_B'}{x_B} = \frac{0.493}{0.8} = 0.616$$

(*d*) Considering liquid A as solvent, the standard state is the pure state at the same conditions as that of solution *i.e.*

$$a_A = \frac{p_A}{p_A^{\circ}} = \frac{p_A}{p_A^{\bullet}} = \frac{605 \text{ mm}}{700 \text{ mm}} = 0.864$$

$$\gamma_A = \frac{a_A}{x_A} = \frac{0.864}{0.8} = 1.08$$

Similarly applying Henry's law of Hypothetical state of $x_A = 1$, $p^0{}_A = 1502.5$ mm so

$$a'_A = \frac{605 \text{ mm}}{1502.5 \text{ mm}} = 0.403$$

$$\gamma'_A = \frac{a_A'}{x_A} = \frac{0.403}{0.8} = 0.5$$

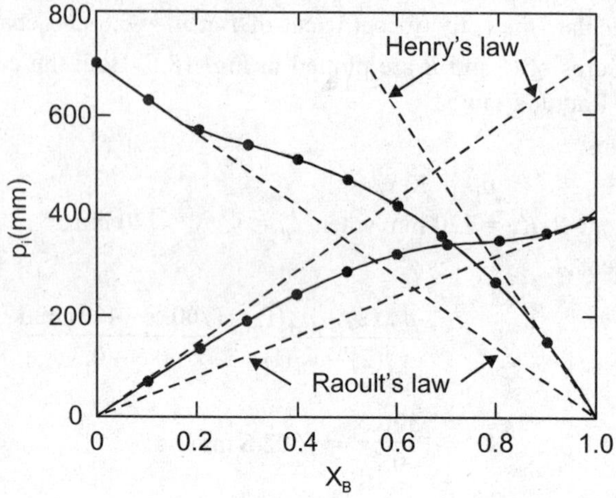

Fig. 18.6

18.7.2. *Application of Gibbs-Duhem Equation*

This method is applicable only if the activity or activity coefficient of one of the components of the binary mixture is known. The equation is

$$\log\frac{a_1}{x_1} - \log\frac{a_1'}{x_1'} = -\int_{x_2'}^{x}\frac{x_2}{x_1}d\log\frac{a_2}{x_2} \qquad \text{...(18.39}a\text{)}$$

for infinite dilute solution *i.e.* x_1' 1, $= x_2' = 0$, $\frac{a_1'}{x_1'} = 1$, so

$$\log\frac{a_1}{x_1} = -\int_0^x\frac{x_2}{x_1}d\log\frac{a_2}{x_2} \qquad \text{...(18.39}b\text{)}$$

where a_i are the activities,x_i are the mole fraction of the two components in a binary mixture. Integrationis carried out graphically *i.e.* $\log\frac{a_2}{x_2}$ is plotted against $\frac{x_2}{x_1}$ and area under the curve will be equal to $\log\frac{a_1}{x_1}$.

Thus a_1 or γ_1 can be calculated.

For Solute: The equation (18.39*b*) is applicable only to solvent. In case of solute the equation reduces to

$$\log\frac{a_2}{x_2} = \int_0^{x_1}\frac{x_1}{x_2}d\log\frac{a_1}{x_1} \qquad \text{...(18.40)}$$

The integration cannot be carried out because for infinite dilute solution *i.e.* for $x_2 = 0$, so $\frac{x_1}{x_2} = \infty$.

The following procedure is followed. Write $\gamma_2 = \frac{a_2}{x_2}$ and the integration is carried over from x_1' to x_1 for fixed value of x_2. Equation (18.40) becomes:

$$\log \frac{\gamma_2}{\gamma_2'} = -\int_{x_1'}^{x_1} \frac{x_1}{x_2} \log \frac{\gamma_1}{\gamma_1'} \qquad ...(18.41)$$

From the graphical method, $\frac{\gamma_2}{\gamma_2'}$ can be calculated at various concentration of x_2.

Now $\frac{\gamma_2}{\gamma_2'}$ is plotted against x_2 and the extrapolation to $x_2 = 0$ gives the value of $\frac{1}{\gamma_2'}$ because

$$\lim_{x_2 \to 0} \frac{a_2}{x_2} = \lim_{x_2 \to 0} \gamma_2 = 1$$

Once the value of γ_2' is known the value of γ_2 at any concentration can be obtained from the respective ratio.

***Exercise* 18.13:** The activity of solute B as a function of x_2 is given as follows, calculate the activity of solvent A at $x_2 = 0.8$.

x_2	**0.1**	**0.2**	**0.3**	**0.4**	**0.5**	**0.6**	**0.7**	**0.8**
a_2	0.18	0.35	0.45	0.6	0.713	0.814	0.863	0.888
$\frac{a_2}{x_2}$	1.8	1.75	1.5	1.5	1.426	1.357	1.233	1.11
$\log \frac{a_2}{x_2}$	0.255	0.243	0.176	0.176	0.154	0.132	0.091	0.045
$\frac{x_2}{x_1}$	0.111	0.25	0.429	0.667	1.0	1.5	2.33	4.0

For the given value of x_2 and a_2; $\frac{a_2}{x_2}$, $\log \frac{a_2}{x_2}$ and $\frac{x_2}{x_1}$ are calculated and tabulated in the above table, $\log \frac{a_2}{x_2}$ is plotted against $\frac{x_2}{x_1}$ in Fig. 18.7 and the area under the curve up to $x_2 = 0.8$ is -0.475. From equation (18.39*b*)

$$\log \frac{a_1}{x_1} = -0.475$$

or

$$\frac{a_1}{x_1} = 0.335$$

and

$$a_1 = (0.335)\,(0.2) = 0.067$$

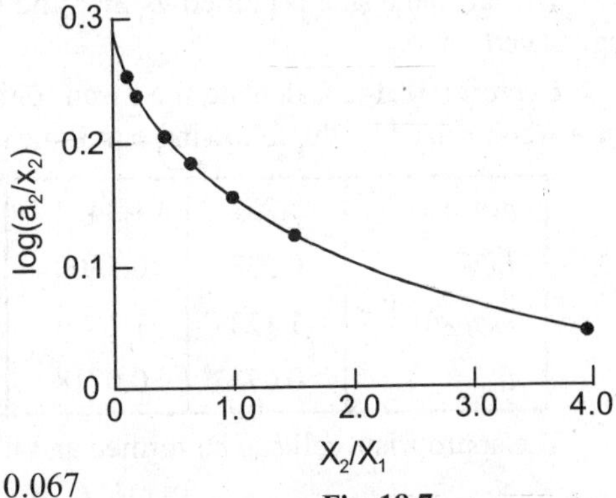

Fig. 18.7

18.7.3. *Measurement of Activity by EMF Method*

The activities of non-electrolytes or electrolytes can be calculated by setting an appropriate galvanic cells and measuring the e.m.f. at various concentrations.

The e.m.f. of the cell is given by

$$E = E^0 - \frac{RT}{nF} \ln a \qquad ...(18.42)$$

where E^0 is the standard e.m.f. of the cell, n the number of electrons, F the Faraday constant. If E^0 is not known, the above equation can be written as

$$E = E^0 - \frac{RT}{nF} \ln m - \frac{RT}{nF} \ln \gamma \qquad ...(18.43)$$

for non-electrolytes. It can be rearranged as

$$E + \frac{RT}{nF} \ln m = E^0 - \frac{RT}{nF} \ln \gamma \qquad ...(18.44)$$

The left-hand side is plotted against m and the extrapolation to $m = 0$ gives E^0, because as $m \to 0$, $\gamma \to 1$ so $\ln \gamma = 0$. Once E^0 is known, γ and a can be calculated at any concentration.

The equation (18.43) for electrolytes can be written as

$$E = E^0 - \frac{RT\nu}{nF} \ln m_\pm - \frac{RT\nu}{nF} \ln \gamma_\pm \qquad ...(18.45)$$

or

$$E + \frac{RT\nu}{nF} \ln m_\pm = E^0 - \frac{RT\nu}{nF} \ln \gamma_\pm \qquad ...(18.46)$$

where m_+ and γ_+ are mean ionic molality and activity coefficient of the electrolyte. The mean ionic molality can also be replaced by the molality of the electrolyte

$$m_\pm^\nu = (\nu_+^{\nu+} \nu_-^{\nu-}) m^\nu \qquad ...(18.47)$$

i.e.

$$E + \frac{RT\nu}{nF} \ln m + \frac{RT}{nF} \ln (\nu_+^{\nu+} \nu_-^{\nu-}) = E^0 - \frac{RT\nu}{nF} \ln \gamma_\pm \qquad ...(18.48)$$

The left-hand side is plotted vs $m^{1/2}$ and extrapolation to $m = 0$ give E^0 and thus $\gamma_\pm$ and $a_\pm$ can be calculated.

***Exercise* 18.14:** Calculate the mean ionic activity and mean ionic activity coefficient of HBr at $m = 1.262 \times 10^{-4}$ by the following e.m.f. measurements:

$m \times 10^4$	1.262	1.34	1.775	2.066	4.172	6.7	10.994
E(V)	0.533	0.53001	0.51616	0.50792	0.47211	0.44804	0.4228
$\sqrt{m} \times 10^2$	1.123	1.16	1.33	1.44	2.04	2.59	3.32
E'	0.0717	0.0718	0.0724	0. 0720	0.0723	0.0726	0.0728

The appropriate cell can be formed and it is

$$\text{Pt} \mid H_2 \text{ (1 atm)} \mid \text{HBr}(m) \mid \text{AgBr}(s) \mid \text{Ag}$$

The electrode reactions are:

$$\frac{1}{2}H_2(g) \rightarrow H^+(aq) + e^-$$
$$AgBr(s) + e^- \rightarrow Ag(s) + Br^-(aq)$$

Over all reaction is

$$AgBr(s) + \frac{1}{2}H_2(g) \rightarrow H^+(aq) + Br^-(aq) + Ag(s)$$

Since HBr is a 1:1 electrolyte, equation (18.47) gives $m_+ = m$. As $a_{H_2} = 1$, the equation (18.46) at 25°C can be written as

$$E' + 0.1183 \log m = E^0 - 0.1183 \log \gamma_+(HBr)$$

or

$$E' = E + 0.1183 \log m = E^0 - 0.1183 \log \gamma_+(HBr) \quad ...(18.49)$$

E' is calculated by substituting the respective value of E and m. The data is calculated and listed in the table along with the $\sqrt{m}$. The E' is plotted against m in Fig. 18.8. The extrapolation to $m = 0$ gives $E^0 = 0.0711$ V. Now using equation (18.48) and E^0, γ_+(HBr) at 1.262×10^{-4} M solution can be obtained as

$$\log \gamma_+ = \frac{1}{0.1138}[0.0711 - 0.533 - 0.1138 \log(1.262 \times 10^{-4})]$$
$$= -0.16$$

or

$$\gamma_+ = 10^{-0.16} = 0.692$$

Thus

$$a_\pm = \gamma_+ m_\pm = (0.692)(1.262 \times 10^{-4}) = 0.873 \times 10^{-4}$$

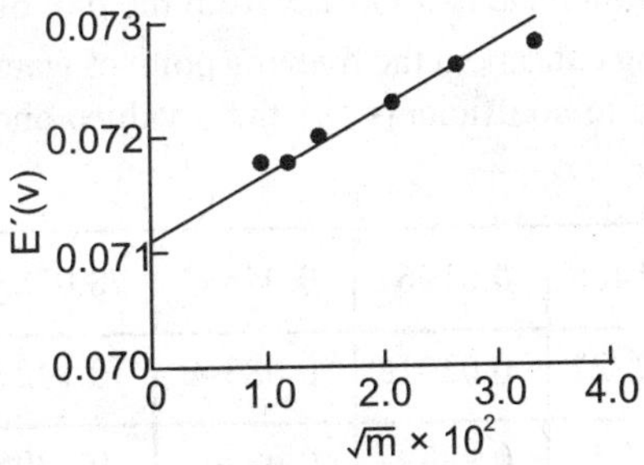

Fig. 18.8 *Plot of E′ vs* $\sqrt{m}$

18.7.4. *By Colligative Properties*

The activities of the solvent can be calculated with the help of equations:

(*a*) *Freezing point depression*

$$\ln a = \frac{\Delta \bar{H}^0_{fus}}{R}\left(\frac{1}{T_0} - \frac{1}{T}\right) \quad ...(18.50)$$

where T_0 and T are the freezing points of the solvent without and with solute.

(*b*) *Boiling point elevation*

$$\ln a = \frac{\Delta \bar{H}^0_{vap}}{R}\left(\frac{1}{T}-\frac{1}{T_0}\right) \qquad ...(18.51)$$

where T and T_0 are the boiling points of solvent with and without solute.

(*c*) *Osmotic pressure*

$$\ln a = -\frac{\bar{V}^0 \pi}{RT} \qquad ...(18.52)$$

where π is the osmotic pressure and $\bar{V}^0$ is the molar volume of the solvent.

(*ii*) ***For solute:*** The activity coefficient of the solute with the freezing point depression can be calculated from

$$\ln \gamma_2 = -j - \int_0^m \left(\frac{j}{m}\right) \mathrm{dm} \qquad ...(18.53a)$$

where $$\theta = K_f m(1-j) \text{ for non-electrolytes} \qquad ...(18.53b)$$

and $$\theta = v K_f m(1-j) \qquad ...(18.54)$$

for electrolytes. $\theta = T_0 - T$, depression in the freezing point, K_f, the molal freezing point constant, m is the molality of solute, v the sum of the ions when a molecule electrolyte dissociates and $(1-j)$ is called osmotic coefficient. The equation (18.53 *a*) for electrolytes is

$$\ln \gamma_{+} = -j - \int_0^m \left(\frac{j}{m}\right) dm \qquad ...(18.55)$$

The integral is evaluated by graphical method *i.e.* form the plot of j/m vs m.

***Exercise* 18.15:** The following data about the freezing point of water, in presence of *n*-propylalcohol is observed. Calculate (*a*) osmotic coefficients (*b*) the j values and (*c*) the activity coefficient of *n*-propylalcohol at $m = 0.15$.

m	0.00919	0.01468	0.01967	0.04542	0.07832	0.1102	0.17406
θ	0.01702	0.02721	0.03618	0.08342	0.1427	0.2013	0.3168
$1-j$	0.9967	0.9975	0.9899	0.9884	0.9805	0.9830	0.9794
j	0.0033	0.0025	0.0101	0.0116	0.0195	0.017	0.0205
$\frac{j}{m}$	0.36	0.17	0.51	0.255	0.249	0.154	0.118

The values of $1-j$ at various m values are calculated by using equation (18.53 *b*) and $k_f = 1.8582$ for $m = 0.00919$

$$1 - j = \frac{0.01702}{(1.8582)(0.00919)}$$

$$= 0.9967$$

Similarly other values of $(1-j)$ are calculated and the j/m is plotted against m in Fig. 18.9. The area under the curve up to $m = 0.15$ *i.e.*

$$\int_0^m \frac{j}{m}\,dm = 0.036, j = 0.0195$$

so using equation (18.53*a*)

$$\log \gamma_2 = -0.0195 - 0.0360 = 0.0555$$

$$\gamma_2 = 0.88$$

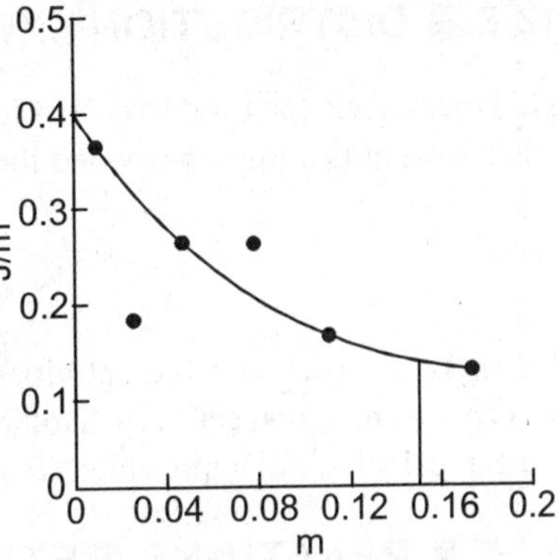

Fig. 18.9

***Exercise* 18.16:** Calculate the activity coefficient of H_2O at 0°C with the following data m (propylalcohol) = 0.13146 mol kg^{-1}, $T = -0.239$°C, $\Delta\bar{H}^0_{\text{fus}} = 6.01$ kJ mol^{-1}

Using equation (18.50)

$$\log a = \frac{6.01\times10^3 \text{ J mol}^{-1}}{8.314 \text{ J K}^{-1} \text{ mol}^{-1}}\left(\frac{1}{273} - \frac{1}{272.761}\right)\text{K}^{-1}$$

$$= \frac{(6.01\times10^3)(-0.239)}{(8.314)(273)(272.761)} = -2.32 \times 10^{-3}$$

$$a = \exp(-2.32 \times 10^{-3}) = 0.998$$

$$n \text{ (propylalcohol)} = 0.13146 \text{ mol}, \; n\,(H_2O) = \frac{1000 \text{ g}}{18 \text{ g mol}^{-1}} = 55.556 \text{ mol}$$

$$x_{H_2O} = \frac{55.556 \text{ mol}}{(55.556 + 0.13146) \text{ mol}} = 0.9976$$

Using equation (18.30) $\gamma_{H_2O} = \dfrac{0.998}{0.9976} = 1.0004$

***Exercise* 18.17:** The boiling point of water containing $x_{NH_4NO_3} = 0.1$ is 104.8°C at 760 mm pressure. If $\Delta\bar{H}^0_{\text{vap}} = 40.656$ kJ mol^{-1}, calculate the activity coefficient of water at 100°C.

Using equation (18.51)

$$\ln a = \frac{40656 \text{ J mol}^{-1}}{8.314 \text{ J K}^{-1} \text{ mol}^{-1}}\left(\frac{1}{377.8} - \frac{1}{373}\right)\text{K}^{-1}$$

$$= \frac{(40656)(-4.8)}{(8.314)(377.8)(373)} = -0.1666$$

$$a = e^{-0.1666} = 0.847$$

Using equation (18.30*a*)

$$\gamma = \frac{0.847}{0.9} = 0.941$$

18.8. DISTRIBUTION LAW

This method can be used to determine the activity of solute in one solvent if the activity of the solute in other solvent is known, provided the two solvents are completely immiscible with each other, *i.e.*

$$(a_2)_B = (a_2)_A \frac{(f_2^0)_A}{(f_2^0)_B} \qquad ...(18.56)$$

i.e. $(a_2)_A$ and $(a_2)_B$ are the activities, $(f_2^0)_A$ and $(f_2^0)_B$ are the standard states fugacities of solute in solvents *A* and *B* respectively. In other words, if we know the Henry's constants of solute in two solvents, we can calculate $(a_2)_B$ provided $(a_2)_A$ is known.

18.9. RELATIONS BETWEEN ACTIVITY, MEAN ACTIVITY, ACTIVITY COEFFICIENT, MEAN ACTIVITY COEFFICIENT ETC. OF IONS IN ELECTROLYTE

If a and γ, $a_\pm$ and $\gamma_\pm$, a_+ and γ_+ and a_- and γ_- are the activity and activity coefficient, mean ionic activity and mean ionic activity coefficient of an electrolyte, activity and activity coefficient of positive and negative ions respectively, then

$$a = a_\pm^v \qquad ...(18.57)$$

$$a_\pm^v = a_+^{v+} a_-^{v-} \qquad ...(18.58)$$

$$\gamma_\pm^v = \gamma_+^{v+} \gamma_-^{v-} \qquad ...(18.59)$$

$$m_\pm^v = (v_+^{v+} v_-^{v-}) m^v \qquad ...(18.47)$$

***Exercise* 18.18:** Prove the relation 18.58.

Consider an electrolyte which is completely dissociated and each 1 mol dissociates into v_+ and v_- mol of positive and negative ions respectively. Then $v = v_+ + v_-$ is the total mol of ions produced by 1 mol of electrolyte. If there is n mol of electrolyte then $n_+ = v_+ n$ and $n_- = v_- n$. Also the free energy of the system will be

$$G = n_s\mu_s + n_+\mu_+ + n_-\mu_-$$
$$= n_s\mu_s + n(v_+\mu_+ + v_-\mu_-) \qquad ...(18.60a)$$

where n_s is the number of mol of solvent and μ's are the chemical potential of respective species. If μ is the chemical potential of an electrolyte then

$$G = n_s\mu_s + n\mu \qquad ...(18.60b)$$

Comparing equations (18.60*a*) and (18.60*b*)

$$\mu = v_+\mu_+ + v_-\mu_-$$

Further mean ionic chemical potential $\mu_\pm$ is defined as

$$v\mu_\pm = v_+\mu_+ + v_-\mu_- \qquad ...(18.60c)$$

Similar to the equation (18.29), the chemical potential of various species can be defined as

$$\mu = \mu^0 + RT \ln a \qquad ...(18.60d)$$

$$\mu_\pm = \mu_\pm^0 + RT \ln a_\pm \qquad ...(18.60e)$$

$$\mu_+ = \mu_+^0 + RT \ln a_+ \qquad ...(18.60f)$$

$$\mu_- = \mu_-^0 + RT \ln a_- \quad \text{...(18.60g)}$$

Using equations (18.60*c*), (18.60*e*), (18.60*f*) and (18.60*g*) one gets

$$\nu\mu_\pm^0 + RT\nu \ln a_\pm = \nu_+\mu_+^0 + \nu_+ RT \ln a_+ + \nu_-\mu_-^0 + \nu_- RT \ln a_- \quad \text{...(18.60h)}$$

Similar to equation (18.60*c*), the equation for standard states also exists as

$$\nu\mu_\pm^0 = \nu_+\mu_+^0 + \nu_-\mu_-^0$$

Equation (18.60*h*) becomes

$$RT \ln a_\pm^\nu = RT \ln a_+^{\nu+} \cdot a_-^{\nu-}$$

or

$$a_\pm^\nu = a_+^{\nu+} \cdot a_-^{\nu-}$$

18.10. DEBYE-HÜCKEL LIMITING LAW

The mean ionic activity coefficient, calculated with the help of Debye-Hückel theory, is given by

$$\ln \gamma_\pm = \frac{-e^2 \kappa Z_+ Z_-}{2Dk_b T(1+\kappa a_0)} \quad \text{...(18.61)}$$

and κ (kappa) is defined as

$$\kappa = \left[\frac{8\pi N_{av} e^2}{1000\, D\, k_b T\mu}\right]^{1/2} \quad \text{...(18.62)}$$

where N_{av} is Avogadro's number, *e* is the electronic charge, Z_+ and Z_- are the valences of positive and negative ion, *D* is the dielectric constant of medium, k_b is the Boltzmann's constant, a_0 is the mean distance of closest approach of ions and μ is the ionic strength of solution, given by

$$\mu = \frac{1}{2}\sum_i C_i Z_i^2 \quad \text{...(18.63)}$$

where C_i is the concentration of ions in mol kg^{-1} and Z_i is the valence of ion *i*. The summation is carried over all the ions present in the solution.

If $\kappa a_0 \ll 1$, the equation (18.61) reduces to, after substituting the κ,

$$\log \gamma_\pm = -\frac{1}{2.303}\left[(2\pi N_{av})^{1/2}\left(\frac{e^2}{10\, k_b\, TD}\right)^{3/2}\right] Z_+ Z_- \mu^{1/2} \quad \text{...(18.64)}$$

This is known as "Debye-Hückel limiting law". It can be written as

$$\log \gamma_\pm = -A\, Z_+ Z_- \mu^{1/2} \quad \text{...(18.65)}$$

For water at 298 K the value of $A = 0.509$, so

$$\log \gamma_+ = -0.509\, Z_+ Z_- \mu^{1/2} \quad \text{...(18.66)}$$

***Exercise* 18.19:** Calculate the ionic strength of the following solutions:

(*i*) 50 mL of 0.2 M KNO_3, 20 mL of 0.15 M K_2SO_4, 30 mL of 0.05 M $Cu(NO_3)_2$.

(*ii*) 60 mL of 0.1 M KCl, 20 mL of 0.2 M K_2SO_4 and 20 mL of H_2O.

(*i*) (1) The total volume of the solution after mixing becomes 100 mL. In the mixture the molarity of KNO_3 is $\frac{0.2 \times 50}{100} = 0.1$ M

(2) of $K_2SO_4 = \frac{20 \times 0.1}{100} = 0.02$ M and

(3) of $Cu(NO_3)_2 = \frac{30 \times 0.5}{100} = 0.015$ M

Each molecule of K_2SO_4 and $Cu(NO_3)_2$ will give 2 ions of K^+ and NO_3^- respectively so their concentration will be double in the solution. Using equation (18.63)

$$\mu = \frac{1}{2}\,[[K^+]Z_{K^+}^2 + [NO_3^-]Z_{NO_3}^{2-} + [K^+]\,Z_{K^+}^2 + [SO_4^{2-}]Z_{SO_4^{2-}}^2 + [Cu^{2+}]Z_{Cu^{2+}}^2 + [NO_3^-]Z_{NO_3^-}^2]$$

$$= \frac{1}{2}[(0.1)(+1)^2 + (0.1)(-1)^2 + (2 \times 0.02)(+1)^2 + (0.02)(-2)^2 + (0.015)(+2)^2 + (2 \times 0.015)(-1)^2]$$

$$= 0.205$$

(*ii*) Total volume of the solution after mixing is 100 mL. In the mixture the molarity of (1) KCl solution

$$= \frac{0.1 \times 60}{100} = 0.06 \text{ M and (2) } K_2SO_4 = \frac{0.2 \times 20}{100} = 0.04 \text{ M}$$

So

$$\mu = \frac{1}{2}\,[([K^+]\,Z_{K+}^2 + [Cl^-]\,Z_{Cl^-}^2) + ([K^+]\,Z_{K+}^2 + [SO_4^{2-}]\,Z_{SO_4^-}^2)]$$

Since water contributes no ions.

$$\mu = \frac{1}{2}(0.06)(+1)^2 + (0.06)(-1)^2 + (2 \times 0.04)(+1)^2 + (0.04)(-2)^2]$$

$$= 0.18$$

***Exercise* 18.20:** (*a*) Calculate the activity coefficients of Cu^{2+} and NO_3^- ions in 2.0×10^{-3} M solution of $Cu(NO_3)_2$, (*b*) using the above data calculate the mean ionic activity coefficients, (*c*) calculate the mean ionic concentration, and (*d*) mean ionic activity.

(*a*) The ionic strength of 2.0×10^{-3} M $Cu(NO_3)_2$ solution is

$$\mu = \frac{1}{2}\,[(2.0 \times 10^{-3})(2)^2 + (2 \times 2.0 \times 10^{-3})(-1)^2] = 6.0 \times 10^{-3}$$

The activity coefficient of any ion by Debye-Hückel's limiting law is given as

$$\log \gamma_i = -0.509\, Z_i^2\, \mu^{1/2} \quad \text{...(18.67)}$$

$$\log \gamma_{Cu^{2+}} = -(0.509)(2^2)(6.0 \times 10^{-3})^{1/2} = -0.1577$$

$$\gamma_{Cu^{2+}} = .6950$$

and

$$\log \gamma_{NO_3^-} = -(0.509)(-1)^2(6.0 \times 10^{-3})^{1/2} = -0.0394$$

$$\gamma_{NO_3^-} = 0.9133$$

(*b*) Here $v = 2 + 1 = 3$, $v_+ = 1$, $v_- = 2$

Using equation (18.60)

$$\gamma_\pm^3 = (0.6950)^1\,(0.9133)^2$$
$$\gamma_\pm = [(0.6950)\,(0.9133)^2]^{1/3} = 0.8341$$

(*c*) Using equation (18.47)

$$m_\pm = [(2.0 \times 10^{-3})^1\,(2 \times 2.0 \times 10^{-3})^2]^{1/3}$$
$$= 3.17 \times 10^{-3}\ \text{M}$$

(*d*) $a_\pm = \gamma_\pm\, m_\pm = (0.8341)\,(3.17 \times 10^{-3})$
$= 2.64 \times 10^{-3}$

18.11. ACIDS AND BASES

No chemistry course is complete without the study of acid-base chemistry and thus forms an important class of electrolytes. Equilibrium between acid-base is very important and wide spread. The precise balance of their concentrations or pH in our bodies is necessary for proper function of enzymes, maintenance of osmotic pressure and so on. A deviation from the normal pH value, by as small an amount as one-tenth of a unit can cause disease or even death. So, in this section, we will briefly discuss the importance of acid-base equilibrium, buffers and amino acids.

18.11.1. *Definition of Acid and Base*

According to Brønsted, acid is a substance that can donate proton and base is a substance that can accept proton. These kinds of substances are known as Brønsted acids. For example

$$HCl(aq) + H_2O\,(l) \rightarrow H_3O^+(aq) + Cl^-(aq)$$

Here HCl donates proton to water molecule, thus HCl is known as Brønsted acid and water is a Brønsted base. On the other hand, H_3O^+ is called conjugate acid of base H_2O and Cl^- is conjugate base of acid HCl. Further, if an acid is a strong acid (HCl), its conjugate base (Cl^-) will be a weak base, whereas weak acid has a strong conjugate base. Other examples can be:

$$CH_3COOH(aq) + H_2O(l) \rightarrow CH_3OO^-(aq) + H_3O^+(aq)$$
$$NH_3(aq) + H_2O(l) \rightarrow NH_4^+(aq) + OH^-(aq)$$

Here acetic acid is a weak acid and acetate ion will be a strong conjugate base, whereas ammonia is a weak base and ammonium ion will be relatively stronger conjugate acid. On the other hand, metallic hydroxides, *e.g.* NaOH, KOH, does not contain proton to donate but when dissolved in water dissociate to give Na^+, K^+ and OH^- ions. The ions OH^- accept proton to form water and thus can be called base.

Definition of Brønsted acid-base can be extended to substances which contain protons and do not need separate substance to exchange protons, *e.g.*

$$NH_3 + NH_3 \rightarrow NH_4^+ + NH_2^-$$

There are many similar examples, like H_2O, H_2SO_4 etc.

18.11.2. *Lewis Acid-Base*

Broader definition of acid base was provided by Gilbert Lewis. According to Lewis, a substance which donates electron pair is called *Lewis base* and a substance which accepts a lone pair is called *Lewis acid*. Lewis acid-base not only covers the molecules which contain proton but also those molecules which do not contain protons. For example

$$BF_3 + NH_3 \rightarrow F_3B : NH_3$$

$$BF_3 + F^- \rightarrow BF_4^-$$

$$Cu^{2+} + 6CN^- \rightarrow [Cu(CN)_6]^{4-}$$

In these examples BF_3 and Cu^{2+} are Lewis acids and the other molecules or ions are called Lewis bases.

18.11.3. *Water As:Both Acid and Base*

Although water is a weak electrolyte, a poor conductor of electricity, it undergoes dissociation to a small extent

$$H_2O(l) + H_2O(l) \rightarrow H_3O^+(aq) + OH^-(aq)$$

The reaction can be called *auto-ionisation* of water. $H_3O^+(aq)$, called *hydronium ion*, is a conjugate acid of the base H_2O, whereas OH^- is a conjugate base of water acid. Evidence also suggests that besides H_3O^+ ion, more complex structures such as $H_9O_4^+$ are also present. The structure of such complex species is:

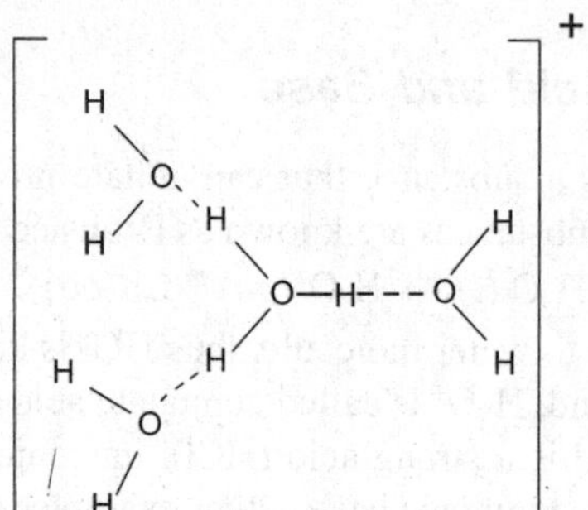

The auto-ionisation of water can be written simply as:

$$H_2O(l) \rightleftarrows H^+(aq) + OH^-(aq) \qquad ...(18.68)$$

and the thermodynamic dissociation constant can be written as

$$K_w = \frac{a_{H^+} a_{OH^-}}{a_{H_2O}} \qquad ...(18.69)$$

Taking $a_{H_2O} = 1$, equation (18.69) reduces to

$$K_w = a_{H^+} a_{OH^-} \qquad ...(18.70)$$

or in terms of concentrations

$$K_w = [H]^+[OH]^- \qquad ...(18.71)$$

On the same lines it can be mentioned that K_w is also temperature dependent and some values of K_w as a function temperature is given below:

T (K)	273	298	313	373
K_w	0.12×10^{-14}	1.0×10^{-14}	2.9×10^{-14}	5.4×10^{-13}

If the solution is neutral, $[H^+] = [OH^-] = 10^{-7}$ mol L^{-1} at 298 K. In acidic solution $[H^+] > [OH^-]$ and in case of basic solution $[H^+] < [OH^-]$. The pH of the solution is defined

$$pH = -\log a_{H^+} \qquad ...(18.72)$$

and in terms of concentration, pH is defined as

$$pH = -\log[H^+] \qquad ...(18.73)$$

Equation (18.73) is valid for dilute solutions at low ionic strengths ($[H^+] \leq 0.1$ M and $\mu < 0.1$). In the similar manner pOH⁻ can also defined as

$$pOH = -\log[OH^-] \qquad ...(18.74)$$

Further taking negative logarithm of both sides of equation (18.70), we get

$$-\log K_w = -\log a_{H^+} - \log a_{OH^-} \qquad ...(18.75)$$

$$pK_w = pH + pOH \qquad ...(18.76)$$

Thus at 298 K, equation (18.76) can be written as

$$pH + pOH = 14 \qquad ...(18.77)$$

and for neutral solution, pH = pOH = 7.0. The solution is acidic if pH < 7.0 and solution is basic if pH > 7.0. It may be mentioned here that the pH range for all practical purposes lies between 1 and 14. Negative pH commonly known as Hammett's scale (H_0), as well as, positive pH (greater than 14, $H\_$) do exist.

***Exercise* 18.21:** Calculate the value of $K_w \sim 1 \times 10^{-14}$: (*i*) Assuming the solution to be ideal and (*ii*) Taking into account the mean ionic activity coefficient of H^- and OH^- ions.

(*i*) Equilibrium constant for the dissociation of water, *i.e.*,

$$H_2O\,(l) \rightleftharpoons H^+(aq) + OH^-(aq) \qquad ...(1)$$

$$K_w = \frac{a_{H^+}a_{OH^-}}{a_{H_2O}} = \frac{[H^+]_e\gamma_{H^+}[OH^-]_e\gamma_{OH^-}}{a_{H_2O}}$$

$$= [H^+]_e[OH^-]_e\gamma_{\pm}^2 \qquad ...(2)$$

For the water as solvent, the reference function is the mole fraction, with $X_{H_2O} = 1$, the reference state. For pure water, therefore we can take $a_{H_2O} = 1$. Similarly the activity of the solute is also close to unity, since it is an uncharged species. We also know that

$$K_w = e^{-\Delta\mu^0/RT} \qquad ...(3)$$

where

$$\Delta\mu_0 = \mu^0_{H_3O^+} + \mu^0_{HO^-} - \mu^0_{H_2O}$$

$$= 0 - 157.24 \text{ kJ mol}^{-1} - (-237.13 \text{ kJ mol}^{-1})$$

$$= 79.89 \text{ kJ mol}^{-1}$$

So at 298 K,

$$K_w = e^{-\Delta\mu^0/RT} = e^{-79890 \text{ J mol}^{-1}/(8.314 \text{ J K}^{-1} \text{ mol}^{-1})(298\text{K})}$$

$$= 1.01 \times 10^{-14} \qquad ...(4)$$

If $[H^+]_e = m$, $[OH^-]_e = m$, thus equation (4) can be written as

$$K_w = [H^+]_e[OH^-]_e\gamma_\pm^2 = m^2\gamma_\pm^2 = 1.01 \times 10^{-14} \qquad ...(5)$$

For dilute solutions, $\gamma_\pm^2 = 1.0$, equation (5) reduces to

$$K_w = 1.01 \times 10^{-14}$$

(*ii*) Knowing $[H^+]_e = [OH^-]_e = m$, from equation (5), the value of $[H^+] = 1 \times 10^{-7}$ mol L^{-1}
Value of $\gamma +$ can be obtained from equation (18.65). The ionic strength, μ will be

$$\mu = \frac{1}{2}[H^+]Z_{H^+}^2 + [OH^-]Z_{OH^-}^2 = \frac{1}{2}(10^{-7})(1)^2 + (10^{-7})(1)^2 = 10^{-7} \text{ mol L}^{-1}$$

$$\log\gamma_\pm = -0.509(1)(1)\sqrt{10^{-7}} = -1.61 \times 10^{-4}$$

or

$$\gamma_\pm = 0.9996 \simeq 1.0$$

Since the mean ionic activity of the ions H^+ and OH^- is so closed to unity, we take the ionic product of water $K_w = 10^{-14}$.

Now we will take up some examples where the concentrations of the ions are affected when the activities are taken into account.

18.11.4. *Dissociation of Acids and Bases*

To the first approximation, strong acids and strong bases dissociate completely and equilibrium can be written as (for example HNO_3, or H_2SO_4 etc.):

$$HNO_3(aq) + H_2O(l) \rightarrow NO_3^-(aq) + H_3O^+(aq)$$
$$H_2SO_4(aq) + H_2O(l) \rightarrow H_2SO_4(aq) + H_3O(aq)$$

On the other hand, dissociation of weak acid or base is not complete and its thermodynamic equilibrium constant can be written as:

$$HA(aq) + H_2O(l) \rightleftarrows H_3O^+(aq) + A^-(aq)$$

$$K_a = \frac{a_{H_3O^+}a_{A^-}}{a_{HA}a_{H_2O}} \qquad ...(18.78)$$

where *a* represents the activity of respective species. In most of the aqueous solutions the amount of water hardly changes, *i.e.* remains nearly constant (one litre, 1000 g, 55.56 mol) by the dissociation process. Therefore water can be considered in its standard state, (pure liquid) and activity of this kind of liquid will be unity, *i.e.* $a_{H_2O} = 1$. Thus equation (18.78) reduces to.

$$K_a = \frac{a_{H_3O^+}a_{A^-}}{a_{HA}} = \frac{[H^+]\gamma_{H^+}[A^-]\gamma_{A^-}}{[HA]\gamma_{HA}} \qquad ...(18.79)$$

Because HA is an uncharged species, we can take $\gamma_{HA} = 1$ and using equation (18.59), *i.e.* $\gamma_\pm^2 = \gamma^+\gamma^-$, equation (18.79) reduces to

$$K_a = \frac{a_{H_3O^+} a_{A^-}}{a_{HA}} = \frac{[H^+][A^-]\gamma_\pm^2}{[HA]} \quad ...(18.80)$$

If the acid is a very weak acid and the concentration of the ionic species are less than 0.05 M, mean ionic activity coefficient can be neglected. Thus equation (18.80) reduces to

$$K_a = \frac{[H^+][A^-]}{[HA]} \quad ...(18.81)$$

Similar to the selation between hydrogen ion concentration and ph, another way of defining dissociation constant is

$$pK_a = -\log K_a \quad ...(18.81a)$$

The larger the value of K_a, or smaller the value of pK_a stronger the acid will be. Another way to measure the strength of the acid is to calculate its percent dissociation, defined by

$$\text{Percent dissociation} = \frac{[H^+]_{eq}}{[HA]_0} \quad ...(18.82)$$

where $[H^+]_{eq}$ is the concentration of proton at equilibrium and $[HA]_0$ is the initial concentration of acid. Dissociation constants are compiled in Table 18.4.

TABLE 18.4. Dissociation constants of some weak acids and their pK_a values at 298 K

Acid	K_a	pK_a
HF	7.1×10^{-4}	3.15
HCN	4.9×10^{-10}	9.31
HNO_2	4.5×10^{-4}	3.35
H_2S	5.7×10^{-8} ($K_{a,1}$)	7.24 (pKa,1)
HS^-	1.2×10^{-15} ($K_{a,2}$)	14.92 (pKa,2)
H_2CO_3	4.2×10^{-7} ($K_{a,1}$)	6.38 (pKa,1)
HCO_3^-	7.1×10^{-11} ($K_{a,2}$)	10.32 (pKa,2)
H_2SO_4	Very high	–
HSO_4^-	1.3×10^{-2} ($K_{a,2}$)	1.89 (pKa,2)
H_3BO_3 $B(OH)_3(aq) + H_2O(l) \rightleftarrows B(OH)_4^- + H^+$	7.3×10^{-10}	9.14
H_3PO_4	7.5×10^{-3} ($K_{a,1}$)	2.13 (pKa,1)
$H_2PO_4^-$	6.2×10^{-8} ($K_{a,2}$)	7.21 (pKa,2)
HPO_4^{2-}	4.8×10^{-13} ($K_{a,3}$)	12.32 (pKa,3)
CH_3COOH	1.75×10^{-5}	4.76
C_6H_5COOH	6.30×10^{-5}	4.20
HCOOH	1.77×10^{-4}	3.75
$ClCH_2COOH$	1.36×10^{-3}	2.87
C_6H_5OH	1.30×10^{-10}	9.89
HOOCCOOH (oxalic acid)	6.5×10^{-2} ($K_{a,1}$)	1.19 (pKa,1)
	6.1×10^{-5} ($K_{a,2}$)	4.21 (pKa,2)

$(CH_2COOH)_2$ (succinic acid)	6.4×10^{-5} ($K_{a,1}$) 2.7×10^{-6} ($K_{a,2}$)	4.19 (pKa,1) 5.57 (pKa,2)
$C_6H_8O_6$ (ascorbic acid)	8×10^{-5} ($K_{a,1}$) 1.6×10^{-12} ($K_{a,2}$)	4.1 (pKa,1) 11.79 (pKa,2)
$(CHCOOH)_2$ (fumaric acid)	9.3×10^{-4} ($K_{a,1}$) 3.4×10^{-5} ($K_{a,2}$)	3.03 (pKa,1) 4.47 (pKa,2)
$CH_3CH(OH)CH_2COOH$ (lactic acid)	1.39×10^{-4}	3.86
$HOOCCH(OH)CH_2COOH$ (malic acid)	4×10^{-4} ($K_{a,1}$) 9×10^{-6} ($K_{a,2}$)	3.40 (pKa,1) 5.50 (pKa,2)
$HOOCCH_2CH(OH)COOHCH_2COOH$ (citric acid)	8.7×10^{-4} ($K_{a,1}$) 1.8×10^{-5} ($K_{a,2}$) 4.0×10^{-6} ($K_{a,3}$)	3.06 (pKa,1) 4.75 (pKa,2) 5.40 (pKa,3)

Dissociation of bases can be treated in the same manner as we done for acids, *i.e.*

$$NH_3 + H_2O(l) \rightleftharpoons NH_4^+(aq) + OH^-(aq)$$

On the same lines as we have carried out he dissociation of acids, the dissociation constants of bases can be written as:

$$K_b = \frac{a_{NH_4^+}a_{OH^-}}{a_{NH_3}a_{H_2O}} = \frac{a_{NH_4^+}a_{OH^-}}{a_{NH_3}} = \frac{[NH_4^+][OH^-]\gamma_\pm^2}{[NH_3]} \quad ...(18.83)$$

Here we have set $a_{H_2O} = 1$. In terms of concentration, equation (18.83) can be written as:

$$K_b = \frac{[NH_4^+][OH^-]}{[NH_3]} \quad ...(18.84)$$

Similar to pK_a, pK_b can be defined as

$$pK_b = -\log K_b \quad ...(18.84a)$$

Basicity constants of few bases are compiled in Table 18.5.

TABLE 18.5. Basicity constants of few bases at 298 K

Base	K_b	pK_b	Base	K_b	pK_b
Ammonia	1.8×10^{-5}	4.75	Morphine	7.4×10^{-7}	6.13
Aniline	3.80×10^{-10}	9.42	Nicotine	7×10^{-7}	6.2
Caffeine	4.1×10^{-4}	3.39	Novocaine	7×10^{-6}	5.2
Cocaine	2.57×10^{-6}	5.39	Pyridine	1.71×10^{-9}	8.77
Ceatine	1.92×10^{-11}	10.72	Quinine	1.1×10^{-6} (K_{b1}) 1.35×10^{-10} (K_{b2})	5.96 (pK_{b1}) 9.87 (pK_{b2})
Ethylamine	5.6×10^{-4}	3.25	Strychnine	1.0×10^{-6} (K_{b1}) 2×10^{-12} (K_{b2})	6.0 (pK_{b1}) 11.7 (pK_{b2})
Methylamine	4.38×10^{-5}	3.36	Urea	1.5×10^{-14}	13.82
Trimethyl ammonia	6.25×10^{-10}	9.20			

***Exercise* 18.22:** What is the molarity of $[H^+]$ ions on a 0.01 M solution of benzoic acid (C_6H_5COOH), and in presence of 0.1 M NaCl. $K_a = 6.30 \times 10^{-5}$?

(*i*) Without considering the activities into account, the solution will give us the approximate concentration of ions. Writing the dissociation equation, we get

$$C_6H_5COOH(aq) + H_2O(l) \rightleftarrows H_3O^+(aq) + C_6H_5COO^-(aq) \quad ...(1)$$

$$K_a = \frac{a_{H_3O^+} a_{A^-}}{a_{HA}} = \frac{[H^+][A^-]\gamma_\pm^2}{[HA]} \quad ...(2)$$

where $[A^-] = [C_6H_5COO^-]$ and $\gamma_\pm$ is the mean ionic coefficient of proton and benzoate ions. Assuming $\gamma_\pm = 1$, and $[C_6H_5COO^-] = [H^+]$, equation (2) reduces to

$$K_a = \frac{[H^+]^2}{[HA]}$$

or
$$[H^+] = \sqrt{K_a[HA]} \quad ...(3)$$

Since the acid is weak acid, we can assume that the concentration of the acid will be close to the initial concentration, *i.e.* 0.01 M. Substituting the values of the parameters, we get

or
$$[H^+] = \sqrt{6.30 \times 10^{-5}(0.01)} = 7.937 \times 10^{-4}\,M$$

(*ii*) The ionic strength of 0.1 M NaCl

$$\mu = \frac{1}{2}[Na^+]Z^2_{Na^+} + [Cl^-]Z^2_{Cl^-} = \frac{1}{2}(0.1)(1)^2 + (0.1)(-1)^2 = 0.1 \text{ mol L}^{-1}$$

$$\log \gamma_\pm = -0.509(1)(1)\sqrt{0.1} = -0.161$$

or
$$\gamma_\pm = 0.69$$

$$6.30 \times 10^{-5} = \frac{[H^+]^2(0.69)^2}{0.01} = \frac{[H^+]^2(0.69)^2}{0.01} = 47.61[H^+]^2$$

or
$$[H^+]^2 = \frac{6.30 \times 10^{-5}}{47.61} = 1.3233 \times 10^{-6}$$

or
$$[H^+] = 1.15 \times 10^{-3}\,M$$

In presence of NaCl the dissociation constant of benzoic acid increased and this phenomenon is called as the *salting in*.

***Exercise* 18.23:** Consider a very weak acid having $K_a = 6.4 \times 10^{-11}$. Calculate the magnitude of hydrogen ion concentration in 0.001 M acid solution.

This problem is little more complex than exercise (18.22), *i.e.* $C_a \times K_a = 0.001 \times 6.4 \times 10^{-11} = 6.4 \times 10^{-14}$. This factor is not very different from the value of $K_w = 10^{-14}$. Thus we have to include the contribution from both the acids. The equilibria of both the acids are:

$$H_2O \rightleftharpoons H^+ + OH^- \quad (1)$$
$$HA \rightleftharpoons H^+ + A^- \quad (2)$$

Let the concentration of $[H^+]$ from water is m_1 and from HA is m_2. The value of dissociation constants of both the acids can be written as:

$$K_w = [H^+][OH^-] = (m_1 + m_2)(m_1) = 10^{-14} \quad (1)$$

$$K_a = 6.4 \times 10^{-11} = \frac{[H^+][OH^-]}{C_0 - [m_2]} = \frac{[m_1 + m_2][m_2]}{0.001 - [m_2]} = \frac{[m_1 + m_2][m_2]}{0.001} \quad (2)$$

The final term has been written assuming m_2 is too small as compared to 0.001. The value of m_1 and m_2 can be solved as, *i.e.* divide (1) by (2), we get

$$\frac{(m_1 + m_2)(m_1)(0.001)}{(m_1 + m_2)m_2} = \frac{10^{-14}}{6.4 \times 10^{-11}}$$

$$\frac{m_1}{m_2} = \frac{10^{-14}}{6.4 \times 10^{-14}} = \frac{1}{6.4} \text{ or } m_2 = 6.4\, m_1 \quad (3)$$

Substituting the value of m_2 in equation (2), we get

$$(m_1 + 6.4\, m_1)m_1 = 10^{-14}$$

$$m_1^2 = \frac{10^{-14}}{7.4} = 0.135 \times 10^{-14} \quad \text{or } m_1 = \sqrt{0.135 \times 10^{-14}} = 0.3676 \times 10^{-7}\,\text{M}$$

$$m_2 = 6.4 \times 0.3776 \times 10^{-7} = 2.3527 \times 10^{-7}\,\text{M}$$

Thus the total hydrogen ion concentration is

$$m_1 + m_2 = (0.3676 \times 10^{-7} + 2.3527 \times 10^{-7}) = 2.72\ 10^{-7}\,\text{M}$$

***Exercise* 18.24:** Calculate the hydrogen ion concentration of 0.01 M acetic acid solution containing 0.01 M sodium acetate. K_a for acetic acid = 1.75×10^{-5}.

Sodium acetate is a strong electrolyte and thus will be assumed to be completely dissociated. Dissociation of acetic acid can be written as

$$CH_3COOH \rightleftharpoons CH_3COO^- + H^+$$

Before we calculate the concentration of hydrogen ions in presence of sodium acetate, let us calculate the concentration of hydrogen ions in absence of sodium acetate, *i.e.*

$$K_a = 1.75 \times 10^{-5} = \frac{[H^+][CH_3COO^-]}{[CH_3COOH]} = \frac{(m)(m)}{0.01} = \frac{m^2}{0.01}$$

i.e. $$m = \sqrt{(1.75 \times 10^{-5})(0.01)} = 4.183 \times 10^{-4}\ \text{M}$$

Let amount of acetic acid dissociated be m M, $[H^+] = m$, $[CH_3COO^-] = m + 0.01$, $[CH_3COOH] = 0.01 - m = \sim 0.01$ M.

$$K_a = 1.75 \times 10^{-5} = \frac{[H^+][CH_3COO^-]}{[CH_3COOH]} = \frac{m(m + 0.01)}{0.01} \quad (1)$$

To the first approximation assume that $m << 0.01$ and thus $m + 0.01 \sim 0.01$, equation (1) reduces to:

$$\frac{m(0.01)}{0.01} = m = 1.75 \times 10^{-5}\ \text{M} \quad (2)$$

To get more accurate results, let us solve the quadratic equation in m as:

$$m^2 + 0.01\, m - 1.75 \times 10^{-7} = 0$$

$$m = \frac{-m \pm \sqrt{m^2 - 4ac}}{2a} = \frac{-(0.01) \pm \sqrt{(0.01)^2 - 4(1)(-1.75 \times 10^{-7})}}{2} = \frac{-(0.01) \pm 0.0100349}{2}$$

$$= 1.747 \times 10^{-5} \text{ M} \quad (3)$$

This means that approximation used is not that bad as amount of hydrogen ion concentration is nearly same. The results show that in the presence of common ion, the concentration of other ion decreases. This process is known as *salting out*. This is opposite to what we learnt in exercise 18.22, wherein presence of neutral strong electrolyte the dissolution or dissociation of weak electrolyte increases.

18.11.5. *Relationship between the Dissociation Constant of an Acid and that of a Base*

Consider the dissociation of weak acid acetic acid. Dissociation constant in terms of concentration can be written as:

$$CH_3COOH\,(aq) \rightleftharpoons CH_3COO^-(aq) + H^+(aq)$$

$$K_a = \frac{[H^+][CH_3COO^-]}{[CH_3COOH]} \quad ...(18.85)$$

Conjugate base CH_3COO^-, obtained from the dissolution of sodium acetate reacts with H_2O

$$CH_3COO^-(aq) + H_2O(l) \rightleftharpoons CH_3COOH(aq) + OH^-(aq)$$

$$K_b = \frac{[OH^-][CH_3COOH]}{[CH_3COO^-]} \quad ...(18.86)$$

Using equations (18.85) and (18.86), product of K_a and K_b is given by

$$K_aK_b = \frac{[H^+][CH_3COO^-]}{[CH_3COOH]} \cdot \frac{[OH^-][CH_3COOH]}{[CH_3COO^-]} = [H^+][OH^-]$$

$$K_aK_b = K_w \quad ...(18.87)$$

Equation (18.87) enables us to draw a very important conclusion: *The stronger the acid (the larger the K_a), the weaker will be its conjugate base (the smaller K_b) and vice versa.*

18.11.6. *Salt Hydrolysis*

When salt of strong acid and strong base (say NaCl) is dissolved in water, the solution is neutral. In case if a salt of weak acid and strong base (say CH_3COONa) is dissolved in water, the solution shows basic behaviour and pH increases with increase of the amount of sodium acetate. On the other hand, if a salt of strong acid and weak base (say NH_4Cl) is dissolved in water, the solution shows the acidic behaviour and pH of the solution is less than 7. This behaviour is results from the hydrolysis of salt, which is a reaction of anion, or cation or both with water.

$$CH_3COOH(l) \xrightarrow{H_2O} CH_3COO^-(aq) + H^+(aq)$$

$$K_a = \frac{[CH_3COO^-][H^+]}{[CH_3COOH]} \quad ...(18.88)$$

Similarly for the hydrolysis of sodium acetate, we can write the equation as

$$CH_3COONa(s) \xrightarrow{H_2O} CH_3COO^-(aq) + Na^+(aq)$$

$$CH_3COO^-(aq) + H_2O(l) \rightleftharpoons CH_3COOH(aq) + OH^-(aq)$$

Sodium acetate is a strong electrolyte and thus it will dissociate completely. Since acetic acid is a weak acid, so it will dissociate slowly and thus the equilibrium will shift towards the right, producing large amount of OH^- ions and a basic solution. The equilibrium reaction will be similar to that of a base dissociation, *i.e.*

$$K_b = \frac{[OH^-][CH_3COOH]}{[CH_3COO^-]} = \frac{[OH^-][H^+][CH_3COOH]}{[CH_3COO^-][H^+]} = \frac{K_w}{K_a} \quad ...(18.89)$$

Using equation (18.88) and taking $K_a = 1.75 \times 10^{-5}$, we get

$$K_b = \frac{K_w}{K_a} = \frac{10^{-14}}{1.75 \times 10^{-5}} = 5.7 \times 10^{-10} \quad ...(18.90)$$

We can assume that sodium ions do not react with water.

***Exercise* 18.25:** Calculate the concentration of OH^- ion in 0.02 M sodium formate aqueous solution at 298 K. $K_a = 1.77 \times 10^{-4}$.

Hydrolytic reaction of sodium formate can be written as

$$HCOONa(s) \xrightarrow{H_2O} HCOO^-(aq) + Na^+(aq)$$

$$HCOO^-(aq) + H_2O(l) \rightleftharpoons HCOOH(aq) + OH^-(aq) \quad (1)$$

$$K_b = \frac{[OH^-][HCOOH]}{[HCOO^-]} \quad (2)$$

Contribution to OH^- ion can be from the self dissociation of water and from equation (1). Let m_1 be the OH^- ion concentration from water and m_2 the concentration of OH^- ion from equation (2). So

$[OH^-] = m_1 + m_2$; $[H^+] = m_1$; $[HCOOH] = m_2$; $[HCOO^-] = C_0 - m_2$

Substituting the desired parameters in equation (2), we get:

$$K_h = \frac{[OH^-][HCOOH]}{[HCOO^-]} = \frac{m_2(m_1 + m_2)}{(C_0 - m_2)} \quad (3)$$

Similarly the K_w can be written as by taking $\gamma_\pm(H_2O) = 1$

$$K_w = [H^+][OH^-] = m_1\,(m_1 + m_2) \quad (4)$$

The values of m_1 and m_2 can be obtained by solving equations (3) and (4). From equation (3),

$$m_1 = \frac{K_w}{\sqrt{K_w + K_h C_0}} \quad (5)$$

$$m_2 = \frac{C_0 K_h}{\sqrt{K_w + K_h C_0}} \quad (6)$$

Thus the total concentration of $[OH^-]$ can be obtained by adding $m_1 + m_2$, *i.e.*

$$[OH^-] = m_1 + m_2 = \frac{K_w}{\sqrt{K_w + K_h C_0}} + \frac{C_0 K_h}{\sqrt{K_w + K_h C_0}} = \frac{K_w + K_h C_0}{\sqrt{K_w + K_h C_0}} = \sqrt{K_w + K_h C_0} \quad ...(18.91)$$

K_h can be obtained by using equation (18.90), *i.e.*

$$K_h = \frac{K_w}{K_a} = \frac{10^{-14}}{1.77 \times 10^{-4}} = 5.65 \times 10^{-11} \quad (7)$$

Using equation (18.91) and substituting the values of parameters, we get

$$[OH^-] = \sqrt{K_w + K_h C_0} = \sqrt{10^{-14} + (5.65 \times 10^{-11})(0.02)} = \sqrt{10^{-14} + (113 \times 10^{-14})}$$

$$= 1.067 \times 10^{-6} \text{ M}$$

18.11.7. *Multi-protic Acids or Bases*

From Table 18.4. it is clear that we can represent the individual equilibrium when each proton dissociates. Consider diprotic acid H_2CO_3 and write the dissociation reactions as:

$$H_2CO_3(aq) \rightleftharpoons HCO_3^-(aq) + H^+(aq) \quad ...(18.92)$$

$$HCO_3^-(aq) \rightleftharpoons CO_3^{2-}(aq) + H^+(aq) \quad ...(18.93)$$

The equilibrium constants for each reaction can be written as

$$K_{a1} = \frac{[H^+][HCO_3^-]}{[H_2CO_3]} \quad ...(18.94)$$

$$K_{a2} = \frac{[H^+][CO_3^{2-}]}{[HCO_3^-]} \quad ...(18.95)$$

The value for, each equilibrium constant can be determined. From Table 18.4, it is clear that dissociation constant for second and further dissociation is several orders of magnitude smaller than the preceding ones. Further many examples of complex equilibriums have been discussed in Section 17.8 without taking into account the use of activity coefficients.

18.11.8. *Buffer Solutions*

One of the most important applications of acid-base solutions is buffering. A buffer solution is a solution of (*i*) a weak acid or weak base and (*ii*) its salt; both components must be present. The solution has the ability to resist changes in pH on the addition of small amounts of either an acid or a base. Presence of buffers plays a major role in the functioning of different parts of the body, *e.g.* proper functioning action of enzymes, the balance of osmotic pressure and so on. For example, pH of the blood plasma is about 7.4, whereas that of gastric juice, a fluid produced by glands in the mucosa membrane lining the stomach, is about 1.2. Consider the dissociation of a weak acid HA:

$$HA \rightleftharpoons H^+ + A^- \quad ...(18.96)$$

$$K_a = \frac{[H^+][A^-]}{[HA]} \quad ...(18.97)$$

Taking negative logarithm of both sides and writing in the form as

$$-\log K_a = -\log[H^+] - \log\frac{[A^-]}{[HA]} \quad ...(18.98)$$

or

$$pK_a = pH - \log\frac{[A^-]}{[HA]}$$

or it can written as

$$pH = pK_a + \log\frac{[A^-]}{[HA]} \quad ..(18.99)$$

Equation (18.99) is known as the Henderson-Hasselbalch equation, which is generally written as:

$$pH = pK_a + \log\frac{[\text{Conjugate base}]}{[\text{Acid}]} \quad ...(18.100)$$

Similar equation can be obtained when a buffer is prepared by the addition of weak base and its conjugate acid, *i.e.* equation (18.101).

$$pH = pK_a + \log\frac{[\text{Base}]}{[\text{Conjugate acid}]} \quad ...(18.101)$$

***Exercise* 18.26:** Calculate the number of moles of sodium acetate to be added to 0.015 M solution of aqueous of acetic acid to have the pH = 6.25. pK_a $(CH_3COOH) = 4.76$.

Using equation (18.100) and substitute the desired parameters, we get

$$6.24 = 4.76 + \log\frac{[\text{Base}]}{0.015}$$

or

$$\log\frac{[\text{Base}]}{0.015} = 6.24 - 4.76 = 1.48$$

or

$$\frac{[\text{Base}]}{0.015} = 10^{1.48} = 30.2$$

or

$$[\text{Base}] = 0.015 \times 30.2 = 0.453 \text{ M}.$$

18.11.9. *Effect of Ionic Strength and Temperature on Buffer Solutions*

So far equations (18.100) and (18.101) have derived assuming that the solutions are behaving in an ideal manner and we have used the concentration units. A more accurate solution will be the use of activity instead of concentration. Again taking the example of a weak acid HA, we write K_a as:

$$K_a = \frac{a_{H^+} a_{A^-}}{a_{HA}} \quad ...(18.102)$$

For neutral weak acid, we can replace the activity by concentration and equation (18.102) can be written after rearranging as:

$$a_{H^+} = \frac{K_a[HA]}{\gamma_{A^-}[A^-]} \quad ...(18.103)$$

where $a_{A^-} = \gamma_{A^-}[A^-]$ Taking the negative logarithm of both sides of equation (18.103), we get

$$pH = pK_a + \log\frac{[A^-]}{[HA]} + \log\gamma_{A^-} \quad ...(18.104)$$

Using Debye-Hückel limiting law [equation (18.67)], the activity coefficient can be expressed as

$$\log\gamma_{A^-} = -0.0509(Z^-)^2\sqrt{\mu} \quad ...(18.67)$$

and substituting the value in equation (18.104) we get the modified form of Henderson-Hasselbalch equation:

$$pH = pK_a + \log\frac{[A^-]}{[HA]} - 0.0509(Z^-)^2\sqrt{\mu} \quad ...(18.105)$$

Equation (18.105) consists of four terms. Three terms represent either concentration of the species or ionic strength and thus are independent of temperature. The only term which depend on temperature is the dissociation constant K_a and this in turn depend on the magnitude of the enthalpy of dissociation ($\Delta H°$). Thus knowing the value of $\Delta H°$ one can calculate the value of K_a at different temperatures and the pH values. In general we calculate the pH values at 298 K and then we can calculate pH at the body temperature (between 30° C to 40° C) or at any other temperatures.

18.11.10. *Buffer Capacity*

Buffer capacity measures the effectiveness of a buffer, which is the amount of acid or base that must be added to the buffer to produce a unit change of pH. Thus

$$\beta = \frac{d[B]}{dpH} \quad ...(18.106)$$

where $d[B]$ is the increase (in mol L^{-1}) in concentration of base B and d[pH] is the corresponding increase in pH. Equation (18.106) can be discussed as: Consider that to a buffer we have added small amount of acid. The acid will react with base to reduce its concentration, *i.e.* $d[B]$ will be negative. Similarly the pH of the solution will also decrease, *i.e.* d(pH) will also be negative. Thus equation (18.106) can be written as:

$$\beta = \frac{-d[B]}{-dpH} \quad ...(18.107)$$

In other words, sign of both the quantities are similar and thus the buffer capacity will always be positive value. Just to mention, buffer capacity depends on (*i*) nature of the buffer, (*ii*) the concentration of the acid and conjugate base and (*iii*) the pH.

The quantity which is also important is the *range of buffer* in which it is very useful. In terms of concentrations, the ratio [conjugate base]/[acid] should be between 0.1 and 10. This is due to slow variation of the logarithmic term over this range of ratios. Thus the useful range of buffer should be

$$pH = pK_a \pm 1 \quad ...(18.108)$$

18.12. MEAN ACTIVITY AND ACTIVITY COEFFICIENTS OF ELECTROLYTES

The following methods are specially used to calculate the activity coefficients of electrolytes.

18.12.1. *From Dissociation of a Weak Electrolytes*

Let the weak electrolyte AB dissociates as

$$AB \rightleftharpoons A^+ + B^-$$

$$K_{th} = \frac{(a_{A^+})(a_{B^-})}{a_{AB}} = \frac{[A^+][B^-]}{[AB]} \frac{\gamma_{A^+}\gamma_{B^-}}{\gamma_{AB}} \quad ...(18.109)$$

If α is degree of dissociation and C is the concentration of electrolyte then

$$\frac{[A^+][B^-]}{[AB]} = \frac{\alpha^2 C}{1-\alpha} = K \quad ...(18.110)$$

and if the electrolyte is weak $\gamma_{AB} = 1$ and $\gamma_{A^+}\gamma_{B^-} = \gamma_+^2$ (AB)

Substituting these in equation (18.109) and taking log of both sides and arranging

$$\log K = \log K_{th} - 2 \log \gamma_\pm \quad ...(18.111)$$

α can be calculated from the Arrhenius relation $\left(\alpha = \frac{\Lambda}{\Lambda_0}\right)$

or

$$\alpha = \frac{\Lambda}{\Lambda_e} \text{ where } \Lambda_e = \lambda_e^+ + \lambda_e^-$$

From Debye-Hückel limiting law and at 25°C

$$\log \gamma_\pm = -0.5091\,(Z^+)\,(Z^-)\,\sqrt{\mu} \quad ...(18.112)$$

and

$$\mu = \frac{1}{2}\,\Sigma C_i Z_i^2$$

For the 1:1 electrolyte, $C_A = \alpha C$, $C_B = \alpha C$

So $\mu = \frac{1}{2}\left[C_A Z_A^2 + C_B Z_B^2\right] = \frac{1}{2}\left[\alpha\, C(1)^2 + \alpha\, C(-1)^2\right] = \alpha C$

i.e.

$$\log \gamma_\pm \propto \sqrt{C\alpha}$$

The plot of $\log K$ vs $\sqrt{C}$ will be a straight line at small concentration and extrapolated to $C = 0$ will give K_{th}. Thus the value of $\gamma_\pm$ can be calculated at any concentration knowing K.

***Exercise* 18.27:** Calculate the value of $\gamma_\pm$ for *o*-chlorobenzoic acid from the following data:

$C \times 10^3$ (M)	6.662	3.331	1.666	0.8327	0.4163
Λ	134.6	174.0	218.0	262.6	302.1
Λ_e	375.8	377.5	378.9	380.3	381.1

The above data can be used to calculate α, K as

$C \times 10^3$	6.662	3.331	1.666	0.8327	0.4163
α	0.358	0.461	0.575	0.691	0.801
$K \times 10^4$	13.31	13.13	12.96	12.86	12.54
$\log K$	–2.876	–2.882	–2.887	–2.891	–2.901
$\sqrt{C\alpha}$	0.0487	0.0392	0.0310	0.0240	0.0182

log K is plotted vs $\sqrt{C}$ in Fig. 18.10 and the value of K at $C = 0$ is $12.15 \times 10^{-4} = K_{th}$. Using equation (18.111)

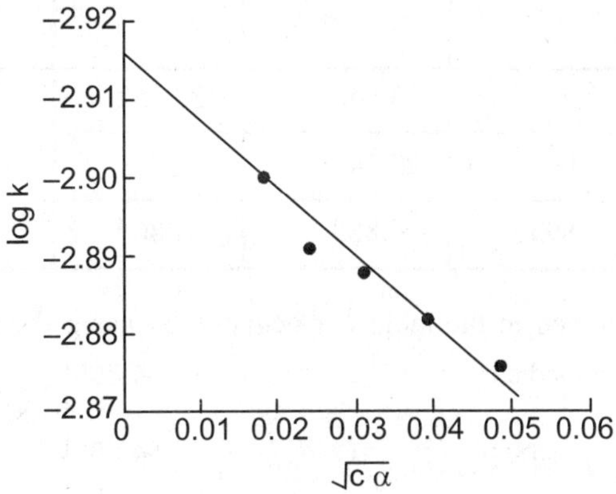

Fig. 18.10

$$\log \gamma_+ = \frac{1}{2} (\log K_{th} - \log K) = \frac{1}{2} \log \frac{K_{th}}{K}$$

or $$\gamma_+ = \left(\frac{K_{th}}{K}\right)^{1/2}$$

Thus for $C = 6.6 \times 10^{-3}$ M, $K = 13.31 \times 10^{-4}$

$$\gamma_+ = \left(\frac{12.15 \times 10^{-4}}{13.31 \times 10^{-4}}\right)^{1/2} = 0.955$$

Similarly for $C = 3.331 \times 10^{-3}$ M, 1.666×10^{-3} M, 0.8327×10^{-3} M and 0.4163×10^{-3} M, γ_+ are 0.961, 0.965, 0.972 and 0.984 respectively.

18.12.2. *Activity Coefficient from Solubility Measurements*

Consider 1:1 electrolyte (AB)

$$AB(s) \rightleftharpoons A^+ + B^-$$

$$K_{th} = (a_{A+})(a_{B-}) = (\gamma_{A+})(\gamma_{B-})[A^+][B^-]$$
$$= \gamma_+^2 (AB)[A^+][B^-] = \gamma_+^2 (AB)(s^2) \quad ...(18.113)$$

Taking log of both sides

$$\log s = \frac{1}{2} \log K_{th} - \log \gamma_+ \quad ...(18.114)$$

where s is the solubility of electrolyte that dissolves in water. According to Debye-Hückel theory log $\gamma_+ \propto \sqrt{\mu}$. So plot log s vs $\sqrt{\mu}$ (where μ is ionic strength of the solution) will be a straight line and at $\mu = 0$ gives K_{th}. Knowing s and K_{th}, γ_+ can be calculated.

***Exercise* 18.28:** The solubility of AgCl is measured in presence of different amounts of HNO_3. Calculate the activity coefficient of AgCl using equation (18.113) and Debye-Hückel limiting law. The given data is

$s \times 10^5$ (mol dm^{-3})	1.280	1.318	1.352	1.387	1.422
concn. of solvent electrolyte (mol dm^{-3})	1.28×10^{-5}	72.33×10^{-5}	286.4×10^{-5}	5.695×10^{-5}	900.9×10^{-5}
$\mu \times 10^5$	2.56	73.648	287.752	570.887	902.322
$\sqrt{\mu} \times 10^3$	5.06	27.14	53.75	75.56	94.99
log s	– 4.893	– 4.880	– 4.869	– 4.858	– 4.847

The ionic strength mentioned in the table is calculated by using the relation $\mu = \frac{1}{2}\Sigma C_i Z_i^2$ for example, the first value is obtained as

$$\mu = \frac{1}{2}[[H^+](Z_{H^+})^2 + [NO_3^-](Z_{NO_3^-})^2 + [Ag^+](Z_{Ag^+})^2 + [Cl^-][Z_{Cl^-}]^2]$$

$$= \frac{1}{2}[(1.28(1)^2 + (1.28)(-1)^2 + (1.28)(1)^2 + (1.28)(1.28)(-1)^2] \times 10^{-5}$$

$$= 2.56 \times 10^{-5} \text{ mol dm}^{-3}$$

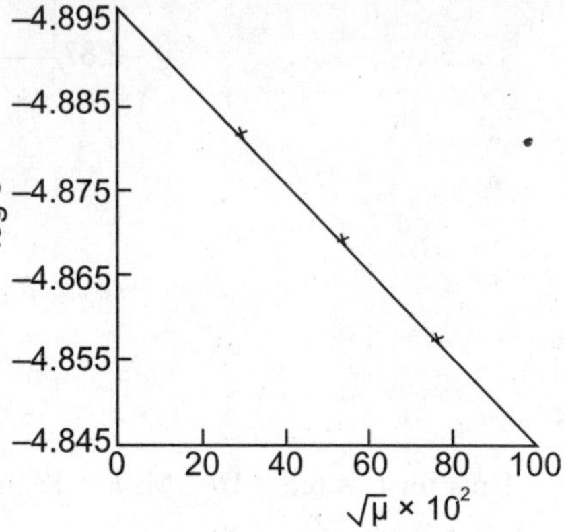

Fig. 18.11

Similarly the other values of μ are calculated. The plot of log s vs $\sqrt{\mu}$ (Fig. 18.11) gives $K_{th} = 1.618 \times 10^{-10}$. The activity coefficient of AgCl at ionic strength 570.9×10^{-5} mol dm^{-3} can be obtained by using the equation (18.114) *i.e.* the value of s at this ionic strength is 1.387×10^{-5} mol dm^{-3}

$$\log \gamma_+ = \frac{1}{2}\log(1.618 \times 10^{-10}) - \log(1.384 \times 10^{-5})$$

$$= -0.0366$$

$$\gamma_+ = 0.919$$

Using equation (18.30*a*)

$$a_\pm = (0.919)(1.387 \times 10^{-5}) = 1.275 \times 10^{-5}$$

18.13. EFFECT OF IONIC STRENGTH ON DISSOCIATION CONSTANT AND SOLUBILITY PRODUCT

When a strong electrolyte is added to a weak electrolyte or to a solution containing sparingly soluble salt, the ionic strength of the solution will change (strong electrolyte will be completely dissociated). Due to this the activity coefficient and thus the dissociation constant of weak electrolyte or the solubility of sparingly soluble salt will change. This can be seen with the exercises 18.29 and 18.30.

***Exercise* 18.29:** The solubility product of $BaSO_4$ is $10^{-10.0}$ mol^2 dm^{-6}. Calculate the concentration of Ba^{2+} ions in (*a*) water (*b*) 0.01 M solution of $NaNO_3$ and (*c*) 0.01 M $BaCl_2$ solution.

The $BaSO_4$ dissociates in water as

$$BaSO_4\,(s) \rightleftharpoons Ba^{2+}(aq) + SO_4^{2-}(aq)$$

and solubility product is

$$K_{sp} = a_{Ba^{2+}} \cdot a_{SO_4^{2-}} = \gamma_{Ba^{2+}} \cdot \gamma_{SO_4^{2-}} \cdot m_{Ba^{2+}} \cdot m_{SO_4^{2-}}$$

$$= \gamma_{\pm}^{\nu} (\nu_+^{\nu+} \nu_-^{\nu-}) m^{\nu^+ + \nu^-}$$

Here $\nu_+ = 1,\ \nu_- = 1$ and $\nu = 1 + 1 = 2$. So

$$K_{sp} = \gamma_{\pm}^2 m^2 \qquad ...(18.115)$$

If in problem (*a*) it is assumed that $\gamma_{\pm} = 1$ So

$$K_{sp} = m^2 \qquad ...(18.116)$$

or

$$m = (K_{sp})^{1/2} = (10^{-10})^{1/2} = 10^{-5} \text{ mol dm}^{-3}$$

So the concentration of Ba^{2+} ions in water is 10^{-5} mol dm^{-3}.

The assumption made that $\gamma_{\pm} = 1$ can be verified with the help of Debye-Hückel limiting law *i.e.* using equation (18.63)

$$\mu = \frac{1}{2}[(Ba^{2+})\, Z_{Ba^{2+}}^2 + (SO_4^{2-})\, Z_{SO_4^{2-}}^2]$$

$$= \frac{1}{2}[(10^{-5})(+2)^2 + (10^{-5})(-2)^2]$$

$$= 4.0 \times 10^{-5} \text{ mol dm}^{-3}$$

Substituting the respective values in equation (18.112)

$$\log \gamma_{\pm} = -(0.509)(2)(2)(4.0 \times 10^{-5})^{1/2} = -0.0129$$

$$\gamma_{\pm} = 0.9943$$

It is quite close to unity.

(*b*) In presence of 0.01 M $NaNO_3$ solution

$$\mu = \frac{1}{2}[(1.0 \times 10^{-2})(1)^2 + (1.0 \times 10^{-2})(-1)^2] = 1.0 \times 10^{-2}$$

The ionic concentration due to $BaSO_4$ is very small and can be neglected. Now the mean ionic activity coefficient of $BaSO_4$ in presence of 0.01M $NaNO_3$ is

$$\log \gamma_{\pm} = -(0.509)(2)(2)(1.0 \times 10^{-2})^{1/2}$$

$$= -2.036 \times 1 \times 10^{-1} = -0.2036$$

$$\gamma_{\pm} = 0.6218$$

Substituting the values of K_{sp} and $\gamma_{\pm}$ in (18.64)

$$m = \frac{(K_{sp})^{1/2}}{\gamma_{\pm}} = \frac{(10^{-10})^{1/2}}{0.6218} \text{ mol dm}^{-3}$$

$$= 1.62 \times 10^{-5} \text{ mol dm}^{-3}$$

Concentration of both ions or solubility of $BaSO_4$ increases,

(*c*) In presence of 0.01 M $BaCl_2$ the ionic strength is

$$\mu = \frac{1}{2}[(1.0 \times 10^{-2})(+2)^2 + (2 \times 1.0 + 10^{-2})(-1)^2]$$

$$= 3.0 \times 10^{-2}$$

Now

$$\log \gamma_+ = -(0.509)(2)(1)(3.0 \times 10^{-2})^{1/2} = -0.1763$$
$$\gamma_+ = 0.6664$$

Now

$$K_{sp} = 10^{-10} = (0.6664)^2 [(SO_4^{2-})\{1.0 \times 10^{-2}\} + (Ba^{2+})]$$

or

$$[Ba^{2+}] = [SO_4^{2-}] \text{ when } BaSO_4 \text{ dissociates.}$$
$$[Ba^{2+}]^2 + 0.01\,[Ba^{2+}] - 2.27 \times 10^{-10} = 0$$

gives

$$[Ba^{2+}] = 0.01\text{M}$$

and

$$[SO_4^{2-}] = \frac{K_{sp}}{Ba^{2+}} = \frac{10^{-10}}{0.01} = 10^{-8}\text{ M}$$

This is called common ion effect.

***Exercise* 18.30:** The dissociation constant of propionic acid is 1.336×10^{-5}. Will the degree of dissociation increase or decrease if 0.01 M KCl solution is added in 0.1 M propionic acid.

Propionic acid dissociates as

$$C_2H_5COOH \rightleftharpoons C_2H_5COO^- + H^+$$

If α is the degree of dissociation

$$K = \frac{\alpha^2 m}{1-\alpha} \cdot \frac{\gamma_\pm^2}{\gamma_{HA}} = \alpha^2 m \,\frac{\gamma_\pm^2}{\gamma_{HA}} \qquad ...(18.117)$$

if α is small

If it is assumed that activity coefficients are unity and α is small, the equation (18.117) becomes

$$K = \alpha^2 m \qquad ...(18.118)$$

or

$$\alpha = \sqrt{\frac{K}{m}} = \left(\frac{1.336 \times 10^{-5}}{0.1}\right)^{1/2} = 1.15 \times 10^{-2}$$

In the presence of 0.01 M KCl, the activity coefficients will be different. Using equation (18.63)

$$\mu = \frac{1}{2}[(0.01)(+1)^2 + (0.01)(-1)^2] = 0.01$$

Here the concentration of H^+ and $C_2H_5COO^-$ will be small and can be neglected.

Using equation (18.112)

$$\log \gamma_+ = -(0.509)(1)(1)(1.0 \times 10^{-2})^{1/2} = -0.0509$$
$$\gamma_+ = 0.8894$$

Substituting the values of γ_+, $\gamma_{HA} = 1$, m and K in (18.117), the α can be calculated as

$$1.336 \times 10^{-5} = (\alpha)^2 (0.1)(0.8894)^2$$

$$\alpha = \left[\frac{1.336 \times 10^{-5}}{(0.1)(0.8894)^2}\right]^{1/2} = 1.303 \times 10^{-2}$$

There is an increase in the degree of dissociation

$$\% \text{ Increase} = \frac{1.303 \times 10^{-2} - 1.15 \times 10^{-2}}{1.15 \times 10^{-2}} \times 100$$

$$= 13.3\%.$$

PROBLEMS

18.1 The densities of water-methanol mixtures of different compositions are as follows:

Wt % CH_3OH	d kg dm^{-3}	Wt % CH_3OH	d kg dm^{-3}	Wt % CH_3OH	d kg dm^{-3}
0	0.9982	50	0.9156	80	0.8469
10	0.9815	55	0.9052	85	0.8340
20	0.9666	60	0.8946	90	0.8202
30	0.9515	65	0.8834	95	0.8062
40	0.9345	70	0.8715	100	0.7917
45	0.9245	75	0.8592		

Calculate the partial molal volume of methanol and water at 2 molal solution by graphical method.

18.2 Calculate the partial molal volume of H_2SO_4 and water at 6 molal solution of H_2SO_4 at 20°C by graphical method.

Wt % H_2SO_4	d kg dm^{-3}	Wt % H_2SO_4	d kg dm^{-3}	Wt % H_2SO_4	d kg dm^{-3}
0	0.9982	55	1.4453	85	1.7841
10	1.0661	60	1.4983	90	1.8144
20	1.1394	65	1.5533	95	1.8337
30	1.2185	70	1.6105	100	1.8305
40	1.3028	75	1.6692		
50	1.3951	80	1.7272		

18.3 Use the data of problems 18.1 and 18.2 to calculate the partial molal volumes of CH_3OH, H_2SO_4 and H_2O, at the desired concentrations, by the method of intercepts.

18.4 Prove the following relations:

(*a*) $\bar{V}_2 = \dfrac{M_2 - \left(\dfrac{\partial d}{\partial C}\right)}{d_1 - C\left(\dfrac{\partial d}{\partial C}\right)}$ (*b*) $\bar{V}_1 = \dfrac{M_1}{d - C\left(\dfrac{\partial d}{\partial C}\right)}$

where d_1 and d are the densities of the solution and of solvent respectively, C is the concentration of solute in mol dm^{-3} and M_2 the molecular weight of solute in g mol^{-1}.

18.5 Use the data of problems 18.1, 18.2 and 18.6 to calculate the partial molal volumes of respective species by using the relations mentioned in problem 18.4.

18.6 Calculate the partial molal volumes of KBr and water at 1 molal solution of KBr at 20°C by graphical method and the following data:

wt % KBr	d kg dm^{-3}	wt % KBr	d kg dm^{-3}	wt % KBr	d kg dm^{-3}
1	1.00599	10	1.08033	20	1.17449
2	1.01385	12	1.09810	24	1.21627
4	1.02984	14	1.11638	28	1.26078
6	1.04623	16	1.13519	30	1.28413
8	1.06305	18	1.15455		

18.7 The total volume (dm^{-3}) of the mixture of compound A with water at 20°C can be expressed as

$$10^3 \times V = 1001.80 + 24.582\ \text{m}^{1/2} + 3.486\ m + 40.452\ \text{m}^{3/2}$$

where m is the molality of the solute. Calculate the partial molal volumes of the solute and water at 1 molal concentration.

18.8 Consider the following processes at 20°C.

(*i*) $H_2O(l)$ → H_2O (2 molal solution of A)
(*ii*) $H_2O(l)$ → H_2O (55% aqueous CH_3OH)
(*iii*) $H_2SO_4(l)$ → H_2SO_4 (35 % aqueous H_2SO_4)

Using the data of problems 18.7, 18.1 and 18.2, compute the volume changes.

18.9 Prove the following relation between apparent partial molal volume and density of the solution

$$\phi V_2 = \frac{1}{C} - \frac{1}{d}\left[\frac{d_1}{C} - M_2 \times 10^{-3}\right]$$

where C is the concentration in mol dm^{-3}, d and d_1 are the densities of the solvent and solution respectively.

18.10 Use the data of problems 18.1, 18.2 and 18.6 to calculate the (*a*) apparent molal volume, (*b*) the partial molal volume of the solute and (*c*) the partial molal volume of the solvent at the respective concentrations.

18.11 Repeat the problem 18.10 by using the relation of ϕV_2 and d, mentioned in problem 18.9.

18.12 The relative apparent molal heat content at 30°C of sucrose in water can be represented by

$$\phi L_2 = 140.2\ \text{m} - 7.08\ \text{m}^2$$

with $m \leq 2.2$ mol kg^{-1}. Calculate the relative molal heat content at 2 molal solution and at 30°C.

18.13 Heat absorbed by dissolving m_2 mol of an electrolyte in 1 kg of water can be represented as

$$\Delta H = 420.6\ \text{m}_2^{1/2} + 220.4\ \text{m}_2 - 654.0\ \text{m}_2^{3/2} + 625.6\ \text{m}_2^2$$

(*a*) Calculate the $\bar{L}_2$ for 0.1 and 0.04 molal solution
(*b*) Calculate $\bar{L}_1$ for above solution.

Activity and Activity Coefficient

18.14 Will the value of activity coefficient be less than unity or greater than unity for liquids with (*a*) positive deviation and (*b*) negative deviation for Raoult's law.

18.15 The partial vapour pressures of water (p_1) and *n*-propyl alcohol (p_2) at 25°C are as follows:

x_2	p_1 (mm)	p_2 (mm)	x_2	p_1 (mm)	p_2 (mm)
0	23.756	0	0.4	21.7	14.2
0.01	23.4	2.68	0.6	17.9	15.5
0.02	23.5	5.05	0.8	13.4	17.8
0.05	23.2	10.8	0.9	8.13	19.4
0.1	22.7	13.2	0.95	4.20	20.8
0. 2	21.8	13.6	1.00	0.0	21.76

(*a*) Plot the partial pressures of water and propyl alcohol as a function of composition and indicate whether there is a positive deviation or negative deviation from Raoult's law.

(*b*) Calculate Henry's constant for both liquids.

(*c*) Calculate the activity coefficient of water at mole fraction$x_2 = 0.6$, considering the standard state as (*i*) pure component (*ii*) Hypothetical state of Henry's law at unit mole fraction of A assuming that vapours are ideal.

(*d*) Calculate the activity coefficient of alcohol at the above composition considering it as solvent.

18.16 Repeat the problem 18.15 with the following data of CS_2 (A) and benzene (B) at 25°C.

$x_B \times 10^2$	p_A (mm)	p_B (mm)	$x_B \times 10^2$	p_a (mm)	p_b (mm)
0	361.1	0	49.86	197.4	53.7
2.11	352.7	3.5	63.02	156.6	63.5
4.68	344.1	7.1	76.63	102.5	75.3
11.53	321.4	16.0	87.13	60.2	83.8
18.24	299.1	24.0	94.19	27.9	89.5
30.02	263.6	35.1	100.00	0.0	94.4

18.17 Repeat the problem 18.15 with the following data of $CHCl_3$ (*A*) and ethyl ether (*B*) at 17°C.

$x_A \times 100$	p_A (mm)	p_B (mm)	$x_A \times 100$	p_A (mm)	p_B (mm)
0	0	397	60	71	95
10	6	354	70	91	60
20	13	304	80	111	32
30	22	251	90	129	14

40	34	196	100	143	0
50	50	143			

18.18 Calculate the activity coefficients of components mentioned in problems 18.15, 18.16 and 18.17 for each concentration, plot them as a function of concentration and calculate the limiting values at infinite dilution.

18.19 From the e.m.f. studies of the cell Ag(s)/AgCl in LiCl (1), Cl_2 (g)/Pt the following data of a_1 is obtained:

x_1(AgCl)	1	0.804	0.69	0.573	0.469	0.252	0.136
a_1	1	0.884	0.791	0.741	0.69	0.529	0.392

Calculate the activity and activity coefficient of LiCl by using Gibbs-Duhem relation at $x_2 = 0.427$.

18.20 From the data of problems 18.15 and 18.16, calculate the activity and activity coefficient of water (considering it as solvent) by Gibb's-Duhem relation.

18.21 If the two liquids have partial miscibility then at equilibrium between two layers of two component system a_1 (layer = 1) = a_1' (layer 2). Further if in one layer component 1 is very slightly miscible show that $\gamma_1 = \dfrac{1}{x_1}$.

18.22 Using the above relation derived in problem 18.21 calculate the activity coefficient of n phenyl alcohol in water if the percentage solubility of alcohol in water is only 0.624.

18.23 Water and ter. butyl alcohol are partially miscible. Calculate the ratio of activity coefficient of ter. butyl alcohol in water rich and alcohol rich phase at 25°C from the following data. Alcoholic rich phase contains 79.64% (by weight) of alcohol and water rich phase 7.31% (by weight) of alcohol.

18.24 Calculate the activity and activity coefficient of potassium in mercury at $x_k = 0.01055$ from the following e.m.f. data of the cell $K(C_2)$ amalgam | KCl soln | $K(C_1)$ amalgam.

x_2 (K)	0.01984 sat.	0.00065	0.01418	0.01055	0.00712	0.0025	0.0013
E(V)	0	– 0.1242	–.0189	–.03279	–.0485	–.0823	–.1006

18.25 The elevation in boiling point of water was noted as 5.26° when mole fraction of the solution containing H_2SO_4 is 0.06. Calculate the activity coefficient of water.

18.26 The water solution containing 0.5 molal glucose solution boils at 100.3°C. Calculate the activity coefficient of water at the boiling point.

18.27 The following data is observed for the boiling point of water when different amount of KCl was added to water:

m (mol kg^{-1})	0.1	0.3	0.5	1.0	2.0	5.0	7.82
$\dfrac{\Delta t_B}{x}$	52.8	52.3	52.3	54.1	55.0	63.0	69.6

Calculate the activity coefficient of KCl at 0.4 molal solution.

18.28 For different concentration of sucrose solution in water, the osmotic coefficient observed is as follows:

m (mol kg^{-1})	$1-j$	m (mol kg^{-1})	$1-j$	m (mol kg^{-1})	$1-j$
0.1	1.0073	0.7	1.0596	2.5	1.2436
0.2	1.0151	1.0	1.0897	3.0	1.2923
0.3	1.0234	1.5	1.1415		
0.5	1.041	2.0	1.1929		

Calculate the activity coefficient of sucrose at 0.4 molal concentration.

18.29 Calculate the mean ionic activity and activity coefficient of acetic acid at 0.03685 M from the following data:

m (M)	0.0737	0.03685	0.01842	0.00921	0.00461	0.0023
Λ	6.086	8.591	12.09	16.98	23.81	33.22
Λ_0	385.7	386.5	386.9	388.1	388.9	389.2

18.30 The thermodynamic equilibrium constant of cynoacetic acid is 35.0×10^{-4}. If the values of Λ and Λ_0 for 0.01487 molar cynoacetic acid are 152.5 and 377.9 Ω^{-1} cm^2 mol^{-1}, calculate $\gamma_\pm$ and $a_\pm$ for this concentration.

18.31 Calculate the mean ionic activity and activity coefficient for 0.06 m HCl from the following e.m.f. data of the cell Pt | H_2 (1 atm) | HCl (m) | AgCl (s) | Ag:

m	0.1238	0.0539	0.0256	0.0134	0.00914	0.00562
E (V)	0.34199	0.38222	0.41824	0.44974	0.4686	0.49257

18.32 The standard state as well as the emf of the cell Ag | Ag_2SO_4 (s) | H_2SO_4(0.1)| H_2(1 atm)| Pt at 25°C are –0.653 V and –0.775 V respectively. Calculate the mean ionic activity coefficient of H_2SO_4. Also use Debye-Hückel's limiting law to calculate $\gamma_\pm$ and compare the two values.

18.33 In the following cell, Ag | AgBr (s) | HBr (0.1) | H_2 (1.0 atm) | Pt, the e.m.f. of the cell at 25°C is 0.22 V. If the solubility product of AgBr is 1×10^{-12} and ϵ^0 (Ag^+/Ag) = 0.7991 V, calculate $\gamma_\pm$ for 0.1 molar HBr.

18.34 The following j values are observed for different concentration of ammonium chloride solutions:

m	0.001	0.002	0.005	0.01	0.02	0.05	0.1	0.2
j	0.014	0.0251	0.0405	0.0411	0.0503	0.0652	0.0787	0.0927

Calculate the activity coefficient of ammonium chloride at 0.01 molal solution.

18.35 The change in the freezing points of water observed for various concentration of KNO_3(molality) in water is given in the data below. Calculate the osmotic coefficient and the activity coefficient of KNO_3 at 0.07 molal solution.

m	θ	m	θ	m	θ
0.001	3.674	0.01	3.587	0.1	3.332
0.002	3.657	0.02	3.536	0.2	3.185
0.005	3.624	0.05	3.438	0.3	3.073

18.36 The solubility of AgCl in various solutions containing different amounts of $Ba(NO_3)_2$ is as follows:

$m \times 10^5$ ($Ba(NO_3)_2$ mol dm^{-3})	0.64	3.615	21.108	70.64	149.9	219.2	308.3
$s \times 10^5$ (mol dm^{-3})	1.28	1.291	1.309	1.339	1.372	1.394	1.421

Calculate the activity coefficients of AgCl at various concentration and also the limiting value.

Debye-Hückel Theory

18.37 Calculate the ionic strength of the solution, formed by mixing 10 mL of 0.017 M $Fe(NO_3)_3$, 40 mL of 0.1 M KNO_3, 10 mL of 0.1 M HNO_3, 10 mL of 0.0096 M $Na_2S_2O_3$, 20 mL of 0.1 M KI and 10 mL of H_2O.

18.38 Calculate the ionic strength of the following solutions:

(*a*) 0.02 M solution of $Cu_3(PO_4)_2$

(*b*) 0.15 M solution of H_3PO_4 if it dissociates

$$H_3PO_4 \rightarrow 3H^+ + PO_4^{3-}$$

completely.

(*c*) 0.2 M solution of acetic acid of the degree of dissociation is 0.60%.

(*d*) 0.1 M solution of $K_3[Fe(CN)_6]$.

18.39 Prove the equations (18.59), (18.47) and (18.58).

18.40 If the mean activity coefficient of 0.05 M H_2SO_4 is 0.34, calculate the activity of the electrolyte as well as mean molality.

18.41 Using the limiting form of Debye-Hückel theory, calculate the activity coefficients of Cu^{2+} and PO_4^{3-} ions as well as the $\gamma_{\pm}$ of 0.02 M solution of $Cu_3(PO_4)_2$.

18.42 Calculate the activity coefficients of H^+ and propionate ions in 0.05 N propionic acid solution if the dissociation constant of propionic acid is 1.34×10^{-5} at 25°C.

18.43 Calculate the γ_i of each ion and $\gamma_{\pm}$ of first electrolyte in the following solutions:

(*i*) 0.02 M KCl, 0.04 $BaCl_2$

(*ii*) 0.01 M $CuSO_4$, 0.03 Na_2SO_4

(*iii*) 0.015 M $LaCl_3$, 0.02 $MgCl_2$

18.44 Calculate the mean ionic molality of the following electrolytes at 0.5 M solutions of (*i*) NaOH (*ii*) H_2SO_4 (*iii*) $Ca(NO_3)_2$ (*iv*) $La(NO_3)_3$.

18.45 Calculate the activity of the electrolytes mentioned in problem (18.44).

Effect of Ionic Strength on Solubility and degree of dissociation

18.46 The solubility product of Ag_2SO_4 at 25°C is $10^{-4.8}$ mol^3 dm^{-9}. Calculate the solubility of Ag_2SO_4 in 0.1, 0.01 and 0.005 M Ca $(NO_3)_2$ solution when the mean activity coefficients of $Ca(NO_3)_2$ at these concentrations are 0.48, 0.71 and 0.77 respectively.

18.47 The solubility product of Fe $(OH)_3$ is $10^{-38.4}$ mol^4 dm^{-12}. Calculate the solubility of Fe $(OH)_3$ in (*a*) water, (*b*) 0.1 M KOH, (*c*) 0.05 M K_2SO_4 and (*d*) 0.5 $Ca(NO_3)_2$ solution.

18.48 The dissociation constant of chloroacetic acid is 1.379×10^3. Calculate the degree of dissociation of the acid in (*i*) 0.1 M solution, (*ii*) 0.1 M solution containing 0.055 M K_2SO_4 and (*iii*) 0.1 M solution containing 0.1 M La $(NO_3)_2$.

18.49 The basicity constant (K_b) of NH_4OH at 25°C is 1.8×10^{-5}. Calculate the degree of dissociation of base in 0.1 M solution using the limiting law of Debye-Hückel theory. Compare the results when one neglects the activity coefficients.

18.50 The dissociation constant of formic acid is 1.8×10^4 at 25°C. Calculate the degree of dissociation when the concentration of formic acid is 0.4 M. If the 0.1 M of $Al_2(SO_4)_3$ is added, calculate the degree of dissociation of formic acid and also when 0.1 M sodium formate is added.

Additional Problems

18.51 Calculate the force between Cu^{2+} and SO_4^{2-} ions (*i*) in vacuum and (*ii*) in water at 298 K.

$\varepsilon\,(H_2O) = 78.54$, $q_{Cu^{2+}} = 3.2 \times 10^{-19}\ C$ and $q_{SO_4^{2-}} = -3.2 \times 10^{-19}\ C$.

18.52 Repeat the problem 18.51 for Fe^{2+} and SO_4^{2-} ions (*i*) in vacuum and (*ii*) in water at 298 K.

$\varepsilon\,(H_2O) = 78.54$, $q_{Cu^{2+}} = 3.2 \times 10^{-19}\ C$ and $q_{SO_4^{2-}} = -3.2 \times 10^{-19}\ C$.

18.53 Calculate the force between Al^{3+} and SO_4^{2-} ions (*i*) in vacuum and (*ii*) in water at 298 K.

$\varepsilon\,(H_2O) = 78.54$, $q_{Al^{3+}} = 4.8 \times 10^{-19}\ C$ and $q_{SO_4^{2-}} = -3.2 \times 10^{-19}\ C$.

18.54 (*a*) We have a binary mixture of components *A* and *B*. Suggests which of the following expressions is incorrect as a representation of the partial molar volume of the component *A* in a two–component solution? Why and how would you correct it?

$$\left(\frac{\partial V_m}{\partial n_A}\right)_{T,p,n_B}, \left(\frac{\partial V_m}{\partial x_A}\right)_{T,p,x_B}$$

(*b*) Let the molar volume of this mixture V_m is given by

$$V_m = 0.43 + 2.8\, x_A x_B + 0.45 x_B(1 - x_A)$$

Calculate the partial molar volume of *A* when $x_A = 0.28$.

18.55 Consider a binary mixture of two components *A* and *B*. The partial molar volumes of liquid *A* and liquid *B* solution at 25°C at a mole fraction of liquid *A* (0.45) are $\bar{V}_A = 0.180$ L mol^{-1} and $\bar{V}_B = 0.115$ L mol^{-1} respectively. (*i*) Calculate the volume of a solution when 1 mole of each component is mixed, (*ii*) If the molar volume of liquids *A* and *B* are 0.165 L mol^{-1} and 0.110 L mol^{-1}, calculate the change in volume on mixing 2 moles of liquid *A* and 1 mole of liquid *B*, (*iii*) Based on these results, deduce the nature of intermolecular forces between liquids *A* and *B*.

18.56 Calculate the heat of solution for KI from the data given below:

	NaCl	NaI	KCl	KI
Lattice energy (kJ mol^{-1})	787	700	716	643
Heat of solution (kJ mol^{-1})	3.8	–5.1	17.1	?

18.57 Consider a binary liquid mixture of *A* and *B* where *A* is non-volatile and *B* is a volatile liquid. The composition of the mixture in term of mole fraction is $\chi_A = 0.05$ and $\chi_B = 0.95$. The vapour pressure of pure liquid *B* at 298 K is 250 mm of Hg and that of the mixture at 298 K is 11.5 mm Hg. Calculate the activity coefficient of liquid *B*.

18.58 The Debye-Hückel limiting law, the Debye radius which defines the size of the ionic radius is given by $1/\kappa$, which is defined by equation (18.62). Calculate the value of Debye radius in a 0.005 m aqueous solution of (*i*) NaCl, (*ii*) Na_2SO_4 and (*ii*) Na_3PO_4.

18.59 For which electrolyte, Debye-Hückel law is more accurate; 1:1 electrolyte or 2 : 2 electrolyte?

18.60 100 g of glucose is dissolved in 900 g of water. If the molar mass of glucose is 180 g mol^{-1}, calculate the activity coefficient of glucose in this solution if the freezing point of water is –0.85 ^{0}C.

18.61 Vapour pressure of pure water at 40°C is 55.3 mm Hg and vapour pressure of glucose-water solution is 49.2 mm Hg at 45°C when 780 g of glucose was dissolved in 900 g of water.

18.62 Calculate the hydrogen ion concentration of an aqueous solution containing 0.05 M formic acid in the (*i*) absence of $Ca(NO_3)_2$ and (*ii*) in presence of 0.05 M $Ca(NO_3)_2$. Dissociation constant of formic acid is 1.36×10^{-4}.

18.63 Repeat the problem 18.62 for the chloroacetic acid ($ClCH_2COOH$). Dissociation constant of chloroacetic acid is 1.36×10^{-3}.

18.64 Repeat the problem 18.62 for an aqueous solution containing 0.2 M HF solution at 298.

$$K_a\,(HF) = 6.31 \times 10^{-4}.$$

18.65 Calculate the [H^+] ion concentration when 0.2 M ammonium carbonate is dissolved in water at

$$298\text{ K. } K_b\,(NH_3) = 1.8 \times 10^{-5}.$$

18.66 0.15 mole of acetic acid and 0.2 mole of formic acid are added to one litre of water. Calculate the hydrogen ion concentration. $K_a\,(CH_3COOH) = 1.75 \times 10^{-5}$, $K_a\,(HCOOH) = 1.77 \times 10^{-4}$.

18.67 Calculate the hydrogen ion concentration for (*i*) 0.15 M and (*ii*) 0.25 M aqueous solution of trimethylammonium chloride at 298 K assuming that this salt dissociates completely in water. $K_b\,(CH_3)_3N = 6.25 \times 10^{-10}$.

18.68 Calculate the values of [H^+], [OH^-], [H_2CO_3], [HCO_3^-] and [CO_3^{2-}]. Given the strength of H_2CO_3 as 1.1×10^{-5}. $pK_a(1) = 6.38$ and $pK_a(2) = 10.32$.

18.69 Repeat the problem 18.67 when 0.2 M and 0.02 M $Ca(NO_3)_2$ is added to the solution.

18.70 Calculate the concentration of Ag^+ ions in a saturated solution of AgCl containing (*i*) 0.2 M NaCl and (*ii*) 0.2 M $NaNO_3$ solution at 298 K. Thermodynamic K_{sp} (AgCl) = 1.6×10^{-10}. Explain the results if they are different.

18.71 Repeat the problem 18.70 for determining the solubility of $BaSO_4$. Thermodynamic K_{sp} = 1.1×10^{-10}. (*i*) in water, (*ii*) 0.2 M $CuSO_4$ and (*iii*) 0.2 M $Cu(NO_3)_2$.

18.72 Indicate which of the species are Brønsted acid or base or both: (*i*) H_2O, (*ii*) NH_3, (*iii*) HCl, (*iv*) HCN, (*v*) OH^-, (*vi*) H_3O^+, (*vii*) CN^-, (*viii*) NH_2^-, (*ix*) NH_4^+ and (*x*) CO_3^{2-}.

18.73 Write the conjugate acids or the conjugate base of the species given in problem (18.72).

18.74 Calculate the pH of the following solutions: (*i*) 10^{-8} M HCl, (*ii*) 10^{-3} M HCl, (*iii*) 10^{-2} M H_2SO_4, (*iv*) 10^{-9} M KOH and (*v*) 10^{-2} M KOH.

18.75 Calculate the dissociation constant of the monoprotic acid which is having 0.025 degree of dissociation for 0.03 M acid.

18.76 Consider the monoprotic acid whose dissociation constant is 7.1×10^{-4}. Calculate the degree of dissociation by (*i*) assuming that acid is ideal, (*ii*) by considering the $\gamma_{\pm} = 0.85$. The concentration of the acid is 0.36 M.

18.77 The ion product of D_2O is 1.35×10^{-15} at 298 K. (*i*) Calculate the value of *pD* where $pD = -\log[D^+]$, (*ii*) For what value of *pD* will a solution be acidic in D_2O and (*iii*) derive a relation between *pD* and *pOD*.

18.78 The observation that although HF is a weak acid its strength increases with concentration. Explain briefly [**Hint:** F^- reacts with HF to form HF_2^-. The equilibrium constant for this reaction is 5.2 at 298 K].

18.79 Consider a solution containing 0.02 M H_3PO_4. Taking the values of different dissociation constants from the Table 18.4, calculate the concentration of H^+, $H_2PO_4^-$, $H_2PO_4^{2-}$ and PO_4^{3-}.

18.80 Repeat the problem 18.63 for 0.4 M H_2CO_3 acid, *i.e.* concentration of H^+, HCO_3^-, H_2CO_3 and CO_3^{2-}.

18.81 Novacaine is a weak base with $K_b = 8.91 \times 10^{-6}$. Calculate the ratio the base to its conjugate acid if the pH of the solution is 8.9.

18.82 Calculate the degree of dissociation of HCN acid at (*i*) 0.2 M and (*ii*) 0.02 M solution.

18.83 Taking the data from Table 18.5 for the K_b of pyridine calculate the amount of pyridine present in 100 mL of the solution.

18.84 Calculate the pH of the solution containing 0.2 M solution sodium acetate.

18.85 Repeat the problem 18.81 for 0.15 M solution of NH_4Cl solution.

18.86 Write the hydrolysis reactions from the following salts. (*i*) KCl, (*ii*) CH_3CH_2COONa, (*iii*) KCN, (*iv*) KF, (*v*) NH_4Cl, (*vi*) C_6H_5COONa, (*vii*) $CaSO_4$, (*viii*) CH_3COONH_4.

18.87 Repeat the problem 18.83 in presence of (*i*) 0.2 M KCl and (*ii*) 0.5 M Na_2SO_4.

18.88 Which of the following combinations will lead to the formation of buffers? (*i*) NH_3/NH_4Cl, (*ii*) $KHSO_4/H_2SO_4$, (*iii*) KCl/HCl, (*iv*) K_2HPO_4/KH_2PO_4, (*v*) Sodium borate/boric acid, (*vi*) Potassium hydrogen phthalate/phthalic acid, (*vii*) Sodium citrate/citric acid.

18.89 25 mL of 0.5 M of sodium citrate and 25 mL of 0.25 M citric acid are mixed together. Calculate the pH of the solution.

18.90 How would you prepare a buffer solution containing CH_3COONa/CH_3COOH of pH = 5.2 and ionic strength (*i*) 0.95 and (*ii*) 0.80.

18.91 Taking the help of Tables 18.4 and 18.5, name the acids or base that will give you the pH of (*i*) 4.0, (*ii*) 7.5 and (*iii*) 8.60.

CHAPTER 19

CHEMICAL KINETICS

19.1. DEFINITIONS

19.1.1. *Rate of Reaction*

Rate is defined as the change of concentration per unit time. The common unit is mol dm $^{-3}$ s^{-1}. In general the rate of reaction increases with increase of concentration and can be expressed as

$$\text{Rate} \propto f(C_1, C_2, ..., C_i) \quad ...(19.1)$$

or

$$\text{Rate} = kf(C_1, C_2, ..., C_i) \quad ...(19.2)$$

where k is rate constant, also called specific rate constant or velocity constant, C_1, C_2... are the concentrations of the reactants and products. For example, in case of general reaction

$$aA + bB + ... \rightarrow pP + qQ + ... \quad ...(19.3)$$

differential rate of the reaction can be expressed in terms of any of the reactants or products.

$$-\frac{1}{a}\frac{[A]}{dt} = -\frac{1}{b}\frac{[B]}{dt} = ... = \frac{1}{p}\frac{d[P]}{dt} = \frac{1}{q}\frac{d[Q]}{dt} = k\,[A]^l[B]^m ... \quad ...(19.4)$$

where $a, b, ..., p, q, ...$ are the stochiometric coefficients of reactants and products, l, m... are the order of the reaction with respect to A, B... . In the above expression it is assumed that volume does not change during the course of reaction. If it does, the above equation gets modified. Further rate of the reaction is in general a function of the concentration of reactants.

19.1.2. *Rate Constant*

It is defined as the rate of reaction when the concentration of each species is unity. Its unit depends on the order of reaction. Physically, k represents a reaction probability averaged overall internal variables, impact parameters and relative speeds.

19.1.3. *Elementary Reaction*

Any reaction which is a one step process is called elementary reaction *e.g.*

$$H + Cl_2 \rightarrow HCl + Cl$$
$$NO + N_2O_5 \rightarrow 3NO_2$$

19.1.4. *Complex Reaction*

A collection of elementary reactions which give the required products or describe the mechanism of a reaction is called complex reaction *e.g.*

$$2N_2O_5 \rightarrow 4NO_2 + O_2$$

can be written in terms of elementary reactions as

$$N_2O_5 \rightarrow NO_2 + NO_3$$
$$NO_2 + NO_3 \rightarrow N_2O_5$$
$$NO_2 + NO_3 \rightarrow NO_2 + O_2 + NO$$
$$NO + N_2O_5 \rightarrow 3NO_2$$

19.1.5. *Molecularity*

Number of molecules taking part in an elementary step is known as molecularity. The order and molecularity of an elementary step are equal. But for complex reactions this may not be so, *e.g.* the molecularity of each elementary reaction, mentioned above, has molecularity 2 and that is also of complex reaction but its order is one. In general molecularity is associated with the elementary reaction.

19.1.6. *Order of Reaction*

The order of a reaction describes the mathematical form in which the experimental results can be represented. The order of a reaction can only be calculated experimentally and can only be predicted if a reaction mechanism is known. Overall order of reaction can be defined as the sum of the exponents for each reactants, whereas the value of exponent for each reactant is known as the order of reaction for that component.

***Exercise* 19.1:** Find molecularity of the following reactions:

(*a*) $NO + N_2O_5 \rightarrow 3NO_2$

(*b*) $2NO + Cl_2 \rightarrow 2NOCl$

(*c*) $Cl + Cl + M \rightarrow Cl_2 + M$

(*d*) $C_6H_5-CH \rightarrow C_6H_5-CH$

$\quad\quad\quad\|\quad\quad\quad\quad\quad\|$

$\quad NC-CH \quad\quad\quad CH-CN$

(*a*) The number of molecules in this reaction are two so the molecularity is two or bimolecular. Similarly reactions (*b*), (*c*) and (*d*) are bimolecular, termolecular and unimolecular respectively.

***Exercise* 19.2:** (*a*) The value of k of certain reaction is 2.14×10^{-4} cm^3 mol^{-1} s^{-1}. What is its value in unit of cm^3 molecule^{-1} s^{-1}?

(*a*) 1 mol of substance contains 6.02×10^{23} molecules so 1 mol^{-1} substance will have $[6.02 \times 10^{23}]^{-1}$ molecules^{-1}.

(*b*) Express k in unit of dm^3 mol^{-1} s^{-1}, $k = 5 \times 10^{-4}$ s^{-1} atm^{-1}.

(*c*) Express k in unit of dm^3 mol^{-1} s^{-1}, $k = 1.25 \times 10^{-11}$ cm^3 mol^{-1} s^{-1}.

Now
$$\begin{aligned} k &= 2.14 \times 10^{-4}\ \text{cm}^3\ \text{mol}^{-1}\ \text{s}^{-1} \\ &= 2.14 \times 10^{-4}\ \text{cm}^3\ (6.02 \times 10^{23}\ \text{molecule})^{-1}\ \text{s}^{-1} \\ k &= 3.55 \times 10^{-28}\ \text{cm}^3\ \text{molecule}^{-1}\ \text{s}^{-1} \end{aligned}$$

(*b*) From ideal gas equation $P = \frac{n}{V}RT$

$$P = CRT \text{ or } C\ (\text{mol dm}^{-3}) = \frac{1\ \text{atm}}{(0.0821\ \text{atm dm}^3\ \text{mol}^{-1}\ \text{K}^{-1})(273\ \text{K})}$$

$$C = 0.0446\ \text{mol dm}^{-3}\ \text{atm}^{-1}$$

i.e. $\quad 1\ \text{atm} = 0.0446\ \text{mol dm}^{-3}$

$$k = (5.0 \times 10^{-4}\ \text{s}^{-1}\ \text{atm}^{-1})\left(\frac{1}{0.0446}\ \text{mol}^{-1}\ \text{dm}^3\ \text{atm}\right)$$

$$= 1.12 \times 10^{-2}\ \text{dm}^3\ \text{mol}^{-1}\ \text{s}^{-1}$$

(*c*) $1\ \text{mol} = 6.02 \times 10^{23}$ molecules and $1\ \text{dm}^3 = 1000\ \text{cm}^3$

or $\quad 1\ \text{mole} = \frac{1}{6.02 \times 10^{23}}$ mol and $1\ \text{cm}^3 = 10^{-3}\ \text{dm}^3$

Now $\quad k = (1.25 \times 10^{-11})\ \text{cm}^3\ \text{mole}^{-1}\ \text{s}^{1}$

$$= (1.25 \times 10^{-11})(10^{-3}\ \text{dm}^3)\left(\frac{1}{6.02 \times 10^{23}}\text{mol}\right)^{-1} s^{-1}$$

$$= (1.25)(6.02)(10^{-14+23})\ \text{dm}^3\ \text{mol}^{-1}\ \text{s}^{-1} = 7.525 \times 10^9\ \text{dm}^3\ \text{mol}^{-1}\ \text{s}^{-1}$$

***Exercise* 19.3:** Write the differential rate equations of the following reactions:

(*a*) $2A + B \rightarrow$

(*b*) $A + 2B + 2C \rightarrow$

(*c*) $3A + B \rightarrow 2C + D + 2E$

The differential rates of the above reactions can be written as (assuming them to be elementary steps)

(*a*) $$-\frac{1}{2}\frac{d[A]}{dt} = -\frac{d[B]}{dt} = k[A]^2[B]$$

(*b*) $$-\frac{d[A]}{dt} = -\frac{1}{2}\frac{d[B]}{dt} = -\frac{1}{2}\frac{d[C]}{dt} = k[A][C]^2[B]^2$$

(*c*) $$-\frac{1}{3}\frac{d[A]}{dt} = -\frac{d[B]}{dt} = \frac{1}{2}\frac{d[C]}{dt} = \frac{d[D]}{dt} = \frac{1}{2}\frac{d[E]}{dt} = k[A]^3[B]$$

19.2. ZERO-ORDER REACTIONS

Zero-order reactions are those whose rate can be written as

$$-\frac{d[A]}{dt} = k \qquad ...(19.5)$$

where k is the zero-order rate constant. The above differential equation can be integrated with initial conditions, $t = 0$, $[A] = [A]_0$

$$[A] = [A]_0 - kt \qquad ...(19.6)$$

or

$$k = \frac{[A]_0 - [A]}{t} \qquad ...(19.6a)$$

Equation (19.5) also tells that the rate of zero-order reactions are independent of concentration of the reactant. The units of zero-order rate constant k is mol L^{-1} s^{-1} (Ms^{-1}).

***Exercise* 19.4:** A reaction is 50% complete in 10 minutes. It is allowed to proceed another 5 minutes. How much of the reaction would be complete at the end of these 15 minutes if the reaction follows zero-order kinetics?

Let $[A]_0$ be 1×10^3 mol dm^{-3}. Half of $[A]_0$ is consumed in 10 minutes so k from equation (19.6a) is

$$k = \frac{(1-1/2)10^3 \text{ mol dm}^{-3}}{10 \text{ min}} = 50 \text{ mol dm}^{-3} \text{ min}^{-1}$$

Now in further 5 minutes *i.e.* after 15 minutes.

$$[A] = [1 \times 10^3 - 50 \times 15] \text{ mol dm}^{-3} = 250 \text{ mol dm}^{-3}$$

19.3. FIRST-ORDER REACTIONS

First-order reactions are those whose rate is directly proportional to the concentration of the reactant *i.e.*

$$-\frac{d[C]}{dt} = k[C] \qquad ...(19.7)$$

which on integration gives

$$\ln [C] = \ln [C]_0 - kt \qquad ...(19.8a)$$

or

$$[C] = [C]_0 e^{-kt} \qquad ...(19.8b)$$

or

$$k = \frac{1}{t} \ln \frac{[C]_0}{[C]} \qquad ...(19.8c)$$

$[C]_0$ is the concentration of the reactant at $t = 0$. For first-order reactions, plot of ln $[C]$ (or log $[C]$) vs t is a straight line. The intercept gives the concentration at $t = 0$ and k can be calculated from the slope. The units of first order sate constant is s^{-1}. The other parameter is the average life time (τ) of the reaction. It is defined as the inverse of the sum of the rate constants (k_i) of all the steps involved in reaction, *i.e.*,

$$\tau = \frac{1}{\sum_i k_i} \qquad ...(19.9)$$

***Exercise* 19.5:** Phosphorescence emission of Acetone-d_6 (0.05 M) in Acetonitrile at 20°C was measured at 450 nm. Calculate the rate constant for the emission and also calculate the average life time of triplet state of acetone from the following data:

t(μ sec)	20	32	40	60	80	100	120	140
Intensity (I) (arbitrary units)	5.5	4.6	4.0	2.9	2.1	1.5	1.05	0.75
log I	0.74	0.66	0.60	0.46	0.33	0.18	0.025	0.12

log I is plotted vs time in Fig. 19.1 and slope obtained is $-0.72 \times 10^4 \text{ s}^{-1}$.

Now $k = -\text{slope} \times 2.303$

$= -(-0.72 \times 10^4 \text{ s}^{-1}) \times 2.303$

$= 1.66 \times 10^4 \text{ s}^{-1}$

Using equation (19.9)

$$\text{Average life of triplet state} = \frac{1}{k}$$

$$= \frac{1}{1.66 \times 10^4 \text{ s}^{-1}} = 60 \text{ μS}$$

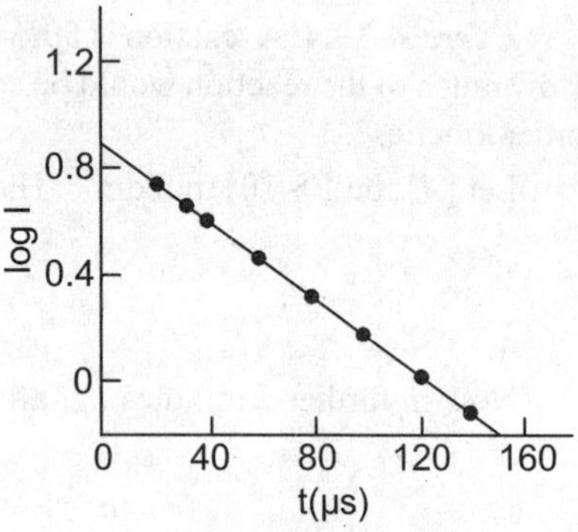

Fig. 19.1. *Plot of log I vs time*

***Exercise* 19.6:** A reaction is 25% complete in 25 minutes. If it follows first-order kinetics, what would be the concentration at the end of another 50 minutes. The initial concentration is 2×10^4 mol dm^{-3}.

Since 25 % of the reactants are consumed after 25 min, the concentration after 25 min will be

$$= 2 \times 10^4 - 2 \times 10^4 \times \frac{25}{100} = 1.5 \text{ mol dm}^{-3}$$

Substituting the values of $[C] = 1.5 \times 10^4$ mol dm^{-3} and $[C]_0 = 2.0 \times 10^4$ mol dm^{-3}, $t = 25$ min in (19.8c)

$$k = \frac{2.303}{(25 \text{ min})} \log \left(\frac{2 \times 10^4 \text{ mol dm}^{-3}}{1.5 \times 10^4 \text{ mol dm}^{-3}} \right) = 1.152 \times 10^{-2} \text{ min}^{-1}$$

Now $[C] = [C]_0 \exp(-kt) = 2 \times 10^4 \exp[(-1.152 \times 10^{-2})(75)]$

The concentration at the end of 75 min = 0.846 mol dm^{-3}.

19.4. SECOND–ORDER REACTIONS

In second-order reactions, the rate is directly proportional to the square of the concentration of one of the reactants or to the product of the concentration raised to the power unity of two of the reactants.

19.4.1. *Case I*

$$2A \rightarrow \text{Products} \qquad \text{...(19.10)}$$

$$-\frac{d[A]}{dt} = k[A]^2$$

which on integration gives

$$\frac{1}{[A]} = \frac{1}{[A]_0} + kt \qquad ...(19.11)$$

where $[A]_0$ is the concentration of the reactant at $t = 0$. Equation (19.11) can also be written as

$$[A] = \frac{[A]_0}{1+[A]_0 k\tau} \qquad ...(19.11a)$$

19.4.2. *Case II*

$$aA + bB \rightarrow \text{Products} \qquad ...(19.11b)$$

where $a \neq b$ and $[A]_0 \neq [B]_0$. The differential rate equation is

$$-\frac{1}{a}\frac{d[A]}{dt} = -\frac{1}{b}\frac{d[B]}{dt} = k\,[A]\,[B] \qquad ...(19.12)$$

and the integrated rate equation is

$$\frac{1}{b[A]_0 - a[B]_0} \ln\left(\frac{[B]_0[A]}{[B][A]_0}\right) = kt \qquad ...(19.13)$$

If $a = b = 1$, equation (19.13) becomes

$$\frac{1}{[A]_0 - [B]_0} \ln\left(\frac{[B]_0[A]}{[B][A]_0}\right) = kt \qquad ...(19.14)$$

Plots of left-hand side of equations (19.13) and (19.14) against t will be straight line. Rate constants can be calculated from the slopes and initial concentrations of the reactants from the intercept.

***Exercise* 19.7:** ClO radical decays as

$$2\,ClO \rightarrow Cl_2 + O_2$$

Calculate the value of the rate constant k from the following data :

t(ms)	0.12	0.62	0.96	1.60	3.20	4.00	5.75
[ClO] × 10^6 (mol dm^{-3})	8.49	8.09	7.10	5.79	5.20	4.77	3.95
1/[ClO] (dm^3 mol^{-1} × 10^6)	0.118	0.124	0.141	0.173	0.192	0.21	0.253

1/[ClO] is plotted against time in Fig. 19.2 and slope is the same as rate constant.

$$\text{Slope} = \frac{(0.246-0.152)\times 10^6 \text{ mol}^{-1}\text{ dm}^3}{(5.5-1.5)\times 10^{-3}\text{ s}} = 2.35 \times 10^7 \text{ dm}^3 \text{ mol}^{-1} \text{ s}^{-1}$$

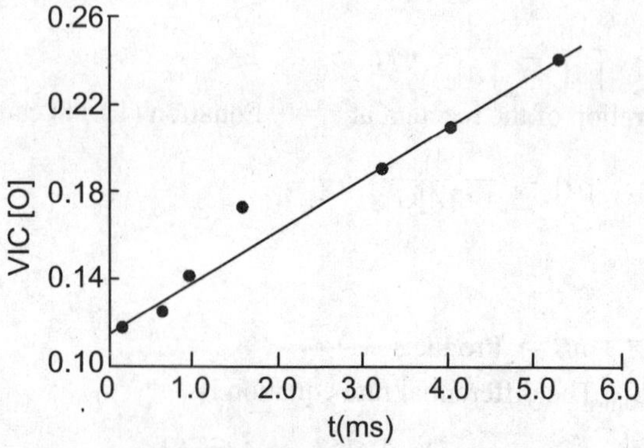

Fig. 19.2

The rate constant $k = 2.35 \times 10^7$ dm^3 mol^{-1} s^{-1}.

***Exercise* 19.8:** In the reaction between bromine atoms and Cl_2O,

$$Br + Cl_2O \rightarrow BrCl + ClO$$

the course of the reaction was followed by measuring the concentration of ClO radicals. If the initial concentrations of bromine atoms and Cl_2O are 12.2×10^{-6} mol dm^{-3} and 24.4×10^{-6} mol dm^{-3}, calculate the rate constant.

Time(μs)	10	20	30	40	50	60	70	80	100
$[ClO] \times 10^6$ (mol dm^{-3})	1.68	2.74	3.66	4.7	5.6	6.19	6.55	7.40	7.84
$\log\left(\frac{[B]_0 - x}{[A]_0 - x}\right)$	0.336	0.360	0.386	0.419	0.455	0.480	0.497	0.55	0.585

The reaction is second order with $a = b = 1$ and $[A]_0 \neq [B]_0$. If x is the concentration of ClO at time t, the concentration of bromine atoms and Cl_2O will be $[A]_0 - x = [A]$ and $[B]_0 - x = [B]$ respectively. The equation (19.14) will be

$$\log \frac{[B]_0 - x}{[A]_0 - x} = \frac{[B]_0 - [A]_0}{2.303} kt + \log \frac{[A]_0}{[B]_0}$$

$\log \frac{[B]_0 - x}{[A]_0 - x}$ is calculated for various values of x and is mentioned in the above table.

$\log \frac{[B]_0 - x}{[A]_0 - x}$ is plotted against t in Fig. 19.3, and slope $= 2.91 \times 10^3$ s^{-1}

$$k = \text{Slope} \times \frac{2.303}{[B]_0 - [A]_0} = \frac{2.91 \times 2.303 \times 10^3}{12.2 \times 10^{-6} \text{ mol dm}^{-3}}$$

$$= 5.49 \times 10^8 \text{ dm}^3 \text{ mol}^{-1} \text{ s}^{-1}$$

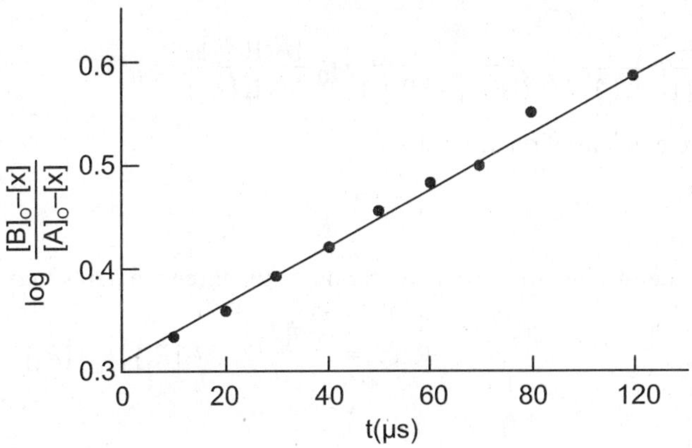

Fig. 19.3 *Plot of log* $\frac{[B]_0 - x}{[A]_0 - x}$ *vs t*

19.5. THIRD-ORDER REACTIONS

In a third-order reaction, three different cases can be considered.

19.5.1. *Case I*

Rate is directly proportional to the cube of the concentration of one of reactant, *i.e.*

$$3R \rightarrow P \qquad ...(19.15)$$

$$-\frac{d[R]}{dt} = k\,[R]^3$$

and integrated equation with $[R] = [R]_0$ at $t = 0$ is

$$\frac{1}{2\,[R]^2} - \frac{1}{2\,[R]_0^2} = kt \qquad ...(19.16a)$$

or

$$k = \frac{1}{2t}\left[\frac{1}{[R]^2} - \frac{1}{[R]_0^2}\right] \qquad ...(19.16b)$$

19.5.2. *Case II*

Rate is proportional to the square of the concentration of one of the reactant and first power of concentration of the second, *i.e.*

$$-\frac{d[R_2]}{dt} = k\,[R_1]^2\,[R_2] \qquad ...(19.17)$$

and integrated form of the equation is

$$\frac{1}{[R_2]_0 - [R_1]_0}\left[\frac{1}{[R_1]} - \frac{1}{[R_1]_0}\right] + \frac{1}{([R_2]_0 - [R_1]_0)^2} \ln \frac{[R_1][R_2]_0}{[R_1]_0 [R_2]} = kt \quad ...(19.18)$$

with $[R_1]_0$ and $[R_2]_0$ are concentration at $t = 0$.

19.5.3. *Case III*

Rate is proportional to the product of the concentration of the three reactants, *i.e.*

$$\text{Rate} = -\frac{d[R_1]}{dt} = k\,[R_1]\,[R_2]\,[R_3] \quad ...(19.19)$$

and integrated form is

$$\frac{1}{([R_1]_0 - [R_2]_0)([R_2]_0 - [R_3]_0)([R_3]_0 - [R_1]_0)}$$

$$\ln\left[\left(\frac{[R_1]_0}{[R_1]}\right)^{[R_2]_0 - [R_3]_0}\left(\frac{[R_2]_0}{[R_2]}\right)^{[R_3]_0 - [R_1]_0}\left(\frac{[R_3]_0}{[R_3]}\right)^{[R_1]_0 - [R_2]_0}\right] = kt \quad ...(19.20)$$

***Exercise* 19.9:** Considering the data of the exercise 19.13, calculate the value of rate constant.

It is found in exercise (19.13) that the rate equation is

$$-\frac{d\,[BrO_3^-]}{dt} = k\,[H^+]\,[BrO_3^-][H_2O_2]$$

Taking anyone of the rows, (*e.g.* take the first one) substitute the values in the above equation and k can be calculated *i.e.*

$$9 \times 10^{-7} \text{ mol dm}^{-3}\, t^{-1} = k\,[0.1 \text{ mol dm}^{-3}]\,[0.00587 \text{ M}]\,[0.36 \text{ M}]$$

$$k = \frac{9 \times 10^{-7} \text{ M}\, t^{-1}}{(0.1)\,(0.00587)\,(0.036)\text{ M}^3} = 4.265 \times 10^{-2} \text{ mol}^{-2} \text{ dm}^6\, t^{-1}$$

Similarly taking the data of last row

$$k = \frac{28 \times 10^{-7} \text{ M}\, t^{-1}}{(0.36 \text{ M})\,(0.00493 \text{ M})\,(0.0363 \text{ M})} = 4.346 \times 10^{-2} \text{ mol}^{-2} \text{ dm}^6\, t^{-1}$$

The two values are quite close to each other. In this way using the data of other rows, the average values of k can be obtained.

19.6. PSEUDO ORDER REACTIONS

In these reactions, the concentrations of one or more than one reactants are in large excess than the other or one of the reactant is acting as catalyst. Due to this concentration of these species will nearly remain same and can be taken as constant. Thus, the order of the reaction will be reduced *e.g.* the hydrolysis of esters catalysed by acids are

$$RCOOR_1 + H_2O \xrightarrow{H^+} RCOOH + R^1OH$$

and order of the reaction is unity if water is in excess.

Exercise: **19.10:** Consider the acid catalysed reaction of carboxylic acid with alcohol as

$$RCOOH + R^1OH \xrightarrow{H^+} RCOOR^1 + H_2O$$

and the rate equation observed for the complete reaction is

$$-\frac{d[A]}{dt} = k[H^+][A][B]$$

where A is acid and B alcohol with $[A]_0 = [B]_0 = 0.2$ mol dm^{-3} and pH = 3, it is found that half life of the above reaction is 50 min. Calculate the true and pseudo rate constant k.

Since H^+ is a catalyst, its concentration remains constant so

$$k_{app} = k[H^+]$$

Further the reaction is first order with respect to A and first order with respect to B, their initial concentrations are equal so, according to equation (19.21)

$$t_{1/2} = \frac{1}{k_{app}[A]_0}$$

$$k_{app} = \frac{1}{(50\,\text{min})(0.2\,\text{mol dm}^{-3})} = 0.1\ \text{mol}^{-1}\ \text{dm}^3\ \text{min}^{-1}$$

$$k_{true} = \frac{k_{app}}{[H^+]} = \frac{0.1\ \text{mol}^{-1}\ \text{dm}^3\ \text{min}^{-1}}{1.0\times10^{-3}\ \text{mol dm}^{-3}}$$

$$= 1.0 \times 10^2\ \text{mol}^{-2}\ \text{dm}^6\ \text{min}^{-1}$$

TABLE 19.1. Half-lives of common radioactive isotopes

Isotope	Decay reaction	$t_{1/2}$
$^3H_1$	$^3H_1 \rightarrow {^3He_2} + {^0\beta_{-1}}$	12.3 yr
$^{14}C_6$	$^{14}C_6 \rightarrow {^{14}N_7} + {^0\beta_{-1}}$	5.73×10^3 yr
$^{24}Na_{11}$	$^{24}Na_{11} \rightarrow {^{24}Mg_{12}} + {^0\beta_{-1}}$	15 hr
$^{32}P_{15}$	$^{32}P_{15} \rightarrow {^{32}S_{16}} + {^0\beta_{-1}}$	14.3 days
$^{35}S_{16}$	$^{35}S_{16} \rightarrow {^{35}S_{17}} + {^0\beta_{-1}}$	88 days
$^{60}Co_{27}$	Emission of γ-rays	5.26 yr
$^{99m}Tc_{43}$	Emission of γ-rays	6 hr
$^{131}I_{53}$	$^{131}I_{53} \rightarrow {^{131}Xe_{54}} + {^0\beta_{-1}}$	8.05 days

Exercise **19.11:** Consider A and B as the two strands of DNA and double helix is formed by the following reaction as

$$A + B \rightarrow AB$$

(*i*) Fraction of strands that dissociate, assuming $[A] = [B] = [A]$, (*ii*) if C_0 is the initial total concentration of the single strand, derive the relation between $[A]_0$ and C_0, (*iii*) if the repeating sequence is one, calculate the value of $[A]_0$ and if there is no repeating sequence, calculate $[A]_0$, (*iv*) find the rate constant of this reaction, (*v*) derive the relation between the fraction (f) and the C_0, (*vi*) if $C_0t_{1/2}$ at the fraction = 1/2 is equal to 2×10^{6} M s, calculate the value of k and (*vii*) assuming that second order rate constant is the same for different single strands, calculate the complexity factor N if $C_0t_{1/2}$ at the fraction = 1/2 is equal to 3×10^3 M s.

(*i*) Recombination of two strands of DNA is a case I of second order reaction. The concentration of *A*, using equation (19.20*a*) can be written as

$$[A] = \frac{[A]_0}{1+[A]_0 kt} \quad ...(19.20a)$$

where $[A]_0$ is initial concentration (in moles of nucleotide per litre) of fragment *A* that is complementary to fragment *B*, [*A*] is the concentration of the single strand *A* at time *t*, also in moles per litre. Fraction of the strands that dissociate is given as:

$$f = \frac{[A]}{[A]_0} = \frac{1}{1+[A]_0 kt} \quad ...(1)$$

(*ii*) If C_0 is the total concentration of all the single strands before recombination (is called re-naturation), then the relation between $[A]_0$ and C_0 is given as

$$[A]_0 = \frac{C_0}{2N} \quad ...(2)$$

where *N* is the smallest repeating sequence, called *complexity* of the nucleotide pairs.

(*iii*) If $N = 1$, equation (2) reduces to (or the relation between C_0 and $[A]_0$)

$$[A]_0 = \frac{C_0}{2} \quad ...(3)$$

Combining equations (1) and (2), we get

$$f = \frac{1}{1+\frac{C_0}{2Nkt}} \quad ...(4)$$

(*iv*) The half-life of the recombination reaction is the amount of time it takes for half of the single strands to renature, so $f = 1/2$ and equation (4) reduces to

$$f = \frac{1}{2} = \frac{1}{1+\frac{C_0}{2Nkt_{1/2}}} = \frac{2Nkt_{1/2}}{2Nkt_{1/2}+C_0}$$

or
$$\frac{1}{2} = \frac{1}{1+\frac{C_0kt_{1/2}}{2N}} = \frac{2N}{2N+C_0k_{1/2}t}$$

or
$$2N + C_0kt_{1/2} = 4N$$

or
$$C_0t_{1/2} = \frac{2N}{k} = \frac{N}{k'} \quad ...(5)$$

where
$$k' = k/2.$$

(*v*) Using equation (5) and substituting the values of parameters, we get

$$k = \frac{2N}{C_0t_{1/2}} = \frac{2 \times 1}{2 \times 10^{-6}\ \text{M s}} = 1 \times 10^6\ \text{M}^{-1}\ \text{s}^{-1}$$

(*vi*)
$$N = \frac{C_0t_{1/2}k}{2} = \frac{(3 \times 10^3\ \text{M s})(10^6\ \text{M}^{-1}\ \text{s}^{-1})}{2} = 1.5 \times 10^9$$

Thus, it can be concluded that this sample of DNA has a very high order of complexity.

19.7. HALF-LIFE

Half-life is defined as the time required when half the concentration of a reactant is consumed. Half-life can be precisely defined if only one kind of reactant is involved but if a reaction is between different kind of reactants, half-life has to be defined with respect to particular reactant. For one component system, half-life is related to its initial concentration by the relation (19.21) except $n = 1$, *i.e.*

$$t_{1/2} = \frac{2^{n-1}-1}{(n-1)[C]_0^{n-1}k} \quad \text{...(19.21)}$$

and for $n = 1$

$$t_{1/2} = \frac{0.69}{k} \quad \text{...(19.22)}$$

i.e. half-life is independent of concentration for first-order reactions whereas in all other reactions, half-life time is inversely proportional to the initial concentration. Half-life times of first-order kinetics is commonly associated with the redioactive decay. Half-life of some commonly occurring isotopes are compiled in Table 19.1. If the reaction is pseudo-order, the half-life is always defined with respect to the concentration of species present in small amounts.

***Exercise* 19.12:** Half-life for disintegration of radium is 1590 years. Calculate the rate constant in s^{-1} and also how many years will be taken for the disintegration of 80%?

$$t_{1/2} = 1590 \text{ years} = 5.014 \times 10^{10} \text{ s}$$

Using equation (19.22)

$$k = \frac{0.69}{5.014 \times 10^{10} \text{ s}} = 1.376 \times 10^{-11} \text{ s}^{-1}$$

Now if $[C]_0 = 100$ mol dm^{-3}, $[C] = 20$ mol dm^{-3} and using equation (19.8*c*)

$$t = \frac{2.303 \times 1590 \text{ years}}{0.69} \log \frac{100}{20} = 3661.8 \text{ years}$$

***Exercise* 19.13:** The reaction $2A + B \rightarrow C + D$ goes to completion and follows the following rate law

$$-\frac{d[A]}{dt} = k[A]^2[B]$$

Fill the gaps left in the following information :

$10^6 \times [A]_0$ (mol dm^{-3})	$[B]_0 \times 10^6$ (mol dm^{-3})	half time (s)
300	40	62.6
300	60	—
5	300	625
10	300	—

In more than one reactant system, the half-life is the time for half of the reactant present in small amount.

In the first set, reactant B is in small amount, the half-life is for reactant B. Further the rate is first-order with respect to B, so the half-life is independent of initial concentration, *i.e.* missing data is 62.6s.

In the second set, the half-life is for reactant A, and rate is second-order with respect to A, the equation 19.21 becomes

$$t_{1/2} = \frac{1}{k[A]_0}$$

Now

$$\frac{(t_{1/2})_1}{(t_{1/2})_2} = \frac{1}{k[A_1]_0} \times \frac{k[A_2]_0}{1}$$

Substituting the values

$$(t_{1/2})_2 = (t_{1/2})_1 \frac{[A_1]_0}{[A_2]_0} = \frac{(625\,\text{s})\,(5\times10^{-6}\ \text{mol dm}^{-3})}{(10\times10^{-6}\ \text{mol dm}^{-3})}$$

$$= 312.5\ \text{s}.$$

19.8. METHODS OF DETERMINING RATE CONSTANT AND ORDER OF REACTION

19.8.1. *Differential Method*

In this method the data is not collected in the form of concentration versus time but expressed as the rate of change of concentration with time against concentration of the reactant. Consider a general rate equation involving two reactants.

$$-\frac{d[R_1]}{dt} = k\,[R_1]^l\,[R_2]^m \qquad ...(19.23)$$

or

$$\log\left(-\frac{d[R_1]}{dt}\right) = \log k + l \log [R_1] + m \log [R_2] \qquad ...(19.24)$$

l (or m) can be evaluated from the slopes of plots between $\log\left(-\frac{d[R_1]}{dt}\right)$ and $\log [R_1]$ (or $\log [R_2]$) keeping $[R_2]$ (or $[R_1]$) fixed. k can be calculated from the intercept of the above plot.

***Exercise* 19.14:** In the reaction given below, the following data was obtained. Calculate the rate constant and order of reaction

$$3H_2O_2 + BrO_3^- \rightarrow Br^- + 3O_2 + 3H_2O$$

$[H^+]$ (M)	$[BrO_3^-]$ (M)	$[H_2O_2]$ (M)	Initial rate $-\frac{d[BrO_3^-]}{dt} \times 10^7$ M t^{-1}
0.10	0.00587	0.036	9
0.10	0.01174	0.036	19

0.120	0.00978	0.0385	20
0.120	0.00978	0.077	40
0.120	0.00489	0.0363	9
0.240	0.00489	0.0363	18
0.240	0.00493	0.0363	18
0.360	0.00493	0.0363	28

Seeing at the data, let the rate expression be

$$-\frac{d[BrO_3^-]}{dt} = R = k[H^+]^l\,[BrO_3^-]^m\,[H_2O_2]^n$$

or

$$\log R = \log k + l\log[H^+] + m\log[BrO_3^-] + n\log[H_2O_2]$$

The values of l, m, n can be calculated by making use of the above equation and selecting the data in such a way that either $[H^+]$ or $[BrO_3^-]$ or $[H_2O_2]$ is varying and the other two are held constant.

From the first two rows and using equation (19.24 *a*) log 19 – log 9 = m [log 0.01174 – log 0.00587]

i.e.

$$m = \frac{\log\left(\frac{19}{9}\right)}{\log\left(\frac{0.01174}{0.00587}\right)} = 1.08$$

Similarly making use of rows 6 and 7, the value of m obtained is also unity.
'n' can be calculated by making use of data of rows 3 and 4 *i.e.*
log 40 – log 20 = n [log 0.077 – log 0.0385]

$$n = \frac{\log\left(\frac{40}{20}\right)}{\log\left(\frac{0.077}{0.0385}\right)} = 1$$

'l' can be obtained by making use of the data of rows 7 and 8

$$\log 28 - \log 18 = l(\log 0.36 - \log 0.24)$$

$$l = \frac{\log\left(\frac{28}{18}\right)}{\log\left(\frac{0.36}{0.24}\right)} = 1.09$$

The data of rows 5 and 6 also shows that the value of l is unity. Thus, the rate equation is

$$R = k[H^+]\,[BrO_3^-][H_2O_2]$$

Now substituting the values of respective quantities

$$k = \frac{R}{[H^+][BrO_3^-]H_2O_2]} = \frac{9\times10^{-7}\text{ M}\,t^{-1}}{(0.1\text{M})(0.00587\text{ M})(0.036\text{M})}$$

$$= 0.0426\text{ M}^{-2}\,t^{-1}$$

Similarly other values of k can be calculated and average can be obtained or graphical method can be used also.

19.8.2. *Integral Method*

This method is a trial-and-error method *i.e.* change of concentration with time is measured and value of k is calculated by using the different integrated equations for different order of reactions. The order of the reaction will be obtained from the equation which gives the consistent value of k. This can be done either analytically or graphically.

Exercise **19.15:** Following data is obtained in the reaction of bromine atoms with Cl_2O. Calculate the order of the reaction with respect to bromine atoms and Cl_2O.

Time (μs)	0	40	70	100
$[Cl_2O] \times 10^3$ (mol m^{-3})	24.4	19.70	17.85	16.56
$[Br] \times 10^3$ (mol m^{-3})	12.2	7.50	5.65	4.36

Let n_1 and n_2 be order of the reaction with respect to Cl_2O and bromine atoms. The values of k are calculated for different combinations of n_1 and n_2 from the above data and is listed in table. It is clear that for $n_1 = n_2 = 1$ the value of k obtained is consistent.

	k		
$n_1 = 0, n_2 = 0$	0.118×10^{-3}	0.062×10^{-3}	0.050×10^{-3}
$n_1 = 1, n_2 = 0$	5.34	3.29	2.33
$n_1 = 0, n_2 = 1$	12.06×10^1	9.44×10^1	8.61×10^1
$n_1 = 2, n_2 = 0$	2.50×10^5	2.33×10^5	2.0×10^5
$n_1 = 0, n_2 = 2$	1.27×10^5	1.47×10^5	1.70×10^5
$n_1 = 1, n_2 = 1$	5.40×10^5	5.30×10^5	5.30×10^5

19.8.3. *Half-life Method*

This method requires the determination of half-life times as a function of concentration. If half-life is independent of concentration, the order of reaction is unity. Otherwise, slope of the plot between log $t_{1/2}$ versus log $[C]_0$ gives the value of order of reaction as in equation

$$t_{1/2} = \frac{2^{n-1}-1}{(n-1)[C]_0^{\,n-1}k} \quad \text{...(19.21)}$$

$$\log t_{1/2} = \log\frac{2^{n-1}-1}{(n-1)\,k} - (n-1)\log[C]_0 \quad \text{...(19.25)}$$

Exercise **19.16:** Following data was recorded in the decay of ClO radicals. Calculate the order of the reaction.

$[ClO]_0 \times 10^3$ (mol m^{-3})	8.7	8.44	7.44	7.39	7.13
*log $[ClO]_0$	0.94	0.926	0.892	0.869	0.853
$t_{1/2}$ (ms)	4.8	4.9	5.4	5.5	5.8
log $t_{1/2}$	0.681	0.690	0.732	0.74	0.763

*Neglecting the common factor 10^{-3}.

Data shows that half-life is not independent of initial concentration of ClO radicals. Plot of log [ClO] vs log $t_{1/2}$ is a straight line (Fig. 19.4) and slope is – 1.

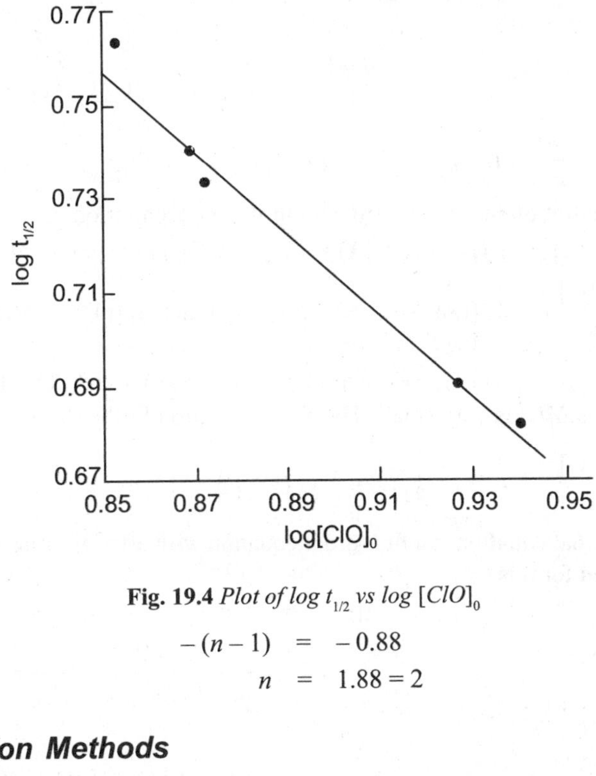

Fig. 19.4 *Plot of log $t_{1/2}$ vs log $[ClO]_0$*

This gives

$$-(n-1) = -0.88$$

$$n = 1.88 = 2$$

19.8.4. *Relaxation Methods*

These methods are used to study fast reactions. In this method the mixture of reactants are disturbed slightly from the equilibrium position with the help of either temperature jump, pressure jump or electrical pulse method. The disturbed system returns to either new equilibrium or old equilibrium and in general follows first order kinetics. The equation describing the behaviour is

$$\Delta[C_i] = \Delta[C_i]_0 e^{-t/\tau} \quad ...(19.26)$$

where $\Delta[C_i]$ and $\Delta[C_i]_0$ are the displacements in the concentration from equilibrium position at time t and $t = 0$, τ is the relaxation time. The differential form can be obtained by differentiating with respect to t *i.e.*

$$\frac{d\,\Delta[C_i]}{dt} = \Delta[C_i]_0 e^{-t/\tau}\left(-\frac{1}{\tau}\right) \qquad ...(19.27)$$

or

$$-\frac{d\,(\Delta[C_i])}{dt} = \frac{1}{\tau}\Delta[C_i] \qquad ...(19.28)$$

***Exercise* 19.17 :** (*a*) Find the relationship between k_1, k_2 and τ for the reaction of type

$$A \underset{k_2}{\overset{k_1}{\rightleftharpoons}} X + Y$$

(*b*) If $[X]_e = [Y]_e = 5 \times 10^{-2}$ M, $\tau = 50$ μs and $[A]_e = 0.1$ M, calculate k_1 and k_2.

In this case the forward reaction is first-order and reverse reaction is second-order. If concentrations of *A, X, Y* at any time *t* are $[A]_0 - [C]$, $[X]_0 + [C]$, $[Y]_0 + [C]$,

then

$$-\frac{d[A]}{dt} = k_1\,[(A)_0 - (C)] - k_2\,[(X)_0 + (C)]\,[(Y)_0 + (C)]$$

At equilibrium $-\dfrac{d[A]}{dt} = 0 = k_1\,[A]_e - k_2\,[X]_e\,[Y]_e$

Now define concentration in reference to equilibrium concentration *i.e.*

$$[A] = [A]_e - \Delta[C],\ [X] = [X]_e + \Delta[C],\ [Y] = [Y]_e + \Delta[C],$$

so

$$\frac{d\,\Delta[C]}{dt} = k_1\,[(A)_e - \Delta(C)] - k_2[(X)_e + \Delta(C)]\,[(Y)_e + \Delta(C)]$$

$$= k_1\,[A]_e - k_2\,[X]_e\,[Y]_e - [k_1 + k_2\,(X)_e + k_2\,(Y)_e]\,\Delta[C]$$

neglecting $k_2\,(\Delta[C])^2$ as $\Delta[C]$ is very small. The above equation further becomes

$$\frac{d\,\Delta[C]}{dt} = -\,[k_1 + k_2\,(X)_e + k_2\,(Y)_e]\,\Delta[C]$$

The above differential equation is a first-order equation with effective rate constant $k = k_1 + k_2\,[X]_e + k_2\,[Y]_e$ and integrated form is

$$\Delta[C] = \Delta[C]_0\,e^{-kt} = \Delta[C]_0{}'\,e^{-t/\tau}$$

where

$$\tau = \frac{1}{k_1 + k_2\,[X]_e + k_2\,[Y]_e} \qquad ...(19.29)$$

(*b*) First calculate *K*

$$K = \frac{[X]_e\,[Y]_e}{[A]_e} = \frac{(5\times10^{-2})(5\times10^{-2})}{(0.1)} = 0.025 = \frac{k_1}{k_2}$$

and

$$k_1 = 0.025\,k_2\ \text{M} \qquad ...(19.30a)$$

Substitute the respective values in (19.29)

$$k_1 + k_2\,(5 \times 10^{-2} + 5 \times 10^{-2}) = \frac{1}{50\times10^{-6}\ \text{s}} = 2 \times 10^4\ \text{s}^{-1}$$

or

$$k_1 + 0.1\,k_2 = 2 \times 10^4 \qquad ...(19.30b)$$

Solving for k_1 and k_2 from equations (19.30*a*) and (19.30*b*) $k_1 = 4.0 \times 10^3\ \text{s}^{-1}$ and $k_2 = 1.6 \times 10^5$ $\text{M}^{-1}\ \text{s}^{-1}$.

19.8.5. *Guggenheim's Method of Analysis*

In certain experiments it becomes necessary to take readings at $t = \propto$ before it can be analysed whereas this method does not involve this reading. The basis for this method is that it requires readings at equal intervals of time. The necessary mathematics is as follows. Consider a first-order reaction and equation (19.8*b*)

$$[C]_1 = [C]_0 e^{-kt} \quad ...(19.8b)$$

$$[C]_2 = [C]_0 e^{-k(t+\Delta t)} \quad ...(19.31a)$$

so

$$[C]_1 - [C]_2 = [C]_0 e^{-kt} (1 - e^{-k\Delta t})$$

or

$$\ln [(C)_1 - (C)_2] = -kt + \ln [C]_0 (1 - e^{-k\Delta t}) \quad ...(19.31b)$$

i.e. $\ln [[C]_1 - (C)_2]$ is plotted vs t and slope gives the value of k.

***Exercise* 19.18:** The following data is observed in a certain reaction $A \rightarrow B$, calculate the value of k by Guggenheim's method:

t(min)	0	5	15	20	35	52	80	100	140
$[C] \times 10^{+2}$ (mol dm^{-3})	7.5	6.75	6.25	5.35	4.25	3.2	2.1	1.5	0.75
t' (min)	0	10	20	30	40	50	60	70	80
$[C] \times 10^2$ (mol dm^{-3})	7.5	6.37	5.35	4.6	3.9	3.3	2.85	2.45	2.1
$[C]_1 - [C]_2$	1.13	1.02	0.75	0.7	0.6	0.45	0.4	0.35	
$\ln [C]_1 - [C]_2)'$	0.12	0.02	–.29	–.36	–.51	–.8	–.916	–1.05	

Concentration is plotted vs time and the values of $[C]$ is obtained from the curve at the interval of 10 min. $[C]_1 - [C]_2$ and $\ln [(C_1) - (C_2)]$ is calculated. The values are given in the above table. $l_n [(C)_1 - (C)_2]$ is plotted vs t.

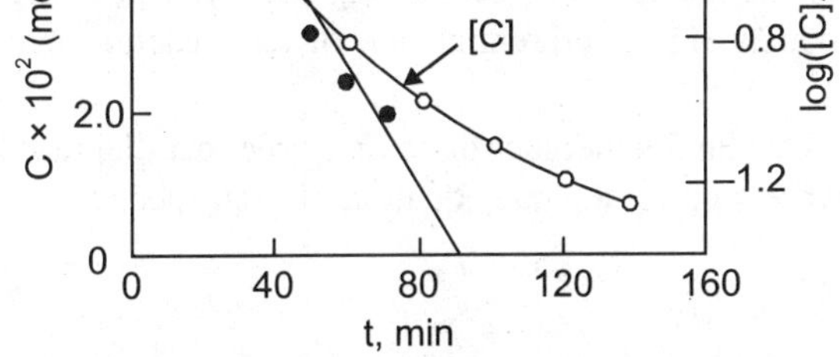

Fig. 19.5 *Plot of concentration (0–0–0) and log ([C]$_1$ – [C]$_2$)(–. –.) vs time*

$$\text{Slope} = \frac{-0.23 + 0.75}{20 - 50} = -0.0173 = -k$$

or $$k = 1.73 \times 10^{-2} \text{ min}^{-1}$$

19.9. COMPLEX REACTIONS

The combination of few elementary reactions are called complex reactions. Since, it is possible that the same reactant or the product formed in one of reaction may react in further reactions, the rate equation for any particular species will be the sum of the rates in all the elementary reactions. If the differential equation involves the rate of formation, then on right-hand side of the equation, positive sign will be for the terms involving formation and negative sign for the removal of particular species.

***Exercise* 19.19:** In the following reaction scheme, write the rate equation for the removal of species *A, B, C* and *D* in differential form.

$$A + B \xrightarrow{k_1} C + D$$

$$C + D \xrightarrow{k_2} A + B$$

$$C + B \xrightarrow{k_3} E + D$$

$$2D \xrightarrow{k_4} F$$

The reactant A is removed in step 1 and reformed in step 2, so

$$-\frac{d[A]}{dt} = k_1 [A][B] - k_2 [C][D]$$

The reactant B is removed in steps 1, 3 and formed in step 2 so

$$-\frac{d[B]}{dt} = k_1 [A][B] - k_2 [C][D] + k_3 [C][B]$$

Similarly for species C and D

$$-\frac{d[C]}{dt} = k_2 [C][D] + k_3 [C][B] - k_1 [A][B]$$

$$-\frac{d[D]}{dt} = k_2 [C][D] - k_1 [A][B] - k_3 [C][B] + 2k_4 [D]^2$$

Mainly the non-chain complex reactions can be classified into three categories.

19.9.1. *Parallel Reactions*

In these reactions, the same reactant may be reacting itself or with other reactant in different ways to form the different products. The preferential rate of any reaction may be varied by changing the temperature or catalyst.

***Exercise* 19.20:** A reactant R undergoes three first-order parallel reactions to give three products P_1, P_2 and P_3. Write the expression for the rate equation of each species.

$$R \xrightarrow{k_1} P_1$$
$$\xrightarrow{k_2} P_2$$
$$\xrightarrow{k_3} P_3$$

Rate of removal of R is

$$-\frac{d[R]}{dt} = k_1[R] + k_2[R] + k_3[R]$$
$$= (k_1 + k_2 + k_3)[R] \quad ...(19.32)$$

and can be integrated to give

$$[R] = [R]_0 e^{-kt} \quad ...(19.33)$$

where $[R]_0$ is concentration of R at $t = 0$ and $k = k_1 + k_2 + k_3$ = Overall first-order rate constant.

Similarly rate expression for products are

$$\frac{d[P_1]}{dt} = -\frac{k_1}{k}\frac{d[R]}{dt} = k_1[R] = k_1[R]_0 e^{-kt}$$

After integration with condition that at time $t = 0$, $[P_1] = [P_1]_0$, we get

$$[P_1] = [P_1]_0 + \frac{k_1}{k}(1 - e^{-kt})[R]_0 \quad ...(19.34a)$$

Similarly

$$[P_2] = [P_2]_0 + \frac{k_2}{k}(1 - e^{-kt})[R]_0 \quad ...(19.34b)$$
$$[P_3] = [P_3]_0 + k_3/k(1 - e^{-kt})[R]_0 \quad ...(19.34c)$$

If $[P_1]_0 = [P_2]_0 = [P_3]_0 = 0$, the equations (19.34$a$), (19.34$b$), (19.34$c$) reduce to

$$[P_1] = \frac{k_1}{k}(1 - e^{-kt})[R]_0 \quad ...(19.35a)$$

$$[P_2] = \frac{k_2}{k}(1 - e^{-kt})[R]_0 \quad ...(19.35b)$$

$$[P_3] = \frac{k_3}{k}(1 - e^{-kt})[R]_0 \quad ...(19.35c)$$

It is clear from equations (19.35a), (19.35b) and (19.35c) that the products are formed in the ratios of these rate constants for respective reactions

$$[P_1] : [P_2] : [P_3] :: k_1 : k_2 : k_3$$

***Exercise* 19.21:** Isopropyl chloride undergoes hydrolysis in aqueous media by two reaction mechanisms. Write expression for calculating k_1 and k_2.

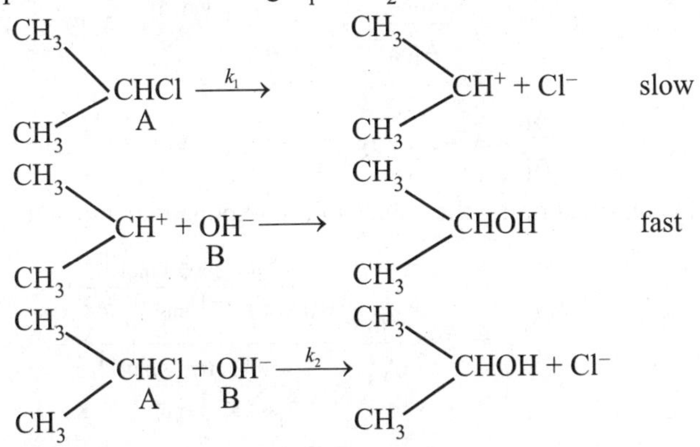

The rate determining step in scheme 1 is first one so the differential rate equation for the total removal of isopropyl chloride will be

$$-\frac{d[A]}{dt} = k_1 [A] + k_2 [A] [B]$$

The simplest procedure is to write it as

$$-\frac{1}{[A]}\frac{d[A]}{dt} = k_1 + k_2 [B] \qquad \text{...(19 36}a\text{)}$$

or

$$-\frac{d(\ln [A])}{dt} = k_1 + k_2 [B] \qquad \text{...(19.36}b\text{)}$$

Plot of $-\dfrac{d\ln [A]}{dt}$ vs $[B]$ will be a straight line. Slope gives k_2, and intercept gives k_1.

***Exercise* 19.22:** Consider the following parallel reactions :

$$A \xrightarrow{k_1} B$$

$$A \xrightarrow{k_2} C$$

The activation energies for the steps (*i*) and (*ii*) are 65.60 kJ mol^{-1} and 35.60 kJ mol^{-1}. If the rate constants $k_1/k_2 = 2$ at 450 K, calculate the value of temperature at which $k_1 = k_2$.

Using Arrehenius equation the ratios of k_1 and k_2 can be written as:

$$\frac{k_1}{k_2} = \frac{A_0^1 e^{-\frac{E_1}{RT}}}{A_0^2 e^{-\frac{E_2}{RT}}} \qquad \text{...(1)}$$

Substituting the respective values of the variables in equation (1), we get

$$\frac{k_1}{k_2} = 2 = \frac{A_0^1 e^{-\frac{65.6 \times 1000 \text{ J mol}^{-1}}{(8.314 \text{ J K}^{-1} \text{ mol}^{-1})(450 \text{ K})}}}{A_0^2 e^{-\frac{35.6 \times 1000 \text{ J mol}^{-1}}{(8.314 \text{ J K}^{-1} \text{ mol}^{-1})(450 \text{ K})}}}$$

$$= \frac{A_0^1 e^{-17.534}}{A_0^2 e^{-9.515}} = \frac{A_0^1}{A_0^2} e^{-8.019} = 3.29 \times 10^{-4} \frac{A_0^1}{A_0^2}$$

$$\frac{A_0^1}{A_0^2} = 6076.28 \qquad \text{...(2)}$$

Using equation (1) and substituting the values of pre-exponential and activation energies, we get

$$\frac{k_1}{k_2} = 1 = \frac{A_0^1}{A_0^2} \frac{e^{-\frac{65.6 \times 1000 \text{ J mol}^{-1}}{(8.314 \text{ J K}^{-1} \text{ mol}^{-1})(T \text{ K})}}}{e^{-\frac{35.6 \times 1000 \text{ J mol}^{-1}}{(8.314 \text{ J K}^{-1} \text{ mol}^{-1})(T \text{ K})}}}$$

$$= 6076.28e^{-\frac{[65.6-35.6]\times 1000 \text{ J mol}^{-1}}{(8.314 \text{ J K}^{-1} \text{ mol}^{-1})(T \text{ K})}}$$

$$= 6076.28e^{-\frac{3608.38}{T}}$$

or
$$= e^{-\frac{3608.38}{T}} = \frac{1}{6076.28} = 1.6458 \times 10^{-4}$$

Taking natural logarithms of both sides, we get

$$-\frac{3608.38}{T} = \ln 1.6458 \times 10^{-4} = -8.712$$

or
$$T = \frac{3608.38}{8.712} = 414 \text{ K}.$$

19.9.2. *Opposing Reactions*

In this category, the reaction procedes on both directions and at equilibrium, the rate of forward direction is equal to that of reverse direction *i.e.* net rate of formation of each species is zero. It is possible to collect the concentration on one side and all the rate constants on other side of the equation. Knowing K (equilibrium constant) and concentration of species involved at time t and at equilibrium, one can calculate rate constants for the forward and backward directions.

***Exercise* 19.23:** Calculate the rate constant k_f, k_r and equilibrium constant K for the following reaction

$$R \underset{k_r}{\overset{k_f}{\rightleftharpoons}} P$$

Let $[R]$ be the concentration of R at time t and that of P is $[P] = [R]_0 - [R]$. So

$$-\frac{d[R]}{dt} = k_f[R] - k_r([R]_0 - [R]) \qquad ...(19.37)$$

At equilibrium $\frac{d[R]}{dt} = 0$, so

$$\frac{[R]_0 - [R]_e}{[R]_e} = \frac{k_f}{k_r} = K$$

Also
$$k_r = k_f \frac{[R]_e}{[R]_0 - [R]_e} \qquad ...(19.38)$$

Substituting k_r in equation (19.37) and rearranging we get

$$-\frac{d[R]}{[R]-[R]_e} = \frac{[R]_0}{[R]_0 - [R]_e} k_f\, dt$$

and integrated form with initial condition $[R] = [R]_0$ at $t = 0$ is

$$\ln \frac{[R]_0 - [R]_e}{[R] - [R]_e} = \frac{[R]_0}{[R]_0 - [R]_e} k_f t \quad ...(19.39)$$

Plot of $\ln \frac{[R]_0 - [R]_e}{[R] - [R]_e}$ vs t will be a straight line passing through origin and slope $= \frac{[R]_0}{[R]_0 - [R]_e} k_f$ knowing $[R]_0$ and $[R]_e$, k_f can be calculated.

Similarly k_r can also be calculated from equation (19.38).

If rate of the reaction is followed by measuring the concentration of product P, then equation (19.39) becomes

$$\ln \frac{[P]_e}{[P]_e - [P]} = \frac{k_f [R]_0 t}{[P]_e} \quad ...(19.40)$$

and also with the help of equation (19.38)

$$\ln \frac{[P]_e}{[P]_e - [P]} = t (k_f + k_r) \quad ...(19.41)$$

***Exercise* 19.24:** Calculate k_f, k_r for the given reaction $R \rightarrow P$. $[R]_0 = 0.15$ M, $[R]_e = 0.086$ M, $[R]_{10} = 0.142$ M, $[R]_{100} = 0.102$ M.

Using equation (19.39) for $t = 10$ s

$$\ln \left(\frac{0.15 - 0.086}{0.142 - 0.086} \right) = \frac{(0.15)\,(10\text{ s})}{0.15 - 0.086} k_f$$

$$k_f = 5.69 \times 10^{-3} \text{ s}^{-1}$$

Similarly for $t = 100$ s

$$\ln \frac{0.15 - 0.086}{0.102 - 0.086} = \frac{(0.15)\,(100\text{ s})}{0.15 - 0.086} k_f$$

$$k_f = 5.91 \times 10^{-3} \text{ s}^{-1}$$

Two values are quite consistent so average value is $k_f = 5.8 \times 10^{-3}$ s^{-1}.

Now

$$[P]_e = 0.15 - 0.086 = 0.064 \text{ M}$$

$$K = \frac{0.064 \text{ M}}{0.086 \text{ M}} = 0.744$$

$$k_r = \frac{k_f}{K} = \frac{5.8 \times 10^{-3} \text{ s}^{-1}}{0.744} = 7.79 \times 10^{-3} \text{ s}^{-1}$$

19.9.3. *Consecutive Reactions*

Consecutive reaction is one in which intermediate formed in one of the steps reacts further to form product. Consider a simplest consecutive reaction

$$R_1 \xrightarrow{k_1} R_2 \xrightarrow{k_2} R_3$$

The concentration of each species at time t, with the initial conditions as

$$[R_1] = [R_1]_0$$

$$[R_2] = [R_3] = 0 \text{ at } t = 0 \text{ are}$$

$$[R_1] = [R_1]_0 e^{-k_1 t} \quad ...(19.42)$$

$$[R_2] = \frac{[R_1]_0\, k_1}{k_2 - k_1}(e^{-k_1 t} - e^{-k_2 t}) \quad ...(19.43)$$

$$[R_3] = [R_1]_0\left(1 - \frac{k_2 e^{-k_1 t} - k_1 e^{-k_2 t}}{k_2 - k_1}\right) \quad ...(19.44)$$

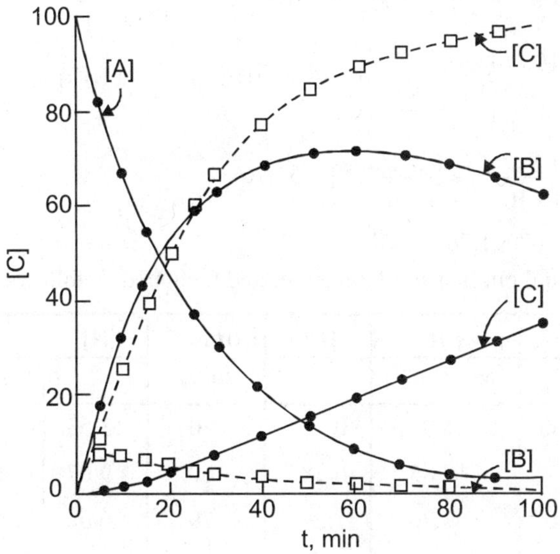

Fig. 19.6 *Plot of concentrations of A = R_1, B = R_2 and C = R_3 vs time for a consecutive reaction*

•–•–• ($k_1 : k_2 :: 1 : 0.15$) □–□–□ ($k_1 : k_2 :: 1 : 10$)

Further as a check at any time t,

$$[R_1]_0 = [R_1] + [R_2] + [R_3] \quad ...(19.45)$$

It can be seen from Fig. 19.6 that concentration of R_1 keeps on decreasing exponentially, concentration of R_2 increases to maximum and then decreases and concentration of R_3 keeps on increasing.

The maximum concentration of R_2 obtained and time required to reach maximum concentration is given by the expressions (19.46) and (19.47) respectively

$$[R_2]_{max} = [R_1]_0\left(\frac{k_2}{k_1}\right)^{k_3/k_1 - k_2} \quad ...(19.46)$$

and $$t_{max} = \frac{\ln(k_1/k_2)}{k_1 - k_2} \quad ...(19.47)$$

***Exercise* 19.25:** Isopropyl benzene is alkylated by *n*-butylamine in presence of hydrogen fluoride in a two step consecutive reaction as

$$\text{Isopropyl-benzene} \xrightarrow{k_1} \text{Isopropyl-sec butyl-benzene} \xrightarrow{k_2} \text{Isopropyl-di–sec butyl benzene}$$

If the initial concentration of isopropyl benzene is 100 mol and relative rate constants $k_1 : k_2 :: 1 : 0.15$, prepare a diagram showing the concentration of each species.

Let $k_1 = 4.0 \times 10^{-2}\ \text{min}^{-1}$, so

$k_2 = 0.15 \times 4 \times 10^{-2} = 0.6 \times 10^{-2}\ \text{min}^{-1}$

As a sample calculations (19.42), (19.43) and (19.45) give for $t = 10$ min.

$$[R_1] = 100\ e^{-4.0\times10^{-2}(10)} = 100\ e^{-0.4} = 67$$

$$[R_2] = \frac{(100)(4\times10^{-2})}{0.6\times10^{-2} - 4.0\times10^{-2}}[e^{-4.0\times10^{-2}(10)} - e^{-0.60\times10^{-2}(10)}]$$

$$= -\frac{4}{3.4\times10^{-2}}[e^{-0.4} - e^{-0.06}] = 31.76$$

$$R_3 = 100 - 67 - 31.76 = 1.24$$

Other data can be found out in a similar manner and tabulated as follows:

time (min)	[A]	[B] (mol dm^{-3})	[C]	time (min)	[A]	[B] (mol dm^{-3})	[C]
0	100	0	0	50	13.53	71.24	15.23
5	81.87	17.85	0.28	60	9.07	71.41	19.52
10	67.00	31.76	1.24	70	6.08	70.15	23.77
15	54.88	42.96	2.16	80	4.08	68.00	27.92
20	44.93	51.48	3.59	90	2.73	65.34	31.92
25	36.79	57.98	5.23	100	1.83	62.41	35.76
30	30.12	62.83	7.05	200	0.0033	35.05	64.95
40	20.19	68.79	11.02				

The above data is plotted in Fig. 19.6. The t_{max} and $[B]_{max}$ obtained from the Fig. 19.6 is 55 min and 71.6% respectively.

***Exercise* 19.26:** With the help of data given in exercise (19.25), calculate $[B]_{max}$ and t_{max}, compare it with values obtained from the Fig. 19.6.

Substituting the respective values in equations (19.46) and (19.47),

$$[B]_{max} = (100\text{ M})\left(\frac{0.15\text{ min}^{-1}}{1.0\text{ min}^{-1}}\right)^{\frac{0.15}{1-0.15}}$$

$$= 71.55\text{ M}$$

$$t_{max} = \frac{\ln\dfrac{1.0}{0.15}}{(1.0-0.15)\text{ min}^{-1}} = 55.80\text{ min}$$

From the Fig. 19.6 the values of $[B]_{max}$ and t_{max} are 71.6% and 55 min respectively. The agreement is good.

***Exercise* 19.27:** Repeat the exercise 19.25 with the values of $k_1 = 4.0 \times 10^{-2}\text{ min}^{-1}$, $k_2 = 4.0 \times 10^{-1}$ min^{-1}.

Using equations (19.42), (19.43) and (19.45) and proceeding in the same manner as in exercise (19.25), the concentrations of species *A*, *B* and *C* after certain time interval are mentioned in the table below and are plotted in Fig. 19.6.

Time (min)	[*A*]	[*B*] (mol dm^{-3})	[*C*]	Time (min)	[*A*]	[*B*] (mol dm^{-3})	[*C*]
0	100	0	0	40	20.19	2.24	77.57
5	81.87	7.59	10.54	50	13.53	1.50	84.97
10	67.00	7.24	25.76	60	9.07	1.00	89.93
15	54.88	6.07	39.05	70	6.08	0.68	93.24
20	44.93	4.99	50.08	80	4.08	0.45	97.47
25	36.79	4.09	59.12	90	2.73	0.30	96.97
30	30.12	3.35	66.53	100	1.83	0.20	97.97

One can see from the plot the concentration of *B* rises to maximum very quickly, decays fast and remains nearly stationary (< 2% for pretty long time). This is only if k_2 is 10 times faster than k_1. If $k_2 \gg k_1$, the stationary state of species *B* (nearly constant concentration of *B*) can be achieved quickly and remains stationary for long time.

19.10. STEADY STATE HYPOTHESIS

The assumption involved in steady state hypothesis is that concentration of reactive intermediates can be assumed to be constant, *i.e.*, $\frac{d[C_i]}{dt} = 0$. *Note that steady state approximation can only be applied to short-lived (or very reactive) species.* The following steps are used to calculate the rate laws in terms of stable species :

(*i*) The differential rate laws are written down for each species.

(*ii*) The differential rate laws of reactive intermediates are put equal to zero and concentrations of reactive intermediates are calculated in terms of stable species.

(*iii*) Steady state concentrations of intermediates calculated in (*ii*) are substituted in the expressions so that the rate laws can be written only in terms of stable species (reactants and products).

***Exercise* 19.28:** In the following reaction mechanism, write down the expression for the rate of formation of *P.*

$$R_1 + R_2 \underset{k_2}{\overset{k_1}{\rightleftharpoons}} I$$

$$I \overset{k_3}{\rightarrow} P$$

I is a reactive intermediate.

Writing the differential rate equation for *I* and equating it to zero.

$$\frac{d[I]}{dt} = k_1[R_1][R_2] - k_2[I] - k_2[I] = 0$$

$$[I] = \frac{k_1[R_1][R_2]}{k_2 + k_3}$$

Now

$$\frac{d[P]}{dt} = k_3[I]$$

$$= \frac{k_3 k_1}{k_2 + k_3}[R_1][R_2] = k[R_1][R_2]$$

where k is the effective rate constant. $k = \dfrac{k_1 k_3}{k_2 + k_3}$

***Exercise* 19.29:** Thermal gas phase reaction between H_2 and Br_2 has the following mechanism

$$Br_2 \overset{k_1}{\rightarrow} 2Br$$

$$Br + H_2 \overset{k_2}{\rightarrow} HBr + H$$

$$H + Br_2 \overset{k_3}{\rightarrow} HBr + Br$$

$$H + HBr \overset{k_4}{\rightarrow} H_2 + Br$$

$$Br + Br \overset{k_5}{\rightarrow} Br_2$$

Write the expression for $\dfrac{d[HBr]}{dt}, \dfrac{d[H]}{dt}$ and $\dfrac{d[Br]}{dt}$.

Assuming H and Br to be reactive intermediates, solve $\dfrac{d[HBr]}{dt}$ in terms of H_2, Br_2 and HBr. What will be pseudo-order of the formation of HBr in the initial stage?

The rate expression for three species are

$$\frac{d[\mathrm{H}]}{dt} = k_2[\mathrm{Br}][\mathrm{H_2}] - k_3[\mathrm{H}][\mathrm{Br_2}] - k_4[\mathrm{H}][\mathrm{HBr}] = 0$$

$$\frac{d[\mathrm{Br}]}{dt} = 2k_1[\mathrm{Br_2}] - k_2[\mathrm{Br}][\mathrm{H_2}] + k_3[\mathrm{H}][\mathrm{Br_2}] + k_4[\mathrm{H}][\mathrm{HBr}] - 2k_5[\mathrm{Br}]^2 = 0$$

$$\frac{d[\mathrm{HBr}]}{dt} = k_2[\mathrm{Br}][\mathrm{H_2}] + k_3[\mathrm{H}][\mathrm{Br}] - k_4[\mathrm{H}][\mathrm{HBr}]$$

Solving the first two equations,

$$[\mathrm{Br}] = \left(\frac{k_1}{k_5}\right)^{1/2} [\mathrm{Br_2}]^{1/2}$$

$$[\mathrm{H}] = k_2\left(\frac{k_1}{k_5}\right) \frac{[\mathrm{Br_2}]^{1/2}[\mathrm{H}]}{k_3\,[\mathrm{Br_2}] + k_4\,[\mathrm{HBr}]}$$

Substituting [H] and [Br] in last equation gives

$$\frac{d\,[\mathrm{HBr}]}{dt} = 2k_2\left(\frac{k_1}{k_5}\right)^{1/2} \frac{[\mathrm{H_2}]\,[\mathrm{Br_2}]^{3/2}}{k_3\,[\mathrm{Br_2}] + k_4\,[\mathrm{HBr}]}$$

At initial stages [HBr] = 0, above equation becomes

$$\frac{d\,[\mathrm{HBr}]}{dt} = 2\frac{k_2}{k_3}\left(\frac{k_1}{k_5}\right)^{1/2} [\mathrm{H_2}]\,[\mathrm{Br_2}]^{1/2} = k\,[\mathrm{H_2}]\,[\mathrm{Br_2}]^{1/2}$$

pseudo-order is 3/2, 1 with respect to H_2 and half with respect to Br_2.

The effective rate constant

$$k = 2\frac{k_2}{k_3}\left(\frac{k_1}{k_5}\right)^{1/2}$$

19.11. EFFECT OF TEMPERATURE ON RATE CONSTANT

Generally rate constant increases with the increase of temperature and its value approximately doubles for every rise of 10°C. The first quantitative relation between k and temperature is due to Arrhenius.

$$k = A \exp\,(-E_a/RT) \qquad \text{...(19.48)}$$

or

$$\ln k = \ln A - \frac{E_a}{RT} \qquad \text{...(19.49)}$$

where A is pre-exponential or frequency factor, E_a is the activation energy, *i.e.*, the molecules must possess this much energy before they react to form products. The above equations require that the unit of E_a should be (energy) ($\mathrm{mol^{-1}}$). The physical meaning of $\mathrm{mol^{-1}}$ is simply that concentration terms used in chemical kinetics are expressed as mol dm^{-3} or mol m^{-3}. Both the factors A and E_a are known as Arrhenius parameters. Plot of log k vs T^{-1} is linear for large number of reactions and over moderate temperature. The relation between the rate constants at two temperatures is

$$\ln \frac{k_2}{k_1} = \frac{E_a}{R}\left[\frac{1}{T_1} - \frac{1}{T_2}\right] \quad ...(19.50)$$

The plot of E (energy) vs the progress of the reaction (reaction coordinate) is shown in Fig. 19.7. The hump in the curve gives activation energy. The difference between the energy of the reactants and products is known as heat of reaction. Thus the activation energy for the backward reaction will be

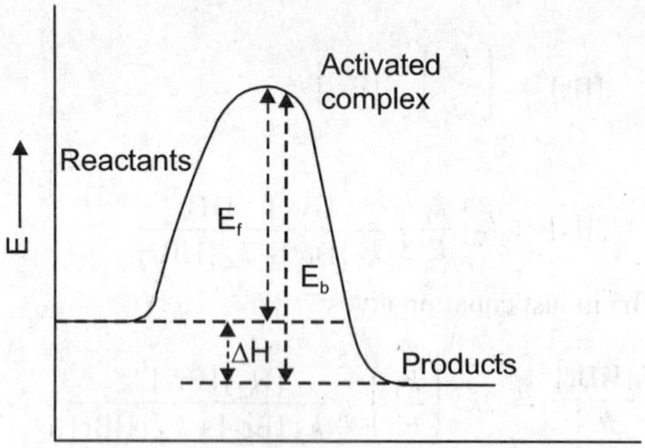

Fig. 19.7 *Plot of E vs reaction coordinates*

$$E_a(r) = E_a(f) - \Delta H \quad ...(19.51)$$

***Exercise* 19.30:** Find the relation between half-life ($t_{1/2}$) and the temperature.

Arrhenius equation gives

$$\ln k = \ln A - \frac{E_a}{RT} \quad ...(19.49)$$

Except for $n = 1$, the relation between half-life and rate constant is

$$t_{1/2} = \frac{2^{n-1} - 1}{k\,(n-1)[C]_0^{\,n-1}} \quad ...(19.21)$$

or

$$\ln t_{1/2} = \ln \frac{2^{n-1} - 1}{(n-1)[C]_0^{\,n-1}} - \ln k \quad ....(19.21a)$$

Combining equations (19.21 *a*) and (19.49)

$$\ln t_{1/2} = \ln \frac{2^{n-1} - 1}{(n-1)[C]_0^{\,n-1}} - \ln A + \frac{E_a}{RT}$$

$$= \ln A' + \frac{E_a}{RT} \quad ...(19.52)$$

where
$$A' = \frac{2^{n-1} - 1}{(n-1)[C]_0^{n-1} A}$$

Relation (19.52) shows that increase of temperature decreases the half-life, as expected. Further plot of $\ln t_{1/2}$ vs $1/T$ is a straight line and slope gives the value of E_a.

***Exercise* 19.31:** In exercise 19.29, the effective rate constant $k' = 2\frac{k_2}{k_3}\left(\frac{k_1}{k_5}\right)^{1/2}$ at the early stage of reaction. Calculate the effective activation energy.

Now

$$k' = 2\frac{k_2}{k_3}\left(\frac{k_1}{k_5}\right)^{1/2}$$

and due to Arrhenius equation

$$k = A \exp(-E_a/RT)$$

Combining the two equations

$$k' = A' \exp(-E_a'/RT) = 2\frac{A_2 \exp(-E_{a(2)}/RT)}{A_3 \exp(-E_{a(3)}/RT)}\left(\frac{A_1 \exp(-E_{a(1)}/RT)}{A_5 \exp(-E_{a(5)}/RT)}\right)^{1/2}$$

$$= 2\frac{A_2 A_1^{1/2}}{A_3 A_5^{1/2}} \exp(-[E_{a(2)} - E_{a(3)} + \tfrac{1}{2}E_{a(1)} - \tfrac{1}{2}E_{a(5)}])/RT$$

So effective activation energy E_a' is

$$E_a' = E_{a(2)} - E_{a(3)} + \tfrac{1}{2}(E_{a(1)} - E_{a(5)})$$

and

$$A' = 2\frac{A_2 A_1^{1/2}}{A_3 A_5^{1/2}}$$

***Exercise* 19.32:** The values of k observed for the reaction $2HI \rightarrow H_2 + I_2$ at 356°C and 443°C are 3.02×10^{-5} mol^{-1} dm^3 s^{-1} and 2.53×10^{-3} mol^{-1} dm^3 s^{-1} respectively. Calculate the $E_{a,f}$ for forward and $E_{a,r}$ for backward reaction if $\Delta H = 16.32$ kJ mol^{-1}.

Using equation (19.50)

$$\ln\left(\frac{2.53 \times 10^{-3} \text{mol}^{-1}\,\text{dm}^3\text{s}^{-1}}{3.02 \times 10^{-5} \text{mol}^{-1}\,\text{dm}^3\text{s}^{-1}}\right) = \frac{E_{a,f}}{(8.314 \text{JK}^{-1}\text{mol}^{-1})}\left[\frac{1}{629} - \frac{1}{716}\right] \text{K}^{-1}$$

$$E_{a,f} = 190.58 \text{ kJ mol}^{-1}$$

$$\Delta H = E_{a,f} - E_{a,r}$$

$$E_{a,r} = E_{a,f} - \Delta H = 190.58 - 16.32 = 174.26 \text{ kJ mol}^{-1}$$

19.12. SIMPLE COLLISION THEORY

According to simple collision theory the rate of reaction is based on (*a*) number of collisions per unit volume per unit time and (*b*) the molecules taking part in collisions must possess enough energy (activation energy) before they can be converted into products. Based on these assumptions, the rate of reaction is

$$\text{rate} = P\left(8\pi k_b T\frac{m_A + m_B}{m_A m_B}\right)^{1/2} d_{AB}^2 n_A n_B e^{-E_a/RT} \qquad ...(19.53a)$$

and rate constant

$$k = P\left(8\pi k_b T\frac{m_A + m_B}{m_A m_B}\right)^{1/2} d_{AB}^2 e^{-E_a/RT}$$

$$k(\text{M}^{-1}\ \text{s}^{-1}) = [10^3 \text{N}_{\text{av}}]\, P\left[8\pi RT\cdot\frac{M_A + M_B}{M_A M_B}\right]^{1/2} d_{AB}^2 e^{-E_a/RT} \qquad ...(19.53b)$$

for unlike molecules and

$$\text{rate} = 2P\left(\frac{\pi k_b T}{m_A}\right)^{1/2} d_{AA}^2 n_A^2 e^{-E_a/RT} \qquad ...(19.54)$$

$$k = 2P\left(\frac{\pi k_b T}{m_A}\right)^{1/2} d_{AA}^2 e^{-E_a/RT} \qquad ...(19.55)$$

$$= 2[10^3 N_{\text{av}}]P\left(\frac{k_b T}{m_A}\right)^{1/2} d_A^2 e^{-E_a/RT} \qquad ...(19.55a)$$

for like molecules. Comparing the above equations with Arrhenius equation the pre-exponential factors for two kinds of molecules are

$$A = P\left(8\pi k_b T\frac{m_A + m_B}{m_A m_B}\right)^{1/2} d_{AB}^2 \qquad ...(19.56)$$

$$A = 2P\left(\frac{\pi k_b T}{m_A}\right)^{1/2} d_{AA}^2 \qquad ...(19.57)$$

where m_A, m_B are the masses of molecules *A* and *B*, d_{AB} is the sum of the radii of the molecules *A* and *B*. *P* is the probability steric factor, less than or equal to unity and takes into account the geometry of molecules or direction of approach of the reacting molecules, n_A, n_B arc the number of molecules per c.c. so the units of *k* are molecule^{-1} c.c. s^{-1} and can be converted to mol^{-1} dm^3 s^{-1} by multiplying it by a factor 6×10^{20} (*i.e.*, $6.02 \times 10^{23} \times 10^{-3}$). Equation (19.55) was derived by involving (*a*) area cross section of the reactants, (*b*) overage relative speeds of the appraoch of reactants and (*c*) fraction of reactants having energies greater than the activation energy. Equation (19.55) can be modified by including the following constraints, *i.e.*,

$$\sigma v(E_r) = \begin{bmatrix} 0 & E_r \le E_a \\ \pi d_{AB}^2 & E_r \ge E_a \end{bmatrix} \qquad ...(19.58)$$

can be written as

$$k = \left(\frac{2}{k_b T}\right)^{\frac{3}{2}}\left(\frac{1}{\mu\pi}\right)^{\frac{1}{2}} \int_{E_a}^{\infty} \pi d_{AB}^2 E_r e^{-Er/kbT} dE_r$$

$$= \pi d_{AB}^2 \left(\frac{8k_bT}{\pi\mu}\right)^{\frac{1}{2}} e^{-Ea/kbT}\left[1+\frac{E_b}{k_bT}\right] \quad ...(19.59)$$

where πd^2_{AB} is a hard sphere collision cross-section.

***Exercise* 19.33:** Calculate A and rate constant for the reaction

$$CH_3 + H_2 \rightarrow CH_4 + H$$

at 200°C if the collision diameter is 3.0×10^{-10} m and it is assumed that activation energy is zero for the above reaction.

Here the two species are different, so equation (19.56) is used. Consider CH_3 as a unit mass of 15 a.m.u. so the reduced mass is

$$\mu = \frac{m_a m_b}{m_a + m_b} = \frac{15 \times 2}{(15+2)\, 6.023 \times 10^{23}} = 0.293 \times 10^{-23} \text{ g mol}^{-1}$$

$$= 2.93 \times 10^{-27} \text{ kg mol}^{-1}$$

$$A = P\left[\frac{(8)\left(\frac{22}{7}\right)(1.381\times10^{-23}\text{J K}^{-1})(473\text{k})}{2.93\times10^{-27}\text{kg mol}^{-1}}\right]^{1/2} (3.0 \times 10^{-10}\text{ m})^2$$

$$= 6.74 \times 10^{-16}\, P \text{ m}^3 \text{ s}^{-1} \text{ mol}^{-1}$$

taking P as unity, $E_a = 0$ (19.52) becomes (19.56)

$$k = (6.74 \times 10^{-13} \text{ dm}^3 \text{ mole}^{-1} \text{ s}^{-1})(6.023 \times 10^{23} \text{ mol}^{-1}) = 4.06 \times 10^{11} \text{ dm}^3 \text{ mol}^{-1} \text{ s}^{-1}$$

***Exercise* 19.34:** If the activation energy of a reaction is 80.9 kJ mol^{-1}, calculate the fraction of molecules at 400°C which have enough energy to react to form products.

As said earlier, those collision will yield products if the energy possessed by the molecules is equal to or greater than activation energy. This is given as

$$\frac{\Delta N}{N_{av}} = e^{-E_a/RT} = e\left(-\frac{80.9\times10^3 \text{ J mol}^{-1}}{(8.314\text{ K J}^{-1}\text{mol}^{-1})(673\text{ K})}\right)$$

$$= e^{-14.459} = 5.257 \times 10^{-7}$$

i.e., fraction of molecules which have enough energy to go for the formation of products is very small, *i.e.*, 5.257×10^{-7}.

19.13. ACTIVATED COMPLEX OR TRANSITION STATE THEORY

The rate or rate constant calculated with the help of activated complex theory is based on (*a*) the reactants are converted to an activated complex before being converted to products, and (*b*) there exists an equilibrium between the activated complex and the reactants. The reaction can be written as

$$A + B \rightleftharpoons AB^* \rightarrow \text{Products} \quad ...(19.60)$$

and rate constant calculated with the help of statistical mechanics for the formation of products is

$$k = \kappa \frac{k_bT}{h} \frac{Q_{\#}}{Q_AQ_B} e^{-E^0_{0}/RT} \quad ...(19.61)$$

where κ is the transmission coefficient, nearly equal to unity, Q_i is the partition function and E^0_0 is the difference in the zero point energies of the activated complex and the reactants. The partition function Q_i is given as

$$Q = Q_{\text{tr}}\, Q_{\text{rot}}\, Q_{\text{vib}}\, Q_{\text{el}} \quad \text{...(19.62)}$$

Each partition function can be calculated by the equations (9.69), (9.61) or (9.65) and (9.55 or 9.57) and (9.49). The number of vibrational degrees of freedom for activated complex is $3N_A + 3N_B - 7$ for non-linear and $3N_A + 3N_B - 6$ for linear activated complex (one degree of freedom is less than the normal molecule as this is converted to translational motion along the reaction coordinate) and $3N_i - 6$ or $3N_i - 5$ are the vibrational degrees of freedom for normal non-linear and linear molecules.

***Exercise* 19.35:** Absolute rate theory relates the dependence of A on temperature ($A \propto T^n$). Calculate the value of exponent for the following reaction :

Atom + Linear molecule → Non-linear molecule

The equation (19.61) can be written as

$$k = k\frac{k_bT}{h}\frac{q_{\text{tr}}^{\#3}\, q_{\text{rot}}^{\#3}\, q_{\text{vib}}^{3N_A+3N_B-7}}{q_{A(\text{tr})}^{3}\, q_{(B)\text{tr}}^{3}\, q_{B(\text{rot})}^{2}\, q_{B_{(\text{vib})}}^{3N_B-1}}$$

Let

$$q_{\text{vib}} = 1,\ q_{\text{tr}} \propto T^{3/2},\ q_{\text{rot}} \propto T \text{ for linear}$$

$$q_{\text{rot}} \propto T^{3/2} \text{ for non-linear}$$

$$A \propto T\frac{T^{3/2}\, T^{3/2}}{T^{3/2}\, T^{3/2}\, T} \quad \text{or } A \propto T^0$$

A is independent of temperature.

***Exercise* 19.36:** If it is assumed that the values of q_{tr}, q_{rot} and q_{vib} for different molecules are equal, calculate the value of A at 298 K for the reaction mentioned in exercise 19.35. $q_{\text{tran}} = 10^8$, $q_{\text{rot}} = 10$, $q_{\text{vib}} = 1$ for each degree of freedom.

Using the equation in exercise 19.36, κ = 1

$$k = \frac{(1.38\times10^{-23}\,\text{J K}^{-1})(298\text{ K})}{(6.62\times10^{-34}\,\text{J s})}\ \frac{(10^8)^3(10)^3(1)}{(10^8)^3(10^8)^3(10)^2(1)}(10^{-3}\times 6.02\times10^{23})$$

$$= 3.74\times10^{10}\ \text{dm}^3\ \text{mol}^{-1}\ \text{s}^{-1}$$

which is quite close to 10^{11} dm^3 mol^{-1} s^{-1} and is a good approximation.

19.14. THERMODYNAMIC FORMULATION OF RATE CONSTANT

The assumption involved in the transition state theory is that there is an equilibrium between the activated complex and the reactants, *i.e.*, for the reaction (19.60).

$$K_c = \frac{C_\#}{C_AC_B} = \frac{k_bT}{h}\frac{q_\#}{q_Aq_B}e^{-E_0/RT} = \frac{k_bT}{h\nu}K_\# \quad \text{...(19.63)}$$

The relation between K_c and $K_\#$ is the same as between $q^\#$ and $q_\#$. The terms $K_\#$ and $q_\#$ must be multiplied by a factor $\frac{k_bT}{h\nu}$, where ν is the frequency of decomposition of the activated complex. $K_\#$ can be calculated with the help of thermodynamics as

$$K_\# = \exp(-\Delta G_\#^0/RT)$$

$$= \exp(S_\#^0/R)\exp(-\Delta H_\#^0/RT) \quad \text{...(19.64)}$$

Thus

$$\text{rate} = \nu C_{\#} = \frac{k_b T}{h} K_{\#}[A][B] = \frac{k_b T}{h} \exp(-\Delta G^0_{\#}/RT)[A][B] \quad ...(19.65)$$

and

$$k = \frac{k_b T}{h} e^{-\Delta G^0_{\#}/RT}$$

$$= \frac{k_b T}{h} e^{\Delta S^0_{\#}/R} e^{-\Delta H^0_{\#}/RT}$$

$$= \frac{k_b T}{h} e^{1-\Delta n} e^{\Delta S^0_{\#}/R} e^{-E_a/RT} \quad ...(19.66)$$

where $\Delta G^0_{\#}$, $\Delta S^0_{\#}$ and $\Delta H^0_{\#}$ are the free energy, entropy and enthalpy of activation, Δn is the change in the number of mol in forming activated complex. Comparing it with Arrhenius equation

$$A = \frac{k_b T}{h} e^{1-\Delta n} e^{\Delta S^0_{\#}/R} \quad ...(19.67)$$

and

$$\Delta H^0_{\#} = E_a - (1 - \Delta n)RT \quad ...(19.68)$$

It must be kept in mind that the unit for standard states of enthalpy and entropy in kinetics in general are mol dm^{-3} since mostly reactants are expressed in mol dm^{-3}. The two standard states expressed as 1 atmosphere and 1 mol dm^{-3} are related as

$$\Delta H^0_p = \Delta H^0_c + \Delta n\, RT \quad ...(19.69)$$

and

$$\Delta S^0_p = \Delta S^0_c + \Delta nR + \Delta nR \ln RT \quad ...(16.70)$$

where Δn is the change in the number of mol of reaction.

***Exercise* 19.37:** Consider the unimolecular reaction cyclopropane → propylene. The value of log A for high pressure limit is 15.17. Calculate the value of $\Delta S_{\#}$ for the reaction. Calculate the contribution of ΔS_{trans} in the above reaction at 298 K.

According to equation (19.67), $\Delta n = 0$

$$A = \frac{k_b T_e}{h} \exp(\Delta S^0_{\#}/R)$$

or

$$\frac{\Delta S^0_{\#}}{R} = 2.303 \log A - 2.303 \log \frac{k_b\, Te}{h}$$

$$\frac{ek_b\, T}{h} = \frac{(1.38 \times 10^{-23}\,\text{J K}^{-1})(298\,\text{K})(2.718)}{(6.62 \times 10^{-34}\,\text{J s})} = 1.70 \times 10^{13}$$

$$\frac{\Delta S^0_{\#}}{R} = 2.303 \log \frac{Ah}{k_b Te} = 2.303 \log \frac{1.479 \times 10^{15}}{1.70 \times 10^{13}} = 4.466$$

$$\Delta S^0_{\#} = (8.314 \text{ J K}^{-1} \text{ mol}^{-1})(4.466) = 37.13 \text{ J K}^{-1} \text{ mol}^{-1}$$

Since the reactant and the product have the same mass $\Delta S^0_{\text{trans}} = 0$. The whole contribution is due to internal degrees of freedom.

***Exercise* 19.38:** The $\Delta S^0_{\#}$ for the reaction $2CH_3 \rightarrow C_2H_6$ is – 155.23 EU mol^{-1}. Calculate the value of A at 400 K for the above reaction.

The entropy change given above is with respect to standard state of 1 atm. Using equation (19.70) $\Delta S^0_{\#}$ can be obtained to

$$\Delta S^0_{\#} = -155.23 - (-1)(8.314 \text{ J K}^{-1} \text{ mol}^{-1}) - (-1)(8.314 \text{J K}^{-1}\text{mol}) \ln(8.314 \times 400)$$

$$= -82.496 \text{ J K}^{-}\text{mol}^{-1}$$

Using equation (19.67)

$$A = \frac{(1.38 \times 10^{-23} \text{ J K}^{-1})(400 \text{ K})}{6.63 \times 10^{-34} \text{ J s}} e^{1-(-1)} e^{-\frac{82.496 \text{ J K}^{-1}\text{mol}^{-1}}{8.314 \text{ J K}^{-1}\text{mol}^{-1}}}$$

$$= 3.02 \times 10^9 \text{ s}^{-1} \text{ mol}^{-1} \text{ dm}^3$$

19.15. PRIMARY SALT EFFECT

The thermodynamic equilibrium constant can be written as

$$K = \frac{a^{\#}}{a_A \, a_B} = \frac{c^{\#}}{c_A \, c_B} \frac{\gamma^{\#}}{\gamma_A \, \gamma_B} \quad \text{...(19.71)}$$

where a_i and γ_i are the activity and activity coefficient for ith species. In gas phase the use of concentration is a valid approximation as the mixture are dilute but not in case of solution and specially ionic solutions. The rate constant for reactions in solutions obtained with the help of transition state theory is

$$k = \frac{k_b \, T}{h} K_{\#} \frac{\gamma_A \, \gamma_B}{\gamma_{\#}} \quad \text{...(19.72)}$$

$$= k_0 \frac{\gamma_A \, \gamma_B}{\gamma_{\#}} \quad \text{...(19.73)}$$

where $k_0 = k$ when $\gamma_i = 1$

According to Debye-Hückle theory, for dilute aqueous solutions at 25°C

$$\log \gamma_i = -Z_i^2 (0.5116) \, \mu^{1/2} \quad \text{...(19.74)}$$

where Z_i is the valence of the ion i and μ is the ionic strength of solution. Combining equations (19.73) and (19.74).

$$\ln k = \ln k_0 + 1.0232 \, Z_A \, Z_B \, \mu^{1/2} \quad \text{...(19.75)}$$

i.e., plot of log k vs $\mu^{1/2}$ is linear and slope is either + ve or – ve depending upon the sign of the charge on the ions.

19.16. GENERAL PRINCIPLES OF CATALYSIS

A *catalyst* is a substance that increases the rate of a reaction without itself being consumed by the process. A reaction in which a catalyst is involved is called *catalysed reaction* and the process is called *catalysis*. Following characteristics are observed in the catalysed reaction.

(*i*) A catalyst lowers the Gibbs energy of activation by providing a different mechanism for the reaction. This mechanism enhances the rate and it applies to both the forward and reverse directions of the reaction.

(*ii*) A catalyst forms an intermediate with the reactant(s) in the initial step of the mechanism and is released in the product-forming step. The catalyst does not appear in the overall reaction.

(*iii*) Regardless of the mechanism and the energetic of a reaction, a catalyst cannot affect the enthalpies or Gibbs energies of the reactants and products. Thus, catalysts increase the rate of approach to the equilibrium, but cannot alter the thermodynamic equilibrium constant.

19.16.1. *Enzyme Catalysis*

Before we discuss the kinetics of enzyme catalysis, let us try to understand few concepts. Most of the enzymes are proteins. An enzyme usually contains one or more active sites where reactions with substrates take place. Fischer proposed a lock and key theory of *enzyme specificity*. According to Fischer the active site in the protein is rigid, similar to a lock. A substrate molecule then has a complimentary structure and functions as a key. Thus, it is known as lock and key mechanism. Later modifications have been suggested in this mechanism, called phenomenon of*cooperativity*. Cooperativity means the binding of a substrate to an enzyme with multiple binding sites can alter the substarte's affinity for binding at its other sites.

As mentioned above that reaction rate increases in enzyme kinetics like any other catalysed reactions. Reaction based on transition state theory, rate constant can be written as

$$k = \frac{\kappa T}{h} e^{-\Delta G^{\dagger}/RT} (M^{1-m}) = \frac{\kappa T}{h} e^{\Delta S^{\dagger}/R} e^{-\Delta H^{\dagger} RT} (M^{1-m}) \qquad ...(19.76)$$

In the above expression (19.76), there are two terms, entropy of activation and enthalpy of activation. The enthalpy of activation is nearly equal to the activation energy (E_a) in the Arrhenius equation. It is known that activation energy decreases in any catalytic reaction and thus would increase the rate constant. It may be mentioned here that this is not the only explanation for the decrease in the rate constant. Entropy of activation also plays a major role in determining the efficiency of enzyme catalysis. Let us compare the enzyme kinetics with bimolecular reactions, *i.e.*

$$A + B \rightarrow AB^{\dagger} \rightarrow \text{Product} \qquad ...(19.77)$$

Now each species *A* and *B*, in the bimolecular reaction, contains three degrees of translational freedom, three degrees of rotational freedom and rest of them are vibrational degrees of freedom. At 298 K, contribution to the entropy of activation due to translational motion is ~120 J K^{-1} mol^{-1}, followed by the rotational motion is 80 J K^{-1} mol^{-1} and nearly 15 J K^{-1} mol^{-1}. When the activated complex is formed, there will be decrease in the three degrees of translational motion and three degrees of rotational degrees of freedom. In other words, there will be a decrease of nearly 200 J K^{-1} mol^{-1} when the activated complex is formed. This loss in entropy is compensated for a small extent by new modes of internal rotation and vibrational motion in the activated complex. When a comparison is made between the theoretical calculations applied to uni-molecular reactions and bimolecular reactions, shows a difference of as much as 3×10^{10} in the $e^{\Delta S^{\dagger}/R}$ term, favouring the unimolecular reaction.

19.16.2. *Enzyme Kinetics: Michaelis-Menton Kinetics*

Michaelis-Menton enzyme kinetics involves the following steps:

$$E + S \underset{k_{-1}}{\overset{k_1}{\rightleftharpoons}} ES \xrightarrow{k_2} P + S \qquad ...(19.78)$$

where E is the enzyme concentration, S stands for substrate concentration, ES is the activated complex formed between E and S, which decomposes to enzyme and product with rate constant k_2. Thus enzyme is regenerated again for fresh participation in the reaction. That is substrate combine with enzyme with a rate constant k_1 in a reversible manner (k_{-1}) to form the complex. The initial rate of formation for the products is given as:

$$\text{v}_0 = \left(\frac{d[P]}{dt}\right)_0 = k_2[ES] \qquad ...(19.79)$$

As usual to depict the rate in terms of stable species, Michaelis-Menton assumed that $k_{-1} \gg k_2$. Due to this the first step {formation of $[ES]$} can be treated as the establishment of a fast equilibrium process. Equilibrium constant K_S for the dissociation process can be written as:

$$K_S = \frac{k_{-1}}{k_1} = \frac{[E][S]}{[ES]} \qquad ..(19.80)$$

The total enzyme concentration $[E]$ can be written as

$$[E]_0 = [E] + [ES] \qquad ...(19.81)$$

Thus,
$$K_S = \frac{([E]_0 - [ES])[S]}{[ES]} \qquad ...(19.82)$$

After rearranging equation (19.82), $[ES]$ can be written as

$$[ES] = \frac{[E]_0[S]}{K_S + [S]} \qquad ...(19.83)$$

Substituting the value of $[ES]$ in equation (19.79), we get

$$\text{v}_0 = \frac{k_2[E]_0[S]}{K_S + [S]} \qquad ...(19.84)$$

Equation (19.84) can be treated in different manners. Let $a = k_2 [E]_0$ and when $K_S \gg [S]$ (*i.e.* at low substrate concentration) equation (19.84) can be written as

$$\text{v}_0 = \frac{k_2[E]_0[S]}{K_S} = \frac{a[S]}{K_S} \qquad ...(19.85)$$

At high substrate concentration $[S] \gg [E]_0$, equation (19.84) can be written as

$$\text{v}_0 = k_2[E]_0 \qquad ...(19.79)$$

Equation (19.79) shows that under these conditions: (*i*) the initial rate is independent of substrate concentration, all the enzyme sites will be saturated or covered will substrate and thus reaction rate becomes independent. (*ii*) The measured initial rate will be maximum (V_{max}) under these conditions, thus

$$V_{max} = k_2[E]_0 \qquad ...(19.79a)$$

(*iii*) When $[S] = K_S$, equation (19.84) reduces to

$$v_0 = \frac{k_2[E]_0[S]}{[S] + [S]} = \frac{k_2[E]_0}{2} = \frac{k_2V_{max}}{2}$$

i.e. when [*S*] equals to K_S, initial rate is half of the maximum initial rate.

***Exercise* 19.39:** Derive the Michaelis-Menton mechanism using steady state hypothesis.

[*ES*] concentration can be derived by using steady state concentration, *i.e.*

$$\frac{d[ES]}{dt} = 0 = k_1[E][S] - k_{-1}[ES] - k_2[ES]$$

or $$= k_1([E]_0 - [S])[S] - k_{-1}[ES] - k_2[ES]$$

After rearranging, [*ES*] can be written as

$$[ES] = \frac{k_1[E]_0[S]}{k_1[S] + k_{-1} + k_2}$$

Substituting the value of [*ES*] in equation (19.79), we get

$$v_0 = k_2[ES] = \frac{k_2k_1[E]_0[S]}{k_1[S] + k_{-1} + k_2} = \frac{k_2[E]_0[S]}{[S] + \dfrac{k_{-1} + k_2}{k_1}} = \frac{k_2[E]_0[S]}{[S] + K_M} \qquad ...(19.86)$$

where $K_M = \dfrac{k_{-1} + k_2}{k_1}$, is called *Michaelis constant*. Comparing equation (19.79) with (19.86), we see that both expressions show the similar dependence on substrate, but $K_S \neq K_M$ unless $k_{-1} \gg k_2$. Using $V_{max} = k_2[E]_0$ equation (19.86), we get

$$v_0 = \frac{V_{max}[S]}{[S] + K_M} \qquad ...(19.87)$$

Analysis of equation (19.87) can be carried out by taking inverse as

$$\frac{1}{v_0} = \frac{1}{V_{max}} + \frac{K_M}{V_{max}[S]} \qquad ...(19.88)$$

Plot of $1/v_0$ *versus* $1/[S]$ gives a straight line. From intercept we can get V_{max} and combining intercept and slope we can get K_M. Similarly equation (19.88) can be arranged by multiplying it with $v_0 V_{max}$, we get

$$V_{max} = v_0 + \frac{v_0K_M}{[S]} \qquad ...(19.89)$$

or $$v_0 = V_M - \frac{v_0K_M}{[S]} \qquad ...(19.89a)$$

Plot of v_0 versus $v_0/[S]$ gives a straight line and plot is called Eadie-Hofstee plot. Slope gives $v_0 K_M$ and intercept as V_M.

PROBLEMS

19.1 In the following hypothetical reactions the differential rate laws are also written. Discuss the nature of each species

(*a*) $A + 2B \rightarrow C + D, \; -\dfrac{d[A]}{dt} = k\,[A]\,[B]$

(*b*) $A + B + C \rightarrow P_1 + P_2 + C, \; -\dfrac{d[A]}{dt} = k\,[A]\,[B]$

(*c*) $2N_2O_5 \rightarrow 2N_2O_4 + O_2, \; -\dfrac{d[N_2O_2]}{dt} = k\,[N_2O_5]$

(*d*) $A + B \rightarrow 2C, \; \dfrac{d[C]}{dt} = k\,[A]\,[B]^{-1}$

(*e*) $2A + B \rightarrow C + D, \; \dfrac{d[D]}{dt} = k\,[A]^2\,[B]\,[D]$

19.2 The second-order rate constant of a reaction is 2.25×10^7 dm^8 mol^{-1} s^{-1} and initial concentrations of the reacting species are 6.0×10^{-4} mol dm^{-3}. What is the initial rate in mol dm^{-3} s^{-1}, in mol cc^{-1} s^{-1} and molecules cc^{-1} s^{-1}?

19.3 The ClO radicals decay by second-order reaction. If the initial concentration is 2.5×10^{-5} mol dm^{-3}, calculate its (*a*) first half-life, (*b*) second half-life and (*c*) concentration of ClO radicals after 4 m. Given $k = 2.25 \times 10^7$ dm^3 mol^{-1} s^{-1}.

19.4 The data of a chemical reaction is plotted as $1/[C]$ vs time and the plot is a straight line. If intercept is 2×10^3 mol^{-1} dm^3 and slope 2×10^{-2} mol^{-1} dm^3 s^{-1}, calculate the half-life of the reaction.

19.5 Consider the reaction $2A + B \rightarrow C + D$. The rate law follows the equation

$$-\frac{1}{2}\frac{d[A]}{dt} = k\,[A]\,[B]$$

The value of k is 4.5×10^{-3} mol^{-1} dm^3 s^{-1}. If the ratio of the initial concentration of A to B is 60 : 40, and total pressure is 1 atm, calculate the concentration of A after 20 min.

19.6 Consider the problem 19.5 and calculate the half-life time for the following concentrations:

(*a*) $[A]_0 = 1.0$ M, $[B]_0 = 0.8$ M

(*b*) $[A]_0 = 0.4$ M, $[B]_0 = 2.0$ M

(*c*) $[A]_0 = 0.4$ M, $[B]_0 = 0.8$ M

19.7 A certain reaction follows the following differential rate law

$$-\frac{d[C]}{dt} = k\,[C]^{3/2}$$

(*a*) Integrate the above equation if initial concentration is $[C]_0$.

(*b*) How would you check the data graphically that rate law follows 3/2 order reaction?

(*c*) Calculate the half-life time.

(*d*) What will be the unit of rate constant?

19.8 The unimolecular decomposition of benzene diazonium chloride is followed by measuring the pressure of nitrogen gas evolved.

(*a*) Reduce the equation 19.8 (*c*) into the form

$$k = \frac{1}{t}\ln\frac{p}{p - p_t}$$

where p is the pressure of total nitrogen evolved and p_t is the pressure of nitrogen at any time

(*b*) calculate the rate constant k from the following data :

$t(s)$	0	240	720	1200	2040	3000	∞
p_t (torr)	0	2.15	5.89	9.00	13.09	16.28	22.62

19.9 Hydrolysis of sucrose in the presence of fixed concentration of H^+ ions (0.14 mol dm^{-3}) follows a pseudo first-order kinetics ($k_1 = k_{bi}[H^+]$). Reaction is followed by measuring the optical rotation (θ) and data obtained is

time(s)	0	120	540	1500	3600	∞
θ (degrees)	32.4	32.2	31.7	30.8	28.9	1.7

Calculate the pseudo first-order rate constant as well as the second-order rate constant (k_{bi}).

19.10 In the hydrolysis of ethyl acetate by NaOH, the course of reaction is followed by titrating NaOH with

$$CH_3COOC_2H_5 + OH^- \rightarrow CH_3COO^- + C_2H_5OH$$

standard acid. If x_0 is the volume of the standard acid needed to titrate OH^- at the beginning of the experiment, x_t is the volume after t sec and x_∞ that equivalent to the excess alkali remaining when all ester has been hydrolysed, then equation (19.9) can be written as

$$k = \frac{2.303}{tx_\infty}\log\frac{x_t(x_0 - x_\infty)}{x_0(x_t - x_\infty)}$$

(*a*) What would you require to convert the above results in unit of $dm^3\ mol^{-1}\ s^{-1}$?

(*b*) Calculate k if following data is available :

$$[\text{Acid}] = 0.04155\ \text{N}, \quad V = 0.100\ dm^3$$

t(s)	0	224	377	629	816	∞
x_i (dm^{-3})	0.062 $(x)_0$	0.0543	0.0506	0.0463	0.0439	0.0331

19.11 The reaction between propionaldehyde and hydrocyanic acid has been studied at 25°C by W.J. Svirbely and J.F. Roth (*J. Amer. Chem. Soc.* 75, 3106 (1953)). The data recorded were as follows

Time (min)	0	3.28	11.12	24.43	40.35	67.22
HCN (mol dm^{-3})		0.0657	0.0619	0.0569	0.0515	0.0463
C_3H_7CHO (mol dm^{-3})		0.0384	0.0346	0.0296	0.0242	0.0190

What is the order of the reaction and the value of the rate constant k? $[HCN]_0 - [C_3H_7CHO]_0 =$ 0.042 mol dm^{-3}.

19.12 In a certain reaction its half life doubles if the initial concentration doubles. What is the order of reaction?

19.13 In the reaction between NO and H_2, the following data is obtained:

Experiment I, p_{H_2} = Constant

p_{NO} (mm)	359	300	152
$-\frac{dp}{dt}$ (mm.s^{-1})	1.50	1.03	0.25

Experiment II, p_{NO} = C onstant

p_{H_2}	289	205	147
$-\frac{dp}{dt}$	1.60	1.10	0.79

Find the order of reaction with respect to H_2 and NO. Also the overall order of reaction.

19.14 Calculate the rate constant k' in the problems 19.8 and 19.9 with the help of Gugenheim's method.

19.15 (*a*) In an experiment, the reaction $A + 2B \rightarrow C + D$ takes place. If the reaction is followed by measuring the rate of formation of D at various original concentrations of A and B, calculate the order of the reaction (*i*) with respect to A (*ii*) with respect of B and (*iii*) total order from the following data:

Rate	Concentration (mol dm^{-3})	
(mol dm^{-3} s^{-1})	[A]	[B]
0.0032	0.10	0.10
0.0096	0.30	0.10
0.0064	0.10	0.20

(*b*) Which of the following mechanisms agrees with the rate law?

(*i*) $A + B \rightarrow E$ (slow) (*ii*) $2A \rightarrow A_2$ (fast)

$E + B \rightarrow C + D$ (fast) $A_2 + B \rightarrow C + D$ (slow)

19.16 The following mechanisms are proposed for the reaction $CO + NO_2 \rightarrow CO_2 + NO$ at low temperature:

(*i*) $$2NO_2 \underset{k_2}{\overset{k_1}{\rightleftharpoons}} N_2O_4 \text{ (fast)}$$

$$N_2O_4 + 2CO \xrightarrow{k_3} 2CO_2 + 2NO \text{ (slow)}$$

(*ii*) $$2NO_2 \xrightarrow{k_1'} NO_3 + NO \text{ (slow)}$$

$$NO_3 + CO \xrightarrow{k_2'} NO_2 + CO_2 \text{ (fast)}$$

Which of the above mechanisms are consistent with the observed rate law?

$$\frac{d[CO_2]}{dt} = k[NO_2]^2$$

19.17 Derive the relation between the relaxation time and the rate constants for the reaction $A \underset{k_2}{\overset{k_1}{\rightleftharpoons}} B$, which is displaced from the equilibrium position by a small amount?

19.18 In a temperature jump experiment on pure water, the relaxation time τ obtained at 25°C is 36μ s. Calculate k_1 and k_2 for the reaction

$$H_2O \underset{k_2}{\overset{k_1}{\rightleftharpoons}} H^+ + OH^-$$

19.19 Derive the relation between the relaxation time and the rate constants k_1 and k_2 for the reactions

$$A + B \underset{k_2}{\overset{k_1}{\rightleftharpoons}} C + D.$$

19.20 Reaction $NOCl + nBuOH \underset{k_2}{\overset{k_1}{\rightleftharpoons}} nBuONO + HCl$ was studied with the help of temperature jump method at 5°C. The following data is obtained. $k_1 = 0.055\times 10^3$ dm^3 mol^{-1} s^{-1}, $K = 0.1$, $[NOCl]_e = 8.1 \times 10^{-2}$ mol dm^{-3} and $[nBuOH]_e = 1.8 \times 10^{-1}$ mol dm^{-3}. Calculate k_2 and τ. (relaxation time for the reaction)

19.21 In the competing reactions

$$A + \text{Reagent} \xrightarrow{k_1} \text{Product}$$

$$B + \text{Reagent} \xrightarrow{k_2} \text{Product}$$

Calculate (*a*) ratio of k_1/k_2 if only 50% of *B* will have been reacted when 94% of *A* has been reacted, (*b*) If the ratio of k_1/k_2 is 100 and 98% of *A* has reacted, find the % of *B* consumed.

19.22 In the oxidation of H_2 at intermediate pressure, hydrogen atom can react with HO_2 radical in three different ways.

$$H + HO_2 \xrightarrow{k_1} H_2 + O_2$$

$$H + HO_2 \xrightarrow{k_2} 2OH$$

$$H + HO_2 \xrightarrow{k_3} H_2O + O$$

After certain time t, the ratio of the products formed are 0.62 : 0.54 : 0.11. Calculate the ratio of the rate constants k_1, k_2 and k_3.

19.23 Bi undergoes β emission (99.96%) and α emission (0.04%). If the half-life is 19.7 min find k_α and k_β.

19.24 Decomposition of HI was studied at 410°C and the data is expressed as the amount of HI decomposed. If the initial concentration of HI is 4.46×10^{-2} mol dm^{-3} and equilibrium concentration 0.938×10^{-2} mol dm^{-3}, calculate k_1 and k_2 from the following data:

$$2HI \underset{k_2}{\overset{k_1}{\rightleftharpoons}} H_2 + I_2$$

t (min)	50	100	150	200
[HI] × 10^3 decomposed (mol dm^{-3})	2.79	4.95	6.16	7.37

19.25 The muta rotation of π-bromonitrocamphor at 285 K was followed by measuring the optical rotation. If the initial angle of rotation (α_0) and angle of rotation at equilibrium position (α_e) are 189° and 31.3° respectively, calculate $k_1 + k_2$ (sum of forward and backward reaction) from the following data:

t (hrs)	3	5	7	24	72
α_i (°)	169.0	156.0`	146.0	84.5	37.3

19.26 The reaction between acetone (A) and HCN (B) can be written as $A + B \underset{k_2}{\overset{k_1}{\rightleftharpoons}} C$. Derive the expression for k in terms of concentration of A and B. Now $[A]_0 = 0.1164$ mol dm^{-3}, $[B]_0 = 0.0758$ mol dm^{-3}, $[A]_e = 0.0772$ mol dm^{-3}, $[B]_e = 0.0366$ mol dm^{-3}. If the concentration of $C(x)$ after 73.23 min is 0.0711 mol dm^{-3}, calculate k_1 and k_2. (W.J. Svirbely and J.F. Roth, *J. Amer. Chem. Soc*, **75**, 3106, 1953.)

19.27 Prove the relations (19.46) and (19.47).

(*a*) Plot the concentration of *A, B, C* as a function of time for $k_1 = 1$ s^{-1}, $k_2 = 0.25$ s^{-1}. Assume $[B]_0 = [C]_0 = 0$ initially.

(*b*) On the same graph, plot the concentration of *A, B, C* for $k_1 = 1$ s^{-1}, and $k_2 = 2 \times 10^2$ s^{-1}. Does this problem tell anything about when steady state approximation is justified?

(*c*) At what time is the concentration of *B* maximised in each of the cases?

19.28 Decomposition of N_2O_5 can be explained by the following mechanism:

$$N_2O_5 \xrightarrow{k_1} NO_2 + NO_3$$

$$NO_2 + NO_3 \xrightarrow{k_2} N_2O_5$$

$$NO_2 + NO_3 \xrightarrow{k_3} NO + O_2 + NO_2$$

$$NO + NO_3 \xrightarrow{k_4} 2NO_2$$

Using steady state approximation for NO and NO_3, show that rate of decomposition of N_2O_5 follows first-order kinetics in N_2O_5. Calculate the effective first-order rate constant.

19.29 Lindemann's mechanism for the unimolecular reactions is

$$A + A \xrightarrow{k_1} A^* + A$$

$$A^* + A \xrightarrow{k_2} A + A$$

$$A^* \xrightarrow{k_3} \text{Products}$$

Calculate the rate of formation of products if A^* is a reactive intermediate. Under what conditions is the rate first order with respect to reactant A?

19.30 Simple Rice-Herzfeld mechanism for the decomposition of acetaldehyde is

$$CH_3CHO \xrightarrow{k_1} CH_3 + CHO$$

$$CHO \xrightarrow{2} CO + H$$

$$H + CH_3CHO \xrightarrow{3} H_2 + CH_3CO$$

$$CH_3 + CH_3CHO \xrightarrow{4} CH_4 + CH_3CO$$

$$CH_3CO \xrightarrow{5} CH_3 + CO$$

$$2CH_3 \xrightarrow{6} C_2H_6$$

If the rate of first step is neglected then the rate of decomposition of acetaldehyde follows 3/2 order. Find the effective rate constant.

19.31 A reaction has the following mechanism:

$$A + B \underset{k_2}{\overset{k_1}{\rightleftharpoons}} C + D$$

$$C + B \xrightarrow{k_3} E + D$$

$$2D \xrightarrow{k_4} F$$

where A and B are reactants, C and D are reactive intermediates, E and F are products. If $r_2 >> r_3$ and using steady state approximation, find (*a*) rate of reaction in terms of A and B, (*b*) partial order with respect to A and B and (*c*) the total order of the reaction.

19.32 The rate of certain reaction at 50°C is three times that at 15°C. Calculate the activation energy. If the heat of the reaction is 50 kJ mol^{-1}, draw the energy *vs* reaction coordinate diagram.

19.33 Westenberg and DeHass (*J. Chem. Phys.*, **50,** 2, 1966) found that rate constant k for the reaction varies with temperature as

$$O + CH_4 \rightarrow CH_3 + OH$$

T (K)	297	363	419	516	605	904
k (dm^3 mol^{-1}s^{-1})	1.26×10^4	9.2×10^4	3.4×10^5	2.5×10^6	1.0×10^7	1.3×10^8

Calculate the activation energy and pre-exponential factor.

19.34 The half-life for a reaction between fixed concentration of reactants varies with temperature as follows:

T (°C)	520	533	555	574	604	620
$t_{1/2}$ (s^{-1})	1288	813	562	477	89.1	49

Calculate the activation energy of this reaction.

19.35 A bottle of milk stored at 30°C sours in 36 hours, stored in a refrigerator at 5°C sours after one week. Assuming the rate constant to be inversely related to the souring time, estimate the activation energy of the chemical reaction involved in souring reaction.

19.36 There are two second-order reactions with rate constants k_1 and k_2 and $E_{a(1)} > E_{a(2)}$, where E_a is the activation energy. The temperature of both the systems is raised from T_1 to T_2. Which of the following relations is correct?

(*a*) $\dfrac{k_1(T_2)}{k_1(T_1)} = \dfrac{k_2(T_2)}{k_2(T_1)}$, (*b*) $\dfrac{k_1(T_2)}{k_1(T_1)} > \dfrac{k_3(T_2)}{k_2(T_1)}$

(*c*) $\dfrac{k_1(T_2)}{k_1(T_1)} < \dfrac{k_2(T_2)}{k_2(T_1)}$, (*d*) none of the above.

19.37 The molecular diameters of H_2 and I_2 are 0.273 nm and 0.5 nm. Calculate the collision frequency of the reaction

$$H_2 + I_2 \rightarrow 2HI$$

at 700 K and 1 atm (1.013×10^5 N m^{-2}). If the activation energy is 170 kJ mol^{-1}, find the rate constant.

19.38 Calculate the rate of reaction, frequency factor and rate constant at 700 K and 1 atm (1.013×10^5 N m^{-2}) for the decomposition of hydrogen iodide if its collision diameter is assumed to 0.35 nm. The activation energy is 184 kJ mol^{-1}.

19.39 The activation energy for a bimolecular reaction $A + B_2 \rightarrow AB + B$ is 20.90 kJ mol^{-1}. What are the temperatures at which 0.15, 0.25 and 0.40% of hard sphere collisions between reactant molecules occur with sufficient relative kinetic energy along the line of centres?

19.40 Calculate the rate constants for the following reactions at 300 K and 1 atm (1.103×10^5 N m^{-2}) for the following reactions:

1. $ClO + ClO \rightarrow Cl_2 + O_2$
2. $2NO_2 \rightarrow 2NO + O_2$
3. $NO + Cl_2 \rightarrow NOCl + Cl$

The hard sphere diameters for the above reactants are:
ClO(0.18 nm), NO_2(0.19 nm), NO(0.14 nm) and Cl_2(0.24 nm) whereas activation energies for the above reactions are 10.45 kJ mol^{-1}, 111.19 kJ mol^{-1}, 84.86 kJ mol^{-1}.

19.41 Absolute rate theory relates the dependence of frequency factor on temperature ($A \propto T^n$), calculate the value of exponent '*n*' for the following reaction

	Reactant		Reactant		Activated complex
	A	+	*B*		*AB*
1.	Atom	+	Atom	$\rightarrow$	Linear

2. Atom + Linear molecule → Linear
3. Atom + Non-linear molecule → Non-linear
4. Linear molecule + Linear molecule → Non-linear

19.42 If it is assumed that the values of translational, rotational and vibrational partition functions for different systems are equal, calculate the value of pre exponential factor for each reaction of problem 19.41 if the value of $kT/h = 10^{13}\ s^{-1}$. Calculate the stearic factor (p) if the collision theory gives the value for A as $10^{12}\ dm^3\ mol^{-3}\ s^{-1}$. Given $q_{trans} \approx 10^8$, $q_{rot} \approx 10$ and $q_{vib} \approx 1$.

19.43 Arrhenius theory assumed that frequency factor is constant, independent of temperature, whereas Collision theory predicts temperature dependence of frequency factor. Show that for a typical reaction, exponential term overweights the change observed in$A$ due to the neglect of temperature. Collision diameter $\sigma = 0.4$ nm, $M = 150$ g mol^{-1} and $E_a = 100$ kJ mol^{-1}. Consider temperature range 200°C to 400°C.

19.44 The entropy of activation observed for the reaction at 300° K is 8 J K^{-1} mol^{-1}. Calculate the frequency factor $H_2 + I_2 \rightarrow 2HI$ and also the rate constant at 300 K if activation energy is 170 kJ mol^{-1}. Compare the value with that obtained in problem 19.37.

19.45 Entropy of activation observed for the reactions at 300 K are 71 J K^{-1} and 46 J K^{-1} mol^{-1} respectively.

(*i*) $CH_3N_2CH_3 \rightarrow C_2H_6 + N_2$

(*ii*) 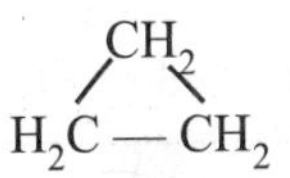 $\rightarrow CH_3CH = CH_2$

Calculate the values of A in each case. Explain the higher value obtained.

19.46 In the ClO radical decay at high pressure the four centred complex has been assumed by Herchbach et al. (*J. Chem. Phys.*, **25**, 736, 1956). The following properties are assigned to the complex, $g = 1$, $I_1 \times I_2 \times I_3 = 2.20 \times 10^{-114}$ cgs units, v's are 1500, 800, 700, 600 and 200 cm^{-1} whereas those of ClO radicals are: $g = 2$, $I = 4.3 \times 10^{-39}$ cgs units, $v = 800$ cm^{-1}. Calculate the frequency factor and compare the results with that obtained in 19.40.

19.47 The collision diameter σ_A for methyl radical recombination is 0.45 nm. If the formation of singlet state only leads to the formation of ethane ($\kappa = 1/4$), calculate the rate constant for the recombination of methyl radicals at 400 K and $E_a = 0$.

19.48 The recombination reaction generally leads to negative entropy of activation whereas dissociation of unimolecular reactions positive. Explain.

19.49 The reaction of hydrogen atom with Cl_2 has a second order rate constant 2.11×10^{11} mol^{-1} dm^3 s^{-1}. If ΔH and ΔS of this reaction are; 188.86 kJ and 14.205 J K^{-1} respectively, calculate the Arrhenius parameters for the reverse reaction.

19.50 A container is having a pure gas at 500°C and a pressure of 0.5 atm. If the molecular diameter is 0.2 nm, calculate the rate of bimolecular collision. If each collision is effective in carrying the chemical reaction, calculate the half-life of the reaction.

19.51 The Arrhenius equations for the cis-trans isomerisation of 2 butene ($CH_3CH = CHCH_3$) and 2 butene nitrile ($CH_3CH = CHCN$) are

$$k(s^{-1}) = 10^{13.8} \exp(-263.5 \text{ kJ mol}^{-1}/RT)$$

and

$$k^{-1}(s^{-1}) = 10^{11} \exp(-214.5 \text{ kJ mol}^{-1}/RT)$$

Calculate the temperature at which $k = k^{-1}$.

19.52 The conversion of compound $A \rightarrow B$ is a first order process and requires activation energy of 60 kJ mol^{-1}. 20% solution of A was heated at 25°C and 20% decomposition was complete in 20 minutes. How much decomposition will be achieved with a 30% solution at 40°C?

19.53 The rate constant for a certain unimolecular reaction is 2.5×10^{-2} s^{-1} at 500°C. If the activation energy is 68 kJ mol^{-1}, calculate the entropy of activation at this temperature.

19.54 The reaction between NO and H_2 is

$$NO + H_2 \rightarrow 1/2N_2 + H_2O$$

If the rate constant at 827°C is 145.5 dm^6 mol^{-2} s^{-1} with activation energy 184 kJ mol^{-1}, calculate the entropy of activation at this temperature.

19.55 A unimolecular reaction is carried out at 127°C. If the activation energy is 150 kJ mol^{-1} and f = 11, calculate the frequency factor for this reaction using Hinshelwood's theory provided $Z_2 = 10^{12}$ dm^3 mol^{-1} s^{-1}.

19.56 In a chemical reaction $Br + H_2 \rightarrow HBr + H$ a linear activated complex is assumed with following characteristics: H–H distance 0.092 nm, H–Br distance 0.159 nm, v's = 1074, 1325 cm^{-1}, g = 2, calculate (a) the moment of inertia of this complex and rotational partition function, (b) vibrational partition function, (c) total partition function of the activated complex, (d) calculate the total partition function for the reactants Br and H_2 taking H–H distance as 0.074 nm, q_{Br} = 4 and fundamental frequency of H_2 and 4395 cm^{-1}, and (e) calculate the value of rate constant if the activation energy of this reaction is 82.5 kJ mol^{-1}.

19.57 Following data is obtained in the reaction of HCN with propionaldehyde:

μ	0.0342	0.0234	0.09	0.2	0.01	0.509
(min^{-1} dm^3 mol^{-1})k	1.23	1.23	1.24	1.25	0.628	0.655

Calculate k_0.

Additional Problems

19.58 Derive equations (19.21) and (19.22).

19.59 Derive equations (19.8c), (19.11) and (19.14).

19.60 Derive equations (19.16b), (19.18) and (19.20).

19.61 Write the differential rate equations in terms of disappearance of reactants and formation of products for the following reactions:

(i) $2O_3 \rightarrow 3O_2$

(ii) $C_2H_4 + H_2 \rightarrow C_2H_6$

(iii) $C_4H_8 + C_2H_4 \rightarrow C_6H_{10}$

(iv) $2AsH_3 \rightarrow 2As + 3H_2$

19.62 In the following first-order reaction: $A \rightarrow B + C$, the half-life of the first order rate law is 45 s. Calculate the fraction of A left in the reaction mixture after 5 min.

19.63 Half-life of the first-order decay of ^{14}C is 5720 years. (i) Calculate the rate constant for the reaction. (ii) The natural abundance of ^{14}C in a given sample was 1.5×10^{-13} mol%. If the mol% ^{14}C at present left in the sample is 0.85×10^{-14}, calculate the age of the sample.

19.64 ^{32}P is used to study the certain aspect of protein metabolism. If the protein sample initially has activity of 10^5 disintegration min^{-1}, calculate the activity after 150 hrs. The half-life time of ^{32}P is 14.28 days.

19.65 The radioactive decay of uranium-238 follows: $^{238}U \rightarrow {}^{236}Pb$. The half-life of this decay is 4.51×10^9 yrs. A sample was analysed for the presence of uranium and lead. It was found that the sample contains 1.25 mg of ^{238}U and 0.452 mg of ^{206}Pb. Calculate the age of the sample assuming that ^{206}Pb is formed only from the decay of uranium and ^{206}Pb does not decay further.

19.66 $^{40}K_{19}$ undergoes radioactive decay by two paths. One of the paths giving argon is used in Geology to date sedimentary rocks. ^{40}K decays by the following two paths are:

$$^{40}K_{19} \rightarrow {}^{40}Ca_{20} + {}^{0}e_{-1} \qquad ...(89.3\%)$$

$$^{40}K_{19} \rightarrow {}^{40}Ar_{18} + {}^{0}e_{-1} \qquad ...(10.73\%)$$

The overall half-life for the decay of ^{40}K is 1.3×109 yrs. Estimate the age of the sedimentary rocks with a ^{40}Ar to ^{40}K ratio of 0.1012.

19.67 The half-life time of the reaction increases from 4 min to 16 min, when the initial concentration of the sample decreases from 20 M to 5 M. Calculate the order of the reaction and its rate constant.

19.68 Scientists working in radioactive materials commonly assume that these materials become harmless after 10 half lives. Calculate the fraction of radioactive material left in the sample after this time.

19.69 Many reactions involving heterogeneous reactions are found to be zero order reactions if the pressure of the gas is greater than 1.0 atm. The rate of the reaction is found to be equal to rate constant. Explain this behaviour.

19.70 The following reaction proceeds slowly in aqueous medium, *i.e.*

$$S_2O_8^{2-} + 2I^- \rightarrow 2SO_4^{2-} + I_2$$

Fe^{3+} ion is found to catalyse this reaction. Given that Fe^{3+} can oxidise I^- and Fe^{2+} can reduce $S_2O_8^{2-}$, write a possible two step mechanism for this reaction. Explain why the uncatalysed reaction?

19.71 In the following reaction: $A(g) + 2B(g) \rightarrow C(g) + 2D(g)$; initial amount of A is very large and amount of B follows the data given below. Calculate the order of reaction with respect to both A and B and as well as the rate constant of this reaction.

Time(s)	0	15	45	75	105	135	195	255
p_B (torr)	1.76	1.31	0.864	0.645	0.515	0.428	0.32	0.257

19.72 In continuity to the opposing reaction discussed in exercise 19.25, consider the following opposing reaction as:

$$A + B = \underset{k_2}{\overset{k_1}{\rightleftharpoons}} P$$

Calculate the values of k_1, k_2 and K.

19.73 In continuity to exercise 19.23, consider only two parallel reactions as

$$A \xrightarrow{k_1} P_1$$

$$A \xrightarrow{k_2} P_2$$

If E_a is the overall activation energy for the reaction of A, it is related to other parameters by the equation as

$$E_a = \frac{k_1 E_1 + k_2 E_2}{k_1 + k_2}$$

where E_1 and E_2 are the activation energies for the reaction 1 and 2 respectively.

19.74 Excited ozone (O_3^*) undergoes following parallel reactions:

$$O_3^* \xrightarrow{k_1} O_3 \text{ (Fluorescence)}$$

$$O_3^* \xrightarrow{k_2} O + O_2 \text{ (Decomposition)}$$

$$O_3^* + M \xrightarrow{k_3} O_3 + M \text{ (Deactivation)}$$

where M is an inert species. Calculate the fraction of exited ozone molecules undergoing deactivation process in terms if rate constants.

19.75 Shullivan showed that the only bimolecular reaction, $H_2 + I_2 \rightarrow 2HI$, also proceeds by two steps mechanism as

$$I_2 \rightleftharpoons 2I$$

$$H_2 + 2I \rightarrow 2HI$$

It is found that rate of formation of HI increases with the intensity of visible radiation. (*i*) Explain why this fact supports the two step mechanism. (*ii*) Explain why the visible light has no effect on the formation of H atoms. (**Hint:** The colour of I_2 vapour is purple, whereas the H_2 is colourless gases).

19.76 In recent times chlorofluorocarbons (CFCs) has not been used as refrigerator materials. It is because it is consuming the ozone layer in the stratosphere. Any leak of CFCs molecule such as $CFCl_3$ is first decomposed by the UV radiations. The chlorine atom then reacts with ozone and reaction scheme is given as:

$$CFCl_3 \rightarrow {}^{\bullet}CFCl_2 + Cl$$

$$Cl + O_3 \rightarrow ClO + O_2$$

$$ClO + O \rightarrow Cl + O_2$$

(*i*) Write the overall reaction for the last two steps,

(*ii*) What are the roles of Cl and ClO?

(*iii*) Why is the fluorine atom is not used in removing the ozone?

19.77 The energy of the activation for the reaction: $2HI \rightarrow H_2 + I_2$ is 180 kJ mol^{-1} at 283° C and the collision diameter for HI is 0.35 nm. Calculate the rate constant of this reaction at this temperature assuming the pressure is 1 atm.

19.78 A mixture of two reactants are present in a container and both undergo first order decomposition reactions. If the half-life times of the two reactants are 25 min and 45 min respectively, calculate time required that the concentration of A to be half of B. Given $[A]_0 = [B]_0$.

19.79 Before we try to solve the radioactivity problem, let us define certain thing, *i.e.* activity. Activity of radioactive sample is defined as the number of nuclear disintegrations per second, which is equal to the first order rate constant times the number of radioactive nuclei presents in the sample. The fundamental unit of radioactivity is curie (Ci), where 1 Ci corresponds to 3.70×10^{10} disintegrations per second and this decay rate is equivalent to 1 g of ^{226}Ra. Calculate the rate constant and half-life for the radium decay. Starting with 1.0 g of radium sample, what is the activity after 500 years? The molar mass of ^{226}Ra is 226.03 g mol^{-1}.

19.80. Consider the following parallel reactions:

$$A \xrightarrow{K_1} B$$

$$A \xrightarrow{K_2} C$$

The activation energies for the steps (*i*) and (*ii*) are 55.0 kJ mol^{-1} and 65.6 kJ mol^{-1}. If the rate constants are equal at 350 K, calculate the value of temperature at which $k_1/k_2 = 3$.

19.81 The rate of electron exchange reaction between naphthalene and its anion $[C_{10}H_8]^-$ at different temperatures are given below:

$$C_{10}H_8 + C_{10}H_8^- \rightleftharpoons C_{10}H_8^- + C_{10}H_8$$

T (K)	273	289	299	307
$k(10^9\ M^{-1}\ s^{-1})$	1.43	1.96	2.40	2.71

(*i*) Calculate the values of E_a, $\Delta H^{\dagger 0}$, $\Delta S^{\dagger 0}$ and $\Delta G^{\dagger 0}$ at 315 K.

(**Source:** Physical chemistry for Biosciences, Richmond Chang.)

19.82 The thermal decomposition of an organic compound was studied by monitoring the progress of the reaction using absorbance data, as given below. Calculate the rate constant of the reaction by assuming that reaction follows first order kinetics and absorbance of the products does not interfere with that of the reactants.

t (*s*)	0	2000	4000	6000	8000	10000	12000	∞
Absorbance (*A*)	1.50	1.26	1.07	0.92	0.81	0.72	0.65	0.40

19.83 Pre-exponential factor and activation energy for the uni-molecular reaction of

$$CH_3NC(g) \rightarrow CH_3CN(g)$$

are 4.0×10^{13} s^{-1} and 272 kJ mol^{-1}. Calculate the values of $\Delta_{\ddagger}^{0}H, \Delta_{\ddagger}^{0}S$ and $\Delta_{\ddagger}^{0}G$.

19.84 In catalysed reactions, both rates of forward and backward reactions decrease. Explain.

19.85 Derive the following relation from equation (19.88):

$$\frac{v_0}{[S]} = \frac{V_{max}}{K_M} - \frac{v_0}{K_M}.$$

19.86 Hydrolysis of compound *A* is catalysed by some enzyme. If the turnover rate number is 25500, calculate how long it will take for the enzyme to leave one molecule of *A*.

19.87 In the above kind of enzyme catalytic reaction, an enzyme has K_M value of 4.5×10^{-5} M. Initial concentration of substrate is 3×10^{-3} M and after 2 minutes it is found that the amount of product formed is 5×10^{-5} M. Calculate the value of V_{max} and the amount of product formed after 6.0 min.

19.88 K_M value for lysozyme is 6.0×10^{-6} M with hexa-N-acetylglucosamine as a substrate. It is assayed at the following substrate concentration: (*i*) 2.5×10^{-7} M, (*ii*) 4.8×10^{-5} M, (*iii*) 6.5×10^{-3} M, (*iv*) 0.061 M. The initial rate measured at 0.061 M was 3.2×10^{-6} M min^{-1}. Calculate the initial rates at the other substrate concentrations.

19.89 The hydrolysis of N-glutaryl-L-phenylalanine-*p*-nitroanilide (GPNA) to *p*-nitroaniline and N-glutaryl-*L*-phenylalanine is catalysed by α-chymotrypsin. The following data are given below:

$[S]$ (10^{-4} M)	2.5	5.0	10.0	15.0
v_0 (10^{-6} M min^{-1})	2.2	3.8	5.9	7.1

$[S]$ = GPNA. Assuming Michaelis-Menton kinetics, calculate the values of V_{max}, K_M and k_2. Another method to treat these data is to plot v_0 versus $v_0/[S]$, which is the Eadie-Hofstee plot. Calculate the values of V_{max}, K_M and k_2 from Eadie-Hofstee treatment, given that $[E]_0 = 4.0 \times 10^{-6}$ *M*. (This problem has been taken from J. A. Hurlbut, T. N. Ball, H. C. Pound and J.L. Graves, *J. Chem. Edu*., 50, 149, 1973).

19.90 The initial rates at various substrate concentrations for an enzyme-catalysed reaction are as follows:

$[S]$ ($\times 10^{-5}$ M)	2.5	4.00	6.00	8.00	16.0	20.0
v_0 ($\times 10^{-6}$ M min^{-1})	38.0	53.4	68.6	80.0	106.8	114.0

(*i*) Does this reaction follow Michaelis-Menton kinetics? (*ii*) Calculate the value of V_{max} of the reaction, (*iii*) Calculate K_M value of the reaction, (*iv*) calculate the initial rates at $[S] = 5.0 \times 10^{-5}$ M and $[S] = 3.00 \times 10^{-1}$ M, (*v*) what is the total amount of product formed during the first 4 min at $[S] = 10.6 \times 10^{-5}$ M? How would an increase in the enzyme concentration by a factor of 2 affect each of the following quantities

K_M, V_{max} and v_0 at $[S] = 5.0 \times 10^{-5}$ M.

(**Source:** Physical Chemistry for the Bioscience, Raymond Chang.)

19.91 Consider the enzyme kinetics with different mechanism as

$$E + S \underset{k_2}{\overset{k_1}{\rightleftharpoons}} ES \underset{k_4}{\overset{k_3}{\rightleftharpoons}} E + P$$

Derive the rate equation v.

19.92 Consider enzyme kinetics involving an inhibitor I and the mechanism involved is

$$E + S \underset{k_{-1}}{\overset{k_1}{\rightleftharpoons}} ES$$

$$E + I \underset{k_{-2}}{\overset{k_2}{\rightleftharpoons}} EI$$

$$ES \xrightarrow{k_3} E + P$$

Derive the rate equation v.

19.93 Consider the hydrogenation of ethylene as given below:

$$C_2H_4\,(g) + H_2\,(g) \rightarrow C_2H_6\,(g)$$

If the diameters of H_2 and C_2H_4 are 0.270 nm and 0.470 nm respectively, calculate the rate constant of this reaction considering hard sphere model at 298 K in units of cm^3 mole^{-1} s^{-1} and dm^3 mol^{-1} s^{-1}.

19.94 In continuation to the problem 19.93 and using the value of *k* obtained therein, it is found that the experimental value of *k* obtained is 3.49×10^{26} dm^3 mol^{-1} s^{-1} for this reaction. Calculate the value of E_a using equation (19.55) and equation (19.59).

Appendix-I
Character Tables of Molecular Symmetry Groups

1. The Non-axial Groups

C_1	E
A	1

C_s	E	σ_h		
A′	1	1	$T_x, T_y; R_z,$	$\alpha_{xx}, \alpha_{yy}, \alpha_{zz}, \alpha_{xy}$
A″	1	–1	$T_z; R_x, R_y,$	α_{yz}, α_{zx}

C_i	E	i		
A_g	1	1	R	α
A_u	1	–1	T	

2. The C_n Groups

C_2	E	C_2		
A	1	1	T_z, R_z	$\alpha_{xx}, \alpha_{yy}, \alpha_{zz}, \alpha_{xy}$
B	1	–1	$T_x, T_y; R_x, R_y$	α_{yz}, α_{zx}

C_3	E	C_3	C_3^2		$\varepsilon = exp\left(\frac{2\pi i}{3}\right)$
A	1	1	1	$T_z; R_z$	$\alpha_{xx} + \alpha_{yy}, \alpha_{zz}$
E	$\begin{cases} 1 \\ 1 \end{cases}$	$\begin{matrix} \varepsilon \\ \varepsilon^* \end{matrix}$	$\begin{matrix} \varepsilon^* \\ \varepsilon \end{matrix}$	$(T_x, T_y); (R_x, R_y)$	$(\alpha_{xx} - \alpha_{yy}, \alpha_{xy}), (\alpha_{yz}, \alpha_{zx})$

C_4	E	C_4	C_2	C_4^3		
A	1	1	1	1	T_z, R_z	$\alpha_{xx}+\alpha_{yy}$, α_{zz}
B	1	−1	1	−1		$\alpha_{xx}-\alpha_{yy}$, α_{xy}
E	$\begin{cases}1\\1\end{cases}$	i $-i$	−1 −1	$-i$ i	(T_x, T_y); (R_x, R_y)	$(\alpha_{yz}, \alpha_{zx})$

C_5	E	C_5^1	C_5^2	C_5^3	C_5^4		$\varepsilon = \exp\left(\frac{2\pi i}{5}\right)$
A	1	1	1	1	1	T_z, R_z	$\alpha_{xx}+\alpha_{yy}$, α_{zz}
E_1	$\begin{cases}1\\1\end{cases}$	ε ε^*	ε^2 ε^{2*}	ε^{2*} ε^2	ε^* ε	(T_x, T_y); (R_x, R_y)	$(\alpha_{yz}, \alpha_{zx})$
E_2	$\begin{cases}1\\1\end{cases}$	ε^2 ε^{2*}	ε^* ε	ε ε^*	ε^{2*} ε^2		$(\alpha_{xx}-\alpha_{yy}, \alpha_{xy})$

C_6	E	C_6^1	C_3	C_2	C_3^2	C_6^5		$\varepsilon = \exp\left(\frac{2\pi i}{6}\right)$
A	1	1	1	1	1	1	T_z, R_z	$\alpha_{xx}+\alpha_{yy}$, α_{zz}
B	1	−1	1	−1	1	−1		
E_1	$\begin{cases}1\\1\end{cases}$	ε ε^*	$-\varepsilon^*$ $-\varepsilon$	−1 −1	$-\varepsilon$ $-\varepsilon^*$	ε^* ε	(T_x, T_y); (R_x, R_y)	$(\alpha_{yz}, \alpha_{zx})$
E_2	$\begin{cases}1\\1\end{cases}$	$-\varepsilon^*$ $-\varepsilon$	$-\varepsilon$ $-\varepsilon^*$	1 1	$-\varepsilon^*$ $-\varepsilon$	ε ε^*		$(\alpha_{xx}-\alpha_{yy}, \alpha_{xy})$

C_7	E	C_7^1	C_7^2	C_7^3	C_7^4	C_7^5	C_7^6		$\varepsilon = \exp\left(\frac{2\pi i}{7}\right)$
A	1	1	1	1	1	1	1	T_z, R_z	$\alpha_{xx}+\alpha_{yy}$, α_{zz}
E_1	$\begin{cases}1\\1\end{cases}$	ε ε^*	ε^2 ε^{2*}	ε^3 ε^{3*}	ε^{3*} ε^3	ε^{2*} ε^2	ε^* ε	(T_x, T_y); (R_x, R_y)	$(\alpha_{xz}, \alpha_{yz})$
E_2	$\begin{cases}1\\1\end{cases}$	ε^2 ε^{2*}	ε^{3*} ε^3	ε^* ε	ε ε^*	ε^3 ε^{3*}	ε^{2*} ε^2		$(\alpha_{xx}-\alpha_{yy}, \alpha_{xy})$
E_3	$\begin{cases}1\\1\end{cases}$	ε^3 ε^{3*}	ε^* ε	ε^2 ε^{2*}	ε^{2*} ε^2	ε ε^*	ε^{3*} ε^3		

C_8	E	C_8^1	C_4	C_2	C_4^3	C_8^3	C_8^5	C_8^7		$\varepsilon = \exp\left(\frac{2\pi i}{8}\right)$
A	1	1	1	1	1	1	1	1	T_z, R_z	$\alpha_{xx} + \alpha_{yy}$, α_{zz}
B	1	–1	1	1	1	–1	–1	–1		
E_1	$\begin{cases}1\\1\end{cases}$	ε / ε^*	i / $-i$	–1 / –1	$-i$ / i	$-\varepsilon^*$ / $-\varepsilon$	$-\varepsilon$ / $-\varepsilon^*$	ε^* / ε	(T_x, R_y) / (R_x, R_y)	α_{xz}, α_{yz}
E_2	$\begin{cases}1\\1\end{cases}$	i / $-i$	–1 / –1	1 / 1	–1 / –1	$-i$ / i	i / $-i$	$-i$ / i		$\alpha_{xx} - \alpha_{yy}$, α_{xy}
E_3	$\begin{cases}1\\1\end{cases}$	$-\varepsilon$ / $-\varepsilon^*$	i / $-i$	–1 / –1	$-i$ / i	ε^* / ε	ε / ε^*	$-\varepsilon^*$ / $-\varepsilon$		

3. The D_n Groups

D_2	E	$C_2(z)$	$C_2(y)$	$C_2(x)$		
A	1	1	1	1		α_{xx}, α_{yy}, α_{zz}
B_1	1	1	–1	–1	T_z, R_z	α_{xy}
B_2	1	–1	1	–1	T_y, R_y	α_{zx}
B_3	1	–1	–1	1	T_x, R_x	α_{yz}

D_3	E	$2C_3$	$3C_2$		
A_1	1	1	1		$\alpha_{xx} + \alpha_{yy}$, α_{zz}
A_2	1	1	–1	T_z; R_z	
E	2	–1	0	$(T_x; T_y)$; $(R_x; R_y)$	$(\alpha_{xx} - \alpha_{yy}, \alpha_{xy})$; $(\alpha_{yz}, \alpha_{zx})$

D_4	E	$2C_4$	$C_4^2 = C_2$	$2C_2'$	$2C_2''$		
A_1	1	1	1	1	1		$\alpha_{xx} + \alpha_{yy}$, α_{zz}
A_2	1	1	1	–1	–1	T_z, R_z	
B_1	1	–1	1	1	–1		$\alpha_{xx} - \alpha_{yy}$
B_2	1	–1	1	–1	1		α_{xy}
E	2	0	–2	0	0	$(T_x; T_y)$; $(R_x; R_y)$	$(\alpha_{yz}, \alpha_{zx})$

D_5	E	$2C_5$	$2C_5^2$	$5C_2$		
A_1	1	1	1	1		$\alpha_{xx} + \alpha_{yy}$, α_{zz}
A_2	1	1	1	–1	T_z; R_z	
E_1	2	2 cos 72°	2 cos 144°	0	(T_x, T_y); (R_x, R_y)	$(\alpha_{yz}, \alpha_{zx})$
E_2	2	2 cos 144°	2 cos 72°	0		$(\alpha_{xx} - \alpha_{yy}, \alpha_{xy})$

D_6	E	$2C_6$	$2C_3$	C_2	$3C'_2$	$3C''_2$		
A_1	1	1	1	1	1	1		$\alpha_{xx} + \alpha_{yy}, \alpha_{zz}$
A_2	1	1	1	1	–1	–1	$T_z; R_z$	
B_1	1	–1	1	–1	1	–1		
B_2	1	–1	1	–1	–1	1		
E_1	2	1	–1	–2	0	0	$(T_x, T_y); (R_x, R_y)$	$(\alpha_{yz}, \alpha_{zx})$
E_2	2	–1	–1	2	0	0		$(\alpha_{xx} - \alpha_{yy}, \alpha_{xy})$

4. The C_{nv} Groups

C_{2v}	E	$C_2(z)$	$\sigma'_v(zx)$	$\sigma_v(yz)$		
A_1	1	1	1	1	T_z	$\alpha_{xx}, \alpha_{yy}, \alpha_{zz}$
A_2	1	1	–1	–1	R_z	α_{xy}
B_1	1	–1	1	–1	$T_x; R_y$	α_{zx}
B_2	1	–1	–1	1	$T_y; R_x$	α_{yz}

C_{3v}	E	$2C_3$	$3\sigma_v$		
A_1	1	1	1	T_z	$\alpha_{xx} + \alpha_{yy}, \alpha_{zz}$
A_2	1	1	–1	R_z	
E	2	–1	0	$(T_x, T_y); (R_x, R_y)$	$(\alpha_{xx} - \alpha_{yy}, \alpha_{xy}), (\alpha_{yz}, \alpha_{zx})$

C_{4v}	E	$2C_4$	$C_4^2 = C_2$	$2\sigma_v$	$2\sigma_d$		
A_1	1	1	1	1	1	T_z	$\alpha_{xx} + \alpha_{yy}, \alpha_{zz}$
A_2	1	1	1	–1	–1	R_z	
B_1	1	–1	1	1	–1		$\alpha_{xx} - \alpha_{yy}$
B_2	1	–1	1	–1	1		α_{xy}
E	2	0	–2	0	0	$(T_x, T_y); (R_x, R_y)$	$(\alpha_{yz}, \alpha_{zx})$

C_{5v}	E	$2C_5$	$2C_5^2$	$5\sigma_v$		
A_1	1	1	1	1	T_z	$\alpha_{xx} + \alpha_{yy}, \alpha_{zz}$
A_2	1	1	1	–1	R_z	
E_1	2	2 cos 72°	2 cos 144°	0	$(T_x, T_y); (R_x, R_y)$	$(\alpha_{yz}, \alpha_{zx})$
E_2	2	2 cos 144°	2 cos 72°	0		$(\alpha_{xx} - \alpha_{yy}, \alpha_{xy})$

C_{6v}	E	$2C_6$	$2C_3$	C_2	$3\sigma_v$	$3\sigma_d$		
A_1	1	1	1	1	1	1	T_z	$\alpha_{xx} + \alpha_{yy}, \alpha_{zz}$
A_2	1	1	1	1	–1	–1	R_z	
B_1	1	–1	1	–1	1	–1		
B_2	1	–1	1	–1	–1	1		
E_1	2	1	–1	–2	0	0	$(T_x, T_y); (R_x, R_y)$	$(\alpha_{yz}, \alpha_{zx})$
E_2	2	–1	–1	2	0	0		$(\alpha_{xx} - \alpha_{yy}, \alpha_{xy})$

C_{1h}	E	σ_h		
A′	1	1	R_z, T_x, T_y	(α_{xx}, α_{yy}, α_{zz}, α_{xy})
A″	1	–1	R_x, R_y, T_z	(α_{yz}, α_{zx})

C_{2h}	E	C_2	i	σ_h		
A_g	1	1	1	1	R_z	α_{xx}, α_{yy}, α_{zz}, α_{xy}
B_g	1	–1	1	–1	R_x, R_y	α_{yz}, α_{zx}
A_u	1	1	–1	–1	T_z	
B_u	1	–1	–1	1	T_x, T_y	

C_{3h}	E	C_3	C_3^2	σ_h	S_3	S_3^5		$\varepsilon = exp\left(\frac{2\pi i}{3}\right)$
A′	1	1	1	1	1	1	R_z	$\alpha_{xx} + \alpha_{yy}$, α_{zz}
E′	{1	ε	ε*	1	ε	ε*	(T_x, T_y)	($\alpha_{xx} - \alpha_{yy}$, α_{xy})
	{1	ε*	ε	1	ε*	ε		
A″	1	1	1	–1	–1	–1	T_z	
E″	{1	ε	ε*	–1	–ε	–ε*	(R_x, R_y)	(α_{yz}, α_{zx})
	{1	ε*	ε	–1	–ε*	–ε		

C_{4h}	E	C_4	C_2	C_4^3	i	S_4^3	σ_h	S_4		
A_g	1	1	1	1	1	1	1	1	R_z	$\alpha_{xx} + \alpha_{yy}$, α_{zz}
B_g	1	–1	1	1	1	–1	1	–1		$\alpha_{xx} - \alpha_{yy}$, α_{xy}
E_g	{1	i	–1	–i	1	i	–1	–i	(R_x, R_y)	α_{yz}, α_{zx}
	{1	–i	–1	i	1	–i	–1	i		
A_u	1	1	1	1	–1	–1	–1	–1	T_z	
B_u	1	–1	1	–1	–1	1	–1	1		
E_u	{1	i	–1	–i	–1	–i	1	i	(T_x, T_y)	
	{1	–i	–1	i	–1	i	1	–i		

C_{5h}	E	C_5	C_5^2	C_5^3	C_5^4	σ_h	S_5	S_5^7	S_5^3	S_5^9		$\varepsilon = exp\left(\frac{2\pi i}{5}\right)$
A′	1	1	1	1	1	1	1	1	1	1	R_z	$\alpha_{xx}+\alpha_{yy}$, α_{zz}
E'_1	1	ε	ε^2	ε^{2*}	ε^*	1	ε	ε^2	ε^{2*}	ε^*	(T_x, T_y)	
	1	ε^*	ε^{2*}	ε^2	ε	1	ε^*	ε^{2*}	ε^2	ε		
E'_2	1	ε^2	ε^*	ε	ε^{2*}	1	ε^2	ε^*	ε	ε^{2*}		$(\alpha_{xx}-\alpha_{yy}, \alpha_{xy})$
	1	ε^{2*}	ε	ε^*	ε^2	1	ε^{2*}	ε	ε^*	ε^2		
A″	1	1	1	1	1	−1	−1	−1	−1	−1	T_z	
E''_1	1	ε	ε^2	ε^{2*}	ε^*	−1	$-\varepsilon$	$-\varepsilon^2$	$-\varepsilon^{2*}$	$-\varepsilon^*$	(R_x, R_y)	α_{yz}, α_{zx}
	1	ε^*	ε^{2*}	ε^2	ε	−1	$-\varepsilon^*$	$-\varepsilon^{2*}$	$-\varepsilon^2$	$-\varepsilon$		
E''_2	1	ε^2	ε^*	ε	ε^{2*}	−1	$-\varepsilon^2$	$-\varepsilon^*$	$-\varepsilon$	$-\varepsilon^{2*}$		
	1	ε^{2*}	ε	ε^*	ε^2	−1	$-\varepsilon^{2*}$	$-\varepsilon$	$-\varepsilon^*$	$-\varepsilon^2$		

C_{6h}	E	C_6	C_3	C_2	C_3^2	C_6^5	i	S_3^5	S_6^5	σ_h	S_6	S_3		$\varepsilon = exp\left(\frac{2\pi i}{6}\right)$
A_g	1	1	1	1	1	1	1	1	1	1	1	1	R_z	$\alpha_{xx}+\alpha_{yy}$, α_{zz}
B_g	1	−1	1	−1	1	−1	1	−1	1	−1	1	−1		
E_{1g}	1	ε	$-\varepsilon^*$	−1	$-\varepsilon$	ε^*	1	ε	$-\varepsilon^*$	−1	$-\varepsilon$	ε^*	R_x, R_y	$(\alpha_{yz}, \alpha_{zx})$
	1	ε^*	$-\varepsilon$	−1	$-\varepsilon^*$	ε	1	ε^*	$-\varepsilon$	−1	$-\varepsilon^*$	ε		
E_{2g}	1	$-\varepsilon^*$	$-\varepsilon$	1	$-\varepsilon^*$	$-\varepsilon$	1	$-\varepsilon^*$	$-\varepsilon$	1	$-\varepsilon^*$	$-\varepsilon$		$(\alpha_{xx}-\alpha_{yy}, \alpha_{xy})$
	1	$-\varepsilon$	$-\varepsilon^*$	1	$-\varepsilon$	$-\varepsilon^*$	1	$-\varepsilon$	$-\varepsilon^*$	1	$-\varepsilon$	ε^*		
A_u	1	1	1	1	1	1	−1	−1	−1	−1	−1	−1	T_z	
B_u	1	−1	1	−1	1	−1	−1	1	−1	1	−1	1		
E_{1u}	1	ε	$-\varepsilon^*$	−1	$-\varepsilon$	ε^*	−1	$-\varepsilon^*$	$-\varepsilon$	1	ε	$-\varepsilon^*$	(T_x, T_y)	
	1	ε^*	$-\varepsilon$	−1	$-\varepsilon^*$	ε	−1	ε	$-\varepsilon^*$	1	ε^*	$-\varepsilon^*$		
E_{2u}	1	$-\varepsilon^*$	$-\varepsilon$	1	$-\varepsilon^*$	$-\varepsilon$	−1	ε^*	ε	−1	ε^*	ε		
	1	$-\varepsilon$	$-\varepsilon^*$	1	$-\varepsilon$	$-\varepsilon^*$	−1	ε	ε^*	−1	ε	ε^*		

5. The D_{nh} Groups

D_{2h}	E	$C_2(z)$	$C_2(y)$	$C_2(x)$	i	$\sigma(xy)$	$\sigma(zx)$	$\sigma(yz)$		
A_g	1	1	1	1	1	1	1	1		α_{xx}, α_{yy}, α_{zz}
B_{1g}	1	1	−1	−1	1	1	−1	−1	R_z	α_{xy}
B_{2g}	1	−1	1	−1	1	−1	1	−1	R_y	α_{zx}
B_{3g}	1	−1	−1	1	1	−1	−1	1	R_x	α_{yz}
A_u	1	1	1	1	−1	−1	−1	−1		
B_{1u}	1	1	−1	−1	−1	−1	1	1	T_z	
B_{2u}	1	−1	1	−1	−1	1	−1	1	T_y	
B_{3u}	1	−1	−1	1	−1	1	1	−1	T_x	

D_{3h}	E	$2C_3$	$3C_2$	σ_h	$2S_3$	$3\sigma_v$		
A'_1	1	1	1	1	1	1		$\alpha_{xx}+\alpha_{yy}, \alpha_{zz}$
A'_2	1	1	–1	1	1	–1	R_z	
E'	2	–1	0	2	–1	0	(T_x, T_y)	$(\alpha_{xx}-\alpha_{yy}, \alpha_{xy})$
A''_1	1	1	1	–1	–1	–1		
A''_2	1	1	–1	–1	–1	1	T_z	
E''	2	–1	0	–2	1	0	(R_x, R_y)	$(\alpha_{yz}, \alpha_{zx})$

D_{4h}	E	$2C_4$	C_2	$2C'_2$	$2C''_2$	i	$2S_4$	σ_h	$2\sigma_v$	$2\sigma_d$		
A_{1g}	1	1	1	1	1	1	1	1	1	1		$\alpha_{xx}+\alpha_{yy}, \alpha_{zz}$
A_{2g}	1	1	1	–1	–1	1	1	1	–1	–1	R_z	
B_{1g}	1	–1	1	1	–1	1	–1	1	1	–1		$\alpha_{xx}-\alpha_{yy}$
B_{2g}	1	–1	1	–1	1	1	–1	1	–1	1		α_{xy}
E_g	2	0	–2	0	0	2	0	–2	0	0	(R_x, R_y)	$(\alpha_{yz}, \alpha_{zx})$
A_{1u}	1	1	1	1	1	–1	–1	–1	–1	–1		
A_{2u}	1	1	1	–1	–1	–1	–1	–1	1	1	T_z	
B_{1u}	1	–1	1	1	–1	–1	1	–1	–1	1		
B_{2u}	1	–1	1	–1	1	–1	1	–1	1	–1		
E_u	2	0	–2	0	0	–2	0	2	0	0	(T_x, T_y)	

D_{5h}	E	$2C_5$	$2C_5^2$	$5C_2$	σ_h	$2S_5$	$2S_5^3$	$5\sigma_v$		
A'_1	1	1	1	1	1	1	1	1		$\alpha_{xx}+\alpha_{yy}, \alpha_{zz}$
A'_2	1	1	1	–1	1	1	1	–1	R_z	
E'_1	2	2 cos 72°	2 cos 144°	0	2	2 cos 72°	2 cos 144°	0	(T_x, T_y)	
E'_2	2	2 cos 144°	2 cos 72°	0	2	2 cos 144°	2 cos 72°	0		$(\alpha_{xx}, -\alpha_{yy}, \alpha_{xy})$
A''_1	1	1	1	1	–1	–1	–1	–1		
A''_2	1	1	1	–1	–1	–1	–1	1	T_z	
E''_1	2	2 cos 72°	2 cos 144°	0	–2	–2 cos 72°	–2 cos 144°	0	(R_x, R_y)	$(\alpha_{yz}, \alpha_{zx})$
E''_2	2	2 cos 144°	2 cos 72°	0	–2	–2 cos 144°	–2 cos 72°	0		

D_{6h}	E	$2C_6$	$2C_3$	C_2	$3C'_2$	$3C''_2$	i	$2S_3$	$2S_6$	σ_h	$3\sigma_d$	$3\sigma_v$		
A_{1g}	1	1	1	1	1	1	1	1	1	1	1	1		$\alpha_{xx}+\alpha_{yy}, \alpha_{zz}$
A_{2g}	1	1	1	1	–1	–1	1	1	1	1	–1	–1	R_z	
B_{1g}	1	–1	1	–1	1	–1	1	–1	1	–1	1	–1		
B_{2g}	1	–1	1	–1	–1	1	1	–1	1	–1	–1	1		
E_{1g}	2	1	–1	–2	0	0	2	1	–1	–2	0	0	(R_x, R_y)	$(\alpha_{yz}, \alpha_{zx})$
E_{2g}	2	–1	–1	2	0	0	2	–1	–1	2	0	0		$(\alpha_{xx}-\alpha_{yy}, \alpha_{xy})$
A_{1u}	1	1	1	1	1	1	–1	–1	–1	–1	–1	–1		
A_{2u}	1	1	1	1	–1	–1	–1	–1	–1	–1	1	1	T_z	
B_{1u}	1	–1	1	–1	1	–1	–1	1	–1	1	–1	1		
B_{2u}	1	–1	1	–1	–1	1	–1	1	–1	1	1	–1		
E_{1u}	2	1	–1	–2	0	0	–2	–1	1	2	0	0	(T_x, T_y)	
E_{2u}	2	–1	–1	2	0	0	–2	1	1	–2	0	0		

D_{8h}	E	$2C_8$	$2C_8^3$	$2C_4$	C_2	$4C'_2$	$4C''_2$	i	$2S_8$	$2S_8^3$	$2S_4$	σ_h	$4\sigma_d$	$4\sigma_v$		
A_{1g}	1	1	1	1	1	1	1	1	1	1	1	1	1	1		$\alpha_{xx}+\alpha_{yy}, \alpha_{zz}$
B_{1g}	1	–1	–1	1	1	1	–1	1	–1	–1	1	1	1	–1		
B_{2g}	1	–1	–1	1	1	–1	1	1	–1	–1	1	1	–1	1		
E_{1g}	2	$\sqrt{2}$	$-\sqrt{2}$	0	–2	0	0	2	$\sqrt{2}$	$-\sqrt{2}$	0	–2	0	0	(R_x, R_y)	$(\alpha_{xz}, \alpha_{yz})$
E_{2g}	2	0	0	–2	2	0	0	2	0	0	–2	2	0	0		$(\alpha_{xx} -$
E_{3g}	2	$-\sqrt{2}$	$\sqrt{2}$	0	–2	0	0	2	$-\sqrt{2}$	$\sqrt{2}$	0	–2	0	0		$\alpha_{yy}, \alpha_{xy})$
A_{1u}	1	1	1	1	1	1	1	–1	–1	–1	–1	–1	–1	–1		
A_{2u}	1	1	1	1	1	–1	–1	–1	–1	–1	–1	–1	1	1	T_z	
B_{1u}	1	–1	–1	1	1	1	–1	–1	1	1	–1	–1	–1	1		
B_{2u}	1	–1	–1	1	1	–1	1	–1	1	1	–1	–1	1	–1		
E_{1u}	2	$\sqrt{2}$	$-\sqrt{2}$	0	–2	0	0	–2	$-\sqrt{2}$	$\sqrt{2}$	0	2	0	0	(T_x, T_y)	
E_{2u}	2	0	0	–2	2	0	0	–2	0	0	2	–2	0	0		
E_{3u}	2	$-\sqrt{2}$	$\sqrt{2}$	0	–2	0	0	–2	$\sqrt{2}$	$-\sqrt{2}$	0	2	0	0		

6. The D_{nd} Groups

D_{2d}	E	$2S_4$	C_2	$2C'_2$	$2\sigma_d$		
A_1	1	1	1	1	1		$\alpha_{xx}+\alpha_{yy}, \alpha_{zz}$
A_2	1	1	1	–1	–1	R_z	
B_1	1	–1	1	1	–1		$\alpha_{xx}-\alpha_{yy}$
B_2	1	–1	1	–1	1	T_z	α_{xy}
E	2	0	–2	0	0	(T_x, T_y); (R_x, R_y)	$(\alpha_{yz}, \alpha_{zx})$

D_{3d}	E	$2C_3$	$3C_2$	i	$2S_6$	$3\sigma_d$		
A_{1g}	1	1	1	1	1	1		$\alpha_{xx}+\alpha_{yy}, \alpha_{zz}$
A_{2g}	1	1	–1	1	1	–1	R_z	$(\alpha_{xx}-\alpha_{yy}; \alpha_{xy})$
E_g	2	–1	0	2	–1	0	(R_x, R_y)	$(\alpha_{yz}, \alpha_{zx})$
A_{1u}	1	1	1	–1	–1	–1		
A_{2u}	1	1	–1	–1	–1	1	T_z	
E_u	2	–1	0	–2	1	0	(T_x, T_y)	

D_{4d}	E	$2S_8$	$2C_4$	$2S_8^3$	C_2	$4C'_2$	$4\sigma_d$		
A_1	1	1	1	1	1	1	1		$\alpha_{xx}+\alpha_{yy}, \alpha_{zz}$
A_2	1	1	1	1	1	–1	–1	R_z	
B_1	1	–1	1	–1	1	1	–1		
B_2	1	–1	1	–1	1	–1	1	T_2	
E_1	2	$\sqrt{2}$	0	$-\sqrt{2}$	–2	0	0	(T_x, T_y)	
E_2	2	0	–2	0	2	0	0		$(\alpha_{xx}-\alpha_{yy}, \alpha_{xy})$
E_3	2	$-\sqrt{2}$	0	$\sqrt{2}$	–2	0	0	(R_x, R_y)	$(\alpha_{xz}, \alpha_{yz})$

D_{5d}	E	$2C_5$	$2C_5^2$	$5C_2$	i	$2S_{10}^3$	$2S_{10}$	$5\sigma_d$		
A_{1g}	1	1	1	1	1	1	1	1		$\alpha_{xx}+\alpha_{yy}, \alpha_{zz}$
A_{2g}	1	1	1	–1	1	1	1	–1	R_z	
E_{1g}	2	2 cos 72°	2 cos 144°	0	2	2 cos 72°	2 cos 144°	0	(R_x, R_y)	$(\alpha_{zx}, \alpha_{yz})$
E_{2g}	2	2 cos 144°	2 cos 72°	0	2	2 cos 144°	2 cos 72°	0		$(\alpha_{xx}-\alpha_{yy}, \alpha_{xy})$
A_{1u}	1	1	1	1	–1	–1	–1	–1		
A_{2u}	1	1	1	–1	–1	–1	–1	1	T_z	
E_{1u}	2	2 cos 72°	2 cos 144°	0	–2	–2 cos 72°	–2 cos 144°	0	(T_x, T_y)	
E_{2u}	2	2 cos 144°	2 cos 72°	0	–2	–2 cos 144°	–2 cos 72°	0		

D_{6d}	E	$2S_{12}$	$2C_6$	$2S_4$	$2C_3$	$2S_{12}^5$	C_2	$6C_2'$	$6\sigma_d$		
A_1	1	1	1	1	1	1	1	1	1		$\alpha_{xx}+\alpha_{yy}, \alpha_{zz}$
A_2	1	1	1	1	1	1	1	–1	–1	R_z	
B_1	1	–1	1	–1	1	–1	1	1	–1		
B_2	1	–1	1	–1	1	–1	1	–1	1	T_z	
E_1	2	$\sqrt{3}$	1	0	–1	$-\sqrt{3}$	–2	0	0	(T_x, T_y)	
E_2	2	1	–1	–2	–1	1	2	0	0		$(\alpha_{xx}-\alpha_{yy}, \alpha_{xy})$
E_3	2	0	–2	0	2	0	–2	0	0		
E_4	2	–1	–1	2	–1	–1	2	0	0		
E_5	2	$-\sqrt{3}$	1	0	–1	$\sqrt{3}$	–2	0	0	(R_x, R_y)	$(\alpha_{yz}, \alpha_{zx})$

7. The S_n Groups

S_4	E	S_4	C_2	S_4^3		
A	1	1	1	1	R_z	$\alpha_{xx}+\alpha_{yy}, \alpha_{zz}$
B	1	–1	1	–1	T_z	$\alpha_{xx}-\alpha_{yy}, \alpha_{xy}$
E	$\begin{cases}1\\1\end{cases}$	i $-i$	–1 –1	$-i$ i	(T_x, T_y); (R_z, R_y)	$(\alpha_{yz}+\alpha_{zx})$

S_6	E	C_3	C_3^2	i	S_6^5	S_6		$S_6 = C_3 \times i,\ \varepsilon = exp\left(\frac{2\pi i}{3}\right)$
A_g	1	1	1	1	1	1	R_z	$\alpha_{xx}+\alpha_{yy}, \alpha_{zz}$
E_g	$\begin{cases}1\\1\end{cases}$	ε ε^*	ε^* ε	1 1	ε ε^*	ε^* ε	(R_x, R_y)	$(\alpha_{xx}-\alpha_{yy}, \alpha_{xy})$; $(\alpha_{yz}, \alpha_{zx})$
A_u	1	1	1	–1	–1	1	T_z	
E_u	$\begin{cases}1\\1\end{cases}$	ε ε^*	ε^* ε	–1 –1	$-\varepsilon$ $-\varepsilon^*$	$-\varepsilon^*$ $-\varepsilon$	(T_x, T_y)	

S_8	E	S_8	C_4	S_8^3	C_2	S_8^5	C_4^3	S_8^7		$\varepsilon = exp\left(\frac{2\pi i}{8}\right)$
A	1	1	1	1	1	1	1	1	R_z	$\alpha_{xx} + \alpha_{yy}, \alpha_{zz}$
B	1	–1	1	–1	1	–1	1	–1	T_z	
E_1	1	ε	i	$-\varepsilon^*$	–1	$-\varepsilon$	$-i$	ε^*	(T_x, T_y);	
	1	ε^*	$-i$	$-\varepsilon$	–1	$-\varepsilon^*$	i	ε	R_x, R_y	
E_2	1	i	–1	$-i$	1	i	–1	$-i$		
	1	$-i$	–1	i	1	$-i$	–1	i		$(\alpha_{xx} - \alpha_{yy}, \alpha_{xy})$
E_3	1	$-\varepsilon^*$	$-i$	ε	–1	ε^*	i	$-\varepsilon$		
	1	$-\varepsilon$	i	ε^*	–1	ε	$-i$	$-\varepsilon^*$		$(\alpha_{zx}, \alpha_{yz})$

8. The Cubic Groups

T	E	$4C_3$	$4C_3^2$	$3C_2$		$\varepsilon = exp\left(\frac{2\pi i}{3}\right)$
A	1	1	1	1		$\alpha_{xx} + \alpha_{yy} + \alpha_{zz}$
E	1	ε	ε^*	1		$(\alpha_{xx} + \alpha_{yy} - 2\alpha_{zz}, \alpha_{xx} - \alpha_{yy})$
	1	ε^*	ε	1		
T	3	0	0	–1	T, R	$\alpha_{xy}, \alpha_{yz}, \alpha_{zx}$

$T_h = T \times i$; T in F_u; R in F_g; α in A_g, E_g, F_g

T_d	E	$8C_3$	$3C_2$	$6S_4$	$6\sigma_d$		
O	E	$8C_3$	$3C_2$	$6C_4$	$6C_2'$		
A_1	1	1	1	1	1		$\alpha_{xx} + \alpha_{yy} + \alpha_{zz}$
A_2	1	1	1	–1	–1		
E	2	–1	2	0	0		$(\alpha_{xx} + \alpha_{yy} - 2\alpha_{zz}, \alpha_{xx} - \alpha_{yy})$
T_1	3	0	–1	1	–1	R; T in O	
T_2	3	0	–1	–1	1	T in T_d	$(\alpha_{xy}, \alpha_{yz}, \alpha_{zx})$

T_h	E	$4C_3$	$4C_3^2$	$3C_2$	i	$4S_6$	$4S_6^5$	$3\sigma_h$		$\varepsilon = exp\left(\frac{2\pi i}{3}\right)$
A_g	1	1	1	1	1	1	1	1		$\alpha_{xx} + \alpha_{yy} + \alpha_{zz}$
E_g	1	ε	ε^2	1	1	ε	ε^2	1		$(\alpha_{xx} + \alpha_{yy} - 2\alpha_{zz}, \alpha_{xx} - \alpha_{yy})$
	1	ε^2	ε	1	1	ε^2	ε	1		
T_g	3	0	0	–1	3	0	0	–1	R	$\alpha_{xy}, \alpha_{yz}, \alpha_{zx}$
A_u	1	1	1	1	–1	–1	–1	–1		
E_u	1	ε	ε^2	1	–1	$-\varepsilon$	$-\varepsilon^2$	–1		
	1	ε^2	ε	1	–1	$-\varepsilon^2$	$-\varepsilon$	–1		
T_u	3	0	0	–1	–3	0	0	1	T	

O_h	E	$8C_3$	$3C_2$	$6C_4$	$6C'_2$	i	$8S_6$	$3\sigma_h$	$6S_4$	$6\sigma_d$		
A_{1g}	1	1	1	1	1	1	1	1	1	1		$\alpha_{xx} + \alpha_{yy} + \alpha_{zz}$
A_{2g}	1	1	1	−1	−1	1	1	1	−1	−1		
E_g	2	−1	2	0	0	2	−1	2	0	0		$(2\alpha_{zz} - \alpha_{xx} - \alpha_{yy}, \alpha_{xx} - \alpha_{yy})$
T_{1g}	3	0	−1	1	−1	3	0	−1	1	−1	R	
T_{2g}	3	0	−1	−1	1	3	0	−1	−1	1		$\alpha_{xy}, \alpha_{yz}, \alpha_{zx}$
A_{1u}	1	1	1	1	1	−1	−1	−1	−1	−1		
A_{2u}	1	1	1	−1	−1	−1	−1	−1	1	1		
E_u	2	−1	2	0	0	−2	1	−2	0	0		
T_{1u}	3	0	−1	1	−1	−3	0	1	−1	1	T	
T_{2u}	3	0	−1	−1	1	−3	0	1	1	−1		

9. The $C_{\infty v}$ and $D_{\infty h}$ Groups for Linear Molecules

$C_{\infty v}$	E	$2C_\infty^\varphi$.........$\infty\sigma_v$		
$A_1 \equiv \Sigma^+$	1	1.........1	T_z	$\alpha_{xx} + \alpha_{yy}, \alpha_{zz}$
$A_2 \equiv \Sigma^-$	1	1.........−1	R_z	
$E_1 \equiv \Pi$	2	2 cos φ 0	(T_x, T_y); (R_x, R_y)	$(\alpha_{yz}, \alpha_{zx})$
$E_2 \equiv \Delta$	2	2 cos 2φ 0		$(\alpha_{xx} - \alpha_{yy}, \alpha_{xy})$
$E_3 \equiv \Phi$	2	2 cos 3φ 0		
....	....			

$D_{\infty h}$	E	$2C_\infty^\varphi$...$\infty\sigma_v$	i	$2S_\infty^\varphi$ ...... ∞C_2		
Σ_g^+	1	1........1	1	1........1		$\alpha_{xx} + \alpha_{yy}, \alpha_{zz}$
Σ_g^-	1	1......−1	1	1......−1	R_z	
Π_g	2	2 cos φ.....0	2	−2 cos φ.....0	(R_x, R_y)	$(\alpha_{yz}, \alpha_{zx})$
Δ_g	2	2 cos 2φ...0	2	2 cos 2φ...0		$(\alpha_{xx} - \alpha_{yy}, \alpha_{zy})$
......	....	....	....	....		
Σ_u^+	1	1........1	−1	−1......−1	T_z	
Σ_u^-	1	1......−1	−1	−1........1		
Π_u	2	2 cos φ......0	−2	2 cos φ...0	(T_x, T_y)	
Δ_u	2	2 cos 2φ....0	−2	−2 cos 2φ...0		
......	....	....	....	....		

Some of the character tables contain complex irreducible representations, they occur in pairs and one candidate of the pair is complex conjugate of the other. Thus frequencies corresponding to two (*i.e.*, Γ and Γ*) representations are accidently degenerate, and hence one frequency corresponding to either of the two representations need to be calculated. The complex representations can also be transformed to real one's by adding the complex conjugate according to the formula: $a_j = \sum N_{R(j)}\, \chi(R)\, \chi^j(R)$ and dividing the resultant real representation by a factor of two. The sign before the real characters of the representation is obtained on the basis of the orthogonality condition, *i.e.*, $a_j = \sum N_{R(j)}\, \chi(R)\, \chi^j(R) = 0$. This is demonstrated here by taking the example of C_3 character table.

$$a_j = 1 \times \varepsilon \times \varepsilon^* + 1 \times \varepsilon^* \times \varepsilon$$

$$= e^{\frac{2\pi i}{3}} \times e^{-\frac{2\pi i}{3}} + e^{-\frac{2\pi i}{3}} \times e^{\frac{2\pi i}{3}} = e^0 + e^0 = 1 + 1 = 2$$

Thus the character of the real representation is 2/2 = 1.

Accordingly we write the character table for the C_3 point group as

C_3	E	C_3	C_3^2
A	1	1	1
E	$\begin{cases} 1 \\ 1 \end{cases}$	$-\frac{1}{2}$ $-\frac{1}{2}$	$-\frac{1}{2}$ $-\frac{1}{2}$

or

C_3	E	C_3	C_3^2
A	1	1	1
E	2	–1	–1

and finally

C_3	E	$2C_3$
A	1	1
E	2	–1

This meets all the qualifications of C_3–character table.

The real character tables corresponding to the character tables possessing complex E representation as worked out on the above pattern are listed below:

C_3	E	$2C_3$
A	1	1
E	2	–1

C_4	E	$2C_4$	C_4^2
A	1	1	1
B	1	–1	1
E	2	0	2

C_5	E	$2C_5^1$	$2C_5^2$
A	1	1	1
E_1	2	2 cos 72°	2 cos 144°
E_2	2	2 cos 144°	2 cos 72°

C_6	E	C_6^1	$C_6^2 \equiv C_3$	$C_6^3 \equiv C''_2$
A	1	1	1	1
B	1	–1	1	–1
E_1	2	1	–1	–2
E_2	2	–1	–1	2

C_{3h}	E	C_3	σ_h	S_3
A′	1	1	1	1
A″	1	1	–1	–1
E′	2	–1	2	–1
E″	2	–1	–2	1

C_{4h}	E	C_4	$C_4^2 \equiv C''_2$	σ_h	S_4	i
A_g	1	1	1	1	1	1
A_u	1	1	1	–1	–1	–1
B_g	1	–1	1	1	–1	–1
B_u	1	–1	1	–1	1	–1
E_g	2	0	–2	–2	0	2
E_u	2	0	–2	2	0	–2

C_{5h}	E	$2C_5$	$2C_5^2$	σ_h	$2S_5^3$	$2S_5^2$
A′	1	1	1	1	1	1
A″	1	1	1	–1	–1	–1
E'_1	2	2 cos 72°	2 cos 144°	2	2 cos 72°	2 cos 144°
E''_1	2	2 cos 72°	2 cos 144°	–2	–2 cos 72°	–2 cos 144°
E'_2	2	2 cos 144°	2 cos 72°	2	2 cos 144°	2 cos 72°
E''_2	2	2 cos 144°	2 cos 72°	–2	–2 cos 144°	–2 cos 72°

C_{6h}	E	C_6	$C_6^2 \equiv C_3$	$C_6^3 \equiv C''_2$	σ_h	S_6	S_3	i
A_g	1	1	1	1	1	1	1	1
A_u	1	1	1	1	–1	–1	–1	–1
B_g	1	–1	1	–1	–1	1	–1	1
B_u	1	–1	1	–1	1	–1	1	–1
E_{1g}	2	1	–1	–2	–2	–1	1	2
E_{1u}	2	1	–1	–2	2	1	–1	–2
E_{2g}	2	–1	–1	2	2	–1	–1	2
E_{2g}	2	–1	–1	2	–2	1	1	–2

T	E	$8C_3$	$3C_2$
A	1	1	1
E	2	–1	2
T	3	0	–1

T_h	E	$8C_3$	$3C_2$	$8S_6$	$3\sigma_h$	i
A_g	1	1	1	1	1	1
A_u	1	1	1	–1	–1	–1
E_g	2	–1	2	–1	2	2
E_u	2	–1	2	1	–2	–2
T_g	3	0	–1	0	–1	3
T_u	3	0	–1	0	–1	–3

APPENDIX-II
CHAIN RULE

The chain rule for differentiation is used to convert the partial differentials of a given function from one set of variables to another set. For example, let there be a function of variables x, y, z, *i.e.* $f(x, y, z)$. The same function can be expressed in terms of variables r, θ, ϕ. Then

$$\frac{\partial f}{\partial x} = \left(\frac{\partial r}{\partial x}\right)_{y,z}\left(\frac{\partial f}{\partial r}\right)_{\theta,\phi} + \left(\frac{\partial \theta}{\partial x}\right)_{y,z}\left(\frac{\partial f}{\partial \theta}\right)_{r,\phi} + \left(\frac{\partial \phi}{\partial x}\right)_{y,z}\left(\frac{\partial f}{\partial \phi}\right)_{r,\theta}$$

$$\frac{\partial f}{\partial y} = \left(\frac{\partial r}{\partial y}\right)_{x,z}\left(\frac{\partial f}{\partial r}\right)_{\theta,\phi} + \left(\frac{\partial \theta}{\partial y}\right)_{x,z}\left(\frac{\partial f}{\partial \theta}\right)_{r,\phi} + \left(\frac{\partial \phi}{\partial y}\right)_{x,z}\left(\frac{\partial f}{\partial \phi}\right)_{r,\theta}$$

$$\frac{\partial f}{\partial z} = \left(\frac{\partial r}{\partial z}\right)_{x,y}\left(\frac{\partial f}{\partial r}\right)_{\theta,\phi} + \left(\frac{\partial \theta}{\partial z}\right)_{x,y}\left(\frac{\partial f}{\partial \theta}\right)_{r,\phi} + \left(\frac{\partial \phi}{\partial z}\right)_{x,y}\left(\frac{\partial f}{\partial \phi}\right)_{r,\theta}$$

ANSWERS

Chapter 1

1.2 m^3; **1.3** (*a*) s^{-1}, (*b*) (*i*) 23.1 J K^{-1} mol^{-1}, (*ii*) –34.45 J K^{-1} mol^{-1}, (*iii*) –91.77 J K^{-1} mol^{-1}; **1.4** (*a*) 9.93 × 10^{-19} J, (*b*) 7.94 × 10^{-19} J, (*c*) 1.986 × 10^{-20} J; **1.6** J mol^{-1}; **1.7** m^{-1}; **1.8** m_{eff} (S.I.) = 7.9774 × 10^2(χ_MT)$^{1/2}$, m_{eff} (cgs) = 2.8279 (χ_MT)$^{1/2}$; **1.9** χ_M(SI) = χ_M (cgs) × 4π × 10^{-6} m^3 mol^{-1}; **1.10** B.M. (cgs) = 5.81 × 10^{-20} esu cm molecule^{-1}, B.M. (SI) = 9.26 × 10^{-24} A m^{-2} molecule^{-1}.

Chapter 2

2.1 (*a*) 19.86 × 10^{-17} J, 4.75 × 10^{-17} cal, 4.92 × 10^{-21} eV; (*b*) 0.132 × 10^{-17} J, 3.16 × 10^{-19} cal, 3.275 × 10^{-23} eV; (*c*) 6.02 × 10^{-19} J, 1.438 × 10^{-19} cal, 1.49 × 10^{-23} eV; (*d*) 3.31 × 10^{-19} J, 7.91 × 10^{-20} cal, 8.198 × 10^{-24} eV; (*e*) 1.986 × 10^{-22} J, 4.75 × 10^{-23} cal, 4.92 × 10^{-27} eV; (*f*) 19.86 × 10^{-26} J, 4.747 × 10^{-26} cal, 4.92 × 10^{-30} eV; (*g*) 1.986 × 10^{-26} J, 0.475 × 10^{-26} cal, 4.92 × 10^{-32} eV; **2.2** 238 nm; **2.3** 1.239 × 10^{15} s^{-1}; **2.4** (*a*) 2.85 × 10^{-60} J molecule^{-1}; (*b*) 5.58 × 10^{-50} J molecule^{-1}; (*c*) 6.91 × 10^{-38} J molecule^{-1}; (*d*) 1.38 × 10^{-20} J molecule^{-1}. **2.5** (*a*) 110.98 × 10^{-16} J; (*b*) 1.87 × 10^{-15} J; (*c*) 2.26 × 10^{-11} J; **2.6** (*a*) 1.877 × 10^{-10} J; (*b*) 1.116 × 10^{-8} J; (*c*) 1.26 × 10^{-8} J; **2.7** (*i*) 7.83 × 10^{24} quanta s^{-1}, (*ii*) 1.96 × 10^{27} quantas s^{-1}; **2.8** 5962 K **2.9** 5.72 ×10^5 m s^{-1}; **2.10** (*a*) 4.34 V, (*b*) 285.8 nm; **2.11** 9.75 × 10^{-15} J; **2.12** (*a*) 5.75 × 10^{14} s^{-1}, (*b*) 6.38 × 10^{-34} J s; **2.13** (*a*) 4.35 × 10^{14} s^{-1}, (*b*) 6.83 × 10^{-34} J s; **2.14** 364.7 nm; **2.15** $\bar{\nu} = -\dfrac{2\pi\mu e^4}{h^3 c}(1/n_2^2 - 1/n_1^2)$; **2.16** 30.12 cm^{-1}; **2.17** (*i*) 3.419 × 10^5 m s^{-1}, (*ii*) 4.38 × 10^5 m s^{-1}, (*iii*) 3.65 × 10^5 m s^{-1}, **2.18** (*i*) 0.845 nm, (*ii*) 1.32 nm, (*iii*) 1.90 nm; **2.19** (*i*) 2467.74 cm^{-1}, (*ii*) 1340.50 cm^{-1}, (*iii*) 3808.25 cm^{-1}; **2.20** (*i*) 2466.7 cm^{-1}, (*ii*) 1339 cm^{-1}, (*iii*) 3806.64 cm^{-1}; **2.21** (*i*) 0.845 nm, (*ii*) 1.319 nm, (*iii*) 1.90 nm; **2.22** (*i*) 1.642 × 10^6 m s^{-1}, 1.314 × 10^6 m s^{-1}, 1.095 × 10^6 m s^{-1}; (*ii*) 93.8 pm, 146.6 pm, 211.1 pm; (*iii*) 22209.66 cm^{-1}, 12064.5 cm^{-1}, 34274.25 cm^{-1}; **2.23** 2467.74 cm^{-1}, 3808.25 cm^{-1}, 4336.7 cm^{-1}; **2.24** $\mu = \dfrac{m_e}{2}$, **2.17** will have the same values, (*ii*) 1.69 pm, 2.64 pm, 3.80 pm, (*iii*) 1233.87 cm^{-1}, 670.25 cm^{-1}, 1904.13 cm^{-1}; **2.25** (*i*) 2.17 will have the same values, (*ii*) 4.54 pm, 7.095 pm, 10.22 pm; (*iii*) 459147.7 cm^{-1}, 249413.43 cm^{-1}, 708563 cm^{-1}; **2.26** 1257.6 kcal, 2829.6 kcal, 5030.4 kcal; **2.27** (*i*) 1.03 × 10^{14} s^{-1}, (*ii*) 5.28 × 10^{14} s^{-1}, (*iii*) 3.06 × 10^{14} s^{-1}; **2.28** (*a*) (*i*) 3551.4 K, (*ii*) 1928.98 K, (*iii*) 5480.1 K, (*b*) (*i*) 3.28 × 10^5 m s^{-1}, (*ii*) 2.42 × 10^5 m s^{-1}, (*iii*) 4.075 × 10^5 m s^{-1}; **2.29** 38.7 pm; **2.30**

0.286 pm; **2.31** 1.99 × 10^{-38} m; **2.33** 2.53 ×10^{-7} m s^{-1}; **2.34** 9.94 × 10^6 V, 0.39 pm; **2.35** 2.04 × 10^{-8} V; **2.36** 0.664 nm, 6.64 nm, 66.4 nm; **2.37** $\lambda = 0.87 \times 10^{-10}$ m, yes; **2.38** 4.39 × 10^{-28} m s^{-1}; **2.39** (*a*) Δp = 1.05 × 10^{-23} kg m s^{-1}, (*b*) it cannot be measured as inaccuracy is more than the magnitude of the momentum; **2.40** Heisenberg uncertainty principle is not valid. **2.41** 3.31 ps; **2.42** 29.6 nm; **2.45** $m^{-3/2}$; **2.46** (*a*) yes, $\frac{-a^2h}{4\pi}$, (*b*) No, as at $x = \infty$, the function goes to ∞, a must be +ve constant; **2.47** wave function will remain same, total energy of each orbital will be increased by V; **2.48** uncertainty principle is violated and probability will be zero; **2.49** (*b*) No; **2.52** (*a*) $\sqrt{1/a}$, (*b*) $\left[\frac{1}{3\pi a_0^3}\right]^{1/2}$, (*c*) $\left[\frac{1}{4\pi(3a_0/2)^7\, 6!}\right]^{1/2}$; **2.53** e^{-ikx}; **2.54** $\hat{H} = -\frac{h^2}{8\pi^2\mu}\frac{d^2}{dx^2} + \frac{1}{2}kx^2$; **2.55** $\psi_0 = \left(\frac{a}{\pi}\right)^{1/4} e^{\frac{-\xi^2}{2}}$, $\psi_1 = \sqrt{2}\left(\frac{a}{\pi}\right)^{1/4} \xi e^{\frac{-\xi^2}{2}}$, $\psi_2 = \sqrt{1/2}\left(\frac{a}{\pi}\right)^{1/4}(2\xi^2 - 1)\, e^{\frac{-\xi^2}{2}}$, $\psi_3 = \sqrt{1/3}\left(\frac{a}{\pi}\right)^{1/4}(2\xi^3 - 3\xi)\, e^{\frac{-\xi^2}{2}}$, **2.56** $E_0 = \frac{1}{2}h\nu$, $E_1 = \frac{3}{2}h\nu$, $E_2 = \frac{5}{2}h\nu$, $E_3 = \frac{7}{2}h\nu$; **2.59** (*a*) $\frac{a}{2}, \frac{a^2}{3}$; **2.60** $\frac{nh^2}{2a}$; **2.61** $\frac{hk}{2\pi i}\left[\frac{\cos 2ak - 1}{\sin 2ak + 2ka}\right]$; **2.63** 0.015; **2.64** (*i*) 1.344 × 10^{-11}, (*ii*) 2.033 × 10^{-10}, (*iii*) 1.86 × 10^{-10}, **2.65** $h/4\pi$; **2.66** $\frac{(v+\frac{1}{2})}{\alpha}$, $(v+\frac{1}{2})\frac{h^2\alpha}{4\pi^2}$, $\frac{h^2}{4\pi^2}(v+\frac{1}{2})^2$; **2.67** (*i*) 2.72 × 10^{14} s^{-1}, (*ii*) 4.53 × 10^{14} s^{-1}; (*iii*) 10.9 × 10^{14} s^{-1}; **2.68** (*i*) g = 3, (*ii*) g = 6, (*iii*) g = 3; **2.69** g = 1; **2.70** (*a*) E_1 and E_2 will have two electrons each, (*b*) *a* = 0.563 nm; **2.71** λ = 1293 nm; **2.73** (*i*) Yes; No **2.79** (*a*) 7.33 × 10^{-19} J, (*b*) 6.18 × 10^{-40} J, ΔE is much smaller in latter than in the former; **2.80** 7.43 × 10^{18} states.

Chapter 3

3.1 (*a*) $\frac{h^2}{8\pi^2 m_e}\left[\frac{\partial^2\psi}{\partial x_1^2} + \frac{\partial^2\psi}{\partial y_1^2} + \frac{\partial^2\psi}{\partial z_1^2}\right] + \frac{h^2}{8\pi^2 M}\left[\frac{\partial^2\psi}{\partial x_2^2} + \frac{\partial^2\psi}{\partial y_2^2} + \frac{\partial^2\psi}{\partial z_2^2}\right] + (E - V)\psi = 0$, (*c*) $\frac{h^2}{8\pi^2(m_e + M)}\left[\frac{\partial^2\psi}{\partial x_c^2} + \frac{\partial^2\psi}{\partial y_c^2} + \frac{\partial^2\psi}{\partial z_c^2}\right] + \frac{h^2}{8\pi^2\mu}\left[\frac{\partial^2\psi}{\partial x^2} + \frac{\partial^2\psi}{\partial y^2} + \frac{\partial^2\psi}{\partial z^2}\right] + (E - V)\psi = 0$; **3.2** $\frac{\partial\psi}{\partial x} = \cos\phi\,\frac{\partial\psi}{\partial\rho} - \frac{\sin\phi}{\rho}\frac{\partial\psi}{\partial\phi}$, $\frac{\partial\psi}{\partial y} = \sin\phi\,\frac{\partial\psi}{\partial r} + \frac{\cos\theta}{r}\frac{\partial\psi}{\partial\theta}$, $\frac{\partial\psi}{\partial z} = \frac{\partial\psi}{\partial z}$, $\nabla^2 = \frac{\partial^2}{\partial\rho^2} + \frac{1}{\rho}\frac{\partial}{\partial\rho} + \frac{1}{\rho^2}\frac{\partial^2}{\partial\phi^2} + \frac{\partial^2}{\partial z^2}$; **3.3** $\frac{\partial}{\partial r} = \sin\theta\frac{\partial}{\partial r} + \frac{\cos\theta}{r}\frac{\partial}{\partial\theta}$, $\frac{\partial}{\partial z} = \cos\theta\,\frac{\partial}{\partial r} - \frac{\sin\theta}{r}\frac{\partial}{\partial\theta}$, $\frac{\partial}{\partial\phi} = \frac{\partial}{\partial\phi}$ $\nabla^2 = \frac{\partial^2}{\partial r^2} + \frac{1}{r}\frac{\partial}{\partial r} + \frac{1}{r^2}\frac{\partial^2}{\partial\theta^2} + \frac{\partial^2}{\partial\phi^2}$; **3.4** $\frac{\partial}{\partial x}$ = sin θ sin

$\phi \dfrac{\partial}{\partial r} + \dfrac{\cos\theta\cos\phi}{r}\dfrac{\partial}{\partial\theta} + \dfrac{\sin\phi}{r\sin\theta}\dfrac{\partial}{\partial\phi}$, $\dfrac{\partial}{\partial y} = \sin\theta\sin\phi\dfrac{\partial}{\partial r} + \dfrac{\cos\theta\sin\phi}{r}\dfrac{\partial}{\partial r} + \dfrac{\cos\phi}{r\sin\theta}\dfrac{\partial}{\partial\phi}$, $\dfrac{\partial}{\partial z} = \cos\theta\dfrac{\partial}{\partial r} - \dfrac{\sin\theta}{r}\dfrac{\partial}{\partial\theta}$, $\nabla^2 = \dfrac{1}{r^2}\dfrac{\partial}{\partial r}\left(r^2\dfrac{\partial}{\partial r}\right) + \dfrac{1}{r^2\sin\theta}\dfrac{\partial}{\partial\theta}\left(\sin\theta\dfrac{\partial}{\partial\theta}\right) + \dfrac{1}{r^2\sin^2\theta}\dfrac{\partial^2}{\partial\phi^2}$; **3.5** $\Theta_{2P}(\theta) = 6\sqrt{2}\sin\theta$ $\cos\phi$, $\Theta_{4f}(\theta) = \left(\dfrac{7}{24}\right)^{1/2}\dfrac{\sin\theta}{6}(210\cos^4\theta - 180\cos^2\theta + 8)$; **3.6** $R_{20} = -\left(\dfrac{Z}{a_0}\right)^{3/2}\left(\dfrac{1}{16}\right)^{1/2}(2\rho - 4)e^{-\rho/2}$,

$$R_{3,1} = \left(\frac{2Z}{3a_0}\right)^{3/2}\frac{1}{(6\times 24^2)^{1/2}}(96 - 24\rho)\rho^3 e^{-\rho/2};$$

3.7 (*i*)

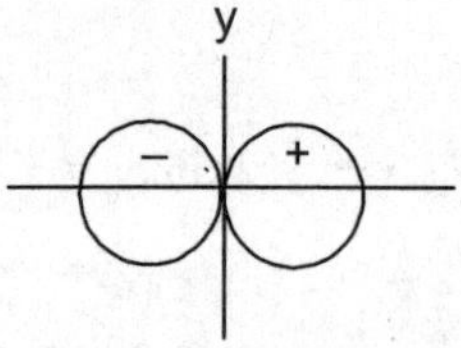

(*ii*)

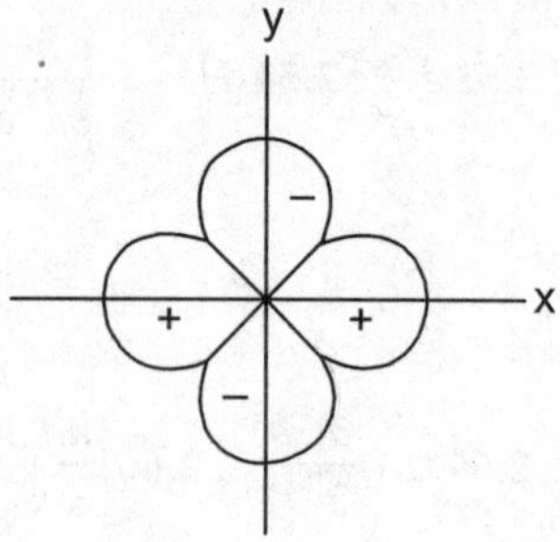

3.8 (*ii*)

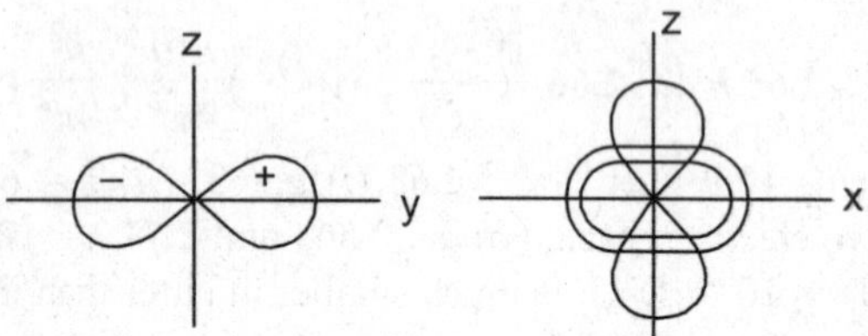

(*iii*)

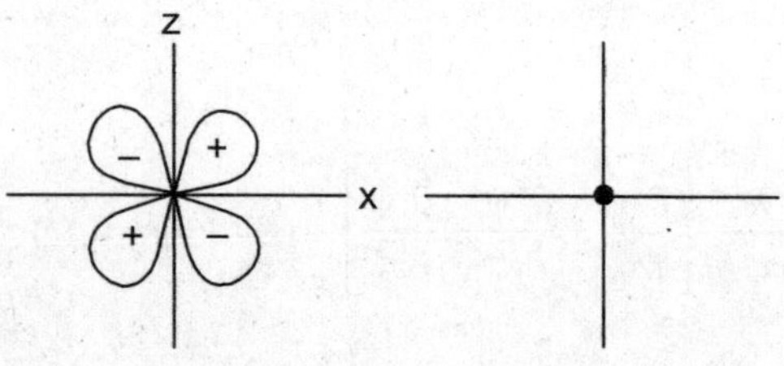

(*iv*)

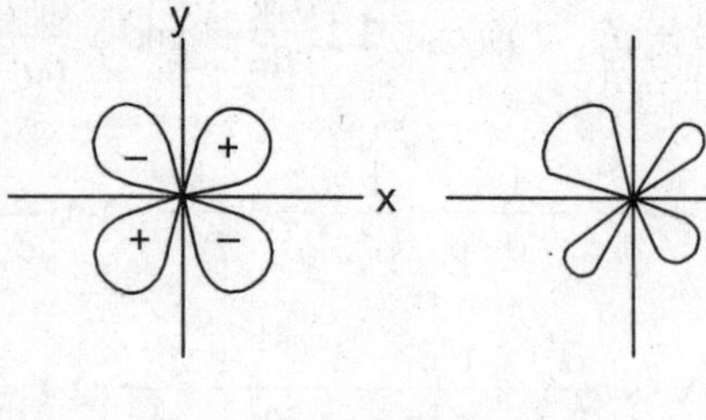

(v)

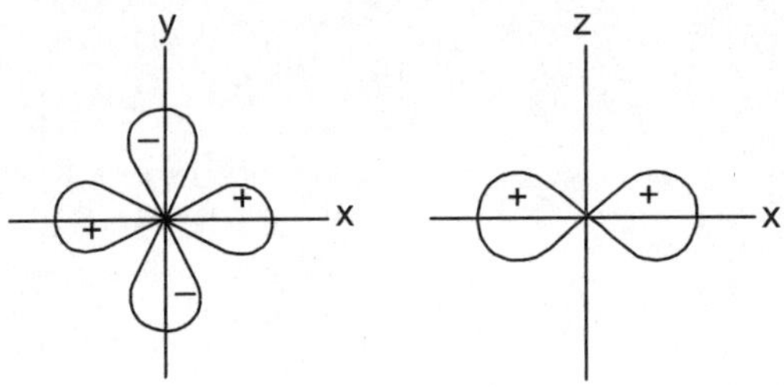

(vi)

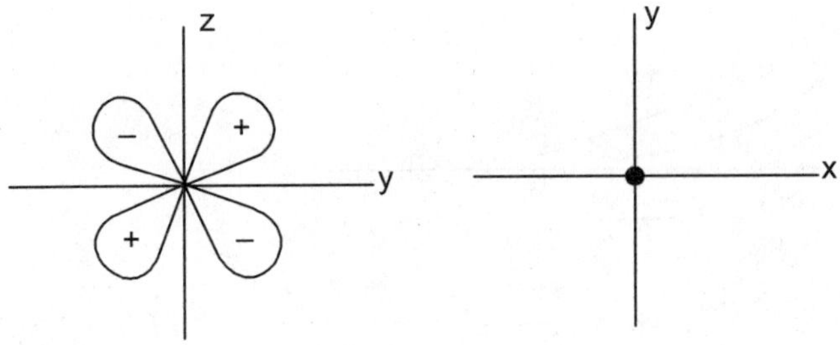

3.9 (i)

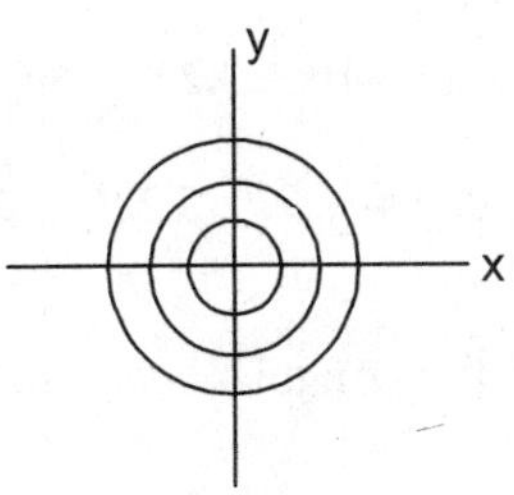

(ii)

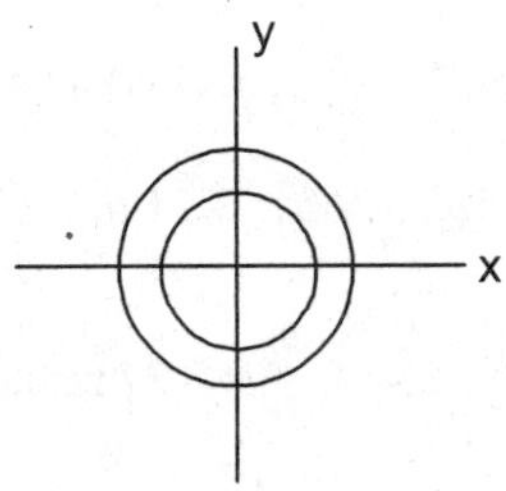

3.10

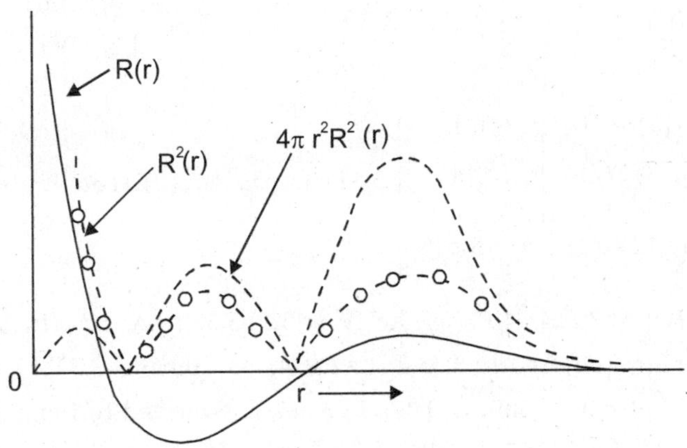

3.11

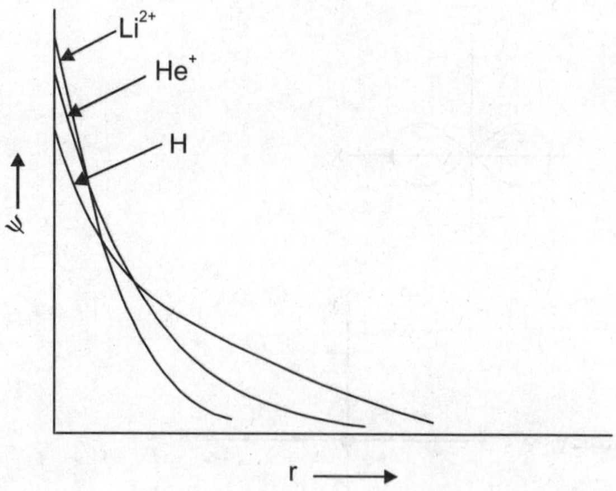

3.12 $a = 18$, $b = -\dfrac{12Z}{a_0}$, $c = \dfrac{4Z^2}{3a_0^2}$, $d = \dfrac{2Z}{6a_0}$, $E_3 = \dfrac{-2\pi^2 \mu z^2 e^4}{9h^2(4\pi\varepsilon_0)^2}$; **3.17** 0.762; **3.18** ~ 1; **3.19** (*a*) 0.9995, (*b*) = 1; **3.20** 3.34×10^{-4}; **3.21** (*i*) 0.0525, (*ii*) 0.97, (*iii*) 2.22×10^{-5}; **3.22** $2.66a_0$; **3.23** (*i*) $6a_0$, $\dfrac{52}{3}a_0^2$, (*ii*) $32a_0^2$, (*iii*) $10.5a_0$, $126a_0^2$; **3.24** no, $r(2p) = 4a_0$, $r(2s) = 0.764a_0$, $2a_0$, $5.236a_0$; **3.25** 106 pm;

3.26 0, $1.414\dfrac{h}{2\pi}$, $2.45\dfrac{h}{2\pi}$, $3.464\dfrac{h}{2\pi}$; **3.27** radial nodes (1, 0, 0, 1, 0, 1) Angular nodes (0, 1, 2, 2, 3, 3);

3.28 (*i*) 1, (*ii*) 4, (*iii*) 16, (*iv*) 36; **3.29** $\psi = 1/\sqrt{2}\begin{vmatrix} \phi_{1s}(1)\alpha(1) & \phi_{1s}(2)\alpha(2) \\ \phi_{1s}(1)\beta(1) & \phi_{1s}(2)\beta(2) \end{vmatrix}$

3.30 $\psi = 1/\sqrt{5}\begin{vmatrix} \phi_{1s}(1)\alpha(1) & \phi_{1s}(2)\alpha(2) \ldots \psi_{1s}(5)\alpha(5) \\ \phi_{1s}(1)\beta(1) & \phi_{1s}(2)\beta(2) \ldots \phi_{1s}(5)\beta(5) \\ \vdots & \\ \phi_{2p}(1)\alpha(1) & \phi_{2p}(2)\alpha(2) \ldots \phi_{2p}(5)\alpha(5) \end{vmatrix}$ **3.31** (*i*) Ar $3d^3 4s^2$, (*ii*) Ar $3d^5 4s^1$, (*iii*) Ar $3d^5 4s^2$,

(*iv*) Ar $3d^8 4s^2$, (*v*) Ar $3d^{10} 4s^1$ (*vi*) Ar $3d^{10} 4s^2 4p^6$, (*vii*) Xe $5s^2 5p^1 5d^1$; **3.32** (*i*) Ar $3d^5$, (*ii*) Ar$3d^4$, (*iii*) Ar $3d^3$, (*iv*) Xe, (*v*) Ar $3d^7$, (*vi*) Ar $3d^6$, (*vii*) Ar $3d^5$; **3.33** $Z_{eff} = 1.034$, $r = 8.09$ nm; **3.34** $Z_{eff} = 0.66$, 2.26, 0.76, 2.26, 2.62 and 1.17, $r = 12.68$ nm, 3.7 nm, 11.0 nm, 14.8 nm, 7.18 nm and 16.1 nm; **3.36** (*i*), (*iii*), (*v*), (*vi*), (*viii*), (*ix*); **3.37** (*i*) ^4F, (*ii*) ^6S, (*iii*) ^2P, (*iv*) ^4S, (*v*) ^3P, (*vi*) ^2P, (*vii*) ^2S, (*viii*) ^6S, (*ix*) ^1S, (X) ^5D; **3.38** (*i*) ^5D, ^3H, ^3G, $^3F(2)$, ^3D, $^3P(2)$, ^1I, $^1G(2)$, ^1F, $^1D(2)$, $^1S(2)$, (*ii*) ^6S, ^4G, ^4F, ^4D, ^4P, ^2I, ^2H, $^2G(2)$, $^2F(2)$, $^2D(3)$, ^2P, ^2S, (*iii*) ^3H, ^3F, ^3P; ^1I, ^1G, ^1D, ^1S; (*iv*) ^4I, ^4G, ^4F, ^4D, ^4S, ^2L, ^2K, ^2I, $^2H(2)$, $^2G(2)$, $^2F(2)$, $^2D(2)$, ^2P and not given in the book; (*v*) ^4F, ^4P, ^2H, ^2G, ^2F, $^2D(2)$, ^2P; **3.39** (*i*) ^3S,^1S, (*ii*) ^3D, ^3P, ^3S, ^1D, ^1P, ^1S, (*iii*) ^3F, ^3D, ^3P, ^1F, ^1D, ^1P, (*iv*) ^4G, $^4F(2)$, $^4D(3)$, $^4P(2)$, ^4S, $^2G(2)$, $^2F(4)$, $^2D(6)$, $^2P(4)$, $^2S(2)$, (*v*) ^4F, ^4D, ^4P, ^2F, ^2D, ^2P, ^2F, ^2D, ^2P; **3.40** (*i*) Kr$4d^4 5s^1$, (*ii*) Kr $4d^5 5s^1$, (*iii*) Kr $4d^7 5s^2$, (*iv*) Xe $4d^{10} 5s^2 5p^6 5d^2$; **3.41** b, c, d, e; **3.43** (*i*) $Z_{eff} = 3.954$, $\mu = 0.046$, (*ii*) 2.956, 0.044, (*iii*) 6.017, –0.017, (*iv*) 4.0,0.0; **3.44** 2.117, 0.883;

3.133, 0.867; 3.9999,0; 6.008, –0.008; **3.45** $\Delta(^2P_{3/2}–^2p_{1/2}) = 57.7$, $\Delta(^2D_{5/2} – ^2D_{3/2}) = 2.3$ cm^{-1}; **3.46** $^3P_0 \leftrightarrow {}^3D_1$, $^3P_1 \leftrightarrow {}^3D_1$, $^3P_1 \leftrightarrow {}^3D_2$, $^3P_2 \leftrightarrow {}^3D_1$, $^3P_2 \leftrightarrow {}^3D_2$, $^3P_2 \leftrightarrow {}^3D_3$.

Chapter 4

4.1 0.706 eV; **4.2** $b = 3.744 \times 10^{10}$ m^{-1}, $a = 5.86 \times 10^{-16}$ J; **4.3** 373.54 kJ mol^{-1}; **4.4** (*a*) $b = 3.126 \times 10^{10}$ m^{-1}, $a = 1.089 \times 10^{-15}$ J, (*b*) 283.72 kJ mol^{-1}; **4.5** 481.54 kJ mol^{-1}; **4.6** 356.56 kJ mol^{-1}; **4.7** equally shared in homonuclear; **4.8** *AB* more ionic, *b* no; **4.9** $\psi_T = [C_1\phi_{1s} + C_2\phi_{2p_z}]^2$; **4.10** $[C_1\phi_{2p_z} + C_2\phi_{3p_z}]^2$; **4.11** (*i*) $C_1 1s_H(1)2_{pz}F(2) + C_2 1s_H(2)2\,p_zF(1) + C_3 1_H(2)ls_H(1) + C_4 2p_zF(1)2p_zF(2)$, (*ii*) $C_1 2p_zF(1)\,3p_zCl(2) + C_2\,2p_zF(2)3p_zCl(1) + C_3\,2_{Pz}F(1)2p_zF(2) + C_4\,3p_zCl(1)\,3p_zCl(2)$; **4.12** (*a*) ionisation energy, (*b*) electron affinity, (*c*) Coulombic energy; **4.13** Attractive Coulombic energy overweighs electron affinity; **4.14** HF; **4.15** (*i*), (*iv*), (*v*) and (*vi*);

4.16 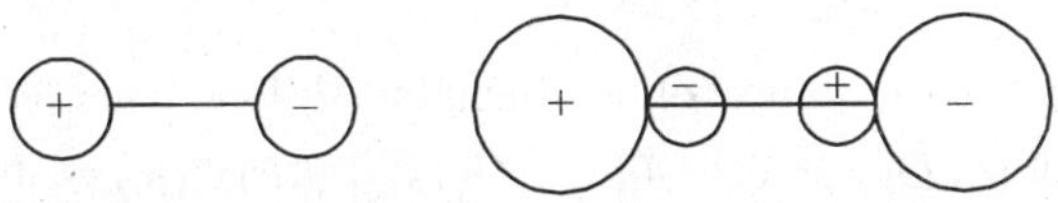

4.17 (*i*) $\sigma_{1s}^2\,\sigma_{1s}^{*2}\,\sigma_{2s}^2\,\sigma_{2s}^{*2}\,\pi_{2p_x}^2\,\pi_{2p_y}^2\,\sigma_{2p_z}^2$ (*ii*) $\sigma_{1s}^2\,\sigma_{1s}^{*2}$, (*iii*) $\sigma_{1s}^2\,\sigma_{1s}^{*2}\,\sigma_{2s}^2\,\sigma_{2s}^{*2}\,\sigma_{2p_x}^2\,\pi_{2p_z}^2\,\pi_{2p_y}^2\,\pi_{2p_y}^{*2}\,\pi_{2p_y}^{*2}$, (*iv*) K(Li) σ_z^2 (*v*) K(C)K(N) $\sigma_z^2\,\sigma_y^{*2}\,\sigma_x^2\,\pi_w^3$; (*vi*) K(O) K(S) L(S) $\sigma_z^2\,\sigma_y^{*2}\,\sigma_x^2\,\pi_w^4\,\pi_v^{*2}$, (*vii*) K(F) K(Cl) L(Cl) $\sigma_z^2\,\sigma_y^{*2}\,\sigma_x^2\,\pi_w^4\,\pi_v^{*4}$, (*viii*) K(O) K(Cl) L(Cl) $\sigma_z^2\,\sigma_y^{*2}\,\sigma_x^2\,\pi_w^4\,\pi_v^{*3}$, (*x*) K(N) K(O) $\sigma_z^2\,\sigma_y^{*2}\,\sigma_x^2\,\pi_w^4\,\pi_v^{*1}$; **4.18** (*i*) 3, (*ii*) 0, (*iii*) 1, (*iv*) 1, (*v*) $\frac{5}{2}$ (*vi*) 2, (*vii*) 1, (*viii*) 1, (*ix*) $\frac{5}{2}$; **4.20** (*a*) (*i*) AB, (*ii*) B_3C, (*iii*) $C \equiv C$, (*v*) D, (*b*) (*i*) K(A) K(B) L(B) M(B) $[C_1 2p_x + C_2 4s]^2$, (*ii*) K(B) K(C) L(B) M(B) $2s^2(C)[C_1p_x(C) + C_2 2s(B)]^2\,[b_1p_y(C) + b_2 4s(B)]^2[C_1p_z(C) + C_2 4s(B)]^2$, (*iii*) K(C) K(C) $2s^2(C)\,2s^2(C)\,[p_x(C) + p_x(C)]^2[p_y(C) + p_y(C)]^2[p_z(C) + p_z(C)]^2$, (*iv*) $1s^2 2s^2 2p^6 3s^2 3p^6$; **4.22** (*a*) OF^+, CN^{-1}, O_2^{2+}, (*b*) OF^+, CN^{-1}, O_2^{2+}; **4.23** $^1\Sigma$, (*iii*) $^1\Sigma$, (*iv*) $^1\Sigma$, (v) $^2\Pi$, (*vi*) $^1\Delta$, $^3\Sigma^-$, $^1\Sigma^+$, (*vii*) $^1\Sigma$, (*viii*) $^2\Pi$, (*ix*) $^2\Pi$; **4.24** (*i*) $^1\Delta$, $^3\Sigma^-$, $^1\Sigma^+$, (*ii*) $^1\Delta$, $^3\Sigma^-$, $^1\Sigma^+$, (*iii*) for O_2 $^1\Delta$, $^1\Sigma^+$ for O_2^- $^2\Pi$, for O_2^{2+} $^1\Sigma$; **4.25** (*i*) $\sigma_{1s}^2\,\sigma_{1s}^{*2}\,\sigma_{2s}^2\,\sigma_{2s}^{*2}\,\pi_{2p_y=2p_z}^4$ (*ii*) $\sigma_{1s}^2\,\sigma_{1s}^{*2}\,\sigma_{2s}^2\,\sigma_{2s}^{*2}\,\pi_{2p_y=2p_z}^2$; **4.26** 3.98, **4.27** 3.76; **4.28** 39.57; **4.29** (*i*) 3.26, (*ii*) 3.09, (*iii*) 2.86; **4.30** 218.28 kJ mol^{-1}; **4.31** 9.198 kJ mol^{-1}; **4.32** (*i*) $C = O \leftrightarrow {}^-C \equiv O^+ \leftrightarrow {}^+C – O^-$, (*ii*) $N = O \leftrightarrow {}^-N \equiv O^+$; **4.33** (*i*) 44.82, (*ii*) 16.75, (*iii*) 11.5, (*iv*) 4.88; **4.34** (*i*) 0.90, (*ii*) 0.45, (*iii*) 0.36, (*iv*) 0.23; **4.35** $e_H = –0.144 \times 10^{-19}$ C, $e_C = –0.856 \times 10^{-19}$C, $e_H = 1.0 \times 10^{-19}$ C; **4.36** 1.458D; **4.37** (*a*) 1.9 D, (*b*) 1.266 D, (*c*) 1.902 D; **4.38** % IC = 69.86, $\lambda = 1.522$, $\chi_{cs} = 1.27$; **4.39** ClO_2, SO_2; **4.40** No, energy of the hybrid orbital is greater than atomic orbital; **4.41** hybrid orbitals, one bond each; **4.42** (*i*) $\psi_{(sp)} = \frac{1}{\sqrt{2}}(s + p)$, $\psi_{(sp)} = \frac{1}{\sqrt{2}}(s – p)$, (*ii*) $\psi_1(sp^2) = \frac{1}{\sqrt{3}}[s + p_x + p_y]$, $\psi_2(sp^2) = \frac{1}{\sqrt{6}}(2s – p_x – p_y)$, $\psi_3(sp^2) = \frac{1}{\sqrt{2}}(p_x – p_y)$ (*iii*) $\psi_1(dsp^2) = \frac{1}{2}(s + \sqrt{2}p_x + d_{x^2-y^2})$, $\psi_2(dsp^2) = \frac{1}{2}(s + \sqrt{2}p_y – d_{x^2-y^2})$, $\psi_3(dsp^2) = \frac{1}{2}(s + d_{x^2-y^2} – \sqrt{2}p_x)$, $\psi_4(dsp^2) = \frac{1}{2}(s – d_{x^2-y^2} – \sqrt{2}p_y)$, (*iv*) $\psi_1(dsp^3) = \frac{1}{\sqrt{5}}s + \frac{1}{\sqrt{2}}p_z + \sqrt{\frac{3}{10}}d_{z^2}$, $\psi_2(dsp^3) = \frac{1}{\sqrt{5}}s – \frac{1}{\sqrt{2}}p_z + \sqrt{\frac{3}{10}}d_{z^2}$, $\psi_3(dsp^3)$

$= \frac{1}{\sqrt{5}}s + \sqrt{\frac{2}{3}}p_x - \sqrt{\frac{4}{30}}d_{z^2}$, $\psi_4(dsp^3) = \frac{1}{\sqrt{5}}s - \frac{1}{\sqrt{6}}p_x - \frac{1}{\sqrt{2}}p_y - \sqrt{\frac{4}{30}}d_{z^2}$, $\psi_5(dsp^3) = \frac{1}{\sqrt{5}}s - \frac{1}{\sqrt{6}}p_x - \frac{1}{\sqrt{2}}p_y + \sqrt{\frac{4}{30}}d_{z^2}$, (*v*) $\psi_1(d^2sp^3) = \frac{1}{\sqrt{6}}s + \frac{1}{\sqrt{2}}p_z + \frac{1}{\sqrt{3}}d_{z^2}$, $\psi_2(d^2sp^3) = \frac{1s}{\sqrt{6}} - \frac{1}{\sqrt{2}}p_z + \frac{1}{\sqrt{3}}d_{z^2}$, $\psi_3(d^2sp^3) = \frac{1}{\sqrt{6}}s + \frac{1}{\sqrt{2}}p_x + \frac{1}{2}d_{x^2-y^3} - \frac{1}{\sqrt{12}}d_{z^2}$, $\psi_4(d^2sp^3) = \frac{1}{\sqrt{6}}s - \frac{1}{\sqrt{2}}p_x + \frac{1}{2}d_{x^2-y^2} - \frac{1}{\sqrt{12}}d_{z^2}$, $\psi_5(d^2sp^3) = \frac{1}{\sqrt{6}}s + \frac{1}{\sqrt{2}}p_x - \frac{1}{2}d_{x^2-y^2} - \frac{1}{\sqrt{12}}d_{z^2}$, $\psi_6(d^2sp^3) = \frac{1}{\sqrt{6}}s - \frac{1}{\sqrt{2}}p_x - \frac{1}{2}d_{x^2-y^2} - \frac{1}{\sqrt{12}}d_{z^2}$,

4.43 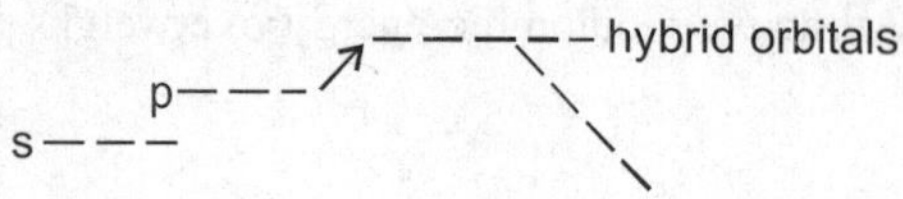

4.44 (*a*) electronegativities of atoms, (*b*) size of atoms, (*c*) bond length, (*d*) lone pairs on central atom; **4.49** In BX_3 octet is not full on B; **4.50** (*a*) $O(1s)^2\,Te^4[O(Te) + H(1s)_\sigma]^4$, (*b*) $C(K)\,(K)[C(sp) + H(1s)_\sigma C-H]^4\,[C(sp) + C(sp)\sigma^2]_{C-C}[C(p_z) + C(p_z), \pi]^2{}_{C-C}[C(p_y) + C(p_y), \pi]^2{}_{C-C}$, (*c*) $N(Tr)^2(Tr)^2[N(Tr) + N(Tr), \sigma)^2{}_{N-N}[N(Tr) + F(p_x), \sigma]^4{}_{N-F}[N(p_z) + N(p_z), \pi]^2$, (*d*) $C(K)Te^4[C(Te) + H(1s),\sigma]^8$, (*e*) $C(K)(K)[C(Tr) + H(1s), \sigma]^8{}_{C-H}[C(Tr) + C(Tr), \sigma]^2{}_{C-C}[C(p_z) + C(p_z), \pi]^2{}_{C-C}$, (*f*) $C(K)C(K)[C(Te) + H(1s), \sigma]^{12}{}_{C-H}[C(Te) + C(Te), \sigma]^2{}_{C-C}$; **4.51** (*i*) $C = 0 \leftrightarrow {}^-C \equiv O^+ \leftrightarrow {}^+C–O^-$, (*ii*) $N = O \leftrightarrow N \equiv O^+$,

(*iii*) 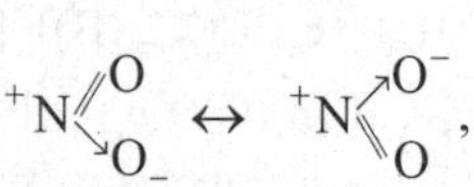, (*iv*) 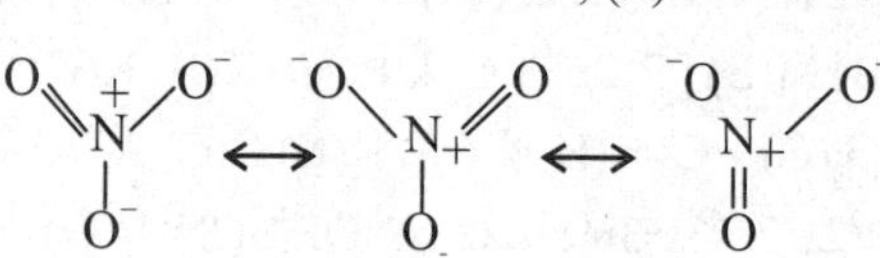,

(*v*) 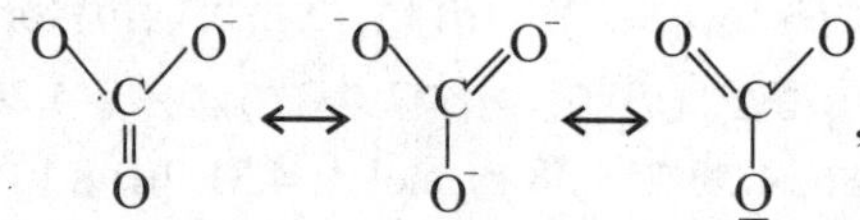,

(*vi*) $^-:\ddot{N} = C = N^+ = N^+ = N^-$, (*vii*) 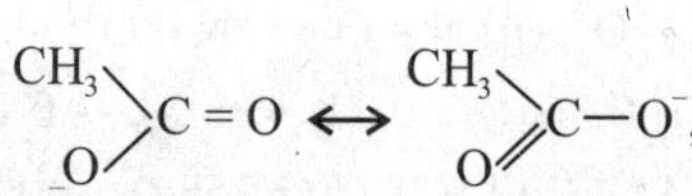, (*viii*) $^-N - N \equiv N \leftrightarrow N \equiv N–N^- \leftrightarrow {}^-N = N^+ = N^-$; **4.52** (*i*) In bond pair S = 0.2, P = 0.8, in lone pairs, S = 0.3, P = 0.7, (*ii*) In bond pair S = 0.264, P = 0.736, in lone pairs, S = 0.236, P = 0.764, (*iii*) same as in $O(Cl)_2$, (*iv*) In lone pair S = 0.313, P = 0.687, bond pair S = 0.229, P = 0.771, (*v*) S = 0.25, P = 0.75, (*vi*) bond pair S = 0.0495, P = 0.905, lone pair, S = 0.8515, P = 0.1485; **4.53** Due to lone pair-lone pair interaction on O; **4.54** (*a*) bond pair in C_2H_6 is replaced by lone pair on $N_2H_5^+$, (*b*) $N_2H_6^{2+}$; **4.55** (*i*) linear, sp^3d, (*ii*) linear, sp^3d, (*iii*) linear, sp^3d, (*iv*) T shape, sp^3d, (*v*) irregular tetrahedron, sp^3d, (*vi*) square planar, sp^3d^2, (*vii*) square planar sp^3d^2, (*viii*) square planar, sp^3d^2, (*ix*) Trigonal bipyramidal, sp^3d, (*x*) square pyramidal, sp^3d^2, (*xi*) trigonal bipyramidal, sp^3d, (*xii*) octahedron, sp^3d^2, (*xiii*) octahedron, sp^3d^2, (*xiv*) pentagonal bipyramidal, sp^3d^3, (*xv*) linear, sp, (*xvi*) triangular pyramidal, sp^3, (*xvii*) tetrahedral, sp^3, (*xviii*) tetrahedral, sp^3, (*xix*) irregular tetrahedral, sp^3d, (*xx*) triangular planar, sp^2, (*xxi*) triangular pyramidal, sp^3, (*xxii*) tetrahedral sp^3, (*xxiii*) tetrahedral, sp^3; **4.56** linear CS_2, C_3, CSTe, $CdCl_2$, $C_2N_2^{2-}$, bent = ClO_2, OCl_2, $SnCl_2$, S_3^{2-}, ONCl; **4.57** (*i*) planar, (*ii*) triangular pyramidal, and for rest molecules; **4.58** (*i*) bent, (*ii*) linear, (*iii*) tetrahedron, (*iv*) triangular bipyramidal, (*v*) octahedron, (*vi*) octahedron, (*viii*) octahedron; **4.59** and **4.60** (*iii*) 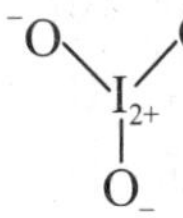 (*iv*) H – C ≡ N, (*v*) 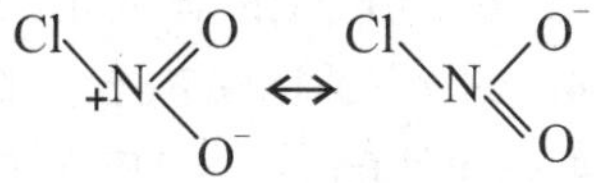 (*vi*) same as IO_3^-,

(*vii*) 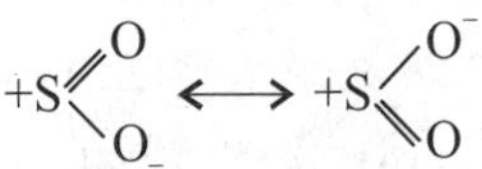 (*viii*) 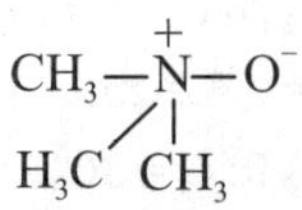 (*ix*) same as SO_2, (*x*) F – N = O, **4.61** (*i*) $E_1 = \alpha + 1.618\beta$, $E_2 = \alpha + 0.618\beta$, $E_3 = \alpha - 0.618\beta$, $E_4 = \alpha - 1.618\beta$, $p_{34} = p_{12} = 0.895$, $p_{34} = p_{12} = 1.895$, $r_{12} = 1.354 \times 10^{-10}$ m, $= r_{34}$, $P_{23} = 1.448$, $r_{23} = 1.438 \times 10^{-10}$ m; **4.62** $\alpha + 2\beta$, $\alpha + \beta$, $\alpha - \beta$, $\alpha - 2\beta$, $p_{12} = p_{23} = p_{34} = p_{41} = 0.5$, $p = 1.5$, $r = 1.428 \times 10^{-10}$ m; **4.63** $\psi_1 = 0.749\phi_1 + 0.393(\phi_2 + \phi_5) + 0.254(\phi_3 + \phi_4)$, $E_1 = \alpha + 2.55\beta$, $\psi_2 = 0.503\phi_1 - 0.089(\phi_2 + \phi_5) - 0.605(\phi_3 + \phi_4)$ $E_2 = \alpha + 1.15\beta$, $\psi_3 = 0.601(\phi_2 + \phi_5) + 0.372(\phi_3 - \phi_4)$, $E_3 = \alpha + 0.618\beta$, $\psi_4 = 0.413\phi_1 - 0.581(\phi_2 + \phi_5) + 0.265(\phi_3 + \phi_4)$, $E_4 = \alpha - 1.20\beta$, $\phi_5 = 0.372(\phi_2 - \phi_5) - 0.601(\phi_3 - \phi_4)$, $E_5 = \alpha - 1.62\beta$, Total energy = $6\alpha + 8.64\beta$, Delocalising energy = 1.64β; **4.64** cyclic complex; **4.65** (*i*), (*v*), (*vi*), (*vii*), (*viii*); **4.66** (*i*) $E_1 = \alpha + \sqrt{2}\beta$, $E_2 = \alpha$, $E_3 = \alpha - \sqrt{2}\beta$, (*ii*) $E_1 = \alpha + \delta\beta$, $E_2 = \alpha + \beta$, $E_3 = \alpha - \beta$; **4.67** (*i*) 337.5 nm; (*ii*) 230.4 nm; **4.68** (*i*) 47379.2 cm^{-1}, 61666.7 cm^{-1}, (*ii*) 21162.51 cm^{-1}, 35571.43 cm^{-1}, **4.69** 3.197 nm; **4.71** $\bar{v} = 1.01 \times 10^6$ cm^{-1}; **4.71** (*i*) low spin, (*ii*) high spin, (*iii*) high spin, (*iv*) low spin, (*v*) high spin, (*vi*) low spin, (*vii*) low spin, (*viii*) high spin, (*ix*) low spin, (*x*) high spin, (*xi*) low spin, (*xii*) high spin, **4.72** (*i*) 2.83 BM, (*ii*) 4.9 BM, (*iii*) 5.92 BM, (*iv*) 2.83 BM, (*v*) 5.92 BM, (*vi*) 0 BM, (*vii*) 0 BM, (*viii*) 4.9 BM, (*ix*) 3.87 BM, (*x*) 2.83 BM, (*xi*) 1.73 BM, (*xii*) 2.83 BM; **4.73** inner field = (*i*), (*iv*), (*vi*), (*vii*), outer field = (*ii*), (*iii*), (*v*), (*viii*), (*ix*), (*x*), (*xi*), (*xii*); **4.74** all square planar; **4.75** tetrahedral will have greater magnetic moment; **4.76** $[Ni(CN)_4]^{2-}$ square planar, $[Ni(NH_3)_4]^{2+}$, tetrahedral; **4.77** 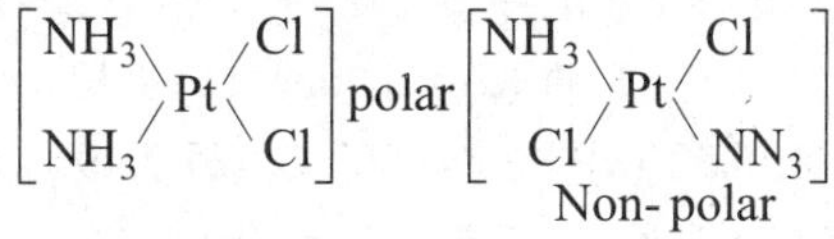

, **4.78** pairing energy is more than crystal field splitting.

Chapter 5

5.1 $\mu = 4.03 \times 10^{-26}$ kg, $I = 1.845 \times 10^{-45}$ kg m^{-2}; **5.2** $v_{1-0} = 9.06 \times 10^9$ s^{-1}, $v_{2-1} = 1.812 \times 10^{10}$ s^{-1}, $v_{3-2} = 2.719 \times 10^{10}$ s^{-1}, $v_{4-3} = 3.624 \times 10^{10}$ s^{-1}; **5.3** $I = 2.646 \times 10^{-47}$ kg m^2, $v_{1-0} = 6.336 \times 10^{11}$ s^{-1}, $v_{2-1} = 12.66 \times 10^{11}$ s^{-1}, $v_{3-2} = 18.99 \times 10^{11}$ s^{-1} $v_{4-3} = 25.32 \times 10^{11}$ s^{-1}; **5.4** $v_{1-0} = 1.02 \times 10^{12}$ s^{-1}, $v_{2-1} = 2.04 \times 10^{12}$ s^{-1}, $v_{3-2} = 3.06 \times 10^{12}$ s^{-1}, $v_{4-3} = 4.08 \times 10^{12}$ s^{-1}; **5.5** 1.28×10^{-10} m; **5.6** 0.106 cm^{-1}; **5.7** $v = 1.247 \times 10^{14}$ s^{-1}, $\epsilon_0 = 4.129 \times 10^{-20}$ J; **5.8** $v = 0.904 \times 10^{14}$ s^{-1}, $\epsilon_0 = 2.99 \times 10^{-20}$ J; **5.9** $w_e x_e = 13.22$ cm^{-1}; **5.10** $v_{N-N} = 0.313 \times 10^{14}$ s^{-1}, $v_{N=N} = 0.537 \times 10^{14}$ s^{-1}, $v_{N \equiv N} = 0.706 \times 10^{14}$ s^{-1}; **5.11**(*a*) $k_{H_2} = 1.90 \times 10^2$ N m^{-1}, $k_{HD} = 1.92 \times 10^2$ N m^{-1}, $k_{D-D} = 1.91 \times 10^2$ N m^{-1}, (*b*) yes; **5.12** $G(1) = 1167.52$ cm^{-1},

$G(2) = 1921.13$ cm^{-1}, $G(3) = 2654.23$ cm^{-1}, $G(4) = 3368.93$ cm^{-1}; **5.13** (*a*) $\bar{\nu} = 793.2$ cm^{-1}, (*b*) 298.2 cm^{-1}; **5.16** 207.06 kJ mol^{-1}; **5.17** (*a*) $w_e = 579.91$ cm^{-1}, $w_e x_e = 10.89$ cm^{-1}, (*b*) $\nu = 28$, $D_e = 7429.45$ cm^{-1}, (*c*) 89.50 kJ mol^{-1}; **5.18** (*a*) $w_e x_e = 0.759$ cm^{-1}, $w_e = 246.52$ cm^{-1}, (*b*) $\nu = 150$, $D_e = 1990.97$ cm^{-1}, (*c*) 239.0 kJ mol^{-1}; **5.19** (*a*) for HF, $\in_0 = 4.129 \times 10^{-20}$ J, $\in(1) = 12.386 \times 10^{-20}$ J, $\in(2) = 16.515 \times 10^{-20}$ J, $\in(3) = 20.644 \times 10^{-20}$ J, for DF $\in(0) = 2.99 \times 10^{-20}$ J, $\in(1) = 8.974 \times 10^{-20}$ J, $\in(2) = 14.957 \times 10^{-20}$ J, $\in(3) = 20.94 \times 10^{-20}$ J; (*b*) ν_{1-0}(HF) $= 1.247 \times 10^{14}$ s^{-1}, ν_{1-0}(DF) $= 0.904 \times 10^{14}$ s^{-1}, $\Delta\nu = 0.343 \times 10^{14}$ s^{-1}; (*c*) For HFF(0) = 0, F(1) = 63 cm^{-1}, F(2) = 126 cm^{-1}, F(3) = 252 cm^{-1}, F(4) = 420 cm^{-1}; For DF, F(0) = 0, F(1) = 22.2 cm^{-1}, F(2) = 66.6 cm^{-1}, F(3) = 133.2 cm^{-1}, F(4) = 222.0 cm^{-1}; (*e*) ν(HF) = 8478 cm^{-1}, $\bar{\nu}$(DF) = 6155 cm^{-1}, difference = 2323 cm^{-1}; **5.20** R(8) = 3048.14 cm^{-1}, P(8) = 2439.79 cm^{-1}; **5.21** 0.149 nm: **5.22** (*a*) $B_{\nu'} = 1.5846$ cm^{-1}, $B_{\nu''} = 1.7466$ cm^{-1}, (*b*) $B_{\nu'} = 1.575$ cm^{-1},. $B_{\nu''} = 1.650$ cm^{-1}, **5.23** $B_{\nu'} = 1.905$ cm^{-1}, $B_{\nu''} = 1.9232$ cm^{-1}; **5.24** 1, 3, 3, 6, 9, 30, 6, 6,15, 3; **5.25** bent; **5.26** linear; **5.27** (*i*) overtone, (*ii*) combination band $\nu = 2\nu_2 + \nu_3$, (*iii*) combination band, $\nu = \nu_1 - \nu_2$, (*iv*) combination band $\nu = \nu_1 + \nu_2 + \nu_3$, (*v*) combination band, $\nu = 2\nu_1 + \nu_4$; **5.28** (*i*) overtone 2646 cm^{-1}, (*ii*) 2939 cm^{-1}, (*iii*) 4439 cm^{-1}, (*iv*) 4262 cm^{-1}, (*v*) 3689 cm^{-1}, (*vi*) 5012 cm^{-1}; **5.29** (*i*) active, (*ii*) active, (*iii*) active, (*iv*) inactive, (*v*) active, (*vi*) inactive, (*vii*) inactive, (*viii*) active, (*ix*) active, (*x*) active: **5.30** (*i*) 14, (*ii*) 16; **5.31** 1406 K; **5.32** J (27°C) = 2, J(127°C) = 3; **5.36** (*i*) $N_1/N_0 = 2.46$, (*ii*) $N_2/N_0 = 2.74$, $N_3/N_0 = 2.11$, $N_4/N_0 = 1.22$, (*ii*) $N_1/N_0 = 1.27$, $N_2/N_0 = 0.38$, $N_8/N_0 = 0.04$, $N_4/N_0 = 0.002$; **5.37** decrease, yes it will be different; **5.38** sunlight; **5.39** Stokes line = 441.3 nm, anti-Stokes line = 435.73 nm; **5.40** Stokes line = 264 nm, anti-Stokes line = 243.98 nm; **5.41** rotational spectrum close to exciting light and vibrational spectrum away from exciting line; **5.42** $\bar{\nu} = 892.6$ cm^{-1}; **5.43** (*i*), (*ii*), (*iv*) and (*viii*) are allowed; **5.44** (*a*) (*i*) will be observed, (*ii*) increase to maxima and decreases, (*iii*) discrete diffuse discrete; **5.45** $D_e = 187.99$ kJ mol^{-1}; **5.46** D_e(CH) = 351.73 kJ mol^{-1}, D_e(CO) = 1084.74 kJ mol^{-1}, D_e(HCl) = 445.17 kJ mol^{-1}, D_e(NO) = 643.26 kJ mol^{-1}; **5.47** $w_e = 707.98$ cm^{-1}, $w_e x_e = 14.99$ cm^{-1}; **5.48** 264.84 kJ mol^{-1}; **5.49** (*a*) $n \to \sigma^*$, $\sigma \to \sigma^*$, (*b*) $\pi \to \pi^*$, (*c*) $n \to \pi^*$, $\sigma \to \sigma^*$, (*d*) $n \to \pi^*$, $\pi \to \pi^*$, (*e*) $n \to \pi^*$, $\pi - \pi^*$; **5.50** (*a*) yes, one, (*b*) 4.305×10^6 Hz; **5.51** B(H) = 23.468×10^3 G, B(D) = 152.939×10^3 G; **5.52** (*a*) 1.73×10^5 G, (*b*) 2.49×10^4 G, (*c*) 8.87×10^4 G; **5.53** CH_3CHO; **5.54** CH_2ClCH_2Cl; **5.55** CH_3CCl_3; **5.56** (*i*) one line, (*ii*) three lines, 1:1:6, (*iii*) two lines 1 : 9, (*iv*) one line, (*v*) two lines one due to OH and other due to CH_3 protons; **5.58** (*i*), (*ii*), (*iv*), (*v*), (*vi*), (*viii*) and (*ix*); **5.59** 21.85%.

Chapter 6

6.1 1040.05 kJ mol^{-1}; **6.2** 5349.23 kJ mol^{-1}; **6.3** 641.63 kJ mol^{-1}; **6.4** 2845.67 kJ mol^{-1}; **6.5** 700.63 kJ mol^{-1}; **6.6** (*a*) 1.64×10^4 kg m^{-3}, 0.82×10^4 kg m^{-3}, (*b*) 2.87×10^{-10} m, 3.52×10^{-10} m, (*c*) 636.99 kJ mol^{-1}, 761.38 kJ mol^{-1}; **6.7** $r_e = 3.39 \times 10^{-10}$ m, $\Delta H = 2598.23$ kJ mol^{-1}; **6.8** (*i*) 655.46 kJ mol^{-1}, (*ii*) 1139.72 kJ mol^{-1}; **6.9** $\Delta H = \dfrac{-N_{av} Z_+ Z_- e^2 M}{4\pi \in_0 r_e}$ [1 – 1/*n*] + 2RT; **6.10** (*i*) 629.03 kJ mol^{-1}, (*ii*) 2503.40 kJ mol^{-1}, (*iii*) 655.66 kJ mol^{-1}, (*iv*) 1077.59 kJ mol^{-1};

6.11

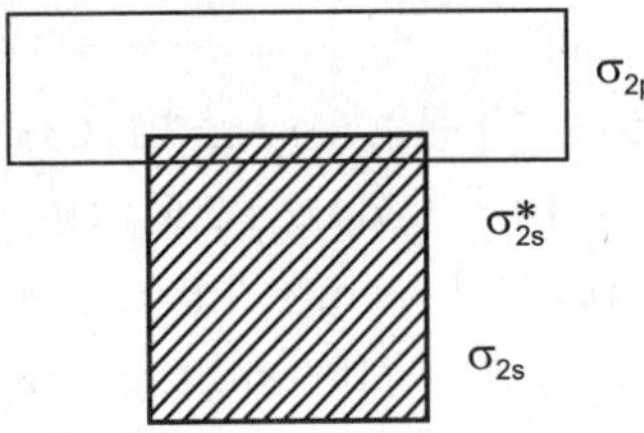

6.12

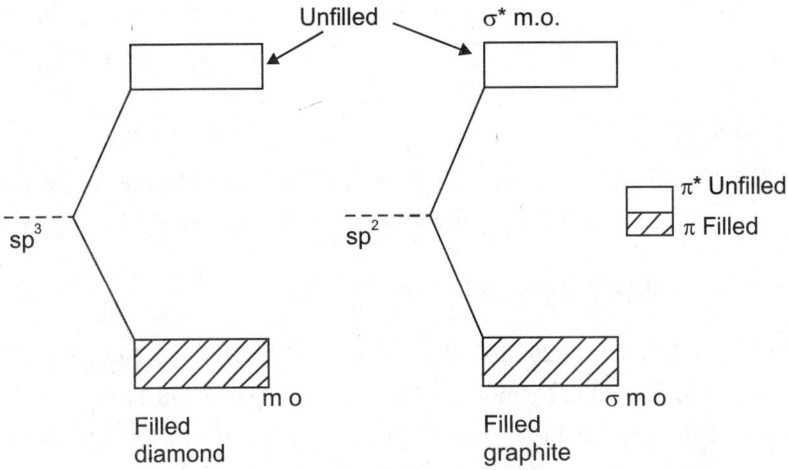

6.13 (*a*) increase of temperature increases molecular vibrations in metals and obstructs the motion of electrons, (*b*) for insulators, electrons from filled m.o. goes to unfilled m.o.; **6.14** electrons go from filled to unfilled m.o.; **6.15** $\bar{C}_v = 0.23$ J mol^{-1} K^{-1}; **6.16** $\bar{C}_v = 0.187$ J mol^{-1} K^{-1}; **6.17** 5.73×10^{-10} m; **6.18** 6.08×10^{-10} m; **6.19** (*a*) 10.67%, (*b*) 6.26%; **6.20** 52.38%; **6.21** 8; **6.22** (0, 0, 0), ($\frac{1}{2}$, $\frac{1}{2}$, 0), (0, $\frac{1}{2}$, $\frac{1}{2}$), ($\frac{1}{2}$, 0, $\frac{1}{2}$), for face centred, (0, 0, 0), ($\frac{1}{2}$, $\frac{1}{2}$, $\frac{1}{2}$) for body centred; **6.23** (*a*) 2.165×10^{-10} m, (*b*) 1.768×10^{-10} m, (*c*) 2.5×10^{-10} m; **6.24** (*a*) $r/R = 0.227$, (*b*) $r/R = 0.414$; **6.25** (*a*) 74.11%, (*b*) 52.38%, (*c*) 68.04%; **6.26** 74.1%; **6.27** AB_3; **6.28** (*a*) XY_2, (*b*) 2.6×10^{-10} m; **6.29** (*a*) Rb^+($\frac{1}{2}$, $\frac{1}{2}$, $\frac{1}{2}$), (*b*) I^-(0, 0, 0), (0, 0, 1), (0, 1, 0), (1, 0, 0), (*b*) $\dfrac{r(Rb^+)}{r(I^-)} = 0.69$, (*c*) 0.42 nm, (*d*) 4.75×10^3 kg m^{-3}; **6.30** (*a*) S^{2-}(0, 0, 0), ($\frac{1}{2}$, $\frac{1}{2}$, 0), ($\frac{1}{2}$, 0, $\frac{1}{2}$), (0, $\frac{1}{2}$, $\frac{1}{2}$), Mg^{2+}($\frac{1}{2}$, 0, 0), (0, $\frac{1}{2}$, 0), (0. 0, $\frac{1}{2}$), ($\frac{1}{2}$, $\frac{1}{2}$, $\frac{1}{2}$), (*b*) 0.35, (*c*) 4, (*d*) 8.53 kg dm^{-3}; **6.31** (*i*) 4.2×10^{-10} m, (*ii*) 3.52×10^{-10} m; **6.32** 1/6; **6.33** (*a*) 97% (*b*) 95% ; **6.34** (*b*) 3.08×10^3 kg m^{-3}, (*c*) 2.598×10^{-10} m, (*d*) CN of A is 8 and that of B is 4; **6.35** (*a*) Na^+($\frac{1}{4}$, $\frac{1}{4}$, $\frac{1}{4}$), ($\frac{1}{4}$, $\frac{1}{4}$, $\frac{3}{4}$), ($\frac{1}{4}$, $\frac{3}{4}$, $\frac{1}{4}$), ($\frac{3}{4}$, $\frac{1}{4}$, $\frac{1}{4}$), ($\frac{3}{4}$, $\frac{3}{4}$, $\frac{1}{4}$), ($\frac{3}{4}$, $\frac{1}{4}$, $\frac{3}{4}$), ($\frac{1}{4}$, $\frac{3}{4}$, $\frac{3}{4}$), ($\frac{3}{4}$, $\frac{3}{4}$, $\frac{3}{4}$), 0^{2-}, (0, 0, 0), (0, $\frac{1}{2}$, $\frac{1}{2}$), ($\frac{1}{2}$, 0, $\frac{1}{2}$), ($\frac{1}{2}$, $\frac{1}{2}$, 0), (*b*) 0.24 nm, (*c*) Na_2O; **6.36** $r_{O^{2-}} = 2.51 \times 10^{-10}$ m; **6.37** 0.212 m; **6.38** $r_{Na^+} = 0.128$ nm; **6.39** 6.41.

Chapter 7

7.1 (*a*) U_L(He – He) = 1.078 × 10^{-23} J, U_L(Ar – Ar) = 3.54 × 10^{-22} J, (*b*) Ar; **7.2** 3.33 kJ mol^{-1}; **7.3** U_{D-D} = – 2.14 × 10^{-26} J, U_{ind} = – 3.96 × 10^{-25} J, U_L = – 5.31 × 10^{-22} J; **7.4** α = 2.6 × 10^{-39} cm^2 V^{-1}; **7.5** 5.65 × 10^{-40} cm^2 V^{-1}; **7.6** 1 poise = 0.1 Pl; **7.7** (*a*) viscous, (*b*) 51.46 N m^{-2}; **7.8** 0.32 m s^{-1}, 0.318 m s^{-1}, 0.030 m s^{-1}, 0.0269 m s^{-1}, 0.0205 m s^{-1}, 0.0115 m s^{-1}; **7.9** 1.327 × 10^{-7} m^3 s^{-1}; **7.10** 31.865 N m^{-2}; **7.11** 0.488; **7.12** 2.54 × 10^{-5} dm^3 s^{-1}; **7.13** 0.578 × 10^{-2} P; **7.14** 3.28 × 10^{-2} P; **7.15** 6.25 kJ mol^{-1}; **7.16** 0.029 m; **7.17** 2 × 10^{-4} m; **7.18** (*a*) 140.86 dynes cm^{-1}, (*b*) 1.50 × 10^{-4} m; **7.19** θ = 72.6°; **7.20** p/p_0 = 10.66.

Chapter 8

8.1 –459.4°F; **8.2** –297.2°F; **8.3** (*i*) 53.44 × 10^3 N m^{-2}, (*iii*) 5.33 × 10^5 dyne cm^2; **8.4** –40°C; **8.5** 241.35 mm; **8.6** $x_{N_2} : x_{O_2}$ :: 0.82 : 0.18; **8.7** h_{N_2} = 6258.5 m, h_{O_2} = 5476.15 m; **8.8** 34.07 mm; **8.9** 443 K; **8.10** 16.67 mm; **8.11** 87.65 K; **8.12** 16.99 psi; **8.13** 1.667 dm^3; **8.14** $p_{eq} = \dfrac{p_A V_A + p_B V_B}{V_B + V_B}$; **8.15** 2.6 atm; **8.16** 24 dm^3; **8.17** 0.57 cm; **8.18** (*a*) 18.20 dm^3, (*b*) 3.53 atm; **8.19** 66.99 g; **8.20** C_6H_6; **8.21** ω(H_2) = 0.038 g, ω(Zn) = 1.24 g; **8.22** M_{av} = 38.21 g mol^{-1}; **8.23** M_{av} = 17.97 g mol^{-1}; **8.24** M_{av} = 28.79 g mol^{-1}; **8.25** *x*(Cl) = 0.621; *d* = 0.596 g dm^{-3}, **8.26** M_B = 113.3 g mol^{-1}; **8.27** (*a*) *p* = 1.41 atm, (*b*) 9.11 mol; **8.28** (*a*) 13.34 dm^3, (*b*) p_{H_2} = 0.615 atm, p_{O_2} = 0.185 atm; **8.29** (*a*) 9.34 × 10^{-28} m^3; (*b*) 7.19 × 10^{-28} m^3, (*c*) 1.52 × 10^{-27} m^3, (*d*) 1.74 × 10^{-28} m^3; **8.30** (*a*) *p* = 23.307 atm, (*b*) p_{Cl_2} = 19.43 atm, p_{CO_2} = 21.96 atm. p_{He} = 25.026 atm, (*c*) $p_i(Cl_2)$ = 24.466 atm, $p_i(CO_2)$ = 24.466 atm = p_i(He); **8.31** V_{CH_4} = 12.133 dm^3, V_{Cl_2} = 11.80 dm^3, V_{CO_2} = 12.02 dm^3, V_{He} = 12.28 dm^3; **8.32** (*i*) *B* = – 0.493 dm^3 mol^{-1}, *C* = 1.83 × 10^{-3} dm^6 mol^{-2}, (*ii*) *B* = – 0.104 dm^3 mol^{-1}, *C* = 1.82 × 10^{-3} dm^6 mol^{-2}, (*iii*) *B* = – 0.209 dm^3 mol^{-1}, *C* = 3.161 × 10^{-3} dm^6 mol^{-2}, (*iv*) *B* = 0.022 dm^3 mol^{-1}, *C* = 5.62 × 10^{-4} dm^6 mol^{-2}; **8.33** (*i*) 641.47 K, (*ii*) 1055.35 K, (*iii*) 1406.73 K, (*iv*) 17.54 K; **8.34** $T_r = \dfrac{27}{8}$; **8.35** (*a*) α = 1/*T*, (*b*) $\alpha = \dfrac{R(\overline{V}-b)}{\overline{V}\left(RT - \dfrac{2a(\overline{V}-b)^2}{\overline{V}^3}\right)}$;

8.37 (*i*) V_c = 0.128 dm^3 mol^{-1}, p_c = 45.595 atm, T_c = 190.0 K, (*ii*) V_c = 0.128 dm^3, p_c = 73.01 atm, T_c = 303.8 K, (*iii*) V_c = 0.1566 dm^3 mol^{-1}, p_c = 76.085 atm, T_c = 416.81 K, (*iv*) V_c = 0.0711 dm^3 mol^{-1}, p_c = 2.25 atm, T_c = 0.65 K; **8.38** (*i*) *b* = 0.0237 dm^3 mol^{-1}, *a* = 0.0342 dm^6 atm mol^{-2}, (*ii*) *b* = 0.0429 dm^3 mol^{-1}, *a* = 3.612 dm^6 mol^{-2} atm; (*iii*) *b* = 0.0562 dm^3 mol^{-1}, *a* = 6.5 dm^6 mol^{-2} atm; (*iv*) *b* = 0.0428 dm^3 mol^{-1}, *a* = 2.257 dm^6 atm mol^{-2}; **8.39** (*i*) 3.438 dm^3; (*ii*) 3.33 dm^3, (*iii*) 3.40 dm^3, (*iv*) V_{ideal} = 3.474 dm^3; **8.40** (*i*) 0.992, (*ii*) 0.965, (*iii*) 0.9826, (*iv*) 1.004 ; **8.41** 89.16 g mol^{-1}; **8.42** 43.34 g mol–^{1}; **8.43** (*a*) $\bar{u}_1 < \bar{u}_2$ (*b*) $N_1 < N_2$, (*c*) $T_1 < T_2$, (*d*) $V_1 < V_2$, (*e*) $p_1 < p_2$; **8.44** (*a*) $\dfrac{n_{O_2}}{n_{N_2}}$ = 0.935, (*b*) 0.17 × 10^{-23} kg m s^{-1}; **8.45**

C_{rms} = 340.79 m s^{-1}, C_{mp} = 278.25 m s^{-1}, $\bar{C}$ = 313.91 m s^{-1}; **8.46** C_{rms} = 411 m s^{-1}, C_{mp} = 335.58 m s^{-1}, $\bar{C}$ = 378.59 m s^{-1}; **8.47** (*a*) 1:0.829, (*b*) 436.36 K; **8.49** (*a*) & (*c*) $\bar{C}$(CO_2, 27°C) = 379.859 m s^{-1}, $\bar{C}$(SO_2, 27°C) = 314.96 m s^{-1}, $\bar{C}$(SO_2, 120°C) = 360.49 m s^{-1}, $\bar{C}$(CO_2, 120°C) = 434.767 m s^{-1}, (*b*) $\bar{C}$(CO_2, 27°C) = 379.669 m s^{-1}, $\bar{C}$(CO_2, 120°C) = 434.59 m s^{-1}, $\bar{C}$(SO_2, 27°C) = 314.80 m s^{-1}, $\bar{C}$(SO_2, 120°C) = 360.31 m s^{-1}; **8.50** 3.06 × 10^{21} molecules; **8.51** 400 mm; **8.52** 1.758 g; **8.53** rate (O_2):rate (SO_2):*r*(SO_3)::3.9530:1.677:1.0; **8.54** *r*(He):*r*(Ne):*r*(Ar): *r*(Kr)::160.4:20.49:5.8:1.0; **8.55** (*a*) 1.115 × 10^{28} molecule m^{-2} s^{-1}, (*b*) 1.852 mol cm^{-2} s^{-1}; **8.56** 0.926 mol cm^{-2} s^{-1}; **8.57** 4.63 mol cm^2 s^{-1}; **8.58** 48.5s; **8.59** (*a*) Z_1:Z_2::1:3, (*b*) temperature same, (*c*) Z_1:Z_2::1.414:1; (*d*) u_1:u_2::1.225:1; **8.60** (*a*) CO_2 more collision; (*b*) Z_1:Z_2 :: 0.3196:1; **8.61** 6.184 mm; **8.62** (*a*) & (*b*) 4.575 × 10^{-8} m, (*c*) 6.078 × 10^{28} collision s^{-1} dm^3; **8.63** (*a*) 18.56 × 10^{-8} m, (*b*) Z_1 = 3.236 × 10^9 mol collision s^{-1} m^{-3}, Z_{11} = 1.23 × 10^9 mol collision s^{-1} m^{-3}; **8.64** Z_{11} = 0.692 × 10^{10} mol collision s^{-1} m^{-3}, *l* = 2.42 × 10^{-7} m; **8.65** 2.37 × 10^{-4} poise; **8.66** 2.043 × 10^{-4} poise; **8.67** 5.6 × 10^{-10} m; **8.68** 4.72 × 10^{-10} m.

Chapter 9

9.1 1/9! (factorial) = 2.756 × 10^{-6}; **9.2** Ω's 30, 60, 15, 120, 60, 20, 90, 30, 1, Ω_T = 426, P_i 0.0704, 0.1408, 0.0352, 0.2817, 0.1408, 0.0469, 0.2113, 0.0704, 0.0025, Ω_{mp} = 0.28; **9.3** Ω's 30, 2160, 1500, 7200, 129600, 20, 3240, 38880, 46656, Ω_T = 229286, P_i, 1.31 × 10^{-4}, 9.42 × 10^{-3}, 6.452 × 10^{-3}, 0.0315, 0.5652, 8.72 × 10^{-5}, 0.0141, 0.1696, 0.2035, **9.5** (*i*) 84, (*ii*) 307.55 Sterling's theorem is applicable for larger system; **9.6** 9; **9.8** $\Omega_{max} = e^{-n'2/2n}$; **9.9** 0.49 ; **9.10** Ω_T = 10910, P_{MP} = 0.66; **9.11** ϵ_i^2 3.33, 3.0, 3.0, 2.33, 2.0, 2.0, 1.66, 1.33,1.0, $\epsilon^2_{ensemble}$ = 2.243; **9.12** 1.92; **9.13** No, as $\frac{N_2}{N_1} \neq \frac{N_3}{N_2}$; **9.14** 290 K; **9.15** N_1/N_0 = 0.9415, N_2/N_0 = 0.547, N_3/N_0 = 2.38 ×10^{-3}; **9.16** N_1/N_0 = 1.412, N_2/N_0 = 1.641, N_3/N_1 = 5.66 × 10^{-3}; **9.20** 414.95 × 10^{23}; **9.21** (*b*) 849.43 K; **9.22** (*b*) 831.3 K, (*c*) 18.38 kJ; **9.23** (*b*) Ω_{max} = 392; **9.24** (*a*) 5.98 × 10^6 N_{av}, (*b*) 11.96 × 10^6 N_{av}; **9.25** (*a*) 1.01 × 10^7 N_{av}, (*b*) 2.05 × 10^8 N_{av} (*c*) 2.53 × 10^7 N_{av}, 5.125 × 10^8 N_{av}; **9.26** (*a*) 0.164 × 10^{-26} kg, (*b*) 0.33 × 10^{-46} kg m^2, (*c*) 24.63, (*d*) 82.1; **9.27** (*i*) 5.47, (*ii*) 5.14; **9.28** (*a*) 3813.22 K, (*b*) N_1/N_0 = 3.018 × 10^{-6}, N_2/N_0 = 9.1726 × 10^{-12}, N_3/N_0 = 2.75 × 10^{-17}, (*c*) 1.0; **9.29** (*a*) 1.096 × 10^{-6}, N_{av}, (*b*) 7.49 × 10^7 N_{av}; **9.30** (*a*) 9.31 × 10^5 N_{av}, (*b*) 6.27 × 10^7 N_{av}; **9.31** (*a*) 3.01 × 10^6 N_{av}, (*b*) 2.08 × 10^8 N_{av}; **9.32** $\bar{E}^0$ = 6.235 kJ mol^{-1}, $\bar{H}^0$ = 8.72 kJ mol^{-1}; **9.33** same as 9.32; **9.34** 198.63 J mol^{-1} K^{-1}; **9.35** (*a*) 169.75 J mol^{-1} K^{-1}, (*b*) 143.33 J K^{-1} mol^{-1}, (*c*) 153.13 J K^{-1} mol^{-1}; **9.36** 20.785 J K^{-1} mol^{-1}; **9.37** 20.785 J K^{-1} mol^{-1}; **9.38** 25.24 J mol^{-1} K^{-1}; **9.39** 161.97 J K^{-1} mol^{-1}; **9.40** ΔH^0(300 K) = – 0.488 kJ, ΔH^0(1000 K) = –4.908 kJ; **9.41** $K_p = \frac{2(m_A + m_B)}{m_A^{\frac{1}{2}}\, m_B^{\frac{1}{2}}} \exp$

$$\left[-\frac{h\nu_{A_2}}{2k_bT}\left(\sqrt{2}\left(\frac{m_A + m_B}{m_B}\right)^{1/2} - \left(\frac{m_A}{m_B}\right)^{1/2} - 1\right)\right]$$

9.42 $K_p = 4.243 \exp\left(-\frac{(h\nu_{H_2})(0.0125)}{k_bT}\right)$; **9.43** K(300 K)

= 8.644 × 10^2, K(1000 K) = 1.34 × 10^2; **9.44** K(300 K) = 10.35, K(1000 K) = 8.02; **9.45** 7.54 × 10^{150}; **9.46** (*a*) K(300 K) = 2.26 × 10^{-20}, K (300 K) = 4.29 × 10^{-18}, (*b*) 1.57 × 10^{-7} = K(100 K), K(1000 K) = 1.25 × 10^{-5}.

Chapter 10

10.1 7.35J; **10.2** 30 kJ mol^{-1}; **10.3** $\frac{1}{2}\alpha A\gamma(L_2^2 - L_1^2)$; **10.4** (*iii*) $q-w$, (*iv*) q_p, (*vi*) $\frac{\Delta H}{T}$, (*vii*) $E + pV$, (*viii*) $H + TS$; **10.6** 35.46 × 10^2 J; **10.7** 49.97 kcal; **10.8** (*a*) 156.8 J, (*b*) ΔT = 0.204 deg; **10.9** (*a*) –240.59 kJ, (*b*) 268.49 kJ, (*c*) –3302.74 kJ, (*d*) –877.32 kJ, (*e*) –1298.39 kJ; **10.10** $C_v(25°C)$ = 66.57 J mol^{-1} deg^{-1}, $C_p(25°)$ = 74.88 J mol^{-1} deg^{-1}, C_p(473 K) = 106.86 J mol^{-1} deg^{-1}, C_v(473 K) = 98.546 J mol^{-1} deg^{-1}; **10.11** 536.72 kJ; **10.12** –888.75 kJ; **10.13** $\bar{C}_p - \bar{C}_v = \frac{R^2T\bar{V}^3}{RT\bar{V}^3 - (\bar{V}-b)^2.2a}$; **10.15** C_v = 6.536 × 10^{-5} J deg^{-1} mol^{-1}, $\bar{C}_v$(500 K) = 12.57 J K^{-1} mol^{-1}; **10.16** $\bar{C}_v$(Cu) = 0.07 J mol^{-1} deg^{-1}, $\bar{C}_v$(Ag) = 0.20 J mol^{-1} deg^{-1}, $\bar{C}_v$(Cu, 300 K) = 23.63 J K^{-1} mol^{-1}, $\bar{C}_v$(Ag, 300 K) = 24.30 J K^{-1} mol^{-1}; **10.17** γ = 2.5 mJ mol^{-1} K^{-2}, Θ_D = 5.035 K, C_v(300 K) = 24.94 J K^{-1} mol^{-1}; **10.18** γ = 3.5 mJ K^{-1} mol^{-1}, Θ_D = 3.37 K; **10.19** for Cu, T = 3.34 K, for Ag T = 1.83 K; **10.20** 8.3 × 10^{-3} K; **10.21** 23.70 J mol^{-1} K^{-1}; **10.22** $\bar{C}_v$(Ag) = 24.48 J K^{-1} mol^{-1}, $\bar{C}_v$(Cu) = 23.79 J K^{-1} mol^{-1}; **10.23** 0.82 J K^{-1} mol^{-1}; **10.24** $\frac{3}{2}R + 1.71\frac{CV^{2.4}}{T}$; **10.25** 465.6 J; **10.26** 12.566 dm^3; **10.27** $Q_{1-2} = \Delta E_{1-2}$ = 25104 J, $Q_{2-3} = W_{2-3}$ = 16440.96 J, Q_{3-1} = – 35080.8 J, W_{3-1} = – 9982.05 J, ΔE_{3-1} = – 25098.75 J; **10.28** $Q_{1-2} = W_{1-2}$ = 1728.45 kJ, $Q_{2-3} = \Delta E_{2-3}$ = 1990.32 kJ, – $W_{3-1} = \Delta E_{3-1}$ = – 1990.30 kJ; **10.29** $Q_{1-2} = \Delta E$ = 81.5 kJ, $\Delta E_{2-3} = W_{2-3}$ = – 81.5 kJ, W_{3-1} = Q_{3-1} = – 29.89 kJ; **10.30** $Q = W$ = 645.61 J; **10.31** (*i*) W = 1164.57 J, $Q_p = Q = \Delta H = \Delta E$ = 11557.84 J, (*ii*) Q = 12721.8 J, ΔE = 11557.23 J, $Q_p = \Delta H$, p_2 = 1.59 atm; **10.32** (*i*) Q = 12773.77 J, ΔE = 11609.20 J; **10.33** (*i*) $Q = W$ = 21.739 kJ, $\Delta H = 0 = \Delta E$, (*ii*) Q = 346.8 kJ, W = 21.749 kJ, $\Delta E = \Delta H$ = 325.06 kJ; **10.34** ΔH = – 17.76 kJ, ΔE = –12.707 kJ, W_T = 0, Q = – 22.813 kJ; **10.35** ΔT = 80729 K; **10.36** p = 0.526 atm; **10.37** V_2 = 86.575 dm^3, V_1 = 0.175 dm^3, W = 8.755 kJ, Q_p = 0.985 kJ, ΔE = – 7.77 kJ; **10.38** W = 3.987 kJ, ΔE = 0.53 kJ; **10.39** (*a*) ΔT = 936.90 deg, (*b*) height has to be increased; **10.40** (*a*) ΔT = 0.036 deg, m = 4.63 g; **10.41** (*i*) $W = -RT \ln\frac{p_2}{p_1} + \frac{C}{2}(p_1^2 - p_1^2)$, (*ii*) $W = RT\ln\frac{\bar{V}_2 - Z}{\bar{V}_1 - Z}$; **10.42** $T^{C_v}(\bar{V} - b)$ = const.; **10.44** $W = \Delta E$ = – 2.65 kJ, ΔH = – 3.74 kJ; **10.45** – $W = \Delta E$ = – 6.095 kJ, ΔH = – 8.508 kJ; **10.46** ΔE = – W = 80.60 kJ, ΔH = 112.56 kJ; **10.47** ΔH = 23.35 kJ, – $W = \Delta E$ = 16.0 kJ; **10.48** (*a*) W = 6.915 kJ = Q, ΔE = 0, (*b*) W = 6.838 kJ, ΔE = 0.11 kJ, Q = 6.948 kJ; **10.49** (*a*) W = – 4.879 kJ, (*b*) W = – 4.853 kJ; **10.50** $W = -\left(p_0 + \frac{1}{\beta}\right)(V_2 - V_1) - \frac{V_2}{\beta}\ln\frac{V_2}{V_1}$; **10.53** 1 deg; **10.54** ΔT = – 24.38 deg; **10.55** (*c*) ΔT = – 0.652 K; **10.59** 0.156 K atm^{-1}; **10.60** ΔT = –173.1 deg.

Chapter 11

11.1 10.573 kJ mol^{-1}; **11.2** 3615.58 J; **11.3** ΔT = 48.46 deg; **11.4** – 3.303.12 kJ mol^{-1}; **11.5** C = 19.25 kJ deg^{-1}; **11.6** (*i*) –1299.61 kJ mol^{-1}, (*ii*) –1191.96 kJ mol^{-1}, (*iii*) –246 kJ mol^{-1}, (*iv*) –764.03 kJ mol^{-1}; **11.7** –6.86 kJ; **11.8** –275.74 kJ mol^{-1}; **11.9** (*a*) 8.4 kJ mol^{-1}, (*b*) –485.77 kJ mol^{-1}, (*c*) –228.1 kJ; **11.10** 707.34 kJ mol^{-1}; **11.11** 156.86 kJ mol^{-1}; **11.12** –276.9 kJ mol^{-1}; **11.13** (*a*) 131.3 kJ, (*b*) –457.33 kJ mol^{-1}, (*c*) –1559.87 kJ, (*d*) –175.98 kJ, (*e*) –3297.17 kJ; **11.14** (*a*) –130.05 kJ; (*b*) –454.85 kJ, (*c*) –1556.75 kJ, (*d*) –173.49 kJ, (*e*) –3300.60 kJ; **11.15** (*a*) –190.43 kJ mol^{-1}, (*b*) –149.1 kJ mol^{-1}, (*c*) 89.4 kJ mol^{-1}, (*d*) –105.22 kJ mol^{1}; **11.16** –344.11 kJ mol^{-1}; **11.17** m = 6.7 g; **11.18** 412.12 kJ mol^{-1}; **11.19** –863.58 kJ mol^{-1}; **11.20** 14.6 kJ mol^{-1}; **11.21** 725.57 kJ; **11.22** –464.41 kJ mol^{-1}; **11.23** –548.1 kJ mol^{-1}; **11.24** –242.67 kJ mol^{-1}; **11.25** 4.44 kJ mol^{-1}; **11.26** 5.27 kJ mol^{-1}; **11.27** –46.19 kJ mol^{-1}; **11.28** difficult to find from this data, too steep curve near $\sqrt{m} = 0$, ΔH_{diff} = 0.3 kJ mol^{-1}; HF; **11.29** ΔH at $\sqrt{m} = 0 = -25$ kJ mol^{-1} HI, ΔH_{diff} = 3.90 kJ mol^{-1}; **11.30** (*a*) 0.31 kJ mol^{-1} HF, (*b*) –2.3 kJ mol^{-1} HI.

Chapter 12

12.1 1; **12.2** 0.214; **12.4** 1.02 kJ; **12.5**

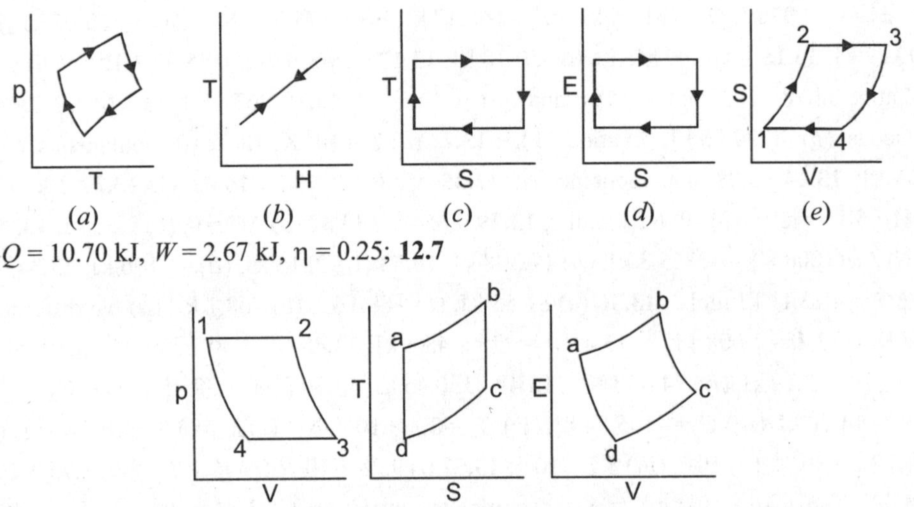

(*a*) (*b*) (*c*) (*d*) (*e*)

12.6 Q = 10.70 kJ, W = 2.67 kJ, η = 0.25; **12.7**

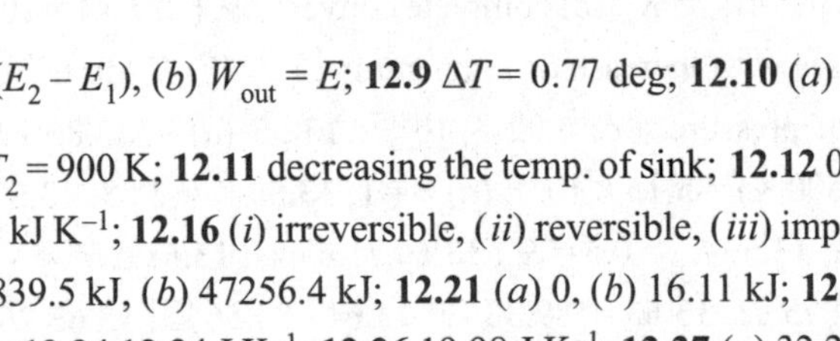

12.8 (*a*) $W_{out} = E - (E_2 - E_1)$, (*b*) $W_{out} = E$; **12.9** ΔT = 0.77 deg; **12.10** (*a*) $T_2^2 = T_1 T_3$, (*b*) $T_2 = \frac{1}{2}(T_1 + T_3)$, (*c*) T_2 = 670.8 K, T_2 = 900 K; **12.11** decreasing the temp. of sink; **12.12** 0.289; **12.13** 804 kJ; **12.14** 1.13 kJ K^{-1}; **12.15** 2.468 kJ K^{-1}; **12.16** (*i*) irreversible, (*ii*) reversible, (*iii*) impossible; **12.19** –689.31 kJ K^{-1} mol^{-1}; **12.20** (*a*) 42839.5 kJ, (*b*) 47256.4 kJ; **12.21** (*a*) 0, (*b*) 16.11 kJ; **12.22** 95.99 J K^{-1}; **12.23** ΔS = 54.39 J K^{-1}, ΔS_{surr} = 0; **12.24** 12.24 J K^{-1}; **12.26** 19.98 J K^{-1}; **12.27** (*a*) 32.29 J K^{-1}, (*b*) ΔS = 0; **12.28** 270.71 J K^{-1} mol^{-1}; **12.29** 5.61 kJ K^{-1}; **12.30** (*a*) ΔV = – 11.164 dm^3, ΔT = 362.88 K, ΔS = – 2.41 J K^{-1}, (*b*) ΔS_{surr} = 0, (*c*) ΔS = 0; **12.31** 24.19 J K^{-1}; **12.32** 40.33 J K^{-1}; **12.33** ΔS_v = 28.73 J K^{-1}, ΔS_p = 40.26 J

K^{-1}; **12.34** 26.24 J K^{-1}; **12.35** (*i*) 31.02 kJ mol^{-1} (*ii*) 35.03 kJ mol^{-1}, (*iii*) 29.36 kJ mol^{-1}; **12.36** (*i*) 109.0 J K^{-1} mol^{-1}, (*ii*) 109.76 J K^{-1} mol^{-1}, (*iii*) 62.22 J K^{-1} mol^{-1}, (*iv*) 73.34 J K^{-1} mol^{-1}; **12.37** –28.29 J K^{-1}; **12.38** 388.49 kJ K^{-1}; **12.39** 3.81 J K^{-1}; **12.40** 2 mC ln $\left[\frac{T_1+T_2}{(T_1T_2)^{1/2}}\right]$; **12.42** (*i*) 2.125 atm, (*ii*) 4.96 atm, (*iii*) 0.163 atm, (*iv*) 580.31 atm; **12.44** 66.98 J K^{-1} mol^{-1}; **12.45** (*b*) –0.207 J K^{-1} mol^{-1}; **12.46** 5.26 J K^{-1}; **12.47** $\Delta S = nR \ln \frac{V_2 - nb}{V_1 - nb}$; **12.48** 10.985 J K^{-1}; **12.51** – 61.48 J K^{-1}; **12.52** 1.379 K^{-1}; **12.54** 1/5 ln 4; **12.55** 264.09 J K^{-1} mol^{-1}; **12.57** imperfect; **12.58** third law is valid; **12.59** (*a*) 12.01 EU, (*b*) –217.23 EU, (*c*) 6.77 EU; **12.60** –128.33 EU; **12.61** (*a*) 201.26 EU, (*b*) –26.29 EU.

Chapter 13

13.1 (*i*) –15.51 kJ, (*ii*) 0, (*iii*) 1.55 kJ, (*iv*) 13.96 kJ, (*v*) 6.20 kJ; **13.2** 8.644 kJ; **13.3** –873.37 kJ; **13.4** ΔE = 2092 J, ΔA =– 12.29 kJ; **13.5** – 864.73 kJ; **13.6** –89.31 kJ; **13.7** –3001.42 kJ; **13.8** (*i*) – 3029.25 kJ, (*ii*) 28.58 kJ; **13.9** (*i*) – 2934.88 kJ, (*ii*) 28.57 kJ, (*iii*) –125.19 kJ; **13.10** (*a*) –33.31 kJ, (*b*) 496.32; **13.11** w = 58.83 kJ, Q = 77.14 kJ, ΔE = 18.31 kJ, ΔH = 77.14 kJ, ΔS = 0.219 kJ, ΔA = – 59.27 kJ, ΔG = –0.44 kJ; **13.12** w = 14.97 kJ, Q = 181.14 kJ, ΔE = 166.17 kJ, ΔS = 175.08 J K^{-1}, ΔG = 533.05 kJ; **13.13** ΔG^0 = – 499.27 kJ; **13.15** – 110.09 kJ; **13.16** –50.18 kJ; **13.17** (*i*) yes, (*ii*) $K_p = 5.99 \times 10^{40}$, (*iii*) $p_{CO_2} = 1.50 \times 10^{40}$ atm; **13.18** (*a*) E and V, (*b*) p and S, (*c*) T and V; **13.19** 263.15 K; **13.20** (*a*) 8.79 kJ, non-spontaneous, (*b*) –1137.45 kJ, (*c*) –562.33 kJ; **13.21** (*a*) $T = 10^4$ K; **13.22** (*i*) spontaneous, solid; **13.23** – 23.65 kJ; **13.24** 95.08, non-spontaneous; **13.25** –5.90 kJ; **13.26** 10.03 kJ; **13.27** 106.04 J; **13.28** (*a*) –313.05 J g mol^{-1}, (*b*) –296.5 J mol^{-1}, **13.29** 5.654 kJ; **13 30** (*a*) ΔG = 0, (*b*) 2.52 kJ; **13.31** 6.32 kJ; **13.32** (*i*) ΔG(1000 K) = –2825.3 kJ, (*ii*) 17.308 kJ; **13.33** (*i*) –2815 kJ, (*ii*) –1.196 kJ; **13.34** –43.08 kJ; **13.35** ΔG = – 43.11 kJ mol^{-1}, **13.36** (*a*) 81.55 kJ, (*b*) –68.1 kJ, (*c*) –68.1 kJ, (*d*) No difference; **13.39** 27.07 J K^{-1}; **13.40** –1.008 J K^{-1}; **13.41** $W = Q$ = –4.86 kJ; **13.48** (*i*) –0.88 kJ mol^{-1}, (*ii*) 1.24 kJ mol^{-1}, (*iii*) ΔG_{mix} = – 22.19 kJ, (*iv*) 74.45 J K^{-1}; **13.49** –130.45 kJ; **13.50** (*i*) 111.38 kJ, (*ii*) G^o(C) = –119.89 kJ, $G^0(C_p)$ = –14.50 kJ, (*iii*) G_f = –135.84 kJ, (*iv*) $K_p = 3.3 \times 10^{19}$, No, (*v*) $K_p = 3.3 \times 10^{19}$, α = 1, (*vi*) same; **13.51** (*i*) 2.3×10^{514}, (*ii*) 10^{-5}, (*iii*) 8.7×10^{21}; **13.52** (*i*) 9.28×10^{511} (*ii*) $K_p = K_c$, (*iii*) 1.75×10^{10}; **13.53** (*a*) –78.68 kJ, (*b*) $6.16 \times 10^{13} = K_p = K_c$, (*c*) complete conversion, (*d*) 2.14×10^9; **13.54** (*a*) –1037.08 kJ, (*b*) –991.9 kJ, (*c*) decreases with increase of temperature, (*d*) K_p, K_c do not change with pressure, K_x increases with increase of pressure, (*e*) 6.92×10^{107}; **13.55** (*a*) –133.85 kJ, (*b*) $K_p = 2.87 \times 10^{28}$, $K_c = 1.43 \times 10^{30}$, (*c*) K_p (500 K) = 5.15×10^{14}; (*d*) x = 1; **13.56** K = 2.58; **13.57**, K = 3.08×10^{45}; **13.58** not possible with this data; **13.59** $K_p = 0.5 = K_x$; **13.60** 71.67 kJ; **13.61** ΔG^0 = – 43.48 kJ, ΔH = 42.07 kJ, ΔS = 4.74 J K^{-1}; **13.62** –4.75 kJ; **13.63** –189.33 kJ; **13.64** –84.56 kJ; **13.65** 26.50 kJ **13.66** 2.05×10^5; **13.67** with increase of pressure (*i*) No effect, (*ii*) backward, (*iii*) forward, (*iv*) backward, (*v*) No effect, (*vi*) forward, (*vii*) backward, (*viii*) forward; **13.68** with increase of temperature, (*i*) backward, (*ii*) backward, (*iii*) backward, (*iv*) forward, (*v*) forward, (*vii*) backward, (*viii*) forward; **13.69** (*i*) α = 6.88

$\times$ 10^{-14}, (*ii*) $\alpha = 1.38 \times 10^{-14}$; **13.70** 4.586×10^{-3}; **13.71** 0.495, $K_p = 0.398$; **13.72** 0.021; **13.73** 8.62 K; **13.74** (*a*) 9.55×10^{-2}, (*b*) $p_{NH_3} = 0.365$ atm, $p_{HCl} = 0.257$ atm; **13.75** 0.3645; **13.76** (*i*) –19.94 kJ, (*ii*) 1.053×10^{-4}; **13.77** 96.88 atm; **13.78** 102.73 atm; **13.79** (*i*) 99.55 atm; (*ii*) 102.66 atm; **13.80** (*i*) 93.24 atm, (*ii*) 99.47 atm; **13.81** $f(CO_2) = 96.43$ atm, $f(CO) = 102.48$ atm, $f(H_2O) = 88.76$ atm, $f(H_2) = 104.1$ atm; **13.82** $f(273 \text{ K}) = 98.14$ atm, $f(373 \text{ K}) = 98.32$ atm; **13.83** 3.59×10^{-3} atm; **13.84** (*i*) 99.3 atm, (*ii*) 95.26 atm; **13.85** (*i*) 106 atm, (*ii*) 101 atm, (*iii*) 97.0 atm, (*iv*) 76 atm; **13.86** (*i*) 102.7 atm, (*ii*) 103.15 atm, (*iii*) 99.0 atm, (*iv*) 86.83 atm; **13.87** 60.97 atm; **13.88** 99.91 atm; **13.89** 8.79 kJ; **13.90** –4.0 kJ; **13.91** $K = 0.906$.

Chapter 14

14.1 604.11 deg; **14.2** 0.78 mm; **14.3** 47.5 kJ mol^{-1}; **14.4** 0.35 K; **14.5** 1.33×10^{-3} kg dm^{-3}; **14.6** 2.97 mm; **14.7** 239.24 K; **14.8** 88.33 kJ mol^{-1}; **14.9** (*a*) 70.78°C, (*b*) 81.17 J K^{-1} mol^{-1}; **14.10** 358.33 K; **14.11** 556.24 K; **14.12** 36.7 kJ mol^{-1}; **14.14** 12.9°C; **14.15** $T = 243$ K, $p = 3.3$ mm; **14.16** 0.7 mm; **14.17** all ideal liquid same vapour pressure; **14.18** (*a*) 8.13 kJ mol^{-1}, (*b*) 48.13 kJ mol^{-1}, (*c*) $T = 356.5$ K; **14.19** $d_A > d_B$; **14.21** 455.5 atm; **14.22** (*a*) 1.004 mm, (*b*) 1.045 mm; **14.23** 437.6 dm^3; **14.24** (*a*) 1, (*b*) 1, (*c*) 2, (*d*) 2, (*e*) 3, (*f*) 3, (*g*) 3; **14.25** (*a*) 2, (*b*) 2, (*c*) 2, (*d*) 3, (*e*) 2, (*f*) 2, (*g*) 3; **14.26** (*a*) 3, (*b*) 3, (*c*) 2, (*d*) 3, (*e*) 1, (*f*) 1, (*g*) 2; **14.27** No; **14.28**

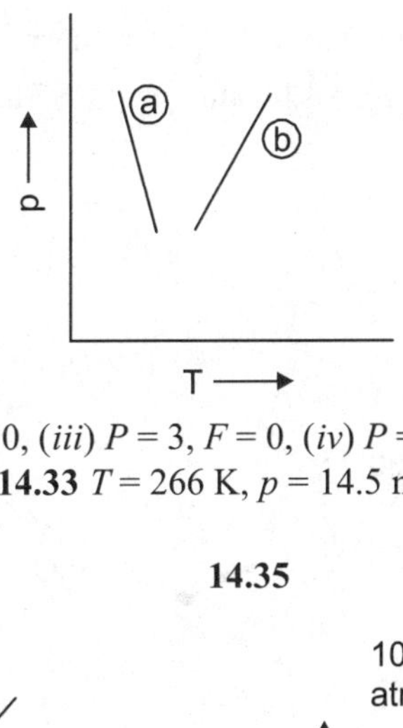

14.29 (*i*) $P = 2, F = 1$, (*ii*) $P = 3, F = 0$, (*iii*) $P = 3, F = 0$, (*iv*) $P = 3, F = 0$, (*v*) $P = 1, F = 2$, (*vi*) $P = 2, F = 1$; **14.32** $T = 282$ K $p = 8.5$ mm; **14.33** $T = 266$ K, $p = 14.5$ mm;

14.34

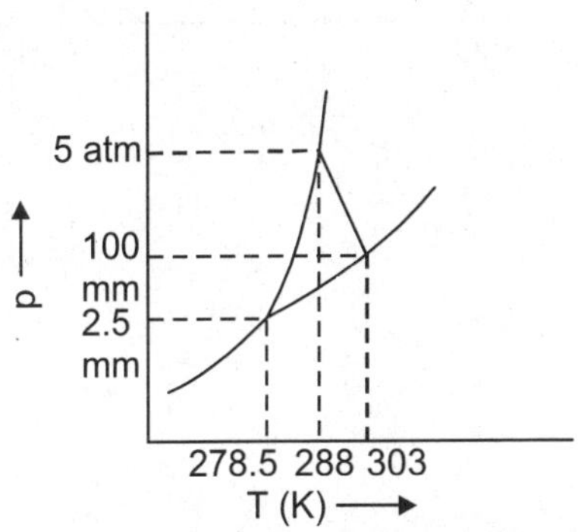

14.35

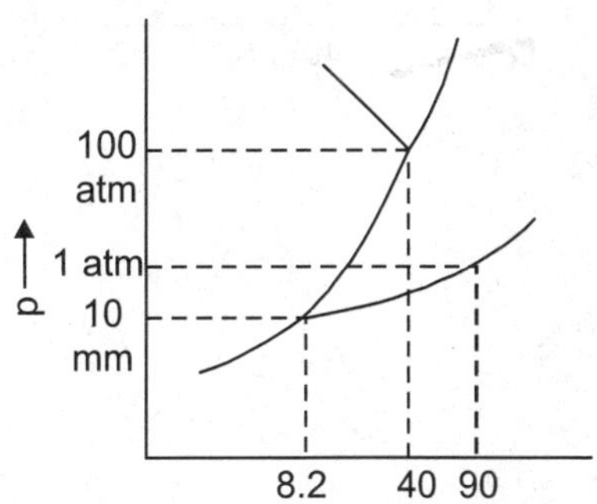

14.36 $T_e = 75°C$, $[B]_e = 49.5\%$; **14.37** $MgZn_2$, **14.39** Mg_2Ge;

14.40

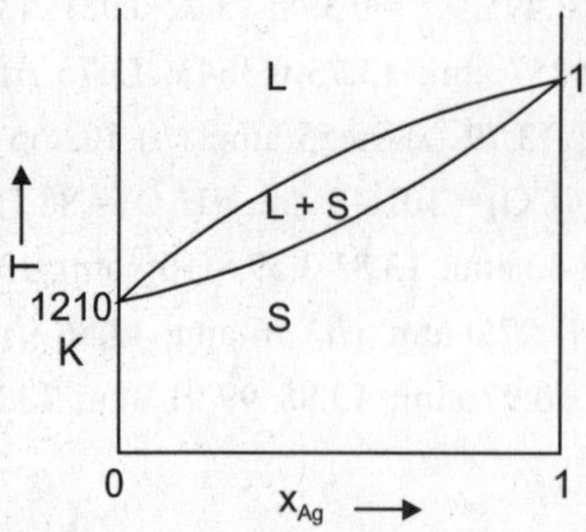

14.42

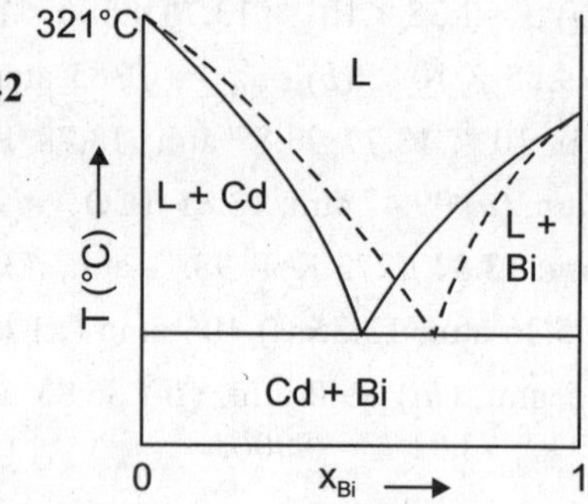

14.43 $x_e(B) = 0.865$, $T_e = 269.5$ K; **14.44** (*a*) 11.24 kJ mol^{-1}, x_e(m xylene) = 0.884, (*b*) 11.899 kJ; **14.45** (*i*) ΔH =16.59 kJ mol^{-1}, (*ii*) 14.98 kJ mol^{-1}; **14.46** T_e = 320 K, x_e = 0.155 (naphthalene); **14.47** (*b*) T = 23°C, $x_B = 0.38$, W_B = 50%, (*d*) [B] = 2.2 g, $[B]_2$ = 17.78 g; **14.48** (*b*) 62°C, W_B = 60.0%, (*d*) $W_B(1)$ = 47.26 g, $W_B(2)$ = 12.74; **14.49** Areas 1, 3, *V*, *L*, *P* = 1, *F* = 2 Areas $L_1 + L_2$, *P* = 2, *F* = 1; **14.50** *W*(1) = 47.27 g, *W*(2) = 12.73 g; **14.53** (*c*) *T* = 53.4 °C, $x(CHCl_3)$ = 0.656; **14.54** (*b*) $y(CHCl_3)$ = 0.11, $x(CHCl_3)$ = 0.19, (*c*) *T* = 64.4°C, $x(CHCl_3)$ = 0.64, (*d*) $(CH_3)_2CO$, (*e*) 60.2°C, (*f*) not possible, one phase, (*g*) $x(CHCl_3)$ = 0.12, $y(CHCl_3)$ = 0.07; **14.55** (*b*) *p* = 337.4 mm, $x(CS_2)$ = 88.47, $y(CS_2)$ = 95.26; (*c*) $K(CS_2)$ = 462.5 , mm, $K(C_6H_6)$ = 222.22 mm, (*d*) $x(CS_2)$ = 0.31, $y(CS_2)$ = 0.65, **14.57** (*a*) y_B = 0.2, *T* = 55°C, pure liquid B, (*b*) Tie line should be such that point divides it in the ratio of 1:3, 58°C, **14.58** (*a*) x_A = 0.8, y_A = 0.46, (*b*) $n_v(A)$ = 0.294, $n_e(A)$ = 0.706, (*c*) P_T vs x_A curve is linear, extend two points x_A = 0.7, P_T = 1 atm, y_A = 0.17, P_T = 1.05 atm, p_B^0 = 1.08 atm, p_A^0 = 0.97 atm.

14.61

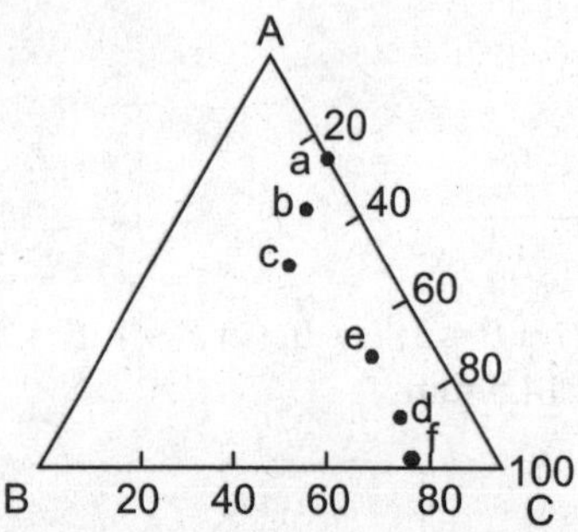

14.62

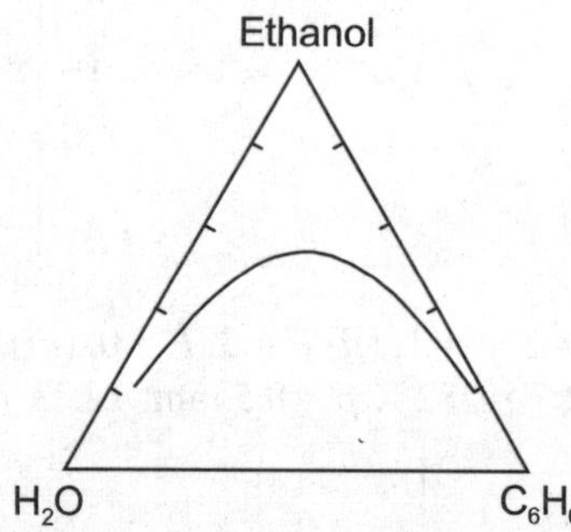

14.63

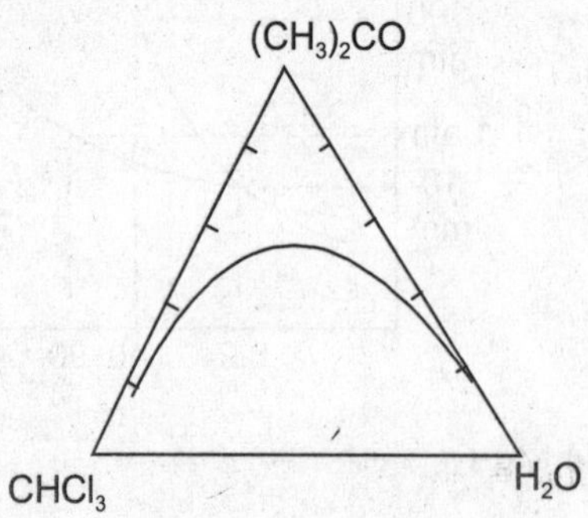

Chapter 15

15.1 0.111 m^{-1}; **15.2** 109.87 m^{-1}; **15.3** 0.011 m; **15.4** $5.1 \times 10^{-6}\ \Omega^{-1}\ m^2\ mol^{-1}$; **15.5** $0.0129\ \Omega^{-1}\ m^2\ mol^{-1}$; **15.6** $137.5 \times 10^{-4}\ \Omega^{-1}\ m^2\ mol^{-1}$; **15.7** 667.1 Ω; **15.8** $547.4 \times 10^{-4}\ \Omega^{-1}\ m^2\ mol^{-1}$; **15.9** $0.033\ \Omega^{-1}\ m^{-1}$; **15.10** (*a*) $\Lambda^0(BaBr_2) - \Lambda^0(BaCl_2) = 2.06 \times 10^{-4}\ \Omega^{-1}\ m^2\ mol^{-1}$, $\Lambda^0(MgBr_2) - \Lambda^0(MgCl_2) = 2.06 \times 10^{-4}\ \Omega^{-1}\ mol^{-1}\ m^2$; **15.11** $0.0524\ \Omega^{-1}\ m^2\ mol^{-1}$; **15.12** $1.9 \times 10^{-2}\ \Omega^{-1}\ mol^{-1}\ m^2$; **15.13** 0.023; **15.14** $404.22 \times 10^{-4}\ \Omega^{-1}\ m^2\ mol^{-1}$; **15.15** $t_{H^+} = 0.829$; **15.16** $t_{Cl^-} = 0.553$; **15.17** $t_{Cl^-} = 0.523$; **15.18** $t_{SO_4^{2-}} = 0.2002$; **15.19** $t_{Ag^+} = 0.473$; **15.20** (*i*) 0.0391 equi dm^{-3}, (*ii*) 0.0397 equi dm^{-3}; **15.21** $\lambda^0_{Na^+} = 50.15 \times 10^{-4}\ \Omega^{-1}\ m^2\ mol^{-1}$, $\lambda^0_{SO_4^{2-}} = 159.53 \times 10^{-4}\ \Omega^{-1}\ m^2\ mol^{-1}$; **15.22** $\lambda^0_{H^+} = 349.88 \times 10^{-4}\ \Omega^{-1}\ mol^{-1}\ m^2$, $\lambda^0_{Cl^-} = 76.28 \times 10^{-4}\ \Omega^{-1}\ m^2\ mol^{-1}$; **15.23** $t_{La^{3+}} = 0.464$, $\lambda^0_{Cl^-} = 77.48 \times 10^{-4}\ \Omega^{-1}\ m^2\ mol^{-1}$, $\lambda_{La^{3+}} = 202.03 \times 10^{-4}\ \Omega^{-1}\ m^2\ mol^{-1}$; **15.24** $\lambda_{Cl^-}(KCl) = 65.77 \times 10^{-4}\ \Omega^{-1}\ m^2\ mol^{-1}$; $\lambda_{Cl^-}(NaCl) = 65.65 \times 10^{-4}\ \Omega^{-1}\ m^2\ mol^{-1}$; **15.26** $U_{H^+} = 3.63 \times 10^{-7}\ m^2\ V^{-1}\ s^{-1}$, $U_{Cl^-} = 7.90 \times 10^{-8}\ m^2\ V^{-1}\ s^{-1}$; **15.27** 42.14 min; **15.28** $U_{Ag^+} = 6.42 \times 10^{-8}\ m^2\ V^{-1}\ s^{-1}$, $U_{Ca^{2+}} = 6.17 \times 10^{-8}\ m^2\ V^{-1}\ s^{-1}$, $U_{Li^+} = 4.01 \times 10^{-8}\ m^2\ V^{-1}\ s^{-1}$, $U_{NH_4^+} = 7.61 \times 10^{-8}\ m^2\ V^{-1}\ s^{-1}$; **15.29** $U_{I^-} = 18.55\ m^2\ V^{-1}\ s^{-1}$; **15.30** 34 min; **15.31** 97.48 g mol^{-1}; **15.32** (*i*) 6.412 deg mol^{-1}, (*ii*) 5.21 deg mol^{-1}; **15.33** (*i*) 3.375, (*ii*) 3.45, (*iii*) 2.8; **15.34** 4.66×10^{-3}; **15.35**,**15.36** and **15.37** Arrhenius law of dissociation is not applicable; **15.38** 2.0×10^{-16}; **15.39** 2.69×10^{-12}; **15.40** $9.72 \times 10^{-6}\ \Omega^{-1}\ m^{-1}$; **15.41** $9.3 \times 10^{-3}\ \Omega^{-1}\ m^{-1}$; **15.42** 1.13×10^{-2}.

Chapter 16

16.1 (*a*) $Fe^{3+} + e^- \rightarrow Fe^{2+}$, $Ti^{4+} + e^- \rightarrow Ti^{3+}$, $Fe^{3+} + Ti^{2+} \rightarrow Fe^{2+} + Ti^{3+}$, (*b*) $Zn^{2+} + 2e^- \rightarrow Zn$, $PbSO_4 + 2e^- \rightarrow Pb + SO_4^{2-}$, $Zn^{2+} + Pb + SO_4^{2-} \rightarrow Zn + PbSO_4$, (*c*) $H^+ + e^- \rightarrow \frac{1}{2}H_2$, $\frac{1}{2}Hg_2Cl_2 + e^- \rightarrow Hg + Cl^-$, $H^+ + Hg + Cl^- \rightarrow \frac{1}{2}H_2 + \frac{1}{2}Hg_2Cl_2$, (*d*) $AgCl + e^- \rightarrow Ag + Cl^-$, $\frac{1}{2}Q + H^+ + e^- \rightarrow \frac{1}{2}QH_2$, $AgCl + \frac{1}{2}QH_2 \rightarrow \frac{1}{2}Ag + H^+ + Cl^-$,

(*e*)
$$\begin{array}{l} CHCOO^- \\ \| \\ CHCOO^- \end{array} + 2H^+ + 2e^- \rightarrow \begin{array}{l} CH_2COO^- \\ | \\ CH_2COO^- \end{array}$$

(*f*) cytochrome (Fe^{3+}) + $e^- \rightarrow$ cytochrome (Fe^{2+}); **16.2** (*a*) Pt | H_2 | H^+; Cu^{2+} | Cu, (*b*) Pt | Cl_2 |Cl^-; Fe^{2+}, Fe^{3+} | Pt. (*c*) Ag | AgCl | Cl^-; Zn^{2+} | Zn, (*d*) Pt | H_2 | H^+; $[Fe(CN)_6]^{3-}$, $[Fe(CN)_6]^{4-}$ | Pt, (*e*) Pt | pyruvate, $2H^+$ lactate; NAD^+, H^+, NADH | Pt; **16.3** (*a*) 2.32×10^{12}, (*b*) 2.12×10^{-14}, (c) 8.56×10^{-10}, (*d*) 7.17×10^{-17}, (*e*) 1.5×10^5, (*f*) 1.23×10^{-20}, (*g*) 4.60×10^{-34}, (*h*) 4.7×10^{11}; **16.4** Ag will not react with I_2; **16.5** (*a*) standard hydrogen electrode, (*b*) Pt | H_2(1 atm) | $H^+(a_1 = 1)$; Fe^{3+}, Fe^{2+} | Pt, (*c*) $Fe^{3+} + H^+ \rightarrow Fe^{2+} + \frac{1}{2}H_2(g)$; **16.6** Pt | $H_2C_2O_4$, H^+; Mn^{2+}, MnO_4^- | Pt, E^0 = 2.0 V; **16.8** (*a*) E^0 = 0.153 V, (*b*) ΔG^0 = – 14764.5 J, (*c*) easy to oxidise Cu to Cu^{2+} than to Cu^+; **16.9** (*a*) 0:007 V, (*b*) –2702 J, (*c*) easy to oxidise Sn to Sn^{2+} than to Sn^{4+}; **16.10** (*a*) 1.468 V, (*b*) 1.348 V, (*c*) ΔG^0 = – 1416.62 kJ, –780.49 kJ, (*d*) easy to oxidise ClO_3^- to Cl_2 than ClO_4^- to $HClO_4$, (*e*) ΔG^0 = – 83.955 kJ, (*f*) = 1.64 V; **16.11** (*a*) 1.51 V, (*b*) MnO_4^{2-} better oxidising than MnO_4^-, (*c*) 1.745 V, (*d*) –327.33 kJ; **16.12** (*a*) $\frac{1}{2}O_2 + H_2O + 2e^- \rightarrow$

$2OH^-$, $2H_2O + 2e^- \rightarrow H_2 + OH^-$, (*b*) $\in^0 = 0.401$ V, (*c*) $E^0 = 1.229$ V; **16.13** $\in^0 = -1.468$ V; **16.14** $\in^0 = 0.037$ V; **16.15** 0.0177 V; **16.16** (*a*) 0.285 V, (*b*) –0.33 V, (*c*) 0.325 V, (*d*) –0.541 V; **16.17** pH = 4.23; **16.18** (*a*) 1.41×10^{11}, (*b*) 2.88×10^9, (*c*) 5.86×10^7, (*d*) 4.92×10^2; **16.19** (*i*) –0.4136 V, (*ii*) –0.4136 V; **16.20** 1.508 V, 1.463 V, 1.439 V; **16.21** (*i*) 0.698 V, (*ii*) 0.462 V; **16.22** Yes; **16.23** (*a*) 0.408 V, (*b*) at pH = 0, $E = -0.42$ V; **16.24** –0.11 V at pH = 0, –0.2282 V at pH = 4; **16.25** $Zn^{2+} = 10^{35.92}$; **16.26** (*a*) anode, $Pb + SO_4^{2-}(C_1) \rightarrow PbSO_4(C_1) + 2e^-$, cathode, $PbSO_4 + 2e^- \rightarrow Pb + SO_4^{2-}(C_2)$, $SO_4^{2-}(C_1) \rightarrow SO_4^{2-}(C_2)$, (*b*) 0.02955 V, (*c*) 0.025 V, **16.27** (*b*) 0.045 V, (*c*) 7.73×10^{-3} V; **16.28** $m_1 = 6.668$; **16.29** 8.895×10^{-3} V; **16.30** $\Delta G^0 = -130\ 332.9$ J, $\Delta S^0 = -125.04$ J K^{-1}, $\Delta H^0 = -167.6$ kJ; **16.31** $\Delta H^0 = (-nF)\ [1.0302 + 9.5 \times 10^{-7}(T - 293^2) - 10^{-9}(2T + 293)(T - 292)^2]$; **16.32** $\Delta G^0 = -68036.36$ J, $\Delta S^0 = -31.266$ J K^{-1}, $\Delta H^0 = -77.35$ KJ; **16.33** –249.28 kJ; **16.34** –176.18 kJ; **16.35** (*a*) 3.31×10^{-3} M; **16.36** 2.39×10^{57}; **16.37** 1.2×10^{-41}; **16.38** (*a*) 8.193, (*b*) no change; **16.39** (*a*) 2.87×10^{-35}, (*b*) 2.12×10^{-4}; **16.40** 4.15×10^{21}; **16.41** 2.45×10^{-70}; **16.42** 5.21×10^{14}; **16.43** 1.82×10^{-37}; **16.44** (*a*) 0.0875 V, (*b*) –0.019 V; **16.45** (*a*) –0.8274 V, (*b*) –0.6744 V; **16.46** (*a*) –0.403 V, (*b*) 0.026; **16.47** 2.92×10^{15}; **16.48** 4.86×10^{-7}.

Chapter 17

17.1 (*a*) 0, (*b*) – 1.7 kJ mol^{-1}, (*c*) 5.71 J K^{-1} mol^{-1}; **17.2** $\Delta S = 82.33$ J K^{-1} mol^{-1}, $\Delta G = -24.7$ kJ mol^{-1}, $\Delta E = \Delta H = 0$; **17.3** $\Delta S = -12.225$ J K^{-1}, $\Delta G = 3667.72$ J; **17.5** $y_{ben} = 0.67$; **17.6** 32 g mol^{-1}; **17.7** $x_A = 0.804$, $y_A = 0.91$; **17.8** negative deviation, $p'_T = 580$ mm, $P_T = 686.9$ mm; **17.9** negative deviation, $P_T^0 = 270$ mm, $P'_T = 193$ mm; **17.10** (*a*) $K_{acetoae} = 428.6$ mm, $K_{CHCl_3} = 371.43$ mm, (*b*) $K_{CHCl_3} = 62.5$ mm, $K_{Et_2CO} = 92.3$ mm; **17.11** $p^2 = p_1^0 p_2^0$; **17.13** $x_B = 0.4$, $y_B = 0.626$; **17.14** (*a*) $p_A^0 = 149.12$ mm, $p_B^0 = 733.92$ mm, (*b*) $y_A = 0.05$; **17.15** $x_A = 0.635$, $p_A^0 = 809.39$ mm, $p_B^0 = 509.69$ mm; **17.16** 22.6 g mol^{-1}; **17.17** 1.34×10^{-3} mol; **17.18** N_2:O_2::0.67:0.33; **17.19** C_2H_6; **17.20** 7.2×10^{-2} mol; **17.21** (*a*) $p_{water} = 23.28$ mm, (*b*) $p = 24.85$ mm; **17.22** $K_p = 3.41 \times 10^{-6}$; **17.23** 0.395 g; **17.24** (*a*) $K = 3.03$, (*b*)$K' = 0.938$; **17.26** $K = 1.01$; **17.28** $K_{AB} = 9.52 \times 10^2$; **17.30** (*a*) 0.058 mol, (*b*) 0.08 mol, (*c*) 0.1192 mol; **17.31** (*a*) 23.29 mm, (*b*) 23.19 mm, (*c*) same as in (*a*); **17.32** $M_{benzoic\ acid}$(water) = 205.6 g mol^{-1}, in benzene 349.3 g mol^{-1}; **17.33** 56 g mol^{-1}; **17.34** (*a*) 17.38 mm, (*b*) more lowering; **17.35** 5.11 K mol^{-1}; **17.36** 446.66 g; **17.37** 66.7 g mol^{-1}; **17.38** 0.9 deg; **17.39** 0.784; **17.40** 0.023 deg; **17.41** not consistent; **17.42** (*i*) 0.614, (*ii*) 0.414, (*iii*) 0.363; **17.43** 5.124; **17.44** degree of association increases, (*b*) K calculated does not mean anything than pure number; **17.45** 6045.2 J mol^{-1}; **17.46** (*a*) $x_A = 0.41$, (*b*) $\Delta \bar{H}_{fus}(A) = 24.90$ kJ mol^{-1}; **17.47** 66.6 g mol^{-1}; **17.48** (*i*) 109.0 g mol^{-1}, (*ii*) 116.43 g mol^{-1}, (*iii*) 121.44 g mol^{-1}, dissociation decrease with concentration; **17.49** (*i*) 1.28, (*ii*) 1.20, (*iii*) 1.15; **17.50** 3.62×10^{-3}; **17.51** (*i*) 106.35°C, (*ii*) 11.54°C, (*iii*) 126.08°C, (*iv*) 145.8°C, (*v*) 165.53°C, (*vi*) 103.0°C; **17.52** 69.5 g mol^{-1}; **17.53** 12.12 kJ mol^{-1}; **17.54** $\Delta T = 2.55$ deg; **17.55** $x_2 = 0.436$; **17.56** 57.37 J K^{-1} mol^{-1}; **17.57** $x_2 = 0.45$; **17.58** $x_2 = 0.92$; **17.59** (*a*) $x_2 = 0.805$, (*b*) $\Delta \bar{H}_{fus} = 5.516$ kJ mol^{-1}; **17.60** 26.37 atm; **17.61** 0.028 atm; **17.62** 93. g mol^{-1}; **17.63** 370 g mol^{-1}; **17.64** 5.46 J mol^{-1}; **17.65** (*i*) 0.35 atm, (*ii*) 19.66 atm; **17.66** (*a*) $[Ba^{2+}] = 0.1$ M, $[K^+] = 0.3$ M, $[Li^+] = 0.25$ M, $[OH^-] = 0.75$ M, (*b*) $[H^+] = 0.45$ M, $[SO_4^{2-}] = 0.3$ M, $[K^+] = 0.375$ M, $[NO_3^-] = 0.225$

M, (*c*) $[Na^+] = 1.47 \times 10^{-3}$ M, $[NO_3^-] = 0.01$ M, $[SO_4^{2-}] = 1.552 \times 10^{-3}$ M, $[Pb^{2+}] = 4.73 \times 10^{-3}$ M, $[K^+] = 3.1 \times 10^{-3}$ M, (*d*) $[Fe^{3+}] = 7.25 \times 10^{-3}$ M, $[ClO_4^-] = 21.75 \times 10^{-3}$ M, $[Ca^{2+}] = 0.31$ M, $[H^+] = 0.25$ M, $[NO_3^-] = 0.87$ M; **17.68** pH = 1.7, **17.69** pH = 2.7, $[F^-] = 0.202$ M, [HF] = 0.598; **17.70** $[H^+] = 5.76 \times 10^{-9}$ M; **17.71** pH = 1.99; **17.72** (*a*) pH = 3.81, (*b*) pH = 3.79; **17.73** (*a*) $[Ba^{2+}] = 10^{-5}$ M, (*b*) $[Ba^{2+}] = 6.25 \times 10^{-10}$ M; **17.74** (*a*) $[Ba^{2+}] = 7.07 \times 10^{-5}$ M, (*b*) $[OH^-] = 5.6 \times 10^{-5}$ M, $[CO_3^{2-}] = 1.47 \times 10^{-5}$ M, $[HCO_3^-] = 2.22 \times 10^{-8}$ M, $[Ba^{2+}] = 3.40 \times 10^{-4}$ M; **17.75** $[H^+] = [OH^-] = 1.0 \times 10^{-7}$ M, $[BaOH^+] = 4.8 \times 10^{-12}$ M, $[HSO_4^-] = 9.8 \times 10^{-11}$ M, $[Ba^{2+}] = [SO_4^{2-}] = 10^{-5}$ M; **17.76** $[Ca^{2+}] = 2.15 \times 10^{-4}$ M, $[F^-] = 4.30 \times 10^{-4}$ M, $[CaF^+] = 9.25 \times 10^{-7}$ M, $[OH^-] = 1.3 \times 10^{-7}$, $[H^+] = 7.8 \times 10^{-8}$, $[HF] = 5.0 \times 10^{-8}$, $[CaOH^+] = 5.6 \times 10^{-10}$ M, **17.77** $[AgCl] = 1.5 \times 10^{-5}$ M, $[Ag(S_2O_2)_2^{3-}] = 7.85 \times 10^{-4}$ M, $[S_2O_3^{2-}] = 4.3 \times 10^{-4}$ M; **17.78** (*a*) $[C_2O_4^{2-}] = 1.66 \times 10^{-12}$ M, (*b*) $[C_2O_4^{2-}] = 1.8 \times 10^{-6}$ M, (*c*) $[Pb^{2+}] = 4.61 \times 10^{-6}$ M; **17.79** $[H^+] = 0.182$ M; **17.80** $[HS^-] = 1.9 \times 10^{-3}$ M, $[Mn^{2+}] = 9.6 \times 10^{-4}$ M, $[MnOH^-] = 3.1 \times 10^{-5}$ M, $[H_2S] = 1.5 \times 10^{-5}$ M, $[OH^-] = 1.3 \times 10^{-5}$ M, $[S^{2-}] = 3.0 \times 10^{-10}$ M, $[H^+] = 7.7 \times 10^{-7}$ M, $[Mn(OH)_3^-] = 1.3 \times 10^{-10}$ M.

Chapter 18

18.1 $\bar{V}_2 = 38.33 \times 10^{-3}$ dm^3 mol^{-1}; **18.2** $\bar{V}_2 = 44.19 \times 10^{-3}$ dm^3 mol^{-1}; **18.3** (*i*) $\bar{V}_2(H_2SO_4) = 43.5 \times 10^{-3}$ dm^3 mol^{-1}, $\bar{V}(H_2O) = 17.8 \times 10^{-3}$ dm^3 mol^{-1}, (*ii*) $\bar{V}_2(CH_3OH) = 38.2 \times 10^{-3}$ dm^3 mol^{-1}, $\bar{V}(H_2O) = 17.8 \times 10^{-2}$ dm^3 mol^{-1}; **18.5** (*i*) $\bar{V}$ $(H_2O) = 16.87 \times 10^{-3}$ dm^3 mol^{-1}, $\bar{V}(H_2SO_4) = 58.10^{-3}$ dm^3 mol^{-1}, (*ii*) $\bar{V}(H_2O) = 18.01 \times 10^{-3}$ dm^3 mol^{-1}, $\bar{V}(CH_3OH) = 36.02 \times 10^{-3}$ dm^3 mol^{-1}: **18.6** (*a*) $\bar{V}_1(H_2O) = 14.41 \times 10^{-3}$ dm^3 mol^{-1}, $\bar{V}(KBr) = 29.5 \times 10^{-3}$dm^3 mol^{-1}, (*b*) $\bar{V}_1(H_2O) = 16.89 \times 10^{-3}$ dm^3 mol^{-1}, $\bar{V}_2(KBr) = 29.78 \times 10^{-3}$ dm^3 mol^{-1}, (*c*) $\bar{V}_1 = 18.15 \times 10^{-3}$ dm^3 mol^{-1}, $\bar{V}_2(KBr) = 29.95 \times 10^{-3}$ dm^3 mol^{-1}; **18.7** $\bar{V}_2 = 76.45 \times 10^{-3}$ dm^3 mol^{-1}, $\bar{V}_1 = 18.03 \times 10^{-3}$dm^3 mol^{-1}; **18.8** (*i*) 156.27×10^{-3} dm^3, (*ii*) 11.50×10^{-3} dm^3, (*iii*) 17.7×10^{-3} dm^3; **18.10** $\bar{V}_2(H_2SO_4) = 43.51 \times 10^{-3}$ dm^3 mol^{-1}, (*ii*) $\bar{V}_2(CH_3OH) = 37.55 \times 10^{-3}$ dm^3 mol^{-1}; **18.11** 37.08×10^{-3} dm^3 mol^{-1}; **18.12** $\bar{L}_2 = 475.84$ J; **18.13** (*a*) $\bar{L}_2 = 439.92$ J at $m_2 = 0.1$, $\bar{L}_2 = 889.35$ J at $m_2 = 0.04$ M, (*b*) $\bar{L}_1 = 0.75$ J at $m_2 = 0.04$ and $\bar{L}_1 = 1.213$ J at $m_2 = 0.1$; **18.14** (*a*) $\gamma_1 > 1$, (*b*) $\gamma_i < 1$; **18.15** (*a*) +ve deviation, (*b*) K (prop. alc.) = 286.66 mm, $K(H_2O) = 94.12$ mm, (*c*) (*i*) $\gamma_{water} = 2.158$, (*ii*) $\gamma_{water} = 0.5$, (*d*) (*i*) $\gamma_{alc} = 1.195$, (*ii*) $\gamma_{alc} = 0.0907$; **18.16** (*a*) +ve deviation, (*b*) $K_{CS_2} = 495$ mm, $K_{ether} = 136$ mm, (*c*) (*i*) $\gamma_{CS_2} = 1.371$, (*ii*) $\gamma_{CS_2} = 1.0$, (*d*) (*i*) 1.089, (*ii*) $\gamma_{ether} = 0.76$; **18.17** (*a*) –ve, (*b*) $K_{CHCl_3} = 52$ mm, $K_{ether} = 124$ mm, (*c*) (*i*) $\gamma_{ether} = 0.598$. (*ii*) $\gamma_{ether} = 1.915$, (*d*) $\gamma_{CHCl_3} = 0.828$, $\gamma_{CHCl_2} = 0.954$; **18.18** γ_{lim} (alc) = 13.2, $\gamma_{lim}(H_2O) = 5.0$, (*ii*) γ_{lim}(ben) = 1.86, $\gamma_{lim}(CS_2) = 1.39$, (*iii*) $\gamma_{lim}(CHCl_3) = 0.39$, γ_{lim}(ether) = 0.31; **18.19** $\gamma_1 = 0.907$; **18.20** $\gamma_{water} = 2.1$; **18.22** $\gamma_{H_2O} = 1.0011$; **18.23** $\gamma'_{alc}/\gamma'_{alc} = 25.68$; **18.24** a = 0.0213; **18.25** $\gamma = 0.886$; **18.26** $\gamma = 0.998$; **18.27** $\gamma = 2.39$; **18.29** $\gamma_\pm = 0.984$; **18.30** $a_\pm = 0.0139$; **18.31** $\gamma_\pm = 0.84$; **18.32** (*a*) $\gamma_\pm = 0.265$, (*b*) $\gamma_\pm = 0.278$; **18.33** $\gamma_\pm = 0.795$; **18.34** $\gamma = 0.877$; **18.35** $\gamma_\pm = 0.757$; **18.36** $\gamma_\pm = 1.0$; **18.37** 0.109 M; **18.38** (*a*) 0.5 M, (*b*) 0.9 M, (*c*) 0.12 M, (*d*) 0.6 M; **18.40** 1.97×10^{-5}; **18.41** $\gamma_\pm = 0.115$; **18.42** $\gamma_\pm = 0.986$; **18.43** (*a*) $\gamma_{Cl^-} = 0.827$, $\gamma_{K^+} = 0.827$, $\gamma_\pm = 0.827$, (b) $\gamma_{SO_4^{2-}} = 0.48$, $\gamma_{Cu^2} = 0.48$, $\gamma_\pm(CuSO_4) = 0.48$, (*iii*) same as (*ii*); **18.44** (*i*) 0.5 M, (*ii*) 0.793 M, (*iii*) 0.793, (*iv*) 1.14 M; **18.45**

(*i*) $\gamma_\pm = 0.698$, (*ii*) $\gamma_\pm = 0.287$, (*iii*) $\gamma_\pm = 0.287$, (*iv*) $\gamma_\pm = 0.071$; **18.46** (*i*) 0.033 M, (*ii*) 0.022 M, (*iii*) 0.0205 M, **18.47** (*a*) $m^0 = 3.48 \times 10^{-10}$ M, (*b*) 4.09×10^{-10} M, (*c*) 5.16×10^{-10} M, (*d*) 5.16×10^{-10} M; **18.48** (*i*) $\alpha^0 = 0.12$, (*ii*) $\alpha = 0.27$, (*iii*) $\alpha = 1.3$; **18.49** $\alpha = 0.0135$, $\alpha^0 = 0.013$; **18.50** (*i*) $\alpha^0 = 0.021$, (*iii*) $\alpha = 6.25 \times 10^{-3}$.

Chapter 19

19.1 (*i*) 1st order in A and B each, second order overall, (*ii*) first order in A and B each, independent of C (catalyst) overall second order, (*iii*) first order in N_2O_5 and overall, (*iv*) first order in A, –negative first order in B and zero overall, (*v*) second order in A, first order in B and D each overall fourth order; **19.2** 8.1×10^{-3} mol cc^{-1} s^{-1}, 4.876×10^{21} mole cc^{-1} s^{-1}; **19.3** (*a*) 1.778 m s, (*b*) 3.556 m s, (*c*) 7.69×10^{-6} mol dm^{-3}; **19.4** 0.1s; **19.5** 15.25%; **19.6** (*a*) 154 s, (*b*) 38.5 s; **19.7** (*a*) $\frac{1}{\sqrt{C}} = \frac{kt}{2} + \frac{1}{\sqrt{C_0}}$, (*b*) slope of the plot log (rate) vs log *C* is –3/2, (*c*) $t_{\frac{1}{2}} = \frac{0.829}{k\sqrt{C_0}}$, (*d*) dm$^{1/2}$ mol$^{-1/2}$ s^{-1}; **19.8** 4.27×10^{-4} s^{-1}; **19.9** 2.4×10^4 dm^3 mol^{-1} s^{-1}; **19.10** (*a*) 5.65×10^{-2} dm^3 mol^{-1} s^{-1}; (*b*) normality of acid and volume of solution taken; **19.11** $k = 0.131$ dm^3 mol^{-1} s^{-1}, 2nd order; **19.12** zero order; **19.13** second order in NO and 1st order in H^+, overall third; **19.14** (*i*) 4.25×10^{-4} s^{-1}; **19.15** (*a*) First order in A and B each, (*b*) mechanism I; **19.16** second mechanism; **19.17** $\tau = \frac{1}{k_1 + k_2}$; **19.18** $k_1 = 2.5 \times 10^{-5}$ s^{-1} $k_2 = 1.39 \times 10^{11}$ dm^3 mol^{-1} s^{-1}; **19.19** $\tau = \frac{1}{k_1([A]_e + [B]_e) + k_2([C]_e + [D]_e)}$; **19.20** 0.0177 s; **19.21** (*a*) 4.644, (*b*) 4% of B consumed; **19.22** 0.62:0.27:0.11; **19.23** $k_\beta = 5.862 \times 10^{-4}$ s^{-1}, $k_\alpha = 2.345 \times 10^{-7}$ s^{-1}; **19.24** 5.55×10^{-5} dm^3 mol^{-1} s^{-1}; **19.25** 0.0482 hr^{-1}; **19.26** $k_1 = 1.794 \times 10^{-3}$ dm^3 mol^{-1} s^{-1}, $k_2 = 1.293 \times 10^{-4}$ s^{-1}; **19.27** (*i*) $t_{max} = 2.66 \times 10^{-2}$ s, $[B]_{max} = 4.8 \times 10^{-3}[A]_0$, (*ii*) $t_{max} = 1.85$ s, $[B]_{max} = 0.63[A]_0$; **19.28** $k_{eff} = \frac{2k_1k_3}{k_2 + 2k_3}$; **19.29** rate $= k_3k_1[A]^2(k_3 + k_2[A])$, rate $= \frac{k_3k_1}{k_2}[A]$ if $k_2[A] \gg k_3$, rate $= k_1[A]^2$ if $k_3 \gg k_2[A]$; **19.30** $k_{eff} = k_4\left(\frac{k_1}{k_6}\right)^{1/2}$; **19.31** (*a*) $k_{eff} = k_4\left(\frac{k.k_3}{k_2k_4}\right)^{2/3}$, (*b*) order with respect to A 2/3, with respect to B 4/3, (*c*) overall 2; **19.32** 24.47 kJ mol^{-1}; **19.33** $E_a = 39.93$ kJ mol^{-1}, $A = 7.94 \times 10^4$; **19.34** 86.47 kJ mol^{-1}; **19.35** 43.15 kJ mol^{-1}; **19.36** b; **19.37** $A = 6.463 \times 10^{10}$ dm^3 mol^{-1} s^{-1}, $k = 1.33 \times 10^{-2}$ dm^3 mol^{-1} s^{-1}; **19.38** $A = 1.39 \times 10^{10}$ dm^3 mol^{-1} s^{-1} $k = 2.58 \times 10^{-4}$ dm^3 mol^{-1} s^{-1}; **19.39** (*a*) 386.6 K, (*b*) 419.6 K, (*c*) 455.3 K; **19.40** (*i*) $A = 3.8 \times 10^9$ mol^{-1} dm^3 s^{-1}, $k = 5.75 \times 10^7$ dm^3 mol^{-1} s^{-1}, (*ii*) $A = 4.487 \times 10^8$ mol^{-1} dm^3 s^{-1}, $k = 2.11 \times 10^{-11}$ mol^{-1} s^{-1} dm^3, (*iii*) $A = 1.30 \times 10^{10}$ mol^{-1} dm^3 s^{-1}, $k = 2.17 \times 10^{-5}$ mol^{-1} dm^3 s^{-1}; **19.41** (*i*) 1/2, (*ii*) –1/2, (*iii*) –1/2, (*iv*)

–1; **19.42** (*i*) 6.02×10^{11} dm^3 mol^{-1} s^{-1}, (*ii*) 6.02×10^{9} dm^3 mol^{-1} s^{-1}, (*iii*) 6.02×10^{9} dm^3 mol^{-1} s^{-1}, (*iv*) 6.02×10^{8} dm^3 s^{-1} mol^{-1}; **19.44** A = 4.46×10^{13} s^{-1}, $k = 1.11 \times 10^{-16}$ mole^{-1} cc^{-1} s^{-1}, **19.45** (*i*) 3.20×10^{16} s^{-1}, (*ii*) 4.30×10^{16} s^{-1}; **19.46** $k = 5.0 \times 10^{7}$ dm^3 mol^{-1} s^{-1}; **19.47** $A = 5.09 \times 10^{10}$ dm^3 mol^{-1} s^{-1}, $k = 1.27 \times 10^{10}$ dm^3 mol^{-1} s^{-1}; **19.49** $A = 1.13 \times 10^{12}$ mol^{-1} dm^3 s^{-1}; **19.50** 5.09 μs; **19.51** 914 K; **19.52** 50.96%; **19.53** $\Delta S = -197.97$ J K^{-1} mol^{-1}; **19.54** $\Delta S = -47.11$ J K^{-1} mol^{-1}; **19.55** 2.46×10^{2} dm^3 mol^{-1} s^{-1}; **19.56** 6.9×10^{-5} dm^3 mol^{-1} s^{-1}; **19.57** 1.21 dm^3 mol^{-1} s^{-1}.

INDEX

A

B

E

I

J

K

L

M

N

O

P

Q

R

S

W

X

Z